CHAPMAN & HALL/CRC COMPUTER and INFORMATION SCIENCE SERIES

Handbook of Computational Molecular Biology

CHAPMAN & HALL/CRC
COMPUTER and INFORMATION SCIENCE SERIES

Series Editor: Sartaj Sahni

PUBLISHED TITLES

HANDBOOK OF SCHEDULING: ALGORITHMS, MODELS, AND PERFORMANCE ANALYSIS
Joseph Y.-T. Leung

THE PRACTICAL HANDBOOK OF INTERNET COMPUTING
Munindar P. Singh

HANDBOOK OF DATA STRUCTURES AND APPLICATIONS
Dinesh P. Mehta and Sartaj Sahni

DISTRIBUTED SENSOR NETWORKS
S. Sitharama Iyengar and Richard R. Brooks

SPECULATIVE EXECUTION IN HIGH PERFORMANCE COMPUTER ARCHITECTURES
David Kaeli and Pen-Chung Yew

SCALABLE AND SECURE INTERNET SERVICES AND ARCHITECTURE
Cheng-Zhong Xu

HANDBOOK OF BIOINSPIRED ALGORITHMS AND APPLICATIONS
Stephan Olariu and Albert Y. Zomaya

HANDBOOK OF ALGORITHMS FOR WIRELESS NETWORKING AND MOBILE COMPUTING
Azzedine Boukerche

HANDBOOK OF COMPUTATIONAL MOLECULAR BIOLOGY
Srinivas Aluru

CHAPMAN & HALL/CRC COMPUTER and INFORMATION SCIENCE SERIES

Handbook of Computational Molecular Biology

Edited by

Srinivas Aluru

Iowa State University

Ames, Iowa, USA

Chapman & Hall/CRC

Taylor & Francis Group

Boca Raton   London   New York

Published in 2006 by
Chapman & Hall/CRC
Taylor & Francis Group
6000 Broken Sound Parkway NW, Suite 300
Boca Raton, FL 33487-2742

International Standard Book Number-10: 1-58488-406-1 (Hardcover)
International Standard Book Number-13: 978-1-58488-406-4 (Hardcover)
Library of Congress Card Number 2005054821

Library of Congress Cataloging-in-Publication Data

Handbook of computational molecular biology / [edited by] Srinivas Aluru.
    p. cm. -- (Chapman & Hall/CRC computer information science ; 9)
   Includes bibliographical references and index.
   ISBN 1-58488-406-1
    1. Computational biology--Handbooks, manuals, etc. 2. Molecular biology--Handbooks, manuals, etc.
I. Aluru, Srinivas. II. Chapman & Hall/CRC computer and information sciences series ; 9.

QH324.2.H357 2005
572.8--dc22   2005054821

Dedication

To my wife,
Maneesha

Preface

Computational Molecular Biology is an exciting and rapidly expanding interdisciplinary research field that is attracting significant attention from both academia and industry. The dynamic and rapidly expanding nature of this young field creates a particular need for appropriate educational material, but also makes it challenging to meet this need. New researchers are routinely attracted to this area and need appropriate training material to learn quickly and begin contributing to the field. Practitioners need access to the latest developments in modeling, analysis, and computational methods. Researchers who are at the forefront of research in one subfield need to become familiar with advancements being made in other subfields to understand how they impact the overall field and to pursue new research directions. There is significant demand for a well trained workforce in this area. The burgeoning of bioinformatics and computational biology graduate programs and faculty hiring in many academic institutions is leading to an explosion in the number of graduate students pursuing computational biology. Yet, the fast pace of discoveries makes it difficult for a single individual to write a textbook providing comprehensive coverage. A number of excellent textbooks and handbooks targeting individual subfields of computational biology — such as sequence analysis, microarrays and phylogenetics — are available. While some textbooks covering multiple areas, or presenting computational biology from unique perspectives such as machine learning or algorithmic paradigms have also been developed, there is no single book that provides a systematic coverage of the breadth of knowledge accumulated in computational biology.

This handbook is designed to serve as a comprehensive introduction to computational molecular biology. The handbook is appropriate for readers with a technical background in computer science — some degree of familiarity with design and analysis of algorithms is required to fully understand the material in the handbook. It is designed to serve as a reference book for researchers and practitioners, a compendium of introductory and advanced material to aid the entry of new researchers into the field, and a resource for classroom instruction of graduate and senior level undergraduate students. The book is divided into eight parts, each part devoted to a particular subfield or a group of related topics. Within most parts, the chapters are designed to reflect the progression from introductory material to advanced material exploring more complex topics or actively pursued research topics of current relevance. The only exception is when a group of relatively independent chapters but with a common theme are grouped into a part. The parts themselves are arranged so that elementary topics and topics which are necessary for understanding other topics precede in the handbook. In this respect, the handbook is designed to be like a textbook — a reader unfamiliar with a subfield covered in a part can read the chapters in a sequence, gain a fairly comprehensive understanding of the subject, and find reference material for further study.

This handbook is truly a community effort, with the chapters authored by leading researchers widely recognized for their research contributions and expertise. Thanks to their contributions, the handbook provides a thorough and researcher's perspective of each topic that would be difficult to achieve in a single author book due to the rapidly growing nature of computational biology. In many cases, the authors went beyond the normal expectations of contributing their individual chapters. Some have offered contributions and provided valuable advice on reorganization of material on some topics. Related chapters have been shared between authors to help interconnect the chapters and better integrate them into the book. While this had lengthened the time for producing the handbook somewhat, it has tremendously helped in improving the quality of the handbook.

Even though the handbook provides a broad coverage of computational molecular biology, it is by no means complete. It is not possible to provide a complete coverage of the subject even in a large handbook such as this. Difficult choices were made to exclude certain topics, limit the coverage of others, and in a few cases a topic is missed because of difficulty in finding a suitable contributor. As a testament to the rapidly growing nature of the field, new and important directions have come up while the handbook was under development. Some of these shortcomings and emerging research topics can be addressed in a revision, and the editor appreciates feedback from readers and experts in this regard.

Organization of the Material

The handbook is organized into eight parts. Part I deals with the fundamental topic of alignments to discover evolutionary relationships between biological sequences. Chapters 1 and 3 deal with the topic of alignments between a pair of sequences, and multiple sequences, respectively. Chapter 2 deals with spliced alignments and their use in gene recognition. The last chapter in this part, Chapter 4, addresses the sensitivity of optimal alignment to changes in the underlying parameter space.

Part II of the book, which can be independently studied from Part I, is devoted to string data structures and their applications in computational biology. Chapter 5 is an introduction to the three widely used string data structures in biology — lookup tables, suffix trees and suffix arrays. Chapter 6 catalogues a number of interesting uses of suffix trees motivated by applications in biology. Finally, Chapter 7 describes enhanced suffix arrays, which are a space-economical way to implement many of the suffix tree based applications.

Part III deals with computational problems arising from genome assembly and expressed sequence tag (EST) clustering. These are important biological applications where fundamental techniques from the first two parts are useful. At a minimum, an understanding of the material in Chapters 1 and 5 is necessary to delve into this part. Chapter 8 presents an overview of the issues involved in shotgun genome assembly and a brief introduction to many programs developed for this purpose. Chapter 9 chronicles the process by which assembly for the public human genome project was carried out, along with the ensuing computational challenges and how they were addressed. Chapter 10 addresses the problem of reducing genome sequencing cost by using known genomes as a guide to assemble the genomes of related organisms. Chapter 11 deals with the construction of physical maps for entire chromosomes. Chapter 12 provides an introduction to expressed sequence tags (ESTs), and methods for EST analysis. Finally, Chapter 13 addresses unifying methodologies for EST clustering and genome assembly with special emphasis on algorithms for addressing large-scale data.

A sampling of computational methods that arise in understanding of genomes, relationships between genomes of various organisms, and the mechanism of genetic inheritance are the subject of Part IV of the handbook. Chapters 14 and 15 are devoted to comparative genomics, a subject of growing importance given the increasing availability of genomes. Chapter 16 addresses computational analysis of alternative splicing, the process by which a single gene codes for multiple mRNAs. Linkage analysis to identify approximate chromosomal locations of genes associated with genetically inherited diseases is the topic of Chapter 17. Chapter 18 explores hyplotype inference, a topic of significant current research interest.

Phylogenetics, or inferring of evolutionary relationships among species and sequences, is the focus of Part V of the handbook. The first chapter, Chapter 19 provides an overview of phylogeny reconstruction and is a gateway for studying the remaining chapters. Chapter 20 deals with the construction of supertrees from trees representing phylogenetic relationships between subsets of species. Recent trends in large-scale phylogenetic analysis are the topic of Chapter 21. Finally, Chapter 22 explores the use of parallel computers for phylogeny reconstruction, in view of the compute-intensive nature of these methods.

Part VI of the handbook is devoted to the study of microarrays and analysis of gene expression to gain insight in gene regulation and inferring gene networks. Chapter 23 presents an overview of the microarray technology and issues in storing, retrieving and annotating microarray data. Chapter 24 is devoted to computational problems arising in designing microarrays. An overview of clustering algorithms for gene expression analysis are presented in Chapter 25. Chapter 26 covers recent advances in the same topic that allow clustering based on subsets of genes and experimental conditions. The last two chapters are on modeling and inferring gene networks. Chapter 27 surveys popular modeling frameworks and explores computational methods for inferring gene networks from large-scale gene expression data. The application of control theory to modeling and analysis of gene networks is the subject of Chapter 28.

Given the central role of proteins in most biological processes and the dependence of their function on structural properties, computational methods related to protein structures constitutes a large subfield of computational biology. Part VII of the handbook presents many approaches for inference and applications of structural information including combinatorial methods, experimental methods and simulation methods. Chapter 29 provides a brief introduction to protein structure and contains an in-depth coverage of prediction of secondary and supersecondary structures. Chapter 30 chronicles early work on protein structure prediction using lattice models which proved useful in understanding the complexity of protein structure prediction. Chapter 31 is devoted to protein structure prediction using nuclear magnetic resonance (NMR) spectroscopy. This subject is continued in Chapter 32 which covers geometric processing methods for reconstructing three dimensional structures from NMR and other experimental approaches. Chapter 33 revisits the problem of homology detection, which is covered at the rudimentary level of sequence alignments in Part I, though what is most often sought is structural homology. This chapter presents the important problem of remote homolog detection and structural homologies using sequence, profile and structure data. Modeling of biomolecular structures through molecular dynamics simulations of atomic level interactions is the subject of Chapter 34.

The last part of the handbook is devoted to algorithmic issues in designing bioinformatic databases and the application of mining techniques to biological data. Chapters 35 and 36 explore the design of storage structures for biological sequence data that support exact matching and approximate matching queries, respectively. Chapter 37 is devoted to the topic of searching for common motifs in sets of biological sequences. Chapter 38 provides an overview of the application of data mining techniques in computational biology with emphasis on mining gene expression data and protein structural data.

Teaching from the Handbook

The handbook can be used as a text to support a variety of courses in computational molecular biology. Most of the handbook can be covered in a two semester course sequence. The sequence can be designed by partitioning the topics, or by covering the introductory material from most parts in the first course followed by coverage of the advanced topics in the second. For example, Chapters 1 and 3 covering pairwise and multiple sequence alignments can be taught as part of the introductory course, while Chapter 2 on spliced alignments and chapter 4 on parametric alignments can be deferred to the advanced course. Such an approach is useful if there are students who will be taking just one course and not follow through with the second. If both courses are expected to be taken by all the students, such as when the course sequence is designed to be part of core curriculum, any appropriate choice for partitioning the topics can be exercised.

The handbook can also support courses in targeted areas related to computational biology. A course in computational genomics can be designed to cover Parts I, III and V of the handbook along with portions of chapters from Part II and selected chapters from Part IV. Part IV can be completely covered by omitting from Parts I, III and V chapters covering advanced material (such as chapters 4 and 10) or those covering specialized topics (such as 11, 13 and 22). Topics in functional genomics can be added by choosing selected chapters from Part VI, perhaps Chapters 22, 24 and 26. A course on string algorithms and their applications to computational biology can be taught by starting with the material in Part II, then covering Chapters 34 and 35, portions of Part I, and supplementing this material with appropriate research papers. Parts VI and VII can be used to support instruction in a course designed around functional genomics and structural biology.

The above suggestions are neither meant to be prescriptive nor are they exhaustive. The editor is interested in feedback on instructional usage of the handbook, and suggestions on improving the contents from that perspective.

I learnt tremendously from the wisdom of authors contributing to the handbook and hope the reader has a similar experience.

Srinivas Aluru
Iowa State University

Acknowledgments

I am most indebted to the authors who contributed their valuable time and expertise in creating the handbook. I was pleasantly surprised by how many of my contacts were met with favorable responses. I am grateful to the contributors for sharing my vision for the handbook, agreeing to my insistence that they write on specific topics I requested, suggesting improvements, and interacting with other contributors as needed. Those wishing to join this exciting field of research have a truly wonderful community to look forward to. If the handbook proves useful to many, the credit squarely belongs to the contributors.

I am grateful to my team of graduate students for their active participation in this project, in particular for reading and presenting the individual chapters to the group as they arrived. The handbook provided wonderful material for our group seminar series. In this regard, I would like to thank Scott Emrich, Benjamin Jackson, Anantharaman Kalyanaraman, Pang Ko, Srikanth Komarina, Mahesh Narayanan, Sarah Orley, and Sudip Seal. I also appreciate their patience in gracefully handling loss of my attention while I focused on the book. My sincere gratitude to Ashraf I. Hamad who helped with initial formatting and Bashar M. Gharaibeh for helping with formatting in later stages, and in working to bring some consistency in listing of references across chapters.

The staff at CRC press have been a pleasure to work with. Bob Stern has been a constant presence in this project from its inception to publication. I appreciate his patience in working with me as I stretched the limits of many deadlines and for allowing me the time to finish the project to my satisfaction. I am grateful to Jessica Vakili who served as the Project Coordinator, and Glenon C. Butler, Jr. for serving as Project Editor and handling proofreading and actual publication of the handbook. Keven Craig and Jonathan Pennell worked on designing the cover for the handbook.

I am greatly indebted to Sartaj Sahni, who is the editor-in-chief of the Chapman & Hall/CRC Computer and Information Science series, of which this handbook is one publication. It was he who initiated this project and I am grateful to him for reposing faith in me to carry out the project, and for his valuable suggestions and help in enlisting some of the authors.

On the personal side, I would like to thank my wife, my two sons, and visiting parents for their love, support, and cheerfully allowing me to spend the time needed in composing the book. I especially appreciate my young son Chaitanya, and Pranav born during this book project, for their understanding and patience at such an early age.

About the Editor

Srinivas Aluru is a Professor and Associate Chair for Research in the Department of Electrical and Computer Engineering at Iowa State University. He serves as the chair of Bioinformatics and Computational Biology graduate program, and is a member of the Laurence H. Baker Center for Bioinformatics and Biological Statistics, and the Center for Plant Genomics at Iowa State. Earlier, he held faculty positions at New Mexico State University and Syracuse University. He received his B. Tech degree in Computer Science from the Indian Institute of Technology, Chennai, India in 1989, and his M.S. and Ph.D. degrees in Computer Science from Iowa State University in 1991 and 1994, respectively.

Dr. Aluru is a recipient of the NSF Career award in 1997, an IBM faculty award in 2002, Iowa State University Young Engineering Faculty Research Award in 2002, and the Warren B. Boast Undergraduate Teaching Award in 2005. He is an IEEE Computer Society Distinguished Visitor from 2004 to 2006. His research interests include parallel algorithms and applications, bioinformatics and computational biology, and combinatorial scientific computing. His contributions to computational biology are in computational genomics, string algorithms, and parallel methods for solving large-scale problems arising in biology. He co-chairs an annual workshop in High Performance Computational Biology (http://www.hicomb.org) and has served as a Guest Editor of the *Journal of Parallel and Distributed Computing* and the *IEEE Transactions on Parallel and Distributed Systems* for special issues on this topic. Dr. Aluru served on the program committees of several conference and workshops, and NSF and NIH panels, in the areas of computational biology, parallel processing and scientific computing.

Contributors

Mohamed Ibrahim Abouelhoda
University of Ulm, Germany

Jonathan Arnold
The University of Georgia, USA

Chandrajit Bajaj
University of Texas at Austin, USA

Suchendra Bhandarkar
University of Georgia, USA

Inna Dubchak
Lawrence Berkeley National Laboratory, Joint Genome Institute, USA

Oliver Eulenstein
Iowa State University, USA

Li M. Fu
University of Florida, USA

Daniel Gusfield
University of California, Davis, USA

Jinling Huang
The University of Georgia, USA

Tamer Kahveci
University of Florida, USA

Mustafa Khammash
University of California, Santa Barbara, USA

Stefan Kurtz
University of Hamburg, Germany

Glenn Martyna
IBM TJ Watson, USA

Richa Agarwala
National Institutes of Health, USA

David Fernández-Baca
Iowa State University, USA

Pierre Baldi
University of California, Irvine, USA

Michael Brudno
University of Toronto, Canada

Ognen Duzlevski
University of Missouri-Columbia, USA

Paolo Ferragina
University of Pisa, Italy

Mikhail S. Gelfand
Russian Academy of Sciences, Russia

William Hart
Sandia National Laboratories, USA

Xiaoqiu Huang
Iowa State University, USA

Anantharaman Kalyanaraman
Iowa State University, USA

Pang Ko
Iowa State University, USA

Guohui Lin
University of Alberta, Canada

Andrei A. Mironov
Russian Academy of Sciences, and Moscow State University, Russia

Srinivas Aluru
Iowa State University, USA

David A. Bader
Georgia Institute of Technology, USA

Catherine Ball
Stanford University, USA

Hui-Hsien Chou
Iowa State University, USA

Scott J. Emrich
Iowa State University, USA

Vladimir Filkov
University of California, Davis, USA

Osamu Gotoh
Kyoto University, Japan

G. Wesley Hatfield
University of California, Irvine, USA

Benjamin N. Jackson
Iowa State University, USA

Laxmikant Kale
University of Illinois at Urbana-champaign, USA

Sameer Kumar
University of Illinois at Urbana-champaign, USA

C. Randal Linder
The University of Texas at Austin, USA

Alexei D. Neverov
Russian Academy of Sciences, Russia

Alantha Newman
Massachusetts Institute of
Technology, USA

James C. Phillips
University of Illinois at
Urbana-champaign, USA

Alejandro A. Schäffer
National Institutes of Health,
USA

Ron Shamir
University of Tel Aviv, Israel

Ambuj Singh
University of California, Santa
Barbara, USA

Steven Skiena
State University of New York at
Stony Brook, USA

Mark Tuckerman
New York University, USA

Xiang Wan
University of Alberta, Canada

Dong Xu
University of
Missouri-Columbia, USA

Zeyun Yu
The University of Texas at
Austin, USA

Gengbin Zheng
University of Illinois at
Urbana-champaign, USA

Enno Ohlebusch
University of Ulm, Germany

**Sanguthevar
Rajasekaran**
University of Connecticut, USA

Klaus Schulten
University of Illinois at
Urbana-champaign, USA

Roded Sharan
University of Tel Aviv, Israel

Mona Singh
Princeton University, USA

Amos Tanay
University of Tel Aviv, Israel

Vamsi Veeramachaneni
Strand Genomics Corporation,
India

Xiu-Feng Wan
University of
Missouri-Columbia, USA

Tetsushi Yada
Kyoto University, Japan

ShinsukeYamada
Waseda University, Japan

Steven Hecht Orzack
Fresh Pond Research institute,
USA

Hana El Samad
University of California, Santa
Barbara, USA

Karlton Sequeira
Rensselaer Polytechnic
Institute, USA

Gavin Sherlock
Stanford University, USA

Robert D. Skeel
Purdue University, USA

Xin Tu
University of Alberta, Canada

Balaji Venkatachalam
Iowa State University, USA

Tandy Warnow
University of Texas at Austin,
USA

Mi Yan
University of New Mexico, USA

Mohammed Zaki
Rensselaer Polytechnic
Institute, USA

Contents

Part V: Phylogenetics

Part VI: Microarrays and Gene Expression Analysis

Part VII: Computational Structural Biology

Part VIII: Bioinformatic Databases and Data Mining

I

Sequence Alignments

Pairwise Sequence Alignment

Benjamin N. Jackson
Iowa State University

Srinivas Aluru
Iowa State University

1.1 Introduction

The discovery of biomolecular sequences and exploring their roles, interplay, and common evolutionary history is fundamental to the study of molecular biology. Three types of sequences fill complementary roles in the cell: DNA sequences, RNA sequences, and protein sequences. DNA sequences are the basis of genetic material and act as the hereditary mechanism, providing the recipe for life. RNA sequences are derived from DNA sequences and play many roles in protein synthesis. Protein sequences carry out most essential processes such as tissue building, catalysis, oxygen transport, signaling, antibody defense, and transcription regulation. The first part of this book will describe the alignment algorithms used to compare these sequences.

For the benefit of the reader unfamiliar with molecular biology, we provide a more detailed introduction to biological sequences. A DNA molecule is composed of simpler molecules known as nucleotides. The nucleotides are differentiated by the differences in their bases — Adenine, Cytosine, Guanine and Thymine, represented by A, C, G, and T, respectively. DNA naturally occurs as a double-stranded helix-shaped molecule, with each nucleotide in one strand pairing with a corresponding nucleotide in the other strand, with A pairing with T and G pairing with C and vice versa. Each strand has a direction, with the two strands having opposite directions. The DNA molecule is represented by the sequence of nucleotides of one strand in that strand's direction. Given one strand, the sequence of the other strand is obtained by reversing the known strand and substituting A for T, C for G, etc. This process is called generating the *reverse complement* and is important when comparing DNA sequences as either strand might be given for the DNA being compared.

Several different terms are used to describe DNA sequences. Each cell in an organism contains the same set of *chromosomes*, which are long DNA sequences. The set of chromosomes in an organism constitutes its *genome*. A *gene* is a contiguous stretch of DNA along a chromosome that codes for a protein or RNA. Genes consist of one or more coding regions called *exons* separated by non-coding regions called *introns*. The terms *promoter, enhancer,* and *silencer* are used to describe DNA sequences involved in regulating gene expression through protein interactions and are often located upstream of the gene. Genes and regulatory regions are often conserved (show high similarity or homology) across species.

An important function of DNA sequences is to code for protein sequences. Like DNA sequences, proteins are also sequences of simpler molecules, in this case amino acids. Amino acids are differentiated by their side chains. There are twenty possible side chains that distinguish the twenty different amino acids found in protein sequences. As with DNA, each of the twenty amino acids is represented by a unique character.

A protein is derived from a gene through an RNA intermediary. Similar to DNA, RNA is a sequence of nucleotides with the base Thymine replaced by Uracil. First, an RNA called *pre-mRNA* containing both exons and introns is copied from the DNA in a process called *transcription*. The introns are excised and the exons are spliced to form an mRNA. The mRNA is then *translated* into an amino acid sequence. A *codon* is three consecutive nucleotides in the mRNA that is translated to an amino acid in the corresponding protein. The mRNA is used as a template to generate an amino acid sequence of one third the length of the coding region. The code mapping the 64 possible codons to the 20 possible amino acids is common to almost all of life. The two step process of transcribing DNA to RNA and translating RNA to protein is popularly known as the central dogma of molecular biology.

Multiple forms of the same gene, known as *alleles*, cause genetic differences between individuals and are responsible for the genetic diversity of a species. Sometimes, variations in alleles lead to undesirable outcomes such as genetic diseases or increased susceptibility to diseases. The differences between alleles are often quite small. Sometimes a single nucleotide change can have a large effect on the resulting protein. DNA sequences are typically modified through insertions, deletions or substitutions. These underlying evolutionary mechanisms provide a starting point for sequence alignment algorithms.

Sequence alignments are intended to discover and illustrate the similarities, differences, or evolutionary relationships between sequences. The algorithms used for sequence comparison vary depending on the types of sequences being compared and the question being asked, giving rise to a variety of sequence alignment algorithms. In this chapter, we will present the basic sequence alignment algorithms, broadly characterized as *global alignment, semiglobal alignment,* and *local alignment*. Global alignment can be used to compare two protein sequences from a closely related gene family, two homologous genes, or two gene alleles. Semiglobal alignment can be used to piece together fragments of DNA from shotgun DNA reads and create a longer inferred sequence, useful in genome assembly. Local alignment can be used as a part of *multiple local alignment*, presented in Chapter 3, to find a common motif among protein sequences or conserved promoter sites in gene sequences. Chapter 2 presents spliced alignments, which are important when aligning DNA with RNA transcripts. Finally, Chapter 4 addresses the characteristics of the problem space, and how changing parameters affect alignment results.

1.2   Global Alignment

The global sequence alignment problem for two sequences is defined as follows. We call the set of unique characters in the input sequences an alphabet Σ. In the case of DNA sequences, that alphabet is $\Sigma = \{a, g, c, t\}$. A string X of length n is a sequence of characters $\langle x_1, x_2, ..., x_n \rangle$ such that $x_i \in \Sigma$. A *prefix* of X is a string of the form $\langle x_1, x_2, ..., x_i \rangle, 1 \leq i \leq n$. A *substring* of X is a string of the form $\langle x_i, x_{i+1}, ..., x_{j-1}, x_j \rangle, 1 \leq i \leq j \leq n$. For example 'aggctga' is a string with substrings 'aggc' and 'gctg', with 'aggc' also being a prefix of 'aggctga'.

A string of characters is the term traditionally used in computer science literature, and it is equivalent to the concept of a sequence in biology. We will use the term string almost exclusively in this chapter. However, it is important to adapt the string algorithms to the specific biological sequences of interest. For example, when comparing DNA sequences, it is important to compare the two input sequences, as well as the reverse complement of one sequence with the other input sequence.

Consider two strings $A = \langle a_1, a_2, ..., a_n \rangle$ and $B = \langle b_1, b_2, ..., b_m \rangle$. Conceptually we wish to create an alignment between the two strings, matching similar regions by aligning each character in string A with a character in string B. Additionally, we can insert gaps in each string (allowing for the possibility of deletions or insertions of sequences of characters). More formally, an alignment between A and B is the production of two new strings of equal length, A_L derived from A and B_L derived from B through insertions of a special gap character '-'. $A_L = \langle a_1, a_2, ..., a_l \rangle$ and $B_L = \langle b_1, b_2, ..., b_l \rangle$, where l is the alignment length, $\max(n, m) \leq l \leq n + m$. Both a_i and b_i may not be gap characters. a_i and b_i are said to be aligned with each other. If a_i is a gap, then b_i is said to be aligned with a gap in A, and vice versa. An example alignment between two strings 'aggctga' and 'agcttg' is shown below.

```
aggct-ga
ag-cttg-
```

The quality of the alignment is measured by its score, which can be thought of as a measure of how similar the two strings are. The score is the summation of the score of each pair of characters a_i and b_i. We will choose a simple scoring function that has roots in our evolutionary model. A character aligned with the same character, a *match*, is given a score α. This corresponds to a conserved character. A character aligned with any other character, a *mismatch*, is given a score β, and corresponds to a substitution. Finally, a character aligned with a gap, a *gap*, is given a score γ, and corresponds to either an insertion or a deletion in one of the strings.

$$score(L) = \sum_{i=1}^{l} score(a_i, b_i)$$

$$score(x, y) = \begin{cases} \alpha & x = y \\ \beta & x \neq y \\ \gamma & x = `-` \text{ or } y = `-` \end{cases}$$

Typically α is positive and γ and β are negative. We will consider the values $\alpha = 2, \beta = -1, \gamma = -1$. Given these values, the example alignment has a total score of 7.

We wish to find an alignment between the two strings that results in the highest score, called an *optimal alignment* between the two strings. A simplistic solution would be to score all possible alignments and chose (from) the highest scoring, but the number of such possibilities is exponential.

1.3   Dynamic Programming Solution

The solution can be sped up using dynamic programming. We see that the problem exhibits an optimal substructure. Consider an optimal alignment L between A and B. If we look at some part of that optimal alignment L' that aligns a substring A' of A with a substring B' of B, we wish to say, for optimal substructure, that L' is an optimal alignment between A' and B'. The proof is simple, using contradiction. If the alignment L' is not optimal, then there exists an alignment L_{new} between A' and B', with $score(L_{new}) > score(L')$. However, L_{new} can be substituted for L' in L, increasing the score of L. Therefore L is not optimal, a contradiction.

We can use the optimal substructure property to solve the problem more efficiently using the following formulation. In order to find the optimal alignment between the two strings, we find the optimal alignment between each prefix A_i of A and each prefix B_j of B, where A_i is the prefix of length i of A and B_j is the prefix of length j of B. Let a_i be the last character in A_i and b_j be the last character in B_j. There are three possibilities that can produce the optimal score.

1. Align a_i with b_j and optimally align A_{i-1} with B_{j-1}.
2. Align a_i with a gap and optimally align A_{i-1} with B_j.
3. Align b_j with a gap and optimally align A_i with B_{j-1}.

We will denote the optimal score of aligning A_i with B_j as $S[i,j]$. Think of a table that records the maximum score of aligning all possible pairs of A_i and B_i. The following recurrence describes how to fill the table using the ideas presented above.

$$S[i,j] = \max \begin{cases} S[i-1,j-1] + \delta(a_i, b_j) \\ S[i,j-1] + \gamma \\ S[i-1,j] + \gamma \end{cases}$$

$$\delta(x,y) = \begin{cases} \alpha & x = y \\ \beta & x \neq y \end{cases}$$

All that remains is to specify the starting conditions. The score of aligning some prefix of A with none of B is the length of that prefix times the gap penalty. Formally, $S[0,j] = \gamma j$ and $S[i,0] = \gamma i$.

A sample table is shown in Figure 1.1. Rows in S correspond to characters in A with the first row corresponding to the empty string. Columns in S correspond to characters in B with the first column corresponding to the empty string. We can initialize the first row and first column of the table as described in the previous paragraph.

Notice that to fill a cell of the table using the recursive definition above, we need to know the value of three other cells — the cell to the north, the cell to the west, and the cell to the northwest. Therefore, if we start to fill the table row by row, from top to bottom and left to right, we will have already filled in these three cells before reaching the current cell.

The amount of time to fill in each cell is constant, so the total time to fill out the table is equal to the number of cells, or $O(nm)$. The space requirement is the same. When the algorithm is finished, the best alignment score is recorded in $S[n,m]$.

a)

S		a	g	g	c	t	g	a
	0	-1	-2	-3	-4	-5	-6	-7
a	-1	2	1	0	-1	-2	-3	-4
g	-2	1	4	3	2	1	0	-1
c	-3	0	3	3	5	4	3	2
t	-4	-1	2	2	4	7	6	5
t	-5	-2	1	1	3	6	6	5
g	-6	-3	0	3	2	5	8	7

b)

S		g	g	g	c	t	g	g	c	g	a
	0	0	0	0	0	0	0	0	0	0	0
a	0	0	0	0	0	0	0	0	0	0	2
g	0	2	2	2	1	0	2	2	1	2	1
g	0	2	4	4	3	2	2	4	3	3	2
c	0	1	3	3	6	5	4	3	6	5	4
g	0	2	3	5	5	5	7	6	5	8	7
g	0	2	4	4	4	4	7	6	5	7	7

FIGURE 1.1: a) The score table for strings "agcttg" and "aggctga" with $\alpha = 2, \beta = -1, \gamma = -1$. The optimal alignment is found by starting at the southeast cell in the table and tracing the path back to the northwest cell. For these strings, more than one alignment produces the score, resulting in more than one possible path. (b) The score table for strings "aggcgg" and "gggctggcga" showing a local alignment. The alignment path through the table is shown with arrows. The optimal alignment is found by searching the table for the maximum value and then tracing a path until reaching a cell with score 0.

The table shown in Figure 1.1 aligns our two sample strings using the parameters $\alpha = 2, \beta = -1, \gamma = -1$. As indicated in the southeast corner cell, the best alignment has a score of 7.

We also wish to construct an alignment corresponding to this score, as it provides information about how the two strings are similar and different, or equivalently it illustrates the homology between the two sequences. We can think of the score in each cell as having a corresponding move, indicating which neighboring cell — north, northwest, or west — was used in producing that cell's score. If we trace these moves from $S[n, m]$ to $S[0, 0]$, called *traceback*, we can construct the alignment. Let's consider cell $S[i, j]$.

1. A diagonal move to $S[i, j]$ corresponds to aligning a_i and b_j.
2. A horizontal move to $S[i, j]$ corresponds to inserting a gap in A after a_i.
3. A vertical move to $S[i, j]$ corresponds to inserting a gap in B after b_j.

One possible way to complete the traceback is to store the moves made for each cell in addition to the score. However, this is unnecessary as the possible moves can be deduced from the score table by considering three cases for each cell.

1. If $S[i, j] - \delta(a_i, b_j) = S[i - 1, j - 1]$, a diagonal move could have been used to reach $S[i, j]$.
2. If $S[i, j] - \gamma = S[i, j - 1]$, a horizontal move could have been used to reach $S[i, j]$.
3. If $S[i, j] - \gamma = S[i - 1, j]$, a vertical move could have been used to reach $S[i, j]$.

Multiple move possibilities imply that there are multiple alignments that produce the optimal score. Figure 1.1 shows that the example strings have more than one alignment that produce a score of 7.

1.4   Semiglobal and Local Alignment

Our dynamic programming solution to the sequence alignment problem resulted in aligning *all* of A with *all* of B. This is called a global alignment, and was first applied in computational biology by Needleman and Wunsch [34]. However, in some cases, a global alignment is not that interesting. Consider the two strings 'agctgctatgataccgacgat' and 'atcata'. An optimal global alignment matches each character perfectly:

```
agctgctatgataccgacgat
a--t-c-at-a----------
```

But a more interesting alignment produces a mismatch and therefore a lower global score, but is much more biologically meaningful. Variations on the global alignment algorithm address this intuition, and were first presented in the context of biological sequences by Smith and Waterman [39].

```
agctgctatgataccgacgat
-------atcata--------
```

1.4.1   Semiglobal Alignment

The first variation is called a *semiglobal*, or *end gaps free* alignment. In this type of alignment, all gaps inserted before or after the string do not affect the score of the alignment. In other words, we are allowed to ignore a prefix of A or a prefix of B but not both. We are also allowed to ignore a suffix of A or a suffix of B but not both. This type of alignment might be interesting if we were assembling a genome from shotgun reads. We would expect high similarity between overlapping ends of two reads, but would not want to incur a penalty for ignoring the non-overlapping ends. Genome assembly is covered more thoroughly in Part III.

In the following discussion, we will use the term *exhausted* to describe the way in which the algorithm uses up characters from A and B. After calculating $S[i,j]$, the algorithm is said to have exhausted the first i characters from A and the first j characters from B.

The semiglobal alignment is achieved through two small modifications to the global alignment algorithm. The first modification addresses inserting gaps at the beginning of a string, or ignoring either a prefix of A or a prefix of B. In a global alignment, we started with an alignment score of 0 only when we had exhausted no characters from both A and B. This corresponded to initializing $S[0,0]$ to 0. Now we wish to be able to start with a score of 0 after ignoring either a prefix of A or a prefix of B. This condition holds as long as we have not exhausted any characters from either A or B, which corresponds to the first row or first column of the table. Therefore we can achieve the result by initializing the first row and column of the table to 0.

The second modification addresses inserting gaps at the end of a string, or ignoring either a suffix of A or a suffix of B. Because we can only ignore a suffix of either A or B, the alignment must exhaust all the characters of either A or B. In terms of the table, this is the case in the last row or column. In a global alignment, we were required to exhaust the characters from both A and B, so the score appeared in the southeast corner of the table. In semiglobal alignment, the score is the maximum over the last row and column of the table.

The traceback of the alignment path through the table proceeds as in global alignment, however the traceback starts at the found maximum and ends at any cell in the first row or column.

1.4.2 Local Alignment

The second variation to global alignment allows even more flexibility than semiglobal alignment. In a *local* alignment, we wish to choose some substring A' of A and some substring B' of B such that A' aligned with B' produces the maximum score. In other words, while for semiglobal alignment we could ignore a prefix of either A or B and a suffix of either A or B, for local alignment we can ignore a prefix and suffix of both A and B.

Possible uses of local alignment include identifying a conserved exon in two genomic sequences and identification of a conserved regulatory region upstream of two genes. We are highly interested in the similar region shared between the two sequences, but are indifferent to remainder of the sequences. In this case, a local alignment would allow us to ignore the parts of the sequence that do not align well, while focusing on the region with the best local similarity.

We create a local alignment by extending the ideas used in semiglobal alignment. Instead of only starting our score at zero in the first row or column (allowing A or B to ignore prefix), we now have the possibility of starting our alignment score at zero in any cell in the table, allowing both A and B to ignore a prefix. This is done by modifying the equation presented in section 1.3.

$$S[i,j] = \max \begin{cases} S[i-1, j-1] + \delta(a_i, b_j) \\ S[i, j-1] + \gamma \\ S[i-1, j] + \gamma \\ 0 \end{cases}$$

We do not allow any cell in the table to take on a negative value. Setting the score of $S[i,j]$ to zero when it would have been negative corresponds to ignoring the prefixes A_i and B_j.

We can deal with ignoring suffixes as an extension of semiglobal alignment as well. Instead of looking for the maximum value over the last row and column, which restricts us to ignoring a suffix of either A or B, we search for the maximum value over the entire table, equivalent to ignoring a suffix of both A and B.

To do a traceback of the local alignment, start at the cell containing the maximum value and traceback until reaching a cell with value 0. An example local alignment with traceback is shown in Figure 1.1. The differences between global, semiglobal, and local alignments are summarized in Table 1.1.

	Global	Semiglobal	Local
Ignore Suffix	no	A or B	A and B
Ignore Prefix	no	A or B	A and B
Reset to Zero	$S[0,0]$	$S[i,0]$, $S[0,j]$	$S[i,j]$
Maximum In	$S[n,m]$	$S[i,m]$, $S[n,j]$	$S[i,j]$

TABLE 1.1   Differences between global, semiglobal, and local alignments

1.5   Space Saving Techniques

We have finished introducing the concept of string alignment. With the basics covered, the rest of the chapter will cover two classes of modifications on this initial concept. The first class are algorithmic improvements, modifications that improve the runtime or space complexity of the algorithms. Space saving techniques, the k-band formulation, and subquadractic alignments fall into this class. We will also cover qualitative modifications such as substitution matrices, normalized alignment, and different gap penalty functions. These modifications often introduce complexities that result in the algorithms taking more time or space, but have the benefit of producing more biologically valid results.

The first modification we discuss is quite important. As mentioned above, the dynamic programming solution described takes $O(nm)$ time and $O(nm)$ space, where n and m are the sizes of the strings. While the $O(nm)$ runtime is quite fast on any reasonably modern computer for string sizes up to a few hundred thousand characters, the space will become a factor before that. Two strings of size 20,000 will require around 1.6 GB of RAM for the dynamic programming table if each cell is a 4 byte integer. For this reason we are interested in reducing the space required to run the algorithm.

We will discuss a technique introduced by Hirschberg [21] that reduces the space requirement from $O(nm)$ to $O(n + m)$ while maintaining the runtime. Obviously, achieving a reduction in space required to hold the table to 160 KB from 1.6 GB is a useful improvement. The space requirement is the theoretical minimum because at the very least the input and output require $O(n + m)$ space to hold the strings themselves. We will first discuss the technique within the context of global alignment.

1.5.1 Preliminaries

If we are looking for the alignment score without producing the alignment, it is easy to reduce the space requirement to $O(n + m)$. Consider filling the table row by row, from top to bottom. To fill any row, we need access to values from the previous and current rows only. It is easy to envision an algorithm which uses two arrays of size m, corresponding to the previous and current rows. When the current row is complete, the two arrays swap roles and the algorithm continues.

If we store only two rows of the table, we cannot proceed with path traceback, because we have lost most of the information needed for this step. While the entire path cannot be found, there is a way to discover a small part of the path; there is enough information in the last two rows to construct the last bit of the path. This observation will serve as the basis for the first naive space optimal algorithm.

As shown in Figure 1.2, the path of the alignment can be described as a list of n intervals, one for each row of the table. Each interval can be defined by its endpoints. Notice that the endpoints of each subsequent interval either overlap or touch through the diagonal. This is because alignment path must move on either a diagonal or a vertical path from row i to row $i - 1$. We will call the left endpoint of row i and the right endpoint of row $i - 1$ the path intersection for row i. We can define the alignment path as $n - 1$ intersections, if we consider that the northwest corner and southwest corner of the matrix will always be endpoints in the interval list.

Each intersection can be discovered using data from only two rows. To construct the entire path, we will build the interval list intersection by intersection. A naive approach finds one intersection per iteration, starting at the bottom of the table and working up to the top. For each iteration, the algorithm calculates the last two rows of the submatrix for A_i and B_j. It uses this information to discover the intersection for row i. This algorithm uses $O(n + m)$ space but runs in $O(n^2 m)$ time.

1.5.2 Using Hirschberg's Recursion

The naive space saving algorithm has a harsh runtime penalty. Hirschberg introduced a divide and conquer approach that was first applied to the biological sequence alignment

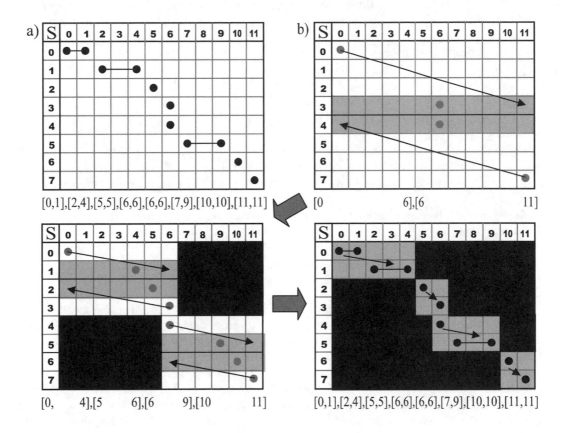

FIGURE 1.2: (a) The alignment path through the table S can be represented as a list of intervals, one per row. (b) Hirschberg's recursion allows us to construct the list in $O(nm)$ time using only $O(n+m)$ space. The black arrows represent the direction the DP algorithm is run on each sub matrix. The gray boxes represent the cells in memory at the end of each step. Partially known intervals are deduced from these cells, as shown by gray dots. In the last stage of the recursion, alignment is run forward in all sub matrices to complete the alignment path.

problem by Myers and Miller [33]. Instead of finding the intersections from bottom to top, we find the intersection in the center of the table first. This will allow us to eliminate more of the table for the next iteration. Refer to Figure 1.2 during the discussion.

To find the intersection for the center row, divide the table in half, with the centerline of the table $t = \lfloor \frac{n}{2} \rfloor + 1$. The top half of the table is $S[0...t-1, 0...m]$ and the bottom half of the table is $S[t...n, 0...m]$. On the top half of the table, run the space-saving algorithm to find the scores along the bottom row of the top half of the table, row $t-1$.

On the bottom half of the table, run the algorithm *backwards*. In other words, initialize the bottom row and right column and run the algorithm from right to left and bottom to top. This is the same as reversing the strings and running the algorithm forwards. The table cell formula is defined as:

$$S[i,j] = \max \begin{cases} S[i+1,j+1] + \delta(a_i, b_j) \\ S[i,j+1] + \gamma \\ S[i+1,j] + \gamma \end{cases}$$

When we finish running the algorithm backwards in the bottom half of the table, we will have scores for the top row of the bottom half of the table, which has index t. Now we will find the move between rows $t-1$ and t (the intersection for t) that produces the maximum score.

More formally, the intersection for t is defined by indexes i in $t-1$ and j in t, $i \le j \le i+1$, that maximize the function:

$$\max_i \begin{cases} S[t-1,i] + S[t,i] + \gamma & \text{j=i} \\ S[t-1,i] + S[t,i+1] + \delta(a_t, b_{i+1}) & \text{j=i+1} \end{cases}$$

Next divide the table into four quadrants. The northeast and southwest quadrants can be ignored, as the optimal path does not travel through them. We must recursively run the algorithm on the northwest and southeast quadrants, defined as $S[0...t-1, 0...i]$ and $S[t...n, j...m]$. The recursion will continue until the number of rows is 1 or 2, at which point the DP algorithm will be run forwards and the optimal path completed.

The problem approximately halves each iteration, because if one splits a rectangle into four quadrants and selects two, the area of the two selected quadrants is half of the original rectangle. The details are left out of this discussion but are easily solved. Because the sum $\sum_{i=0}^{\infty} \frac{1}{2^i} = 2$, the total runtime remains $O(mn)$. The algorithm remembers only $n+1$ cells for each half of the table during each iteration, and as a result the space requirement has been reduced to $O(m+n)$.

The ideas in this algorithm can be extended to both semiglobal and local alignments, as shown by Huang *et al.* [22]. We will consider the case of local alignments, as semiglobal alignments can be handled similarly. Consider the case in which there is exactly one maximum scoring path through the table. The idea can be extended to work with multiple such paths, which we will not consider here.

To solve the problem, we will first find each endpoint of an optimal alignment path, and then run a global alignment on the induced subtable. First run the algorithm forward using the local alignment recursion. As it proceeds, keep track the cell that contains the maximum score. This is the southeast corner of the subtable. Next run the algorithm backwards while keeping track of the cell that gives the maximum score from this direction (the maximum value will be the same). This cell is the northwest corner of the subtable. Run the linear space global alignment algorithm on the subtable defined by these two cells to produce the optimal local alignment.

1.6 Banded Alignment

Imagine that we are studying two orthologous DNA sequences, that is two DNA sequences that are thought to have evolved from the same ancestral sequence. Our two genes code for the same function and the species are evolutionarily close. Therefore, the two sequences are highly similar and are of similar length. Because of this the alignment path between the two sequences will remain close to the main diagonal. A banded alignment makes use of this observation to achieve faster runtime. The idea of a banded alignment was first proposed by Fickett [16].

Consider the subset of inputs in which A and B are highly similar and of the same length n. The k-band algorithms runs in time proportional to the difference between A and B. If A and B are similar enough, then the algorithm's runtime is $O(nk)$ for some small constant k. In the worst case, the runtime is still $O(n^2)$ with an additional constant multiplier of approximately 2.

The k-band algorithm ignores the part of the array distant from the main diagonal during its calculation. Let the value k denote a region of the table called the *k-band*, such that the following is true for the table.

$$S[i,j] \in k\text{-}band \Leftrightarrow |i - j| \leq k$$

For the purpose of the algorithm, all cells outside of the k-band are considered to have a score of $-\infty$, which will cause them to be ignored in the maximum calculation. The algorithm runs as normal, but will only consider those cells within the k-band. Figure 1.3 shows the k-band initialization for $k = 3$. In practice the cells marked $-\infty$ are not actually initialized and do not exist in memory, for this would defeat the purpose of the algorithm. They are shown for clarity.

The end of the algorithm's run will result in an alignment with some score s_k that represents the best alignment under the restriction that the alignment path does not travel

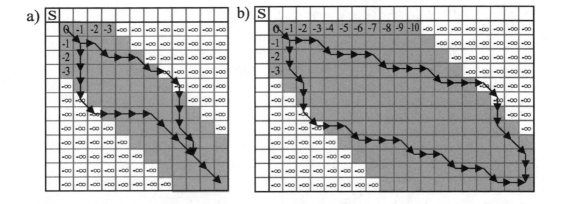

FIGURE 1.3: (a) The k-band initialization for $k = 3$. The k-band is shown in gray. The cells that conceptually take on $-\infty$ do not actually exist in memory. Two hypothetical best paths that travel outside of the k-band are shown. To achieve the best score, all diagonals must represent matches. Hence, the score of these two paths is $2(k+1)\gamma + (n-(k+1))\alpha$. (b) The k-band initialization for global alignment when $k = 3$ and $m > n$. The cost of the hypothetical best path traveling outside of the k-band is $(2(k+1) + m - n)\gamma + (n - (k+1))\alpha$.

outside of the k-band. To show that this alignment is optimal, we consider possible alignments that have paths that travel outside of the k-band. In other words, for some cell on the alignment path, we have $|i - j| > k$. The highest scoring such alignment would have exactly $k + 1$ characters from A aligned with gaps, $k + 1$ characters from B aligned with gaps, and all other $n - (k + 1)$ characters as matches. The score for this alignment would be

$$best_{k+1} = 2(k+1)\gamma + (n - (k+1))\alpha$$

If $s_k >= best_{k+1}$, then the alignment s_k is known to be an optimal alignment, because it beats the best score of any possible alignment that travels outside of the k-band region.

However, if $s_k < best$, then we cannot be sure that s_k is optimal. To solve this problem and maintain our worst case runtime, we double k and rerun the algorithm. In the worst case we keep doubling k until $k \geq n$, at which point all elements in the table are in the k-band. The runtime in the worst case is the sum of all iterations, which is still $O(n^2)$, because the number of elements considered increases exponentially until the number of entries considered constitutes the entire table.

While repeatedly doubling k produces an algorithm that remains asymptotically optimal, it may increase the runtime unnecessarily for some applications. In practice we may wish to only find an alignment if the similarity between two strings is high. If the score of the best alignment is below some minimum threshold, T, then the strings are considered dissimilar and we are no longer interested in finding the alignment. If this is the case, we can choose k such that $best_{k+1} \leq T$ and never have to run the k-band algorithm more than one iteration; if $s_k < best_{k+1}$, then $s_k < T$ and we can report no good alignment. Solving k in terms of T, we have

$$k \geq \frac{T - (n-1)\alpha}{2\gamma - \alpha}$$

The k-band as described for two strings of exactly the same length is very limiting, so we will briefly look at the implications of $|A| = m \neq |B| = n$. For ease of discussion, assume that $m > n$. Now the k-band is redefined.

$$S[i, j] \in k\text{-}band \Leftrightarrow n - m - k \leq i - j \leq k$$

This can be thought of as inserting a parallelogram of width $m - n$ at the center of the k-band defined for $n = m$, as shown in Figure 1.3.

In addition the score $best_{k+1}$ — used as the termination decision and in calculating the k based on the threshold parameter T — is calculated using a more general form of the equation presented for $n = m$.

$$best_{k+1} = (2(k+1) + m - n)\gamma + (n - (k+1))\alpha$$

1.7   Other Gap Penalty Functions

In the algorithms presented thus far, the penalty for aligning a character with a gap has been γ. However, in most biological applications, it does not make sense to penalize gaps in this manner. For example, a single insertion (or deletion) in a DNA sequence typically results in inserting (or deleting, respectively) a string of nucleotides, making multiple consecutive gap characters much more likely than isolated gap characters. The penalty function chosen should reflect this reality.

For this reason, researchers use gap penalties that do not increase linearly with the number of gap characters. In particular, researchers have studied general gap penalty functions [30], affine gap penalty functions [14, 17], and concave and convex gap penalty functions [31, 13, 15].

In terms of computational cost, alignments based on general gap functions cost the most to compute, and in practice they are hardly ever used. We shall discuss them here as motivation for choosing from among simpler functions. Alignments based on convex and concave gap penalty functions can be calculated in $O(nm \log(n+m))$ time. However, affine gap penalty functions offer enough flexibility and can be calculated almost as quickly as linear gap penalty functions. Therefore, these gap penalty functions are almost always used in practice.

1.7.1 General Gap Penalties

Envision a general gap penalty function $\omega(i)$ which is the penalty for inserting a gap of length i. If we allow for this arbitrary gap penalty function, then the runtime increases considerably. An $O(n^2m + nm^2)$ algorithm can be constructed by modifying the recursive definition presented for a linear gap penalty function.

Now, instead of considering a constant number of cells when calculating $S[i,j]$, we must consider $O(i+j)$ cells. This is because, when aligning two suffixes A_i and B_j, one must consider the possibility of aligning a_i with any character b_k $1 < k \leq j$. One must also consider aligning all of A_i with the empty string. Extending this idea to both strings we end up with four possibilities.

1. Align a_i with b_k, $1 < k \leq j$ and A_{i-1} with B_{k-1}.
2. Align b_j with a_k, $1 < k \leq i$ and A_{k-1} with B_{j-1}.
3. Align A_i with a gap
4. Align B_i with a gap

The equation for $S[i,j]$ becomes

$$S[i,j] = \max \begin{cases} \max_{k=1}^{j} S[i-1, k-1] + \delta(a_i, b_k) + \omega(j-k) \\ \max_{k=1}^{i} S[k-1, j-1] + \delta(a_k, b_j) + \omega(i-k) \\ \omega(i) \\ \omega(j) \end{cases}$$

Finally, for global alignment, we initialize the first row and column of the table based on the gap penalty function.

$$S[i,0] = \omega(i)$$

$$S[0,j] = \omega(j)$$

As $O(n+m)$ possibilities are considered for each cell, and there are $O(nm)$ cells, the total runtime of the algorithm is $O(n^2m + nm^2)$.

1.7.2 Affine Gap Penalties

The runtime penalty to allow the flexibility of a general gap penalty function is harsh. It would be nice to find a function with a more complex shape that incurs less of a runtime cost. Using *affine gap penalty* functions allows us to maintain an $O(nm)$ runtime.

An affine gap penalty function has two values. A gap opening penalty g is the cost of starting a new gap. A gap extension penalty h is the cost of extending a gap. The first gap character in an affine gap has a score of $g + h$. Each subsequent gap character in the affine gap has a score of h. Affine gap penalties are widely preferred by biologists because consecutive gap characters likely correspond to a single insertion/deletion event, while an equal number of scattered gaps correspond to as many insertion/deletion events, which is much less probable. For this reason h is often much smaller than g.

To solve the problem in quadratic time, we augment table S with two additional tables, G_A and G_B. $G_A[i, j]$ is the best score of aligning A_i with B_j under the restriction that a_i is aligned with a gap. $G_B[i, j]$ is the best score of aligning A_i and B_j under the restriction that b_j is aligned with a gap. As before, $S[i, j]$ holds the optimal score of aligning A_i and B_j under no restrictions.

There are three possible ways in which the maximum score can arise, with two of the cases consisting of two parts each.

1. a_i is aligned with b_j and A_i is aligned with B_j.

2. a_i is aligned with a gap. In this case, we must consider two sub-cases.

    (a) A_{i-1} is aligned with B_j such that a_{i-1} is not aligned with a gap, and we start a gap.

    (b) A_{i-1} is aligned with B_j such that a_{i-1} is aligned with a gap, and we extend the gap.

3. b_i is aligned with a gap. In this case we must consider two sub-cases.

    (a) A_i is aligned with B_{j-1} such that b_{j-1} is not aligned with a gap, and we start a gap.

    (b) A_i is aligned with B_{j-1} such that b_{j-1} is aligned with a gap, and we extend the gap.

These possibilities are captured and scored in the following equations:

$$S[i, j] = \max \begin{cases} S[i-1, j-1] + \delta(a_i, b_j) \\ G_A[i, j] \\ G_B[i, j] \end{cases}$$

$$G_A[i, j] = \max \begin{cases} S[i-1, j] + g + h \\ G_A[i-1, j] + h \end{cases}$$

$$G_B[i, j] = \max \begin{cases} S[i, j-1] + g + h \\ G_B[i, j-1] + h \end{cases}$$

The number of cells considered in each cell calculation is constant for all tables. Because there are $O(nm)$ number of cells per table, the total runtime of the algorithm is $O(nm)$.

Consider global alignment. We initialize the first row and column of each table such that $S[i, 0] = g + hi$, $S[0, j] = g + hj$, and $S[0, 0] = 0$. $S[n, m]$ contains the optimal alignment score after the algorithm finishes. We can construct the alignment by tracing back the path, starting at position $S[n, m]$. With an extension of the ideas presented for global alignment with linear gap penalties, the traceback can be accomplished without storing any pointers. The details are omitted.

Semiglobal alignment can be handled in a straightforward way. Initialize the first row and column of each table to 0. The maximum value in the last row and column of S is the optimal alignment score. Starting at this position, trace the alignment path back through the table until reaching the first row or column.

Assume that matches are scored positive, while mismatches and gaps are scored negative. Now local alignment with affine gaps is easy, as seen by the observation that every local alignment starts with some character a_k aligned with some b_l and ends with some $a_{k'}$ aligned with some $b_{l'}$. The proof is by contradiction. Assume that an optimal local alignment L starts with a character aligned with a gap. Then there exists a new alignment with higher score constructed removing the first character from L. Therefore, L is not optimal, a contradiction. The same reasoning holds for a gap character at the end of an alignment. As a result of this observation, we can handle local alignment by modifying the equation for $S[i, j]$

$$S[i,j] = \max \begin{cases} S[i-1, j-1] + \delta(a_i, b_j) \\ G_A[i,j] \\ G_B[i,j] \\ 0 \end{cases}$$

The alignment is found by searching for the maximum score in S, corresponding to aligning $a_{k'}$ and $b_{l'}$ by the observation above. The traceback continues until reaching some 0 in S, corresponding to the initial alignment of a_k and b_l.

1.8   Substitution Matrices

In this section we will consider the specific problem of aligning two amino acid sequences and the additional considerations needed in order to produce a biologically meaningful result.

Proteins are sequences of amino acids that fold into an energetically stable shape. The surface of a protein interacts with other proteins and molecules through its shape and chemical properties. It can be the case that proteins with rather different sequences can fold into molecules with similar shapes and properties — and consequently perform the same function. Moreover, a mutation occurring within the DNA sequence corresponding to the protein can result in an amino acid substitution, insertion, or deletion, having varying affects on the protein by affecting the protein's properties.

Some amino acid substitutions might be more acceptable than others. For example, six of the twenty amino acids are hydrophobic, which prefer to face the interior to avoid interacting with water. A substitution within this class of amino acids is more acceptable than a substitution with an amino outside of the class. For this reason, matching a hydrophobic amino acid with another hydrophobic amino acid should be scored higher than matching a hydrophobic amino acid with a hydrophilic one (an amino acid attracted to water).

In section 1.2, we presented a delta function for scoring the alignment of two characters:

$$\delta(x,y) = \begin{cases} \alpha & x = y \\ \beta & x \neq y \end{cases}$$

Now we wish to use a more complex function $\delta : D \times D \to \Re$, where D is the set of 20 amino acids. In practice this function is stored in a 20×20 matrix for use during the execution of an alignment algorithm. Two classes of such matrices called PAM and BLOSUM matrices are used, and we shall look at the origins of both.

1.8.1   PAM Matrices

A biologically valid scoring function δ arises from a complex process that is hard to model analytically. For this reason experimental data has been used to discover appropriate values. Dayhoff *et al.* [10, 11] described an evolutionary model used to interpret experimental data and derive the scoring function δ.

As an organism evolves, mutations will cause changes in the proteins. Those changes that are allowed to remain by an organism are said to be accepted or retained mutations. When comparing two proteins from divergent organisms, one would expect to observe some of these mutations as differences in the amino acid chains for the proteins of the two organisms.

Dayhoff *et al.* built a phylogenetic tree of closely related proteins in an attempt to discover accepted mutations. They accumulated the number of times amino acid i was substituted by amino acid j as one traveled up the phylogenetic tree. This data was stored in a 20×20 matrix, symmetric along the main diagonal, as a transition from i to j was considered a substitution from both i to j and j to i.

Using this data, they calculated the conditional probability of seeing an amino acid j given an amino acid i for all amino acid pairs (i, j). From this the PAM matrix was born. The PAM matrix stands for *Point Accepted Mutation*. From the experimental data, they calculated the probability matrix M. $M[i, j]$ contains the probability that amino acid i will be substituted for amino acid j in one evolutionary unit of time. The PAM evolutionary unit of time is the amount of time it takes for one amino acid in every hundred to undergo an accepted mutation.

Given the M matrix, the matrix M^k is the probability of substituting amino acids in k units of time. The PAMk scoring matrices are derived from the M^k probability matrices using the following equation, where p_j is the probability of a random occurrence of amino acid j.

$$\mathrm{PAM}k[i, j] = 10\,log\frac{M^k[i, j]}{p_j}$$

1.8.2   BLOSUM Matrices

When considering protein sequences that are highly diverged, PAM matrices are not well suited as they were constructed based on closely related proteins with less than 15% difference. The BLOSUM matrices, introduced by Heinikoff and Heinikoff [20] are constructed using an approach that allows comparison of more highly diverged proteins. They are constructed using conserved regions of proteins. These regions, called blocks, give rise to the name BLOSUM, which stands for *BLock SUbstitution Matrix*.

The blocks are found by aligning multiple proteins in protein families (see Chapter 3 for a discussion of multiple alignments). As mentioned before, blocks are regions with a high degree of similarity. Within these regions, it could be the case that certain proteins are nearly identical. For this reason, in calculating the BLOSUMX matrix, multiple proteins that are X percent identical are weighted as one protein. Varying X gives rise to different scoring matrices, labeled BLOSUM30, BLOSUM50, BLOSOM62 and so forth.

The substitution frequencies are calculated based on the enumeration of all pairs of amino acids appearing in each column of the multiple alignment blocks. A value p_{ij} is calculated for amino acids i and j based on this multiple alignment. The details of this score are not easily described without a greater understanding of multiple alignment. If p_i is the probability of seeing amino acid i at random and p_j the probability of seeing amino acid j at random, then the BLOSUM matrix is calculated using the following equation:

	A	R	N	D	C	Q	E	G	H	I	L	K	M	F	P	S	T	W	Y	V
A	4	-1	-2	-2	0	-1	-1	0	-2	-1	-1	-1	-1	-2	-1	1	0	-3	-2	0
R	-1	5	0	-2	-3	1	0	-2	0	-3	-2	2	-1	-3	-2	-1	-1	-3	-2	-3
N	-2	0	6	1	-3	0	0	0	1	-3	-3	0	-2	-3	-2	1	0	-4	-2	-3
D	-2	-2	1	6	-3	0	2	-1	-1	-3	-4	-1	-3	-3	-1	0	-1	-4	-3	-3
C	0	-3	-3	-3	9	-3	-4	-3	-3	-1	-1	-3	-1	-2	-3	-1	-1	-2	-2	-1
Q	-1	1	0	0	-3	5	2	-2	0	-3	-2	1	0	-3	-1	0	-1	-2	-1	-2
E	-1	0	0	2	-4	2	5	-2	0	-3	-3	1	-2	-3	-1	0	-1	-3	-2	-2
G	0	-2	0	-1	-3	-2	-2	6	-2	-4	-4	-2	-3	-3	-2	0	-2	-2	-3	-3
H	-2	0	1	-1	-3	0	0	-2	8	-3	-3	-1	-2	-1	-2	-1	-2	-2	2	-3
I	-1	-3	-3	-3	-1	-3	-3	-4	-3	4	2	-3	1	0	-3	-2	-1	-3	-1	3
L	-1	-2	-3	-4	-1	-2	-3	-4	-3	2	4	-2	2	0	-3	-2	-1	-2	-1	1
K	-1	2	0	-1	-3	1	1	-2	-1	-3	-2	5	-1	-3	-1	0	-1	-3	-2	-2
M	-1	-1	-2	-3	-1	0	-2	-3	-2	1	2	-1	5	0	-2	-1	-1	-1	-1	1
F	-2	-3	-3	-3	-2	-3	-3	-3	-1	0	0	-3	0	6	-4	-2	-2	1	3	-1
P	-1	-2	-2	-1	-3	-1	-1	-2	-2	-3	-3	-1	-2	-4	7	-1	-1	-4	-3	-2
S	1	-1	1	0	-1	0	0	0	-1	-2	-2	0	-1	-2	-1	4	1	-3	-2	-2
T	0	-1	0	-1	-1	-1	-1	-2	-2	-1	-1	-1	-1	-2	-1	1	5	-2	-2	0
W	-3	-3	-4	-4	-2	-2	-3	-2	-2	-3	-2	-3	-1	1	-4	-3	-2	11	2	-3
Y	-2	-2	-2	-3	-2	-1	-2	-3	2	-1	-1	-2	-1	3	-3	-2	-2	2	7	-1
V	0	-3	-3	-3	-1	-2	-2	-3	-3	3	1	-2	1	-1	-2	-2	0	-3	-1	4

TABLE 1.2   The BLOSUM62 matrix.

$$\mathrm{BLOSUM}[i, j] = \frac{1}{\lambda} \log \frac{p_{ij}}{p_i p_j}$$

λ is a scaling factor used to generate scores that can be converted into integers. The BLOSUM62 matrix, the default matrix used for BLAST [3], is shown as Table 1.2.

1.9   Local Alignment Database Search

The dynamic programming algorithm finds the highest scoring alignment between two strings. However, performing a full alignment is often prohibitively slow. For example, if we were to compile a database of protein sequences, we could represent the database as a string D constructed by concatenating each string in the database. If we then attempted to find the optimal local alignment between some query Q and D, the runtime would likely be prohibitive, as the total cost would be the total length of all sequences in the database times the length of the query sequence.

Various approximations for local alignment have been proposed to speed up this basic problem of database search. The first of these was Fasta [25, 26], which we will not discuss here. In 1990, Altschul *et al.* presented the basic local alignment search tool [3] as a method to search protein databases quickly. We will present the second version of their algorithm, published in [27].

The basic idea behind BLAST is that good local alignments contain good ungapped alignments. An ungapped alignment is an alignment not allowing gaps. We wish to find these good ungapped alignments quickly and then extend them to find good local alignments.

Consider a window of size ω. As we move this window along the string, we can see ω characters of the string at a time. The number of unique strings, called ω-mers, of length ω is Σ^{ω}. If the strings are protein sequences, then the alphabet size is 20 and the number of strings we can make of length 3 is $20^3 = 8000$. We will create an index into the database showing all the locations of each ω-mer.

Additionally, we can calculate the score of aligning any ω-mer with any other ω-mer without gaps. Now, for a given word a, there is a set of words S for which $s \in S$ if and only if the score of aligning a with s is above some threshold T.

Now given a query string Q and a database D, we wish to find good local alignments

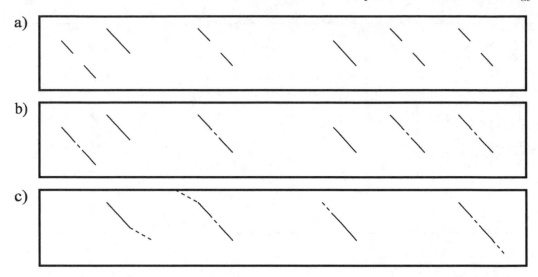

FIGURE 1.4: The BLAST program runs in phases. (a) In one phase pairs of hits are found that lie on the same diagonal of a conceptual dynamic programming table. (b) The region between these hits is aligned without gaps. (c) Finally, the alignment is completed by extending the ends of the alignment for those seed alignments scoring above some threshold.

between Q and D. The BLAST algorithm performs the following steps (see Figure 1.4).

1. The first step is to find hits between Q and the database. Each hit corresponds to some word a in Q matching to some word d in the database such that the score of aligning a and d is above the threshold.
2. From the hits discovered in step one, find those pairs of hits (h_i, h_j) that can be part of the same gapless alignment. That is, they would lie on one diagonal on the dynamic programming table.
3. Perform a gapless extension between these two hits by aligning each character in the query string with the corresponding character in the database. This will produce some alignment score.
4. For those gapless alignments with a score deemed significant, perform a gapped alignment extension from each end of the gapless alignment, such that the total score of the alignment does not drop below some threshold.

BLAST is a popular program for protein database searches, but recently researchers have revisited the problem. An algorithm called DASH (for Diagonal Aggregating Search Heuristic) reports runtimes ten times faster than BLAST with similar sensitivity [18]. Their heuristic extends the idea of BLAST. First they find all words occurring in the same diagonal region. Next, global alignments connect these gapless regions. Finally, they extend the end of the alignments using a global alignment on some part of the dynamic programming table. In addition to the diagonal region heuristic, a key technique they use to improve runtime is to mask those words that occur with high frequency in each sector of the database. This reduces the number of initial database hits.

1.10   Similarity and Distance Measures

This section details some alternate characterizations of the problem found in the bioinformatics field. We have described the solution to the alignment problem as finding the maximum score of an alignment between two strings. This score was the summation of the pairwise scores of each pair of characters involved in the alignment.

$$score(L) = \sum_{i=1}^{l} score(a_i, b_i)$$

The score is considered a measure of the similarity of the two strings, and it easily allows for the extensions into semiglobal and local alignment. However, one interesting result of these extensions is that the scoring system can fail to follow the triangle inequality.

$$score(A, B) + score(B, C) \leq score(A, C)$$

An alternate way of looking at the problem is to define a sequence of elementary operations on a string — insertions, deletions, and substitutions. One can transform A into B through a sequence of these operations $T = \langle t_1, t_2, ...t_n \rangle$. We now assign a cost function $cost(t)$ to the set of operations such that three conditions hold for any strings A, B, C.

1. $dist(A, A) = 0$
2. $dist(A, B) = dist(B, A)$ (symmetry)
3. $dist(A, B) + dist(B, C) \leq dist(A, C)$ (triangle inequality)

Where the $dist(A, B)$, called the *edit distance*, is the minimum sum of the cost of a sequence of operations that transforms A into B.

$$dist = \min_T \sum_i cost(t_i)$$

These requirements impose restrictions on our cost function. For symmetry, we require

1. $cost(insertion) = cost(deletion)$
2. $cost(substitute(a, b)) = cost(substitute(b, a))$

To satisfy the triangle inequality, we require

1. $cost(insertion), cost(deletion), cost(substitute(a, b)) \geq 0$

Distance measures have their uses, as the triangle inequality allows for certain reasoning and analysis that would otherwise be impossible. For example, performance guarantee proofs on approximation algorithms for the computationally expensive problem of multiple alignment are only valid using distance metrics. Multiple alignments are covered in Chapter 3.

An important result in the study of distance and similarity is that for any distance metric used in the alignment problem, one can construct a corresponding similarity metric. That is, finding the minimum distance for some cost function will simultaneously find the maximum score for some similarity function. Smith and Waterman [40] presented a theorem and proof of this assertion, and the idea is fully developed in [38]. The key observation is that for each alignment between two strings A and B, containing a matches, b mismatches, and g gaps, the following equation, known as the *alignment invariant*, is true:

$$n + m = 2(a + b) + g$$

In practice, we construct some scoring scheme based on the cost scheme using an arbitrary constant P.

$$\delta(a, b) = P - cost(sub(a, b))$$

$$\gamma = \frac{P}{2} - cost(insertion)$$

The maximum score and the distance under these valuation schemes are related by the equation:

$$score(A, B) + dist(A, B) = \frac{P(m + n)}{2}$$

Therefore, distances can be quickly calculated from similarity scores.

1.11 Normalized Local Alignment

Assume that we have two DNA sequences that we wish to compare using an alignment algorithm. Importantly, we wish to find regions of high similarity. Local alignment is somewhat suitable for this task, as it will return an alignment between substrings A' and B' that gives the highest scoring alignment.

However, there is a basic problem in the presentation of local alignment, in that the lengths of A' and B' are not taken into account when calculating the score. Therefore an alignment of length 100 and score 51 is considered better than an alignment of length 50 and score 50, although the second alignment has a much higher average score per base. Alexander and Solovyev [2] argued that the local alignment algorithm did not always find the most biologically relevant alignment because it did not consider alignment length.

One can think of post processing local alignments, but the highest scoring local alignment might mask some lower scoring alignment with higher normalized score. Instead, one could individually look at each pair of cells in the global alignment score matrix and compute the normalized alignment score.

$$\max_{i,j,k,l} \frac{S[i,j] - S[k,l]}{(i - k - 1) + (j - l - 1)}, 0 \leq i \leq k \leq n, 0 \leq j \leq l \leq m$$

However, the number of such combinations is $\Theta(n^2 m^2)$, which is expensive.

The problem was explored in [41, 35, 4]. Pevzner *et al.* [5] were the first to provide an $O(nm(\log n))$ algorithm to compute the normalized alignment of some minimum length, and we will present their ideas here. For the sake of brevity, we will discuss the algorithm within the context of linear gap penalties, but the ideas extend easily to affine gap penalties, as shown by Pevzner *et al.*

The score of a best local alignment between two substrings A' and B' is $a\alpha + b\beta + g\gamma$, where a, b, and g are the number of matches, mismatches, and gaps. According to the *alignment invariant* first presented in Section 1.10, we have $n + m = 2a + 2b + g$, where n and m are the lengths of A' and B'. Pevzner proposed to measure the length of the alignment as $n + m + L$ where L is some positive constant. Then the length of some alignment with the score $a\alpha + b\beta + g\gamma$ is $2a + 2b + g + L$.

The best local alignment is found as:

$$LA(A, B) = \max_{(A',B')} a\alpha + b\beta + g\gamma$$

The best normalized local alignment is:

$$NLA(A, B) = \max_{(A', B')} \frac{a\alpha + b\beta + g\gamma}{2a + 2b + g + L}$$

The ideas used in solving the normalized local alignment quickly were introduced by Dinkelbach [12], who developed a general scheme for maximization problems which displayed the following properties:

1. The optimization involves a ratio $\frac{g}{h}$, where g and h are functions
2. The domain of g is equal to the domain of h
3. h is always positive

For our normalized local alignment, we have,

$$\max_{domain} \frac{g(a, b, g)}{h(a, b, g)}$$

Without going into details, we will illustrate some of the main ideas underlying this approach. First, we introduce an alignment called a *parametric local alignment* for some parameter λ.

$$PA(A, B, \lambda) = \max_{domain} g(a, b, g) - \lambda h(a, b, g)$$

$$PA(A, B, \lambda) = \max_{(A', B')} a\alpha + b\beta + g\gamma - \lambda(2a + 2b + g + L)$$

Dinkelbach's interesting result is that the following equation holds:

$$\lambda = NLA(A, B) \Leftrightarrow PA(A, B, \lambda) = 0$$

That is, λ is the score of the best normalized local alignment if and only if the parametric local alignment for λ has a score of zero.

Dinkelbach proposed an iterative search method to find the zero of $PA(A, B, \lambda)$ that has no provable run time but runs well in practice. His ideas are used in the following algorithm. First, initialize lambda by finding the local alignment $LA(A, B)$ and selecting $\lambda = \frac{a\alpha + b\beta + g\gamma}{2a + 2b + g + L}$. Next, repeat two steps until lambda stops changing.

1. Find the parametric local alignment $PA(A, B, \lambda)$.
2. Set λ' to $\frac{a\alpha + b\beta + g\gamma}{2a + 2b + g + L}$, and then set λ to λ'.

This method is faster in practice than the provably optimal alternative. However, if one restricts the values α, β, and γ to rational numbers, one can find the proper λ in $O(\log n)$ time using Megiddo's technique [29], the details of which are omitted here but can be found in [4].

The key to completing the algorithm is to effectively find $PA(A, B, \lambda)$. With some manipulation we see that parametric local alignment can be rewritten in terms of local alignment.

$$PA(A, B, \lambda) = \left(\max_{(A', B')} a(\alpha - 2\lambda) + b(\beta - 2\lambda) + g(\gamma - \lambda) \right) - L\lambda$$

To solve the parametric local alignment, we can first solve local alignment with $\alpha' = \alpha - 2\lambda$, $\beta' = \beta - 2\lambda$, and $\gamma' = \gamma - \lambda$, and then subtract a constant to find the score of the parametric local alignment.

Using this method of solving the parametric local alignment takes the same time and space as local alignment, $O(nm)$ time and $O(n+m)$ space. Therefore, if the scoring scheme is restricted to rational numbers then the time required to complete the normalized local alignment is $O(nm \log n)$. In practice, the Dinkelbach search is known to work equally well.

1.12   Asymptotic Improvements

Normalized local alignment is used to produce a more valid biological result. In the final section of this chapter, we explore some interesting techniques that can be used to reduce the asymptotic runtime complexity of the algorithm. For ease of presentation, assume $\Theta(m) = \Theta(n)$. It might seem at first glance that an $O(n^2)$ solution is as fast as the problem can be solved; however, this is not true. Masek and Paterson [28] were the first to introduce an $O\left(\frac{n^2}{\log n}\right)$ solution. However, their solution was limited in that it did not provide an answer to local alignment problem and required that the scoring method consist of rational numbers only.

Crochemore and Landau presented an algorithm [8] that answered these limitations. Their algorithm makes use of the periodic nature of strings to achieve a runtime of $O\left(\frac{hn^2}{\log n}\right)$, where h is the entropy [9] measure of the strings, varying between 0 and 1. Obviously even when the strings are random, with an entropy of 1, the algorithm shows an asymptotic improvement over $O(nm)$, but strings that are highly repetitive gain a larger improvement.

1.12.1   LZ Parsing of Strings

The algorithm uses a version of Lempel-Ziv parsing [24, 43, 44], which compresses a string by exploiting its repeat structure. The basic idea behind LZ compression is that one can divide a string S into a set of blocks. The blocks are formed in a greedy way, from left to right, using the following formulation. Suppose that we have divided the string into blocks up to position j and block i. In the Lempel-Zip parsing scheme, we will define block $i+1$ using a substring of $S[1...j]$ and a character c. More specifically, we look for the maximal substring $M = S[s...e]$ $(s \le e \le j)$ that matches $S[j+1...k]$. The new block is represented by the triple, $\langle s, l, c \rangle$, where s is the starting index of the substring M, l is the length of the substring M, and c is the character $S[k+1]$, $S[k+1] \ne S[e+1]$.

It has been shown that the number of such blocks for a string of length n is $O\left(\frac{hn}{\log n}\right)$ [23], where again h is the entropy of the string. An example is given in Figure 1.5.

This is the most general version of LZ parsing. The alignment algorithm uses a slightly more restricted version known as LZ78. LZ78 parsing only allows the reuse of complete

```
LZ Parsing
a|g|gg|ga|ac|aacc|
(0,0,a) (0,0,g) (1,1,g) (1,1,a) (0,1,c) (4,3,c)
------------------------------------------------
LZ78 Parsing
a|g|gg|ga|ac|aa|c|c|
(0,a) (0,g) (2,g) (2,a) (1,c) (1,a) (0,c) (1,$)
```

FIGURE 1.5: Example LZ parsings of the string "agggaacacc".

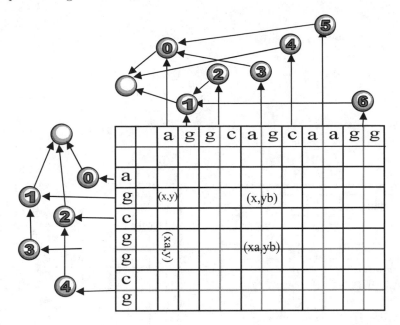

FIGURE 1.6: The dynamic programming table is decomposed into blocks based on each string's LZ78 block decomposition. In addition, two trie indexes are created that capture the structure of this decomposition. For each block (xa, yb), the blocks (x, yb), (xa, y), and (x, y) exist in the submatrix to the left and above block (xa, yb); the block indexes for substrings x and y can be found using the tries.

blocks rather than some arbitrary substring. One nice implication is that each block can be encoded using only two values $\langle i, c \rangle$, where i is the block index and c is the next character. Obviously, this method produces more blocks than the general scheme, but the total number of blocks is the same asymptotically, and the storage per block is less.

1.12.2 Decomposing the Problem

We will create a trie representation of the LZ78 parsing. For more information on tries and specifically suffix trees, refer to Chapter 5. The nodes of the trie correspond to each block in the parsing, and a node's parent corresponds to the block used as the prefix block. Edges point from children to the parents, and correspond to the extending character. The LZ78 parsing of the strings and the construction of the tries takes $O(n)$ time using suffix trees.

Using the block boundaries, as Figure 1.6 shows, we can conceptually divide the table into subtables, which we will also call blocks, as confusion can be avoided through context. Each block G defined for substrings xa of A and yb of B, written as (xa, yb), where x and y are strings and a and b are characters.

The intuition behind the algorithm is that the path information for all cells except for the bottom right cell should have been previously calculated. This is because blocks corresponding to (xa, y), (x, yb), and (x, y) exist in the submatrix above and to the left of the current block. We want to use this observation to do work proportional to the number of cells on block boundaries, which is $O\left(\frac{hn^2}{\log n}\right)$, our goal.

First, consider viewing the alignment problem at the block level, as shown in Figure 1.7.

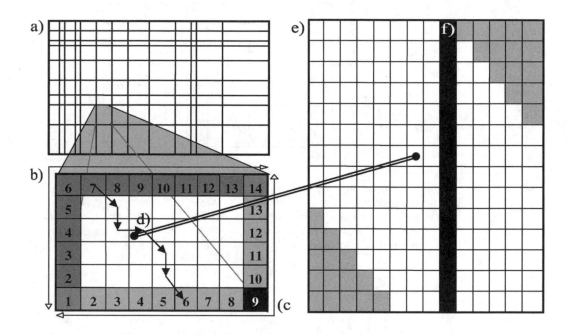

FIGURE 1.7: This figure shows an expanded view of each block in the dynamic programming table (a). The left column and top row are considered input cells for the block (b). The right column and bottom row are considered output cells (c). The total number of input cells or output cells is called p. One can consider every highest scoring path connecting each input cell with each output cell (d), and think of a $p \times p$ square matrix representing the score of each path through the block (e). This path matrix is incomplete as some input and output cells have no paths connecting them (gray). For this block, we only calculate and store one row of this matrix (f), which takes $O(p)$ time and space.

Let G be a block of width w and height h. We define the perimeter size of G as $p = w+h-1$. Note that this is not the same as the classical perimeter of G ($2(w+h)$). We call the left column and the top row the input cells of G and the right column and bottom row the output cells of G. The optimal path between some input cell in and some output cell out has a score $path[in, out]$. We create a $p \times p$ matrix called the path matrix, with the rows corresponding to the input cells and the columns the output cells. This matrix stores all optimal path scores for pairs of input cells and output cells.

For a block G defined as $G = B(xa, yb)$ where x and y are substrings and a and b are characters, we have the following information:

1. The score of all paths from input to output cells except for the bottom right cell, br. There are two reasons why this is the case. First, all paths must move down and to the right. Second, as stated previously, the blocks $B(x, yb), B(xa, y)$, and $B(x, y)$ have already been calculated.

2. The score of a best alignment path from the origin cell to each input cell, $input_i$.

We wish to calculate two things:

1. Scores of optimal paths from each input cell to the bottom right cell, br. This corresponds to one column in the path matrix as shown in figure 1.7. For each input cell in, the score is the maximum of three values:

$$path[in_i, br] = \max \begin{cases} path[in_i, northwest[br]] + \delta(a_i, b_j) \\ path[in_i, west[br]] + \gamma \\ path[in_i, north[br]] + \gamma \end{cases}$$

   Assuming that the three previously calculated path scores can be accessed in constant time, calculating the new path takes constant time for each input cell.

2. The input cell scores for the blocks neighboring our output blocks. This is accomplished by first calculating the output scores for each output cell, $output_j$, the score of the optimal path from the origin cell to the output cell.

$$output_j = \max_i(path[in_i, out_j] + input_i)$$

   The input cell scores can be calculated as the maximum of three values, using the bordering output cell scores.

It appears as if we have not reached our runtime goal. We wish to spend time proportional to the number of perimeter cells p in G. Certainly this is the case in step one, as we use a constant number of operations per input cell. However in step two it appears as if we break this requirement by searching for a max over all input cells for each output cell, which naively appears to take $O(p^2)$ time.

1.12.3 The SMAWK algorithm

It has been shown [1, 37] that the path matrix is *Monge* [32] by showing that $score[a, c] + score[b, d] \le score[a, b]$ for all $a < b, c < d$, which is the concave requirement. In turn, the matrix is *totally monotone* because any Monge matrix is also totally monotone. Again our matrix meets the concave condition of total monotonicity: $score[a, c] \le score[b, c] \Rightarrow score[a, d] \le score[b, d]$.

Aggarwal and Park [1] gave a recursive algorithm, called SMAWK, that solves the problem of finding all row and column maxima in an $n \times n$ totally monotone, full, rectangular matrix in $O(n)$ time. The idea it uses is very simple: as one travels down the matrix from top to bottom, the row maxima must move left to right. Equivalently, the column maxima move top to bottom as one moves between columns left to right.

With this in mind, assume that we know the column maxima for all even columns. We can find the column maxima for all odd columns in $O(n)$ time by searching only the rows between column maxima in adjacent even columns. This alone does not produce the desired runtime; a simple recursion would produce an $O(n \log n)$ bound.

However, given any irregular matrix with more rows R than columns C, it is obvious that only C rows can actually produce column maxima. Using the total monotonicity property, we can find the set of rows producing maxima in $O(R)$ time by eliminating all rows not producing maxima.

We construct a stack s of size $|s|$ that contains the set of rows producing maxima. The top row on the stack will be represented as s_t and the next row down s_{t-1}, and row at position n is denoted as s_n. The stack is initially empty. We will consider the rows from top to bottom. We will place a row r on s only if $r[|s|] < s_t[|s|]$. If this is not the case then we will pop s_t off the stack, because it can contain no maxima. The test is repeatedly applied to r and the top of the stack until the condition is met or the stack becomes empty.

Why can we pop s_t off the stack when $r[|s|] \geq s_t[|s|]$? By total monotonicity, $r[c] \geq s_t[c]$ for all columns $c \geq |s|$. It is also the case that $s_{t-1}[c] \geq s_t[c]$ for all columns $c < |s|$. This can be proved as follows: Assume, for a contradiction, that $s_{t-1}[c] < s_t[c]$ for some $c < |s|$. Then, by total monotonicity, $s_{t-1}[c'] < s_t[c']$ for all $c' > c$. Therefore, $s_{t-1}[|s-1|] < s_t[|s-1|]$. However, by construction, $s_{t-1}[s-1] \geq s_t[s-1]$, a contradiction. Therefore, it must be the case that $s_{t-1}[c] \geq s_t[c]$ for all columns $c < |s|$. Therefore, unless s_t meets the condition, it can be discarded as containing no maxima.

It follows from the proof that row s_n may only contain column maxima for columns $c >= n$. This property is desirable because it bounds the stack size to C, as any rows placed on a stack of size C would not be able to contain any column maxima and can be thrown away. Therefore, when the algorithm is complete the stack contains at most C rows that will contain all column maxima. This set of rows will be fed into the recursion. The runtime of this algorithm is $O(R)$, as each row is pushed and popped off the stack at most once.

Thus we can halve both the row and column size for each recursive step in linear time, and the runtime is given by $O(n + \frac{n}{2} + \frac{n}{4} + ...) = O(n)$. Therefore, the total runtime of the algorithm is linear.

1.12.4 String Alignment in Subquadratic Time

Returning to our problem, we wish to use the SMAWK algorithm to find all column maxima in our path matrix. The algorithm can be adapted for this purpose only after we complete the matrix. The matrix is incomplete because paths must move down and to the right. Therefore some input and output cells cannot be connected by paths. Consequently, the highest scoring path between these cells is undefined. However, we can choose values for the corresponding positions in the matrix that will not result in row or column maximum while maintaining the totally monotone property.

1. For each cell in the upper right triangle, assign the value $-\infty$.
2. For each cell in the lower left triangle, assign the value $-(n + i + j)k$, where k is the maximum possible theoretical score of some path through the block, or $k = |w - |w - h||\alpha + |w - h|\gamma$.

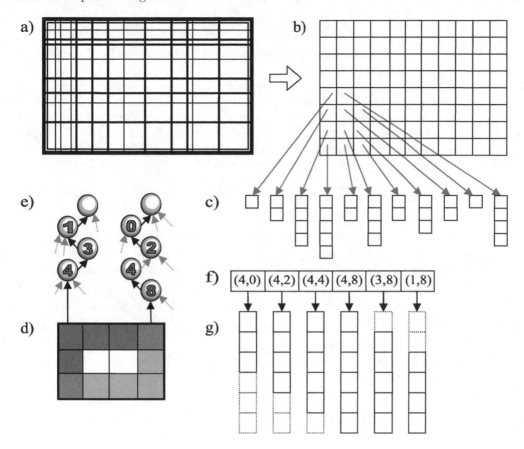

FIGURE 1.8: From the block decomposition of the dynamic programming table (a), we create an index (b) of size $O((\frac{hn}{\log n})^2)$ that points to the score column stored for each block (c). Each block (d) points to its corresponding node in the two trie indexes (e). Travelling up the path of these two trees, we can create a temporary array (f) that collects the rows needed to access the path matrix for this block (g) in constant time. Refer to Figure 1.7 for a description of the path matrix.

Now, let's say that for each row i corresponding to some input cell, we conceptually add $input_i$ to each cell in that row. After this operation, the totally monotone property is maintained as we change all values in each row by the same amount. After the addition, the result for each output cell is found by searching for each column maxima.

While we cannot spend the time to do the addition before running the SMAWK algorithm, we can do the addition for only those cells encountered during the run of the SMAWK algorithm and achieve the same affect. Therefore, we can find the needed maxima for all columns in time proportional to the number of rows and columns, which is the desired result.

Finally, we need to be able to find the path matrix values in constant time. However, we only have direct access to one column of the path matrix, the one that we constructed for this block. We need an indexing scheme that allows us to find the rows for all other output cells in constant time per cell. As shown in Figure 1.8, we will create a two dimensional array of size $O\left(\left(\frac{hn}{\log n}\right)^2\right)$ of pointers to the column stored for each block. This matrix in indexed by the tries we constructed for each string corresponding to the LZ decomposition of the string.

On the trie of each string, there is a path from the node for our current block to the root. Using the block IDs stored on the trie along this path as indexes into the two dimensional array, we can find the columns for the blocks used as prefixes of the current block in $O(1)$ time per block. There are p blocks of interest, each pointing to one column of our path matrix. To allow access to the path matrix values in constant time during the execution of the SMAWK algorithm, we create a temporary array of pointers to each column of the path matrix.

This is the last detail needed to finish the algorithm. In summary, the following steps are done for each block G with perimeter size p:

1. The column of the path matrix corresponding to connecting all input cells to the bottom right cell is constructed in $O(p)$ time using the columns for the blocks (x, yb), (x, y), and (xa, y), which can be found in constant time using the tries.

2. The path matrix for the block is constructed by constructing a temporary array of size p pointing to the columns of the matrix. This can be done in $O(p)$ time.

3. The output scores for this matrix are compiled using the SMAWK algorithm to find the column maxima. This also takes $O(p)$ time.

4. The input scores for the next block are calculated using the output scores from surrounding blocks, taking $O(p)$ time.

Therefore the total time for the algorithm is proportional to the number of perimeter cells, which as stated previously is $O\left(\frac{hn^2}{\log n}\right)$.

1.12.5 Space Requirements

Using Hirschberg's technique, we used $O(n + m)$ space and $O(nm)$ time to produce the alignment. In this section, we have described how to reduce the time required by the algorithm, but there is some expense. There is no known way in which to achieve subquadratic time and linear space in the same algorithm.

There is no published way to reduce the space bound without sacrificing flexibility, and we can find the space bound through a direct list of those data structures needed to solve the problem.

1. Two trie indexes corresponding to the block decomposition of our strings and two linear indexes into these trees, one for each row and column, as shown in Figure 1.6. This takes $O(\frac{hn}{\log n})$ space.

2. The block index structure corresponding to the block decomposition of S, as shown in Figure 1.8 (b). This takes $O\left(\left(\frac{hn}{\log n}\right)^2\right)$ space.

3. Input and output scores for each block, as shown in Figure 1.7 (b) and (c), taking p space per block, for a total of $O\left(\frac{hn^2}{\log n}\right)$ space.

4. One path matrix column for each block, as shown in Figure 1.8 (c), which takes p space per block, for a total of $O\left(\frac{hn^2}{\log n}\right)$ space.

Therefore, the total space complexity is the same as the runtime complexity, $O\left(\frac{hn^2}{\log n}\right)$.

1.13 Summary

In this chapter, we have provided a thorough presentation of fundamental techniques used to find the homology between a pair of DNA or protein sequences. While the basic alignment technique is simple to understand, the diversity of related problems quickly leads to new problem formulations and the resulting semiglobal, local, and banded alignments. We covered many advanced topics, including space saving techniques, normalized alignment, and subquadratic time alignment. Still, this chapter only represents an introduction to the field of alignments. The upcoming chapters in this part will expand on the ideas presented here and introduce a breadth of new formulations and solutions.

References

[1] A. Aggarwal and J. Park. Notes on searching in multidimensional monotone arrays. In *Proceedings of the 29th Annual Symposium on Foundations of Computer Science*, pages 497–512, 1988.

[2] N.N. Alexander and V.V. Solovyev. Statistical significance of ungapped alignments. In *Proceedings of the Pacific Symposium on Biocomputing*, pages 463–472, 1998.

[3] S.F. Altschul, W. Gish, W. Miller, and M. Myers *et al.* Basic local alignment search tool. *Journal of Molecular Biology*, 215(3):403–410, 1990.

[4] A.N. Arslan and O. Egecioglu. Efficient algorithms for normalized edit distance. *Journal of Discrete Algorithms*, 1(1):3–20, 2000.

[5] A.N. Arslan, O. Egecioglu, and P.A. Pevzner. A new approach to sequence comparison: normalized sequence alignment. *Bioinformatics*, 17(4):327–337, 2001.

[6] C. Branden and J. Tooze. *Introduction to Protein Structure*. Garland Publishing, 1991.

[7] T.H. Cormen, C.E. Leiserson, R.L. Rivest, and C. Stein. *Introduction to Algorithms*. MIT Press, 2001.

[8] M. Crochemore, G.M. Landau, and Z. Ziv-Ukelson. A sub-quadratic sequence alignment algorithm for unrestricted cost matrices. In *Proceedings of the 13th Annual ACM-SIAM Symposium on Discrete Algorithms*, pages 679–688, 2002.

[9] M. Crochemore and W. Rytter. *Text Algorithms*. Oxford University Press, 1994.

[10] M.O. Dayhoff and R.M. Schwartz. Matrices for detecting distant relationships. In M.O. Dayhoff, editor, *Atlas of Protein Structure*, pages 353–358. National Biomedical Reasearch Foundataion, 1979.

[11] M.O. Dayhoff, R.M. Schwartz, and B.C. Orcutt. A model of evolutionary change in proteins. In M.O. Dayhoff, editor, *Atlas of Protein Structure*, pages 345–352. National Biomedical Research Foundation, 1979.

[12] W. Dinkelbach. On nonlinear fractional programming. *Management Science*, 13:492–498, 1967.

[13] D. Eppstein. Sequence comparison with mixed convex and concave costs. *Journal of Algorithms*, 11(1):85–101, 1990.

[14] D. Eppstein, Z. Galil, R. Giancarlo, and I. Italiano. Sparse dynamic programming I: linear cost functions. *Journal of the ACM*, 39(3):519–545, 1992.

[15] D. Eppstein, Z. Galil, R. Giancarlo, and I. Italiano. Sparse dynamic programming II: convex and concave cost functions. *Journal of the ACM*, 39(3):546–567, 1992.

[16] J. Fickett. Fast optimal alignment. *Nucleic Acids Research*, 12(1):175–179, 1984.

[17] Z. Galil and R. Giancarlo. Speeding up dynamic programming with applications to

molecular biology. Technical Report 110–87, Columbia University Department of Computer Science, 1987.

[18] P. Gardner-Stephen and G. Knowles. DASH: localising dynamic programming for order of magnitude faster, accurate sequence alignment. In *Proceedings of the 2004 IEEE Computational Systems Bioinformatics Conference*, pages 732–733, 2004.

[19] D. Gusfield. *Algorithms on Strings, Trees and Sequences: Computer Science and Computational Biology*. Cambridge University Press, 1997.

[20] S. Henikoff and J.G. Henikoff. Amino acid substitution matrices from protein blocks. *Proceedings of the National Acadamy of Sciences of the U.S.A*, 89(22):10915–10919, 1992.

[21] D.S. Hirschberg. A linear space algorithm for computing maximal common subsequences. *Communications of the ACM*, 18(6):341–343, 1975.

[22] X. Huang, R.C. Hardison, and W. Miller. A space-efficient algorithm for local similarities. *Computer Applications in Biosciences*, 6(4):373–381, 1990.

[23] J. Kärkkäinen and E. Ukkonen. Lempel-Ziv parsing and sublinear-size index structures for string matching. In *Proceedings of the 3rd South American Workshop on String Processing*, pages 141–155, 1996.

[24] A. Lempel and J. Ziv. On the complexity of finite sequences. *IEEE Transactions on Information Theory*, 22(1):783–795, 1976.

[25] D.J. Lipman and W.R. Pearson. Rapid and sensitive protein similarity searches. *Science*, 227(4693):1435–1441, 1985.

[26] D.J. Lipman and W.J. Wilbur. Rapid similarity searches of nucleic acid and protein data banks. *Proceedings of the National Academy of Sciences of the U.S.A.*, 80(3):726–730, 1983.

[27] D.J. Lipman, J. Zhang, R.A. Schffer, and S.F. Altschul *et al.* Gapped BLAST and PSI-BLAST: a new generation of protein database search programs. *Nucleic Acids Research*, 25:3389–3402, 1997.

[28] W.J. Masek and M.S. Paterson. A faster algorithm for computing string edit distances. *Journal of Computer and System Sciences*, 20(1):18–31, 1980.

[29] N. Megiddo. Combinatorial optimization with rational objective functions. *Mathematics of Operations Research*, 4(4):414–424, 1979.

[30] J. Meidanis. Distance and similarity in the presence of nonincreasing gap-weighting functions. In *Proceedings of the 2nd South American Workshop on String Processing*, pages 27–37, 1995.

[31] W. Miller and E.W. Myers. Sequence comparison with concave weighting functions. *Bulletin of Mathematical Biology*, 50(2):97–120, 1988.

[32] G. Monge, Deblai, and Rembai. *Memoires del l'Academie des Sciences*, 1781.

[33] E.W. Myers and W. Miller. Optimal alignments in linear space. *Computer Applications in the Biosciences*, 4(1):11–17, 1988.

[34] S.B. Needleman and C.D. Wunsch. A general method applicable to the search for similarities in the amino acid sequence of two proteins. *Journal of Molecular Biology*, 48:443–453, 1970.

[35] B.J. Oommen and K. Zhang. The normalized string editing problem revisited. *IEEE Transactions on Pattern Analysis and Machine Intelligence*, 18(6):669–672, 1996.

[36] D. Sankoff and J.B. Kruskal. *Time Warps, String Edits and Macromolecules: the Theory and Practice of Sequence Comparison*. Addison Wesley, 1983.

[37] J.P. Schmidt. All highest scoring paths in weighted grid graphs and their application to finding all approximate repeats in strings. *SIAM Journal on Computing*, 27(4):972–992, 1998.

[38] C. Setubal and J Meidanis. *Introduction to Computational Biology*, chapter 3, pages

47–104. PWS Publishing, 1997.

[39] T.F. Smith and M.S. Waterman. Identification of common molecular subsequences. *Journal of Molecular Biology*, 147:195–197, 1981.

[40] T.F. Smith, M.S. Waterman, and W.M. Fitch. Comparative biosequence metrics. *Journal of Molecular Evolution*, 18:38–46, 1981.

[41] E. Vidal, A. Marzal, and P. Aibar. Fast computation of normalized edit distances. *IEEE Transactions on Pattern Analysis and Machine Intelligence*, 17(9):899–902, 1995.

[42] M.S. Waterman. *Mathematical methods for DNA sequences*. CRC Press, 1991.

[43] J. Ziv and A. Lempel. A universal algorithm for sequential data compression. *IEEE Transactions on Information Theory*, 23(3):337–343, 1977.

[44] J. Ziv and A. Lempel. Compression of individual sequences via variable rate coding. *IEEE Transactions on Information Theory*, 24(5):530–536, 1978.

2

Spliced Alignment and Similarity-based Gene Recognition

Alexei D. Neverov
State Scientific Center GosNIIGenetika

Andrei A. Mironov
State Scientific Center GosNIIGenetika and Moscow State University

Mikhail S. Gelfand
State Scientific Center GosNIIGenetika and Russian Academy of Sciences

2.1 Introduction

Algorithms for gene recognition can be divided into two major groups: statistical algorithms that use differing features of protein-coding and non-coding DNA, and algorithms utilizing similarity to ESTs or homologous genes and proteins. For a long time this distinction was almost absolute, although recently the boundary becomes blurred. As shown in [14], statistical programs show reasonable sensitivity, but their specificity strongly depends on the length of intergenic regions. False exons predicted in long spacers considerably decrease the specificity of predictions. One of the main reasons for that is the lack of good statistical models for gene boundaries.

Similarity-based gene recognition algorithms solve the reverse problem: reconstruction of the exon-intron structure of a gene given a spliced product of this gene (in the simplest case) or a homologous gene. These algorithms are highly specific, and their performance depends mainly on the level of similarity between the gene and the homolog. A serious drawback of

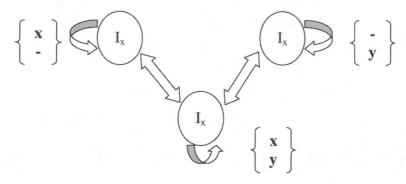

FIGURE 2.1: State transition diagram for a hidden Markov model for alignment of two sequences. X and Y denote the sequence being aligned. The HMM states are: MATCH (M), INSERTION IN SEQUENCE X, (I_x), and INSERTION IN SEQUENCE Y, (I_y). State M emits a pair of matched symbols in X and Y; states I_x and I_y emit pairs deletion–symbol.

such algorithms is that they cannot be applied to genes that have no known homologs. A special case is identification of genes by spliced alignment of genomic sequences containing homologous genes. It is based on the observation that protein-coding regions generally evolve at a slower rate than non-coding regions, and the pattern of mutations is different (e.g. single nucleotide insertions and deletions that would disrupt the reading frame are avoided). This approach is limited only by its inability to predict genes specific to genomes or very narrow taxons.

The *BLAST* family of programs [2] allows one to find regions of local similarity between two sequences, e.g. protein and DNA. Such regions often correspond to exons, but in the general case it is difficult to determine the exon boundaries exactly using *BLAST* alone. There exist several algorithms for fast identification of homologous sequences using *BLAST* with subsequent accurate mapping of exon-intron boundaries [10], [17], [28]. A clear advantage of such programs is the speed, as they are based on fast search over indexed databases. A serious drawback is that they do not guarantee finding the optimal structure. The latter is overcome by the spliced alignment technique that uses variations of the dynamic programming algorithm. The spliced alignment problem was stated in [11], [25].

2.1.1   Representation of Alignments by Hidden Markov Models

This section contains a brief introduction to Hidden Markov Models (HMMs) and their application to sequence alignment. It is necessary, as the HMM language is used in many spliced alignment algorithms, and, in particular, serves as a base for combining statistical and similarity-based approaches.

As it has been noted above, the problem of finding the optimal spliced alignment of two sequences is solved by dynamic programming algorithms. These algorithms are generalized by the Viterbi algorithm for finding the optimal hidden Markov chain [9]. An HMM is defined by an ordinary Markov matrix, or a diagram of transitions between states. In a hidden chain, the sequences of states cannot be observed directly. For instance, in the alignment problem, the observable variables are these sequences, whereas the hidden states are MATCH, INSERTION IN THE FIRST SEQUENCE, and INSERTION IN THE SECOND SEQUENCE (Figure 2.1). Each sequence of hidden states corresponds to an alignment. Beside transition probabilities, each state is characterized by probabilities of emission of the observed symbols.

In the alignment problem, the state MATCH generates a pair of symbols in both sequences, whereas the INSERTION states generate symbols in only one sequence. We will deal with the problem of finding the optimal sequence of hidden states that would maximize the probability of the observed sequences. The Viterbi algorithm is an algorithm for finding the optimal path in an oriented acyclic graph representing all variants of chains of the hidden states.

A more complicated hidden Markov model allowed for merging alignment and statistical gene recognition [6, 29, 20].

2.1.2 Generalized Hidden Markov Models

A standard hidden Markov model generates in each state a single symbol of the observed sequence. In the alignment case, at most one symbol per sequence is generated. If some state allows for transitions into itself, the duration of this state is described by a geometric distribution. Consider the model of exon-intron structure. The tail of the intron length distribution can be well approximated by the geometric distribution, but the latter does not describe well the distribution of the exon lengths. The generalized Markov models allow one to model states with arbitrary durations. Such generalized HMM was successfully applied for statistical gene recognition in the *GENSCAN* program [6].

2.2 Spliced Alignment of DNA with ESTs and cDNA

2.2.1 Statement of the Spliced Alignment Problem

Accumulation of the data about expressed genes in the form of protein, mRNA and EST sequences allowed for development of similarity-based approaches to gene recognition. Since formation of mature mRNA in eukaryotes is preceded by splicing of introns, identification of genes by similarity to processed products of gene expression should take into account existence of introns in the genomic sequences: unlike exons, introns cannot be aligned with the processed product. There exist two statements of the spliced alignment problem. The so-called *block problem* was suggested in [11].

Consider the genomic sequence as a string in the alphabet {A, C, G, T}, and the product of gene expression as a string in the same (in the case of mRNA or EST) or a different (in the case of protein) alphabet. The correspondence between the DNA and product alphabets is defined by a substitution matrix. The given DNA string is supplied by a set of (overlapping) subwords, that is, candidate exons. The goal is to find the set of subwords, whose concatenation is most similar to the expression product according to the substitution matrix. In the most general case, candidate exons correspond to any substring between AG and GT dinucleotides. Then the spliced alignment problem can be formulated as the so-called *site problem*, that is a generalization of the global spliced alignment [26] with introns treated as deletion of a specific type [25].

The *EST_GENOME* algorithm [25] allows one to align EST to a genomic sequence. The algorithm considers two types of introns in the genomic sequence: proper introns bounded by AG–GT dinucleotides (CT–AC if the complementary strand is considered), and *splices*, that is, deletions in the genomic sequence that do not require fixed dinucleotides at the termini, and have a smaller penalty than introns. Two types of intron-like deletions are needed for additional flexibility that allows the algorithm to recover short exons.

Let $W(m, n)$ be the weight of the alignment of the EST segment $(1, \ldots, m)$ and DNA segment $(1, \ldots, n)$, and let $W_{best}(m)$ be the weight of the best alignment of the EST segment $(1, \ldots, m)$ with $K(m)$ being the corresponding genomic position. Then

$$W_{m,n} = \max \begin{cases} W(m-1,n) - D \\ W(m-1,n-1) + M(m,n) \\ W(m,n-1) - D \\ W_{best}(m) - \triangle \\ 0 \end{cases}$$

where D denotes the "standard" deletion penalty in the genomic and EST sequences, $M(m,n)$ is the weight of matching symbols at positions m and n, and

$$\triangle = \begin{cases} \triangle_{splice} \\ \triangle_{intron} \end{cases}$$

where $\triangle = \triangle_{intron}$ is the intron penalty if $(K(m),n)$ is a pair of splicing sites, and $\triangle = \triangle_{splice}$ is the splice penalty otherwise; we set

$$(W_{best}(m), K(m)) = (W(m,n),n)) \, if \, W_{best}(m) < W(m,n)$$

Despite the drawbacks caused by using fixed intron penalty, the approach of this algorithm became quite popular. The running time of the algorithm is $O(MN)$, where M and N are the lengths of the EST and genomic sequences respectively.

2.2.2   The Use of HMM to Set the Intron Penalty

Probabilistic interpretation of the spliced alignment algorithm created a convenient way to combine the statistical gene recognition and sequence alignment. Use of the statistical models was necessary when the similarity between the aligned sequences was low and insufficient to exactly define the exon boundaries from alignment alone. Initially the problem was solved by filtering the candidate splicing sites and candidate exons. In the block variant of the spliced alignment problem, implemented in *Procrustes* [11], a set of candidate exons is filtered at a preliminary statistics-based step [24]. Other possibilities are filtering of candidate splicing sites, as in *Pro-EST* [22] and *Pro-Gene* [27]. One more algorithm of this family, *Pro-Frame* uses the fact that the similarity between the genomic and protein sequence exists at only one side of a true splicing site [23]. Finally, in a probabilistic setting of *GeneSeqer* [33], the intron probability depends on the probability of corresponding candidate sites determined by a specific statistics-based module *SplicePredictor*.

2.2.3   Determination of the Exon-Introns Structure of a Gene by Spliced Alignment with ESTs from Another, Related Gene

GeneSeqer aligns a DNA sequence with EST from a related gene, e.g., an orthologous gene from a different species [33]. It is intended for the annotation of plant genomes, where the number of ETSs for a genome might be rather low, as EST and genome sequencing projects do not always cover same organisms.

The HMM formalism is used for the alignment, and as the similarity between the aligned sequences may be rather low, additional information is needed to precisely map splicing sites. As mentioned above, initially site probabilities were set by a statistics-based model. In a subsequent study [5] sites were scored using the generalized Bayesian likelihood. Each

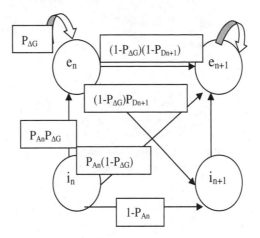

FIGURE 2.2: State transition diagram for a hidden Markov model implementing spliced alignment of EST and genomic sequence [33]. The transition probabilities τ are shown above the arrows. Each position n in the genomic sequence is ascribed EXON (e_n) or INTRON (i_n) state. Notation: $P_{\triangle G}$ is the deletion probability, P_{Dn} and P_{An} are the probabilities that n-th nucleotide in the DNA sequence is, respectively, first or last position in an intron.

candidate sites is classified to one of the following seven categories: true site in the reading frame $0, 1, 2$; false site in the reading frame $0, 1, 2$; and false site in non-coding region.

Consider a genomic sequence G of length N and a EST sequence C of length M. Represent the spliced alignment by an HMM with two states, EXON e_n and INTRON i_n, where $n = 1, \ldots, N - 1$ a position in the genomic sequence. The transition diagram is shown in Figure 2.2. The transition probabilities at the arcs are denoted by τ, for instance, τ_{e_{n-1}, e_n} is the probability to remain in the EXON state moving to the next genomic position. Probabilities of transitions between the EXON and INTRON states are estimated via the candidate site probabilities computed by *SplicePredictor*. Denote by n and m current genomic and EST positions respectively, let $S(n, m)$ be an alignment of sequences $G_1 G_2 \ldots G_n$ and $C_1 C_2 \ldots C_m$, and let $Z = z_1 z_2 z3 \ldots z_k$ be the sequence of hidden states; $\max m, n \leq l \leq m + n$. The maximum probability is computed using a standard formula

$$
\begin{aligned}
P &= \max[E_m^n, I_m^n] \\
E_m^n &= \max P(Z = z_1 z_2 \ldots z_l, z_l = e_n, S(m, n)) \\
I_m^n &= \max P(Z = z_1 z_2 \ldots z_l, z_l = i_n, S(m, n)) \\
E_0^n &= I_0 = 1 \\
E_m^0 &= 1 \\
I_m^0 &= 0 \\
n &= 0, 1, \ldots, N \\
m &= 0, 1, \ldots, M
\end{aligned}
$$

where the recursions for computing the probabilities of the sequence of states to end at position n of an exon E_m^n or an intron I_m^n are as follows:

$$E_m^n = \max \left\{ \begin{array}{l} \max[E_m^{n-1}\tau_{e_{n-1},e_n}, I_m^{n-1}\tau_{i_{n-1},e_n}]P_{e_n} \left(\begin{array}{c} G_n \\ - \end{array} \right), \\[2ex] \max[E_{m-1}^{n-1}\tau_{e_{n-1},e_n}, I_{m-1}^{n-1}\tau_{i_{n-1},e_n}]P_{e_n} \left(\begin{array}{c} G_n \\ C_m \end{array} \right), \\[2ex] \max[E_{m-1}^n\tau_{e_n e_n}, I_{m-1}^n\tau_{i_n e_n}]P_{e_n} \left(\begin{array}{c} - \\ C_m \end{array} \right) \end{array} \right\}$$

$$I_m^n = \max\{E_m^{n-1}\tau_{e_{n-1},i_n}, I_m^{n-1}\tau_{i_{n-1},i_n}\}$$

Since the transition probabilities depend on the site probabilities, the probability of the intron-type deletion implicitly depends on its sites.

2.2.4 Determination of the Exon-Introns Structure of a Gene by Spliced Alignment with EST Clusters

Spliced alignment with multiple ESTs is useful for determining the complete gene structure, finding alternatively spliced isoforms, and mapping gene termini. In early programs, e.g *Pro-EST* [22], it was done by spliced alignment of genomic sequences with pre-computed EST contigs. However, this approach is limited by assumptions used in construction of these contigs.

A more robust way to use the EST information is spliced alignment with individual ESTs with simultaneous construction of complete, alternative exon-intron structures. *GeneSeqer* [5] uses a decision tree to merge fragments of the exon-intron structure from individual spliced alignments. Another program, *TAP*, aligns EST to the genome using an empirical fast algorithm *sim4* [10] described in more detail below, and uses the following procedure for construction of alternative exon-intron structures.

ESTs aligned with identity exceeding 92% are ascribed to the DNA chain using database annotation and additional verification by analysis of invariant dinucleotides at the intron termini. 3′-ESTs are used to find the polyadenylation sites: such a site is defined either by a cluster of at least three ESTs or by a polyadenylated EST if the alignment contains a canonical site AATAAA or ATTAAA, whereas the genomic sequence does not contain a polyA-run. Pairs of splicing sites corresponding to introns can be in one of three possible relationships: continuous (belong to one alignment), transitive (belong to alignments overlapping in an exon), or conflicting. The algorithm constructs a matrix of such relationships between site pairs; an element of this matrix is the number of ESTs confirming the given relationships. This matrix is used to construct the path of the highest total weight, the next one, etc. The highest scoring path corresponds to the most represented isoform.

2.2.5 Clustering of cDNA (mRNA)

Two main approaches for clustering of full-length cDNAs are pairwise comparison of cDNA sequences and comparison with the genomic sequences. The former approach is applied when the genomic sequence is not available. In both cases the dynamic programming algorithm is too slow for mass comparisons: in the former case, too many pairwise comparisons are needed, whereas in the latter case, the genome sequence is too large.

The following filtering procedure was used in [31]. The local filter identifies an exactly coinciding fragment in two cDNAs whose length exceeds a given threshold. The global filter finds an ordered set of coinciding fragments whose total length exceeds a threshold. The program uses the *EST_GENOME* algorithm [25] modified so as to allow for pairwise comparison of cDNAs. This is achieved by using zero weights of matched nucleotides, penalties for external and internal deletions, and fixed penalties for deletions longer than

40 nucletoides, the latter corresponding to retained introns and other differences caused by alternative splicing.

The score of the optimal alignment of a cDNA pair assumed to be generated by alternative splicing of one pre-mRNA transcript, should be below some fixed threshold. The thresholds for the local and global filters are determined dependent on the alignment parameters. The local filter is implemented using a hash table. Construction of the hash table requires time proportional to the database size, whereas the search time for all cDNAs having a common word is proportional to the word length. The global filter uses a modification of the algorithm for construction of the maximal chain of common words in two sequences, whose complexity is $O(MN)$, where M and N are the lengths of the compared sequences [16], [15]. The authors suggested a modification whose run time is $O(N + KM)$, where $M - K$ is the minimum allowed word length.

2.2.6 Heuristic Algorithms of EST-DNA Spliced Alignment

The complexity of the spliced alignment algorithms is proportional to the product of the sequence lengths. Such algorithms guarantee finding the optimal alignment, but they are too slow for database search. A family of *BLAST*-like algorithms were developed for the latter purpose: *sim4* [10], *Spidey* [35], *BLAT* [17], *Squall* [28]. Such algorithms start with sensitive database similarity search aimed at decrease of the number of sequences requiring exact alignment. The database can be a genomic sequence and the query can be a EST or a protein, or vice versa, the database can be a set of ESTs and the query can be a fragment of the genomic sequence. Fast current algorithms do total alignment of human ESTs (3.73×10^6 fragments of total length 1.75×10^9) against the human genome (2.88×10^9 nucleotides). Heuristic spliced ailgnent algorithms do not guarantee finding the optimal alignment, but they are sufficiently specific and sensitive. The reason for that is that very similar sequences are aligned (more than 90% identity). If the similarity level is lower, the quality of predictions drops dramatically.

sim4 [10] aligns EST to DNA using the following strategy. Pairs of segments with maximum similarity are determined using a *BLAST*-like procedure: coinciding words of length 12 are found and then extended to form local similarity segments. A set of aligned segment pairs that could represent a gene is formed. The start and end positions of these segments should form increasing sequences in the EST and genomic DNA, and the offset of diagonals representing the segments in the alignment matrix should be either almost coinciding or sufficient to accommodate an intron. To determine the exon boundaries, pairs from almost coinciding diagonals are merged and their projections to the genomic sequence form exon cores. If projections of the exon cores to the EST sequence overlap, the common part of the cores is cut so as to form an intron with canonical GT–AG dinucleotides at the intron boundaries (or CT–AC if the EST is complementary to the gene strand). If the exon cores do not overlap, they are extended until the EST projections of the corresponding diagonals overlap. The intersection point is adjusted so as to define an intron with the canonical dinucleotides. If this procedure fails, a search for shorter matching segments is performed in the area between the cores.

BLAT [17] is intended for identification and fast alignment of very similar sequences, in particular, human ESTs with the human genome, and the human and mouse genomes.

Again, a search for highly similar fragments is performed first. Several variants of local similarity regions are defined: exact match of the length exceeding the threshold, an inexact match with at most one mismatching position, a chain of shorter exact matches within a given interval off one diagonal in the alignment matrix. The parameters are selected by considering the probability of a match in two sequences of the given identity so that to maximize the specificity at a fixed sensitivity level of 99%. The alignment procedure differs for EST-genome and protein-genome comparisons.

To construct the EST-genome alignments the obtained alignment segments are extended as in *sim4*; to fill the remaining gaps the procedure of the previous paragraph is applied iteratively with more liberal thresholds. The exon boundaries are selected so as to satisfy the GT–AG rule whenever possible. To construct a protein-DNA alignment, the initial fragments are extended without deletions, and then an oriented graph is constructed, whose vertices are alignment fragments, and the arcs connect vertices if the end of the fragment corresponding to the out-vertex is upstream of the start of the fragment corresponding to the in-vertex in both sequences.

Squall [28] also is intended for the EST alignment with complete genomes. Like the previous two programs, a fast search using a hash table of the large genome sequence is used to identify initial exact alignments. They are then extended and the remaining gaps are filled using the dynamic programming algorithm [12], if the lengths of the unaligned region are similar in the EST and genomic sequence. Otherwise it is assumed that the genomic sequence contains an intron. Otherwise hanging EST end is aligned to DNA fragments of the same length at both termini of the unaligned region, and the intron position is selected so as to satisfy the GT–AG rule whenever possible.

It should be noted that all algorithms of this type have a number of common problems, the most important of which is the possibility of missing short exons.

2.3 Protein-DNA Spliced Alignment

2.3.1 Block Problem

The block variant of the spliced alignment problem is solved by the *Procrustes* algorithm [11]. Consider a set of *blocks* (candidate exons) in the DNA sequence. The aim is to find a chain of exons with the highest similarity to a given protein. Denote by $W_{m,n,b}$ the weight of the optimal alignment ending in position n of block b in the genomic sequence and m of the protein sequence, and let $W_{m,n,best}$ be the weight in position m of the best block that has ended upstream of n.

The following recursions are used to find the optimal alignment:

$$
\begin{aligned}
W_{m,n,best} &= \max(W_{m,n-1,best}, W_{m,n,b}) & \text{if } n \text{ is the last position of block } b; \\
W_{m,n,b} &= W_{m,n,best} & \text{if } n \text{ is the first position of block } b; \\
W_{m,n,b} &= \max \begin{cases} W_{m-1,n,b} - D, \\ W_{n,n-1,b} - D, \\ W_{m-1,n-1,b} - M(a_m, b_n) \end{cases} & \text{if } n \text{ is the first position of block } b
\end{aligned}
$$

Here D denotes the deletion penalty. Consider blocks having the same left end (blocks **A** and **B** in Figure 2.3), and call the largest of such blocks a *covering* block. For such blocks the alignment matrix lines in the intersection region are filled with the $W_{m,n,b}$ values only once, and thus the complexity of the algorithm linearly depends on the number of covering blocks K_0. The average covering block length is at worst proportional to the genomic sequence length N. At the end of each block it is necessary to update the values $W_{m,n,best}$. If K be the number of all blocks, this procedure requires KM steps. Thus the overall complexity of the algorithm is $O(NMK_0 + MK)$.

The quality of predictions depends on the degree of similarity between the gene and the related protein and on the number of candidate exons [24]. There exist extensions of the basic algorithm that take into account splicing statistics (see above) and thus allow one to

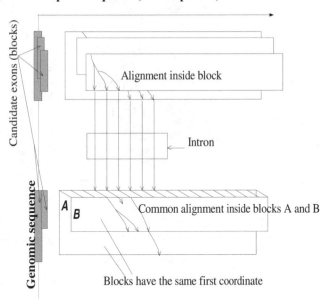

FIGURE 2.3: Dynamic programming graph for the block algorithm of spliced alignment *Procrustes* [11].

use low-similarity homologous proteins [34], [33]. Another problem is the necessity to filter candidate exons, as true exons might be lost. Moreover, statistical filtering of candidate exons becomes impossible when the genomic DNA has been sequenced with errors, especially single nucleotide indels leading to frameshifts. This problem is addressed by introduction of a special type of indels in the genomic sequence, as, e.g., in *Pro-Frame* [23].

The site variant of the spliced alignment problem is used. The transition graph is shown in Figure 2.4. The FRAMESHIFT transitions are used only if the previous vertex has not resulted from a frameshift. The INTRON transitions are allowed only if the corresponding genomic position contains an acceptor splicing site.

The algorithm allows for correction of errors corrupting splicing sites. This situation is likely if the alignment contains large indels in the region adjacent to an intron. The correction procedure identifies a corrupted site in the neighborhood of the indel. It is based on the analysis of short words via the conjecture that a correct donor site has the following property: sequences are similar upstream of a correct donor site, but not similar downstream of such a site. Thus a site is tentatively accepted, despite the lack of the GT dinucleotide, if there are much more conciding words upstream of it, compared to the downstream region. A similar procedure is used for acceptor sites.

2.3.2 Using Distant Homologs

Intuitively it is clear that the quality of homology-based gene predictions depends on the similarity level. Such dependence was considered in [24]. The structure of a human gene was predicted using a set of proteins from mammals, vertebrates, animals, plants, and even prokaryotes. If the similarity level between the predicted and homologous proteins exceeded 60%, it could be guaranteed that the prediction quality measured as the correlation between the predicted and true genes was at least 80%, and the average correlation was 97%. The

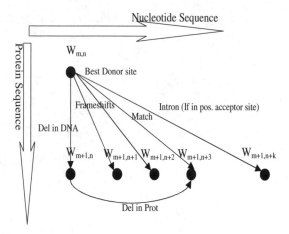

FIGURE 2.4: Dynamic programming graph for spliced alignment tolerant to errors in the genomic sequence *Pro-Frame*[23]. $W_{m,n}$ denotes the optimal weights in the graph vertices.

average correlation in the case of 30% identity exceeded 93%, so such predictions could still be considered as reliable. Using more distant homologs requires additional information, e.g. about the splicing sites, as in [34], see above. Another possibility is to use statistics of protein-coding regions.

Using distant homologous proteins makes it necessary to account for a possibility of long indels. Their length (in nucleotides) should be divisible by three, and thus they are different from indels of one and two nucleotides, as in *Pro-Frame* [23]. A spliced alignment algorithm from [12] accounts for long indels by a bounded affine penalty represented by an affine function of the indel length until the latter exceeds some threshold, and a constant for longer indels. Indels whose length is not divisible by three are additionally penalized. The intron penalty linearly depends on the site scores. Additionally, coding potential of candidate exons is taken into account. Although no explicit probabilistic model is introduced, this is the first step towards combining similarity-based and statistical algorithms.

GeneWise uses a hidden Markov model for protein-genome alignment (`www.sanger.ac. uc/\\Software/Wise2`). It combines site probabilities, a model of the intronic polypyrimidine run, accounts for sequencing errors (frameshifts) and codons interrupted by introns.

2.4 Using Local Similarity Identified by BLAST

Hidden Markov Models that Take into Account Protein Homologs

GenomeScan [37] is a modification of a well-known statistical HMM-based program *GENSCAN* [6]. It uses information about local similarity between a genomic fragment and proteins from a database to increase scores of candidate exon-intron structures that agree with these data and to penalize disagreements.

One of the main problems of statistics-based programs such as *GENSCAN* is low specificity in long genomic fragments containing large intergenic regions. The reason for that is the lack of statistical models suitable for recognition of gene boundaries. Existence of protein homologs identified, e.g. by *blastX*, allows one to change probabilities of hidden states, increasing probabilities of exons containing local similarities. Moreover, if the similarity extends to the protein termini, gene boundaries also can be identified.

Exon-intron structure conforms to *blastX* results if an aligned segment lies within an exon. As in general aligned segments do not have well-defined termini, an exon should cover only the point of maximum local similarity (*central point*). Let Ω be the structure summarizing all similarity data for a genomic sequence G (co-ordinate of the central point, its reading frame and p-value). Denote by Z_Ω the set of exon-intron structures conforming to Ω. As not all local similarities reflect homology, denote artificial local similarity by A_Ω, and denote real homology by H_Ω. Clearly, the probability of artificial similarity $P(A)$ should depend on p-value, and the following heuristics is used: $P(A) = (p-value)^{\frac{1}{r}}$, where r is a small integer. Let $P_\Omega = P(H_\Omega)$ be the probability that the observed similarity is due to homology; then $P(A) = 1 - P_\Omega$. Now, probability of an exon-intron structure z_i is

$$P(z_i, G \mid \Omega) = \begin{cases} (\frac{P_\Omega}{P(Z_\Omega)} + (1 - P_\Omega))P(z_i, G) & \text{if } (z_i \in Z_\Omega) \\ (1 - P_\Omega)P(z_i, G) & \text{if } (z_i \notin Z_\Omega) \end{cases}$$

where $P(z, G)$ is the *GENSCAN* probability (that is, disregarding the similarity data). Note that in all cases

$$(\frac{P_\Omega}{P(Z_\Omega)} + (1 - P_\Omega)) \geq and(1 - P_\Omega) \leq 1.$$

Thus probabilities of exon-intron structures conforming to the similarity data increase, whereas the contradicting structures are penalized.

2.4.1 Gene Recognition by Comparison of Genomic Sequences Containing Homologous Genes

In general, protein-coding exons are more conserved that non-coding DNA, and there exists a large group of algorithms using this fact: CEM [3], ROSETTA [4], TWINSCAN [18], SGP-1 [36], SGP-2 [30].

SGP-2 [30] uses TblastX to identify genomic fragments containing homologous genes. Statistics-based scores of exons covered by similar segments are increased, and then the optimal chain of exons is found using GENEID [13].

TWINSCAN [18] uses local alignment constructed by WU-BLAST [19], [blast.wustl.edu]. The alignment results for each nucleotide are recoded as a sequence of special symbols: MATCH, MISMATCH, UNALIGNED. Hidden states of the GENSCAN HMM are modified so that both the genomic sequence and the conservation sequence are modeled. Let z be a hidden state, *e.g.* EXON, $G_{i,j}$ be a genomic subsequence, $1 \leq i < j \leq N$ (N is the subsequence length), and let $C_{i,j}$ be the corresponding conservation subsequence. Then, assuming independence,

$$P(G_{i,j}, C_{i,j}|z) = P(G_{i,j}|z)P(C_{i,j}|z).$$

The first term, the DNA sequence probability, is computed as in GENSCAN. The second term, the conservation sequence probability is described by a Markov chain of the fifth order.

SGP-1 [36] uses similarity data to identify conserved splicing sites. Genomic sequences are aligned using a local similarity search algorithm, TblastX or simply blastN. Candidate exons are defined as aligned fragments with matching sites at both boundaries and used for construction of exon-intron structures.

A similar scheme is used in CEM [3]. In some neighborhood of local similarities identified by TblastX all possible pairs of candidate donor and acceptor splicing sites are considered. An acyclic graph is constructed, whose vertices are these sites, and arcs are exons and introns. Each exon arc should satisfy the reading frame given by the local alignment. The predicted exon-intron structure corresponds to the path of the maximal weight constructed by dynamic programming.

2.4.2 DNA-DNA Spliced Alignment

The algorithms considered in the previous paragraph use local similarity to identify candidate exons. They assume conservation of the exon-intron structure of the aligned genes, and, like all heuristic algorithms, do not guarantee finding the optimal solution. Here we consider programs that use more technically rigorous algorithms.

Spliced alignment of genomic sequences was implemented in Pro-Gene [27]. Denote the genomic sequences containing homologous genes by X and Y. Candidate splicing sites, start and stop codons are identified by standard profile search. Denote the weight of the optimal alignment of subsequences $X_{1,m}$ and $Y_{1,n}$ by $W(m,n)$. Let $\mu_m(n)$, $\nu_n(m)$ and λ_{mn} be the weights of the optimal alignment ending, respectively, at a donor site in position m in X, at a donor site in position n in Y, and in a pair of donor sites in both sequences (resp. formulas (a), (b), (c) below).

$$
\begin{aligned}
\mu_m(n) &= \max\{W(i,n)|i < m, i \text{ is a donor position in } X\}, &\text{(a)}\\
\nu_n(m) &= \max\{W(m,j)|j < n, j \text{ is a donor position in } Y\}, &\text{(b)}\\
\lambda_{(mn)} &= \max\{W(i,j)|i < m, i \text{ is a donor position in } X, j < n, j \text{ is a donor} \\
& \qquad \text{position in } Y\}. &\text{(c)}
\end{aligned}
$$

The recursion for computing the weight $W(m,n)$ of the optimal alignment is

$$
W(m,n) = \max \begin{cases}
W(m\text{-}3, n\text{-}3) + M(x_m, y_n) & \text{(matching codons } [x_m x_{m+1} x_{m+2}] \text{ and } [y_n y_{n+1} y_{n+2}]) \\
W(m, n\text{-}3)\text{-}D & \text{(deletion in } X \text{ against codon } [y_n y_{n+1} y_{n+2}] \text{ in } Y) \\
W(m\text{-}3, n)\text{-}D & \text{(deletion in } Y \text{ against codon } [x_m x_{m+1} x_{m+2}] \text{ in } X) \\
\mu_m(n\text{-}3)\text{-}\triangle & \text{if } m \text{ is an acceptor position (intron in } X) \\
\mu_m(n)\text{-}\triangle\text{-}D & \text{if } m \text{ is an acc. pos. (intron in } X, \text{ deletion in } Y) \\
\nu_n(m\text{-}3)\text{-}\triangle & \text{if } n \text{ is an acceptor position (intron in } Y) \\
\nu_n(m)\text{-}\triangle\text{-}D & \text{if } n \text{ is an acc. pos. (intron in } Y, \text{ deletion in } X) \\
\lambda_{mn}\text{-}2\triangle & \text{if } m, n \text{ are acceptor positions (introns in } X \text{ and } Y)
\end{cases}
$$

The notation in the above recursion is as follows: $D = D_{gap}$ (the deletion opening penalty) or $D = D_{del}$ (deletion extension penalty); $\triangle$ is the fixed intron penalty, $M(A,B)$ is the score for matching codons A and B. Note that there is no need to compute the values of λ, μ, ν for each acceptor site, as they change only at donor sites. The dynamic programming graph is shown in Figure 2.5.

2.4.3 Pairwise Generalized Hidden Markov Chains

Further development leads to the creation of algorithms that combine similarity and statistics in a single procedure. A pairwise generalized hidden Markov model suggested in a theoretical study [29] aligns two genomic sequences or a genomic sequence with a spliced product, EST or protein. This model was implemented in DOUBLESCAN [20] and SLAM [1].

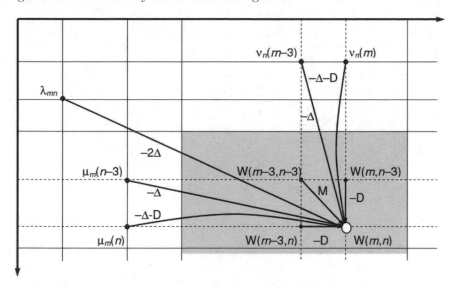

FIGURE 2.5: Dynamic programming graph for spliced alignment of two genomic sequences `Pro-Gene` [27]. Axes: positions in sequences X and Y. All possible incoming arcs for vertex (m,n) are shown. The region of constancy of λ, μ, ν is shadowed.

`DOUBLESCAN` [20] predicts exon-intron structures of homologous genes that may differ by intron insertion/loss. A genomic sequence may contain incomplete genes or several genes. The HMM states classify nucleotides in both genomic sequences as ALIGNED EXONIC, INSERTED (exonic non-aligned), INTRONIC and INTERGENIC (Figure 2.6). A simplified state diagram is shown in Figure 2.7. The HMM does not have generalized states with a given distribution of length. Splicing sites are described by hidden states whose probabilities are ascribed by statistical models implemented in an external procedure. A similar approach was used also in `GeneSeqer` [33], see above Figure 2.2. Modeling of splicing sites is used to identify 5'- and 3'-untranslated exons and thus to improve mapping of gene termini. It is easy to incorporate statistical modeling of exons and introns, although the original algorithm does not do that. Reconstruction of the exon-intron structure is done using the classical Viterbi algorithm and a heuristic `blastN`-based algorithm that divides the original alignment matrix into intersecting subspaces between regions of local similarity and thus to a much faster procedure.

Exon-intron structure of orthologous human and mouse genes is conserved in 86% cases, and most differences are caused by single

intron insertions or deletions [21]. This fact is used in `Projector` that maps an annotated exon-intron structure to a genomic fragment containing an orthologous gene from another genome.

2.4.4 Prediction of Alternative Splicing by HMM Sampling

Current estimates of the prevalence of alternative splicing in the human genome range from 30% through 60%. The Viterbi algorithm allows one to determine the optimal sequence of hidden states in an HMM, whereas suboptimal exon-intron structures also might be biologically relevant. The crucial resource in HMM algorithms is memory. A fast and memory-efficient non-deterministic algorithm for finding suboptimal solution in HMMs was suggested in [7]. It is based on a procedure that allows one to select a path with the

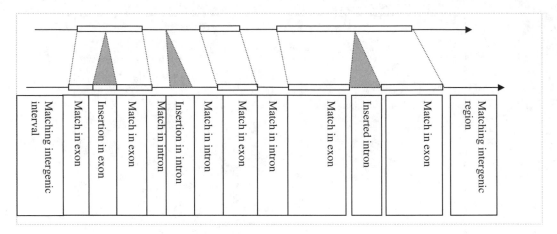

FIGURE 2.6: An example of alignment of homologous genes by DOUBLESCAN [20]. Horizontal lines with arrows denote two genomic sequences containing homologous genes. Exons are shown by empty rectangles. States of the hidden Markov chain are shown below in vertical rectangles.

probability proportional to its weight by the time $O(L)$, where L is the path length (the number of arcs).

The algorithm for finding k paths has the time complexity $O(NT + kT)$ and the memory complexity $O(NT)$, where T is the sequence length, and N is the number of hidden states in the model. Another variant has lower memory complexity $O(N)$, but higher time complexity $O(kNT \log(T))$. For a pairwise HMM applied to sequences of lengths T and U, the time complexity $O(kNTU)$ is prohibitively large. An algorithm with time complexity $O(NTU + kN(T + U))$ and memory complexity $O(N(T + U))$, developed in [7], was implemented as a separate module in SLAM. Most of generated k paths coincide with the optimal exon-intron structure, but there is a probability to observe a suboptimal path. An example of identification of conserved alternative splicing in the human and mouse genome was described, but no systematic application was presented.

2.4.5 Gene Recognition in Several Sequences by a Gibbs Sampler

Two problems in combining statistical and similarity-based approaches are: slow performance (compared to ordinary spliced alignment algorithms), the need for *a priori* setting of alignment and recognition parameters, difficulties in distinguishing coding and conserved regulatory regions, and sensitivity to genome rearrangements changing the order of genes and their strand location. These problems are addressed by the Gibbs sampler approach where multiple genomic sequences $y^1, \ldots, y^n$ containing orthologous genes are considered as realizations of one HMM [8]. The algorithm maximizes the log-likelihood

$$\max_{\theta}[\log P(y^1, \ldots, y^n | \theta)] = \max_{\theta}[\sum_{j=1}^{n} \log P(y^j | \theta)]$$

over all sets of parameters θ and all sequences of hidden states $z^1, \ldots, z^n$. The main technical trick is the separation of steps sampling the hidden states from the distribution $p(z^i | z^1, \ldots, z^{i-1}, z^{i+1}, \ldots, z^n; \theta; y^1, \ldots, y^n)$, $(1 \leq i \leq n)$ and the parameters, from the distribution $p(\theta | z^1, \ldots, z^n; y^1, \ldots, y^n)$ instead of sampling from the joint posterior distribution

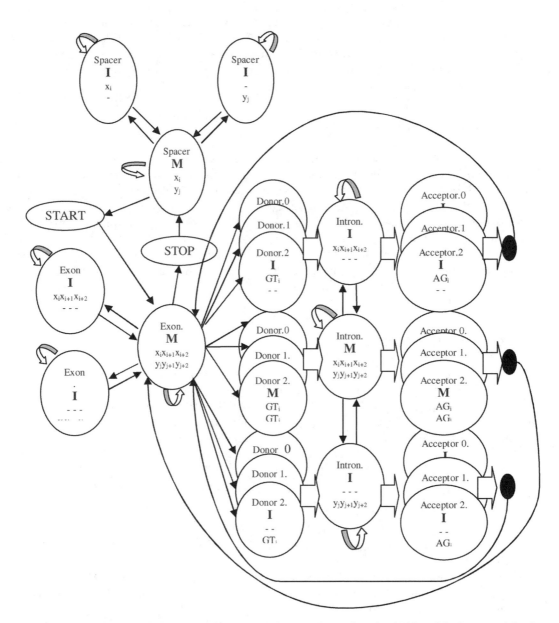

FIGURE 2.7: Simplified diagram of states and transitions for the hidden Markov model of DOUBLESCAN [20]. X and Y are the aligned sequences. M and I are the MATCH and INDEL states, respectively.

$p(z^1, \ldots, z^n; \theta | y^1, \ldots, y^n)$.

Gene recognition is performed as follows: genes are predicted separately in each sequence, then candidate genes identified in each sequence are compared to each other. Weight of an exon is defined as the sum of scores of candidate genes to which this exon belongs. Parameters such as oligonucleotide frequencies in exons and introns, distribution of exon lengths etc. are determined simultaneously with mapping of exons. This approach was applied in the frame of the NISC project aimed at sequencing of some genome regions in multiple mammalian genomes [32]. The method developed in [8] proved to be highly sensitive (89%) and specific (90%).

2.5 Conclusions

It is clear that the similarity-based algorithms are a powerful tool of gene recognition. Various approaches have been developed, applicable in different experimental situations. The most recent developments are (i) algorithms combining statistical and similarity-based approaches, that use similarity data when available, but rely on statistics when predicting the exon-intron structure of orphan, genome-specific genes having no or only distant homologs; (ii) EST-based algorithms that use multiple, alternatively spliced ESTs or ESTs from other species; (iii) algorithms based on spliced alignment of genomic sequences, and specifically, multiple spliced alignment. The most promising directions of research seem to be using multiple genomes at different level of taxonomic relatedness, and prediction of alternative splicing.

Acknowledgements

We are grateful to Alexander Favorov and Vsevolod Makeev for many useful discussions. This study was supported by grants from the Ludwig Institute of Cancer Research (CRDF RB0-1268), the Howard Hughes Medical Institute (55000309), the Russian Fund of Basic Research ($04 - 04 - 49440$), and programs "Molecular and Cellular Biolog" and "Origin and Evolution of the Biosphere" of the Russian Academy of Sciences.

References

[1] M. Alexandersson, S. Cawley, and L. Pachter. SLAM: cross-species gene finding and alignment with a generalized pair hidden markov model. *Genome Research*, 13:495–502, 2003.

[2] S.F. Altschul, T.L. Madden, A.A. Schaffer, and J. Zhang *et al.* Gapped BLAST and PSI-BLAST: a new generation of protein database search programs. *Nucleic Acids Research*, 1997:3389–3402, 1997.

[3] V. Bafna and D.H. Huson. The conserved exon method for gene finding. *Proc Int Conf Intell Syst Mol Biol*, 8:3–12, 2000.

[4] S. Batzoglou, L. Pachter, J.P. Mesirov, and B. Berger *et al.* Human and mouse gene structure: comparative analysis and application to exon prediction. *Genome Research*, 10:950–958, 2000.

[5] V. Brendel, L. Xing, and W. Zhu. Gene structure prediction from consensus spliced alignment of multiple ESTs matching the same genomic locus. *Bioinformatics*, 20:1157–1169, 2004.

[6] C. Burge and S. Karlin. Prediction of complete gene structures in human genomic DNA. *Journal of Molecular Biology*, 268:78–94, 1997.

[7] S.L. Cawley and L. Pachter. HMM sampling and applications to gene finding and alternative splicing. *Bioinformatics*, 19:36–41, 2003.

[8] S. Chatterji and L. Pachter. Multiple organism gene finding by collapsed gibbs sampling. *Proceedings of the Eighth Annual International Conference on Computational Molecular Biology (RECOMB 2004)*, pages 187–193, 2004.

[9] R. Durbin, S. Eddy, A. Krogh, and G. Mitchison. Biological sequence analysis. *Cambridge University Press, Cambridge, UK*, 1998.

[10] L. Florea, G. Hartzell, Z. Zhang, and G.M. Rubin *et al.* A computer program for aligning a cDNA sequence with a genomic DNA sequence. *Genome Research*, 8:967–974, 1998.

[11] M.S. Gelfand, A.A. Mironov, and P.A. Pevzner. Gene recognition via spliced sequence alignment. *Proceedings of the National Academy of Sciences USA*, 93:9061–9066, 1996.

[12] O. Gotoh. Homology-based gene structure prediction: simplified matching algorithm using a translated codon (tron) and improved accuracy by allowing for long gaps. *Bioinformatics*, 16:190–202, 2000.

[13] R. Guigo. Assembling genes from predicted exons in linear time with dynamic programming. *Journal of Computational Biology*, 5:681–702, 1998.

[14] R. Guigo, P. Agarwal, J.F. Abril, and M. Burset *et al.* An assessment of gene prediction accuracy in large DNA sequences. *Genome Research*, 10:1631–1642, 2000.

[15] D. Gusfield. Algorithms on strings, trees, and sequences: Computer science and computational biology. *Cambridge University Press, Cambridge, UK.*, 1997.

[16] D.S. Hirschberg. A linear space algorithm for computing maximal common subsequences. *Communications of the ACM*, 18:341–343, 1975.

[17] W.J. Kent. BLAT–the BLAST-like alignment tool. *Genome Research*, 12:656–664, 2002.

[18] I. Korf, P. Flicek, D. Duan, and M.R. Brent. Integrating genomic homology into gene structure prediction. *Bioinformatics*, 17:140–148, 2001.

[19] R. Lopez, V. Silventoinen, S. Robinson, and A. Kibria *et al.* WU-Blast2 server at the european bioinformatics institute. *Nucleic Acids Research*, 31:3795–3798, 2003.

[20] I.M. Meyer and R. Durbin. Comparative ab initio prediction of gene structures using pair HMMs. *Bioinformatics*, 18:1309–1318, 2002.

[21] I.M. Meyer and R. Durbin. Gene structure conservation aids similarity based gene prediction. *Nucleic Acids Research*, 32:776–783, 2004.

[22] A.A. Mironov, J.W. Fickett, and M.S. Gelfand. Frequent alternative splicing of human genes. *Genome Research*, 9:1288–1293, 1999.

[23] A.A. Mironov, P.S. Novichkov, and M.S. Gelfand. Pro-Frame: similarity-based gene recognition in eukaryotic DNA sequences with errors. *Bioinformatics*, 17:13–15, 2001.

[24] A.A. Mironov, M.A. Roytberg, P.A. Pevzner, and M.S. Gelfand. Performance-guarantee gene predictions via spliced alignment. *Genomics*, 51:332–339, 1998.

[25] R. Mott. EST_GENOME: a program to align spliced DNA sequences to unspliced genomic DNA. *Comput. Appl. Biosci.*, 13:477–478, 1997.

[26] S.B. Needleman and C.D. Wunsch. A general method applicable to the search for similarities in the amino acid sequence of two proteins. *Journal of Molecular Biology*, 48:443–453, 1970.

[27] P.S. Novichkov, M.S. Gelfand, and A.A. Mironov. Gene recognition in eukaryotic DNA by comparison of genomic sequences. *Bioinformatics*, 17:1011–1018, 2001.

[28] J. Ogasawara and S. Morishita. A fast and sensitive algorithm for aligning ESTs to

the human genome. *J. Bioinform. Comput. Biol.*, 1:363–386, 2003.

[29] L. Pachter, M. Alexandersson, and S. Cawley. Applications of generalized pair hidden markov models to alignment and gene finding problems. *Journal Comput. Biol.*, 9:369–399, 2002.

[30] G. Parra, P. Agarwal, J.F. Abril, and T. Wiehe *et al.* Comparative gene prediction in human and mouse. *Genome Research*, 13:108–117, 2003.

[31] T. Shibuya, H. Kashima, and A. Konagaya. Efficient filtering methods for clustering cDNAs with spliced sequence alignment. *Bioinformatics*, 20:29–39, 2004.

[32] J.W. Thomas, J.W. Touchman, R.W. Blakesley, and G.G. Bouffard *et al.* Comparative analyses of multi-species sequences from targeted genomic regions. *Nature*, 424:788–793, 2003.

[33] J. Usuka and V. Brendel. Gene structure prediction by spliced alignment of genomic DNA with protein sequences: increased accuracy by differential splice site scoring. *Journal of Molecular Biology*, 297:1075–1085, 2000.

[34] J. Usuka, W. Zhu, and V. Brendel. Optimal spliced alignment of homologous cDNA to a genomic DNA template. *Bioinformatics*, 16:200–211, 2000.

[35] S.J. Wheelan, D.M. Church, and J.M. Ostell. Spidey: a tool for mRNA-to-genomic alignments. *Genome Research*, 11:1952–1957, 2001.

[36] T. Wiehe, S. Gebauer-Jung, T. Mitchell-Olds, and R. Guigo. SGP-1: prediction and validation of homologous genes based on sequence alignments. *Genome Research*, 11:1574–1583, 2001.

[37] R.-F. Yeh, L.P. Lim, and C.B. Burge. Computational inference of homologous gene structures in the human genome. *Genome Research*, 11:803–816, 2001.

3

Multiple Sequence Alignment

Osamu Gotoh
Kyoto University

Shinsuke Yamada
Waseda University

Tetsushi Yada
Kyoto University

3.1 Biological Background

All genes in contemporary organisms on earth have relatives that originated in a common ancestor. The genes or gene products (proteins and RNAs) derived from a common ancestor are said to be homologous to one another, and comprise a family or a superfamily. The homologous relationships may also apply to individual nucleotides or amino acids (collectively called residues) in a family of genes or proteins. Multiple sequence alignment (MSA) is aimed at reproducing these homologous relationships among individual residues in a set of gene or protein sequences. Mutations (substitutions, deletions, or insertions) that have occurred during the evolutionary process make the inter-residue homologous relationships obscure and sometimes barely detectable. Hence, it is vital to obtain a reliable MSA from a set of remotely related sequences. On the other hand, the sequence data to be analyzed are accumulating at an increasing rate, and therefore, fast and reliable MSA methods are earnestly desired. The main purpose of this chapter is to introduce a variety of computational methods to tackle this difficult problem. We tried to cover both theoretical and practical approaches. However, because the relevant area is too wide to be entirely covered in this chapter, we concentrate most of our attention on the global alignment of protein sequences. Some recent reviews and book chapters on these topics, the reader is referred to [9, 16, 17, 34, 76, 72].

Before we discuss the methodological parts, the interconnected relationships among MSA and various fields of molecular and computational biology will be discussed briefly. Today, the most frequently used tool in the field of molecular computational biology is certainly (approximate) pairwise sequence alignment (PSA), which drives routine homology search [5, 80]. Although the usefulness of PSA is obvious, MSA has several advantages over PSA.

(1) In general, the quality of alignment is improved when multiple related sequences are aligned simultaneously compared with the case when only two members are aligned. This is particularly true when the members have diverged considerably. The high quality of alignment is a prerequisite for homology modeling [93, 113], which is currently the sole practical

method for predicting reliable protein tertiary structures from sequences. Moreover, most modern methods for detecting subtle resemblance in protein sequences/structures rely on profile-to-sequence or profile-to-profile comparison, where MSA provides the most informative resources for the construction of sensitive profiles [82, 117].

(2) MSA also significantly improves the accuracy of the *de novo* prediction of higher-order structures of proteins and RNAs. For example, all the current sensible methods for predicting protein secondary structures rely on MSA to gain 5% or higher accurate prediction compared to their single-sequence counterparts [58, 89, 25, 90, 92].

(3) The conservation/variation pattern along an MSA indicates functionally important sites common to a family or those specific to a subset of sequences within that family. Information about conservation/variation becomes more relevant when coupled with information about the protein tertiary structure. The evolutionary trace method is a successful realization along this strategy [67, 121]. Phylogenetic footprint is another example of a promising approach to decipher signals hidden in genomic sequences [104, 24].

(4) MSA is an indispensable step for the reliable reconstruction of a phylogenetic tree from contemporary protein or nucleotide sequences. There are two major methods for tree reconstruction, *i.e.* the distance-matrix methods and the character-based methods [23, 75]. The former include the unweighted pair group with arithmetic mean (UPGMA) method [100] and the neighbor-joining (NJ) method [91]. Although evolutionary distances may be calculated by other means, *e.g.*, by PSA, the most reliable estimates are obtained from an MSA. On the other hand, the latter include the maximum parsimony (MP) method [18] and maximum likelihood (ML) method [22], both of which are designed to use an MSA as the input. Further details on use of MSA in phylogenetics can be found in Chapter 19 of the handbook.

(5) Although not extensively discussed here, MSA makes a central contribution to the shotgun sequencing procedure to obtain a consensus from error-prone individual reads and to assemble fragments into contigs [11, 46] (see also Chapter 8). MSA is also useful in the design of probes or PCR primers of desired specificity and sensitivity. In summary, MSA plays a central role in the investigation of the so-called FESS relationships, *i.e.*, the relationships among function, evolution, sequence, and structure of a family of genes or proteins, as emphasized in [34].

3.2   Definitions and Properties

3.2.1   Definition

An alignment of a set of sequences $(\bar{s}_1, \bar{s}_2, \ldots, \bar{s}_N)$ represents correspondence among elements, $s'_{ni}s$, one taken from each sequence, allowing for the absence of correspondents but keeping the order within each sequence. (The upper bar indicates a row vector.) When $\bar{s}_n$ is a DNA sequence, $\bar{s}_n$ is a string over the alphabet $\Sigma = \{A, C, G, T\}$. Likewise, when $\bar{s}_n$ is a protein sequence, $s_{ni} \in \Sigma = \{$the 20 amino acids$\}$. We call an alignment a pairwise sequence alignment (PSA) when $N = 2$, and a multiple sequence alignment (MSA) when $N \geq 3$. Alignment may also refer to the process for obtaining an object. In this context, pairwise alignment may involve more than two sequences that are originally present in two separate groups of sequences. An alignment is usually represented by a rectangular character matrix $\mathcal{A} = \{a_{nl}\}(1 \leq n \leq N, 1 \leq l \leq L)$ over $\Sigma' = \Sigma \cup \{-\}$, where a dash '-' denotes a null implying the absence of the correspondent. A run of one or more contiguous nulls in a row is called a gap or an indel (insertion-deletion). As a long gap may be produced by a single evolutionary event, gaps and nulls refer to related yet distinct entities and are clearly discriminated in this chapter. An alignment matrix $\mathcal{A}$ must satisfy the following two conditions.

A) When all nulls are removed, a row of $\mathcal{A}$, $\bar{\mathbf{a}}_n$, is reduced to $\bar{\mathbf{s}}_n$.

B) There must not be a column consisting of only nulls. Because of these conditions, the following inequality holds,

$$\max_{1 \leq n \leq N} | \bar{\mathbf{s}}_n | \leq L \leq \sum_{1 \leq n \leq N} | \bar{\mathbf{s}}_n |, \tag{3.1}$$

where $|\bar{\mathbf{s}}_n|$ indicates the length of the sequence $\bar{\mathbf{s}}_n$. The relationship between the elements of $\bar{\mathbf{s}}'_n s$ and $\bar{\mathbf{a}}'_n s$ is conveniently expressed by an 'index matrix' $\mathcal{R} = \{r_{nl}\}(1 \leq n \leq N, 1 \leq l \leq L)$ of the same size as that of $\mathcal{A}$, where $a_{nl} =$ '-' if $r_{nl} = 0$, or otherwise, $a_{nl} = s_{nr_{nl}}$. Complementarily to $\mathcal{R}$, we may introduce the 'gap state matrix' $\mathcal{Q} = \{q_{nl}\}(1 \leq n \leq N, 1 \leq l \leq L)$, where $q_{nl} = 0$ if $a_{nl} \neq$ '-' or otherwise, q_{nl} denotes the number of contiguous nulls counted from the left up to the position (n, l). Table 3.1 columns a-c show examples of matrices $\mathcal{A}, \mathcal{R}$, and $\mathcal{Q}$, respectively.

3.2.2 Representation

Within a computer program, the $\mathcal{R}$ and $\mathcal{Q}$ matrices may be stored in a more compact format than those shown in Table 3.1b and c. Table 3.1d and e show two alternative ways. In the format of Table 3.1d, the position, $l - 1$, preceding a gap and its length are recorded in a pair. In the last format, the lengths of a run of non-null residues and a gap are shown alternately, where gap lengths are represented by negative values for clarity [20]. These representations contain essentially the same information and are easily inter-convertible. The merit of using these compact representations is two-fold. First, we can save space to store alternative MSAs constructed from the same set of sequences. Second, we can easily insert or remove gaps without massive memory operations during a progressive procedure or an iterative one (see below).

	(a)	(b)	(c)	(d)	(e)
Carp:	MAYPMQL-FQ	1234567089	0000000100	(7,1;10,0)	(7,-1,2)
Hawk:	MAH--QLGF-	1230045670	0001200001	(3,2;9,1)	(3,-2,4,-1)
Lion:	MAN-SQLGFQ	1230456789	0001000000	(3,1;10,0)	(3,-1,6)
hline					

TABLE 3.1   Various ways to represent an MSA. (a) Ordinary alignment matrix. (b) Index matrix. (c) Gap state matrix. (d) Gap position and length. (e) Numbers of contiguous non-null (positive) and null characters (negative).

For a human interface, the format of an MSA shown in Table 3.1a is accessible only when its length is less than the width of a conventional terminal or printer. To display longer alignments, two types of formats are in use. The first is 'sequential,' in which sequences are written one after another and delineated by a special character or a line. Null characters are embedded accordingly so that the MSA is reproduced upon reloading. The most widely used sequential format is the FASTA format, in which a line starting with a 'greater than' symbol '>' delineates sequences. Although computers can easily recognize the sequential format, human eyes cannot. Hence, the most current MSA programs output the results in an 'interleaved' format, in which blocks of alignment, such as Table 3.1a with a fixed width, are arranged one after another until the entire alignment is exhausted. Table 3.2 shows an example of the MSF format that is most widely used in the community for exchanging

MSF:	177	Type: P	Check:	7844	..

Name: ggewa3	oo Len: 177	Check:	4939	Weight:	0.5375
Name: ggwn2c	oo Len: 177	Check:	9604	Weight:	0.5375
Name: ggicea	oo Len: 177	Check:	1667	Weight:	0.2897
Name: ggice7	oo Len: 177	Check:	3998	Weight:	0.2897
Name: gggaa	oo Len: 177	Check:	7028	Weight:	0.2795
Name: gglmf	oo Len: 177	Check:	608	Weight:	0.2795

//

```
ggewa3    K.......KQ   CGVLEGLKVK   SEWGRAYG..   SGHDREAFS.   QAIWRATFAQ
ggwn2c    D.......TC   CSIEDRREVQ   ALWRSIWSA.   EDTGRRTLIG   RLLFEELFEI
ggicea    VA..TPAMPS   MTDAQVAAVK   GDWEKI....   KGSGVEILY.   .....FFLNK
ggice7    ........AP   LSADQASLVK   STWAQV....   RNSEVEILA.   .....AVFTA
gggaa     .........S   LSAAEADLAG   KSWAPVFANK   NANGADFLV.   .....ALFEK
gglmf     PIVDSGSVAP   LSAAEKTKIR   SAWAPVYSNY   ETSGVDILV.   .....KFFTS

ggewa3    VPESRSLFKR   VHGDHTSD..   ...PAFIAHA   ERVLGGLDIA   ISTLD...QP
ggwn2c    DGATKGLFKR   VNVDDTHS..   ...PEEFAHV   LRVVNGLDTL   IGVLG...DS
ggicea    FPGNFPMFKK   L.GNDLAA.A   KGTAEFKDQA   DKIIAFLQGV   IEKLG..SDM
ggice7    YPDIQARFPQ   FAGKDVAS.I   KDTGAFATHA   GRIVGFVSEI   IALIGNESNA
gggaa     FPDSANFFAD   FKGKSVAD.I   KASPKLRDVS   SRIFTRLNEF   VNDAA...NA
gglmf     TPAAQEFFPK   FKGMTSADQL   KKSADVRWHA   ERIINAVNDA   VASMD...DT

ggewa3    ATLKEELDHL   QVQH.EGR.K   IPDNYFDAFK   TAILHVVAAQ   LGERCYSNNE
ggwn2c    DTLNSLIDHL   AEQH.KARAG   FKTVYFKEFG   KALNHVLPEV   AS..CFNPEA
ggicea    GGAKALLNQL   GTSH.KAM.G   ISQAQFNEFR   QALTELLGNL   GF..GGNIGA
ggice7    PAVQTLVGQL   AASH.KAR.G   ISQAQFNEFR   AGLVSYVSSN   VAWNAAAESA
gggaa     GKMSAMLSQF   AKEH.VGF.G   VGSAQFENVR   SMFPGFVASV   A.....APPA
gglmf     EKMSMKLRDL   SGKHAKSF.Q   VDPQYFKVLA   AVIADTVA..   .........A

ggewa3    EIHDAIACDG   FARVLPQVLE   RGIKGHH
ggwn2c    WNH...CFDG   LVDVISHRID   G......
ggicea    WNA...TVDL   MFHVIFNALD   GTPV...
ggice7    WTA...GLDN   IFGLLFAAL.   .......
gggaa     GAD...AWTK   LFGLIIDALK   AAGK...
gglmf     GDA...GFEK   LMSMICILLR   SAY....
hline
```

TABLE 3.2 An example of MSA represented in the MSF format.

MSA data.

3.2.3 Objective Functions

In PSA, an alignment score is defined as the sum of similarity values $S_2(a_{1l}, a_{2l})$ given to all matched pairs, a_{1l} and $a_{2l}(1 \leq l \leq L)$, and penalty values, $g(k)$, assigned to gaps of length k. This scoring system well accords with the underlying evolutionary process, and has been almost invariably adopted in current PSA tools with minor variations in terms of the choice of similarity values and gap penalty functions. The situation becomes much more complicated when we wish to design an objective function that evaluates the goodness of an MSA. Given an MSA of N sequences, $\mathcal{A}$, the objective function $H_N(\mathcal{A})$ generally takes the form:

$$H_N(\mathcal{A}) = \sum_{1 \leq l \leq L} \{S_N(\mathbf{a}_l) - G_N(*_l)\}, \tag{3.2}$$

where $S_N(\mathbf{a}_l)$ indicates the similarity score assigned to the column vector $\mathbf{a}_l$, $G_N(*_l)$ indicates the gap penalty given to column l, and the subscript N indicates the number of sequences involved. Several scoring systems for evaluating $H_N(\mathcal{A})$ are summarized below. Note that the term concerning $G_N(*_l)$ is not defined in some systems.

(1) Maximum parsimony (MP). Historically, MP was the first scoring system introduced in MSA algorithms [94]. Given an evolutionary tree of input sequences, $S_N(\mathbf{a}_l)$ is defined as the minimum possible number of mutations (substitutions, and single-letter insertions and deletions) for realizing $\mathbf{a}_l$. More generally, each mutation may be weighted according

to its type, and $S_N(\mathbf{a}_l)$ is defined as the minimum sum of the weights. In either case, a dynamic programming algorithm calculates $S_N(\mathbf{a}_l)$ in $O(N \cdot | \Sigma' |)$ [95]. Hein [40, 41] proposed a method that simultaneously performs the alignment and reconstruction of ancestral sequences under a given tree. An MSA obtained with this scoring system is called a tree alignment. The procedures for tree alignment are reviewed in [84].

(2) Maximum likelihood (ML). Similar to MP, ML is a major principle of phylogenetic tree reconstruction from sequence data [22, 57]. Of all the objective functions considered here, the ML score is probably the soundest from a biological point of view. Unfortunately, however, the computational costs are too high, and no practical MSA program based on this principle has been developed so far. Most objective functions considered here may be regarded as (extensively) simplified approximations of ML devised to circumvent this computational difficulty.

(3) Sum-of-pairs (SP) score. The SP scoring system [73, 13] is the most popular in MSA programs. Let $\bar{\mathbf{a}}_m$ and $\bar{\mathbf{a}}_n$ be two distinct row vectors extracted from alignment $\mathcal{A}$. Let $H_2(\bar{\mathbf{a}}_m, \bar{\mathbf{a}}_n)$ be the alignment score calculated in the same way as ordinary pairwise alignment with $S_2(a_{ml}, a_{nl})(1 \leq l \leq L)$ and $g(k = |q_{ml} - q_{nl}|)$, where matching nulls are ignored. For $S_2(a_{ml}, a_{nl})$, amino acid substitution matrices, such as PAMn [15], BLOSUMn [42], PET [50], and Gonnet [27], are commonly used directly or after normalization. For gap penalties, any functional forms may be acceptable, although affine functions of the form $g(k) = uk + v$ (u and v are non-negative constants) are the most popular. Then, the SP score is defined as:

$$H_{\mathrm{SP}}(\mathcal{A}) = \sum_{1 \leq m < n \leq N} H_2(\bar{\mathbf{a}}_m, \bar{\mathbf{a}}_n). \tag{3.3}$$

Although this row-wise definition suggests $O(LN^2)$ calculation steps, there exists a virtually $O(LN)$ algorithm if an affine gap penalty is adopted [31].

(4) Weighted sum-of-pairs (WSP) score. WSP [4] is an extension of the SP score where pairs of sequences are weighted so that they contribute differently to the overall alignment score such that:

$$H_{\mathrm{WSP}}(\mathcal{A}) = \sum_{1 \leq m < n \leq N} w_{m,n} H_2(\bar{\mathbf{a}}_m, \bar{\mathbf{a}}_n). \tag{3.4}$$

The weights $\{w_{m,n}\}$ play two superficially conflicting roles. First, closer pairs are given larger weights, because these pairs are expected to be aligned more reliably. Second, pairs between "dense" members with many close relatives are down-weighted, because they represent redundant information. The rationale-2 weights proposed by Altschul *et al.* [4] are ideal in these respects. The weights obtained by the three-way method [32] closely approximate the rationale-2 weights, and facilitate the profile-based efficient calculation of $H_{\mathrm{WSP}}(\mathcal{A})$. Wheras these 'pair-weights' are specifically assigned to sequence pairs, most other weights proposed so far [21, 43, 99, 108, 115] (reviewed in [16]) are assigned to individual sequences. With a set of these 'sequence weights' $\{w_m\}$, a pair weight is expressed as $w_{m,n} = w_m w_n$. Weighting sequences or sequence pairs are certainly beneficial [33, 112], and WSP scoring systems are widely adopted in recently developed high-performance MSA programs (see below). However, the relative performance of various weighting systems for MSA has not been extensively examined.

(5) Star alignment (SA) score. SA is a special case of tree alignment in which the tree topology is star-like. At the center of the tree, one of the input sequences or an inferred consensus sequence is placed. Denoting this center sequence by $\bar{\mathbf{c}}$, we obtain the SA score by:

$$H_{\mathrm{SA}}(\mathcal{A}) = \sum_{1 \leq n \leq N} H_2(\bar{\mathbf{c}}, \bar{\mathbf{a}}_n), \tag{3.5}$$

where $H_2(\bar{\mathbf{c}}, \bar{\mathbf{a}}_n)$ is the alignment score between $\bar{\mathbf{c}}$ and $\bar{\mathbf{a}}_n$, similarly computed as $H_2(\bar{\mathbf{a}}_m, \bar{\mathbf{a}}_n)$ in SP. For convenience, we define $H_2(\bar{\mathbf{c}}, \bar{\mathbf{a}}_n) = 0$ when $H_2(\bar{\mathbf{c}} = \bar{\mathbf{a}}_n)$. The SA scoring system has been used in the context of MSA with a guaranteed error bound [39, 83]. Besides this theoretical interest, however, SA scores are rarely used in practice because they are less informative than (W)SP (SP or WSP) scores.

(6) Maximum entropy difference (MED) or maximum relative entropy (MRE). If we ignore the evolutionary relatedness among sequences to be aligned, we can assume that residues within a column $\mathbf{a}_l$ are independent. Then, it is possible to evaluate $S_N(\mathbf{a}_l)$ by the extent to which the residue frequencies of $\mathbf{a}_l$ deviate from random expectations [97]. The MED score is defined as [26]:

$$H_{\mathrm{MED}}(\mathcal{A}) = \sum_{1 \leq l \leq L} \left\{ \sum_{c \in \Sigma} p_l(c) \ln(p_l(c)) - \sum_{c \in \Sigma} q(c) \ln(q(c)) \right\}, \tag{3.6}$$

where $p_l(c)$ and $q(c)$ are the relative frequencies of residue c observed in $\mathbf{a}_l$ and in the background, respectively. The MRE score is defined analogously as:

$$H_{\mathrm{MRE}}(\mathcal{A}) = \sum_{1 \leq l \leq L} \left\{ \sum_{c \in \Sigma} p_l(c) \ln(p_l(c)/q(c)) \right\}. \tag{3.7}$$

Entropy-based scores are often used for locating conserved columns within an MSA [96]. MED or MRE loses too much information as regards evolutionary relatedness and similarity among amino acid residues, and is not widely accepted as an objective function of global MSA.

(7) Maximum consistency (MC). This scoring scheme is indirect in the sense that an evolutionary process is not explicitly modeled, but implicitly incorporated in the pairwise alignments whose mutual consistency is maximized in the MSA procedure. We will discuss this scoring system in more detail in subsection 3.3.3 below.

Compared to the similarity score $S_N(\mathbf{a}_l)$, less attention has been paid to the gap penalty function $G_N(*_l)$. In many theoretical formulations, the contiguity of nulls in a row is ignored, which enables the $G_N(*_l)$ term to be absorbed into $S_N(\mathbf{a}_l)$ by taking a null character as an element of the extended alphabet set Σ'. For PSA, this is equivalent to using a 'proportional' gap penalty function of the form $g(k) = uk$, where k denotes the length of the gap and u is a positive constant known as the 'gap-extension penalty.' In reality, affine gap penalty functions of the form $g(k) = uk + v$ for $k > 0$ [28] are almost invariably used as noted above, where the constant term v denotes the so-called 'gap-open penalty.' An affine gap penalty function can be naturally incorporated into the SP, WSP or SA scoring system, in which $G_N(*_l)$ is explicitly formulated as a function of $\mathbf{q}_{l-1}$ and $\mathbf{a}_l, G(\mathbf{q}_{l-1}, \mathbf{a}_l)$ [30]. On the other hand, no established method has been reported for evaluating $G_N(*_l)$ in other MSA scoring schemes. The easy incorporation of affine gap penalties is one of the reasons why the SP or WSP scoring system has been most widely used in practice, and has proven to yield satisfactory results.

3.3   Algorithms

3.3.1   Exact Methods

N-dimensional Dynamic Programming

To evaluate the relative goodness of each alignment, $\mathcal{A}$, we have introduced an objective function, $H_N(\mathcal{A})$. The optimal alignment $\mathcal{A}^*$ is then defined as the one that has the best score among all the possible alternatives, *i.e.*, $\mathcal{A}^* = \text{argmax}_{\text{allpossible}\mathcal{A}}\{H_N(\mathcal{A})\}$. Given N sequences of a geometric average length of $\bar{L}$, a rough estimate of the number of possible distinct alignments among these sequences is $(N\bar{L})!/(\bar{L}!)^N \approx N^{N\bar{L}}$, which is tremendous even for a small N and a moderate $\bar{L}$, say $\bar{L} \approx 100$.

For PSA, the computational explosion can be avoided by using the well-known dynamic programming (DP) algorithm, which can rigorously and efficiently find the alignment(s) with the optimal score among all the possible alternatives by fulfilling the following recursion for $1 \leq i \leq I$ and $1 \leq j \leq J$ [120], where $I = |\bar{s}_1|$ and $J = |\bar{s}_2|$:

$$H_{i,j} = \max\left\{H_{i,j}^1, H_{i,j}^2, H_{i,j}^3\right\} \tag{3.8}$$

$$H_{i,j}^1 = H_{i-1,j-1} + S_2(s_{1i}, s_{2j}) \tag{3.9}$$

$$H_{i,j}^2 = \max_{1 \leq k \leq i}\left\{H_{i-k,j} - g(k)\right\} \tag{3.10}$$

$$H_{i,j}^3 = \max_{1 \leq k \leq j}\left\{H_{i,j-k} - g(k)\right\}. \tag{3.11}$$

It is natural to extend the above algorithm to N dimensions. Let $\mathbf{x} = \{x_n\}(1 \leq n \leq N, 0 \leq x_n \leq |\bar{s}_n|)$ be a coordinate indicating a node in the DP graph, and $\mathbf{b} = \{b_n\}(b_n \in \{0,1\})$ be a bit vector indicating an edge that joins adjacent nodes. Temporally, we will use a proportional gap penalty for simplicity. Then, the DP recursive relation is written as:

$$H_N(\mathbf{x}) = \max_{\mathbf{b}}\left\{H_N(\mathbf{x} - \mathbf{b}) + S_N(\mathbf{a})\right\}, \tag{3.12}$$

where $\mathbf{a} = \{a_n\}, a_n = s_{xn}$ if $b_n = 1, a_n = $ '-' if $b_n = 0$, and the elements of $\mathbf{b}$ take all possible combinations of 0 and 1 except for $\mathbf{b} = (0,0,\ldots,0)^{\text{T}}$. Most objective functions discussed in the previous section take $\text{O}(N)$ computational steps to evaluate a score for each column. There are $\prod_{1 \leq n \leq N}(|\bar{s}_n| + 1) \approx \bar{L}^N$ nodes in the DP graph, and we must examine $2^N - 1$ configurations (partial alignments) at each node to find the optimal path through the graph. Thus, the overall computation takes $\text{O}(N(2\bar{L})^N)$ steps using $\text{O}(\bar{L}^N)$ memory. In fact, it is known that MSA problems with reasonable objective functions are all NP-hard [118, 119, 51, 49]. If we use an affine or a more general gap penalty function, computational complexity is further increased. For the simplest case of an SP scoring system with $N = 3$ and an affine gap penalty function, we must consider 13 different configurations and seven types of state transitions, as shown in Tables 3.3 (a) and (b). Table 3.3 (c) shows the state transitions together with the number of gaps that open in association with the transition. For a large N, we would have to consider $\sim [N/(e \ln 2)]^N \sqrt{N}$ configurations and $2^N - 1$ types of transition [2]. Without additional speeding up techniques discussed in the following sub-subsections, $N = 3$ is the upper limit of applicability of straightforward DP algorithms.

(a)

Configuration	1	2	3	4	5	6	7	8	9	10	11	12	13
Seq1	*	*	*	*	.*	.*	-	-	*-	-	*-	--	--
Seq2	*	*	-	-	*-	--	*	*	.*	-	--	.*	*-
Seq3	*	-	*	-	--	*-	*	-	--	*	.*	*-	.*

(b)

Transition	1	2	3	4	5	6	7
Seq1	*	*	*	*	-	-	-
Seq2	*	*	-	-	*	*	-
Seq3	*	-	*	-	*	-	*

(c)

t\c	1	2	3	4	5	6	7	8	9	10	11	12	13
1	1,0	1,0	1,0	1,0	1,0	1,0	1,0	1,0	1,0	1,0	1,0	1,0	1,0
2	2,2	2,0	2,2	2,1	2,0	2,1	2,2	2,1	2,0	2,2	2,2	2,1	2,2
3	3,2	3,2	3,0	3,1	3,1	3,0	3,2	3,2	3,2	3,1	3,0	3,2	3,1
4	4,2	5,1	6,1	4,0	5,0	6,0	4,2	5,2	5,1	6,2	6,1	5,2	6,2
5	7,2	7,2	7,2	7,2	7,2	7,2	7,0	7,1	7,1	7,1	7,1	7,0	7,0
6	8,2	9,1	8,2	9,2	9,1	9,2	12,1	8,0	9,0	12,2	12,2	12,0	12,1
7	10,2	10,2	11,1	11,2	11,2	11,1	13,1	13,2	13,2	10,0	11,0	13,1	13,0

TABLE 3.3 Thirteen states (a) and seven transitions (b) used in a three-way DP algorithm. An asterisk, a dash, and a dot indicate (runs of) non-null, null, and either character, respectively. (c) Transition table used for counting gap-opening penalty in the three-way DP alignment. The configuration and the transition type are encoded by a number, as shown in (a) and (b), respectively. The first number in each cell indicates the resulting configuration induced by the transition type of the row from the original configuration of the column. The second number indicates the number of gaps that open in association with the transition. For example, transition type 7 converts state 3 into state 11 whereby one new gap opens.

MSA

MSA is an implementation of an N-dimensional dynamic programming algorithm with a restricted search space, called the Carrillo-Lipman algorithm [13, 38, 68]. It reduces the search space by using upper bounds estimated from the information of a provisional MSA $\mathcal{A}^{\#}$ and optimal pairwise sequence alignments. Note that *MSA* optimizes an objective function by not maximizing a similarity score but minimizing a transformation cost.

The objective function of *MSA* is the WSP score:

$$C(\mathcal{A}) = \sum_{1 \leq m < n \leq N} w_{m,n} C(\mathcal{A}_{m,n}).$$

The pair-weights are calculated using the rationale-1 method of Altschul *et al.* [4]. $C(\mathcal{A}_{m,n})$ is the cost of the induced pairwise alignment between $\bar{\mathbf{a}}_m$ and $\bar{\mathbf{a}}_n$. Obviously,

$$L(S) \leq C(\mathcal{A}^*) \leq C(\mathcal{A}^{\#}),$$

where $\mathcal{A}^*$ is the optimal alignment of $S = \{\bar{\mathbf{s}}_n\}$. $L(S)$ is the sum of the cost of optimal pairwise alignments, that is,

$$L(S) = \sum_{1 \leq m < n \leq N} w_{m,n} C^*(\bar{\mathbf{s}}_m, \bar{\mathbf{s}}_n),$$

where $C^*(\bar{\mathbf{s}}_m, \bar{\mathbf{s}}_n)$ is the minimum cost between $\bar{\mathbf{s}}_m$ and $\bar{\mathbf{s}}_n$. For any S and $1 \leq p < q \leq N$,

$$C(\mathcal{A}^*) - L(S) = \sum_{1 \leq m < n \leq N} w_{m,n} \left\{ C(\mathcal{A}^*_{m,n}) - C^*(\bar{\mathbf{s}}_m, \bar{\mathbf{s}}_n) \right\} \qquad (3.13)$$

$$\geq w_{p,q} \left\{ C(\mathcal{A}^*_{p,q}) - C^*(\bar{\mathbf{s}}_p, \bar{\mathbf{s}}_q) \right\}. \qquad (3.14)$$

Hence,

$$w_{p,q} C(\mathcal{A}_{p,q}^*) \leq C(\mathcal{A}^\#) - L(S) + w_{p,q} C^*(\bar{\mathbf{s}}_p, \bar{\mathbf{s}}_q), \tag{3.15}$$

because $C(\mathcal{A}^*) \leq C(\mathcal{A}^\#)$. $C(\mathcal{A}^\#)$ is calculated using a simple progressive method.

In order to calculate $C(\mathcal{A}^*)$, the right-hand side of (3.15) is performed as an upper bound for (p, q)-plane to reduce the search space. The upper bound, however, is usually large, because $O(N^2)$ terms are discarded to obtain (3.15). Although *MSA* can use this upper bound, it uses by default a cost of an induced pairwise alignment from $\mathcal{A}^\#$ as a heuristic upper bound for a plane. This heuristic upper bound is able to align more sequences than the upper bound by (3.15), although optimality of an alignment cannot be guaranteed any longer.

To determine the search space, *MSA* first determines the admissible points on $\binom{N}{2}$ planes. An admissible point (i, j) satisfies $C_{m,n}(i, j) \leq U_{m,n}$, where $U_{m,n}$ is an upper bound for (m, n)-plane, that is, a cost of an induced pairwise alignment between $\bar{\mathbf{a}}_m^\#$ and $\bar{\mathbf{a}}_n^\#$. $C_{m,n}(i, j)$ is the minimum cost between $\bar{\mathbf{s}}_m$ and $\bar{\mathbf{s}}_n$ via (i, j), which is the sum of the minimum costs from $(0, 0)$ to (i, j) and from $(| \bar{\mathbf{s}}_m |, | \bar{\mathbf{s}}_n |)$ to $(i + 1, j + 1)$. Because only the admissible points are candidates for those contributing to an optimal alignment, the search space is the intersection of the admissible points on every plane.

Before describing the algorithm for finding an optimal alignment, we address two data structures (an open list and a cell) and a gap cost of *MSA*. The open list Q stores a pointer to a cell that is going to be visited. A cell consists of a previous node *prev*, a current node *curr* and a cost from the start node to *curr*, *cost*. The open list is implemented as a priority queue. Since *MSA* uses the well-known Dijkstra's algorithm, all costs including substitution matrix elements are converted to non-negative.

MSA uses a quasi-natural gap cost, which is an approximate affine gap cost. The use of the affine gap cost (or the so-called natural gap cost) is impractical, because large memory and much computation are required, as mentioned above. The quasi-natural gap cost penalizes gap opens based on the two adjacent edges $\mathbf{u} = \mathbf{x} \to \mathbf{y}$ and $\mathbf{v} = \mathbf{y} \to \mathbf{z}$ on a path, where $\mathbf{x}, \mathbf{y}$ and $\mathbf{z}$ indicate nodes. Let v_m be the mth element of $\mathbf{z} - \mathbf{y}$. If $v_m = 0$, v_m means a null; otherwise, v_m denotes the z_mth residue of $\bar{\mathbf{s}}_m$. Table 3.4 shows the rule for penalizing gap opens. This rule penalizes no less than the actual number of gap opens. When $(u_m, u_n) = (0, 0)$ and (v_m, v_n) is either $(0, 1)$ or $(1, 0)$, an existing gap may extend at $\mathbf{z}$. However, this rule always assigns a gap open penalty in such cases, *i.e.*, the quasi-natural gap cost adopts the pessimistic view. Let $C(\mathbf{u}, \mathbf{v})$ be the transformation cost associated with the move from node $\mathbf{x}$ to $\mathbf{z}$ along edges $\mathbf{u}$ and $\mathbf{v}$. Then,

$$C(\mathbf{u}, \mathbf{v}) = \sum_{1 \leq m < n \leq N} w_{m,n} \{c(t_m, t_n) + g(u_m, u_n, v_m, v_n)\}, \tag{3.16}$$

where $t_k = z_k \cdot v_k$. $c(a, b)$ is the transformation cost between a and b. If either a or b is null, $c(a, b)$ is the gap-extension penalty. If both a and b are null, $c(a, b) = 0$. $g(u_m, u_n, v_m, v_n)$ is the gap-open penalty based on Table 3.4.

Using the upper bound $U = \sum_{1 \leq m < n \leq N} U_{m,n}$, *MSA* tries to find an optimal alignment within the reduced search space. It first extracts and removes a cell u with the minimum cost from the open list. Then, each admissible node (say $\mathbf{z}$) adjacent to $u.curr$ is examined. If Q does not have cell v such that $v.prev = u.curr$ and $v.curr = \mathbf{z}$, such cell v with cost $U + 1$ is inserted into Q. If $u.cost + C(u, v) < v.cost$, $v.cost$ is replaced by $u.cost + C(u, v)$. After every admissible node adjacent to $u.curr$ is checked, another cell is extracted from the open list, and these procedures are repeated until *curr* of the extracted cell is the end node $(| \bar{\mathbf{s}}_1 |, \ldots, | \bar{\mathbf{s}}_N |)^{\mathrm{T}}$.

$\binom{v_m}{v_n}$	$\binom{u_m}{u_n}$			
	$\binom{-}{-}$	$\binom{-}{*}$	$\binom{*}{-}$	$\binom{*}{*}$
$\binom{-}{-}$	0	1	1	0
$\binom{-}{*}$	0	0	1	0
$\binom{*}{-}$	0	1	0	0
$\binom{*}{*}$	0	1	1	0

TABLE 3.4 The rule for penalizing gap opens. An element of an edge is expressed by a symbol, $-$ for a null and $*$ for a residue. A gap open penalty is imposed on a cell with the value of unity.

Note that *MSA* does not always find an optimal alignment subject to the constraints; it may obtain an alignment whose WSP is greater than U. If this happens, the user must manually increase the upper bounds to find an optimal alignment. In addition, the number of sequences to be aligned is limited to single figures even when the heuristic upper bounds are used. Therefore, *MSA* cannot be used for aligning many sequences.

Several methods [47, 66, 98] use the A^* algorithm for aligning more sequences than *MSA* can. These A^*-based methods adopt essentially the same strategy as *MSA*. The major difference lies in the way of calculating the cost assigned to a move from node $\mathbf{y}$ to $\mathbf{z}$. Let $C(\mathbf{y}, \mathbf{z})$ be the cost from $\mathbf{y}$ to $\mathbf{z}$. The modified cost $C'(\mathbf{y}, \mathbf{z})$ is defined as

$$C'(\mathbf{y}, \mathbf{z}) = C(\mathbf{y}, \mathbf{z}) + L(\mathbf{y} \to \mathbf{t}) - L(\mathbf{z} \to \mathbf{t}), \qquad (3.17)$$

where $\mathbf{t} = (|\ \bar{\mathbf{s}}_1\ |, \ldots, |\ \bar{\mathbf{s}}_N\ |)^{\mathrm{T}}$ is the end node and $L(\mathbf{y} \to \mathbf{t})$ denotes the lower bound of the alignment costs for subsequences $s_{m y_m}, \ldots, s_{m|\bar{\mathbf{s}}_m|}$. $L(\mathbf{z} \to \mathbf{t})$ is defined in a similar manner. The use of $C'(\mathbf{y}, \mathbf{z})$ decreases more the number of admissible nodes than *MSA* does, and hence speeds up the computation. *GSA* [66] achieves further performance improvements. It uses better lower bounds estimated from three-way alignments instead of pairwise alignemnts, and dynamically updates an upper bound during the alignment process. However, it is still difficult to align many sequences within a reasonable time. Reinert *et al.* [88] have improved the scalability of their A^*-based algorithm by combining it with the divide and conquer algorithm (DCA) [111, 102], a heuristics for finding anchor points for global MSA from a set of PSAs.

COSA

COSA is an implementation of the integer linear programming (ILP) method instead of N-dimensional dynamic programming [1]. ILP maximizes an objective function $\sum w_x \cdot x$ subject to some constraints, where x is an integer variable to be optimized and w_x is a weight associated with x. An important feature of *COSA* is that it can accept *any* gap costs, such as affine, convex or position-specific gap cost. We briefly describe the algorithm of *COSA*. Note that *COSA* maximizes the similarity score rather than minimizes a transformation cost.

To represent an alignment, two types of variables are used: alignment and gap variables. Both variables take a value of either 0 or 1. An alignment variable denotes whether or not a pair of residues is aligned, whereas a gap variable represents whether or not a segment of a sequence is aligned with a gap of another sequence. Let $a(s_{mi}, s_{nj})$ be an alignment variable for a residue pair of s_{mi} and s_{nj}, and $g_q(s_{pi}, s_{pj})$ be a gap variable for a segment $s_{pi}, \ldots, s_{pj}$ and a sequence $\bar{\mathbf{s}}_q$. $a(s_{mi}, s_{nj}) = 1$ means that residues s_{mi} and s_{nj} are aligned, and $g_q(s_{pi}, s_{pj}) = 1$ means that a segment $s_{pi}, \ldots, s_{pj}$ is aligned with a gap inserted in $\bar{\mathbf{s}}_q$. The total number of these variables is $\sum_{1 \leq m < n \leq N} |\bar{\mathbf{s}}_m||\bar{\mathbf{s}}_n| + N(N-1) \sum_{1 \leq n \leq N} \binom{|\bar{\mathbf{s}}_n|+1}{2}$.

Four constraints are required to calculate an optimal alignment by ILP. First, each residue of a sequence must either correspond to a residue of another one, or be within a region represented by a gap variable. Second, there should not be any incompatibly aligned residue pairs. For example, if $a(s_{pi}, s_{qj}) = a(s_{pl}, s_{qk}) = 1$ with ($i < l$ and $k < j$), then the corresponding residue pairs are incompatible. Third, the regions of two gap variables should not overlap and there must be at least one aligned residue pair between these regions. When a convex gap cost is used, this constraint is satisfied automatically. Fourth, the transitivity of three alignment variables has to be satisfied. Specifically, if $a(s_{pi}, s_{qj}) = 1$ and $a(s_{pi}, s_{rk}) = 1$, then $a(s_{qj}, s_{rk})$ must be 1.

Alignment and gap variables can be thinned out in advance, as most of them are unlikely to take the value of 1. The idea of reducing the number of variables is essentially the same as that of the search space determination of *MSA*. A heuristic alignment is first constructed to obtain a lower bound L. Then, an upper bound U_v is calculated for every variable v. If $U_v \leq L$, v is set to 0 and therefore removed. Assuming that v is associated with sequences $\bar{s}_m$ and $\bar{s}_n$, $U_v = C_v^*(\bar{s}_m, \bar{s}_n) + \sum_{(p,q) \neq (m,n)} C^*(\bar{s}_p, \bar{s}_q)$, where $C_v^*(\bar{s}_m, \bar{s}_n)$ is the score of the optimal alignment between $\bar{s}_m$ and $\bar{s}_n$ such that $v = 1$, and $C^*(\bar{s}_p, \bar{s}_q)$ is the score of the optimal PSA between $\bar{s}_p$ and $\bar{s}_q$.

Because solving an ILP is NP-complete, *COSA* adopts a cutting plane algorithm. This algorithm solves a linear programming (LP), called a relaxation of an ILP, which is obtained by omitting integer constrains of the ILP. Each variable of the relaxation then takes a value between 0 and 1. If a solution of the relaxation is integral, it is also the solution of the ILP. Otherwise, a cutting plane is calculated from the solution of the relaxation and added to the constraints of the relaxation. A cutting plane is a linear constraint that does not exclude an optimal solution of the ILP. This procedure is repeated until an integer solution is obtained. *COSA* provides another way to construct alignments in an exact way. Similar to *MSA*, however, it cannot align many sequences.

3.3.2 Progressive Methods

Strategy

Progressive methods heuristically construct alignments using a two-dimensional dynamic programming method that aligns two groups of sequences (that is, two alignments) to obtain a single alignment. The outline of the progressive methods is as follows.

1. Calculate all possible $\binom{N}{2}$ pairwise alignments to obtain a distance matrix.
2. Construct a guide tree from the matrix.
3. Progressively align groups of sequences following the branching order in the tree.

Because we usually assume that the sequences to be aligned are phylogenetically related, a guide tree is calculated and then an alignment is constructed according to the branch order in the tree. The guide tree is usually constructed by a distance matrix method, such as UPGMA [100] or the NJ method [91]. Progressive methods begin by aligning the most similar pair of sequences. Then, they align the two sequences (or a sequence and an alignment, or two alignments) that are the second most similar. These procedures are repeated until all sequences are aligned into a single alignment. Note that once constructed, intermediate alignments remain unchanged; once alignment errors occur, they cannot be corrected. Therefore, the alignment order directly affects the quality of the final result. Although progressive methods can rapidly calculate large alignments, their accuracies tend to be inferior to those obtained by more elaborate methods.

ClustalW

ClustalW [107] is the most widely used program of the progressive methods. It features the use of position-specific gap penalties. In this subsection, we provide the detailed algorithm of *ClustalW* along the outline presented above.

The distance between sequences $\bar{s}_m$ and $\bar{s}_n$, $D_{m,n}$, is defined as $D_{m,n} = 1 - I_{m,n}$, where $I_{m,n}$ is the degree of sequence identity between $\bar{s}_m$ and $\bar{s}_n$. *ClustalW* calculates $I_{m,n}$ using the dynamic programming algorithm of Myers and Miller [74]. These $D_{m,n}$ are not corrected for multiple substitutions. In this phase, *ClustalW* uses the Gonnet250 amino-acid substitution matrix [27]. After obtaining the distance matrix, a guide tree is constructed by the NJ method [91].

Next, *ClustalW* calculates sequence weights. To this end, *ClustalW* places the root in the guide tree, because the weights [108] are set proportional to the average sum of branch lengths from each sequence (leaf) to the root. The root is placed where the average sum of branch lengths of both sides of the root should be equal. Using the guide tree and the sequence weights, *ClustalW* constructs a multiple alignment along with the branching order of the tree. However, if there are divergent sequences such that the sequence identities are less than 30%, these sequences are not aligned until the other sequences have been aligned.

To align group $\mathcal{A}$ of length I with group $\mathcal{B}$ of length J, *ClustalW* conducts four steps as shown below. First, a substitution matrix is chosen depending on the average sequence identity between $\mathcal{A}$ and $\mathcal{B}$, which is defined as $\sum_{m=1}^{M} \sum_{n=1}^{N} I_{m,n}/MN$. The ranges of distances and substitution matrices used are 35-100%, Gonnet80; 25-35%, Gonnet120 (if $\min(M,N) < 100$, Gonnet250); and 0-25%, Gonnet160 (if $\min(M,N) < 100$, Gonnet350). The minimal element of the raw matrix is subtracted from each matrix element to convert all the elements into non-negative.

Second, *ClustalW* calculates an initial gap open penalty.* The initial gap open penalty $v^{\mathcal{AB}}$ between $\mathcal{A}$ and $\mathcal{B}$ is given by $v^{\mathcal{AB}} = v \cdot \bar{\mu}_s \lambda l(I,J)$, where v is the default gap open penalty. $\bar{\mu}_s$ and λ are an average score of the substitution matrix other than the diagonal elements and a constant specific to the matrix, respectively. $\bar{\mu}_s$ and λ reflect the concept that an optimum gap penalty depends on a substitution matrix and sequence similarities. $l(I,J)$ is a factor adjusting for the unequal lengths between I and J.

Third, position-specific gap penalties are calculated. The basic idea of position-specific gap penalties is to decrease gap penalties on gap-containing regions but to increase them near such regions. The algorithm for calculating position-specific gap penalties consists of three steps:

1. If $\mathbf{c}_l$ includes nulls, $u_l^{\mathcal{C}} \leftarrow u/2$ and $v_l^{\mathcal{C}} \leftarrow v^{\mathcal{AB}} \cdot G_l/N$, and the calculation is terminated.

2. If $\mathbf{c}_l$ has residues on hydrophilic stretches, $v_l^{\mathcal{C}} \leftarrow v^{\mathcal{AB}}/2$.
   Otherwise, $v_l^{\mathcal{C}} \leftarrow \sum_{n=1}^{N} r(c_{nl})/N$.

3. If $\mathbf{c}_l$ is within T columns from a gap-containing region, $v_l^{\mathcal{C}} \leftarrow v_l^{\mathcal{C}} \cdot (2 + 2 \times (T-d)/T)$.

$u_l^{\mathcal{C}}$ and $v_l^{\mathcal{C}}$ are the gap-extension penalty and the open penalty at column $\mathbf{c}_l$. $v^{\mathcal{AB}}$ is the initial gap open penalty between groups $\mathcal{A}$ and $\mathcal{B}$ and u is the gap extension-penalty. G_l is the number of nulls on $\mathbf{c}_l$. $r(c_{nl})$ is a residue-specific gap factor of c_{nl}. T is set to 4. A hydrophilic stretch consists of five or more consecutive hydrophilic residues. If a column having no gap is on a hydrophilic stretch, the gap open penalty of this column is decreased,

---

*As of *ClustalW* version 1.83, the default gap extension penalty is not modified.

because hydrophilic stretches indicate loop regions. If there is neither null nor a residue on a hydrophilic stretch at the column, residue-specific gap factors are used for the calculation. A residue-specific gap factor of residue r indicates how likely a gap occurs adjacent to r.

After that, *ClustalW* aligns $\mathcal{A}$ with $\mathcal{B}$ based on the substitution matrix and position-specific gap penalties. For calculation of the score, *ClustalW* employs the modified algorithm of Myers and Miller so that it can deal with position-specific gap penalties [106].

Although the basic idea for calculating position-specific gap penalties is simple, its algorithm is somewhat complicated and requires much computation, compared with the usual affine gap penalty. As a result, iterative refinement methods, such as *MAFFT* [52] and *MUSCLE* [20], are even faster than *ClustalW*. These iterative methods are generally more accurate than *ClustalW*.

POA

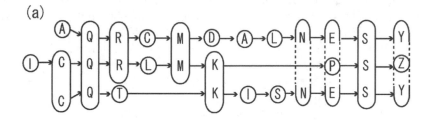

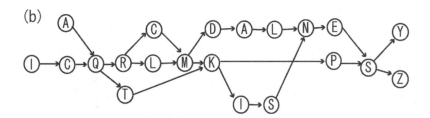

FIGURE 3.1: An example of partially ordered alignment (POA).

ClustalW and most of the iterative refinement methods discussed below convert a column of MSA into a profile vector before two MSAs are subjected to group-to-group alignment. This conversion is irreversible and inevitably leads to loss of information. A directed acyclic graph (DAG) , on the other hand, can reversibly represent an MSA in a form as compact as a profile (Figure 3.1). The nodes (residues) in a DAG are partially ordered, implying that a node in a branch (insertion or mismatch) is not necessarily defined to be located before or after the nodes in another branch. If the underlying evolutionary process involves recombination, alternative splicing, or domain shuffling, a DAG is a more realistic representation than the profile representation of an MSA. The partial order alignment (*POA*) method proposed by Lee and coworkers [65, 35] takes advantage of this feature of the DAG. Pairwise alignment between a DAG and a usual sequence or between DAGs is performed with a network alignment algorithm [61], an ordinary DP algorithm extended to a branch-and-merge structure in an additional dimension. Hein [40] was the first to use this data structure in an MSA procedure, *i.e.*, tree alignment based on the MP principle. Instead of MP or SP, *POA* adopts a surprisingly simple scoring system; at a meeting point of several branches, the path that gives the best score at that point is adopted. For example, since different kinds of residues are represented by multi-furcating branches, the matching score between gapless columns $\mathbf{a}_i$ and $\mathbf{b}_j$ in the original MSAs under comparison is evaluated as: $S_2(\mathbf{a}_i, \mathbf{b}_j) = \max_{a \in \mathbf{a}_i, b \in \mathbf{b}_j} S_2(a, b)$, where a and b denote residue type included in $\mathbf{a}_i$ and $\mathbf{b}_j$, respectively, and $S_2(a, b)$ is the corresponding substitution matrix element (Blosum80 by default). Thus, although the DAG structure may hold full information about the original MSA, the alignment procedure of *POA* actually utilizes only part of the information. *POA* can be run with either of the two types of progressive procedures, sequential (sequences are processed in consecutive order) or tree-based. A few performance tests have indicated that the tree-based *POA* with this 'cheap' scoring scheme well competes with more sophisticated MSA programs, such as *ClustalW*. This finding is instructive because it urges us to reconsider the best balance between the complexity of the theoretical model and the actual performance of MSA programs.

The jumping alignment algorithm [101] shares a property with *POA* in the process of MSA vs. sequence alignment, *i.e.*, only one of the sequences within the input MSA contributes to the pairwise alignment score. This 'representative' sequence may vary column by column, although such a transition is penalized. The penalty signifies horizontal continuity of sequence conservation, whereas such horizontal information is lost in profile or DAG representations. The jumping alignment algorithm does not seem to have been applied to the construction of MSAs, but may be useful not only for recognizing potential recombinations but also for reducing noise associated with conventional profiles.

3.3.3   Consistency-based Methods

General Concepts

The concept of consistency among a set of pairwise alignments was introduced by Gotoh [29] to look for 'anchor points,' *i.e.*, local MSAs plausibly embedded in an optimal global MSA. Consider three sequences, $\bar{\mathbf{s}}_1, \bar{\mathbf{s}}_2, \bar{\mathbf{s}}_3$, and three PSAs between them, $\mathcal{A}_2(\bar{\mathbf{s}}_1, \bar{\mathbf{s}}_2), \mathcal{A}_2(\bar{\mathbf{s}}_1, \bar{\mathbf{s}}_3)$ and $\mathcal{A}_2(\bar{\mathbf{s}}_2, \bar{\mathbf{s}}_3)$. Each PSA can be represented by an undirected bipartite graph (V, E), where a vertex corresponds to a residue and an edge, $e(s_{mi}, s_{nj})$, corresponds to a matched pair ($m \neq n \in [1, 2, 3], s_{mi} \in \bar{\mathbf{s}}_m, s_{nj} \in \bar{\mathbf{s}}_n$). The set of edges, $E(\bar{\mathbf{s}}_m, \bar{\mathbf{s}}_n)$, is called trace [60]. In this trace formulation, the degree of each vertex is either 0 or 1 depending on whether it matches a null or a residue. In Gotoh's formulation [29], on the other hand, a vertex corresponds not to a residue itself but to a joint between them or either end of a sequence, and every vertex belongs to one or more edge (Figure 3.2 c). The latter formulation can take care of gapped alignments more precisely than the trace formulation. For the sake of cohesion with other methods, however, we consider here that each vertex corresponds to a residue (Figure 3.2 b). If there exist three edges $e(s_{1i}, s_{2j}) \in E(\bar{\mathbf{s}}_1, \bar{\mathbf{s}}_2), e(s_{2j}, s_{3k}) \in E(\bar{\mathbf{s}}_1, \bar{\mathbf{s}}_3)$ and $e(s_{1j}, s_{3k}) \in E(\bar{\mathbf{s}}_1, \bar{\mathbf{s}}_3)$, the triple edges and vertices are said to be consistent (Figure 3.2 d, thick bars). A set of residues belonging to contiguous consistent edges form a consistently aligned region (Figure 3.2 e). Essentially the same formulation is used for finding weakly conserved regions among a set of local MSAs (blocks) [62]. For $N > 3$, the above argument applies to every combination of three sequences in $\{\bar{\mathbf{s}}_n\}$ ($1 \leq n \leq N$). When consistency holds for any combination of three vertices from $\{s_{ni}\}$, the vertex set is considered to be consistent. Alternatively, $\{s_{ni}\}$ may be considered consistent if every s_{ni} participates in at least one consistent triple vertex. The computational complexity is O($N^3 \bar{L}$) with either definition. Vingron and Pevzner [114] considered intermediate cases in which each edge

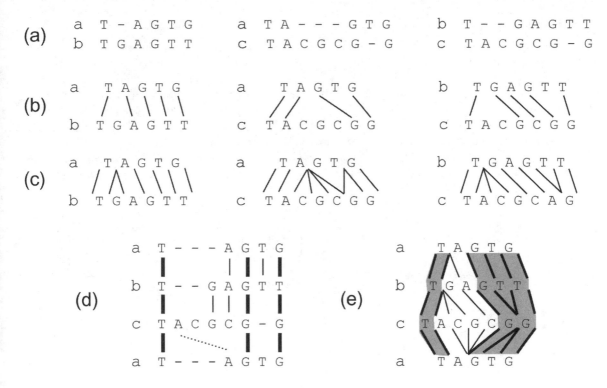

FIGURE 3.2: Consistency of pairwise alignments between three sequences $\bar{a}, b$, and $\bar{c}$. (a) Alignments. (b) Traces. (c) Edges of bipartite graphs representing the alignments of (a). (d) Trace edges realized (solid bars) and unrealized (dotted bar) in an MSA. Consistent edges are shown by thick bars. (e) Another formalization of the consistently aligned regions (shaded areas). In (d) and (e), sequence $\bar{a}$ is shown in duplicate to enhance viewing.

$e(s_{mi}, s_{nj})$ belongs to at least $K(1 \le K \le N - 2)$ consistent triplets, although the vertexes are derived not from PSAs but from dot-matrix plots between every pair of sequences.

Maximum Weight Trace Problem

The maximum weight trace problem was formulated by Kececioglu based on a motivation similar to that discussed above [54], namely, given a set of PSAs, find an MSA that is closest to all PSAs in the set. The input to the maximum weight trace problem is a set of edges, that are obtained from a set of pairwise optimal sequence alignments in a special case. The algorithm finds an MSA in which the maximum number of input edges (or more generally, maximum sum of weights associated with the edges) is realized (Figure 3.2 d). Because the MSA must satisfy the two conditions A and B mentioned in subsection 3.2.1, the maximum weight trace problem is not trivial, and in fact proven to be NP-hard [54]. Kececioglu proposed a branch-and-bound algorithm somewhat similar to the *MSA* program described above [54]. Recently, Kececioglu et al. [53] reformulated the problem, and proposed an ILP algorithm to solve it with a branch-and-cut technique (3.3.1). Because of the intrinsic difficulty, the applicability of even this new approach does not seem to much exceed that of the *MSA*.

T-Coffee

The *T-Coffee* algorithm developed by Notredame *et al.* [78] is based on a more practical use of consistency information derived from a set of PSAs, compared to the strategies discussed above. Assume that we are given N sequences $\bar{s}_n (1 \leq n \leq N)$ and $KN(N-1)/2$ PSAs between them, $\mathcal{A}_2^k(\mathbf{s}_m, \mathbf{s}_n)(1 \leq k \leq K)$, in the form of a bipartite graph, where K denotes the total number of PSAs obtained with different PSA methods (*e.g.*, global and local) for each pair of sequences. The primary library consists of residue pairs $(s_{mi}, s_{ni}) \in E^k(\bar{s}_m, \bar{s}_n)$, with associated weights. The weight assigned to such a residue pair, $w^k(s_{mi}, s_{ni}) = w^k(\bar{s}_m, \bar{s}_n)$, is the percent sequence identity of $\mathcal{A}_2^k(\bar{s}_m, \bar{s}_n)$. If $(s_{mi}, s_{ni}) \notin E^k(\bar{s}_m, \bar{s}_n), w^k(s_{mi}, s_{ni})$ is 0. When the same residue pair appears in the set of PSAs more than once, the associated weights are summed up to yield the primary weight $W_p(s_{mi}, s_{nj}) = \sum_{1 \leq k \leq K} w^k(s_{mi}, s_{nj})$.

Note that $W_p(s_{mi}, s_{ni}) = 0$, if (s_{mi}, s_{ni}) is absent in any $E^k(\mathbf{s}_m, \mathbf{s}_n)$. The primary library is nearly equivalent to the input edges used in the special case of the maximum weight trace problem mentioned above. The edges in a primary library occupy only a sparse subset of all the edges to be examined by a complete MSA procedure. *T-Coffee* extends the library by adding residue pairs that are 'indirectly' matched. For example, if $(s_{mi}, s_{pk}) \in E^k(\bar{s}_m, \bar{s}_p)$ and $(s_{nj}, s_{pk}) \in E^k(\bar{s}_n, \bar{s}_p)$ for a triplet $(m, n, p; 1 \leq m, n, p \leq N)$, the residue pair (s_{mi}, s_{nj}) is said to be indirectly matched, and added to the 'extended library' even if $(s_{mi}, s_{nj}) \notin E^k(\bar{s}_m, \bar{s}_n)$. The weight to an indirectly matched pair (s_{mi}, s_{nj}) mediated by s_{pk} in a third sequence $\bar{s}_p$ is defined as $w^k(s_{mi}, s_{nj}; s_{pk}) = \min\{w^k(s_{mi}, s_{pk}), w^k(s_{nj}, s_{pk})\}$. Now, the total weight given to a residue pair (s_{mi}, s_{nj}) is obtained by

$$W(s_{mi}, s_{nj}) = \sum_{1 \leq k \leq K} \left\{ w^k(s_{mi}, s_{nj}) + \sum_{p \neq m, p \neq n} w^k(s_{mi}, s_{nj}; s_{pk}) \right\}. \tag{3.18}$$

T-Coffee uses these weights in place of an ordinary score matrix, such as PAMn or Blosum$n$, in the DP-based pairwise alignment of single or pre-aligned groups of sequences without imposing any gap penalties. Otherwise, *T-Coffee* adopts the typical progressive alignment strategy. The distance matrix and the guide tree are constructed in the same manner as those of *ClustalW*. The major advantage of *T-Coffee* over *ClustalW* and other ordinary progressive methods is that information about alignments between all pairs of sequences is condensed in the weights $W(s_{mi}, s_{nj})$, which are utilized even at the very beginning of the progressive procedure. Another advantage of *T-Coffee* is that several different sources of alignment information, *e.g.*, that obtained from global and local PSAs, are mixed to compute a residue-pair weight. The computational complexity is O($N^2\bar{L}^2$) for pairwise alignment, O($N^3\bar{L}$) for library construction, and O($N\bar{L}^2$) for the progressive alignment.

DIALIGN

The notion of 'consistency' implied in the *DIALIGN* algorithm [71, 70] differs from that mentioned above, but simply means that two ungapped segment pairs (diagonals or fragments), $\mathbf{f} = \mathcal{A}_2(a_i a_{i+1} \cdots a_{i+k-1}, b_j b_{j+1} \cdots b_{j+k-1})$ and $\mathbf{f}' = \mathcal{A}_2(a_{i'} a_{i'+1} \cdots a_{i'+k-1}, b_{j'} b_{j'+1} \cdots b_{j'+k-1})$, are arranged in the order of $\mathbf{f} \leq \mathbf{f}'$ or $\mathbf{f}' \leq \mathbf{f}$, where $\mathbf{f} \leq \mathbf{f}'$ holds if $i + k \leq i'$ and $j + k \leq j'$, and vice versa. To avoid confusion, we will use the term 'compatible' instead of 'consistent' here to refer to such situations. The idea of the *DIALIGN* algorithm is somewhat related to that of the maximum weight trace problem, although units used in an alignment process are fragments with positive weights (significant fragments) rather than individual matched pairs. The objective function is the sum of weights of the fragments that are involved in the final alignment, where no penalty is imposed on the gaps. The residues that are not included in these fragments remain unaligned. Hence, *DIALIGN* tends to produce global to more local alignments with a decrease in similarity of sequences under comparison.

For PSA, *DIALIGN*, as well as most standard procedures, uses a DP algorithm. At each node (i, j) of recursion, *DIALIGN* examines $\min(i, j)$ possible segment pairs that end at (i, j), indicating $O(\bar{L}^3)$ overall computational steps. By restricting the fragment sizes below a fixed value K $(= 40$ by default), the computational complexity is reduced to $O(K\bar{L}^2)$, although it is still considerably greater than that of simpler DP algorithms that leave highly divergent parts unaligned [3, 45]. The most crucial step of the *DIALIGN* algorithm is the evaluation of the weight for a fragment $\mathbf{f}$, $w(\mathbf{f})$. When a fragment of length k has an alignment score of $s(\mathbf{f}) = \sum_{0 \leq l \leq k} S_2(a_{i+l}, b_{j+l})$, $w(\mathbf{f})$ is defined as $w(\mathbf{f}) = -\log P(k, s(\mathbf{f}))$, where $P(k, s(\mathbf{f}))$ denotes the probability of observing by chance one with an alignment score $\geq s(\mathbf{f})$ among $(|\bar{\mathbf{a}}| - k + 1)(|\bar{\mathbf{b}}| - k + 1)$ pairs of segments of length k each having random sequences. Since the expected value for $P(k, s(\mathbf{f}))$ is close to 1, a majority of the fragments have negative weights and are excluded from the list of candidates.

For MSA, *DIALIGN* adopts a greedy strategy. First, all combinations of input sequences are aligned as described above to yield the initial list of significant fragments. The weight for each fragment is recalculated in a similar fashion to that used in the construction of the extended library in the *T-Coffee* program, except that the supplementary weights are derived from indirectly aligned segment pairs. The fragments in the initial list are examined in the order of their weight values, and moved to the second list as long as they are compatible with all the fragments already present in the second list. After all the fragments are examined, the process is repeated again from the first member remaining in the initial list until no member in the initial list is compatible with those in the second list.

As the above procedure suggests, *DIALIGN* is most effective for detecting local alignments among distantly related sequences or sequences composed of several domains [63]. The local nature is favorable for searching exons or regulatory elements in genomic sequences [105]. On the other hand, *DIALIGN* is too expensive to align many, say $N > 100$, sequences, because the theoretical computational complexity is $O(N^4 \bar{L}^2)$, which is spent for recalculating weight values. It might be wise to introduce a hierarchical structure so that the *DIALIGN* algorithm is applied to a set of prealigned groups of sequences [81].

3.3.4 Iterative Methods

It is commonly recognized that the major drawback of progressive methods rests on the lack of an appropriate procedure to correct earlier errors when more sequences are added later on. Many ideas have been proposed to overcome this drawback, and most of the proposed methods have adopted some kind of iterative procedure [10, 12, 14, 103]. An extensive review of these methods has been published [34]; thus, we discuss only relatively recent results here. MSA methods based on a hidden Markov model (HMM) may be considered as a variant of this strategy, but will be discussed separately in subsection 3.3.6, since HMM methods have their own mathematical background.

General Strategy

The basic strategy of iterative refinement methods is summarized as follows. Given a preliminary MSA, the alignment is divided into a few (usually two) groups. Columns consisting of null characters only are depleted from each group so that the condition B in subsection 3.2.1 is satisfied. After the two groups are optimally aligned, the total score must never be lower than that of the original alignment. By repeating the process for various ways of division, the overall alignment is gradually improved and ultimately reaches convergence. Several variations exist in the way and the order of division, as reviewed by Hirosawa *et al.* [44]. Recent methods that show good performance adopt similar strategies. (1) The initial alignment is obtained by a progressive method. (2) Division into two groups is guided by a tree that is constructed by a distance matrix method from the initial alignment. The order of division is either random or predetermined. (3) Sequences or sequence pairs are weighted. (4) Some heuristics are used for locating anchor points to accelerate overall calculation. Three representative methods are introduced below.

Prrn

The heart of iterative refinement methods is group-to-group pairwise alignment, with which the overall alignment score is improved. When we use a proportional gap penalty function, a DP algorithm solves the problem straightforwardly. However, the worst-case computational complexity is dramatically increased when we use an affine gap penalty under the (W)SP scoring system, as Kececioglu and Starrett have recently proven the problem to be NP-complete with respect to the total number, $M + N$, of sequences in the two groups [55]. In practice, Gotoh has devised very efficient algorithms that solve the problem in time complexity nearly independent of $M + N$ [30, 31]. Two key ideas have made the algorithm feasible. First, the so-called candidate list paradigm [69] extends the usual DP procedures without loss of rigor but with only a moderate increase in time and space complexity. Second, the data structure of 'generalized profiles' facilitates exact and efficient calculation of affine gap penalties. Note that the natural gap penalties are imposed rather than the quasi-natural gap costs [2] used in *MSA* [68].

Let $\mathcal{A}_i = \mathbf{a}_1 \mathbf{a}_2 \cdots \mathbf{a}_i$ and $\mathcal{B}_j = \mathbf{b}_1 \mathbf{b}_2 \cdots \mathbf{b}_j$ be the prefixes of the groups of sequences, $\mathcal{A}$ and $\mathcal{B}$, to be aligned. A set of candidates at each node (i, j) of the DP procedure correspond to distinct configurations of alignment between $\mathcal{A}_i$ and $\mathcal{B}_j$. By mutual competition, only those candidates that have the possibility to contribute to the final optimal alignment are retained. In the earlier versions of *Prrn/Prrp*, four criteria, which Kececioglu and Starrett call extremal pruning, were used to prune candidates. Since Ver. 3.0 [34], a single criterion, the dominance pruning [55], has been adopted. The efficiencies of the extremal and dominance pruning are virtually equivalent, whereas the dominance pruning can be coded significantly more compactly. Incidentally, from this version, *Prrp* for protein sequences was merged into a single program *Prrn* that had been used to align nucleotide sequences only.

Another unique feature of *Prrn* is the use of a doubly nested randomized iterative strategy. In the inner loop of this strategy, the tree-partitioned iterative refinement of MSA is performed with a set of weights given to all pairs of sequences. These pair-weights are calculated from an unrooted tree by the 'three-way' method [32]. The tree used for partitioning and calculation of weights is obtained by a distance matrix method, UPGMA or the NJ method. The distance matrix is, in turn, obtained from an MSA. Thus, MSA, tree, and pair-weights are mutually interdependent. *Prrn* repeats the iteration until this triad becomes mutually consistent [33, 34].

From a practical point of view, the *Prrn* algorithm is somewhat over-luxuriant. For example, the rigorous group-to-group pairwise alignment algorithm (Algorithm D in [30]) may be replaced by a cheaper one (Algorithm B or C in [30]) without significant loss of accuracy, as assessed with a method discussed in section 3.4. *MAFFT* and *MUSCLE* discussed below follow this idea.

MAFFT

The features of *MAFFT* [52] are rapid construction of the guide tree and fast search for anchor points by means of the fast Fourier transform (FFT) method [87]. *MAFFT* can construct accurate alignments even faster than *ClustalW*. *MAFFT* differs from *Prrn* in three major respects: (1) preparation of the initial alignment, (2) the method for detecting anchor points, and (3) treatment of the gap-open penalty. These differences are described in detail below.

MAFFT constructs an initial alignment using a progressive method *twice*. The first phase aligns sequences based on a roughly estimated guide tree. A modified method of Jones *et al.* [50] is used for calculating distance. The distance between $\bar{s}_m$ and $\bar{s}_n$, $D_{m,n}$, is obtained by:

$$D_{m,n} = 1 - \frac{T_{m,n}}{\min(T_{m,n}, T_{m,n})}, \qquad (3.19)$$

where $T_{m,n}$ is the number of K-mer segments ($K = 6$ in *MAFFT*) shared by $\bar{s}_m$ and $\bar{s}_n$. In the calculation of $T_{m,n}$, the 20 amino acids are grouped into six categories depending on their physico-chemical properties. This method requires only $O(\bar{L})$ computational steps for each sequence pair provided that the 6-mer frequencies are precomputed for all sequences, which requires $O(N\bar{L})$. Hence, the total computation requires $O(N^2\bar{L})$. By contrast, the dynamic programming method requires $O(\bar{L}^2)$ for each pair, and $O(N^2\bar{L}^2)$ in total. Thus, much computation time can be saved by the K-mer method. *MAFFT* uses a slightly modified version of the UPGMA method [100] for the construction of the guide tree.

The second phase uses the progressive method again. The second phase differs from the first phase in the guide tree, which is reconstructed from the distance matrix estimated from the MSA obtained by the first phase. The distance $D_{m,n}$ is defined as $D_{m,n} = -\log I_{m,n}$, where $I_{m,n}$ is the degree of sequence identity between $\bar{a}_m$ and $\bar{a}_n$. The second phase alignment is likely to be more accurate than that of the first phase, since the new guide tree is expected to be more reliable.

Next, we explain the methods of FFT preprocessing and the group-to-group sequence alignment algorithm used in *MAFFT*. The FFT preprocessing method finds anchor points that vertically divide a group into several disjoint sections. In this procedure, correlations between two groups are rapidly calculated with the FFT algorithm, and 20 positional lags (diagonals) with the highest correlation scores are identified. These diagonals are then searched for high-scoring segment pairs with average matching scores per column exceeding a threshold.

The correlation between groups $\mathcal{A}$ and $\mathcal{B}$ with positional lag k, $\rho(k)$, is defined as

$$\rho(k) = \rho_v(k) + \rho_p(k), \qquad (3.20)$$

where

$$\rho_v(k) = \sum_{1 \leq i \leq \min(I, J-k)} \{v_{\mathcal{A}}(i)v_{\mathcal{B}}(i+k)\}. \qquad (3.21)$$

$\rho_v(k)$ denotes the correlation of the volume component. The correlation of the polarity component $\rho_p(k)$ is defined in a similar way. $v_{\mathcal{C}}(j)$ and $p_{\mathcal{C}}(j)$ are the weighted sum of the volume values and the polarity values for the j-th column of $\mathcal{C} \in \{\mathcal{A}, \mathcal{B}\}$. *MAFFT* obtains the weights in the same way as *ClustalW* in the progressive phase, and by the three-way method [32] in the iterative phase. The calculation of $\rho(k)$ requires $O(\bar{L} \log \bar{L})$ computation used in the FFT procedure to obtain $\rho_v(k)$ and $\rho_p(k)$.

To determine the positions of high-scoring segment pairs in each diagonal, *MAFFT* uses a sliding window method with the window size of 30. If successive high-scoring segment pairs

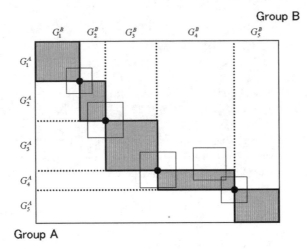

FIGURE 3.3: FFT preprocessing. Open squares with thin lines denote high-scoring segment pairs detected. In this example, four of the five high-scoring segment pairs are included in the optimal compatible assembly. Both groups are cut into five sections at the midpoint of each high-scoring segment pair indicated by a filled circle. After division, each pair of G_i^A and G_i^B is aligned. Since only the shaded regions are examined, *MAFFT* attains significantly rapid calculation.

overlap, they are merged. Potential noises are removed by a sparse DP algorithm that finds the optimal combination of compatible high-scoring segment pairs. The midpoints of the high-scoring segments in this optimal combination serve as the boundaries by which each group is divided into several sections (Figure 3.3). Then, each pair of sections is aligned by the group-to-group sequence alignment algorithm described below.

MAFFT uses the WSP-type objective function for group-to-group sequence alignment, but the gap-opening penalty differs from that of the natural WSP scoring system. The gap-opening penalty assigned to a gap that opens or closes opposite to two successive columns $\mathbf{c}_p$ and $\mathbf{c}_q$ is defined as:

$$v\left[\{1 - g_s^{\mathcal{C}}(p)\} + \{1 - g_e^{\mathcal{C}}(q)\}\right]/2, \tag{3.22}$$

where v is the basic gap-opening penalty. $g_s^{\mathcal{C}}(p)$ is the number of gaps starting at $\mathbf{c}_p$ in $\mathcal{C}$ and $g_e^{\mathcal{C}}(q)$ is the number of gaps ending at $\mathbf{c}_q$. *MAFFT* also uses a normalized substitution matrix. A score of substitution matrix $S_2(a, b)$ is normalized as

$$\bar{S}_2(a, b) = \frac{S_2(a, b) - \mu_2}{\mu_1 - \mu_2} + u, \tag{3.23}$$

where $\mu_1 = \sum_{a \in \Sigma} f_a S_2(a, a)$ and $\mu_2 = \sum_{a, b \in \Sigma} f_a f_b S_2(a, b)$. u is a parameter with $u \ll 1$, corresponding to a gap extension penalty. f_a is the stationary composition of amino acid a derived from the substitution matrix. An average score per position between segments satisfies $u \leq \mu_{\bar{s}} \leq 1 + u$. *MAFFT* uses an ordinary two-dimensional DP for aligning two groups.

Note that *MAFFT* is not always faster than *MAFFT without* FFT preprocessing; since FFT preprocessing may not identify high-scoring segment pairs among highly divergent sequences, the search space may not be sufficiently reduced to compensate for the cost of the preprocessing. In addition, the alignment accuracies of *MAFFT* are slightly decreased because FFT preprocessing reduces the search space without guaranteeing optimality.

MUSCLE

Recently, a new program called *MUSCLE* [20] has been reported. *MUSCLE* uses nearly the same strategy as *MAFFT*. *MUSCLE* differs from *MAFFT* in that it adopts an oligomer counting [19] method instead of FFT preprocessing for the detection of anchor points, which requires $O(L)$ computation. Moreover, instead of the classical profiles [37] used in *Prrn* or *MAFFT*, *MUSCLE* adopts a log expectation scoring scheme to evaluate the $S_2(\mathbf{a}, \mathbf{b})$ term for group-to-group alignment. The log expectation score between profile columns $\mathbf{a}$ and $\mathbf{b}$ is defined as

$$(1 - a[-])(1 - b[-]) \sum_{x,y \in \Sigma} \log \left(a[x]b[y]\frac{p_{x,y}}{p_x p_y} \right), \qquad (3.24)$$

where $a[-]$ and $a[x]$ ($b[-]$ and $b[y]$) are frequencies of null and residue x (y) of profile column $\mathbf{a}$ ($\mathbf{b}$), respectively. $p_{x,y}$ is a joint probability between residues x and y, and p_z is a background probability of residue z. Both $p_{x,y}$ and p_z are derived from the probabilistic substitution matrix of amino acids.

The gap penalty of *MUSCLE* is basically the same as that of *MAFFT*, but *MUSCLE*, like *ClustalW*, adjusts gap penalties according to the hydrophobicity of the surrounding residues.

3.3.5   Stochastic Methods

Simulated Annealing

A class of strategies, known as stochastic algorithms, have often been used for solving such complex combinatorial optimization problems as MSA. Two representative methods in this class are simulated annealing (SA) and genetic algorithm (GA). The process of SA consists of a series of two kinds of operations: (i) change in the current state and (ii) acceptance or rejection of the change. Not only locally favorable changes but also unfavorable changes are accepted at a certain probability. The probability is controlled by the "temperature" and the energy gap between the old and new states. In MSA, each state change is realized by a move of a gap or a segment of gaps [48, 56]. The energy is calculated in the same way as an SP score (the sign is inverted), although any other scoring systems may be appropriate. The intensive computational power required by SA is often shared by distributed CPUs, each of which takes charge of distinct temperatures. Gibbs sampling algorithm for local MSA [64] is based on essentially the same principle as that of the SA for global MSA, except that each state corresponds to an ungapped local MSA of a fixed length.

Genetic Algorithm

Several instances of the application of GA to the MSA problem are found in the literature [77, 79, 122]. Although the general strategy of these methods is common, the fitness function and the actual procedures of mutation and crossover considerably differ from one another. The most orthodox settings are those of Notredame and Higgins, in which a mutation corresponds to a move of a gap, a crossover corresponds to a recombination between alignments at a consistent cut site, and fitness corresponds to a WSP score. As a tool for standard M-SA, GA is not as efficient as progressive or iterative strategies. However, stochastic methods including GA and SA are more flexible in terms of the scoring scheme. One promising application is the alignment of RNA sequences considering conserved features in both primary and secondary structure levels. Another possible application is protein structure-sequence alignment under some constraints, such as experimentally determined distances between a subset of residue pairs.

3.3.6 Hidden Markov Models

A hidden Markov model (HMM)[85] is a probabilistic model that is well suited for many tasks in bioinformatics, although it has been mostly developed for speech recognition since the early 1970s. One of the most popular uses of the HMM in bioinformatics is as a 'probabilistic profile' of an MSA of a protein/DNA family, which is called a profile HMM[59, 8]. A profile HMM resembles a profile [36], but is a full probabilistic model, whereas a profile has a non-probabilistic structure. The main contribution of the profile HMM is that it treats gaps in a probabilistic manner. A profile HMM can be constructed from a set of aligned/unaligned sequences of a protein/DNA family and is used for searching a database for other members of the family. Further, a profile HMM is capable of producing an MSA of a sequence family.

Profile HMMs

Figure 3.4 shows an example of the basic type of a profile HMM. The edges not shown have zero transition probability. In profile HMMs, a set of match (M_i), insert (I_i) and delete (D_i) states stretches from left to right. The number of stretches is called the 'length' of a profile HMM, and i is called the 'position' in a profile HMM. State B is the *begin* state, so that the two-step process, namely a repetitive series of transition and emission, always starts here. A transition never moves to the left, so that with time, the current state gradually moves to the right, eventually ending at the *end* state E. When this state is reached, the two-step process ends. A match or delete state is never visited more than once. A set of emission symbols of the profile HMM consists of the twenty amino acids or four nucleotides with one null symbol denoted by δ. Delete states emit δ with a probability of one. Each match and insert state has its own emission distribution over the twenty amino acids or four nucleotides, but cannot emit δ, that is, only a delete state can emit δ, and each delete state emits only δ.

The profile HMM is capable of producing various sets of sequences by changing the distributions of transition and emission probabilities. Let us consider two extreme cases. If the emission probabilities for the match and insert states are uniform over all symbols, the profile HMM will produce random sequences that do not have much in common except their lengths. At the other extreme, if each state emits one specific symbol with a probability of one, and if the transitions from M_i to M_{i+1} have a probability of one, then the profile

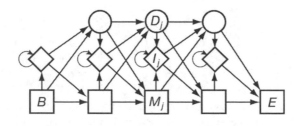

FIGURE 3.4: The structure of a profile HMM.

HMM will always produce the same sequence. Somewhere in between these two cases, the parameters of the profile HMM can be set so that it produces sequences that are similar, thus producing what can be thought of as a 'family' of sequences. Each choice of parameters produces a different family. It is also possible that the similarity is high in some positions of the sequences produced and low in others. This will happen if the emission distributions of some match states are concentrated on a few symbols, whereas those of the others are approximately uniform over all the symbols. It is frequently observed that a set of protein/DNA sequences of a family has some regions of higher conservation and other regions of lower conservation. These indicate that a profile HMM can describe the sequence features of a protein/DNA family.

If we adopt a gap symbol '-' instead of δ, the profile HMM is capable of producing a set of sequences that looks like an MSA. Let us follow the two-step process for producing MSA-like sequences. A gap always opens when the transition from a match/insert state to a delete state occurs, and a gap always extends when the transition from one delete state to another delete state occurs. Each transition has its own probability. This indicates that the profile HMM involves a framework of the affine gap. The difference from the standard affine gap is that the gap extension penalty of the profile HMM can vary with position, whereas that of the standard affine gap is always constant. The transitions concerned with an insert state completely correspond to the standard affine gap. The transition from a match state to an insert state corresponds to the gap open penalty, and the transition from an insert state to itself corresponds to the gap extension penalty. As the latter transition always moves to itself, it is guaranteed to have the same penalty. In the profile HMM, the transition probabilities and the emission distribution can differ along the entire positions, because each transition has its own probability and each state has its own distribution. On the other hand, an MSA of a protein/DNA family generally has a mosaic structure consisting of well-conserved and less-conserved regions. Gaps are rarely observed in the former and frequently observed in the latter. This indicates that the profile HMM is sufficiently flexible to describe the varying features observed in an MSA of a protein/DNA family.

This is, however, a drawback of the profile HMM as well, because the flexibility largely depends on the number of parameters. In the case of a protein family, the profile HMM shown in Figure 3.3.6 has approximately $49L$ parameters ($9L$ transition probabilities and $40L$ emission probabilities) in total, where L is the length of the profile HMM. L is on the order of a few hundred, resulting immediately in the profile HMMs with more than 10,000 free parameters. This can be a problem when only a few sequences are available in a family, which is not an uncommon situation in the early stages of genome projects, and overestimation of the parameters easily occurs. Therefore, careful estimation of the parameters by introducing pseudocounts and sequence weights is essential (see below for details).

Deriving profile HMMs from MSAs

We discuss here how to construct a profile HMM when an MSA is given. Suppose that well-conserved columns are shown in the MSA. The choice of length of the profile HMM corresponds to the decision of which MSA columns to assign to the match and delete states, and on which MSA columns to assign to the insert states. In many cases, the profile HMM, whose length is the same as the number of the conserved columns, is prepared, and the i-th conserved column is assigned to the i-th match and delete states (M_i and D_i) in the profile HMM. The successive non-conserved columns after the i-th conserved column are assigned to the i-th insert state (I_i). These assignments lead us to a simple procedure for estimating the parameters of the profile HMM, because these imply that we identify the two-step processes through the profile HMM for producing each MSA sequence, that is, we just count up the number of times each transition or emission occurs during the processes, and assign probabilities according to

$$a_{kl} = \frac{A_{kl}}{\sum_{l'} A_{kl'}} \quad \text{and} \quad b_k(c) = \frac{B_k(c)}{\sum_{c'} B_k(c')}$$

where k and l are indices over states, a_{kl} is the transition probability from k to l, and $b_k(c)$ is the emission probability that the symbol c is emitted from k. A_{kl} and $B_k(c)$ are the corresponding frequencies. This procedure corresponds to the ML estimation of the parameters, that is, the estimated parameters maximize the joint probability that the profile HMM produces the given MSA. To avoid overestimation of the parameters, the emission distributions of the insert states are often set to be the average composition of the residues over the MSA, because the significant biases of the corresponding distributions are not observed in many cases. In the case that well-conserved columns in the MSA are not shown, we will use a heuristic rule for deciding which columns should be assigned to which states. A simple rule that works well is that columns that are more than half gap symbols should be assigned to insert states.

A clear problem with the ML estimation is that if there are only a few sequences then some transitions or emissions may not be observed in the MSA. This will frequently occur. However, it is quite likely that if there are more sequences then the corresponding transitions or emissions will be observed in the MSA. We should estimate the parameters with consideration of such a situation. A method to deal with this problem is to add pseudocounts to the observed counts. Pseudocounts are some fake imagined data based on our prior knowledge of protein/DNA sequences to represent all the other things that might happen. A very simple and much-used pseudocount method is to add a constant to all the counts. When the constant is one, this is called 'Laplace's rule'. A slightly sophisticated method is to add a quantity proportional to the background probability, e.g. the average residue composition over the MSA. These methods have the appealing feature, that is, the effect of pseudocount becomes significant and the estimated parameter values approximate our prior knowledge if very little data is available. At the other extreme, the effect of pseudocount becomes insignificant and the parameter estimation approximates the ML estimation if a large amount of data is available. See [16] for detailed information of pseudocount methods.

There are often some sequences that are very closely related to each other in the given MSA. Since some of the information from these sequences is shared, we should not give them each the same weight in the parameter estimation as a single sequence that is more highly diverged from all the others. Many weighting methods are based on building a tree relating the sequences. Since sequences in a family are related by an evolutionary tree, it is very natural to try to construct this tree and use it when estimating the independent contribution of each of the given sequences, down-weighting sequences that have only recently diverged. See [16] for detailed information of tree construction and weight computation.

Deriving profile HMMs from unaligned sequences

We show here how to construct a profile HMM when only one set of protein/DNA sequences is given. A method for constructing the profile HMM is summarized as follows. First, we choose the length of the profile HMM and initialize the parameters. Then, we estimate the profile HMM parameters with the Baum-Welch algorithm or other alternatives.

Since a profile HMM has a repeating linear structure of match/insert/delete states, the only decision that must be made in choosing an initial architecture for the Baum-Welch estimation is the length of a profile HMM. A commonly used rule is to set the length to be the average length of the given sequences (or to set it based on prior knowledge of an expected MSA).

It is important to initialize the parameters of a profile HMM carefully, because the Baum-Welch estimation finds local optima and not global optima. The profile HMM should be encouraged to use sensible transitions; for example, transitions into match states should be large compared to other transition probabilities. Moreover, we should start the Baum-Welch estimation from several different sets of the initial parameters to check whether all converge to approximately the same optimum. We have to introduce some randomness in the choice of the initial parameters. A more sophisticated method is to sample the initial parameters from the Bayesian Dirichlet prior distributions [16, 7] over parameters. An alternative method is as follows: we can initialize the profile HMM with frequencies derived from the prior, use this profile HMM to produce a small number of random sequences, and then use these counts as data to estimate an initial profile HMM. It is known that the natural probabilistic priors on HMM parameters are Dirichlet distributions.

In the case that only one set of protein/DNA sequences is given, we cannot directly perform the ML estimation of the parameters, because there are many two-step processes for producing each given sequence. The Baum-Welch algorithm solves this problem by introducing a framework of expectation-maximization (EM) algorithm. Let us consider a simple case that a sequence $s = (s_1, \ldots, s_{|s|})$ is given. First, the Baum-Welch algorithm computes the forward and backward variables on the basis of the initial parameters. The forward variable $\alpha_i(k)$ is the joint probability that the HMM produces the subsequence $s_1, \ldots, s_k$ ($1 \leq k \leq |s|$) ending at state i. The backward variable $\beta_i(k)$ is the joint probability that the HMM produces the subsequence $s_{k+1}, \ldots, s_{|s|}$ beginning from state i. The forward and backward variables are computed for all i and k recursively, and we obtain $\sum_i \alpha_i(|s|)$ and $\sum_i \beta_i(0)$ which are the joint probabilities that the HMM produces the sequence. The algorithms computing the forward and backward variables are called the forward and backward algorithms, respectively. The structure of these algorithms is similar to that of dynamic programming. Next, the algorithm computes the expected counts of transitions and emissions from sequence s on the basis of the forward and backward variables. Finally, the algorithm re-estimates the parameters from the expected counts, and the parameters are overwritten by the re-estimated ones. The algorithm repeats the above steps until the parameter values converge. We can easily extend the algorithm in order to treat a set of sequences. See [85] for a detailed description of the algorithm.

The parameter estimation of a profile HMM is done by a straightforward application of the standard Baum-Welch algorithm described above. There are, however, major differences: each state in the profile HMM has at most three entering transitions (see Figure 3.3.6), and the delete states cannot emit any symbols. Let M_0 and M_{L+1} denote the begin and end states, respectively. Below we give the algorithms for profile HMMs.

Forward algorithm for profile HMMs

Initialization:

$$\alpha_{M_0}(0) = 1$$

Recursion:

$$\alpha_{M_i}(k) = b_{M_i}(s_k)[\alpha_{M_{i-1}}(k-1)a_{M_{i-1}M_i} + \alpha_{I_{i-1}}(k-1)a_{I_{i-1}M_i} + \alpha_{D_{i-1}}(k-1)a_{D_{i-1}M_i}]$$
$$\alpha_{I_i}(k) = b_{I_i}(s_k)[\alpha_{M_i}(k-1)a_{M_iI_i} + \alpha_{I_i}(k-1)a_{I_iI_i} + \alpha_{D_i}(k-1)a_{D_iI_i}]$$
$$\alpha_{D_i}(k) = \alpha_{M_{i-1}}(k)a_{M_{i-1}D_i} + \alpha_{I_{i-1}}(k)a_{I_{i-1}D_i} + \alpha_{D_{i-1}}(k)a_{D_{i-1}D_i}$$

Termination:

$$\alpha_{\mathrm{M}_{L+1}}(|s|+1) \;=\; \alpha_{\mathrm{M}_L}(|s|)a_{\mathrm{M}_L\mathrm{M}_{L+1}} + \alpha_{\mathrm{I}_L}(|s|)a_{\mathrm{I}_L\mathrm{M}_{L+1}} + \alpha_{\mathrm{D}_L}(|s|)a_{\mathrm{D}_L\mathrm{M}_{L+1}}$$

Backward algorithm for profile HMMs

Initialization:

$$
\begin{aligned}
\beta_{\mathrm{end}}(|s|+1) &= 1 \\
\beta_{\mathrm{M}_L}(|s|) &= a_{\mathrm{M}_L\mathrm{M}_{L+1}} \\
\beta_{\mathrm{I}_L}(|s|) &= a_{\mathrm{I}_L\mathrm{M}_{L+1}} \\
\beta_{\mathrm{D}_L}(|s|) &= a_{\mathrm{D}_L\mathrm{M}_{L+1}}
\end{aligned}
$$

Recursion:

$$
\begin{aligned}
\beta_{\mathrm{M}_i}(k) &= \beta_{\mathrm{M}_{i+1}}(k+1)a_{\mathrm{M}_i\mathrm{M}_{i+1}}b_{\mathrm{M}_{i+1}}(s_{k+1}) \\
&\quad + \beta_{\mathrm{I}_i}(k+1)a_{\mathrm{M}_i\mathrm{I}_i}b_{\mathrm{I}_i}(s_{k+1}) + \beta_{\mathrm{D}_{i+1}}(k)a_{\mathrm{M}_i\mathrm{D}_{i+1}} \\
\beta_{\mathrm{I}_i}(k) &= \beta_{\mathrm{M}_{i+1}}(k+1)a_{\mathrm{I}_i\mathrm{M}_{i+1}}b_{\mathrm{M}_{i+1}}(s_{k+1}) \\
&\quad + \beta_{\mathrm{I}_i}(k+1)a_{\mathrm{I}_i\mathrm{I}_i}b_{\mathrm{I}_i}(s_{k+1}) + b_{\mathrm{D}_{i+1}}(k)a_{\mathrm{I}_i\mathrm{D}_{i+1}} \\
\beta_{\mathrm{D}_i}(k) &= \beta_{\mathrm{M}_{i+1}}(k+1)a_{\mathrm{D}_i\mathrm{M}_{i+1}}b_{\mathrm{M}_{i+1}}(s_{k+1}) \\
&\quad + \beta_{\mathrm{I}_i}(k+1)a_{\mathrm{D}_i\mathrm{I}_i}b_{\mathrm{I}_i}(s_{k+1}) + \beta_{\mathrm{D}_{i+1}}(k)a_{\mathrm{D}_i\mathrm{D}_{i+1}}
\end{aligned}
$$

where L is the length of the profile HMM.

The forward and backward variables can then be combined to re-estimate the transition and emission probability parameters as follows:

Re-estimation equations for profile HMMs

Expected transition counts from sequence s:

$$
\begin{aligned}
A_{\mathrm{S}_i\mathrm{M}_{k+1}} &= \frac{1}{P(s)}\sum_k \alpha_{\mathrm{S}_i}(k)a_{\mathrm{S}_i\mathrm{M}_{i+1}}b_{\mathrm{M}_{i+1}}(s_{k+1})\beta_{\mathrm{M}_{i+1}}(k+1) \\
A_{\mathrm{S}_i\mathrm{I}_i} &= \frac{1}{P(s)}\sum_k \alpha_{\mathrm{S}_i}(k)a_{\mathrm{S}_i\mathrm{I}_i}b_{\mathrm{I}_i}(s_{k+1})\beta_{\mathrm{I}_i}(k+1) \\
A_{\mathrm{S}_i\mathrm{D}_{i+1}} &= \frac{1}{P(s)}\sum_k \alpha_{\mathrm{S}_i}(k)a_{\mathrm{S}_i\mathrm{D}_{i+1}}\beta_{\mathrm{D}_{i+1}}(k)
\end{aligned}
$$

Expected emission counts from sequence s:

$$
\begin{aligned}
B_{\mathrm{M}_i}(c) &= \frac{1}{P(s)}\sum_{k|s_k=c} \alpha_{\mathrm{M}_i}(k)\beta_{\mathrm{M}_i}(k) \\
B_{\mathrm{I}_i}(c) &= \frac{1}{P(s)}\sum_{k|s_k=c} \alpha_{\mathrm{I}_i}(k)\beta_{\mathrm{I}_i}(k)
\end{aligned}
$$

where $S_i \in \{M_i, I_i, D_i\}$ and $P(s) = \alpha_{M_{L+1}}(|s|+1)$.

A framework of stochastic search algorithm is introduced into the Baum-Welch algorithm to avoid converging local optima. The most common stochastic algorithm is SA. The important point of SA is that it can escape local optima because of the stochastic choice of temporal solution. A similar effect can be obtained by adding noise to the expected counts computed in the Baum-Welch algorithm. See [16] for a detailed description of the algorithm. The expected counts become an index to the length adaptation of the profile HMM [59]. We can see how much a certain transition is used by the given sequences from the expected counts. The usage of a state is the sum of counts for all symbols in the state. If a certain match state is rarely used, we should remove the match state together with the corresponding insert and delete states, because the match state is redundant and should be absorbed in an insert state. Similarly, if a certain insert state is frequently used, we should create a new match state at the corresponding position together with corresponding new insert and delete states, because the insert state absorbs much sequence and should be expanded. Although this approach is *ad hoc*, it works well in practice.

MSAs with given profile HMMs

Once a profile HMM is prepared according to the methods described above, it enables us to produce an MSA in a systematic manner. Suppose that a set of sequences of a protein/DNA family and a profile HMM for the family are given. A method for producing an MSA of the sequences is summarized as follows: for each sequence, we use the Viterbi algorithm for determining a two-step process (a path) through the profile HMM that is most likely to produce that sequence, and then we build an MSA from the paths.

The Viterbi algorithm finds the most likely path through a profile HMM by introducing a framework of dynamic programming. Let us consider a simple case that a sequence $s = (s_1, \ldots, s_{|s|})$ is given. The Viterbi algorithm computes for the Viterbi variable $v_i(k)$, which is the joint probability of the most likely path that the profile HMM produces the subsequence $s_1, \ldots, s_k$ $(1 \leq k \leq |s|)$ ending at state i. The Viterbi algorithm also computes the traceback variable $\psi_i(k)$, that stores an entering transition of state i for computing $v_i(k)$. The Viterbi and traceback variables are computed for all i and k recursively. Then, $\sum_i v_i(|s|)$, which is the joint probability that the HMM produces the sequence from the most likely path, is obtained, and the algorithm backtracks the traceback variables from $\psi_{\hat{i}}(|s|)$, where $\hat{i} = \text{argmax}_i \psi_i(|s|)$, to find the most likely path through the HMM. See [85] for a detailed description of the Viterbi algorithm.

Finding the most likely path through a profile HMM is done by a straightforward application of the standard Viterbi algorithm described above. There are, however, major differences: each state in the profile HMM has at most three entering transitions (see Figure 3.3.6), and the delete states cannot emit any symbols. Moreover, the log-odds score is adopted as the Viterbi variable, and all the products in the algorithm are converted into sums by the log transformation. Let M_0 and M_{L+1} denote the begin and end states, respectively. Below we give the algorithm for profile HMMs.

Viterbi algorithm for profile HMMs

Initialization:

$$v_{M_0}(0) = 0$$

Recursion:

$$
v_{M_i}(k) = \log \frac{b_{M_i}(s_k)}{q_{s_k}} + \max \begin{cases} v_{M_{i-1}}(k-1) + \log a_{M_{i-1}M_i} \\ v_{I_{i-1}}(k-1) + \log a_{I_{i-1}M_i} \\ v_{D_{i-1}}(k-1) + \log a_{D_{i-1}M_i} \end{cases}
$$

$$
v_{I_i}(k) = \log \frac{b_{I_i}(s_k)}{q_{s_k}} + \max \begin{cases} v_{M_j}(k-1) + \log a_{M_i I_i} \\ v_{I_j}(k-1) + \log a_{I_i I_i} \\ v_{D_j}(k-1) + \log a_{D_i I_i} \end{cases}
$$

$$
v_{D_i}(k) = \max \begin{cases} v_{M_{i-1}}(k) + \log a_{M_{i-1}D_i} \\ v_{I_{i-1}}(k) + \log a_{I_{i-1}D_i} \\ v_{D_{i-1}}(k) + \log a_{D_{i-1}D_i} \end{cases}
$$

Termination:

$$
v_{M_{L+1}}(|s|+1) = \max \begin{cases} v_{M_L}(|s|) + \log a_{M_L M_{L+1}} \\ v_{I_L}(|s|) + \log a_{I_L M_{L+1}} \\ v_{D_L}(|s|) + \log a_{D_L M_{L+1}} \end{cases}
$$

where q_c is the average composition of residue c over the given sequences. The equations relevant to the traceback variables are not shown.

The Viterbi algorithm finds a path through a profile HMM that is most likely to produce a sequence, and this corresponds to aligning the sequence to the profile HMM. The construction of an MSA requires computing such a Viterbi alignment for each sequence. Residues aligned to the same match state are aligned in columns. Indels are then inserted appropriately for insertions and deletions. This implies an important difference between profile HMM induced MSAs and conventional MSAs, which will be more clearly shown by an example. Consider the protein sequences NTPFS, NCYDFLS and NKYLS. Suppose that the profile HMM has length 4 and the most likely paths of the sequences are $M_1\ I_1\ I_1\ M_2\ D_3\ M_4$, $M_1\ I_1\ I_1\ I_1\ M_2\ M_3\ M_4$ and $M_1\ I_1\ M_2\ M_3\ M_4$, respectively (the begin and end states are not shown). Then, we can obtain the alignment of these paths by aligning positions that were generated by the same match state.

$$
\begin{array}{ccccccc}
M_1 & I_1 & I_1 & & M_2 & D_3 & M_4 \\
M_1 & I_1 & I_1 & I_1 & M_2 & M_3 & M_4 \\
M_1 & I_1 & & & M_2 & M_3 & M_4
\end{array}
$$

This leads to the following MSA.

$$
\begin{array}{ccccccc}
N & T & P & & F & - & S \\
N & C & Y & D & F & L & S \\
N & K & & & Y & L & S
\end{array}
$$

This method can give ambiguous results in some cases. In the above example, it is not clear how to align the TP from the first sequence to the CYD from the second sequence and the K from the third sequence, whereas the other residues are obviously aligned. Such ambiguous residues are emitted from the insert states. A profile HMM does not attempt to align these residues, because the insert state residues usually represent parts of the sequences that are atypical, unconserved, and not meaningfully alignable. In contrast, many other MSA algorithms align the whole sequences, regardless of which parts of the sequence are meaningfully alignable or not. Another advantage of this method is that it allows the sequences themselves to guide the MSA, rather than having a precomputed substitution matrix and gap penalties. Thus, less bias is introduced.

	no. of alignments	characteristics of alignments
Reference 1	82	phylogenetically equi-distant
Reference 2	23	including orphan sequences
Reference 3	12	equi-distant families
Reference 4	12	long terminal insertions
Reference 5	12	long internal insertions
Reference 6	13	repeats
Reference 7	8	transmembrane
Reference 8	10	circular permutations
hline		

TABLE 3.5 BAliBASE version 2 contents [6]. References 6-8 were added upon the release of version 2.

3.4 Methods for Assessment

To assess alignment accuracies of programs, we must prepare a reference data set and evaluate the accuracies using the data set. The best-known reference data set is BAliBASE [6, 109]. In the latest version (BAliBASE2), references are divided into eight categories depending on the nature of the structural alignments (Table.3.5). In addition, BAliBASE defines the core segments for each alignment. The core segments of an alignment represent explicitly the alignable regions within it. Recently published data sets, such as OXBench [86], PREFAB [20] and SABmark [116], contain more references than BAliBASE.

The most widely used measures for evaluating MSAs are sum-of-pairs and column scores [110]. The sum-of-pairs score (SPS) is defined as the proportion of correctly aligned pairs:

$$SPS = \frac{\sum_{i=1}^{I} SP_i^t}{\sum_{j=1}^{J} SP_j^r},\qquad(3.25)$$

where I and J are the numbers of columns of test and reference alignment, respectively. SP_i^t is defined as:

$$SP_i^t = \sum_{1 \leq m < n \leq N} p_i(m, n).$$

If aligned residue pair a_{mi} and a_{ni} of the test alignment also exists in the reference alignment, $p_i(m, n) = 1$; otherwise, $p_i(m, n) = 0$. SP_j^r is the total number of aligned pairs in the reference MSA. The column score (CS) represents the proportion of correctly aligned columns:

$$CS = \frac{\sum_{i=1}^{I} C_i}{J}.\qquad(3.26)$$

If the column of the test alignment is identical to the ith column of the reference, $C_i = 1$; otherwise, $C_i = 0$. Both SPS and CS consider not the magnitude of alignment error but the correctness of an alignment. The measure recently proposed by Raghava *et al.* [86] takes the magnitude of error into consideration. In any case, the quality of reference alignments critically affects the evaluation results. A measure without reference alignments, called APDB [86], has also been proposed. The idea of APDB is to evaluate the goodness of structural superposition induced by the test alignment.

The performance of a program may be represented by the mean or median of the distribution of scores. These values, however, should be assessed with care, because the distributions are possibly asymmetric. Instead of means or medians, non-parametric statistical tests, such as the Wilcoxon matched pair signed rank test and the Friedman test, are often used for assessing the relative performance of programs. The Wilcoxon matched pair signed rank test asks whether there is a significant difference in accuracy between the MSAs produced by two programs. By contrast, the Friedman test examines whether all programs achieve equivalent performance. If there is a significant difference among the programs, the Friedman test can also be used for assessing the difference between two methods. The Wilcoxon matched pair signed rank test is generally more discriminative than the Friedman test, because the latter assesses relative relationships whereas the former considers the absolute score differences.

3.5 Summary

MSA is an old yet highly active area in computational molecular biology. With the rapid progress in genome projects, a huge amount of sequence data have been accumulated and MSA is undoubtedly one of the most powerful computational tools for drawing the functional implicationsngwerful computational tools tocussed above from these data. For a long time, progressive methods (subsection 3.3.2) were the sole practical approach to solving large MSA problems. This situation has been changed by the recent progress in iterative refinement strategies (subsection 3.3.4). If only moderate evolutionary changes, such as substitutions and short indels, are involved, current iterative methods may produce good alignments of the order of 10^3 protein sequences within reasonable time. On the other hand, more drastic evolutionary changes, such as the insertion or deletion of long segments, recombination, and domain shuffling, are not well modeled by the current objective functions, as discussed in subsection 3.2.3. The adequate combination of local and global similarities must be incorporated in the alignment procedure. The methods discussed in subsection 3.3.3 steer toward this direction, although much remains to be studies. The MSA of nucleic acid sequences is another area that requires in-depth investigations. The three most important problems are the MSA of structural RNAs, the MSA of regulatory elements on genomic sequences, and the MSA of whole genome sequences. The various ideas discussed in this chapter may be applied to these problems by appropriate adaptations.

References

[1] E. Althaus, A. Caprara, H.P. Lenhof, and K. Reinert. Multiple sequence alignment with arbitrary gap costs: Computing an optimal solution using polyhedral combinatorics. *Bioinformatics*, Vol. 18, Suppl. 2:S4–S16, 2002.

[2] S.F. Altschul. Gap costs for multiple sequence alignment. *J. Theor. Biol.*, 138:297–309, 1989.

[3] S.F. Altschul. Generalized affine gap costs for protein sequence alignment. *Proteins*, 32:88–96, 1998.

[4] S.F. Altschul, R.J. Carroll, and D.J. Lipman. Weights for data related by a tree. *J. Mol. Biol.*, 207:647–653, 1989.

[5] S.F. Altschul, T.L. Madden, A.A. Schaffer, and J. Zhang *et al.* Gapped BLAST and PSI-BLAST: a new generation of protein database search programs. *Nucleic Acids Res.*, 25:3389–3402, 1997.

[6] A. Bahr, J.D. Thompson, J.C. Thierry, and O. Poch. BAliBASE (Benchmark Alignment dataBASE): enhancements for repeats, transmembrane sequences and circular permutations. *Nucleic Acids Res.*, 29:323–326, 2001.

[7] P. Baldi and S. Brunak. *Bioinformatics: The Machine Learning Approach*. The MIT Press, second edition, 2001.

[8] P. Baldi, Y. Chauvin, T. Hunkapiller, and M.A. McClure. Hidden markov models of

biological primary sequence information. *Proc. Natl. Acad. Sci. USA*, 91:1059–1063, 1994.

[9] G.J. Barton. Protein sequence alignment techniques. *Acta Crystallogr. D Biol. Crystallogr.*, 54:1139–1146, 1998.

[10] G.J. Barton and M.J.E. Sternberg. A strategy for the rapid multiple alignment of protein sequences. confidence levels from tertiary structure comparisons. *J. Mol. Biol.*, 198:327–337, 1987.

[11] S. Batzoglou, D.B. Jaffe, K. Stanley, and J. Butler *et al.* ARACHNE: a whole-genome shotgun assembler. *Genome Res.*, 12:177–189, 2002.

[12] M.P. Berger and P.J. Munson. A novel randomized iterative strategy for aligning multiple protein sequences. *Comput. Appl. Biosci.*, 7:479–484, 1991.

[13] H. Carrillo and D. Lipman. The multiple sequence alignment problem in biology. *SIAM J. Appl. Math.*, 48:1073–1082, 1988.

[14] F. Corpet. Multiple sequence alignment with hierarchical clustering. *Nucleic Acids Res.*, 16:10881–10890, 1988.

[15] M.O. Dayhoff, R.M. Schwartz, and B.C. Orcutt. A model of evolutionary change in proteins. In *Atlas of protein sequence and structure*, volume Vol. 5, Suppl. 3, pages 345–352. National Biomedical Research Foundation, Silver Spring, ML, 1978.

[16] R. Durbin, S. Eddy, A. Krogh, and G. Mitchison. *Biological sequence analysis: Probabilistic models of proteins and nucleic acids.* Cambridge University Press, Cambridge, 1998.

[17] L. Duret and S. Abdeddaim. Multiple alignments for structural, functional, or phylogenetic analyses of homologous sequences. In *Bioinformatics: Sequence, structure and databanks*, pages 51–76. Oxford University Press, Oxford, 2000.

[18] R.V. Eck and M.O. Dayhoff. *Atlas of protein sequence and structure.* National Biomedical Research Foundation, Springs, MD, 1966.

[19] R.C. Edgar. Local homology recognition and distance measures in linear time using compressed amino acid alphabets. *Nucleic Acids Res.*, 32:380–385, 2004.

[20] R.C. Edgar. MUSCLE: a multiple sequence alignment method with reduced time and space complexity. *BMC Bioinformatics*, 5:113, 2004.

[21] J. Felsenstein. Maximum-likelihood estimation of evolutionary trees from continuous characters. *Am. J. Hum. Genet.*, 25:471–492, 1973.

[22] J. Felsenstein. Evolutionary trees from DNA sequences: a maximum likelihood approach. *J. Mol. Evol.*, 17:368–376, 1981.

[23] J. Felsenstein. Inferring phylogenies from protein sequences by parsimony, distance, and likelihood methods. *Methods Enzymol.*, 266:418–427, 1996.

[24] J.W. Fickett and W.W. Wasserman. Discovery and modeling of transcriptional regulatory regions. *Curr. Opin. Biotechnol.*, 11:19–24, 2000.

[25] D. Frishman and P. Argos. Seventy-five percent accuracy in protein secondary structure prediction. *Proteins*, 27:329–335, 1997.

[26] M. Gerstein and R.B. Altman. Average core structures and variability measures for protein families: application to the immunoglobulins. *J. Mol. Biol.*, 251:161–175, 1995.

[27] G.H. Gonnet, M.A. Cohen, and S.A. Benner. Exhaustive matching of the entire protein sequence database. *Science*, 256:1443–1445, 1992.

[28] O. Gotoh. An improved algorithm for matching biological sequences. *J. Mol. Biol.*, 162:705–708, 1982.

[29] O. Gotoh. Consistency of optimal sequence alignments. *Bull. Math. Biol.*, 52:509–525, 1990.

[30] O. Gotoh. Optimal alignment between groups of sequences and its application to

multiple sequence alignment. *Comput. Appl. Biosci.*, 9:361–370, 1993.

[31] O. Gotoh. Further improvement in methods of group-to-group sequence alignment with generalized profile operations. *Comput. Appl. Biosci.*, 10:379–387, 1994.

[32] O. Gotoh. A weighting system and algorithm for aligning many phylogenetically related sequences. *Comput. Appl. Biosci.*, 11:543–551, 1995.

[33] O. Gotoh. Significant improvement in accuracy of multiple protein sequence alignments by iterative refinement as assessed by reference to structural alignments. *J. Mol. Biol.*, 264:823–838, 1996.

[34] O. Gotoh. Multiple sequence alignment: algorithms and applications. *Adv. Biophys.*, 36:159–206, 1999.

[35] C. Grasso and C. Lee. Combining partial order alignment and progressive multiple sequence alignment increases alignment speed and scalability to very large alignment problems. *Bioinformatics*, 20:1546–1556, 2004.

[36] M. Gribskov, R. Lüthy, and D. Eisenberg. Profile analysis. *Methods Enzymol.*, 183:146–159, 1990.

[37] M. Gribskov, A.D. McLachlan, and D. Eisenberg. Profile analysis: detection of distantly related proteins. *Proc. Natl. Acad. Sci. USA*, 84:4355–4358, 1987.

[38] S.K. Gupta, J.D. Kececioglu, and A.A. Schaffer. Improving the practical space and time efficiency of the shortest-paths approach to sum-of-pairs multiple sequence alignment. *J. Comput. Biol.*, 2:459–472, 1995.

[39] D. Gusfield. Efficient methods for multiple sequence alignment with guaranteed error bounds. *Bull. Math. Biol.*, 55:141–154, 1993.

[40] J. Hein. A new method that simultaneously aligns and reconstructs ancestral sequences for any number of homologous sequences, when the phylogeny is given. *Mol. Biol. Evol.*, 6:649–668, 1989.

[41] J. Hein. Unified approach to alignment and phylogenies. *Methods Enzymol.*, 183:626–645, 1990.

[42] S. Henikoff and J.G. Henikoff. Amino acid substitution matrices from protein blocks. *Proc. Natl. Acad. Sci. USA*, 89:10915–10919, 1992.

[43] S. Henikoff and J.G. Henikoff. Position-based sequence weights. *J. Mol. Biol.*, 243:574–578, 1994.

[44] M. Hirosawa, Y. Totoki, M. Hoshida, and M. Ishikawa. Comprehensive study on iterative algorithms of multiple sequence alignment. *Comput. Appl. Biosci.*, 11:13–18, 1995.

[45] X. Huang and K.M. Chao. A generalized global alignment algorithm. *Bioinformatics*, 19:228–233, 2003.

[46] X. Huang and A. Madan. CAP3: A DNA sequence assembly program. *Genome Res.*, 9:868–877, 1999.

[47] T. Ikeda and H. Imai. Enhanced A* algorithms for multiple alignments: optimal alignments for several sequences and k-opt approximate alignments for large cases. *Theor. Comp. Sci.*, 210:341–374, 1999.

[48] M. Ishikawa, T. Toya, M. Hoshida, and K. Nitta *et al.* Multiple sequence alignment by parallel simulated annealing. *Comput. Appl. Biosci.*, 9:267–273, 1993.

[49] T. Jiang and L. Wang. Algorithmic methods for multiple sequence alignment. In *Current topics in computational molecular biology*, Computational molecular biology, pages 71–110. The MIT Press, Cambridge, 2002.

[50] D.T. Jones, W.R. Taylor, and J.M. Thornton. The rapid generation of mutation data matrices from protein sequences. *Comput. Appl. Biosci.*, 8:275–282, 1992.

[51] W. Just. Computational complexity of multiple sequence alignment with SP-score. *J. Comput. Biol.*, 8:615–623, 2001.

[52] K. Katoh, K. Misawa, K. Kuma, and T. Miyata. MAFFT: a novel method for rapid multiple sequence alignment based on fast fourier transform. *Nucleic Acids Res.*, 30:3059–3066, 2002.

[53] J. Kececioglu, H.-P. Lenhof, K. Mehlhorn, and P. Mutzel *et al.* A polyhedral approach to sequence alignment problems. *Disc. Appl. Math.*, 104:143–186, 2000.

[54] J.D. Kececioglu. The maximum weight trace problem in multiple sequence alignment. *Lecture Notes Comp. Sci.*, 684:106–119, 1993.

[55] J.D. Kececioglu and D. Starrett. Aligning alignments exactly. In *Proceedings of the 8th ACM Conference on Computational Molecular Biology, 85-96, 2004*, 2004.

[56] J. Kim, S. Pramanik, and M.J. Chung. Multiple sequence alignment using simulated annealing. *Comput. Appl. Biosci.*, 10:419–426, 1994.

[57] H. Kishino, T. Miyata, and M. Hasegawa. Maximum likelihood inference of protein phylogeny and the origin of chloroplasts. *J. Mol. Evol.*, 31:151–160, 1990.

[58] A. Kloczkowski, K.L. Ting, R.L. Jernigan, and J. Garnier. Combining the GOR V algorithm with evolutionary information for protein secondary structure prediction from amino acid sequence. *Proteins*, 49:154–166, 2002.

[59] A. Krogh, M. Brown, I. S. Mian, and K. Sjölander *et al.* Hidden markov models in computational biology: Applications to protein modeling. *J. Mol. Biol.*, 235:1501–1531, 1994.

[60] J.B. Kruskal. An overview of sequence comparison. In *Time warps, string edits, and macromolecules: The theory and practice of sequence comparison*, pages 1–44. Addison-Wesley, Reading, MA, 1983.

[61] J.B. Kruskal and D. Sankoff. An anthology of algorithms and concepts for sequence comparison. In *Time warps, string edits, and macromolecules: The theory and practice of sequence comparison*, pages 265–310. Addison-Wesley, Reading, MA, 1983.

[62] V. Kunin, B. Chan, E. Sitbon, and G. Lithwick *et al.* Consistency analysis of similarity between multiple alignments: prediction of protein function and fold structure from analysis of local sequence motifs. *J. Mol. Biol.*, 307:939–949, 2001.

[63] T. Lassmann and E.L. Sonnhammer. Quality assessment of multiple alignment programs. *FEBS Lett.*, 529:126–130, 2002.

[64] C.E. Lawrence, S.F. Altschul, M.S. Boguski, and J.S. Liu *et al.* Detecting subtle sequence signals: A Gibbs sampling strategy for multiple alignment. *Science*, 262:208–214, 1993.

[65] C. Lee, C. Grasso, and M.F. Sharlow. Multiple sequence alignment using partial order graphs. *Bioinformatics*, 18:452–464, 2002.

[66] M. Lermen and K. Reinert. The practical use of the A* algorithm for exact multiple sequence alignment. *J. Comput. Biol.*, 7:655–671, 2000.

[67] O. Lichtarge, H.R. Bourne, and F.E. Cohen. An evolutionary trace method defines binding surfaces common to protein families. *J. Mol. Biol.*, 257:342–358, 1996.

[68] D.J. Lipman, S.F. Altschul, and J.D. Kececioglu. A tool for multiple sequence alignment. *Proc. Natl. Acad. Sci. USA*, 86:4412–4415, 1989.

[69] W. Miller and E.W. Myers. Sequence comparison with concave weighting functions. *Bull. Math. Biol.*, 50:97–120, 1988.

[70] B. Morgenstern. DIALIGN 2: improvement of the segment-to-segment approach to multiple sequence alignment. *Bioinformatics*, 15:211–218, 1999.

[71] B. Morgenstern, K. Frech, A. Dress, and T. Werner. DIALIGN: finding local similarities by multiple sequence alignment. *Bioinformatics*, 14:290–294, 1998.

[72] D.W. Mount. Multiple sequence alignment. In *Bioinformatics: Sequence and genome analysis*, pages 163–225. Cold Spring Harbor Laboratory Press, Cold Spring

Harbor, New York, second edition, 2004.

[73] M. Murata, J.S. Richardson, and J.L. Sussman. Simultaneous comparison of three protein sequences. *Proc. Natl. Acad. Sci. USA*, 82:3073–3077, 1985.

[74] E.W. Myers and W. Miller. Optimal alignments in linear space. *Comput. Appl. Biosci.*, 4:11–17, 1988.

[75] M. Nei and S. Kumar. *Molecular evolution and phylogenetics*. Oxford University Press, Oxford, 2000.

[76] C. Notredame. Recent progress in multiple sequence alignment: a survey. *Pharmacogenomics*, 3:131–144, 2002.

[77] C. Notredame and D.G. Higgins. SAGA: sequence alignment by genetic algorithm. *Nucleic Acids Res.*, 24:1515–1524, 1996.

[78] C. Notredame, D.G. Higgins, and J. Heringa. T-Coffee: A novel method for fast and accurate multiple sequence alignment. *J. Mol. Biol.*, 302:205–217, 2000.

[79] C. Notredame, L. Holm, and D.G. Higgins. COFFEE: an objective function for multiple sequence alignments. *Bioinformatics*, 14:407–422, 1998.

[80] W.R. Pearson. Comparison of methods for searching protein sequence databases. *Protein Sci.*, 4:1145–1160, 1995.

[81] J. Pei, R. Sadreyev, and N.V. Grishin. PCMA: fast and accurate multiple sequence alignment based on profile consistency. *Bioinformatics*, 19:427–428, 2003.

[82] D. Petrey, Z. Xiang, C.L. Tang, and L. Xie *et al.* Using multiple structure alignments, fast model building, and energetic analysis in fold recognition and homology modeling. *Proteins*, Vol. 53, Suppl. 6:430–435, 2003.

[83] P. Pevzner. *Computational molecular biology: An algorithmic approach*. Computational molecular biology. The MIT Press, Cambridge, MA, 2000.

[84] A. Phillips, D. Janies, and W. Wheeler. Multiple sequence alignment in phylogenetic analysis. *Mol. Phylogenet. Evol.*, 16:317–330, 2000.

[85] L.R. Rabiner. A tutorial on hidden markov models and selected applications in speech recognition. *Proc. IEEE*, 77:257–286, 1989.

[86] G.P. Raghava, S.M. Searle, P.C. Audley, and J.D. Barber *et al.* OXBench: a benchmark for evaluation of protein multiple sequence alignment accuracy. *BMC Bioinformatics*, 4:47, 2003.

[87] S. Rajasekaran, X. Jin, and J.L. Spouge. The efficient computation of position-specific match scores with the fast fourier transform. *J. Comput. Biol.*, 9:23–33, 2002.

[88] K. Reinert, J. Stoye, and T. Will. An iterative method for faster sum-of-pairs multiple sequence alignment. *Bioinformatics*, 16:808–814, 2000.

[89] S.K. Riis and A. Krogh. Improving prediction of protein secondary structure using structured neural networks and multiple sequence alignments. *J. Comput. Biol.*, 3:163–183, 1996.

[90] B. Rost and C. Sander. Progress of 1D protein structure prediction at last. *Proteins*, 23:295–300, 1995.

[91] N. Saitou and M. Nei. The neighbor-joining method: A new method for reconstructing phylogenetic trees. *Mol. Biol. Evol.*, 4:406–425, 1987.

[92] A.A. Salamov and V.V. Solovyev. Prediction of protein secondary structure by combining nearest-neighbor algorithms and multiple sequence alignments. *J. Mol. Biol.*, 247:11–15, 1995.

[93] A. Sali, J.P. Overington, M.S. Johnson, and T.L. Blundell. From comparisons of protein sequences and structures to protein modelling and design. *Trends Biochem. Sci.*, 15:235–240, 1990.

[94] D. Sankoff. Minimal mutation trees of sequences. *SIAM J. Appl. Math.*, 78:35–42, 1975.

[95] D. Sankoff and R. Cedergren. Simultaneous comparison of three or more sequences related by a tree. In *Time warps, string edits, and macromolecules: The theory and practice of sequence comparison.* Addison-Wesley, Reading, MA, 1983.

[96] T.D. Schneider and R.M. Stephens. Sequence logos: a new way to display consensus sequences. *Nucleic Acids Res.*, 18:6097–6100, 1990.

[97] T.D. Schneider, G.D. Stormo, L. Gold, and A. Ehrenfeucht. Information content of binding sites on nucleotide sequences. *J. Mol. Biol.*, 188:415–431, 1986.

[98] T. Shibuya and H. Imai. New flexible approaches for multiple sequence alignment. *J. Comput. Biol.*, 4:385–413, 1997.

[99] P.R. Sibbald and P. Argos. Weighting aligned protein or nucleic acid sequences to correct for unequal representation. *J. Mol. Biol.*, 216:813–818, 1990.

[100] P.H.A. Sneath and R.P. Sokal. *Numerical taxonomy.* Freeman, San Francisco, CA, 1973.

[101] R. Spang, M. Rehmsmeier, and J. Stoye. A novel approach to remote homology detection: jumping alignments. *J. Comput. Biol.*, 9:747–760, 2002.

[102] J. Stoye, V. Moulton, and A.W. Dress. DCA: an efficient implementation of the divide-and-conquer approach to simultaneous multiple sequence alignment. *Comput. Appl. Biosci.*, 13:625–626, 1997.

[103] S. Subbiah and S.C. Harrison. A method for multiple sequence alignment with gaps. *J. Mol. Biol.*, 209:539–548, 1989.

[104] D.A. Tagle, B.F. Koop, M. Goodman, and J.L. Slightom *et al.* Embryonic epsilon and gamma globin genes of a prosimian primate (*galago crassicaudatus*). nucleotide and amino acid sequences, developmental regulation and phylogenetic footprints. *J. Mol. Biol.*, 203:439–455, 1988.

[105] L. Taher, O. Rinner, S. Garg, and A. Sczyrba *et al.* AGenDA: homology-based gene prediction. *Bioinformatics*, 19:1575–1577, 2003.

[106] J.D. Thompson. Introducing variable gap penalties to sequence alignment in linear space. *Comput. Appl. Biosci.*, 11:181–186, 1995.

[107] J.D. Thompson, D.G. Higgins, and T.J. Gibson. CLUSTAL W: improving the sensitivity of progressive multiple sequence alignment through sequence weighting, position-specific gap penalties and weight matrix choice. *Nucleic Acids Res.*, 22:4673–4680, 1994.

[108] J.D. Thompson, D.G. Higgins, and T.J. Gibson. Improved sensitivity of profile searches through the use of sequence weights and gap excision. *Comput. Appl. Biosci.*, 10:19–29, 1994.

[109] J.D. Thompson, F. Plewniak, and O. Poch. BAliBASE: a benchmark alignment database for the evaluation of multiple alignment programs. *Bioinformatics*, 15:87–88, 1999.

[110] J.D. Thompson, F. Plewniak, and O. Poch. A comprehensive comparison of multiple sequence alignment programs. *Nucleic Acids Res.*, 27:2682–2690, 1999.

[111] U. Tönges, S.W. Perrey, J. Stoye, and A.W. Dress. A general method for fast multiple sequence alignment. *Gene*, 172:GC33–41, 1996.

[112] W.S. Valdar. Scoring residue conservation. *Proteins*, 48:227–241, 2002.

[113] C. Venclovas. Comparative modeling in CASP5: progress is evident, but alignment errors remain a significant hindrance. *Proteins*, Vol. 53, Suppl. 6:380–388, 2003.

[114] M. Vingron and P.A. Pevzner. Multiple sequence comparison and consistency on multipartite graphs. *Adv. Appl. Math.*, 16:1–22, 1995.

[115] M. Vingron and P.R. Sibbald. Weighting in sequence space: A comparison of methods in terms of generalized sequences. *Proc. Natl. Acad. Sci. USA*, 90:8777–8781, 1993.

[116] I.V. Walle, I. Lasters, and L. Wyns. SABmark - a benchmark for sequence alignment

that covers the entire known fold space. *Bioinformatics*, 2004.

[117] G. Wang and Jr. Dunbrack, R.L. Scoring profile-to-profile sequence alignments. *Protein Sci.*, 13:1612–1626, 2004.

[118] L. Wang and T. Jiang. On the complexity of multiple sequence alignment. *J. Comput. Biol.*, 1:337–348, 1994.

[119] H.T. Wareham. A simplified proof of the NP- and MAX SNP-hardness of multiple sequence tree alignment. *J. Comput. Biol.*, 2:509–514, 1995.

[120] M.S. Waterman, T.F. Smith, and W.A. Beyer. Some biological sequence metrics. *Adv. Math.*, 20:367–387, 1976.

[121] H. Yao, D.M. Kristensen, I. Mihalek, and M.E. Sowa *et al.* An accurate, sensitive, and scalable method to identify functional sites in protein structures. *J. Mol. Biol.*, 326:255–261, 2003.

[122] C. Zhang and A.K. Wong. A genetic algorithm for multiple molecular sequence alignment. *Comput. Appl. Biosci.*, 13:565–581, 1997.

4

Parametric Sequence Alignment

David Fernández-Baca
Iowa State University

Balaji Venkatachalam
Iowa State University

4.1 Introduction

The optimum solution to any sequence alignment problem depends on the choice of a number of parameters, such as the penalties for mismatches and gaps. Even slight changes in the parameter values can result in a complete structural change in the optimum alignment. *Parametric sequence alignment* explores the effect of parameter variation on the optimum alignment over a *range* of parameter values, to assist in making the right choice for the given circumstances. Several specific problems are of interest. For instance, in *sensitivity analysis*, the goal is to assess the robustness of an optimum alignment by determining the amount by which the current parameter values can be perturbed without altering the optimality of the alignment. The goal in *inverse alignment* is to locate the parameter values for which the optimum alignment exhibits certain desired features, for example, a certain number of matches or gaps. These and other questions are the subject of this chapter, which is organized around four interrelated topics: combinatorial complexity, parametric search, construction, and parametric problems in hidden Markov models of sequence alignment. We now give an overview of these issues; more precise definitions can be found in the next section.

Under the linear scoring schemes used in practice, the parameter space decomposes into a finite set of convex and polyhedral *optimality regions*, each of which is a maximal connected subset of parameters within which essentially one alignment is optimum. The number of regions of the decomposition is the *combinatorial complexity* of the decomposition; determining this quantity is one of the basic problems in parametric analysis. Because of the shared properties of commonly-used scoring schemes, it is possible to establish general bounds on the combinatorial complexity. These bounds and their applications are presented in 4.3.

Parametric search problems ask to find parameter settings satisfying certain specified conditions. The inverse alignment problem described above falls in this category. *Ray shooting* is a parametric search problem that arises as a subtask in sensitivity analysis: Given a point in the parameter space, an alignment optimal at that point, and a ray emanating from the point, the question is to find the point along the ray at which the alignment ceases to be optimal. Techniques for parametric search are reviewed in 4.4.

The *construction problem* is to build the parameter space decomposition induced by the alignment of two strings. As discussed in 4.2, this is equivalent to building the *upper envelope* of a set of linear score functions. The time needed to build the decomposition clearly depends on the combinatorial complexity of the output produced. Approaches for construction are reviewed in 4.5.

Hidden Markov models offer an alternative view of sequence alignment scoring schemes. Here, the evolution of two sequences from a common ancestor is viewed as a stochastic process where unobservable mutations, insertions, and deletions take place according to certain probabilities, which are the parameters of the model. The goal is to infer the most likely evolutionary process for the given sequences. Parametric problems arise naturally in this context, since the models depend highly on the underlying probabilities. The relationship between these questions and the parametric problems of 4.3 – 4.5 is studied in 4.6.

The chapter closes in 4.7 with some historical notes, comments, and additional references.

4.2   Preliminaries

The study of parametric sequence alignment relies on certain basic geometric notions, together with the properties of scoring schemes. We review both of these next, after which we formally define parametric sequence alignment and introduce some of the basic problems in parametric analysis.

4.2.1   Geometric Notions

DEFINITION 4.1   Let H be a set of d-variate real-valued functions. The *upper envelope* of H is the function defined as:

$$\mathcal{U}_H(\lambda) = \max_{h \in H} h(\lambda). \tag{4.1}$$

That is, $\mathcal{U}_H$ is the point-wise maximum of the functions in H (see Figure 4.1). The *lower envelope* of H is defined as

$$\mathcal{L}_H(\lambda) = \min_{h \in H} h(\lambda). \tag{4.2}$$

That is, $\mathcal{L}_H$ is the point-wise minimum of the functions in H.

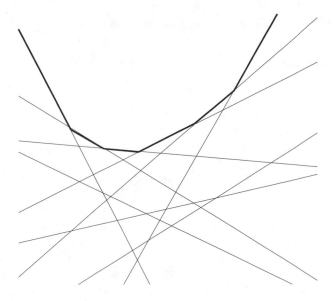

FIGURE 4.1: A collection of linear functions, with their upper envelope shown in bold.

PROPOSITION 4.1   Let H be a collection of d-variate linear functions. Then, $\mathcal{U}_H$ ($\mathcal{L}_H$) is a piecewise-linear convex (concave) function.

In the one-dimensional case, the upper envelope of a collection of linear functions takes on a particularly simple form: it is a chain of line segments of successively larger slopes. The parameter values at which the slope changes are called *breakpoints*. (See Figure 4.1.)

DEFINITION 4.2   Let H be a set of d-variate real-valued functions. The *maximization diagram* (*minimization diagram*) of H is a data structure that consists of two parts:

(i) The subdivision of $\mathbb{R}^d$ induced by $\mathcal{U}_H$ ($\mathcal{L}_H$); that is, the projection of the faces of dimensions 0 through d of $\mathcal{U}_H$ ($\mathcal{L}_H$) onto $\mathbb{R}^d$.

(ii) For each d-face f of the subdivision, the unique function $h \in H$ such that $\mathcal{U}_H(\lambda)$ ($\mathcal{L}_H(\lambda)$) equals $h(\lambda)$ for all $\lambda \in f$. Face f is the *optimality region* for function h.

PROPOSITION 4.2   Let H be a collection of d-variate linear functions. Then, the maximization (minimization) diagram of H is a decomposition of $\mathbb{R}^d$ into convex polyhedral regions.

Observe that in the one-dimensional case, the maximization diagram yields a subdivision of the real line into a sequence of intervals. The boundary between each successive pair of intervals is a breakpoint of the upper envelope.

4.2.2   Scoring Schemes

Every sequence alignment method is based on some scheme that assigns numeric scores to alignments. The score of an alignment $\mathcal{A}$ is a function of $\mathcal{A}$ itself and a user-specified *parameter vector* $\lambda = (\lambda_1, \ldots, \lambda_d) \in \mathbb{R}^d$, which gives the weights of the various features of an alignment (for example, the cost of mismatches and gaps).

DEFINITION 4.3   The *optimum alignment problem* is to compute

$$Z(\lambda) = \max_{\mathcal{A}} \ score(\mathcal{A}, \lambda) \tag{4.3}$$

for some fixed λ, where *score* is the scoring function and $\mathcal{A}$ ranges over all alignments of the input sequences. The alignment $\mathcal{A}^*$ that maximizes (4.3) is called the *optimum alignment at* λ. Function Z is the *optimum score function*.

Throughout the rest of this chapter the input sequences are denoted by S_1 and S_2; their lengths are n and m, respectively, with $n \le m$. The value of the maximum score of an alignment between S_1 and S_2 is sometimes called the *similarity score* (or simply the *similarity*) of S_1 and S_2. Intuitively, the higher the similarity between S_1 and S_2, the closer the two sequences are. Note that there are cases where the objective is to find an alignment of *minimum* score. These are sometimes called *minimum edit distance problems* and the value of the minimum score is called the *distance* between the sequences [28]. In this chapter, all results are for similarity scoring; however, these results can be translated to minimization problems. Indeed, it is often possible to establish correspondences between similarity- and distance-based scoring schemes [28].

DEFINITION 4.4   An *evaluator* is a procedure that given a pair of input strings and a parameter vector $\lambda^{(0)}$, computes an optimum alignment $\mathcal{A}^{(0)}$ at $\lambda^{(0)}$.

Note that an evaluator also allows us to compute the value of $Z(\lambda^{(0)})$, which equals $score(\mathcal{A}^{(0)}, \lambda^{(0)})$.

DEFINITION 4.5   A scoring scheme is *linear* if, for every alignment $\mathcal{A}$, $score(\mathcal{A}, \lambda)$ is a linear function of the parameter vector λ.

Essentially all scoring schemes used in sequence alignment are linear. For example, in global alphabet-independent scoring, the score is a linear function of four parameters,

$$score(\mathcal{A}, \alpha, \beta, \gamma, \delta) = \alpha w - \beta x - \gamma y - \delta z, \tag{4.4}$$

where w, x, y, z are the number of matches, mismatches, indels, and gaps in $\mathcal{A}$, and $\alpha, \beta, \gamma, \delta$ are the respective weights. An evaluator for this problem is any optimum global alignment algorithm, such as the standard $O(nm)$ dynamic programming procedure [28].

In global alphabet-dependent scoring, we are given a $|\Sigma| \times |\Sigma|$ matrix $M = [\mu_{ab}]$, where Σ is the alphabet and μ_{ab} is the score for aligning characters a and b. Then,

$$score(\mathcal{A}, M, \gamma, \delta) = \sum_{a,b \in \Sigma} \mu_{ab} w_{ab} - \gamma y - \delta z, \tag{4.5}$$

where w_{ab} denotes the number of times characters a and b are aligned in $\mathcal{A}$, and y, z, γ, δ are as in Equation (4.4). As in the alphabet-independent case, an optimum alignment can be found in $O(nm)$ time for any fixed parameter vector.

The next definition captures some of the characteristics of commonly used scoring schemes.

DEFINITION 4.6   A scoring scheme is *feature-based* if there exists a (many to one) mapping f from alignments to a subset F of $\mathbb{Z}^d$ and the score of any alignment $\mathcal{A}$ as a function of the parameters depends only on $f(\mathcal{A})$. Set F is the *feature set* of the problem; $p = f(\mathcal{A})$ is the *feature vector* of $\mathcal{A}$ and each coordinate of p is called a *feature*. A feature-based scoring scheme is *simple* if for any alignment $\mathcal{A}$, the score of $\mathcal{A}$ can be expressed as

$$score(\mathcal{A}, \lambda) = p \cdot \lambda, \qquad (4.6)$$

where $p \in \mathbb{Z}^d$ is the feature vector of $\mathcal{A}$ and $\lambda \in \mathbb{R}^d$ is the parameter vector.

In all applications considered in this chapter, the size of the feature set is bounded as a function of the sequence lengths. This is explored in greater depth in 4.3.

The feature vector of an alignment $\mathcal{A}$ represents discrete characteristics of $\mathcal{A}$. To illustrate this, observe that the alphabet-independent scheme of Equation (4.4) is feature-based and simple. The feature vector of an alignment in this case is $p = (w, x, y, z)$. Each of p's features is bounded by the total number of characters in the input strings, and the parameter vector $\lambda = (\alpha, \beta, \gamma, \delta)$ assigns a weight to each feature. The alphabet-dependent scoring scheme of Equation (4.5) is also feature-based and simple. In this case, there are $|\Sigma|^2 + 2$ features, given by the vector

$$p = (w_{a_1 a_1}, \ldots, w_{a_1 a_{|\Sigma|}}, \ldots, w_{a_{|\Sigma|} a_1}, \ldots, w_{a_{|\Sigma|} a_{|\Sigma|}}, y, z). \qquad (4.7)$$

Each feature is at most equal to the total number of characters in the input strings.

The following is an example of an alphabet-dependent scoring scheme that is feature-based but not simple [31]. Let the scoring matrix $M = [\mu_{ab}]$ be fixed, but assign different weights to matches and mismatches as shown below.

$$score(\mathcal{A}, \alpha, \beta, \gamma, \delta) = \alpha \sum_{a \in \Sigma} \mu_a w_{aa} + \beta \sum_{a, b \in \Sigma} \mu_{ab} w_{ab} - \gamma y - \delta z \qquad (4.8)$$

The feature vector in this case is given by Equation (4.7), but there are now only four parameters — α, β, γ, and δ — instead of $|\Sigma|^2 + 2$.

Definition 4.6 does not capture all scoring schemes used in practice. For example, the definition does not cover schemes where the cost of a gap is some constant (the gap penalty) times the logarithm of the gap length [28]. Even though this method does not fit within our framework, it nevertheless still exhibits some of the characteristics of feature-based scoring, since, for fixed sequence lengths, the number of gaps and their various lengths can only assume a finite set of integer values. Note also that while the dependence on the gap length is non-linear, the scoring scheme is itself linear in the weight given to gaps.

4.2.3 Parametric Sequence Alignment

Parametric sequence alignment studies the effect of varying the parameter vector on the optimum alignment. Its central object of study is Z, the optimum score function of Definition 4.3. Under linear similarity scoring, Z is the upper envelope of a set of linear functions

$$H = \{score(\mathcal{A}, \lambda) : \mathcal{A} \text{ is optimal at some } \lambda^{(0)} \in \mathbb{R}^d\}.$$

(For minimum edit distance problems, Z is a lower envelope.) For brevity, we refer to the maximization diagram of H as the maximization diagram of Z. By Propositions 4.1 and 4.2, the maximization diagram of Z decomposes the parameter space, $\mathbb{R}^d$, into convex polyhedral optimality regions. This is illustrated in Figure 4.2, which shows the maximization diagram of the optimum score function for global alphabet-independent alignment between two sequences as the mismatch penalty β and the indel penalty γ are varied across the

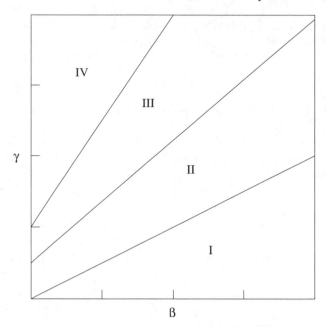

FIGURE 4.2: Decomposition of the parameter space induced by sequences S_1 = BAABBB and S_2 = ABBAAA. The corresponding optimum alignments are $\mathcal{A}_{\mathrm{I}}$ = (BAABBB $---,---$ A $-$ BBAAA), $\mathcal{A}_{\mathrm{II}}$ = (BAABB $--$ B, $--$ ABBAAA), $\mathcal{A}_{\mathrm{III}}$ = (BAAB $-$ BB, $-$ ABBAAA), $\mathcal{A}_{\mathrm{IV}}$ = (BAABBB, ABBAAA).

positive quadrant, while each match gets a (constant) reward of one and the gap penalty is zero. There are exactly four optimality regions, whose corresponding alignments are shown. Note that in general, each optimality region may have several co-optimal alignments.

The structure of the maximization diagram of Figure 4.2 is particularly simple: any vertical cross section encounters the same series of alignments when going from bottom to top. A representative slice is shown in Figure 4.3, which displays the optimal score function $Z(\gamma)$ for the alignment problem of Figure 4.2, when β is fixed at one. The picture shows how, indeed, $Z(\gamma)$ is the upper envelope of the score functions of various alignments.

The simplicity of Figure 4.2 is due to the scoring scheme, and is analyzed in more detail in 4.3. Other scoring schemes yield more intricate structures, as shown in Figure 4.4, taken from [31]. The figure gives the maximization diagram for the score of the optimum global alignment of immunoglobulins FABVL and FABVH, as a function of indel and gap penalties. The scoring here is alphabet-dependent, using the PAM250 matrix [15] with a constant of 8 added to each entry; gaps at the end of an alignment are assigned a score of zero.

4.2.4 Some Basic Parametric Problems

We now list five basic problems encountered in parametric sequence analysis, regardless of the scoring system used. The first of these is central to many parametric analysis problems:

PROBLEM 1 *(Ray shooting)* Given a parameter vector $\lambda^{(0)} \in \mathbb{R}^d$, an optimum alignment $\mathcal{A}^{(0)}$ at $\lambda^{(0)}$ and a ray ρ originating at $\lambda^{(0)}$, find the last point on ρ at which $\mathcal{A}^{(0)}$ is optimal.

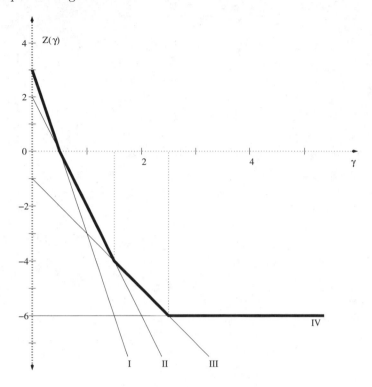

FIGURE 4.3: The optimum score function for the two sequences of Figure 4.2, when the mismatch penalty is fixed at one. Each line segment is the score of one optimum alignment (labeled I–IV to correspond to Figure 4.2); the Z function is shown in bold.

By suitable re-parametrization Problem 1 can be converted to a one-parameter problem: find the largest $\lambda^* > \lambda^{(0)}$ such that $\mathcal{A}^{(0)}$ is optimal at λ^* (thus, λ^* is a breakpoint of the optimum score function or infinity).

The next problem is a multidirectional version of the preceding one. It addresses the robustness of alignments to changes in parameter settings.

PROBLEM 2 *(Sensitivity analysis)* Given a parameter vector $\lambda^{(0)} \in \mathbb{R}^d$ and an optimum alignment $\mathcal{A}^{(0)}$ at $\lambda^{(0)}$, find the largest subset F of $\mathbb{R}^d$ such that $\mathcal{A}^{(0)}$ is optimal for all $\lambda \in F$.

Problem 2 is equivalent to computing a complete description of the optimality region containing a given parameter vector. The next problem asks to find *all* the optimality regions.

PROBLEM 3 *(Parameter space decomposition)* Construct the maximization diagram of $Z(\lambda)$.

In applications, the maximization diagram is only constructed for a subset of $\mathbb{R}^d$. For instance, in two-parameter problems, this subset may be the positive quadrant. In other

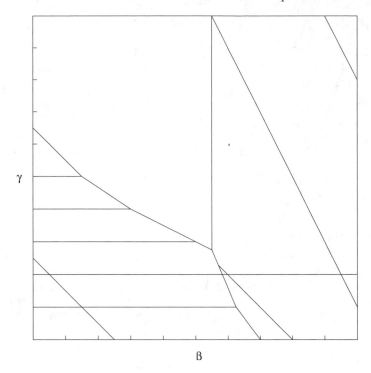

γ

β

FIGURE 4.4: Decomposition of the parameter space induced by two protein sequences.

cases, we may only be interested in constructing the maximization diagram within a (hyper-) rectangle in $\mathbb{R}^d$.

The time needed to construct the maximization diagram of the optimum score function depends heavily on the number of regions in the diagram. Thus, closely related to Problem 3 is the following question.

PROBLEM 4 *(Combinatorial complexity)* Establish bounds on the number of optimality regions in the maximization diagram of Z as a function of the lengths of the sequences.

One way to determine parameter settings that are likely to produce biologically meaningful results is to use a set of sample alignments (produced, perhaps, by manually editing computer-generated alignments [38]) as a training set. This motivates our final problem.

PROBLEM 5 *(Inverse optimization)* Given a *reference alignment* $\mathcal{A}^{(0)}$, find a parameter vector $\lambda^* \in \mathbb{R}^d$ that minimizes $Z(\lambda) - score(\mathcal{A}^{(0)}, \lambda)$.

Problem 4 is, in a sense, the most basic of the five and is thus studied first, in 4.3. Problems 1 and 5 are parametric search problems, which are discussed in 4.4. Problem 2 is also treated in 4.4, where it is shown that it can be solved through ray shooting. Problem 3 is discussed in 4.5. The relationship between Problems 1–5 and hidden Markov models of sequence alignment is examined in 4.6.

4.3 Combinatorial Complexity

The combinatorial complexity of the optimum score function is the number of optimality regions in its maximization diagram. This quantity impacts the time needed to solve various parametric alignment problems. The most obvious example is the construction problem, since the work needed to build the maximization diagram clearly depends directly on the number of regions in the decomposition (see Section 4.5). The combinatorial complexity also affects the time needed for parametric search (see Section 4.4).

For feature-based scoring schemes, the number of distinct score functions is trivially bounded by the product of the sizes of the ranges of the features. We summarize this as follows.

THEOREM 4.1 *Let $\mathcal{P}$ be a d-parameter alignment problem with feature-based scoring, such that for every $i \in [d]$, feature i can assume one of n_i distinct values. Then, the maximization diagram of the optimum score function of $\mathcal{P}$ has $O(\prod_{i=1}^{d} n_i)$ optimality regions.*

The above result can give quick estimates on the number of optimality regions, since individual features — such as the number of matches, mismatches, gaps, or indels — are typically linear in the lengths of the input strings (see 4.3.1).

For small d and simple feature-based scoring, the combinatorial properties of lattice polytopes can be used to improve the bound of Theorem 4.1 as follows [21, 42].

THEOREM 4.2 *Let $\mathcal{P}$ be a d-parameter alignment problem with simple feature-based scoring, such that for every $i \in [d]$, feature i can assume integer values in an interval of length n_i. Then, the maximization diagram of the optimum score function of $\mathcal{P}$ has*
$$O\left(\left(\prod_{i=1}^{d} n_i\right)^{(d-1)/(d+1)}\right) \text{ optimality regions.}$$

In the remainder of this section, we present some applications of these results. Our goal is not to be exhaustive, but to give the reader an idea of how the theorems can be used.

4.3.1 Global Alignment

We begin with alphabet-independent scoring. The score of a global alignment of two strings, Equation (4.4) of 4.2.2, is a linear function of four parameters, $\alpha, \beta, \gamma, \delta$. The feature vector is (w, x, y, z) and each feature is bounded by the length of the strings as follows [29]:

$$0 \leq w, x \leq n, \qquad m - n \leq y \leq m + n, \qquad 0 \leq z \leq 2n + 1. \tag{4.9}$$

That is, the range of values for each feature is $O(n)$. Therefore, applying Theorem 4.1, we obtain an upper bound of $O(n^4)$ on the number of optimality regions. This is reduced to $O(n^{12/5})$ by applying Theorem 4.2. In this case as in others, it is possible to improve on these estimates significantly by exploiting dependencies between features. We give the details below.

The reward for matches, α, is always assumed to be positive. Thus, we can divide through by α and re-parameterize appropriately, to obtain the following equivalent score function

$$score(\mathcal{A}, \beta, \gamma, \delta) = w - \beta x - \gamma y - \delta z \tag{4.10}$$

The number of matches can be eliminated from the score function through the following observation. Since every match and every mismatch involves exactly two characters from each of the input strings and every indel involves exactly one character from one of the strings, we have that [29]

$$2w + 2x + y = m + n. \tag{4.11}$$

Solving for w and substituting in Equation (4.10), we obtain [29, 19]

$$score(\mathcal{A}, \alpha, \beta, \gamma, \delta) = \frac{n+m}{2} - (\beta + 1)x - \left(\gamma + \frac{1}{2}\right)y - \delta z. \tag{4.12}$$

Observe that since the term $(n+m)/2$ is common to all score functions, it can be eliminated, allowing us to redefine the score, after suitable re-parametrization [29, 19], as

$$score(\mathcal{A}, \beta, \gamma, \delta) = -\beta x - \gamma y - \delta z. \tag{4.13}$$

We consider two cases. First, suppose the gap penalty δ is zero, and thus $score(\mathcal{A}, \beta, \gamma, \delta) = -\beta x - \gamma y$. The feature vector is therefore (x, y). Using Equation (4.9) and Theorem 4.2 yields a bound of $O(n^{2/3})$ on the number of optimality regions.

For the problem of global alignment with varying gap penalty $\delta > 0$, there are three parameters and the feature vector is (x, y, z). Using Equation (4.9) and Theorem 4.2 yields a bound of $O(n^{3/2})$ on the number of optimality regions.

REMARK 4.1   Equation (4.13) implies that global alphabet-independent alignment with zero gap penalty is effectively a one parameter problem in γ, in the sense that the optimal alignments encountered as γ is varied are exactly the same regardless of the value of β, as long as it is positive. Thus, the simple structure of the maximization diagram in Figure 4.2 is not accidental. By a similar reasoning, for $\beta > 0$ global alphabet-independent alignment with non-zero gap penalty is effectively a two parameter problem in γ and δ.

We now turn to alphabet-dependent scoring, where the score function is given by Equation (4.5). The number of parameters $d = |\Sigma|^2 + 2$ and the feature vector is given by Equation (4.7). The values of y and z in this vector obey the bounds of expression (4.9). For every $a, b \in \Sigma$, $0 \le w_{ab} \le n$, since the number of times any two characters in the alphabet can be paired in an alignment cannot exceed the number of characters in the shorter string. Thus, Theorem 4.2 gives a bound of $n^{O(|\Sigma|^2)}$ optimality regions, which is polynomial if the alphabet size is fixed.

Consider now the four-parameter alphabet-dependent scoring scheme of Equation (4.8) of 4.2.2. Since the scheme is not simple, we use Theorem 4.1 to bound the number of optimality regions. The feature vector in this case has $\Theta(|\Sigma|^2)$ coordinates (the same as in the previous example), each of which is at most n. This yields a bound of $n^{O(|\Sigma|^2)}$ on the combinatorial complexity of Z.

4.3.2   Local Alignment

A *local alignment* of two strings S_1, S_2 is a global alignment between two substrings S_1', S_2', of S_1, S_2, respectively [28]. Thus, in the alphabet-independent case, the score function takes on the same form as Equation (4.10). The bounds of inequalities (4.9) on the ranges of w, x, y, and z still hold. However, since local alignments involve *substrings* of S_1 and S_2, Equation (4.11) becomes an inequality, preventing us from eliminating w [29].

When the gap penalty is zero, there are three parameters, with feature vector (w, x, y). Hence, Theorem 4.2 gives a bound of $O(n^{3/2})$ on the number of optimality regions. When the gap penalty is allowed to vary, we have four parameters and feature vector (w, x, y, z), giving a bound of $O(n^{12/5})$. For alphabet-dependent scoring, the same arguments as in the preceding subsection give an upper bound of $n^{O(|\Sigma|^2)}$ on the number of optimality regions.

4.3.3 Multiple Sequence Alignment

Since Theorem 4.1 and 4.2 depend only on the form of the scoring scheme, they can be applied to problems beyond pairwise alignment, including a variety of multiple alignment problems. The key idea is to aggregate the features of the induced pairwise alignments into a feature vector for the multiple alignment as a whole. As an example, we consider sum-of-pairs scoring; further applications are given in [21, 19].

A *multiple alignment* $\mathcal{A}$ of the set of sequences $S_1, \ldots, S_k$, $k \geq 2$, is obtained by inserting spaces in each string to obtain k strings $S'_1, \ldots, S'_k$ of the same length. For every pair $i, j \in [k]$, $i \neq j$, alignment $\mathcal{A}$ induces a pairwise alignment $\mathcal{A}_{ij}$ between strings S_i and S_j, obtained by taking S'_i and S'_j and striking out any column containing two spaces. Assume that we have chosen some scoring mechanism for pairwise alignments that depends on a parameter vector λ. Then, the *sum-of-pairs* (*SP*) score of $\mathcal{A}$ is given by

$$score(\mathcal{A}, \lambda) = \sum_{i<j} score(\mathcal{A}_{ij}, \lambda). \tag{4.14}$$

We now examine in some detail the case where global alphabet-independent scoring is used to score pairwise alignments; we then summarize the results for other pairwise scoring schemes.

Let w, x, y, z be, respectively, the total number of matches, mismatches, indels, and gaps in all pairwise alignments induced by $\mathcal{A}$; for example, $w = \sum_{i<j} w_{ij}$, where w_{ij} is the number of matches in the induced pairwise alignment $\mathcal{A}_{ij}$. The score can be again expressed as in Equation (4.4), and the feature vector is (w, x, y, z). Assuming for simplicity that all strings are of the same length n, and using the fact that the features of the pairwise alignments obey Equation (4.9), each aggregate feature is within a range of size $O(k^2 n)$. Furthermore, we can eliminate w as we did for pairwise alignment, leaving us with two parameters for the case where gaps have zero weight and three for the case where the weight is positive [19]. Therefore, Theorem 4.2 gives an upper bound of $O(n^{2/3} k^{4/3})$ for the first case, and a bound of $O(n^{3/2} k^3)$ for the second.

For local alignment under alphabet-independent scoring, we obtain, combining the arguments above with those of the previous subsection, bounds of $O(n^{3/2} k^3)$ and $O(n^{12/5} k^{24/5})$ regions with zero and variable gap penalties, respectively. Under alphabet-dependent scoring, for either the global or local alignment, we obtain bounds of $(nk)^{O(|\Sigma|^2)}$.

4.4 Parametric Search

The term *parametric search* refers to any question that involves finding a parameter vector λ^* that satisfies some specified property. Problems 1 and 5 of 4.2.4 fall in this category. In both cases, λ^* is a vertex of the maximization diagram of Z: For ray shooting, after re-parametrization, the λ^* we seek is a breakpoint of Z. For inverse optimum alignment, the point λ^* that minimizes the difference between $Z(\lambda)$ and $score(\mathcal{A}^{(0)}, \lambda)$ can always be chosen to be a vertex of the maximization diagram of Z.

We review four methods of parametric search: bisection search, Newton's method, gradient descent, and Megiddo's method. All of these operate by generating a set of candidate values for λ^*, using them to narrow down the search by invoking either an evaluator (Definition 4.4) or an oracle.

DEFINITION 4.7   An *oracle* for a one-dimensional parametric search problem is a procedure that, given a parameter value $\hat{\lambda}$ determines whether or not $\hat{\lambda}$ is less than or equal to the parameter value being sought.

Oracles and evaluators are often related. To illustrate this, consider the following ray-shooting problem that is used as a sample application of three of the methods presented here.

> **Indel penalty sensitivity analysis in global alignment.** Let $A^{(0)}$ be an optimum global alignment for some given indel penalty $\gamma^{(0)}$. Assuming that the gap penalty is zero and that the reward for matches and the mismatch penalty are each one, find the largest indel penalty $\gamma^* \geq \gamma^{(0)}$ such that $A^{(0)}$ is optimal for every $\gamma \in [\gamma^{(0)}, \gamma^*]$.

Note that, by Remark 4.1, no generality is lost by the above choices for the weights of matches and mismatches. As seen in 4.2.2, an evaluator for this problem is the standard dynamic programming algorithm for optimum global alignment. An oracle for the problem must determine whether a given $\hat{\gamma} \geq \gamma^{(0)}$ is less than or equal to γ^*. To test this, first use the evaluator to find an optimum alignment $\hat{A}$ when the indel penalty equals $\hat{\gamma}$. Next, compare $score(A^{(0)}, \hat{\gamma})$ and $score(\hat{A}, \hat{\gamma})$. If they are equal, then $\hat{\gamma} \leq \gamma^*$. Otherwise (since $\hat{A}$ is optimum at $\hat{\gamma}$), the only possibility is that $score(A^{(0)}, \hat{\gamma}) < score(\hat{A}, \hat{\gamma})$, and thus $\hat{\gamma} > \gamma^*$.

The rest of this section is organized as follows. In 4.4.1–4.4.4, we give an overview of bisection search, Megiddo's method, Newton's method, and gradient descent. This is followed by applications of parametric search to inverse alignment (4.4.5) and sensitivity analysis (4.4.6).

4.4.1   Bisection Search

Bisection search for one-parameter problems is easy to describe. Suppose λ^* (which, by assumption, is a breakpoint) is known to lie in some interval $\mathcal{I}$ on the real line. Repeatedly halve $\mathcal{I}$, taking the left or right half depending on the outcome of an oracle call at the midpoint. The search stops when $\mathcal{I}$ is too small to contain more than one breakpoint; the sole breakpoint that remains in $\mathcal{I}$ must be λ^*. We use the indel penalty sensitivity analysis problem to illustrate this technique. We show that the properties of the scoring function imply a logarithmic bound on the number of halving steps (and, therefore, the number of oracle calls) required. Similar ideas can be used to prove the efficiency of the bisection search in other applications.

The goal is to locate the first breakpoint γ^* of Z that follows $\gamma^{(0)}$. For this, we

(i) choose a sufficiently large search interval $\mathcal{I}$,
(ii) repeatedly bisect $\mathcal{I}$ until it has at most one breakpoint of Z, but still contains γ^*, and
(iii) locate γ^* within $\mathcal{I}$.

The oracle for the problem has already been described. Its running time is $O(nm)$ (the work to compute an optimum alignment using dynamic programming). It remains to explain the implementation of steps (i)–(iii). As usual, let n and m be the lengths of the sequences, $n \leq m$.

Consider step (i). Let $A^{(1)}, A^{(2)}, \ldots$ denote the series of optimal alignments along the interval $\mathcal{I}$. Let w_i, x_i, y_i denote the number of matches, mismatches, and indels in $A^{(i)}$. Let $\Delta w_i = w_{i+1} - w_i$, $\Delta x_i = x_{i+1} - x_i$, $\Delta y_i = y_{i+1} - y_i$ and let $\gamma^{(i)}$ be the breakpoint where

$\mathcal{A}^{(i)}$ and $\mathcal{A}^{(i+1)}$ are co-optimal. Then,

$$\gamma^{(i)} = \frac{\Delta w_i - \Delta x_i}{\Delta y_i}. \tag{4.15}$$

By integrality, $\Delta y_i \geq 1$ and, by inequalities (4.9), $w_i \leq n$. Hence, $\gamma^{(i)} \leq n$ for all i. Thus, our search can be restricted to the interval $\mathcal{I} = (\gamma^{(0)}, n]$.

Consider step (ii). For any two successive breakpoints $\gamma^{(i)}$, $\gamma^{(i+1)}$ of Z,

$$\gamma^{(i+1)} - \gamma^{(i)} = \frac{(\Delta w_{i+1} - \Delta x_{i+1})\Delta y_i - (\Delta w_i - \Delta x_i)\Delta y_{i+1}}{\Delta y_{i+1}\Delta y_i}. \tag{4.16}$$

By (4.9), $\Delta y_i \leq 2n$. Since the left-hand side of Equation (4.16) must be positive and the various Δ terms are integers, the numerator must be at least 1. Thus, $\gamma^{(i+1)} - \gamma^{(i)} \geq 1/(4n^2)$. Therefore, in step (ii) we stop as soon as the length of the search interval drops below $1/(4n^2)$.

After step (ii) is complete, we know that γ^* must lie in the interval $\mathcal{I} = (\gamma^{(0)}, \gamma^{(1)}]$, within which Z has at most one breakpoint. To locate γ^* within $\mathcal{I}$ (step (iii)), do as follows. First, compute the optimal alignment $\mathcal{A}^{(1)}$ at $\gamma^{(1)}$. There are two cases:

Case 1: There is no breakpoint inside $\mathcal{I}$, and therefore there are no breakpoints beyond $\gamma^{(0)}$. This is true if either $score(\mathcal{A}^{(0)}, \gamma^{(0)}) = score(\mathcal{A}^{(1)}, \gamma^{(0)})$ or $score(\mathcal{A}^{(0)}, \gamma^{(1)}) = score(\mathcal{A}^{(1)}, \gamma^{(1)})$. In this case, return $\gamma^* = +\infty$.

Case 2: There is exactly one breakpoint inside $\mathcal{I}$, which must be the value λ^* being sought. In this case, return the value γ^* such that $score(\mathcal{A}^{(0)}, \gamma^*) = score(\mathcal{A}^{(1)}, \gamma^*)$.

The number of bisection steps is $O(\log n)$, each requiring $O(nm)$ time. The final step requires computing one optimum alignment plus $O(1)$ additional work. The total time is therefore $O(nm \log n)$.

While the details above are specific to sensitivity analysis, similar ideas can be used for other search problems, such as inverse optimal alignment (see 4.4.5 and [50]). Extensions to two-parameter problems are possible. In this case, instead of maintaining an interval, we maintain a polygonal region and, instead of splitting an interval through the middle, we split the current polygonal region by a line through its centroid (see [50]).

4.4.2   Megiddo's Method

Megiddo's method [36, 37] provides a precise relationship between the complexity of solving a parametric problem and the complexity of the problem's fixed-parameter version. Here we discuss the one-parameter version of Megiddo's method; generalizations to any fixed number of parameters are explained elsewhere [9, 3].

In what follows λ denotes a scalar parameter. Let the value being sought be denoted by λ^*, which is known to to be greater than or equal to some value $\lambda^{(0)}$. Like bisection search, Megiddo's method generates a sequence of test values that are used to reduce the search interval with the aid of the oracle. The key difference is that the test values are generated by simulating the execution of an algorithm for the underlying fixed-parameter problem. This algorithm must be of a certain kind.

DEFINITION 4.8   An algorithm is *piecewise linear* if each value it computes is a linear combination of the input parameters.

Any reasonable dynamic programming algorithm is piecewise linear. For example, consider the standard (table-based) dynamic programming algorithm for global alignment with zero gap penalty. We argue that each entry of the dynamic programming table is a linear combination of α, β and γ. This claim is trivially true for the first row and column of the table. Now assume the claim is true for every entry (i', j') such that (i', j') is lexicographically smaller than (i, j). Entry (i, j) is the maximum of three entries of the table, each with index (i', j') lexicographically smaller than (i, j), plus α, minus β, or minus γ. Therefore, entry (i, j) is itself a linear combination of α, β and γ.

Megiddo's method simulates the execution of a piecewise linear algorithm $\mathcal{B}$ for the underlying fixed-parameter problem in order to find $\mathcal{B}$'s execution path at λ^*. Instead of manipulating numbers, the simulation manipulates linear functions of λ. This is possible because every value v manipulated by $\mathcal{B}$ can be represented symbolically as $v(\lambda) = p_v + q_v \lambda$. Megiddo's method maintains an interval $\mathcal{I} = [\lambda^{(0)}, \lambda^{(1)})$ that is updated so that the following invariant holds after i steps of $\mathcal{B}$ have been simulated:

$$\lambda^* \in \mathcal{I} \text{ and the first } i \text{ steps of } \mathcal{B}\text{'s execution path are the same for every } \lambda \in \mathcal{I}. \quad (4.17)$$

Initially, $\lambda^{(1)} = +\infty$. Suppose a certain number of $\mathcal{B}$'s steps have been simulated. To simulate the next step, proceed as follows.

- If the step is an arithmetic operation, execute it symbolically to obtain a new linear function of λ. We make the mild assumption that symbolic execution of an operation only increases its running time by a constant factor.

- If the step is a comparison between two numbers $u(\lambda) = p_u + q_u \lambda$ and $v(\lambda) = p_v + q_v \lambda$, compute $\hat{\lambda}$ such that $u(\hat{\lambda}) = v(\hat{\lambda})$. If no such $\hat{\lambda}$ exists, u and v are either identical or one is larger than the other for all λ. In either case, the outcome of the comparison can easily be determined and the step can be executed. If $\hat{\lambda}$ exists, invoke the oracle to determine the position of $\hat{\lambda}$ relative to λ^*. The outcome of the call determines the outcome of the comparison between u and v at λ^*. If $\hat{\lambda} \le \lambda^*$, set $\lambda^{(0)} = \max(\lambda^{(0)}, \hat{\lambda})$. Otherwise, set $\lambda^{(1)} = \min(\lambda^{(1)}, \hat{\lambda})$

At the end of the simulation, we have an interval $\mathcal{I}$ such that for any $\lambda \in \mathcal{I}$, algorithm $\mathcal{B}$ always executes the same way. Therefore Z has no breakpoints in $\mathcal{I}$ and, hence, $\lambda^* = \lambda^{(0)}$.

THEOREM 4.3 *Let $\mathcal{P}$ be a parametric search problem that has an oracle that runs in worst-case time b. Suppose that there exists a piecewise linear algorithm to evaluate $Z(\lambda)$ that executes t steps in the worst case. Then, $\mathcal{P}$ can be solved in time $O(t \cdot b)$.*

For example, in the indel penalty sensitivity problem, t and b are both $O(nm)$. Thus, by Theorem 4.3, Megiddo's method yields a $O(n^2 m^2)$ algorithm for the problem, which is considerably slower than bisection search. This can be improved to $O(nm \operatorname{polylog} n)$ by simulating a *parallel* alignment algorithm instead of a sequential one (see [37, 29] for details), but this is at the expense of a considerably more involved procedure.

4.4.3 Newton's Method

Newton's classic zero-finding method can be adapted for ray shooting (Problem 1). Recall that the question is as follows: Given a parameter vector $\lambda^{(0)} \in \mathbb{R}^d$, an optimum alignment $\mathcal{A}^{(0)}$ at $\lambda^{(0)}$, and a ray ρ originating at $\lambda^{(0)}$, find the last point λ^* on the ray such that $\mathcal{A}^{(0)}$ is optimal at λ^*. Without loss of generality, assume that the problem has been reparameterized so that λ is a scalar. Furthermore, we restrict the search for λ^* to a finite interval $\mathcal{I} = (\lambda^{(0)}, \lambda^{(1)}]$. This is not a limitation in practice, since $\lambda^{(1)}$ can always be chosen to be large enough (an example of this is, in fact, given in 4.4.1).

The key observation is that if $\lambda^* < \lambda^{(1)}$, then $\mathcal{A}^{(0)}$ is co-optimal with some other alignment at λ^*. This leads to the following version of Newton's method, adapted for piecewise linear functions.

Algorithm NEWTON

Input: An interval $\mathcal{I} = (\lambda^{(0)}, \lambda^{(1)}]$, an optimum alignment $\mathcal{A}^{(0)}$ at $\lambda^{(0)}$, and an evaluator for Z.

Output: The largest value $\lambda^* \in \mathcal{I}$ such that $\mathcal{A}^{(0)}$ is optimal at λ^*.

1. Compute an optimal alignment $\mathcal{A}^{(1)}$ at $\lambda^{(1)}$.
2. Set $i = 1$.
3. While $score(\mathcal{A}^{(0)}, \lambda^{(i)}) < score(\mathcal{A}^{(i)}, \lambda^{(i)})$, do the following steps:
    (a) Let $\lambda^{(i+1)}$ be λ-value such that $score(\mathcal{A}^{(0)}, \lambda) = score(\mathcal{A}^{(i)}, \lambda)$.
    (b) Set $i = i + 1$.
4. Return $\lambda^* = \lambda^{(i)}$.

The execution of NEWTON is illustrated in Figure 4.5. The convexity and piecewise linearity of Z imply that the $\lambda^{(i)}$ values form a decreasing sequence and that at all times $\lambda^{(i)} \geq \lambda^*$. At termination, we must have $score(\mathcal{A}^{(0)}, \lambda^{(i)}) = score(\mathcal{A}^{(i)}, \lambda^{(i)})$, which implies that $\lambda^{(i)} = \lambda^*$. Also, the successive alignments computed by the algorithm must have distinct score functions. This leads to the following result.

THEOREM 4.4 *Algorithm* NEWTON *correctly solves the ray shooting problem. The number of evaluations it requires is at most equal to the number of optimality regions of the maximization diagram.*

For feature-based scoring schemes we can invoke Theorem 4.2: If each feature is in the same integer range of size N, then the number of evaluations of Z required by the algorithm is $O(N^{d(d-1)/(d+1)})$, where d is the number of features. For example, in the indel penalty sensitivity analysis problem, the $O(n^{2/3})$ bound on the number of regions (see 4.3.1) implies that only that many evaluations are needed in the worst case, resulting in a $O(n^{5/3}m)$ bound on the search time.

4.4.4 Gradient Descent

Gradient descent (also called *steepest descent*) is a numerical method to obtain the minimum of a function within a given interval [40, 44]. The method is iterative, generating a sequence of points that converges to a minimum. If the current point is not minimum, the algorithm chooses the next point by moving some distance in the direction opposite to the direction of the gradient. The intuition is that advancing in that direction should reduce the value of the function. More formally, assume that the function $F : \mathbb{R}^d \to \mathbb{R}$ to be minimized is continuously differentiable and that $\lambda^{(t)}$ is the current (non-optimum) point. The next point in the sequence is given by

$$\lambda^{(t+1)} = \lambda^{(t)} - \theta \nabla F(\lambda^{(t)}),$$

where θ is a scalar, which denotes the step distance, and $\nabla F(\lambda^{(t)})$, the *gradient* of F at $\lambda^{(t)}$, is the vector whose elements are partial derivatives with respect to the d dimensions.

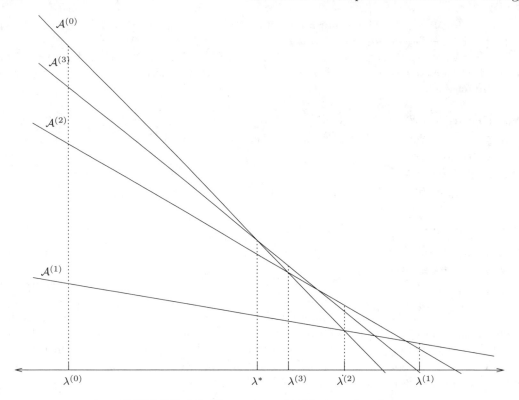

FIGURE 4.5: Newton's method for ray shooting.

That is,

$$\nabla F = \left(\frac{\partial F}{\partial \lambda_1}, \ldots, \frac{\partial F}{\partial \lambda_d} \right).$$

To apply this approach to piecewise linear functions, which are not everywhere differentiable, we need a new concept.

DEFINITION 4.9   Let F be a function $F : \mathbb{R}^d \to \mathbb{R}$. A vector $s \in \mathbb{R}^d$ is a *subgradient of F at $\lambda^{(0)} \in \mathbb{R}^d$* if for all $\lambda \in \mathbb{R}^d$

$$F(\lambda) \geq F(\lambda^{(0)}) + s \cdot (\lambda - \lambda^{(0)}).$$

The collection of subgradients at $\lambda^{(0)}$ is called the *sub-differential at $\lambda^{(0)}$*, and is denoted by $\partial F(\lambda^{(0)})$.

It can be shown that $\partial F(\lambda^{(0)}) \neq \emptyset$ at all points $\lambda^{(0)}$ [40]. Subgradients play the role of gradients in searching for the minimizer of functions that are not everywhere differentiable. In particular, it can be shown that $\lambda^{(0)}$ is optimal if and only if $0 \in \partial F(\lambda^{(0)})$ [40].

The subgradient algorithm is as follows:

Algorithm SUBGRADIENT
Input: A point $\lambda^{(0)} \in \mathbb{R}^d$, a sequence $\theta^{(0)}, \theta^{(1)}, \ldots$ of real numbers, and a procedure for computing a sub-gradient of function F at any point.
Output: A value λ^* at which $F(\lambda)$ is minimum.

1. Compute a subgradient $s^{(0)} \in \partial F(\lambda^{(0)})$.

2. Set $t = 0$.

3. While $s^{(t)} \neq 0$, do the following steps:

   (a) Let $\lambda^{(t+1)} = \lambda^{(t)} - \theta^{(t)} s^{(t)}$

   (b) Choose a subgradient $s^{(t+1)} \in \partial F(\lambda^{(t+1)})$.

   (c) Set $t = t + 1$

4. Return $\lambda^* = \lambda^{(t)}$.

The procedure to compute a subgradient depends on the given problem; we explain how to find a subgradient for inverse alignment in the next subsection. In practice it may be difficult to determine if $0 \in \partial F(\lambda^{(t)})$, since only one subgradient is computed at any point. One way to handle this is by terminating the algorithm if the function has not decreased by a certain amount after some number of iterations. We note that the convergence and running time of algorithm SUBGRADIENT depend on the choice of the $\theta^{(i)}$ sequence [40]. Although the algorithm is fast in practice, it is not in general possible to establish combinatorial bounds on its running time.

4.4.5 Parametric Search and Inverse Sequence Alignment

As defined in 4.2.4, the inverse optimal alignment (Problem 5) is a parametric search problem whose goal is to find a parameter vector λ^* that minimizes the function $F(\lambda)$ defined as

$$F(\lambda) = Z(\lambda) - score(\mathcal{A}^{(0)}, \lambda),$$

where $\mathcal{A}^{(0)}$ is a reference alignment and, as usual, $Z(\lambda)$ is the optimal score function. Since Z is piecewise linear and convex and $score(\mathcal{A}^{(0)}, \lambda)$ is a linear function, F is also piecewise linear and convex. In fact, the decomposition of the parameter space induced by F is identical to that induced by Z. By convexity, the solution λ^* to the problem can always be chosen to be a vertex of the maximization diagram of Z. We discuss how to solve the inverse optimal alignment problem through bisection search, Megiddo's method and gradient descent. Newton's method can also be adapted to solve this problem [46].

For bisection search and Megiddo's method, the key is to implement the oracle. In the one-parameter case, we can determine if a given $\hat{\lambda}$ is greater than λ^* by computing the optimum alignment $\hat{\mathcal{A}}$ at $\hat{\lambda}$. Then $\hat{\lambda} > \lambda^*$ if and only if $score(\hat{\mathcal{A}}, \lambda) - score(\mathcal{A}^{(0)}, \lambda)$ has a positive slope. Generalizations to more parameters are discussed in [3, 50].

To apply gradient descent, we need a means to compute a sub-gradient in $\partial F(\lambda^{(t)})$. This can be done as follows. Let $\mathcal{A}^{(t)}$ be an optimum alignment at $\lambda^{(t)}$. The function $score(\mathcal{A}^{(t)}, \lambda) - score(\mathcal{A}^{(0)}, \lambda)$ has the form $a_0 + \sum_{i=1}^{d} a_i \lambda_i$. Then, the vector $(a_1, \ldots, a_d)$ is a sub-gradient at $\lambda^{(t)}$. Algorithm SUBGRADIENT of the previous section can now be used to obtain the inverse optimal value.

4.4.6 Ray Shooting and Sensitivity Analysis

The sensitivity analysis problem (4.2.4, Problem 2) can be solved by repeated ray shooting. Using the notation of 4.2.4, let $\lambda^{(0)}$ be a given point in the parameter space and let $\mathcal{A}^{(0)}$ be an optimal alignment at that point. The problem is to find the maximal region around $\lambda^{(0)}$ where $\mathcal{A}^{(0)}$ is optimal. In the one-parameter case, this translates into finding an interval around $\lambda^{(0)}$, which can be done by shooting two rays from $\lambda^{(0)}$. The first, in the negative direction, yields a point $\lambda^{(1)}$; the other, in the positive direction, yields a point $\lambda^{(2)}$. The

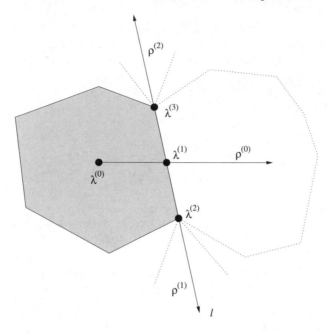

FIGURE 4.6: Ray shooting to determine an edge of the region of optimality of alignment $\mathcal{A}^{(0)}$.

interval $[\lambda^{(1)}, \lambda^{(2)}]$ is the maximal region of $\mathbb{R}^1$ within which alignment $\mathcal{A}^{(0)}$ is optimal. We describe how to extend the idea to two-parameter problems, where the optimality regions are polygons. Extensions to higher dimensions are possible (see the notes in 4.7).

Let F be the region of optimality of $\mathcal{A}^{(0)}$. The first step is to choose an arbitrary ray $\rho^{(0)}$ emanating from $\lambda^{(0)}$ and shoot a ray to find the point $\lambda^{(1)}$ along $\rho^{(0)}$ that intersects the boundary of F. Ray shooting is assumed to be adapted to yield an alignment $\mathcal{A}^{(1)}$ that is co-optimal at $\lambda^{(1)}$. Let l be the line defined by the intersection of $score(\mathcal{A}^{(0)}, \lambda)$ and $score(\mathcal{A}^{(1)}, \lambda)$. Then l contributes a segment e to the boundary of F. From $\lambda^{(1)}$ shoot two rays $\rho^{(1)}$ and $\rho^{(2)}$ in opposite directions along l, to find the end points of edge e. The process is illustrated in Figure 4.6.

To find the remaining edges, repeat the above steps with other rays emanating from $\lambda^{(0)}$. Each new ray must be in a direction away from any previously discovered edges of the boundary of F. To ensure of this, one can use the data structure depicted in Figure 4.7. The solid lines there represent the edges already discovered. Edges with common endpoints are joined as these endpoints are identified. Chains of known edges are linked to each other by dashed lines, indicating unknown regions of the boundary. Each successive ray generated by the algorithm goes between the endpoints of such a region. The discovery of a new edge fills in part of the missing information for that portion of the boundary.

There are two special cases. One occurs when $\mathcal{A}^{(0)}$ remains optimal along the entire length of the current ray. To handle this, it is convenient to assume that the parameter space is enclosed within a large rectangular interval. We can then use one of the boundary edges of the interval as a boundary of the region. The other case is when the ray $\rho^{(t)}$ goes through a vertex of the optimality region. Then $\mathcal{A}$ remains optimal only along at most one the two rays along the boundary line, allowing us to recognize the vertex.

We need three ray searches to find each edge of F. This gives us a bound of $O(e)$ ray searches to determine a polygon of e edges.

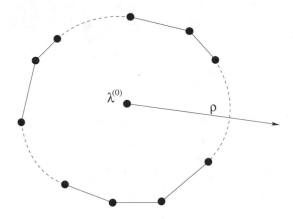

FIGURE 4.7: Ray shooting data structure.

4.5 Constructing the Parameter Space Decomposition

Constructing the maximization diagram of the optimum score function requires producing all faces of dimensions zero through d of the maximization diagram, along with the incidence relationships between faces, and the optimum alignments associated with all faces of dimension d. The construction problem can be solved repeatedly applying the ray-shooting idea used for sensitivity analysis (4.4.6) to generate all regions in some order [31]. For two-parameter problems this method runs in $O(t \cdot f + f^2)$ time, where t and f are, respectively, the time to do a single evaluation of Z and the number of faces in the maximization diagram [31]. Another approach is to *lift* the execution of a dynamic programming optimum alignment algorithm so that instead of computing an alignment for a single parameter value, it computes alignments for *all* values simultaneously [41]. To do so, the lifted algorithm takes advantage of the piecewise linearity of dynamic programming algorithms, operating on linear functions instead of real numbers. In particular, sums of real values become sums of piecewise linear functions and the max operator on numbers is replaced by the upper envelope operator on piecewise linear functions [41].

In the remainder of this section, we present a simple and intuitive algorithm that constructs Z through a series of evaluations of the optimum solution at various points in the parameter space [16, 21, 20]. In effect, the algorithm generates a sequence of score functions whose upper envelope converges to Z. For two parameters, the method has the same running time as above-mentioned procedures, and naturally generalizes to any number of parameters.

4.5.1 The One-Parameter Case

The one-parameter case gives much of the intuition behind the higher-dimensional generalization. The method presented here constructs the maximization diagram of a one-parameter problem within a given interval $\mathcal{I} = [\lambda^{(r)}, \lambda^{(l)}]$. Throughout its execution, the algorithm updates a function W, which is an increasingly closer approximation to Z, while maintaining the following invariant:

W is the upper envelope of a subset of $\{\mathit{score}(\mathcal{A}, \lambda) : \mathcal{A} \text{ is optimal at some } \lambda \in \mathcal{I}\}$ (4.18)

Thus, W is a piecewise linear convex function such that $W(\lambda) \le Z(\lambda)$ for all $\lambda \in \mathcal{I}$; W is represented by its maximization diagram. A breakpoint $\lambda^{(0)}$ of W is said to be *verified* if it is known to also be a breakpoint of Z and $W(\lambda^{(0)}) = Z(\lambda^{(0)})$; otherwise, the breakpoint is *unverified*. At termination, all breakpoints are verified and $W = Z$.

Algorithm 1PARAMCONSTRUCT

Input: An interval $\mathcal{I} = [\lambda^{(l)}, \lambda^{(r)}]$ and an evaluator for Z.

Output: The maximization diagram of Z.

1. Compute optimal alignments $\mathcal{A}^{(l)}$ and $\mathcal{A}^{(r)}$ at $\lambda^{(l)}$ and $\lambda^{(r)}$.
2. Let W be the upper envelope of $score(\mathcal{A}^{(l)}, \lambda)$ and $score(\mathcal{A}^{(r)}, \lambda)$.
3. If W has no breakpoints inside $\mathcal{I}$, then return W and stop.
4. Otherwise, declare the intersection point $\lambda^{(m)}$ of $score(\mathcal{A}^{(l)}, \lambda)$ and $score(\mathcal{A}^{(r)}, \lambda)$ to be unverified.
5. Iterate the following steps until W has no unverified breakpoints.
   (a) Choose any unverified breakpoint $\lambda^{(0)}$ of W.
   (b) Evaluate Z at $\lambda^{(0)}$; let $\mathcal{A}^{(0)}$ be the alignment returned.
   (c) If $Z(\lambda^{(0)}) = W(\lambda^{(0)})$, declare $\lambda^{(0)}$ to be verified.
   (d) Otherwise, the new W is the upper envelope of current W and $score(\mathcal{A}^{(0)}, \lambda)$. Any new breakpoints created in the process are declared to be unverified.
6. Return W

The execution of this algorithm is illustrated in Figure 4.8. At termination, all breakpoints of W are also breakpoints of Z; thus, since invariant (4.18) holds throughout the execution, we must have $W(\lambda) = Z(\lambda)$ for all $\lambda \in \mathcal{I}$. To count the number of evaluations, observe that there can be at most one evaluation at any λ that is not a breakpoint of Z and at most two evaluations at any λ that is a breakpoint of Z. Each update to W in step (5d) takes $O(1)$ time. We summarize the analysis as follows [16, 26].

THEOREM 4.5   *Algorithm* 1PARAMCONSTRUCT *correctly computes the maximization diagram of Z within interval $\mathcal{I}$ in time $O(t(b+1))$, where b is the number of breakpoints of Z that lie within $\mathcal{I}$ and t is the time for an evaluation of Z.*

For example, consider the global alignment problem with zero gap penalty. A single evaluation in this case takes time $t = O(nm)$. By Remark 4.1, 4.3, while there are formally three parameters, α, β, and γ, there is effectively only one, γ. Thus, we can use 1PARAMCONSTRUCT to build the maximization diagram of Z, which in this case has $O(n^{2/3})$ regions (see 4.3.1), for a total bound of $O(n^{5/3}m)$.

4.5.2   Generalization to Higher Dimensions

Algorithm 1PARAMCONSTRUCT can be extended to two parameters in a natural way [20]. The two-parameter algorithm constructs the maximization diagram within a rectangular interval $\mathcal{I} = [\lambda_1^{(0)}, \lambda_1^{(1)}] \times [\lambda_2^{(0)}, \lambda_2^{(1)}]$ by repeatedly invoking an evaluator to obtain optimal alignments at various points within $\mathcal{I}$. Throughout its execution, the algorithm maintains a function W satisfying invariant (4.18). W is represented by its maximization diagram — in this case a subdivision of the plane into convex polyhedral regions — whose vertices are classified as either *verified* or *unverified*, depending on whether or not $W(\lambda) = Z(\lambda)$

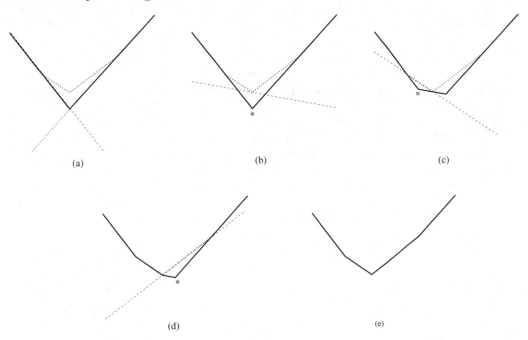

FIGURE 4.8: Executing 1PARAMCONSTRUCT: (a) Initial configuration. The solid curve is the upper envelope of $score(\mathcal{A}^{(l)}, \lambda)$ and $score(\mathcal{A}^{(r)}, \lambda)$; the dotted curve is $Z(\lambda)$. (b–d) A series of evaluations and updates. An asterisk is shown next to the breakpoint being verified; the corresponding score function is shown as a dashed line. (e) The final result. Not shown is the verification of the three intermediate breakpoints of W.

at those points. The two-parameter algorithm is similar to 1PARAMCONSTRUCT, the main difference being that wherever that procedure refers to a breakpoint, the new procedure deals with a vertex. As in 1PARAMCONSTRUCT, any new vertices created during an update of W are initialized as unverified. To initialize the procedure, we evaluate Z at the four corners of $\mathcal{I}$ and take the upper envelope of the resulting functions as the original value of W.

Essentially the same arguments used for 1PARAMCONSTRUCT can be used to justify the correctness of the two-parameter algorithm. There are at most two evaluations of Z for each vertex of the maximization diagram and at most one evaluation for each edge and face. Since the total number of vertices and edges is linearly related to the number f of faces of the maximization diagram, the number of evaluations required is $O(f)$. The only significant complication in handling two parameters instead of one arises when updating W in the analogue of step (5d): In the two-parameter case, replacing W by the upper envelope of W and $score(\mathcal{A}^{(0)}, \lambda_1, \lambda_2)$ requires time proportional to the number of faces of the maximization diagram of W, not constant time (see [20]). Indeed, by geometric duality [34], the question is equivalent to updating the lower convex hull of a set of points after adding a new point [11]. We summarize the analysis below.

THEOREM 4.6 *For two-parameter problems, the maximization diagram of Z restricted to a rectangular interval $\mathcal{I}$ can be computed in time $O(t \cdot f + f^2)$, where f is the number of* *faces of Z that lie within $\mathcal{I}$ and t is the time for an evaluation of Z.*

For example, consider the global alignment problem with non-zero gap penalties. By Remark 4.1, 4.3, this is effectively a two-parameter problem in γ, the indel penalty, and δ, the gap extension penalty and the number of optimality regions is $f = O(n^{3/2})$ (see 4.3.1). Using the standard $O(nm)$ dynamic programming algorithm for evaluating Z, Theorem 4.6 implies a construction time of $O(n^{5/2}m)$.

Going beyond two parameters requires essentially no new concepts. The maximization diagram of Z can again be built by successive evaluations at the vertices of the maximization diagram of a function W that represents the approximation to Z built so far. As a result, the algorithm produces a sequence of hyperplanes, each of which is the score function of an alignment that is optimum at some point in the region of interest. Each new hyperplane is used to incrementally update the current estimate of Z. The technique is reminiscent of the methods used to determine the shape of a convex polyhedron through a series of hyperplane probes [13, 14]. Indeed, these results can be used to prove the following.

THEOREM 4.7 *Let d denote the number of parameters, let f and v be, respectively, the number of facets (d-dimensional faces) and vertices (0-dimensional faces) of the maximization diagram of Z within a hyperrectangle $\mathcal{I}$ of interest, and let t be the time needed for a single evaluation. Then, the maximization diagram of Z within $\mathcal{I}$ can be computed in $O(t \cdot (f + dv))$ time, plus the time needed to construct the upper envelope of all the score functions generated during the computation.*

The last part of the statement of the above theorem requires elaboration. As for one and two parameters, the problem of incrementally building the upper envelope of a set of d-variate linear functions is dual to constructing the lower convex hull of a set of points in $\mathbb{R}^{d+1}$, a problem for which an extensive literature exists (see [49]). In particular, this body of research addresses many of the subtle implementation issues encountered in such computations, such as representation and numerical stability [5, 25].

The time needed to construct an upper envelope depends heavily on the complexity of the output produced, as measured by its total number of faces of dimensions 0 through d. This number increases exponentially with the dimension. The *Upper Bound Theorem* [35] states that the total complexity is $\Theta(N^{\lfloor(d+1)/2\rfloor})$, where N is the number of actual hull vertices. Duality therefore implies an upper bound of $\Theta(f^{\lfloor(d+1)/2\rfloor})$ for the complexity of the maximization diagram of Z, where f is the number of regions. This limits the number of parameters that can be handled in practice.

4.6   Parametric Problems in Hidden Markov Models of Sequence Alignment

Hidden Markov models (HMMs) provide a rigorous mathematical framework for the scoring schemes used in sequence alignment. An HMM represents a system that undergoes a series of unobservable state transitions, during which a sequence of observable values is generated. One of the main questions in HMMs is to infer the state sequence, based on the observations. Given to us are the allowed state transitions, represented by a directed graph, together with the respective transition probabilities, which are the parameters of the model. The values of these probabilities have a profound effect on the model's accuracy.

Our goal is to show the close connection between parametric sequence analysis and parameter choice in HMMs of sequence evolution. To this end, the rest of this section is organized as follows. We first provide an overview of Markov and hidden Markov models. Next, we introduce the pair HMM, which models the evolutionary relationship between two sequences. We then present the basic parametric problems in HMMs and show that, through the log transform, many of these reduce to the linear problems of 4.2.4. For concreteness, we focus on the pair HMM for *global* alignment. HMMs for local alignment, multiple alignment, and other sequence analysis problems also exist [15]; analogous parametric issues arise for these models [41].

4.6.1 Hidden Markov Models

A *Markov model* is a system that can be in one of a finite number of distinct *states*, numbered 1 through N. The relationship between states is represented by a directed graph whose nodes are the states and where there is an edge from state i to state j if and only if a transition from i to j is allowed. The state of the system changes at discrete instants of time $t = 1, 2, \ldots$; the state of the system at time i is denoted by x_i. A key property of Markov models is that x_i depends only on x_{i-1}. That is,

$$\Pr(x_i | x_{i-1}, \ldots, x_1) = \Pr(x_i | x_{i-1}).$$

The value of $\Pr(x_i = l | x_{i-1} = k)$ is called the *transition probability* from state k to state l and is denoted s_{lk}. Note that the transition probabilities must satisfy

$$s_{ij} \geq 0 \qquad \text{and} \qquad \sum_{j=1}^{N} s_{ij} = 1.$$

A two-state Markov model, along with its transition probabilities, is shown in Figure 4.9.

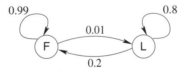

FIGURE 4.9: State transition diagram for a simple Markov model.

In a *hidden Markov model*, each state of the system emits a symbol, chosen from some finite alphabet Σ. The emission probability of symbol b at state i is denoted by $e_i(b)$. The sequence of symbols emitted by the system is called the *observed sequence*; the corresponding sequence of states is called the *emission path* or *path* for short. Such a model is called "hidden," because an observer can see the emitted sequence, but not the emission path. The joint probability of state sequence $x = x_1 \ldots x_n$ and the emitted sequence $b = b_1 \ldots b_n$ is given by

$$\Pr(b, x) = \prod_{i=1}^{n} s_{x_{i-1} x_i} e_{x_i}(b_i), \tag{4.19}$$

where $s_{x_0 x_1}$ gives the probability distribution of the initial state. A given observed sequence may have been generated by different paths, each of which has a probability given by Equation (4.19). One of the fundamental problems in HMMs is the *inference problem*:

given an observed sequence, find the path that is most likely to have generated it. This path depends on the values of the transition probabilities, which are the parameters of the system.

As an example, consider again the Markov model depicted in Figure 4.9. The system can be used to model the behavior of an occasionally dishonest casino, where, unbeknownst to the players, a loaded die is occasionally used. There are two hidden states in the system, F and L, indicating whether a fair or a loaded die is being tossed. Each state emits symbols from the same set $\Sigma = \{1, 2, 3, 4, 5, 6\}$ (the result of tossing the die), but the emission probabilities are different: $e_F(i) = 1/6$ for $i \in \Sigma$; while $e_L(i) = 1/10$ for $i \in \Sigma - \{6\}$ and $e_L(6) = 1/2$. While players do not know the state of the system, they can observe the sequence of numbers produced. The inference problem for this model is to determine the sequence of dice (fair or loaded) that were tossed, based on the sequence of numbers produced.

It is often convenient to allow some states of an HMM to be *silent*; that is, to emit no symbol. For instance, in the previous example, it can be useful to define an additional silent "begin" state, from which the system can go to the F or L state.

4.6.2   The Pair HMM

The *pair HMM* [15] is a probabilistic model of the evolution of two sequences from a common ancestor through a series of mutations, insertions and deletions. The transition diagram of the model is illustrated in Figure 4.10. There are five states: B and E are the begin state and end state respectively, M is the match/mismatch state, I and D are the insertion and deletion states, respectively. The first two states are silent. For the remaining states, the single-symbol emission strategy of ordinary HMMs is extended to allow for the generation of *two* sequences S_1 and S_2 instead of one. A D state emits a single symbol to be put in S_1, an I state emits a symbol to be put in S_2, and an M state emits two symbols, one for S_1, the other for S_2.

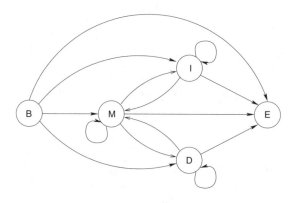

FIGURE 4.10: The pair HMM

There is a one-to-one correspondence between global alignments of two sequences S_1 and S_2 and emission paths that generate the sequences in the pair HMM. Consider any path that emits S_1, S_2. The path begins at state B and ends at state E. In between, successive columns in the alignment correspond to successive nodes on the path: a node is an M if the column aligns two characters (equal or not); it is an I if the first row is a space and the second a character from S_2; it is a D if the first row is a character and the second a space. For example, for sequences $S_1 = $ AGCT and $S_2 = $ ACTC, the path corresponding to alignment

$$
\begin{array}{ccccc}
\text{A} & \text{G} & - & \text{C} & \text{T} \\
\text{A} & \text{C} & \text{T} & \text{C} & -
\end{array}
\qquad (4.20)
$$

is BMMIMDE. Similarly, given any alignment $\mathcal{A}$ between S_1 and S_2, once can construct a path where each successive state between B and E corresponds to one column of $\mathcal{A}$. Because of this bijection between alignments and paths, the two terms are used interchangeably in what follows.

Note that the diagram of Figure 4.10 has no transitions between states I and D. While there is no mathematical reason to disallow these transitions, their absence prevents the generation of alignments where insertions are immediately followed by deletions, which are typically considered undesirable.

To fully specify the pair HMM, in addition to the state transition diagram, one must also specify 13 state transition probabilities (for the edges of Figure 4.10), and $|\Sigma|^2 + 2|\Sigma|$ emission probabilities, where Σ is the emission alphabet; these numeric values are the parameters of the system. The emission probabilities for state D are given by a function $e_D(a)$, $a \in \Sigma$. Similarly, the emission probabilities for I are given by $e_I(a)$, where $a \in \Sigma$. For state M, the probabilities are given by $e_M(a, b)$, $a, b \in \Sigma$. The bound of $|\Sigma|^2 + 2|\Sigma| + 13$ on the total number of parameters is in general an overestimate. For example, it is common to assume that $e_I(a) = e_D(a)$ for all $a \in \Sigma$ or even that $e_I(a) = e_I(b)$ for all a, b.

Given the various probability values, it is straightforward to obtain the probability of an alignment through Equation (4.19). For example, the probability of alignment (4.20) is given by

$$
s_{\text{BM}} \cdot e_M(\text{A}, \text{A}) \cdot s_{\text{MM}} \cdot e_M(\text{G}, \text{C}) \cdot s_{\text{MI}} \cdot e_I(\text{T}) \cdot s_{\text{IM}} \cdot e_M(\text{C}, \text{C}) \cdot s_{\text{MD}} \cdot e_D(\text{T}) \cdot s_{\text{DE}}.
$$

4.6.3 Parametric Problems in HMMs

We now examine some of the parametric problems that arise in the pair HMM and then show their connection to the issues presented in 4.2.4.

Perhaps the most basic question in the pair HMM is the following fixed parameter problem.

PROBLEM 6 *(Maximum likelihood alignment)* Given fixed transition and emission probabilities and sequences S_1, S_2, find the path $\mathcal{A}^*$ in the pair HMM that is most likely to have emitted S_1, S_2. That is, compute

$$
\mathcal{A}^* = \arg\max_{\mathcal{A}} \ \Pr(S_1, S_2, \mathcal{A}), \qquad (4.21)
$$

where $\mathcal{A}$ ranges over all paths that generate S_1, S_2.

The next two questions address the robustness of the model, an issue that arises when there is uncertainty about the correct parameter choices.

PROBLEM 7 *(Sensitivity analysis)* Let $\mathcal{A}^{(0)}$ be a most likely alignment for S_1 and S_2 for some parameter vector $\lambda^{(0)}$. Find the set of all parameter values such that $\mathcal{A}^{(0)}$ is the most likely alignment for S_1 and S_2.

PROBLEM 8 *(Parametric inference)* Find the most likely alignment between two given sequences S_1 and S_2 for every possible choice of transition and emission probabilities.

A different problem arises when trying to train the alignment algorithm to produce a desired result.

PROBLEM 9 *(Inverse optimization)* Given strings S_1 and S_2 and an alignment $\mathcal{A}$ for them, find transition and emission probabilities for which $\mathcal{A}$ is the most likely alignment.

The pair HMM can generate *any* pair of sequences over alphabet Σ; each pair can be generated in a variety of ways. Motivated by this observation, our final problem is as follows.

PROBLEM 10 *(Full probability estimation)* Given sequences S_1, S_2, compute the probability that the sequences are generated by the pair HMM. That is, compute

$$\Pr(S_1, S_2) = \sum_{\mathcal{A}} \Pr(S_1, S_2, \mathcal{A}), \qquad (4.22)$$

where $\mathcal{A}$ ranges over all paths that emit S_1, S_2.

Problems 6–9 are related to the basic parametric analysis problems of 4.2.4 through the *log transform*. To explain this, we examine Problem 6.

The *Viterbi algorithm* finds the most likely alignment $\mathcal{A}^*$. We now describe its main ideas (see [15] for details). For $p \in \{1, 2\}$, let $S_{p1}, \ldots, S_{pn_i}$ be the characters in sequence S_p. For $q \in \{M, I, D\}$, let $C_q(i, j)$ be the probability of the most likely alignment between subsequences $S_{11} \ldots S_{1i}$ and $S_{21} \ldots S_{2j}$ beginning at state B and ending at state q. Assume for simplicity that $s_{Bq} = s_{qE} = \tau$ for all $q \in \{M, I, D\}$. Then, the $C_q(i, j)$'s can be calculated using the following recurrences.

$$C_M(i, j) = e_M(S_{1i}, S_{2j}) \max \begin{cases} s_{MM} \cdot C_M(i-1, j-1) \cdot \\ s_{IM} \cdot C_I(i-1, j-1) \\ s_{DM} \cdot C_D(i-1, j-1) \end{cases} \qquad (4.23)$$

$$C_D(i, j) = e_D(S_{1i}) \max \begin{cases} s_{MD} \cdot C_M(i-1, j) \\ s_{DD} \cdot C_D(i-1, j) \end{cases} \qquad (4.24)$$

$$C_I(i, j) = e_I(S_{2j}) \max \begin{cases} s_{MI} \cdot C_M(i, j-1) \\ s_{II} \cdot C_I(i, j-1) \end{cases} \qquad (4.25)$$

The probability of the most likely alignment between S_1 and S_2 is given by

$$\tau^2 \max\{C_M(n_1, n_2), C_D(n_1, n_2), C_I(n_1, n_2)\}. \qquad (4.26)$$

Viterbi's algorithm solves the recurrences by a method that is structurally identical to the well-known dynamic programming algorithm for sequence alignment, except that multiplication is used instead of addition. In fact, we can make the two algorithms identical through the *log transform*, which replaces probabilities by their logarithms, thus enabling us to deal with sums, instead of products. Define $u_{..} = \log s_{..}$ and $v_{..} = \log e_{...}$. In the log domain, the probability of alignment (4.20) above is

$$u_{BM} + v_M(A, A) + u_{MM} + v_M(G, C) + u_{MI} + v_I(T) + u_{IM} + v_M(C, C) + u_{MD} + v_D(T) + u_{DE}.$$

Problem 6 is equivalent to solving

$$\mathcal{A}^* = \arg \max_{\mathcal{A}} \ \log \Pr(S_1, S_2, \mathcal{A}). \qquad (4.27)$$

In the log domain, recurrences (4.23)–(4.25) become identical to those of the dynamic programming algorithm for computing optimum alignments. Therefore, the algorithms for finding a maximum likelihood alignment and for finding a maximum score alignment are essentially the same.

The log transform can be used parametrically, yielding expressions that are linear in the parameters (which are logs of probabilities). Thus, the parametric inference problem (Problem 8) is equivalent to the construction problem (Problem 3). Similarly, Problems 7 and 9 are equivalent to Problems 2 and 5, respectively. Thus, all the techniques described in the previous sections apply to these HMM problems as well.

Problem 10 is the one problem in our list that does not benefit from the log transform. Nevertheless, it can be solved by the standard *forward algorithm* for the HMMs, which is essentially the Viterbi method, except that we add probabilities instead of taking maximums [15]. The full probability for a pair of sequences is a polynomial in the probability values and thus it is itself a parametric expression.

4.7 Notes and Comments

Parametric issues in optimization, especially linear programming, have been studied since the 1950s. *Parametric linear programming*, where the coefficients of the objective function are variable, was initially formulated by Gass and Saaty [24]. In the terminology of this chapter, Gass and Saaty presented a simplex-based algorithm for one-dimensional sensitivity analysis of parametric linear programs. The method can be used for construction and search as well. The combinatorial complexity of parametric linear programming was studied by Murty [39], who showed that there exists a parametric linear program with n variables and $2n$ constraints where there are 2^n basic feasible solutions, each of which is a unique optimal solution for some suitably chosen value of the parameter. Parametric versions of various optimization problems have been studied and bounds for various problems have been established. A sampling of the parametric optimization problems considered in the literature includes network flows [23], stable marriage [30], matroid optimization [12], and scheduling [32].

Many of the approaches discussed in this chapter are specializations of more general techniques to sequence comparison. The geometric definitions and results of Section 4.2.1 are adapted from Agarwal and Sharir's text [2]. Megiddo's method of parametric search technique originally appeared in [36] as a method for solving optimization problems with rational objective functions. An improvement based on simulating parallel algorithms instead of sequential ones is also due to Megiddo [37]. The application of Megiddo's method to sensitivity analysis was first investigated by Gusfield [26, 27]. Ray shooting is an important problem in computer graphics, where it is used to detect and remove hidden surfaces and in computing the intersection of polyhedra; Agarwal and Matoušek describe these and other geometric applications of parametric search in [1] (see also Salowe's survey [48]). Newton's zero-finding algorithm and the gradient descent method are classical algorithms that can be traced back to Newton and Cauchy respectively. Radzik [47] describes the application of Newton's method to solve fractional combinatorial optimization problems. The gradient descent method for optimization is well known and discussed in many textbooks [44, 40]. Polyak [43] was among the first to study the subgradient method's theoretical aspects. Held and Karp [33] were the first to apply the method to mathematical programming problems.

Parametric sequence comparison was first considered by Fitch and Smith [22], who studied the effect of varying the gap penalty on the optimum alignment of two sample sequences. By careful analysis, they showed that there are 7 and 11 different optimal alignments (optimality regions) for their sample pair when end gaps are weighted and unweighted, respectively. Waterman et al. [52] proposed a systematic way of finding the optimality regions. Vingron and Waterman [51] studied the implications of parameter choice through a series of case studies. Independently of Waterman et al.'s work, Gusfield et al. [29] formally defined parametric alignment and gave the first bounds on the number of regions. Among their results is the $O(n^{2/3})$ on the number of optimality regions for global alignment with zero gap penalty presented in 4.3.1. Fernández-Baca et al. [18] prove that this bound is tight when the alphabet size is unbounded [18]; in fact, it is the only combinatorial complexity bound for parametric sequence comparison known to be exact. The best known lower bound when the alphabet size is bounded is $\Omega(\sqrt{n})$ [18]. The properties of parametric alignment problems with feature-based scoring schemes were first investigated by Fernández-Baca et al. [19], who obtained combinatorial bounds for several problems, such as multiple sequence alignment and phylogeny construction, by observing that they all have a similar integer parametric nature. Tighter bounds (Theorem 4.2) are due to Pachter and Sturmfels [42] and Fernández-Baca and Venkatachalam [21].

Gusfield [26] attributes the one-parameter construction algorithm of 4.5.1 to Eisner and Severance [16]. The two-parameter construction algorithm presented in 4.5.2 is due to Fernández-Baca and Srinivasan [20]. Zimmer and Lengauer [53] used this algorithm in their parametric sequence alignment software. The techniques for reconstructing multidimensional convex geometric objects through probing, upon which Theorem 4.7 is based, were developed by Dobkin et al. [13, 14], who extended the work on two-dimensional probing by Cole and Yap [10].

Gusfield and Stelling's publicly-available XPARAL system [31] implements the ray-shooting approach (using Newton's method) for two-parameter sensitivity analysis described in 4.4.6 and applies it to construct the maximization diagram for two-parameter alignment problems under a wide variety of scoring functions. While, in principle, ray shooting can be used for sensitivity analysis (and, hence, construction) for any number of parameters, there appears to be no reference to this in the literature. One way to achieve this generalization is to use the probing idea mentioned above; the probes here are ray-shooting queries, instead of evaluations. Each probe returns a supporting hyperplane of the region being generated. The results of [13, 14] imply a number of queries proportional to the number of vertices and facets of the region.

Pachter and Sturmfels [42, 41] describe an implementation of the lifting algorithm for the construction problem mentioned at the beginning of 4.5. Their software relies on Gawrilow and Joswig's `polymake` tool [25].

XPARAL solves the inverse alignment problem using the gradient descent method described in 4.4.4. Sun et al. [50] give efficient algorithms for inverse sequence alignment with and without gaps that exploit the properties of feature-based scoring. Other inverse parametric optimization problems are studied by Eppstein [17].

For general background on hidden Markov models, a good starting point is Rabiner's survey article [45]. HMMs were first used for sequence alignment by Borodovsky et al. [6, 7, 8]. Durbin et al. [15] and Baldi and Brunak [4] give good introductions to the application of HMMs to sequence alignment and bioinformatics in general. Pachter and Sturmfels [42, 41] build a mathematical theory of statistical models for biological applications and show connections between parametric analysis and statistical models.

References

[1] P. Agarwal and J. Matoušek. Ray shooting and parametric search. *SIAM Journal on Computing*, 22:794–806, 1993.

[2] P. K. Agarwal and M. Sharir. *Davenport-Schinzel Sequences and their Geometric Applications.* Cambridge University Press, Cambridge–New York–Melbourne, 1995.

[3] R. Agarwala and D. Fernández-Baca. Weighted multidimensional search and its application to convex optimization. *SIAM Journal on Computing*, 25:83–99, 1996.

[4] P. Baldi and S. Brunak. *Bioinformatics: The machine learning approach.* Adaptive Computation and Machine Learning. MIT Press, Cambridge, MA, 2nd edition, 2001.

[5] C.B. Barber, D.P. Dobkin, and H. Huhdanpaa. The quickhull algorithm for convex hulls. *ACM Transactions on Mathematical Software*, 22(4):469–483, 1996.

[6] M. Borodovsky, Yu. Sprizhitsky, E. Golovanov, and A. Alexandrov. Statistical patterns in primary structures of functional regions in the E. coli genome: I. Oligonucleotide frequencies analysis. *Molecular Biology*, 20:826–833, 1986.

[7] M. Borodovsky, Yu. Sprizhitsky, E. Golovanov, and A. Alexandrov. Statistical patterns in primary structures of functional regions in the E. coli genome: II. Non-homogeneous markov models. *Molecular Biology*, 20:833–840, 1986.

[8] M. Borodovsky, Yu. Sprizhitsky, E. Golovanov, and A. Alexandrov. Statistical patterns in primary structures of functional regions in the E. coli genome: III. Computer recognition of coding regions. *Molecular Biology*, 20:1145–1150, 1986.

[9] E. Cohen and N. Megiddo. Maximizing concave functions in fixed dimension. In P. M. Pardalos, editor, *Complexity in Numerical Optimization*, pages 74–87. World Scientific, Singapore, 1993.

[10] R. Cole and C. Yap. Shape from probing. *Journal of Algorithms*, 8:19–38, 1987.

[11] M. de Berg, M. van Kreveld, M. Overmars, and O. Schwarzkopf. *Computational Geometry: Algorithms and Applications.* Springer-Verlag, Berlin, 2nd edition, 2000.

[12] T. Dey. Improved bounds on planar k-sets and related problems. *Discrete and Computational Geometry*, 19(3):373–382, 1998.

[13] D. Dobkin, H. Edelsbrunner, and C.K. Yap. Probing convex polytopes. In *Proceedings of the 18th Annual ACM Symposium on Theory of Computing*, pages 424–432, 1986.

[14] D. Dobkin, H. Edelsbrunner, and C.K. Yap. Probing convex polytopes. In Cox and Wilfong, editors, *Autonomous Robot Vehicles*, pages 328–341. Springer-Verlag, 1990.

[15] R. Durbin, S.R. Eddy, A. Krogh, and G. Mitchison. *Biological Sequence Analysis: Probabilistic Models of Proteins and Nucleic Acids.* Cambridge University Press, 1998.

[16] M.J. Eisner and D.G. Severance. Mathematical techniques for efficient record segmentation in large shared databases. *Journal of the Association for Computing Machinery*, 23:619–635, 1976.

[17] D. Eppstein. Setting parameters by example. *SIAM J. Computing*, 32(3):643–653, 2003.

[18] D. Fernández-Baca, T. Seppäläinen, and G. . Bounds for parametric sequence comparison. *Discrete Applied Mathematics*, 118:181–198, 2002.

[19] D. Fernández-Baca, T. Seppäläinen, and G. Slutzki. Parametric multiple sequence alignment and phylogeny construction. *Journal of Discrete Algorithms*, 2:271–287, 2004. Special issue on Combinatorial Pattern Matching, R. Giancarlo and D. Sankoff, eds.

[20] D. Fernández-Baca and S. Srinivasan. Constructing the minimization diagram of a two-parameter problem. *Operations Research Letters*, 10:87–93, 1991.

[21] D. Fernández-Baca and B. Venkatachalam. Parametric analysis, duality, and lattice polytopes. unpublished manuscript, May 2004.

[22] W.M. Fitch and T.F. Smith. Optimal sequence alignments. *Proceedings of the National Academy of Sciences USA*, 80:1382–1386, 1983.

[23] G. Gallo, M.D. Grigoriades, and R.E. Tarjan. A fast parametric maximum flow algorithm and applications. *SIAM Journal on Computing*, 18:30–55, 1989.

[24] S.I. Gass and T. Saaty. The computational algorithm for the parametric objective function. *Naval Research and Logistics Quarterly*, 2:39–45, 1955.

[25] E. Gawrilow and M. Joswig. `polymake`: an approach to modular software design in computational geometry. In *Proceedings of the 17th Annual Symposium on Computational Geometry*, pages 222–231. ACM Press, 2001.

[26] D. Gusfield. Sensitivity analysis for combinatorial optimization. Technical Report UCB/ERL M80/22, University of California, Berkeley, May 1980.

[27] D. Gusfield. Parametric combinatorial computing and a problem in program module allocation. *Journal of the Association for Computing Machinery*, 30(3):551–563, 1983.

[28] D. Gusfield. *Algorithms on Strings, Trees, and Sequences: Computer Science and Computational Biology*. Cambridge University Press, Cambridge–New York–Melbourne, 1997.

[29] D. Gusfield, K. Balasubramanian, and D. Naor. Parametric optimization of sequence alignment. *Algorithmica*, 12:312–326, 1994.

[30] D. Gusfield and R.W. Irving. Parametric stable marriage and minimum cuts. *Information Processing Letters*, 30:255–259, 1989.

[31] D. Gusfield and P. Stelling. Parametric and inverse-parametric sequence alignment with XPARAL. In Russell F. Doolittle, editor, *Computer methods for macromolecular sequence analysis*, volume 266 of *Methods in Enzymology*, pages 481–494. Academic Press, 1996.

[32] N.G. Hall and M.E. Posner. Sensitivity analysis for scheduling problems. *J. of Scheduling*, 7(1):49–83, 2004.

[33] M. Held and R.M. Karp. The traveling salesman problem and minimum spanning trees: part II. *Mathematical Programming*, 6:6–25, 1971.

[34] E. Herbert. *Algorithms in Combinatorial Geometry*. Springer-Verlag, Heidelberg, 1987.

[35] P. McMullen. The maximum number of faces of a convex polytope. *Mathematika*, 17:179–184, 1970.

[36] N. Megiddo. Combinatorial optimization with rational objective functions. *Math. Oper. Res.*, 4:414–424, 1979.

[37] N. Megiddo. Applying parallel computation algorithms in the design of serial algorithms. *Journal of the Association for Computing Machinery*, 30(4):852–865, 1983.

[38] D. Mount. *Bioinformatics: Sequence and Genome Analysis*. Cold Spring Harbor Press, Cold Spring Harbor, New York, 2001.

[39] K. Murty. Computational complexity of parametric linear programming. *Math. Programming*, 19:213–219, 1980.

[40] G.L. Nemhauser and L.A. Wolsey. *Integer and Combinatorial Optimization*. Wiley-Interscience Series in Discrete Mathematics and Optimization. John Wiley & Sons, 1988.

[41] L. Pachter and B. Sturmfels. Parametric inference for biological sequence analysis. *Proceedings of the National Academy of Sciences USA*, 101(46):16138–16143, 2004.

[42] L. Pachter and B. Sturmfels. Tropical geometry of statistical models. *Proceedings of the National Academy of Sciences USA*, 101(46):16132–16137, 2004.

[43] B.T. Polyak. A general method for solving extremal problems. *Soviet. Math. Dokl.*, 8:593–597, 1967.

[44] W.H. Press, B.P. Flannery, S.A. Teukolsky, and W.T. Vetterling. *Numerical Recipes: The Art of Scientific Computing*. Cambridge University Press, Cambridge (UK) and New York, 2nd edition, 1992.

[45] L.R. Rabiner. A tutorial on hidden Markov models and selected applications in speech recognition. *Proceedings of the IEEE*, 77(2):257–286, 1989.

[46] T. Radzik. *Algorithms for some linear and fractional combinatorial optimization problems*. Department of Computer Science, Stanford University, Stanford, CA 94305, August 1992.

[47] T. Radzik. Newton's method for fractional combinatorial optimization. In *33rd Annual Symposium on Foundations of Computer Science*, pages 659–669, Pittsburgh, PA, October 1992. IEEE.

[48] J. Salowe. Parametric search. In J.E. Goodman and J. O'Rourke, editors, *Handbook of Discrete and Computational Geometry*, chapter 37, pages 683–696. CRC Press LLC, Boca Raton, FL, 1997.

[49] R. Seidel. Convex hull computations. In J.E. Goodman and J. O'Rourke, editors, *Handbook of Discrete and Computational Geometry*, chapter 19, pages 361–376. CRC Press LLC, Boca Raton, FL, 1997.

[50] F. Sun, D. Fernández-Baca, and W. Yu. Inverse parametric sequence alignment. *Journal of Algorithms*, 53(1):36–54, 2004.

[51] M. Vingron and M.S. Waterman. Sequence alignment and penalty choice: Review of concepts, case studies, and implications. *J. Mol. Biol.*, 235:1–12, 1994.

[52] Michael S. Waterman, M. Eggert, and E. Lander. Parametric sequence comparisons. *Proceedings of the National Academy of Sciences USA*, 89:6090–6093, 1992.

[53] R. Zimmer and Th. Lengauer. Fast and numerically stable parametric alignment of biosequences. In *Proceedings of RECOMB 97, Santa Fe, NM*, pages 344–353. ACM Press, 1997.

II

String Data Structures

5

Lookup Tables, Suffix Trees and Suffix Arrays

Srinivas Aluru
Iowa State University

Pang Ko
Iowa State University

5.1 Introduction

Fundamental string data structures, and their myriad applications in computational molecular biology are the focus of this part of the handbook. Sequence alignments and string data structures form the twin foundations for many applications in computational genomics. The utility of string data structures stems from the fact that at a basic level, various types of DNA and RNA sequences, and protein sequences can be modeled as strings — DNA as strings over the alphabet {A,C,G,T}, RNA as strings over the alphabet {A,C,G,U}, and proteins as strings over an alphabet of size 20 corresponding to the 20 amino acid residues. While simplistic, modeling of biological sequences as mere strings serves as a sufficient level of abstraction for a plethora of applications.

Given the large volume of sequence data that many computational biology applications must deal with, proper organization of the data to facilitate fast access is important to achieve desirable run-times. From this perspective, string data structures serve the same purpose for biological sequence data as binary search trees serve for ordered numeric data, and quadtrees serve for spatial data.

String data structures are ideal for uncovering exact matching patterns in sequences. Due to evolutionary mechanisms which alter biomolecular sequences, errors introduced by experimental processes, and many other factors that permit variations — such as the degeneracy of genetic code, protein sequences with some sequence similarity showing significant structural similarity — one is rarely interested in exact matches as an end in itself. Despite this, exact matches play a role because they are typically fast — requiring linear time as opposed to the quadratic time of alignment algorithms. As an example, consider the task of finding good local alignments between a query sequence and a database consisting of tens to hundreds of millions of sequences. It is computationally expensive to do as many pairwise local alignments. If we are interested in a pairwise alignment only if it exhibits significant homology, such an alignment should also contain regions of exact matches. For instance, if an aligning region of $100bp$ length contains at most 4 positions of difference, there should be an exact match of length at least 20 in this region. Exact matches can be used as a filter to eliminate large number of pairs that would not yield a good local alignment by performing alignments only on pairs that have an exact matching region larger than a determined threshold. It is in this spirit that many problems related to exact matches find applications in computational biology. String data structures are also useful when performing approximate matches where only a small number of differences are permitted.

In this chapter, we provide a detailed introduction to the three most frequently used string data structures in computational molecular biology — lookup tables, suffix trees and suffix arrays. The focus of this chapter will be on algorithms for constructing these data structures, which tend to be somewhat complex in the case of suffix trees and suffix arrays. We will also explore the relationships between these data structures. Chapter 6 provides several illustrations of biological applications where suffix trees play a central role. A number of new research results on solving biological applications using the more space efficient suffix array data structure, and its augmented variants, are presented in Chapter 7.

5.2   Lookup Tables

Lookup table is a simple data structure that records the positions of occurrences of substrings of a prespecified length in one or more strings. Lookup tables are used in a number of important bioinformatic tools including such popular programs as BLAST [1, 2] for database searches, and CAP3 [15] for genome assembly.

We use the following notation throughout the chapter: Let s be a string over the alphabet Σ. $|s|$ denotes the size of s, $s[i]$ denotes the ith character of s, and $s[i..j]$ denotes the substring $s[i]s[i+1]\ldots s[j]$. Let w denote a prespecified length, sometimes referred to as window-size. The lookup table is an array LT of size $|\Sigma|^w$, corresponding to the $|\Sigma|^w$ possible substrings of length w. Let $f : \Sigma \to \{0, 1, \ldots |\Sigma| - 1\}$ be the one-to-one function such that $f(c) = j - 1$ if c is the jth lexicographically smallest character. For the purpose of the lookup table, any arbitrary ordering of the characters can be taken as lexicographic ordering. Using f, a substring of length w can be treated as a w digit number in a base $|\Sigma|$ system, and converted to its decimal equivalent. We use the notation $F(\alpha)$ to denote the decimal number corresponding to a w-long substring α.

Each entry in the lookup table LT points to a linked list of specific locations within the input set of strings where the substring corresponding to the index for the entry occurs. The lookup table for the DNA sequence CATTATTAGGA with $w = 2$ is shown in Figure 5.1. It is constructed by using the mapping A→ 0, C→ 1, G→ 2 and T→ 3. The substring TA corresponds to the index $(30)_4 = 12$. The entry at index 12 indicates that the substring TA occurs in the string starting at positions 4 and 7.

Let s be a string of length n. It is easy to construct the lookup table for s in $O(|\Sigma|^w + n)$ time. First, create and initialize each entry to a null list in $O(|\Sigma|^w)$ time. Then, insert

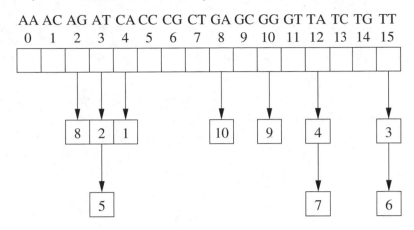

FIGURE 5.1: Lookup table for the DNA sequence CATTATTAGGA with $w = 2$. The table contains 16 entries, corresponding to the 16 different nucleotide sequences of length 2.

substrings one at a time. First compute $index = F(s[1..w])$ in $O(w)$ time. Insert position 1 in the linked list corresponding to $LT[index]$. Using the identity

$$F(s[k+1..k+w+1]) = \left(F(s[k..k+w] - f(s[k])|\Sigma|^{w-1}\right) \times |\Sigma| + f(s[k+w+1])$$

$F(s[k+1..k+w+1])$ can be computed from $F(s[k..k+w])$ in $O(1)$ time. As each starting position $1 \ldots n-w+1$ occurs in a linked list, the total size of all linked lists is $O(n)$ (typically $n >> w$). Thus, the size of the lookup table data structure is $O(|\Sigma|^w + n)$. The lookup table can be easily generalized to a set of strings. Let $\mathcal{S} = \{s_1, s_2, \ldots, s_k\}$ be a set of k strings of total length N. To create the corresponding lookup table, substrings from each of the strings are inserted in turn. A location in a linked list now consists of a pair giving the string number and the position of the substring within the string. The space and run-time required for constructing the lookup table is $O(|\Sigma|^w + N)$.

The size of the lookup table depends exponentially on the window-size w. To achieve space usage that is linear in the input data size, the value of w should be no greater than $\log_{|\Sigma|} N$. A window-size of 10 for DNA sequences assuming a 4-letter alphabet results in a lookup table with $2^{20} > 1$ million entries.

Lookup table is conceptually a very simple data structure to understand and implement. Once the lookup table for a database of strings is available, given a query string of length w, all occurrences of it in the database can be retrieved in $O(w + k)$ time, where k is the number of occurrences. The main problem with this data structure is its dependence on an arbitrary predefined substring length w. If the query string is of length $l > w$, the lookup table does not provide an efficient way of retrieving all occurrences of the query string in the database. Nevertheless, lookup tables are widely used in many bioinformatic programs due to their simplicity and ease of use.

5.3 Suffix Trees and Suffix Arrays

5.3.1 Basic Definitions and Properties

Suffix trees and suffix arrays are versatile data structures fundamental to string processing applications. Let s' denote a string over the alphabet Σ. Let $\$ \notin \Sigma$ be a unique termination character, and $s = s'\$$ be the string resulting from appending $\$$ to s'. Let $suff_i = s[i]s[i+1]\ldots s[|s|]$ be the suffix of s starting at i^{th} position. The suffix tree of s, denoted $ST(s)$ or simply ST, is a compacted trie of all suffixes of string s. Let $|s| = n$. It has the following properties:

1. The tree has n leaves, labeled $1 \ldots n$, one corresponding to each suffix of s.
2. Each internal node has at least 2 children.
3. Each edge in the tree is labeled with a substring of s.
4. The concatenation of edge labels from the root to the leaf labeled i is $suff_i$.
5. The labels of the edges connecting a node with its children start with different characters.

The paths from root to the leaves corresponding to the suffixes $suff_i$ and $suff_j$ coincide up to their longest common prefix, at which point they bifurcate. If a suffix of the string is a prefix of another longer suffix, the shorter suffix must end in an internal node instead of a leaf, as desired. It is to avoid this possibility that the unique termination character is added to the end of the string. Keeping this in mind, we use the notation $ST(s')$ to denote the suffix tree of the string obtained by appending $\$$ to s'. Throughout this chapter, '$\$$' is taken to be the lexicographically smallest character.

As each internal node has at least 2 children, an n-leaf suffix tree has at most $n-1$ internal nodes. Because of property (5), the maximum number of children per node is bounded by $|\Sigma|+1$. Except for the edge labels, the size of the tree is $O(n)$. In order to allow a linear space representation of the tree, each edge label is represented by a pair of integers denoting the starting and ending positions, respectively, of the substring describing the edge label. If the edge label corresponds to a repeat substring, the indices corresponding to any occurrence of the substring may be used. The suffix tree of the string CATTATTAGGA is shown in Figure 5.2. For convenience of understanding, we show the actual edge labels.

The suffix array of $s = s'\$$, denoted $SA(s)$ or simply SA, is a lexicographically sorted array of all suffixes of s. Each suffix is represented by its starting position in s. $SA[i] = j$ iff $suff_j$ is the ith lexicographically smallest suffix of s. The suffix array is often used in conjunction with an array termed Lcp array, containing the lengths of the longest common prefixes between every consecutive pair of suffixes in SA. We use $lcp(\alpha, \beta)$ to denote the longest common prefix between strings α and β. We also use the term lcp as an abbreviation for the term *longest common prefix*. $Lcp[i]$ contains the length of the lcp between $suff_{SA[i]}$ and $suff_{SA[i+1]}$, i.e., $Lcp[i] = |lcp(suff_{SA[i]}, suff_{SA[i+1]})|$. As with suffix trees, we use the notation $SA(s')$ to denote the suffix array of the string obtained by appending $\$$ to s'. The suffix and Lcp arrays of the string CATTATTAGGA are shown in Figure 5.2.

Let v be a node in the suffix tree. Let $path\text{-}label(v)$ denote the concatenation of edge labels along the path from root to node v. Let $string\text{-}depth(v)$ denote the length of $path\text{-}label(v)$. To differentiate this with the usual notion of depth, we use the term $tree\text{-}depth$ of a node to denote the number of edges on the path from root to the node. Note that the length of the longest common prefix between two suffixes is the string depth of the lowest common ancestor of the leaf nodes corresponding to the suffixes. A repeat substring of string S is *right-maximal* if there are two occurrences of the substring that are succeeded by different characters in the string. The path label of each internal node in the suffix tree corresponds to a right-maximal repeat substring and vice versa.

Let v be an internal node in the suffix tree with path-label $c\alpha$ where c is a character and α is a (possibly empty) string. Therefore, $c\alpha$ is a right-maximal repeat, which implies that α is also a right maximal repeat. Let u be the internal node with path label α. A pointer from node v to node u is called a *suffix link*; we denote this by $SL(v) = u$. Each suffix $suff_i$

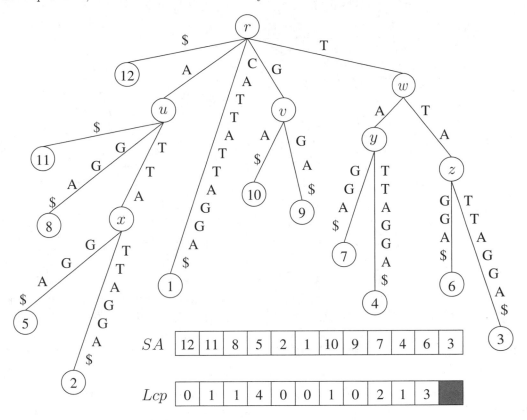

FIGURE 5.2: Suffix tree, suffix array and *Lcp* array of the string CATTATTAGGA. The suffix links in the tree are given by $x \to z \to y \to u \to r$, $v \to r$, and $w \to r$.

in the subtree of v shares the common prefix $c\alpha$. The corresponding suffix $suff_{i+1}$ with prefix α will be present in the subtree of u. The concatenation of edge labels along the path from v to leaf labeled i, and along the path from u to leaf labeled $i + 1$ will be the same. Similarly, each internal node in the subtree of v will have a corresponding internal node in the subtree of u. In this sense, the entire subtree under v is contained in the subtree under u.

Every internal node in the suffix tree other than the root has a suffix link from it. Let v be an internal node with $SL(v) = u$. Let v' be an ancestor of v other than the root and let $u' = SL(v')$. As *path-label*(v') is a prefix of *path-label*(v), *path-label*(u') is also a prefix of *path-label*(u). Thus, u' is an ancestor of u. Each proper ancestor of v except the root will have a suffix link to a distinct proper ancestor of u. It follows that *tree-depth*$(u) \geq$ *tree-depth*$(v) - 1$.

Suffix trees and suffix arrays can be generalized to multiple strings. The generalized suffix tree of a set of strings $\mathcal{S} = \{s_1, s_2, \ldots, s_k\}$, denoted $GST(\mathcal{S})$ or simply GST, is a compacted trie of all suffixes of each string in $\mathcal{S}$. We assume that the unique termination character $\$$ is appended to the end of each string. A leaf label now consists of a pair of integers (i, j), where i denotes the suffix is from string s_i and j denotes the starting position of the suffix in s_i. Similarly, an edge label in a GST is a substring of one of the strings. An edge label is represented by a triplet of integers (i, j, l), where i denotes the string number, and j and l denote the starting and ending positions of the substring in s_i. For convenience of

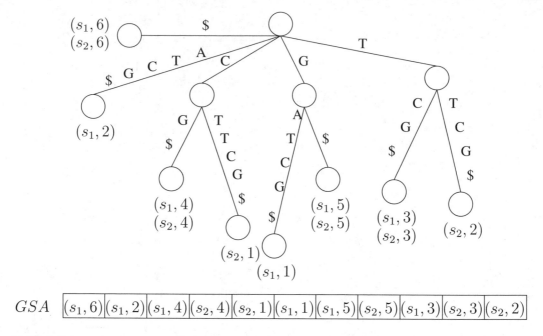

GSA $(s_1,6)$ $(s_1,2)$ $(s_1,4)$ $(s_2,4)$ $(s_2,1)$ $(s_1,1)$ $(s_1,5)$ $(s_2,5)$ $(s_1,3)$ $(s_2,3)$ $(s_2,2)$

FIGURE 5.3: Generalized suffix tree and generalized suffix array of strings GATCG and CTTCG.

understanding, we will continue to show the actual edge labels. Note that two strings may have identical suffixes. This is compensated by allowing leaves in the tree to have multiple labels. If a leaf is multiply labeled, each suffix should come from a different string. If N is the total number of characters (including the $ in each string) of all strings in $\mathcal{S}$, the GST has at most N leaf nodes and takes up $O(N)$ space. The generalized suffix array of $\mathcal{S}$, denoted $GSA(\mathcal{S})$ or simply GSA, is a lexicographically sorted array of all suffixes of each string in $\mathcal{S}$. Each suffix is represented by an integer pair (i, j) denoting suffix starting from position j in s_i. If suffixes from different strings are identical, they occupy consecutive positions in the GSA. For convenience, we make an exception for the suffix $ by listing it only once, though it occurs in each string. The GST and GSA of strings GATCG and CTTCG are shown in Figure 5.3.

Suffix trees and suffix arrays can be constructed in time linear to the size of the input. Suffix trees are very useful in solving a plethora of string problems in optimal run-time bounds. Moreover, in many cases, the algorithms are very simple to design and understand. For example, consider the classic pattern matching problem of determining if a pattern P occurs in text T over a constant sized alphabet. Note that P occurs starting from position i in T iff P is a prefix of $suff_i$ in T. Thus, whether P occurs in T or not can be determined by checking if P matches an initial part of a path from root to a leaf in $ST(T)$. Traversing from the root matching characters in P, this can be determined in $O(|P|)$ time, independent of the size of T. As another application, consider the problem of finding a longest common substring of a pair of strings. Once the GST of the two strings is constructed, all that is needed is to identify an internal node with the largest string depth that contains at least one leaf from each string. Applications of suffix trees in computational molecular biology are explored in great detail in the next chapter. Suffix arrays are of interest because they require much less space than suffix trees, and can be used to solve many of the same problems. Such methods are explored in Chapter 7. In this chapter, we concentrate on linear time construction algorithms for suffix trees and suffix arrays.

5.3.2 Suffix Trees vs. Suffix Arrays

In this section, we explore linear time construction algorithms for suffix trees and suffix arrays. We also show how suffix trees and suffix arrays can be derived from each other in linear time. We first show that the suffix array and Lcp array of a string can be obtained from its suffix tree in linear time. Define lexicographic ordering of the children of a node to be the order based on the first character of the edge labels connecting the node to its children. Define lexicographic depth first search to be a depth first search of the tree where the children of each node are visited in lexicographic order. The order in which the leaves of a suffix tree are visited in a lexicographic depth first search gives the suffix array of the corresponding string. In order to obtain lcp information, the string-depth of the current node during the search is remembered. This can be easily updated in $O(1)$ time per edge as the search progresses. The length of the lcp between two consecutive suffixes is given by the smallest string-depth of a node visited between the leaves corresponding to the two suffixes.

Given the suffix array and the Lcp array of a string s ($|s\$| = n$), its suffix tree can be constructed in $O(n)$ time. This is done by starting with a partial suffix tree for the lexicographically smallest suffix, and repeatedly inserting subsequent suffixes from the suffix array into the tree until the suffix tree is complete. Let T_i denote the compacted trie of the first i suffixes in lexicographic order. The first tree T_1 consists of a single leaf labeled $SA[1] = n$ connected to the root with an edge labeled $suff_{SA[1]} = \$$.

To insert $SA[i+1]$ into T_i, start with the most recently inserted leaf $SA[i]$ and walk up $(|suff_{SA[i]}| - |lcp(suff_{SA[i]}, suff_{SA[i+1]})|) = ((n - SA[i] + 1) - Lcp[i])$ characters along the path to the root. This walk can be done in $O(1)$ time per edge by calculating the lengths of the respective edge labels. If the walk does not end at an internal node, create an internal node. Create a new leaf labeled $SA[i+1]$ and connect it to this internal node with an edge. Set the label on this edge to $s[SA[i+1] + Lcp[i]..n]$. This creates the tree T_{i+1}. The procedure is illustrated in Figure 5.4. It works because no other suffix inserted so far shares a longer prefix with $suff_{SA[i+1]}$ than $suff_{SA[i]}$ does. To see that the entire algorithm runs in $O(n)$ time, note that inserting a new suffix into T_i requires walking up the rightmost path in T_i. Each edge that is traversed ceases to be on the rightmost path in T_{i+1}, and thus is never traversed again. An edge in an intermediate tree T_i corresponds to a path in the suffix tree ST. When a new internal node is created along an edge in an intermediate tree, the edge is split into two edges, and the edge below the newly created internal node corresponds to an edge in the suffix tree. Once again, this edge ceases to be on the rightmost path and is never traversed again. The cost of creating an edge in an intermediate tree can be charged to the lowest edge on the corresponding path in the suffix tree. As each edge is charged once for creating and once for traversing, the total run-time of this procedure is $O(n)$.

Finally, the Lcp array itself can be constructed from the suffix array and the string in linear time [20]. Let R be an array of size n such that $R[i]$ contains the position in SA of $suff_i$. R can be constructed by a linear scan of SA in $O(n)$ time. The Lcp array is computed in n iterations. In iteration i of the algorithm, the longest common prefix between $suff_i$ and its respective right neighbor in the suffix array is computed. The array R facilitates locating an arbitrary suffix $suff_i$ and finding its right neighbor in the suffix array in constant time. Initially, the length of the longest common prefix between $suff_1$ and its suffix array neighbor is computed directly and recorded. Let $suff_j$ be the right neighbor of $suff_i$ in SA. Let l be the length of the longest common prefix between them. Suppose $l \geq 1$. As $suff_j$ is lexicographically greater than $suff_i$ and $s[i] = s[j]$, $suff_{j+1}$ is lexicographically greater than $suff_{i+1}$. The length of the longest common prefix between them is $l - 1$. It follows that the length of the longest common prefix between $suff_{i+1}$ and its right neighbor in the suffix

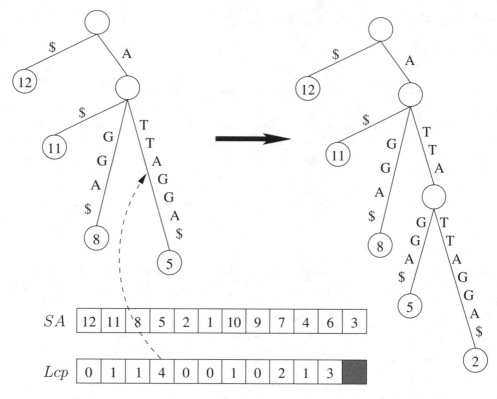

FIGURE 5.4: The construction of a suffix tree from the corresponding suffix and Lcp arrays. The example illustrates the insertion of $suff_2$ in the partial tree resulting from previously inserting the first four suffixes in the suffix array. The length of the *lcp* between the last inserted suffix and the new suffix gives the number of characters under the root and along the rightmost path at which the new leaf is inserted.

array is $\geq l-1$. To determine its correct length, the comparisons need only start from the lth characters of the suffixes.

To prove that the run time of the above algorithm is linear, charge a comparison between the rth character in suffix $suff_i$ and the corresponding character in its right neighbor suffix in SA to the position in the string of the rth character of $suff_i$, i.e., $i+r-1$. A comparison made in an iteration is termed successful if the characters compared are identical, contributing to the longest common prefix being computed. Because there is one failed comparison in each iteration, the total number of failed comparisons is $O(n)$. As for successful comparisons, each position in the string is charged only once for a successful comparison. Thus, the total number of comparisons over all iterations is linear in n.

In light of the above discussion, a suffix tree and a suffix array can be constructed from each other in linear time. Thus, a linear time construction algorithm for one can be used to construct the other in linear time. In the following sections, we explore such algorithms. Each algorithm is interesting in its own right, and exploits interesting properties that could be useful in designing algorithms using suffix trees and suffix arrays.

In suffix tree and suffix array construction algorithms, three different types of alphabets are considered — a constant or fixed size alphabet ($|\Sigma|(1)$), integer alphabet ($\Sigma = \{1, 2, \ldots, n\}$), and arbitrary alphabet. Suffix trees and suffix arrays can be constructed in linear time for both constant size and integer alphabets. The constant alphabet size case covers DNA and protein sequences in molecular biology. The integer alphabet case is interesting because a string of length n can have at most n distinct characters. Furthermore, some algorithms use a recursive technique that would generate and require operating on strings over integer alphabet, even when applied to strings over a fixed alphabet.

5.4 Linear Time Construction of Suffix Trees

Let s be a string of length n including the termination character \$. Suffix tree construction algorithms start with an empty tree and iteratively insert suffixes while maintaining the property that each intermediate tree represents a compacted trie of the suffixes inserted so far. When all the suffixes are inserted, the resulting tree will be the suffix tree. Suffix links are typically used to speedup the insertion of suffixes. While the algorithms are identified by the names of their respective inventors, the exposition presented does not necessarily follow the original algorithms and we take the liberty to comprehensively present the material in a way we feel contributes to ease of understanding.

5.4.1 McCreight's Algorithm

McCreight's algorithm inserts suffixes in the order $suff_1, suff_2, \ldots, suff_n$. Let T_i denote the compacted trie after $suff_i$ is inserted. T_1 is the tree consisting of a single leaf labeled 1 that is connected to the root by an edge with label $s[1..n]$. In iteration i of the algorithm, $suff_i$ is inserted into tree T_{i-1} to form tree T_i. An easy way to do this is by starting from the root and following the unique path matching characters in $suff_i$ one by one until no more matches are possible. If the traversal does not end at an internal node, create an internal node there. Then, attach a leaf labeled i to this internal node and use the unmatched portion of $suff_i$ for the edge label. The run-time for inserting $suff_i$ is proportional to $|suff_i| = n - i + 1$. The total run-time of the algorithm is $\Sigma_{i=1}^{n}(n - i + 1) = O(n^2)$.

In order to achieve an $O(n)$ run-time, suffix links are used to significantly speedup the insertion of a new suffix. Suffix links are useful in the following way — Suppose we are inserting $suff_i$ in T_{i-1} and let v be an internal node in T_{i-1} on the path from root to leaf labeled $(i - 1)$. Then, $path\text{-}label(v) = c\alpha$ is a prefix of $suff_{i-1}$. Since v is an internal node, there must be another suffix $suff_j$ $(j < i - 1)$ that also has $c\alpha$ as prefix. Because $suff_{j+1}$ is previously inserted, there is already a path from the root in T_{i-1} labeled α. To insert $suff_i$ faster, if the end of path labeled α is quickly found, comparison of characters in $suff_i$ can start beyond the prefix α. This is where suffix links will be useful. The algorithm must also construct suffix links prior to using them.

LEMMA 5.1 Let v be an internal node in $ST(s)$ that is created in iteration $i - 1$. Let $path\text{-}label(v) = c\alpha$, where c is a character and α is a (possibly empty) string. Then, either there exists an internal node u with $path\text{-}label(u) = \alpha$ or it will be created in iteration i.

Proof As v is created when inserting $suff_{i-1}$ in T_{i-2}, there exists another suffix $suff_j$ $(j < i - 1)$ such that $lcp(suff_{i-1}, suff_j) = c\alpha$. It follows that $lcp(suff_i, suff_{j+1}) = \alpha$. The tree T_i already contains $suff_{j+1}$. When $suff_i$ is inserted during iteration i, internal node u with path-label α is created if it does not already exist.

The above lemma establishes that the suffix link of a newly created internal node can be established in the next iteration.

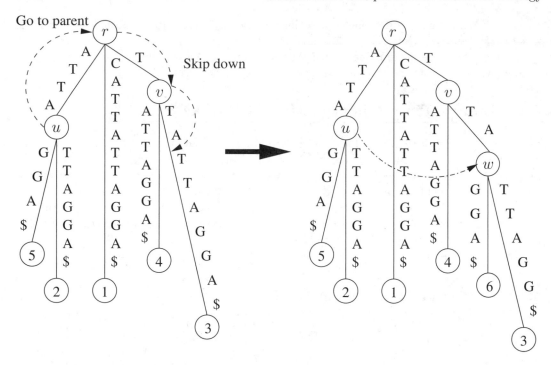

FIGURE 5.5: Illustration of suffix tree construction using McCreight's algorithm on the string CATTATTAGGA. The tree to the left is the compacted trie of suffixes 1 through 5. The process of inserting the next suffix $suff_6$ is shown in the figure.

The following procedure is used when inserting $suff_i$ in T_{i-1}. Let v be the internal node to which $suff_{i-1}$ is attached as a leaf. If v is the root, insert $suff_i$ using character comparisons starting with the first character of $suff_i$. Otherwise, let $path\text{-}label(v) = c\alpha$. If v has a suffix link from it, follow it to internal node u with path-label α. This allows skipping the comparison of the first $|\alpha|$ characters of $suff_i$. If v is newly created in iteration $i-1$, it would not have a suffix link yet. In that case, walk up to parent v' of v. Let β denote the label of the edge connecting v' and v. Let $u' = SL(v')$ unless v' is the root, in which case let u' be the root itself. It follows that $path\text{-}label(u')$ is a prefix of $suff_i$. Furthermore, it is guaranteed that there is a path below u' that matches the next $|\beta|$ characters of $suff_i$. Traverse $|\beta|$ characters along this path and either find an internal node u or insert an internal node u if one does not already exist. In either case, set $SL(v) = u$. Continue by starting character comparisons skipping the first $|\alpha|$ characters of $suff_i$.

This procedure is illustrated in Figure 5.5 for the string CATTATTAGGA. The tree to the left is the compacted trie after $suff_1$, $suff_2$, $suff_3$, $suff_4$ and $suff_5$ are inserted. To insert $suff_6$, consider the internal node u under which it is inserted as a leaf. Since u did not exist previously but was created during the insertion of $suff_5$, it does not have a suffix link yet. Therefore, walk up the 4 character edge to the parent of u to take a suffix link. However, the parent is the root r itself, and no suffix link is taken. To insert $suff_6$, walk down $4 - 1 = 3$ characters by only comparing one character per edge label and skipping edges at the rate of constant time per edge. At this position, create a new internal node w and set $SL(u) = w$. Continue to insert $suff_6$ below w.

The above procedure requires two different types of traversals — one in which it is known that there exists a path below that matches the next $|\beta|$ characters of $suff_i$ (type I), and the other in which it is unknown how many subsequent characters of $suff_i$ match a path below (type II). In the latter case, the comparison must proceed character by character until a mismatch occurs. In the former case, however, the traversal can be done by spending only $O(1)$ time per edge irrespective of the length of the edge label. At an internal node during such a traversal, the decision of which edge to follow next is made by comparing the next character of $suff_i$ with the first characters of the edge labels connecting the node to its children. However, once the edge is selected, the entire label or the remaining length of β must match, whichever is shorter. Thus, the traversal can be done in constant time per edge, and if the traversal stops within an edge label, the stopping position can also be determined in constant time.

The insertion procedure during iteration i can now be described as follows: Start with the internal node v to which $suff_{i-1}$ is attached as a leaf. If v has a suffix link, follow it and perform a type II traversal to insert $suff_i$. Otherwise, walk up to v's parent, take the suffix link from it unless it is the root, and perform a type I traversal to either find or create the node u which will be linked from v by a suffix link. Continue with a type II traversal below u to insert $suff_i$.

LEMMA 5.2   The total time spent in type I traversals over all iterations is $O(n)$.

Proof   A type I traversal is performed by walking down along a path from root to a leaf in $O(1)$ time per edge. Each iteration consists of walking up at most one edge, following a suffix link, and then performing downward traversals (either type II or both type I and type II). Recall that if $SL(v) = u$, then $tree\text{-}depth(u) \geq tree\text{-}depth(v) - 1$. Thus, following a suffix link may reduce the depth in the tree by at most one. It follows that the operations that may cause moving to a higher level in the tree cause a decrease in depth of at most 2 per iteration. As both type I and type II traversals increase the depth in the tree and there are at most n levels in ST, the total number of edges traversed by type I traversals over all the iterations is bounded by $3n$.

LEMMA 5.3   The total time spent in type II traversals over all iterations is $O(n)$.

Proof   In a type II traversal, a suffix of the string $suff_i$ is matched along a path in T_{i-1} until there is a mismatch. When a mismatch occurs, an internal node is created if there does not exist one already. Then, the remaining part of $suff_i$ becomes the edge label connecting leaf labeled i to the internal node. Charge each successful comparison of a character in $suff_i$ to the corresponding character in the original string s. Note that a character that is charged with a successful comparison is never charged again as part of a type II traversal. Thus, the total time spent in type II traversals is $O(n)$.

The above lemmas prove that the total run-time of McCreight's algorithm is $O(n)$.

5.4.2   Ukkonen's Algorithm

Ukkonen's suffix tree construction algorithm is also a linear time algorithm but with an important on-line property: The algorithm reads the input string one character at a time and maintains a suffix tree of the prefix of the string seen so far. As before, let s be a string of length n including the terminal '$\$$' character. The algorithm constructs a series of trees $T_1, T_2, \ldots, T_n$, where T_i is the suffix tree of $s[1..i]$. After constructing T_i, the algorithm reads $s[i+1]$ and updates T_i to create T_{i+1}. The total run-time spent by the time the algorithm constructs T_i is $O(i)$, even though the time spent in transitioning from one tree to the next is not necessarily constant.

When considering the string $s[1..i]$, a suffix of it may be repeated elsewhere in it because the unique terminal symbol is only at $s[n]$. Hence, a compacted trie of all suffixes of $s[1..i]$ may not have each suffix represented by a path that ends in a leaf. Therefore, we relax the definition of suffix trees by requiring that a downward path from the root corresponding to each suffix exist but not necessarily end in a leaf node. Such a tree is called *implicit suffix tree*. This would not pose any problem as the implicit suffix tree for $s[1..n]$ is the same as $ST(s)$ due to the terminal symbol $s[n] = \$$.

Ukkonen's algorithm employs a few additional ideas in conjunction with those already illustrated under McCreight's algorithm. Consider the prefix $s[1..i]$. We now use the notation $suff_k$ to denote the suffix starting from position k in the current string, i.e., $s[k..i]$. Let j be the position such that $suff_j = s[j..i]$ is the longest suffix of $s[1..i]$ that occurs elsewhere in it. Observe that the compacted trie of just suffixes $suff_1, suff_2, \ldots, suff_{j-1}$ is the same as the compacted trie of all suffixes $suff_1, suff_2, \ldots, suff_i$ of $s[1..i]$. In the implicit suffix tree of $s[1..i]$, the paths corresponding to first $j-1$ suffixes end in leaves and the paths corresponding to the remaining suffixes end otherwise.

Consider building T_{i+1} from T_i. Viewed naively, this requires extending all suffixes in T_i with the newly added character $s[i+1]$, and finally inserting a new suffix corresponding to the last character. Let j' be the position such that $s[j'..i+1]$ is the longest suffix of $s[1..i+1]$ that occurs elsewhere in it. Clearly, $j' \geq j$, where $s[j..i]$ is the longest suffix of $s[1..i]$ that occurs elsewhere in it. In creating T_{i+1}, the suffixes of $s[1..i+1]$ can be considered in three categories:

1. Suffixes $suff_i \ldots suff_{j-1}$: The corresponding suffixes in T_i already end in a leaf and they all need to be extended by the newly read character $s[i+1]$. Instead of explicitly doing this, this is implicitly achieved by assuming that all leaf labels (the labels of edges incident to leaves) end at the current end of the string.

2. Suffixes $suff_j \ldots suff_{j'-1}$: These suffixes are inserted in turn using ideas presented in McCreight's algorithm. This will be dealt in more detail later.

3. Suffixes $suff_{j'} \ldots suff_{i+1}$: We need not bother about these suffixes as the compacted trie of the suffixes in the two categories above automatically accounts for these suffixes also.

Observe that work is required only for inserting suffixes in category 2 above. The suffixes that are processed under category 2 in creating T_{i+1} from T_i, will become category 1 suffixes in subsequent tree constructions and are never worked on again. As the trees $T_1 \ldots T_n$ are constructed in Ukkonen's algorithm, each suffix is inserted as a category 2 suffix exactly once. Taken together, these suffix insertions can be thought of as similar to McCreight's suffix insertions. Essentially the same techniques will give linear run-time for these suffix insertions.

Consider T_i and let $s[j..i]$ be the longest suffix of $s[1..i]$ that is repeated elsewhere in it. This is actually realized while attempting to insert $s[j..i]$ while transitioning to T_i. The entire suffix $s[j..i]$ would be found within an already existing root to leaf path. This is a signal that T_i is already constructed. As we transition to T_{i+1}, the first suffix to insert is $s[j..i+1]$. Note that we are already at the end of the downward path from the root corresponding to $s[j..i]$. If the path can continue with $s[i+1]$, there are no category 2 insertions that need to be made. Otherwise, an internal node v is created at the end of $s[j..i]$ unless such a node v already exists, and a leaf attached to it using the edge label that start at $s[i+1]$ and ends at the current end of the string. Then, the next suffix is inserted as in McCreight's algorithm by taking suffix link from v if v was present beforehand, or by walking up to v's parent and taking suffix link from it if v was newly created and hence missing a suffix link from it. This process of inserting consecutive suffixes is carried out until one finds a suffix $s[j'..i+1]$ that is already represented, or the last suffix of the current string $s[i+1]$ is inserted.

The mechanism for moving from one suffix to the next is identical to the process described in McCreight's algorithm. Simply walk up from the current insertion point until the first node with a suffix link is reached, take the suffix link, walk down using type I traversal for the guaranteed portion of the match, and continue with type II traversal from that point on. Suffix links are also created during the execution of the algorithm as in McCreight's. Another way to view this algorithm is in terms of two shifting pointers j and i. The first pointer points to a suffix being inserted and the second pointer points to the current end of the string. If the suffix needs to be inserted, j is incremented by 1 to insert the next suffix. If the suffix is already found, i is incremented by 1 to switch to the next larger prefix of the string. As we advanced one of the pointers by 1 and the total length of the string is n, the number of steps before the two pointers sweep all the indices is at most $2n$. All of the suffix insertions together take $O(n)$ time, giving the algorithm a linear run-time.

McCreight's algorithm and Ukkonen's algorithm are suitable for constant sized alphabets. The dependence of the run-time and space for storing suffix trees on the size of the alphabet $|\Sigma|$ is as follows: A simple way to allocate space for internal nodes in a suffix tree is to allocate $|\Sigma| + 1$ pointers for children, one for each distinct character with which an edge label may begin. With this approach, the edge label beginning with a given character, or whether an edge label exists with a given character, can be determined in $O(1)$ time. However, as all $|\Sigma| + 1$ pointers are kept irrespective of how many children actually exist, the total space is $O(|\Sigma|n)$. If the tree is stored such that each internal node points only to its leftmost child and each node also points to its next sibling, if any, the space can be reduced to $O(n)$, irrespective of $|\Sigma|$. With this, searching for a child connected by an edge label with the appropriate character takes $O(|\Sigma|)$ time. Thus, McCreight's algorithm can be run in $O(n)$ time using $O(n|\Sigma|)$ space, or in $O(n|\Sigma|)$ time using $O(n)$ space. It is possible to obtain $O(n \log |\Sigma|)$ time with $O(n)$ space using an ordered list of pointers at each internal node. However, this is unlikely to be faster in practice, especially for small alphabet sizes such as for DNA and proteins.

5.4.3 Generalized Suffix Trees

The above linear time algorithms can be easily adapted to build the generalized suffix tree for a set $\mathcal{S} = \{s_1, s_2, \ldots, s_k\}$ of strings of total length N in $O(N)$ time. A simple way to do this is to construct the string $S = s_1\$_1 s_2\$_2 \ldots s_k\$_k$, where each $\$_i$ is a unique string termination character that does not occur in any string in $\mathcal{S}$. Using a linear time algorithm, $ST(S)$ can be computed in $O(N)$ time. This differs from $GST(\mathcal{S})$ in the following way: Consider a suffix $suff_j$ of string s_i in $GST(\mathcal{S})$. The corresponding suffix in $ST(S)$ is $s_i[j..|s_i|]\$_i s_{i+1}\$_{i+1} \ldots s_k\$_k$. Let v be the parent of the leaf representing this suffix in $ST(S)$. As each $\$_i$ is unique and $path\text{-}label(v)$ must be a common prefix of at least two suffixes in S, $path\text{-}label(v)$ must be a prefix of $s_i[j..|s_i|]$. Thus, by simply shortening the edge label below v to terminate at the end of the string s_i and attaching a common termination character $\$$ to it, the corresponding suffix in $GST(\mathcal{S})$ can be generated in $O(1)$ time. Additionally, all suffixes in $ST(S)$ that start with some $\$_i$ should be removed and replaced by a single suffix $\$$ in $GST(\mathcal{S})$. Note that the suffixes to be removed are all directly connected to the root in $ST(S)$, allowing easy $O(1)$ time removal per suffix. Thus, $GST(\mathcal{S})$ can be derived from

$ST(S)$ in $O(N)$ time.

Instead of first constructing $ST(S)$ and shortening edge labels of edges connecting to leaves to construct $GST(S)$, the process can be integrated into the tree construction itself to directly compute $GST(S)$. We will explain this in the context of using McCreight's algorithm. When inserting the suffix of a string, directly set the edge label connecting to the newly created leaf to terminate at the end of the string, appended by $. As each suffix that begins with $\$_i$ in $ST(S)$ is directly attached to the root, execution of McCreight's algorithm on S will always result in a downward traversal starting from the root when a suffix starting from the first character of a string is being inserted. Thus, we can simply start with an empty tree, insert all the suffixes of one string using McCreight's algorithm, insert all the suffixes of the next string, and continue this procedure until all strings are inserted. To insert the first suffix of a string, start by matching the unique path in the current tree that matches with a prefix of the string until no more matches are possible, and insert the suffix by branching at this point. To insert the remaining suffixes, continue as described in constructing the tree for one string.

This procedure immediately gives an algorithm to maintain the generalized suffix tree of a set of strings in the presence of insertions and deletions of strings. Insertion of a string is the same as executing McCreight's algorithm on the current tree, and takes time proportional to the length of the string being inserted. To delete a string, we must locate the leaves corresponding to all the suffixes of the string. By mimicking the process of inserting the string in GST using McCreight's algorithm, all the corresponding leaf nodes can be reached in time linear in the size of the string to be deleted. To delete a suffix, examine the corresponding leaf. If it is multiply labeled, it is enough to remove the label corresponding to the suffix. It it has only one label, the leaf and edge leading to it must be deleted. If the parent of the leaf is left with only one child after deletion, the parent and its two incident edges are deleted by connecting the surviving child directly to its grandparent with an edge labeled with the concatenation of the labels of the two edges deleted. As the adjustment at each leaf takes $O(1)$ time, the string can be deleted in time proportional to its length.

Suffix trees were invented by Weiner [29], who also presented the first linear time algorithm to construct them for a constant sized alphabet. McCreight's algorithm is a more space-economical linear time construction algorithm [26]. A linear time on-line construction algorithm for suffix trees was invented by Ukkonen [28]. In fact, our presentation of McCreight's algorithm also draws from ideas developed by Ukkonen. A unified view of these three suffix tree construction algorithms is studied by Giegerich and Kurtz [10]. Farach [6] presented the first linear time algorithm for strings over integer alphabets. The algorithm recursively constructs suffix trees for all odd and all even suffixes, respectively, and uses a clever strategy for merging them. The complexity of suffix tree construction algorithms for various types of alphabets is explored in [7].

5.5 Linear Time Construction of Suffix Arrays

Suffix arrays were proposed by Manber and Myers [25] as a space-efficient alternative to suffix trees. While suffix arrays can be deduced from suffix trees, which immediately implies any of the linear time suffix tree construction algorithms can be used for suffix arrays, it would not achieve the purpose of economy of space. Until recently, the fastest known direct construction algorithms for suffix arrays all required $O(n \log n)$ time, leaving a frustrating gap between asymptotically faster construction algorithms for suffix trees, and asymptotically slower construction algorithms for suffix arrays, despite the fact that suffix trees contain all the information in suffix arrays. This gap is successfully closed by a number of researchers in 2003, including Kärkkäinen and Sanders [19], Kim *et al.* [21], and Ko and Aluru [22, 23]. All three algorithms work for the case of integer alphabet. Given the simplicity and/or space efficiency of some of these algorithms, it is now preferable to construct suffix trees via the construction of suffix arrays.

5.5.1 Kärkkäinen and Sanders' Algorithm

Kärkkäinen and Sanders' algorithm is the simplest and most elegant algorithm to date to construct suffix arrays, and by implication suffix trees, in linear time. The algorithm also works for the case of an integer alphabet. Let s be a string of length n over the alphabet $\Sigma = \{1, 2, \ldots, n\}$. For convenience, assume n is a multiple of three and $s[n+1] = s[n+2] = 0$. The algorithm has the following steps:

1. Recursively sort the $\frac{2}{3}n$ suffixes $suff_i$ with $i \bmod 3 \neq 0$.
2. Sort the $\frac{1}{3}n$ suffixes $suff_i$ with $i \bmod 3 = 0$ using the result of step (1).
3. Merge the two sorted arrays.

To execute step (1), first perform a radix sort of the $\frac{2}{3}n$ triples $(s[i], s[i+1], s[i+2])$ for each $i \bmod 3 \neq 0$ and associate with each distinct triple its rank $\in \{1, 2, \ldots, \frac{2}{3}n\}$ in sorted order. If all triples are distinct, the suffixes are already sorted. Otherwise, let $suff_i'$ denote the string obtained by taking $suff_i$ and replacing each consecutive triplet with its corresponding rank. Create a new string s' by concatenating $suff_1'$ with $suff_2'$. Note that all $suff_i'$ with $i \bmod 3 = 1$ ($i \bmod 3 = 2$, respectively) are suffixes of $suff_1'$ ($suff_2'$, respectively). A lexicographic comparison of two suffixes in s' never crosses the boundary between $suff_1'$ and $suff_2'$ because the corresponding suffixes in the original string can be lexicographically distinguished. Thus, sorting s' recursively gives the sorted order of $suff_i$ with $i \bmod 3 \neq 0$.

Step (2) can be accomplished by performing a radix sort on tuples $(s[i], rank(suff_{i+1}))$ for all $i \bmod 3 = 0$, where $rank(suff_{i+1})$ denotes the rank of $suff_{i+1}$ in sorted order obtained in step (1).

Merging of the sorted arrays created in steps (1) and (2) is done in linear time, aided by the fact that the lexicographic order of a pair of suffixes, one from each array, can be determined in constant time. To compare $suff_i$ ($i \bmod 3 = 1$) with $suff_j$ ($i \bmod 3 = 0$), compare $s[i]$ with $s[j]$. If they are unequal, the answer is clear. If they are identical, the ranks of $suff_{i+1}$ and $suff_{j+1}$ in the sorted order obtained in step (1) determines the answer. To compare $suff_i$ ($i \bmod 3 = 2$) with $suff_j$ ($i \bmod 3 = 0$), compare the first two characters of the two suffixes. If they are both identical, the ranks of $suff_{i+2}$ and $suff_{j+2}$ in the sorted order obtained in step (1) determines the answer.

The run-time of this algorithm is given by the recurrence $T(n) = T\left(\lceil \frac{2n}{3} \rceil\right) + O(n)$, which results in $O(n)$ run-time. Note that the $\frac{2}{3} : \frac{1}{3}$ split is designed to make the merging step easy. A $\frac{1}{2} : \frac{1}{2}$ split does not allow easy merging because when comparing two suffixes for merging, no matter how many characters are compared, the remaining suffixes will not fall in the same sorted array, where ranking determines the result without need for further comparisons. Kim *et al.*'s linear time suffix array construction algorithm is based on a $\frac{1}{2} : \frac{1}{2}$ split, and the merging phase is handled in a clever way so as to run in linear time. This is much like Farach's algorithm for constructing suffix trees [6] by constructing suffix trees for even and odd positions separately and merging them. Both the above linear time suffix array construction algorithms partition the suffixes based on their starting positions in the string. A more detailed account of Kärkkäinen and Sanders' Algorithm including pseudocode and an example suffix array construction, along with application of this algorithm to construct suffix arrays on disks can be found in Chapter 35.

s	M	I	S	S	I	S	S	I	P	P	I	\$
Type	L	S	L	L	S	L	L	S	L	L	L	L/S
Pos	1	2	3	4	5	6	7	8	9	10	11	12

FIGURE 5.6: The string CATTATTAGGA\$ and the types of its suffixes.

5.5.2 Ko and Aluru's Algorithm

A completely different way of partitioning suffixes based on the lexicographic ordering of a suffix with its right neighboring suffix in the string is used by Ko and Aluru to derive a linear time algorithm [22, 23]. Consider a string s of size n over the alphabet $\Sigma = \{1 \ldots n\}$. As before, we use '\$' to denote the last character of s, considered unique and lexicographically the smallest. For strings α and β, we use $\alpha \prec \beta$ to denote that α is lexicographically smaller than β.

A high level overview of the algorithm is as follows: The suffixes are classified into two types, S and L. Suffix $suff_i$ is of type S if $suff_i \prec suff_{i+1}$, and is of type L if $suff_{i+1} \prec suff_i$. The last suffix $suff_n$ is classified as both type S and type L. The positions of the type S suffixes partition the string into a set of substrings. We substitute each of these substrings by its rank among all the substrings and produce a new string s'. The suffixes of the new string are then recursively sorted. The suffix array of s' gives the lexicographic order of all type S suffixes. The lexicographic order of all suffixes can be deduced from this order.

The first step of the algorithm is to classify suffixes into types S and L. Consider $suff_i$ $(i < n)$.

- If $s[i] < s[i+1]$, $suff_i$ is of type S.
- If $s[i] > s[i+1]$, $suff_i$ is of type L.
- If $s[i] = s[i+1]$, find the smallest $j > i$ such that $s[j] \neq s[i]$. If $s[j] > s[i]$, then $suff_i, suff_{i+1}, \ldots, suff_{j-1}$ are of type S. Otherwise, they are all of type L.

Thus, all suffixes can be classified using a left to right scan of s in $O(n)$ time. The type of each suffix of the string CATTATTAGGA\$ is shown in Figure 5.6.

LEMMA 5.4   A type S suffix is lexicographically greater than a type L suffix that begins with the same first character.

Proof   Let $suff_i$ be type S and $suff_j$ be type L such that $s[i] = s[j] = c$. We can write $suff_i = c^k c_1 \alpha$ and $suff_j = c^l c_2 \beta$, where c^k and c^l denote the character c repeated for $k, l > 0$ times, respectively, $c_1 > c$, $c_2 < c$, and α and β are (possibly empty) strings.

  Case 1: If $k < l$, c_1 is compared to a character c in c^l. Then $c_1 > c \Rightarrow suff_j \prec suff_i$.

  Case 2: If $k > l$, c_2 is compared to a character c in c^k. Then $c > c_2 \Rightarrow suff_j \prec suff_i$.

  Case 3: If $k = l$ then c_1 is compared to c_2. Since $c_1 > c$ and $c > c_2$, then $c_1 > c_2 \Rightarrow suff_j \prec suff_i$.

It follows that in the suffix array of s, among all suffixes that start with the same character, the type S suffixes appear after the type L suffixes.

Let A be an array containing all suffixes of s, not necessarily in sorted order. Let B be an array of all suffixes of type S, sorted in lexicographic order. Using B, the lexicographic

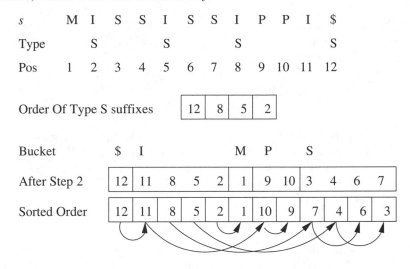

FIGURE 5.7: Illustration of how to obtain the sorted order of all suffixes, from the sorted order of type S suffixes of the string CATTATTAGGA\$.

sorted order of all suffixes of s can be computed as follows:

1. Bucket all suffixes of s according to their first character in array A in $O(n)$ time.

2. Scan B from right to left. For each suffix encountered in the scan, move the suffix to the current end of its bucket in A, and advance the current end by one position to the left. More specifically, the move of a suffix in array A to a new position should be taken as swapping the suffix with the suffix currently occupying the new position. After completion of the scan, all type S suffixes are in their correct positions in A. The time taken is $O(|B|)$, which is bounded by $O(n)$.

3. Scan A from left to right. For each entry $A[i]$, if $suff_{A[i]-1}$ is a type L suffix, move it to the current front of its bucket in A, and advance the front of the bucket by one. This takes $O(n)$ time. At the end of this step, A contains all suffixes of s in sorted order.

In Figure 5.7, the suffix pointed by the arrow is moved to the current front of its bucket when the scan reaches the suffix at the origin of the arrow. The following lemma proves the correctness of the procedure in Step 3.

LEMMA 5.5 In step 3, when the scan reaches $A[i]$, suffix $suff_{A[i]}$ is already in its sorted position in A.

Proof By induction on i. To begin with, the smallest suffix in s must be of type S and hence in its correct position $A[1]$. By inductive hypothesis, assume that $A[1], A[2], \ldots, A[i]$ are the first i suffixes in sorted order. We now show that when the scan reaches $A[i+1]$, then the suffix in it, i.e., $suff_{A[i+1]}$ is already in its sorted position. Suppose not. Then there exists a suffix referenced by $A[k]$ ($k > i+1$) that should be in $A[i+1]$ in sorted order, i.e., $suff_{A[k]} \prec suff_{A[i+1]}$. As all type S suffixes are already in correct positions, both $suff_{A[k]}$ and $suff_{A[i+1]}$ must be of type L. Because A is bucketed by the first character of the suffixes prior to step 3, and a suffix is never moved out of its bucket, $suff_{A[k]}$ and $suff_{A[i+1]}$ must begin with the same character, say c. Let $suff_{A[i+1]} = c\alpha$ and $suff_{A[k]} = c\beta$. Since $suff_{A[k]}$ is type L, $\beta \prec suff_{A[k]}$. From $suff_{A[k]} \prec suff_{A[i+1]}$, $\beta \prec \alpha$. Since $\beta \prec suff_{A[k]}$, and the correct sorted position of $suff_{A[k]}$ is $A[i+1]$, β must occur in $A[1] \ldots A[i]$. Because $\beta \prec \alpha$, $suff_{A[k]}$ should have been moved to the current front of its bucket before $suff_{A[i+1]}$. Thus, $suff_{A[k]}$ can not occur to the right of $suff_{A[i+1]}$, a contradiction.

So far, we showed that if all type S suffixes are sorted, then the sorted position of all suffixes of s can be deduced in $O(n)$ time. A similar result can also be obtained by sorting all suffixes of type L: First bucket all suffixes of s based on their first characters into an array A. Scan the sorted order of type L suffixes from left to right and determine their correct positions in A by moving them to the current front of their respective buckets. Subsequently, scan A from right to left and when $A[i]$ is encountered, if $suff_{A[i]-1}$ is of type S, move it to the current end of its bucket. Since the suffix array of s can be deduced either from sorting all type S suffixes, or from sorting all type L suffixes, it is advantageous to choose the type which has fewer suffixes. Without loss of generality, assume there are fewer type S suffixes. We now show how to recursively sort these suffixes.

Define position i of s to be a type S position if $suff_i$ is of type S, and similarly to be a type L position if $suff_i$ is of type L. The substring $s[i..j]$ is called a type S substring if both i and j are type S positions, and every position in between is a type L position.

Our goal is to sort all type S suffixes in s. To do this we first sort all the type S substrings. The sorting generates buckets where all the substrings in a bucket are identical. The buckets are numbered using consecutive integers starting from 1. We then generate a new string s' as follows: Scan s from left to right and for each type S position in s, write the bucket number of the type S substring starting from that position. This string of bucket numbers forms s'. Observe that each type S suffix in s naturally corresponds to a suffix in the new string s'. In Lemma 5.6, we prove that sorting all type S suffixes of s is equivalent to sorting all suffixes of s'. We sort s' recursively.

We first show how to sort all the type S substrings in $O(n)$ time. Consider the array A, consisting of all suffixes of s bucketed according to their first characters. For each suffix $suff_i$, define its *S-distance* to be the distance from its starting position i to the nearest type S position to its left (excluding position i). If no type S position exists to the left, the *S-distance* is defined to be 0. Thus, for each suffix starting on or before the first type S position in s, its *S-distance* is 0. The type S substrings are sorted as follows (illustrated in Figure 5.8):

1. For each suffix in A, determine its *S-distance*. This is done by scanning s from left to right, keeping track of the distance from the current position to the nearest type S position to the left. While at position i, the *S-distance* of $suff_i$ is known and this distance is recorded in array $Dist$. The *S-distance* of $suff_i$ is stored in $Dist[i]$. Hence, the *S-distances* for all suffixes can be recorded in linear time.

2. Let m be the largest *S-distance*. Create m lists such that list j ($1 \le j \le m$) contains all the suffixes with an *S-distance* of j, listed in the order in which they appear in array A. This can be done by scanning A from left to right in linear time, referring to $Dist[A[i]]$ to put $suff_{A[i]}$ in the correct list.

3. We now sort the type S substrings using the lists created above. The sorting is done by repeated bucketing using one character at a time. To begin with, the bucketing based on first character is determined by the order in which type S suffixes appear in array A. Suppose the type S substrings are bucketed according to their first $j - 1$ characters. To extend this to j characters, we scan list j. For

s	M	I	S	S	I	S	S	I	P	P	I	$
Type		S		S			S					S
Pos	1	2	3	4	5	6	7	8	9	10	11	12

A | 12 | 2 5 8 11 | 1 | 9 10 | 3 4 6 7 |

Step 1. Record the S-distances

Pos	1	2	3	4	5	6	7	8	9	10	11	12
Dist	0	0	1	2	3	1	2	3	1	2	3	4

Step 2. Construct S-distance Lists

1 | 9 | 3 6 |

2 | 10 | 4 7 |

3 | 5 8 11 |

4 | 12 |

Step 3. Sort all type S substring

Original

| 12 | 2 5 8 |

Sort according to list 1

| 12 | 8 | 2 5 |

Sort according to list 2

| 12 | 8 | 2 5 |

Sort according to list 3

| 12 | 8 | 2 5 |

Sort according to list 4

| 12 | 8 | 2 5 |

FIGURE 5.8: Illustration of the sorting of type S substrings of the string CATTATTAGGA$.

each suffix $suff_i$ encountered in the scan of a bucket of list j, move the type S substring starting at $s[i - j]$ to the current front of its bucket, then move the current front to the right by one. After a bucket of list j is scanned, new bucket boundaries need to be drawn between all the type S substrings that have been moved, and the type S substrings that have not been moved. Because the total size of all the lists is $O(n)$, the sorting of type S substrings only takes $O(n)$ time.

The sorting of type S substrings using the above algorithm respects lexicographic ordering of type S substrings, with the following important exception: If a type S substring is the prefix of another type S substring, the bucket number assigned to the shorter substring will be larger than the bucket number assigned to the larger substring. This anomaly is designed on purpose, and is exploited later in Lemma 5.6.

As mentioned before, we now construct a new string s' corresponding to all type S substrings in s. Each type S substring is replaced by its bucket number and s' is the sequence of bucket numbers in the order in which the type S substrings appear in s. Because every type S suffix in s starts with a type S substring, there is a natural one-to-one correspondence between type S suffixes of s and all suffixes of s'. Let $suff_i$ be a suffix of s and $suff'_{i'}$ be its corresponding suffix in s'. Note that $suff'_{i'}$ can be obtained from $suff_i$ by replacing every type S substring in $suff_i$ with its corresponding bucket number. Similarly, $suff_i$ can be obtained from $suff'_{i'}$ by replacing each bucket number with the corresponding substring and removing the duplicate instance of the common character shared by two consecutive type S substrings. This is because the last character of a type S substring is also the first character of the next type S substring along s.

LEMMA 5.6   Let $suff_i$ and $suff_j$ be two suffixes of s and let $suff'_{i'}$ and $suff'_{j'}$ be the corresponding suffixes of s'. Then, $suff_i \prec suff_j \Leftrightarrow suff'_{i'} \prec suff'_{j'}$.

Proof   We first show that $suff'_{i'} \prec suff'_{j'} \Rightarrow suff_i \prec suff_j$. The prefixes of $suff_i$ and $suff_j$

corresponding to the longest common prefix of $suff'_{i'}$ and $suff'_{j'}$ must be identical. This is because if two bucket numbers are the same, then the corresponding substrings must be the same. Consider the leftmost position in which $suff'_{i'}$ and $suff'_{j'}$ differ. Such a position exists and the characters (bucket numbers) of $suff'_{i'}$ and $suff'_{j'}$ in that position determine which of $suff'_{i'}$ and $suff'_{j'}$ is lexicographically smaller. Let k be the bucket number in $suff'_{i'}$ and l be the bucket number in $suff'_{j'}$ at that position. Since $suff'_{i'} \prec suff'_{i'}$, it is clear that $k < l$. Let α be the substring corresponding to k and β be the substring corresponding to l. Note that α and β can be of different lengths, but α cannot be a proper prefix of β. This is because the bucket number corresponding to the prefix must be larger, but we know that $k < l$.

Case 1: β is not a prefix of α. In this case, $k < l \Rightarrow \alpha \prec \beta$, which implies $suff_i \prec suff_j$.

Case 2: β is a proper prefix of α. Let the last character of β be c. The corresponding position in s is a type S position. The position of the corresponding c in α must be a type L position. Since the two suffixes that begin at these positions start with the same character, the type L suffix must be lexicographically smaller then the type S suffix. Thus, $suff_i \prec suff_j$.

From the one-to-one correspondence between the suffixes of s' and the type S suffixes of s, it also follows that $suff_i \prec suff_j \Rightarrow suff'_{i'} \prec suff'_{j'}$.

From the above lemma, the sorted order of the suffixes of s' determines the sorted order of the type S suffixes of s. Hence, the problem of sorting the type S suffixes of s reduces to the problem of sorting all suffixes of s'. Note that the characters of s' are consecutive integers starting from 1. Hence the suffix sorting algorithm can be recursively applied to s'.

If s has fewer type L suffixes than type S suffixes, the type L suffixes are sorted using a similar procedure − Call $s[i..j]$ a type L substring if both i and j are type L positions, and every position in between is a type S position. Now sort all the type L substrings and construct the corresponding string s' obtained by replacing each type L substring with its bucket number. Sorting s' gives the sorted order of type L suffixes.

Thus, the problem of sorting the suffixes of a string s of length n can be reduced to the problem of sorting the suffixes of a string s' of size at most $\lceil \frac{n}{2} \rceil$, and $O(n)$ additional work. This leads to the recurrence $T(n) = T\left(\lceil \frac{n}{2} \rceil\right) + O(n)$, resulting in $O(n)$ run time. The algorithm can be made to run in only $2n$ words plus $1.25n$ bits for strings over constant alphabet [23]. Algorithmically, Kärkkäinen and Sanders' algorithm is akin to mergesort and Ko and Aluru's algorithm is akin to quicksort.

It may be more space efficient to construct a suffix tree by first constructing the corresponding suffix array, deriving the Lcp array from it, and using both to construct the suffix tree. For example, while all direct linear time suffix tree construction algorithms depend on constructing and using suffix links, these are completely avoided in the indirect approach. Furthermore, the resulting algorithms have an alphabet independent run-time of $O(n)$ while using only the $O(n)$ space representation of suffix trees. This should be contrasted with the $O(|\Sigma|n)$ run-time of either McCreight's or Ukkonen's algorithms.

5.6 Space Issues

Suffix trees and suffix arrays are space efficient in an asymptotic sense because the memory required grows linearly with input size. However, the actual space usage is of significant concern, especially for very large strings. For example, the human genome can be represented as a large string over the alphabet $\Sigma = \{A,C,G,T\}$ of length over 3×10^9. Because of linear dependence of space on the length of the string, the exact space requirement is easily characterized by specifying it in terms of the number of bytes per character. Depending on the number of bytes per character required, a data structure for the human genome may fit in main memory, may need a moderate sized disk, or might need a large amount of secondary storage. This has significant influence on the run-time of an application as access to secondary storage is considerably slower. It may also become impossible to run an application for large data sizes unless careful attention is paid to space efficiency.

Consider a naive implementation of suffix trees. For a string of length n, the tree has n leaves, at most $n - 1$ internal nodes, and at most $2n - 2$ edges. For simplicity, count the space required for each integer or a pointer to be one word, equal to 4 bytes on most current computers. For each leaf node, we may store a pointer to its parent, and store the starting index of the suffix represented by the leaf, for $2n$ words of storage. Storage for each internal node may consist of 4 pointers, one each for parent, leftmost child, right sibling and suffix link, respectively. This will require approximately $4n$ words of storage. Each edge label consists of a pair of integers, for a total of at most $4n$ words of storage. Putting this all together, a naive implementation of suffix trees takes $10n$ words or $40n$ bytes of storage.

Several techniques can be used to considerably reduce the naive space requirement of 40 bytes per character. Many applications of interest do not need to use suffix links. Similarly, a pointer to the parent may not be required for applications that use traversals down from the root. Even otherwise, note that a depth first search traversal of the suffix tree starting from the root can be conducted even in the absence of parent links, and this can be utilized in applications where a bottom-up traversal is needed. Another technique is to store the internal nodes of the tree in an array in the order of their first occurrence in a depth first search traversal. With this, the leftmost child of an internal node is found right next to it in the array, which removes the need to store a child pointer. Instead of storing the starting and ending positions of a substring corresponding to an edge label, an edge label can be stored with the starting position and length of the substring. The advantage of doing so is that the length of the edge label is likely to be small. Hence, one byte can be used to store edge labels with lengths < 255 and the number 255 can be used to denote edge labels with length at least 255. The actual values of such labels can be stored in an exceptions list, which is expected to be fairly small. Using several such techniques, the space required per character can be roughly cut in half to about 20 bytes [24].

A suffix array can be stored in just one word per character, or 4 bytes. Most applications using suffix arrays also need the *Lcp* array. Similar to the technique employed in storing edge labels on suffix trees, the entries in *Lcp* array can also be stored using one byte, with exceptions handled using an ordered exceptions list. Provided most of the *lcp* values fit in a byte, we only need 5 bytes per character, significantly smaller than what is required for suffix trees. Further space reduction can be achieved by the use of compressed suffix trees and suffix arrays and other data structures [8, 11]. However, this often comes at the expense of increased run-time complexity.

5.7 Lowest Common Ancestors

Consider a string s and two of its suffixes $suff_i$ and $suff_j$. The longest common prefix of the two suffixes is given by the path label of their lowest common ancestor. If the string-depth of each node is recorded in it, the length of the longest common prefix can be retrieved from the lowest common ancestor. Thus, an algorithm to find the lowest common ancestors quickly can be used to determine longest common prefixes without a single character comparison. In this section, we describe how to preprocess the suffix tree in linear time and be able to answer lowest common ancestor queries in constant time [4].

5.7.1 Bender and Farach's *lca* algorithm

Let T be a tree of n nodes. Without loss of generality, assume the nodes are numbered $1 \ldots n$. Let $lca(i, j)$ denote the lowest common ancestor of nodes i and j. Bender and Farach's algorithm performs a linear time preprocessing of the tree and can answer *lca* queries in constant time.

Let E be an Euler tour of the tree obtained by listing the nodes visited in a depth first search of T starting from the root. Let L be an array of level numbers such that $L[i]$ contains the tree-depth of the node $E[i]$. Both E and L contain $2n - 1$ elements and can be constructed by a depth first search of T in linear time. Let R be an array of size n such that $R[i]$ contains the index of the first occurrence of node i in E. Let $RMQ_A(i, j)$ denote the position of an occurrence of the smallest element in array A between indices i and j (inclusive). For nodes i and j, their lowest common ancestor is the node at the smallest tree-depth that is visited between an occurrence of i and an occurrence of j in the Euler tour. It follows that

$$lca(i, j) = E[RMQ_L(R[i], R[j])]$$

Thus, the problem of answering *lca* queries transforms into answering range minimum queries in arrays. Without loss of generality, we henceforth restrict our attention to answering range minimum queries in an array A of size n.

To answer range minimum queries in A, do the following preprocessing: Create $\lfloor \log n \rfloor + 1$ arrays $B_0, B_1, \ldots, B_{\lfloor \log n \rfloor}$ such that $B_j[i]$ contains $RMQ_A(i, i + 2^j)$, provided $i + 2^j \leq n$. B_0 can be computed directly from A in linear time. To compute $B_l[i]$, use $B_{l-1}[i]$ and $B_{l-1}[i + 2^{l-1}]$ to find $RMQ_A(i, i + 2^{l-1})$ and $RMQ_A(i + 2^{l-1}, i + 2^l)$, respectively. By comparing the elements in A at these locations, the smallest element in the range $A[i..i + 2^l]$ can be determined in constant time. Using this method, all the $\lfloor \log n \rfloor + 1$ arrays are computed in $O(n \log n)$ time.

Given an arbitrary range minimum query $RMQ_A(i, j)$, let k be the largest integer such that $2^k \leq (j - i)$. Split the range $[i..j]$ into two overlapping ranges $[i..i + 2^k]$ and $[j - 2^k..j]$. Using $B_k[i]$ and $B_k[j - 2^k]$, a smallest element in each of these overlapping ranges can be located in constant time. This will allow determination of $RMQ_A(i, j)$ in constant time. To avoid a direct computation of k, the largest power of 2 that is smaller than or equal to each integer in the range $[1..n]$ can be precomputed and stored in $O(n)$ time. Putting all of this together, range minimum queries can be answered with $O(n \log n)$ preprocessing time and $O(1)$ query time.

The preprocessing time is reduced to $O(n)$ as follows: Divide the array A into $\frac{2n}{\log n}$ blocks of size $\frac{1}{2} \log n$ each. Preprocess each block such that for every pair (i, j) that falls within a block, $RMQ_A(i, j)$ can be answered directly. Form an array B of size $\frac{2n}{\log n}$ that contains the minimum element from each of the blocks in A, in the order of the blocks in A, and record the locations of the minimum in each block in another array C. An arbitrary query $RMQ_A(i, j)$ where i and j do not fall in the same block is answered as follows: Directly find the location of the minimum in the range from i to the end of the block containing it, and also in the range from the beginning of the block containing j to index j. All that remains is to find the location of the minimum in the range of blocks completely contained between i and j. This is done by the corresponding range minimum query in B and using C to find the location in A of the resulting smallest element. To answer range queries in B, B is preprocessed as outlined before. Because the size of B is only $O\left(\frac{n}{\log n}\right)$, preprocessing

B takes $O\left(\frac{n}{\log n} \times \log \frac{n}{\log n}\right) = O(n)$ time and space.

It remains to be described how each of the blocks in A is preprocessed to answer range minimum queries that fall within a block. For each pair (i, j) of indices that fall in a block, the corresponding range minimum query is precomputed and stored. This requires computing $O(\log^2 n)$ values per block and can be done in $O(\log^2 n)$ time per block. The total run-time over all blocks is $\frac{2n}{\log n} \times O(\log^2 n) = O(n \log n)$, which is unacceptable. The run-time can be reduced for the special case where the array A contains level numbers of nodes visited in an Euler Tour, by exploiting its special properties. Note that the level numbers of consecutive entries differ by $+1$ or -1. Consider the $\frac{2n}{\log n}$ blocks of size $\frac{1}{2} \log n$. Normalize each block by subtracting the first element of the block from each element of the block. This does not affect the range minimum query. As the first element of each block is 0 and any other element differs from the previous one by $+1$ or -1, the number of distinct blocks is $2^{\frac{1}{2} \log n - 1} = \frac{1}{2}\sqrt{n}$. Direct preprocessing of the distinct blocks takes $\frac{1}{2}\sqrt{n} \times O(\log^2 n) = o(n)$ time. The mapping of each block to its corresponding distinct normalized block can be done in time proportional to the length of the block, taking $O(n)$ time over all blocks.

Putting it all together, a tree T of n nodes can be preprocessed in $O(n)$ time such that *lca* queries for any two nodes can be answered in constant time. We are interested in an application of this general algorithm to suffix trees. Consider a suffix tree for a string of length n. After linear time preprocessing, *lca* queries on the tree can be answered in constant time. For a given pair of suffixes in the string, the string-depth of their lowest common ancestor gives the length of their longest common prefix. Thus, the longest common prefix can be determined in constant time, without resorting to a single character comparison! This feature is exploited in many suffix tree algorithms.

5.7.2   Suffix Links from Lowest Common Ancestors

Suppose we are given a suffix tree and it is required to establish suffix links for each internal node. This may become necessary if the suffix tree creation algorithm does not construct suffix links but they are needed for an application of interest. For example, the suffix tree may be constructed via suffix arrays, completely avoiding the construction and use of suffix links for building the tree. The links can be easily established if the tree is preprocessed for *lca* queries.

Mark each internal node v of the suffix tree with a pair of leaves (i, j) such that leaves labeled i and j are in the subtrees of different children of v. The marking can be done in linear time by a bottom-up traversal of the tree. To find the suffix link from an internal node v (other than the root) marked with (i, j), note that $v = lca(i, j)$ and $lcp(suff_i, suff_j) = path\text{-}label(v)$. Let $path\text{-}label(v) = c\alpha$, where c is the first character and α is a string. To establish a suffix link from v, node u with path label α is needed. As $lcp(suff_{i+1}, suff_{j+1}) = \alpha$, node u is given by $lca(i+1, j+1)$, which can be determined in constant time. Thus, all suffix links can be determined in $O(n)$ time. This method trivially extends to the case of a generalized suffix tree.

5.8   Conclusions

In this chapter, we focused on linear time construction algorithms for the three most important data structures used in computational biology — lookup tables, suffix trees, and suffix arrays. Some references for further study on this topic are provided in the References section. Compressed suffix arrays, which are briefly mentioned in Section 5.6 can be stored in $O(n)$ bits; Hon *et al.* provided the first linear time construction algorithm [14] for this data structure. In recent years, the size of biological databases has grown rapidly. This generated considerable interest in constructing and maintaining suffix trees and suffix arrays in secondary storage [3, 17, 27]. For a more detailed study of string data structures on secondary storage, the reader is referred to Chapter 35 of the handbook. Some biological applications are data and compute intensive, e.g. genome assembly of complex eukaryotic organisms and clustering large scale expressed sequence tag data. Parallelism is increasingly being used to solve such problems effectively (for example, see [16, 18]). Farach *et al.* [7], Futamura *et al.* [9] and Hariharan [13] have all studied the construction of suffix arrays or suffix trees in parallel environments.

The next two chapters explore in detail how suffix trees and suffix arrays are being used to support applications in computational biology. A comprehensive treatise of suffix trees, suffix arrays and string algorithms can be found in the textbooks by Gusfield [12], and Crochemore and Rytter [5].

Acknowledgements

This work was supported in part by the U.S. National Science Foundation under IIS-0430853.

References

[1] S.F. Altschul, W. Gish, W. Miller, and E.W. Myers *et al.* Basic local alignment search tool. *Journal of Molecular Biology*, 215(3):403–410, 1990.

[2] S.F. Altschul, T.L. Madden, A.A. Schäffer, and J. Zhang *et al.* Gapped BLAST and PSI-BLAST: A new generation of protein database search programs. *Nucleic Acids Research*, 25:3389–3402, 1997.

[3] S.J. Bedathur and J.R. Haritsa. Engineering a fast online persistent suffix tree construction. In *Proc. 20th International Conference on Data Engineering*, pages 720–731, 2004.

[4] M.A. Bender and M. Farach-Colton. The LCA problem revisited. In *Proc. 4th Latin American Theoretical Informatics Symposium*, pages 88–94, 2000.

[5] M. Crochemore and W. Rytter. *Jewels of Stringology*. World Scientific Publishing Company, Singapore, 2002.

[6] M. Farach. Optimal suffix tree construction with large alphabets. In *Proc. 38th Annual Symposium on Foundations of Computer Science*, pages 137–143. IEEE, 1997.

[7] M. Farach-Colton, P. Ferragina, and S. Muthukrishnan. On the sorting-complexity of suffix tree construction. *Journal of the ACM*, 47(6):987–1011, 2000.

[8] P. Ferragina and G. Manzini. Opportunistic data structures with applications. In *Proc. 41th Annual Symposium on Foundations of Computer Science*, pages 390–398. IEEE, 2000.

[9] N. Futamura, S. Aluru, and S. Kurtz. Parallel suffix sorting. In *Proc. 9th International Conference on Advanced Computing and Communications*, pages 76–81, 2001.

[10] R. Giegerich and S. Kurtz. From Ukkonen to McCreight and Weiner: A unifying view of linear-time suffix tree construction. *Algorithmica*, 19:331–353, 1997.

[11] R. Grossi and J.S. Vitter. Compressed suffix arrays and suffix trees with applications to text indexing and string matching. In *Proc. 32nd annual ACM symposium on*

Theory of computing, pages 397–406. ACM, 2000.

[12] D. Gusfield. *Algorithms on Strings Trees and Sequences.* Cambridge University Press, New York, New York, 1997.

[13] R. Hariharan. Optimal parallel suffix tree construction. *Journal of Computer and System Sciences*, 55(1):44–69, 1997.

[14] W.K. Hon, K. Sadakane, and W.K. Sung. Breaking a time-and-space barrier in constructing full-text indices. In *Proc. 44th Annual IEEE Symposium on Foundations of Computer Science*, pages 251–260, 2003.

[15] X. Huang and A. Madan. CAP3: A DNA sequence assembly program. *Genome Research*, 9(9):868–877, 1999.

[16] X. Huang, J Wang, S Aluru, and S.P. Yang *et al.* Pcap: a whole-genome assembly program. *Genome Research*, 13(9):2164–2170, 2003.

[17] E. Hunt, M.P. Atkinson, and R.W. Irving. Database indexing for large DNA and protein sequence collections. *The VLDB Journal*, 11(3):256–271, 2002.

[18] A. Kalyanaraman, S. Aluru, V. Brendel, and S. Kothari. Space and time efficient parallel algorithms and software for EST clustering. *IEEE Transactions on Parallel and Distributed Systems*, 14(12):1209–1221, 2003.

[19] J. Kärkkäinen and P. Sanders. Simpler linear work suffix array construction. In *Proc. 30th International Colloquium on Automata, Languages and Programming*, pages 943–955, 2003.

[20] T. Kasai, G. Lee, H. Arimura, and S. Arikawa *et al.* Linear-time longest-common-prefix computation in suffix arrays and its applications. In *Proc. 12th Annual Symposium, Combinatorial Pattern Matching*, pages 181–192, 2001.

[21] D.K. Kim, J.S. Sim, H. Park, and K. Park. Linear-time construction of suffix arrays. In *Proc. 14th Annual Symposium, Combinatorial Pattern Matching*, pages 186–199, 2003.

[22] P. Ko and S. Aluru. Space-efficient linear-time construction of suffix arrays. In *Proc. 14th Annual Symposium, Combinatorial Pattern Matching*, pages 200–210, 2003.

[23] P. Ko and S. Aluru. Space efficient linear time construction of suffix arrays. *Journal of Discrete Algorithms*, 3:143–156, 2005.

[24] S. Kurtz. Reducing the space requirement of suffix trees. *Software - Practice and Experience*, 29(13):1149–1171, 1999.

[25] U. Manber and G. Myers. Suffix arrays: a new method for on-line search. *SIAM Journal on Computing*, 22:935–948, 1993.

[26] E.M. McCreight. A space-economical suffix tree construction algorithm. *Journal of the ACM*, 23:262–272, 1976.

[27] S. Tata, R.A. Hankins, and J.M. Patel. Practical suffix tree construction. In *Proc. 13th International Conference on Very Large Data Bases*, pages 36–47, 2004.

[28] E. Ukkonen. On-line construction of suffix-trees. *Algorithmica*, 14:249–60, 1995.

[29] P. Weiner. Linear pattern matching algorithms. In *Proc. 14th Symposium on Switching and Automata Theory*, pages 1–11. IEEE, 1973.

6

Suffix Tree Applications in Computational Biology

Pang Ko
Iowa State University

Srinivas Aluru
Iowa State University

6.1 Introduction

In recent years the volume of biological data has increased exponentially. Concomitantly, the speed with which such data is generated has increased as well. It is now possible to sequence a bacterial genome in a single day. Thus efficient data structures are needed to archive and retrieve biological data. Furthermore, this explosion of data has increased the need to analyze a large amount of data in a reasonable time. With the availability of complete genomes, researchers have begun to compare whole genomes [4, 11, 26, 27]. This further increases the scale of problems addressed, and algorithms that worked well for smaller scale problems are either insufficient or inappropriate. For example, dynamic programming techniques worked well to identify the matching regions between two genes. However, heuristics must be applied when we try to identify highly conserved regions between two genomes in reasonable time and space. Suffix trees can serve as an efficient data structure to analyze DNA and protein sequences. They can also be used to provide exact matches efficiently, which many heuristics depend on.

Computationally, both DNA and protein sequences can be modeled as strings of characters. But unlike natural languages where there are well-defined sentence structures and word boundaries, DNA and protein sequences have no such properties. This makes the traditional approaches of using inverted tables and hash tables less appealing. Suffix trees and generalized suffix trees, the multiple string variant of suffix trees, can be used to solve a number of computational biology related problems in optimal space and time. In this chapter we examine several applications of suffix trees in computational biology. For the most part, our focus will be on solving problems motivated by real applications in molecular biology. In many cases, the algorithms presented here are part of actual bioinformatic software programs developed, illustrating the practical role of suffix trees in computational biology research. We use the same terminology as in the previous chapter where suffix trees and suffix arrays are introduced.

6.2   Basic Applications

In this section, we provide a brief introduction to the pattern matching capabilities of suffix trees. Although pattern matching by itself may not directly correspond to many computational biology applications, it is a basic building block upon which many suffix tree algorithms are founded. Besides, the underlying ideas are frequently used as components within more complicated algorithms, and in some cases they are modified and used in software with vastly different objectives.

6.2.1   Pattern Matching

Given a pattern P and a text T, the pattern matching problem is to find all occurrences of P in T. Let $|P| = m$ and $|T| = n$. Typically, $n >> m$. Moreover, T remains fixed in many applications and the query is repeated for many different patterns. For example, T could be an entire database of DNA sequences and P denotes a substring of a query sequence for homology (similarity) search. Thus, it is beneficial to preprocess the text T so that queries can be answered as efficiently as possible.

The pattern matching problem can be solved in optimal $O(m + k)$ time using $ST(T)$, where k is the number of occurrences of P in T. Suppose P occurs in T starting from position i. Then, P is a prefix of $suff_i$ in T. It follows that P matches the path from root to leaf labeled i in ST. This property results in the following simple algorithm: Start from the root of ST and follow the path matching characters in P, until P is completely matched or a mismatch occurs. If P is not fully matched, it does not occur in T. Otherwise, each leaf in the subtree below the matching position gives an occurrence of P. The positions can be enumerated by traversing the subtree in time proportional to the size of the subtree. As the number of leaves in the subtree is k, this takes $O(k)$ time. If only one occurrence is of interest, the suffix tree can be preprocessed in $O(n)$ time such that each internal node contains the label of one of the leaves in its subtree. Thus, the problem of whether P occurs in T or the problem of finding one occurrence can be answered in $O(m)$ time.

6.2.2   Approximate Pattern Matching

The simpler version of approximate pattern matching problem is as follows: Given a pattern P ($|P| = m$) and a text T ($|T| = n$), find all substrings of length $|P|$ in T that match P with at most k mismatches. To solve this problem, first construct the GST of P and T. Preprocess the GST to record the string-depth of each node, and to answer *lca* queries in constant time. For each position i in T, we will determine if $T[i..i+m-1]$ matches P with at most k mismatches. First, use an lca query $lca((P,1),(T,i))$ to find the largest substring from position i of T that matches a substring from position 1 of P. Suppose the length of this longest exact match is l. Thus, $P[1..l] = T[i..i+l-1]$, and $P[l+1] \neq T[i+l]$. Count this as a mismatch and continue by finding $lca((P,l+2),(T,i+l+1))$. This procedure is continued until either the end of P is reached or the number of mismatches crosses k. As each lca query takes constant time, the entire procedures takes $O(k)$ time. This is repeated for each position i in T for a total run-time of $O(kn)$.

Now, consider the more general problem of finding the substrings of T that can be derived from P by using at most k character insertions, deletions or substitutions. To solve this problem, we proceed as before by determining the possibility of such a match for every starting position i in T. Let $l = string\text{-}depth(lca((P,1),(T,i)))$. At this stage, we consider three possibilities:

1. Substitution $-$ $P[l+1]$ and $T[i+l]$ are considered a mismatch. Continue by finding $lca((P,l+2),(T,i+l+1))$.
2. Insertion $-$ $T[i+l]$ is considered an insertion in P after $P[l]$. Continue by finding $lca((P,l+1),(T,i+l+1))$.
3. Deletion $-$ $P[l+1]$ is considered a deletion. Continue by finding $lca((P,l+2),(T,i+l))$.

After each lca computation, we have three possibilities corresponding to substitution, insertion and deletion, respectively. All possibilities are enumerated to find if there is a sequence of k or less operations that will transform P into a substring starting from position i in T. This takes $O(3^k)$ time. Repeating this algorithm for each position i in T takes $O(3^k n)$ time.

The above algorithm always uses the longest exact match possible from a given pair of positions in P and T before considering the possibility of an insertion or deletion. To prove the correctness of this algorithm, we show that if there is an approximate match of P starting from position i in T that does not use such a longest exact match, then there exists another approximate match that uses only longest exact matches. Consider an approximate match that does not use longest exact matches. Consider the leftmost position j in P and the corresponding position $i+l'$ in T where the longest exact match is violated. i.e., $P[j] = T[i+l']$ but this is not used as part of an exact match. Instead, an insertion or deletion is used. Suppose that an exact match of length r is used after the insertion or deletion. We can come up with a corresponding approximate match where the longest match is used and the insertion/deletion is taken after that. This will either keep the number of insertions/deletions the same or reduce the count. Thus, if the value of k is small, the above algorithms provide a quick and easy way to solve the approximate pattern matching problem. For sophisticated algorithms with better run-times, see [9, 30].

6.3 Restriction Enzyme Recognition Sites

Restriction endonucleases are enzymes that recognize a particular pattern in a DNA sequence and cleave the DNA at or near the recognition site. The enzyme typically cuts both strands of double stranded DNA and the recognition sequence is often a short sequence that is identical on both the strands. Recall that due to opposite directionality of the two strands, the sequences are read in opposite directions relative to each other. Thus, the recognition sequence is what is called a *complemented palindrome*; by reversing the sequence and using complementary substitutions A $\leftrightarrow$ T, and C $\leftrightarrow$ G, one would obtain the sequence itself. As an example, the restriction enzyme *SwaI* recognizes the site ATTTAAAT and cleaves it in the center of the pattern. The restriction enzyme *BamHI* detects the sequence GGATCC and cleaves it after the first base (and similarly after the first base in the complementary strand sequence; the first base in the complementary strand is paired with the last base in the original strand). Most restriction enzymes are derived from bacteria and are named after the organism in which they are first discovered. Restriction enzymes play a defense role by cleaving foreign DNA. The DNA of the host organism is protected by mythelation of its own recognition sites, which makes it immune to restriction enzyme activity.

Consider the problem of finding all complemented palindromic sequences in a given long DNA sequence. We focus on the problem of identifying all maximal complemented palindromes, as all other palindromes are contained in them. Formally, a substring $s[i..j]$ of a string s of length n is called a *maximal complemented palindrome* of s, if $s[i..j]$ is a complemented palindrome and $s[i-1]$ and $s[j+1]$ are not complementary bases, or $i = 1$, or $j = n$. Note that a maximal complemented palindrome must necessarily be of even length. For a palindrome of length $2k$, define the center to be the position between characters k and $k+1$ of the palindrome. The palindrome is said to be of radius k. Starting from the center, a complemented palindrome is a string that reads the same in both directions subject to complementarity. Observe that each maximal palindrome in a string must have a distinct center. As the number of possible centers for a string of length n is $n-1$, the total number of maximal palindromes of a string is $n-1$. All such palindromes can be identified in linear time using the following algorithm.

Let s^r denote the reverse complement of string s. Construct a GST of the strings s and s^r and preprocess the GST to record string depths of internal nodes and for answering lca queries. The maximal even length palindrome centered between $s[i]$ and $s[i+1]$ is given by the length of the longest common prefix between $suff_{i+1}$ of s and $suff_{n-i+1}$ of s^r. This is computed as the string-depth of $lca((s, i+1), (s^r, n-i+1))$ in constant time. Thus all maximal complemented palindromes can be recognized in $O(n)$ time. An example to illustrate this algorithm is presented in Figure 6.1. The figure shows the generalized suffix tree of the DNA sequence TAGAGCTCA and its reverse complement TGAGCTCTA.

6.4   Detection of RNAi Elements

RNA interference (RNAi) is a process that utilizes a double stranded RNA (dsRNA) molecule to inhibit the expression of a particular gene by binding to its mRNA. This process was first discovered by Fire *et al.* [14]. Since then, RNAi has been used as an alternative to gene knockout experiments. Unlike traditional experiments where a gene is permanently removed from the genome, researchers can choose when and where to introduce the dsRNA. This gives biologists greater flexibility in experimental design. It has been shown that RNAi is also used in cells as a way to regulate gene expression, and as a defense mechanism against viruses.

Unlike DNA molecules which are double stranded helixes, RNA molecules are usually single stranded and have a secondary structure that sometimes has important functions. Like all RNA molecules, naturally occurring RNAi elements are also produced by transcription from a corresponding genomic sequence. The transcription produces an RNA molecule that contains a sequence and its reverse complement separated by a short sequence. The reverse complementarity causes the sequences to bind to each other with the short sequence in the middle forming a stem-loop-stem structure. Cleaving of this structure results in dsRNA (see Figure 6.2). The resulting dsRNA will then interact with target messenger RNAs (mRNAs) to prevent them from being translated into proteins, thus controlling gene expression.

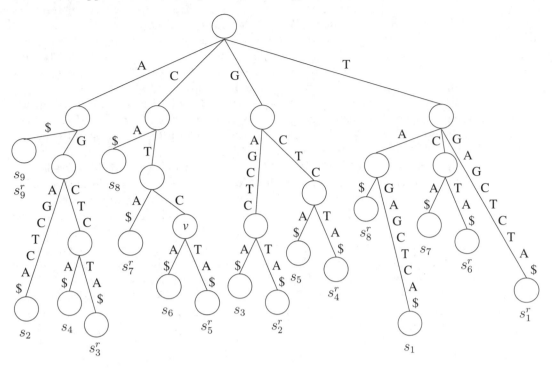

FIGURE 6.1: The generalized suffix tree of the DNA sequence $s = \text{TAGAGCTCA}$ and its reverse complement $s^r = \text{TGAGCTCTA}$. For $i = 5$, $v = lca((s,6),(s^r,5))$, revealing the maximal complemented palindrome GAGCTC.

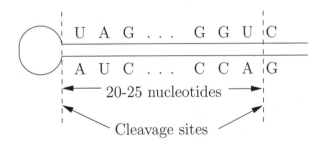

FIGURE 6.2: An example of RNAi element — the stem-loop-stem structure forming a double stranded RNA (dsRNA) that is usually about 20 nucleotides in length.

Horesh *et al.* [19] present a suffix tree based algorithm to detect RNAi elements from a genomic sequence. Here we present how suffix trees can be used to detect such patterns, while avoiding the more complex details necessary for accurate biological modeling.

We can identify RNAi elements in a genome by identifying substrings s_1 and s_2 of the same length (about 20 to 25 nucleotides) that are reverse complements of each other, and separated by a substring s_3 of length $l \leq k$. The parameter k is used to avoid detecting substring and reverse complement pairs separated by great distances. To do this, first a generalized suffix tree is built for the input genomic sequence, and its reverse complement. Let

$s_1 s_3 s_2$ be a substring in the genomic sequence such that s_1 and s_2 are reverse complements of each other, and s_3 is the loop. Then in the generalized suffix tree we just constructed, there is an internal node v that contains two leaves in its subtree corresponding to $suff_i$ and $suff_{n-i-l}$ of the reverse complement sequence, where $l \leq k$ is the length of s_3 (k is a pre-selected threshold); and v is the lowest common ancestor of the two leaves.

The straightforward solution in this case is to traverse the entire tree; for each internal node at a string depth of about 20 to 25, find pairs of leaves that satisfy the criterion mentioned above. However for each leaf representing some suffix $suff_i$ in the original sequence, checking whether there is a leaf corresponding to suffix $suff_{n-i-l}$ of the reverse complement sequence is not easy. Done naively, this could take $O(n^2)$ time, because for each leaf we need to scan all the leaves in the subtree to check whether a suitable counterpart exists.

Here we make the observation that a post-order traversal of the suffix tree induces a complete ordering of all the suffixes. Each suffix in the original sequence and the reverse complement sequence can be associated with a rank, which can be easily obtained by a traversal of the suffix tree. For each internal node v two values ℓ_v and $\ell\ell_v$ are calculated. The value ℓ_v is the number of leaves in the subtree rooted at v. The value $\ell\ell_v$ is the number of leaves in the entire tree to the left of v, i.e., the number of leaves visited in a post-order traversal before visiting v. The ℓ values for all internal nodes can be calculated using post-order traversal as follows: When internal node v is visited, add up ℓ_u of all the nodes u, such that u is a child of v. If a node w is a leaf node then $\ell_w = 1$. To calculate the $\ell\ell$ values, define $\ell\ell_{root} = 0$. Then for each internal node v, $\ell\ell_v = \ell\ell_u + \sum_w \ell_w$, where u is the parent of v and w ranges over all the siblings to the left of v. In other words, the number of leaves to the left of a node v is the number of leaves to the left of its parent plus the total number of leaves in all the subtrees rooted at its siblings to the left. The $\ell\ell$ values can be calculated using a pre-order traversal of the tree.

Consider an internal node v whose string depth is in the target range, say 20 to 25, and a suffix $suff_i$ in the subtree rooted at v. To check whether there is a leaf corresponding to suffix $suff_{n-i-l}$ of the reverse complement sequence in the same subtree, one can scan the ranks of suffix $suff_{n-i-1}$ to suffix $suff_{n-i-k}$ of the reverse complement sequence. Suppose the rank of suffix $suff_{n-i-j}$ is between $\ell\ell_v$ and $\ell\ell_v + \ell_v$, then we know that this suffix is in the subtree rooted by v. If the ith suffix of the original sequence and the $(n-i-j)$th suffix of the reverse complement sequence appear in the subtrees of two different children of v, then v is the lowest common ancestor of the two leaves. The path label of v is a potential dsRNA sequence.

This algorithm takes $O(nk)$ time, where n is the length of the genomic sequence, and k is the maximum length allowed for the stem-loop-stem structure. A more biologically sensible model can be used to take into account the fact that the two strands need not be identical, either because it is enough to have high sequence similarity, or due to potential sequencing errors. An algorithm allowing mismatches on the two strands can also be found in [19]. However, due to the complexity of the model, the algorithm is close to a brute force algorithm.

6.5 Sequence Clustering and Assembly

DNA sequence clustering and assembling overlapping DNA sequences are vital to knowledge discovery in molecular biology. Part III of this handbook is devoted to assembly and clustering applications, and the reader will once again find that suffix trees are used in some of the algorithms presented in that part. In this section, we discuss two suffix tree related problems that are motivated by applications in clustering and assembly. The problems presented in this section are rather artificial, as they are applicable only in the case input data does not contain any errors or genetic variations. Nevertheless, these problems will serve to develop a basic understanding of some of these applications, and suffix tree based algorithms for real clustering and assembly applications can be found in Part III.

6.5.1   Sequence Containment

One problem that is encountered in sequence clustering and assembly applications is redundancy in the input data. Consider a set $S = \{s_1, s_2, \ldots, s_k\}$ of DNA sequences. We wish to identify sequences that are completely contained in other sequences and remove them. In the absence of sequencing errors and other types of variations (such as the DNA sequences being derived from different individuals who may have natural genetic variations), this can be abstracted as the *string containment* problem. Given a set $S = \{s_1, s_2, \ldots, s_k\}$ of strings of total length N, the string containment problem is to identify each string that is a substring of some other string. This problem can be easily solved using suffix trees in $O(N)$ time. First, construct the $GST(S)$ in $O(N)$ time. To find if a string s_i is contained in another, locate the leaf labeled $(s_i, 1)$. If the label of the edge connecting the leaf to its parent is labeled with the string '$', s_i is contained in another string. Otherwise, it is not. This can be determined in $O(1)$ time per string.

6.5.2   Suffix-Prefix Overlaps

The suffix-prefix overlap problem arises in genome assembly problems. At the risk of oversimplification, the problem of genome assembly is to construct a long, target DNA sequence from a large sampling of much shorter fragments of it. This procedure is carried out to extend the reach of DNA sequencing, which can be directly carried only for DNA sequences hundreds of nucleotides long. The first step in assembling the many fragments is to detect pairs of fragments that show suffix-prefix overlaps; i.e., identify pairs of fragments such that the suffix of one fragment in the pair overlaps the prefix of the other fragment in the pair. The suffix-prefix overlaps are then used to assemble the fragments into longer DNA sequences.

Suppose we are given a set of strings $S = \{s_1, s_2, \ldots, s_k\}$ of total length N. In the absence of sequencing errors, the suffix-prefix overlap problem is to identify, for each pair of strings (s_i, s_j), the longest suffix of s_i that is a prefix of s_j. This problem can be solved using $GST(S)$ in optimal $O(N + k^2)$ time. Consider the longest suffix α of s_i that is a prefix of s_j. In $GST(S)$, α is an initial part of the path from the root to leaf labeled $(s_j, 1)$ that culminates in an internal node. A leaf that corresponds to a suffix from s_i should be a child of the internal node, with the edge label '$'. Moreover, it must be the deepest internal node on the path from root to leaf $(s_j, 1)$ that has a suffix from s_i attached in this way. The length of the corresponding suffix-prefix overlap is given by the string depth of the internal node.

Let M be a $k \times k$ output matrix such that $M[i, j]$ should contain the length of the longest suffix of s_i that overlaps a prefix of s_j. The matrix is computed using a depth first search (DFS) traversal of $GST(S)$. During the DFS traversal, k stacks $A_1, A_2, \ldots, A_k$ are maintained, one for each string. The top of the stack A_i contains the string depth of the deepest node along the current DFS path that is connected with edge label '$' to a leaf corresponding to a suffix from s_i. If no such node exists, the top of the stack contains zero. Each stack A_i is initialized by pushing zero onto an empty stack, and is maintained during the DFS as follows: When the DFS traversal visits a node v from its parent, check to see if v is attached to a leaf with edge label '$'. If so, for each i such that string s_i contributes a suffix labeling the leaf, *push string-depth(v)* on to stack A_i. The string depth of the current node can be easily maintained during the DFS traversal. When the DFS traversal leaves the node v to return back to its parent, again identify each i that has the above property and *pop* the top element from the corresponding stack A_i.

The output matrix M is built one column at a time. When the DFS traversal reaches a leaf labeled $(j, 1)$, the top of stack A_i contains the longest suffix of s_i that matches a prefix of s_j. Thus, column j of matrix M is obtained by setting $M[i, j]$ to the top element of stack S_i. To analyze the run-time of the algorithm, note that each *push* (similarly, *pop*) operation on a stack corresponds to a distinct suffix of one of the input strings. Thus, the total number of *push* and *pop* operations is bounded by $O(N)$. The matrix M is filled in $O(1)$ time per element, taking $O(k^2)$ time. Hence, all suffix-prefix overlaps can be identified in optimal $O(N + k^2)$ time.

The above solutions for sequence containment and suffix-prefix overlap problems are not useful in practice because they assume a perfect input free of errors and genetic variations. In practice, one is interested in detecting strong homologies rather than exact matches. The reader interested in how suffix trees can be used for such applications is referred to Chapter 13.

6.6   Whole Genome Alignments

With the availability of multiple genomes, the field of comparative genomics is gaining increasing attention. By comparing the genomic sequences of two closely related species, one can identify potential genes, coding regions, and other genetic information preserved during evolution. On the other hand, by comparing the genomic sequences of distantly related species, one might be able to identify genes that are most likely vital to life. Several programs have been developed to identify such "local" regions of interest [1, 4, 7, 28]. An important problem in comparative genomics is whole genome comparison, i.e., a global or semi-global alignment of two genomes. This allows researchers to understand the genomic differences between the two species. This is particularly useful in comparing two strains of the same virus or bacteria, or even two versions of the assembly of the genome of the same species. We describe a suffix tree based approach for whole genome alignments, as utilized in the popular whole genome alignment tool *MUMmer*, developed by Delcher *et al.* [11, 12]. A suffix array based solution for the same problem is presented in the next chapter.

The *MUMmer* program is based on the identification of maximal unique matches (MUMs). A maximal match between strings s_1 and s_2 is a pair of matching substrings $s_1[i..i+k] = s_2[i'..i'+k] = \alpha$, that cannot be extended in either direction, i.e. $s_1[i-1] \neq s_2[i'-1]$ and $s_1[i+k+1] \neq s_2[i'+k+1]$. A maximal unique match implies that the pair of matching substrings is not only maximal, but also unique; i.e., the substring α is maximal, and occurs exactly once in each s_1 and s_2. A long MUM is very likely to be in the optimal alignment of two sequences. The program has the following stages:

1. Find all MUMs between the two sequences.
2. Find the longest sequence of MUMs, that occur in the same order in either sequence.
3. Align the regions between the MUMs.

To illustrate the use of suffix trees in whole genome alignment, we focus on identification of MUMs, a step that utilizes suffix trees. Given two strings s_1 and s_2, assume the last characters of s_1 and s_2 are $\$_1$ and $\$_2$, respectively, characters that do not occur anywhere else in both strings. First build the $GST(\{s_1, s_2\})$ of the two strings. Let $suff_i^j$ denote

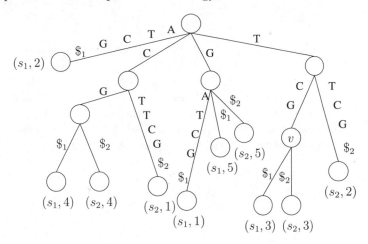

FIGURE 6.3: The generalized suffix tree of sequences GATCG$\$_1$ and CTTCG$\$_2$. Each leaf label is a tuple identifying the string number, followed by the position of the corresponding suffix within the string. In this example, the path-label of node v, is the MUM TCG.

the suffix of string s_j starting at position i. $Lcp(suff_i^1, suff_{i'}^2)$ is a MUM if and only if it is unique in both sequences and $s_1[i-1] \neq s_2[i'-1]$. Let u be the internal node with path label $lcp(suff_i^1, suff_{i'}^2)$. Then the uniqueness part implies u must have exactly two children, the leaves corresponding to suffixes $suff_i^1$ and $suff_{i'}^2$. To ensure left maximality, we need to compare the left characters of the two suffixes under u to make sure that they are unequal. Thus all internal nodes corresponding to MUMs can be identified in a traversal of the $GST(\{s_1, s_2\})$. An example of MUM identification is shown in Figure 6.3.

The space required for the algorithm can be considerably reduced by building the suffix tree of only one string, say s_1, and streaming the other string s_2 to identify MUMs [12]. The algorithm works by considering all suffixes of s_2 starting from $suff_1^2$.

- Find the longest possible match in the suffix tree for $suff_1^2$, the first suffix of string s_2. This is done by traversing from the root of the suffix tree and matching consecutive characters of $suff_1^2$ until no further matches are possible.

- If the match ends inside the edge label between an internal node and a leaf node, then check the left character in s_1 of the suffix corresponding to the leaf and the left character of the suffix from s_2. If they are not the same, then the match is reported.

- After finishing with the first suffix, the same method can be repeated with the second suffix $suff_2^2$. But instead of starting from the root and matching the suffix, suffix links are used to shortcut the process. Let u be the last internal node encountered while matching the previous suffix. Then, take the suffix link from u to, say, u'. Suppose the previous suffix match ended l characters away from node u. It is guaranteed that these l characters will match a path below u'. Therefore, these characters can be matched at the rate of constant time per edge using the same technique as employed in McCreight's suffix tree construction algorithm (Chapter 5). Once the end of these l characters is reached, further matching will continue by examining the subsequent characters of $suff_2^2$ one by one.

- Repeat the process for all suffixes of the second string.

The above algorithm correctly reports all the maximal matches between the two strings. However uniqueness is not preserved for the second string, because we do not actually insert the suffixes of the second string. For example, if we build the suffix tree for the string ATGACGGTCCT$_1$, and subsequently stream the second string ATGATGAG$_2$, then the substring ATGA will be reported twice. This streaming algorithm also runs in $O(n)$ time, because building the suffix tree for the first string takes $O(|s_1|)$ time, and streaming of the second string is equivalent to inserting all suffixes of it using McCreight's algorithm, which takes $O(|s_2|)$ time.

6.7 Tandem Repeats

Tandem repeats — segments of short DNA repeated multiple times consecutively — are believed to play a role in regulating gene expression. Tandem repeats also have a much higher rate of variation then the rest of the genome (in terms of the number of copies), and this makes them ideal markers to distinguish one individual from another.

A tandem repeat can consist of anywhere from two to hundreds of repetitions. If we have the ability to detect a 2-repeat tandem sequence, it can be used to deduce tandem sequences with more repeats. Therefore, the problem is modeled by defining a tandem repeat to be a string $\beta = \alpha\alpha$, i.e., a consecutive occurrence of two copies of the same string α. Tandem repeats are further divided into two sub-categories, primitive and non-primitive. String β is called a primitive tandem repeat if it does not contain another tandem repeat. For example, strings aa and $abab$ are primitive tandem repeats, while $aaaa$ is not a primitive tandem repeat. A 2-repeat tandem sequence is sometimes referred to as a square. When a substring α repeats more than twice consecutively, it is sometimes referred to as a tandem array. A tandem repeat/array in string s given by $\beta = \alpha^k = s[i..i + k|\alpha| - 1]$, where $|\alpha|$ is the length of the substring α, is represented as a triple (i, α, k). We can also represent a tandem repeat $\beta = \alpha\alpha$ as a tuple $(i, 2|\alpha|)$. We use the notation that best suits the situation we are describing.

Detection of tandem repeats is a well-studied problem in computational biology. Crochemore presented an algorithm that computes all occurrences of primitive tandem repeats in $O(n \log n)$ time [3, 10]. On the other hand, all occurrences of tandem repeats (both primitive and non-primitive) can be found in $O(n \log n + occ)$ [24, 29], where occ is the number of occurrences of tandem repeats in the string. We first present a simple $O(n \log n + occ)$ algorithm due to Stoye and Gusfield [29].

6.7.1 Stoye and Gusfield's $O(n \log n)$ Algorithm

Consider a tandem repeat in string s starting at position i of the form $s[i..2|\alpha| + i - 1] = \alpha\alpha$, and $\alpha = a\gamma$, where a is the first character of α and γ is the remainder of α. If character $s[2|\alpha| + i] = x \neq a$, then in the suffix tree there is an internal node v at string depth $|\alpha|$, and suffix $suff_i$ and suffix $suff_{|\alpha|+i-1}$ will be in the subtrees of two different children of v. Since the two suffixes branch, the tandem repeat is referred to as a branching tandem repeat. An example is shown in Figure 6.4.

If a tandem repeat $(i, a\gamma, 2)$ is not a branching tandem repeat, then $(i + 1, \gamma a, 2)$ is also a tandem repeat. However, $(i + 1, \gamma a, 2)$ may not be a branching tandem repeat either. This property of non-branching tandem repeats is easy to see; if $s[i..2|\alpha| + i - 1] = a\gamma a\gamma$ is a non-branching tandem repeat, then $s[i + 1..2|\alpha| + i] = \gamma a\gamma a$ is a tandem repeat. We say that $(i, a\gamma, 2)$ is on the left of $(i + 1, \gamma a, 2)$, while $(i + 1, \gamma a, 2)$ is on the right of $(i, a\gamma, 2)$. In

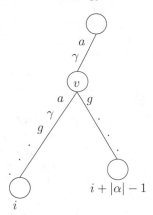

FIGURE 6.4: An example of a branching tandem repeat. $(i, \alpha, 2)$ is a branching tandem repeat, node v is at string depth $|\alpha\gamma|$, and suffixes $suff_i$ and $suff_{|\alpha|+i-1}$ branch from node v.

Figure 6.5, both tandem repeats starting at positions i and $i + 1$ are non-branching, while the tandem repeat starting at position $i + 2$ is a branching tandem repeat. We refer to tandem repeat $(i, a\gamma, 2)$ as a left rotation of tandem repeat $(i + 1, \gamma a, 2)$; right rotation is similarly defined.

FIGURE 6.5: An example of a non-branching tandem repeat. Tandem repeats starting at positions i and $i + 1$ are both non-branching, while the tandem repeat starting at position $i + 2$ is branching. It is easy to see from this example each non-branching tandem repeat is to the left of another tandem repeat.

The non-branching tandem repeats that are next to each other can be considered a chain, with a branching tandem repeat at the end of the chain. Therefore, by locating all branching tandem repeats, and detecting the non-branching tandem repeats to their left, all tandem repeats can be identified. Hence, we focus on identifying all the branching tandem repeats.

A naive algorithm to identify branching tandem repeats is as follows:

1. For each internal node v, collect all the leaves in the subtree rooted by v in a list $\ell\ell(v)$.
2. Let α be the path label of v. Each leaf represents a suffix $suff_i$, and for each suffix $suff_i$ in $ll(v)$ check if $suff_{|\alpha|+i}$ is in $\ell\ell(v)$.
3. If so check if character $s[i]$ is the same as character $s[2|\alpha| + i]$. If so, $(i, \alpha, 2)$ is a branching tandem repeat.

If we can identify whether a suffix $suff_j$ is in $\ell\ell(v)$ in constant time, then the algorithm runs in $O(n^2)$ time. If we number all leaf nodes according to the order they are encountered in a post-order traversal, then leaves in the subtree under any internal node are consecutive. We can mark this range for each internal node v by storing the number of the first leaf, i.e., the leftmost leaf in the subtree; and the last leaf, i.e., the rightmost leaf. Suppose the leaf that represents $suff_i$ is the jth leaf we encounter in the post-order traversal, then in a separate array R we store j in $R[i]$. Therefore, we can identify whether a suffix $suff_i$ is in $\ell\ell(v)$ by checking if the value stored in $R[i]$ lies in the range of node v.

Since for each suffix $suff_i$ there can be $O(n)$ internal nodes on the path from the root to the leaf, the naive algorithm runs in $O(n^2)$ time. However, we can reduce this runtime to $O(n \log n)$ using the knowledge that the two suffixes in a branching tandem repeat are under different children of the node whose path label is the tandem repeat. So we can check the leaves under all but one child, and all branching tandem repeats can be identified. This is because if a branching tandem repeat has a leaf under the child we did not check, then the other leaf must be in a child we did check.

Let node v' be a child of node v that has the most leaves of all of node v's children. We define $\ell\ell'(v) = \ell\ell(v) - \ell\ell(v')$, and modify the naive algorithm by checking all leaves in $\ell\ell'(v)$ instead of $\ell\ell(v)$. Suppose a suffix $suff_i$ is in both $\ell\ell'(v)$ and $\ell\ell'(u)$, where u is a child of v. Then $|\ell\ell'(u)| \leq \frac{|\Sigma|-1}{|\Sigma|} \ell\ell'(v)$ where $|\Sigma|$ is the size of the alphabet; i.e., the number of leaves in $|\ell\ell'(u)|$ is at most $\frac{|\Sigma|-1}{|\Sigma|}$ times the number of leaves in $\ell\ell'(v)$. So a suffix can be in at most $\log_{|\Sigma|/(|\Sigma|-1)} n$ number of lists, resulting in an $O(n \log n)$ time algorithm.

After locating all branching tandem repeats we can find all the non-branching tandem repeats. Suppose that $(i, \alpha, 2)$ is a branching tandem repeat. If $s[i-1] = s[i + 2|\alpha| - 1]$ then $(i - 1, \delta, 2)$ is a non-branching tandem repeat, where $\delta = s[i - 1..i + 2|\alpha| - 2]$. So for each branching tandem repeat $(i, \alpha, 2)$, we check if its left rotation is a tandem repeat, if so we check the left rotation of this new tandem repeat until it is no longer true. This yields an $O(n \log n + occ)$ runtime algorithm, where occ is the total number of tandem repeats.

6.7.2   Stoye and Gusfield's $O(n)$ Algorithm

In 1998, Fraenkel and Simpson [15] proved that for a string $|s| = n$ there are at most $O(n)$ different types of tandem repeats. Tandem repeats $\beta = \alpha\alpha$ and $\gamma = \delta\delta$ are of different types if and only if $\alpha \neq \delta$. Since all occurrences of tandem repeats can be found from the knowledge of all the types of tandem repeats, it is of interest to find the latter. The set of all types of tandem repeats of a string s is also referred to as the vocabulary of s. Gusfield and Stoye designed a linear time algorithm to identify these [18].

String decomposition

The linear time tandem repeat identification algorithm uses Lempel-Ziv string decomposition, illustrated in Figure 6.6. At some stage during the execution of the algorithm, let i be the first position that is not in any block. Find a position $j < i$ that maximizes $|lcp(suff_i, suff_j)|$. Then mark the next block to be of length $\max\{1, |lcp(suff_i, suff_j)|\}$ starting from the ith position. This procedure is continued until the whole string is decomposed into blocks. An example of the Lempel-Ziv decomposition is shown in Figure 6.7.

This decomposition can be easily obtained using a suffix tree. Given a string s first build its suffix tree $ST(s)$. Then in a postorder traversal of the tree, mark each internal node u with the index of the smallest suffix in its subtree. As the postorder traversal visits all children of an internal node u before visiting u itself, node u is marked with the smallest of the numbers marking its children.

To create the decomposition, start by traversing along the path from root to leaf labeled

---

FIGURE 6.6: Lempel-Ziv Decomposition

Procedure Lemple_Ziv_Decomposition(S)

        $blocks \leftarrow \emptyset$

        $block\_start \leftarrow 1$

        $block\_end \leftarrow 1$

        While $block\_end < |s|$ **do**

            Let $block\_len = \max\{1, \max_{k=1}^{block\_start-1} |lcp(suff_k, suff_{block\_start})|\}$

            $block\_end \leftarrow block\_start + block\_len - 1$

            $blocks \leftarrow blocks \cup (block\_start, block\_end)$

            $block\_start \leftarrow block\_end + 1$

      end while

  end procedure

FIGURE 6.7: An example of the Lempel-Ziv decomposition of a string, each number under the block corresponds to the block number.

$suff_1$ in $ST(s)$. The traversal will continue only if the next node along the path is marked with a number smaller than the current position in the string. Continue the traversal until we cannot go any further, and this is the end of the block. Repeat this process by starting at the next position in the string and the root of $ST(s)$.

Using the string given in Figure 6.7 as an example, there is a node u with edge label a from the root of the suffix tree. This node is marked with 1, because $suff_1$ is in its subtree. When we start at position 1, we cannot go to node u because while its edge label is a, its marker is not less than 1. So the end of the block starting at position 1 is 1. The procedure is continued starting at position 2. It is easy to see that this algorithm produces the correct result, and its run time is $O(n)$.

Leftmost-covering set

Since we are only interested in discovering the vocabulary of tandem repeats, and not all their occurrences, it suffices to discover the leftmost occurrence of each type of tandem repeat. Recall that a non-branching tandem repeat is on the left of another tandem repeat with equal length, and this series of consecutive equal length tandem repeats forms a chain. Let (i, l) and (j, l) be two tandem repeats in such a chain. We say that (i, l) covers (j, l) if and only if $i < j$. A set of tandem repeats is a leftmost-covering set if and only if the leftmost occurrence of each type of tandem repeat is covered by a tandem repeat in the set.

Figure 6.8 shows an example of the leftmost-covering set. Tandem repeats $(1, 8)$, $(2, 2)$, $(2, 8)$, $(3, 8)$, $(4, 2)$, $(7, 6)$, $(11, 8)$ are the leftmost tandem repeats of their types. But tandem repeats $(2, 8)$ and $(3, 8)$ are covered by $(1, 8)$, so the leftmost-covering set is $\{(1, 8), (2, 2), (4, 2), (7, 6), (11, 8)\}$. Also note that this is the minimal leftmost-covering set, i.e., no other leftmost-covering set has fewer elements. However, in general a leftmost-covering set need not be minimal.

LEMMA 6.1 The leftmost occurrence of any tandem repeat type must span at least two

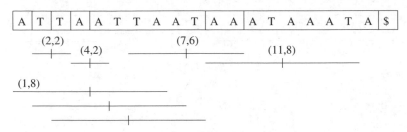

FIGURE 6.8: An example of leftmost-covering set. The leftmost occurrence of each tandem repeat is marked. The leftmost-covering set is $\{(1,8),(2,2),(4,2),(7,6),(11,8)\}$

blocks of the Lempel-Ziv decomposition.

Proof   Let $\beta = \alpha\alpha = S[i..2|\alpha|+i-1]$ be the leftmost occurrence of a type of tandem repeat. If β spans only one block in the Lempel-Ziv decomposition, then by definition of the Lemple-Ziv decomposition, there must exist $suff_j, j < i$, such that $lcp(suff_j, suff_i) \geq 2|\alpha|$. Then, $s[j..2|\alpha|+j-1] = \beta$ must be an earlier occurrence of that type of tandem repeat.

LEMMA 6.2   The second half of any tandem repeat must not span more than two blocks of the Lempel-Ziv decomposition.

Proof   If the second half of a tandem repeat spans more than two blocks of the Lempel-Ziv decomposition, then one block of the decomposition must lie completely inside the second half of the tandem repeat. But by the definition of Lempel-Ziv decomposition, this is impossible. If a block starts at position k of the second half of a tandem repeat, then the suffix starting at position k of the first half of the tandem repeat is sufficient to propel the block to the end of the tandem repeat.

COROLLARY 6.1   By Lemma 6.2, if a block of the Lemple-Ziv decomposition starts at a character that is part of the second half of a tandem repeat, then this block will last until at least the end of the second half of the tandem repeat.

From Lemmas 6.1 and 6.2, one of the following situations must occur for the leftmost occurrence of any tandem repeat type.

- There is a block starting at the same position as the start of the second half of a leftmost tandem repeat.
- There is a block starting after the start of the second half of a leftmost tandem repeat.
- There is no block starting on or after the first character of the second half of a leftmost tandem repeat.

Each of the three cases described above can be split into two sub-cases, based on whether there is another block contained in the left half of the tandem repeat or not. Figure 6.9 illustrates these six cases. Stoye and Gusfield presented two algorithms that will detect all tandem repeats with structures illustrated in Figure 6.9. The two algorithms are run for each block B of the Lempel-Ziv decomposition. Let h be the starting position of the current

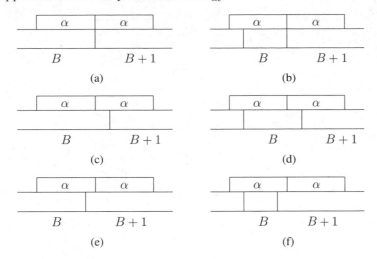

FIGURE 6.9: An enumeration of the possible cases

block, and h_1 be the starting position of the next block. Let $lcp\_r(suff_i, suff_j)$ denote the lcp in the reverse direction starting at positions i and j, i.e., the longest common suffix of prefixes ending at i and j. This can be easily calculated by reversing the string and building the suffix tree for it, along with the usual lcp algorithm.

To see how the first algorithm (Figure 6.10) works, suppose that block B starts at the ith character in the first half of the tandem repeat. Then k will eventually reach the ith character in the second half of the tandem repeat. At this point both k_1 and k_2 will be non-zero, and the length of the tandem repeat is $2k$. This corresponds to cases (b), (d), (e), and (f). On the other hand, the algorithm in Figure 6.11 starts from the ith character in the second half of the tandem repeat, and tries to detect the ith character in the first half; this detects cases (a), and (c).

The above two algorithms take $O(n)$ time because each block is processed once by Backward Detection (see Figure 6.11), and twice by Forward Detection (see Figure 6.10). Each position of the block takes constant time to process by each algorithm. Therefore, the total time is $O(n)$ so far. Also note that since the algorithm runs in $O(n)$ time, the

---

FIGURE 6.10: Forward Detection

Procedure Forward_Detection()

        for $k \leftarrow 1, |B| + |B + 1|$
            $q \leftarrow h + k$
            $k_1 \leftarrow lcp(S_q, S_h)$
            $k_2 \leftarrow lcp\_r(S_{q-1}, S_{h-1})$
            if $k_1 + k_2 \geq k$ and $k_1, k_2 > 0$
                if $\max(h - k_2, h - k + 1) + k < h_1$
                    Output $(\max(h - k_2, h - k + 1), 2k)$
                end if
            end if
        end for
    end procedure

FIGURE 6.11: Backward Detection

Backward_Detection()
       for $k \leftarrow 1, |B|$
          $q \leftarrow h_1 - k$
          $k_1 \leftarrow lcp(S_q, S_{h_1})$
          $k_2 \leftarrow lcp\_r(S_{q-1}, S_{h_1-1})$
          If $k_1 + k_2 \geq k$ and $k_1 > 0$
              if $\max(q - k_2, q - k + 1) + k < h_1$
                  Output $(\max(q - k_2, q - k + 1),\ 2k)$
              end if
          end if
       end for
end procedure

---

number of tandem repeats reported is also $O(n)$. However the result may not be a minimal leftmost-covering set, i.e., some of the tandem repeats reported are either not the leftmost occurrence of its type, or are covered by other tandem repeats in the set, or both.

We have successfully computed a leftmost-covering set, and would now like to mark the tandem repeats in this set in the suffix tree. We begin by first sorting all the tandem repeats by their beginning position, and then by their length (from longest to shortest). This way all the tandem repeats starting from position i are next to each other and ranked according to their length. All such tandem repeats that start form position i are associated with the leaf node v, representing $suff_i$. We call this list of tandem repeats $p(v)$.

Let u be the parent of v, and let k be the string depth of node u. For each tandem repeat $(i, l) \in p(v)$, if $l \geq k$ then mark the position on the edge from u to v or on node u itself, and continue until $l < k$. Since $p(v)$ is sorted the amount of work is proportional to the number of tandem repeats processed. After all the children of node u are processed, then we need to calculate the list $p(u)$. It is not possible to merge all the lists of the children, because this will take $O(n)$ time for each node, and $O(n^2)$ total time.

Each node is labeled with the number of the suffix that has the smallest index in its subtree, i.e., we label node v with i if and only if $j > i$ for each suffix $suff_j$ in v's subtree. To compute $p(u)$, we simply adopt $p(v)$ where v is the child with the smallest label.

LEMMA 6.3 By adopting the list of the child with the smallest label, all the tandem repeats in the leftmost-covering set will be marked.

Proof By induction, assume that all the tandem repeats in the leftmost-covering set under a node u are marked correctly. This is true for internal nodes whose children are all leaf nodes, which serves as the base case. Now we show that the edge e between u and its parent v is marked correctly. Suppose that (i, l) is a part of the leftmost-covering set, and that a position on e should be marked as a result. Then (i, l) must be the first occurrence of that type of tandem repeat. Thus $suff_i$ is the first suffix with that type of tandem repeat as a prefix. Therefore (i, l) is an entry in $p(w)$, where w is the child with the smallest label.

Once the leftmost-covering set is marked in the suffix tree, any tandem repeat is covered by one of the tandem repeats in this set. Let $\beta = \alpha\alpha = a\gamma$, where a is a character. If there is a tandem repeat to its right with the same length, then this tandem repeat must be of the form γa. To mark this tandem repeat, if an internal node v has the path label $a\gamma$, one can travel from $a\gamma$ to γ in the suffix tree by using the suffix link from v. Otherwise, let u be the parent of v, and $a\gamma$ lies inside the edge label of the edge between u and v. Then, first go up to node u and travel to node u' using the suffix link from u. Then, travel down in the suffix tree. Note that for each edge encountered, every character of the edge label need not be compared. One can simply compare the first character, and move down by the length of the edge label. This marks all types of tandem repeats.

Although this compare-and-skip method allows us to traverse each edge in constant time, the number of the edges in the traversal could be large, and result in a non-linear time algorithm. In order to calculate how many times an edge is traversed in the algorithm, we first state the theorem presented in Fraenkel and Simpson [15].

THEOREM 6.1   *Each position i in string s can be the starting position of at most two rightmost occurrences of tandem repeats.*

From the above theorem we can deduce the following.

LEMMA 6.4   For each edge e between node u and node v, there can be at most two marked positions each being the endpoint of some tandem repeat.

Proof   Suppose that an edge e between node u and its child node v has more than two marked positions. Let suffix $suff_i$ be the rightmost suffix in string s under node v, i.e., for all $suff_k$ in the subtree rooted at node v, $k < i$. Then position i is the starting position of the rightmost occurrence of more than two types of tandem repeats, a contradiction.

LEMMA 6.5   Each edge is traversed no more than $O(|\Sigma|)$ times in marking all the tandem repeat types.

Proof   Let node u be the parent of node v, let u' be the internal node reachable from u using the suffix link labeled c, let v' be the internal node reachable from v using the suffix link labeled c. Since there is an edge between u and v, then there is a path between u' and v'; we call this a suffix link induced path. Let edge e be an edge on this suffix link induced path. By Lemma 6.4 there are only two marked positions between nodes u and v. As a result e will be traversed at most twice in order to mark the tandem repeats that are right rotations of the two tandem repeats ending between nodes u and v. Furthermore, any edge e can only be on $|\Sigma|$ number of suffix link induced paths. Thus each edge e is traversed $O(|\Sigma|)$ times.

By Lemma 6.5 each edge is traversed at most $O(|\Sigma|)$ times. Since there are $O(n)$ edges, the total runtime of the algorithm is $O(|\Sigma|n)$. For constant size alphabet, the runtime is $O(n)$.

6.8   Identification of Promoters and Regulatory Sequences

Gene expression, the process by which a gene is transcribed into corresponding mRNA sequences, is aided by promoters and other regulatory sequences usually located upstream of the transcribed portion of the gene. The upstream region typically consists of several important short subsequences, usually 4-10 nucleotides long, that play a role as binding sites for transcription factors. It is known that these sequences are often conserved between similar genes, and also genes that are similarly expressed. The problem of identifying multiple unknown patterns with flexible distance constraints between them is in general known as *structured motif identification problem*. By extracting potential motifs of regulatory sites in gene upstream regions, biologists can gain valuable insight into gene expression regulation. It is natural to use a suffix tree to identify motifs in DNA sequences, because of its suitability to find common substrings in multiple sequences. Marsan and Sagot [25] proposed algorithms to solve the sequence motif identification problem. We present a simplified version by focusing on identification of two patterns. For a more detailed treatment of motif identification problems the reader is referred to [25] and to Chapter 37 of this handbook.

Given a set of m DNA sequences each corresponding to the upstream region of a gene, if a nucleotide sequence of length k is found upstream in all the sequences, then this sequence is a possible motif. This is a simplified view of sequence motifs, because of the following: 1) Not all the m genes may have similar function, so that they might have different motifs. 2) Not all the upstream regions will have an identically common sequence due to evolution, and random mutations. 3) All the motifs should be a similar distance away from the gene. For example, if a sequence occurs 20 base pairs upstream from a gene, while the exact sequence occurs 1000 base pairs upstream from another gene, then it is more likely to be a coincidence than an actual motif. 4) It is possible that the set of m DNA sequence have the same subsequence upstream by chance, therefore we should restrict the motif to be more complicated than one short exact match.

We consider the two pattern motif problem: $((\beta_1, \beta_2), (d_{min}, d_{max}))$ is a motif if there is a subset of q sequences out of all the m input sequences that have substring matches β_1 and β_2, and the two substrings are at least d_{min} away from each other, and at most d_{max} away from each other. This definition can be relaxed, such that we do not need exact matches to β_1 and β_2, but allow a few mismatches. We can restrict the definition by setting a length k for β_1 and β_2.

Build a generalized suffix tree for all the m input sequences. Augment each internal node v of the suffix tree with a boolean array $sequences_v$ of size m, such that $sequences_v[i]$ is set to 1 if and only if a suffix from sequence i is a leaf in the subtree rooted at node v. We also augment each internal node v with a counter $count_v$, such that $count_v$ is the number of 1's in $sequences_v$. Then all the motifs can be identified by a tree traversal. Let p be a position inside the edge label of the edge (u, v) where node u is the parent of node v. If the string depth from the root of the suffix tree to p is between $2k + d_{min}$ and $2k + d_{max}$ and $count_v \geq q$, then the concatenation of all edge labels from the root to p is a potential motif. All potential motifs can be generated in $O(mn)$ time, where m is the number of sequences and n is the total length of all the sequences.

Suppose we would like to consider substrings with e number of mismatches as well. Then we can generate all strings of length k and test if a particular string s_i can be β_1 of the motif. Then we consider all paths beginning at the root of the suffix tree that are e mismatches away from s_i. To find the number of sequences similar to s_i, combine all $sequences_v$ arrays with a logical OR and count the number of 1's in the array. Figure 6.12 shows a generalized suffix tree of AGTACG$\$_1$ and ACGTCA$\$_2$. Suppose the pattern is AGT, and one mismatch is allowed, then the path AGT and CGT will be found. After β_1 is found, find downward paths below the position corresponding to the end of β_1, with lengths between d_{min} and d_{max}, and search for β_2. In the case of the example in Figure 6.12, assume $d_{min} = 0$, $d_{max} = 1$ and allow m_2 to be length 2. Then we can identify the motif $((AGT,CA),(0,1))$ in strings AGTACG$\$_1$ and ACGTCA$\$_2$.

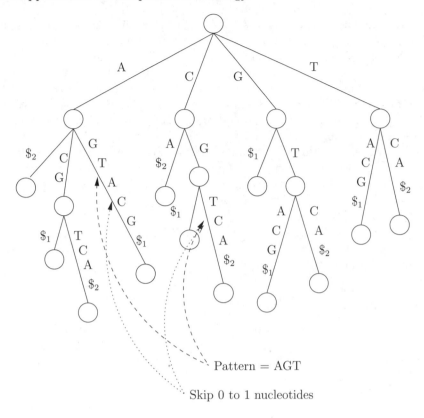

FIGURE 6.12: The generalized suffix tree of AGTACG$_1$ and ACGTCA$_2$. The search for the pattern AGT, allowing one mismatch yields the paths AGT and CGT. Then we skip 0 or 1 nucleotides and try to identify the other part of the motif. We then find $((AGT,CA),(0,1))$ as a motif common to the strings.

6.9 Oligonucleotide Selection

Microarrays are useful in measuring the concentration levels of a target set of DNA sequences. They are based on the concept that two DNA sequences exhibiting complementarity hybridize to each other. If the sequences of the target DNA molecules are known, we can choose a unique oligonucleotide (a short DNA sequence) called a *probe* for each target DNA molecule and attach the probes to the microarray. It is important that the probe be unique in the sense that it hybridizes to only its intended target DNA. To measure the concentration of the target DNA molecules in a solution, they are separated into single stranded molecules, colored with a fluorescent dye, and allowed to hybridize with the fixed probes on the microarray. By using a laser to detect the fluorescence at each microarray spot, the intensity can be used to estimate the concentration of the target DNA molecule. DNA microarrays are commonly used to simultaneously measure the expression levels of tens of thousands of genes of an organism. They have also been used to detect the concentration levels of microorganisms by designing unique probes based on their genomic sequences.

The design of oligonucleotides is challenging because the probes must each be unique to a target sequence. Furthermore, a DNA sequence can hybridize to a probe that it does not match exactly. To account for this, we must select a set of probes such that each probe is unique up to k differences. Because of the hybridization process, if the two probes differ in the first or last k nucleotides and the remaining nucleotides are same, unintended hybridizations are still likely because hybridization can happen to the common part. Therefore, it is best to have the differing positions distributed evenly throughout the probe. Kurtz *et al.* [21] developed an algorithm to design probes as one of the many applications of their repeat finding software REPuter [22]. Subsequently, Kaderali and Schliep [20] have proposed a more complex model for probe design by further screening the unique sequences using their hybridization temperature. In this section we present the approach by Kurtz *et al.* to illustrate how suffix trees can be used in probe design. For a thorough treatment of probe design, the reader is referred to Chapter 24 of the handbook.

For ease of understanding, we restrict ourselves to the problem of designing probes for two target sequences S_1 and S_2. If the probes are too short, the sequences cannot be distinguished from each other; longer probes are harder to manufacture. To model this, let ℓ_{min} and ℓ_{max} be the minimum and maximum allowable length of the probes, respectively. As mentioned above, the probes should also include at least k mismatches, distributed as evenly as possible throughout the probes.

6.9.1 Maximal k-mismatch repeat

In order to design the probes we first look at the maximal k-mismatch repeat problem. Two substrings $s_1[i_1..j_1]$ and $s_2[i_2..j_2]$ are said to be a k-mismatch repeat if we can obtain one from the other by exactly k character replacements, i.e., they mismatch at exactly k positions. A k-mismatch repeat is said to be maximal if we cannot extend it at either end without incurring an extra mismatch.

To identify maximal k-mismatch repeats, first a generalized suffix tree is built for the two target sequences s_1 and s_2. Traverse the tree and mark each internal node u as mixed, if and only if u is the $lca(w_1, w_2)$ where w_1 and w_2 are leaves from s_1 and s_2 respectively. All the mixed internal nodes can be found in $O(n)$ time, where $n = |s_1| + |s_2|$. For each node u we maintain two Boolean values m_1 and m_2; m_1 is set to true if and only if there is a leaf corresponding to a suffix of s_1 in the subtree rooted at u, or u is a leaf node and corresponding to a suffix of s_1; m_2 is similarly defined. This can be done in $O(n)$ time with one post-order traversal of the tree. Then an internal node u is mixed if and only if m_1 at v_1 is true and m_2 at v_2 is true, where v_1 and v_2 are two of u's children. The mixed nodes can be identified with another post-order traversal in $O(n)$ time. In fact the two post-order traversals can be combined into one without changing the asymptotic run-time.

Suppose we are interested in maximal k-mismatch repeats of length at least l. This implies that the maximal k-mismatch repeat has a maximal exact match of length at least $\frac{l}{k+1}$. For each internal node v of string depth at least $\frac{l}{k+1}$ and marked mixed, identify a pair of leaves w_1 and w_2 such that $lca(w_1, w_2) = v$, where w_1 corresponds to a suffix of s_1, and w_2 corresponds to a suffix of s_2. This can be done by a bottom up traversal of the tree. For each node maintain a list of all the leaves of s_1 — call this $list_1$, and another list of all the leaves of s_2 — call this $list_2$. For a leaf node one of the lists is empty and the other has exactly one element. For an internal node v, all distinct pairs of leaves can be generated by choosing an element from $list_1$ of one of the children and an element from $list_2$ of another child. After all pairs of leaves are generated, $list_1$ for v can be constructed by joining all the $list_1$'s children, and $list_2$ for v can be constructed in the same manner. This step can be done in $O(n)$ space and $O(n + occ)$ time, where occ is the number of pairs generated. The space for the suffix tree is $O(n)$ and the total size of the lists is also $O(n)$. While the bottom up traversal takes only $O(n)$ time, the number of pairs generated can be $\Theta(n^2)$ in the worst case.

For each pair of leaves w_1 and w_2 generated, find the length of $lcp(w_1, w_2)$. Let $s_1[i_1..j_1]$ and $s_2[i_2..j_2]$ be the two substrings in s_1 and s_2, respectively, corresponding to $lcp(w_1, w_2)$. It is clear that $s_1[j_1 + 1] \neq s_2[j_2 + 1]$, because the *lcp* would be longer otherwise. Let $suff^1_{j_1+2}$ be the $(j_1 + 2)$th suffix of S_1, and $suff^2_{j_2+2}$ be the $(j_2 + 2)$th suffix of S_2. If $lcp(suff^1_{j_1+2}, suff^2_{j_2+2})$ is of length r, then substrings $s_1[i_1..j_1 + 1 + r]$ and $s_2[i_2..j_2 + 1 + r]$ are a maximal 1-mismatch repeat. We can repeat this procedure to find the maximal k-mismatch repeat by extending to the right and/or to the left. Given a pair of leaves as seed, we can identify a maximal k-mismatch repeat in $O(k)$ time, because finding each required *lcp* takes only constant time with preprocessing for *lca*. We can then check if the maximal k-mismatch repeat is within the specified length constraints.

6.9.2 Oligonucleotide design

From the algorithm presented above, we can generate all the maximal k-mismatch repeats and check if their length l is within l_{min} and l_{max} in $O(n + occ \cdot k)$ time, where occ is the number of pairs generated. Note that a maximal k-mismatch cannot be extended on either side without incurring an extra mismatch. However, the maximal k-mismatch can be shortened on either side if necessary by deleting nucleotides at either end without going as far as the the first mismatch position. This flexibility can be used to increase the chance of finding a k-mismatch probe within the specified length constraints. The probe selection algorithm for two sequences that is presented here can be extended to more than two sequences, with the same run-time of $O(n + occ \cdot k)$, where n is the total length of all the sequences. In the worst case, the number of pairs generated is $\sum_{i=1}^{k} \sum_{i<j} n_i \cdot n_j$, where n_i and n_j are the length of sequence i and j, respectively.

6.10 Protein Database Classification and Peptide Inference

6.10.1 Protein sequence database classification

As previously mentioned, the volume of biological sequence data has increased exponentially in recent years. To better facilitate the analysis of this data, efficient indexing is needed. Also due to high throughput sequencing, it is no longer efficient or even feasible to have researchers manually process the large number of data generated each day, and update sequence databases. With these two goals in mind, an automated process was designed and implemented by Gracy and Argos [16, 17] to classify an entire protein sequence database. In this section we present their process, and the role played by suffix trees.

Since our knowledge of the protein structure and their function is limited, data mining methods are used to discover similarities between individual proteins in a protein family or cluster. However, in order to apply these data mining methods, the protein sequences in the database must first be classified into homologous families. In order to achieve this goal, very similar protein sequences are first classified together by a composition similarity search, where a compositional vector is calculated for each protein sequence based on the number of amino acids and dipeptides. Then a pairwise composition distance is calculated by finding the L_1 distance of the compositional vectors associated with each pair of protein sequences. Only pairs with distances smaller than a threshold are selected into a family.

The composition distance gives a good starting point for further, more sensitive comparisons. From each family identified in the previous step, one protein sequence is selected as a representative. The goal of this step is to further reduce the number of families by grouping together representative sequences. To accomplish this goal efficiently, regions of local similarity need to be identified and used as an anchor. The local similarities in this case are equal length subsequences that share comparable prefixes. To identify these equal length comparable prefixes a generalized suffix tree is built for all the selected sequences. Then the nodes of the suffix tree are visited by a depth-first traversal. For each node v encountered, all the nodes u of equal depth are found and a similarity score is calculated for the path labels of each pair of nodes uv. The identified anchors are then extended to both ends. If the resulting match exceeds a cut-off point then the two protein families are potentially similar, and further checks are performed.

6.10.2   Experimental Interpretation

Background

Tandem mass spectrometry can be used to identify protein sequences. When a large number of experiments are conducted, it is impractical to interpret each experimental output manually. This tedious and repetitive process can be done by sequence database searches. Due to the large data size, an efficient database is needed to effectively identify potential candidates.

In protein studies, the first step is usually the identification of the protein sequence. A protein sequence is first digested with an enzyme to produce short peptides. A mass spectrometry (MS) is then used to measure the mass-to-charge (m/z) ratios of the resulting peptides and these ratios are used in further selecting peptides of interest. The selected peptides are fragmented by a pass through a collision cell, in a step referred to as collision-induced dissociation (CID). At this point the peptide is broken into shorter peptides and individual amino acids. Another mass spectrometry (MS) measures the m/z ratio of the resulting amino acids. This procedure allows us to deduce the mass of the amino acids in the peptide if the charges are known.

However, this does not directly tell us the sequence of the peptide. In order to find the sequence of the peptide, a database of possible peptides must be searched to produce the best answer. Further complicating this process, we may not know which enzyme is used to produce the initial peptides, so the leading/trailing amino acid is not known. Post-translational modification could change the mass of a peptide, which will also have to be taken into consideration during the construction or the search of the index. So clearly the goal is to build an indexing structure and a search routine so that the best interpretation of a tandem mass spectrometry experiment can be found quickly in the database. In this section we will study the indexing and searching method proposed by Lu and Chen [23].

NC-spectrum graph

Given a mass spectrograph as an input, a NC-spectrum graph is constructed using the algorithm by Chen *et al.* [8]. Suppose there are k peaks in the mass spectrograph, then the peptide is broken into k fragments $I_1, \ldots, I_k$ with masses denoted by $w_1, \ldots, w_k$, respectively. Figure 6.13(a) shows an example mass spectrograph with four peaks.

The NC-spectrum graph with $2k + 2$ vertices is created on the real number line. Let $m = 2k + 1$, vertex z_0 and z_m correspond to zero mass, and the total mass of the peptide W, respectively. For each peak I_j, two vertices z_j and z_{m-j} are added to the graph, one at position w_j, and the other at position $W - w_j$, respectively. This is the same as assuming a fragment is either a prefix or a suffix of the peptide. Obviously, a fragment could be a substring in the middle of the original peptide. However, since we do not always know which enzyme was used to digest the protein, we lack the start and end points. Thus assuming a fragment is a substring in the middle of a peptide does not help us deduce the actual

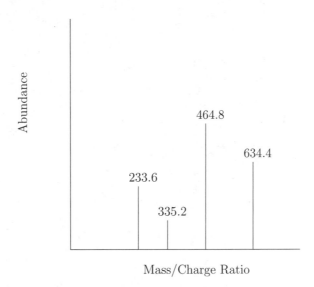

(a) Mass spectrograph of a hypothetical peptide.

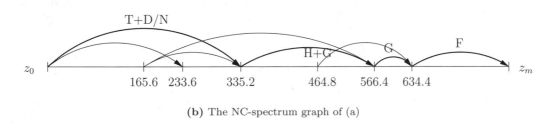

(b) The NC-spectrum graph of (a)

FIGURE 6.13: The total mass of the peptide is 800 amu, and the bolded edges forms a possible peptide. The symbol '+' means 'concatenation', and the symbol '/' means 'disjunction'. For example, the express 'T+D/N' means 'TD' or 'TN'.

sequence, and it will complicate the search effort because we do not know where to place it on the line. After the vertices are fixed on the line, edges are added to the graph. Suppose $z_i < z_j$ are two vertices in the graph. If the difference between z_i and z_j corresponds to the mass of some amino acid, an edge is drawn from z_i to z_j and is labeled with that amino acid.

Figure 6.13(b) shows an example of the NC-spectrum graph corresponding to the mass spectrograph of Figure 6.13(a). This NC-spectrum graph has only eight vertices instead of the ten vertices. This is because the peaks corresponding to molecular weights 335.2 and 464.8 map to the same two vertices on the line. It is easy to see that the concatenation of all the edge labels of a path from vertex z_0 to z_m in the NC-spectrum graph corresponds to a peptide. This peptide is one of the many peptides that could have produced the mass spectrograph.

The Peptide Inference Algorithm

Given a database of proteins, the goal is to identify a set of good candidate peptides from the database for a particular mass spectrograph. First a generalized suffix tree is constructed, consisting of all the proteins in the database. Then a depth first traversal is done on the

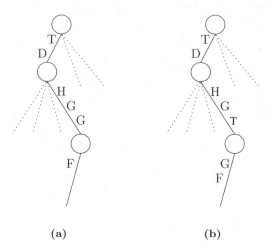

(a)   (b)

FIGURE 6.14: Suppose the NC-spectrum graph in Figure 6.13 (b) is used in our search. (a) A path that matches the peptide TDHGGF completely, when F is encountered in the search. The peptide TDHGGF is returned as a candidate peptide. (b) A path with an insertion T in the middle (between the two G's), the peptide is also returned as a candidate.

suffix tree by referring to the NC-spectrum graph. Start at the root of the suffix tree r, and vertex z_0 of the NC-spectrum graph. Let (r, u) be an edge in the suffix tree with edge label l_{ru}. We can map this edge onto the NC-spectrum graph by starting from z_0 and following the appropriate edge on the graph. Continue this until a position in the suffix tree is reached such that the corresponding path in the NC-spectrum graph reaches the last vertex z_m, or the path in the NC-spectrum graph can no longer be extended. In the first case the concatenation of the edge labels of the suffix tree from the root r to the current position is a candidate peptide. In the second case the current position in the suffix tree cannot yield a possible match. In this case, backtrack in the suffix tree and the NC-spectrum graph by taking a different edge in the suffix tree and continue; see Figure 6.14(a).

In order to account for experimental errors, the search can be relaxed by allowing errors. If the search is at node u in the suffix tree and vertex z_j in the NC-spectrum graph, but there is no outgoing edge from u in the suffix tree that has the same label as the edge from z_j we are interested in, we could skip one character from u and check if the edge label is available. For example, in Figure 6.14(b), we are searching for the peptide TDHGGF in the suffix tree. The prefix TDHG is located, however the next amino acid is T instead of G. This T is skipped and the search routine tries to locate the remaining sequence GF, which comes after T. So the peptide TDHGTGF is returned as a candidate. If a match cannot be found by skipping one character in the suffix tree, more characters can be skipped depending on the quality of the spectrograph.

This algorithm allows searching for all possible candidates in a protein database using $O(n + |G|)$ space, where n is the size of the protein database, and $|G|$ is the size of the NC-spectrum graph. Since we would like to return all the possible candidates in the protein database, a complete traversal of the suffix tree may be necessary. This takes $O(n)$ time. As the entire suffix tree may need to be traversed in the worst case, it is advantageous to use the linked list implementation of children of internal nodes to save space without increasing the worst case run-time.

After all the candidate peptides are located, a probability can be associated with each pep-

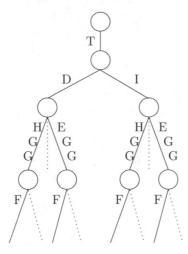

FIGURE 6.15: Post-translational modifications. Consider D can be modified to be I, H can be modified to E, and all other amino acids have no allowed modifications.

tide, and the peptide with the highest matching probability is output. Careful readers may notice we have not yet addressed the problem of post-translational modification. This can be done by modifying the search algorithm. Keep a table of all possible post-translational modifications such that for each amino acid a_i, a list of all its possible modifications are stored. During the depth first traversal of the suffix tree, instead of simply referring to the NC-spectrum graph to decide whether a path in the suffix tree is a candidate, each amino acid in the edge label of the NC-spectrum graph is also substituted with all its possible modifications to check if a path in the suffix tree can be a possible candidate. For example, in Figure 6.15, the only two modifications are from D to I and H to E. Suppose the path label in the NC-spectrum graph is TDHGGF, then the path TDHGGF will generate a match, and because of the modifications the paths TDEGGF, TIHGGF, and TIEGGF will also be considered as matches.

6.11 Conclusions

Suffix trees and its variants are used in many applications in computational biology. This chapter provides a diverse, but by no means exhaustive, sample of the many applications in which suffix trees have been used. Variants of suffix trees have also been developed for use in Markov models. These include prediction suffix trees [6] and probabilistic suffix trees [2, 5, 13].

Acknowledgements

This work was supported in part by the U.S. National Science Foundation under IIS-0430853.

References

[1] S.F. Altschul, T.L. Madden, A.A. Schäffer1, and J. Zhang *et al.* Gapped blast and psi-blast: a new generation of protein database search programs. *Nucleic Acids Research*, 25(17):3389–3402, 1997.

[2] A. Apostolico and G. Bejerano. Optimal amnesic probabilistic automata or how to learn and classify proteins in linear time and space. *Journal of Computational Biology*, 7(3):381–393, 2000.

[3] A. Apostolico and F.P. Preparata. Optimal off-line detection of repetitions in a string. *Theoretical Computer Science*, 22:297–315, 1983.

[4] S. Batzoglou, L. Pachter, J.P. Mesirov, and B. Berger *et al.* Human and mouse gene structure: Comparative analysis and application to exon prediction. *Genome Research*, 10(7):950–958, 2000.

[5] G. Bejerano, Y. Seldin, H. Margalit, and N. Tishby. Markovian domain fingerprinting: Statistical segmentation of protein sequences. *Bioinformatics*, 17(10):927–934, 2001.

[6] G. Bejerano and G. Yona. Variations on probabilistic suffix trees: Statistical modeling and prediction of protein families. *Bioinformatics*, 17(1):23–43, 2001.

[7] N. Bray, I. Dubchak, and L. Pachter. Avid: A global alignment program. *Genome Research*, 13(1):97–102, 2003.

[8] T. Chen, M.Y. Kao, M. Tepel, and J. Rush *et al.* A dynamic programming approach to de novo peptide sequencing via tandem mass spectrometry. In *Proc. 11th annual ACM-SIAM symposium on Discrete algorithms*, pages 389–398, 2000.

[9] R. Cole and R. Hariharan. Approximate string matching: A simpler faster algorithm. *SIAM Journal on Computing*, 31, 2002.

[10] M. Crochemore. An optimal algorithm for computing the repetitions in a word. *Information Processing Letters*, 12(5):244–250, 1981.

[11] A.L. Delcher, S. Kasif, R.D. Fleischmann, and J. Peterson *et al.* Alignment of whole genomes. *Nucleic Acids Research*, 27(11):2369–2376, 1999.

[12] A.L. Delcher, A. Phillippy, J. Carlton, and S.L. Salzberg. Fast algorithms for large-scale genome alignment and comparison. *Nucleic Acids Research*, 30(11):2478–2483, 2002.

[13] E. Eskin, W.S. Noble, and Y. Singer. Protein family classification using sparse markov transducers. *Journal of Computational Biology*, 10(2):187–213, 2003.

[14] A. Fire, S. Xu, M.K. Montgomery, and S.A. Kostas *et al.* Potent and specific genetic interference by double-stranded rna in caenorhabditis elegans. *Nature*, 391(6669):806–811, 1998.

[15] A.S. Fraenkel and J. Simpson. How many squares can a string contain? *Journal of Combinatorial Theory, Series A*, 82(1):112–120, 1998.

[16] J. Gracy and P. Argos. Automated protein sequence database classification. i. integration of compositional similarity search, local similarity search, and multiple sequence alignment. *Bioinformatics*, 14(2):164–173, 1998.

[17] J. Gracy and P. Argos. Automated protein sequence database classification. ii. delineation of domain boundaries from sequence similarities. *Bioinformatics*, 14(2):174–187, 1998.

[18] D. Gusfield and J. Stoye. Linear-time algorithms for finding and representing all tandem repeats in a string. *Journal of Computer and System Sciences*, 69(4):525–

546.

[19] Y. Horesh, A. Amir, S. Michaeli, and R. Unger. A rapid method for detection of putative rnai target genes in genomic data. *Bioinformatics*, 19(Suppl. 2):73ii–80ii, September 2003.

[20] L. Kaderali and A. Schliep. Selecting signature oligonucleotides to identify organisms using DNA arrays. *Bioinformatics*, 18(10):1340–1349, 2002.

[21] S. Kurtz, J.V. Choudhuri, E. Ohlebusch, and C. Schleiermacher *et al.* REPuter: the manifold applications of repeat analysis on a genomic scale. *Nucleic Acids Research*, 29(22):4633–3642, 2001.

[22] S. Kurtz and C. Schleimermacher. REPuter: fast computation of maximal repeats in complete genomes. *Bioinformatics*, 15(5):426–427, 1999.

[23] B. Lu and T. Chen. A suffix tree approach to the interpretation of tandem mass spectra: applications to peptides of non-specific digestion and post-translational modifications. *Bioinformatics*, 19(Suppl. 2):ii 113–ii 121, 2003.

[24] M.G. Main and R.J. Lorentz. An O(n log n) algorithm for finding all repetitions in a string. *Journal of Algorithms*, 5(3):422–432, 1984.

[25] L. Marsan and M.F. Sagot. Algorithms for extracting structured motifs using a suffix tree with an application to promoter and regulatory site consensus identification. *Journal of Computational Biology*, 7(3):345–362, 2000.

[26] B. Morgenstern. DIALIGN2: Improvement of the segement-to-segment approach to mulitple sequence alignment. *Bioinformatics*, 15(3):211–218, 1999.

[27] B. Morgenstern, O. Rinner, S. Abdeddaïm, and D. Haase *et al.* Exon discovery by genomic sequence alignment. *Bioinformatics*, 18(6):777–787, 2002.

[28] S. Schwartz, Z. Zhang, K.A. Frazer, and A. Smit *et al.* Pipmaker – a web server for aligning two genomic dna sequences. *Genome Research*, 10(4):577–586, 2000.

[29] J. Stoye and D. Gusfield. Simple and flexible detection of contiguous repeats using a suffix tree. In *Proc. 9th Annual Symposium, Combinatorial Pattern Matching*, pages 140–152, 1998.

[30] E. Ukkonen. Approximate string-matching over suffix trees. In *Proc 4th Annual Symposium, Combinatorial Pattern Matching*, pages 228–242, 1993.

7

Enhanced Suffix Arrays and Applications

Mohamed I. Abouelhoda
University of Ulm

Stefan Kurtz
University of Hamburg

Enno Ohlebusch
University of Ulm

7.1   Introduction

The suffix tree is undoubtedly one of the most important data structures in string processing. This is particularly true if the sequences to be analyzed are very large and do not change. An example of prime importance from the field of bioinformatics is genome analysis, where the sequences under consideration are whole genomes (the human genome, for example, contains more than $3 \cdot 10^9$ base pairs).

The suffix tree of a sequence S is an index structure that can be computed and stored in $O(n)$ time and space [34], where $n = |S|$. Once constructed, it can be used to efficiently solve a "myriad" of string processing problems [3]. Table 7.1 shows the applications discussed in this chapter. These applications can be classified into the following kinds of tree traversals:

- a bottom-up traversal of the complete suffix tree
- a top-down traversal of a subtree of the suffix tree
- a traversal of the suffix tree using suffix links

Therefore, Table 7.1 also shows which kind of traversal is used for the respective application.

While suffix trees play a prominent role in algorithmics, they are not as widespread in actual implementations of software tools as one should expect. There are two major reasons for this:

(i) Although being asymptotically linear, the space consumption of a suffix tree is quite large; even recently improved implementations of linear time constructions still require 20 bytes per input character in the worst case; see, e.g., [22].

Application	Type of tree traversal		
	bottom-up	top-down	suffix-links
supermaximal repeats	√		
maximal repeats	√		
maximal unique matches	√		
maximal multiple exact matches	√		
tandem repeats	√		
tandem repeats (brute force)		√	
exact pattern matching		√	
maximal exact matches (space efficient)		√	√

TABLE 7.1   The suffix tree applications discussed here and the kinds of traversals they require.

(ii) In most applications, the suffix tree suffers from a poor locality of memory reference, which causes a significant loss of efficiency on cached processor architectures, and renders it difficult to store in secondary memory.

These problems have been identified in several large scale applications like the repeat analysis of whole genomes [23] and the comparison of complete genomes [11, 16].

More space efficient data structures than the suffix tree exist. The most prominent one is the *suffix array*, which was introduced by Manber and Myers [27] and independently by Gonnet et al. [12] under the name PAT array. The suffix array requires only $4n$ bytes in its basic form and it can be constructed in $O(n)$ time in the worst case by first constructing the suffix tree of S; see [14]. Recently, it was shown independently and contemporaneously in [17, 19, 20] that a direct linear time construction of the suffix array is possible. However, the suffix array has less structure than the suffix tree, so that it is not clear that (and how) an algorithm using a suffix tree can be replaced with an algorithm based on a suffix array. In this chapter, we will show that the problems listed in Table 7.1 can also be solved with suffix arrays plus additional information. The algorithms presented here are not only more space efficient than previous ones, but they are also faster and easier to implement. In many applications the above-mentioned "additional information" consists of the longest common prefix (lcp) information. Kasai et al. [18] coined the name *virtual suffix tree* for a suffix array enhanced with the lcp information. However, there are also other applications that cannot be solved with this virtual suffix tree data structure. Thus, we will use the generic name *enhanced suffix array* for data structures consisting of the suffix array and additional tables representing the required information.

In Section 7.3, we treat applications (computation of supermaximal repeats and maximal unique matches) that are solely based on the properties of the enhanced suffix array.

In Section 7.4, we will introduce the concept of *lcp-interval tree*. The lcp-interval tree of an enhanced suffix array is only conceptual (i.e., it is not really built) but it allows us to simulate all kinds of suffix tree traversals very efficiently. As examples of a bottom-up traversal, we will show how to compute all maximal repeated pairs of a string as well as all maximal multiple exact matches of a set of strings. These applications use the suffix array and the lcp-table, both of which can be stored in $4n$ bytes. In order to compute tandem repeats efficiently, one further needs the inverse suffix array; see Section 7.6.

In Section 7.5, we consider the exact pattern matching problem, which consists of finding all occurrences of a pattern P of length m in a string S of length n. We suppose that the enhanced suffix array of S has been constructed. It is well-known that the exact pattern matching problem can be solved by a binary search in the suffix array: A decision query "Is P a substring of S?" can be answered in $O(m \log n)$ time. Manber and Myers [27] showed how this can be improved to $O(m + \log n)$ running time using an additional table. This result is, however, only of theoretical interest, and therefore we only describe the

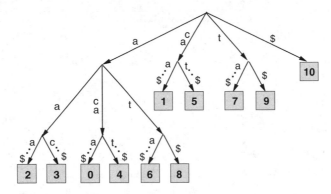

FIGURE 7.1: The suffix tree for $S = acaaacatat$.

$O(m \log n)$-method. Furthermore, we will show how to answer decision queries in optimal $O(m)$ time and how to find all z occurrences of a pattern P in optimal $O(m + z)$ time. These results are achieved by using the basic suffix array enhanced with the lcp-table and an additional table, called the child-table, that requires $4n$ bytes.

A space efficient algorithm for computing maximal exact matches of two strings (genomes) is presented in Section 7.7. Because it uses suffix links — a well-known feature of suffix trees — we will show how to incorporate the concept of suffix links into enhanced suffix arrays. To this end, we further enhance the suffix array with an additional table, called the suffix link table, that stores the left and right boundaries of suffix link intervals. This table can be stored in $8n$ bytes.

We would like to point out that in practice both the lcp-table and the child-table can be stored in n bytes, whereas the suffix link table requires $2n$ bytes; see [2] for implementation details. Experiments revealed that this space reduction entails no loss of performance; see [2].

7.2   Basic Notions

Let Σ be a finite ordered *alphabet*. Σ^* is the *set of all strings over* Σ. We use Σ^+ to denote the set $\Sigma^* \setminus \{\varepsilon\}$ of non-empty strings. Let S be a string of length $|S| = n$ over Σ. To simplify analysis, we suppose that the size of the alphabet is a constant, and that $n < 2^{32}$. The latter implies that an integer in the range $[0, n]$ can be stored in 4 bytes. We assume that the special sentinel symbol $ is an element of Σ (which is larger then all other elements) but does not occur in S. $S[i]$ denotes the *character at position* i in S, for $0 \leq i < n$. For $i \leq j$, $S[i \ldots j]$ denotes the *substring* of S starting with the character at position i and ending with the character at position j. The substring $S[i \ldots j]$ is also denoted by the *pair* (i, j) *of positions*.

A *suffix tree* for the string S is a rooted directed tree with exactly $n + 1$ leaves numbered 0 to n. Each internal node, other than the root, has at least two children and each edge is labeled with a nonempty substring of $S$$. No two edges out of a node can have edge-labels beginning with the same character. The key feature of the suffix tree is that for any leaf i, the concatenation of the edge-labels on the path from the root to leaf i exactly spells out the string S_i, where $S_i = S[i \ldots n - 1]$$ denotes the i-th nonempty suffix of the string $S$$, $0 \leq i \leq n$. Figure 7.1 shows the suffix tree for the string $S = acaaacatat$.

The *suffix array* suftab of the string S is an array of integers in the range 0 to n,

| | | | childtab | | | suflink | | | | |
	i	suftab	lcptab	1.	2.	3.	l	r	suftab^{-1}	bwttab	$S_{\text{suftab}[i]}$
	0	2	0		2	6			2	c	$aaacatat\$$
	1	3	2				0	5	6	a	$aacatat\$$
	2	0	1	1	3	4	0	10	0		$acaaacatat\$$
	3	4	3				6	7	1	a	$acatat\$$
	4	6	1	3	5				3	c	$atat\$$
	5	8	2				8	9	7	t	$at\$$
	6	1	0	2	7	8			4	a	$caaacatat\$$
	7	5	2				0	5	8	a	$catat\$$
	8	7	0	7	9	10			5	a	$tat\$$
	9	9	1				0	10	9	a	$t\$$
	10	10	0	9					10	t	$\$$

FIGURE 7.2: Suffix array of the string $S = acaaacatat$ enhanced with the lcp-table, the child-table, the suffix link table, the inverse suffix array, and the Burrows and Wheeler table bwttab. The child-table and the suffix link table will be explained later.

specifying the lexicographic ordering of the $n + 1$ suffixes of the string $S\$$. That is, $S_{\text{suftab}[0]}, S_{\text{suftab}[1]}, \ldots, S_{\text{suftab}[n]}$ is the sequence of suffixes of $S\$$ in ascending lexicographic order; see Figure 7.2. The suffix array requires $4n$ bytes.

The *lcp-table* lcptab is an array of integers in the range 0 to n. We define $\text{lcptab}[0] = 0$ and $\text{lcptab}[i]$ to be the length of the longest common prefix of $S_{\text{suftab}[i-1]}$ and $S_{\text{suftab}[i]}$, for $1 \leq i \leq n$. Since $S_{\text{suftab}[n]} = \$$, we always have $\text{lcptab}[n] = 0$. The lcp-table can be computed as a by-product during the construction of the suffix array (see,e.g., [17]), or alternatively, in linear time from the suffix array [18]. The lcp-table requires $4n$ bytes in the worst case.

The *inverse suffix array* suftab^{-1} is a table of size $n + 1$ such that $\text{suftab}^{-1}[\text{suftab}[q]] = q$ for any $0 \leq q \leq n$. suftab^{-1} can be computed in linear time from the suffix array and needs $4n$ bytes.

The table bwttab contains the *Burrows and Wheeler transformation* [8] known from data compression. It is a table of size $n + 1$ such that for every $i, 0 \leq i \leq n$, $\text{bwttab}[i] = S[\text{suftab}[i] - 1]$ if $\text{suftab}[i] \neq 0$. $\text{bwttab}[i]$ is undefined if $\text{suftab}[i] = 0$. The table bwttab is stored in n bytes and constructed in one scan over the suffix array in $O(n)$ time.

7.3 Computation of Supermaximal Repeats and Maximal Unique Matches

Motivation: Repeat Analysis

Repeat analysis plays a key role in the study, analysis, and comparison of complete genomes. In the analysis of a single genome, a basic task is to characterize and locate the repetitive elements of the genome. In the comparison of two or more genomes, a basic task is to find similar subsequences of the genomes. As we shall see later, this problem can also be reduced to the computation of certain types of repeats of the string that consists of concatenated genomes.

The repetitive elements of the human genome can be classified into two large groups: dispersed repetitive *DNA* and tandemly repeated *DNA*. Dispersed repetitions vary in size and content and fall into two basic categories: transposable elements and segmental duplications

[25]. Transposable elements belong to one of the following four classes: SINEs (short interspersed nuclear elements), LINEs (long interspersed nuclear elements), LTRs (long terminal repeats), and transposons. Segmental duplications, which might contain complete genes, have been divided into two classes: chromosome-specific and trans-chromosome duplications [30]. Tandemly repeated *DNA* can also be classified into two categories: simple sequence repetitions (relatively short k-mers such as micro and minisatellites) and larger ones, which are called blocks of tandemly repeated segments. While bacterial genomes usually do not contain large parts of redundant *DNA*, a considerable portion of the genomes of higher organisms is composed of repeats. For example, 50% of the 3 billion basepairs of the human genome consist of repeats. Repeats also comprise 11% of the mustard weed genome, 7% of the worm genome and 3% of the fly genome [25]. Clearly, one needs extensive algorithmic support for a systematic study of repetitive *DNA* on a genomic scale. The algorithms for this task usually use the suffix tree to locate repetitive structures such as maximal or supermaximal repeats; see [14]. In this section we show how to locate all supermaximal repeats, while Section 7.4 treats maximal repeated pairs. Let us recall the definitions of these notions.

A pair of substrings $R = ((i_1, j_1), (i_2, j_2))$ is a *repeated pair* if and only if $(i_1, j_1) \neq (i_2, j_2)$ and $S[i_1 \ldots j_1] = S[i_2 \ldots j_2]$. The length of R is $j_1 - i_1 + 1$. A repeated pair $((i_1, j_1), (i_2, j_2))$ is called *left maximal* if $S[i_1 - 1] \neq S[i_2 - 1]^1$ and *right maximal* if $S[j_1 + 1] \neq S[j_2 + 1]$. A repeated pair is called *maximal* if it is left and right maximal. A substring ω of S is a *(maximal) repeat* if there is a (maximal) repeated pair $((i_1, j_1), (i_2, j_2))$ such that $\omega = S[i_1 \ldots j_1]$. A *supermaximal repeat* is a maximal repeat that does not occur as a substring of any other maximal repeat.

The lcp-Intervals of an Enhanced Suffix Array

We start this subsection with the introduction of the first essential concept of this chapter, namely lcp-intervals. Then we will derive two new algorithms that solely exploit the properties of lcp-intervals. The algorithms are much simpler than the corresponding ones based on suffix trees.

DEFINITION 7.1   An interval $[i..j]$, $0 \leq i < j \leq n$, is an *lcp-interval of lcp-value* ℓ if

1. $\mathsf{lcptab}[i] < \ell$,
2. $\mathsf{lcptab}[k] \geq \ell$ for all k with $i + 1 \leq k \leq j$,
3. $\mathsf{lcptab}[k] = \ell$ for at least one k with $i + 1 \leq k \leq j$,
4. $\mathsf{lcptab}[j + 1] < \ell$.

We will also use the shorthand ℓ-interval (or even ℓ-$[i..j]$) for an lcp-interval $[i..j]$ of lcp-value ℓ. Every index k, $i + 1 \leq k \leq j$, with $\mathsf{lcptab}[k] = \ell$ is called ℓ-index. The set of all ℓ-indices of an ℓ-interval $[i..j]$ will be denoted by $\ell Indices(i, j)$. If $[i..j]$ is an ℓ-interval such that $\omega = S[\mathsf{suftab}[i] \ldots \mathsf{suftab}[i] + \ell - 1]$ is the longest common prefix of the suffixes $S_{\mathsf{suftab}[i]}$, $S_{\mathsf{suftab}[i+1]}, \ldots, S_{\mathsf{suftab}[j]}$, then $[i..j]$ is called the ω-interval. The size of an interval $[i..j]$ is $j - i + 1$.

---

[1] This definition has to be extended to the cases $i_1 = 0$ or $i_2 = 0$, but throughout the chapter we do not explicitly state boundary cases like these.

As an example, consider the table in Figure 7.2. $[0..5]$ is a 1-interval because $\mathsf{lcptab}[0] = 0 < 1$, $\mathsf{lcptab}[5+1] = 0 < 1$, $\mathsf{lcptab}[k] \geq 1$ for all k with $1 \leq k \leq 5$, and $\mathsf{lcptab}[2] = 1$. Furthermore, 1-$[0..5]$ is the a-interval and $\ell Indices(0,5) = \{2,4\}$. We shall see later that lcp-intervals correspond to internal nodes of the suffix tree.

Computation of Supermaximal Repeats

Next, we present an algorithm that computes all supermaximal repeats of a string. The reader is invited to compare our simple algorithm with the suffix-tree based algorithm of [14, page 146].

DEFINITION 7.2   An ℓ-interval $[i..j]$ is called a *local maximum* in the lcp-table if $\mathsf{lcptab}[k] = \ell$ for all $i+1 \leq k \leq j$.

For instance, in the lcp-table of Figure 7.2, the local maxima are the intervals $[0..1]$, $[2..3]$, $[4..5]$, $[6..7]$, and $[8..9]$.

LEMMA 7.1   A string ω is a supermaximal repeat if and only if there is an ℓ-interval $[i..j]$ such that

- $[i..j]$ is a local maximum in the lcp-table and $[i..j]$ is the ω-interval.
- the characters $\mathsf{bwttab}[i], \mathsf{bwttab}[i+1], \dots, \mathsf{bwttab}[j]$ are pairwise distinct.

Proof   "if": Since ω is a common prefix of the suffixes $S_{\mathsf{suftab}[i]}, \dots, S_{\mathsf{suftab}[j]}$ and $i < j$, it is certainly a repeat. The characters $S[\mathsf{suftab}[i] + \ell], S[\mathsf{suftab}[i+1] + \ell], \dots, S[\mathsf{suftab}[j] + \ell]$ are pairwise distinct because $[i..j]$ is a local maximum in the lcp-table. By the second condition, the characters $\mathsf{bwttab}[i], \mathsf{bwttab}[i+1], \dots, \mathsf{bwttab}[j]$ are also pairwise distinct. It follows that ω is a maximal repeat and that there is no repeat in S which contains ω. In other words, ω is a supermaximal repeat.

"only if": Let ω be a supermaximal repeat of length $|\omega| = \ell$. Furthermore, suppose that $\mathsf{suftab}[i], \mathsf{suftab}[i+1], \dots, \mathsf{suftab}[j], 0 \leq i < j \leq n$, are the consecutive entries in suftab such that ω is a common prefix of $S_{\mathsf{suftab}[i]}, S_{\mathsf{suftab}[i+1]}, \dots, S_{\mathsf{suftab}[j]}$ but neither of $S_{\mathsf{suftab}[i-1]}$ nor of $S_{\mathsf{suftab}[j+1]}$. Because ω is supermaximal, the characters $S[\mathsf{suftab}[i] + \ell], S[\mathsf{suftab}[i+1] + \ell], \dots, S[\mathsf{suftab}[j] + \ell]$ are pairwise distinct. Hence $\mathsf{lcptab}[k] = \ell$ for all k with $i+1 \leq k \leq j$. Furthermore, $\mathsf{lcptab}[i] < \ell$ and $\mathsf{lcptab}[j+1] < \ell$ hold because otherwise ω would also be a prefix of $S_{\mathsf{suftab}[i-1]}$ or $S_{\mathsf{suftab}[j+1]}$. All in all, $[i..j]$ is a local maximum of the array lcptab and $[i..j]$ is the ω-interval. Finally, the characters $\mathsf{bwttab}[i], \mathsf{bwttab}[i+1], \dots, \mathsf{bwttab}[j]$ are pairwise distinct because ω is supermaximal.

The preceding lemma does not only imply that the number of supermaximal repeats is smaller than n, but it also suggests a simple linear time algorithm to compute all supermaximal repeats of a string S.

Motivation: Comparison of Whole Genomes

To date (September 2005), Genbank contains "complete" genomes for more than 1,000 viruses, over 250 microbes, and 36 eukaryota. This abundance of complete genomic *DNA*-sequences has boosted the field of comparative genomics. Comparative studies are emerging

---

FIGURE 7.3: Computation of the supermaximal repeats of string S

---

find all local maxima in the lcp-table of S
for each local maximum $[i..j]$ in the lcp-table of S **do**
    if bwttab$[i]$, bwttab$[i+1]$, . . . , bwttab$[j]$ are pairwise distinct characters
    then report the string $S[\mathsf{suftab}[i] \ldots \mathsf{suftab}[i] + \mathsf{lcptab}[i] - 1]$ as supermaximal repeat

---

as a powerful tool for the identification of genes and regulatory elements. Comparing the genomes of related species gives us new insights into the complex structure of organisms at the *DNA*-level and protein-level.

The first step in the comparison of genomes is to produce an alignment, i.e., a colinear arrangement of sequence similarities. Alignment of nucleic or amino acid sequences has been one of the most important methods in sequence analysis. Nowadays, many sophisticated algorithms are available for aligning sequences with similar regions. These require to score all possible alignments (typically, the score is the sum of the similarity/identity values for the aligned symbols, minus a penalty for the introduction of gaps), in conjunction with a dynamic programming method to find optimal or near-optimal alignments according to this scoring scheme (see, e.g., [29]). The dynamic programming algorithms run in time proportional to the product of the lengths of the sequences. Hence they are not suitable for aligning entire genomes. Recently, several genome alignment programs have been developed, all using an anchor-based method to compute an alignment (for an overview see [9]). The anchor-based methods are composed of the following three phases:

(1) Computation of all potential anchors (exact or approximate matches).

(2) Computation of an optimal colinear sequence of non-overlapping potential anchors: these are the anchors that form the basis of the alignment.

(3) Closure of the gaps in between the anchors by applying the same method recursively — yielding a divide and conquer method — or, e.g., by a standard dynamic programming method.

In this chapter, we will focus on phase (1) and explain several algorithms to compute exact matches. The first one is used in the software tool *MUMmer* [11]. It is based on a maximal unique match (*MUM*) decomposition of two genomes S_1 and S_2. The implementation of *MUMmer* uses the suffix tree of $S_1 \# S_2$ to compute *MUM*s in $O(n)$ time and space, where $n = |S_1 \# S_2|$ and $\#$ is a symbol neither occurring in S_1 nor in S_2.

Computation of Maximal Unique Matches

The space consumption of the suffix tree has been identified to be a major problem in the computation of the maximal unique match decomposition of two large genomes; see [11]. We will solve this problem by using the suffix array enhanced with the lcp-table.

DEFINITION 7.3 Given two strings S_1 and S_2, a *MUM* is a string that occurs exactly once in S_1 and once in S_2, and is not contained in any longer such string.

LEMMA 7.2 Let $\#$ be a unique separator symbol not occurring in S_1 and S_2 and let $S = S_1 \# S_2$. The string u is a *MUM* of S_1 and S_2 if and only if u is a supermaximal repeat in S such that

1. there is only one maximal repeated pair $((i_1, j_1), (i_2, j_2))$ with $u = S[i_1 \ldots j_1] = S[i_2 \ldots j_2]$,

2. $j_1 < p < i_2$, where $p = |S_1|$ is the position of # in S.

Proof "if": It is a consequence of conditions (1) and (2) that u occurs exactly once in S_1 and once in S_2. Because the repeated pair $((i_1, j_1), (i_2, j_2))$ is maximal, u is a *MUM*.

"only if": If u is a *MUM* of the sequences S_1 and S_2, then it occurs exactly once in S_1 (say, $u = S_1[i_1 \ldots j_1]$) and once in S_2 (say, $u = S_2[i_2 \ldots j_2]$), and is not contained in any longer such sequence. Clearly, $((i_1, j_1), (p + 1 + i_2, p + 1 + j_2))$ is a repeated pair in $S = S_1 \# S_2$, where $p = |S_1|$. Because u occurs exactly once in S_1 and once in S_2, and is not contained in any longer such sequence, it follows that u is a supermaximal repeat in S satisfying conditions (1) and (2).

The first version of *MUMmer* [11] computed *MUMs* with the help of the suffix tree of $S = S_1 \# S_2$. Using an enhanced suffix array, this task can be done more time and space economically as follows.

FIGURE 7.4: Computation of maximal unique matches of two strings S_1 and S_2

find all local maxima in the lcp-table of $S = S_1 \# S_2$
for each local maximum $[i..j]$ in the lcp-table of S **do**
  if $i + 1 = j$ **and** bwttab$[i] \neq$ bwttab$[j]$ **and** suftab$[i] < p <$ suftab$[j]$
  then report the string $S[$suftab$[i] \ldots$ suftab$[i] +$ lcptab$[i] - 1]$ as *MUM*

The algorithms that compute supermaximal repeats and *MUMs* require tables suftab, lcptab, and bwttab, but do not access the input sequence. More precisely, instead of the input string, we use table bwttab without increasing the total space requirement. This is because the tables suftab, lcptab, and bwttab can be accessed in sequential order, thus leading to an improved cache coherence and in turn considerably reduced running time; see [2] for details and experimental results.

7.4 Computation of Maximal Repeats and Maximal (Multiple) Exact Matches

Next, we introduce the second essential concept of this chapter — the lcp-interval tree.

DEFINITION 7.4 An m-interval $[l..r]$ is said to be *embedded* in an ℓ-interval $[i..j]$ if it is a subinterval of $[i..j]$ (i.e., $i \leq l < r \leq j$) and $m > \ell$.[2] The ℓ-interval $[i..j]$ is then called the interval *enclosing* $[l..r]$. If $[i..j]$ encloses $[l..r]$ and there is no interval embedded in $[i..j]$ that also encloses $[l..r]$, then $[l..r]$ is called a *child interval* of $[i..j]$.

---

[2]Note that we cannot have both $i = l$ and $r = j$ because $m > \ell$.

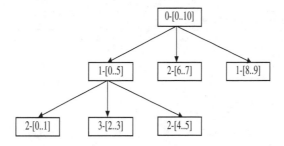

FIGURE 7.5: The lcp-interval tree of the string $S = acaaacatat$.

This parent-child relationship constitutes a conceptual (or virtual) tree which we call the *lcp-interval tree* of the suffix array. The root of this tree is the 0-interval $[0..n]$; see Figure 7.5. The lcp-interval tree is basically the suffix tree without leaves (more precisely, there is a one-to-one correspondence between the nodes of the lcp-interval tree and the internal nodes of the suffix tree). These leaves are left implicit in our framework, but every leaf in the suffix tree, which corresponds to the suffix $S_{\text{suftab}[l]}$, can be represented by a *singleton interval* $[l..l]$. The parent interval of such a singleton interval is the smallest lcp-interval $[i..j]$ with $l \in [i..j]$. For instance, continuing the example of Figure 7.2, the child intervals of $[0..5]$ are $[0..1]$, $[2..3]$, and $[4..5]$.

In Figure 7.6, the lcp-interval tree is traversed in a bottom-up fashion by a linear scan of the lcp-table, while storing needed information on a stack. We stress that the lcp-interval tree is not really built: whenever an ℓ-interval is processed by the generic function *process*, only its child intervals have to be known. These are determined solely from the lcp-information, i.e., there are no explicit parent-child pointers in our framework. In Figure 7.6, the elements on the stack are lcp-intervals represented by quadruples $\langle lcp, lb, rb, childList \rangle$, where lcp is the lcp-value of the interval, lb is its left boundary, rb is its right boundary, and $childList$ is a list of its child intervals. Furthermore, $add([c_1, \ldots, c_k], c)$ appends the element c to the list $[c_1, \ldots, c_k]$ and returns the result.

In this chapter, several problems will be solved merely by specifying the function *process* called in line 8 of Figure 7.6.

Computation of Maximal Repeated Pairs

The computation of maximal repeated pairs plays an important role in the analysis of a genome. The algorithm of Gusfield [14, page 147] computes maximal repeated pairs of a sequence S of length n in $O(|\Sigma|n + z)$ time, where z is the number of maximal repeated pairs. This running time is optimal. To the best of our knowledge, Gusfield's algorithm was first implemented in the *REPuter*-program [23], based on space efficient suffix trees described in [22]. The software tool *REPuter* uses maximal repeated pairs as seeds for finding degenerate (or approximate) repeats. In this section, we show how to implement Gusfield's algorithm using enhanced suffix arrays. This considerably reduces the space requirements, thus removing a bottle neck in the algorithm. As a consequence, much larger genomes can be searched for repetitive elements. As in the algorithms in Section 7.3, the implementation requires tables suftab, lcptab, and bwttab, but does not access the input sequence. The accesses to the three tables are in sequential order, thus leading to an improved cache coherence and in turn to a considerably reduced running time; see [2].

We begin by introducing some notation: Let $\perp$ stand for the undefined character. We

---

FIGURE 7.6: Traverse and process the lcp-interval tree

$lastInterval := \perp$
$push(\langle 0, 0, \perp, [\,] \rangle)$
for $i := 1$ **to** n **do**
    $lb := i - 1$
    while $\mathsf{lcptab}[i] < top.lcp$
       $top.rb := i - 1$
       $lastInterval := pop$
       $process(lastInterval)$
       $lb := lastInterval.lb$
       if $\mathsf{lcptab}[i] \leq top.lcp$ **then**
          $top.childList := add(top.childList, lastInterval)$
          $lastInterval := \perp$
    if $\mathsf{lcptab}[i] > top.lcp$ **then**
       if $lastInterval \neq \perp$ **then**
          $push(\langle \mathsf{lcptab}[i], lb, \perp, [lastInterval] \rangle)$
          $lastInterval := \perp$
       else $push(\langle \mathsf{lcptab}[i], lb, \perp, [\,] \rangle)$

---

assume that it is different from all characters in Σ. Let $[i..j]$ be an ℓ-interval and $u = S[\mathsf{suftab}[i]] \ldots \mathsf{suftab}[i] + \ell - 1]$. Define $\mathcal{P}_{[i..j]}$ to be the set of positions p such that u is a prefix of S_p, i.e., $\mathcal{P}_{[i..j]} = \{\mathsf{suftab}[r] \mid i \leq r \leq j\}$. We divide $\mathcal{P}_{[i..j]}$ into disjoint and possibly empty sets according to the characters to the left of each position: For any $a \in \Sigma \cup \{\perp\}$ define

$$\mathcal{P}_{[i..j]}(a) = \begin{cases} \{0 \mid 0 \in \mathcal{P}_{[i..j]}\} & \text{if } a = \perp \\ \{p \mid p \in \mathcal{P}_{[i..j]}, p > 0, \text{ and } S[p-1] = a\} & \text{otherwise} \end{cases}$$

The algorithm computes position sets in a bottom-up traversal. In terms of an lcp-interval tree, this means that the lcp-interval $[i..j]$ is processed only after all child intervals of $[i..j]$ have been processed.

Suppose $[i..j]$ is a singleton interval, i.e., $i = j$. Let $p = \mathsf{suftab}[i]$. Then $\mathcal{P}_{[i..j]} = \{p\}$ and

$$\mathcal{P}_{[i..j]}(a) = \begin{cases} \{p\} & \text{if } p > 0 \text{ and } S[p-1] = a \text{ or } p = 0 \text{ and } a = \perp \\ \emptyset & \text{otherwise} \end{cases}$$

Now suppose that $i < j$. For each $a \in \Sigma \cup \{\perp\}$, $\mathcal{P}_{[i..j]}(a)$ is computed step by step while processing the child intervals of $[i..j]$. These are processed from left to right. Suppose that they are numbered, and that we have already processed q child intervals of $[i..j]$. By $\mathcal{P}^q_{[i..j]}(a)$ we denote the subset of $\mathcal{P}_{[i..j]}(a)$ obtained after processing the q-th child interval of $[i..j]$. Let $[i'..j']$ be the $(q+1)$-th child interval of $[i..j]$. Due to the bottom-up strategy, $[i'..j']$ has been processed and hence the position sets $\mathcal{P}_{[i'..j']}(b)$ are available for any $b \in \Sigma \cup \{\perp\}$.

The interval $[i'..j']$ is processed in the following way: First, maximal repeated pairs are output by combining the position set $\mathcal{P}^q_{[i..j]}(a)$, $a \in \Sigma \cup \{\perp\}$, with position sets $\mathcal{P}_{[i'..j']}(b)$, $b \in \Sigma \cup \{\perp\}$. In particular, $((p, p + \ell - 1), (p', p' + \ell - 1))$, $p < p'$, are output for all $p \in \mathcal{P}^q_{[i..j]}(a)$ and $p' \in \mathcal{P}_{[i'..j']}(b)$, $a, b \in \Sigma \cup \{\perp\}$ and $a \neq b$.

It is clear that u occurs at positions p and p'. Hence $((p, p + \ell - 1), (p', p' + \ell - 1))$ is a repeated pair. By construction, only those positions p and p' are combined for which the characters immediately to the left, i.e., at positions $p - 1$ and $p' - 1$ (if they exist), are different. This guarantees left-maximality of the output repeated pairs.

The position sets $\mathcal{P}^q_{[i..j]}(a)$ were inherited from child intervals of $[i..j]$ that are different from $[i'..j']$. Hence the characters immediately to the right of u at positions $p+\ell$ and $p'+\ell$ (if they exist) are different. As a consequence, the output repeated pairs are maximal.

Once the maximal repeated pairs for the current child interval $[i'..j']$ have been output, the union $\mathcal{P}^{q+1}_{[i..j]}(e) := \mathcal{P}^q_{[i..j]}(e) \cup \mathcal{P}_{[i'..j']}(e)$ is computed for all $e \in \Sigma \cup \{\bot\}$. That is, the position sets are inherited from $[i'..j']$ to $[i..j]$.

In Figure 7.6, if the function *process* is applied to an lcp-interval, then all its child intervals are available. Hence the maximal repeated pair algorithm can be implemented by a bottom-up traversal of the lcp-interval tree. To this end, the function *process* in Figure 7.6 outputs maximal repeated pairs and further maintains position sets on the stack (which are added as a fifth component to the quadruples). The bottom-up traversal requires $O(n)$ time.

There are two operations performed when processing an lcp-interval $[i..j]$. Output of maximal repeated pairs by combining position sets and union of position sets. Each combination of position sets means to compute their Cartesian product. This delivers a list of position pairs, i.e., maximal repeated pairs. Each repeated pair is computed in constant time from the position lists. Altogether, the combinations can be computed in $O(z)$ time, where z is the number of repeats. The union operation for the position sets can be implemented in constant time, if we use linked lists. For each lcp-interval, we have $O(|\Sigma|)$ union operations. Since $O(n)$ lcp-intervals have to be processed, the union and add operations require $O(|\Sigma|n)$ time. Altogether, the algorithm runs in $O(|\Sigma|n + z)$ time.

Let us analyze the space consumption of the algorithm. A position set $\mathcal{P}_{[i..j]}(a)$ is the union of position sets of the child intervals of $[i..j]$. If the child intervals of $[i..j]$ have been processed, the corresponding position sets are obsolete. Hence it is not required to copy position sets. Moreover, we only have to store the position sets for those lcp-intervals which are on the stack used for the bottom-up traversal of the lcp-interval tree. So it is natural to store references to the position sets on the stack together with other information about the lcp-interval. Thus the space required for the position sets is determined by the maximal size of the stack. Since this is $O(n)$, the space requirement is $O(|\Sigma|n)$. In practice, however, the stack size is much smaller. Altogether the algorithm is optimal, since its space and time requirement is linear in the size of the input plus the output.

Computation of Maximal (Multiple) Exact Matches

In this section, we come back to the problem of aligning genomic sized sequences. In Section 7.3, we have seen that the software tool *MUMmer* uses *MUMs* as potential anchors in the first phase of its anchor-based method. Delcher et al. [11] wrote: "The crucial principle behind this step is the following: if a long, perfectly matching sequence occurs exactly once in each genome, it is almost certain to be part of the global alignment." It has been argued in [16] that the restriction to *MUMs* as anchors seems unnecessary because exact matches occurring more than once in a genome may also be meaningful. This consideration lead to the notion of maximal exact matches.

DEFINITION 7.5 An *exact match* between two strings S_1 and S_2 is a triple (l, p_1, p_2) such that $p_1 \in [0, |S_1| - l]$, $p_2 \in [0, |S_2| - l]$, and $S_1[p_1 \ldots p_1 + l - 1] = S_2[p_2 \ldots p_2 + l - 1]$. An exact match is *left maximal* if $S[p_1 - 1] \neq S[p_2 - 1]$ and *right maximal* if $S[p_1 + l] \neq S[p_2 + l]$. A *maximal exact match* (*MEM*) is a left and right maximal exact match.

Of course, computing maximal exact matches between two strings S_1 and S_2 boils down

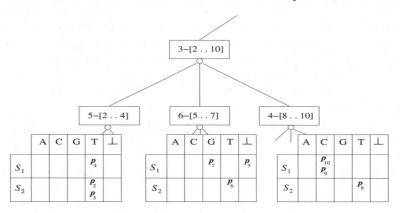

FIGURE 7.7: The position sets of a part of an lcp-interval tree. p_i denotes the position suftab$[i]$.

to computing maximal repeated pairs of the string $S = S_1 \# S_2$. This is made precise in the following lemma.

LEMMA 7.3   Let $\#$ be a unique separator symbol not occurring in the strings S_1 and S_2 and let $S = S_1 \# S_2$. (l, p_1, p_2) is a *MEM* if and only if $((p_1, p_1 + l - 1), (p_2, p_2 + l - 1))$ is a maximal repeated pair of the string S such that $p_1 + l - 1 < p < p_2$, where $p = |S_1|$ is the position of $\#$ in S.

Therefore, one can use the maximal repeated pair algorithm with the following modification to compute maximal exact matches. The position sets are divided into two disjoint and possibly empty subsets: One subsets contains all positions that correspond to S_1 (these are smaller than $|S_1|$). Another subset contains all positions that correspond to S_2 (these are greater than $|S_1|$). Let $\mathcal{P}_{[i..j]}(S_1, a)$ denote the set of all positions $p \in \mathcal{P}_{[i..j]}$ such that p corresponds to a position in S_1 and $S[p - 1] = a \in \Sigma \cup \{\bot\}$. $\mathcal{P}_{[i..j]}(S_2, a)$ is defined analogously. (Figure 7.7 shows an example.) To compute maximal exact matches, the Cartesian product is build from each position set $\mathcal{P}_{[i..j]}(S_1, a)$, $a \in \Sigma \cup \{\bot\}$ and the position sets $\mathcal{P}_{[i..j]}(S_2, b)$, where $a \neq b \in \Sigma \cup \{\bot\}$. It is not difficult to see that this modification does not affect the time and space complexity of the algorithm.

However, if one uses *MEMs* instead of *MUMs* in a global alignment tool, then one is faced with the problem that the number of *MEMs* can be very large. (This is because the number of *MEMs* is determined by a Cartesian product of position sets.) Needless to say that this phenomenon occurs especially in the comparison of highly repetitive genomes. Clearly, *MEMs* that occur too many times should be excluded from the set of potential anchors. In our opinion, the exact definition of "too many times" should be left to the user of the global alignment tool. In other words, it should be a parameter of the program. The next definition makes this precise.

DEFINITION 7.6   Suppose (l, p_1, p_2) is a *MEM* and let $u = S_1[p_1 \ldots p_1 + l - 1]$ be the corresponding sequence. Define the set

$$MP_u = \{(p_1', p_2') \mid (l, p_1', p_2') \text{ is a } MEM \text{ and } u = S_1[p_1' \ldots p_1' + l - 1]\}$$

of position pairs where the *MEMs* with sequence u start. Given a threshold $t \in \mathbb{N}$, the

MEM (l, p_1, p_2) is called

- *infrequent in S_1* if $r_1 := |\{p_1' \mid (p_1', p_2') \in MP_u\}|$ satisfies $r_1 \leq t$,
- *infrequent in S_2* if $r_2 := |\{p_2' \mid (p_1', p_2') \in MP_u\}|$ satisfies $r_2 \leq t$,
- *infrequent* if $r := |MP_u|$ satisfies $r \leq t$.

It is our next goal to calculate the values r, r_1, and r_2. To this end, the following notation will be useful. For a position set $\mathcal{P}$, let $C_{\mathcal{P}}(S_1, a) = |\mathcal{P}(S_1, a)|$ and $C_{\mathcal{P}}(S_1) = \sum_{a \in \Sigma \cup \{\perp\}} C_{\mathcal{P}}(S_1, a)$ (the values $C_{\mathcal{P}}(S_2, a)$ and $C_{\mathcal{P}}(S_2)$ are defined similarly). The value $C_{\mathcal{P}}(S_1)$ represents the number of repeats in S_1 from which the *MEMs* are derived. One can also impose constraints on this value.

For any lcp-interval $[i..j]$ that contains k child intervals, the value r can be calculated according to the following formula, where $q, q' \in [1, k]$:

$$r = \frac{1}{2} \sum_{a \in \Sigma \cup \{\perp\}} \sum_{q \neq q'} C_{\mathcal{P}^q_{[i..j]}}(S_1, a) \cdot (C_{\mathcal{P}^{q'}_{[i..j]}}(S_2) - C_{\mathcal{P}^{q'}_{[i..j]}}(S_2, a))$$

In the example of Figure 7.7, the calculation of the value r for the interval $[2..10]$ yields $r = (0 * 1 + 0 * 1) + (2 * 2 + 2 * 1) + (1 * 2 + 1 * 1) + (1 * 0 + 1 * 0) + (1 * 2 + 1 * 1) = 12$.

For any lcp-interval $[i..j]$ that contains k child intervals, the value r_1 can be calculated by the formula:

$$r_1 = \sum_{a \in \Sigma \cup \{\perp\}} \sum_{q \in [1, k]} C'_{\mathcal{P}^q_{[i..j]}}(S_1, a)$$

where

$$C'_{\mathcal{P}^q_{[i..j]}}(S_1, a) = \begin{cases} 0, & \text{if } \forall q' \in [1, k], q' \neq q : C_{\mathcal{P}^{q'}_{[i..j]}}(S_2) = C_{\mathcal{P}^{q'}_{[i..j]}}(S_2, a) \\ C_{\mathcal{P}^q_{[i..j]}}(S_1, a) & \text{otherwise} \end{cases}$$

In the example of Figure 7.7, we have $r_1 = 0 + 2 + 1 + 0 + 1 = 4$. It is obvious how the algorithm for the computation of *MEMs* has to be modified to compute infrequent *MEMs*. Choosing an appropriate threshold on r (or a threshold on the combination of r_1, r_2, $C_{\mathcal{P}}(S_1)$, and $C_{\mathcal{P}}(S_2)$), the user can fine tune the computation of infrequent *MEMs*. Thus the concept of infrequent *MEMs* provides a reasonable compromise between *MUMs* and *MEMs*.

We would also like to comment on methods for multiple genome comparisons. As already mentioned, the *DNA* sequences of entire genomes are being determined at a rapid rate. Nowadays, it is quite common for a project to sequence the genome of an organism that is very closely related to another completed genome. For example, the genomes of several strains of the bacteria *E. coli* and *S. aureus* have already been completely sequenced. A global alignment of the genomes may help, for example, in understanding why a strain of a bacterium is pathogenic or resistant to antibiotics while another is not. Current software tools for multiple alignment of genomic sized sequences build a multiple alignment from pairwise alignments; see [28, 6, 7, 32]. However, if the organisms under consideration are closely related, then it makes sense to build a multiple alignment directly from exact matches that occur in each genome. The software tool *MGA* [16] is capable of aligning three or more closely related genomes. In the first phase of its anchor-based method, all *maximal multiple exact matches* (*multiMEMs*) longer than a specified minimum length are computed. The notion of a *multiMEM* is the natural extension of maximal exact matches to more than two genomes. Roughly speaking, a *multiMEM* is a sequence that occurs in all genomes to be aligned and cannot simultaneously be extended to the left or right in every genome. The maximal repeated pair algorithm can also be modified such that it computes (infrequent) *multiMEMs*. A detailed description of the resulting algorithm can be found in [24].

7.5 Exact Pattern Matching

In this section we consider the exact pattern matching problem. We suppose that the suffix array for S is available. We consider two methods: A practical method based on binary searches in the suffix array (running in $O(m \log n)$ time), and an optimal method requiring only $O(m)$ time.

A Method Based on Binary Searches in the Suffix Array

Consider a pattern P of length m. To find all occurrences of P in S, we need to find all the suffixes of S that have P as a prefix. Since the suffix array for S stores the suffixes of $S\$$ in lexicographic order, all these suffixes are consecutive in the suffix array. Using a binary search, one can therefore compute the leftmost (i.e., smallest) position $l, 0 \leq l \leq n$, in the suffix array such that P is a prefix of $S_{\mathsf{suftab}[l]}$. If such an l does not exist, then P does not occur as a substring of S. Otherwise, one computes the rightmost (i.e., largest) position $r, 0 \leq r \leq n$, in the suffix array such that P is a prefix of $S_{\mathsf{suftab}[r]}$. Given l and r, all occurrences of P in S are computed as follows: Output $\mathsf{suftab}[j]$ for all $j, l \leq j \leq r$. This obviously requires $O(r - l)$ time.

Figure 7.8 presents a pseudo-code implementation for this method. In particular, the function *findleftmost* delivers the leftmost position l and the function *findrightmost* delivers the rightmost position r, given some boundaries l_0 and r_0 in the suffix array and some value h_0, such that all suffixes in the range l_0 to r_0 have a common prefix of length at least h_0. To reduce the number of character comparisons, the length of the longest common prefix of all suffixes between the current boundaries is computed from h_l and h_r. These two values are the lengths of the longest common prefix of P with $S_{\mathsf{suftab}[l]}$ and $S_{\mathsf{suftab}[r]}$, respectively. Hence $S_{\mathsf{suftab}[mid]}$ and P have a common prefix of length $\min\{h_l, h_r\}$, where $mid = \lfloor (l + r)/2 \rfloor$ is the midpoint between the current boundaries l and r. The longest common prefix can be skipped when comparing $S_{\mathsf{suftab}[mid]}$ and P using the function *compare*; see Figure 7.8.

The function *compare* performs the comparison of a suffix of $S\$$ with some string w of length m. More precisely, let $0 \leq i \leq n$ and $v = S_{\mathsf{suftab}[i]}$ and suppose that w and v have a common prefix of length q. Then $compare(w, m, i, q)$ delivers a pair (c, f_c) such that the following holds:

- c is the length of the longest common prefix of w and v. (Note that $c \geq q$.)
- If w is a prefix of v, i.e., $c = m$ holds, then $f_c = 0$.
- Otherwise, if w is not a prefix of v, then $w[c] \neq v[c]$. There are two cases

    - If $w[c] < v[c]$, then $f_c = -1$.
    - If $w[c] > v[c]$, then $f_c = 1$.

The function *sasearch* calls *findleftmost* and *findrightmost*. If the left boundary l is smaller than or equal to the right boundary r, then all start positions of the suffixes between these boundaries are reported. If (h_0, l_0, r_0) is set to $(0, 0, n)$, then both searches take $O(m \log_2 n)$ time. That is, the running time of *sasearch* is $O(m \log_2 n + z)$ where z is the number of occurrences of P in S.

Note that we do not always have to start the binary searches with the boundaries 0 and n. One can divide the suffix array into buckets such that each bucket contains all suffixes

---

FIGURE 7.8: Search all occurrences of P in S using the suffix array for S.

---

function *findleftmost*(P, m, h, l, r)
$(h_l, f_l) := compare(P, m, l, h)$
if $f_l \leq 0$ **then**
    return l
$(h_r, f_r) := compare(P, m, r, h)$
if $f_r > 0$ **then**
    return $r + 1$
while $r > l + 1$ **do**
    $mid := \lfloor (l + r)/2 \rfloor$
    $(c, f_c) := compare(P, m, mid, \min\{h_l, h_r\})$
    if $f_c \leq 0$ **then**
        $(h_r, r) := (c, mid)$
    else
        $(h_l, l) := (c, mid)$
return r

function *findrightmost*(P, m, h, l, r)
$(h_l, f_l) := compare(P, m, l, h)$
if $f_l < 0$ **then**
    return $- 1$
$(h_r, f_r) := compare(P, m, r, h)$
if $f_r \geq 0$ **then**
    return r
while $r > l + 1$ **do**
    $mid := \lfloor (l + r)/2 \rfloor$
    $(c, f_c) := compare(P, m, mid, \min\{h_l, h_r\})$
    if $f_c \geq 0$ **then**
        $(h_l, l) := (c, mid)$
    else
        $(h_r, r) := (c, mid)$
return l

function *compare*(w, m, i, q)
$v := S_{\mathsf{suftab}[i]}$
$c := q$
while $c < \min\{m, |v|\}$ **do**
    if $w[c] < v[c]$ **then**
        return $(c, -1)$
    else
        if $w[c] > v[c]$ **then**
            return $(c, 1)$
        else
            $c := c + 1$
if $c = m$ **then**
    return $(c, 0)$
else
    return $(c, -1)$

function *sasearch*(P, m, h_0, l_0, r_0)
$l := findleftmost(P, m, h_0, l_0, r_0)$
$r := findrightmost(P, m, h_0, l_0, r_0)$
if $l \leq r$ **then**
    for $j := l$ **to** r **do**
        print "`match at pos`" $\mathsf{suftab}[j]$

---

having the same prefix of length d, where d is a given parameter. Formally, one precomputes a table bcktab_d storing for each string $u \in \Sigma^*$ of length d a pair $\mathsf{bcktab}_d(u) := (l_d, r_d)$ of integer values. l_d and r_d are the leftmost and rightmost positions of all suffixes in the suffix array having prefix u. Table bcktab_d can be precomputed in $O(n)$ time in one scan over tables suftab and lcptab. Given a string u of length d, the pair $\mathsf{bcktab}_d(u)$ can easily be accessed by computing a unique integer code for u in the range 0 to $|\Sigma|^d - 1$. This requires $O(d)$ time.

Now suppose $m \geq d$ and let $(l_d, r_d) := \mathsf{bcktab}_d(P[0 \ldots d - 1])$. If $l_d > r_d$, then P does not occur in S. Otherwise, all suffixes of S having prefix $P[0 \ldots d - 1]$ occur between the boundaries l_d and r_d. Moreover, all suffixes between these boundaries have a common prefix of length d. Hence one can start the binary search with these boundaries and with prefix length d to search for the remaining suffix $P[d \ldots m - 1]$ of P. In this way, the overall search time for P using *sasearch* becomes $O(d + (m - d) \log_2(r_d - l_d + 1))$ with an additional space consumption of $O(|\Sigma|^d)$ for table bcktab_d. In practice one chooses d such that bcktab_d can be stored in n bytes, so that the space requirement only increases by 25%.

The algorithm presented here can be improved to $O(m + \log n)$ by employing an extra table containing the length of the longest common prefix for each pair of suffix array boundaries which may occur in a binary search; for details, see [27]. However, the asymptotic improvement does not lead to an improved running time in practice.

An Optimal Method with Additional Tables

In this subsection we will demonstrate how to answer decision queries "Is P a substring of S?" in optimal $O(m)$ time. The same method allows one to find all z occurrences of a pattern P in optimal $O(m + z)$ time. To achieve this time complexity, one must be able to determine, for any ℓ-interval $[i..j]$, all its child intervals in constant time. We achieve this goal by enhancing the suffix array with the lcp-table and an additional table: the child-table childtab; see Figure 7.2. The child-table is a table of size $n + 1$ indexed from 0 to n and each entry contains three values: *up*, *down*, and *nextℓIndex*. Each of these three values requires 4 bytes in the worst case, but it is possible to store the same information in only one byte; see [2] for details. Formally, the values of each childtab-entry are defined as follows (we assume that $\min \emptyset = \max \emptyset = \bot$):

$$\begin{aligned}
\text{childtab}[i].up &= \min\{q \in [0..i-1] \mid \text{lcptab}[q] > \text{lcptab}[i] \text{ and} \\
&\qquad \forall k \in [q+1..i-1] : \text{lcptab}[k] \geq \text{lcptab}[q]\} \\
\text{childtab}[i].down &= \max\{q \in [i+1..n] \mid \text{lcptab}[q] > \text{lcptab}[i] \text{ and} \\
&\qquad \forall k \in [i+1..q-1] : \text{lcptab}[k] > \text{lcptab}[q]\} \\
\text{childtab}[i].nextℓIndex &= \min\{q \in [i+1..n] \mid \text{lcptab}[q] = \text{lcptab}[i] \text{ and} \\
&\qquad \forall k \in [i+1..q-1] : \text{lcptab}[k] > \text{lcptab}[i]\}
\end{aligned}$$

In essence, the child-table stores the parent-child relationship of lcp-intervals. Roughly speaking, for an ℓ-interval $[i..j]$ whose ℓ-indices are $i_1 < i_2 < \cdots < i_k$, the childtab$[i].down$ or childtab$[j+1].up$ value is used to determine the first ℓ-index i_1. The other ℓ-indices $i_2, \ldots i_k$ can be obtained from childtab$[i_1].nextℓIndex$, $\ldots$ childtab$[i_{k-1}].nextℓIndex$, respectively. Once these ℓ-indices are known, one can determine all the child intervals of $[i..j]$ according to the following lemma.

LEMMA 7.4 Let $[i..j]$ be an ℓ-interval. If $i_1 < i_2 < \cdots < i_k$ are the ℓ-indices in ascending order, then the child intervals of $[i..j]$ are $[i..i_1 - 1]$, $[i_1..i_2 - 1], \ldots, [i_k..j]$ (note that some of them may be singleton intervals).

Proof   Let $[l..r]$ be one of the intervals $[i..i_1 - 1]$, $[i_1..i_2 - 1], \ldots, [i_k..j]$. If $[l..r]$ is a singleton interval, then it is a child interval of $[i..j]$. Suppose that $[l..r]$ is an m-interval. Since $[l..r]$ does not contain an ℓ-index, it follows that $[l..r]$ is embedded in $[i..j]$. Because lcptab$[i_1] = $ lcptab$[i_2] = \cdots = $ lcptab$[i_k] = \ell$, there is no interval embedded in $[i..j]$ that encloses $[l..r]$. That is, $[l..r]$ is a child interval of $[i..j]$. Finally, it is not difficult to see that $[i..i_1 - 1]$, $[i_1..i_2 - 1], \ldots, [i_k..j]$ are all the child intervals of $[i..j]$, i.e., there cannot be any other child interval.

As an example, consider the enhanced suffix array in Figure 7.2. The 1-$[0..5]$ interval has the 1-indices 2 and 4. The first 1-index 2 is stored in childtab$[0].down$ and childtab$[6].up$. The second 1-index is stored in childtab$[2].nextℓIndex$. Thus, the child intervals of $[0..5]$ are $[0..1]$, $[2..3]$, and $[4..5]$. Before we show in detail how the child-table can be used to determine the child intervals of an lcp-interval in constant time, we address the problem of building the child-table efficiently.

Construction of the Child-Table

The childtab can be computed in linear time by a bottom-up traversal of the lcp-interval tree (Figure 7.6) as follows. At any stage, when the function *process* is applied to an ℓ-interval $[i..j]$, all its child intervals are known and have already been processed (note that $[i..j] \neq [0..n]$ must hold). Let $[l_1..r_1], [l_2..r_2], \ldots, [l_k..r_k]$ be the k child intervals of $[i..j]$, stored in its *childList*. If $k = 0$, then $[i..j]$ is a leaf in the lcp-interval tree. In this case, the ℓ-indices of $[i..j]$ are the indices $i + 1, i + 2, \ldots, j$. Otherwise, if $k > 0$, then the ℓ-indices of $[i..j]$ are the indices $l_2, \ldots, l_k$ plus all those indices from $[i..j]$ that are not contained in any of the child intervals (these indices correspond to singleton intervals). In our example from Figure 7.5, when the function *process* is applied to the 1-interval $[0..5]$, its child intervals are $[0..1]$, $[2..3]$, and $[4..5]$; hence 2 and 4 are the 1-indices of $[0..5]$. Let $i_1, \ldots, i_p$ be the ℓ-indices of $[i..j]$, in ascending order. The first ℓ-index i_1 is assigned to childtab$[j + 1].up$ and childtab$[i].down$. The other ℓ-indices $i_2, \ldots, i_p$ are stored in the *nextℓIndex* field of childtab$[i_1]$, childtab$[i_2]$, ..., childtab$[i_{p-1}]$, respectively. Finally, one has to determine the 0-indices $j_1, \ldots, j_q$ (in ascending order) of the interval $[0..n]$ (which remained on the stack) and store them in the *nextℓIndex* field of childtab$[j_1]$, childtab$[j_2]$, ..., childtab$[j_{q-1}]$. The resulting child-table is shown in Figure 7.2, where the fields 1, 2, and 3 of the childtab denote the *up*, *down*, and *nextℓIndex* field. However, the space requirement of the child-table can be reduced in two steps. First, the three fields *up*, *down*, and *nextℓIndex* of the childtab can be stored in one field of 4 bytes. Second, only one byte is used in practice; see [2] for details.

Determining Child Intervals in Constant Time

Given the child-table, the first step to locate the child intervals of an ℓ-interval $[i..j]$ in constant time is to find the first ℓ-index in $[i..j]$, i.e., the minimum of the set $\ell Indices(i, j)$. This is possible with the help of the *up* and *down* fields of the child-table:

LEMMA 7.5 For every ℓ-interval $[i..j]$, the following statements hold:

1. $i < $ childtab$[j + 1].up \leq j$ or $i < $ childtab$[i].down \leq j$.
2. childtab$[j + 1].up$ stores the first ℓ-index in $[i..j]$ if $i < $ childtab$[j + 1].up \leq j$.
3. childtab$[i].down$ stores the first ℓ-index in $[i..j]$ if $i < $ childtab$[i].down \leq j$.

Proof (1) First, consider index $j + 1$. Suppose lcptab$[j + 1] = \ell'$ and let I' be the corresponding ℓ'-interval. If $[i..j]$ is a child interval of I', then lcptab$[i] = \ell'$ and there is no ℓ-index in $[i + 1..j]$. Therefore, childtab$[j + 1].up = \min \ell Indices(i, j)$, and consequently $i < $ childtab$[j + 1].up \leq j$. If $[i..j]$ is not a child interval of I', then we consider index i. Suppose lcptab$[i] = \ell''$ and let I'' be the corresponding ℓ''-interval. Because lcptab$[j + 1] = \ell' < \ell'' < \ell$, it follows that $[i..j]$ is a child interval of I''. We conclude that childtab$[i].down = \min \ell Indices(i, j)$. Hence, $i < $ childtab$[i].down \leq j$.
(2) If $i < $ childtab$[j + 1].up \leq j$, then the claim follows from childtab$[j + 1].up = \min\{q \in [i + 1..j] \mid$ lcptab$[q] > $ lcptab$[j + 1]$, lcptab$[k] \geq $ lcptab$[q] \ \forall k \in [q + 1..j]\} = \min\{q \in [i + 1..j] \mid$ lcptab$[k] \geq $ lcptab$[q] \ \forall k \in [q + 1..j]\} = \min \ell Indices(i, j)$.
(3) Let i_1 be the first ℓ-index of $[i..j]$. Then lcptab$[i_1] = \ell > $ lcptab$[i]$ and for all $k \in$

$[i+1..i_1-1]$ the inequality $\mathsf{lcptab}[k] > \ell = \mathsf{lcptab}[i_1]$ holds. Moreover, for any other index $q \in [i+1..j]$, we have $\mathsf{lcptab}[q] \geq \ell > \mathsf{lcptab}[i]$ but *not* $\mathsf{lcptab}[i_1] > \mathsf{lcptab}[q]$.

Once the first ℓ-index i_1 of an ℓ-interval $[i..j]$ is found, the remaining ℓ-indices $i_2 < i_3 < \cdots < i_k$ in $[i..j]$, where $1 \leq k \leq |\Sigma|$, are obtained successively from the *nextℓIndex* field of $\mathsf{childtab}[i_1], \mathsf{childtab}[i_2], \ldots, \mathsf{childtab}[i_{k-1}]$. It follows that the child intervals of $[i..j]$ are the intervals $[i..i_1-1], [i_1..i_2-1], \ldots, [i_k..j]$; see Lemma 7.4. The pseudo-code implementation of the following function *getChildIntervals* takes a pair (i,j) representing an ℓ-interval $[i..j]$ as input and returns a list containing the pairs $(i, i_1-1), (i_1, i_2-1), \ldots, (i_k, j)$.

FIGURE 7.9: *getChildIntervals*, applied to an lcp-interval $[i..j] \neq [0..n]$.

     $\mathsf{intervalList} = [\,]$
     if $i < \mathsf{childtab}[j+1].up \leq j$ **then**
        $i_1 := \mathsf{childtab}[j+1].up$
     else $i_1 := \mathsf{childtab}[i].down$
     $add(\mathsf{intervalList}, (i, i_1-1))$
     while $\mathsf{childtab}[i_1].nextℓIndex \neq \bot$ **do**
        $i_2 := \mathsf{childtab}[i_1].nextℓIndex$
        $add(\mathsf{intervalList}, (i_1, i_2-1))$
        $i_1 := i_2$
     $add(\mathsf{intervalList}, (i_1, j))$

The function *getChildIntervals* runs in time $O(|\Sigma|)$. Since we assume that $|\Sigma|$ is a constant, *getChildIntervals* runs in constant time. Using *getChildIntervals* one can simulate every top-down traversal of a suffix tree on an enhanced suffix array. To this end, one can easily modify the function *getChildIntervals* to a function *getInterval* which takes an ℓ-interval $[i..j]$ and a character $a \in \Sigma$ as input and returns the child interval $[l..r]$ of $[i..j]$ (which may be a singleton interval) whose suffixes have the character a at position ℓ. Note that all the suffixes in $[l..r]$ share the same ℓ-character prefix because $[l..r]$ is a subinterval of $[i..j]$. If such an interval $[l..r]$ does not exist, *getInterval* returns $\bot$. Clearly, *getInterval* has the same time complexity as *getChildIntervals*.

With the help of Lemma 7.5, it is also easy to implement a function $getlcp(i,j)$ that determines the lcp-value of an lcp-interval $[i..j]$ in constant time as follows.

FIGURE 7.10: Function $getlcp(i,j)$

---

     if $i < \mathsf{childtab}[j+1].up \leq j$
     then return $\mathsf{lcptab}[\mathsf{childtab}[j+1].up]$
     else return $\mathsf{lcptab}[\mathsf{childtab}[i].down]$

Answering Queries in Optimal Time

Now we are in a position to show how enhanced suffix arrays can be used to answer decision queries of the type "Is P a substring of S?" in optimal $O(m)$ time. Moreover, enumeration queries of the type "Where are all z occurrences of P in S?" can be answered in optimal $O(m + z)$ time, totally independent of the size of S.

FIGURE 7.11: Answering decision queries.

$c := 0$
$queryFound := True$
$(i, j) := getInterval(0, n, P[c])$
while $(i, j) \neq \perp$ and $c < m$ and $queryFound = True$
  if $i \neq j$ **then**
    $\ell := getlcp(i, j)$
    $min := \min\{\ell, m\}$
    $queryFound := S[\mathsf{suftab}[i] + c \ldots \mathsf{suftab}[i] + min - 1] = P[c \ldots min - 1]$
    $c := min$
    $(i, j) := getInterval(i, j, P[c])$
  else $queryFound := S[\mathsf{suftab}[i] + c \ldots \mathsf{suftab}[i] + m - 1] = P[c \ldots m - 1]$
if $queryFound$ **then**
  $report(i, j)$ /* the P-interval */
else $print$ "pattern P not found"

The algorithm starts by determining with $getInterval(0, n, P[0])$ the lcp or singleton interval $[i..j]$ whose suffixes start with the character $P[0]$. If $[i..j]$ is a singleton interval, then pattern P occurs in S if and only if $S[\mathsf{suftab}[i] \ldots \mathsf{suftab}[i] + m - 1] = P$. Otherwise, if $[i..j]$ is an lcp-interval, then we determine its lcp-value ℓ by the function $getlcp$. Let $\omega = S[\mathsf{suftab}[i] \ldots \mathsf{suftab}[i] + \ell - 1]$ be the longest common prefix of the suffixes $S_{\mathsf{suftab}[i]}$, $S_{\mathsf{suftab}[i+1]}, \ldots, S_{\mathsf{suftab}[j]}$. If $\ell \geq m$, then pattern P occurs in S if and only if $\omega[0 \ldots m-1] = P$. Otherwise, if $\ell < m$, then we test whether $\omega = P[0 \ldots \ell - 1]$. If not, then P does not occur in S. If so, we search with $getInterval(i, j, P[\ell])$ for the ℓ'- or singleton interval $[i'..j']$ whose suffixes start with the prefix $P[0 \ldots \ell]$ (note that the suffixes of $[i'..j']$ have $P[0 \ldots \ell-1]$ as a common prefix because $[i'..j']$ is a subinterval of $[i..j]$). If $[i'..j']$ is a singleton interval, then pattern P occurs in S if and only if $S[\mathsf{suftab}[i'] + \ell \ldots \mathsf{suftab}[i'] + m - 1] = P[\ell \ldots m - 1]$. Otherwise, if $[i'..j']$ is an ℓ'-interval, let $\omega' = S[\mathsf{suftab}[i'] \ldots \mathsf{suftab}[i'] + \ell' - 1]$ be the longest common prefix of the suffixes $S_{\mathsf{suftab}[i']}, S_{\mathsf{suftab}[i'+1]}, \ldots, S_{\mathsf{suftab}[j']}$. If $\ell' \geq m$, then pattern P occurs in S if and only if $\omega'[\ell \ldots m-1] = P[\ell \ldots m-1]$ (or equivalently, $\omega[0 \ldots m-1] = P$). Otherwise, if $\ell' < m$, then we test whether $\omega[\ell \ldots \ell' - 1] = P[\ell \ldots \ell' - 1]$. If not, then P does not occur in S. If so, we search with $getInterval(i', j', P[\ell'])$ for the next interval, and so on.

Enumerative queries can be answered in optimal $O(m + z)$ time as follows. Given a pattern P of length m, we search for the P-interval $[l..r]$ using the preceding algorithm. This takes $O(m)$ time. Then we can report the start position of every occurrence of P in S by enumerating $\mathsf{suftab}[l], \ldots, \mathsf{suftab}[r]$. In other words, if P occurs z times in S, then reporting the start position of every occurrence requires $O(z)$ time in addition.

We would like to mention that the very recent results concerning $RMQs$ [19, 5, 31] can

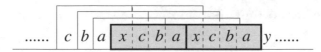

FIGURE 7.12: Chain of non-branching tandem repeats *axcb*, *baxc*, and *cbax*, derived by successively shifting a window one character to the left, starting from a branching tandem repeat *xcba*.

be used to obtain a different method to simulate top-down traversals of a suffix tree.

7.6 Computation of Tandem Repeats

As already mentioned at the beginning of Section 7.3, repeats play an important role in molecular biology. If the repeated segments are occurring adjacent to each other, then we speak of tandem repeats. Large tandem repeats can span hundreds of thousands of base pairs. There are two types of large tandem repeats: Those that contain genes and those that do not. If a large tandem repeat contains a gene, then it also contains a second copy of that gene. These genes are called paralogous genes. (By contrast, orthologous genes result from speciation rather than duplication.) An example of such tandemly duplicated genes is the human TRGV Locus, which is a region that contains nine repeated genes [26]. Examples of large tandem repeats that do not contain genes are those that are located in or near the centromeres and telomeres of the chromosomes, the so-called (*macro*) *satellites*.

Let us recall the mathematical definition of tandem repeats. A substring of S is a *tandem repeat* if it can be written as $\omega\omega$ for some nonempty string ω. An *occurrence* of a tandem repeat $\omega\omega = S[p\ldots p + 2|\omega| - 1]$ is represented by the pair $(p, |\omega|)$. Such an occurrence $(p, |\omega|)$ is *branching* if $S[p + |\omega|] \neq S[p + 2|\omega|]$.

There is an abundance of papers dealing with the efficient computation of tandem repeats; see, e.g., [33] for references. It is known that tandem repeats of a string S can be computed in $O(n)$ time in the worst case; see [15, 21]. Because these algorithms are quite complicated, we will present simpler algorithms, albeit with non-optimal worst case time complexities.

Stoye and Gusfield [33] described how all tandem repeats can be derived from branching tandem repeats by successively shifting a window to the left; see Figure 7.12. For this reason, we restrict ourselves to the computation of all branching tandem repeats.

A Brute Force Method

The simplest method to find branching tandem repeats is to process all lcp-intervals by top-down traversals of the lcp-interval tree. For a given ω-interval $[i..j]$ one checks whether there is a child interval $[l..r]$ (which may be a singleton interval) of $[i..j]$ such that $\omega\omega$ is a prefix of $S[\mathsf{suftab}[q]]$ for each $q \in [l..r]$. Using the child-table, such a child interval can be detected in $O(|\omega|)$ time (if it exists) with the algorithm of the previous section. Without the child-table, such a child interval can be found in $O(|\omega| \log(j - i))$ time by the algorithm of Manber and Myers [27] described in Section 7.5. This algorithm searches for ω in $[i..j]$. It turns out that the running time of the brute force algorithm is $O(n^2)$ (take, e.g., $S = a^n$). However, the expected length of the longest repeated subword is $O(\log n)$ according to [4]. As a consequence, in practice the brute force method is faster and more space efficient than other methods; see the experimental results in [1].

The Optimized Basic Algorithm

The *optimized basic algorithm* of [33] computes all branching tandem repeats in $O(n \log n)$ time. It is based on a traversal of the suffix tree, in which each branching node is annotated by its leaf list, i.e., by the set of leaves in the subtree below it. The leaf list of a branching node corresponds to an lcp-interval in the lcp-interval tree. As a consequence, it is not difficult to implement the optimized basic algorithm via a traversal of the lcp-interval tree. For didactic reasons, we start with the *basic algorithm* of [33], which is justified by the following lemma.

LEMMA 7.6 Let $\omega = S[p \ldots p+\ell-1]$, where $\ell = |\omega| > 0$, and let $[i..j]$ be the ω-interval. The following statements are equivalent.

1. (p, ℓ) is an occurrence of a branching tandem repeat.
2. $i \leq \mathsf{suftab}^{-1}[p] \leq j$ and $i \leq \mathsf{suftab}^{-1}[p+\ell] \leq j$ and $S[p+\ell] \neq S[p+2\ell]$.

Proof (1) $\Rightarrow$ (2): If (p, ℓ) is an occurrence of the tandem repeat ω, then $\omega = S[p \ldots p+\ell-1] = S[p+\ell \ldots p+2\ell-1]$. Since ω is the longest common prefix of $S_{\mathsf{suftab}[i]}, \ldots, S_{\mathsf{suftab}[j]}$, it follows that $i \leq \mathsf{suftab}^{-1}[p] \leq j$ and $i \leq \mathsf{suftab}^{-1}[p+\ell] \leq j$. Furthermore, $S[p+\ell] \neq S[p+2\ell]$ because (p, ℓ) is branching.
(2) $\Rightarrow$ (1): If $i \leq \mathsf{suftab}^{-1}[p] \leq j$ and $i \leq \mathsf{suftab}^{-1}[p+\ell] \leq j$, then the longest common prefix of S_p and $S_{p+\ell}$ has length at least ℓ. Thus (p, ℓ) is an occurrence of a tandem repeat. Since $S[p+\ell] \neq S[p+2\ell]$, this occurrence is branching.

FIGURE 7.13: Basic algorithm for the computation of tandem repeats

---

for each ℓ-interval $[i..j]$ with $\ell > 0$ **do**
    for $q := i$ **to** j **do**
        $p := \mathsf{suftab}[q]$
        if $i \leq \mathsf{suftab}^{-1}[p+\ell] \leq j$ **and** $S[p+\ell] \neq S[p+2\ell]$
        then report (p, ℓ) as an occurrence of a branching tandem repeat

---

The basic algorithm finds all occurrences of branching tandem repeats in time proportional to the sum of the sizes of all ℓ-intervals, which is $O(n^2)$ (take, e.g., $S = a^n$). A simple modification of the basic algorithm yields the optimized basic algorithm of [33] as follows. We saw that if (p, ℓ) is an occurrence of the tandem repeat ω, then $\mathsf{suftab}^{-1}[p]$ and $\mathsf{suftab}^{-1}[p+\ell]$ are elements of the ω-interval $[i..j]$. The modification relies on the following observation: If the occurrence is branching, then $\mathsf{suftab}^{-1}[p]$ and $\mathsf{suftab}^{-1}[p+\ell]$ belong to different child intervals of $[i..j]$. Thus, we can omit all indices of one child $[l..r]$ of $[i..j]$ in the second for-loop of the basic algorithm, provided that for each $q \in [i..j] \setminus [l..r]$ we do not only look forward from $p := \mathsf{suftab}[q]$ (i.e., consider $p + \ell$) but we also look backward from it (i.e., we must also consider $p - \ell$). This is made precise in the next algorithm.

Figure 7.6 can be implemented by a bottom-up traversal of the lcp-interval tree. Then, for each ℓ-interval, a child interval of maximum size can be determined in constant time. Since the largest child interval is always excluded in the second for-loop of Figure 7.6, the

---

FIGURE 7.14: Optimized basic algorithm for the computation of tandem repeats

---

for each ℓ-interval $[i..j]$ **do**
    determine the child interval $[l..r]$ of maximum size among all children of $[i..j]$
    for $q \in [i..j] \setminus [l..r]$ **do**
        $p := \mathsf{suftab}[q]$
        if $i \leq \mathsf{suftab}^{-1}[p+\ell] \leq j$ and $S[p+\ell] \neq S[p+2\ell]$
        then report (p, ℓ) as an occurrence of a branching tandem repeat
        if $i \leq \mathsf{suftab}^{-1}[p-\ell] \leq j$ and $S[p-\ell] \neq S[p+\ell]$
        then report $(p-\ell, \ell)$ as an occurrence of a branching tandem repeat

---

algorithm runs in $O(n \log n)$ time; see [33]. Figure 7.6 requires the tables lcptab, suftab, and suftab^{-1} plus some space for the stack used during the bottom-up traversal of the lcp-interval tree.

It is possible to further improve Figure 7.6 by exploiting the fact that $S[p] = S[p + |\omega|]$ for an occurrence $(p, |\omega|)$ of a branching tandem repeat $\omega\omega$. Namely, if $p + |\omega| = \mathsf{suftab}[q]$ for some q in the ω-interval $[i..j]$, then p must occur in the child interval $[l_a..r_a]$ storing the suffixes of S which have ωa as a prefix, where $a = S[\mathsf{suftab}[i]] = S[p]$. The interested reader is referred to [1] for details and for experimental results.

7.7 Space Efficient Computation of Maximal Exact Matches

It is our next goal to compute the maximal exact matches of two genomes in a very space efficient manner. Because the algorithm is based on suffix links, we show how to incorporate them into our framework. Let us first recall the definition of suffix links. In the following, we denote a node u in the suffix tree by $\overline{\omega}$ if and only if the concatenation of the edge-labels on the path from the root to u spells out the string ω. It is a property of suffix trees that for any internal node $\overline{a\omega}$, there is also an internal node $\overline{\omega}$. A pointer from $\overline{a\omega}$ to $\overline{\omega}$ is called a *suffix link*.

DEFINITION 7.7   Let $S_{\mathsf{suftab}[i]} = a\omega$. If index j, $0 \leq j < n$, satisfies $S_{\mathsf{suftab}[j]} = \omega$, then we denote j by $\mathsf{link}[i]$ and call it the suffix link (index) of i.

The suffix link of i can be computed with the help of the inverse suffix array as follows.

LEMMA 7.7   If $\mathsf{suftab}[i] < n$, then $\mathsf{link}[i] = \mathsf{suftab}^{-1}[\mathsf{suftab}[i] + 1]$.

Proof   Let $S_{\mathsf{suftab}[i]} = a\omega$. Since $\omega = S_{\mathsf{suftab}[i]+1}$, $\mathsf{link}[i]$ must satisfy $\mathsf{suftab}[\mathsf{link}[i]] = \mathsf{suftab}[i] + 1$. This immediately proves the lemma.

Under a different name, the function link appeared already in [13].

DEFINITION 7.8   Given ℓ-interval $[i..j]$, the smallest lcp-interval $[l..r]$ satisfying the inequality $l \leq \mathsf{link}[i] < \mathsf{link}[j] \leq r$ is called the *suffix link interval* of $[i..j]$.

Suppose that the ℓ-interval $[i..j]$ corresponds to an internal node $\overline{a\omega}$ in the suffix tree. Then there is a suffix link from node $\overline{a\omega}$ to the internal node $\overline{\omega}$. The following lemma states that node $\overline{\omega}$ corresponds to the suffix link interval of $[i..j]$.

LEMMA 7.8 Given the $a\omega$-interval ℓ-$[i..j]$, its suffix link interval is the ω-interval, which has lcp-value $\ell - 1$.

Proof Let $[l..r]$ be the suffix link interval of $[i..j]$. Because the lcp-interval $[i..j]$ is the $a\omega$-interval, $a\omega$ is the longest common prefix of $S_{\mathsf{suftab}[i]}, \ldots, S_{\mathsf{suftab}[j]}$. Consequently, ω is the longest common prefix of $S_{\mathsf{suftab}[\mathsf{link}[i]]}, \ldots, S_{\mathsf{suftab}[\mathsf{link}[j]]}$. It follows that ω is the longest common prefix of $S_{\mathsf{suftab}[l]}, \ldots, S_{\mathsf{suftab}[r]}$, because $[l..r]$ is the smallest lcp-interval satisfying $l \le \mathsf{link}[i] < \mathsf{link}[j] \le r$. That is, $[l..r]$ is the ω-interval and thus it has lcp-value $\ell - 1$.

Construction of the Suffix Link Table

In order to incorporate suffix links into the enhanced suffix array, we proceed as follows. In a preprocessing step, we compute for every ℓ-interval $[i..j]$ its suffix link interval $[l..r]$ and store the left and right boundaries l and r at the first ℓ-index of $[i..j]$. The corresponding table, indexed from 0 to n is called suffix link table and denoted by suflink; see Figure 7.2 for an example. Note that the lcp-value of $[l..r]$ need not be stored because it is known to be $\ell - 1$. Thus, the space requirement for suflink is $2 \cdot 4n$ bytes in the worst case. To compute the suffix link table suflink, the lcp-interval tree is traversed in a breadth first left-to-right manner. For every lcp-value encountered, we hold a list of intervals of that lcp-value, which is initially empty. Whenever an ℓ-interval is computed, it is appended to the list of ℓ-intervals; this list is called ℓ-list in what follows. In the example of Figure 7.5, this gives

0-list: $[0..10]$
1-list: $[0..5], [8..9]$
2-list: $[0..1], [4..5], [6..7]$
3-list: $[2..3]$

Note that the ℓ-lists are automatically sorted in increasing order of the left-boundary of the intervals and that the total number of ℓ-intervals in the ℓ-lists is at most n. For every lcp-value $\ell > 0$ and every ℓ-interval $[i..j]$ in the ℓ-list, we proceed as follows. We first compute $\mathsf{link}[i]$ according to Lemma 7.7. Then, by a binary search in the $(\ell - 1)$-list, we search in $O(\log n)$ time for the interval $[l..r]$ such that l is the largest left boundary of all $(\ell - 1)$-intervals with $l \le \mathsf{link}[i]$. This interval is the suffix link interval of $[i..j]$. Finally, we determine in constant time the first ℓ-index of $[i..j]$ according to Lemma 7.5 and store l and r there. Because there are less than n lcp-intervals and for each interval the binary search takes $O(\log n)$ time, the preprocessing phase requires $O(n \log n)$ time. Table suftab^{-1} and the ℓ-lists require $O(n)$ space, but they are only used in the preprocessing phase and can be deleted after the computation of the suffix link table.

Theoretically, it is possible to compute the suffix link intervals in time $O(n)$ via the construction of the suffix tree. By avoiding the binary search over the ℓ-lists and reducing the problem of computing the suffix link intervals to the problem of answering range minimum queries, it is also possible to give a linear time algorithm without intermediate construction of the suffix tree; see [2] for details.

Space Efficient Computation of MEMs for two Genomes

In a previous section we have described an algorithm that uses the enhanced suffix array of the concatenation of S_1 and S_2 to compute *MEMs*. The following algorithm only needs the enhanced suffix array of S_1. It computes all *MEMs* of length at least ℓ by matching S_2 against the enhanced suffix array of S_1. The matching process delivers substrings of S_2 represented by *locations* in the enhanced suffix array. This notion is defined as follows: Suppose that u occurs as a substring of S_1 and consider the enhanced suffix array of S_1. Let ω be the maximal prefix of u such that there is an ω-interval $[l..r]$ and $u = \omega t$ for some string t. We distinguish between two cases:

- If $t = \varepsilon$, then $u = \omega$ and $[l..r]$ is defined to be the *location of u*.
- If $t \neq \varepsilon$, then $([l..r], t)$ is defined to be the *location of u*.

The *location of u* is denoted by $loc(u)$.

The enhanced suffix array represents all suffixes $S_1[i \ldots |S_1| - 1]\$$ of $S_1\$$. The algorithm processes S_2 suffix by suffix from longest to shortest. In the jth step, for $0 \leq j \leq |S_2| - 1$, the algorithm processes suffix $R_j = S_2[j \ldots |S_2| - 1]$ and computes the locations of two prefixes minp_j and maxp_j of R_j defined as follows:

- maxp_j is the longest prefix of R_j that is a substring of S_1.
- minp_j is the prefix of maxp_j of length $\min\{\ell, |\text{maxp}_j|\}$.

If $|\text{minp}_j| < \ell$, then less than ℓ of the first characters of R_j match a substring of S_1. If $|\text{minp}_j| = \ell$, then at least the first ℓ characters of R_j match a substring of S_1. Now determine an lcp-interval $[l'..r']$ as follows: If the location of minp_j is the lcp-interval $[l..r]$ then let $[l'..r'] := [l..r]$. If the location of minp_j is of the form $([l..r], at)$ for some lcp-interval $[l..r]$, some character a, and some string t, then let $[l'..r'] := getInterval([l..r], a)$. By construction, at least one suffix in the lcp-interval $[l'..r']$ matches at least the first ℓ characters of R_j. To extract the *MEMs*, $[l'..r']$ is traversed in a depth first order. The depth first traversal maintains for each ω-interval encountered the length of the longest common prefix of ω and R_j. Each time a leaf-interval $[l''..l'']$ is visited, one first checks whether $S_1[i - 1] \neq S_2[j - 1]$, where $i = \text{suftab}[l'']$. If this is the case, then $(\bot, i, j)$ is a left maximal exact match and one determines the length c of the longest common prefix of $S_{\text{suftab}[i]}$ and R_j. By construction, $c \geq \ell$ and $S_1[i + c] \neq S_2[j + c]$. Hence (c, i, j) is a *MEM*. Now consider the different steps of the algorithm in more detail:

Computation of $loc(\text{minp}_j)$: For $j = 0$, one computes $loc(\text{minp}_j)$ by greedily matching $S_2[0 \ldots \ell - 1]$ against the suffix array, using the algorithms described in Section 7.5. For each $j, 1 \leq j \leq |S_2| - 1$, one considers two cases: (a) follow the suffix link of $loc(\text{minp}_{j-1})$, if this is an lcp-interval. (b) follow the suffix link of $[l..r]$ if $loc(\text{minp}_{j-1})$ is of the form $([l..r], w)$. This shortcut via the suffix link leads to an lcp-interval on the path from the root of the lcp-interval tree to $loc(\text{minp}_j)$. Starting from this location, one matches the next characters of R_j. The method is similar to the matching-statistics computation of [10], and one can show that its overall running time for the computation of all $loc(\text{minp}_j)$, $0 \leq j \leq |S_2| - 1$, is $O(|S_2|)$.

Computation of $loc(\text{maxp}_j)$: Starting from $loc(\text{minp}_j)$ one computes $loc(\text{maxp}_j)$ by greedily matching $S_2[|\text{minp}_j| \ldots |S_2| - 1]$ against the enhanced suffix array of S_1. To facilitate the computation of longest common prefixes, one keeps track of the list of lcp intervals on the path from $loc(\text{minp}_j)$ to $loc(\text{maxp}_j)$. This list is called the *match path*. Since $|\text{maxp}_{j-1}| \geq 1$ implies $|\text{maxp}_j| \geq |\text{maxp}_{j-1}| - 1$, we do not always have to match the edges of the lcp-interval tree completely against the corresponding substring of S_2. Instead, to reach $loc(\text{maxp}_j)$, one rescans most of the edges by only looking at the first character of the edge label to determine the appropriate edge to follow. Thus the total time for this step is $O(|S_2| + \alpha)$ where α is the total length of all match paths. α is upper bounded by the total size β of the subtrees below $loc(\text{minp}_j)$, $0 \leq j \leq |S_2| - 1$. β is upper bounded by the number z_r of right maximal exact matches between S_1 and S_2. Hence the running time for this step of the algorithm is $O(|S_2| + z_r)$.

Depth first traversal: This maintains an *lcp*-stack which stores for each visited lcp-interval, say the ω-interval $[l..r]$, a pair of values (*onmatchpath*, *lcpvalue*), where the boolean value *onmatchpath* is true if and only if $[l..r]$ is on the match path, and *lcpvalue* stores the length of the longest common prefix of ω and R_j. Given the match path, the *lcp*-stack can be maintained in constant time for each branching node visited. For each leaf-interval $[l..l]$ visited during the depth first traversal, the *lcp*-stack allows to determine in constant time the length of the longest common prefix of $S_{\text{suftab}[l]}$ and R_j. As a consequence, the depth first traversal requires time proportional to the size of the traversed subtree. As exploited above, this is bounded by the number of right maximal matches in the traversed subtree. Thus the total time for all depth first traversals of the subtrees below $loc(\text{minp}_j)$, $0 \leq j \leq |S_2| - 1$, is $O(z_r)$.

Altogether, the algorithm described here runs in $O(|S_1|+|S_2|+z_r)$ time and $O(|S_1|)$ space, where z_r is the number of right maximal matches. Table 7.2 shows the values computed when running the algorithm for two concrete sequences.

Acknowledgements

M.I.A. and E.O. were supported by DFG-grant Oh 53/4-1. S.K. was supported in part by DFG-grant Ku 1257/3-1.

References

[1] M.I. Abouelhoda, S. Kurtz, and E. Ohlebusch. The Enhanced Suffix Array and its Applications to Genome Analysis. In *Proceedings of the Second Workshop on Algorithms in Bioinformatics*, pages 449–463. Lecture Notes in Computer Science 2452, Springer-Verlag, 2002.

[2] M.I. Abouelhoda, S. Kurtz, and E. Ohlebusch. Replacing Suffix Trees with Enhanced Suffix Arrays. *Journal of Discrete Algorithms*, 2:53–86, 2004.

[3] A. Apostolico. The Myriad Virtues of Subword Trees. In *Combinatorial Algorithms on Words, Springer Verlag*, pages 85–96, 1985.

[4] A. Apostolico and W. Szpankowski. Self-Alignments in Words and Their Applications. *Journal of Algorithms*, 13:446–467, 1992.

[5] M.A. Bender and M. Farach-Colton. The LCA Problem Revisited. In *Latin American Theoretical Informatics*, pages 88–94, 2000.

[6] N. Bray, I. Dubchak, and L. Pachter. AVID: A global alignment program. *Genome Research*, 13:97–102, 2003.

[7] M. Brudno, C.B. Do, G.M. Cooper, and M.F. Kim *et al.* LAGAN and Multi-LAGAN: Efficient Tools for large-scale Multiple Alignment of Genomic DNA. *Genome Research*, 13:721–731, 2003.

[8] M. Burrows and D.J. Wheeler. A Block-Sorting Lossless Data Compression Algorithm. Research Report 124, Digital Systems Research Center, 1994.

[9] P. Chain, S. Kurtz, E. Ohlebusch, and T.R. Slezak. An Applications-Focused Review

j	$minp_j$:$loc(minp_j)$	remainder of matchpath	depth first traversal	linking locations
0	ac:([0..5],c)		S(0) C(0,0) M(2) S(4) C(4,0) M(2)	ac:([0..5],c)→c:([0..10],c)
1	c:([0..10],c)			c:([0..10],c)→:[0..10]
2	t:[8..9]			t:[8..9]→:[0..10]
3	ta:([8..9],a)		S(7) C(7,3) M(2)	ta:([8..9],a)→a:[0..5]
4	aa:[0..1]	aaaca:([0..1],aca)	S(2) C(2,4) M(5) S(3) C(3,4) M(2)	aa:[0..1]→a:[0..5]
5	aa:[0..1]	aaca:([0..1],ca)	S(2) C(2,5) M(2) S(3) C(3,5)	aa:[0..1]→a:[0..5]
6	ac:([0..5],c)	[2..3] acaaac:([2..3],aac)	S(0) C(0,6) M(6) S(4) C(4,6)	ac:([0..5],c)→c:([0..10],c)
7	ca:[6..7]	caaac:([6..7],aac)	S(1) C(1,7) S(5) C(5,7)	ca:[6..7]→a:[0..5]
8	aa:[0..1]	aaac:([0..1],ac)	S(2) C(2,8) S(3) C(3,8) M(2)	aa:[0..1]→a:[0..5]
9	aa:[0..1]	aac:([0..1],c)	S(2) C(2,9) M(2) S(3) C(3,9)	aa:[0..1]→a:[0..5]
10	ac:([0..5],c)		S(0) C(0,10) M(2) S(4) C(4,10)	ac:([0..5],c)→c:([0..10],c)
11	c:([0..10],c)			

TABLE 7.2 Computation of MEMs of length $\geq \ell = 2$ between $S_1 = acaaacatat$ and $S_2 = acttaaacaaact$, using the enhanced suffix array of S_1 (see Figure 7.2). For each $j \in [0, |S_2| - \ell]$, we show the values $minp_j$ and $loc(minp_j)$ separated by a colon. Furthermore, for the case that $|minp_j| \geq \ell$, we show the remainder of the matchpath, i.e., all elements, except for $loc(minp_j)$. The fourth column shows the actions performed during the depth first traversal of the subtree below $minp_j$. $S(i)$ means that the suffix $S_1[i \ldots n - 1]\$$ of S_1 is visited, where $n = |S_1|$. $C(i, j)$ means that $S_1[i \ldots n - 1]\$$ and R_j are checked for left maximality. If this is the case, then $M(q)$ means that a *MEM* of length q is output. The final column shows the link which is followed to obtain a location representing a prefix of $minp_{j+1}$. Consider the situation for $j = 6$. Then $minp_j = ac$ is obtained by scanning $R_j = acaaact$ starting at the location $[0..5]$ of a. Further scanning $aaact$ starting at location $loc(ac) = ([0..5], c)$ visits the intermediate lcp-interval $[2..3]$ end ends at $loc(acaaac) = ([2..3], aac) = maxp_j$. The depth first traversal of subtree below the lcp-interval $[2..3]$ visits the suffixes of S_1 starting position 0 and 4, respectively. For both suffixes, it is checked whether their left context is different than the left context of R_j. This is the case for $S_1[0 \ldots n - 1]\$$, but not for $S_1[4 \ldots n - 1]\$$. The length 6 of the longest common prefix of $S_1[0 \ldots n - 1]\$$ is determined from the depth of $maxp_j$.

of Comparative Genomics Tools: Capabilities, Limitations and Future Challenges. *Briefings in Bioinformatics*, 4(2):105–123, 2003.

[10] W.I. Chang and E.L. Lawler. Sublinear Approximate String Matching and Biological Applications. *Algorithmica*, 12(4/5):327–344, 1994.

[11] A.L. Delcher, S. Kasif, R.D. Fleischmann, and J. Peterson *et al.* Alignment of Whole Genomes. *Nucleic acids research*, 27:2369–2376, 1999.

[12] G. Gonnet, R. Baeza-Yates, and T. Snider. New Indices for Text: PAT trees and PAT arrays. In W. Frakes and R. A. Baeza-Yates, editors, *Information Retrieval: Algorithms and Data Structures*, pages 66–82. Prentice-Hall, Englewood Cliffs, NJ, 1992.

[13] R. Grossi and J.S. Vitter. Compressed Suffix Arrays and Suffix Trees with Applications to Text Indexing and String Matching. In *ACM Symposium on the Theory of Computing*, pages 397–406. ACM Press, 2000.

[14] D. Gusfield. *Algorithms on Strings, Trees, and Sequences*. Cambridge University Press, New York, 1997.

[15] D. Gusfield and J. Stoye. Linear Time Algorithms for Finding and Representing all the Tandem Repeats in a String. Report CSE-98-4, Computer Science Division, University of California, Davis, 1998.

[16] M. Höhl, S. Kurtz, and E. Ohlebusch. Efficient Multiple Genome Alignment. *Bioinformatics*, 18(Suppl. 1):S312–S320, 2002.

[17] J. Kärkkäinen and P. Sanders. Simple Linear Work Suffix Array Construction. In *Proceedings of theInternational Colloquium on Automata, Languages and Programming*, pages 943–955. Lecture Notes in Computer Science 2719, Springer Verlag, 2003.

[18] T. Kasai, G. Lee, H. Arimura, and S. Arikawa *et al.* Linear-Time Longest-Common-Prefix Computation in Suffix Arrays and its Applications. In *Proceedings of the 12th Annual Symposium on Combinatorial Pattern Matching*, pages 181–192. Lecture Notes in Computer Science 2089, Springer-Verlag, 2001.

[19] D.K. Kim, J.S. Sim, H. Park, and K. Park. Linear-Time Construction of Suffix Arrays. In *Proceedings of the 14th Annual Symposium on Combinatorial Pattern Matching (CPM)*, volume 2676 of *LNCS*, pages 186–199. Springer-Verlag, 2003.

[20] P. Ko and S. Aluru. Space Efficient Linear Time Construction of Suffix Arrays. In *Proceedings of the 14th Annual Symposium on Combinatorial Pattern Matching (CPM)*, volume 2676 of *LNCS*, pages 200–210. Springer-Verlag, 2003.

[21] R. Kolpakov and G. Kucherov. Finding Maximal Repetitions in a Word in Linear Time. In *Symposium on Foundations of Computer Science*, pages 596–604. IEEE Computer Society, 1999.

[22] S. Kurtz. Reducing the Space Requirement of Suffix Trees. *Software—Practice and Experience*, 29(13):1149–1171, 1999.

[23] S. Kurtz, J.V. Choudhuri, E. Ohlebusch, and C. Schleiermacher *et al.* REPuter: The Manifold Applications of Repeat Analysis on a Genomic Scale. *Nucleic acids research*, 29(22):4633–4642, 2001.

[24] S. Kurtz and S. Lonardi. Computational Biology. In D.P. Mehta and S. Sahni, editor, *Handbook on Data Structures and Applications*. CRC Press, 2004.

[25] E.S. Lander, L.M. Linton, B. Birren, and C. Nusbaum *et al.* Initial Sequencing and Analysis of the Human Genome. *Nature*, 409:860–921, 2001.

[26] M.P. LeFranc, A. Forster, and T.H. Rabbitts. Rearrangement of two distinct T-cell Gamma-Chain Variable-Region genes in human DNA. *Nature*, 319(6052):420–422, 1986.

[27] U. Manber and E.W. Myers. Suffix Arrays: A New Method for On-Line String Searches. *SIAM Journal on Computing*, 22(5):935–948, 1993.

[28] B. Morgenstern. A Space Efficient Algorithm for Aligning Large Genomic Sequences. *Bioinformatics*, 16:948–949, 2000.

[29] S.B. Needleman and C.D. Wunsch. A General Method Applicable to the Search for Similarities in the Amino-Acid Sequence of Two Proteins. *Journal of Molecular Biology*, 48:443–453, 1970.

[30] C. O'Keefe and E. Eichler. The Pathological Consequences and Evolutionary Implications of Recent Human Genomic Duplications. In *Comparative Genomics*, pages 29–46. Kluwer Press, 2000.

[31] K. Sadakane. Succinct Representations of lcp Information and Improvements in the Compressed Suffix Arrays. In *Proceedings of ACM-SIAM SODA*, pages 225–232, 2002.

[32] S. Schwartz, Z. Zhang, K.A. Frazer, and A. Smit *et al.* PipMaker–a web server for aligning two genomic DNA sequences. *Genome Research*, 10(4):577–586, 2000.

[33] J. Stoye and D. Gusfield. Simple and Flexible Detection of Contiguous Repeats Using a Suffix Tree. *Theoretical Computer Science*, 270(1-2):843–856, 2002.

[34] P. Weiner. Linear Pattern Matching Algorithms. In *Proceedings of the 14th IEEE Annual Symposium on Switching and Automata Theory*, pages 1–11, 1973.

III

Genome Assembly and EST Clustering

8

Computational Methods for Genome Assembly

Xiaoqiu Huang
Iowa State University

8.1 Introduction

Advances in genomics are driven by genome sequencing projects. The goal of a genome sequencing project for an organism is to determine the genome sequence of the organism. Only short sequences of up to 1000 base pairs (bp) can be directly produced by sequencing machines. However, genomes are huge; bacterial genomes are a few million base pairs (Mb) in size, animal genomes can be a few billion base pairs (Gb) in size, and plant genomes can be tens of Gb in size. Thus long genome sequences have to be constructed from short sequences, which is called fragment assembly or genome assembly.

Whole-genome shotgun sequencing (WGS) is an efficient strategy for producing a draft genome sequence of an organism. In this strategy, multiple copies of the genome are broken into pieces. Both ends of every piece are read by automated sequencing machines to produce two short sequences called reads, one from each strand of the piece. The size of the piece is measured, which is the distance between the two reads on the genome. The two reads along with their distance and orientation information are referred to as a read pair. Figure 8.1 illustrates an application of the shotgun sequencing procedure in determination of the sequence of a small DNA segment. Because of space restriction, the example in Figure 8.1 is not realistic. An example involving real sequences is available at (`http://genome.cs. mtu.edu/cap/data/`).

A read is a word of length up to 1000 over the alphabet $\{A, C, G, T, N\}$, where N represents an ambiguous base. Each letter at a position of the read is called a base. The left and right ends of a read are called the $5'$ and $3'$ ends of the read. Errors of missing a base, adding a base, and misreading a base are occasionally made by sequencing machines in generation of bases. More errors occur at the $3'$ end of a read. The accuracy of a read is represented by a sequence of numbers called quality values, one quality value per base. The quality value of a base is $q = -10 \log p$, where p is the estimated error probability for

A

```
5'      AGCTTTGTGGGGGAGAAAGTGGATGAGGAGGGGCTGAAGAAGCTGATGGG      3'
3'      TCGAAACACCCCCTCTTTCACCTACTCCTCCCCGACTTCTTCGACTACCC      5'
```

B

```
            s1.b                          s2.b
5'      agctttgtgggGGAGAAAGTGGAT|gaggaggggctgAAGAAGCTGATGGG      3'
3'      TCGAAACACCCCctctttcaccta|CTCCTCCCCGACttcttcgactaccc      5'
                    s1.g                          s2.g

            s3.b                       s4.b
5'      AGCTT|tgtgggggagaAAGTGGATGAGG|aggggctgaagaAGCTGATGGG      3'
3'      TCGAA|ACACCCCCTCTTtcacctactcc|TCCCCGACTTCTTcgactaccc      5'
                    s3.g                          s4.g
```

C

```
s1.b:  5'   agctttgtggg    3'      s1.g:  5'   atccactttctc   3'
s2.b:  5'   gaggaggggctg   3'      s2.g:  5'   cccatcagcttctt 3'
s3.b:  5'   tgtgggggaga    3'      s3.g:  5'   cctcatccact    3'
s4.b:  5'   aggggctgaaga   3'      s4.g:  5'   cccatcagc      3'
```

D

```
s1.b+ agctttgtggg
s3.b+      tgtgggggaga
s1.g-           gagaaagtggat
s3.g-              agtggatgagg
s2.b+                  gaggaggggctg
s4.b+                     aggggctgaaga
s2.g-                            aagaagctgatggg
s4.g-                               gctgatggg
      ----------------------------------------------------
c 5'   AGCTTTGTGGGGGAGAAAGTGGATGAGGAGGGGCTGAAGAAGCTGATGGG   3'
```

FIGURE 8.1: An application of the shotgun DNA sequencing procedure in determination of
a small DNA segment. (A) A DNA segment of complementary strands. (B)
Generation of reads. Two copies of the DNA segment are cut, at positions indicated
by vertical lines, into pieces. Two reads, shown in lower case letters, are produced
from a piece, with one read from the 5' end of each strand of the piece. (C) A list
of reads in given orientation. Each read is conventionally written with its 5' end on
the left and its 3' end on the right. (D) A contig of reads. The contig is represented
by an alignment of reads. The reads in the contig are ordered and oriented with
respect to one strand of the DNA segment. The names of the reads are shown
on the left, where the '+' sign indicates that the read is in given orientation and
the '-' sign indicates that the read is in reverse orientation, that is, the read is the
reverse complement of the read in given orientation. The consensus sequence of
the contig is shown on the bottom.

the base. For example, a quality value of 30 corresponds to an error probability of 1 in 1000. For each read, its base sequence and its quality value sequence are produced by a base-calling program from traces of digital signals from a sequencing machine [8, 7]. The input to an assembly program consists of a file of base sequences, a corresponding file of quality value sequences, and a file of read pairs.

In addition to sequencing errors, reads occasionally have two other types of errors. First, a read may contain a contaminant at its ends, where the contaminant is a short sequence of bases from sources other than the genome of the organism. Most of the contaminants can be found and removed by comparing the sequences of reads with known contaminant sequences. However, the remaining contaminants have to be addressed by the assembly program. Second, a read may be a concatenation of two regions that are far apart on the genome, which is called a chimeric read. Chimeric reads have to be addressed by the assembly program.

The most critical issue in assembly is how to deal with repetitive regions of the genome, which are highly similar in sequence. Most of the major errors in assembly are due to repetitive regions. This issue must be addressed by the assembly program.

A number of whole-genome assembly programs have recently been developed: Celera Assembler [22], ARACHNE [3, 17], RePS [28], JAZZ [2], Phusion [21], PCAP [16], and Atlas [12]. The PaCE program was originally developed for clustering and assembly of EST sequences [18] and was later adapted for assembly of the maize genome [6]. Those programs are based on the experiences of previous sequence assembly programs [26, 23, 13, 10, 19, 27, 15, 20, 5, 24].

In this book chapter, we first describe a general whole-genome assembly algorithm. Then we present special features of existing whole-genome assembly programs. We conclude with comments on the efficiency and accuracy of existing assembly programs, and on future developments in this area.

8.2 A General Whole-Genome Assembly Algorithm

We start with definitions of a few terms used in an assembly algorithm. An overlap between two reads is an overlapping alignment of two reads with the maximum score. An overlapping alignment of two reads consists of base matches, base mismatches, and gaps. Note that every base of the two reads is on the overlapping alignment. A gap involving an end of a read is called terminal. Base matches are given a positive score, whereas base mismatches and internal gaps are given negative scores. Terminal gaps are given a score of 0. The score of an overlapping alignment is the sum of scores of each match, each mismatch, and each gap. A contig is a list of overlapping reads that are ordered and oriented with respect to a region of the target genome. A contig is represented by a multiple alignment of reads and a consensus sequence. A scaffold is a list of contigs that are ordered and oriented with respect to the target genome. A scaffold is represented by an ordered and oriented list of contig consensus sequences. Figure 8.2 provides examples of overlaps, contigs, and scaffolds.

A general whole-genome assembly algorithm works in three major phases. In the first phase, overlaps between reads are computed and highly repetitive regions of reads are identified. The overlap computation step and the repeat identification step are performed alternately. A region of a read is identified to be highly repetitive if it occurs in many overlaps. Once a highly repetitive region of a read is found, no overlap involving the region is computed. A non-highly repetitive region is called a unique region. In the second phase, the 5′ and 3′ clipping positions of reads are determined based on unique overlaps. Poor end regions of each read are removed. Unique overlaps are ranked in a decreasing order

A

5'   AGCTTTGTGGGGGAGAAAGTGGATGAGGAGGGGCTGAAGAAGCTGATGGG   3'
3'   TCGAAACACCCCCTCTTTCACCTACTCCTCCCCGACTTCTTCGACTACCC   5'

B

                              t1.b
5'   AGCTTT|gtggggggagaaagtggatGAGGAGGGGCTGAAGAAG|CTGATGGG   3'
3'   TCGAAA|CACCCCCTCTTTCACCTACTCctccccgacttcttc|GACTACCC   5'
                                                    t1.g

                            t2.b
5'   AGCTTTGTGGGGGAGA|aagtggatgaGGAGGGGCTGAAGAAGCTGATG|GG   3'
3'   TCGAAACACCCCCTCT|TTCACCTACTCCTCCCCGgacttcttcgactac|CC   5'
                                                  t2.g

C

t1.b+   gtggggggagaaagtggat
t2.b+   aagtggatga
t1.g-   gaggggctgaagaag
t2.g-   ctgaagaagctgatg
         --------------------   ---------------------
c1+ 5'   GTGGGGGAGAAAGTGGATGA   3'
c2- 5'   GAGGGGCTGAAGAAGCTGATG   3'

D

t2.g+   catcagcttcttcag
t1.g+   cttcttcagccccctc
t2.b-   tcatccactt
t1.b-   atccactttctcccccac
         ---------------------   ---------------------
c2+ 5'   CATCAGCTTCTTCAGCCCCTC   3'
c1- 5'   TCATCCACTTTCTCCCCCAC   3'

FIGURE 8.2: Examples of overlaps, contigs, and scaffolds. (A) A double stranded DNA segment. (B) Generation of reads. Two copies of the DNA segment are cut into pieces. Two reads are produced from a piece. (C) A scaffold of two contigs. Each contig consists of two reads with an overlap. The two contigs are ordered and oriented with respect to the top strand of the DNA segment. (D) The reverse complement of the scaffold in part C. The two contigs are ordered and oriented with respect to the bottom strand of the DNA segment.

of overlap strengths. Reads are assembled into contigs by processing the unique overlaps in the decreasing order. Corrections to contigs are made based on read pairs. Corrections include breaking a contig in the middle and joining broken contig pieces. Contigs are linked into scaffolds with read pairs. Corrections to scaffolds are made based on read pairs. In the third phase, a multiple alignment is constructed for each contig and a consensus sequence is generated from the alignment. The alignments and consensus sequences are reported. Below we describe each phase in detail, as designed and implemented for our whole genome assembler PCAP, to provide a concrete illustration.

8.2.1 Overlap computation and repeat identification

The whole set of reads is partitioned into subsets of similar sizes. The subsets are compared with the whole set in parallel with each comparison of a subset with the whole set performed on a different processor. The goal of each comparison is to compute overlaps between reads in the subset and reads in the whole set and to find repetitive regions of reads in the subset.

The whole set S is stored only on a hard disk accessible by all processors. Every subset and its data structures are stored in the main memory of a processor. To compare S with the subset, the reads in S are considered one at a time, where the current read in S is brought into the main memory and compared with the subset.

The set S is compared with the subset twice. In the first comparison, repetitive regions of reads in the subset are identified. In the second comparison, overlaps between unique regions of reads in the subset and any regions of reads in S are computed. Consider the first comparison. A region of a read is repetitive if it is highly similar to regions of many reads. Repetitive regions of reads in the subset are identified by computing the coverage arrays of reads in the subset. The coverage array of a read is an integer array of the same length, where the value at a position of the array is the number of overlaps between the read and other reads that cover the position. A region of a read is repetitive if the values at every position of the corresponding region of the coverage array are greater than a repeat coverage cutoff.

Identification of repetitive regions of reads depends on computation of overlaps involving the reads. However, it may not be computationally feasible to compute all overlaps because there are a huge number of overlaps between repetitive regions. Our strategy is to alternate computation of overlaps and identification of repetitive regions. Initially, some overlaps are computed. Then repetitive regions are identified based on the overlaps. The repetitive regions are processed so that no overlap involving any of the repetitive regions is computed again. The two-step procedure is performed many times until all repetitive regions are identified.

A lookup table for a word length w is constructed for the subset [14]. Given any word of length w, the lookup table is used to locate each occurrence of the word in reads of the subset. The reads in S are compared, one at a time, with the subset through the lookup table. Let f be the current read in S. The lookup table is used to find each read g in the subset such that the reads f and g have two close word matches. For each pair of close word matches, the left word match is extended into a high-scoring segment pair (HSP). HSPs from the same read in the subset are combined into high-scoring chains. For each read g in the subset with a chain of score greater than a cutoff, an overlap between the reads f and g is computed by a banded Smith-Waterman algorithm. The coverage array of the read g is updated over the corresponding region of g in the overlap.

After every ns reads from S are compared, where ns is the number of reads in the subset, the coverage arrays of reads in the subset are used to find new repetitive regions of reads in the subset. The current lookup table is replaced by a new lookup table, where all known repetitive regions are excluded in the construction of the new lookup table. After all reads from S are compared with the subset, the repetitive regions of reads in the subset are reported in a file. A final lookup table is constructed for the unique regions of reads in the subset.

In the second comparison, the reads in S are compared with the subset through the final lookup table. Overlaps between reads in S and reads in the subset are computed. The overlaps are reported in a file. Because the final lookup table covers only the unique regions of reads in the subset, all overlaps computed in the second comparison involve unique regions of reads in the subset.

8.2.2   Construction of contigs and scaffolds

The second phase consists of three major steps. First, each overlap is evaluated based on the depths of coverage of the two regions in the overlap. Second, poor ends of every read are identified and trimmed. Chimeric reads are identified and removed. Third, reads are assembled into contigs based on unique overlaps. Contigs are corrected and linked into scaffolds based on read pairs. Scaffolds are also corrected based on read pairs. Those steps are described in detail below.

For each read, the depths of coverage by overlaps for each position of the read are computed, where the depth of coverage of a position of the read is the number of times the position occurs in an overlap. The score of each overlap is adjusted based on the depths of coverage of the regions in the overlap such that an overlap with a larger adjusted score is more likely to be true. The adjustment is performed by multiplying the overlap score by the smaller of the average coverage scores of the two regions in the overlap. Let *repcocut* denote the repeat coverage cutoff. The coverage score of a position of a read is the logarithm of the ratio of *repcocut* to the depth of coverage of the position. The average coverage score of a region of f_x is the sum of coverage scores of each position in the region divided by the length of the region.

Poor ends of each read f are located and removed by computing the 5' and 3' clipping positions of f. The quality values of f and overlaps involving f are used in the computation. The quality values of f are used to determine 5' and 3' ranges for potential 5' and 3' clipping positions of f, whereas overlaps are used to select the 5' and 3' clipping positions of f in the 5' and 3' ranges.

Chimeric reads are identified based on their chimeric character. A chimeric read consists of two pieces from different parts of the genome. A pair of similar regions between two reads ends (starts) with an overhang if the regions of the reads after (before) the similar regions are sufficiently long and different. A pair of similar regions between a chimeric read and a real read often ends or starts with an overhang. A read is identified as a chimeric read if it has an internal position such that all the overlaps involving the read start or end around the position with an overhang. Chimeric reads are not considered in construction of contigs.

Contigs are constructed by processing overlaps with adjusted scores greater than a cutoff. Initially, each read is a contig by itself. The overlaps are ranked in a decreasing order of their adjusted scores. Then the overlaps are considered one by one in the order for construction of contigs. The overlap being considered is called the current overlap. For the current overlap between two reads, if the reads are in different contigs and the two contigs have an overlap consistent with the current overlap, then the two contigs are merged into a larger contig. Otherwise, no action is performed for the current overlap. This process is repeated until all the overlaps are considered. The computation is performed on one processor with enough memory to hold the overlaps and contigs.

Read pairs are used to find and break misjoins in contigs, and to make additional joins for contigs. A read pair is satisfied if the two reads are in a contig, the upstream read is in forward orientation, the downstream read is in reverse orientation, and the distance between the reads is within the given range. Otherwise, the read pair is unsatisfied. An overlap is unused if the overlap is not consistent with any contig. A group of unsatisfied read pairs support an unused overlap if, after corrections are made to the current set of contigs, all the read pairs in the group are satisfied with respect to the resulting set of contigs.

There are three steps for making corrections to contigs. In step 1, each read pair is evaluated with respect to the current set of contigs. Unsatisfied read pairs are partitioned into groups such that all read pairs in a group support an unused overlap. In step 2, an unused overlap supported by the largest number of unsatisfied read pairs is selected for consideration. If the number of unsatisfied read pairs supporting the unused overlap is sufficiently larger than the number of satisfied read pairs against the unused overlap, then corrections are made to the set of contigs so that the overlap is used in the resulting set of contigs. If no correction is made for the selected overlap, then other unused overlaps are considered until corrections are made or no more overlap is available for selection. In step 3, if no correction is made in step 2, then the process terminates. Otherwise, steps 1 and 2 are repeated.

Read pairs are used to order and orient contigs into scaffolds as follows. Initially, each contig is a scaffold by itself. Unsatisfied read pairs are partitioned into groups such that all read pairs in a group link a pair of scaffolds. The groups of unsatisfied read pairs are considered in a decreasing order of their sizes. For the current group of unsatisfied read pairs, if the number of read pairs in the group is sufficiently large, the read pairs link two scaffolds, and the two scaffolds can be combined by using the read pairs in the group, then the two scaffolds are combined into a larger scaffold. Otherwise, no action is performed for the current group.

Read pairs are also used to make corrections to scaffolds. The algorithm for making corrections to contigs based on read pairs can be extended into an algorithm for making corrections to scaffolds. After corrections are made to scaffolds, the scaffolds are arranged in a decreasing order of sizes, which are referred to as scaffold 0, scaffold 1, scaffold 2, etc. Then the scaffolds are partitioned into m groups, where for $0 \leq k < m$, group k consists of scaffolds q with $k = q \bmod m$. This partition ensures that the groups are balanced in scaffold sizes.

8.2.3   Generation of consensus sequences

From now on, the m groups of scaffolds are processed in parallel, with each group of scaffolds on a separate processor. For each scaffold, a set of repetitive reads that are linked by read pairs to unique reads in the scaffold is identified. For each gap in the scaffold, a subset of repetitive reads that may fall into the gap are selected from the set for the scaffold. An attempt is made to close the gap with the subset of repetitive reads. After all gaps in the group of scaffolds are considered for closure, a consensus sequence is generated for each contig and a list of ordered and oriented contig consensus sequences is reported for each scaffold.

Generation of a consensus sequence for a contig is based on a multiple alignment of reads in the contig, which is constructed as follows. The reads in the contig are sorted in an increasing order of their positions in the contig. A multiple alignment is constructed by repeatedly aligning the current read with the current alignment, with the resulting alignment as the current alignment for the next iteration. The reads in the contig are considered one by one in the order. In iteration 1, the current alignment is empty and the current read becomes the current alignment for iteration 2. For each column of the final multiple alignment, a weighted sum of quality values is calculated for each base type. The base type with the largest sum of quality values is taken as the consensus base for the column.

Consider aligning one read with the current alignment of reads. Only a $3'$ portion of the alignment that gets changed is used for alignment. The $3'$ portion, called a block, is replaced by the resulting alignment. A profile of average quality values is constructed for the block. For each column of the block, there are seven average quality values: five for substitution, one for deletion, and one for insertion. Of the five values for substitution, four are for the four regular base types, and one for the ambiguous base type. The average quality value for a base type is the signed sum of quality values of each base in the column divided by the number of bases in the column. The average quality values of the block and the quality values of the read are used to weight match and difference scores. A global alignment of the block and the read with the maximum score is computed in linear space, where $3'$ gaps are not penalized.

8.3   Existing Whole-Genome Assembly Programs

We are aware of eight existing programs for assembly of large genomes: Atlas, ARACHNE, Celera Assembler, JAZZ, PCAP, Phusion, PaCE, and RePS. The programs are based on two similar paradigms: overlap-layout-consensus and overlap-clustering-assembly. In the overlap-layout-consensus paradigm, the layouts of contigs are constructed by processing all available overlaps. In the overlap-clustering-assembly paradigm, the reads are clustered by processing all available overlaps, and the clusters are individually assembled with a small-genome assembly program. The small-genome assembly program is based on the overlap-layout-consensus paradigm. The ARACHNE, Celera Assembler, JAZZ, and PCAP programs are in the overlap-layout-consensus group, while Atlas, Phusion, PaCE, and RePS are in the overlap-clustering-assembly group. Below we comment on the special features of the methods used in the eight assembly programs.

8.3.1   Atlas

The Atlas genome assembly system is a suite of programs for processing sequence reads from both clone-by-clone (CBC) and WGS libraries. It follows the overlap-clustering-assembly paradigm, where the clustering step is aided by CBC sequence reads. Initially, reads are trimmed to remove low quality ends and to remove vector sequences and other contaminants.

In the overlap step, a fast method is used to find pairs of reads with a potential overlap. Any pair of reads sharing a rare word of length 32 is considered to have a potential overlap. To find rare words of length 32 in efficient space, a small table is constructed to record all words of length 32 that occurs at least R times in the trimmed WGS reads. Note that all words of length 32 that are not in the table are rare words. The fast method is controlled by two word frequency parameters R' and Y with $R \leq R' \leq Y$. If a pair of reads share a word of length 32 with frequency less than or equal to Y, then an overlap between the reads is computed by a banded alignment algorithm [4]. The banded alignment algorithm is seeded by a rarest word of length 32 between the reads, where words of length 32 with frequency less than R' are considered to be of the equal frequency. All WGS-WGS and BAC-WGS overlaps are computed. Overlaps with an adjusted score less than a cutoff are rejected.

In the clustering step, WGS reads are distributed to BAC clusters as follows. Initially, each BAC cluster consists only of BAC reads. For each BAC read, best N BAC-WGS overlaps at each end of the read are selected and the WGS reads involved in the overlaps are added to the BAC cluster. Additional WGS reads are added to BAC clusters based on the best N WGS-WGS overlaps of each WGS read in the BAC cluster. WGS reads that are linked by read pairs to WGS reads in BAC clusters are added to the BAC clusters.

In the assembly step, each enriched BAC (eBAC) cluster is assembled with Phrap [10], where full-length reads are provided to Phrap. Misjoins in contigs are detected and split through use of read pairs. Then for each BAC cluster, contigs are linked into scaffolds by using read pairs. Next contigs of BAC clones called bactigs are constructed and checked based on BAC mapping information, BAC scaffolds, and comparison with the human and mouse genomes as follows. Pairs of eBACs with a potential overlap are found based on WGS reads, read pairs, and BAC mapping data. The eBAC scaffolds of overlapping BAC clones usually share WGS reads or are linked by read pairs. Overlapping BAC clones usually have similar fingerprint contig (FPC) patterns. For each pair of eBACs with a potential overlap, an overlap between the eBACs is computed by aligning the consensus sequences of the eBAC scaffolds with BLASTZ [25]. The alignment must be end to end. BAC overlaps that are inconsistent with read pairs or mapping data are rejected. BAC clones with consistent overlaps are merged into bactigs. Misjoins in bactigs are found and corrected by comparison with independently generated maps, and the human and mouse genomes. Finally, a consensus sequence for each bactig is produced with Phrap in a piece-by-piece manner.

The Atlas system is special in that it supports the combined approach to a genome sequencing project. It uses BAC reads and BAC maps extensively to produce an accurate genome assembly. It uses a small table to remember the unique words of length 32 by keeping a count only for each repetitive word. The Atlas package has been used to produce a rat genome assembly at Baylor College of Medicine.

8.3.2   ARACHNE

The ARACHNE program follows the overlap-layout-consensus paradigm. Initially, reads are trimmed to remove low quality ends and to remove vector sequences and other contaminants. In the overlap computation phase, highly repetitive regions of reads are identified based on high coverage of word matches of length k, where k is set to 24. A word of length k is highly repetitive if the number of its occurrences in reads is greater than a cutoff. Otherwise, the word is unique. A region of a read is highly repetitive if every position in the region is covered by a highly repetitive word. For each pair of reads with one or more unique word matches, an overlap between the reads is first computed by a BLAST-like method and then refined by a banded alignment algorithm. Sequencing errors in a read are identified and corrected based on quality values of the read and a multiple alignment of the read with other reads. An overlap is rejected if there are a sufficient number of differences at bases of high quality values.

In the contig and scaffold construction phase, reads with consistent overlaps are assembled into contigs. If a read f has overlaps with reads g and h, but reads g and h do not have an implied overlap, then the reads have inconsistent overlaps. Contigs with a high depth of coverage are identified as repetitive and not used in construction of scaffolds. Scaffolds are constructed by an iterative procedure. In each iteration, scaffolds are merged by processing read pair links in a decreasing order of scores. Then corrections are made to scaffolds based on read pairs.

In the consensus generation phase, a consensus sequence for a contig is constructed based on a multiple alignment of reads and the quality values of reads in the contig. The generation starts with the leftmost read in the contig, which is the current read, and proceeds base by base to the right end of the contig. If the current base of the current read is of high quality value and is confirmed by bases of other reads on the multiple alignment, then the current base becomes the consensus base. Otherwise, another read with high quality values in the neighborhood and with bases confirmed by other reads on the multiple alignment is selected as the current read.

Most of the assembly tasks in ARACHNE are performed on one processor, whereas an initial processing of input files is multithreaded. The ARACHNE algorithm has two special features. First, sequencing errors in reads are identified and corrected. Second, an iterative method is used to construct scaffolds, which allows the algorithm to correct errors made in previous steps. The ARACHNE program has been used in mouse, chimpanzee, dog, and fungal genome projects.

8.3.3 Celera Assembler

Celera Assembler is the first whole-genome assembly program to handle genomes of at least 100 Mb. The program is based on the overlap-layout-consensus paradigm. Initially, reads are trimmed to remove low quality ends and to remove vector sequences and other contaminants. Ribosomal and heterochromatic regions of reads are masked through a comparison with a database of ribosomal and heterochromatic sequences. Highly repetitive regions of reads are identified through a comparison with a database of repetitive sequences.

Overlaps between reads are computed by the seed-and-extend technique of the BLAST program [1]. The computation is performed in parallel on many processors. Then reads with consistent overlaps are merged into contigs. Contigs that are consistently linked by read pairs are ordered and oriented into scaffolds. Gaps in scaffolds are filled by placing lonely contigs into the gaps in three increasingly more aggressive steps. Step 1 places into a gap every lonely contig that is linked to contigs around the gap by at least two read pairs. Step 2 places into a gap every lonely contig that is linked to a contig on one side of the gap by one read pair. In addition, it is required that the gap be completely covered by a list of overlapping contigs including the lonely contig. Step 3 fills each gap with a list of overlapping contigs, where each overlap between adjacent contigs on the list is of high quality. Finally, a multiple alignment of reads is constructed for each contig and a consensus sequence is generated from the alignment.

Celera Assembler is the first program to demonstrate that whole-genome shotgun sequencing is efficient. There are two innovations in the program. First, the regions of reads are classified into unique and repetitive regions, and unique regions are assembled into contigs called unitigs. Second, read pairs are used to link unitigs into scaffolds. The two innovations are fundamental to the success of the whole-genome assembler. Celera Assembler has been used in fly, human, mouse, and dog genome projects.

8.3.4 JAZZ

The JAZZ program follows the overlap-layout-consensus paradigm. Initially, reads are trimmed to remove low quality ends and to remove vector sequences and other contaminants. In the overlap step, pairs of reads with at least 10 unique word matches are identified. For each pair of reads, an overlap between the reads is computed by the banded Smith-Waterman algorithm. Overlaps with a percent identity below a cutoff are rejected. In the layout step, unique reads with consistent overlaps are merged into contigs. Layout errors in contigs are identified and corrected in an iterative process. Contigs are linked into scaffolds through use of read pairs. In the consensus step, for each contig, a subset of reads that spans the entire contig is selected. Then a reference sequence for the contig is constructed by concatenating high-quality regions of the reads in the subset. Next the reads in the contig are aligned to the reference sequence. A consensus sequence for the contig is generated from the alignment by quality-weighted voting.

The use of 10 short word matches, instead of one long word match, as a condition for finding pairs of reads with a potential overlap, is unique to JAZZ. The consensus method in JAZZ is similar to the method in Phrap [10]. The JAZZ program has been used in fish, tree, and frog genome projects at Joint Genome Institute.

8.3.5 PCAP

The PCAP program is based on the general whole-genome assembly algorithm in the last section. The algorithm has three special features. First, an extension of a word match into an HSP is triggered by two close word matches, instead of one longer word match. This feature allows the algorithm to find overlaps with more sequencing errors. Second, repetitive regions of reads are identified based on high coverage by chains of HSPs, instead of words. A chain is an ordered list of HSPs, where an HSP comes from a word match. Third, poor ends of reads are clipped during the assembly. This feature allows the algorithm to use those reads whose contaminants are not detected in a preassembly processing.

The PCAP program takes as input a number of compressed files of base sequences and an equal number of compressed files of quality value sequences, along with a file of read pairs. The input files are in a common file system accessible by all processors. The overlap and consensus steps are performed in parallel by multiple jobs with each job on a separate processor. Every job running on a processor copies each compressed file to a local disk of the processor, uncompresses the file on the local disk, and reads the uncompressed file on the local disk. This feature significantly reduces the I/O load on the common file system as there can be over 100 jobs running on different processors simultaneously. The PCAP program produces assembly results in the .ace format, which can be viewed in the Consed autofinishing package [9].

The PCAP program was initially developed and evaluated on a mouse whole-genome data set of 30 million reads and a human chromosome 20 data set of 1.7 million reads. The PCAP program has been used by Washington University Genome Sequencing Center in St. Louis in chimpanzee, chicken, fruit fly, and fungal genome projects. The program has been used on a maize data set of 1.6 million reads.

8.3.6 Phusion

The Phusion program partitions the entire set of reads into clusters such that each cluster can be handled by the existing assembly program Phrap [10]. Initially, reads are trimmed to remove low quality ends and to remove vector sequences and other contaminants. The clustering step is performed through a histogram analysis. For every word of length k, where k is set to 17, the histogram shows the number of occurrences of the word in the reads. A word is repetitive if the number of its occurrences in the reads is greater than a cutoff D and is unique otherwise. Two reads are related if they share at least M unique words. Two reads are transitively related if they are related, or if there is another read such that it is transitively related to each of the two reads. The reads are partitioned into clusters such that all reads in a cluster are transitively related. Values for the parameters D and M are selected such that each cluster can be handled by Phrap.

The initial family of clusters are refined by repeatedly performing assembly of clusters by Phrap and making changes to clusters based on assembly results. In each iteration, each cluster of reads along with the reads linked to the cluster by read pairs are assembled independently by Phrap. Note that adding the reads that are linked to the cluster results in inclusion of reads in multiple contigs. Read pairs are used to make corrections to contigs. A new family of clusters are constructed based on contigs. This process ends with a final set of contigs.

In the final set of contigs, different contigs contain common reads. Additional corrections, based on the following rule, are made to contigs to remove common reads. If two contigs share a read, but are different over other regions, then the read is removed from the smaller of the two contigs. After removal of all common reads, overlapping contigs that are linked by read pairs are merged.

Contigs are linked into scaffolds in an increasing order of gap sizes. Pairs of contigs linked by at least two read pairs are sorted in an increasing order of their gap sizes. Initially, each contig is a scaffold by itself. Then scaffolds are combined into longer scaffolds by processing pairs of contigs one at a time.

The Phusion algorithm has two special features. First, clusters are constructed in an iterative process. The iterative process allows the algorithm to correct errors made in previous steps. Second, scaffolds are constructed in an increasing order of contig gap sizes, instead of read pair strengths. This feature can add into scaffolds more small contigs that are located in gaps between large contigs. The Phusion program has been used at Sanger Institute in mouse, worm, and zebrafish genome projects.

8.3.7   PaCE

The PaCE assembly program is based on the overlap-clustering-assembly paradigm. The PaCE program runs on a parallel computer with multiple processors. Initially, reads are trimmed to remove low quality ends and to remove vector sequences and other contaminants. Repetitive regions of reads are masked by comparison with a repeat database.

In the clustering phase, a generalized suffix tree of masked reads [11] is constructed in parallel on multiple processors, with each processor saving part of the tree in its local memory. The tree is used to generate maximal word matches between reads in a decreasing order of word match lengths. Initially, each read is a cluster by itself. The maximal word matches of lengths greater than a cutoff are processed one at a time in the decreasing order. For the current word match between reads f and g, if the reads are already in the same cluster, then no action is performed. Otherwise, an overlap between the reads is computed. If the percent identify of the overlap is greater than a cutoff, then the cluster containing read f and the cluster containing read g are combined into one cluster.

In the assembly phase, each cluster of unmasked reads is assembled by CAP3 [15]. Consensus sequences are linked into scaffolds based on matches to protein and cDNA sequences.

The PaCE program has a special feature. A generalized suffix tree is constructed for finding pairs of reads with potential overlaps. The suffix tree data structure allows the algorithm to produce overlaps in order of their strengths with a strongest overlap first. The PaCE program has been used on a maize data set of about one million reads.

8.3.8   RePS

The RePS program appears to be a simplified version of the Phusion program. A word of length 20 is repetitive if the number of its occurrences in the reads is greater than a cutoff. All repetitive words in the reads are masked. Then the reads are clustered with BLAST [1]. Next the clusters are distributed among the processors. The clusters are independently assembled with Phrap on each processor. Gaps between contigs that are due to repeat masking are closed by placing fully masked reads that are linked by read pairs to reads in the contigs. Finally contigs are linked into scaffolds by using read pairs. The RePS program has been used in a rice genome project.

8.4 Efficiency and Accuracy of Existing Programs

It is difficult to compare the efficiency and accuracy of the whole-genome assembly programs discussed above for three reasons. First, the assembly programs require different powerful computational facilities for a genome of a few Gb in size. It is difficult to find the computational facilities that can accommodate all the assembly programs. Each program was developed on a powerful platform that was available to the program developers. Second, the assembly programs require different types of input data sets and have unique features to deal with their input data sets. For example, some programs require that all contaminants in reads be removed, whereas other programs can take full-length reads because of their ability to clip poor read ends. As another example, some programs can use a complete genome sequence to aid in assembly of a closely related genome. It is difficult to produce standard input data sets for comparison. Third, the assembly programs are continuously improved to meet the needs of whole-genome sequencing projects. It is difficult to select the latest version of each assembly program for comparison.

We provide information on the efficiency and accuracy of PCAP as an example. An input data set for a genome of a few Gb in size consists of 20 to 40 million reads. Both base and quality value sequences are required for each read. It is recommended that clearly low quality ends of reads, such as ends consisting mostly of bases of quality values at most 9, be trimmed for efficiency. It is desirable to remove contaminants in reads. An input set of clone and subclone read pairs is necessary for construction of scaffolds. A draft assembly on the input data sets can be produced with PCAP in a week on 75 processors with 30 Gb of main memory for one processor and 4 Gb of main memory for each of the other processors. The draft assembly may contain a few global misassemblies, where a global assembly involves the joining of two regions that are far away on the genome. The draft assembly covers unique regions of the genome as well as highly repetitive regions that are shorter than the lengths of reads.

Genome assembly projects are so difficult that no one existing assembly program is able to handle all situations perfectly. Currently each major genome center is supported by a different team of assembly program developers to address effectively and quickly the needs of the center in genome projects. Multiple assembly programs will continue to be improved in the near future.

Continued improvements to the existing programs are necessary in several directions. First, the existing programs lack an ability to assemble highly repetitive regions of the genome when those regions are longer than sequence reads. This problem is especially serious for plant genomes with a very high percentage of repeats. Second, the number of global and local misassemblies need to be reduced through the generation and use of read pairs with different distances. Third, errors in contig consensus sequences at the base level need to be reduced by developing improved algorithms. Fourth, the existing programs lack an ability to address problems caused by polymorphism. True overlaps between polymorphic reads have a sufficient number of differences at bases of high-quality values and are therefore rejected as false overlaps. Polymorphic reads end up in separate contigs and scaffolds. Thus a genome assembly is fragmented for regions with a high rate of polymorphism.

A modular open-source (AMOS) assembler is currently under development by the AMOS consortium (http://www.cs.jhu.edu/~genomics/AMOS). The open-source and modular features of this project make it easy for any person to make contributions to the development of the AMOS assembler. When computers in ordinary labs are powerful enough to process tens of millions of reads, many labs will be able to participate in the AMOS project by making improvements to the assembler and testing the improved assembler on their computers.

Acknowledgements

I am grateful to Srinivas Aluru for inviting me to write this book chapter. I thank Shiaw-Pyng Yang, LaDeana Hillier, and Asif Chinwalla for suggestions on improvements to PCAP. X.H. is supported by NIH grants R01 HG01502 and R01 HG01676.

References

[1] S.F. Altschul, W. Gish, W. Miller, and E.W. Myers *et al.* Basic local alignment search tool. *Journal of Molecular Biology*, 215(3):403–410, 1990.

[2] S. Aparicio, J. Chapman, E. Stupka, and N. Putnam *et al.* Whole-genome shotgun assembly and analysis of the genome of *fugu rubripes*. *Science*, 297(5585):1301–1310, 2002.

[3] S. Batzoglou, D. Jaffe, K. Stanley, and J. Butler *et al.* ARACHNE: A whole-genome shotgun assembler. *Genome Research*, 12(1):177–189, 2002.

[4] K. M. Chao, W.R. Pearson, and W. Miller. Aligning two sequences within a specified diagonal band. *Computer Applications in the Biosciences*, 8(5):481–487, 1992.

[5] T. Chen and S.S. Skiena. A case study in genome-level fragment assembly. *Bioinformatics*, 16(6):494–500, 2000.

[6] S.J. Emrich, S. Aluru, Y. Fu, and T.J. Wen *et al.* A strategy for assembling the maize (*zea mays l.*) genome. *Bioinformatics*, 20(2):140–147, 2004.

[7] B. Ewing and P. Green. Base-calling of automated sequencer traces using phred. II. error probabilities. *Genome Research*, 8(3):186–194, 1998.

[8] B. Ewing, L. Hillier, M.C. Wendl, and P. Green. Base-calling of automated sequencer traces using phred. I. accuracy assessment. *Genome Research*, 8(3):175–185, 1998.

[9] D. Gordon, C. Abajian, and P. Green. Consed: A graphical tool for sequence finishing. *Genome Research*, 8(3):195–202, 1998.

[10] P. Green. Technical report. http://www.phrap.org.

[11] D. Gusfield. *Algorithms on Strings, Trees, and Sequences: Computer Science and Computational Biology*. Cambridge University Press, New York, 1997.

[12] P. Havlak, R. Chen, K.J. Durbin, and A. Egan *et al.* The atlas genome assembly system. *Genome Research*, 14(4):721–732, 2004.

[13] X. Huang. A contig assembly program based on sensitive detection of fragment overlaps. *Genomics*, 14(1):18–25, 1992.

[14] X. Huang. *Bio-sequence comparison and applications*, pages 45–69. Current Topics in Computational Molecular Biology. MIT Press, Cambridge, 2002.

[15] X. Huang and A. Madan. CAP3: A DNA sequence assembly program. *Genome Research*, 9(9):868–877, 1999.

[16] X. Huang, J. Wang, S. Aluru, and S.P. Yang *et al.* PCAP: A whole-genome assembly program. *Genome Research*, 13(9):2164–2170, 2003.

[17] D.B. Jaffe, J. Butler, S. Gnerre, and E. Mauceli *et al.* Whole-genome sequence assembly for mammalian genomes: ARACHNE 2. *Genome Research*, 13(1):91–96, 2003.

[18] A. Kalyanaraman, S. Aluru, S. Kothari, and V. Brendel. Efficient clustering of large EST data sets on parallel computers. *Nucleic Acids Research*, 31(11):2963–2974, 2003.

[19] J.D. Kececioglu and E.W. Myers. Combinatorial algorithms for DNA sequence assembly. *Algorithmica*, 13(1):7–51, 1995.

[20] S. Kim and A.M. Segre. AMASS: A structured pattern matching approach to shotgun

sequence assembly. *Journal of Computational Biology*, 6(2):163–186, 1999.

[21] J.C. Mullikin and Z. Ning. The phusion assembler. *Genome Research*, 13(1):81–90, 2003.

[22] E.W. Myers, G.G. Sutton, A.L. Delcher, and I.M. Dew *et al.* A whole-genome assembly of *drosophila*. *Science*, 287(5461):2196–2204, 2000.

[23] H. Peltola, H. Soderlund, and E. Ukkonen. SEQAID: A DNA sequence assembling program based on a mathematical model. *Nucleic Acids Research*, 12(1):307–321, 1984.

[24] P.A. Pevzner, H. Tang, and M.S. Waterman. An eulerian path approach to DNA fragment assembly. *Proceedings of the National Academy of Sciences USA*, 98(17):9748–9753, 2001.

[25] S. Schwartz, W.J. Kent, A. Smit, and Z. Zhang *et al.* Human-mouse alignments with BLASTZ. *Genome Research*, 13(1):103–107, 2003.

[26] R. Staden. A new computer method for the storage and manipulation of DNA gel reading data. *Nucleic Acids Research*, 8(16):3673–3694, 1980.

[27] G.G. Sutton, O. White, M.D. Adams, and A.R. Kerlavage. TIGR assembler: A new tool for assembling large shotgun sequencing projects. *Genome Science and Technology*, 1(1):9–19, 1995.

[28] J. Wang, G.K. Wong, P. Ni, and Y. Han *et al.* RePS: A sequence assembler that masks exact repeats identified from the shotgun data. *Genome Research*, 12(5):824–831, 2002.

9

Assembling the Human Genome

Richa Agarwala
National Institutes of Health

9.1   Introduction

The sequence of the human genome is over 3 billion basepairs (bp) long. Current sequencing technology can only identify sequence of 500 – 700 bp templates generated from pieces of the genome. The process of deciphering the sequence of a genome from the sequence of its orders of magnitude smaller pieces and any other additional information we may have about the genome is called *assembling* the genome. For a genome with genetic variation between individuals (e.g. the human genome where the genome of even monozygotic twins differs after somatic and/or epigenetic DNA changes) and DNA for sequencing taken from usually a very small number of individuals, an assembled sequence for a genome can only represent some of the variation present in the population. However, as most of the genome is common between two individuals in a population (e.g. $\approx 99\%$ between two humans), the assembled genome does serve as a reference for the population.

For the human genome, there were two major sequencing initiatives undertaken, namely, publicly funded Human Genome Project (HGP) and privately funded effort at Celera Genomics. HGP used DNA from a few individuals (with exact number not mentioned by HGP but put at 13 by popular press [15]) and Celera used DNA from 5 individuals: two white men, one black woman, one Asian woman, and one Hispanic woman. This chapter reviews various challenges faced by the human genome assemblers as we progressed from the days when there were a handful of well understood regions on the human genome to the day we have finished sequence for essentially the entire genome. We show how the balance of techniques used for assembling data generated by HGP shifted from biology to computer science to computational science as properties of the data available for assembly changed. We also summarize significantly different techniques used by the effort at Celera

Genomics for producing and assembling the human genome. Discussions comparing and contrasting efforts of HGP to Celera Genomics by the principals of these two efforts can be found in [120, 89, 36, 121, 4, 66]. The National Center for Biotechnology Information's (NCBI) role in assembling the human genome using the data produced by HGP and our perspective on assemblies we produced in context of the assemblies discussed in the two landmark publications by HGP [33] and by Celera Genomics [115] is also presented in this chapter. Publication [33] also included an article [1] comparing the Celera assembly with the assembly produced by NCBI for the data available at the time.

This chapter is not a discussion on the scope, achievements, or implications of HGP. However, to understand the changes in underlying data for assembly and various assemblies that were made available, we present the basic timeline of main events relevant for human genome assemblies from the inception of HGP in 1990 to the delivery of an essentially finished human genome in April 2003.

The initial goals for HGP were stated in 1990 [39] by the U.S. Department of Energy and the National Institutes of Health, building upon the work done by the Committee on Mapping and Sequencing the Human Genome [6, 122, 22]. The main components of the project were:

- to build physical and genetic maps (defined later) for human and smaller sized model organisms,
- to enhance available technology for sequencing [67, 84, 102, 104, 87, 108, 98, 83, 28, 113] and mapping [125, 85, 16, 20, 43, 92, 124, 81] and to develop new technology,
- to sequence human genome and model organisms,
- to develop informatics infrastructure, and
- to be answerable to the ethical, legal, and social issues arising from HGP.

Not as a part of HGP but appropriately enough, software for finding whether two sequences are significantly similar, namely, BLAST was also developed in 1990 [7, 8].

Physical locations on a chromosome that can be uniquely identified by some mechanism are called *markers*. If the mechanism of identification is through inheritance, the marker is more specifically called a *genetic marker*. Some other methods of identification include restriction enzyme cutting sites, expressed regions of DNA, tags around marker that identify it uniquely (referred to as sequence tag sites), etc. A *map* places markers relative to each other on a chromosome. Two types of maps, namely, physical maps and genetic maps are defined in the next section. The first set of physical maps covering an entire human chromosome, namely the Y chromosome [118, 50], and a genetic map for the human genome [126] were published in 1992. In the decade that followed, significant increase in mapping activity provided much better coverage of chromosomes by physical maps [29, 64, 27, 99, 17], generated markers using radiation hybrids [111, 91], and produced more genetic markers and maps [19, 23, 114, 40, 18, 76]. Efforts were also made to integrate different types of maps [42, 37, 110, 74, 5]. Some physical maps were made by cloning pieces of genome sequence in Yeast Artificial Chromosomes (YAC) where a YAC is a vector that can be used for generating replicated copies of DNA fragments. Just like a YAC, a Bacterial Artificial Chromosome (BAC) is also a vector. We will give some details for how a BAC is created and used in the next section.

On the sequencing front, it became clear that while YACs were useful for large scale continuity in map construction, they were poor substrates for DNA sequencing, with the situation made worse by YAC libraries tending to have high rates of chimerism and deletions [56, 105]. Pieces of genome that can be cloned using BACs have average size of only about 120 to 250 kilobases (Kb) compared to 350 to > 1000 Kb for YACs but they were found to be more stable, have reduced cloning biases, and could be easily purified into DNA suitable for DNA sequencing [107, 73, 93]. Sequencing of human genome for HGP was done using BACs.

The initial goals for HGP were revised in 1993 [31]. This revision made explicit that one of the goals of HGP is to identify genes within maps and sequences. A strategy to find expressed genes by expressed sequence tags (EST) had been proposed earlier in 1991 [3]. Complementary DNA (cDNA) microarrays were added to the arsenal of biologists studying gene expression patterns in 1995 [103]. EST, cDNA, and mRNA are different types of *transcripts* that are subsequences from portion of the genome having a gene. We will show in this chapter that transcripts, BAC ends, plasmids, and mapping information are not only of use to biologists for their studies, they are also useful in assembling the genome.

In Feb 1996, members of the International Human Genome Sequencing Consortium (I-HGSC), a group of 20 international centers (`http://www.genome.gov/11006939`) doing sequencing for HGP, met for the first time in Bermuda to discuss the sequencing process. Six major groups were funded to start large-scale sequencing of the human genome and six pilot projects were initiated to sequence the ends of BAC clones. IHGSC also decided to release, within 24 hours [12], any sequence data that met minimum standards to members of the International Nucleotide Sequence Database Collaboration, namely the DNA Database of Japan (DDBJ), the European Molecular Biology Laboratory (EMBL), and Genbank at NCBI. With data getting updated daily, NCBI started semi-automatic assembly and manual curation of finished regions using the sequence overlap information included in GenBank submission for BACs, if any.

New goals following up on five year report of 1993 were published in 1998 [32]. In the same year, Celera Genomics was formed with the goal of sequencing human genome in 3 years [116] using a whole genome shotgun approach [117] that would first be used for sequencing an organism with much smaller genome, namely *Drosophila* [88], in collaboration with Drosophila genome project [2]. Shortly thereafter, HGP changed their strategy of carefully selecting and completely sequencing each BAC from a tiling path (TPF) of BACs for each chromosome that cover the genome in a minimal fashion to one where more (possibly redundant) BACs are sequenced to a lower coverage, known as BAC in draft stage [30], without necessarily knowing the place where BAC fits on current TPF. The change in strategy was partly in response to concerns that a human genome sequence might not be in the public domain at about the same time it was expected to be in private domain, but mostly because of the realization that producing a draft genome is good enough to answer many scientific questions and can speed up the process of finishing the genome. This resulted in a drastic acceleration in number of draft BACs being submitted to GenBank without an up-to-date accompanying TPF for the genome. In a truly international fashion, TPFs were maintained at EMBL and available to all members of IHGSC at all times but it was clear to everyone that the concern expressed in [30] that sequencing can outpace mapping of clones on TPF was a valid concern. Note that the term TPF is usually overloaded to mean both the TPF for a chromosome as well as the set of TPF's for all chromosomes in the genome.

The mission of NCBI is not only to be the central national repository of biological data but is also to develop new information technologies to aid in the understanding of data. With the drastic increase in number of draft BACs in GenBank without an up-to-date accompanying TPF, simply presenting draft BACs to users without an attempt to assemble them together to the best of our ability would have been insufficient. Therefore, NCBI embarked on the project of developing software to assemble BACs that did not depend on presence of a TPF. Each assembly published by NCBI is associated with a "data freeze" referred to by a *Build* number. A data freeze takes a snapshot at a particular time of the information in GenBank needed for assembly and by other steps of the genome pipeline [75]. Assembly for a build only utilizes BACs, transcripts, maps, etc. present in the corresponding freeze. TPF available from HGP at the time of freeze was also included in the freeze. UCSC developed GigAssembler [72] that is dependent on a TPF and used data freezes carried out at NCBI. Groups at NCBI and UCSC using data for their pipeline from the same data freeze aided in:

- both groups being involved in evaluating assemblies produced by the other,
- pointing out problems present in the TPF, and
- filling gaps in the TPF for which BACs were present in GenBank.

We will show later that the assembly software at NCBI utilized maximum transcript information available that did not contradict sequence overlap information of BACs. The TPF being produced by HGP and utilized by UCSC was based on published maps as well as the map data that was being generated by HGP. Maps have granularity that is coarser than that of a gene but they give a long range correct placement of BACs including BACs in regions of duplication. Hence, it was not surprising that NCBI had a better assembly for some regions around a gene and UCSC had a better assembly for some regions with duplication. Some independent comparisons of assemblies of the human genome produced at NCBI, UCSC, and Celera Genomics at various stages can be found in [1, 106, 101, 26, 79].

Joint statement announcing first draft genome at the White House on June 26, 2000 and publications by HGP [33] and Celera Genomics [115] was a major milestone. The assembly discussed in the publication by HGP utilized GigAssembler. Finished sequence for chromosome 21 [60] and chromosome 22 [44] was available and it was clear that in a matter of another couple of years, that will be the case for all chromosomes. With focus shifting on finishing the draft BACs and no major discrepancies left between the BAC order generated by NCBI during assembly and TPF produced by HGP, NCBI adapted their assembly software to directly utilize TPFs as a source of reliable information. This removed the main difference between assemblies produced by NCBI and UCSC and made the two assemblies very similar. UCSC stopped doing human genome assembly after August 2001 (`http://genome.ucsc.edu/FAQ/FAQreleases\#release2`) but continued to serve as evaluators of assemblies produced at NCBI.

The announcement of essentially finished human genome sequence by HGP was made on the 50th anniversary of the discovery of the double helical structure of DNA [123] and before the 60th anniversary of the paper [9] demonstrating that genetic material is DNA. The stitches to be made for getting assembled finished genome from BACs are specified by the sequencing centers; similar to the overlap information provided in the initial stage when a handful of finished BACs were available. The sequence following the provided recipe is generated at NCBI as part of the NCBI genome pipeline. Other notable efforts at assembling (not necessarily human) genome sequences include [95, 57, 71, 112, 62, 24, 65, 97, 96, 11, 119, 68, 86, 25, 63].

In Section 9.2, biological information useful for assembling genomes is presented. Readers who are familiar with meaning of terms BACs, shotgun, transcripts, plasmids, Bacends, and physical maps can skip this section. In Section 9.3, computer science terms and techniques necessary for describing the assembly process are presented. Readers who are familiar with meaning of terms interval graph, clique tree, lexicographic breadth first search, and greedy algorithm can skip this section. In Section 9.4, four major stages of evolution for data produced by HGP alluded to earlier in this section and changes needed in NCBI's assembly process as a result are discussed. We briefly mention our current role in the process that yields finished assembly for the human genome available at our web site, among others.

Section 9.5 presents an overview of the whole genome shotgun assembly process and a brief summary of the data and techniques utilized by Celera Genomics to assemble the human genome. We conclude by presenting a brief discussion of what we may expect to change in human genome assemblies in near future.

9.2 Biological Data

In this section the relationship of the human genome to various pieces of biological data that can be used as hints for assembly is presented. The primer on molecular genetics [41] published by DOE human genome program is an excellent introduction to biological terms we assume readers are familiar with, so are not covered in this chapter. For a fun and easy to read introduction to genetic engineering, the author personally likes [109] but for someone interested in more details, we recommend [78] and Part IV of [58]. A primer for more global overview of bioinformatics and description of terms and techniques in non-technical language can be found at (http://www.ncbi.nlm.nih.gov/About/primer/index.html).

9.2.1 Chromosome maps

Chromosome maps give the order of markers along a chromosome. There are two types of chromosome maps: physical and genetic.

Physical maps decipher order of markers along a chromosome using techniques that are based on the actual nucleotide base pair distance between the markers. If we have finished sequence for a genome, then distance between markers and their order can be computed precisely. However, if we don't have finished genome sequence, then techniques like fluorescence in situ hybridization (FISH) mapping, mapping of sequence tagged sites (STS) on YAC clones, and radiation hybrid (RH) mapping are used for creating physical maps.

Genetic maps decipher order of markers along a chromosome using techniques that are based on the relationship of their genetic linkage where markers closer to each other segregate together through meiosis in a family more so than markers that are farther apart. For segregation to be observed, markers need to have more than one allele. Different types of markers used for constructing genetic maps are: restriction fragment length polymorphisms (RFLP), variable number tandem repeats (VNTR), microsatellites with alleles coming from length variation of simple di-, tri-, and tetra- nucleotide repeats, and single nucleotide polymorphisms (SNP).

The order of markers produced by genetic linkage is same as the order in physical maps, except in the unusual situation where there is a polymorphism involving inversion, but there is no constant scale factor that relates the physical and genetic distances. Genetic distances vary between males and females but physical distances do not.

9.2.2 Plasmids

Plasmids are circular double stranded DNA molecules that are separate from the chromosomal DNA but have the ability to replicate independently from the chromosomal DNA. Plasmids that contain one or more genes capable of providing antibiotic resistance and a pair of promoters sandwiching the site where a sequence of interest, called *insert*, can be added are used as *cloning vectors* for replicating inserts as follows: (i) ligate insert into plasmid, (ii) insert modified plasmid into bacteria using a *transformation* process that creates pores in the bacterial cell wall membrane through which the plasmid can enter the cell, (iii) provide medium for bacteria to replicate, (iv) apply antibiotic to replicated bacteria to kill

Abbr.	Library name	Cloning vector	Source	Enzyme
RP11	RPCI human BAC library 11 (segments 1 to 4)	pBACe3.6	blood	EcoRI
RP11	RPCI human BAC library 11 (segment 5)	pTARBAC1	blood	MboI
RP13	RPCI human BAC library 13	pBACe3.6	blood	EcoRI/MboI/DpnII
CH14	CHORI 14 human male BAC library	pTARBAC2	blood	EcoRI
CH15	CHORI 15 human female BAC library	pTARBAC1.3	blood	MboI
CTA	CalTech human BAC library A	pBeloBACII	987 SK cells	HindIII
CTB	CalTech human BAC library B	pBeloBACII	987 SK cells	HindIII
CTC	CalTech human BAC library C	pBeloBACII	sperm	HindIII
CTD	CalTech human BAC library D	pBeloBACII	sperm	HindIII/EcoRI

TABLE 9.1   Cloning vector, source of human DNA, and fragmentation mechanism used for some BAC libraries

all bacteria that do not contain the plasmid, (v) isolate plasmid, and (vi) extract insert for sequencing. In principle, the final insert extracted is identical to the insert injected, but mutations do occur during DNA replication.

Length of plasmids is from 1 to 250 Kb and maximum amount of insert sequence that a plasmid can take in practice is $\approx$ 15 Kb although there is no theoretical limit on insert size. *Paired plasmid reads* are sequence reads of about 500 bp from both ends of insert. Directed sequencing of just the insert ends is feasible because of the presence of promoter on each side of the insertion site.

9.2.3   BACs

A Bacterial Artificial Chromosome (BAC) is a synthetic oligonucleotide (as against naturally occurring plasmids) that can (i) hold up to 350 Kb of insert sequence, (ii) serve as a vector that is incorporated into a bacterial cell, and (iii) multiply when bacteria replicates [107]. That is, the molecule is constructed to function like a plasmid but can take much bigger inserts than a plasmid can. A bacterial colony produced by replication contains clones of the insert sequence. The term *BAC clone* refers to the copy of insert sequence present in the replicated bacteria. However, the term *BAC* is overloaded to mean sequence of the insert and not the vector.

A collection of BAC clones generated as part of one experiment or project is called a *BAC library*. Libraries may be for the whole genome or may be chromosome specific. To produce a BAC library, the genome (or chromosome) is cleaved by restriction enzymes or sheared into smaller segments up to 350 kilobases long. Segments are separated by size using gel electrophoresis and BAC clones made for each segment. Different BAC libraries use different cloning vector and/or different restriction enzymes. See Table 9.1 for information on some human BAC libraries.

BACs are sequenced by (i) cleaving or shearing clones into smaller fragments, (ii) sequencing fragments, and (iii) assembling the sequence of fragments into a sequence for the BAC. The process of splitting big segments into smaller pieces is called *shotgun*. The number of times the DNA in a region of genome is sequenced is called the *depth of coverage* for that region. Depending on the depth of coverage for a BAC, fragments sequenced so far, and complexity of the BAC leading to problems in assembling the fragments, a BAC can be in one of the following four *phases*:

Phase 3: BACs in phase 3 are *finished* and they have a single contiguous sequence for the whole BAC. A finished BAC is guaranteed to be 99.99% accurate and typically has 9 or 10 as the depth of coverage.

Phase 2: BACs in phase 2 are *draft* and the sequence of such a BAC is in a few *fragments* but all fragments are ordered and oriented with respect to each other.

That is, we have the sequence for the BAC from one end to the other but sequence for some portions is missing. Draft BACs in phase 2 have 8 or 9 as the depth of coverage.

Phase 1: BACs in phase 1 are draft BACs where the sequence of the BAC is in several fragments and relative order and orientation for all fragments is not known. That is, we only have fragments that we know belong to a BAC but have little idea about how the sequence of BAC relates to the fragments. Draft BACs in phase 1 have only 4 or 5 as the depth of coverage. Hence, they are often referred to as "working" draft BACs.

Phase 0: BACs in phase 0 are sequenced to a very small depth of coverage of about 0.5 to 1 for the purpose of low pass screening, say, to determine whether the BAC looks redundant with respect to a BAC that has already been sequenced to a better coverage or not.

HGP created a minimal tiling path of BACs for each chromosome by mapping BACs to chromosomes. A BAC sequenced by HGP may not be present on the tiling path of any chromosome if there was not enough mapping information available to map the BAC to a single location on the genome or if the BAC was found to be redundant with a BAC already present at the location where it was mapped.

9.2.4 WGS sequences and mate pairs

When the whole genome is sheared into segments of a set of pre-defined length, typically 1 Kb to 10 Kb, and cloned into a plasmid vector for sequencing, the fragments so produced are called *whole genome shotgun* sequences. A pair of sequence reads from both ends of a segment of known size is called a *mate pair*. It is critical for whole genome shotgun libraries to have uniform size, to be nonchimeric, to represent the genome uniformly, and for mate pairing to be accurate as whole genome shotgun assembly software needs to make these assumptions. The whole genome shotgun sequencing and assembly process does not necessarily require creation of chromosome maps but does require assembly software that can assemble millions of sequence fragments in an acceptable amount of time, say, a few weeks for the human genome assembly. The strategy also requires high-throughput DNA sequencers that enable the sequencing of hundreds of thousands of bases per day.

For the human genome, sequencing was not considered a hurdle but much of the concern and skepticism among scientists of the whole genome shotgun strategy for sequencing human genome was for the assembly. Many considered that it was impossible to expect whole genome shotgun assembly software to correctly assemble the sequence of the human genome, because it contains millions of repetitive DNA sequences. The best known whole genome shotgun assembly software at the time was developed at The Institute for Genome Research (TIGR) [112] for the assembly of *Haemophilus influenza* genome [49] that is much smaller and less repetitive than the human genome. Scientists at Celera Genomics were able to develop software [88] that assembled the human genome sequenced by whole genome shotgun. The software for assembling the sequence of the human genome was engineered from the software they developed for *Drosophila*.

9.2.5 Transcripts

Messenger RNA (mRNA) is a single stranded nucleotide sequence generated by *transcribing* a portion of genomic DNA corresponding to a gene that is later *translated* into a protein. As mRNA is unstable outside the cell, enzymes called reverse transcriptases are used to

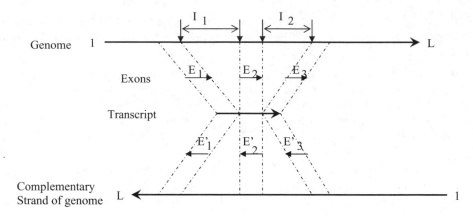

FIGURE 9.1: Relationship between genome and transcripts

convert it into complementary DNA (cDNA) that is stable. Expressed sequence tags (EST) are sequence of a few hundred nucleotides (usually 200 to 500 bp) generated by sequencing portions of cDNA.

Eukaryotic genes are often not contiguous stretches of genome but have some subsequences removed. Portions of the genome that are removed are called *introns*, portions that are retained in a gene are called *exons*, and the process that removes introns is called *splicing*.

For example, figure 9.1 shows a transcript that has three exons E_1, E_2, E_3, two introns I_1, I_2, and the relationship between sequence of the transcript and the genome. Note that both the order of exons as well as the orientation of exons is same in the transcript as on one strand of the genome. That is, each exon is a subsequence from one strand of the genome sequence and the order and orientation of subsequences corresponding to exons of a transcript is same in a transcript as it is on that strand of the genome. Relationship between the transcript and the complementary strand of the genome reverses both the order and the orientation of exons, as shown by E_3' followed by E_2' followed by E_1' where E_i' is reverse complement of E_i.

Chromosome maps, mate pairs, plasmid pair reads, and transcripts place restrictions on what constitutes a correct assembly and can, therefore, be used while doing assembly for restricting the search space. Transcripts with more than one exon (and similarly plasmid pair reads and mate pairs) can be used to order and orient two different fragments (or cluster of overlapping fragments) if both fragments happen to overlap with the same transcript because the order and orientation of exons gives the order and orientation of the fragments with respect to the genome.

9.3   Computer Science Terms and Techniques

Algorithms and graph theory terms used in this chapter, but not covered in this section can be found in [48, 34]. Very briefly, a *graph* is a mathematical representation for objects and a relationship between pairs of objects. The set of objects represented in a graph are called its *nodes* or *vertices*. If two objects are related, they are *joined* by an *edge*. The *degree* of a node is the number of edges joining it to other nodes in the graph. A node is *singleton* if it has no edges. A graph is a *directed* graph if its edges are directed edges; otherwise, it is an *undirected* graph. A *path* from node u to node v in graph G is an alternating sequence of nodes and edges of G, beginning with u and ending with v, such that no node is repeated and every edge joins the nodes immediately preceding it and following it. The *length* of a path is the number of edges in the sequence defining the path. A *cycle* is a path that begins and ends at the same node. If a graph has no cycles, it is called *acyclic*. If G and H are graphs then H is a *subgraph* of G if and only if the set of nodes and edges in H is a subset of the set of nodes and edges in G. A graph is *connected* if there is a path, ignoring the direction of edges if graph is directed, between every pair of nodes in the graph; otherwise the graph is said to be have more than one *connected component*.

Assembling a genome such that the sequence produced agrees with what is already known about the genome can be reformulated as a few *optimization problems* [94, 128] as presented in next section. The general structure of an optimization problem is to determine the values of n problem variables $x_1, \cdots, x_n$, so as to minimize an objective function $F(x_1, x_2, \cdots, x_n)$ subject to lying within a region G which includes any constraints on the variables. The constraints define all feasible solutions, and the problem is normally expressed as the determination of values $(x_1, x_2, \cdots, x_n)$

$$to \ minimize \ or \ maximize \ F(x_1, x_2, \cdots, x_n)$$
$$subject \ to \ (x_1, x_2, \cdots, x_n) \in G.$$

An assignment of values that minimizes or maximizes the function over the whole of the feasible region is called an *optimal solution*. Strictly speaking, it is a *global optimum*, but there are usually a number of *local optimum*. A local optimum minimizes the function within a local region which is a subset of the feasible region defined with respect to a candidate solution.

There are a variety of available search procedures for solving optimization problems. However, if the problem is *NP-complete* [52], which (simply stated) means that it is extremely unlikely to have an algorithm for the problem that grows polynomially with the size of input and guarantees a global optimal solution, one looks for *approximation algorithms* or *heuristics*. Approximation algorithms guarantee that solution reported by the algorithm is no worse than a certain factor compared to optimal. Even if an approximation algorithm is known, a heuristic that does not give any guarantee on the solution can be faster or may empirically "tend" to produce better results than an approximation algorithm. Heuristics typically find a local optimum but can get "lucky" and produce global optima, which is also why heuristics work well for problems that do not have many local optimums.

We now present some information about an optimization problem known as *maximum interval subgraph*, a technique called *lexicographic breadth first search (LBFS)* for navigating graphs and its extension *LBFS** that detects interval graphs, and a greedy heuristic for finding *directed acyclic subgraph* of a directed graph. These tools and techniques will be used in the next section.

9.3.1   Maximum interval subgraph

A graph G is called an *interval graph* if it is possible to assign an interval on real line to each node such that two nodes are connected in the graph by an edge if and only if their corresponding intervals overlap. It has been shown that an interval graph can not have following two *obstructions* present in it:

1. *Chordless cycle:* A cycle with more than three nodes without a *chord* where a chord is an edge that connects two nodes in the cycle that are not adjacent in the cycle. A graph without any chordless cycles is called a *chordal* graph.
2. *Asteroidal triple:* An independent set of three vertices such that each pair is joined by a path that avoids the neighborhood of the third.

In other words, the class of interval graphs is the subset of the class of chordal graphs that do not have an asteroidal triple.

The problem of finding a maximum interval subgraph is to find a subgraph of the input graph that is an interval graph and has the maximum number of edges. In the weighted version of maximum interval subgraph, we look for a subgraph where sum of weights of all edges in the solution is maximum. Verifying whether a graph is an interval graph or not has very efficient solutions [14, 77, 35, 61], but finding maximum interval subgraph of a given graph is an NP-complete problem [55].

A couple of different characterizations of interval graphs that we will use later are as follows:

Node ordering [90]: A graph is an interval graph if and only if there exists a linear order $<$ on the nodes such that for every choice of u, v, w with $u < v$ and $v < w$, if there is an edge between u and w, then there exists an edge between u and v.

Maximal clique ordering [54]: A graph is an interval graph if and only if its maximal cliques can be linearly ordered in such a way that for every vertex in the graph the maximal cliques to which it belongs occur consecutively in the linear order.

A few more definitions needed to state the facts used later in the chapter are given next.

Minimal separator: A set of nodes S is a *separator* for a pair of vertices a and b if removal of all nodes in S separates a and b into two distinct components. If no proper subset of S is a separator for a and b, then S is called a *minimal a, b separator*.

Clique graph: A *clique graph* for a chordal graph represents each maximal clique of the chordal graph by a node, has an edge joining node for clique A and node for clique B if and only if $A \cap B$ is a minimal separator for each node $a \in A - B$ and $b \in B - A$. The weight of the edge joining the node for clique A and the node for clique B is the size of $A \cap B$. Number of maximal cliques in a chordal graph is linear in the size of the graph and clique graph for a chordal graph can be constructed in polynomial time [51].

Clique tree: A *clique tree* for a chordal graph G is a subgraph of clique graph C for G such that it is a tree and for each node a in G, the set of nodes in C that have a as an element induce a subtree. It was shown in [51] that a spanning tree of a clique graph for a chordal graph is a clique tree if and only if it is a maximum weight spanning tree. Figure 9.6 illustrates the transformation of a chordal graph in a clique graph and two maximum weight spanning trees with different leafage.

Note that the linear ordering of maximal cliques as described in the maximal clique ordering characterization of interval graphs above admits that there exists a clique tree for interval graphs which is a Hamiltonian path in the clique graph. We will use this observation for developing a heuristic for the maximum interval subgraph problem.

9.3.2   LBFS and LBFS*

LBFS [100] is a breadth first search procedure that breaks ties lexicographically using labels where a label for a node is the set of neighbors already visited. In the beginning, all nodes have an empty set as label and no number. If there are N nodes in the graph, the algorithm for finding LBFS has N iterations for assigning a node number and updating labels. In the ith iteration, an unnumbered node with highest label is chosen to receive number $N - i + 1$ and all its unnumbered neighbors get $N - i + 1$ added to their label. LBFS was designed to find a special ordering of nodes in chordal graphs called a *perfect elimination ordering*. Note that in LBFS, there may be nodes tied at each iteration; any such ties are broken arbitrarily.

LBFS* is a four sweep LBFS strategy for interval graph recognition proposed by Corneil et. al. [35] that utilizes the characterization of interval graphs in terms of a linear order of nodes and breaks ties intelligently based on previous sweeps. The first sweep is the original LBFS algorithm in which there is an arbitrary choice made when nodes are tied. The next two sweeps decide ties by the highest index in the previous sweep. The last sweep decides among tied vertices by choosing two candidates based on the previous two sweeps and then deciding between them depending on their and their neighbor's edge relationships with nodes that come after the set of tied vertices. Corneil et. al. showed that the ordering produced by the final sweep is a node ordering (as defined in the previous subsection) if and only if the graph is an interval graph.

Both LBFS and LBFS* are linear time algorithms.

9.3.3 Directed acyclic subgraph and a greedy heuristic

A directed acyclic subgraph can be recognized in linear time by using *topological sort*. However, if the given directed graph has cycles and we wish to find a directed acyclic subgraph by removing minimum number of edges, the problem is called *feedback edge set* problem and is known to be NP-complete [70, 52] even on directed graphs with total vertex in-degree and out-degree smaller than 3 [53] but is polynomially solvable on planar directed graphs [82] and on undirected graphs (by using maximum weight spanning tree algorithms as the resulting graph will have to be a tree). The problem is APX-hard [69]. The best approximation algorithm known makes the set approximable within $O(\log |V| \log \log |V|)$ [47] and requires solving a linear program. Although the complementary problem of finding a maximum acyclic subgraph can be approximated by a ratio even smaller than 2 [13, 59], the constant factor is with respect to the size of the input graph and not with respect to the "problem spots" giving us the cycles.

Heuristic methods do not guarantee to find optimal solutions, but they do find what one hopes are 'near-optimal' solutions. Some heuristics for feedback edge set problem can be found in [46, 45]. In the weighted version of the feedback edge set problem with weights on edges, most heuristics try to either find a light edge to remove from each cycle or find edges that are in a large number of cycles even if the weight of some of those edges is high. A heuristic that mixes the two approaches was presented by Finocchi et. al. [38] using the local-ratio principle [10].

The weighted directed graph for which we are interested in finding a maximal directed acyclic subgraph has

- nodes that represent sequence of fragments or sequence of clusters of overlapping fragments or their reverse complements,
- edges that connect a pair of nodes when the sequence represented by two nodes overlap
  - different fragments of a phase 2 BAC, or
  - different exons of a transcript T, or
  - different ends of a paired plasmid

- direction on edge as deduced from the order of fragments/exons/ends, and
- weight on edges that gives preference to phase 2 BAC orientation over mRNA, mRNA over EST, and EST over paired plasmid reads.

For example, if we have three fragments A, B, C and a transcript with three exons as shown in figure 9.1 such that A overlaps with E_1, B overlaps with E_2' and C overlaps with E_3, then the graph can have nodes N_A, N_B', N_C where N_i represents node for sequence i and N_i' represents node for sequence of reverse complement of i, a directed edge from N_A to N_B' and another directed edge from N_B' to N_C or, equivalently, the graph can have nodes N_A', N_B, N_C', a directed edge from N_C' to N_B and another directed edge from N_B to N_A'. We chose to use the heuristic of removing light edges as the weight reflects quality of information in the directed graph we produced. The heuristic was implemented as a greedy heuristic where edges were sorted by weight and added from maximum weight to the minimum as long as adding the next edge did not create a cycle in the current partial solution or attempted to add a node whose sequence is reverse complement of the sequence of a node already present in the current partial solution.

9.4 BAC Data and Assembly

Sequence data produced by IHGSC evolved in four main stages (Figure 9.2), with each stage requiring data driven changes in algorithms for assembling the data. BACs meeting the minimum requirements for HGP were submitted to the high throughput genomic sequence (HTGS) section of GenBank. BACs used for human genome assembly were the phase 1, phase 2, and phase 3 present in HTGS. That is, phase 0 BACs were ignored for assembly. Four stages of sequence data were as follows:

Till July 1999: As per the original plan of HGP, BACs submitted to HTGS in early stages were almost all finished BACs carefully selected to cover the region being sequenced. The center submitting a BAC would sometimes also submit instructions on how to join the BAC with its predecessor/successor. NCBI would verify the joins and either make the join if supporting data agreed or report back if there were problems.

July 1999 to January 2000: In this stage, we saw that GenBank had a balance between amount of finished and draft sequence with total amount of sequence being only a fraction of the genome size. The primary responsibility for genome assembly from this stage onwards for genome pipeline at NCBI was shouldered by the author, with upstream and downstream processing done by other members of the genome pipeline group at NCBI. The software for assembly, described later in this section, came up with a tiling path for BACs and decided on joins to make while preserving the finished regions that were curated following the process of the previous stage.

January 2000 to October 2001: Following the change in strategy adopted by HGP in October 1999, this stage saw an explosive growth in sequence coverage of the human genome by draft BACs. The software developed in the previous stage to come up with a tiling path order for BACs and make joins that minimize the conflicts in all supporting data was modified to handle a high number of draft BACs.

October 2001 to April 2003: The final stage of assemblies at NCBI saw a balance between amount of finished and draft sequence with HGP proposing a tiling path order for BACs. NCBI continued with the approach of defining an overall tiling

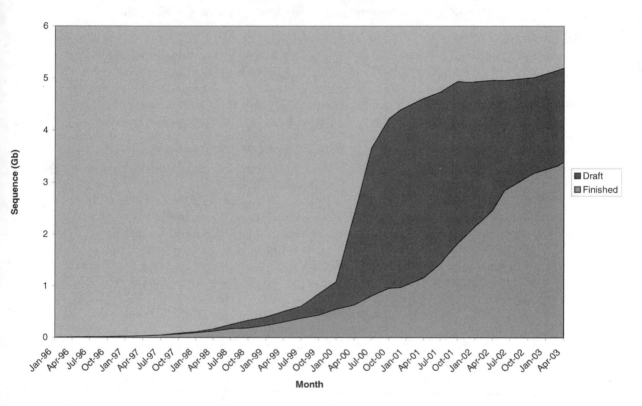

FIGURE 9.2: Growth of draft and finished sequence in High Throughput Genome Sequencing section of GenBank

path order but took the tiling path order proposed by HGP unless it had some evidence that suggested otherwise.

The recipe for joining BACs to produce the essentially finished human genome sequence in April 2003 was given by IHGSC. Ongoing finishing efforts of IHGSC are aimed at closing gaps present in this sequence. Revisions to the assembly are also expected to come with complete instructions on how to join BACs. The sequence following the given recipe is generated at NCBI as part of the NCBI genome pipeline.

The strategy employed by NCBI to assemble the human genome in all four stages was what is typically called "*overlap–layout–assemble*". Specifically, the steps involved were as follows:

Containment   Dovetail

FIGURE 9.3: Consistent overlaps showing containment or dovetail

Overlap generation: Starting from the BLAST hits between fragments of pairs of BACs, merge hits and find *consistent* overlaps between them where a hit is consistent if it either has one fragment *contained* in the other fragment or has a *dovetail* with the other fragment (Figure 9.3). Different parameters determined the consistency depending on the phase of BACs in the overlap. For two finished BACs to be considered consistent, they were allowed to have a tail of at most 50 bp whereas draft BACs were allowed a tail of 1 Kb. Among the set of consistent overlaps, only those that satisfied a minimum length and average percent identity criteria were considered for assembly. The criteria used were that for two finished BACs, it was sufficient to have overlap length of $\geq$ 95 bp at $\geq$ 99% whereas overlap involving a draft BAC needed at least one fragment to have $\geq$ 2500 bp at $\geq$ 98%. BACs that could be assigned to a chromosome based on the mapping information available were marked as such while rest were marked as unassigned. Overlaps between BACs being placed on different chromosomes by map information were removed.

BAC layout: Reformulate the problem of assembling a genome as "given a set of BACs and overlap information between BACs, subdivide BACs into disjoint sets and order the BACs in each set such that all the BACs in a set are unassigned or assigned to the same chromosome and subdivison and ordering is most truthful to the maximum amount of good quality overlap information." This reformulation makes a basic assumption that the false negative rate in overlap information is negligible and can be ignored. The order of BACs in a set is the tiling path for that set. Note that if we have BACs for all regions of the genome, we have one tiling path for each chromosome. Otherwise, regions of the genome not represented in any BAC create a *gap* in the sequence that can be assembled for the genome and split the tiling path for the chromosome into smaller tilings paths. Placement of all tiling paths on chromosomes with appropriate orientation for each tiling path is the BAC layout for the genome.

For producing BAC layout, a graph is created where BACs are represented as nodes, overlaps as edges, a weight is given to each edge depending on the phase of BACs, overlap length, and percent identity of the overlap. The solution desired is the maximum weight interval subgraph. Each connected component in the subgraph has the set of BACs and overlaps to produce a tiling path. Each tiling path results in making one contiguous sequence (*contig*) so we sometimes use the term contig to mean the tiling path for the set of BACs that make that contig.

Create melds: For each pair of fragments joined by overlaps kept in contig, a breakpoint is chosen and sequences joined. This results in one contiguous sequence for a contig if all BACs in the contig are finished but may result in a set of contiguous sequences, called *melds*, if there is a draft BAC such that some of its fragments do not become part of same meld by the overlaps with fragments from other BACs in the contig.

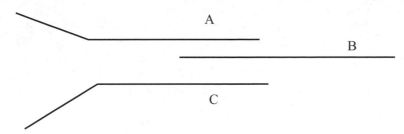

FIGURE 9.4: Pairwise consistent overlaps that have inconsistent multiple overlap

Order and orient melds: After sequence joins are made, if a contig has more than one meld, it is sometimes possible to order and orient melds using information provided by HGP for phase 2 BACs, transcripts, and paired plasmid reads. This is achieved by making a directed graph as described in the previous section with each meld or its reverse complement being represented by a node and directed edges being derived from the orientation information available. Desired solution is a directed acyclic graph that maximizes weight of edges kept in the solution. The translation of directed acyclic graph into a linear order of melds aims to minimize the difference between expected length of BACs and the length of their extents on contig. The linear order of melds is made into a contig by putting a 100 bp gap between consecutive melds.

As three pairwise consistent overlaps for three BACs may not necessarily result in a single multiple alignment for all three BACs (see figure 9.4), it is not necessarily true that the layout step is completed before the melding step but can instead be interleaved. The decision to have BAC layout stage completely precede melding stage or to interleave the two steps depends on the quality of BACs that in turn dictates the quality of overlaps for pairs of BACs we observe. The contigs produced were oriented and placed on chromosome by using map information. We now present how above steps were carried out for each stage.

9.4.1 A few finished regions

Each BAC has a clone name that identifies its microtitre plate address (plate number, row, and column) prefixed by a library abbreviation. For example, the BAC with clone name RP4-740C4 is from RP4 library, plate number 740, row C and column 4. When a sequence for a BAC is submitted to GenBank, it is assigned an accession and version number 1. Revised submission of a clone already present in GenBank receives a new version number but the same accession. The NCBI Clone Registry (http://www.ncbi.nlm.nih.gov/genome/clone/) provides links between clones and their accession numbers. It integrates information about genomic clones and libraries, sequence data, genomic position, and distributor information. It was also a place for genome sequencing groups to register their intent to sequence specific clones and advise the database of changes in their sequencing status.

When BACs were submitted and overlap with the predecessor and/or successor clone was known, the submittor sometimes also submitted this information to GenBank. The following steps were sufficient to do the assembly when GenBank had mostly finished BACs covering relatively small portion of the genome:

1. *Relate clone names and accessions* using clone registry. This was needed because information in GenBank entries was in terms of clone names but sequences were assigned an accession and version number in GenBank.

2. *Verify overlaps* specified in GenBank entries using sequence alignment tools like BLAST. The verification was to find out if the overlap was present and if it was, then to verify that the overlap is a consistent overlap.

3. *Make a join* if a consistent overlap is found by deciding *breakpoint* at which to switch from one BAC to next. A breakpoint could be one of the two endpoints of the consistent overlap.

For example, GenBank entry for AL513531.15 (clone RP11-361M21) says,

> "*The true left end of clone RP4-740C4 is at 1997 in this sequence. The true right end of clone RP3-395M20 is at 2000 in this sequence.*"

We find that clones RP4-740C4 and RP3-395M20 have accessions AL513477 and AL139246, respectively. BLAST between AL513531.15 and AL513477.21 does indeed show a consistent overlap starting at first basepair of AL513477.21 to region starting at basepair 1997 of AL513531.15. BLAST between AL513531.15 and AL139246.20 gives a consistent overlap ending at basepair 2000 of AL513531.15. However, the portion of AL139246.20 that is overlapping is the first 2 Kb on reverse strand. Hence, the instructions in GenBank entry for AL513531.15 are correct but the sequence submitted for AL139246.20 is the reverse complement of what was expected by submitter of AL513531.15. As the overlaps could be verified, above instruction in GenBank results in making an assembly of the region that includes AL139246.20 from 2001 bp to 154736 bp reverse complemented followed by AL513531.15 from 1 bp to 1996 bp followed by AL513477.21 from 1 bp to 106323 bp.

Factors other than sequence submitted in reverse complement that complicate the process described above include discrepant information submitted within a clone or between two clones that are supposedly adjacent. For example, the entry for AL139246.20 (clone RP3-395M20) says,

> "*The true left end of clone RP4-755G5 is at 90075 in this sequence. The true right end of clone RP4-755G5 is at 90078 in this sequence. The true right end of clone RP11-361M21 is at 135438 in this sequence.*"

Clearly, a clone cannot have its left end and right end in the same clone without it being redundant! Furthermore, as shown in the previous example, the overlap between AL513531.15 and AL139246.20 is first 2 Kb of both sequences; not one where overlap starts at 135438 bp of AL513531.15. It is likely that sequence of clones were revised without the information being synchronized with other clones that referred to them resulting in information getting out-of-sync.

This stage did not need the order and orient step as all BACs were finished. Also, information specified by submittor overrode parameters we had for keeping an overlap. An example of a situation where overriding parameters was necessary is when a submittor specified that the only overlap for 2 BACs is a 6 bp restriction site, as in case of last 6 bp of AC002088.2 (clone CTB-13P7) overlapping with first 6 bp of AC002124.2 (clone CTB-18O1) in GenBank entry for AC002088.2. Hence, the assembly in this stage was driven more by biology than by computational methods.

It should be noted that the algorithm used in this first stage for finding contigs that assembles finished BACs as per submittors direction continued to produce contigs for all the remaining stages as well and provided a set of contigs, referred to as *ngtable* contigs, to keep "as is" in any solution produced by the software in subsequent stages but could be extended at either end by other BACs.

9.4.2 Balance of draft and finished regions: Fraction of genome represented

A natural advantage of finished BACs over draft BACs is that the false positive rate in consistent overlaps with some modest minimum requirements is negligible. Hence, in the second stage, where number of BACs were still rather modest with almost half of all BACs being finished, it was plausible to attempt using computer science techniques to try and produce a solution to instances of the NP-complete maximum interval subgraph problem before melds were created. The heuristic used for finding a solution to maximum interval subgraph was not provably optimal but probably achieved solutions that were pretty close to optimal as measured by the total number of edges in the solution kept at high *level* where the level of an edge is high for a consistent overlap between a pair of finished BACs, medium for a consistent overlap between a finished BAC and a draft BAC, and low for a consistent overlap between a pair of draft BACs. Having a curated set of overlaps to keep from the previous stage also simplified the problem.

The heuristic used for producing a solution to the layout problem was as follows:

Step 1: Create a graph with each BAC represented by a node and each consistent overlap represented by an edge where an overlap between a pair of BACs A and B results in an edge between nodes that represent A and B. We use the length, percent identity, and level of an edge to mean the length, percent identity, and level of the overlap it represents.

Step 2: Remove all edges for any node with degree more than 20. The number of nodes made singletons by removing all its edges due to high degree was typically less than 100. In the data set illustrated in tables 9.2 and 9.3, the graph has 9219 nodes. The maximum degree of any node in this graph was 102 and there was only one such node, there were 36 nodes with degree in range 51 to 100, 59 nodes with degree 21 to 50, and remaining 9123 nodes had degree at most 20. Do remaining steps for each connected component of the graph.

Step 3: Using LBFS*, find whether the graph is an interval graph. If the graph is not an interval graph, it has at least one chordless cycle and/or an asteroidal triple. Repeat steps 4 to 6 while any obstruction exists. Steps 4 to 6 need to be done repeatedly because deleting a chordless cycle can result in creation of an asteroidal triple while deleting an asteroidal triple can result in creation of a chordless cycle. For example, in figure 9.5, we initially have chordless cycle A, B, C, D but no asteroidal triple. Deleting edge (A, B) breaks the chordless cycle but creates asteroidal triple A, B, F (also A, B, G). Deleting edge (C, E) removes asteroidal triple A, B, F (but not A, B, G) and creates a chordless cycle C, D, E, F.

Step 4: While a chordless cycle exists in the graph, delete the worst edge from every chordless cycle where the goodness of an edge is determined by first its level, then its percent identity, and last by its overlap length. Cycles were not chosen in any special order. Use LBFS* to see if any asteroidal triples are present in the graph. If they are, do steps 5 and 6; otherwise, go to step 7.

Step 5: Since there is no chordless cycle in the graph, we have a chordal graph. Find the clique graph of the chordal graph [51] and a maximum weight spanning tree for the clique graph with low leafage. Note that the problem of finding the maximum weight spanning tree with minimum leafage is NP-complete (as one can easily reduce finding Hamiltonian path to it). Some results on upper and lower bounds on leafage of a chordal graph can be found in [80]. We find a

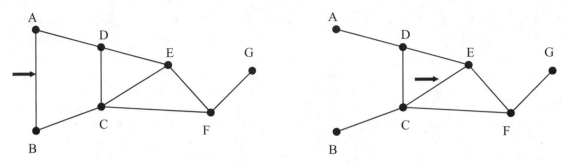

FIGURE 9.5: Deleting an edge from a chordless cycle resulting in an asteroidal triple and vice versa

maximum weight spanning tree using Prim's algorithm [34] and reduce leafage by iteratively finding a pair of edges *new* and *old* such that *new* is currently not in the solution, *old* is in the solution, and swapping *old* by *new* results in a maximal weight spanning tree with lower leafage. Since reduction in leafage is a requirement, it is sufficient to consider only those edges as candidates for *new* that are connected to a node that is a leaf in the current solution and to restrict finding candidates for *old* to the edges that are part of the cycle that is created if *new* is added to the current solution. Intuitively, finding a maximum weight spanning tree with low leafage for the clique graph approximates finding a Hamiltonian path and any node in the solution with degree more than 2 is "near" an asteroidal triple.

Step 6: Find the set of nodes S in the clique tree found in the previous step that have degree more than 2. For each node $A \in S$, find a neighbor B of A such that the weight of edge (which is same as the size of separator) connecting A to B is minimum. We developed three different heuristics for removing asteroidal triples. *Delete clique*, represented as -clique, deletes edges in original graph to remove neighbor from separator. That is, all edges connecting $B - (A \cap B)$ to $A \cap B$ are deleted for every pair of A and B found above. *Delete edge*, represented as -edge, finds the worst edge in all the ones that are candidates for deletion in -clique and only deletes that single edge. *Delete middle*, represented as -middle, deletes single worst edge for each pair A and B.

Step 7: Translate each clique path into a tiling path for a contig using LBFS*. Since not every BAC is assigned to a chromosome, it is possible to have connected components in the solution that has BACs from more than one chromosome. Check overall solution and break components into smaller components so that each component has BACs from at most one chromosome.

It is easy to see that among the three heuristics described in Step 6, -clique is fastest but creates more contigs, -edge is slow as it attempts to preserve as many good edges as possible but may delete more bad edges than -clique, and -middle tries to take advantage of both -clique and -edge by deleting worst edge of "likely" independent triples in each iteration. Results in Table 9.2 show that all three variations keep a very high percentage of edges at high levels. Results in Table 9.3 show that while -clique kept more total number of edges than -edge, it has those edges divided in higher number of contigs as was expected. We used -edge heuristic for all our processing. We present the results of -clique and -middle that are much more aggressive than -edge in deleting edges to show that graphs in this stage had a

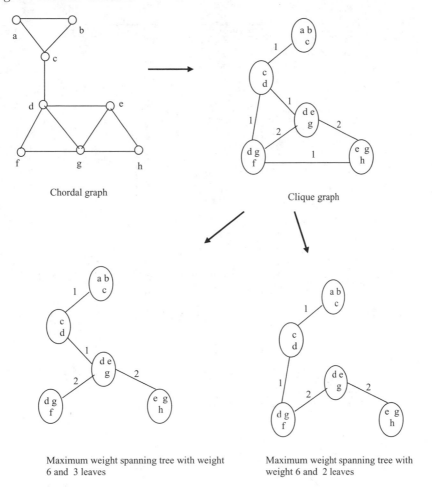

FIGURE 9.6: Transformation of chordal graph to clique graph and two maximum weight spanning trees of clique graph

phase	consistent	> 20 removed	-edge	-middle	-clique
Finished-Finished	2425	2424	2345	2327	2318
Finished-Draft	3249	1994	1439	1433	1549
Draft-Draft	4843	1702	1060	1048	1277
Total	10517	6120	4844	4808	5144

TABLE 9.2 Count of overlaps between finished and draft BACs at different stages of the algorithm and different heuristics

few "problem spots" while most of the graph was preserved by all reasonable heuristics.

9.4.3 Explosive growth in draft genome

With explosive growth in number of draft BACs, maximum interval subgraph was no longer necessarily a good representation of contig layout because the chances of multiple pairwise consistent overlaps resulting in a single multiple alignment for all BACs was much lower (see figure 9.4). In other words, it was no longer practical to have layout step separate from

Size	> 20 removed	-edge	-middle	-clique
1384, 117, 90, 88, 47	1 each			
113		1	1	
79, 44		1 each		
73, 43, 26			1 each	
71, 67, 49, 48, 36, 24				1 each
42	1	1	1	2
40		2	1	
35		1	2	1
33		1	2	
30		1		1
27		1	1	
25	1		2	2
23		1	1	
22		1	2	2
21				2
20		1	2	
19		3	1	1
18	2	4	4	4
17		3	3	
16		2	1	4
15		3	3	3
14	4	7	4	8
13	4	7	8	6
12	7	7	6	18
11	7	12	12	12
10	7	16	15	12
9	10	19	17	19
8	20	31	27	30
7	18	30	32	30
6	29	45	45	39
5	57	72	68	70
4	99	119	122	111
3	223	245	253	239
2	600	659	662	653
1	3951	4061	4071	4109
Total	5045	5358	5372	5384

TABLE 9.3   Distribution of connected component sizes with different heuristics

the melding step. This stage as well as the next one did not have an explicit layout step but did the layout in conjunction with melding. Steps taken were as follows:

Step 1: Create a contig for each *ngtable* contig and for every BAC not in ngtable contigs. A contig representing a draft BAC had a meld for each fragment and a gap of 100 bp between melds.

Step 2: Using map information for BACs in a contig, assign a lower and upper map index to the contig for each map. Maps used were Généthon and Marshfield linkage maps, and Whitehead, Stanford G3, Genemap99 GB4 , and Genemap99 G3 radiation hybrid maps. Maps were scaled such that one scaled unit represented approximately one million basepairs. This allowed us to apply same threshold to all maps for determining whether two contigs are close to each other or not. Conditions for saying that contig A is close to contig B based on a map were (i) $low(A) \leq high(B) + 1.25$ and (ii) $high(A) \geq low(B) - 1.25$, where $low(X)[high(X)]$ is the minimum [maximum] scaled unit for any BAC in contig X.

Step 3: Map overlaps in BAC coordinates to contig coordinates. Retain overlaps between two contigs only if it is consistent with map information where an overlap was considered consistent with map information if (i) at least two maps put the contigs near each other, or (ii) no map placed the two contigs apart, or (iii) there was at most one map placing contigs apart and one map placing them together.

Step 4: Assign weight to overlaps by using the formula:
$$weight = (pid - PF) \times (\min(len, LF)/100)$$

where *pid* is the percent identity of the overlap, PF is the percent factor set at 96, *len* is overlap length in bp, and LF is length factor set at 1 Kb. Minimum percent identity for any overlap used for making contigs was 97%. Sort the overlaps first by level and then by weight from highest to lowest.

Step 5: Starting from the sorted list of overlaps generated in Step 4, find the first overlap that can merge a pair of contigs and merge them. An overlap between two contigs may not result in a merge if one contig is phase 2, other contig is finished or phase 2, and adding the overlap will result in expanding a contig beyond 1.4 times average length of a BAC taken to be 250 Kb. If no contigs can be merged, go to Step 7.

Step 6: Remove overlap between contigs A and B that were merged in Step 5 to all other contigs X in the current solution. Remap all overlaps between contig X and A and X and B to and overlap between X and contig resulting from merging A and B. If there are any consistent overlaps, go back to step 4.

Step 7: Compute the *warping* of each BAC by dividing the length of BAC on contig by the length of sequence for BAC. If there are warped BACs with factor more than 1.4, then attempt to reduce their warping by removing melds consisting of a single fragment of size at most 2500 bp and by removing melds that are at least 80% contained in another meld in same contig when removing such melds results in reducing warping for the BAC.

Step 8: Order and orient melds in a contig using the directed cyclic subgraph greedy heuristic described in previous section.

9.4.4 Balance of draft and finished regions: Full genome represented

The assembly for the final stage again had the situation where we had a balance of draft and finished regions but the number of draft sequences was high enough that it was not practical to have separate layout and melding steps. Additional changes in data at this stage to be incorporated by the assembly software were as follows:

- We had a TPF that could be used as a reliable source of information for finding pairs of BACs that are close to each other on the genome.
- We had BACs that were marked as having sequence for some variations (referred to as *haplotype*) being studied in some small portions of the genome that were different from what was to be present in the assembly for the whole genome.
- Some chromosomes were finished and recipe for assembling these chromosomes was provided by IHGSC.

The following four modifications to the algorithm of previous stage were sufficient to incorporate all the data changes:

1. Create an instance of the assembly problem that is haplotype specific. This is achieved by partitioning set of BACs such that each partition has BACs with same haplotype. Assemble each haplotype separately.
2. Add TPF as a map with scaling factor of 4 as average size of a BAC is 250,000 bp and one scaled unit should represent 1 million basepairs. With scaling factor 4 and conditions for consistency with respect to a map being $low(A) \leq high(B) + 1.25$ and $high(A) \geq low(B) - 1.25$, a BAC at index N on TPF is considered consistent with respect to TPF map with only those BACs that are present on TPF and have indices $N - 5 \ldots N + 5$.

3. Upgrade the level of an overlap by 6 if the overlap is consistent with TPF map. This ensures that all overlaps consistent with TPF map are considered before considering an overlap involving a contig with no BACs on TPF.

4. Delete assembly for finished chromosomes and replace by the recipe given by HGP for that chromosome.

Above changes were designed to reflect that TPF is reliable but the assembly software can fill gaps in TPF if it finds BACs that can do so.

Running time for the assembly software in all stages after the data freeze was less than an hour on a single processor workstation. Data freeze includes generation of ngtable contigs and running BLAST for all BACs against all other BACs and transcripts. However, these are done incrementally every day and simply frozen when we wish to have a new Build and hence not counted towards the time taken for assembly.

9.5   WGS Data and Assembly

This section outlines the whole genome shotgun data assembly process and presents a brief summary of data and assembly specifics for the Celera Genomics human genome assembly. Details for the Celera Genomics human genome assembly are clearly presented in [115].

In WGS sequencing, each nucleotide of DNA in the genome is covered numerous times in fragments of about 500 bp corresponding to a shotgun read. Mate pairs made from libraries of different lengths are generated to provide order and orientation at different interval sizes. As in the case of sequencing a single BAC, repeats and missing sequence lead to several clusters instead of one cluster. These clusters are called contigs in case of WGS assembly and are called fragments in case of a BAC. Using overlaps to mate pairs and the length of library to which the mate pair belongs gives relative placement and distance information between two contigs in WGS assembly, thus, creating a scaffold consisting of more than one contig with an educated guess for the distance between consecutive contigs in the scaffold. Scaffolds are placed on chromosome using map information. An important design decision for a whole genome shotgun assembler is how it manipulates regions of repeats. Masking repeats before computing fragment overlaps results in missing true overlaps but allowing repeats to be present for overlap computation results in overcollapsing of regions.

A major advantage of WGS sequencing is that there is no need to produce BAC libraries and mapping BACs to the chromosome (although we showed in previous section that BAC based assembly can also be done if BACs are not mapped to chromosomes) making it much faster than BAC based sequencing. Disadvantages of WGS sequencing are that sequence coverage needs to be high, quality of data needs to be very good, and the assembly process is harder and time consuming with major problem being collapse of repeats. Different whole genome programs usually differ in how they process repeats, overcome repeat induced overcollapse, and reduce running time. For example, PCAP uses BLASTs 2-hit model for deciding whether to compute an overlap between a pair of fragments or not while Celera Genomics assembler relies on its high performance computational facility and software engineering techniques to deliver assembly in a few weeks.

For the human genome, Celera Genomics made plasmid libraries with insert size 2, 10, and 50 Kb and had 5.11 fold genome coverage from reads they generated from these plasmid libraries. An additional 2.9 fold coverage was achieved by shredding of data produced by HGP into 550 bp segments and 104,018 BAC end-sequence pairs [115]. Celera Genomics' state of the art sequencing facility with ABI 3700 fully automated capillary array sequencer and a high performance computational facility produced almost 14.9 billion bases corresponding to 5.11 fold coverage for a 2.9 Gb human genome in less than a year. They were able to carry out one run of their assembler in about 6 weeks real time and about 20,000 CPU hours with the largest machine having 64 Gb RAM. The design decision for repeats in Celera Genomics assembler was to introduce repeat regions slowly by first creating overlaps between unique sequence and detecting if any of the clusters so produced are over-collapsed by computing log-odds ratio of the probability that the distribution of fragment start points in the cluster is representative of a correct versus an overcollapsed cluster of two repeat copies.

9.6   Conclusion

This chapter presents how human genome assembly has progressed from inception of the Human Genome Project, its acceleration during the middle stages at the same time it met a companion in Celera Genomics, to the day we have a finished sequence for the human genome. We presented that different methods are appropriate for assembling different types of data and what those methods are.

In the future, NCBI expects to get revisions from HGP for the essentially finished assembly that reduces amount of gaps remaining. Celera Genomics has employed their updated software to produce a new assembly [66] that they have recently submitted to GenBank but with the change of direction in business plans of Celera Genomics, it looks unlikely that they will do any more updates in future. Ideally, we would like to see one assembly that can take advantage of clone based assembly for repeat regions that are hard to handle by the WGS approach and WGS assembly for regions that are hard to clone and therefore missing from the clone based assembly. Lessons learned with the dual approach that can take cost advantage of shotgun approach and control of clone by clone approach are likely to yield new approaches for sequencing large genomes [127, 21] in the future.

Acknowledgement

Thanks to Dr. David Lipman and Dr. James Ostell for providing clear direction and motivation needed over the years especially when the role of NCBI in HGP beyond GenBank became unclear to the author. The user support group at NCBI deserves praise for answering most questions from users coping with different assemblies.

References

[1] J. Aach, M.L. Bulyk, G.M. Church, and J. Comander *et al.* Computational comparison of two draft sequences of the human genome. *Nature*, 409(6822):856–859, 2001.

[2] M.D. Adams, S.E. Celniker, R.A. Holt, and C.A. Evans *et al.* The genome sequence of drosophila melanogaster. *Science*, 287(5461):2185–2195, 2000.

[3] M.D. Adams, J.M. Kelley, J.D. Gocayne, and M. Dubnick *et al.* Complementary DNA sequencing: expressed sequence tags and human genome project. *Science*, 252(5013):1651–1656, 1991.

[4] M.D. Adams, G.G. Sutton, H.O. Smith, and E.W. Myers *et al.* The independence of our genome assemblies. *Proceedings of the National Academy of Sciences USA*, 100(6):3025–3026, 2003.

[5] R. Agarwala, D.L. Applegate, D. Maglott, and G.D. Schuler *et al.* A fast and scalable radiation hybrid map construction and integration strategy. *Genome Research*, 10(3):350–364, 2000.

[6] B.M. Alberts, D. Botstein, and S. Brenner *et al.* Report of the committee on mapping and sequencing the human genome. Technical report, National Academy of Science, Washington D.C., 1988.

[7] S.F. Altschul, W. Gish, W. Miller, and E.W. Myers *et al.* Basic local alignment search tool. *Journal Molecular Biology*, 215(3):403–410, 1990.

[8] S.F. Altschul, T.L. Madden, A.A. Schäffer, and J. Zhang *et al.* Gapped BLAST and PSI-BLAST: a new generation of protein database search programs. *Nucleic Acids Research*, 25(17):3389–3402, 1997.

[9] O.T. Avery, C.M. MacLeod, and M. McCarty. Studies on the chemical nature of the substance inducing transformation of pneumococcal types. induction of transformation by a desoxyribonucleic acid fraction isolated from pneumococcus type III. *The Journal of experimental medicine*, 79(1):137–158, 1944.

[10] R. Bar-Yehuda and S. Even. A local-ratio theorem for approximating the weighted vertex cover problem. *Annals of Discrete Mathematics*, 25:27–46, 1985.

[11] S. Batzoglou, D.B. Jaffe, K. Stanley, and J. Butler *et al.* Arachne: A whole-genome shotgun assembler. *Genome Research*, 12(1):177–189, 2002.

[12] D.R. Bentley. Genomic sequence information should be released immediately and freely in the public domain. *Science*, 274(5287):533–534, 1996.

[13] B. Berger and P.W. Shor. Approximation algorithms for the maximum acyclic subgraph problem. In *Proc. First Annual ACM-SIAM Symposium on Discrete Algorithms*, pages 236–243, 1990.

[14] K.S. Booth and G.S. Lueker. Testing for the consecutive ones property, interval graphs and graph planarity using PQ-tree algorithms. *Journal of Computer and System Sciences*, 13(3):335–379, 1976.

[15] S. Borenstein. The human genome project. Technical report, 2000. http://www.freep.com/news/nw/gene27_20000627.htm.

[16] D. Botstein, R.L. White, M. Skolnick, and R.W. Davis *et al.* Construction of a genetic linkage map in man using restriction fragment length polymorphisms. *American Journal of Human Genetics*, 32(3):314–331, 1980.

[17] G.G. Bouffard, J.R. Idol, V.V. Braden, and L.M. Iyer *et al.* A physical map of human chromosome 7: An integrated YAC contig map with average STS spacing of 79 kb. *Genome Research*, 7(7):673–692, 1997.

[18] K.W. Broman, J.C. Murray, V.C. Sheffield, and R.L. White *et al.* Comprehensive human genetic maps: individual and sex-specific variation in recombination. *American Journal of Human Genetics*, 63(3):861–869, 1998.

[19] K.H. Buetow, J.L. Weber, S. Ludwigsen, and T. Scherpbier *et al.* Integrated human genome-wide maps constructed using the CEPH reference panel. *Nature Genetics*, 6(4):391–393, 1994.

[20] D.T. Burke, G.F. Carle, and M.V. Olson. Cloning of large segments of exogenous DNA into yeast by means of artificial chromosome vectors. *Science*, 236(4803):806–812, 1987.

[21] W.-W. Cai, R. Chen, R.A. Gibbs, and A. Bradley *et al.* A clone-array pooled shotgun strategy for sequencing large genomes. *Genome Research*, 11(10):1619–1623, 2001.

[22] C.R. Cantor. Orchestrating the human genome project. *Science*, 248(4951):49–51, 1990.

[23] Cooperative Human Linkage Center. A comprehensive human linkage map with centimorgan density. *Science*, 265(5181):2049–2054, 1994.

[24] T. Chen and S.S. Skiena. A case study in genome-level fragment assembly. *Bioinformatics*, 16(6):494–500, 2000.

[25] V. Choi and M. Farach-Colton. Barnacle: an assembly algorithm for clone-based sequences of whole genomes. *Gene*, 320:165–176, 2003.

[26] S.L. Christian, J. McDonough, C.-Y. Liu, and S. Shaikh *et al.* An evaluation of the assembly of an approximately 15-mb region on human chromosome 13q32-q33 linked to bipolar disorder and schizophrenia. *Genomics*, 79(5):635–658, 2002.

[27] I.M. Chumakov, P. Rigault, E.S. Lander, and C. Bellanne *et al.* A YAC contig map of the human genome. *Nature*, 377(6547 Suppl):175–297, 1995.

[28] A.S. Cohen, D.R. Najarian, and B.L. Karger. Separation and analysis of DNA sequence reaction products by capillary gel electrophoresis. *Journal of Chromatography*, 516(1):49–60, 1990.

[29] D. Cohen, I. Chumakov, and J. Weissenbach. A first-generation physical map of the human genome. *Nature*, 366(6456):698–701, 1993.

[30] F.S. Collins. Contemplating the end of the beginning. *Genome Research*, 11(5):641–643, 2001.

[31] F.S. Collins and D. Galas. A new five-year plan for the u.s. human genome project. *Science*, 262(5130):43–46, 1993.

[32] F.S. Collins, A. Patrinos, E. Jordan, and A. Chakravarti *et al.* New goals for the U.S. human genome project: 1998-2003. *Science*, 282(5389):682–689, 1998.

[33] International Human Genome Sequencing Consortium. Initial sequencing and analysis of the human genome. *Nature*, 409(6822):860–921, 2001.

[34] T.H. Cormen, C.E. Leiserson, R.L. Rivest, and C. Stein *et al. Introduction to Algorithms.* MIT Press and McGraw-Hill, second edition, 2001.

[35] D.G. Corneil, S. Olariu, and L. Stewart. The ultimate interval graph recognition algorithm? In *Proc. Ninth Annual ACM-SIAM Symposium on Discrete Algorithms*, pages 175–180, 1998.

[36] N.R. Cozzarelli. Revisiting the independence of the publicly and privately funded drafts of the human genome. *Proceedings of the National Academy of Sciences USA*, 100(6):3021, 2003.

[37] P. Deloukas, G.D. Schuler, and G. Gyapay *et al.* A physical map of 30,000 human genes. *Science*, 282(5389):744–746, 1998.

[38] C. Demetrescu and I. Finocchi. Combinatorial algorithms for feedback problems in directed graphs. *Information Processing Letters*, 86(3):129–136, 2003.

[39] U.S. Department of Energy and National Institutes of Health. Understanding our genetic inheritance: The U.S. human genome project, the first five years: Fiscal years 1991-1995. Technical report, 1990. DOE/ER-0452P, NIH Publication No. 90-1590, http://www.ornl.gov/sci/techresources/Human_Genome/project/5yrplan.

[40] C. Dib, S. Faure, C. Fizames, and D. Samson *et al.* A comprehensive genetic map of the human genome based on 5,264 microsatellites. *Nature*, 380(6570):152–154, 1996.

[41] DOE human genome program. Primer on molecular genetics. Technical report, U.S. Department of Energy, Office of energy research, Washington DC, 1992.

[42] N.A. Doggett, L.A. Goodwin, J.G. Tesmer, and L.J. Meincke *et al.* An integrated physical map of human chromosome 16. *Nature*, 377(6547 Suppl):335–365, 1995.

[43] H. Donis-Keller, P. Green, C. Helms, and S. Cartinhour *et al.* A genetic linkage map of the human genome. *Cell*, 51(2):319–337, 1987.

[44] I. Dunham, N. Shimizu, B.A. Roe, and S. Chissoe *et al.* The DNA sequence of human chromosome 22. *Nature*, 402(6761):489–495, 1999.

[45] P. Eades and X. Lin. A new heuristic for the feedback arc set problem. *Australian Journal of Combinatorics*, 12:15–26, 1995.

[46] P. Eades, X. Lin, and W.F. Smyth. A fast and effective heuristic for the feedback arc set problem. *Information Processing Letters*, 47(6):319–323, 1993.

[47] G. Even, J. Naor, B. Schieber, and M. Sudan *et al.* Approximating minimum feedback sets and multi-cuts in directed graphs. In *Proceedings of the 4th International IPCO Conference on Integer Programming and Combinatorial Optimization*, Lecture Notes In Computer Science, pages 14–28, London, UK, 1995. Springer Verlag.

[48] S. Even. *Graph Algorithms*. Computer Science Press, Rockville, Maryland, 1979.

[49] R.D. Fleischmann, M.D. Adams, O. White, and R.A. Clayton *et al.* Whole-genome random sequencing and assembly of *Haemophilus influenzae Rd*. *Science*, 269(5223):496–512, 1995.

[50] S. Foote, D. Vollrath, A. Hilton, and D.C. Page *et al.* The human Y chromosome: overlapping DNA clones spanning the euchromatic region. *Science*, 258(5079):60–66, 1992.

[51] P. Galinier, M. Habib, and C. Paul. Chordal graphs and their clique graphs. In *Workshop on Graph-Theoretic Concepts in Computer Science*, 1995.

[52] M.R. Garey and D.S. Johnson. *Computers and Intractability: A Guide to the Theory of NP-Completeness*. W. H. Freeman and co., San Francisco, California, 1979.

[53] F. Gavril. Some NP-complete problems on graphs. In *Proceedings of the 11th conference on Information Sciences and Systems*, pages 91–95, 1977.

[54] P.C. Gilmore and A.J. Hoffman. A characterization of comparability graphs and of interval graphs. *Canadian Journal of Mathematics*, 16(99):539–548, 1964.

[55] P.W. Goldberg, M.C. Golumbic, H. Kaplan, and R. Shamir *et al.* Four strikes against physical mapping of DNA. *Journal of Computational Biology*, 2(1):139–152, 1995.

[56] E.D. Green, H.C. Riethman, J.E. Dutchik, and M.V. Olson *et al.* Detection and characterization of chimeric yeast artificial-chromosome clones. *Genomics*, 11(3):658–669, 1991.

[57] P. Green. PHRAP and cross_match. http://www.phrap.org.

[58] C.M. Grisham and R.H. Garrett. *Biochemistry*. Brooks Cole, second edition, 1998.

[59] R. Hassin and S. Rubinstein. Approximations for the maximum acyclic subgraph problem. *Information Processing Letters*, 51(3):133–140, 1994.

[60] M. Hattori, A. Fujiyama, T.D. Taylor, and H. Watanabe *et al.* The DNA sequence of human chromosome 21. *Nature*, 405(6784):311–319, 2000.

[61] W.-L. Hsu and T.-H. Ma. Fast and simple algorithms for recognizing chordal comparability graphs and interval graphs. *SIAM Journal on Computing*, 28(3):1004–1020, 1999.

[62] X. Huang and A. Madan. CAP3: A DNA sequence assembly program. *Genome Research*, 9(9):868–877, 1999.

[63] X. Huang, J. Wang, S. Aluru, and S.P. Yang *et al.* Pcap: A whole-genome assembly program. *Genome Research*, 13(9):2164–2170, 2003.

[64] T.J. Hudson, L.D. Stein, S.S. Gerety, and J. Ma *et al.* An STS-based map of the human genome. *Science*, 270(5244):1945–1954, 1995.

[65] D.H. Huson, K. Reinert, S.A. Kravitz, and K.A. Remington *et al.* Design of a compartmentalized shotgun assembler for the human genome. *Bioinformatics*, 17 Suppl 1:S132–139, 2001.

[66] S. Istrail, G.G. Sutton, L. Florea, and A.L. Halpern *et al.* Whole-genome shotgun assembly and comparison of human genome assemblies. *PNAS*, 101(7):1916–1921, 2004.

[67] D.A. Jackson, R.H. Symons, and P. Berg. Biochemical method for inserting new genetic information into DNA of simian virus 40: circular *sv40* DNA molecules con-

taining lambda phage genes and the galactose operon of escherichia coli. *Proceedings of the National Academy of Sciences USA*, 69(10):2904–2909, 1972.

[68] D.B. Jaffe, J. Butler, S. Gnerre, and E. Mauceli *et al.* Whole-genome sequence assembly for mammalian genomes: Arachne 2. *Genome Research*, 13(1):91–96, 2003.

[69] V. Kann. *On the Approximability of NP-complete Optimization Problems.* PhD dissertation, Royal Institute of Technology, Stockholm, Department of Numerical Analysis and Computing Science, 1992.

[70] R.M. Karp. Reducibility among combinatorial problems. In R.E. Miller and J.W. Thatcher, editors, *Complexity of computer computations*, pages 85–103. Plenum Press, 1972.

[71] J.D. Kececioglu and E.W. Myers. Combinatorial algorithms for DNA sequence assembly. *Algorithmica*, 13:7–51, 1995.

[72] W.J. Kent and D. Haussler. Assembly of the working draft of the human genome with *GigAssembler. Genome Research*, 11(9):1541–1548, 2001.

[73] U.-J. Kim, B.W. Birren, T. Slepak, and V. Mancino *et al.* Construction and characterization of a human bacterial artificial chromosome library. *Genomics*, 34(2):213–218, 1996.

[74] I.R. Kirsch and T. Ried. Integration of cytogenetic data with genome maps and available probes: present status and future promise. *Seminars in hematology*, 37(4):420–428, 2000.

[75] P. Kitts. *Genome Assembly and Annotation Process*, chapter 14 in The NCBI handbook, http://www.ncbi.nlm.nih.gov/books/bookres.fcgi/handbook/ch14d1.pdf. 2002.

[76] A. Kong, D.F. Gudbjartsson, J. Sainz, and G.M. Jonsdottir *et al.* A high-resolution recombination map of the human genome. *Nature Genetics*, 31(3):241–247, 2002.

[77] N. Korte and R.H. Möhring. An incremental linear time algorithm for recognizing interval graphs. *SIAM Journal on Computing*, 18(1):68–81, 1989.

[78] B. Lewin. *Genes VIII.* Prentice Hall, first edition, 2003.

[79] S. Li, G. Cutler, J.J. Liu, and T. Hoey *et al.* A comparative analysis of hgsc and celera human genome assemblies and gene sets. *Bioinformatics*, 19(13):1597–1605, 2003.

[80] I.-J. Lin, T.A. McKee, and D.B. West. Leafage of chordal graphs. *Discussiones Mathematicae Graph Theory*, 18(1):23–48, 1998.

[81] M. Litt and J.A. Luty. A hypervariable microsatellite revealed by in vitro amplification of a dinucleotide repeat within the cardiac muscle actin gene. *American Journal of Human Genetics*, 44(3):397–401, 1989.

[82] C.L. Lucchesi. *A minimax equality for directed graphs.* PhD dissertation, University of Waterloo, Department of Numerical Analysis and Computing Science, 1966.

[83] J.A. Luckey, H. Drossman, A.J. Kostichka, and D.A. Mead *et al.* High speed DNA sequencing by capillary electrophoresis. *Nucleic Acids Research*, 18(15):4417–4421, 1990.

[84] A.M. Maxam and W. Gilbert. A new method for sequencing DNA. *Proceedings of the National Academy of Sciences USA*, 74(2):560–564, 1977.

[85] V.A. McKusick and F.H. Ruddle. The status of the gene map of the human chromosomes. *Science*, 196(4288):390–405, 1977.

[86] J.C. Mullikin and Z. Ning. The phusion assembler. *Genome Research*, 13(1):81–90, 2003.

[87] K. Mullis, F. Faloona, S. Scharf, and R. Saiki *et al.* Specific enzymatic amplification of DNA in vitro: the polymerase chain reaction. *Cold Spring Harbor symposia on quantitative biology*, 51(Part 1):263–273, 1986.

[88] E.W. Myers, G.G. Sutton, A.L. Delcher, and I.M. Dew *et al.* A whole-genome assembly of *Drosophila*. *Science*, 287(5461):2196–2204, 2000.

[89] E.W. Myers, G.G. Sutton, H.O. Smith, and M.D. Adams *et al.* On the sequencing and assembly of the human genome. *Proceedings of the National Academy of Sciences USA*, 99(7):4145–4146, 2002.

[90] S. Olariu. An optimal greedy heuristic to color interval graphs. *Information processing letters*, 37(1):21–25, 1991.

[91] M. Olivier, A. Aggarwal, J. Allen, and A.A. Almendras *et al.* A high-resolution radiation hybrid map of the human genome draft sequence. *Science*, 291(5507):1298–1302, 2001.

[92] M. Olson, L. Hood, C. Cantor, and D. Botstein *et al.* A common language for physical mapping of the human genome. *Science*, 245(4925):1434–1435, 1989.

[93] K. Osoegawa, P.Y. Woon, B. Zhao, and E. Frengen *et al.* An improved approach for construction of bacterial artificial chromosome libraries. *Genomics*, 52(1):1–8, 1998.

[94] C.H. Papadimitriou and K. Steiglitz. *Combinatorial Optimization: Algorithms and Complexity*. Prentice-Hall, Englewood Cliffs, New Jersey, 1982.

[95] H. Peltola, H. Soderlund, and E. Ukkonen. SEQAID: a DNA sequence assembling program based on a mathematical model. *Nucleic Acids Research*, 12(1):307–321, 1984.

[96] P.A. Pevzner and H. Tang. Fragment assembly with double-barreled data. *Bioinformatics*, 17(Suppl. 1):S225–S233, 2001.

[97] P.A. Pevzner, H. Tang, and M.S. Waterman. An eulerian path approach to DNA fragment assembly. *Proceedings of the National Academy of Sciences USA*, 98(17):9748–9753, 2001.

[98] J.M. Prober, G.L. Trainor, R.J. Dam, and F.W. Hobbs *et al.* A system for rapid DNA sequencing with fluorescent chain-terminating dideoxynucleotides. *Science*, 238(4825):336–341, 1987.

[99] S. Qin, N.J. Nowak, J. Zhang, and S.N. Sait *et al.* A high-resolution physical map of human chromosome 11. *Proceedings of the National Academy of Sciences USA*, 93(7):3149–3154, 1996.

[100] D.J. Rose, R.E. Tarjan, and G.S. Leuker. Algorithmic aspects of vertex elimination on graphs. *SIAM Journal on Computing*, 5(2):266–283, 1976.

[101] E.C. Rouchka, W. Gish, and D.J. States. Comparison of whole genome assemblies of the human genome. *Nucleic Acids Research*, 30(22):5004–5014, 2002.

[102] F. Sanger, S. Nicklen, and A.R. Coulson. DNA sequencing with chain-terminating inhibitors. *Proceedings of the National Academy of Sciences USA*, 74(12):5463–5467, 1977.

[103] M. Schena, D. Shalon, R.W. Davis, and P.O. Brown *et al.* Quantitative monitoring of gene expression patterns with a complementary DNA microarray. *Science*, 270(5235):467–470, 1995.

[104] D.C. Schwartz and C.R. Cantor. Separation of yeast chromosome-sized DNAs by pulsed field gradient gel electrophoresis. *Cell*, 37(1):67–75, 1984.

[105] L. Selleri, J.H. Eubanks, M. Giovannini, and G.G. Hermanson *et al.* Detection and characterization of chimeric yeast artificial chromosome clones by fluorescent in situ suppression hybridization. *Genomics*, 14(2):536–541, 1992.

[106] C.A.M. Semple, S. W. Morris, D.J. Porteous, and K.L. Evans *et al.* Computational comparison of human genomic sequence assemblies for a region of chromosome 4. *Genome Research*, 12(3):424–429, 2002.

[107] H. Shizuya, B. Birren, U.-J. Kim, and V. Mancino *et al.* Cloning and stable maintenance of 300-kilobase-pair fragments of human DNA in escherichia coli using an

f-factor-based vector. *Proceedings of the National Academy of Sciences USA*, 89(18):8794–8797, 1992.

[108] L.M. Smith, J.Z. Sanders, R.J. Kaiser, and P. Hughes *et al.* Fluorescence detection in automated DNA sequence analysis. *Nature*, 321(6071):674–679, 1986.

[109] W.H. Sofer. *Introduction to Genetic Engineering*. Butterworth-Heinemann, 1991.

[110] H.H. Stassen and C. Scharfetter. Integration of genetic maps by polynomial transformations. *American Journal of Medical Genetics*, 96(1):108–113, 2000.

[111] E.A. Stewart, K.B. McKusick, A. Aggarwal, and E. Bajorek *et al.* An STS-based radiation hybrid map of the human genome. *Genome Research*, 7(5):422–433, 1997.

[112] G.G. Sutton, O. White, M.D. Adams, and A. Kerlavage *et al.* TIGR assembler: A new tool for assembling large shotgun sequencing projects. *Genome science and technology*, 1:9–19, 1995.

[113] H. Swerdlow, S.L. Wu, H. Harke, and N.J. Dovichi *et al.* Capillary gel electrophoresis for DNA sequencing. laser-induced fluorescence detection with the sheath flow cuvette. *Journal of Chromatography*, 516(1):61–67, 1990.

[114] The Utah Marker Development Group. A collection of ordered tetranucleotide-repeat markers from the human genome. *American Journal of Human Genetics*, 57(3):619–628, 1995.

[115] J.C. Venter, M.D. Adams, E.W. Myers, and P.W. Li *et al.* The sequence of the human genome. *Science*, 291(5507):1304–1351, 2001.

[116] J.C. Venter, M.D. Adams, G.G. Sutton, and A.R. Kerlavage *et al.* Genomics: Shotgun sequencing of the human genome. *Science*, 280(5369):1540–1542, 1998.

[117] J.C. Venter, H.O. Smith, and L. Hood. A new strategy for genome sequencing. *Nature*, 381(6581):364–366, 1996.

[118] D. Vollrath, S. Foote, A. Hilton, and L.G. Brown *et al.* The human Y chromosome: a 43-interval map based on naturally occurring deletions. *Science*, 258(5079):52–59, 1992.

[119] J. Wang, G. K.-S. Wong, P. Ni, and Y. Han *et al.* Reps: A sequence assembler that masks exact repeats identified from the shotgun data. *Genome Research*, 12(5):824–831, 2002.

[120] R.H. Waterston, E.S. Lander, and J.E. Sulston. On the sequencing of the human genome. *Proceedings of the National Academy of Sciences USA*, 99(6):3712–3716, 2002.

[121] R.H. Waterston, E.S. Lander, and J.E. Sulston. More on the sequencing of the human genome. *Proceedings of the National Academy of Sciences USA*, 100(6):3022–3024, 2003.

[122] J.D. Watson. The human genome project: past, present, and future. *Science*, 248(4951):44–49, 1990.

[123] J.D. Watson and F.H. Crick. Molecular structure of nucleic acids; a structure for deoxyribose nucleic acid. *Nature*, 171(4356):737–738, 1953.

[124] J.L. Weber and P.E. May. Abundant class of human DNA polymorphisms which can be typed using the polymerase chain reaction. *American Journal of Human Genetics*, 44(3):388–396, 1989.

[125] M.C. Weiss and H. Green. Human-mouse hybrid cell lines containing partial complements of human chromosomes and functioning human genes. *Proceedings of the National Academy of Sciences USA*, 58(3):1104–1111, 1967.

[126] J. Weissenbach, G. Gyapay, C. Dib, and A. Vignal *et al.* A second-generation linkage map of the human genome. *Nature*, 359(6398):794–801, 1992.

[127] M.C. Wendl, M.A. Marra, L.W. Hillier, and A.T. Chinwalla *et al.* Theories and applications for sequencing randomly selected clones. *Genome Research*, 11(2):274–

280, 2001.

[128] L.A. Wolsey and G.L. Nemhauser. *Integer and Combinatorial Optimization.* Wiley-Interscience, New York, New York, first edition, 1999.

Comparative Methods for Sequence Assembly

Vamsi Veeramachaneni
Strand Genomics Corporation

10.1 Introduction

The past few years have seen remarkable progress in the field of DNA sequencing. Some of the landmark achievements are the sequencing of several key experimental organisms including a yeast (*Saccharomyces cerevisiae*) [16], the multicellular organism *Caenorhabditis elegans* [44], the plant *Arabidopsis thaliana* [15], the fruitfly *Drosophila melanogaster* [1, 32] and the completion of the draft sequence of the human genome [52, 27]. At present, the complete genome sequences of over 200 species (mostly bacteria) are available in the Genomes section of GenBank.

The GenBank sequence database [4], a repository of all publicly available DNA sequences, continues to grow at an exponential rate, doubling in size approximately every 14 months. As of release 146, February 2005, GenBank contained over 46.8 billion nucleotide bases from 42.7 million different sequences. Complete genomes represent a growing portion of the database, with over 70 of the 229 complete microbial genomes available in April, 2005 in GenBank deposited over the past year.

However, current costs associated with large-scale sequencing remain a limiting factor for genome sequencing efforts [17]. For example, establishing approximately one-fold sequence coverage of a mammalian genome requires the generation of roughly 6 million sequence reads and at present costs about US $10–20 million (and so, a working draft sequence with approximately five-fold coverage would cost US $50–100 million). As a result, even though more than 140,000 different organisms are represented in GenBank, it is clear that only a handful of different genomes can realistically be sequenced in a comprehensive fashion, at least by the available methods.

In this chapter, we describe methods that use existing genomic sequences as a guide to assembling the genomic sequences of related organisms. The underlying assumption in all these methods is that genomic organization tends to be evolutionarily conserved between species — an idea that was first postulated in the early 1900s [7, 19]. Recent studies that characterize the nature and extent of sequence conservation, and some comparative genomics methods that exploit the presence of conserved segments are summarized in Section 10.2. Section 10.3 outlines the main computational problems encountered in hierarchical sequencing — physical mapping and contig ordering. Comparative approaches that have been used to solve the mapping problem are summarized in 10.4.1. In section 10.4.2, we introduce the main idea behind the comparative methods for contig ordering presented in greater detail in the rest of the chapter.

10.2   Conserved Segments in Chromosomes

During evolution, inter- and intra- chromosomal rearrangements such as fission, fusion, reciprocal translocation, transposition and inversion disrupt the order of genes along the chromosome, as depicted in Figure 10.1. However, gene order remains fixed in the segment between any two neighboring breakpoints resulting from the rearrangements. These segments with conserved gene order (*conserved segments*) tend to become shorter with time as they are disrupted by new events.

Comparative mapping studies have shown that closely related organisms share large conserved chromosome segments while more distantly related species exhibit shorter conserved segments [37, 11]. Many studies have shown that large stretches of DNA are conserved in mammalian species as divergent as human and fin whales [33, 41, 42, 55] and that gene order is also conserved to a significant extent [35, 46]. For instance, the cat genome can be reorganized into the human genome by as few as 13 translocation steps [36]. However, exceptions to the slow rate of rearrangements occur in several mammalian orders. Rodent species, particularly mouse and rat, show rapid pattern of change, with some 180-280 conserved segments shared between human and mouse [18, 27] and 60 shared between mouse and rat [34].

The availability of genome sequences of multiple organisms has allowed researchers to compute the extent of sequence conservation in homologous regions more precisely. One rather surprising discovery was that a large number of actively conserved regions are unlikely to be genes [9]. Overall it is estimated that 5% of the mammalian genome is under purifying selection [53]. This is a much higher proportion than can be explained by protein coding sequence alone. To better describe unique conserved regions, Mural *et al.* [31] use the term *syntenic anchor* to refer to short stretches of DNA in different species that show significant sequence match to each other but to no other region in either genome[1]. While some of the anchors correspond to conserved exons, the majority do not. In a comparison of the human and mouse genomes, an analysis of the distribution of syntenic anchors mapped to chromosome 16 of mouse revealed that only 56% are within gene boundaries (34% correspond to exons) and that the density of anchors along the chromosome was unaffected

---

[1]The terms *orthologous landmark* [53] and *clusters of orthologous bases (COBS)* [10] have also been used in similar contexts.

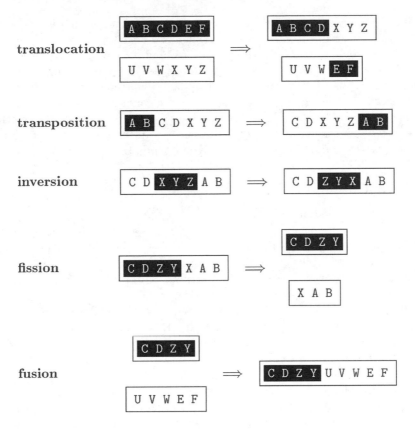

FIGURE 10.1: Chromosomal rearrangements. Colors are used to highlight the changes in gene order. Assuming that the rearrangements take place in the order shown, the final chromosomal segment contains 4 conserved segments — (C D), (Z Y), (U V W) and (E F) — from the initial chromosomes.

by gene density. Over 50% of the syntenic anchors occurred in runs of at least 128 in a row with the same order and orientation in both the genomes with no interleaving anchors. Similar results based on a three-way comparison between human, mouse and dog [10], and a 12 species comparison of regions orthologous to a 1.8 MB section of human chromosome 7 confirm the wide-spread presence of syntenic anchors [49].

Comparative genomics methods have been used to discover regulatory elements [20, 29], construct comparative gene maps [48, 26], refine annotation [5], predict gene structure based on sequence conservation at corresponding locations in two sequences [25, 12, 30], and identify orthologs based on conserved gene order [45]. In this chapter we focus on how comparative sequencing approaches can take advantage of the presence of the large number of syntenic anchors distributed across genomes.

10.3   Hierarchical Sequencing

Hierarchical (map based) sequencing and whole genome shotgun (WGS) sequencing [54] are the two most common strategies for the sequencing of large, complex genomes. The relative merits of these approaches have been summarized in [17]. The initial large scale sequencing projects (including those for the *S. cerevisiae*, *C. elegans* and *A. thaliana* genomes) were guided by carefully constructed clone maps which are an integral part of the hierarchical approach. While mapped BAC clones continue to play a fundamental role in obtaining a finished sequence, the fast and relatively less expensive WGS sequencing method has come to be an important component of many genome-scale sequencing projects. Mixed strategies are being used in the sequencing of the mouse and rat genomes. In this chapter, we discuss only the hierarchical approach since it is more amenable to comparative methods. The two phases in this approach and the computational problems involved are described in this section. The next section describes comparative methods that tackle the same problems.

10.3.1　Physical mapping

The process of physical mapping starts with the breaking of the DNA molecule into small pieces (e.g., with restriction enzymes). Multiple copies of each fragment are created by incorporating the fragment into a self-replicating host. The self-replication process then creates large number of copies (clones) of the fragment.

Over the past several years, the bacterial artificial chromosome (BAC) has emerged as the vector system of choice for cloning. The BAC cloning system allows the isolation of genomic DNA fragments that are reasonably large (80 to >200 kilobases [kb]). At the same time, BAC clones are less prone to artifacts than other available vector systems for large-fragment cloning, e.g., cosmids and yeast artificial chromosomes (YACs). In the rest of the chapter, we often refer to clones as BAC clones. However, the methods discussed are often applicable to other types as well.

As a result of the cloning process, biologists obtain a *clone library* consisting of thousands of clones (each representing a short DNA fragment) from the same DNA molecule. After a clone library is constructed biologists want to *order* the clones since information regarding the location of clones is lost in the construction of the clone library. The problem of determining placement of clones along the DNA molecule to construct a *clone map* is also called the *physical mapping* problem.

The main idea in physical mapping is to describe each clone using an easily determined fingerprint, which can be thought of as a set of "key words" for the clone. In addition to being inexpensive to determine, a good clone fingerprinting scheme should satisfy two criteria — clones with substantial overlap should have similar fingerprints and non-overlapping clones should have distinct fingerprints with high probability. Two commonly used fingerprints are

- the sizes of the fragments created when a clone is cut by a particular restriction enzyme
- the list of probes that hybridize to a clone

The computational problems underlying the construction of a clone map based on these fingerprints are NP-hard [38]. However, in practice, these fingerprints allow biologists to distinguish between overlapping and non-overlapping clones and to reconstruct the order of the clones (*physical map*).

10.3.2　Shotgun sequencing

Minimally overlapping clones that form a tiling path across the target DNA fragment are selected from the clone map to form a *sequence ready clone map*. Individual clones from this map are sequenced using shotgun sequencing — a method that has been the mainstay of all large scale sequencing techniques since the 1980s. In this method, multiple copies of

```
5'  →   AAAAGCTTCTAGAACCACTGTAGGAGGTACAAGATGCTCCTGAGAACTCAGTAGAGGTGG   →   3'
3'  ←   TTTTCGAAGATCTTGGTGACATCCTCCATGTTCTACGAGGACTCTTGAGTCATCTCCACC   ←   5'
```

```
              f₁                     f₂                       f₃
5'  →   AAAAGCTTCTAGAACCACTGTAGGAGGTACAAGATGCTCCTGAGAACTCAGTAGAGGTGG   →   3'
3'  ←   TTTTCGAAGATCTTGGTGACATCCTCCATGTTCTACGAGGACTCTTGAGTCATCTCCACC   ←   5'
              f₄                     f₅
```

FIGURE 10.2: A sample order/orient problem instance. Sequencing the DNA fragment shown in the upper half of the figure results in five contigs $f_1, \ldots, f_5$ with unknown order and orientation. The figure in the lower half shows the locations of the contigs in the original fragment.

the DNA fragment to be sequenced are created by cloning and then randomly partitioned into smaller fragments. Approximately 500-700 consecutive basepairs from the ends of these fragments can be determined using variations of the Sanger method [40]. These sequences are called *reads*. Overlapping *reads* are assembled into *contigs*, i.e., presumably contiguous sections of the genomic sequence. The ideal outcome of this step is a single contig that represents the entire sequence of the original DNA fragment. More commonly, the result is a set of contigs with unknown order and orientation (strand information).

For example, consider the double stranded DNA fragment shown in Figure 10.2. Since strands are read in the $5' \rightarrow 3'$ direction, the fragment is represented by the sequence $s = \text{AAAAG} \cdots \text{GTGG}$ or its reverse complement $s^R = \text{CCAC} \cdots \text{CTTTT}$. If the result of shotgun sequencing is the set of contigs $\{f_1, f_2, f_3, f_4, f_5\}$, we would like to order the contigs along one of the strands. We call this the *contig order/orient problem*. In this instance there are two solutions depending on the strand along which we choose to place the contigs – $\langle f_1, f_4^R, f_2, f_5^R, f_3 \rangle$ and $\langle f_3^R, f_5, f_2^R, f_4, f_1^R \rangle$.

In the case of hierarchical sequencing, the clone map imposes a partial order on the sets of all contigs obtained i.e., if $C_1 C_2 \ldots C_n$ is the sequence ready clone map for a DNA fragment X, then it follows that when the resulting contigs are correctly ordered along X, all contigs from clone C_i will appear to the left of all contigs from clone C_j for any $i < j$. Thus, the problem reduces to that of ordering and orienting contigs from individual clones separately. This can usually be accomplished by consulting a detailed list of markers.

However, since the physical mapping process is not present in the whole genome shotgun (WGS) sequencing strategy, an alternative method is used. Pairs of reads, called *mates*, are sequenced from the ends of long inserts randomly sampled from the genome. Since the lengths of these long inserts can be measured accurately, the presence of mates in different contigs serves to order and orient the contigs and give the approximate distance between them. The result of this assembly process is, therefore, a collection of *scaffolds*, where each scaffold is a set of contigs that are ordered, oriented and positioned with respect to each other.

10.4 Comparative Methods in Sequencing

10.4.1 Construction of physical maps

A few hundred thousand clones are needed to make a library that provides an adequate representation of a mammalian-sized genome. This makes both the fingerprinting techniques

discussed in the previous section relatively expensive to implement for the complete characterization of an entire library. BAC end sequencing, in which probes are created based on
the sequence obtained by reading the ends of BACs, is presently nearly three times more
expensive than sequencing other substrates because of the difficulty in obtaining good yields
of BAC DNA for sequencing etc. Furthermore, appropriate DNA preparation is difficult
and automated preparation systems have not been developed.

As for restriction enzyme-based fingerprinting, only in the last few years has the methodology become reliable enough at high throughput levels to generate the amount of high
quality data needed to build maps that are accurate enough and of high enough resolution
to guide the choice of clones for sequencing. The costs of constructing a fingerprint-based
map have decreased somewhat recently, but remain high. For example, for 200,000 clones,
end sequencing currently costs at least $1.5 million, while fingerprinting costs more than $2
million.

The availability of a draft human genome sequence [52, 27] has radically improved the
ability to perform comparative mapping with related species. Specifically, the locations
and sequences of genes in the human genome can be used to guide map construction in
other species through the computational detection of orthologous sequences. Some recent
methods that use the finished genomic sequence of one species, S_1, to assemble a physical
map of clones of a second related species S_2 are described below:

- *Required:* Locations and sequences of genes in species S_1.
  Method: The gene sequences of S_1 are used to query databases containing DNA
  sequences of species S_2. Orthologous S_2 sequences found by the search are used
  to design S_2 specific hybridization probes. The probes are used to screen a S_2
  clone library. The resulting hybridization data is used to infer overlaps among
  clones and to construct a clone map. This approach is well suited to produce
  sequence ready maps that are known to contain a gene(s) and are homologous to
  a specific section of the S_1 genome. Note that the designed probes can also be
  used in assembling clone maps of any species S_3 that is very close to species S_2.
  This is especially helpful if very little sequence data is available for S_3 compared
  to S_2.
  More than 3,800 mouse BACs representing approximately 40% of the mouse
  genome that is homologous to human chromosome 7 were assembled in this manner by this method [48]. In addition, the same probes were used to isolate clones
  from a rat clone library [47]. A map spanning 90% of mouse genomic sequence
  homologous to human chromosome 19 was also created in a similar manner [24].
  The main limitation of the above comparative mapping approach is the requirement for an extensive DNA sequence resource for the second species S_2, such
  as large collections of expressed-sequence tags (ESTs), BAC-end sequences, or
  whole-genome shotgun sequences.
- *Required:* Genomic sequence of species S_1.
  Method: S_2 BAC end sequences (BESs) are determined by sequencing the ends
  of clones from a S_2 BAC clone library. If possible, BESs that comprise mainly of
  repetitive elements are discarded. Matches between the BESs and the S_1 genomic
  sequence are identified using standard sequence comparison programs and BESs
  with non-unique matches or no matches are eliminated. A clone for which both
  BESs have high quality matches is mapped onto the location on S_1 bounded by
  the BESs match sites if the distance between the match sites is proportional to

the length of the cloned fragment [2]. Clone overlaps are inferred in a natural manner when clones are mapped to overlapping locations on the S_1 genomic sequence. Clones, whose paired BESs match sites on different chromosomes or sites separated by intervals disproportional to the clone length, probably straddle genomic rearrangement breakpoints and cannot be mapped easily.

The main problem with this method is that the presence of repetitive sequences reduces the effective amount of DNA available for BESs similarity search by nearly half in the case of mammalian BAC clones. Species-specific sequence and highly diverged sequence further reduce the number of usable BESs. This means that when this method is applied to distantly related species only a small fraction of BACs in the library might be successfully positioned. To obtain a sequence ready clone map covering an entire chromosome(s), it may be necessary to start with a high coverage clone library.

In a recent application of this method [14], chimp BESs were compared with the human genome RefSeq contigs. 67% of the BESs which aligned with high similarity were used to map 24,580 (out of 64,116) BAC clones creating a physical map that covered 48.6% of the human genome. In more distantly related species, the numbers are a bit lower. In the construction of a human-cattle comparative map that covers 32% of the human genome [28], only 1,242 (5%) out of 24,855 clones had both ends matching human sequence with significant similarity. This approach has also been used in conjunction with the traditional restriction enzyme finger-printing method for assembling the human-mouse homology clone map [18]. Pooled Genomic Indexing (PGI), a variation of this method that uses shotgun sequence reads (instead of BESs) and a pooled array of clones to minimize the number of distinct sequencing operations is currently being used to create a clone map for Rhesus macaque [21].

- *Required:* Genomic sequences for two species S_1, S_2.

  Method: Orthologous genomic sequences from the two species are masked for repetitive elements and then aligned. Regions with high sequence conservation are identified and a short candidate probe sequence is selected from the conserved region. Unique sequences from this set which are separated by 30-40kb are selected to serve as hybridization probes. These probes can be used to screen libraries from different species and the hybridization data can be used for the parallel construction of sequence-ready clone maps in multiple species.

  Note that these probe sequences are not necessarily from genic regions – many studies have shown a significant fraction of conserved regions is present in intergenic and intronic regions [8, 13]. By choosing probe sequences based only on the extent of cross-species conservation and distribution along the genome, this approach maximizes the chance of generating a tiling map that spans gene poor regions. This approach was used to create sequence-ready clone maps for multiple mammalian species (chimpanzee, baboon, cat, dog, cow, and pig) [23].

---

[2]Spacing between adjacent syntenic anchors varies from species to species. Syntenic anchors are more closely spaced in mouse than in human, reflecting the 14% smaller overall genome size [53].

10.4.2 Ordering and orienting contigs

In the absence of high resolution marker maps or mate pair information, we can still infer some order/orient relationships by comparing the conserved regions present in contigs of two organisms that are close in evolutionary terms. This process was performed manually by Pletcher *et al.* [39] to order and orient mouse contigs derived from low-pass (2.2x) sequencing. Figure 10.3 illustrates the sort of inference that is possible.

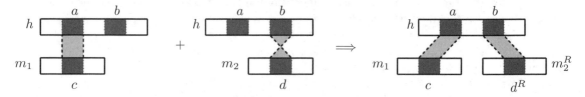

FIGURE 10.3: Use of sequence comparison to orient/order contigs. Contig h (say, of human) includes region a, which aligns with region c in contig m_1 (say, of mouse). Also, another region of h, denoted b, aligns with d^R, the reverse complement of region d of mouse contig m_2. We infer that m_1 precedes m_2^R, relative to the orientation in which h is given. Note that the distance between m_1 and m_2^R *cannot* be inferred from such comparisons.

In Section 10.5 we formulate a general version of the order-orient problem as an optimization problem and summarize the complexity results shown in [50]. Algorithms for simpler versions of the problem which are more frequently encountered in practice are discussed in Section 10.6.

10.5 Consensus Sequence Reconstruction (CSR) problem

We model the problem of determining order/orient relationships from alignments between contigs as follows. Data consists of a set of "h-contigs" and a set of "m-contigs", where each contig is simply an ordered list of conserved regions. We use $\sigma(a, b)$ to denote the score of the alignment between a and b, where a or a^R is a conserved region of an h-contig and b or b^R is a conserved region of an m-contig. An example of a permissible data set consists of contigs $h_1 : \langle a, b, c \rangle$, $h_2 : \langle d \rangle$, $m_1 : \langle s, t \rangle$, $m_2 : \langle u, v \rangle$ and the alignment scores $\sigma(a, s) = 4$, $\sigma(a, t) = 1$, $\sigma(b, t^R) = 3$, $\sigma(c, u) = 5$, $\sigma(d, t) = \sigma(d, v^R) = 2$. See Fig. 10.4.

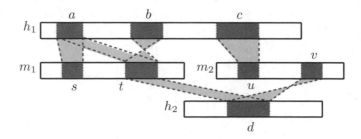

FIGURE 10.4: Picture of a sample set of data, discussed in the text.

Alignments involving conserved regions in contig h_1 may serve to orient and order several m-contigs relative to each other. Some of these m-contigs may in turn orient and order h_1 relative to additional h-contigs, and so on. This leads to an "island"[3] of contigs that are oriented and ordered relative to one another. With ideal data, this process would partition the set of contigs into islands, such that inter-island order/orient relationships cannot be determined from the alignments. In reality, the set of given alignments is frequently inconsistent with any proposed orientation and ordering of the contigs. Simple examples are shown in Fig. 10.5. More complex examples arise in practice when regions have been shuffled by evolutionary processes, when incorrect alignments are computed, and when contigs are incorrectly assembled from reads.

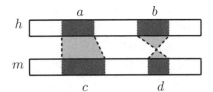 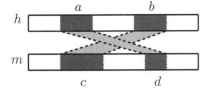

FIGURE 10.5: Two potential inconsistencies among alignments between contigs. In the first example, contig h contains regions a and b, where a aligns with region c of contig m, and b aligns with d^R, where d is another region of m. The $a-c$ alignment supports the current orientation, while the $b-d$ alignment calls for reversal of m. The second example violates our requirement that aligning regions be in the same order in the two sequences.

Our goal is to determine orientations and an order for each of the two sets of contigs that, possibly together with deletions of some of the conserved regions, gives two equal-length and consistently ordered lists of conserved regions showing high overall similarity. Ideally, this would mean maximizing the sum of the scores σ. For a simple example, consider the dataset given several paragraphs above. We can delete (i.e., ignore) b and t, reverse h_2 and place it after h_1 (giving $\langle a, c, d^R \rangle$), then place m_1 before m_2 in their given orientation (giving $\langle s, u, v \rangle$), which yields the score $\sigma(a, s) + \sigma(c, u) + \sigma(d^R, v) = 4 + 5 + 2 = 11$. See Fig. 10.6 for a picture of the solution.

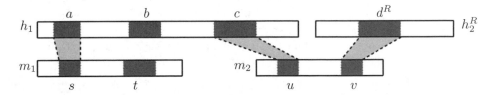

FIGURE 10.6: Solution to the orient/order problem of Fig. 10.4. All alignments pictured in Fig. 10.4 that are inconsistent with this layout have been discarded.

---

[3]While similar to scaffolds [54], islands present a different combinatorial problem because they involve fragments of different species, do not imply any distance information and cannot overlap with other islands.

Note that once orientations and an order of the contigs are chosen, it is easy to decide how sites should be deleted to maximize the score—this is simply the classic problem of aligning two lists of symbols. (Here, however, each symbol of the "sequence" denotes a conserved region, rather than an individual nucleotide.) The difficulty lies with determining an optimal set of orient/order operations.

In results described elsewhere ([50]) we have shown that no polynomial-time algorithm can be guaranteed to orient and order the contigs so as to always maximize the resulting score. In fact, even if we make a number of simplifying assumptions, such as

- each conserved region is involved in precisely one alignment (e.g., for each a, $\sigma(a, b) > 0$ for just one b),
- there is only one h-contig and
- each m-contig has only two conserved regions,

the problem of computing an optimal set of orient/order operations is MAX-SNP hard.

In particular, this result implies that for any existing heuristic one can generate data such that the heuristic result will be far from the correct one. This poses a challenge that can be addressed in two ways -

- Characterize the types of data that would "fool" the heuristic. Any time the heuristic is used, show that the input does not contain data with these "bad" properties.
- Find an algorithm that is designed on different principles and compare the two outcomes.

While the first option is preferable, it is difficult to formalize. Approximation algorithms offer alternative design principles to greedy heuristics and have been described in detail in [50]. In the next section, we show three different ways for tackling a simpler version of the CSR problem.

10.6   The 1-CSR Problem

In this section, we assume that one of the genomes is assembled i.e., we have a set of m-contigs and a single h-contig, h. We consider the problem of ordering and orienting the m-contigs based on their alignments with h. Our formulation, described in Section 10.6.1, is well suited to handle real data sets generated by local alignment tools in a natural manner. We describe a 2-approximation algorithm in Section 10.6.2 and the exact branch-and-bound algorithm by Veeramachaneni *et al.* [51] in Section 10.6.4. Finally, we describe a heuristic that uses both the approximation algorithm and the exact algorithm as subroutines in Section 10.6.5. Guidelines for the effective use of the heuristic are summarized at the end of the section.

10.6.1   Problem Statement

One can use a local alignment tool such as BLAST to compare all the m-contigs with h. We assume that each local alignment that is found by the comparison program identifies a region of a m-contig, m_i (or its reverse complement m_i^R), a region of h, and the alignment score. We model these alignments as *hits*. More formally,

Definition 10.1 *A hit, p, aligns interval $[b_m, e_m]$ of some contig m_l^d (where d is blank or R) with interval $[b_h, e_h]$ of contig h with score s (see Figure 10.7). We use the notation*

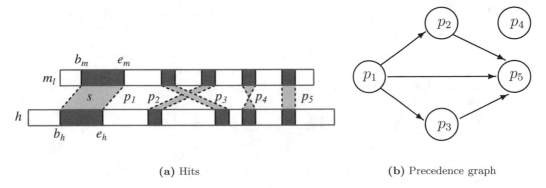

(a) Hits   (b) Precedence graph

FIGURE 10.7: The hit precedence graph restricted to hits owned by m_l is shown on the right. Hit p_4 is inconsistent with all other hits and appears as an isolated vertex.

- $owner(p) = m_l$
- $dir(p) = d$
- $m\text{-}interval(p) = [b_m(p), e_m(p)] = [b_m, e_m]$
- $h\text{-}interval(p) = [b_h(p), e_h(p)] = [b_h, e_h]$
- $\sigma(p) = s$

We view the input to our problem as a set of hits D. Our solution, a subset of D, will represent a collection of "consistent alignments". The definitions that follow formalize this notion.

Definition 10.2 *We say that hit p precedes hit q ($p \prec q$) if*

- $owner(p) = owner(q)$
- $dir(p) = dir(q)$
- $e_m(p) < b_m(q)$ and $e_h(p) < b_h(q)$

Also, $p \preceq q$ if $p \prec q$ or $p = q$.

Definition 10.3 *The* hit precedence graph *is the weighted directed acyclic graph $G = (D, E, w)$ where*

- $\forall p, q \in D, (p, q) \in E$ iff $p \prec q$
- $\forall p \in D, w(p) = \sigma(p)$

Definition 10.4 *In the hit precedence graph $G = (D, E, w)$*

- *a path, P, is a sequence of vertices $\langle p_1, \ldots, p_n \rangle$ s.t. $\forall i < n, (p_i, p_{i+1}) \in E$.*
- *the weight of P, $w(P)$, is the sum of the weights of the vertices in P*
- *$maxpath(p, q)$ is the maximum weight path that starts at p and ends at q*

Using dynamic programming we can easily compute $maxpath(p, q)$, for every pair of hits p, q which satisfy $p \prec q$.

Definition 10.5 *If $p \preceq q$, we define match $\mathbf{p} = match(p, q)$ with the following attributes:*

- $owner(\mathbf{p}) = owner(p)$
- $dir(\mathbf{p}) = dir(p)$

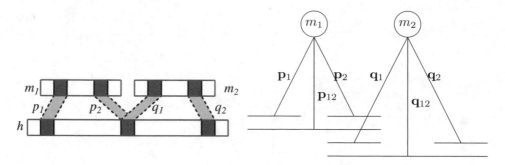

(a) Hit based representation   (b) Match based representation

FIGURE 10.8: The match based representation shown on the right displays owners of matches (as circles) and the *h-interval*s of matches as horizontal lines. $\mathbf{p}_1 = match(p_1), \mathbf{p}_{12} = match(p_1, p_2)$ etc.

- $h\text{-}interval(\mathbf{p}) = [b_h(p), e_h(q)]$
- $m\text{-}interval(\mathbf{p}) = [b_m(p), e_m(q)]$
- $\sigma(\mathbf{p}) = w(maxpath(p, q))$

For a set of matches A, we can extend the above notation in a straight forward manner — using addition for numerical values, union otherwise. In particular,

- $owner(A) = \{owner(\mathbf{p}) \mid \mathbf{p} \in A\}$
- $h\text{-}interval(A) = \{h\text{-}interval(\mathbf{p}) \mid \mathbf{p} \in A\}$
- $\sigma(A) = \sum_{\mathbf{p} \in A} \sigma(\mathbf{p})$

Definition 10.6 *Two matches* $\mathbf{p}$, $\mathbf{q}$ *are compatible if*

- $owner(\mathbf{p}) \neq owner(\mathbf{q})$
- $h\text{-}interval(\mathbf{p}) \cap h\text{-}interval(\mathbf{q}) = \phi$

A set of matches is compatible if the matches are pairwise compatible.

Intuitively, adding a match $\mathbf{p}$ to a solution represents the anchoring of the m-contig $owner(\mathbf{p})$ to the location $h\text{-}interval(\mathbf{p})$ on h. The compatibility definition given above ensures that each m-contig is placed at atmost one location along h and that no two m-contigs are placed at the same location.

Our problem is the following: given D, find a compatible set of matches with maximum score. It follows from Definition 10.5 that we can compute in polynomial time the set of all matches defined by D. So we can reduce our problem to a simpler one: given a set of matches $\mathcal{M}$ find the compatible subset with maximum score. Following the notation used in the previous section, we call this the 1-CSR problem. Note that once we define the problem in terms of matches, we no longer need to consider the orientation of a m-contig nor the $m\text{-}interval$s of its hits. We, therefore, use the simpler representation shown in Figure 10.8 for 1-CSR problem instances.

10.6.2 Approximation algorithm

We can reduce 1-CSR to a more abstract Interval Selection Problem, ISP for short, where we are given set A of integer intervals and a non-negative profit function $p : [1, k] \times A \to R^+$.

The task is to select at most one interval of A for each $i \in [1, k]$, so that the selected intervals are disjoint and the sum of profits is maximal. ISP was studied in the context of scheduling[4] by Bar-Noy *et al.* [2], who described an algorithm with ratio 2. Later Berman and DasGupta [6] described a *Two Phase Algorithm* (TPA) that obtains ratio 2. A version of the TPA suitable for the 1-CSR problem is shown in Figure 10.9.

THEOREM 10.1 *TPA is a 2-approximation algorithm for the 1-CSR problem that runs in time $O(n \log n)$, where $n = |\mathcal{M}|$.*

FIGURE 10.9: Two-phase algorithm for 1-CSR.

(* definitions *)

> $\mathcal{M}$ is sorted so that $e_h(\mathbf{p})$s are non-decreasing
> **S** is an initially empty stack that stores (*value, match*) *pairs*;
> RegionTotal(b, e) is the sum of *values* of those $(v, \mathbf{p})$ *pairs* on **S**
>   that have $e_h(\mathbf{p}) \in [b, e]$;
> OwnerTotal(m_l, c) is the sum of *values* of those $(v, \mathbf{p})$ *pairs* on **S**
>   that have $e_h(\mathbf{p}) \in [1, c]$ and *owner*$(\mathbf{p}) = m_l$;
> DONE $\leftarrow n + 1$;
> **for** (each m_l) InSoln$[m_l] \leftarrow$ false;

(* evaluation phase *)

> **for** (each $\mathbf{p}$ from $\mathcal{M}$)
> { $v \leftarrow \sigma(\mathbf{p}) - $ RegionTotal$(h\text{-}interval(\mathbf{p})) - $ OwnerTotal$(owner(\mathbf{p}), b_h(\mathbf{p}) - 1)$;
>   **if** ($v > 0$)
>   push$((v, \mathbf{p}),$ **S**);
> }

(* selection phase *)

> **while** (**S** is not empty)
> { $(v, \mathbf{p}) \leftarrow$ pop(**S**);
>   **if** (not InSoln$[owner(\mathbf{p})]$ and $e_h(\mathbf{p}) < $ DONE)
>   insert $\mathbf{p}$ into the solution,
>   InSoln$[owner(\mathbf{p})] \leftarrow$ true, DONE $\leftarrow b_h(\mathbf{p})$;
> }

---

[4]More general versions of ISP are considered with different A_i for each $i \in [1, k]$.

10.6.3 Upper and lower bounds

Every pair of matches in a valid solution satisfies two properties — different owners and disjoint *h-intervals*. If we relax the first property, the problem reduces to that of computing the weighted maximum independent set in the interval graph $G = (V, E, w)$, where

- $V = h\text{-}interval(\mathcal{M})$
- $E = \{(v_1, v_2) \in V \times V \mid v_1 \cap v_2 \neq \phi\}$
- $\forall \mathbf{p} \in \mathcal{M}, w(h\text{-}interval(\mathbf{p})) = \sigma(\mathbf{p})$

The weighted maximum independent set problem in intervals graphs can be solved in $O(n \log n)$ time [22]. We let $MIS(\mathcal{M})$ denote the optimal solution to this relaxed version of the 1-CSR problem.

REMARK 10.1   Let $\mathcal{M}$ be a 1-CSR instance, $TPA(\mathcal{M})$ the solution returned by the Two Phase Algorithm (TPA) and $Opt(\mathcal{M})$ the optimal solution.

1. $\sigma(TPA(\mathcal{M})) \leq \sigma(Opt(\mathcal{M})) \leq \sigma(MIS(\mathcal{M}))$
2. If each contig participates in exactly one match then $\sigma(MIS(\mathcal{M})) = \sigma(Opt(\mathcal{M}))$

It follows from the previous remark that $lb(\mathcal{M})$, $ub(\mathcal{M})$ defined as $\sigma(TPA(\mathcal{M}))$, $\sigma(MIS(\mathcal{M}))$ respectively are valid lower and upper bounds on $\sigma(Opt(\mathcal{M}))$.

10.6.4 Exact algorithm

We can solve a 1-CSR problem instance by examining all compatible sets of matches and selecting the set with highest score. However, explicitly enumerating all elements of the search space of feasible solutions could take exponential time. We use the classic branch-and-bound method to eliminate from consideration parts of the search space where no optimal solution can exist.

To apply the branch-and-bound method we

- divide the search space of the problem to create smaller sub-problems
- compute upper and lower bounds for each sub-problem and if possible eliminate it from further consideration

The method starts by considering the original problem with the complete search space – this is the root problem. The search space is then divided into two parts. The problems aimed at finding the optimal solution for each part become the children of the root search node. Applying the division procedure recursively, we generate a tree of subproblems.

Given a 1-CSR problem instance, $\mathcal{M}$, we can represent a node, u, of the tree as a pair (S, C), where

- $S, C \subseteq \mathcal{M}, S \cap C = \phi$
- S is a compatible set of matches
- each match of C is individually compatible with S i.e. $\forall a \in C, S \cup \{a\}$ is a feasible solution

Let $\mathcal{U}$ represent the set of all feasible solutions. The node u represents the subset R_u of solutions that satisfy

$$R_u = \{S' \in \mathcal{U} \mid S \subseteq S' \subseteq S \cup C\}$$

For a node $u = (S, C)$ our division procedure create subproblems v, w as follows:

1. let a be the match in C with maximum score
2. let $\text{incompat}(a) = \{b \in \mathcal{M} \mid a \text{ is incompatible with } b\}$
3. $v = (S \cup \{a\}, C - \text{incompat}(a)), w = (S, C - \{a\})$

Note that $R_v = \{S' \in R_u \mid a \in S'\}$ while $R_w = \{S' \in R_u \mid a \notin S'\}$. Since $R_u = R_v \cup R_w$ this division procedure ensures that we do not overlook any feasible solutions.

To generate all feasible solutions, we can start with a root node $(\phi, \mathcal{M})$ and apply the recursive procedure until all leaf nodes are of the form (S, ϕ). The key to avoid generating some subtrees whose leaves contain only sub-optimal solutions is the following observation: if (S', ϕ) is a leaf in a subtree rooted at (S, C) then $\sigma(S') \leq \sigma(S) + ub(C)$. So, if $\sigma(S) + ub(C) \leq \sigma(B)$ where B is some known feasible solution, we can prune the entire subtree rooted at (S, C).

The actual implementation uses a queue data structure to carry out a breadth first traversal of the search tree and is described in Figure 10.10. The queue Q can become exponentially long if few sub-problems are pruned. Therefore, in an implementation of the algorithm one might want to abort the computation if the queue length exceeds a certain threshold.

FIGURE 10.10: Pseudo-code for the branch-and-bound algorithm.

---

function Exact($\mathcal{M}$)
    $B \leftarrow TPA(\mathcal{M})$ is a feasible solution
    Initialize queue Q with the root node $(\phi, \mathcal{M})$
    while (Q is not empty)
    {   let $u = (S, C)$ be the first element of Q;
        if ($\sigma(S) + lb(C) > \sigma(B)$)
            $B \leftarrow S \cup TPA(C)$;
        if ($\sigma(S) + ub(C) > \sigma(B)$)
            compute v, w the children of u;
            add v, w to the queue Q;
    }
    return B
}

10.6.5 Heuristic

In practice, it may be infeasible to apply the exact algorithm to large problem instances. In this section we describe criteria to

1. eliminate matches that cannot be in any optimal solution
2. select matches that are guaranteed to be in some optimal solution

These techniques can be used till saturation (no more progress can be made) and the remaining problem can be solved by using *TPA* or by greedily selecting compatible matches. In our tests, we found that the majority of the solution is obtained using the above criteria. We, therefore, obtain a better result than that obtained by using *TPA* alone.

Removing dominated matches

Definition 10.7 *A contig is considered* single *if it is the owner of exactly one match of* $\mathcal{M}$[5]. *Let singles*$(\mathcal{M})$ *denote the matches of* $\mathcal{M}$ *that are owned by singles.*

Because we keep eliminating matches, $\mathcal{M}$ will stand for the set of matches that are not yet eliminated. Definitions based on $\mathcal{M}$ are thus dynamic and need to be recomputed when $\mathcal{M}$ changes.

Definition 10.8 *A match* $\mathbf{p}$ *is* dominated *by a compatible set of matches A if*

- $\forall \mathbf{q} \in A$, *owner*$(\mathbf{q})$ *is either a single or the same as owner*$(\mathbf{p})$
- $\forall \mathbf{q} \in A$, *h-interval*$(\mathbf{q}) \subseteq$ *h-interval*$(\mathbf{p})$
- $\sigma(\mathbf{p}) < \sigma(A)$

A match is dominated if there exists a compatible set of matches that dominates it.

LEMMA 10.1   The optimal solution does not contain any dominated matches.

Proof.   Let S be any optimal solution. Suppose $\mathbf{p} \in S$ is dominated by a compatible set of matches A. Then it is easy to verify that $S - \{\mathbf{p}\} \cup A$ is a compatible set of matches with score greater than $\sigma(S)$. $\square$

Definition 10.9 *Let A be any set of matches. Then A* restricted to (b, e) *is defined as* $A(b, e) = \{\mathbf{p} \in A \mid$ *h-interval*$(\mathbf{p}) \subseteq [b, e]\}$.

Let $S = Opt(singles(\mathcal{M})) = MIS(singles(\mathcal{M}))$. Then it is easy to see that a match $\mathbf{p} \in \mathcal{M}$ is dominated if

- $\sigma(\mathbf{p}) < \sigma(S(b_h(\mathbf{p}), e_h(\mathbf{p})))$ or
- $\exists \mathbf{q} \in \mathcal{M}$ that satisfies
    1. *owner*$(\mathbf{q}) = $ *owner*$(\mathbf{p})$,
    2. *h-interval*$(\mathbf{q}) \subseteq$ *h-interval*$(\mathbf{p})$, and
    3. $\sigma(\mathbf{p}) < \sigma(S(b_h(\mathbf{p}), b_h(\mathbf{q}) - 1)) + \sigma(\mathbf{q}) + \sigma(S(e_h(\mathbf{q}) + 1, e_h(\mathbf{p})))$

Since the number of matches can be quadratic in the number of hits, say n, the method of removing dominated matches suggested by the second condition above has $O(n^4)$ time complexity. In practice, few contigs have more than 10 hits and the time taken is almost linear in n. In later stages of the heuristic, whenever the set of singles changes, one can use the simpler check for dominated matches stated in the first condition.

Checking for local solutions

Here we formulate a sufficient condition for recognizing that a solution to a sub-problem is contained in an optimal solution for the full problem.

Definition 10.10 *The* overlap components *of* $\mathcal{M}$ *are the connected components of the graph* $G = (\mathcal{M}, E)$ *where*

$$(\mathbf{p}, \mathbf{q}) \in E \ \textit{iff} \ \textit{h-interval}(\mathbf{p}) \cap \textit{h-interval}(\mathbf{q}) \neq \phi$$

---

[5]As we remove matches from consideration, an increasing number of contigs will become single.

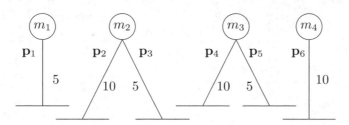

FIGURE 10.11: A 1-CSR problem instance with 3 overlap components $C_1 = \{\mathbf{p}_1, \mathbf{p}_2\}$, $C_2 = \{\mathbf{p}_3, \mathbf{p}_4\}$ and $C_3 = \{\mathbf{p}_5, \mathbf{p}_6\}$.

An overlap-closed set *is the union of one or more overlap components.*

Definition 10.11 *Contig m_k is local to a set of matches A if*

$$\forall \mathbf{p} \in \mathcal{M}, owner(\mathbf{p}) = m_k \Longrightarrow \mathbf{p} \in A$$

Also, we say that A has a local solution *if all matches in $Opt(A)$ have owners that are local to A.*

LEMMA 10.2   If an overlap-closed set C has a local solution then

$$Opt(\mathcal{M}) = Opt(C) \cup Opt(\mathcal{M} - C) \tag{10.1}$$

Proof.   Clearly $\sigma(Opt(\mathcal{M})) \leq \sigma(Opt(C)) + \sigma(Opt(\mathcal{M} - C))$. To prove equality it suffices to show that $Opt(C) \cup Opt(\mathcal{M} - C)$ is a feasible solution. Let $\mathbf{p} \in Opt(C), \mathbf{q} \in Opt(\mathcal{M} - C)$ be any two matches. Then,

- $owner(\mathbf{p}) \neq owner(\mathbf{q})$ because $owner(\mathbf{p})$ is local to C
- $h\text{-}interval(\mathbf{p}) \cap h\text{-}interval(\mathbf{q}) = \phi$ because the overlap component containing $\mathbf{p}$ is a subset of C

By Definition 10.6, $\mathbf{p}$ is compatible with $\mathbf{q}$. Since $\mathbf{p}$, $\mathbf{q}$ were chosen arbitrarily from $Opt(C)$, $Opt(\mathcal{M} - C)$ it follows that $Opt(C) \cup Opt(\mathcal{M} - C)$ is a compatible set of matches. $\qquad\square$

Lemma 10.2 suggests that we can solve a 1-CSR instance by repeatedly identifying overlap-closed sets with local solutions and solving them using the Exact algorithm described earlier in Section 10.6.4. For example, consider the 1-CSR problem instance $\mathcal{M} = \{\mathbf{p}_1, \ldots, \mathbf{p}_6\}$ shown in Figure 10.11. The optimal solutions to the overlap components are $Opt(C_1) = \{\mathbf{p}_2\}$, $Opt(C_2) = \{\mathbf{p}_4\}$ and $Opt(C_3) = \{\mathbf{p}_6\}$. C_3 is the only overlap component with a local solution. But once this sub-problem is solved and the matches of C_3 are removed, $\mathcal{M}$ reduces to $C_1 \cup C_2$. Now, C_2 has a local solution and $Opt(C_2)$ can be added to the optimal solution. Finally, we are left only with C_1 and then $Opt(C_1)$ can also be added to the optimal solution. Function LocalSolutions shown in Figure 10.12 generalizes this approach.

We first call the function LocalSolutions with the set of overlap components. When the function terminates, none of the remaining overlap components has a local solution. At this point, we can try to form larger overlap-closed sets and check if these larger sets have local solutions. In our implementation, we simply collapse connected components of the graph into large overlap-closed sets, and use the LocalSolution procedure again. However, this

FIGURE 10.12: Procedure for identifying overlap-closed sets with local solutions and solving them separately. S is the optimal solution computed and $V(G)$ consists of overlap-closed sets whose matches form the residual problem.

function LocalSolutions($\mathcal{C}$)
{ $\mathcal{C}$ is a collection of disjoint overlap-closed sets;
    Compute the directed graph G where
        $V(G) \leftarrow \mathcal{C}$;
        for (each $C \in V(G)$)
        {   **if** ($Opt(C)$ cannot be computed by algorithm Exact)
                $(C, C) \in E(G)$;
            else
                $\forall C' \neq C,\ (C, C') \in E(G)$ iff $owner(Opt(C)) \cap owner(C') \neq \phi$;
        }

    while (some $C \in \mathcal{C}$ has out-degree 0)
    { $S \leftarrow S \cup Opt(C)$;
        remove C with all incident edges from G;
    }
    return $(S, V(G))$
}

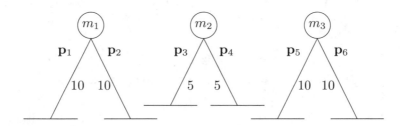

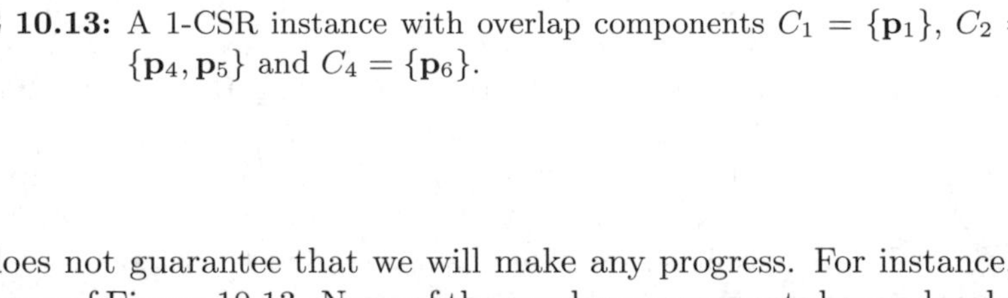

FIGURE 10.13: A 1-CSR instance with overlap components $C_1 = \{p_1\}$, $C_2 = \{p_2, p_3\}$, $C_3 = \{p_4, p_5\}$ and $C_4 = \{p_6\}$.

method does not guarantee that we will make any progress. For instance, consider the 1-CSR instance of Figure 10.13. None of the overlap components have a local solution, and the graph G at the end of the procedure has 4 edges – $(C_1 \rightarrow C_2), (C_2 \rightarrow C_1), (C_3 \rightarrow C_4), (C_4 \rightarrow C_3)$. But once we collapse the connected components, we still do not have any vertices with out-degree 0: we now have a graph with edges $((C_1 \cup C_2) \rightarrow (C_3 \cup C_4)), ((C_3 \cup C_4) \rightarrow (C_1 \cup C_2))$.

    We interrupt the merging process when the optimal solutions to the overlap-closed sets can no longer be computed by the branch-and-bound algorithm. At this point, we can choose compatible matches greedily or use TPA to find an approximate solution to the residual problem.

Measuring solution quality

Let $\mathcal{M}$ be the problem instance and $\mathcal{M}' \subseteq \mathcal{M}$ be some overlap-closed set with a local solution S'. Then it follows from Equation 10.1 that

$$
\begin{aligned}
\sigma(Opt(\mathcal{M})) &= \sigma(Opt(\mathcal{M}')) + \sigma(Opt(\mathcal{M} - \mathcal{M}')) \\
&\leq \sigma(S') + ub(\mathcal{M} - \mathcal{M}')
\end{aligned}
$$

Since $\sigma(S') + ub(\mathcal{M} - \mathcal{M}')$ is an upper bound on $\sigma(Opt(\mathcal{M}))$, the quality of any solution S relative to this upper bound is given by

$$
q = \frac{\sigma(S)}{\sigma(S') + ub(\mathcal{M} - \mathcal{M}')} \times 100
$$

A high quality score indicates that a solution may be hard to improve. On the other hand, near-optimal solutions may sometimes have low quality scores if the sub-problem $\mathcal{M} - \mathcal{M}'$ is large.

10.6.6   Results

A data set consisting of the sequence of human chromosome 22 (hsa22) and 700,000 mouse contigs from the *arachne.3* [3] assembly was used to test the performance of the heuristic. 45,115 contigs with high scoring BLASTZ [43] hits to hsa22 sequence were selected and the hits were used to generate 112,554 matches. After discarding 40,200 (35%) dominated matches, the remaining matches formed 14,000 overlap components. A substantial portion of the problem was solved by using the LocalSolutions procedure along with our merging method for forming larger overlap-closed sets. The residual problem consisting of 1,416 overlap components was solved using TPA. The overall solution quality was 95%. By contrast, the solution generated by an alternate greedy algorithm that orders contigs based on their highest hit score had quality 84%, and the solution obtained by simply using TPA had quality 90%. By comparing with the assembled mouse genome sequences, we were able to verify that solutions with higher quality imposed an ordering on the contigs that was closer to that found in the assembled sequence.

The brief results presented here are only to give an indication of how effective the different steps of the heuristic are at reducing the size of the problem and solving sub-problems. The different parameters that can affect the final ordering generated by the heuristic are discussed in the next section.

10.6.7   Discussion

The actual performance of our heuristic is influenced by several parameters — the degree of sequence conservation between the two species, the quality and size of the contigs, the presence of duplicated regions and gene families, the choice of an alignment tool etc. Some issues that need to be considered before using the heuristic are described below.

Organisms are closely related

In the case of closely related species such as human and chimp, even a greedy algorithm will find solutions with high score. In such cases, one can still use the heuristic to estimate the upper bound and, thus, gauge the quality of the greedy solution. A solution with score within 98-99% of the optimum should be expected for these data sets.

Filtering hits

Large overlap components may occur when multiple hits share the same *h-interval*. By examining the genomic sequence corresponding to such an *h-interval* and, if possible, any associated annotation, one can determine if the region contains a member of a gene-family or an unmasked repetitive element. Based on such analysis we may choose to discard all or some of the participating hits. Eliminating such hits may increase the number of overlap components and this in turn allows a larger fraction of the solution to be computed in an optimal manner.

Note that even if the only hits remaining in the problem instance correspond to syntenic anchors, the problem remains NP-hard (see Section 10.5). In other words, elimination of hits does not change the complexity of the underlying problem but simply makes the problem more tractable.

Assembled reference genome has multiple chromosomes

A simplistic (but wrong) way out is to solve multiple 1-CSR problem instances — one for each chromosome. Unfortunately, this may result in a *m*-contig appearing in more than one 1-CSR solution. Such ambiguous mappings of *m*-contigs are likely because mammalian genomes consist of several gene families some of which are spread across multiple chromosomes. Contigs with high scoring alignments to one member of the gene family are likely to also align with other members of the family.

A second method, which preserves the underlying assumption of the heuristic viz. all the *m*-contigs belong to regions homologous to the *h*-contig, is to create a single artificial *h*-contig which represents the concatenation of all the chromosomes in some arbitrary order. Filtering methods discussed above could be used to restrict the size of the resulting problem instance.

Accommodating external information

To extend an existing partial solution, we can simply discard all matches in the problem instance that are incompatible with the partial solution and then run the heuristic on the residual problem.

10.7 Conclusions

The number of species with publicly available sequence data is expected to grow rapidly in the next few years. However, it is clear that not all species genomes will be fully sequenced. In most cases, sequence data will remain in the form of numerous contigs. In this chapter, we described some methods for ordering and orienting contigs based on alignments with contigs of a related species. While the underlying problem is NP-hard, the heuristic described in the previous section is expected to perform well on real data sets.

References

[1] M.D. Adams, S.E. Celniker, R.A. Holt, and C.A. Evans *et al.* The genome sequence of *drosophila melanogaster*. *Science*, 287(5461):2185–95, 2000.

[2] A. Bar-Noy, S. Guha, J. Naor, and B. Schieber. Approximating the throughput of multiple machines in real-time scheduling. *Proc. 31st ACM STOC*, pages 622–631,

1999.

[3] S. Batzoglou, D.B. Jaffe, K. Stanley, and J. Butler *et al.* ARACHNE: a whole-genome shotgun assembler. *Genome Res*, 12(1):177–89, 2002.

[4] D.A. Benson, I. Karsch-Mizrachi, D.J. Lipman, and J. Ostell *et al.* GenBank: update. *Nucleic Acids Res*, 32 Database issue:D23–6, 2004.

[5] C.M. Bergman, B.D. Pfeiffer, D.E. Rincon-Limas, and R.A. Hoskins *et al.* Assessing the impact of comparative genomic sequence data on the functional annotation of the *drosophila* genome. *Genome Biol*, 3(12):RESEARCH0086, 2002.

[6] P. Berman and B. DasGupta. Multi-phase algorithms for throughput maximization for real-time scheduling. *Journal Combinatorial Optimization*, 4(3):307–323, 2000.

[7] W. Castle and W. Wachter. Variation of linkage in rats and mice. *Genetics*, 9:1–12, 1924.

[8] P. Dehal, P. Predki, A.S. Olsen, and A. Kobayashi *et al.* Human chromosome 19 and related regions in mouse: conservative and lineage-specific evolution. *Science*, 293(5527):104–11, 2001.

[9] E.T. Dermitzakis, A. Reymond, R. Lyle, and N. Scamuffa *et al.* Numerous potentially functional but non-genic conserved sequences on human chromosome 21. *Nature*, 420(6915):578–82, 2002.

[10] Kirkness E.F., V. Bafna, A.L. Halpern, and S. Levy *et al.* The dog genome: survey sequencing and comparative analysis. *Science*, 301(5641):1898–903, 2003.

[11] J. Ehrlich, D. Sankoff, and J.H. Nadeau. Synteny conservation and chromosome rearrangements during mammalian evolution. *Genetics*, 147(1):289–96, 1997.

[12] P. Flicek, E. Keibler, P. Hu, and I. Korf *et al.* Leveraging the mouse genome for gene prediction in human: from whole-genome shotgun reads to a global synteny map. *Genome Res*, 13(1):46–54, 2003.

[13] K.A. Frazer, J.B. Sheehan, R.P. Stokowski, and X. Chen *et al.* Evolutionarily conserved sequences on human chromosome 21. *Genome Res*, 11(10):1651–9, 2001.

[14] A. Fujiyama, H. Watanabe, A. Toyoda, and T.D. Taylor *et al.* Construction and analysis of a human-chimpanzee comparative clone map. *Science*, 295(5552):131–4, 2002.

[15] The *Arabidopsis* Genome Initiative. Analysis of the genome sequence of the flowering plant *arabidopsis thaliana*. *Nature*, 408:796–815, 2000.

[16] A. Goffeau, R. Aert, M.L. Agostini-Carbone, and A. Ahmed *et al.* The yeast genome directory. *Nature*, 387:S1–S105, May 1997.

[17] E.D. Green. Strategies for the systematic sequencing of complex genomes. *Nature Reviews Genetics*, 2(8):573–83, 2001.

[18] S.G. Gregory, M. Sekhon, J. Schein, and S. Zhao *et al.* A physical map of the mouse genome. *Nature*, 418(6899):743–50, 2002.

[19] J. Haldane. The comparative genetics of color in rodents and carnivora. *Biol. Rev. Camb. Philos. Soc*, 2:199–212, 1927.

[20] R.C. Hardison, J. Oeltjen, and W. Miller. Long human-mouse sequence alignments reveal novel regulatory elements: a reason to sequence the mouse genome. *Genome Res*, 7(10):959–66, 1997.

[21] A.R. Jackson, M. Csûrös, E. Sodergren, and A. Milosavljevic. Pooled genomic indexing of Rhesus macaque (*macaca mulatta*) BACs. In preparation.

[22] Y. H. Ju and Y. T. Chuan. An efficient algorithm for finding a maximum weight 2-independent set on interval graphs. *Information Processing Letters*, 43(5):229–235, 1992.

[23] Thomas. J.W., A. B. Prasad, T.J. Summers, and S.Q. Lee-Lin *et al.* Parallel construction of orthologous sequence-ready clone contig maps in multiple species. *Genome*

Res, 12(8):1277–85, 2002.

[24] J. Kim, L. Gordon, P. Dehal, and H. Badri *et al.* Homology-driven assembly of a sequence-ready mouse BAC contig map spanning regions related to the 46-Mb gene-rich euchromatic segments of human chromosome 19. *Genomics*, 74(2):129–41, 2001.

[25] I. Korf, P. Flicek, D. Duan, and M.R. Brent. Integrating genomic homology into gene structure prediction. *Bioinformatics*, 17 Suppl 1:S140–8, 2001.

[26] A.E. Kwitek, P.J. Tonellato, D. Chen, and J. Gullings-Handley *et al.* Automated construction of high-density comparative maps between rat, human, and mouse. *Genome Res*, 11(11):1935–43, 2001.

[27] E.S. Lander, L.M. Linton, B. Birren, and C. Nusbaum *et al.* Initial sequencing and analysis of the human genome. *Nature*, 409(6822):860–921, 2001.

[28] D.M. Larkin, A. Everts-van der Wind, M. Rebeiz, and P. A. Schweitzer *et al.* A cattle-human comparative map built with cattle BAC-ends and human genome sequence. *Genome Res*, 13(8):1966–72, 2003.

[29] Y. Liu, X.S. Liu, L. Wei, and R.B. Altman *et al.* Eukaryotic regulatory element conservation analysis and identification using comparative genomics. *Genome Research*, 14(3):451–458, 2004.

[30] J.E. Moore and J.A. Lake. Gene structure prediction in syntenic DNA segments. *Nucleic Acids Res*, 31(24):7271–9, 2003.

[31] R.J. Mural, M.D. Adams, E.W. Myers, and H.O. Smith *et al.* A comparison of whole-genome shotgun-derived mouse chromosome 16 and the human genome. *Science*, 296(5573):1661–71, 2002.

[32] E.W. Myers, G.G. Sutton, A.L. Delcher, and I.M. Dew *et al.* A whole-genome assembly of *drosophila*. *Science*, 287(5461):2196–204, 2000.

[33] W.G. Nash and S.J. O'Brien. Conserved regions of homologous G-banded chromosomes between orders in mammalian evolution: carnivores and primates. *Proceedings of the National Academy of Sciences USA*, 79(21):6631–5, 1982.

[34] S. Nilsson, K. Helou, A. Walentinsson, and C. Szpirer *et al.* Rat-mouse and rat-human comparative maps based on gene homology and high-resolution zoo-FISH. *Genomics*, 74(3):287–98, 2001.

[35] R.J. Oakley, M.L. Watson, and M.F. Seldin. Construction of a physical map on mouse and human chromosome 1: Comparison of 13 Mb of mouse and 11 Mb of human DNA. *Human Molecular Genetics*, 1:616–620, 1992.

[36] S.J. O'Brien, M. Menotti-Raymond, W.J. Murphy, and W.G. Nash *et al.* The promise of comparative genomics in mammals. *Science*, 286(5439):458–62, 479–81, 1999.

[37] A.H. Paterson, T.H. Lan, K.P. Reischmann, and C. Chang *et al.* Toward a unified genetic map of higher plants, transcending the monocot-dicot divergence. *Nat Genet*, 14(4):380–2, 1996.

[38] P. Pevzner. *Computational Molecular Biology*. The MIT Press, Cambridge, Massuchusetts, 2000.

[39] M.T. Pletcher, T. Wiltshire, D.E. Cabin, and M. Villanueva *et al.* Use of comparative physical and sequence mapping to annotate mouse chromosome 16 and human chromosome 21. *Genomics*, 74(1):45–54, 2001.

[40] F. Sanger, S. Nicklen, and A.R. Coulson. DNA sequencing with chain-terminating inhibitors. *Proc Natl Acad Sci U S A*, 74(12):5463–7, 1977.

[41] J.R. Sawyer and J.C. Hozier. High resolution of mouse chromosomes: banding conservation between man and mouse. *Science*, 232(4758):1632–5, 1986.

[42] H. Scherthan, T. Cremer, U. Arnason, and H.U. Weier *et al.* Comparative chromosome painting discloses homologous segments in distantly related mammals. *Nature Genetics*, 6(4):342–7, 1994.

[43] S. Schwartz, Z. Zhang, K.A. Frazer, and A. Smit *et al.* PipMaker–a web server for aligning two genomic DNA sequences. *Genome Res*, 10(4):577–86, 2000.

[44] The *C. elegans* Sequencing Consortium. Genome sequence of the nematode c. elegans: a platform for investigating biology. *Science*, 282(5396):2012–2018, December 1998.

[45] L. D. Stein, Z. Bao, D. Blasiar, and T. Blumenthal *et al.* The genome sequence of *caenorhabditis briggsae*: A platform for comparative genomics. *PLoS Biol*, 1(2):E45, 2003.

[46] L. Stubbs, E.M. Rinchik, E. Goldberg, and B. Rudy *et al.* Clustering of six human 11p15 gene homologs within a 500-kb interval of proximal mouse chromosome 7. *Genomics*, 24(2):324–32, 1994.

[47] T.J. Summers, J.W. Thomas, S.Q. Lee-Lin, and V.V. Maduro *et al.* Comparative physical mapping of targeted regions of the rat genome. *Mamm Genome*, 12(7):508–12, 2001.

[48] J.W. Thomas, T.J. Summers, S.Q. Lee-Lin, and V.V. Maduro *et al.* Comparative genome mapping in the sequence-based era: early experience with human chromosome 7. *Genome Res*, 10(5):624–33, 2000.

[49] J.W. Thomas, J.W. Touchman, R.W. Blakesley, and G.G. Bouffard *et al.* Comparative analyses of multi-species sequences from targeted genomic regions. *Nature*, 424(6950):788–93, 2003.

[50] V. Veeramachaneni, P. Berman, and W. Miller. Aligning two fragmented sequences. *Disc. Appl. Mathematics*, 127(1), 2003.

[51] V. Veeramachaneni, Z. Zhang, P. Berman, and W. Miller. Comparative assembly of genome shotgun data. In preparation.

[52] J.C. Venter, M.D. Adams, E.W. Myers, and P.W. Li *et al.* The sequence of the human genome. *Science*, 291(5507):1304–51, 2001.

[53] R.H. Waterston, K. Lindblad-Toh, E. Birney, and J. Rogers *et al.* Initial sequencing and comparative analysis of the mouse genome. *Nature*, 420(6915):520–62, 2002.

[54] J. L. Weber and E. W. Myers. Human whole-genome shotgun sequencing. *Genome Res*, 7(5):401–9, 1997.

[55] J. Weinberg and R. Stanyon. Chromosome painting in mammals as an approach to comparative genomics. *Current Opinion in Genetics and Development*, 5:792–797, 1995.

11

Information Theoretic Approach to Genome Reconstruction

Suchendra Bhandarkar
The University of Georgia

Jinling Huang
The University of Georgia

Jonathan Arnold
The University of Georgia

11.1 Introduction

Creating maps of entire chromosomes, which could then be used to reconstruct the chromosome's DNA sequence, has been one of the fundamental problems in genetics right from its very inception [47]. These maps are central to the understanding of the structure of genes, their function, their transmission and their evolution. Chromosomal maps fall into two broad categories — *genetic maps* and *physical maps*. Genetic maps represent an ordering of genetic markers along a chromosome where the distance between two genetic markers is related to their recombination frequency. Genetic maps are typically of low resolution i.e., 1 to 10 million base pairs (Mb). Lander and Green [32] pioneered the use of computational techniques for the assembly of genetic maps with many markers. While genetic maps enable a scientist to narrow the search for genes to a particular chromosomal region, it is a physical map that ultimately allows the recovery and molecular manipulation of genes of interest.

A physical map is defined as an ordering of distinguishable (i.e., sequenced) DNA fragments called *clones* or *contigs* by their position along the entire chromosome where the clones may or may not contain genetic markers. The physical mapping problem is therefore one of reconstructing the order of clones and determining their position along the chromosome. A physical map has a much higher resolution than a genetic map of the same chromosome i.e., 10 to 100 thousand base pairs (Kb). Physical maps have provided fundamental insights into gene development, gene organization, chromosome structure, recombination and the role of sex in evolution and have also provided a means for the recovery and molecular manipulation of genes of interest.

11.1.1   The Physical Mapping Protocol

The physical mapping protocol essentially determines the nature of clonal data and the probe selection procedure. The physical mapping protocol used in our work is the one based on *sampling without replacement* [19]. This protocol is simple, adaptable and relatively inexpensive, and has been used successfully in physical mapping projects of several fungal genomes under the Fungal Genome Initiative at the University of Georgia [3].

The protocol that generates the probe set $\mathcal{P}$ and the clone set $\mathcal{C}$ is an iterative procedure which can be described as follows. Let $\mathcal{C}^i$ and $\mathcal{P}^i$ be the clone set and the probe set respectively at the ith iteration. The initial clone set $\mathcal{C}^0$ consists of all the clones in the library whereas the initial probe set $\mathcal{P}^0 = \phi$. The clones in $\mathcal{C}^0$ are designed to be of the same length and to be overlapping so that each clone samples a fragment of the chromosome and coverage of the entire chromosome is made possible. At the ith iteration a clone c is chosen at random from $\mathcal{C}^i$ and added to $\mathcal{P}^i$. Clone c is hybridized to all the clones in $\mathcal{C}^i$. The subset of clones $\mathcal{C}^c$ that hybridize to clone c are removed from $\mathcal{C}^i$ so that $\mathcal{C}^{i+1} = \mathcal{C}^i - \mathcal{C}^c$. Note that $c \in \mathcal{C}^c$ since a clone hybridizes to itself. The hybridization experiment entails extracting complementary DNA from both ends of a probe, washing the DNA over the arrayed plate and recording all clones in the library to which the DNA attaches (i.e., hybridizes). The above procedure is halted at the kth iteration when $\mathcal{C}^k = \phi$. The final probe set is given by $\mathcal{P} = \mathcal{P}^k$ and the clone set by $\mathcal{C} = \mathcal{C}^0 - \mathcal{P}^k$. In the absence of errors, the probe set $\mathcal{P}$ represents a maximal nonoverlapping subset of $\mathcal{C}^0$ that covers the entire length of the chromosome.

The clone-probe overlap pattern is represented in the form of a binary hybridization matrix H where H_{ij} denotes the hybridization of the ith clone $\in \mathcal{C}$ to the jth probe $\in \mathcal{P}$. $H_{ij} = 1$ if the ith clone $\in \mathcal{C}$ hybridizes to the jth probe $\in \mathcal{P}$ and $H_{ij} = 0$ otherwise. If the probes in $\mathcal{P}$ were ordered with respect to their position along a chromosome, then by selecting from H a common overlapping clone for each pair of adjacent probes in $\mathcal{P}$ a minimal set of clones and probes that covers the entire chromosome (i.e., a minimal tiling) could be obtained. Note that a common overlapping clone between two adjacent probes would hybridize to both probes. The minimal tiling in conjunction with the sequencing of each individual clone/probe in the tiling and a sequence assembly procedure that determines the overlaps between successive sequenced clones/probes in the tiling [29] could then be used to reconstruct the DNA sequence of the entire chromosome.

In reality, the hybridization experiments are rarely error-free. The hybridization matrix H could be expected to contain false positives and false negatives. H_{ij} would be a false positive if $H_{ij} = 1$ (denoting hybridization of the ith clone with the jth probe) when in fact $H_{ij} = 0$. Conversely, H_{ij} would be a false negative if $H_{ij} = 0$ when in fact $H_{ij} = 1$. Other sources of error include chimerism where a single clone samples two or more distinct regions of a chromosome, deletions where certain regions of the chromosome are not sampled during the cloning process and repeats where a clone samples a region of the chromosome with repetitive DNA structure. In this chapter, we confine ourselves to errors in the form of false positives and false negatives. Since the clones (and probes) in the mapping projects at the University of Georgia that use the aforementioned protocol are generated using cosmids which makes them sufficiently small (around 40 Kb), chimerism and deletions do not pose a serious problem. However, repeats do pose a problem but are not explicitly addressed here; rather they are treated as multiple isolated incidences of false positives.

11.1.2 Computation of a Physical Map

Several techniques exist for computation of physical maps from contig libraries. These techniques are specific to an experimental protocol and the type of data collected, for example, mapping by nonunique probes [17], mapping by unique probes [16, 22, 25], mapping by unique endprobes [11], mapping using restriction fragments [18, 26], mapping using radiation-hybrid data [6, 46] and optical mapping [28, 34, 40]. Likewise, several computation techniques based on deterministic optimization and stochastic optimization in the context of physical mapping have been reported. Examples of stochastic optimization algorithms include simulated annealing [16, 17, 13, 39], and the random cost algorithm [48] whereas those of deterministic optimization algorithms include linear programming [25], integer programming [11], integer linear programming with polyhedral combinatorics [12] and semidefinite programming [10]. Various statistical analyses of the aforementioned physical mapping techniques have also been reported in the literature [4, 5, 33, 41, 14, 49, 50].

In this chapter we describe a physical mapping approach based on the information theoretic concept of maximum likelihood reconstruction of signals that have been corrupted by noise when transmitted through a communications channel. We model the chromosome as the original signal, the hybridization experiments as the process of transmission through a communications channel, the hybridization errors as the noise introduced by the communications channel, the hybridization matrix as the observed corrupted signal at the receiving end of the communications channel and the desired physical map as the reconstructed signal. In particular, we describe the maximum likelihood (ML) estimation-based approach to physical map construction proposed in [7, 30] which determines the probe ordering and inter-probe spacings that maximize the probability of occurrence of the experimentally observed hybridization matrix H under a probabilistic model of hybridization errors consisting of false positives and false negatives. The estimation procedure involves a combination of discrete and continuous optimization where determining the probe ordering entails discrete (i.e., combinatorial) optimization whereas determining the inter-probe spacings for a particular probe ordering entails continuous optimization. The problem of determining the optimal probe ordering is intractable and can be shown to be isomorphic to the classical NP-hard Traveling Salesman Problem (TSP) [20]. Moreover, the ML objective function is non-linear in the probe ordering thus rendering the classical linear programming or integer linear programming techniques inapplicable. However, for a given probe ordering, determining the optimal inter-probe spacings is shown to be a tractable problem that is solvable using gradient descent-based search techniques.

In this chapter we present three stochastic combinatorial optimization algorithms for computation of the optimal probe ordering based on simulated annealing (SA), large step Markov chains (LSMC) and the genetic algorithm (GA). The computation of the optimal inter-probe spacings for a specified probe ordering is shown to be best achieved by the conjugate gradient descent algorithm. We propose a two-tier parallelization strategy for efficient implementation of the ML estimation-based physical mapping algorithm. The upper level represents parallel discrete optimization using the aforementioned stochastic combinatorial optimization algorithms whereas the lower level comprises of parallel conjugate gradient descent search. The resulting parallel algorithms are implemented on a cluster of shared-memory symmetric multiprocessors (SMPs). The conjugate gradient descent search algorithm is parallelized using shared-memory multithreaded programming on a single SMP whereas the stochastic combinatorial optimization algorithm is implemented on the SMP cluster using the distributed-memory Message Passing Interface (MPI) environment [42]. Convergence, speedup and scalability characteristics of the parallel algorithms are compared, analyzed and discussed.

11.2 The Maximum Likelihood Reconstruction of a Physical Map

The probe ordering problem can be formally stated as follows. Given a set $\mathcal{P} = \{P_1, P_2, \ldots, P_n\}$ of n probes and a set $\mathcal{C} = \{C_1, C_2, \ldots, C_k\}$ of k clones generated using the protocol described in Section 11.1.1, and the $k \times n$ clone-probe hybridization matrix H containing both false positives and false negatives with predefined probabilities, reconstruct the correct ordering $\Pi = (\pi_1, \pi_2, \ldots, \pi_n)$ of the probes and also the correct spacings $Y = (Y_1, Y_2, \ldots, Y_n)$ between the probes. The ordering Π is a permutation of $(1, \ldots, n)$ that gives the labels (indices) of the probes in left-to-right order across the chromosome. In the inter-probe spacings vector Y, Y_1 denotes the space between the left end of the first probe P_{π_1} and the left end of the chromosome, and Y_i the spacing between the right end of probe $P_{\pi_{i-1}}$ and the left end of probe P_{π_i} (where $2 \leq i \leq n$). The spacing between the right end of probe P_{π_n} and the right end of the chromosome is given by $Y_{n+1} = N - nM - \sum_{i=1}^{n} Y_i$ where N is length of the chromosome and M is the length of each clone/probe. Note that the protocol described in Section 11.1.1 requires that all probes and clones be of the same length.

The problem as stated above is ill-posed as defined by Hadamard [23] since the underlying constraints do not imply a unique solution. In the absence of errors, any probe ordering $(\pi_1, \pi_2, \ldots, \pi_n)$ with inter-probe spacings $Y = (Y_1, Y_2, \ldots, Y_n)$ that satisfies the constraint $N \geq nM + \sum_{i=1}^{n} Y_i$ is a feasible solution. In the presence of errors the problem is formulated as one of determining a probe ordering and an inter-probe spacings vector that maximize the likelihood of the observed hybridization matrix H given predefined probabilities for false positives and false negatives.

11.2.1 Mathematical Notation

The mathematical notation used in the formulation of the maximum likelihood estimator is given below:

N : length of the chromosome (in base pairs),

M : length of a clone/probe (in base pairs),

n : number of probes,

k : number of clones,

ρ : probability of false positive,

η : probability of false negative,

$H = ((h_{i,j}))_{1 \leq i \leq k, \, 1 \leq j \leq n}$: clone-probe hybridization matrix, where

$$h_{i,j} = \begin{cases} 1 & \text{if clone } C_i \text{ hybridizes with probe } P_j \\ 0 & \text{otherwise,} \end{cases}$$

H_i : ith row of the hybridization matrix (also termed as the binary hybridzation signature of the ith clone)

$\Pi = (\pi_1, \ldots, \pi_n)$: permutation of $\{1, 2, \ldots, n\}$ which denotes the probe labels in the ordering when scanned from left to right along the chromosome,

$p_i = \sum_{j=1}^{n} h_{i,j}$: number of 1's in H_i,

$P = \sum_{i=1}^{k} p_i$: total number of 1's in H,

$Y = (Y_1, Y_2, \ldots, Y_n)$: vector of inter-probe spacings where Y_i is the spacing between the right end of $P_{\pi_{i-1}}$ and the left end of P_{π_i} ($2 \leq i \leq n$), and Y_1 is the spacing between the left end of P_{π_1} and the left end of the chromosome, and

$\mathcal{F} \subseteq \mathcal{R}^n$: set of feasible inter-probe spacings $Y = \{Y_1, \ldots, Y_n\}$ such that $Y_i \geq 0$, $1 \leq i \leq n$ and $N - nM - \sum_{i=1}^{n} Y_i \geq 0$.

Under the assumptions that (i) the false positive and false negative errors at different positions along the clonal hybridization signature H_i are independent of each other, and (ii) the clones $\in \mathcal{C}$ are independently distributed along the chromosome i.e., each row of H is independent of the other rows, the probability of observing a hybridization matrix H for a given probe ordering Π and inter-probe spacing vector Y is given by:

$$P(H \mid \Pi, Y) = \prod_{i=1}^{k} C_i \left\{ R_i - \sum_{j=1}^{n+1} (a_{i,\pi_j} - 1)(a_{i,\pi_{j-1}} - 1) \min(Y_j, M) \right\} \tag{11.1}$$

where

$$a_{i,j} = \begin{cases} \frac{\eta}{(1-\rho)} & \text{if } h_{i,j} = 0 \text{ and } j = 1, \ldots, n \\ \frac{(1-\eta)}{\rho} & \text{if } h_{i,j} = 1 \text{ and } j = 1, \ldots, n \\ 0 & \text{otherwise,} \end{cases}$$

$$C_i = \frac{\rho^{p_i}(1-\rho)^{(n-p_i)}}{N-M}, \tag{11.2}$$

and

$$R_i = N - nM + M \sum_{j=1}^{n-1} a_{i,\pi_j} a_{i,\pi_{j+1}}. \tag{11.3}$$

The detailed derivation of equation (11.1) can be found in [7].

The goal therefore is to determine Π and Y that maximize $P(H \mid \Pi, Y)$ as given in equation (11.1), that is determine $(\hat{\Pi}, \hat{Y})$ where

$$(\hat{\Pi}, \hat{Y}) = \arg\max_{(\Pi, Y)} P(H \mid \Pi, Y) \tag{11.4}$$

Alternatively we could consider the negative log-likelihood (NLL) function $f(\Pi, Y)$ given by

$$\begin{aligned} f(\Pi, Y) &= -\ln P(H \mid \Pi, Y) \\ &= C - \sum_{i=1}^{k} \ln \left\{ R_i - \sum_{j=1}^{n+1} (a_{i,\pi_j} - 1)(a_{i,\pi_{j-1}} - 1) \min(Y_j, M) \right\} \end{aligned} \tag{11.5}$$

where C is a constant given by

$$C = k \ln(N - M) - P \ln \frac{\rho}{(1-\rho)} - nk \ln(1-\rho) \tag{11.6}$$

and $\pi_0 = \pi_{n+1} = 0$. Since $\ln x$ is a monotonically increasing function of x for all $x > 0$, it follows that

$$(\hat{\Pi}, \hat{Y}) = \arg\max_{(\Pi, Y)} P(H \mid \Pi, Y) = \arg\min_{(\Pi, Y)} f(\Pi, Y) \tag{11.7}$$

Let $\Pi^R = (\pi_n, \ldots, \pi_1)$. It is easy to verify that

$$P(H \mid \Pi, Y) = P(H \mid \Pi^R, Y) \tag{11.8}$$

This means that the likelihood as a function of Π is unique up to reversals which implies that it is possible to recover the ordering of the probes uniquely only up to reversals.

11.3   Computation of the Maximum Likelihood Estimate

Computing the values of $\hat{\Pi}$ and $\hat{Y}$ (equation (11.7)) involves a two stage procedure:

Stage 1: We first determine the optimal spacing $\hat{Y}_\Pi$ for a given probe ordering Π i.e., determine $\hat{Y}_\Pi = (\hat{Y}_1, \ldots, \hat{Y}_n)$ such that for a given Π,

$$f(\Pi, \hat{Y}_\Pi) = \min_Y f(\Pi, Y) = \min_Y f_\Pi(Y) \qquad (11.9)$$

Here the minimum is taken over all feasible solutions Y that satisfy the constraints $Y_i \geq 0$; $i = 1, \ldots, n$ and $\sum_{i=1}^n Y_i \leq N - nM$.

Stage 2: We determine $\hat{\Pi}$ for which,

$$f(\hat{\Pi}, \hat{Y}_{\hat{\Pi}}) = \min_\Pi f(\Pi, \hat{Y}_\Pi) = \min_\Pi f_{\hat{Y}_\Pi}(\Pi) \qquad (11.10)$$

Here the minimum is taken over all Π where Π is a permutation of $\{1, \ldots, n\}$.

A region $\mathcal{D} \subseteq \mathcal{R}^n$ is deemed to be convex if for any pair of points p, $q \in \mathcal{D}$, all points along the line segment $\alpha p + (1 - \alpha)q \in \mathcal{D}$ where $0 \leq \alpha \leq 1$. A function $h : \mathcal{D} \mapsto \mathcal{R}$ defined on a convex set $\mathcal{D}$ is deemed convex if for all points $\alpha p + (1 - \alpha)q \in \mathcal{D}$ where $0 \leq \alpha \leq 1$, $h(\alpha p + (1 - \alpha)q) \leq \alpha h(p) + (1 - \alpha)h(q)$. Furthermore, a region $\mathcal{D} \subseteq \mathcal{F}$ is considered *good* if for all $Y \in \mathcal{D}$, $Y_i \neq M$, $1 \leq i \leq n + 1$. The significance of a good region is that $f_\Pi(Y)$ is differentiable within it. It can be shown that $f_\Pi(Y)$ is convex in every good convex region $\mathcal{D}$ and therefore possesses a unique local minimum which is also a global minimum [7, 30]. Consequently this minimum can be reached using continuous local search-based techniques such as gradient descent (i.e., steepest descent) search or conjugate gradient descent search [15].

Consider the four disjoint subregions $\mathcal{F}_{+1,+1}$, $\mathcal{F}_{+1,-1}$, $\mathcal{F}_{-1,+1}$ and $\mathcal{F}_{-1,-1}$ within $\mathcal{F}$ where

$$\mathcal{F}_{a,b} \triangleq \{Y \in \mathcal{F} : aY_1 \leq aM; Y_i \leq M, 2 \leq i \leq N; bY_{n+1} \leq bM\} \qquad (11.11)$$

Each of these regions is convex since they result from the intersection of half spaces. Also, it can be shown that since the derivative of $f_\Pi(Y)$ is defined in the interior of each subregion, each subregion is good [7]. Note that we can define the derivative on the boundary of each subregion $\mathcal{F}_{a,b}$, $a, b \in \{-1, +1\}$, based on the direction in which the boundary point is approached. Thus, by selecting a starting point in each of the subregions (or as many subregions as possible without violating any feasibility constraints), one can compute a local minimum for $f_\Pi(Y)$ in each of the subregions and select the minimum of these local minima to be the global minimum of $f_\Pi(Y)$ [30].

Gradient Descent Search

We illustrate the computation of $\hat{Y}_\Pi$ in each of the subregions $\mathcal{F}_{a,b}$ using the steepest descent (SD) search technique. The SD search technique is a simple iterative procedure which consists of three steps: (i) determine the initial value of Y, (ii) compute the downhill gradient at Y and (iii) update the current value of Y using the computed value of the downhill gradient. Steps (ii) and (iii) are repeated until the gradient vanishes. The point at which the gradient vanishes is considered to be the desired local minimum. In practice, the SD search procedure is halted when the magnitude of the gradient is less than a prespecified threshold.

The initial value of $Y = (Y_1, \ldots, Y_n)$ in each of the subregions $\mathcal{F}_{a,b}$ can be determined by assigning $\frac{N-nM}{n+1}$ to each of Y_i's, i.e., distributing the spacings equally. Having obtained the starting value for $\hat{Y}$, we compute the gradient of the negative log likelihood function. The negative of the local gradient (i.e., the downhill gradient) represents the direction along which the function decreases the most rapidly. The local downhill gradient of $f_\Pi(Y)$ is given by

$$
\begin{aligned}
-\nabla f(\Pi, \hat{Y}) &= -(\frac{\partial f(\Pi, Y)}{\partial Y_1}, \dots, \frac{\partial f(\Pi, Y)}{\partial Y_n})\mid_{Y=\hat{Y}} \\
&= (U_1, \dots, U_n)\mid_{Y=\hat{Y}},
\end{aligned}
\tag{11.12}
$$

The current value of $\hat{Y} = \hat{Y}_{old}$ is updated by moving along the downhill gradient direction $U = -\nabla f(\Pi, \hat{Y})|_{\hat{Y}=\hat{Y}_{old}}$. The new value of $\hat{Y} = \hat{Y}_{new}$ is given by

$$
\hat{Y}_{new} = \hat{Y}_{old} + sU.
\tag{11.13}
$$

We attempt to find an optimal value of s, say s^* such that

$$
f(\Pi, \hat{Y} + s^*U) = \min_s f(\Pi, \hat{Y} + sU).
\tag{11.14}
$$

Having obtained the value of s^*, then the new inter-probe spacings are given by

$$
\hat{Y}_{new} = \hat{Y}_{old} + s^*U.
\tag{11.15}
$$

Our specific problem is that of constrained minimization of $f_\Pi(Y)$ in a good convex region $\mathcal{F}_{a,b}$, $a, b \in \{-1, +1\}$. The downhill gradient U at a given point Y on the boundary of $\mathcal{F}_{a,b}$ may be outside the region. In this case, we need to proceed along a direction U' that is directed inside $\mathcal{F}_{a,b}$ along which $f_\Pi(Y)$ decreases, albeit at a slower rate. Since the boundaries of $\mathcal{F}_{a,b}$ where $a, b \in \{-1, +1\}$ are hyperplanes, we determine, at the boundary point Y, all the boundary constraints (hyperplanes) $\Gamma_1, \Gamma_2, \dots, \Gamma_r$ that are violated by the downhill gradient U. Let $\mathbf{T}_i(U)$ denote the projection of U on Γ_i. Then the resulting direction U' is given by successively projecting U on each of the violating hyperplanes as follows

$$
U' - \mathbf{T}_r(\mathbf{T}_{r-1}(\dots(\mathbf{T}_1(U))\dots)
\tag{11.16}
$$

If $U' = 0$ then the current point Y is the local minimum of $f_\Pi(Y)$ in $\mathcal{F}_{a,b}$, $a, b \in \{-1, +1\}$. If $U' \neq 0$ then the non-violating hyperplanes are used to determine upper and lower bounds on the value of s denoted by $[s_{low}, s_{high}]$. Since the function $f_\Pi(Y)$ is convex with respect to s in $\mathcal{F}_{a,b}$ where $a, b \in \{-1, +1\}$, the bisection method [44] can be used to determine s^*. The SD search procedure is terminated when either U or U' vanishes depending on which situation is encountered first.

One of the problems with the SD search is that its convergence rate is sensitive to the starting point and the shape of the solution landscape [15]. The SD search typically takes several small steps while descending a narrow valley in the solution landscape thus resulting in a slow convergence rate [43]. The conjugate gradient descent (CGD) search, on the other hand, is known to be one of the fastest in the class of gradient descent search-based optimization methods [24]. The CGD search procedure is very similar to the SD procedure except that different directions are followed while minimizing the objective function. Instead of consistently following the local downhill gradient direction (i.e., the direction of steepest descent), a set of n mutually orthonormal (i.e., conjugate) direction vectors are generated from the downhill gradient vector where n is the dimensionality of the solution space. The orthonormality condition ensures that the minimization can proceed along any given direction vector independently of the other direction vectors. The CGD search procedure

FIGURE 11.1: Serial CGD search algorithm

Conjugate Gradient Descent Algorithm:

Phase 1: *Initialization*
Start with an initial guess of $Y = Y_i$;
Calculate gradient $G = \nabla f(\Pi, Y_i)$;
$G = G_1 = G_2 = -G$;

Phase 2: *Iterative Refinement*
while (1)
beginwhile

Project G ;

if $(|G| < \epsilon)$ break;

Bracket the minimum along the direction G;

Minimize along the direction G:
Find the optimal s^* such that $f(\Pi, Y_i + s^*G) = \min_s f(\Pi, Y_i + sG)$;

$Y_{i+1} = Y_i + s^*G$; $\Delta f = f(\Pi, Y_i) - f(\Pi, Y_{i+1})$; $Y_i = Y_{i+1}$;

if $(\Delta f < \epsilon)$ break;

Compute gradient $G = \nabla f(\Pi, Y_i)$; $g_1 = (G + G_2) \cdot G$; $g_2 = G_2 \cdot G_2$; $g_3 = g_1/g_2$; $G_2 = -G$;
$G = G_1 = G_2 + g_3 G_1$;

endwhile

Phase 3: *Output Result*
$Y = Y_i$;

guarantees convergence to a local minimum of a quadratic function within n steps [15] making it one of the fastest in the class of gradient descent search-based optimization methods [24]. The CGD search algorithm depicted in Figure 11.1 is based on the one presented in [44] with suitable adaptations (similar to the ones for the SD search algorithm described above) to take into account the fact that the solution space of the inter-probe spacings is constrained.

11.3.1 Computation of $\hat{\Pi}$

Determining the optimal clone ordering $\hat{\Pi}$, is an intractable problem that entails a combinatorial search through the discrete space of all possible permutations of $\{1, \ldots, n\}$. We present three stochastic discrete optimization algorithms for this purpose, Simulated Annealing (SA), Large Step Markov Chain (LSMC) and the Genetic Algorithm (GA). In particular, we chose to augment the classical GA with the stochastic hill climbing capability of the LSMC algorithm. We provide a brief description of these three algorithms.

Simulated Annealing

A single iteration of the SA algorithm consists of three basic phases: (i) perturb, (ii) evaluate, and (iii) decide. In the perturb phase, the current probe ordering Π_i is systematically perturbed to yield another candidate probe ordering Π_j by reversing the ordering within a block of probes where the endpoints of the block are chosen at random. This perturbation is referred to as a *2-opt* heuristic in the context of the TSP [35]. In the evaluate phase, the function $f(\Pi_j, \hat{Y}_{\Pi_j})$ (equation (11.9)) is computed. In the decide phase, the new can-

FIGURE 11.2: SA algorithm for computing the probe ordering and inter-probe spacings

Simulated Annealing Algorithm:

1. Choose a random order of probes, Π and a sufficiently high value of the temperature T. Compute $f(\Pi, \hat{Y}_\Pi)$.

2. Choose a random block within the ordering Π.

3. Perform a block reversal and call the new ordering Π'.

4. Compute $f(\Pi', \hat{Y}_{\Pi'})$.

5. If $f(\Pi', \hat{Y}_{\Pi'}) < f(\Pi, \hat{Y}_\Pi)$, then replace the existing solution $(\Pi, \hat{Y}_\Pi)$ by the new solution $(\Pi', \hat{Y}_{\Pi'})$,

   Else if $f(\Pi', \hat{Y}_{\Pi'}) \geq f(\Pi, \hat{Y}_\Pi)$, then

   Generate a random number x that is uniformly distributed in the interval $[0, 1]$.

   If $x < \exp(-(f(\Pi', \hat{Y}_{\Pi'}) - f(\Pi, \hat{Y}_\Pi)/T)$ then, replace the existing solution $(\Pi, \hat{Y}_\Pi)$ by the new solution $(\Pi', \hat{Y}_{\Pi'})$,

   Else retain the existing solution $(\Pi, \hat{Y}_\Pi)$.

6. Repeat steps 2-5 for a given value of T until D reorderings have been attempted or S successful reorderings have been recorded, whichever comes first. An ordering is deemed successful if it lowers the objective function $f(\Pi, \hat{Y}_\Pi)$.

7. Check if the convergence criterion has been met. If yes, stop; if not, reduce T using the annealing schedule and go to step 2.

didate solution Π_j is accepted with probability p which is computed using the Metropolis function [37]:

$$
p = \begin{cases} 1 & \text{if } f(\Pi_j, \hat{Y}_{\Pi_j}) < f(\Pi_i, \hat{Y}_{\Pi_i}) \\ \exp\left(-\frac{f(\Pi_j, \hat{Y}_{\Pi_j}) - f(\Pi_i, \hat{Y}_{\Pi_i})}{T}\right) & \text{if } f(\Pi_j, \hat{Y}_{\Pi_j}) \geq f(\Pi_i, \hat{Y}_{\Pi_i}) \end{cases} \tag{11.17}
$$

or using the Boltzmann function $B(T)$ [1]:

$$
p = B(T) = \frac{1}{1 + \exp\left(\frac{f(\Pi_j, \hat{Y}_{\Pi_j}) - f(\Pi_i, \hat{Y}_{\Pi_i})}{T}\right)} \tag{11.18}
$$

at a given value of temperature T, whereas Π_i is retained with probability $(1 - p)$.

Both, the Metropolis function and the Boltzmann function, ensure that SA generates an asymptotically ergodic Markov chain of solution states at a given temperature value. Geman and Geman [21] have shown that logarithmic annealing schedules of the form $T_k = R/\log k$ for some value of $R > 0$ ensure asymptotic convergence to a global minimum with unit probability in the limit $k \to \infty$. The convergence criterion in our case was the fact that the value of the objective function had not changed for the past k successive annealing steps. However, we used a geometric annealing schedule of the form $T_{k+1} = \alpha T_k$ where α is a value less than but close to 1. We used a value of $\alpha = 0.95$ in all of our experiments. Although the aforementioned geometric annealing schedule does not ensure strict asymptotic convergence to a global optimum as the logarithmic annealing schedule does, it is much faster and has been found to yield very good solutions in practice [45]. Figure 11.2 gives the outline of the serial SA algorithm.

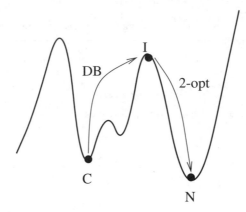

C: Current locally optimal solution

I: Intermediate Solution

N: New locally optimal solution

DB: double bridge perturbation

2-opt: exhaustive 2-opt search

FIGURE 11.3: Exploration of the solution space by the LSMC algorithm.

Large Step Markov Chain Algorithm

A single iteration of the LSMC algorithm (like the SA algorithm) also consists of the perturb, evaluate, and decide phases. The major difference between the SA and the LSMC algorithms arises from the fact that the classical SA algorithm performs a strictly *local* perturbation in the perturb phase whereas the LSMC algorithm performs a non-local perturbation followed by an exhaustive local search. The LSMC algorithm combines the stochastic decision function with an exhaustive local search using the 2-opt heuristic. The current solution at every stage is guaranteed to be locally optimal under the 2-opt heuristic. In the perturb phase, the current solution (which is locally optimal) is subject to a non-local perturbation termed as a *double-bridge kick* [36] which results in a transition to a non-local point in the search space. An exhaustive local 2-opt search is performed starting from this new point yielding a new local optimum. The choice between the new local optimum and the current solution is then made using the Metropolis decision function or the Boltzmann decision function as in the case of the SA algorithm. Figure 11.3 depicts the search strategy employed by the LSMC algorithm.

The exhaustive local search using the 2-opt heuristic would, strictly speaking, entail the evaluation of the objective function $f(\Pi, Y)$ after each 2-opt perturbation. This would cause the LSMC algorithm to be computationally extremely intensive. As an effective compromise, the exhaustive local search is performed using a modified objective function. The column in the hybridization matrix H corresponding to a given probe could be considered as a binary hybridization signature of that probe. The modified objective function $f_D(\Pi)$ computes the sum of the Hamming distances between the binary hybridization signatures of successive probes in a given probe ordering Π. The local minimum of $f_D(\Pi)$ is sought using the 2-opt heuristic. Since the modified objective function $f_D(\Pi)$ is much easier to compute than the original objective function $f(\Pi, Y)$, the exhaustive local search is very fast. Note, that whereas the SA algorithm samples the entire search space, the LSMC algorithm samples only the space of locally optimal solutions. The Metropolis decision function or the Boltzmann decision function in the case of the LSMC algorithm is annealed

FIGURE 11.4: LSMC algorithm for computing the probe ordering and inter-probe spacings

Large Step Markov Chain (LSMC) Algorithm:

1. Choose a random order of probes, Π and a sufficiently high value of the temperature T.

2. Perform an exhaustive local search using the 2-opt heuristic and the Hamming distance objective function on Π to yield new probe order Π'. Compute $f(\Pi', \hat{Y}_{\Pi'})$

3. Perform a non-local double-bridge perturbation on Π' to yield a new probe order Π''.

4. Perform an exhaustive local search using the 2-opt heuristic and the Hamming distance objective function on Π'' to yield new probe order Π'''. Compute $f(\Pi''', \hat{Y}_{\Pi'''})$.

5. If $f(\Pi''', \hat{Y}_{\Pi'''}) < f(\Pi', \hat{Y}_{\Pi'})$, then replace the existing solution $(\Pi', \hat{Y}_{\Pi'})$ by the new solution $(\Pi''', \hat{Y}_{\Pi'''})$,
   Else if $f(\Pi''', \hat{Y}_{\Pi'''}) \geq f(\Pi', \hat{Y}_{\Pi'})$, then

   Generate a random number x that is uniformly distributed in the interval $[0, 1]$.

   If $x < \exp(-(f(\Pi''', \hat{Y}_{\Pi'''}) - f(\Pi', \hat{Y}_{\Pi'})/T)$ then, replace the existing solution $(\Pi', \hat{Y}_{\Pi'})$ by the new solution $(\Pi''', \hat{Y}_{\Pi'''})$,

   Else retain the existing solution $(\Pi', \hat{Y}_{\Pi'})$.

6. Repeat steps 2-5 for a given value of T until D reorderings have been attempted or S successful reorderings have been recorded, whichever comes first. An ordering is deemed successful if it lowers the objective function $f(\Pi, \hat{Y}_{\Pi})$.

7. Check if the convergence criterion has been met. If yes, stop; if not, reduce T using the annealing schedule and go to step 2.

in a manner similar to the SA algorithm. Figure 11.4 gives the outline of the serial LSMC algorithm. The LSMC algorithm starting from an initial solution, also generates, in the limit, an ergodic Markov chain of solution states which asymptotically converges to a stationary Boltzmann distribution [1]. The Boltzmann distribution asymptotically converges to a globally optimal solution when subject to the annealing process [21].

The Genetic Algorithm

The Genetic Algorithm (GA) [38] begins with an initial ensemble or population of candidate solutions (typically represented by bit strings) and iterates through a number of generations before reaching a locally optimal solution. In each iteration or generation, the solutions in the current population are subject to the genetic operators of selection, crossover and mutation. The selection operator selects two candidate solutions from the current population with probability in direct proportion to their fitness values using a *roulette-wheel* procedure. The fitness function is the negative of the NLL objective function in equation (11.5) so that solutions with lower objective function values have higher fitness values and conversely. During the crossover operation, the solutions selected by the roulette-wheel procedure are treated as parental chromosomes and child chromosomes are generated by exchanges of parental chromosomal segments. This mimics the phenomenon of recombination in biological chromosomes. The purpose of crossover is to enable large-scale exploration of the search space [38]. The child chromosomes resulting from the crossover operator are then subject to the mutation operator which represents a random local (i.e., 2-opt) perturbation in the solution space. The new generation of chromosomes, thus created using the aforementioned genetic operators, replaces the existing population. The population replacement is repeated for several generations of the GA until the fitness value of the best chromosome in the population has not changed over the past k generations.

In our implementation of the GA, the heuristic crossover operator proposed by Jog *et al.* [27] for the TSP was improvised and incorporated. The probes in the ordering are treated as nodes in an undirected graph. The edges between the nodes in the graph are weighted by the Hamming distances between the binary hybridization signatures of the corresponding probes. The heuristic crossover can be described as follows:

1. Choose a start node from one of the selected chromosomes for crossover.

2. Compare two edges leaving from the start node between the two parents and choose the shorter edge; if the shorter edge leads to an illegal sequence, choose the other edge; if both edges introduce illegal sequences, choose an edge from the remaining nodes that has the shortest Hamming distance from the start node.

3. Choose the new node as the start node and repeat step 2 until a complete sequence is generated.

The heuristic crossover described above was slightly improvised in the context of our problem. If the start node is close to the end of the sequence, the crossover will not be effective in most cases where the two parental chromosomes are similar. For example, if the start node is selected at a point after which sequences of both parental chromosomes are the same, no actual exchange of parental chromosomal segments will result from the application of the heuristic crossover operator. To avoid this situation, we start from the beginning of the sequence whenever a node close to the end of the sequence is selected as the start node. This often leads to a much improved child in a single crossover operation.

In the absence of a hill climbing mechanism, the GA exhibits a slow convergence rate [38]. The incorporation of *deterministic* hill climbing into the GA typically results in premature convergence to a local optimum [8]. Our previous experience with the GA has shown that incorporation of a stochastic hill climbing mechanism greatly improves the asymptotic convergence of the GA to a near-globally optimal solution [8]. As a consequence, the GA was enhanced with the incorporation of a *stochastic* hill climbing search similar to that of the LSMC algorithm resulting in a GA-LSMC hybrid algorithm.

In the case of the GA-LSMC hybrid algorithm, the mutation operator is implemented using the non-local double-bridge perturbation followed by an exhaustive local 2-opt search. The mutation rate is set dynamically based on the genetic variation present in the current population. During the early stages of the GA, since genetic variations in the population are relatively high, application of the heuristic crossover operator usually creates better offspring. Therefore the mutation rate is kept relatively low. During the later stages of the GA, when the genetic variation is depleted due to repeated selection and crossover, the mutation rate is increased to introduce more variation in the population.

Two slightly different versions of the GA-LSMC hybrid algorithm were designed and implemented. In the first version, the population replacement strategy is deterministic, i.e., the less fit parent is replaced by the child chromosome if the child chromosome has a lower NLL objective function (i.e., higher fitness function) value. In the second version, the population replacement strategy is stochastic, i.e., the the less fit parent is replaced by the child chromosome with probability p_r computed using the Boltzmann function:

$$p_r = B(T) = \frac{1}{1 + \exp\left(\frac{f(\mathbf{x}_c) - f(\mathbf{x}_p)}{T}\right)} \tag{11.19}$$

where $f(\mathbf{x}_c)$ and $f(\mathbf{x}_p)$ are the NLL objective function values associated with the child chromosome and parent chromosome, respectively. Note that both versions of the GA

FIGURE 11.5: Outline of the GA-LSMC hybrid algorithm with stochastic replacement
The GA-LSMC Hybrid Algorithm with Stochastic Replacement:

1. Create an initial population of locally optimal solutions using the *double-bridge* perturbation and an exhaustive local 2-opt search; (This ensures that all the members in the initial population are locally optimal solutions)

2. While not converged do

    (a) Select two parents using the roulette wheel selection procedure;

    (b) Apply the heuristic crossover to the selected parents to create an offspring S;

    (c) Perform exhaustive local 2-opt search on S to yield a new locally optimum solution S';

    (d) With probability p_m perform a mutation on S' using a double-bridge perturbation followed by exhaustive local 2-opt search to yield a new locally optimum solution S^*;

    (e) Evaluate the NLL objective function at S^* (if mutation is performed) or at S' (if mutation is not performed) using conjugate gradient descent search;

    (f) Compute $\Delta f = f(S^*) - f(P)$ (if mutation is performed) or $\Delta f = f(S') - f(P)$ (if mutation is not performed) where P is the less fit of the two parents;

    (g) In the case of stochastic replacement: Replace P with S^* (if mutation is performed) or S' (if mutation is not performed) with probability p_r computed using the Boltzmann function $p_r = \frac{1}{1+\exp\left(\frac{\Delta f}{T}\right)}$;

    In the case of deterministic replacement: If $\Delta f < 0$ replace P with S^* (if mutation is performed) or S' (if mutation is not performed);

    (h) Update p_m;

    (i) In the case of stochastic replacement, update the temperature using the annealing function $T = A(T)$;

    (j) Check for convergence;

3. Output the best solution in the population as the final solution;

incorporate the hill climbing mechanism of the LSMC algorithm. Figure 11.5 depicts the GA-LSMC hybrid algorithm.

After a sufficient number of generations, the solutions in the population represent locally optimal solutions to the problem. The final solution to the optimization problem is chosen from the ensemble of locally optimal solutions. Since a larger solution space is sampled, the final solution from the GA-LSMC algorithm is potentially better than the one obtained from randomized local neighborhood search algorithms such as SA or the LSMC algorithm. However, since an ensemble of solutions has to be analyzed in each generation or iteration, the execution time per iteration is higher.

11.4 Parallel Computation of the Maximum Likelihood Estimate

Computation of the ML estimate entails two levels of parallelism corresponding to the two stages of optimization discussed in the previous section:

Level 1: Parallel computation of the optimal inter-probe spacing $\hat{Y}_\Pi$ for a given probe ordering Π (equation (11.9)). This entails parallelization of the gradient descent search procedure for constrained optimization in the *continuous* domain.

Level 2: Parallel computation of the optimal probe ordering (equation (11.10)). This entails parallelization of SA, LSMC or the GA for optimization in the *discrete* domain.

The two levels of parallel computation were implemented on a cluster of of SMPs which constitutes a hybrid platform comprising of both, shared memory and distributed memory. Each SMP is a shared-memory multiprocessor whereas the cluster of SMPs constitutes a distributed-memory computing resource. The parallel computation of the optimal inter-probe spacing $\hat{Y}_\Pi$ for a given probe ordering Π was deemed to be well suited for data parallelism using shared-memory multithreaded programming [2]. The distributed-memory message-passing paradigm using the Message Passing Interface (MPI) software environment [42] was deemed to be better suited for parallelization of SA, the LSMC algorithm and the GA.

11.4.1   Parallel SA and LSMC Algorithms

Since a candidate solution in the serial SA or LSMC algorithm can be considered to be an element of an asymptotically ergodic first-order Markov chain of solution states, we formulated and implemented two models of parallel SA (pSA) and parallel LSMC (pLSMC) algorithms based on the distribution of the Markov chain of solution states on the individual processors as described below:

- The Non-Interacting Local Markov chain (NILM) pSA and pLSMC algorithms.
- The Periodically Interacting Local Markov chain (PILM) pSA and pLSMC algorithms.

In the NILM pSA and NILM pLSMC algorithms, each SMP or node in a distributed-memory multiprocessor system runs an independent and asynchronous version of the serial SA or LSMC algorithm. In essence, there are as many Markov chains of solution states as there are physical SMPs within the system. At each temperature value, each SMP iterates through the perturb-evaluate-accept cycle concurrently (but asynchronously) with all the other SMPs. The perturbation function uses a parallel random number generator to generate the Markov chains of solution states. By assigning a distinct seed to each SMP at the start of execution, the independence of the Markov chains in different SMPs is ensured. The evaluation and decision functions are executed concurrently on the solution state within each SMP. On termination, the best solution is selected from among all the solutions available on the individual SMPs. The NILM model is essentially that of multiple independent searches.

The PILM pSA and PILM pLSMC algorithms are similar to their NILM counterparts except that just before the parameter T is updated, the best candidate solution from among those in all the processors is selected and duplicated on all the other processors. The goal of this synchronization procedure is to focus the search in the more promising regions of the solution space. The PILM model is essentially that of multiple periodically interacting searches as described in B(iii) above.

In the case of all the four algorithms, NILM pSA, NILM pLSMC, PILM pSA and PILM pLSMC, a **master process** acts as the overall controlling process and runs on one of the SMPs within the MPI cluster. The master process spawns child processes on each of the other SMPs within the MPI cluster, broadcasts the data subsets needed by each child process, collects the final results from each of the child processes and terminates the child processes. The master process, in addition to the above mentioned functions, also runs its own version of the SA or LSMC algorithm just as any of its child processes. In the case of the PILM pSA/pLSMC algorithms, before the parameter T is updated, the master process collects the results from each child process along with its own result, broadcasts the best result to all the child processes and also replaces its own result with the best result. The master process updates its temperature using the annealing schedule and proceeds with its local version of the SA or LSMC algorithm. On convergence, the master process collects the final results from each of the child processes along with its own, selects the best result as the final solution and terminates the child processes.

Each of the child processes in the PILM pSA/pLSMC algorithms receives the initial parameters from the master process and runs its local version of the SA or LSMC algorithm. At the end of each annealing step, each child process conveys its result to the master process, receives the best result thus far from the master process and replaces its result with the best result thus far before proceeding with the next annealing step. Upon convergence each child process conveys its result to the master process.

11.4.2 Parallel Genetic Algorithm

The approach to parallelizing the GA is based on partitioning the population amongst the available processors [9]. Each processor is responsible for searching for the best solution within its subpopulation. This is tantamount to performing multiple concurrent searches within the search space [9]. The parallel GA (pGA) was also implemented using the master-slave model with MPI on a distributed-memory platform comprising of a network of SMPs. The master process runs on one of the SMPs within the SMP cluster. The master process reads in the problem data, creates the initial population, spawns slave processes on all the SMPs (including its own) and divides the initial population amongst the slave processes. Each slave process runs a serial version of the GA on its subpopulation concurrently with the other slave processes. The slave processes periodically send the solutions within their subpopulation to the master process. The master process on receipt of the solutions from all the slave processes, checks for convergence, mixes the solutions at random and redistributes the population amongst the slave processes. This periodic mixing and redistribution of the population prevents a slave process from premature convergence to a local optimum after having exhausted all the genetic variation within its subpopulation. The master process deems the pGA to have converged if the best solution in the overall population has not changed over a certain number of successive generations.

11.4.3 Parallel Gradient Descent Search

Due to its inherent sequential nature, we deemed data parallelism to be the appropriate parallelization scheme for the CGD search algorithm. Our previous experience showed that the speedup of a data parallel implementation of the CGD search procedure on a distributed-memory multiprocessor did not scale well with an increasing number of processors [7]. This was attributed to the high inter-processor communication overhead in a distributed-memory environment. Consequently, a data parallel implementation of the CGD search algorithm on a shared-memory multiprocessor using multithreaded programming was deemed more suitable.

In our current implementation, the Y and G vectors are distributed amongst the processors within a single SMP. Each processor performs the required operations on its local Y_{loc} and G_{loc} subvectors concurrently with the other processors. Here, $|Y_{loc}| = |Y|/N_p$ and $|G_{loc}| = |G|/N_p$ where N_p is the number of processors within a single SMP. Implementation of the parallel algorithm follows the Master/Slave model, where both the master and slaves are implemented using IEEE POSIX threads [2]. The slave threads are responsible for most of the computation. Coordination and synchronization among the slave threads are carried out by the master thread.

Inter-thread synchronization is realized using data types mutex and semaphore from the POSIX thread (Pthread) library [2] and the barrier function implemented by us. Mutex is used to ensure that a critical section is executed atomically. Semaphore is used to coordinate the order of execution between the master thread and slave threads. Barrier is employed so that no thread can proceed any further until all the threads have reached the same point in their execution. This would prevent certain threads from updating the global variables when some other threads are still using them. Two types of barriers have been used in our implementation. One of the barrier types is used for coordinating the execution of the slave threads. This is useful when the computation is conducted entirely by the slave threads without any need for coordination by the master thread. The other barrier type is used in the computation by slave threads when the coordination and synchronization by the master thread are necessary. Each slave thread is bound to a processor so that the time spent on switching threads between processors is reduced to a minimum. The master thread is not bound to any processor since the time used by the master thread for coordination and synchronization is insignificant compared to the slave threads.

In the case of the pLSMC algorithm, the exhaustive local search using the 2-opt heuristic was also parallelized on a shared-memory SMP using multithreaded programming. However, instead of a data parallel implementation as in the case of the CGD search procedure, a control parallel scheme for the exhaustive local search was designed. The control parallel scheme entails multiple independent searches performed by multiple threads each bound to a distinct processor. For a given ordering of n clones, an exhaustive local search using the 2-opt heuristic would result in generation and evaluation of $O(n^2)$ distinct perturbations [35]. In the control parallel scheme, the generation and evaluation of distinct perturbations in carried out concurrently by multiple threads, each thread bound to a distinct processor within an SMP. The control parallel scheme follows the master-slave model wherein the master thread spawns multiple slave threads, binds each slave thread to a distinct processor within the SMP, assigns each slave thread a distinct partition of the search space (i.e., space of $O(n^2)$ distinct perturbations), collects the final result from each of the slave threads and designates the best result as the locally optimal solution under the 2-opt heuristic.

11.4.4 A Two-tier Approach for Parallel Computation of the ML Estimator

In order to ensure a modular, flexible and scalable implementation, two tiers of parallelism were incorporated in the computation of the ML estimator. The finer or lower level of parallelism pertains to the computation of $\hat{Y}$ for a given probe ordering Π using the parallel multithreaded CGD search algorithm. The coarser or upper level of parallelization pertains to the computation of $\hat{\Pi}$ using SA, the LSMC algorithm or the GA.

At the coarser level, the user has a choice of using either the parallel SA (pSA), parallel LSMC (pLSMC) or the parallel GA (pGA). The parallelization of the CGD search algorithm at the finer level is transparent to pSA, pLSMC or the pGA at the coarser level, i.e., the communication and control scheme for the parallel evolutionary methods is independent of that of the parallel multithreaded CGD search algorithm. For example, one could use the serial or parallel version of the CGD search algorithm (or, for that matter, any other serial or parallel algorithm for continuous optimization at the finer level) without having to make any changes to pSA, pLSMC or the pGA at the coarser level and vice versa.

A parallel multithreaded CGD search process is embedded within each of the pSA, pLSMC

TABLE 11.1   Specifications of the simulated
clone-probe hybridization data

Data Set	n	k	N	M	ρ	η
1	50	300	180	3	2%	2%
2	100	650	850	7	2%	2%
3	200	1300	1480	7	2%	2%
4	500	3250	3700	7	2%	2%

and pGA processes. When the parallel multithreaded CGD search procedure is invoked from within the master or slave pSA, pLSMC or pGA process, a new set of CGD threads is spawned on the available processors within an SMP. The master CGD thread runs on the same processor as the parent pSA, pLSMC or pGA process (master or child). The master and slave CGD threads cooperate to evaluate and optimize the objective function $f_\Pi(Y)$. Having optimized $f_\Pi(Y)$, the master and slave CGD threads are terminated by the parent pGA process and the corresponding processors are available for future computation.

The two-tier approach to parallelization of the ML estimator can be seen to induce a logical tree-shaped interconnection network on the available processors within the SMP cluster. The first level is the collection of SMPs that comprise the cluster. These SMPs run the (master and slave) pSA, pLSMC or pGA processes. The second level is the collection of processors within each SMP that run the master and slave CGD threads spawned by the pSA, pLSMC or pGA process running on that SMP. The processors that run the CGD threads are logically connected to the processor running the parent pSA, pLSMC or pGA process but are independent of the processors running other pSA, pLSMC or pGA processes.

11.5   Experimental Results

The parallel algorithms were implemented on a dedicated cluster comprising 8 nodes where each node is a shared-memory symmetric multiprocessor (SMP) running SUN Solaris-x86. Each SMP comprises four 700MHz Pentium III Xeon processors with 1 MB cache per processor and 1GB of shared memory. The programs were tested with simulated clone-probe hybridization data [7] as well as real data from *cosmid2* ($n = 109$, $k = 2046$) and *cosmid3* ($n = 111$, $k = 1937$) of the fungal genome *Neurospora crassa* made available by the Fungal Genome Resource, Department of Genetics, The University of Georgia. The simulated data used had the specifications outlined in Table 11.1. The simulated data were generated using a program described in [7] which generates clonal data of a given length with the left endpoints of the clones and probes uniformly distributed along the length of a simulated chromosome.

The pGA was implemented with the following parameters: the population size N_{pop} was chosen to be 40 for all the tests, the initial temperature T_{init} was chosen to be 1 for the pGA with stochastic replacement, the annealing factor α in the geometric annealing schedule $T_{next} = \alpha \cdot T_{prev}$ was chosen to be 0.95, the maximum number of trials max_trials before the population was mixed was chosen to be the population size N_{pop} and the convergence criterion used was the fact that no member in the population was replaced for two successive generations (i.e., the population remained the same for two successive generations). Note that this a stricter convergence criterion than the one that requires only the best solution in the population to be unchanged over a certain number of successive generations. The mutation probability p_m was set dynamically as follows: for a population replacement rate greater than 70% of N_{pop}, p_m was set to 0, for a population replacement rate between 30% and 70% of N_{pop}, the p_m was set to 0.1, and for a population replacement rate less than 70% of N_{pop}, p_m was set to 0.2. Thus, the mutation probability is kept low when the population replacement rate is sufficiently high. The crossover operation is the primary mechanism for exploration of the search space and for maintaining genetic variation in the population in this case. When the genetic variation in the population has depleted after repeated selection and crossover operations, resulting in a low population replacement rate, the mutation probability is gradually raised to introduce more genetic variation into the population.

The pSA and pLSMC algorithms were implemented with following parameters: the initial value for the temperature was chosen to be 1.0, the maximum number of iterations D for each annealing step was chosen to be $100 \cdot n$. The current annealing step was terminated when the maximum number of iterations was reached or when the number of successful perturbations equaled $10 \cdot n$ (i.e., 10% of the maximum number of iterations) whichever was encountered first. The annealing factor α in the geometric annealing function was chosen to be 0.95. The algorithm was terminated (and deemed to have reached a global optimum) when the number of successful perturbations in any annealing step equaled 0. When comparing the various parallel versions of the SA and LSMC algorithms, the product of the number of SMPs i.e., N_{SMP} and the maximum number of iterations D performed by a processor in a single annealing step was kept constant i.e., $D = (100 \cdot n)/N_{SMP}$. This ensured that the overall workload remained constant as the number of SMPs was varied thus enabling one to examine the scalability of the speedup and efficiency of the algorithms for a given problem size with increasing number of processors. In the NILM pSA and NILM pLSMC algorithms, each process was independently terminated when the number of successful perturbations in any annealing step for that process equaled 0. In the PILM pSA and PILM pLSMC algorithms, each process was terminated when the number of successful perturbations in an annealing step equaled 0 for *all* the processes i.e., the master process and all the child processes. This condition was checked during the synchronization phase at the end of each annealing step.

The parallel multithreaded CGD (MTCGD) search algorithm was tested on the simulated data set in Table 11.1 and on real data derived from *cosmid2* ($n = 109, k = 2046$) of the fungal genome *Neurospora crassa*. Figure 11.6 shows the speedup of the MTCGD search algorithm for varying N_p (i.e., number of processors within an SMP). As can be seen, the payoff in the parallelization of the CGD search algorithm is better realized for larger values of n (i.e., larger problem sizes). For $n = 200$ and $n = 500$, the best efficiency figures were seen to be in the range 93%–95% whereas for $n = 50$ the best efficiency figure were about 85%. For a given problem size, the efficiency was seen to decrease with an increasing number of processors N_p within the SMP. This is expected since the overhead imposed by barrier synchronization, mutex locks and semaphores increasingly dominates the overall execution time as N_p is increased for a given problem size.

The parallel multithreaded exhaustive local search algorithm was also tested on the simulated data set in Table 11.1. Figure 11.7 shows the speedup of the multithreaded exhaustive local search algorithm for varying N_p (i.e., number of processors within an SMP). For a given problem size, the efficiency was seen to decrease with an increasing number of processors N_p within the SMP, but not appreciably. This can be attributed to the fact that although there is a certain amount of overhead involved in creation of multiple slave threads and binding them to distinct processors, there is no frequent synchronization amongst the slave threads or between the master thread and the slave threads. Since each slave thread performs an independent search of the space of 2-opt perturbations, the only synchronization needed is at the beginning and end of the search process. This is in contrast to the parallel multithreaded CGD search algorithm where the synchronization amongst the slave threads or between the master thread and the slave threads is more frequent.

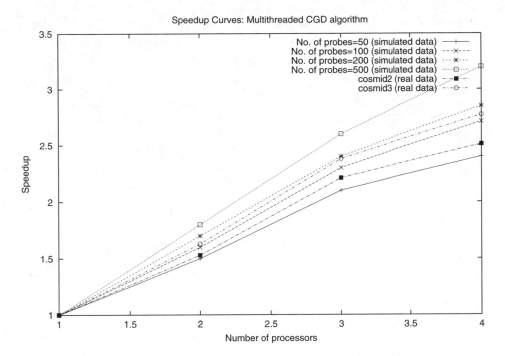

FIGURE 11.6: Speedup curves for the MTCGD algorithm on a single SMP.

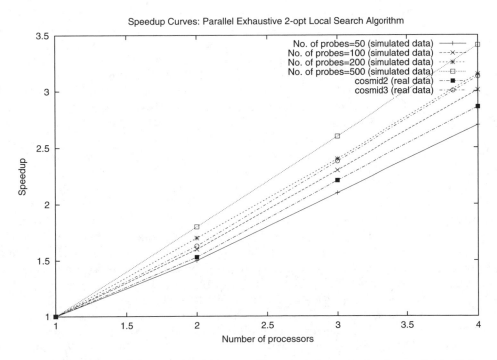

FIGURE 11.7: Speedup curves for the multithreaded exhaustive local search algorithm on a single SMP.

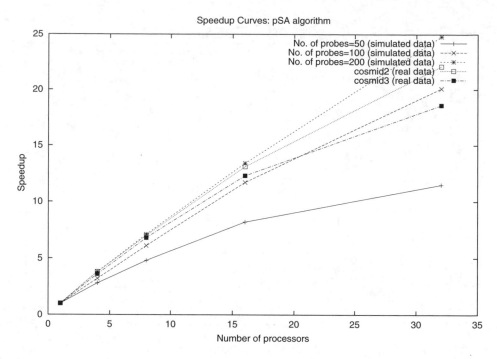

FIGURE 11.8: Speedup curves for the PILM pSA algorithm on an SMP cluster.

Figure 11.8 shows the speedup of the PILM pSA algorithm on the simulated and real data sets. Likewise, Figure 11.9 shows the speedup of the PILM pLSMC algorithm on the simulated and real data sets. The PILM versions of the pSA and pLSMC algorithms were observed to exhibit superior performance when compared to their NILM counterparts. This was expected since the PILM pSA and PILM pLSMC algorithms focus on the more promising regions of the search space via periodic synchronization. As can be observed, the pSA and pLSMC algorithms exhibit consistent and scalable speedup with an increasing total number of processors $N_{proc} = N_p \cdot N_{SMP}$. As expected, the speedup scales better with increasing number of processors for larger values of the number of probes n (i.e., larger problem sizes). Overall, the pSA algorithm was seen to scale better than the PLSMC algorithm. The reason for this is that the serial LSMC algorithm is much faster than the serial SA algorithm [7]. This implies that the pLSMC algorithm is better suited for larger problem instances than the pSA algorithm.

Figure 11.10 shows the speedup of the parallel GA-LSMC (pGA-LSMC) algorithm with stochastic replacement. The performance of the pGA-LSMC algorithm with stochastic replacement was observed to be superior to the pGA-LSMC algorithm with deterministic replacement in that the former was observed to yield better solutions with a smaller convergence time in most cases. This is in conformity with the results from our earlier work [8] which showed that the incorporation of a stochastic hill climbing mechanism (which, in the case of the pGA, is manifested in the form of stochastic replacement using the Boltzmann function), does improve the convergence rate and the ability of the pGA to explore more promising regions of the search space, thus resulting in a better final solution. As in the case of the pSA and pLSMC algorithms, the speedup of the pGA-LSMC algorithm was seen to scale better with increasing number of processors for larger values of n (i.e., larger problem sizes).

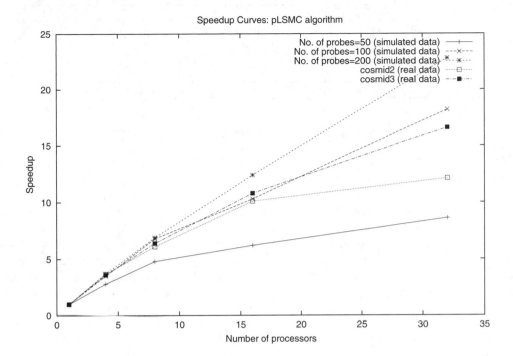

FIGURE 11.9: Speedup curves for the PILM pLSMC algorithm on an SMP cluster.

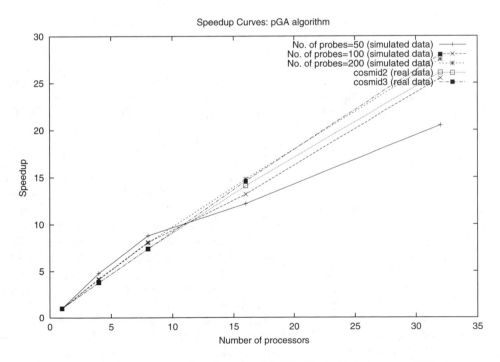

FIGURE 11.10: Speedup curves for the pGA-LSMC algorithm on an SMP cluster.

TABLE 11.2   Comparison of the serial SA, LSMC and GA-LSMC algorithms

Data Set	GA-LSMC			LSMC			SA		
	T (sec)	NLL	C	T (sec)	NLL	C	T (sec)	NLL	C
$n = 50, k = 300$	3325	1548.39	4	10746	1624.09	12	15076	1665.37	12
$n = 100, k = 650$	10919	4262.8	9	7459	4297.5	14	6265	4288.64	13
$n = 200, k = 1300$	181311	11159.48	9	105893	11515.13	24	31013	11574.75	27
cosmid2	27962	12731.37	-	34704	12757.55	-	108499	12949.99	-
cosmid3	14922	12584.62	-	30183	12501.88	-	45533	13212.85	-

T: Execution time, NLL: NLL fuction value, C: Number of contigs recovered

The pGA-LSMC algorithm was also observed to exhibit superlinear speedup in some instances. This can be attributed to two causes, population caching and the stochastic nature of the search process in the serial and parallel GA. Population caching is due to the fact that with a larger number of processors, the population per processor decreases to the point where it can reside entirely in cache. When the number of processors is small, the caching effect can overcome the inter-processor communication and synchronization overhead resulting in superlinear speedup. Also, due to the inherently stochastic nature of the selection, crossover and mutation operations in the GA, the manner in which the search space is traversed by the serial GA and the parallel GA could be entirely different. The difference in the manner of the search tree traversal has been known to cause instances of superlinear speedup in the case of other well known combinatorial search algorithms (both, stochastic and deterministic) such as SA, LSMC and branch-and-bound [31].

The efficiency values for all versions of all the three parallel algorithms (PSA, pLSMC and pGA-LSMC) are observed to be higher for larger problem instances (larger values of n and k) for a given value of N_{proc}. The efficiency values also show an overall declining trend for increasing values of N_{proc} for a given problem instance for all versions of all the three parallel algorithms. These observations are in conformity with the general expectations regarding the performance of these parallel stochastic optimization algorithms.

The results of the serial GA-LSMC hybrid algorithm with stochastic replacement were compared with those of the serial SA and serial LSMC algorithms (Table 11.2). The serial GA-LSMC hybrid algorithm was seen to consistently yield lower NLL values compared to the SA and LSMC algorithms on both, artificial and real data. The only exception is the real data set *cosmid3* where the GA-LSMC hybrid algorithm yielded a slightly higher NLL function value (less than 1% difference) than the LSMC algorithm but with much shorter execution time (less than half). Table 11.2 also shows the number of probe suborderings (i.e., contigs) recovered by the GA-LSMC, SA and LSMC algorithms on the synthetic data sets. In an ideal case, one should be able to recover the true probe ordering as a single contig. In reality, this is unlikely due to the presence of hybridization errors. In a realistic scenario, the physical mapping algorithm would be expected to recover probe suborderings that could then be manually manipulated (via translation and probe order reversal) to yield the final probe order. The fewer and longer these probe suborderings or contigs, the less intensive the subsequent manual editing to recover the desired probe ordering. The GA-LSMC algorithm was seen to consistently yield a physical map with fewer and longer contigs (i.e., fewer contig breaks) than the SA and LSMC algorithms suggesting that the GA-LSMC algorithm is capable of yielding solutions of higher quality. This clearly showed that the heuristic crossover in conjunction with population sampling is a powerful mechanism for exploration of the search space in the case of the GA. Note that whereas the SA and LSMC algorithms possess stochastic hill climbing capabilities, they do not possess the capability for large-scale exploration of the search space via population sampling and the crossover operator as does the GA.

11.6   Conclusions and Future Directions

In this chapter we presented an information theoretic approach to genome reconstruction. Information theoretic reconstruction approaches based on the maximum likelihood (ML) model or the Bayesian maximum a posteriori (MAP) model have been used extensively in image and signal reconstruction. In this chapter, we described a maximum likelihood (ML) estimation-based approach to physical map reconstruction under a probabilistic model of hybridization errors consisting of false positives and false negatives. The ML estimator optimizes a likelihood function defined over the spacings and orderings of probes under an experimental protocol wherein clones of equal length are hybridized to a maximal subset of non-overlapping equal-length clones termed as probes. The estimation procedure was shown to involve a combination of continuous and discrete optimization; the former to determine a set of optimal inter-probe spacings for a given probe ordering and the latter to determine the optimal probe ordering. The conjugate gradient descent (CGD) search procedure was used to determine the optimal spacings between probes for a given probe ordering. The optimal probe ordering was determined using stochastic combinatorial optimization procedures such as Simulated Annealing (SA), the Large Step Markov Chain (LSMC) algorithm and the Genetic Algorithm (GA). In particular, the incorporation of stochastic hill climbing into the traditional GA was shown to result in a GA-LSMC hybrid algorithm with convergence behavior superior to the traditional GA.

The problem of ML estimation-based physical map reconstruction in the presence of errors is a problem of high computational complexity thus providing the motivation for parallel computing. A two-level parallelization strategy was proposed wherein the CGD search procedure was parallelized at the lower level and SA, the LSMC algorithm or the GA was simultaneously parallelized at the higher level. The parallel algorithms were implemented on a networked cluster of shared-memory symmetric multiprocessors (SMPs) running the Message Passing Interface (MPI) environment. The cluster of SMPs offered a hybrid of shared-memory (within a single SMP) and distributed-memory (across the SMP cluster) parallel computing. A shared-memory data parallel approach where the components of the gradient vector are distributed amongst the individual processors within an SMP was deemed more suitable for the parallelization of the CGD search procedure. A distributed-memory control parallel scheme where individual SMPs perform noninteracting or periodically interacting searches was deemed more suitable for the parallelization of SA, the LSMC algorithm and the GA.

Our experimental results on simulated clone-probe data showed that the payoff in data parallelization of the CGD procedure was better realized for large problem sizes (i.e., large values of n and k). A similar trend was observed in the case of the parallel versions of SA, the LSMC algorithm and the GA. In the case of the parallel GA, superlinear speedup was observed in some instances which could be attributed to population caching effects. The parallel exhaustive local 2-opt search algorithm was observed to be the the most scalable in terms of speedup. This was expected since the parallel exhaustive local 2-opt search algorithm entails minimal interprocessor communication and synchronization overhead. Overall, the experimental results were found to be in conformity with expectations based on formal analysis.

Future research will investigate extensions of the ML function that encapsulate errors due to repeat DNA sequences in addition to false positives and false negatives. The current MPI implementation of the ML estimator is targeted towards a homogeneous distributed processing platform such as a network of identical SMPs. Future research will explore and address issues that deal with the parallelization of the ML estimator on a heterogeneous distributed processing platform such as a network of SMPs that differ in processing speeds, a scenario that is more likely to be encountered in the real world. Other combinatorial optimization techniques such as those based on Lagrangian-based global search and tabu search will also be investigated. Extensions of the ML estimation-based approach to a Bayesian MAP estimation-based approach, where the ordering of clones/probes containing genetic markers, as inferred from a genetic map, are used as a prior distribution, will also be investigated.

Acknowledgments

This research was supported in part by an NRICGP grant by the US Department of Agriculture (Grant Award No. USDA GEO-2002-03590) and by a Research Instrumentation grant by the National Science Foundation (Grant Award No. NSF EIA-9986032).

References

[1] E.H.L. Aarts and K. Korst. *Simulated Annealing and Boltzman Machines: A Stochastic Approach to Combinatorial Optimization and Neural Computing*. Wiley, New York, NY, 1989.

[2] G. Andrews. *Foundations of Multithreaded, Parallel, and Distributed Programming*. Addison Wesley Pub. Co., Reading, MA, 2000.

[3] J. Arnold. Editorial. *Fungal Genetics and Biology*, 21:254–257, 1997.

[4] R. Arratia, E.S. Lander, S. Tavare, and M.S. Waterman. Genomic mapping by anchoring random probes: a mathematical analysis. *Genomics*, 11:806–827, 1991.

[5] D.J. Balding. Design and analysis of chromosome physical mapping experiments. *Philos. Trans. Roy. Soc. London Ser. B.*, 334:329–335, 1994.

[6] A. Ben-Dor and B. Chor. On constructing radiation hybrid maps. In *Proc. ACM Conf. Computational Molecular Biology*, pages 17–26, 1997.

[7] S.M. Bhandarkar, S.A. Machaka, S.S. Shete, and R.N. Kota. Parallel computation of a maximum likelihood estimator of a physical map. *Genetics, special issue on Computational Biology*, 157(3):1021–1043, March 2001.

[8] S.M. Bhandarkar and H. Zhang. Image segmentation using evolutionary computation. *IEEE Trans. Evolutionary Computation*, 3(1):1–21, April 1999.

[9] E. Cantu-Paz. *Efficient and Accurate Parallel Genetic Algorithms*. Kluwer Academic Publishers, Boston, MA, November 2000.

[10] B. Chor and M. Sudan. A geometric approach to betweenness. In *Proc. European Symp. Algorithms*, volume 979, pages 227–237. Springer–Verlag Lecture Notes in Computer Science, 1995.

[11] T. Christof, M. Jünger, J.D. Kececioglu, and P. Mutzel *et al.* A branch–and–cut approach to physical mapping of chromosomes by unique end probes. *Journal of Computational Biology*, 4(4):433–447, 1997.

[12] T. Christof and J.D. Kececioglu. Computing physical maps of chromosomes with non-overlapping probes by branch–and–cut. In *Proc. ACM Conf. Computational Molecular Biology*, pages 115–123, Lyon, France, April 1999.

[13] A. J. Cuticchia, J. Arnold, and W. E. Timberlake. The use of simulated annealing in chromosome reconstruction experiments based on binary scoring. *Genetics*, 132:591–601, 1992.

[14] D.S. Greenberg D.B. Wilson and C.A. Phillips. Beyond islands: runs in clone–probe matrices. In *Proc. ACM Conf. Computational Molecular Biology*, pages 320–329,

1997.

[15] C.N. Dorny. *A Vector Space Approach to Models and Optimization.* R.E. Krieger Publishing Company, Huntington, NY, 1980.

[16] D.K. Weisser F. Alizadeh, R.M. Karp and G. Zweig. Physical mapping of chromosomes using unique probes. In *Proc. ACM–SIAM Conf. Discrete Algorithms*, pages 489–500, 1994.

[17] L.A. Newberg F. Alizadeh, R.M. Karp and D.K. Weisser. Physical mapping of chromosomes: a combinatorial problem in molecular biology. *Algorithmica*, 13(1/2):52–76, 1995.

[18] D.P. Fasulo, T. Jiang, R.M. Karp, and R. Settergren *et al.* An algorithmic approach to multiple complete digest mapping. In *Proc. ACM Conf. Computational Molecular Biology*, pages 118–127, 1997.

[19] Y.X. Fu, W.E. Timberlake, and J. Arnold. On the design of genome mapping experiments using short synthetic oligonucleotides. *Biometrics*, 48:337–359, 1992.

[20] M.S. Garey and D.S. Johnson. *Computers and Intractability: A Guide to the Theory of NP–Completeness.* W.H. Freeman, New York, NY, 1979.

[21] S. Geman and D. Geman. Stochastic relaxation, gibbs distribution and the bayesian restoration of images. *IEEE Trans. Pattern Analysis and Machine Intelligence*, 6(6):721–741, 1984.

[22] D.S. Greenberg and S. Istrail. Physical mapping by STS hybridization: algorithmic strategies and the challenge of software evaluation. *Journal Computational Biology*, 2(2):219–273, 1995.

[23] H. Hadamard. *Lectures on the Cauchy Problem in Linear Partial Differential Equations.* Yale University Press, New Haven, CT, 1923.

[24] M. Hestenes and E. Stiefel. Methods of conjugate gradients for solving linear systems. *Journal Research of National Bureau of Standards*, 49:409–436, 1954.

[25] M. Jain and E.W. Myers. Algorithms for computing and integrating physical maps using unique probes. *Journal Computational Biology*, 4(4):449–466, 1997.

[26] T. Jiang and R.M. Karp. Mapping clones with a given ordering or interleaving. In *Proc. ACM–SIAM Conf. Discrete Algorithms*, pages 400–409, 1997.

[27] P. Jog, J.Y. Suh, and D. Van Gucht. The effects of population size heuristic crossover and local improvement on a genetic algorithm for the traveling salesman problem. In *Proc. Intl. Conf. Genetic Algorithms*, pages 110–115, Fairfax, VA, June 1989.

[28] R.M. Karp and R. Shamir. Algorithms for optical mapping. In *Proc. ACM Conf. Computational Molecular Biology*, pages 117–124, 1998.

[29] J.D. Kececioglu and E.W. Myers. Combinatorial algorithms for DNA sequence assembly. *Algorithmica*, 13:7–51, 1995.

[30] J.D. Kececioglu, S.S. Shete, and J. Arnold. Reconstructing distances in physical maps of chromosomes with nonoverlapping probes. In *Proc. ACM Conf. Computational Molecular Biology*, pages 183–192, Tokyo, Japan, April 2000.

[31] T.H. Lai and S. Sahni. Anomalies in parallel branch and bound algorithms. *Communications of the ACM*, 27(6):594–602, June 1984.

[32] E.S. Lander and P. Green. Construction of multi-locus genetic linkage maps in humans. *Proceedings of the National Academy of Sciences*, 84:2363–2367, 1987.

[33] E.S. Lander and M.S. Waterman. Genomic mapping by fingerprinting random clones: a mathematical analysis. *Genomics*, 2:231–239, 1988.

[34] J.K. Lee, V. Dancik, and M.S. Waterman. Estimation for restriction sites observed by optical mapping using reversible-jump Markov chain Monte Carlo. In *Proc. ACM Conf. Computational Molecular Biology*, pages 147–152, 1998.

[35] S. Lin and B. Kernighan. An effective heuristic search algorithm for the traveling

salesman problem. *Operations Research*, 21:498–516, 1973.

[36] O. Martin, S.W. Otto, and E.W. Felten. Large-step Markov chains for the traveling salesman problem. *Complex Systems*, 5(3):299–326, 1991.

[37] N. Metropolis, A. Rosenbluth, M. Rosenbluth, and A. Teller *et al.* Equation of state calculations by fast computing machines. *Journal Chemical Physics*, 21:1087–1092, 1953.

[38] M. Mitchell. *An Introduction to Genetic Algorithms*. MIT Press, Cambridge, MA, 1996.

[39] R. Mott, A.V. Grigoriev, E. Maier, and J.D. Hoheisel *et al.* Algorithms and software tools for ordering clone libraries: application to the mapping of the genome *s. pombe*. *Nucleic Acids Research*, 21(8):1965–1974, 1993.

[40] S. Muthukrishnan and L. Parida. Towards constructing physical maps by optical mapping: an effective, simple, combinatorial approach. In *Proc. ACM Conf. Computational Molecular Biology*, pages 209–219, 1997.

[41] D.O. Nelson and T.P. Speed. Statistical issues in constructing high resolution physical maps. *Statistical Science*, pages 334–354, 1994.

[42] P. Pacheco. *Parallel Programming with MPI*. Morgan Kaufmann Publishers, San Francisco, CA, 1996.

[43] E. Polak. Optimization: Algorithms and consistent approximations. *Applied Mathematical Sciences*, 124, 1997.

[44] W.H. Press, B.P. Flannery, S.A. Teukolsky, and W.T. Vetterling. *Numerical Recipes in C*. Cambridge University Press, New York, NY, 1988.

[45] F. Romeo and A. Sangiovanni-Vincentelli. A theoretical framework for simulated annealing. *Algorithmica*, 6:302–345, 1991.

[46] D. Slonim, L. Kruglyak, L. Stein, and E. Lander. Building human genome maps with radiation hybrids. In *Proc. ACM Conf. Computational Molecular Biology*, pages 277–286, 1997.

[47] A.H. Sturtevant. The linear arrangement of six sex-linked factors in it Drosophila as shown by their mode of association. *The Journal of experimental zoology*, 14:43–49, 1913.

[48] Y. Wang, R.A. Prade, J. Griffith, and W.E. Timberlake *et al.* A fast random cost algorithm for physical mapping. *Proceedings of the National Academy of Sciences*, 91:11094–11098, 1994.

[49] M. Xiong, H.J. Chen, R.A. Prade, and Y. Wang *et al.* On the consistency of a physical mapping method to reconstruct a chromosome *in vitro*. *Genetics*, 142(1):267–284, 1996.

[50] M.Q. Zhang and T.G. Marr. Genome mapping by nonrandom anchoring: a discrete theoretical analysis. In *Proc. ACM Conf. Computational Molecular Biology*, volume 90, pages 600–604, 1993.

Expressed Sequence Tags: Clustering and Applications

Anantharaman Kalyanaraman
Iowa State University

Srinivas Aluru
Iowa State University

12.1 Introduction

When committees advising on the sequencing of complex genomes met in the late 80s, they estimated a wait of 12 to 15 years before the entire human genome sequence could be made available. While the size of the genome to be sequenced was estimated around 3 billion *bp*, rough estimates placed the number of genes to be about a few tens of thousands spanning only about 3% of the entire genome [1]. The fact that the gene space is relatively small spawned a new drive in seeking alternative sequencing methods to quickly discover the gene space without having to wait for the completion of the entire genome project. Invented in early 1990s [1], high throughput *cDNA* sequencing is one such technique in which double stranded DNA molecules called *complementary DNA* (or *cDNA*) molecules are synthesized from messenger RNA (or mRNA) libraries collected from the cells of living tissues. Later, these cDNA clones are read in a single-pass from either end, resulting in a subsequence of the original clone called an *Expressed Sequence Tag* or simply *EST*. As with any other sequencing technologies, this sequencing procedure is also vulnerable to errors. Nevertheless, the simplicity of the procedure has proved instrumental in its proliferation and has provided a cost-effective means to support high-throughput sequencing of transcribed portions of the genome.

EST projects were originally undertaken to provide a head start in the gene discovery process much earlier to genome sequencing efforts. Their scope, however, broadened as more applications of EST data were discovered. They remain a chief source of information for discovering genes even after a genome is sequenced. EST based techniques form one of the most popular classes of gene discovery methods (alongside *ab initio* and comparative techniques). EST data represent the set of all genes transcribed in a living cell under the experimental conditions provided and are so used to decipher the *transcriptome* of an organism, i.e., a catalog of all transcribed genes. Grouping ESTs based on their putative gene sources and counting the resulting number of non-redundant "clusters" is still one of the primary techniques to estimate the number of genes in an organism. ESTs are used in gene-expression studies based on the following main idea: genes showing varying levels of expression under different experimental conditions generate proportionately varying concentrations of ESTs in the sequenced data pools. ESTs are also applied in designing oligos for microarray chips and as sequence tags in physical mapping projects.

Ever since the development of high throughput cDNA sequencing technologies, EST databases have dominated the publicly available sequence repositories, and are continuing to grow at an unprecedented rate. The GenBank repository hosted by the National Center for Biotechnology Information (NCBI) has a database called dbEST for storing ESTs and making them widely accessible [7]. Formed in 1992, this division now has the largest collection of publicly available ESTs, with over 12 billion base pairs sequenced from more than 740 organisms; this occupies about 8% of the entire GenBank database. As more new ESTs continue to populate these databases at a rapid rate, there have been many persistent efforts in the development and application of automated utilities that analyze these data sets and extract useful information from them. Since EST databases have a mixed composition, i.e., with ESTs from numerous genes and their transcripts, the first and primary challenge that these analysis utilities face is to "intelligently" partition the ESTs by their putative sources. This problem is particularly hard because the desired source information is almost never available directly from the sequence databases, and needs to be deduced through alternative means.

The purpose of this chapter is to introduce the readers to ESTs, their abundant presence in sequence databases and their multifarious uses, and to develop an appreciation of both the computational challenges and biological relevance typical of the clustering utilities developed for the purpose of mining desired useful information from the EST data. To this effect, we present a detailed description of ESTs, their generation (in Section 12.2), availability (in Section 12.4) and their numerous applications (in Section 12.3), and a compendium of various clustering techniques that have been developed and applied for drawing biological inferences from various EST data sets (in Section 12.5). One of the main challenges that often surfaces while dealing with EST data is the computational capability to handle very large sequence data sets, and we defer our discussion on such challenges and algorithmic solutions for large scale data to Chapter 13.

12.2　Sequencing ESTs

Expressed sequence tags are single pass sequences obtained from mRNA libraries in the following manner (see Figure 12.1 for an illustration): Cells of living tissues are subjected to certain biological conditions of interest and the mRNAs that result from the transcribed genes within are extracted through experimental procedures such as those described in [18]. The isolated mRNAs are then subjected to reaction with reverse transcriptase, an enzyme that converts a single stranded mRNA molecule to a double stranded cDNA molecule. Due to underlying experimental limitations, however, this procedure may not complete on the entire mRNA thereby resulting in partial length cDNAs. In order to provide sufficient coverage over the entire mRNA, multiple and possibly redundant such partial length cDNAs are generated and each is cloned using a cloning vector. Using primers targeted for known

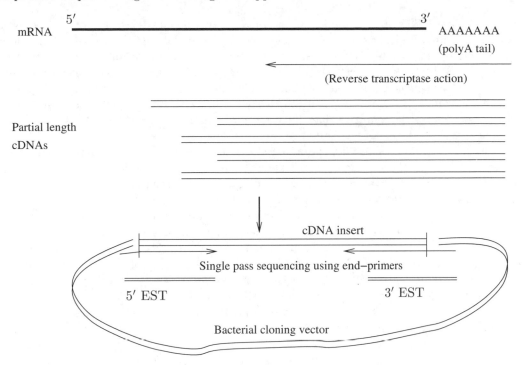

FIGURE 12.1: Illustration of the EST sequencing procedure. An mRNA library when subjected to reaction with the reverse transcriptase enzyme generates partial length cDNAs, which are then cloned using bacterial vectors and sequenced from ends to result in ESTs.

vector sequences near the ends of the inserts, the nucleotide sequence of each inserted clone is then read over a single pass from either end, resulting in fragments called ESTs that are about 500 *bp* long. Because this procedure may oversample the end regions of a cDNA clone, the untranslated regions at the ends of the corresponding mRNA may also get proportionately over-represented. If it is desirable to avoid such bias, sequencing is started from random locations on the cDNA insert using randomly located sequences as primers, or through application of restriction enzymes, breaking the cDNA insert before its shreds are sequenced from their ends.

Cost-effective high throughput sequencing of ESTs has largely been facilitated by the simplicity of the single pass sequencing technology. Nevertheless, the technology does not always generate accurate sequences. Bases are sometimes misread or ambiguously interpreted resulting in low-quality sequences. It is also possible, although rare, that two cDNA sequences representing two distinct mRNA sequences are spliced together resulting in an artifact known as a *chimeric cDNA*. When cloned and sequenced, the resulting ESTs could contain portions from either cDNA, potentially confounding their subsequent analysis. Sequencing error rates are typically in the range of 1-2% with present technology and dealing with these errors is generally deferred to later stages of sequence analysis. We will discuss techniques to deal with these errors and on the capacity to distinguish them from naturally-occurring variations in Section 12.5 in the context of clustering methods.

During sequencing, the two ESTs that originate from the ends of a cDNA insert are sometimes tagged with the clone identifier and stored in the header of the EST sequences in the database. This auxiliary information proves valuable in later stages of the sequence analysis – pairs of ESTs having the same clone identifier are labeled "mate pairs" (or "clone pairs") and are immediately associated with a common source transcript, obviating the need to compute additional evidence to establish their relationship. Mate pair information is not unique to EST sequencing; it is also common in genomic sequencing techniques that involve sequencing from a clone insert. Also available sometimes with EST sequence data are "trace data" that contain the quality values for each base position of the ESTs. Such trace data are measures of sequence quality and are valuable during analysis.

Genes express differentially depending on the tissue they reside and the subjected experimental conditions. Consequently, EST data generated by conventional sequencing techniques have ESTs from overly expressed genes in a proportionately higher concentration than from sparsely expressed genes. Such non-uniformity may be desirable if the ESTs are used in gene expression related studies; otherwise, not only is the effort spent in sequencing multiple ESTs covering the same regions unnecessary, but also such non-uniformity may add significant challenges to the computational methods for EST analysis. For example, one unique EST per gene is sufficient for estimating the number of genes in an organism, while oversampling may significantly increase the computation as a function of the number of ESTs represented per gene. To alleviate this problem, many variations to the original sequencing technique have been invented and these methods can be classified into two groups: normalization and subtractive hybridization. Normalization achieves a balance in the cDNA population within a cDNA library [63, 80], while subtractive hybridization reduces overly represented cDNA population by selectively removing sequences shared across cDNA libraries [24, 27, 74, 75, 87]. For a survey of these two methods see [4].

12.3 Applications of ESTs

12.3.1 Transcriptome and Gene Discovery

One of the earliest identified merits of EST data is in discovering genes with expression evidence [1, 8]. A sequencing experiment can trigger the expression of multiple genes in a target cell/tissue, and so the resulting EST data is a segmented representation of the transcribed portions of all these expressed genes. Thus, if we could "meaningfully" partition an EST collection into "bins" or "clusters" such that the ESTs derived from the same transcript or gene are clustered together, then we would have reverse-engineered the process that sequenced the ESTs in the first place, and the set of clusters would correspond to the portion of the transcriptome represented in the underlying sequence data. However, such EST-to-source mapping is not a readily available information and one of the main challenges in clustering is its inference from other information contained within the sequence data:

- Pairwise overlaps: Any two ESTs that cover a common segment within their gene source are expected to show a significant sequence overlap in the corresponding region(s). Therefore, by partitioning ESTs obtained from an organism using pairwise overlap information is expected to "cluster" the ESTs that were originally derived from the same gene source. Furthermore, if it is possible to assemble the ESTs in each cluster consistent with the pairwise assembly, then the resulting supersequence is likely to correspond to the mRNA transcript that gave rise to the set of ESTs in that cluster. The UniGene project undertaken by NCBI is a typical example of clustering ESTs by gene source [68]; and the Gene Index project undertaken by The Institute of Genome Research (TIGR) clusters by transcript source [69].

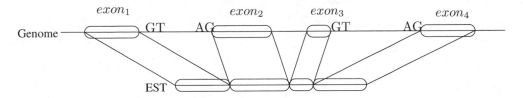

FIGURE 12.2: Spliced alignment of an EST with the corresponding genomic sequence. Applications include gene discovery and gene structure prediction.

- Protein evidence: ESTs sequenced from an organism can be "compared" against a database of known protein sequences, and those that significantly overlap with a common protein are likely to have originated from the same protein-coding transcript. This method of aligning ESTs with a protein database could be used to quickly discover protein-coding genes within an organism that have known protein products. An example application of this approach can be found in [93].

- Alignment with genomic DNA: Aligning ESTs with genomic DNA using a technique called spliced alignment identifies the coding regions represented by the ESTs along the genomic DNA. If the genomic sequence is a gene, and if an EST exhibits a "good" spliced alignment with it, then it is highly likely that the EST originated from this gene. The spliced alignment technique can also be used to discover new genes in long genomic sequences — by aligning each EST along the genomic stretches (potentially the whole genome if available), and identifying those regions that align well and unambiguously with ESTs. The same procedure can also be applied to identify the structure of a known gene. The boundaries of the aligning regions can be used as a good starting point for locating intronic and exonic boundaries within a gene, as shown in Figure 12.2. The spliced alignment technique and its applications are described in detail in Chapter 2.

Given the high costs associated with whole genome projects, the genomes of many organisms of interest are unlikely to be sequenced. In many cases, biologists still depend on EST data to help them with building transcriptomes and gene lists. Numerous transcriptome projects have benefited from EST databases in the past [6, 13, 14, 15, 60]. EST based gene discovery and transcriptome construction projects, however, are not guaranteed to cover the gene space entirely — i.e., genes that are not transcribed during sequencing will be missed subsequently by an EST based discovery process. For example, in Berkeley Drosophila Genome Project, only about 70% of the genes were covered by the cDNA/EST based gene discovery [82].

12.3.2 Gene Annotation and Alternative Splicing

ESTs can be used to annotate the structure of a gene through spliced alignment techniques [32, 73, 90]. Exonic and intronic boundaries within expressed genes can be marked using the alignment pattern of a gene sequence with an EST derived from it. EST based gene annotation has been a vibrant research area [3, 9, 36, 38, 60, 77, 95, 100].

Alternative splicing is the characteristic by which the transcriptional process alternatively selects or excludes exons and introns along a gene. This combinatorial process provides the biological means for a gene to transcribe for multiple mRNAs. Alternative splicing is a well-studied concept and is the subject of Chapter 16. As ESTs are derived from

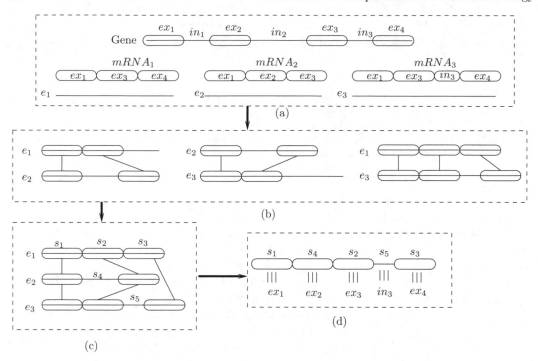

FIGURE 12.3: Deducing alternative splicing events of a gene through pairwise alignments between ESTs derived from it. (a) The unknown truth: A gene, represented as a series of exons (ex) and introns (in), transcribes for three different mRNAs. A subsequent sequencing generates three ESTs, e_1, e_2 and e_3, one from each mRNA, (b) The three ESTs are aligned pairwise allowing for insertion gaps in both sequences, (c) A layout of overlapping ESTs induced by the pairwise alignments (each unique shared segment is marked from s_1 to s_5), and (d) The shared segments are ordered to reflect their relative ordering along their putative gene source.

mRNAs, they provide a means to discover alternative splicing events of the underlying genes [12, 41, 54, 55, 56]. Figure 12.3 illustrates an example case, where an inference that the depicted gene transcribes for three different spliced variants (mRNAs) can be drawn through EST vs. EST alignments. The figure shows a rather optimistic scenario where the ESTs contain all information required to deduce the total order of the shared segments among them (Figure 12.3(d)). In practice, such information may be insufficient and as a result only a partial order can be inferred. For example, if only two mRNAs were available to start with, such that they can be represented as $ex_1.ex_2.ex_4$ and $ex_1.ex_3.ex_4$ respectively, then it is impossible to deduce the relative ordering between the segments ex_2 and ex_3 along the gene, regardless of the number of ESTs sequenced from the mRNAs. For a detailed discussion of algorithms and applications in alternative splicing studies, see Chapter 16.

12.3.3 Alternative Polyadenylation

Polyadenylation occurs during transcription and is the process by which an mRNA sequence is terminated at its $3'$ end. At the terminated end, the transcription process appends a repeat sequence of the nucleotide adenine (termed as a "polyA tail"), which plays important roles in the mRNA's stability and translation initiation. Alternate choice of polyadenylation sites results in corresponding variations at the mRNA ends and is considered an important post-transcriptional regulatory mechanism. ESTs sequenced from the 3' ends of the mRNAs are used to determine alternate polyadenylation sites in genes [31]. ESTs are first clustered and assembled into sequences representing the underlying mRNA transcript. While assembling, polyA discrepancies are detected in positions having additional evidence of conserved motifs for polyadenylation sites such as the hexamer AAUAAA, which are then recorded as possible sites of alternate polyadenylation.

12.3.4 Estimating Gene Counts

What is the number of genes in the human genome? Many research efforts have been in pursuit of an answer for this famous question for more than a decade now [20, 22, 25, 28, 46, 65, 83], and yet no consensus! Of the many techniques used to assess the number, one of the oldest is to approximate the answer to the count of unique human EST "clusters". EST based estimates ran up to over 100,000 genes only a few years ago [46]. However, around the same time another EST based estimate placed it around only 35,000 [25]. Most recently, the International Human Genome Sequencing Consortium announced a surprisingly smaller number between 20,000 and 25,000 [22]. While this is believed to be the most reliable estimate on the number of protein-coding genes [83], the discrepancies in various estimates have been a cause of concern over both EST based and other counting approaches [65].

EST based counting could result in an underestimate because not all genes in an organism may provide expression evidence. However, the gene count could also be inflated if ESTs from different alternatively spliced variants of the same gene or alternative polyadenylations of the same transcript are misplaced in different clusters. According to a study by Jean-Michel Claverie [20] on 82,000 UniGene clusters, alternative splicing and polyadenylation reduces the number of unique "genes" but still there are about 46,000 clusters that do not have protein-coding evidence. Comparing this with the most recent estimation of 20,000-25,000 protein-coding genes, this suggests that a majority of the EST based gene clusters represent non-protein (or just mRNA) coding genes.

12.3.5 Gene Expression Studies

Before the advent of the microarray technology, gene expression and co-regulation related studies were primarily dependent on EST data. During a sequencing experiment, the number of ESTs derived from an expressed gene is correlated to its expression level under the experimental conditions. In 1995, a technique called Serial Analysis of Gene Expression (SAGE) was developed based on the above philosophy [92]. For examples of EST based gene expression studies, see [26, 50]. For a review on different approaches to differential gene expression studies including EST based analysis, see [17]. In addition to expression profiling, ESTs are also used to design oligos for microarray chips [42, 99].

12.3.6 Single Nucleotide Polymorphisms

Single Nucleotide Polymorphisms (SNPs) are the most abundant class of genetic variation occurring almost every 1,200 *bp* along the human genome. SNPs are studied for mapping complex genetic traits. SNPs that occur in coding and regulatory sequences could alter the expression pattern or even the transcriptional behavior of the gene. SNPs have also been identified as causes for various diseases [21]. Such SNPs can be identified as nucleotide variations in assembled ESTs [30, 51, 67, 96]. However, these variations need to be distinguished from those variations seen among ESTs from paralogous genes, or occurring in ESTs due to sequencing errors; otherwise the SNP identification process may result in false predictions. This is usually accomplished by observing a probabilistic distribution that also takes into account the quality values of nucleotides in question. A large database of all identified SNPs is maintained by the NCBI (http://www.ncbi.nlm.nih.gov/projects/SNP/) and is called dbSNP [78]. Although a majority of the SNPs in this database are that of human and mouse, the database is open to SNPs from any species and occurring anywhere within its genome.

12.3.7 Physical Mapping

Sequence-Tagged Sites (STSs) are sequences that map to unique locations on a genome and are therefore used for physical mapping [61]. As ESTs are derived from coding regions along the genome, they can also serve the purpose of STSs as long as the underlying coding sequences themselves map to unique genomic locations. ESTs also have the added advantage of directly mapping to gene-rich regions [1, 10, 29, 52].

12.4 EST Databases

The sizes of publicly available EST databases continue to grow explosively. The dbEST portion (http://www.ncbi.nlm.nih.gov/dbEST/index.html) of the NCBI GenBank is used as a public repository for storing ESTs and full-length cDNAs generated by numerous sequencing efforts. The GenBank team, in its latest release (release 143) [5], reports that the dbEST database contains over 23.4 million ESTs spanning over 12 billion nucleotides, making it the largest public EST data repository. Also, the number of ESTs increased about 29% over the past year (2004) . About 740 organisms are represented in this database, and as of August 2005, human ESTs dominate the pool with about 6.1 million sequences, followed by mouse ESTs with 4.3 million sequences. Among plants, *Triticum aestivum* (wheat) has about 600,000 ESTs, and *Zea mays* (maize) has over 550,000 ESTs.

Besides dbEST, there are other EST databases that are smaller in size and have sequences generated for special purposes. The rat EST project at University of Iowa is one such project that has about 300,000 ESTs available (http://ratest.eng.uiowa.edu). RIKEN (http://www.riken.jp/engn/) hosts full-length cDNAs of many species. Its Mouse Gene Encyclopedia Project that aims at sequencing full-length enriched cDNA clones from various mouse tissues and inferring their chromosomal locations sequenced about 21,000 full-length mouse cDNAs [9]. Full-length cDNAs are also available for Arabidopsis [76] and human [62] catering towards functional annotation projects. The Prostate Expression Database (http://www.pedb.org/) [59], designed for the study of prostate gene expression in normal and disease states, contains ESTs and full-length cDNAs from over 40 human prostate cDNA libraries. The Cancer Genome Anatomy Project (http://cgap.nci.nih.gov/), an ongoing initiative by the National Cancer Institute to understand the anatomy and the functional genomics behind cancerous growths also sequences large collections of ESTs and full-length cDNAs from cancerous tissues, and uses them for gene expression profiling. ESTs are also sequenced as part of genome projects in order to catalog and annotate the genes of the species. For example, the Berkeley Drosophila Genome Project [71, 81] obtained a collection of about 370,000 *Drosophila* ESTs and full-length cDNAs from various tissues and life stages of the species for subsequent use in the project. For a larger but hardly comprehensive list of on-going EST projects, see Table 12.1.

Organism	Institution and Website Information
Arabidopsis	The Institute of Genomic Research www.tigr.org/tdb/e2k1/ath1
Barley	Clemson University Genomics Institute www.genome.clemson.edu/projects/barley
Cotton	Clemson University www.genome.clemson.edu/projects/cotton
Chicken	University of Manchester www.chick.umist.ac.uk
Drosophila	Drosophila Genome Center www.fruitfly.org/EST
Honeybee	University of Illinois, Urbana-Champaign titan.biotec.uiuc.edu/bee/honeybee_project.htm
Human	Washington University & Merck genomeold.wustl.edu/est/index.php?human_merck=1
Maize	Iowa State University www.maizegdb.org/documentation/mgdp/est/index.php schnablelab.plantgenomics.iastate.edu/research/genomics/htp_est
Mouse	Washington University genomeold.wustl.edu/est/index.php?mouse=1
Phaseolus coccineus	University of California, Los Angeles estdb.biology.ucla.edu/PcEST
Pig	Iowa State University pigest.genome.iastate.edu
Protist	University of Montreal megasun.bch.umontreal.ca/pepdb/pep.html
Rat	University of Iowa ratest.eng.uiowa.edu
Trypanosoma cruzi	Uppsala University www.genpat.uu.se/tryp/tryp.html
Xenopus tropicalis	Sanger Institute www.sanger.ac.uk/Projects/X_tropicalis
Zebrafish	Washington University zfish.wustl.edu

TABLE 12.1 A partial list of active EST projects with their URLs as of August 2005.

12.5 EST Clustering

EST clustering algorithms are at the core of most EST analysis projects and have been under continued research and development for improving both their quality and efficiency. Almost all applications described in Section 12.3 engage a clustering mechanism prior to extracting any useful information from EST data. In this section, we describe the EST clustering problem, algorithmic solutions, and ongoing efforts that apply these clustering utilities for maintaining cluster databases for ESTs derived from many organisms.

12.5.1 The Problem and Challenges

The main goal of clustering ESTs is to group ESTs based on their gene or transcript source. Clustering algorithms rely on several "patterns" expected of ESTs originating from the same source to identify those patterns in input set, and deduce relationships before arriving at the final clustering. These patterns include pairwise overlaps, and homology to a common protein or a common region in the genome.

Overlap Detection

Pairwise overlaps can be detected by computing alignments and other related measures. Many algorithms and techniques exist for computing alignments (see Chapter 1): BLAST [2] and its numerous variants (http://www.ncbi.nlm.nih.gov/BLAST/) or dynamic programming techniques [32, 33, 57, 79, 97]. The choice of an appropriate overlap detection scheme is dictated by the goal of clustering. If the goal is to cluster ESTs based on mRNA source, then a global or a semi-global alignment computation is sought because it detects suffix-prefix overlap expected out of two ESTs derived from an overlapping region on the mRNA transcript. However, if clustering by gene source is desired, then in addition to a suffix-prefix type of alignment there is a need to detect overlaps between ESTs derived from different alternatively spliced transcripts of the same gene. This is modeled as finding good local alignments corresponding to the regions containing shared exons. Similar dynamic programming techniques have been developed for detecting homology of ESTs with other types of biological sequences — homology between an EST and a protein/amino acid sequence is computed using cDNA-protein alignment techniques such as in [97], while homology with genomic regions can be computed as spliced alignments (described in Chapter 2). Overlaps can also be detected without computing alignments, as measures that compute word frequencies [11, 89]. While computing these measures may not model the problem accurately for sequence errors and expected patterns in overlaps, these techniques are usually sought as quicker alternatives to alignment based methods.

A fairly straightforward approach to clustering is to first choose the overlap detection scheme, run it on each pair of input sequences, and in the process form the clusters using only those pairs with a significant overlap. The main issue with this naive approach is that its scalability is limited by the quadratic increase in the number of pairs. This can be further aggravated by the high computation cost associated with detecting each overlap. For instance, the run-time complexity for aligning two sequences through a standard dynamic programming approach is proportional to the product of their lengths. Thus, a prime challenge in devising clustering algorithms is to be able to significantly reduce the run-time spent in detecting overlaps, and still obtain correct clustering that would have resulted had all pairs been considered. There are two independent ways of achieving this reduction: (i) reduce the cost of each pair computation by opting for a less rigorous and/or approximate method instead of aligning two sequences, and (ii) device faster methods to detect sequence pairs in advance that exhibit significant promise for a good alignment and then perform rigorous alignment only on those selected "promising pairs".

The nature of sampling inherent in EST data can add significantly to the computational complexity of the clustering process. Even if one were to devise a scheme that intelligently discards all non-overlapping pairs from overlap computation, the number of genuinely overlapping ESTs may still be overwhelming in practice. This is because the sequencing procedure may oversample the ends of the mRNA transcripts (giving them a deep coverage) while undersampling their mid-regions. The result is what we see in Figure 12.4, i.e., a vertical tiling of ESTs on a source mRNA transcript. Thus the number of genuinely overlapping pairs could grow at a quadratic rate as a function of the number of ESTs covering each transcript, which could be very high for transcripts arising from over-expressed genes. This raises a critical issue when dealing with large inputs containing hundreds of thousands to millions of ESTs, especially limiting the applicability of those software packages designed to handle only uniformly sampled sequence data (e.g., fragment assemblers).

Sequencing Errors and Artifacts

With current technology, even though the error rates are as low as 1-2%, it is important for a clustering algorithm to handle such errors in order to guarantee a high prediction accuracy. Handling errors such as an incorrectly interpreted, or included, or excluded nucleotide in a sequence is typically deferred to the overlap detection phase — by modeling such errors as mismatches, insertions and deletions in alignments. There are other types of errors and artifacts that can be detected at an earlier stage and such errors are typically handled in a "preprocessing" step prior to overlap detection:

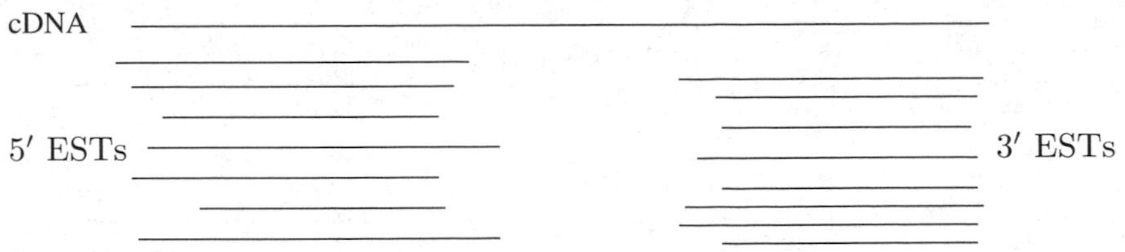

FIGURE 12.4: Non-uniform sampling of mRNA resulting from the EST sequencing procedure.

- During sequencing, the ESTs may get contaminated with the vector sequences adjoining the cDNA clones. These sequences are easy to detect because they are part of the known vector DNA sequence and are expected to occur at ends of ESTs. During preprocessing, such sequences are detected and removed.

- The sequencing procedure may also have ambiguously read some bases and may have marked such bases with low quality values. In the resulting ESTs, these bases are marked with special characters such as 'N' or 'X', so that they can be treated accordingly by a subsequent overlap detection scheme.

- ESTs derived from 3' ends of an mRNA usually retain portions of the mRNA's polyA tail. The presence of such polyA tails in ESTs may be of interest only to alternative polyadenylation related studies. In other studies, such regions are uninformative and if retained may only result in false overlaps. Thus as part of preprocessing, these polyA tails are trimmed off the ends of the ESTs.

- ESTs can also sometimes contain portions of chimeric cDNA clones and accurately detecting such sequencing artifacts is typically hard and deferred until later stages of analysis. If the genome sequence of the organism in question is available, then the results of aligning ESTs against the genome using spliced alignment techniques can reveal such artifacts as ESTs that have their segments aligning to two or more entirely distant locations along the genome. Their detection becomes much harder in the absence of the genome sequence. A common method is to flag those ESTs that "bridge" two otherwise distinct non-overlapping sets of ESTs. The problem with this approach is, however, that there could also be ESTs that genuinely bridge two ends of an mRNA transcript, and therefore this scheme may result in false labeling of such ESTs as with chimeric origins. The number of ESTs in the two distinct sets being bridged can also be taken into account to reduce the chance of a false prediction.

Natural Variations

If a pair of sequences overlap significantly but with a few mismatches and/or indels in their underlying best alignment(s), then there are two ways to explain such disagreements: (i) the underlying sequencing procedure incorrectly read the bases on one of the sequences, or (ii) the two ESTs being compared are from alleles or paralogous genes that have these natural variations because of mutations or single nucleotide polymorphisms. The choice between these two possibilities is made by looking at more than one overlapping pair at a time. For example, of the 10 overlapping sequences shown in Figure 12.5a, only one has a nucleotide that is different from the corresponding nucleotides in the other 9 sequences, indicating the high likelihood of a sequencing error that caused the variation in the singled out sequence. Figure 12.5b shows a different case where such a disagreement is equally distributed among

FIGURE 12.5: Overlap layout suggesting a case of a (a) sequencing error, and (b) natural variation.

the 10 sequences indicating that it is likely the result of a natural variation i.e., that the sequences were extracted from two different gene paralogues or polymorphic genes. The underlying assumption is that the probability of such a variation occurring at the same position evenly across multiple overlapping ESTs is too low to have occurred. If available, base quality values can also be used to enrich the quality of the detection mechanism.

12.5.2 Algorithms, Software and Clustering Projects

For more than a decade now, numerous software programs have been used for clustering ESTs. However, some of these programs were originally developed for solving the related problem of genome assembly. Both genome assembly and EST clustering rely on detecting overlapping pairs of sequences. While clustering only requires computing a partition based on significant overlaps, assemblers go one step further to build representative sequences called "contigs" for each resulting cluster. Nevertheless, the problem of clustering ESTs have other unique complications caused by alternatively spliced or polyadenylated variants, non-uniformly sampled multiple gene sources, etc. For example, it may not be appropriate to generate one contig sequence for a cluster that has ESTs from different alternatively spliced variants of the same gene. For these reasons, considerable effort has been spent in recent years on developing algorithms and software directed towards the EST clustering problem. The use of fragment assemblers is restricted to projects that cluster ESTs by their source transcripts.

In this section, we present an overview of algorithms in both classes. Among the assembler class of algorithms, we will discuss three programs, CAP3 [37], Phrap [34] and TIGR Assembler [84], which are popular among EST clustering community [47]; although in principle, any fragment assembly software can be used for clustering ESTs to the same effect. (For a survey on fragment assembly algorithms and software, see Chapter 8.) Among the algorithms designed specifically for EST clustering, we will discuss UniGene [68], STACK [19, 53], UIcluster [64], TGICL [66], PaCE [39] and xsact [48]. All EST clustering algorithms have three main phases: preprocessing, overlap detection, and cluster formation. The preprocessing step involves removing and/or masking uninformative portions of the ESTs as discussed in Section 12.5.1. It is the techniques used in overlap detection and cluster formation phases that differentiates one clustering algorithm from another.

TIGR Assembler

The TIGR Assembler [84] is one of the oldest fragment assembler programs and has been used in various EST clustering projects [58, 70, 72, 85]. The algorithm is as follows: Given an input of n ESTs, the overlap detection phase "evaluates" all $\binom{n}{2}$ pairs — for each pair, the algorithm identifies all fixed-length ($\sim$ 10 bp) exact matches and then considers only those "promising pairs" that have convincingly long stretches of such matches for further alignment computations. From the aligned pairs, the algorithm selects only those pairs with a satisfactory sequence similarity over the overlapping regions. The clusters are then formed by initially assigning one unique cluster for every sequence ("or seeds") that has "very small number" of overlaps and then iteratively merging clusters by considering the pairs in the decreasing order of their overlap quality. Storing and sorting overlaps implies an $O(n^2)$ space complexity. The run-time is dominated by the cost to align all the promising pairs identified by the algorithm.

Phrap

Phrap [34] starts its overlap detection phase by building a list of sequence pairs with fixed-length matches and then sorting the list such that all matches of the same pair are consecutively placed. For each such "promising pair", it computes an alignment band centered around the diagonal containing all matches and then computes a best alignment using a banded version of the Smith-Waterman technique [79]. If there are many matches, then the band of diagonals is made wider to include all the word matches. Using only those pairs with a band score above a certain desired threshold, a layout of overlaps is then constructed and subsequently a contig is constructed from the layout using only the portions of ESTs that have a high sequence quality (or "quality value"). Because of storing and sorting pairs with fixed-length matching substrings, this algorithm has a space complexity of $O(n^2)$. Even though this is the worst-case complexity, the likelihood of such a quadratic requirement is high for EST data because of the underlying non-uniformity in sampling, as shown in Figure 12.4. The run-time complexity is dominated by the cost to align all the promising pairs identified by the algorithm.

One of the unique features of Phrap is its strong dependence on quality values, as the contigs are not built on consensus but using the quality values of the bases in each position of the layout. As ESTs are typically not accompanied with quality values, the use of Phrap is known to generate inaccurate assemblies of ESTs (i.e., ESTs from different genes falsely collapsed to form one contig) [47]. The Phrap software is typically used in conjunction with two other helper packages — Phred for reading in quality trace files along with sequence files and assigning quality values for each input base, and Consed for viewing, editing, and finishing sequence assemblies created using Phrap.

CAP3

In the overlap detection phase, the CAP3 [37] algorithm detects pairs that show "promising" characteristics for good alignment, without having to enumerate all pairs, as follows: concatenate all input ESTs into one long string using a special delimiter character and then quickly identify high scoring chains of "segment pairs" within each sequence against the concatenated string. This is implemented through a lookup table approach, similar to the method in BLAST [2]. A "segment pair" is an alignment without gaps and is initially computed by looking at all exact matches of a specified fixed-length and extending these matches as far as possible in either direction. Only those pairs that have a chain score greater than a specified threshold value are later considered for global alignment computation [57]. The alignments are then considered in the decreasing order of their scores and an "overlap-layout" is constructed using the order and orientation of each aligning pair. In this greedy process, inconsistencies due to violating alignments can be resolved in favor of the higher scoring alignments. Such inconsistencies may arise from overlapping ESTs derived from different spliced variants of the same gene. The final step is to compute a multiple sequence alignment from each overlap-layout component, thereby resulting in a consensus "contig" corresponding to a putative cDNA transcript from which the composite ESTs were originally derived. The space complexity is worst-case quadratic because of storing and sorting all the promising pairs. The dominant run-time cost is that of aligning all the promising pairs.

An independent study conducted by researchers at TIGR [47] rated CAP3 better than Phrap and TIGR Assembler in the accuracy of the predicted EST clusters, in the absence of any base quality information. CAP3 has been applied in many EST clustering projects including the TIGR Gene Indices project [69].

UniGene

The UniGene project [68] undertaken by the NCBI is an initiative towards clustering all GenBank ESTs by organisms and by individual gene sources, i.e., ESTs from different spliced variants of the same gene are also clustered together. The UniGene clustering scheme performs incremental daily processing of ESTs submitted to the dbEST database, computed as BLAST alignments of each new EST with the contents of all individual clusters. Care is taken that each cluster contains at least one EST derived from the 3′ terminus of the source mRNA transcript. This is ascertained by the presence of a polyA tail in the corresponding EST(s) (which is not removed as part of its preprocessing step). Even though the run-time of the UniGene method is worst-case quadratic in the number of ESTs, incremental processing in batches allows for quick updating of clusters as new sequences are added to the database.

The entire UniGene cluster database is accessible online at the URL http://www.ncbi.nlm.nih.gov/UniGene/. As of September 2005, the database contains over 700,000 gene-oriented sequence clusters representing over 50 organisms, with the human and mouse collections leading the chart with 53,100 and 42,555 UniGene clusters respectively [94].

STACK

STACK (Sequence Tag Alignment and Consensus Knowledgebase) [19, 53] is one of the first EST clustering programs and was developed to achieve tissue-specific clustering that groups ESTs by transcript source. The underlying algorithm performs simple all-versus-all pairwise comparisons with the overlap between each pair detected through a word-multiplicity measure called d^2, a distance measure to assess sequence dissimilarities. Subsequently, the pairs with significantly small distances are used to form the clusters by an agglomerative approach called d^2_cluster [86], as follows: initially, each input EST occupies a cluster of its own, and as the program progresses each significant overlap merges the corresponding clusters forming a supercluster. This mechanism achieves what is called a transitive closure clustering, in which two entirely different sequences are brought together because of a common third sequence with which each share a good overlap. This implies that the clustering procedure results in grouping ESTs by their putative gene sources. Each cluster is post-processed by the Phrap assembler to build transcript assemblies. During clustering, additional information such as clone pairs and tissue-specific information specified by the user are incorporated into the clusters. The STACK algorithm has a run-time complexity that is proportional to the product of $\binom{n}{2}$ and the time taken to compute d^2 measure.

Because of its simplicity, the STACK algorithm is also easily parallelized [16]. The all-pairs work is distributed evenly across processors, and the clustering results are collected and recorded serially by one processor. For an example of STACK's application, see [91].

TGICL

The TGICL clustering software [66] was developed by TIGR. The algorithm achieves clustering by performing an all-versus-all pairwise alignment but using a greedy alignment algorithm called *megablast* [98]. The advantage of using *megablast* is that it provides a significant speedup ($\sim$ 10 times) while aligning two "similar" sequences over its dynamic programming counterparts, although the run-time increases as the similarity decreases. For large clusters of ESTs, this approach is expected to be beneficial when compared to dynamic programming techniques, because of the high source coverage expected of EST data. Because of its simplicity, this algorithm can also be easily parallelized. Post-clustering, CAP3 is used for assembling the sequences of each cluster into putative transcript(s).

TIGR maintains a large database of clustered ESTs called the "TIGR Gene Indices". The initiative is towards maintaining a compendium of transcriptomes of several organisms. The database has clusters built for ESTs collected from over 32 animal species including the human and mouse, and over 33 plant species including wheat and maize.

UIcluster

The UIcluster method [64] was originally developed for clustering 3' generated ESTs into 3' transcripts. The algorithm is based on the following incremental approach: Initially, each EST is in its own cluster. At any point of execution, a list of "representative ESTs" is maintained for each cluster (typically its longest EST(s)). A global hash table is constructed by preprocessing all input ESTs, such that it indexes all fixed-length (<16 *bp*) substrings within all ESTs. The ESTs are then considered one at a time. For a given EST, all clusters with at least one representative EST that has at least a specified number of fixed-length matches are identified. Alignment computations are then performed between the input EST and each of the representative ESTs identified in each cluster. The input EST (and its cluster) is then merged into one of the clusters containing the best overlapping representative, provided that best alignment(s) pass the specified similarity threshold; otherwise the clusters are left intact. The space complexity is proportional to the size of input plus the size of the global hash table. The worst case run-time complexity is proportional to $O(n^2)$ multiplied by the average cost to align two sequences; for large clusters, the run-time is likely to be close to the worst-case behavior, as all potential cluster merges are evaluated through alignments with ESTs considered one at a time.

A parallel version of the UIcluster algorithm has also been developed [88]. The input ESTs and the initial set of clusters are evenly partitioned across processors, and each processor constructs the hash table for its local portion of the ESTs. The algorithm then performs one parallel step for each input EST, in which the sequential algorithm is run locally on each processor and the cluster to which the EST has to be merged with is decided through a collective communication at the end of the step. There are two main issues with this parallel approach: (i) the number of parallel steps is proportional to the number of input ESTs, independent of the number of processors used, and (ii) the speedup achieved in each step is determined by the processor with the most number of alignments to compute. The program is extensively used on clustering rat ESTs (http://ratest.eng.uiowa.edu).

PaCE

The PaCE (for "<u>Pa</u>rallel <u>C</u>lustering of <u>E</u>STs") algorithm [39, 40] was primarily developed to cluster large-scale EST data using parallel processing. The salient features of the algorithm are as follows: 1) Promising pairs are detected based on "maximal matches", i.e., variable-length matches that cannot be extended on either side to result in a longer match. The advantage over algorithms based on fixed-length (say w bp) matches is that this algorithm can detect a long match of length l directly instead of detecting it as a chain of $l - w + 1$ shorter matches. The generalized suffix tree data structure [35] is used for pair generation. 2) The clustering scheme is agglomerative, i.e., each EST is initially in a cluster of its own, and as promising pairs are evaluated for alignments the clusters are updated. 3) The promising pairs are generated on-demand in decreasing order of their maximal match lengths without requiring to sort or store these pairs, and the clustering algorithm selects only those pairs for alignment evaluation that have not been clustered together. This greedy heuristic provides an effective means to significantly reduce the number of pairs considered for alignment work. For example, studies with 168,200 Arabidopsis ESTs show that of the ~7 million pairs generated as promising pairs only under 1.5 million were aligned, achieving about 80% reduction in the workload. The worst case space complexity is linear in the input size, while the run-time is dominated by the cost of aligning those promising pairs selected by the heuristic. Parallelization studies have demonstrated a linear speedup with processor size.

More details of this algorithm and its applications are presented in Chapter 13. The PaCE software is used to cluster and maintain over 41 plant EST collections (largest being that of about 590,000 wheat ESTs) in the PlantGDB database [23, 73] (www.plantgdb.org).

xsact

Concurrent to the development of PaCE, Malde *et al.* [48] developed *xsact*, which is a serial program for EST clustering that also generates promising pairs based on maximal matches. The *xsact* algorithm first constructs a generalized suffix array on the input ESTs. This is achieved by recursively sorting prefixes, similar to the approach in [49] — this algorithm was developed prior to the development of linear time algorithms for directly constructing suffix arrays [43, 44, 45]. The algorithm then detects each pair with a maximal match of length l bp, multiple ($\sim l$) times, but reports only one instance of it to the alignment module. The pairs are generated in no particular order, and all reported pairs are aligned. Only those alignments which satisfy a specified similarity threshold are stored. The pairs are then sorted in decreasing order of their alignment scores and considered in that order for cluster merges. The space complexity of the algorithm is dominated by the number of pairs that have satisfactory alignments, which within each generated EST cluster is worst case quadratic in its number of ESTs. The run-time is dominated by the cost to align all promising pairs reported.

Given the complexities of EST data, ensuring both high quality and high efficiency in a clustering method is the primary challenge faced by an algorithm designer. Even though run-time intensive, alignment based modeling provides the most accurate means to account for sequencing errors, mutations, SNPs, chimeras, alternatively spliced variants, etc. An approach that performs an all-versus-all alignment comparisons, therefore, has enough information to generate an accurate clustering based on overlap information. However, computing alignments for all pairs of ESTs is not a scalable solution. One way to overcome this problem is to resort to faster methods of detecting overlaps. A more efficient approach is to reduce the alignment work without sacrificing quality. In order to reduce the number of pairs to be considered for alignment, the most popular technique used is to preprocess the ESTs such that the algorithm can identify those pairs that show significant promise for a potential good overlap. If two sequences have a good overlap, then they are also expected to contain correspondingly "long" exact matches. Thus, there are algorithms that identify pairs with fixed-length or variable-length exact matches. While both schemes reduce the number of aligned pairs, the variable-length schemes can provide a quicker route to generate these pairs. For example, after a linear time preprocessing, it is possible to generate a pair with a long maximal match in amortized constant time rather than generating it as a chain of many smaller fixed-length matches. Further savings in run-time can be achieved by taking advantage of current clustering at any stage of the algorithm and discarding those promising pairs that have both sequences clustered.

12.6 Conclusions

EST data is highly resourceful and valuable in the advancement of molecular and functional genomics. Numerous genome projects have directly benefited by analyzing EST data towards gene discovery, gene annotation and structure prediction, SNP identification, physical mapping, and gene counting studies. ESTs are also used extensively in gene expression studies and alternative splicing and polyadenylation studies. Clustering methods play a key role in analyzing EST data. While the purpose of these methods is to extract interesting biological inferences from the EST data, there is a growing emphasis on devising computationally efficient methods. EST databases have been growing at alarming rates because of cost-effective high-throughput sequencing strategies. In addition, clustering methods face other difficulties imposed by characteristics that are inherent in EST collections such as non-uniform sampling, alternatively spliced variants, differential gene expression, ESTs from paralogous gene sources, etc. Considerable effort is therefore being spent on developing clever algorithmic techniques that use high-performance computing resources for efficiently analyzing large EST data, and such methods are the subject of the next chapter.

Acknowledgements

This work was supported in part by the U.S. National Science Foundation under ACR-0203782.

References

[1] M.D. Adams, J.M. Kelley, J.D. Gocayne, and M. Dubnick *et al.* Complementary DNA sequencing: expressed sequence tags and human genome project. *Science*, 252(5013):1651–1656, 1991.

[2] S.F. Altschul, W. Gish, W. Miller, and E.W. Myers *et al.* Basic local alignment search tool. *Journal of Molecular Biology*, 215:403–410, 1990.

[3] L.C. Bailey, D.B. Searls, and G.C. Overton. Analysis of EST-Driven Gene Annotation in Human Genomic Sequence. *Genome Research*, 8(4):362–376, 1998.

[4] M.F. Baldo, G. Lennon, and M.B. Soares. Normalization and subtraction: Two approaches to facilitate gene discovery. *Genome Research*, 6:791–806, 1996.

[5] D.A. Benson, I. Karsch-Mizrachi, D.J. Lipman, and J. Ostell *et al.* GenBank. *Nucleic Acids Research*, 3:D34–D38, 2005.

[6] M. Boguski and G. Schuler. ESTablishing a human transcript map. *Nature Genetics*, 10(11):369–371, 1995.

[7] M.S. Boguski, T.M. Lowe, and C.M. Tolstoshev. dbEST - database for "expressed sequence tags". *Nature Genetics*, 4(4):332–333, 1993.

[8] M.S. Boguski, C.M. Tolstoshev, and D.E. Bassett Jr. Gene discovery in dbEST. *Science*, 265(5181):1993–1994, 1994.

[9] H. Bono, T. Kasukawa, M. Furuno, and Y. Hayashizaki *et al.* FANTOM DB: database of Functional Annotation of RIKEN Mouse cDNA Clones. *Nucleic Acids Research*, 30(1):116–118, 2002.

[10] K.P. Brady, L.B. Rowe, H. Her, and T.J. Stevens *et al.* Genetic mapping of 262 loci derived from expressed sequences in a murine interspecific cross using single-strand conformational polymorphism analysis. *Genome Research*, 7(11):1085–1093, 1997.

[11] J. Burke, D. Davison, and W.A. Hide. d2_cluster: A validated method for clustering EST and full-length cDNA sequences. *Genome Research*, 9(11):1135–1142, 1999.

[12] J. Burke, H. Wang, W. Hide, and D.B. Davison. Alternative gene form discovery and candidate gene selection from gene indexing projects. *Genome Research*, 8(3):276–290, 1998.

[13] A.A. Camargo, H.P.B. Samaia, E. Dias-Neto, and D.F. Simao *et al.* From the Cover: The contribution of 700,000 ORF sequence tags to the definition of the human transcriptome. *Proceedings of the National Academy of Sciences USA*, 98(21):12103–12108, 2001.

[14] P. Carninci, K. Waki, T. Shiraki, and H. Konno *et al.* Targeting a Complex Transcriptome: The Construction of the Mouse Full-Length cDNA Encyclopedia. *Genome Research*, 13(6):1273–1289, 2003.

[15] H. Caron, B. Schaik, M. Mee, and F. Baas *et al.* The human transcriptome map: Clustering of highly expressed genes in chromosomal domains. *Science*, 291(5507):1289–1292, 2001.

[16] J.E. Carpenter, A. Christoffels, Y. Weinbach, and W.A. Hide. Assessment of the parallelization approach of d2_cluster for High Performance Sequence Clustering. *Journal of Computational Chemistry*, 23(7):755–757, 2002.

[17] J.P. Carulli, M. Artinger, P.M. Swain, and C.D. Root *et al.* High throughput analysis of differential gene expression. *Journal of Cellular Biochemistry*, 72(S30-31):286–296, 1999.

[18] P. Chomczynski and N. Sacchi. Single-step method of RNA isolation by acid guanidinium thiocyanate-phenol-chloroform extraction. *Analytical Biochemistry*, 162(1):156–159, 1987.

[19] A. Christoffels, A.V. Gelder, G. Greyling, and R. Miller *et al.* STACK Sequence Tag Alignment and Consensus Knowledgebase. *Nucleic Acids Research*, 29(1):234–238, 2001.

[20] J. Claverie. What if there are only 30,000 human genes? *Science*, 291(5507):1255–1257, 2001.

[21] F.S. Collins, M.S. Guyer, and A. Chakravarti. Variations on a theme: cataloging human DNA sequence variation. *Science*, 278(5343):1580–1581, 1997.

[22] International Human Genome Sequencing Consortium. Finishing the euchromatic sequence of the human genome. *Nature*, 431:931–945, 2004.

[23] Q. Dong, S.D. Schlueter, and V. Brendel. PlantGDB, plant genome database and analysis tools. *Nucleic Acids Research*, 32:D354–D359, 2004.

[24] J.R. Duguid and M.C. Dinauer. Library subtraction of in vitro cDNA libraries to identify differentially expressed genes in scrapie infection. *Nucleic Acids Research*, 18(9):2789–2792, 1990.

[25] B. Ewing and P. Green. Analysis of expressed sequence tags indicates 35,000 human genes. *Nature Genetics*, 25:232–234, 2000.

[26] R.M. Ewing, A.B. Kahla, O. Poirot, and F. Lopez *et al.* Large-Scale Statistical Analyses of Rice ESTs Reveal Correlated Patterns of Gene Expression. *Genome Research*, 9(10):950–959, 1999.

[27] J. Fargnoli, N.J. Holbrook, and A.J. Fornace Jr. Low-ratio hybridization subtraction. *Analytical Biochemistry*, 187(2):364–373, 1990.

[28] C. Fields, M.D. Adams, O. White, and J.C. Venter. How many genes in the human genome? *Nature Genetics*, 7:345–346, 1994.

[29] A.K. Fridolfsson, T. Hori, A.K. Winter, and M. Fredholm *et al.* Expansion of the pig comparative map by expressed sequence tags (EST) mapping. *Mammalian Genome*, 8(12):907–912, 1997.

[30] K. Garg, P. Green, and D.A. Nickerson. Identification of candidate coding region single nucleotide polymorphisms in 165 human genes using assembled expressed sequence tags. *Genome Research*, 9(11):1087–1092, 1999.

[31] D. Gautheret, O. Poirot, F. Lopez, and S. Audic *et al.* Alternate Polyadenylation in Human mRNAs: A Large-Scale Analysis by EST Clustering. *Genome Research*, 8(5):524–530, 1998.

[32] M S. Gelfand, A. Mironov, and P.A. Pevzner. Gene recognition via spliced alignment. *Proceedings of the National Academy of Sciences USA*, 93(17):9061–9066, 1996.

[33] O. Gotoh. An improved algorithm for matching biological sequences. *Journal of Molecular Biology*, 162(3):705–708, 1982.

[34] P. Green. Phrap - the assembler. *http//www.phrap.org*, 1994, (Accessed Oct 2005).

[35] D. Gusfield. *Algorithms on strings, trees and sequences Computer Science and Computational Biology.* Cambridge University Press, Cambridge, London, 1997.

[36] X. Huang, M.D. Adams, H. Zhou, and A.R. Kerlavage. A tool for analyzing and annotating genomic sequences. *Genomics*, 46:37–45, 1997.

[37] X. Huang and A. Madan. CAP3: A DNA sequence assembly program. *Genome Research*, 9(9):868–877, 1999.

[38] J. Jiang and H.J. Jacob. EbEST: An Automated Tool Using Expressed Sequence Tags to Delineate Gene Structure. *Genome Research*, 8(3):268–275, 1998.

[39] A. Kalyanaraman, S. Aluru, V. Brendel, and S. Kothari. Space and time efficient parallel algorithms and software for EST clustering. *IEEE Transactions on Parallel and Distributed Systems*, 14(12):1209–1221, 2003.

[40] A. Kalyanaraman, S. Aluru, S. Kothari, and V. Brendel. Efficient clustering of large EST data sets on parallel computers. *Nucleic Acids Research*, 31(11):2963–2974, 2003.

[41] Z. Kan, E.C. Rouchka, W.R. Gish, and D.J. States. Gene Structure Prediction and Alternative Splicing Analysis Using Genomically Aligned ESTs. *Genome Research*, 11(5):889–900, 2001.

[42] T. Kapros, A.J. Robertson, and J.H. Waterborg. A simple method to make better probes for short DNA fragments. *Molecular Biotechnology*, 2(1):95–98, 1994.

[43] J. Karkkainen and P. Sanders. Simple linear work suffix array construction. *Lecture Notes in Computer Science*, 2719:943–955, 2003.

[44] D.K. Kim, J.S. Sim, H. Park, and K. Park. Linear-time construction of suffix arrays. *Lecture Notes in Computer Science*, 2676:186–199, 2003.

[45] P. Ko and S. Aluru. Space efficient linear time construction of suffix arrays. *Lecture Notes in Computer Science*, 2676:200–210, 2003.

[46] F. Liang, I. Holt, G. Pertea, and S. Karamycheva *et al.* Gene index analysis of the human genome estimates approximately 120,000 genes. *Nature Genetics*, 25:239–240,

2000.

[47] F. Liang, I. Holt, G. Pertea, and S. Karamycheva *et al.* An optimized protocol for analysis of EST sequences. *Nucleic Acids Research*, 28(18):3657–3665, 2000.

[48] K. Malde, E. Coward, and I. Joassen. Fast sequence clustering using a suffix array algorithm. *Bioinformatics*, 19(10):1221–1226, 2003.

[49] U. Manber and E. Myers. Suffix arrays a new method for on-line string searches. *SIAM Journal on Computing*, 22(5):935–948, 1993.

[50] M. Mao, G. Fu, J.S. Wu, and Q.H. Zhang *et al.* Identification of genes expressed in human CD34+ hematopoietic stem/progenitor cells by expressed sequence tags and efficient full-length cDNA cloning. *Proceedings of the National Academy of Sciences USA*, 95(14):8175–8180, 1998.

[51] G.T. Marth, I. Korf, M.D. Yandell, and R.T. Yeh *et al.* A general approach to single-nucleotide polymorphism discovery. *Nature Genetics*, 23:452–456, 1999.

[52] L.C. McCarthy, J. Terrett, M.E. Davis, and C.J. Knights *et al.* A first-generation whole genome-radiation hybrid map spanning the mouse genome. *Genome Research*, 7(12):1153–1161, 1997.

[53] R.T. Miller, A.G. Christoffels, C. Gopalakrishnan, and J. Burke *et al.* A Comprehensive Approach to Clustering of Expressed Human Gene Sequence: The Sequence Tag Alignment and Consensus Knowledge Base. *Genome Research*, 9(11):1143–1155, 1999.

[54] A.A. Mironov, J.W. Fickett, and M.S. Gelfand. Frequent alternative splicing of human genes. *Genome Research*, 9(12):1288–1293, 1999.

[55] B. Modrek and C. Lee. A genomic view of alternative splicing. *Nature genetics*, 30:13–19, 2002.

[56] B. Modrek, A. Resch, C. Grasso, and C. Lee. Genome-wide detection of alternative splicing in expressed sequences of human genes. *Nucleic Acid Research*, 29(13):2850–2859, 2001.

[57] S.B. Needleman and C.D. Wunsch. A general method applicable to the search for similarities in the amino acid sequence of two proteins. *Journal of Molecular Biology*, 48:443–453, 1970.

[58] M.A Nelson, S. Kang, E.L. Braun, and M.E. Crawford *et al.* Expressed sequences from conidial, mycelial, and sexual stages of *Neurospora crassa*. *Fungal Genetics and Biology*, 21:348–363, 1997.

[59] P.S. Nelson, C. Pritchard, D. Abbott, and N. Clegg. The human (PEDB) and mouse (mPEDB) Prostate Expression Databases. *Nucleic Acids Research*, 30(1):218–220, 2002.

[60] Y. Okazaki, M. Furuno, T. Kasukawa, and J. Adachi *et al.* Analysis of the mouse transcriptome based on functional annotation of 60,770 full-length cDNAs. *Nature*, 420:563–573, 2002.

[61] M. Olson, L. Hood, C. Cantor, and D. Botstein. A common language for physical mapping of the human genome. *Science*, 245(4925):1434–1435, 1989.

[62] T. Ota, Y. Suzuki, T. Nishikawa, and T. Otsuki *et al.* Complete sequencing and characterization of 21,243 full-length human cDNAs. *Nature Genetics*, 36:40–45, 2004.

[63] S.R. Patanjali, S. Parimoo, and S.M. Weissman. Construction of a uniform-abundance (normalized) cDNA library. *Proceedings of the National Academy of Sciences USA*, 88(5):1943–1947, 1991.

[64] K. Pedretti. Accurate, parallel clustering of EST (gene) sequences. *Masters' Thesis, University of Iowa*, 2001.

[65] E. Pennisi. Gene counters struggle to get the right answer. *Science*, 301(5636):1040–

1041, 2003.

[66] G. Pertea, X. Huang, F. Liang, and V. Antonescu *et al.* TIGR Gene Indices clustering tool (TGICL) a software system for fast clustering of large EST datasets. *Bioinformatics*, 19(5):651–652, 2003.

[67] L. Picoult-Newberg, T.E. Ideker, M.G. Pohl, and S.L. Taylor *et al.* Mining SNPs From EST Databases. *Genome Research*, 9(2):167–174, 1999.

[68] J.U. Pontius, L. Wagner, and G.D. Schuler. UniGene a unified view of the transcriptome. *The NCBI Handbook*, 2003.

[69] J. Quackenbush, F. Liang, I. Holt, and G. Pertea *et al.* The TIGR gene indices reconstruction and representation of expressed gene sequences. *Nucleic Acids Research*, 28(1):141–145, 2000.

[70] S.D. Rounsley, A. Glodek, G. Sutton, and M.D. Adams *et al.* The construction of Arabidopsis expressed sequence tag assemblies. *Plant Physiology*, 112:1177–1183, 1996.

[71] G.M. Rubin, L. Hong, P. Brokstein, and M. Evans-Holm *et al.* A *Drosophila* Complementary DNA Resource. *Science*, 287(5461):2222–2224, 2000.

[72] Y. Satou, L. Yamada, Y. Mochizuki, and N. Takatori *et al.* A cDNA resource from the Basal Chordate *Ciona intestinalis*. *Genesis*, 33:153–154, 2002.

[73] S.D. Schlueter, Q. Dong, and V. Brendel. GeneSeqer@PlantGDB: gene structure prediction in plant genomes. *Nucleic Acids Research*, 31(13):3597–3600, 2003.

[74] D.W. Schmid and C. Girou. Cloning of cDNA derived from mRNA of the electric lobe of Torpedo marmorata and selection of putative cholinergic-specific sequences. *Journal of Neurochemistry*, 48(1):307–312, 1987.

[75] C.W. Schweinfest, K.W. Henderson, J.R. Gu, and S.D. Kottaridis *et al.* Subtraction hybridization cDNA libraries from colon carcinoma and hepatic cancer. *Genetic Analysis : Techniques and Applications*, 7(3):64–70, 1990.

[76] M. Seki, P. Carninci, Y. Nishiyama, and Y. Hayashizaki *et al.* High-efficiency cloning of *Arabidopsis* full-length cDNA by biotinylated CAP trapper. *Plant Journal*, 15(5):707–720, 1998.

[77] M. Seki, M. Narusaka, A. Kamiya, and J. Ishida *et al.* Functional annotation of a full-length Arabidopsis cDNA collection. *Science*, 296(5565):141–145, 2002.

[78] S.T. Sherry, M.H. Ward, M. Kholodov, and J. Baker *et al.* dbSNP: the NCBI database of genetic variation. *Nucleic Acids Research*, 29(1):308–311, 2001.

[79] T.F. Smith and M.S. Waterman. Identification of common molecular subsequences. *Journal of Molecular Biology*, 147:195–197, 1981.

[80] M.B. Soares, M.F. Bonaldo, P. Jelene, and L. Su *et al.* Construction and characterization of a normalized cDNA library. *Proceedings of the National Academy of Sciences USA*, 91(20):9228–9232, 1994.

[81] M. Stapleton, J. Carlson, P. Brokstein, and C. Yu *et al.* A Drosophila full-length cDNA resource. *Genome Biology*, 3(12):research0080.1–0080.8, 2002.

[82] M. Stapleton, G. Liao, P. Brokstein, and L. Hong *et al.* The Drosophila Gene Collection: Identification of Putative Full-Length cDNAs for 70% of D. melanogaster Genes. *Genome Research*, 12(8):1294–1300, 2002.

[83] L.D. Stein. Human genome: End of the beginning. *Nature*, 431:915–916, 2004.

[84] G. Sutton, O. White, M. Adams, and A. Kerlavage. TIGR assembler: A new tool for assembling large shotgun sequencing projects. *Genome Science and Technology*, 1:9–19, 1995.

[85] C. Ton, D.M. Hwang, A.A. Dempsey, and H. Tang *et al.* Identification, characterization, and mapping of expressed sequence tags from an embryonic zebrafish heart *cdna* library. *Genome Research*, 10(12):1915–1927, 2000.

[86] D.C. Torney, C. Burks, D. Davison, and K.M. Sirotkin. *Computers and DNA*. Addison-Wesley, New York, 1990.

[87] G.H. Travis and J.G. Sutcliffe. Phenol emulsion-enhanced DNA-driven subtractive cDNA cloning: isolation of low-abundance monkey cortex-specific mRNAs. *Proceedings of the National Academy of Sciences USA*, 85(5):1696–1700, 1988.

[88] N. Trivedi, J. Bischof, S. Davis, and K. Pedretti *et al*. Parallel creation of non-redundant gene indices from partial mRNA transcripts. *Future Generation Computer Systems*, 18:863–870, 2002.

[89] E. Ukkonen. Approximate string-matching with q-grams and maximal matches. *Theoretical Computer Science*, 92(1):191–211, 1992.

[90] J. Usuka, W. Zhu, and V. Brendel. Optimal spliced alignment of homologous cDNA to a genome database ZmDB. *Bioinformatics*, 16(3):203–211, 2000.

[91] V. VanBuren, Y. Piao, D.B. Dudekula, and Y. Qian *et al*. Assembly, verification, and initial annotation of the NIA mouse 7.4K cDNA clone set. *Genome Research*, 12(12):1999–2003, 2002.

[92] V.E. Velculescu, L. Zhang, B. Vogelstein, and K.W. Kinzler. Serial analysis of gene expression. *Science*, 270(5235):484–487, 1995.

[93] R.E. Verdun, N.D. Paolo, T.P. Urmenyi, and E. Rondinelli *et al*. Gene Discovery through Expressed Sequence Tag Sequencing in *Trypanosoma cruzi*. *Infection and Immunity*, 66(11):5393–5398, 1998.

[94] D.L. Wheeler, T. Barrett, D.A. Benson, and S.H. Bryant *et al*. Database resources of the National Center for Biotechnology Information. *Nucleic Acids Research*, 33:D39–D45, 2005.

[95] C.W. Whitfield, M.R. Band, M.F. Bonaldo, and C.G. Kumar *et al*. Annotated expressed sequence tags and cDNA microarrays for studies of brain and behavior in the honey bee. *Genome Research*, 12(4):555–566, 2002.

[96] Z. Ye and J.M. Parry. The discovery and confirmation of single nucleotide polymorphisms in the human p53R2 gene by EST database analysis. *Mutagenesis*, 17(5):361–364, 2002.

[97] Z. Zhang, W.R. Pearson, and W. Miller. Aligning a DNA sequence with a protein sequence. *Journal of Computational Biology*, 4(3):339–349, 1997.

[98] Z. Zhang, S. Shwartz, L. Wagner, and W. Miller. A greedy algorithm for aligning DNA sequences. *Journal of Computational Biology*, 7(1-2):203–214, 2000.

[99] T. Zhu and X. Wang. Large-Scale Profiling of the Arabidopsis Transcriptome. *Plant Physiology*, 124:1472–1476, 2000.

[100] W. Zhu, S.D. Schlueter, and V. Brendel. Refined annotation of the *Arabidopsis thaliana* genome by complete Expressed Sequence Tag mapping. *Plant Physiology*, 132:469–484, 2003.

13

Algorithms for Large-Scale Clustering and Assembly of Biological Sequence Data

Scott J. Emrich
Iowa State University

Anantharaman Kalyanaraman
Iowa State University

Srinivas Aluru
Iowa State University

13.1   Introduction

Each living cell contains its own "cookbook" called a *genome* that contains recipes called *genes* required for biological processes and phenotypic characteristics. In the past, advancements in molecular biology largely focused on the study of individual genes. For example, unusual phenotypes have been used to isolate corresponding genomic regions using classical genetic mapping techniques. When applied to traditional genetics, this experiment is called a *forward screen*. Genome projects have since led to a paradigm shift in this field — after a complete book is obtained, recipes are deciphered using *reverse screens*. Under this paradigm unusual phenotypes can be generated by disabling a predicted gene and observing possible effects. Because the availability of a complete genome also accelerates the design and analysis of high-throughput experiments, such as those involving microarrays, data about large collections of genes are also now more quickly available.

The challenging aspect of genome projects, however, is their cost in both financial and computational resources. This is a consequence of current experimental limitations; biochemical procedures collectively known as *sequencing* are capable of accurately reading only hundreds of bases from a DNA molecule ($\approx 500 - 1000$ bp). To extend the reach of sequencing to entire genomes, numerous short fragments are sequenced from randomly distributed locations of a larger sequence. These fragments are then combined to form the original sequence through the computational process of *assembly*.

Note that if the larger sequence is the original genome, this fragmentation process is called *whole-genome shotgun* (WGS) sequencing, which was used to complete the earliest genome sequence of bacteria phage λ [26] and smaller genomes such as the bacterial sequences completed in the mid to late 1990s. Concurrent advances in high-throughput sequencing technology and assembly algorithms over the past decade have led to the assembly of increasingly larger and more complex genomes. Nearly twenty years after the completion of the first genome sequence, drafts of the much larger human genome — over sixty-two thousand times larger than λ — were reported by both public [6] and private [30] consortiums and is the subject of Chapter 9. Many genomes are under consideration for large-scale sequencing in the near future, and we now have genome projects either completed or in-progress for Arabidopsis, chimpanzee, dog, fruit fly, mosquito, mouse, nematode, pig, rat, rice, sea urchin, and zebrafish among others. Many of these eukaryotic projects, along with almost every bacterial genome sequenced, either used shotgun sequencing alone or in tandem with other methods.

Despite rapid advances in hardware speeds and memory capacities over the same period, assembling the tens of millions of fragments typical of such large genome projects places enormous demands on computational resources. For example, Celera Genomics estimated that it would take tens of thousands of CPU hours and approximately 600 GB of RAM to assemble the $\sim$3,000 million base human genome sequence based on their previous fruit fly assembly [30]. To reduce these computational requirements, Celera used ten 4-processor SMP clusters with 4 GB memory each along side a 16-processor NUMA machine with 64 GB shared memory, and engineered an incremental method that reduced the memory required to 28 GB before using 20,000 total CPU hours to assemble their 27.27 million shotgun fragments. Other large assembly projects have also been carried out in a similar fashion — by specialized teams running software developed for serial assembly on high-end workstations with tens of gigabytes of main memory while using rudimentary means to partition the workload for carrying out the assembly incrementally and/or concurrently.

A primary contributor to the high memory and run-time requirements of these conventional assembly techniques is the approach taken. As if assembling a massive jigsaw puzzle, fragments are pieced together using the information preserved between overlapping fragments derived from neighboring regions. In the case of human genome assembly above, a dominant part of the computation time was attributed to identifying such overlaps, which was parallelized to complete in 4-5 days (10,000 total CPU hours). While overlap computation poses run-time concerns, it also dictates the memory requirements of conventional assembly techniques because overlaps are stored before generating the final assembly.

Under the uniform sampling that is used for genomes sequenced by WGS, the number of overlaps is expected to be proportional to the number of sequenced fragments; therefore, the memory utilized scales linearly even though it might be on the order of hundreds of gigabytes for large genome sequences such as human. This scenario is changing, however, since alternative strategies [24, 32] were used to selectively sequence gene-rich regions of the maize genome. The non-uniform sampling involved in these approaches imposes a much higher memory requirement because it scales quadratically; on the other hand, the ability to efficiently handle such data will be important in sequencing other important plants including crops such as wheat and barley.

Although the invention of gene-enriched genome sequencing is a relatively new development, biologists have been selectively sequencing genes since the early 1990s by isolating messenger RNA (mRNA) within cells, which corresponds to a subset of the expressed portion of the genome. One popular technique, called Expressed Sequence Tag (EST) sequencing, generates single-pass sequences that typically correspond to either the start or the end of mRNA molecules. Because these data are transcribed by the natural gene-processing mechanisms within cells, they encapsulate valuable information for studies related to gene expression, gene identification and alternative splicing. Processing EST data also depends on overlap information preserved between sequences originating from the same gene source, which is similar to genome assembly. Unlike conventional genome sequence data, however, non-uniform sampling is especially rampant in ESTs because of differential expression behavior exhibited by the underlying cells, posing even greater computational challenges.

To scale up to larger EST collections, most users typically preprocess these data using a *clustering* algorithm that partitions these sequences according to gene source. Each resulting cluster can then be further analyzed to serve the desired objectives, as discussed in Chapter 12. For example, the TGICL program [22] was able to preprocess 1.7 million ESTs of an unspecified species for assembly using 1 hour on a PVM cluster with 20 Pentium III nodes (20 total CPU hours) using a fast alignment scheme. Subsequent assembly completed the next day based on these less accurate clusters. For most clustering applications, however, the compromise in quality for efficiency could either generate incorrect results or require a longer assembly post-processing step, but these can be offset by more accurate alignment based clustering performed on larger distributed memory machines. For example, our own experiments with 726,988 rat ESTs using the PaCE software [14] required 108 CPU hours; however, this result was obtained in only 2 hours using 54 processors of a commodity cluster. Given this result, we can estimate that processing the largest human EST collection that contains over 6 million ESTs would require nearly 900 CPU hours if we assume a linear increase in run-time. A quadratic increase, on the other hand, would require an estimated 7,350 CPU hours, or 5.7 days on the same commodity cluster.

This chapter focuses on methods developed for performing large-scale clustering and assembly that adequately address the computational challenges posed by both uniformly and non-uniformly sampled sequence data. By taking a unified approach to solving the problems of clustering and assembly (in Section 13.2), we describe a clustering algorithm with a space complexity that scales linearly with input size along with parallel processing capabilities (in Sections 13.3 through 13.5) that provide the memory required for generating large-scale assemblies. We demonstrate the effectiveness of this approach in the context of on-going maize genome assembly efforts, and then discuss its application in large-scale EST clustering (in Section 13.6). Finally, we lay foundations for developing a generalized framework that broadens the scope of these ideas toward performing other types of important sequence analyses in parallel.

13.2   A Unified Approach to Clustering and Assembly

Many assemblers follow the overlap-layout-consensus paradigm, in which pairs of overlapping fragments are used to build long contiguous stretches of the genome called *contigs*. Although a jigsaw puzzle is not a perfect analogy, we will use it to illustrate the computational difficulties associated with the overlap computation modules of traditional sequence assemblers. Under uniform random sampling it is expected that each region of the genome is sampled a constant number of times, which is called the *coverage* of the genome. Using the analogy of a jigsaw puzzle, this means that for every piece there will be a fixed number of neighboring pieces. For example, imagine a puzzle of an eclectic town with many uniquely colored houses; each of these houses could yield a single group, one per color, of related pieces. If these houses are uniformly sized, it can be expected that these groups have approximately the same number of pieces. Now suppose that we can no longer assume a uniform sample. In this case we may have to test all pairs of pieces because we cannot ignore the possibility the same region is oversampled. Relating this to the analogy, this is equivalent to trying to assemble a jigsaw puzzle consisting of an ocean; this task will be very time-intensive due to the substantial number of potentially valid combinations of any two "blue" pieces.

Just like assembly, clustering algorithms determine sequence relatives based upon overlaps. The difference between these two problems is illustrated in the following example. During clustering, if A overlaps with B, and B overlaps with C, we can deduce that A is related to C *without* having to compare A and C. Assemblers, however, must directly determine if A overlaps with C because of the need to determine that all relationships are consistent to accurately reconstruct the original genome sequence. Intuitively, it follows that a solution to clustering is easier to compute than its assembly counterpart — as it is typically easier to group pieces together than it is to fit them together correctly.

With this distinction in mind, we now provide a unified view of the two problems using clustering as the initial step. This approach capitalizes on the following observation: most sampling approaches either directly or indirectly miss certain regions of the original sequence. Sequence clustering provides a method for reducing a large assembly problem into many, but smaller, assembly problems [7, 20]. For example, gene enrichment should result in a large number of genomic "islands" that contain most of the genic regions interspersed by non-sampled repetitive regions. A *cluster-then-assemble* approach would therefore first partition the input sequences into corresponding genomic islands and then assemble the individual islands using a serial assembler of choice. This approach shifts the burden of addressing the memory and run-time needs of large-scale sequence analysis to the clustering phase, while benefiting from and not duplicating the painstakingly built-in biological expertise present in conventional assemblers.

There are two additional advantages of this paradigm. First, clustering limits the subsequent peak memory usage to the memory required to assemble the largest subproblem. Second, breaking the problem into clusters allows trivial parallelization; each cluster can be individually processed on a different processor. Both of these properties facilitate the generation of assemblies that are consistent with what would have been generated by any conventional assembly program, except that the maximum solvable problem size is larger and speed is significantly enhanced. In order for this statement to be true, however, we must require that any overlap considered significant by the assembler is also considered acceptable by the clustering algorithm; therefore, in practice the overlap criteria are less stringent than those used during assembly to ensure that the result is the same as if we ran the assembler on the entire dataset.

This strategy is also applicable in conventional genome assembly projects, the rationale for which is as follows. The sampling redundancy — typically between five and seven — can be plugged into the Lander-Waterman equation [17] to determine the expected number of clusters that result from random sampling. A real-life example is the Celera human genome assembly that resulted in 221,036 contigs each spanning ∼11.7 Kbp on an average and the longest contig spanning ∼1.2 million bp (i.e., only 0.48% of the longest chromosome). Even assembling a small genome like that of *N. meningitis* (∼2.18 Mbp) generated 149 contigs with an average length of 14 Kbp [23]. This happens because gaps invariably occur while sequencing, both because of sample size as well as the reality that certain genomic regions are difficult to sample for a variety of reasons.

13.3   Clustering

In this section, we describe sequence clustering methods with special emphasis on their relevance to large-scale sequence data. For generality, we will not assume anything specific about the type of sequence data being clustered — they can be ESTs, cDNAs, genomic sequences, protein sequences, or other types of biological sequences that can be computationally represented as strings over a fixed alphabet. Overlaps are considered to be the only computable information for clustering purposes. Although there may be other auxiliary information available to supplement the clustering decision process, we will defer discussion of such details until the end of this section.

Overlaps can be modeled in two different ways: alignment based and exact match based. Alignment-based modeling of overlaps provides a high degree of flexibility in accounting for sequencing errors and natural variations (e.g., Single Nucleotide Polymorphisms or SNPs). This benefit, however, comes at a high price — computing an optimal alignment between two sequences through standard dynamic programming techniques requires run-time proportional to the product of their lengths. Exact match based modeling, on the other hand, allows for quicker albeit less accurate means to compute overlaps. For example, counting the number of fixed-length matches between two sequences could be used to measure their overlap. While this can be computed in linear run-time, the number itself does not provide much information besides providing a lower bound for the edit distance between the sequences. For this reason, alignment based modeling is generally biologically more meaningful, and we therefore restrict our discussion to clustering methods that model overlaps by computing alignments.

The Sequence Clustering Problem: Let $S = \{s_1, s_2, \ldots, s_n\}$ denote the set of n input sequences over an alphabet Σ. Two sequences $s_i, s_j \in S$ are said to be *related* if either s_i and s_j show a "significant overlap", or $\exists s_k \in S$ to which both s_i and s_j are *related*. The problem of sequence clustering is to partition S such that $\forall s_i, s_j \in S$, s_i and s_j are in the same subset (or "cluster") if and only if s_i and s_j are related.

The above formulation is generic enough to accommodate any preferred alignment method. For instance, in the context of genome assembly, two sequences can be considered to have a significant overlap if there is a good suffix-prefix alignment between them. In the context of EST clustering, if the underlying objective is to cluster together sequences derived from the same gene, then overlaps can be detected as a chain of local alignments.

The above formulation of clustering is also sometimes referred to as *transitive closure clustering* because the defined relationship between sequences is transitive in nature. Thus it is possible to have two entirely distinct sequences in the same cluster simply because there is a third sequence to which both are related. An example in the context of genome assembly is shown in Figure 13.1a. This formulation, however, does not guarantee that the sequences in the same cluster conform to a consistent overlap layout. An example of such an inconsistent layout is illustrated in Figure 13.1b. The task of resolving such inconsistencies is typically deferred to later stages post-clustering.

13.3.1   Promising Pairs

A simple clustering approach involves evaluating all pairs of sequences for significant overlaps, and using the results to generate clusters. Such an approach is expensive for large n because it requires the computation of $\Theta(n^2)$ alignments.

One way to reduce overall computation is to use quicker methods of computing align-

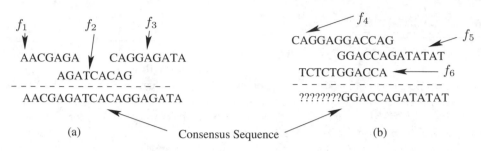

FIGURE 13.1:   Examples to show the effect of transitive closure clustering in the context of genome assembly. Part (a) shows a case where three fragments that are clustered together based on suffix-prefix alignments have a consistent overlap layout despite the fact that two of the fragments, f_1 and f_3, have no significant suffix-prefix alignment. Part (b) shows a complementary case where transitive closure clustering produces an inconsistent overlap layout.

ments, such as *megablast* [34] (as used in *TGICL* [22]) or computing a d^2 measure [29] (as discussed in Chapter 12). A more effective strategy is to reduce the number of alignments computed without affecting the clustering result. Given the low frequency of disagreements resulting from modern sequencing errors and minimal natural variations, sequences with significant aligning regions will contain long exact matches, while the converse is not necessarily true. This observation can be exploited by restricting alignment computation to only pairs containing exact matches of length at least w. Such pairs are termed *promising pairs* and can be identified by constructing a lookup table that indexes substrings of length w (see Chapter 5). In practice, the value of w is usually less than 12 because the size of the lookup table is exponential in w, even though the low error rates may allow higher values. For example, using a value of 12 implies $4^{12} = 16$ million entries to store (assuming the DNA alphabet); while an expected error rate of 2% over a 100 bp long aligning region allows a value up to 33. Another downside of this approach is that a long exact match of length l reveals itself as $(l - w + 1)$ matches of length w.

The PaCE clustering method [14] uses a different mechanism that indexes and identifies promising pairs based on variable-length exact matches. This mechanism is used in conjunction with a greedy heuristic to more effectively reduce the number of alignments computed without sacrificing clustering quality. Furthermore, this method has a worst-case space complexity that is linear in input size, irrespective of the nature of the underlying sequence data. The PaCE clustering method was originally developed for clustering ESTs in parallel, but is also being applied to cluster large-scale genomic data. This clustering method is described below.

13.4   A Space-optimal Heuristic-based Method for Clustering

Consider the following approach to clustering : Initially, each sequence is considered to be in a cluster of its own. As the algorithm progresses, all promising pairs are generated but aligned only if they currently do not belong to the same cluster. If two sequences from two different clusters show a significant overlap, the corresponding clusters are combined. Otherwise, the clusters are left intact, and so the alignment effort is considered wasted. The process of merging is continued until there are no more promising pairs remaining, resulting in the final set of clusters. Because the clustering process can be viewed as a forest of subtrees with a total of n leaves and an internal node for every merge, the overall number of merges is limited by $n - 1$ in the worst-case. This also indicates that there is a linear number of alignments that could lead to the final solution. While such pairs cannot be predicted in advance, their early identification causes clusters to merge sooner, which in turn significantly reduces the number of future alignments because pairs that are already part of the same cluster are never aligned. This hypothesis forms the basis of the following heuristic method.

A "maximal match" between a pair of sequences is an exact match that cannot be extended on either side to result in a longer match. The greedy heuristic generates promising pairs on-demand in non-increasing (henceforth, "decreasing" for ease of exposition) order of the maximal match lengths, and considers them for alignment computation in the same order as dictated by the above clustering scheme. The maximal match length provides an effective mechanism to differentiate among promising pairs — longer the maximal match, higher the likelihood the overlap between the pair is significant. Therefore, evaluating pairs in decreasing order of their maximal match lengths is expected to result in early cluster merges, subsequently reducing the number of alignments required. Furthermore, an on-demand generation of pairs allows them to be considered for alignment as they are generated, obviating the need to store them. Thus the space complexity of the algorithm is determined by the space requirement of the pair generation algorithm.

13.4.1 On-Demand Promising Pair Generation

Ideally, each promising pair should be generated only once. A given pair of strings, however, may have multiple distinct maximal matches, or a given match could be maximal in multiple pairs of locations between the same two strings. See Figure 13.2 for an illustration. One way to avoid generating multiple copies of the same pair in such cases is to record a pair the first time it gets generated and ignore any future reoccurrence of the pair. This simple scheme, however, requires storing all generated pairs, potentially requiring $\Theta(n^2)$ memory. As a compromise, the following algorithm operates in linear space and generates each pair at least once and at most as many times as the number of distinct maximal matches in it. For example, in Figure 13.2 the algorithm will generate (s_1, s_2) exactly once, while (s_3, s_4) is generated at least once and at most twice.

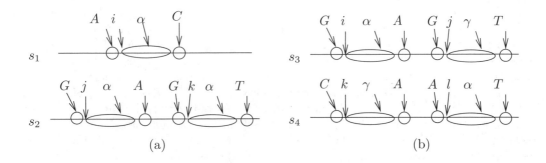

FIGURE 13.2: Examples showing two cases of maximal matches. (a) A match α is maximal in two pairs of locations (i,j) and (i,k) between s_1 and s_2. (b) Two maximal matches α and γ exist between s_3 and s_4.

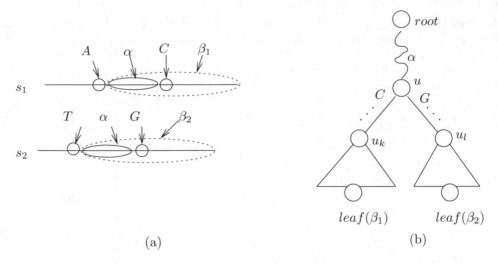

(a)   (b)

FIGURE 13.3: An example showing a maximal match between two strings and its representation
in the corresponding GST. (a) A maximal match α between s_1 and s_2; (b) An
internal node u with path-label α in the GST. The child nodes u_1 and u_2 are
roots to subtrees where the leaf nodes corresponding to the suffixes β_1 and β_2
are located respectively.

Notation

Without loss of generality, assume that S is a set of DNA sequences over the alphabet $\Sigma = \{A, C, G, T\}$. Henceforth, the words "sequence" and "string" are used interchangeably. Let $|s|$ denote the length of string s, and $N = \sum_{i=1}^{n} |s_i|$. A prefix of a string is said to preceded by λ, a null character. A match α between two strings is said to be *left-maximal* (alternatively, *right-maximal*) if the characters that immediately precede (alternatively, follow) α in the two strings are different or if α is a prefix (alternatively, a suffix) of either string. Thus α is a maximal match if it is both left- and right-maximal.

Let G denote the Generalized Suffix Tree (GST) constructed over all suffixes of all strings in S. A special terminal character '$\$$' is appended to each input string in S to ensure there exists a leaf node for every suffix of each string. Let u denote a node in G and $subtree(u)$ denote the set of all nodes in u's subtree (including u). Let *path-label* of u be the string obtained by concatenating all edge labels from the root to u; if u is a leaf node, the terminal character '$\$$' is excluded in its path-label. Let *string-depth* of a node u denote the length of its path-label. Let β be an input string's suffix and $leaf(\beta)$ denote the leaf node with path-label β. Let $leaf\text{-}set(u) \subseteq S$ represent the set of input strings that have path-label(u) as a substring. The $leaf\text{-}set(u)$ of a node u can be partitioned into $|\Sigma| + 1$ sets called *lsets* of u: $\ell_A(u)$, $\ell_C(u)$, $\ell_G(u)$, $\ell_T(u)$ and $\ell_\lambda(u)$. By definition, if a string is in $\ell_c(u)$ (for $c \in \Sigma \cup \{\lambda\}$), then it has a suffix β such that $leaf(\beta) \in subtree(u)$ and β is preceded by c. Observe that such a partition need not be unique. For instance, a string s could have two suffixes β and β' such that $leaf(\beta)$ and $leaf(\beta')$ both are in $subtree(u)$ while β and β' are preceded by two different characters, say C and G in s. Then s could be either in $\ell_C(u)$ or $\ell_G(u)$, and either choice will work for the algorithm.

The Algorithm

If two strings s_1 and s_2 share a right-maximal match α that starts at positions i and j in s_1 and s_2 respectively, then the GST built over the two strings contains an internal node u with path-label α. Also the leaf nodes corresponding to the suffixes starting at i of s_1 and j of s_2 are in subtrees of two different children of u. Therefore, (s_1, s_2) can be generated at u provided the suffixes are also left-maximal. An illustration is provided in Figure 13.3. The *lsets* are useful in checking for left-maximality. The algorithm is as follows.

FIGURE 13.4: Algorithm for generating promising pairs. The operator "$<$" when applied to characters, denotes lexicographic order.

Pair Generation based on Maximal Matches
GeneratePairsFromLeaf(Leaf Node: u)
    1. Compute the *lsets* at u by scanning its labels.
    2. Compute:
$$P_u = \bigcup_{(c_i,c_j)} \ell_{c_i}(u) \times \ell_{c_j}(u), \ \forall (c_i, c_j) \text{ s.t., } c_i < c_j \text{ or } c_i = c_j = \lambda$$

GeneratePairsFromInternalNode(Internal Node: u)
    1. Traverse all *lsets* of all children $u_1, u_2, \ldots, u_q$ of u.
        If a string is present in more than one *lset* **then**
            all but one occurrence of it are removed.
    2. Compute:
$$P_u = \bigcup_{(u_k,u_l)} \bigcup_{(c_i,c_j)} \ell_{c_i}(u_k) \times \ell_{c_j}(u_l), \ \forall (u_k, u_l), \ \forall (c_i, c_j) \text{ s.t.,}$$
$$1 \le k < l \le q, \ c_i \ne c_j \text{ or } c_i = c_j = \lambda$$
    3. Create all *lsets* at u by computing :
        for each $c_i \in \Sigma \cup \{\lambda\}$
$$\ell_{c_i}(u) = \bigcup_{u_k} \ell_{c_i}(u_k), \ 1 \le k \le q$$

The GST G for S is first constructed using a linear time algorithm (see Chapter 5). The nodes in G with string-depth $\ge w$ are then sorted in decreasing order of string-depth. Because string-depth of any node in a GST is bounded by the length of the longest string in S, radix sorting is used to run in linear time. Once sorted, the nodes are processed for pair generation in that order.

The algorithms for generating pairs from leaf and internal nodes are given in Figure 13.4. Let P_u denote the set of pairs generated at node u. If u is a leaf node, the *lsets* are computed directly from its labels: For every pair of different characters in $\Sigma \cup \{\lambda\}$, a Cartesian product of the corresponding *lsets* is computed. In addition, a Cartesian product of $\ell_\lambda(u)$ with itself is also computed. P_u is the union of these Cartesian products.

If u is an internal node, the *lsets* of the children of u are traversed to first eliminate multiple occurrences of the same string in *lsets* of different children. The purpose of this elimination is to avoid generating the same pair multiple times under a given node. Note that the resulting *lsets* at a child of u may no longer represent a partition. Next, for every pair of *lsets* corresponding to different characters under every pair of child nodes, a Cartesian product of their *lsets* is computed. In addition, a Cartesian product of ℓ_λ under each child node with ℓ_λ under every other child node is computed. P_u is the union of these Cartesian products. After pairs are generated, the *lsets* at u are computed from the *lsets* of its children — by taking a union of all *lsets* corresponding to the same character. Because multiple occurrences of the same string were eliminated before, the *lsets* at u will constitute a partition of leaf-set(u).

Traversing *lsets* of all child nodes to eliminate multiple occurrences of a string can be implemented to run in time proportional to the sum of the cardinalities of those *lsets*. A global array $M[1 \ldots n]$, one entry for each input string, is maintained. Let u be an internal node currently being processed. The first time a string s_i is encountered, $M[i]$ is marked with u's identifier. Any future occurrence of s_i under any of u's child nodes is detected as a duplicate occurrence by directly checking $M[i]$. A linked list implementation of the *lsets* allows the union in *Step* 3 of *GeneratePairsFromInternalNode* to be computed using $O(|\Sigma|^2)$ concatenation operations. This restricts the overall space required to store *lsets* to $O(N)$. The assumed arbitrary orderings of the characters in $\Sigma \cup \{\lambda\}$ and the child nodes are to avoid generating a pair at u in both of its forms: (s, s') and (s', s).
In summary, if v is a leaf,

$$P_u = \{(s, s') \mid s \in \ell_{c_i}(u), s' \in \ell_{c_j}(u), c_i, c_j \in \Sigma \cup \{\lambda\}, ((c_i < c_j) \vee (c_i = c_j = \lambda))\}$$

and if u is an internal node,

$$P_u = \{(s, s') \mid s \in \ell_{c_i}(u_k), s' \in \ell_{c_j}(u_l), c_i, c_j \in \Sigma \cup \{\lambda\}, k < l, ((c_i \neq c_j) \vee (c_i = c_j = \lambda))\}$$

The following lemmas prove the correctness and run-time characteristics of the algorithm:

LEMMA 13.1   Let u be a node with path-label α. A pair (s, s') is generated at u only if α is a maximal match between s and s'.

Proof   At a leaf node u, all pairs of strings represented in its *lsets* are automatically right-maximal by definition. If the algorithm generates a pair (s, s') at u, it is because the strings are either from *lsets* representing different characters or from the *lset* representing λ. In either case, α is a maximal match between s and s'. For an internal node u, the algorithm generates a pair (s, s') only if (i) s and s' are from *lsets* either representing different characters or λ, and (ii) s and s' are from *lsets* of two different children of u. The former ensures α is left-maximal; the latter ensures α is right-maximal. Thus α is a maximal match of s and s'. Figure 13.3 illustrates an example case at an internal node.

COROLLARY 13.1   The number of times a pair is generated is at most the number of distinct maximal common substrings of the pair.

Proof   Follows directly from Lemma 1 and the fact that a pair is generated at a node at most once. The latter is true because for any internal node the algorithm retains only one occurrence of a string before generating pairs; whereas for any leaf node there can be at most one occurrence of any string in its *lsets*. While this bounds the maximum number of times a pair is generated, a pair may not be generated as many times.

Note that the converse of Lemma 13.1 need not necessarily hold because the elimination of multiple occurrences of strings while processing an internal node may remove the corresponding occurrences that would otherwise lead to the generation of a pair at a given node. This could, however, happen only if the same pair was generated elsewhere because of a longer maximal match. In other words, it is guaranteed that each promising pair is generated at least once, as proved below.

LEMMA 13.2 Each promising pair is generated at least once.

Proof Let (s, s') be a promising pair. Consider α, a largest maximal match of length $\geq w$ between s and s'. This implies that there exists either a leaf or an internal node u with path-label α. Also there exist suffixes β and β' of s and s' respectively such that $leaf(\beta), leaf(\beta') \in subtree(u)$ and α is a prefix of both β and β' satisfying left- and right-maximal properties. Thus if u is a leaf node, then $s \in l_{c_1}(u)$ and $s' \in l_{c_2}(u)$ such that $c_1 \neq c_2$ or $c_1 = c_2 = \lambda$, implying that the algorithm will generate the pair at u. If u is an internal node, then the fact that α is a largest maximal match ensures that s and s' will occur once, even after the duplicate elimination process at u, in the $lsets$ of different children and the $lsets$ will correspond either to different characters or to λ. Thus the algorithm will generate the pair at u.

LEMMA 13.3 The algorithm runs in time proportional to the number of pairs generated plus $O(N)$. The space complexity of the algorithm is $O(N)$.

Proof Each node at string-depth $\geq w$ is processed exactly once. At an internal node, the initial elimination process reduces the total size of $lsets$ of all its children by at most a factor of $(|\Sigma| + 1)$. This is because a string is present in at most one $lset$ of each child node and the number of children is bounded by $(|\Sigma| + 1)$. The total size of all the $lsets$ of all the children after duplicate elimination is bounded by the number of pairs generated at the node. Taken together, this implies that the cost of the elimination process is bounded by a constant multiple of the number of pairs generated at the node (assuming $|\Sigma|$ is a constant).

The space complexity of the GST data structure is $O(N)$. The space required by $lsets$ is proportional to the total number of $lset$ entries to be stored at all the leaf nodes, which is $O(N)$. This is because $lsets$ at internal nodes are constructed from $lsets$ of their children and so do not require additional space.

13.4.2 The Clustering Algorithm

The overall clustering algorithm can be divided into a preprocessing phase followed by a clustering phase, as shown in Figure 13.5. In the preprocessing phase, the GST for all n input sequences is constructed and its nodes are sorted based on their string-depths. This phase takes linear time and space. The clustering phase begins by initializing n clusters, one for each input sequence. Promising pairs are then generated one at a time, and considered in the same order for alignment on-the-fly. Managing clusters involves two types of operations: finding the current cluster that contains a given sequence; and combining two clusters into one. As these operations are performed on disjoint sets, the union-find data structure [28] is used for managing the clusters. This enables each operation to cost an amortized run-time given by the inverse of Ackermann's function, a constant for all practical purposes. While the space complexity of the clustering phase is $O(N)$, its run-time is proportional to the number of promising pairs generated plus the alignment computation cost for the pairs selected by the clustering mechanism.

Asymptotically, the run-time is likely to be dominated by the time spent in computing alignments, a fact that is corroborated by our experiments on large collections of genomic

FIGURE 13.5: The sequential clustering algorithm. Steps 1 and 2 are collectively called the "preprocessing phase" and the remainder of the algorithm is called "clustering phase". $Find(s_i)$ returns the set containing s_i and $Union(C_p, C_q)$ performs a union of the two sets.

Cluster()
Input: Set $S = \{s_1, s_2, \ldots s_n\}$ **of** n **sequences**
Output: A partition $C = \{C_1, C_2, \ldots C_m\}$ **of** S, $1 \leq m \leq n$
1. $G \leftarrow$ Construct the Generalized Suffix Tree of S
2. Radix sort nodes in G with string-depth $\geq w$ in decreasing order of string-depth.
3. Initialize clusters:
    $$C \leftarrow \{\{s_i\} \mid 1 \leq i \leq n\}$$
4. For each node u in the sorted list
    REPEAT
        $(s_i, s_j) \leftarrow$ generate next pair from u
        $C_p \leftarrow Find(s_i)$
        $C_q \leftarrow Find(s_j)$
        IF $C_p \neq C_q$ THEN
            $Overlap \leftarrow align(s_i, s_j)$
            IF $Overlap$ is significant THEN
                $Union(C_p, C_q)$
    UNTIL no more pairs at u
5. Output C

fragments and ESTs (see Section 13.6). The alignment computation run-time can be reduced by using the maximal match information that caused a pair to be generated to "anchor" its alignment as shown in Figure 13.6. By anchoring, it is only required to compute alignment over the two flanking extensions, thereby saving the alignment run-time. Further savings can be achieved by extending this idea to include multiple maximal matches as part of the same anchor, and in addition computing alignment over a band of diagonals [8] within each area of the table not covered by the anchored maximal matches. Anchoring may, however, produce a sub-optimal alignment, as it is possible that none of the optimal alignments have all the anchored maximal matches in their paths.

In practice, partial clustering information may be available through alternative means for a subset of input sequences prior to clustering. For example, it may be known that two sequences were derived from two ends of the same clone. Such "mate" information can be incorporated into the clustering algorithm by initializing the clusters such that all mates are already clustered.

Space Requirement

While the space complexity is $O(N)$, the constant of proportionality is that of what is required to store the GST on all input sequences. Since there are at most N leaf nodes in the GST, the total number of nodes is limited to $2 \times N - 1$. Because the above algorithm does a bottom-up traversal of the tree, in which a parent is visited only after all its children are visited, the tree can be implemented as follows: The nodes are stored in an array in the depth-first traversal order. Each node in the tree stores its string-depth, a pointer to its *lsets* and a pointer to the rightmost leaf in its subtree. A leaf node's pointer points to itself. Given an internal node, all its children can be accessed as follows: The node immediately next to it in the array is its leftmost child. Its right sibling can be obtained by tracing the

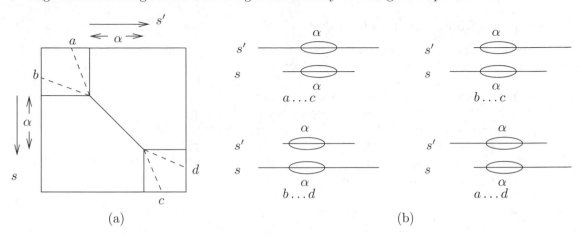

(a)   (b)

FIGURE 13.6: Using maximal match information while computing alignments, assuming an alignment based modeling. (a) Dynamic programming table showing the extension of a maximal match α, at both its ends. The alignment is said to be "anchored" at α. (b) Four possible overlap patterns resulting from suffix-prefix alignment computation and their corresponding optimal paths in the table.

array entry next to its rightmost pointer entry. If a node's rightmost pointer points to the same as its parent's, then it is the rightmost child of its parent. The *lsets* need N entries, one for each suffix in the input. An additional array of at most $2 \times N - 1$ entries is required to store the node identifiers in sorted order of their string-depths.

Our implementation meeting the above storage requirements has a worst case constant of ~ 40 bytes for every input base. Because DNA sequences are double stranded, a sequence should be considered both in its forward and reverse complemented form for overlaps. This doubles the constant to ~ 80 for every input base. As an example, on a set of shotgun sequences sampled at 8x coverage over a megabase long genomic stretch (i.e., for an input size of 8 megabases), this implementation requires 640 MB in the worst case.

13.5   Parallel Clustering

Extending the reach of clustering to gigabases of sequence data requires tens of gigabytes of memory. In the absence of parallel software, a common strategy is to run a sequential code but on a shared memory machine with tens of gigabytes of memory; in the event the memory requirement is still not satisfied, the input is first partitioned among multiple such machines and the code is run concurrently on each before circulating the local portion to other machines to affect their clustering. Furthermore, the alignment workload is distributed among multiple processors within the same machine to reduce the overall run-time. This strategy suffices although it requires durations of days to weeks. In contrast, the PaCE parallel clustering methodology, to the best of our knowledge, is the only "truly parallel" solution with a demonstrated linear scalability over hundreds to thousands of processors. Its parallel algorithms are based on the heuristic described in the previous section. In what follows, we describe the details of its underlying parallel algorithms and implementation.

13.5.1   Parallel Construction of GST

The preprocessing phase requires the construction of a distributed representation of the GST in parallel. There are algorithms for constructing suffix tress in parallel under CR-CW/CREW PRAM models of computing [1, 12]. However, due to the unrealistic assumptions underlying the PRAM model, a direct implementation of these algorithms is not practically useful. The following is an alternative algorithm.

Let p denote the number of processors. Initially, the input set S is partitioned across p processors, such that each processor gets $O(\frac{N}{p})$ of the input. Through a linear scan, each processor partitions the suffix positions of the local input into $|\Sigma|^k$ buckets based on their first $k \leq w$ characters. The suffix positions are then globally redistributed such that those belonging to the same bucket are grouped in the same processor, while maintaining $O(\frac{N}{p})$ suffix positions per processor after redistribution. The value of k should be carefully chosen. While its value is upper bounded by w, a small value may result in too few buckets to distribute among processors, or too many ($>> \frac{N}{p}$) suffixes per bucket. For example, a value of 10 would generate 1 million buckets for distribution among processors assuming DNA alphabet.

Once buckets are assigned to processors, each processor constructs one subtree for all suffixes in each bucket. Note that a sequential suffix tree construction algorithm cannot be used for this purpose, because all suffixes of a string need not be present in the same bucket, unless the string is repetition of a single character. A depth-first construction of each subtree is achieved as follows. First, partition all suffixes in a bucket into at most $|\Sigma|$ sub-buckets based on their first characters. This partitioning is then recursively applied on each sub-bucket until all suffixes separate from one another, or their characters exhausted. As this involves scanning each character of a suffix at most once, the run-time is $O(\frac{N \times \ell}{p})$, where ℓ is the average length of a sequence. Despite not being optimal, this algorithm is practically efficient mainly because ℓ is only about 500–800 bases (independent of n).

There is, however, one catch. Because each subtree is constructed in a depth-first manner, this algorithm assumes the availability of all characters corresponding to its suffixes in the local memory. Storing characters of all suffixes assigned to a processor is not a viable solution, because these suffixes can span n sequences in the worst case. There are two ways to address this problem without affecting the method's scalability. If the underlying machine supports scalable parallel I/O, then each processor can construct its subtrees in batches such that the strings required to be loaded in the memory per batch is $O(\frac{N}{p})$, and acquire those strings before each batch from disk. If the underlying communication network provides a faster alternative, then the following communication-based approach can be pursued: Evenly partition the input among processors' memory and then construct subtrees in batches as in the previous scheme, except that before each batch, strings are acquired from processors that have them by remembering the initial distribution of input. In either approach, further optimizations can be done by masking the latency time with the data transfer time. In practice, disk latencies and data transfer rates are in the order of milliseconds and microseconds, respectively (in the absence of parallel I/O); where as communication latencies and transfer rates are in the order of microseconds and tens of nanoseconds, respectively.

Note that the resulting set of trees over all processors represents a distributed collection of subtrees of the GST for S, except for the top portion consisting of nodes with string-depth $< k$. Once the GST is constructed, the nodes within each processor are locally sorted using radix sort as mentioned in the previous section.

13.5.2   Parallel Clustering Phase

The parallel implementation of the clustering phase follows the master-worker paradigm. A single master processor maintains the set of clusters, selects pairs generated by worker processors for alignment based on the current state of clusters, and then distributes them to worker processors. The worker processors first build the GST in parallel. After sorting its nodes locally, each worker processor then generates promising pairs, sends them to the master processor for evaluation, and computes alignments assigned by the master processor. Instead of communicating one pair at a time, pairs are communicated in batches of size that is just large enough to offset the communication latency but small enough to take advantage of the latest clustering — a value that is empirically determined. The master processor is a necessary intermediary that regulates the work flow and balances the load of alignment work distribution. The space complexity is $O(\frac{N}{p})$ with the same constant of proportionality as our sequential approach. More details about the parallel implementation can be found in [14].

13.6   Applications

Many non-traditional assembly approaches would benefit from this parallel clustering framework. Comparative genome sequencing, which uses a complete genome to anchor a low redundancy skim of a related genome, could easily be performed using this approach. As an example of its application, Marguiles and colleagues [19] proposed that most mammalian genomes can be scaffolded on the completed human genome sequences and therefore would provide both a "road map" for initial sequence assembly as well as an important information resource for those interested in deciphering genes and their functions. Another group of researchers, Katari *et al.* [15], have used low redundancy sequencing of *Brassica oleracea* to identify previously unknown genes in the important model plant *Arabidopsis thaliana*.

The rest of this section focuses on applications to plant genome assembly and EST clustering. In particular, we discuss the design and development of a parallel genome assembly framework we developed in the context of ongoing national efforts spearheaded by U.S National Science Foundation (NSF), Department of Energy (DOE) and Department of Agriculture (USDA) to sequence and assemble the maize genome. Previous approaches have been very useful in identifying gene sequences within plants and are being applied to wheat, sorghum and pine. We will discuss the various sequencing strategies being used, the various biological idiosyncrasies in the maize genome that warrant novel sequencing approaches, and the advantages of using the parallel cluster-then-assemble assembly paradigm on these non-uniformly sampled sequence data.

13.6.1   Maize Genome Assembly

Maize (*i.e.*, corn) is both economically important and a principal model organism for the majority of world food production. In addition to its estimated size of 2.5–3 billion bases [2], which makes it comparable in size to the human genome, maize genome sequencing and assembly is considered particularly challenging because highly similar repetitive sequences span an estimated 65-80% of this genome. Maize is also highly interesting to biologists; it is believed that the complex domestication performed by Native Americans in tandem with modern crop improvement has generated an immense amount of genomic diversity as evidenced by the vast phenotypic diversity in different maize species (Figure 13.7). Some researchers believe that the differences between two maize varieties revealed by genome

FIGURE 13.7: Examples of phenotypic diversity found in different maize lines. Photo courtesy of Lois Girton and Patrick Schnable, Iowa State University

sequencing could be more interesting than any previous comparison, e.g., human and mouse, because they are the same species and therefore must have occurred recently. For this reason, along with the rich scientific history of maize, the U.S. NSF, DOE and USDA have decided to jointly fund large-scale genome sequencing efforts.

Maize Genome Project: An Introduction

Since its role in the rediscovery of Mendel's laws of inheritance [25], maize has had an integral part in furthering the overall understanding of biology. For example, Barbara McClintock's work on maize led to the seminal discovery of transposable elements, which later won her a Nobel Prize. A strong reason for this, and other discoveries, is the community of maize geneticists who have used this species as a model not only for food crops but also for plants in general over the past century. This is in stark contrast to the model plant Arabidopsis that has only recently been used to elucidate biological processes and a strong reason why maize genome sequencing is a high priority for the plant research community.

Over the past few thousand years, the domestication and spread of maize throughout the Americas has led to an immense source of phenotypic variation including kernel color, cob size and other important nutritional traits such as sugar content. It was once believed that *colinearity*, or the preservation of the positions of genes in related species, within the cereal crops (e.g., rice, wheat, barley and maize) would facilitate comparative mapping and discovery of economically important genes faster than traditional approaches. Therefore, draft sequences of the much smaller rice genome (430 million bases) were completed [10] and an international sequencing project was begun. Based on incoming sequence data from smaller intervals of these species, however, biologists now believe that the genomic and evolutionary differences between maize and rice — and even between multiple maize subspecies [4] — are unique and interesting enough to warrant genome sequencing. The inbred line B73, which was developed at Iowa State University and is the genetic ancestor of commercially important lines, was the initial choice for sequencing with the goal of

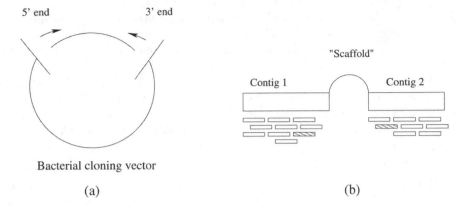

FIGURE 13.8: Illustration of clone pair information. (a) Larger pieces of the target genome, typically thousands of nucleotides, are sub-cloned into a bacterial vector and sequenced from both ends; (b) these paired fragments (shaded) allow the determination of the order and orientation of disjoint contigs based upon the unsequenced region in the middle of the cloning vector.

sequencing other maize lines as and when the resources become available.

Unfortunately, the predominance of repeats within this genome make it unlikely that a cheaper shotgun sequencing approach alone will be able to accurately reconstruct the maize genome given current experimental and computational limitations. Because the highly similar repeats, however, are expected to be scattered throughout the maize genome, breaking the genome into smaller chunks (e.g., Bacterial Artificial Chromosomes, or BACs) should reduce their effect on maize genome assembly. As such, an incremental sequencing approach would increase assembly correctness along with reducing the computational work required at the expense of performing thousands of smaller "mini" genome assemblies as opposed to one large assembly when using a whole genome shotgun approach.

An alternative strategy to sequencing the maize genome, *genome reduction*, consists of filtering out the repeats sometime during the shotgun sequencing project. This would therefore achieve the advantages of BAC-based sequencing with regard to correctness with the added bonus of reduced sequencing costs. To meet the immediate goal of deciphering the 10-15% of the genome [3] outside the highly similar repeat content in this manner, biologists have designed special experimental techniques that bias a shotgun-like sequencing approach toward the gene-rich regions in plants [21, 31]. As of April 2005, a total of 3.1 million sequences totaling over 2.5 billion nucleotides have been generated and deposited in public repositories for experimenting with these different sequencing strategies.

If genome reduction works, why should assembly teams bother with traditional shotgun sequencing? The simple answer is that although highly complex, the portion not captured by these biological filters contains a small fraction of "real" maize genes that may be important [9]. The same is true even for BAC-based sequencing; currently, about 6–8% of all maize genes are absent from the current maize map and therefore would not be captured during sequencing. The availability of unfiltered shotgun data, therefore, allows an unbiased sampling to occur that does not miss any of these potentially important regions.

Another application of genomic shotgun data, which has been used for multiple mammalian genomes, is the construction of inferred physical relationships using data often called "clone pairs" or "mate pairs" (Figure 13.8). The concept is simple: pairs of uniformly sized pieces derived from a larger piece can be used to infer the existence of an unsampled region of the genome during assembly. Moreover, this information can also be used to estimate gaps between contigs and ensure that the placement of component fragments is in line with the expected size. To use these data, most modern genome projects now use a predominant clone size, e.g., 4–5 thousand bases, augmented by larger clones called *fosmids* based on the experimental protocol. Although these longer clones are more useful for ordering and orienting the various contigs into larger structures often called "scaffolds", at present they are subject to a much higher experimental error rate using current technology.

Non-uniform sequencing strategies

The stated goal of the NSF, DOE and USDA joint effort is explicit in its special emphasis on identifying and locating all genes and their associated regulatory regions along the genome. Note that unlike previous sequencing projects, such as the human genome project, the emphasis has shifted from knowing the complete genome to a large collection of genomic contigs whose order and orientation along maize chromosomes is known. Most of this rationale is driven by the highly complex nature of the maize genome and the fact that only a few hundred million bases will be useful to most plant biologists.

There are two primary genome reduction sequencing strategies that have been successfully tried on maize and are now being applied to other plants including wheat, pine and sorghum. The first strategy, *Methyl Filtration* (MF) [24], discards the portions of the genome that are methylated. The second strategy, *High-C_0t* sequencing (HC) [32], utilizes hybridization kinetics to isolate low-copy fractions of a genome. Each of these techniques is explained in detail below.

Methylation occurs at certain nucleotides and is important in multiple biological process- es. One such effect, especially in plants, is the silencing, or turning off of transcription, of certain regions of the genome. In particular, it has been shown that retrotransposons and other repetitive sequences in maize, which spread based on a transcription mechanism, tend to be predominantly silenced by methylation. By sampling the unmethylated regions, as done by the sequencing strategy, the resulting sequences should mostly originate from the gene-rich stretches of the genome. The interesting aspect of this sequencing approach is that it uses a special strain of *E. coli* that is able to select against even a single methyl group. Therefore, the protocol is the same as performed in traditional sub-cloning reactions with a special bacterial host.

The HC sequencing approach is somewhat more complex than MF because it relies on biochemical filtration, but it will be applicable in most eukaryotic genome projects. Repet- itive sequences hybridize more often in a heterogeneous mixture of single-stranded genomic DNA fragments; consequently, removing double-stranded sequences after some elapsed time enriches for lower-copy sequences. Consider, as an example, a bag of marbles. Suppose this bag has more green marbles than grey ones, say by a ratio of 4:1 and our game consists of reaching into the bag and pulling out two marbles at random. If the colors match, we discard them; otherwise, we place all green-grey pairs in a pile that will eventually be placed back into the bag. It should be clear that in the beginning of this exercise we will remove many more green pairs (64% chance) than grey pairs (4% chance). Moreover, the only way a marble survives this game is if it finds an opposite-colored partner. It follows that our original population, therefore, can be normalized from a ratio of 4:1 to 1:1 upon completion by simply using the pile that should be placed back into the bag. Even if we stop this exercise early, such that not all of the marbles have been processed, or place heterogeneous pairs back into the bag (i.e., solution) we still enrich for the underrepresented sequences because the overrepresented sequences preferentially get removed from the game. Note,

Sequence Type	Before Cleaning		After Cleaning	
	Number of Sequences	Total length (in millions)	Number of Sequences	Total length (in millions)
MF	411,654	335	349,950	288
HC	441,184	357	427,276	348
BAC	1,132,295	964	425,011	307
WGS	1,138,997	870	405,127	309
Total	3,124,130	2,526	1,607,364	1,252

TABLE 13.1   Maize genomic sequence data types and size statistics: Methyl-filtrated (MF), High-C_0t (HC), Bacterial Artificial Chromosome derived (BAC), and Whole Genome Shotgun (WGS).

however, that HC selection may also remove non-repetitive sequences as the ratio within the solution approaches equilibrium. HC filtration therefore uses multiple time intervals that both maximize low-copy sequence recovery while minimizing the loss of highly similar genic sequences (e.g., gene families).

Cluster-then-assemble as applied on maize sequence data

Initial work on maize has focused on the biological idiosyncrasies involved in sequencing the maize genome from the perspective of capturing genes. Although it is not yet clear which additional portions of the maize genome are being sampled by enrichment strategies, it is apparent that a substantial number of genes and non-genes are being over-sampled relative to traditional sequencing approaches [7, 27].

A summary of currently available maize data broken into different categories is provided in Table 13.1. As of April 2005, there are a total of 3,124,130 sequences totaling about 2.5 billion bp. Of these, 852,838 are MF and HC sequences, which is about 27% of the total data; 1,138,997 are WGS sequences constituting about 36%; and the remaining sequences are derived from BACs.

The cluster-then-assemble parallel framework used for these maize data consists of three main phases [7] — sequence cleaning, clustering, and assembly. An illustration is provided in Figure 13.9. Raw sequences obtained from sequencing strategies can also be contaminated with foreign DNA elements known as *vectors*, which are removed using the program *Lucy* [5]. This step is not yet parallelized. An efficient masking procedure is important because unmasked repeats cause spurious overlaps that cannot be resolved in the absence of paired fragments spanning multiple length scales. Such data may become available from future sequencing projects; however, repeat masking would remove a substantial number of these random shotgun sequences because of the large repetitive fraction in the maize genome (Table 13.1).

After vector screening and repeat masking procedures, the input is clustered by the PaCE clustering software in parallel. Sequences in each resulting cluster are then assembled using the CAP3 assembler, though any assembler can be substituted here. The assembly process was parallelized by first partitioning the PaCE clusters among compute nodes and then running the CAP3 serially on each cluster.

Generation of a partial maize genome assembly

Even though genome enrichment artificially places gaps into the original genome sequence, the highly repetitive nature of the maize genome may lead to excessive merges of unrelated regions based on common repeats. To solve this problem, a modified transitive closure clustering algorithm was used in order to locate *statistically-defined repeats*, or SDRs, that could later be used to mask the repetitive sequences and thus eliminate the formation of large clusters [7]. Interestingly, an Interpolated Markov Model, which is often used for locating genic regions within genome sequences, was also useful to expand the original, high-confidence SDRs to other regions of the genome.

Once the database of repetitive sequences was initialized in August 2003, accurate clustering and assembly of 730,974 non-uniformly sampled sequences was performed in 4 hours on 64 nodes of a Pentium III 1.1 GHz cluster. The next partial assembly, which incorporated an improved repeat database along with other biological enhancements, assembled a total of 879,523 sequences in under 2 hours on the same machine in October 2003. These results imply that a substantial savings in run-time can be achieved by identifying repeats and not using these sequences as seeds for promising pairs. Both the August and October data sets were on the order of 600–700 million bases post-masking.

Because both sets were relatively small compared to the data described in Table 13.1, their memory requirements both fit into the 64 GB available on the Pentium III cluster. To meet the estimated memory requirement of ~100 GB on the entire maize collection (i.e., 1,252 million bases times 80 bytes per input base) and to accelerate the clustering process, a total of 1,024 dual-processor nodes of the IBM BlueGene/L supercomputer — each with 512 MB RAM and 700 MHz CPUs — were used. This allowed the clustering to complete in 2 hours. In comparison with the October 2003 clustering, this clustering computed almost ten times as many alignments in just the same parallel run-time, using 16 times as many, but less powerful processors.

Figure 13.10a shows the run-times of the preprocessing and clustering phases of PaCE as a function of processor size on a data size of 500 million bp. As shown, these phases scale almost linearly until 1,024 processors. Even though the preprocessing phase takes more time than the clustering phase for this data size, the clustering phase is asymptotically dominant. For example, on the entire 1,252 million bp data set, the preprocessing phase took 13 minutes while the clustering phase took 102 minutes.

Figure 13.10b shows the number of promising pairs generated as a function of input size. The results also indicate the effectiveness of the clustering heuristic in significantly reducing the alignment work. For the 1,252 million bp data set, about 40% of the pairs generated are aligned. However, less than 1% of the pairs aligned are accepted, indicating the presence of numerous repetitive elements that may have survived masking.

The most recent assembly based upon the BlueGene/L clustering described above completed in under 8.5 hours on 40 processors of the Pentium III cluster and resulted in a total of 163,390 maize genomic islands (or contigs) formed by two or more input sequences, and 536,377 singletons (i.e., sequences that did not assemble with any other sequence). On

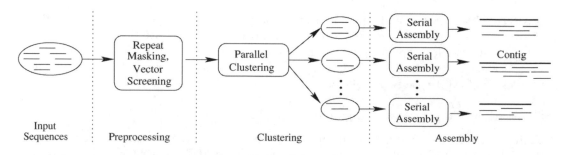

FIGURE 13.9: A parallel framework for genome assembly. PaCE is used for parallel clustering, and CAP3 is used for assembly. The output is a set of sequences representing contiguous stretches of the genome, or *contigs* (shown in thick lines).

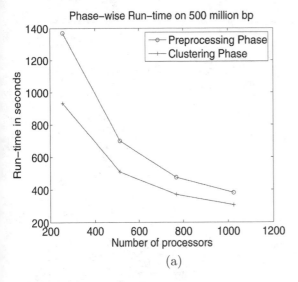
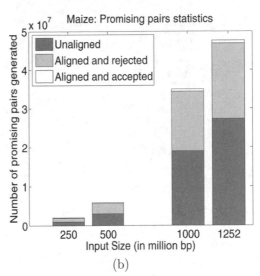

FIGURE 13.10: (a) Phase-wise run-times for PaCE clustering as a function of the processor size on 500 million bases of maize input, and (b) The number of pairs generated, aligned, and accepted as a function of input size.

an average, each cluster assembled into 1.1 contigs; given that the CAP3 assembly is performed with a higher stringency, this result indicates the high specificity of our clustering method and its utility in breaking the large assembly problem into disjoint pieces of easily-manageable sizes for conventional assemblers. The overall size of our contigs is about 268 million bp, which is roughly 10% of the entire maize genome. Upon validation using independent gene-finding techniques, we confirmed that our contigs span a significant portion ($\sim 96\%$) of the estimated gene-space. The average number of input sequences per contig is 6.55 sequences, while the maximum is 2,435. To more accurately assess non-uniformity within these data, coverage throughout the entire maize assembly was analyzed. Given the 1X coverage of the maize genome in this analysis, it is unexpected, and interesting, that the mean observed coverage was 3.24 in this gene-rich assembly. Moreover, 1.34 million bases of this assembly have sequence coverage of 25 or higher and may correspond to unmasked repeats and/or biases from the gene-enrichment approach. The results of our assembly, named Maize Assembled Genomic Islands (MAGIs) [7, 9], can be graphically viewed at (http://www.plantgenomics.iastate.edu/maize). An example of a genomic island is shown in Figure 13.11.

13.6.2 EST Clustering

ESTs are sequenced experimentally from mRNAs, which are transcribed from genes. The number of ESTs derived from a gene is a direct indication of its level of activity under such experimental conditions. ESTs can also be used to identify alternative splicing behavior of genes. Intronic and exonic boundaries can be identified by aligning the ESTs to their corresponding genes. ESTs collected during an experiment may have originated from many gene sources, and clustering them according to their sources provides a good starting point for deriving gene-specific information from them. ESTs, their applications, and an overview on different clustering methods are presented in Chapter 12. In this section, we discuss the computational challenges in clustering large EST data.

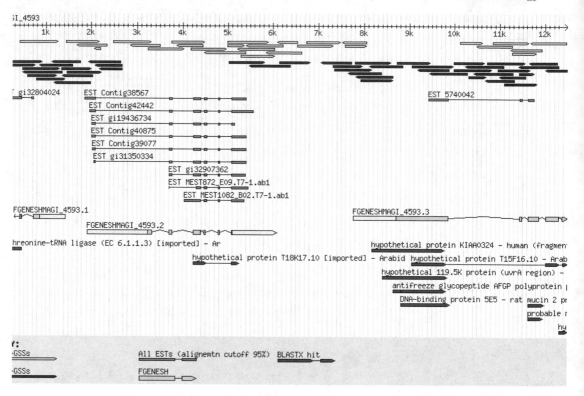

FIGURE 13.11: **(See color insert following page 20-4.)** Graphical representation of MAGI-3.1-4593. The first two rows above, as illustrated in the key, correspond to HC and MF sequences, respectively. The next two rows correspond to genes predicted using either EST-based or *ab initio* prediction approaches and include introns. The last row are annotated protein matches. It follows that this single contig shows a case where there are three genes on a genomic island; however, notice how the sampling sources differ in the different intervals above. Some regions are only captured using HC reads (3–5KB) while others are only captured using MF (8–10KB).

Besides the usual complications introduced by sequencing errors, EST data present additional challenges to the clustering process. EST data may contain sequences derived from alternatively spliced transcripts derived from the same gene. Computationally, such splicing events can be detected by first grouping ESTs based on common gene source and then by common transcript source. Two ESTs are likely from the same gene source if they have a chain of significant local alignments between them, where each local alignment indicates a shared exon. Alternatively, a pair is likely from the same transcript if they have a good suffix-prefix alignment. Because fragment assembly software also detect such suffix-prefix alignments, they can be used for clustering ESTs by transcript source. EST data may also contain introns retained during the transcription process, which can be computationally detected by treating them as special cases of alternative splicing. EST data may also have originated from different genes of the same gene family or may have Single Nucleotide Polymorphisms (SNPs), and the general strategy to identify such cases is to first group based on sequence similarity and then look for desired patterns among sequences within each cluster.

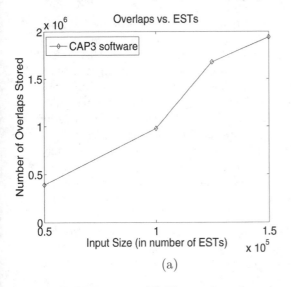
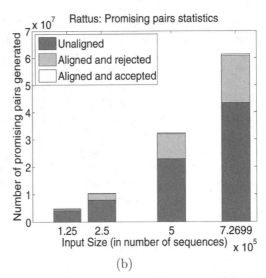

(a)   (b)

FIGURE 13.12: (a) The number of overlaps stored by the CAP3 assembly software as a function of input size on a subset of *Rattus sp.* (rat) ESTs. The peak memory usage reached 2 GB for 150,000 ESTs. (b) The number of pairs generated, aligned, and accepted by the PaCE clustering software as a function of input size for the entire EST data publicly available in GenBank for *Rattus sp.* (or rat) consisting of 726,988 ESTs. The ratio of number of pairs unaligned to the total number of pairs generated signifies the percentage work that is saved by the clustering heuristic.

While it is possible to handle the above properties of EST data by devising specialized variation of alignment methods post-clustering, there are two more properties prevalent in EST data — non-uniform sampling across gene sources and within a gene source — that make their clustering particularly hard especially with increasing data sizes. The number of ESTs derived from a gene depends on its level of activity, and because some genes may be expressed more than others, the resulting EST collection may show highly diverse sampling levels. In addition, the underlying sequencing mechanism may differentially cover a cDNA clone. For example, it is possible that the 3′ end of a clone is covered by hundreds to thousands of ESTs, while the 5′ end is covered by less than tens of ESTs.

Either case of non-uniform sampling implies a worst case quadratic number of "truly" overlapping ESTs to be dealt with during clustering, and may therefore limit the scalability of approaches that store such overlaps. Experimentation on different sized subsets of rat (*Rattus sp.*) ESTs downloaded from GenBank using CAP3 assembler (a software that is widely used for clustering EST data and also highly acclaimed for its clustering quality [18]) show that for an input increase from 100,000 to 150,000 ESTs the number of overlaps stored almost doubled, as illustrated in Figure 13.12a. The peak memory usage for the 150,000 data is 2 GB. Also note that the number of overlaps stored means that at least as many alignments were computed by the software. This quadratic increase in storage is expected of CAP3 or any other assembly software when applied to ESTs because they were originally developed for the assembly problem assuming uniform sampling.

In order to provide readers a perspective of these scaling issues associated with software developed for EST clustering, we experimented on the entire rat collection comprising of 726,988 ESTs and its subsets (data downloaded as of May 2005, from GenBank) using the PaCE software. PaCE clustered the entire collection in 2 hours on 27 dual-processor

nodes of a myrinet cluster with 3.06 GHz Xeon processors and 2 GB RAM each. The peak
memory consumed per processor was ~600 MB. Using a minimum exact match length (w)
of 40 bp, the number of promising pairs generated was ~60 million, of which the greedy
heuristic selected only ~18 million (30%) for alignment. Less than 1% of the aligned pairs
were "accepted" i.e., contributed to merging of clusters, indicating the presence of numerous
exact matches between ESTs that do not overlap significantly. Figure 13.12b shows similar
statistics for other subsets of the rat EST collection. The increase in the number of promising
pairs generated with input size indicates the expected worst-case quadratic behavior.

Because of cost-effectiveness, ESTs are sequenced in large quantities. The largest publicly
available EST database is the dbEST repository maintained by the GenBank division of the
National Center for Biotechnology Information (`http://www.ncbi.nlm.nih.gov/dbEST/`).
As of May 2005, this database has over 26.8 million sequences obtained from 869 organisms.
Among the animals, the human collection is the largest with 6,057,790 ESTs followed by
mouse with 4,334,174 ESTs. Among the plants, the largest is that of *Triticum aestivum*
(wheat) with 589,455 ESTs followed by *Zea mays* (maize) with 452,984 ESTs.

Among publicly funded EST clustering projects, the UniGene project undertaken by
NCBI is an excellent resource that creates and maintains clusters of ESTs based on gene
sources, with a goal of representing the "transcriptome" (or the set of all mRNA transcripts)
in different species. The UniGene clustering procedure is incremental — existing clusters
are updated with every new increment of EST data, thereby making their method applicable
to situations in which input sequences become available over an extended period of time.
As of May 2005, the UniGene collection contained clustering for EST data obtained from
52 organisms, the largest being that of human with over 53,000 clusters.

The Institute for Genomic Research (TIGR) maintains the TIGR Gene Indices (TGI),
by forming EST clusters (obtained using the TGICL software) and a subsequent assembly
(obtained using CAP3). Gene indices are available for 31 animals, 30 plants and many
microbial species.

Another public repository is PlantGDB, a NSF-funded resource for plant comparative genomics (`http://www.plantgdb.org`) that publishes and maintains clusters for plant ESTs.
The project curates clustering results of EST data from about 50 plant species, with sizes of
individual collections ranging from a few hundred ESTs to as large as 554,859 for *Triticum
aestivum* (wheat). The PaCE software is used for clustering purposes, and the resulting clusters are assembled using CAP3 assembly software. The results can be viewed at
(`http://www.plantgdb.org/ESTCluster/progress.php`).

13.7 Towards a Generic Platform for Large-scale Sequence Analysis

In Sections 13.4 and 13.5, we described a clustering-based framework for performing large-
scale genome assembly and EST clustering projects. This framework can be extended into a
more generic platform for conducting other types of large-scale sequence analysis in parallel,
by providing a new capability for analyzing multiple sequence types during the same analysis
experiment. In this section, we discuss various potential applications that motivate the need
to develop such a generalized framework, and then describe a clustering-based approach that
can be tuned with "rules" based on the outcome desired by the end-user.

Public repositories such as GenBank have genes, full-length cDNAs, ESTs, protein and
amino acid sequences, and genomic fragments that can be classified into many sub-types
based on their sources and size ranges (e.g., full-length assembled BACs, whole genome
shotgun sequences, fosmids, known repetitive elements, etc.). Providing the capability to

handle multiple sequence types and integrating information across multiple sequence types will extend the reach of analysis experiments. Listed below are some potential applications that can benefit directly from a platform that supports this capability.

- **Gene identification, structure prediction and alternative splicing studies:** In a genome project, genes can be located within contigs by finding portions that have expression evidence. This can be achieved by identifying the contigs that align "well" with ESTs, cDNAs, and/or proteins that are derived from same or closely related organisms. Once a gene is identified through this alignment process, its intronic and exonic boundaries can also be predicted by marking the boundaries of the alignments. Furthermore, if many ESTs align to the same genic region but not necessarily at same segments of it, a subsequent assembly of the corresponding ESTs can be used to identify the different alternatively spliced forms of the gene.

- **Scaffolding:** In genome projects, aligning expressed sequences with contigs can also provide scaffolding information at the level of a gene. Due to lack of coverage, a gene may be split across multiple contigs. Since expressed data such as ESTs, cDNAs and protein sequences represent genic regions but without their introns, they may have originated from distant exons of the same gene and so can be used to order and orient the contigs containing any of their exons.

  Assuming that the order of most genic and inter-genic regions are conserved across closely related species (as mentioned in Chapter 10), genome projects can also benefit from scaffolding information already computed in previous genome projects. Treating the contigs from two different genome projects, one currently underway and another completed, as two "types", a partial order and orientation of contigs in the first type can be induced from the already computed information of their counterparts in the second type.

- **Gene discovery:** Novel expressed genes in organisms for which the genomes have not yet been sequenced can be predicted by clustering its ESTs and cDNAs and then identifying those clusters that have protein evidence.

- **Incremental clustering:** In sequencing projects, sequences are typically available over a period of time. The goal of incremental clustering is to take advantage of the results of previous clustering and assembly efforts, and to compute only those changes that are induced in the output by the new increment — by treating the old and new data as two different types.

- **Polymorphism detection:** Expressed data extracted from normal and diseased tissues often exhibit Single Nucleotide Polymorphisms (SNPs). Detecting such polymorphic events between sequences from different tissues (types) is an important step in identifying the genetic basis of certain inherited diseases. Similarly, polymorphic events can be used to differentiate among paralogous genes (genes in the same organism that are highly similar).

- **Repeats and vector cleaning:** Sequences can contain foreign DNA elements or repetitive elements that confound the assembly process by indicating "false" overlaps. Given a set of known repeats and vector sequences, the task of locating them in the input sequence data before the assembly process is similar to the task of finding contained substrings or subsequences (to allow for a few errors) between the input and the repeat/vector sequences.

Type I	Type II	Alignment Method
genomic DNA	genomic DNA	semi-global
genomic DNA	cDNA/EST	spliced
cDNA/EST	cDNA/EST	semi-global, local alignment
Amino acid	EST/cDNA	DNA protein alignment without intronic insertions
Amino acid	genomic DNA	DNA protein alignment accounting for introns

TABLE 13.2 A sample rule-table showing the appropriate alignments to compute for corresponding pairs of types. Types I and II denote the types of any two sequences that can be considered for alignment.

13.7.1 A Rule-Based Approach

Considerable work has already been done towards finding and establishing relationships between sequences of different types, although at a sequence to sequence level — the central theme of them all being "related sequences show a corresponding sequence homology". Overall, these methods are specialized alignment techniques that take two sequences and account for the patterns induced by their types, in a way that reflects their original biological origins. For example, aligning a genomic fragment with a cDNA/EST sequence is computed as a spliced alignment that accounts for the intronic insertions that are spliced out during transcription, a subject of focus in Chapter 2. Aligning a genomic fragment with a protein sequence with the goal of gene structure prediction is a well studied problem [11, 13, 16], while aligning a cDNA sequence with a protein sequence is the subject of [33]. Furthermore, there are programs such as BLAST and its myriad varieties used for making local alignment based inferences.

When it comes to performing cross-type analyses between sets of sequences, a popular strategy is to run an appropriate alignment program between all sequences of one type against all sequences of another. While this simple protocol is capable of producing desired biological results, it is computationally a brute-force approach, limiting the scale of data to which such an approach can be applied. Therefore, developing a more efficient platform for conducting large-scale cross-type sequence analysis can accelerate the process of biological inferences drawn from sequence data. The rule-based approach proposed below, which targets efficient solutions for the applications including those listed in Section 13.7, is a good starting point towards developing a more generic platform.

Suppose the user inputs sequence data with information on their types. Along side, the user provides a set of "rules" stating the combination of types to be evaluated for overlap and their corresponding alignment methods. A sample set of such rules is provided in Table 13.2.

Based on the input sequence types and the rules specified by the user, the algorithm first preprocesses the input so that all are represented in one alphabet, either DNA or amino acid characters but not both, and then constructs a string data structure such as the generalized suffix tree or generalized suffix array on this transformed input. For example, if the input comprises genomic fragments and protein sequences, the genomic fragments can be translated to their six reading frames (three each in forward and reverse orientations), and a GST can be constructed over the amino acid alphabet on the translated genomic fragments and the input protein sequences. An exception to this approach is incremental clustering, in which a GST corresponding to the older data can be retrieved and updated with the new increment. The next step is to generate promising pairs and "consider" them for alignment computation consistent with the rules. Variations of the PaCE clustering heuristic that selects alignment workload are sought for this purpose. Performance optimization can be achieved by taking advantage of the rules — if the rules allow only for overlaps to be computed across types, it is needless to generate promising pairs from within the same type of sequences, and so the subtrees or their equivalents with only one type of sequence in it are not constructed, saving both run-time and space.

The efficiency of this rule-based approach lies in the choice of an appropriate definition for promising pairs and in devising efficient mechanisms to generate such pairs in parallel. Measures other than maximal matches may suit different clustering objectives. For example, while detecting single nucleotide polymorphic events in an input, a solution that allows for single character mismatches is preferable over another that allows only for maximal matches. In applications involving cross-species analysis, sequence relatives are expected to show more divergence among themselves. In such cases, allowing up to a fixed number of gaps and mismatches in addition to long matches is likely to increase the chances of a promising pair of genomic fragments passing an alignment test. This can be achieved by developing methods that detect bounded approximate matches within multiple sequences in parallel.

13.8 Conclusions

Molecular biology continues to shift toward data-driven hypotheses. Small questions related to a specific pathway or process are still undertaken; however, these experiments are often aided and sometimes greatly improved based upon the availability of complete genome sequences. To accomplish this goal, an intimate alliance has been formed between traditional biologists and computer scientists to process and analyze these large-scale sequence data resources. In this chapter, we have discussed two of these problems, namely genome assembly and EST clustering.

One of these solutions, the human genome sequence, was a massive endeavor and a great accomplishment. Maize, which is discussed in depth in this chapter, will be another accomplishment and will usher in the application of technology being used to improve human health to improve economically important crops. While the complexity involved in many of these projects is not likely to be more complex than the human and maize genome, these approaches will require a significant amount of computational resources. In particular, the memory requirements of traditional genome assemblers require special consideration or the need to redesign them. The parallel clustering framework presented in this chapter and other previous parallelization approaches have begun this shift to high performance computers that can effectively address these immense problems without a complete overhaul of well-tested assembly techniques.

As the benefit of genomics-driven biology becomes more and more important during every day molecular biology, there is a push to provide even rudimentary sequences of the genes within these species until time and resources can be allocated to completing the entire genome. In this manner genome reduction and EST sequencing represent two valuable approaches that maximize value given minimal resources. Unfortunately, both of these sequencing approaches are subject to non-uniformity that further aggravate the computational resources needed to solve these problems. We believe that a cluster-then-assemble, as presented in this chapter, is a viable solution and demonstrated its usefulness towards the creation of Maize Assembled Genomic Islands (MAGIs) along with in large-scale EST clustering.

There appear to be two intriguing challenges based upon the ideas we have explored within this chapter. First, as the amount and availability of diverse large-scale sequencing data becomes available to bioinformaticians and computational biologists, can the scientific community devise accurate and appropriate algorithms and experiments to extract as much information as possible on diverse topics including molecular evolution, functional genomics, and comparative biology to name only a few. Second, would the sequencing capabilities ever improve to the point that genome assembly may well become a trivial step in the course of personalized medicine? If not, could the current developing understanding on haplotypes and other natural variations within the human species facilitate the design of diagnostic tests based on technology like microarrays? These questions remain to be answered.

Acknowledgements

This work was supported in part by the U.S. National Science Foundation under ACR-0203782 and CNS-0130861.

References

[1] A. Apostolico, C. Iliopoulos, G.M. Landau, and B. Schieber *et al.* Parallel construction of a suffix tree with applications. *Algorithmica*, 3:347–365, 1988.

[2] K. Arumuganathan and E.D. Earle. Nuclear DNA content of some important plant species. *Plant Molecular Biology Reporter*, 9:208–219, 1991.

[3] J.L. Bennetzen, V.L. Chandler, and P.S. Schnable. National Science Foundation-sponsored workshop report. Maize genome sequencing project. *Plant Physiology*, 127:1572–1578, 2001.

[4] S. Brunner, K. Fengler, M. Morgante, and S. Tingey *et al.* Evolution of DNA sequence homologies among maize inbreds. *Plant Cell*, 17:343–360, 2005.

[5] H. Chou, G.G. Sutton, A. Glodek, and J. Scott. Lucy - a sequence cleanup program. In *Proc. Tenth Annual Genome Sequencing and Annotation Conference*, 1998.

[6] International Human Genome Sequencing Consortium. Initial sequencing and analysis of the human genome. *Nature*, 409:860–921, 2001.

[7] S.J. Emrich, S. Aluru, Y. Fu, and T. Wen *et al.* A strategy for assembling the maize (*Zea mays* L.) genome. *Bioinformatics*, 20:140–147, 2004.

[8] J. Fickett. Fast optimal alignment. *Nucleic Acids Research*, 12(1):175–179, 1984.

[9] Y. Fu, S. J. Emrich, L. Guo, and T. Wen *et al.* Quality assessment of Maize Assembled Genomic Islands (MAGIs) and large-scale experimental verification of predicted novel genes. *Proceedings of the National Academy of the Sciences USA*, 102:12282–12287, 2005.

[10] S.A. Goff, D. Ricke, T.H. Lan, and G. Presting *et al.* A draft sequence of the rice genome (*Oryza sativa* L. ssp. *japonica*). *Science*, 296:92–100, 2002.

[11] O. Gotoh. Homology-based gene structure prediction simplified matching algorithm using a translated codon (tron) and improved accuracy by allowing for long gaps. *Bioinformatics*, 16(3):190–202, 2000.

[12] R. Hariharan. Optimal parallel suffix tree construction. *Journal of Computer and System Sciences*, 55(1):44–69, 1997.

[13] X. Huang and J. Zhang. Methods for comparing a DNA sequence with a protein sequence. *Computer Applications in the Biosciences*, 12(6):497–506, 1996.

[14] A. Kalyanaraman, S. Aluru, V. Brendel, and S. Kothari. Space and time efficient parallel algorithms and software for EST clustering. *IEEE Transactions on Parallel and Distributed Systems*, 14(12):1209–1221, 2003.

[15] M.S. Katari, V. Balija, R.K. Wilson, and R.A.Martienssen *et al.* Comparing low coverage random shotgun sequence data from *Brassica oleracea* and *Oryza sativa*

genome sequence for their ability to add to the annotation of *Arabidopsis thaliana*. *Genome Research*, 15(4):496–504, 2005.

[16] P. Ko, M. Narayanan, A. Kalyanaraman, and S. Aluru. Space-conserving optimal DNA-protein alignment. *Proc. IEEE Computational Systems Bioinformatics Conference*, pages 80–88, 2004.

[17] E.S. Lander and M.S. Waterman. Genomic mapping by fingerprinting random clones: a mathematical analysis. *Genomics*, 2:231–239, 1988.

[18] F. Liang, I. Holt, G. Pertea, and S. Karamycheva *et al.* An optimized protocol for analysis of EST sequences. *Nucleic Acids Research*, 28(18):3657–3665, 2000.

[19] E.H. Margulies, J.P. Vinson, W. Miller, and D.B. Jaffe *et al.* An initial strategy for the systematic identification of functional elements in the human genome by low-redundancy comparative sequencing. *Proceedings of the National Academy of Sciences USA*, 102(13):4795–4800, 2005.

[20] J.C. Mullikin and Z. Ning. The Phusion assembler. *Genome Research*, 13(1):81–90, 2003.

[21] L.E. Palmer, P.D. Rabinowicz, A.L. O'Shaughnessy, and V.S. Balija *et al.* Maize genome sequencing by methylation filtration. *Science*, 302(5653):2115–7, 2003.

[22] G. Pertea, X. Huang, F. Liang, and V. Antonescu *et al.* TIGR Gene Indices clustering tool (TGICL) a software system for fast clustering of large EST datasets. *Bioinformatics*, 19(5):651–652, 2003.

[23] P.A. Pevzner, H. Tang, and M.S. Waterman. An Eulerian path approach to DNA fragment assembly. *Proceedings of the National Academy of Sciences USA*, 98:9748–9753, 2001.

[24] P.D. Rabinowicz, K. Schutz, N. Dedhia, and C. Yordan *et al.* Differential methylation of genes and retrotransposons facilitates shotgun sequencing of the maize genome. *Nature Genetics*, 23:305–308, 1999.

[25] M. M. Rhoades. The early years of maize genetics. *Annual Reviews in Genetics*, 18:1–29, 1984.

[26] F. Sanger, A.R. Coulson, G.F. Hong, and D.F. Hill *et al.* Nucleotide sequence of bacteriophage lambda DNA. *Journal of Molecular Biology*, 162:729–773, 1982.

[27] N.M. Springer and W.B. Barbazuk. Utility of different gene enrichment approaches toward identifying and sequencing the maize gene space. *Plant Physiology*, 136:3023–3033, 2004.

[28] R.E. Tarjan. Efficiency of a good but not linear set union algorithm. *Journal of the ACM*, 22(2):215–225, 1975.

[29] D.C. Torney, C. Burks, D. Davison, and K.M. Sirotkin. *Computers and DNA*. Addison-Wesley, New York, 1990.

[30] J.C. Venter, M.D. Adams, E.W. Myers, and P.W. Li *et al.* The sequence of the human genome. *Science*, 291:1304–1351, 2001.

[31] C.A. Whitelaw, W.B. Barbazuk, G. Pertea, and A.P. Chan *et al.* Enrichment of gene-coding sequences in maize by genome filtration. *Science*, 302:2118–2120, 2003.

[32] Y. Yuan, P.J. SanMiguel, and J.L. Bennetzen. High-C_0t sequence analysis of the maize genome. *The Plant Journal*, 34:249–255, 2003.

[33] Z. Zhang, W.R. Pearson, and W. Miller. Aligning a DNA sequence with a protein sequence. *Journal of Computational Biology*, pages 339–49, 1997.

[34] Z. Zhang, S. Shwartz, L. Wagner, and W. Miller. A greedy algorithm for aligning DNA sequences. *Journal of Computational Biology*, 7:203–214, 2000.

IV

Genome-Scale Computational Methods

14

Comparisons of Long Genomic Sequences: Algorithms and Applications

Michael Brudno
University of Toronto

Inna Dubchak
Lawrence Berkeley National Laboratory

14.1 Introduction

Comparing genomic sequences across related species is a fruitful source of biological insight, because functional elements such as exons tend to exhibit significant sequence similarity due to purifying selection, whereas regions that are not functional tend to be less conserved. The first step in comparing genomic sequences is to *align* them – that is, to map the letters of one sequence to those of the others. There are several categories of alignments: *local alignments* identify local similarities between regions of each sequence, *global alignments* find a mapping between all the letters of the sequences. Alignments can be either *pairwise*, between two sequences, or *multiple* that compare several sequences. The main challenge in developing algorithms for genomic alignment is that these must be fast enough to deal with megabase long sequences and gigabase long genomes, but also accurately map individual base pairs.

While generating alignments is difficult computationally, visualization of alignments also presents challenges, such as how to enable users to interact with the data and the processing programs in the context of enormous datasets. Visualization frameworks should be easy to understand by a biologist and provide insight into the mutations that a particular region has undergone. Finally, alignments are useful only if they help shed light on the important functional elements in the genomic sequence. In this chapter, after a detailed discussion of algorithms used to construct genomic alignments and methods to visualize them we give a short overview of several algorithms that use an alignment to improve predictions of transcription factor binding sites.

14.2   Local Alignment

Local alignment is the basic problem of finding similar fragments in two sequences, regardless of the order and location of these similarities. Consequently local alignments allow one to identify rearrangements between two sequences, and are suitable for aligning draft sequences. The original local alignment algorithm is the Smith-Waterman [62] dynamic programming approach. This algorithm, however, runs in time proportional to the product of the lengths of the sequences. As this is impractical for comparing two long genomic intervals there has been extensive work since the mid 1980s on development of fast approaches for local alignment of genomic sequences. While the details of all of these approaches are different, most share some overriding paradigms. In particular, almost all algorithms start with seed generation – the location of short, exact or nearly exact matches between two sequences. Because this can be accomplished quickly by indexing one of the two sequences in an appropriate data structure, such as a lookup table or some variant of a suffix tree, these seeds help to reduce the search area of the local alignment algorithm to just the regions that are likely to be similar. Once generated, nearby seeds may be joined together: the presence of several seeds close to each other is a stronger evidence of homology than a single seed. Finally the individual seeds (or groups of seeds) are extended to find regions that did not match exactly but are still conserved. These three steps form the basis of most local alignment algorithms for DNA sequences; the individual algorithms differ in how they solve each of the steps.

14.2.1   Seed generation

Perhaps the simplest way to generate seeds between two sequences is a straight lookup table technique: all k-long words (k-mers) of one sequence (the database) are indexed in a table, and the k-mers of the other sequence (the query) are used to retrieve from the lookup table the locations at which the particular k-mer of the query is present in the database sequence. This approach was used in the two first (and perhaps the best known) local aligners for long sequences, FASTA [53] and BLAST [1, 2]. An alternative approach for seed generation is to use a suffix tree or one of its variants (such as Aho-Corasick automaton) to search for the seeds. This approach suffers from higher constant overhead, in terms of both the memory requirements and the running time, but allows for the use of longer k-mer matches as it takes advantage of sparseness of longer seeds: there are 4^k possible DNA words of length k, and only a small fraction of them may be present in a particular sequence. This makes suffix tree approaches preferable when one is searching for longer seeds, while direct lookup tables are preferable for shorter ones. Instead of searching for k-long words that match exactly between the two sequences it is possible to use degenerate words as seeds, that is, words where a certain number of the letters are allowed not to match. These seeds allow for a higher sensitivity than exact matching k-mer seeds, but are more computationally intensive to generate. These seeds can be found in one of several ways, all of which, however, lead to an exponential running time increase in the number of degeneracies: it is possible to create indices where all of the possible degenerate positions are absent (i.e. index all possible 7 bases within an 8-mer). This leads to each word being represented only once in an index, but the number of indices grows as $\binom{k}{n}$ where n is the number of degeneracies allowed. Alternatively, one can index each word at all of its degenerate locations. This way there is only one index, but each word is indexed in many places, resulting in a blowup in the amount of memory required for the index. A popular alternative to both of these techniques is known as a "spaced" seed, which was initially implemented in the PatternHunter program [46]. These seeds are similar to degenerate seeds, in that they allow certain positions to not match, but these positions are pre-specified. For example a 110101 seed requires 4 (1st, 2nd, 4th and 6th) out of 6 positions to match. The other two positions may not match. This pattern is referred to as a (4,6) spaced seed. Because the degenerate positions are known ahead of time it is possible to use just a single index to look up all seeds. PatternHunter seeds have been shown to be preferable to regular fixed-length k-mers because two adjacent k-mers will no longer share $k-1$ positions, reducing the correlation between adjacent words. If in the case of exact matching seeds it is sufficient to introduce a mutation every $k-1$ positions in order to prevent the algorithm from finding a seed, this is not sufficient in the case of spaced seed. In fact, for the optimal (9,15) seed 111001010011011 it is necessary to have a mutation at least every 7 base pairs to prevent a single seed from being found.

PatternHunter seeds are preferable to exact matching k-mers when the stretch of similarity between the two sequences is long; their efficacy drops off when one is comparing sequences with very short (< 30 bases) stretches of homology. Another way to generate seeds is to use k-mers extended to the first mismatch, which are called max-mers or maximal extended matches (MEMs) of a certain minimal length. Max-mers can be found in strictly linear time in the sequence length by using a suffix tree (see chapter 5). While max-mers do not offer a sensitivity improvement over k-mers (wherever there is a k-mer there is also a max-mer and vice-versa) they have the advantages of returning only a single seed for every stretch of exact matches, no matter how long. When one is comparing very close sequences, such as two primates or two strains of bacteria, this will lead to a reduced number of seeds that need to be analyzed. However, when one is comparing more distant sequences (such as humans and mice) most of the matches tend to be short, and MEMs do not offer a significant reduction in the number of seeds, while the extra overhead of finding MEMs makes k-mers or PatternHunter seeds preferable. Additional theoretical work on seeds has included vector seeds [11] and random projections [16], but these have not been used in any alignment program.

14.2.2 Joining of neighboring seeds

While one seed can be a good indicator of homology, several smaller seeds located "near" each other can be an even better indicator. This idea has been used by many programs, starting with the first heuristic local aligner FASTA, that required m seeds of length k within a certain window to start a local alignment. It has also been common practice in BLAST and several other programs to search for two nearby seeds (spaced by a few base pairs) before starting an extension. These approaches are commonly implemented by creating a lookup table or hash table for all of the diagonals of the dynamic programming matrix, numbering all of the diagonals from 1 to X+Y for two sequences of length X and Y. Once a new seed is found, its diagonal is looked up the table. If there is already a seed linked to the diagonal, and the seed is close enough to the new one, the two seeds are joined

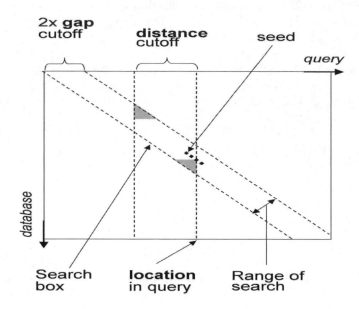

FIGURE 14.1: The CHAOS algorithm: The seed shown can be chained to any seed which lies inside the *search box*. All seeds located less then *distance bp* from the *current location* are stored in a skip list, in which we do a range query for seeds located within a *gap* cutoff from the diagonal on which the current seed is located. The seeds located in the grey areas are not available for chaining to make the algorithm independent of sequence order

together.

An alternative approach is used in the CHAOS program [12]. While similar to FASTA, it allows for groups of seeds to be on nearby diagonals. Each seed, once found, is stored in a skip list (a probabilistically balanced binary tree), indexed by the difference of its indices in the two sequences (diagonal number). For each new seed a range query is done in the skip list, searching for previously stored seeds which have a diagonal number within the permitted gap penalty from the diagonal at which the current seed is. The distance criterion is enforced by removing from the skip list any seed that is too far away from the position at which the new seeds are generated. This finds the possible previous seeds with which the current one can be chained. The highest scoring chain is picked, and it can then be further extended by future seeds (see Figure 14.1).

14.2.3   Ungapped and gapped extension

Once the seeds or groups of seeds are found, it is common practice to do some form of extension of the seeds in order to find the boundary of the homology. For this the most common approach is to do gap-free extensions which were first introduced in the original BLAST program. Here one keeps on adding one letter from each sequence to the already existing alignment, keeping track of the total score. The score goes up if the two letters match, and goes down otherwise. Once the score has fallen significantly lower than the maximum the extension is stopped and the alignment with the maximum score is returned. This process is commonly referred to as BLAST, or ungapped extension and is used in all of the early versions of the BLAST algorithm, as well as the CHAOS program. The alternative to this approach is a Smith-Waterman extension, where one is allowed to introduce gaps around the seeds. This process is much more time consuming, and is most useful when comparing distant genomic sequences. It is used in the BLASTZ alignment program [56], where gapped extensions are only triggered for ungapped alignments meeting a particular scoring threshold.

14.3   Global Alignment

Global alignments find the correspondence between two strings end-to-end by building a monotonically increasing map between the letters of each sequence. This produces a more accurate alignment at the base-pair level when the conserved features (for example genes) are in the same order in the compared species. The original global alignment algorithm is Needleman-Wunsch [51], which requires time proportional to the product of the lengths of the aligned sequences. This algorithm is too inefficient for comparing megabase long genomic sequences. Faster and more accurate methods have been developed recently: DI-ALIGN [50, 12], MUMmer [22, 23], GLASS [4], WABA [39], AVID [9] and LAGAN [13]. All these methods rely on an anchoring approach. It is worth noting that the anchors in global alignment serve the same purpose as the seeds in local alignment – to reduce the inherently quadratic search space. The overall approach followed by all of the tools mentioned above (as well as several others) can be summarized as 1) generate the fragments (local stretches of similarity) between the sequences, 2) resolve the set of fragments into the highest scoring consistent set using the sparse dynamic programming approach or some alternative, and 3) run a thorough global alignment algorithm either between the anchors, or in some region around them.

14.3.1   Finding potential anchors

The next chapter (chaining) shows how to construct the highest scoring consistent chains from fragments. When these algorithms are applied in the context of global alignment, the immediate question is how these fragments should be found, and what is a meaningful fragment. Perhaps the most straightforward method is to use k-mers as these fragments, as was done in the GLASS alignment program. Because individual k-mers are not a reliable guide to homology they were supplemented with a short extension by running the Needleman-Wunsch algorithm in a 12 × 12 window around each k-mer. Simultaneously the authors of MUMmer program suggested the use of maximum unique matches (MUMs) as the fragments used for the chaining algorithm. MUMs are maximum exact matching strings between two sequences that appear exactly once in each of the sequences. The fact that the MUM is a unique word in each of the sequences, reduces the probability of a false positive match. The disadvantage of this approach is its inability to find anchors between divergent genomes, where the maximal exact matches are usually too short to be unique. These two ideas were combined in the AVID program, which searches for maximal exact (not necessarily unique) matches, and does 12 × 12 Needleman-Wunsch windows around each match to verify its quality.

The most recent approach for generating the fragments has been the use of full local alignments. The first implementation of this idea was in the DIALIGN/CHAOS combination [12], where CHAOS local alignments were used to narrow down the search space of the DIALIGN program. CHAOS was subsequently used in a similar way in the LAGAN program, while BLAST local alignments were used in the ORCA program (Arenillas and Wasserman, unpublished).

Other approaches have leveraged biological knowledge in the process of anchor selection, with two prominent examples being the WABA [39] and CONREAL [5] programs, which search for anchors that are likely protein coding regions and transcription factor binding sites, respectively. In the case of WABA, the anchors between two sequences are "wobblemers", or k-mers where every third position is allowed not to match. Wobblemers are equivalent to PatternHunter seeds of the form 110110110...110. Allowing the third base pair to mutate more accurately reflects the conservation pattern of protein coding sequences as mutations in the third base pair of a codon are likely to be "silent": they will not change the amino acid being coded. This allows the wobblemers to be an effective method for anchoring protein regions. The CONREAL program, on the other hand, searches for potential transcription factor binding sites using the TRANSFAC suite of programs [47]. Two potential binding sites for the same transcription factor in the two sequences are used as the potential anchor points. While both WABA and CONREAL have been shown to be successful for their main purpose, that is alignment of protein coding and promoter regions, respectively, they are not general purpose alignment tools. Using a model of conservation rooted in a biological phenomenon usually does not work as well at modeling a different type of conservation: WABA will not work well for promoters and CONREAL will not align gene coding regions accurately.

14.3.2   Building a consistent set of anchors

The simplest way to find a consistent set of local alignment hits so that they can be used as anchor points is to use a greedy approach, where the strongest match (one with highest score using some scoring function) is accepted first, and every subsequent match is included if it does not conflict with any of the previous ones. This, however, leads to the problem of incorrect anchor points when several slightly weaker similarities may be ignored because of one strong one. Still, several alignment programs, including OWEN [54], use this approach.

Eppstein and colleagues [26] showed that it was possible to find the highest scoring consistent set of fragments in time $O(nlogn)$ via a sparse dynamic programming chaining procedure. For practical reasons it is not uncommon to do several rounds of anchoring: at the first step just the most confident fragments are fixed, a second pass fixes the less confident ones, etc. This approach, commonly known as hierarchical anchoring is useful because it allows one to search for lower levels of conservation: using a small seed size while looking at long sequences would lead to a very large number of seeds that are difficult to process. Because fewer anchors are found in each pass, hierarchical anchoring allows for an accurate analysis of each potential anchor.

14.3.3   Filling in the gaps

Once the set of anchors is fixed it is possible to reduce the search space of the final, slow global alignment algorithm to just the areas consistent with the anchors; usually Needleman-Wunsch is used, though the CHAOS/DIALIGN combination uses the DIALIGN program. Both k-mer and max-mer matches can be used as "hard" anchors: because these consist only of matches with no gaps, the optimal alignment has to go exactly over them. By comparison, local alignments are not reliable hard anchors, as a local alignment may have small errors. For this reason LAGAN implemented a more flexible anchoring scheme, where the anchor is not fixed, but rather noted, and the global alignment is required to go near

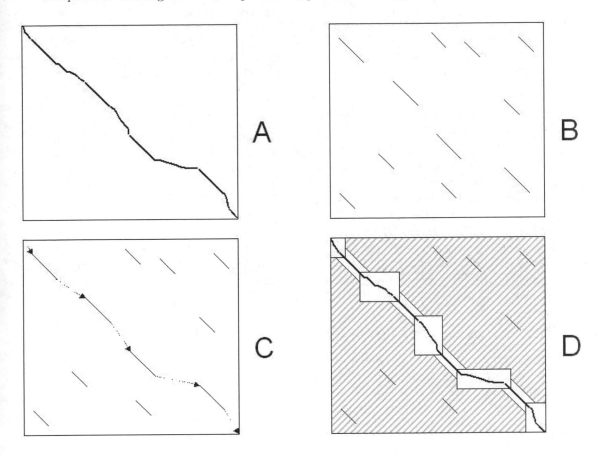

FIGURE 14.2: General scheme for most global alignment algorithms: (a) optimal map between the sequences (target) (b) potential anchors are found (c) highest scoring increasing subset of them is located (d) dynamic programming is done in the limited area. (From [13]).

the anchor, though not necessarily exactly over it (Figure 14.2).

14.4 Multiple Global Alignment

Similarity across large evolutionary distances can reveal conserved, and likely important biological features [30, 35, 65]. More recently it has been shown that it is unnecessary to compare distant species: it suffices to compare many close ones, and through the process of phylogenetic shadowing it is possible to separate the conserved regions from the neutral ones [7]. Comparative analysis, however, depends on *multiple alignments*. Multiple alignments have been shown to be more powerful than pairwise ones both in that they give a higher resolution of conserved regions [17] and a higher overall accuracy alignment [13]. Multiple alignments also allow for estimates of local rates of evolution that give quantitative measures of the strength of evolutionary constraints and the importance of functional elements [61, 63, 19]. Multiple alignments are considerably more difficult to compute than pairwise alignments: the running time scales as the product of the lengths of all the sequences.

Formally, the problem is NP-complete under several formulations [68, 8]. For this reason heuristic approaches are usually applied. Below we will first discuss the problem of scoring a multiple alignment, and then talk about two possible ways of obtaining a multiple alignment, the first via multi-way anchoring and the second via progressive alignment.

14.4.1   Scoring a multiple alignment

Perhaps the most basic, and at the same time the most non-trivial issue with multiple alignment is how to score it. The most common method used is the sum-of-pairs scoring, where the score for a particular column is set to just the sum of all the pairwise substitution and gap events. Alternatively it is possible to use a consensus model: for every column one finds the most likely character and penalizes divergence from the character. Another approach is to measure entropy in the column (see [25] for a summary). It is also possible to combine these approaches: LAGAN, for example, uses sum-of-pairs scoring for matches and mismatches and consensus for gaps. The problem of accurately scoring a multiple alignment remains very much an open one. Some innovative solutions have been published recently, such as T-COFFEE [52], and a full probabilistic framework for modeling multiple sequence alignments was implemented in HANDEL [34].

14.4.2   MGA and DIALIGN alignment algorithms

In order to generate global alignments of long genomic sequences it is necessary to generate anchor points between several sequences. For this only one truly multiple method has been suggested: multi-MEMs, that is multiple exact matches between all of the sequences being aligned. Because it is possible to find the highest scoring consistent chain in an arbitrary number of sequences in sub-quadratic time (see next chapter) these multi-MEMS can be used to reduce the search space for a multiple alignment algorithm. In between the anchor points it is necessary to run a sensitive multiple sequence alignment method to align the individual base pairs. This approach was first implemented in the MGA [33] program. The drawback to the MGA approach is that the requirement of each anchor being present in all sequences is too strict when one is comparing distant sequences. Because it is hard to find enough anchor points, the running time of the procedure becomes prohibitive. Consequently the MGA aligner, while being the first true global multiple alignment method for long genomic sequences is suitable only for comparing very close homologs, such as different strains of a bacterium.

Another approach for construction of multiple alignments from pairwise ones was implemented in the DIALIGN program [50]. The DIALIGN alignment consists of segments of ungapped homology (diagonals). To create a multiple alignment the set of pairwise diagonals is sorted according to the weights of the diagonals in a greedy way. Diagonals are incorporated one by one into the multiple alignment starting with the diagonal of maximum weight, provided they are not contradictory with the diagonals already incorporated. Diagonals contradictory with the growing set of consistent diagonals are rejected. The result of this selection process is a consistent set of diagonals – i.e., a multiple alignment.

14.4.3   Progressive alignment

The most common approach used for multiple alignment of several sequences is a progressive strategy, which constructs a multiple alignment by successive applications of a pairwise alignment algorithm. The best known system based on progressive alignment is perhaps CLUSTALW [66]. Some other systems include MULTALIGN [3], MULTAL [64], and PRRP

[29]. The basic idea behind this approach is that pairwise alignment techniques can be generalized from alignment of two sequence to alignment of two profiles, that is sequences where each position consists of some fraction of A's, C's, T's and G's rather than an individual letter. Because an alignment can be thought of as a profile (the gap can, for instance, be treated as a fifth character) it is possible to generate a multiple alignment via a bottom-up traversal of the phylogenetic tree, where after generating the alignments corresponding to a node's left and right children, one aligns these to get an alignment for the node itself.

Anchoring an alignment of two profiles

The progressive approach described in the previous paragraph can be used to build an alignment of two profiles in time proportional to the product of their lengths. This, however, is too slow for long sequences, and anchoring approaches have been used to reduce the running time of this problem in a very similar manner to the pairwise anchoring problem described above. The MLAGAN multiple alignment program uses all pairwise local alignments between the sequences to generate the set of anchor points. For example, given the sequences X and Y, the alignment between them X/Y and a third sequence Z, the anchors between X/Y and Z are computed as follows: First, all anchors in the rough global maps between X and Z, and between Y and Z, are mapped to their coordinates in the X/Y alignment and become anchors between X/Y and Z, with score equal to their original score. Second, each anchor between X and Z that overlaps an anchor between Y and Z, is reweighed with score equal to $(s1 + s2)\, I/U$, where $s1$, $s2$ are the scores of the (X, Z) and (Y, Z) anchors, respectively, I is the length of intersection, and U is the length of union of the anchors (summed in X/Y and Z). The rough global map between X/Y and Z is the highest scoring consistent chain of these anchors. An alternative method was suggested by the authors of the MAVID program [10]: they use phylogenetic methods to predict the likely ancestor of two sequences, and use the sequence of the ancestor as the representative of the sequences in the progressive steps.

14.5   Alignment of Sequences with Rearrangements

One common way in which the genome evolves is through rearrangement (shuffling) of blocks of DNA. Some of the most common rearrangement events are inversions (a block of DNA changes direction, but not location in the genome), translocations (a piece of DNA moves to a new location in the genome), and duplications (two copies of a block of DNA appear where there was one previously). Of the two main methods for alignment, global alignment, which shows how one sequence can be transformed into another using a combination of the simple edits, and local alignment, which identifies local similarities between regions of sequences, neither handles rearrangement events satisfactorily. Global alignment algorithms do not handle these events at all: the map between the two sequences that a global alignment algorithm creates must be monotonically increasing. While local alignment methods are able to identify homology in the presence of rearrangements between two sequences, they do not suggest how the two sequences could have evolved from their common ancestor. Also, in the case where both sequences have n paralogs (copies) of a particular gene or feature, local aligners return n^2 local alignments between all of the pairs, whereas a simple global alignment more clearly reflects the evolutionary process.

Varre and colleagues proposed a distance metric between two DNA sequences that models various rearrangement events [67]. Their algorithm, called Tnt1, builds a second sequence from an initially empty string using insertions and copying of blocks from the first sequence.

The distance between the two strings is defined to be the Kolmogorov complexity of the program that builds the second sequence. This algorithm has several shortcomings, the most notable being its inability to handle simple edit operations. It was also too slow to have practical applications to genomic sequences. The first programs for alignment of long genomic sequences with rearrangements were Shuffle-LAGAN [14] for pairwise sequences and Mauve [21] for multiple sequences. These are now discussed in turn.

14.5.1   Pairwise alignment with Shuffle-LAGAN

The Shuffle-LAGAN (S-LAGAN algorithm) was built on the LAGAN global alignment framework but allowed for rearrangements using a novel chaining technique. The S-LAGAN algorithm consists of three distinct stages. During the first stage the local alignments between the two sequences are found using the CHAOS tool. Second, the maximal scoring subset of the local alignments under certain gap penalties is picked to form a *1-monotonic conservation map*. It is the structure of this map that makes S-LAGAN different from standard anchored global aligners. Finally, the local alignments in the conservation map that can be part of a common global alignment are joined into *maximal consistent subsegments*, which are aligned using the LAGAN global aligner.

Building the 1-Monotonic Conservation Map

Most tools for rapid global alignment start with a set of local alignments, which they resolve into a "rough global map" – the set of anchors described in section 3.2. The rough global map must be non-decreasing in both sequences. In order to allow S-LAGAN to catch rearrangements, this assumption is relaxed to allow the map to be non-decreasing in only one sequence, without putting any restrictions on the second sequence. This is called a 1-monotonic conservation map.

To build this map we first sort all of the local alignments based on their coordinates in the base genome. For every next alignment we chain it to the previous one that gives the highest overall score subject to the affine chaining penalties. The penalty enforced depends on whether the previous alignment is on the same or different strand than the previous one, and whether it is before or after it in the coordinates of the second sequence (roughly speaking the four cases correspond to regular gap (same strand, after), inversion (different strand, after), translocation (same strand, before), and inverted translocation (different strand, before). By using the Eppstein-Galil sparse dynamic programming algorithm we can reduce the running time of this chaining procedure to $O(nlogn)$ from $O(n^2)$. The resulting highest scoring chain is 1-monotonic (strictly increasing in the base genome, but without any restrictions on the second genome order). The 1-monotonic chain can capture all rearrangement events besides duplications in the second genome.

Aligning Consistent Subsegments

Two local alignments are considered to be consistent if they can both be a part of a global alignment. Once we have a 1-monotonic conservation map it is straight-forward to generate the maximal consistent subsegments of the map by simply sorting all of the local alignments in the 1-monotonic map by their coordinates in the 1-monotonic sequence, taking the first alignment to be the start of a consistent subsegment, and adding additional local alignments while they are all consistent. As soon as an alignment is found to be inconsistent with the current subsegment, we start a new subsegment. Every consistent subsegment is extended to the nearest adjacent local alignment, so as to include areas of homology that did not fall into the local alignment, and are aligned using LAGAN. The overlap between adjacent consistent subsegments is resolved by doing a linear passes through the alignments, and finding the optimal breakpoint at which to end the first alignment and start the second one.

14.5.2 Multiple alignment with Mauve

The general problem of aligning multiple genomes that have undergone recombination events such as translocation, inversion, and duplication-loss remains an open problem. One early method that has shown promise has been implemented in the Mauve genome alignment package [21]. Like the other methods described in this chapter, Mauve uses a seeding procedure to generate candidate anchors, chains these anchors, and finally computes a progressive multiple alignment between anchors. Unlike previous methods, Mauve does not build a single consistent set of anchors, but rather builds one consistent set of anchors for every collinear segment of the genome sequences. Each consistent set of anchors is referred to as a Locally Collinear Block (LCB). LCBs are bounded on either side by a breakpoint: a change in LCB order or orientation among a pair of sequences.

The original Mauve algorithm used a seed-and-extend technique to generate multi-MUMs which were used as candidate anchors. Because multi-MUMs must be exact, unique matches occurring in every genome the original anchoring method had limited sensitivity. Current releases of Mauve use a spaced seed pattern to match multiple sequences simultaneously. The inexact matching method substantially improves anchoring sensitivity. Given a set of potential anchors that match in all of the genomes being aligned, Mauve uses a greedy breakpoint elimination method to filter out matches due to paralogy and random similarity. The greedy breakpoint elimination method repeats three core steps. (1) Use breakpoint analysis [45] to identify breakpoints in the anchor order, yielding LCBs (2) Calculate the weight o f each LCB as the sum of its constituent anchor lengths (3) Identify the lowest weight LCB. If it has weight < MinWeight then delete its anchors and return to step 1; otherwise end.

In the original Mauve paper, the genome alignment algorithm was applied to a group of nine closely related Enterobacteria, identifying numerous genome rearrangements and sites of differential gene content. Currently over 300 bacterial genomes have been finished and many more are nearly complete. As genome sequencing continues, we expect that automated methods for aligning multiple genomes with rearrangements will be tantamount to understanding the evolutionary forces shaping gene and genome function.

14.6 Whole Genome Alignment

The availability of the whole-genome human, mouse, and later rat assemblies presented for the first time the challenge of building multiple alignments of several large genomes. The problem of aligning genomes is more difficult than that of aligning sequences not only because of the size of the problem – a mammalian genome has about three billion basepairs – but also because of necessity of find the orthologous blocks, matching areas between the genomes, in which to apply alignment algorithms. Finding these blocks between two species computationally is a non-trivial task. Local alignment tools find a lot of high scoring matching segments, in particular the orthologous segments, but in addition they identify many paralogous relationships, or even false positive alignments resulting from simple sequence repeats and other artifacts [18]. The initial approaches for whole genome comparison developed for human and mouse genomes were based either on local alignment [55, 46, 6], or on a local/global technique, where stretches of one genome are mapped onto

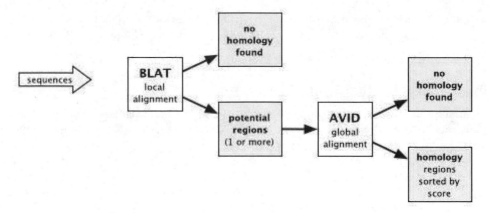

FIGURE 14.3: General computational scheme of tandem local/global genome alignment from (from [20]). The scheme used to aligns individual contigs, supercontigs, or long fragments of assemblies.

the others by a local aligner, and then the homology is confirmed and refined by a global one [20].

14.6.1   Local alignment on a whole genome scale

Perhaps the most straightforward approach to aligning two whole genomes is to do all-by-all local alignment. This is a challenging task due to the large amount of computation required, and is also difficult due to the problem of setting a threshold for individual local alignments: if the threshold is set too low many false positive local alignments could be found. If it is set too high true positive alignments are missed. Additionally classical local alignment methods do not consider whether a particular local alignment falls into a larger syntenic region. This leads to difficulties with local alignments that are either the result of repeats that were not masked, or with paralogous copies. Nevertheless local alignments were initially used for whole genome comparisons as they were better able to consider the rearrangements that are present between two large mammalian genomes. The initial local alignmets of the human and mouse genomes was done with BLASTZ [55] and PatternHunter [46]. For the human/mouse/rat three way alignments MULTIZ, a multiple sequence version of BLASTZ was used [6].

14.6.2   Local-global tandem approach

A computational strategy for finding and aligning orthologous regions that combines advantages of both local and global alignment techniques was first applied to the comparative analysis of the mouse and human genomes [20] and then expanded to the human, mouse and rat genomes [15]. In this technique the mouse genome is split up into contigs of 250 Kbp. The potential human homologs for each contig are found using the BLAT aligner [36]. The human sequence is then extended around the BLAT anchor, and aligned to the mouse contig using a global aligner (Figure 14.3). If the BLAT hits fall on both strands, then the aligner is called both with the original mouse contig and a reverse-complemented copy, making it possible to catch some inversions. This procedure has been expanded to three way alignment for comparing the human, mouse and rat genomes: First, the mouse and rat genomes are aligned using the BLAT program for approximate mapping followed by global alignment of selected regions. This step results in a set of mouse-rat *multi-contigs* (global alignments of rat contigs and mouse genomic sequence) as well as the remaining unaligned sequences. Second, the multi-contigs are aligned to human using the union of all available BLAT local alignments from mouse to human and from rat to human; mouse or rat sequences that could not be aligned to the other rodent are also aligned to human. The local/global tandem approach combines advantages of local and global alignment schemes in order to obtain both specificity (with respect to identifying only orthologous alignments) and sensitivity (in terms of coverage of genomic features of interest). At the same time the method is highly dependent on the parameters used at the local alignment stage, and it has difficulty aligning sequences that have undergone micro-rearrangements: small changes in the order of conserved elements due to inversions, duplications, or translocations of very short (sometimes only tens of base pairs) pieces of DNA.

14.7   Visualization

After obtaining alignments of two or more genomic sequences the next step is to analyze the level of overall homology, distribution of highly conserved elements and other comparative features. Visualization of results is a critical component of a comparative sequence analysis since manual examination of alignment on the scale of megabase long genomic regions is not efficient. Alignment-browsing systems should identify regions that exhibit properties suggestive of a particular biological function, for example well-conserved segments within an alignment, or matching the consensus sequence for a specific transcription factor binding site [49].

14.7.1   Visualization of pairwise alignments

There are several publicly available visualization tools for long pairwise DNA alignments. PIPMaker [55] represents the level of conservation in ungapped regions of BLASTZ local alignment as horizontal dashes called percent identity plots or pips. VISTA [28, 24, 48] displays comparative data in the form of a curve, where conservation is calculated in a sliding window of a gapped global alignment.

PipMaker and the companion server MultiPipMaker (`http://bio.cse.psu.edu`, [57]) visualize BLASTZ [56] local alignments. PipMaker visualizes the local alignments in two different formats: Percent identity plots (pips) and a dot plots. Pips present a compact, understandable display of local alignments on long genomic regions [32]. The program plots the position (in the base sequence) and percent identity of each gap-free segment of the alignments. The top horizontal axis is automatically marked with the positions of repeats and exons. The positions of CpG islands are also computed and displayed along the horizontal axis.

VISTA system is fundamentally based on global alignments, and its plot is generated by moving a user-specified window over the entire alignment and calculating the percent identity over the window at each base pair. The X-axis represents the base sequence; the Y-axis represents the percent identity. If the user supplies an annotation file, genes and exons are marked above the plot. Conserved segments with percent identity X and length Y are defined to be regions in which every contiguous sub-segment of length Y was at least X% identical to its paired sequence. These segments are merged to define the conserved regions. Conserved regions calculated by using user-submitted cutoffs are highlighted under

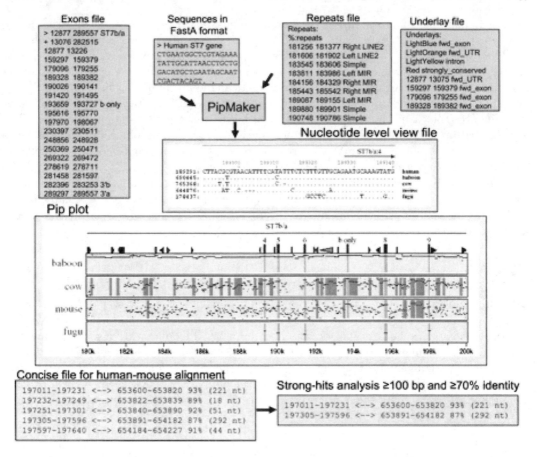

FIGURE 14.4: adapted from Frazer et al., 2003. [27] PipMaker: input and output files. The Pip plot shown is a subregion of the human ST7 interval compared with the orthologous baboon, cow, mouse, or fugu sequences. Each panel represents a pairwise comparison between human sequence and that of the indicated species

the curve, with different colors indicating a conserved non-coding regions, exons and UTRs. Figures 14.4 and 14.5 from the review of Frazer and coauthors ([27]) show the PipMaker and VISTA input and output.

Currently the selection of a particular visualization tool is mostly defined by the type of alignment used for the analysis. It is important to note, that as alignment algorithms become more sophisticated, it is becoming harder to distinguish between local and global alignment tools. For example, a chaining option for BLASTZ [56] allows for the extraction of global alignments from BLASTZ local alignments, and similarly Shuffle-LAGAN [14] is a hybrid *glocal* aligner that explicitly deals with rearrangements between sequences. Thus visualization methods also will have to become universal.

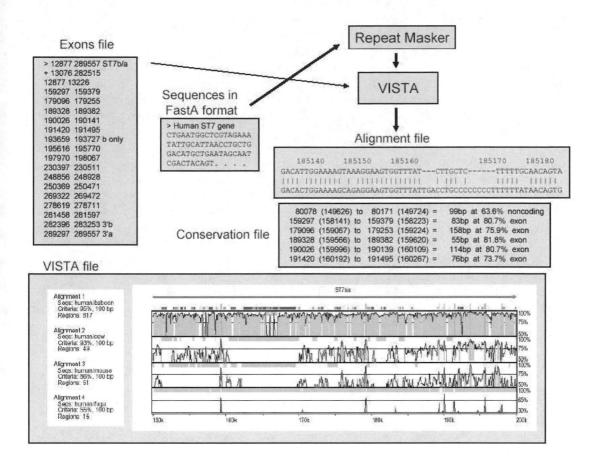

FIGURE 14.5: adapted from Frazer et al., 2003 [27]. VISTA: input and output files. The VISTA plot shown here is also a subregion of the human ST7 interval compared with the orthologous baboon, cow, mouse, or fugu sequences. Conserved sequences represented as peaks [noncoding (red) and coding (blue)] are shown relative to their positions in the human genome (horizontal axes), and their percent identities (50%-100%) are indicated on the vertical axes.

14.7.2 Visualization of multiple alignments

Both the VISTA and PipMaker approaches, described above, support visualization of a multiple alignment by projecting the alignment to a particular base sequence and in effect visualizing pairwise alignments between this base sequence and any number of homologous genomic intervals. This approach, however, only shows a part of the multiple alignment; it will be missing any similarity between fragments of two sequences other that the base that is not present in the base genome. For example, if there is a multiple alignment of human, mouse and rat with human used as the base, areas of conservation between mouse and rat that are not present in human will not be displayed. Full visualization of a multiple alignment is a difficult and largely unsolved problem, and currently is an area of active research.

The first tool to support visualization of multiple alignments was SynPlot [31]. The SynPlot graphical output includes a similarity profile of the long-range alignment together with a diagrammatic representation of both loci. Unlike PIPMaker and VISTA, SynPlot uses an alignment as a base coordinate, so the positions of all features in the individual sequences are mapped to the alignment coordinates. Feature files generated during annotation contain the positions of exons and repeat elements and can be directly imported into the graphical output. Therefore, the SynPlot output conveys comparative gene structure, repeat patterns (plus any other user-defined patterns), and relative sequence homology in a single linear plot. Its main drawback is that the single plot does not allow the user to distinguish the source of the similarity within the multiple alignment: a strongly conserved region in 3 of 5 species that is absent in the other two would look very similar to a weakly conserved region in all 5 of them. A recently developed program from the VISTA family, Phylo-VISTA (short for Phylogenetic VISTA, Shah et al., 2004)[58], uses the phylogenetic relationship as a guide to display and analyze the level of conservation across internal tree nodes. Using the entire multiple alignment, not a reference sequence, as a base in the x-axis allows for additional capabilities in visualization, such as presentation of comparative data together with available annotations for all sequences and computation of a measure of similarity for any node of the tree. The phylogenetic relationship among species is important for building and analyzing multiple alignments, thus visualizing sequence alignment data while taking phylogenetic trees into account is important to make the results easy to interpret.

14.7.3   Whole-genome visualization of alignments

The algorithmic challenge of whole genome alignment has been accompanied by user interface challenges, such as how to visualize information related to enormous datasets and how to enable users to interact with the data and the processing programs.

The principle of selecting a whole genome as a base sequence is utilized on the whole-genome scale in the UCSC genome browser (`http://genome.ucsc.edu` [38]) and the VISTA browser (`http://pipeline.lbl.gov` [20, 15]). These tools provide complementary information for a number of genomes including human, mouse, rat, drosophila. The UCSC browser represents annotations as a series of horizontal "tracks" over the genome sequence. Each track displays a particular type of annotation, such as Genscan gene predictions, m-RNA alignments, interspersed repeats, and others. There are two types of tracks to display comparative data such as alignments and various statistical measures of alignments. The first is a curve, the other is a conventional UCSC genome browser block-based display. The curve track called "Conservation" shows a measure of evolutionary similarity in multiple species based on the phylogenetic Hidden Markov Model (phylo-HMM) [59, 60] and MULTI-Z alignments [6] of human, chimpanzee, mouse, rat, and chicken whole-genome assemblies. Unlike the "Conservation" track, other comparative genomics tracks show particular fragments of alignments as boxes. Thus, "Chained Blastz" shows genomic alignment of different assemblies to the base sequence. Another track, "Alignment Net" [37] shows the best chain for every part of the base genome. There are some other comparative block tracks, such as tight subset of best alignments, differential view of the human/chimp alignment, and others.

The VISTA Browser is a Java applet for interactively visualizing results of alignment of entire genomes in the VISTA format on the scale of whole chromosomes along with annotations [15]. The user may select any genome as the reference or base, and display the level of conservation between this reference and the sequences of another species in a particular interval. The browser has a number of options, such as zoom, extraction of a region to be displayed, user-defined parameters for conservation level, and options for selecting sequence elements to study. A VISTA display is also implemented as a custom track linked to the

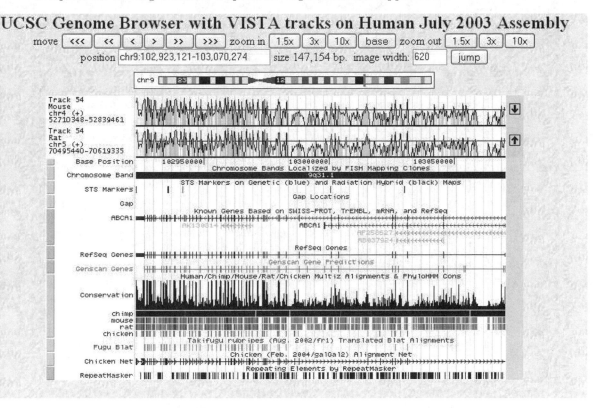

FIGURE 14.6: **(See color insert following page xxx.)** UCSC browser [36] with custom-built VISTA tracks showing conservation between the human chr. 9 interval aligned with orthologous mouse and rat sequences.

UCSC Browser. Figure 14.6 shows the UCSC Browser display of conservation with added VISTA conservation tracks and a control panel as a custom module accessible through the Berkeley Genome Browser.

14.8 Applications of Alignments

Identifying transcriptional and other regulatory elements represents a significant challenge in annotating the genomes of higher vertebrates because these elements are usually very short (5 to 20 bp in length) and have low information content. Adding comparative sequence analysis methods allowed for improving and refining signal searches. These methods help filter computational predictions to reduce noise of false positive predictions at the price of some decrease in sensitivity. Here we briefly describe two approaches. One of them uses functional site prediction information together with comparative data to refine and improve predictions, while the other utilizes comparative and co-expression data.

Clues for identifying sequences involved in the complex regulatory networks of eukaryotic genes are provided by the presence of transcription factor binding sites (TFBS) motifs, the clustering of such binding site motifs, and the conservation of these sites between species. rVISTA [43, 44] takes advantage of all these established strategies to enhance the detection of functional transcriptional regulatory sequences controlling gene expression through its ability to identify evolutionarily conserved and clustered TFBSs. Although the identification of conserved TFBSs on a small genomic interval can be achieved by phylogenetic footprinting [32, 40], a strength of the rVISTA algorithm is its ability to efficiently analyze large genomic intervals and potentially whole genomes. The clustering modules and the user-defined customization of visualized sites make this a further useful tool for the investigation of TFBSs. To take advantage of combining sequence motif recognition and multiple sequence alignment of orthologous regions in an unbiased manner, rVISTA analysis proceeds in four major steps: (1) identification of transcription factor binding sites (TFBS) matches in the individual sequences (the program uses available position weight matrices in the TRANSFAC database and independently locates all TFBS matches in each sequence), (2) identification of globally aligned noncoding TFBSs, (3) calculation of local conservation extending upstream and downstream from each orthologous TFBS, and (4) visualization of individual or clustered noncoding TFBSs. Available sequence annotations are used to identify aligned TFBS matches in noncoding genomic intervals.

Another tool, Consite [41] uses the same principle of combining predictions of TFBS and sequence conservation information. This program used a good quality binding profiles collection of TFBS developed by the authors and provided an efficient graphical web application to visualize results of the analysis. It is also worth mentioning here that CONREAL algorithm, described earlier, is both an alignment algorithm and a tool for finding TFBS. By using potential TFBSs as anchor points CONREAL attempts to more accurately align these. This leads to a decrease in the false positive rate, achieved in the same way as in the rVISTA program.

A novel approach for finding transcription factor binding sites in a genome is to use both conservation information within a species and information about coregulation of genes in the same genome. While the latter allows for ab initio prediction of TFBSs for which no known motifs exist, the former leads to a reduction in the false positive rate. Two programs that use this idea are PhyloCon and CompareProspector.

PhyloCon (Phylogenetic Consensus) [69] is based on the Consensus algorithm previously established by the same group and takes into account both conservation among orthologous genes and co-regulation of genes within a species. This algorithm first aligns conserved regions of orthologous sequences into multiple sequence alignments, or profiles, and then compares profiles representing co-regulated genes. Motifs emerge as common regions in these profiles, and a greedy approach is used to search for common subprofiles in the sequences. PhyloCon thus integrates knowledge of co-regulation of genes in a single species and sequence conservation across multiple species to improve performance of motif finding.

Similar ideas were used in CompareProspector [42]. This program takes as input the upstream sequences of a group of genes that are known or predicted to be coregulated, as well as local sequence conservation calculated based on alignments with orthologous sequences. CompareProspector uses a Gibbs sampling approach to search for motifs in the input sequences, biasing the search toward conserved regions by integrating sequence conservation into the posterior probability in the sampling process. CompareProspector was tested on two data sets from humans using human-mouse comparisons and two data sets from *Caenorhabditis elegans* using *C. elegans* – *C. briggsae* comparisons, and demonstrated the power of comparative genomics-based biased sampling in eukaryotic regulatory element identification.

14.9   Conclusion

In this chapter we have tried to illustrate the standard methodologies that have been used for creation and visualization of alignments of genomic sequence. We would like to emphasize that Comparative Genomics is a vibrant, growing field, and that this chapter represents a snapshot in time. We are bound to have missed covering some programs and algorithms, for example because they appeared too late for us to include them (most of this chapter was written in the summer of 2004). Some parts, especially individual programs and particular visualization techniques may become dated very quickly, however we are hopeful that the readers will find our synopsis of the underlying methodologies helpful.

Acknowledgments

The section on Mauve was contributed by Aaron Darling. We would like to thank Katya Permiakova for reading an early draft of the chapter and making many useful comments. We would like to thank our co-authors on many of the papers on which this chapter is partially based, especially Serafim Batzoglou, Greg Cooper, Chuong (Tom) Do, Burkhard Morgenstern, Alex Poliakov, Arend Sidow, and many others. Their ideas have contributed to our understanding of this area and helped define the general directions of this chapter. We are also grateful to the biologists in the Genomics Division of Lawrence Berkeley National Laboratory, especially Edward Rubin and Kelly Frazer (now at Perlegen Sciences) who introduced us to many challenging problems in comparative genomics.

References

[1] S.F. Altschul, W. Gish, W. Miller, and E.W. Myers *et al*. Basic local alignment search tool. *Journal of Molecular Biology*, 215:403–410, 1990.

[2] S.F. Altschul, T.L. Madden, A.A. Schäffer, and J. Zhang *et al*. Gapped BLAST and PSI-BLAST: a new generation of protein database search programs. *Nucleic Acids Res.*, 25(17):3389–3402, September 1997.

[3] G.J. Barton and M.J. Sternberg. A strategy for the rapid multiple alignment of protein sequences. Confidence levels from tertiary structure comparisons. *J Mol Biol*, 198(2):327–337, Nov 1987.

[4] S. Batzoglou, L. Pachter, J.P. Mesirov, and B. Berger *et al*. Human and mouse gene structure: comparative analysis and application to exon prediction. *Genome Res*, 10(7):950–958, Jul 2000.

[5] E. Berezikov, V. Guryev, R.H.A. Plasterk, and E. Cuppen. CONREAL: conserved regulatory elements anchored alignment algorithm for identification of transcription factor binding sites by phylogenetic footprinting. *Genome Res*, 14(1):170–178, Jan 2004.

[6] M. Blanchette, W.J. Kent, C. Riemer, and L. Elnitski *et al*. Aligning multiple genomic sequences with the threaded blockset aligner. *Genome Res.*, 14(4):708–715, 2004.

[7] D. Boffelli, J. McAuliffe, D. Ovcharenko, and K.D. Lewis *et al*. Phylogenetic shadowing of primate sequences to find functional regions of the human genome. *Science*, 299(5611):1391–1394, Feb 2003.

[8] P. Bonizzoni and G.D. Vedova. The complexity of multiple sequence alignment with SP-score that is a metric. *Theor. Comput. Sci.*, 259(1-2):63–79, 2001.

[9] N. Bray, Dubchak I., and Pachter L. AVID: a global alignment program. *Genome Res*, 13(1):97–102, 2003.

[10] N. Bray and L. Pachter. MAVID: constrained ancestral alignment of multiple sequences. *Genome Res.*, 14(4):693–699, 2004.

[11] B. Brejova, D. Brown, and T. Vinar. Vector seeds: an extension to spaced seeds allows substantial improvements in sensitivity and specificity. In G. Benson and R. Page, editors, *Algorithms and Bioinformatics: 3rd International Workshop (WABI)*, volume 2812 of *Lecture Notes in Bioinformatics*, pages 39–54, Budapest, Hungary, September 2003. Springer.

[12] M. Brudno, M. Chapman, B. Gottgens, and S. Batzoglou *et al.* Fast and sensitive multiple alignment of large genomic sequences. *BMC Bioinformatics*, 4(1):66, 2003.

[13] M. Brudno, C.B. Do, G.M. Cooper, and M.F. Kim *et al.* LAGAN and Multi-LAGAN: efficient tools for large-scale multiple alignment of genomic DNA. *Genome Res*, 13:721–731, 2003.

[14] M. Brudno, S. Malde, A. Poliakov, and C.B. Do *et al.* Glocal alignment: finding rearrangements during alignment. *Bioinformatics*, 19 Suppl 1:54–62, 2003. Evaluation Studies.

[15] M. Brudno, A. Poliakov, A. Salamov, and G.M. Cooper *et al.* Automated whole-genome multiple alignment of rat, mouse, and human. *Genome Res*, 14(4):685–692, 2004.

[16] J. Buhler. Provably sensitive indexing strategies for biosequence similarity search. *Journal of Computational Biology*, 10(3/4):399–417, 2003.

[17] M.A. Chapman, I.J. Donaldson, J. Gilbert, and D. Grafham *et al.* Analysis of multiple genomic sequence alignments: a web resource, online tools, and lessons learned from analysis of mammalian SCL loci. *Genome Res*, 14(2):313–318, Feb 2004.

[18] R. Chen, J.B. Bouck, G.M. Weinstock, and R.A. Gibbs. Comparing vertebrate whole-genome shotgun reads to the human genome. *Genome Res*, 11:1807–1816, 2001.

[19] G.M. Cooper, M. Brudno, E.D. Green, and S. Batzoglou *et al.* Quantitative estimates of sequence divergence for comparative analyses of mammalian genomes. *Genome Res*, 13(5):813–820, May 2003.

[20] O. Couronne, A. Poliakov, N. Bray, and T. Ishkhanov *et al.* Strategies and tools for whole-genome alignments. *Genome Res*, 13(1):73–80, Jan 2003.

[21] A.C.E. Darling, B. Mau, F.R. Blattner, and N.T. Perna. Mauve: multiple alignment of conserved genomic sequence with rearrangements. *Genome Res*, 14(7):1394–1403, Jul 2004.

[22] A.L. Delcher, S. Kasif, R.D. Fleischmann, and J. Peterson *et al.* Alignment of whole genomes. *Nucleic Acids Research*, 27(11):2369–2376, 1999.

[23] A.L. Delcher, A. Phillippy, J. Carlton, and S.L. Salzberg. Fast algorithms for large-scale genome alignment and comparison. *Nucleic Acids Res*, 30(11):2478–2483, Jun 2002.

[24] I. Dubchak, M. Brudno, L.S. Pachter, and G.G. Loots *et al.* Active conservation of noncoding sequences revealed by 3-way species comparisons. *Genome Research*, 10:1304–1306, 2000.

[25] R. Durbin, S.R. Eddy, A. Krogh, and G. Mitchison. *Biological sequence analysis: probabilistic models of proteins and nucleic acids.* Cambridge Univ. Press, 2000. Durbin.

[26] D. Eppstein, Z. Galil, R. Giancarlo, and G.F. Italiano. Sparse dynamic programming I: linear cost functions. *J. ACM*, 39(3):519–545, July 1992.

[27] K.A. Frazer, L. Elnitski, D.M. Church, and I. Dubchak *et al.* Cross-species sequence comparisons: a review of methods and available resources. *Genome Res*, 13(1):1–12,

Jan 2003.

[28] K.A. Frazer, L. Pachter, A. Poliakov, and E.M. Rubin *et al.* VISTA: computational tools for comparative genomics. *Nucleic Acids Res.*, 32:W273–9, July 2004. Web Server issue.

[29] O. Gotoh. Significant improvement in accuracy of multiple protein sequence alignments by iterative refinement as assessed by reference to structural alignments. *J Mol Biol*, 264(4):823–838, Dec 1996.

[30] B. Gottgens, L.M. Barton, M.A. Chapman, and A.M. Sinclair. Transcriptional regulation of the stem cell leukemia gene (SCL)–comparative analysis of five vertebrate SCL loci. *Genome Res*, 12(5):749–759, May 2002. Letter.

[31] B. Gottgens, J.G. Gilbert, L.M. Barton, and . Grafham *et al.* Long-range comparison of human and mouse SCL loci: localized regions of sensitivity to restriction endonucleases correspond precisely with peaks of conserved noncoding sequences. *Genome Res.*, 11:87–97, 2001.

[32] R.C. Hardison, J. Oeltjen, and W. Miller. Long human-mouse sequence alignments reveal novel regulatory elements: a reason to sequence the mouse genome. *Genome Res*, 7(10):959–966, Oct 1997.

[33] M. Hohl, S. Kurtz, and E. Ohlebusch. Efficient multiple genome alignment. *Bioinformatics*, 18 Suppl 1:312–320, 2002. Evaluation Studies.

[34] I. Holmes and W.J. Bruno. Evolutionary HMMs: a Bayesian approach to multiple alignment. *Bioinformatics*, 17(9):803–820, Sep 2001.

[35] M. Kellis, N. Patterson, M. Endrizzi, and B. Birren *et al.* Sequencing and comparison of yeast species to identify genes and regulatory elements. *Nature*, 423(6937):241–254, May 2003.

[36] J. Kent. BLAT - the BLAST-like alignment tool. *Genome Res.*, 12:656–664, 2002.

[37] W.J. Kent, R. Baertsch, A. Hinrichs, and W. Miller *et al.* Evolution's cauldron: duplication, deletion, and rearrangement in the mouse and human genomes. *Proc Natl Acad Sci U S A*, 100(20):11484–11489, Sep 2003.

[38] W.J. Kent, C.W. Sugnet, T.S. Furey, and K.M. Roskin *et al.* The human genome browser at UCSC. *Genome Res.*, 12(6):996–1006, 2002.

[39] W.J. Kent and A.M. Zahler. Conservation, regulation, synteny, and introns in a large-scale C. briggsae-C. elegans genomic alignment. *Genome Res*, 10(8):1115–1125, Aug 2000.

[40] W. Krivan and W.W. Wasserman. A predictive model for regulatory sequences directing liver-specific transcription. *Genome Res*, 11(9):1559–1566, Sep 2001.

[41] B. Lenhard, A. Sandelin, L. Mendoza, and P. Engstrom *et al.* Identification of conserved regulatory elements by comparative genome analysis. *Journal of Biology*, 2(2):13, May 2003. Epub.

[42] Y. Liu, X.S. Liu, L. Wei, and R.B. Altman *et al.* Eukaryotic regulatory element conservation analysis and identification using comparative genomics. *Genome Res.*, 14(3):451–458, 2004.

[43] G. Loots, I. Ovcharenko, L. Pachter, and I. Dubchak *et al.* rVISTA for comparative sequence-based discovery of functional transcription factor binding sites. *Genome. Res.*, 12:832–839, 2002.

[44] G.G. Loots and I. Ovcharenko. rVISTA 2.0: evolutionary analysis of transcription factor binding sites. *Nucleic Acids Res.*, 32:W217–21, 2004. Web Server issue.

[45] Blanchette M., G. Bourque, and D. Sankoff. Breakpoint Phylogenies. *Genome Inform Ser Workshop Genome Inform*, 8:25–34, 1997. JOURNAL ARTICLE.

[46] B. Ma, J. Tromp, and M. Li. PatternHunter: faster and more sensitive homology search. *Bioinformatics*, 18(3):440–445, 2002.

[47] V. Matys, E. Fricke, R. Geffers, and E. Gossling *et al.* TRANSFAC: transcriptional regulation, from patterns to profiles. *Nucleic Acids Res*, 31(1):374–378, Jan 2003.

[48] C. Mayor, M. Brudno, J.R. Schwartz, and A. Poliakov *et al.* VISTA: visualizing global DNA sequence alignments of arbitrary length. *Bioinformatics*, 16:1046–1047, 2000.

[49] W. Miller. Comparison of genomic DNA sequences: solved and unsolved problems. *Bioinformatics*, 17:391–397, 2001.

[50] B. Morgenstern, K. Frech, A. Dress, and T. Werner. DIALIGN: finding local similarities by multiple sequence alignment. *Bioinformatics*, 14(3):290–294, 1998.

[51] S.B. Needleman and C.D. Wunsch. A general method applicable to the search for similarities in the amino acid sequence of two proteins. *J Mol Biol*, 48(3):443–453, Mar 1970.

[52] C. Notredame, D.G. Higgins, and J. Heringa. T-Coffee: A novel method for fast and accurate multiple sequence alignment. *J Mol Biol*, 302(1):205–217, Sep 2000.

[53] W.R. Pearson. Rapid and sensitive sequence comparison with FASTP and FASTA. *Methods Enzymol*, 183:63–98, 1990.

[54] M.A. Roytberg, A.Y. Ogurtsov, S.A. Shabalina, and A.S. Kondrashov. A hierarchical approach to aligning collinear regions of genomes. *Bioinformatics*, 18(12):1673–1680, 2002.

[55] S. Schwartz, L. Elnitski, M. Li, and M. Weirauch *et al.* Nisc comparative sequencing program: Multipipmaker and supporting tools: alignments and analysis of multiple genomic dna sequences. *Nucleic Acids Res.*, 31:3518–3524, 2003.

[56] S. Schwartz, W.J. Kent, A. Smit, and Z. Zhang *et al.* Human-mouse alignments with blastz. *Genome Res.*, 13(1):103–107, April 2003.

[57] S. Schwartz, Z. Zhang, K.A. Frazer, and A. Smit *et al.* PipMaker-a web server for aligning two genomic DNA sequences. *Genome Res.*, 10(4):577–586, 2000.

[58] N. Shah, O. Couronne, L.A. Pennacchio, and M. Brudno *et al.* Phylo-VISTA: interactive visualization of multiple DNA sequence alignments. *Bioinformatics*, 20(5):636–643, Mar 2004. Evaluation Studies.

[59] A. Siepel and D. Haussler. Combining phylogenetic and hidden markov models in biosequence analysis. *In Proceedings of the Seventh Annual International Conference on Computational Molecular Biology (RECOMB 2003)*, pages 277–286, 2003.

[60] A. Siepel and D. Haussler. *Statistical Methods in Molecular Evolution*, chapter Phylogenetic hidden Markov models. Springer, 2004. in press.

[61] A.L. Simon, E.A. Stone, and A. Sidow. Inference of functional regions in proteins by quantification of evolutionary constraints. *Proc Natl Acad Sci U S A*, 99(5):2912–2917, Mar 2002.

[62] T.F. Smith and M.S. Waterman. Identification of common molecular subsequences. *J Mol Biol*, 147(1):195–197, Mar 1981.

[63] K. Sumiyama, C.B. Kim, and F.H. Ruddle. An efficient cis-element discovery method using multiple sequence comparisons based on evolutionary relationships. *Genomics*, 71(2):260–262, Jan 2001.

[64] W.R. Taylor. A flexible method to align large numbers of biological sequences. *J Mol Evol*, 28(1-2):161–169, Dec 1988.

[65] J.W. Thomas, J.W. Touchman, R.W. Blakesley, and G.G. Bouffard *et al.* Comparative analyses of multi-species sequences from targeted genomic regions. *Nature*, 424(6950):788–793, Aug 2003.

[66] J.D. Thompson, D.G. Higgins, and T.J. Gibson. CLUSTAL W: improving the sensitivity of progressive multiple sequence alignment through sequence weighting, position-specific gap penalties and weight matrix choice. *Nucleic Acids Res*, 22(22):4673–4680, Nov 1994.

[67] J.S. Varre, J.P. Delahaye, and E. Rivals. Transformation distances: a family of dissimilarity measures based on movements of segments. *Bioinformatics*, 15(3):194–202, Mar 1999.

[68] L. Wang and T. Jiang. On the complexity of multiple sequence alignment. *Journal of Computational Biology*, 1(4):337–348, 1994.

[69] T. Wang and G.D. Stormo. Combining phylogenetic data with co-regulated genes to identify regulatory motifs. *Bioinformatics*, 19(18):2369–2380, December 2003.

15

Chaining Algorithms and Applications in Comparative Genomics

Enno Ohlebusch
University of Ulm

Mohamed I. Abouelhoda
University of Ulm

15.1 Motivation: Comparison of Whole Genomes

The output of sequence data from worldwide sequencing centers has been rising at an exponential rate for the past decade or two. The first two publications of microbial whole genome sequencing projects appeared in 1995. To date (September 2004), there are 294 published complete genomes as well as 740 prokaryotic and 532 eukaryotic ongoing genome sequencing projects.

Comparative genomics is concerned with comparing genomic sequences to each other. If the organisms under consideration are closely related (that is, if no or only a few genome rearrangements have occurred) or one compares regions of conserved synteny (regions in which orthologous genes occur in the same order), then global alignments can be used for the prediction of genes and regulatory elements. This is because coding regions are relatively well preserved, while non-coding regions tend to show varying degree of conservation. Non-coding regions that do show conservation are thought important for regulating gene expression, maintaining the structural organization of the genome and possibly have other, yet unknown functions. Several comparative sequence approaches using alignments have recently been used to analyze corresponding coding and non-coding regions from different species, although mainly between human and mouse; see, e.g., [8, 44]. These approaches are based on software-tools for aligning two [18, 45, 8, 32, 19, 10] or multiple genomic DNA-sequences [38, 26, 11, 12, 43]; see [14] for a review. Here we focus on comparing two genomes and briefly outline how the methods described in this chapter can be extended

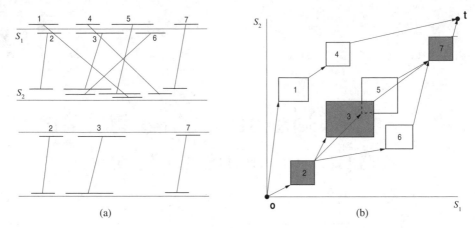

FIGURE 15.1: Given a set of fragments (upper left figure), an optimal global chain of colinear non-overlapping fragments (lower left figure) can be computed, e.g., by computing an optimal path in the graph in (b) (in which not all edges are shown).

to the comparison of multiple genomes. This is of utmost importance because there is an immediate need for "reliable and automatic software for aligning three or more genomic sequences"[37]. To cope with the sheer volume of data, most of the software-tools use an anchor-based method that is composed of three phases:

1. computation of fragments (segments in the genomes that are similar),
2. computation of a highest-scoring global chain of colinear non-overlapping fragments: these are the anchors that form the basis of the alignment,
3. alignment of the regions between the anchors.

This chapter deals, among other things, with algorithms for solving the combinatorial chaining problem of the second phase: finding a highest-scoring global chain of colinear non-overlapping fragments, where gaps between the fragments are penalized. Roughly speaking, two fragments are *colinear* if the order of their respective segments is the same in both genomes. In the pictorial representation of Figure 15.1(a), two fragments are colinear if the lines connecting their segments are non-crossing (in Figure 15.1, for example, the fragments 2 and 3 are colinear, while 1 and 6 are not). Two fragments *overlap* if their segments overlap in one of the genomes (in Figure 15.1, for example, the fragments 1 and 2 are overlapping, while 2 and 3 are non-overlapping). For didactic reasons, the presentation of the global chaining algorithm is separated into two sections. In Section 15.3, we will explain a global chaining algorithm that neglects gap costs, i.e., gaps between the fragments are not penalized at all. An application of this algorithm is cDNA mapping and a variant is an algorithm that solves the heaviest increasing subsequence problem. In Section 15.4, the algorithm is modified in two steps, so that it can deal with certain gap costs. It is worth mentioning that a related technique used to solve the global chaining problem is known under the name *sparse dynamic programming*. It was invented by Eppstein et al. [20] and independently by Myers and Huang [40] in solving a different problem. In our opinion, the approach presented here is conceptually simpler than those of [20, 40].

Global alignments are valuable in comparing genomes of closely related species. For diverged genomic sequences, however, a global alignment strategy is likely predestined to failure for having to align unrelated regions in an end-to-end colinear approach. This is because one expects that many genome rearrangements occurred during evolution. *Genome rearrangements* comprise large scale duplications, *transpositions* (where a section of the genome is excised and inserted at a new position in the genome, without changing orientation), and *inversions* (where a section of the genome is excised, reversed in orientation, and re-inserted). In this case, either local alignments are the strategy of choice or one must first identify syntenic regions, which then can be individually aligned. However, both alternatives are faced with obstacles. Current local alignment programs suffer from a huge running time, while the problem of automatically finding syntenic regions requires a priori knowledge of all genes and their locations in the genomes — a piece of information that is often not available. (It is beyond the scope of this chapter to discuss the computational difficulties of gene prediction and the accurate determination of orthologous genes.) In Section 15.5, we will show that a variant of our global chaining algorithm can be used to solve the problem of automatically finding local regions of similarity in large genomic DNA sequences. As in the anchor-based global alignment method, one first computes fragments between the two genomes. In the second phase, instead of computing a highest-scoring global chain of colinear non-overlapping fragments, we compute significant local chains — chains whose score exceeds a user-defined threshold. Under stringent thresholds, significant local chains of colinear non-overlapping fragments represent candidate regions of conserved synteny. If one aligns these individually, one gets good local alignments. Because we have to deal with inversions, however, we also chain fragments between the first genome and the reverse complement of the second genome in a seperate run of the chaining algorithm. We would like to point out that the automatic identification of regions of similarity is a first step toward an automatic detection of genome rearrangements; see [1].

Section 15.5 also deals with other variations of the chaining problem. For example, apart from its global 2-dimensional chaining option, the software tool PipMaker [45] provides an alternative method — the single coverage option — that "selects a highest-scoring set of alignments such that any position in the first sequence can appear in one alignment, at most." As we shall see, this single coverage problem can be solved by a 1-dimensional global chaining algorithm. Interestingly, the same algorithm can be used to compute the transformation distance between two genomes as defined by Varré et al. [50]. The transformation distance problem addresses the question of how two sequences may be most economically derived from one another by various genome rearrangements.

We conclude Section 15.5 by considering a generalization of the chaining problem, in which the colinearity requirement is removed. This generalized problem is to find a set of non-overlapping fragments such that the amount of sequence covered by the fragments is maximized. In computational biology, a solution to the problem can be used to determine the relative degrees of completion of two sequencing projects. To be more precise, in the comparison of the assemblies of two genomes, one tries to maximize their common regions regardless of the order of the assembled segments. This interesting generalization of the chaining problem is equivalent to finding a maximum weight independent set in a certain intersection graph. It has been shown to be NP-complete by Bafna et al. [6]. Therefore, one has to resort to approximation algorithms for solving it. As an alternative, one could use the approach of Halpern et al. [25]. The key idea of their solution is to minimally "refine" (subdivide) fragments until all overlaps between the refined fragments are "resolved", i.e., the projections of any two refined fragments onto each sequence are either disjoint or identical.

In Section 15.6, we outline the extension of the global and local chaining algorithms to the higher-dimensional case. This makes it possible to compare multiple genomes simultaneously. Among other things, most of the chaining algorithms described in this chapter are implemented in the program CHAINER [2].

It is worth mentioning that chaining algorithms are also useful in other bioinformatics applications such as comparing restriction maps [40]; see also [24, Section 16.10].

15.2   Basic Definitions and Concepts

For $1 \leq i \leq 2$, $S_i = S_i[1 \dots n_i]$ denotes a string of length $|S_i| = n_i$. In our application, S_i is the DNA sequence of a genome. $S_i[l_i \dots h_i]$ is the substring of S_i starting at position l_i and ending at position h_i. A *fragment* f consists of two pairs $beg(f) = (l_1, l_2)$ and $end(f) = (h_1, h_2)$ such that the strings (henceforth also called segments) $S_1[l_1 \dots h_1]$ and $S_2[l_2 \dots h_2]$ are "similar". If the segments are exact matches, i.e., $S_1[l_1 \dots h_1] = S_2[l_2 \dots h_2]$, then we speak of *exact fragments*. Examples of exact fragments are maximal unique matches as used in MUMmer [18, 19], maximal exact matches as used in MGA [26] and AVID [10], and exact k-mers as used in GLASS [8]. In general, however, one may also allow substitutions (yielding fragments as in DIALIGN [39] and LAGAN [12]) or even insertions and deletions (as the BLASTZ-hits [44] that are used in PipMaker [45]). Each fragment f has a positive weight (denoted by $f.weight$) that can, for example, be the length of the fragment (in case of exact fragments) or its similarity score.

A fragment f can be represented by a rectangle in $\mathbb{R}^2$ with the lower left corner $beg(f)$ and the upper right corner $end(f)$, where each coordinate of the corner points is a non-negative integer. To fix notation, we recall the following concepts. For any point $p \in \mathbb{R}^2$, let $p.x_1$ and $p.x_2$ denote its coordinates (we will also sometimes use $p.x$ and $p.y$ instead of $p.x_1$ and $p.x_2$). A rectangle, whose sides are parallel to the axes, is the Cartesian product of two intervals $[l_1 \dots h_1]$ and $[l_2 \dots h_2]$ on distinct coordinate axes, where $l_i < h_i$ for $1 \leq i \leq 2$. A rectangle $[l_1 \dots h_1] \times [l_2 \dots h_2]$ will also be denoted by $R(p, q)$, where $p = (l_1, l_2)$ and $q = (h_1, h_2)$ are the lower left and the upper right corner, respectively.

In what follows, we will often identify the point $beg(f)$ or $end(f)$ with the fragment f. This is possible because we assume that all fragments are known from the first phase of the anchor-based approach described in Section 15.1 (so that every point can be annotated with a tag that identifies the fragment it stems from). For example, if we speak about the score of a point $beg(f)$ or $end(f)$, we mean the score of the fragment f. For ease of presentation, we consider the origin $\mathbf{0} = (0, 0)$ and the terminus $\mathbf{t} = (|S_1| + 1, |S_2| + 1)$ as fragments with weight 0. For these fragments, we define $beg(\mathbf{0}) = \bot$, $end(\mathbf{0}) = \mathbf{0}$, $beg(\mathbf{t}) = \mathbf{t}$, and $end(\mathbf{t}) = \bot$, where $\bot$ stands for an undefined value.

DEFINITION 15.1   We define a binary relation $\ll$ on the set of fragments by $f \ll f'$ if and only if $end(f).x_i < beg(f').x_i$ for $1 \leq i \leq 2$. If $f \ll f'$, then we say that f *precedes* f'.

Note that $\mathbf{0} \ll f \ll \mathbf{t}$ for every fragment f with $f \neq \mathbf{0}$ and $f \neq \mathbf{t}$.

DEFINITION 15.2   A *chain* of colinear non-overlapping fragments (or chain for short) is a sequence of fragments $f_1, f_2, \dots, f_\ell$ such that $f_i \ll f_{i+1}$ for all $1 \leq i < \ell$. The *score* of C is $score(C) = \sum_{i=1}^{\ell} f_i.weight - \sum_{i=1}^{\ell-1} g(f_{i+1}, f_i)$, where $g(f_{i+1}, f_i)$ is the cost (penalty) of connecting fragment f_i to f_{i+1} in the chain. We will call this cost *gap cost*.

DEFINITION 15.3   Given m weighted fragments and a gap cost function, the *global fragment-chaining problem* is to determine a chain of highest score (called *optimal global chain* in the following) starting at the origin $\mathbf{0}$ and ending at terminus $\mathbf{t}$.

The global fragment-chaining problem was previously called fragment alignment problem [51, 20]. A direct solution to this problem is to construct a weighted directed acyclic graph $G = (V, E)$, where the set V of vertices consists of all fragments (including 0 and t) and the set of edges E is characterized as follows: There is an edge $f \to f'$ with weight $f'.weight - g(f', f)$ if $f \ll f'$; see Figure 15.1(b). An optimal global chain of fragments corresponds to a path of maximum score from vertex 0 to vertex t in the graph. Because the graph is acyclic, such a path can be computed as follows. Let $f'.score$ be defined as the maximum score of all chains starting at 0 and ending at f'. $f'.score$ can be expressed by the recurrence: $0.score = 0$ and

$$f'.score \quad = \quad f'.weight + \max\{f.score - g(f', f) : f \ll f'\} \qquad (15.1)$$

A dynamic programming algorithm based on this recurrence takes $O(|V| + |E|)$ time provided that computing gap costs takes constant time. Because $|V| + |E| \in O(m^2)$, computing an optimal global chain takes quadratic time and linear space; see [33, 16]. This graph-based solution works for any number of genomes and for any kind of gap cost. It has been proposed as a practical approach for aligning biological sequences, first for two sequences by Wilbur and Lipman [51] and for multiple sequences by Sobel and Martinez [47]. However, the $O(m^2)$ time bound can be improved by considering the geometric nature of the problem. In order to present the material systematically, we first give a chaining algorithm that neglects gap costs. Then we will modify this algorithm in two steps, so that it can deal with certain gap costs.

15.3 A Global Chaining Algorithm without Gap Costs

15.3.1 The Basic Chaining Algorithm

Because our algorithm is based on orthogonal range-searching for a maximum, we have to recall this notion. Given a set S of points in $\mathbb{R}^2$ with associated score, a *range maximum query* $\texttt{RMQ}(p, q)$ asks for a point of maximum score in $R(p, q)$.

LEMMA 15.1 If the gap cost function is the constant function 0 and $\texttt{RMQ}(0, beg(f') - \vec{1})$ (where $\vec{1}$ denotes the vector $(1, 1)$) returns the end point of fragment f, then we have $f'.score = f'.weight + f.score$.

Proof This follows immediately from recurrence (15.1).

We will further use the line-sweep paradigm to construct an optimal chain. Suppose that the start and end points of the fragments are sorted w.r.t. their x_1 coordinate. Then, processing the points in ascending order of their x_1 coordinate simulates a line that sweeps the points w.r.t. their x_1 coordinate. If a point has already been scanned by the sweeping line, it is said to be *active*; otherwise it is said to be *inactive*. During the sweeping process, the x_1 coordinates of the active points are smaller than the x_1 coordinate of the currently scanned point s. According to Lemma 15.1, if s is the start point of fragment f', then an optimal chain ending at f' can be found by an $\texttt{RMQ}$ over the set of active end points of fragments. Since $p.x_1 < s.x_1$ for every active end point p (w.l.o.g., start points are handled before end points, hence the case $p.x_1 = s.x_1$ cannot occur), the $\texttt{RMQ}$ need not take the first coordinate into account. In other words, the $\texttt{RMQ}$ is confined to the range $R(0, s.x_2 - 1)$, so that the dimension of the problem is reduced by one. To manipulate the point set during the sweeping process, we need a data structure D that stores the end points of fragments and efficiently supports the following two operations: (1) activation and (2) RMQ over the set of active points. The following algorithm is based on such a data structure D, which will be defined later.

FIGURE 15.2: 2-dimensional chaining of m fragments

Sort all start and end points of the m fragments in ascending order w.r.t. their x_1 coordinate and store them in the array points; because we include the end point of the origin and the start point of the terminus, there are $2m + 2$ points. Store all end points of the fragments (ignoring their x_1 coordinate) as inactive (in the 1-dimensional) data structure D.

for $i := 1$ **to** $2m + 2$
    if points$[i]$ is the start point of fragment f' **then**
        $q :=$ RMQ$(0,$ points$[i].x_2 - 1)$
        determine the fragment f with $end(f) = q$
        $f'.prec := f$
        $f'.score := f'.weight + f.score$
    else   /$\star$ points$[i]$ is end point of a fragment f' $\star$/
        activate points$[i].x_2$ in D /$\star$ activate with score $f'.score$ $\star$/

In the algorithm, $f'.prec$ denotes the preceding fragment of f' in a chain. It is an immediate consequence of Lemma 15.1 that Algorithm 15.2 (Note: for Algorithm 15.2, see Figure 15.2 etc.) finds an optimal chain. One can output this chain by tracing back the *prec* pointers from the terminus to the origin. The complexity of the algorithm depends of course on how the data structure D is implemented.

Answering RMQ with Activation Efficiently

To answer RMQ with activation, we use the *priority search tree* devised by McCreight [35]. Let S be a set of m one dimensional points. For ease of presentation, assume that no two points have the same coordinate ([35] shows how to proceed if this is not the case). The priority search tree of S is a minimum-height binary search tree T with m leaves, whose ith leftmost leaf stores the point in S with the ith smallest coordinate. Let $v.L$ and $v.R$ denote the left and right child, respectively, of an interior node v. To each interior node v of T, we associate a canonical subset $C_v \subseteq S$ containing the points stored at the leaves of the subtree rooted at v. Furthermore, v stores the values h_v and p_v, where h_v denotes the largest coordinate of any point in C_v and p_v denotes the point of highest score (priority) in C_v that has not been stored at a shallower depth in T. If such a point does not exist, p_v is undefined; see Figure 15.3. In essence, the priority search tree of S is a variant of a range tree and a heap (i.e., priority queue). It can be built in $O(m \log m)$ time and $O(m)$ space.

In Algorithm 15.2, storing all end points of the fragments as inactive boils down to constructing the priority search tree of the end points, where each point stored at a leaf has score $-\infty$ and p_v is undefined for each interior node v.

In a priority search tree T, a point q can be activated with priority *score* in $O(\log m)$ time as follows. First, update the value of $q.score$ by $q.score := score$. Second, update the p_v values by a traversal of T that starts at the root. Suppose node v is visited in the traversal. If v is a leaf, there is nothing to do. If v is an interior node with $p_v.score \le q.score$, then

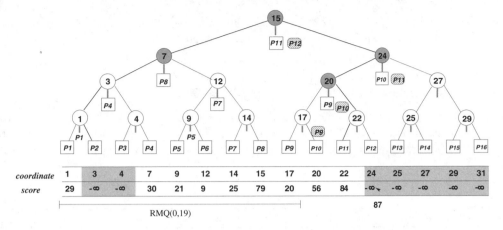

FIGURE 15.3: Priority search tree: The points with score $-\infty$ are inactive; the others are active. The value $h_{v.L}$ in every interior node v is the coordinate that separates the points in its left subtree $v.L$ from those occurring in its right subtree $v.R$. The p_v value of v is depicted as a "child" between $v.L$ and $v.R$. If it is missing, then p_v is undefined. The colored nodes are visited in answering the RMQ$(0, 19)$. The hatched boxes contain the modified p_v values when point p_{12} is activated with score 87.

swap q and p_v (more precisely, exchange the values of the variables q and p_v); otherwise retain q and p_v unchanged. Then determine where the procedure is to be continued. If $q \leq h_{v.L}$, proceed recursively with the left child $v.L$ of v. Otherwise, proceed recursively with the right child $v.R$ of v.

A range maximum query RMQ$(0, q)$ can be answered in $O(\log m)$ time as follows. We traverse the priority search tree starting at the root. During the traversal, we maintain a variable *max_point* that stores the point of highest score in the range $R(0, q)$ seen so far (initially, *max_point* is undefined). Suppose node v is visited in the traversal. If v is a leaf storing the point p_v, we proceed as in case (1) below. If v is an interior node, we distinguish the cases (1) $p_v \leq q$ and (2) $p_v \not\leq q$. In case (1), if $p_v.score \geq max\_point.score$, we update *max_point* by *max_point* $:= p_v$. In case (2), if $q \leq h_{v.L}$, we recursively proceed with the left child $v.L$ of v; otherwise, we recursively proceed with both children of v.

Another example of a data structure that supports RMQ with activation is the kd-tree [9]. In essence, the one-dimensional kd-tree coincides with the priority search tree, but the operations on kd-trees are accelerated in practice by various programming tricks; see [9].

An Alternative Data Structure

The priority search tree is a semi-dynamic data structure, in the sense that points are not really inserted. The advantage of using semi-dynamic data structures in Algorithm 15.2 will become clear in Section 15.6. There it is shown that the approach can be naturally extended to the higher-dimensional case. Here, however, we can also use any other one dimensional *dynamic* data structure D to answer RMQ with activation, provided that it supports the operations[1]

---

[1] In the dynamic case, the sentence preceding the for-loop in Algorithm 15.2 must be deleted.

- *insert(q)*: if q is not in D, then put q into D; otherwise update its satellite information, i.e., the fragment corresponding to q
- *delete(q)*: remove q from D
- *predecessor(q)*: gives the largest element $\leq q$ in D
- *successor(q)*: gives the smallest element $> q$ in D

To answer $\mathrm{RMQ}(0, q)$ boils down to computing *predecessor(q)* in D, and Algorithm 15.4 shows how to activate a point q in D. Note that the operations *predecessor(q)* and *successor(q)* are always well-defined if we initialize the data structure D with the origin and the terminus point.

FIGURE 15.4: Implementation of the operation *activate* in the data structure D

---

if $(q.score > predecessor(q).score)$ **then**
    insert(q)
    while $(q.score > successor(q).score)$
        delete$(successor(q))$

Note that the following invariant is maintained: If $0 \leq q_1 < q_2 < \cdots < q_\ell \leq n$ are the entries in the data structure D, then $q_1.score \leq q_2.score \leq \cdots \leq q_\ell.score$.

Many data structures supporting the aforementioned operations are known. For example, the priority queues devised by van Emde Boas [49, 48] and Johnson's improvement [29] support the operations in time $O(\log \log N)$ and space $O(N)$, provided that every q satisfies $1 \leq q \leq N$. The space requirement can be reduced to $O(n)$, where n denotes the number of elements stored in the priority queue; see [36]. Recall that a fragment corresponds to segments of the strings S_1 and S_2 that are similar. W.l.o.g., we may assume that $n_1 = |S_1| \leq |S_2| = n_2$ (otherwise, we swap the sequences). In Algorithm 15.2, sorting all start and end points of the m fragments in ascending order w.r.t. their x_1 coordinate by counting sort (see, e.g., [16]) takes $O(n_1)$ time. Since Algorithm 15.2 employs at most $O(m)$ priority queue operations, each of which takes time $O(\log \log n_1)$, the overall time complexity of this implementation is $O(n_1 + m \log \log n_1)$. If the fragments are already ordered as in the heaviest increasing subsequence problem (see below), the worst case time complexity reduces to $O(m \log \log n_1)$. Using Johnson's data structure [29], Eppstein et al. [20] showed that their sparse dynamic programming algorithm solves the problem in $O(n_1 + n_2 + m \log \log \min(m, n_1 n_2/m))$ time. However, as noted by Chao and Miller [15], the data structure employed to obtain this theoretical efficiency is unusable in practice. With a practical data structure, the complexity becomes $O(m \log m)$; see also [30, 24]. Moreover, in most applications m is relatively small compared to n_1, so that it is advantageous to sort the start and end points of the m fragments in $O(m \log m)$ time. Then the usage of AVL trees (see, e.g., [5]), red-black trees (see, e.g., [16]), or any other practical data structure that supports the above-mentioned operations in $O(\log m)$ time, gives an $O(m \log m)$ time and $O(m)$ space implementation of Algorithm 15.2.

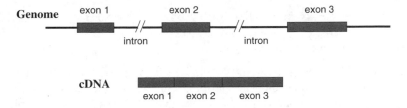

FIGURE 15.5: (See color insert following page 20-4.) cDNA mapped to a genomic sequence.

15.3.2 Applications

Global Alignment

As already mentioned in Section 15.1, software-tools that compute global alignments of large sequences use anchor-based methods to cope with the huge amount of data. These are composed of three phases:

1. computation of fragments,
2. computation of an optimal global chain of colinear non-overlapping fragments,
3. alignment of the regions between the fragments (the anchors) in the computed chain (by applying the same method recursively with less stringent parameters or by using another alignment program).

Obviously, Algorithm 15.2 solves the second phase in sub-quadratic time. The incorporation of gap costs (see Section 15.4) turns the algorithm into a practical tool for computing global alignments of whole genomes. Its extension to the higher-dimensional case makes it possible to compare multiple genomes simultaneously; see Section 15.6.

cDNA Mapping

cDNA (complementary or copy DNA) is DNA obtained from mRNA (messenger RNA) through reverse transcription. The cDNA consists only of the exons of the expressed gene because the introns (which are common in eukaryotes) have been spliced out. The problem of cDNA mapping is to find the gene (and its exon/intron structure) on the genome from which the cDNA was derived; see Figure 15.5. A precise mapping allows further analyses such as finding regulatory elements.

The global fragment-chaining algorithm 15.2 can be used for cDNA mapping. As cDNA lacks the introns that are contained in the DNA sequence from which it was derived, gaps should not be penalized in this application. To avoid that an exon is chained to a spurious match that is very far away, one can add a gap constraint when connecting fragments. To be more precise, the start point $beg(f)$ of a fragment f is connected to the end point of a highest-scoring fragment that lies within a (user-defined) gap distance δ. This gap constraint can be incorporated into Algorithm 15.2 by restricting the range $R(\vec{0}, beg(f) - \vec{1})$ of each RMQ to the range $R(beg(f) - \delta, beg(f) - \vec{1})$, that is, the RMQ$(\vec{0}, beg(f) - \vec{1})$ is replaced with RMQ$(beg(f) - \delta, beg(f) - \vec{1})$.

It has been observed by Shibuya and Kurochkin [46] that if one uses maximal exact matches as fragments, then the corresponding segments may slightly overlap in the cDNA sequence. Thus, they developed a variant of Algorithm 15.2 that can deal with overlaps. The worst case running time of this variant is still $O(m \log m)$.

It is worth mentioning that there are several heuristic algorithms for cDNA mapping.

Algorithms that heuristically map cDNA to a genome include sim4 [22] and BLAT [31].

Longest/Heaviest Increasing Subsequence

The software tools MUMmer [18, 19] and LAGAN [12] compute a heaviest increasing subsequence to find a chain of colinear non-overlapping fragments. In contrast to Algorithm 15.2, however, the computed chain is not necessarily optimal. MUMmer, for example, uses maximal unique matches (MUMs) as fragments, where the length of a MUM yields the weight of the corresponding fragment. Since MUMs are unique in both sequences, every fragment has exactly one segment in the genomic sequence S_1 and one in S_2. The fragments are sorted according to their start position in S_1 and then they are numbered consecutively. Thus, the numbers of the segments in S_1 give the sequence $1, 2, \ldots, m$, while the numbers of the corresponding segments in S_2 yield a permutation π of the sequence $1, 2, \ldots, m$. MUMmer outputs a heaviest increasing subsequence (see below) of π as an optimal global chain. It is easy to see that the resulting chain of MUMs may contain overlapping MUMs, which in turn may lead to inconsistencies (i.e., it may not be possible to find an alignment that is consistent with all selected MUMs). MUMmer takes an ad hoc approach to handle this: It simply removes the overlapping parts from the MUMs.

Let us briefly recall the heaviest increasing subsequence problem. Given a sequence $A = A[1 \ldots m]$ over some linearly ordered alphabet Σ, the *longest increasing subsequence* (LIS) problem is to find a longest subsequence of A that is strictly increasing. For ease of presentation, we assume that $\Sigma = \{1, \ldots, n\}$. If every $A[i]$ has a weight $A[i].weight$, then the *heaviest increasing subsequence* (HIS) problem is to find a strictly increasing subsequence of A such that the sum of the weights of its elements is maximal (among all strictly increasing subsequences). Clearly, the LIS problem is a special case of the HIS problem in which each $A[i]$ has weight 1. There is an abundance of papers on the LIS problem and the closely related longest common subsequence (LCS) problem; we refer the interested reader to [24] for references. If we write the sequence A in the form $(1, A[1]), (2, A[2]), \ldots, (m, A[m])$ and view the pair $(i, A[i])$ as the ith fragment, then it becomes obvious that the HIS problem is a special case of the 2-dimensional fragment-chaining problem. Therefore, the following specialization of Algorithm 15.2 solves the HIS problem. Note that the first three statements of the for-loop correspond to the processing of start points, while the remaining statements correspond to the processing of end points.

FIGURE 15.6: Computation of a heaviest increasing subsequence

---

for $i := 1$ **to** m
    $(j, A[j]) := predecessor(A[i] - 1)$
    $(i, A[i]).prec := (j, A[j])$
    $(i, A[i]).score := (i, A[i]).weight + (j, A[j]).score$
    if $((i, A[i]).score > predecessor(A[i]).score)$ **then**
       $insert((i, A[i]))$
       while $((i, A[i]).score > successor(A[i]).score)$
          $delete(successor(A[i]))$

Algorithm 15.6 takes $O(m \cdot \min\{\log m, \log \log n\})$ time and $O(m)$ space, because no sorting is required in this application. Note that Algorithm 15.6 is a little different from the original

```
ACCXXXX___AGG          ACCXXXXAGG
ACC____YYYAGG          ACCYYY_AGG
```

FIGURE 15.7: Alignments based on the fragments ACC and AGG w.r.t. gap cost g_1 (left) and g_∞ (right), where X and Y are anonymous characters.

HIS algorithm devised by Jacobson and Vo [28].

15.4 Incorporating Gap Costs into the Algorithm

In the previous section, fragments were chained without penalizing the gaps in between them. In this section we modify the algorithm, so that it can take gap costs into account. The usual gap costs used in alignments correspond to gap costs in the L_∞ metric; see right side of Figure 15.7. For didactic reasons, however, we first consider gap costs in the L_1 metric.

15.4.1 Costs in the L_1 Metric

We first handle the case in which the cost for the gap between two fragments is the distance between the end and start point of the two fragments in the L_1 metric. For two points $p, q \in \mathbb{R}^2$, this distance is defined by

$$d_1(p,q) = \sum_{i=1}^{2} |p.x_i - q.x_i|$$

and for two fragments $f \ll f'$ we define $g_1(f', f) = d_1(beg(f'), end(f))$. If an alignment of two sequences S_1 and S_2 shall be based on fragments and one uses this gap cost, then the characters between the two fragments are *deleted/inserted*; see left side of Figure 15.7.

The problem with gap costs in our approach is that an RMQ does not take the cost $g(f', f)$ from recurrence (15.1) into account, and if we would explicitly compute $g(f', f)$ for every pair of fragments with $f \ll f'$, then this would yield a quadratic time algorithm. Thus, it is necessary to express the gap costs implicitly in terms of weight information attached to the points. We achieve this by using the *geometric cost* of a fragment f, which we define in terms of the terminus point t as $gc(f) = d_1(\text{t}, end(f))$.

LEMMA 15.2   Let f, $\tilde{f}$, and f' be fragments such that $f \ll f'$ and $\tilde{f} \ll f'$. Then the inequality $\tilde{f}.score - g_1(f', \tilde{f}) > f.score - g_1(f', f)$ holds true if and only if the inequality $\tilde{f}.score - gc(\tilde{f}) > f.score - gc(f)$ holds.

Proof

$$
\begin{aligned}
&\quad \tilde{f}.score - g_1(f', \tilde{f}) > f.score - g_1(f', f) \\
&\Leftrightarrow \tilde{f}.score - \sum_{i=1}^{2}(beg(f').x_i - end(\tilde{f}).x_i) > f.score - \sum_{i=1}^{2}(beg(f').x_i - end(f).x_i) \\
&\Leftrightarrow \tilde{f}.score - \sum_{i=1}^{2}(\text{t}.x_i - end(\tilde{f}).x_i) > f.score - \sum_{i=1}^{2}(\text{t}.x_i - end(f).x_i) \\
&\Leftrightarrow \tilde{f}.score - gc(\tilde{f}) > f.score - gc(f)
\end{aligned}
$$

The second equivalence follows from adding $\sum_{i=1}^{2} beg(f').x_i$ to and subtracting $\sum_{i=1}^{2} \text{t}.x_i$ from both sides of the inequality. Figure 15.8 illustrates the lemma.

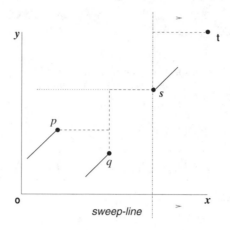

FIGURE 15.8: Points p and q are active end points of the fragments f and $\tilde{f}$. The start point s of fragment f' is currently scanned by the sweeping line and t is the terminus point.

Because t is fixed, the value $gc(f)$ is known in advance for every fragment f. Therefore, Algorithm 15.2 needs only two slight modifications to take gap costs into account. First, we replace the statement $f'.score := f'.weight + f.score$ with

$$f'.score := f'.weight + f.score - g_1(f', f)$$

Second, if points$[i]$ is the end point of f', then it will be activated with $f'.priority := f'.score - gc(f')$. Thus, an RMQ will return a point of highest priority instead of a point of highest score.

The next lemma implies the correctness of the modified algorithm.

LEMMA 15.3 If the range maximum query $\mathrm{RMQ}(0, beg(f') - \vec{1})$ returns the end point of fragment $\tilde{f}$, then we have $\tilde{f}.score - g_1(f', \tilde{f}) = \max\{f.score - g_1(f', f) : f \ll f'\}$.

Proof If the range maximum query $\mathrm{RMQ}(0, beg(f') - \vec{1})$ returns the end point of fragment $\tilde{f}$, then $\tilde{f}.priority = \max\{f.priority : f \ll f'\}$. Since $f.priority = f.score - gc(f)$ for every fragment f, it is an immediate consequence of Lemma 15.2 that $\tilde{f}.score - g_1(f', \tilde{f}) = \max\{f.score - g_1(f', f) : f \ll f'\}$.

15.4.2 Costs in the L_∞ Metric

In this section we consider the gap cost associated with the L_∞ metric. The distance between two points $p, q \in \mathbb{R}^2$ in the L_∞ metric is $d_\infty(p, q) = \max_{i \in \{1,2\}} |p.x_i - q.x_i|$, or equivalently,

$$d_\infty(p, q) = \begin{cases} |p.x_1 - q.x_1| & \text{if } |p.x_1 - q.x_1| \geq |p.x_2 - q.x_2| \\ |p.x_2 - q.x_2| & \text{if } |p.x_2 - q.x_2| \geq |p.x_1 - q.x_1| \end{cases}$$

Furthermore, the gap cost of connecting two fragments $f \ll f'$ in the L_∞ metric is defined by $g_\infty(f', f) = d_\infty(beg(f'), end(f))$. If an alignment of two sequences S_1 and S_2 shall be based on fragments and one uses the gap cost g_∞, then the characters between the two fragments

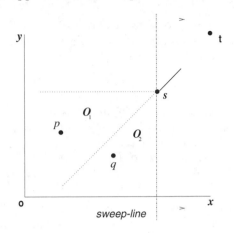

FIGURE 15.9: The first quadrant of point s is divided into two octants.

are *replaced* as long as possible and the remaining characters are *deleted* or *inserted*; see right side of Figure 15.7.

In order to compute the score of a fragment f' with $beg(f') = s$, the following definitions are useful. The *first quadrant* of point $s \in \mathbb{R}^2$ consists of all points $p \in \mathbb{R}^2$ with $p.x_1 \leq s.x_1$ and $p.x_2 \leq s.x_2$. We divide the first quadrant of s into regions O_1 and O_2 by the straight line $x_2 = x_1 + (s.x_2 - s.x_1)$. The *first octant* O_1 of s consists of all points p in the first quadrant of s satisfying $s.x_1 - p.x_1 \geq s.x_2 - p.x_2$, these are the points lying above or on the straight line $x_2 = x_1 + (s.x_2 - s.x_1)$; see Figure 15.9. The *second octant* O_2 consists of all points q satisfying $s.x_2 - q.x_2 \geq s.x_1 - q.x_1$, these are the points lying below or on the straight line $x_2 = x_1 + (s.x_2 - s.x_1)$. Then $f'.score = f'.weight + \max\{v_1, v_2\}$, where $v_i = \max\{f.score - g(f', f) : f \ll f' \text{ and } end(f) \text{ lies in octant } O_i\}$ for $i \in \{1, 2\}$.

However, our chaining algorithms rely on RMQ, and these work only for orthogonal regions, not for octants. For this reason, we will make use of the *octant-to-quadrant* transformations of Guibas and Stolfi [23]. The transformation $T_1 : (x_1, x_2) \mapsto (x_1 - x_2, x_2)$ maps the first octant to a quadrant. More precisely, point $T_1(p)$ is in the first quadrant of $T_1(s)$ if and only if p is in the first octant of point s.[2] Similarly, for the transformation $T_2 : (x_1, x_2) \mapsto (x_1, x_2 - x_1)$, point q is in the second octant of point s if and only if $T_2(q)$ is in the first quadrant of $T_2(s)$. By means of these transformations, we can apply the same technique as in the previous section. We just have to define the geometric cost properly. In the first octant O_1, the geometric cost gc_1 of a fragment f is $gc_1(f) = \mathsf{t}.x_1 - end(f).x_1$, while in second octant O_2 we define $gc_2(f) = \mathsf{t}.x_2 - end(f).x_2$.

LEMMA 15.4 Let f, $\tilde{f}$, and f' be fragments such that $f \ll f'$ and $\tilde{f} \ll f'$. If $end(f)$ and $end(\tilde{f})$ lie in the octant O_i of $beg(f')$, then $\tilde{f}.score - g_\infty(f', \tilde{f}) > f.score - g_\infty(f', f)$ if and only if $\tilde{f}.score - gc_i(\tilde{f}) > f.score - gc_i(f)$.

Proof Similar to the proof of Lemma 15.2.

---

[2] Observe that the transformation may yield points with negative coordinates, but it is easy to overcome this obstacle by an additional transformation (a translation). Hence we will skip this minor problem.

In Section 15.4.1 there was only one geometric cost gc, but here we have to take two different geometric costs gc_1 and gc_2 into account. To cope with this problem, we need two data structures D_1 and D_2, where D_i stores the set of points

$$\{T_i(end(f)) : f \text{ is a fragment}\}$$

If we encounter the end point of fragment f' in Algorithm 15.2, then we activate point $T_1(end(f'))$ in D_1 with priority $f'.score - gc_1(f')$ and point $T_2(end(f'))$ in D_2 with priority $f'.score - gc_2(f')$. If we encounter the start point of fragment f', then we launch two range maximum queries, namely $\mathtt{RMQ}(0, T_1(beg(f') - \vec{1}))$ in the data structure D_1 and $\mathtt{RMQ}(0, T_2(beg(f') - \vec{1}))$ in D_2. If the first $\mathtt{RMQ}$ returns $T_1(end(f_1))$ and the second returns $T_2(end(f_2))$, then f_i is a fragment of highest priority in D_i such that $T_i(end(f_i)) \ll T_i(beg(f'))$, where $1 \le i \le 2$. Because a point p is in the octant O_i of point $beg(f')$ if and only if $T_i(p)$ is in the first quadrant of $T_i(beg(f'))$, it follows that f_i is a fragment such that its priority $f_i.score - gc_i(f_i)$ is maximal in octant O_i. Therefore, according to Lemma 15.4, the value $v_i = f_i.score - g(f', f_i)$ is maximal in octant O_i. Hence, if $v_1 > v_2$, then we set $f'.prec = f_1$ and $f'.score := f'.weight + v_1$. Otherwise, we set $f'.prec = f_2$ and $f'.score := f'.weight + v_2$.

For gap costs in the L_∞-metric, the chaining algorithm runs in $O(m \log m \log \log m)$ time and $O(m \log m)$ space because of the two-dimensional $\mathtt{RMQ}$s required for the transformed points; see Section 15.6. This is in sharp contrast to gap costs in the L_1-metric, where we merely need one-dimensional $\mathtt{RMQ}$s. We would like to point out that the sparse dynamic programming method of Eppstein et al. [20] can solve the chaining problem for gap costs in the L_∞-metric (and for the sum-of-pairs gap cost introduced in [41]) for the special case $k = 2$ in $O(n_1 + n_2 + m \log \log \min(m, n_1 n_2/m))$ time, where $n_1 = |S_1|$ and $n_2 = |S_2|$. As noted by Myers and Miller [41], however, it seems that their approach cannot be extended to the case $k > 2$. By contrast, our method can naturally be extended to the higher-dimensional case; see Section 15.6.

15.5 Variations

15.5.1 Local Chains

Chao and Miller [15] extended the sparse dynamic programming algorithm of Eppstein et al. [20] such that it delivers any desired number of highest-scoring chains at a slight increase in asymptotic time complexity. In this section we will extend our algorithm in the same direction.

In the previous sections, we have tackled the global chaining problem, which asks for an optimal chain starting at the origin 0 and ending at terminus t. However, in many applications (such as searching for local similarities in genomic sequences) one is interested in chains that can start and end with arbitrary fragments. If we remove the restriction that a chain must start at the origin and end at the terminus, we get the local chaining problem; see Figure 15.10.

DEFINITION 15.4 Given m weighted fragments and a gap cost function g, the *local fragment-chaining problem* is to determine a chain of highest score. Such a chain will be called *optimal local chain*.

Note that if g is the constant function 0, then an optimal local chain must also be an optimal global chain, and vice versa. Our solution to the local chaining problem is a

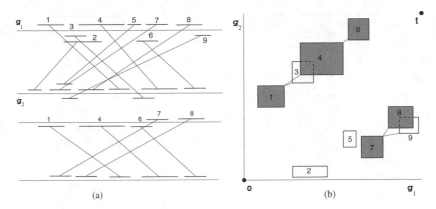

FIGURE 15.10: Computation of local chains of colinear non-overlapping fragments. The optimal local chain is composed of the fragments 1, 4, and 6. Another significant local chain consists of the fragments 7 and 8.

variant of the global chaining algorithm. For ease of presentation, we will use gap costs corresponding to the L_1 metric (see Section 15.4.1), but the approach also works with gap cost in the L_∞-metric (see Section 15.4.2).

DEFINITION 15.5 Let $f'.score = \max\{score(C) : C \text{ is a chain ending with } f'\}$. A chain C ending with f' and satisfying $f'.score = score(C)$ will be called *optimal chain ending with* f'.

LEMMA 15.5 The following equality holds:

$$f'.score = f'.weight + \max(\{0\} \cup \{f.score - g_1(f', f) : f \ll f'\}) \tag{15.2}$$

Proof Let $C' = f_1, f_2, \ldots, f_\ell, f'$ be an optimal chain ending with f', that is, $score(C') = f'.score$. Because the chain that solely consists of fragment f' has score $f'.weight \geq 0$, we must have $score(C') \geq f'.weight$. If $score(C') = f'.weight$, then $f.score - g_1(f', f) \leq 0$ for every fragment f that precedes f', because otherwise it would follow $score(C') > f'.weight$. Hence equality (15.2) holds in this case. So suppose $score(C') > f'.weight$. Clearly, $score(C') = f'.weight + score(C) - g_1(f', f_\ell)$, where $C = f_1, f_2, \ldots, f_\ell$. It is not difficult to see that C must be an optimal chain that is ending with f_ℓ because otherwise C' would not be optimal. Therefore, $score(C') = f'.weight + f_\ell.score - g_1(f', f_\ell)$. If there were a fragment f that precedes f' such that $f.score - g_1(f', f) > f_\ell.score - g_1(f', f_\ell)$, then it would follow that C' is not optimal. We conclude that equality (15.2) holds. $\qquad\blacksquare$

With the help of Lemma 15.5, we obtain an algorithm that solves the local chaining problem.

It is not difficult to verify that we can use the techniques of the previous sections to solve the local fragment-chaining problem in the same time and space complexities as the global fragment-chaining problem.

We stress that Algorithm 15.11 can easily be modified, so that it can report all chains whose score exceeds some threshold T (in Algorithm 15.11, instead of determining a fragment $\tilde{f}$ of highest score, one determines all fragments whose score exceeds T). Such chains

---

FIGURE 15.11: Finding an optimal local chain based on RMQ

---

for every fragment f' **do begin**
    determine $\hat{f}$ such that $\hat{f}.score - g_1(f', \hat{f}) = \max\{f.score - g_1(f', f) : f \ll f'\}$
    $max := \max(\{0\} \cup \{\hat{f}.score - g_1(f', \hat{f})\})$
    if $max > 0$ **then** $f'.prec := \hat{f}$ **else** $f'.prec := NULL$
    $f'.score := f'.weight + max$
end
determine a fragment $\tilde{f}$ such that $\tilde{f}.score = \max\{f.score : f$ is a fragment $\}$
report an optimal local chain by tracing back the pointers from $\tilde{f}.prec$ until a fragment f
with $f.prec = NULL$ is reached

---

will be called *significant local chains*; see Figure 15.10. In this case, however, an additional problem arises: Several chains can share one or more fragments, so that the output can be quite complex. To avoid this, local chains are partitioned into equivalence classes by the starting fragment of the chain [27, 15]. Two local chains belong to the same class if and only if they begin with the same fragment. Instead of all chains in an equivalence class, only one highest-scoring chain is reported as a representative of that class. In Figure 15.10, for example, the chains 1,4,6 and 7,8 are reported.

15.5.2 1-dimensional Chaining

The Single Coverage Problem

As noted by Schwartz et al. [45] the chaining option of PipMaker (which solves the global 2-dimensional chaining problem) "should be used only if the genomic structure of the two sequences are known to be conserved; otherwise a duplication might not be detected." PipMaker also provides an alternative method — the single coverage option — that "selects a highest-scoring set of alignments such that any position in the first sequence can appear in one alignment, at most." In our terminology, this problem is to find a set of fragments of maximum coverage of the x-axis without overlapping on that axis. The following definitions make this precise.

DEFINITION 15.6 We define a binary relation $\ll_x$ on the set of fragments by $f \ll_x f'$ if and only if $end(f).x < beg(f').x$.

DEFINITION 15.7 Given a set $\mathcal{F}$ of m fragments, the *single coverage problem* is to find a subset $\mathcal{F}' = \{f_1, f_2, \ldots, f_\ell\}$ of $\mathcal{F}$ such that the elements of $\mathcal{F}'$ do not overlap on the x-axis (i.e., either $f_i \ll_x f_j$ or $f_j \ll_x f_i$) and $\sum_{i=1}^{\ell}(end(f_i).x - beg(f_i).x)$ is maximized.

Schwartz et al. [45] did not describe PipMaker's algorithm for solving the single coverage problem, but they demonstrated the value of the single coverage option in a comparison of the β-globin gene clusters of human (six genes) and chicken (four genes).
The single coverage problem can be solved as follows. For any 2-dimensional fragment $f = (beg(f), end(f))$, let I_f be the interval $[beg(f).x, \ldots, end(f).x]$, i.e., I_f is the projection of f to the x-axis. Then, given a set $\mathcal{F} = \{f_1, f_2, \ldots, f_m\}$ of 2-dimensional fragments, a solution to the single coverage problem can be found by solving the global 1-dimensional

chaining problem for the set $\{I_{f_1}, I_{f_2}, \ldots, I_{f_m}\}$. The latter can be done in $O(m \log m)$ time by the following algorithm; see also [24].

FIGURE 15.12: Global chaining of m intervals (1-dimensional fragments)

Sort all start and end points of the m intervals in ascending order and store them in the array points;

$I_{max} := 0$

for $i := 1$ **to** $2m$

    if points[i] is the start point of interval I **then**

        $I.prec := I_{max}$

        $I.score := I.weight + I_{max}.score$

    else   /⋆ points[i] is end point of an interval ⋆/

        determine the interval I with $end(I) = $ points[i]

        if $I.score > I_{max}.score$ **then** $I_{max} := I$

In the algorithm, I_{max} denotes the interval that has highest score among all intervals already scanned. Consequently, upon termination of the algorithm, it is the last interval in an optimal global chain. Recall that $I.prec$ denotes a field that stores the preceding interval of I in a chain. Hence we obtain an optimal global chain by following the back-pointers from I_{max} to the origin 0. Note that the for-loop in Algorithm 15.12 requires only $O(m)$ time.

The Transformation Distance

The transformation distance between two genomes as defined in [50] can also be computed by solving the single coverage problem. The transformation distance problem addresses the question of how two sequences may be most economically derived from one another by various genome rearrangements. Varré et al. [50] applied the graph-based $O(m^2)$ time solution described in Section 15.2 to solve this problem, but Algorithm 15.12 obviously provides a better solution.

It should be noted that if one uses exact fragments weighted by their length, then the following gap cost g_x is (implicitly) incorporated in Algorithm 15.12. Given 2-dimensional fragments $f \ll_x f'$, the gap cost $g_x(f', f)$ is the distance between the intervals $I_{f'}$ and I_f on the x-axis, i.e., $g_x(f', f) = beg(f').x - end(f).x$. Because a solution to the single coverage problem maximizes the coverage of the x-axis, it automatically minimizes the overall amount of gaps between the intervals on the x-axis. Note that it is also possible to explicitly incorporate other gap costs into Algorithm 15.12 by using the approach of Section 15.4. Details are left to the reader.

Using the sparse dynamic programming technique of Eppstein et al. [20], Brudno et al. [13] showed that it is also possible to incorporate 2-dimensional gap costs into the algorithm without changing its $O(m \log m)$ time complexity.

15.5.3 A More General Problem

The problems dealt with in previous sections required colinearity of the fragments to be selected. Now we drop this requirement, that is, we are now interested in an optimal

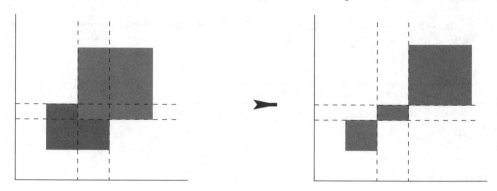

FIGURE 15.13: Two fragments $f_i = (beg(f_i), end(f_i))$ and $f_j = (beg(f_j), end(f_j))$ overlapping on both axes are refined into the three fragments $(beg(f_i), beg(f_j))$, $(beg(f_j), end(f_i))$, and $(end(f_i), end(f_j))$.

set of non-overlapping fragments. For example, in determining the relative degrees of completion of two sequencing projects, an interesting question is "how much of each sequence is not represented in the other, under the restriction that positions in each sequence can be involved in at most one selected match" [25]. In other words, one wants to find a set of non-overlapping fragments such that the amount of sequence covered by the fragments is maximized. The following definition gives a precise formulation.

DEFINITION 15.8   Given a set $\mathcal{F}$ of m fragments, the corresponding *MWIS problem* is to find a subset $\mathcal{F}' = \{f_1, f_2, \ldots, f_\ell\}$ of $\mathcal{F}$ such that the elements of $\mathcal{F}'$ are pairwise non-overlapping and the amount of sequence covered by the fragments, i.e., $\sum_{i=1}^{\ell}(end(f_i).x - beg(f_i).x) + \sum_{i=1}^{\ell}(end(f_i).y - beg(f_i).y)$, is maximized.

In the terminology of graph theory, the preceding problem is to find a maximum weight independent set (MWIS) in the following kind of intersection graph (called 2-union graph): For every fragment f_i there is a vertex labeled f_i with weight $f_i.weight$ in the graph and there is an undirected edge connecting vertices f_i and f_j if and only if f_i and f_j overlap.

Recall that an *independent set* (IS) of a graph $G = (V, E)$ is a subset $V' \subseteq V$ of vertices such that each edge in E is incident on at most one vertex in V'. The independent set problem is to find an independent set of maximum size. If each vertex has a weight as in our problem, then the *maximum weight independent set* problem is to find an independent set of maximum weight. By a reduction from 3-SAT, Bafna et al. [6] showed that the MWIS problem for fragments is NP-complete. (They also provided ideas for approximation algorithms.) Even worse, this problem was recently shown to be APX-hard; see [7]. A maximization problem is called APX-hard if there exists some constant $\epsilon > 0$ such that it is NP-hard to approximate the problem within a factor of $(1 - \epsilon)$.

Halpern et al. [25] studied an interesting variation of the preceding problem, which they called *Maximal Matched Sequence Problem* (MMSP). Given a set $\mathcal{F}$ of fragments, the MM-SP is to compute a set $\mathcal{F}'$ of non-overlapping fragments that are all subfragments of the fragments in $\mathcal{F}$ such that the amount of sequence covered by the fragments is maximized. Halpern et al. [25] showed that this problem can be solved optimally in polynomial time. The key idea of their solution is to minimally "refine" (subdivide) fragments until all overlaps between the refined fragments are "resolved", i.e., the projections of any two refined fragments onto each sequence are either disjoint or identical; see Figures 15.13 and 15.14.

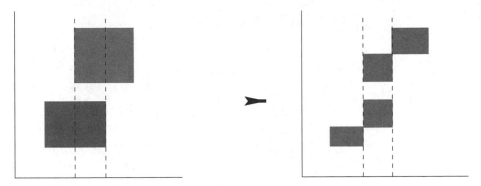

FIGURE 15.14: Two fragments overlapping on the x-axis but not on the y-axis are subdivided into four refined fragments.

For the details which are quite involved, and for experimental results, we refer to [25].

15.6 Higher-dimensional Chaining

As mentioned in Section 15.2, the graph based approach also solves the chaining problem for more than two genomes. However, the running time of this chaining algorithm is quadratic in the number m of fragments. This can be a serious drawback if m is large. To overcome this obstacle, Zhang et al. [53] presented an algorithm that constructs an optimal chain using space division based on kd-trees, a data structure known from computational geometry [9]. However, a rigorous analysis of the running time of their algorithm is difficult because the construction of the chain is embedded in the kd-tree structure. Another chaining algorithm, devised by Myers and Miller [41], is based on the line-sweep paradigm and uses orthogonal range-searching supported by range trees instead of kd-trees. The algorithm presented below improves the sub-quadratic worst-case time complexity $O(m \log^k m)$ of Myers and Miller's algorithm.

A big advantage of Algorithm 15.2 is that it can be naturally extended to higher dimensions. In order to chain k-dimensional fragments, the 1-dimensional data structure D in Algorithm 15.2 has to be replaced with a $(k-1)$-dimensional data structure that supports range maximum queries and activation. In the following, we will outline two such data structures for dimension $d = k - 1$.

Our orthogonal range-searching data structures are based on *range trees*, which are well-known in computational geometry. Given a set S of m d-dimensional points, its range tree can be built as follows (see, e.g., [4, 42]). For $d = 1$, the range tree of S is a minimum-height binary search tree or an array storing S in sorted order. For $d > 1$, the range tree of S is a minimum-height binary search tree T with m leaves, whose ith leftmost leaf stores the point in S with the ith smallest x_1-coordinate. To each interior node v of T, we associate a canonical subset $C_v \subseteq S$ containing the points stored at the leaves of the subtree rooted at v. For each v, let l_v (resp. h_v) be the smallest (resp. largest) x_1 coordinate of any point in C_v and let $C_v^* = \{(p.x_2, \ldots, p.x_d) \in \mathbb{R}^{d-1} : (p.x_1, p.x_2, \ldots, p.x_d) \in C_v\}$. The interior node v stores l_v, h_v, and a $(d-1)$-dimensional range tree constructed on C_v^*. For any fixed dimension d, the data structure can be built in $O(m \log^{d-1} m)$ time and space.

Given a set S of points in $\mathbb{R}^d$, a *range query* (RQ) asks for all points of S that lie in a hyper-rectangle $R(p, q)$. A range query RQ(p, q) for the hyper-rectangle $R(p, q) = [l_1 \ldots h_1] \times [l_2 \ldots h_2] \times \cdots \times [l_d \ldots h_d]$ can be answered as follows. If $d = 1$, the query can be answered in

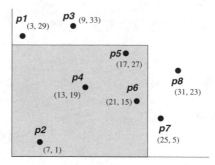

query rectangle [0 .. 22]x[0 .. 28]

FIGURE 15.15: A set of points and a query rectangle.

$O(\log m)$ time by a binary search. For $d > 1$, we traverse the range tree starting at the root. Suppose node v is visited in the traversal. If v is a leaf, then we report its corresponding point if it lies inside $R(p, q)$. If v is an interior node, and the interval $[l_v \dots h_v]$ does not intersect $[l_1 \dots h_1]$, there is nothing to do. If $[l_v \dots h_v] \subseteq [l_1 \dots h_1]$, we recursively search in the $(d-1)$-dimensional range tree stored at v with the hyper-rectangle $[l_2 \dots h_2] \times \cdots \times [l_d \dots h_d]$. Otherwise, we recursively visit both children of v. This procedure takes $O(\log^d m + z)$ time, where z is the number of points in the hyper-rectangle $R(p, q)$.

In Section 15.3.1, we have seen that the priority search tree can be used for 1-dimensional RMQ with activation. For $d \geq 2$ dimensions, we modify the d-dimensional range tree by replacing the range tree in the last dimension with a priority search tree. It is not difficult to show that this data structure supports range maximum queries of the form RMQ$(0, q)$ and activation operations in $O(\log^d m)$ time. Thus, the usage of this data structure yields a chaining algorithm for $k > 2$ sequences that runs in $O(m \log^{k-1} m)$ time. However, we can do even better.

The technique of *fractional cascading* [52] saves one log-factor in answering range queries (in the same construction time and using the same space as the original range tree). Here, we will recall this technique for range queries of the form RQ$(0, q)$ because we want to modify it to answer range maximum queries of the form RMQ$(0, q)$ efficiently. For ease of presentation, we consider the case $d = 2$. In this case, the range tree is a binary search tree (called *x-tree*) of arrays (called *y-arrays*). Let v be a node in the *x-tree* and let $v.L$ and $v.R$ be its left and right children. The *y-array* A_v of v contains all the points in C_v sorted in ascending order w.r.t. their y coordinate. Every element $p \in A_v$ has two downstream pointers: The left pointer $Lptr$ and the right pointer $Rptr$. The left pointer $Lptr$ points to the largest (i.e., rightmost) element q_1 in $A_{v.L}$ such that $q_1 \leq p$ ($Lptr$ is a *NULL* pointer if such an element does not exist). In an implementation, $Lptr$ is the index with $A_{v.L}[Lptr] = q_1$. Analogously, the right pointer $Rptr$ points to the largest element q_2 of $A_{v.R}$ such that $q_2 \leq p$. Figure 15.16 shows an example of this structure.

Locating all the points in a rectangle $R(0, (h_1, h_2))$ is done in two stages. In the first stage, a binary search is performed over the *y-array* of the root node of the *x-tree* to locate the rightmost point p_{h_2} such that $p_{h_2}.y \in [0 \dots h_2]$. Then, in the second stage, the *x-tree* is traversed (while keeping track of the downstream pointers) to locate the rightmost leaf p_{h_1} such that $p_{h_1}.x \in [0 \dots h_1]$. During the traversal of the *x-tree*, we identify a set of nodes which we call *canonical nodes* (w.r.t. the given range query). The set of canonical nodes is the smallest set of nodes $v_1, \dots, v_\ell$ of *x-tree* such that $\biguplus_{j=1}^{\ell} C_{v_j} = $ RQ$(0, (h_1, \infty))$. ($\biguplus$ denotes disjoint union.) In other words, $P := \biguplus_{j=1}^{\ell} A_{v_j} = \biguplus_{j=1}^{\ell} C_{v_j}$ contains every point $p \in S$ such

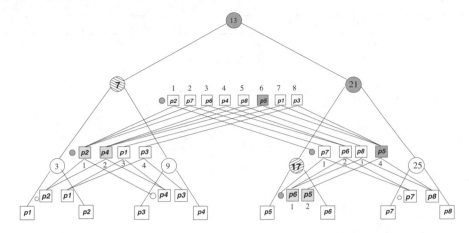

FIGURE 15.16: Range tree with fractional cascading for the points in Fig. 15.15. The colored nodes are visited for answering the range query of Fig. 15.15. Hatched nodes are the canonical nodes. The small circles refer to NULL pointers. In this example, $p_{h_1} = p_6$ and $p_{h_2} = p_5$. The colored elements of the *y-arrays* of the canonical nodes are the points in the query rectangle of Fig. 15.15. The value $h_{v.L}$ in every internal node v is the x coordinate that separates the points in its left subtree from those occurring in its right subtree.

that $p.x \in [0 \ldots h_1]$. However, not every point $p \in P$ satisfies $p.y \in [0 \ldots h_2]$. Here, the downstream pointers come into play. As already mentioned, the downstream pointers are followed while traversing the *x-tree*, and to follow one pointer takes constant time. If we encounter a canonical node v_j, then the element e_j, to which the last downstream pointer points, partitions the list A_{v_j} as follows: Every e that is strictly to the right of e_j is not in $R(0, (h_1, h_2))$, whereas all other elements of A_{v_j} lie in $R(0, (h_1, h_2))$. For this reason, we will call the element e_j the *split* element. It is easy to see that the number of canonical nodes is $O(\log m)$. Moreover, we can find all of them and the split elements of their *y-arrays* in $O(\log m)$ time; cf. [52]. Therefore, the range tree with fractional cascading supports 2-dimensional range queries in $O(\log m + z)$ time, where z is the number of points in the rectangle $R(0, q)$. For dimension $d > 2$, it takes time $O(\log^{d-1} m + z)$.

In order to answer RMQ with activation efficiently, we will further enhance every *y-array* that occurs in the fractional cascading data structure with a priority queue as described in [48, 29]. Each of these queues is (implicitly) constructed over the rank space of the points in the *y-array* (because the *y-arrays* are sorted w.r.t. the y dimension, the rank of an element *y-array*$[i]$ is its index i). The rank space of the points in the *y-array* consists of points in the range $[1 \ldots N]$, where N is the size of the *y-array*. The priority queue supports the operations *insert(r)*, *delete(r)*, *predecessor(r)*, and *successor(r)* in time $O(\log \log N)$, where r is an integer in the range $[1 \ldots N]$; see Section 15.3.1. Algorithm 15.17 shows how to activate a point q in the 2-dimensional range tree and Algorithm 15.18 shows how to answer an RMQ$(0, q)$.

Note that in the outer while-loop of Algorithm 15.17, the following invariant is maintained: If $0 \le i_1 < i_2 < \cdots < i_\ell \le n$ are the entries in the priority queue attached to A_v, then $A_v[i_1].score \le A_v[i_2].score \le \cdots \le A_v[i_\ell].score$.

Algorithm 15.18 gives pseudo-code for answering RMQ$(0, q)$, but we would like to first describe the idea on a higher level. In essence, we locate all canonical nodes $v_1, \ldots, v_\ell$ in D for the hyper-rectangle $R(0, q)$. For any v_j, $1 \le j \le \ell$, let the r_jth element be

---

FIGURE 15.17: Activation of a point q in the data structure D

$v :=$ root node of the *x-tree*
find the rank (*index*) r of q in A_v by a binary search
while $(v \neq \bot)$
   if $(A_v[r].score > A_v[predecessor(r)].score)$ **then**
      $insert(r)$ into the priority queue attached to A_v
      while$(A_v[r].score > A_v[successor(r)].score)$
         $delete(successor(r))$ from the priority queue attached to A_v
   if $(A_v[r] = A_{v.L}[A_v[r].Lptr])$ **then**
      $r := A_v[r].Lptr$
      $v := v.L$
   else
      $r := A_v[r].Rptr$
      $v := v.R$

---

the split element in A_{v_j}. We have seen that $\uplus_{j=1}^{\ell} A_{v_j}$ contains every point $p \in S$ such that $p.x \in [0 \ldots q.x]$. Now if r_j is the index of the split element of A_{v_j}, then all points $A_{v_j}[i]$ with $i \leq r_j$ are in $R(0, q)$, whereas all other elements $A_{v_j}[i]$ with $i > r_j$ are not in $R(0, q)$. Since Algorithm 15.17 maintains the above-mentioned invariant, the element with highest score in the priority queue of A_{v_j} that lies in $R(0, q)$ is $q_j = predecessor(r_j)$ (if r_j is in the priority queue of A_{v_j}, then $q_j = r_j$ because $predecessor(r_j)$ gives the largest element $\leq r_j$). We then compute $max\_score := \max\{A_{v_j}[q_j].score : 1 \leq j \leq \ell\}$ and return $max\_point = A_{v_i}[q_i]$, where $A_{v_i}[q_i].score = max\_score$.

FIGURE 15.18: RMQ$(0, q)$ in the data structure D

$v :=$ root node of the *x-tree*
$max\_score := -\infty$
$max\_point := \bot$
find the rank (*index*) r of the rightmost point p with $p.y \in [0 \ldots q.y]$ in A_v
while $(v \neq \bot)$
   if $(h_v.x \leq q.x)$ **then** /⋆ v is a canonical node ⋆/
      $tmp := predecessor(r)$ in the priority queue of A_v
      $max\_score := \max\{max\_score, A_v[tmp].score\}$
      if $(max\_score = tmp.score)$ **then** $max\_point := A_v[tmp]$
   else if $(h_{v.L}.x \leq q.x)$ **then** /⋆ $v.L$ is a canonical node ⋆/
      $tmp := predecessor(A_v[r].Lptr)$ in the priority queue of $A_{v.L}$
      $max\_score := \max\{max\_score, A_{v.L}[tmp].score\}$
      if $(max\_score = tmp.score)$ **then** $max\_point := A_{v.L}[tmp]$
      $r := A_v[r].Rptr$
      $v := v.R$
   else
      $r := A_v[r].Lptr$
      $v := v.L$

Because the number of canonical nodes is $O(\log m)$ and any of the priority queue operations takes $O(\log \log m)$ time, answering a 2-dimensional range maximum query takes $O(\log m \log \log m)$ time. Since every point occurs in at most $\log m$ priority queues, there are at most $m \log m$ delete operations. Hence the total time complexity of activating m points is $O(m \log m \log \log m)$.

THEOREM 15.1   *Given $k > 2$ genomes and m fragments, an optimal global chain (without gap costs) can be found in $O(m \log^{k-2} m \log \log m)$ time and $O(m \log^{k-2} m)$ space.*

Proof   In Algorithm 15.2, the points are first sorted w.r.t. their first dimension and the RMQ with activation is required only for $d = k - 1$ dimensions. For $d \geq 2$ dimensions, the preceding data structure is implemented for the last two dimensions of the range tree, which yields a data structure D that requires $O(m \log^{d-1} m)$ space and $O(m \log^{d-1} m \log \log m)$ time for m range maximum queries and m activation operations. Consequently, one can find an optimal chain in $O(m \log^{k-2} m \log \log m)$ time and $O(m \log^{k-2} m)$ space.

Details on the incorporation of gap costs into the higher-dimensional chaining algorithm can be found in [3]. (The same worst-case complexities hold for the local fragment-chaining algorithm; see [1].) There, it is shown that the global fragment-chaining problem for m fragments and $k > 2$ genomes can be solved in

- $O(m \log^{k-2} m \log \log m)$ time and $O(m \log^{k-2} m)$ space for gap costs in the L_1 metric,
- $O(k! \, m \log^{k-1} m \log \log m)$ time and $O(k! \, m \log^{k-1} m)$ space for gap costs in the L_∞ metric and also for the sum-of-pairs gap cost as introduced in [41].

If the kd-tree is used instead of the range tree, for m fragments and $k > 2$ genomes, the algorithms take

- $O((k-1) \, m^{2 - \frac{1}{k-1}})$ time and $O(m)$ space for gap costs in the L_1 metric,
- $O(k \, m^{2 - \frac{1}{k}})$ time and $O(m)$ space for gap costs in the L_∞ metric,
- $O(k! \, m^{2 - \frac{1}{k}})$ time and $O(m)$ space for the sum-of-pairs gap cost.

This is because answering *one* d-dimensional range query with the kd-tree takes $O(dm^{1 - \frac{1}{d}})$ time in the worst case; see [34]. Moreover, for small k, a collection of programming tricks can speed up the running time in practice; see [9].

We would like to conclude by pointing out an interesting connection between the global chaining problem and the maximum weight independent set (MWIS) problem in trapezoid graphs. Trapezoid graphs were introduced by Dagan et al. [17] and solutions to the maximum clique problem or the minimum coloring problem for trapezoid graphs are important in channel routing problems in VLSI design. Felsner et al. [21] showed that a maximum weight independent set in a k-trapezoid graph with n vertices can be computed in $O(n \log^{k-1} n)$ time and $O(n \log^{k-2} n)$ space provided that its k-dimensional box representation is given. Interestingly, the solution presented in this chapter solves the MWIS problem for k-trapezoid graphs in $O(n \log^{k-2} n \log \log n)$ time and $O(n \log^{k-2} n)$ space. That is, it improves the time complexity by a factor $\frac{\log n}{\log \log n}$.

Acknowledgement

The authors were supported by DFG-grant Oh 53/4-1 and thank Stefan Kurtz for his comments on the manuscript.

References

[1] M.I. Abouelhoda and E. Ohlebusch. A local chaining algorithm and its applications in comparative genomics. In *Proc. 3rd Workshop on Algorithms in Bioinformatics*, volume 2812 of *Lecture Notes in Bioinformatics*, pages 1–16, Berlin, 2003. Springer-Verlag.

[2] M.I. Abouelhoda and E. Ohlebusch. CHAINER: Software for comparing genomes. In *12th International Conference on Intelligent Systems for Molecular Biology/3rd European Conference on Computational Biology*, 2004. Short paper available at `http://www.iscb.org/ismbeccb2004/short%20papers/19.pdf`.

[3] M.I. Abouelhoda and E. Ohlebusch. Chaining algorithms for multiple genome comparison. *Journal of Discrete Algorithms*, 3:321–341, 2005.

[4] P. Agarwal. Range searching. In J.E. Goodman and J. O'Rourke, editors, *Handbook of Discrete and Computational Geometry*, chapter 31, pages 575–603. CRC Press LLC, 1997.

[5] A.V. Aho, J.E. Hopcroft, and J.D. Ullman. *Data Structures and Algorithms*. Addison-Wesley, 1983.

[6] V. Bafna, B. Narayanan, and R. Ravi. Nonoverlapping local alignments (weighted independent sets of axis-parallel rectangles). *Discrete Applied Mathematics*, 71:41–53, 1996.

[7] R. Bar-Yehuda, M.M. Halldórsson, J. Naor, and H. Shachnai. Scheduling split intervals. In *Proc. ACM-SIAM Symposium on Discrete Algorithms*, pages 732–741, 2002.

[8] S. Batzoglou, L. Pachter, J.P. Mesirov, and B. Berger *et al.* Human and mouse gene structure: Comparative analysis and application to exon prediction. *Genome Research*, 10:950–958, 2001.

[9] J.L. Bentley. K-d trees for semidynamic point sets. In *Proc. 6th Annual ACM Symposium on Computational Geometry*, pages 187–197, 1990.

[10] N. Bray, I. Dubchak, and L. Pachter. AVID: A global alignment program. *Genome Research*, 13:97–102, 2003.

[11] N. Bray and L. Pachter. MAVID multiple alignment server. *Nucleic Acids Res.*, 31:3525–3526, 2003.

[12] M. Brudno, C.B. Do, G.M. Cooper, and M.F. Kim *et al.* LAGAN and Multi-LAGAN: Efficient tools for large-scale multiple alignment of genomic DNA. *Genome Research*, 13(4):721–731, 2003.

[13] M. Brudno, S. Malde, A. Poliakov, and C.B. Do *et al.* Glocal alignment: Finding rearrangements during alignment. *Bioinformatics*, 19:i54–i62, 2003.

[14] P. Chain, S. Kurtz, E. Ohlebusch, and T. Slezak. An applications-focused review of comparative genomics tools: Capabilities, limitations and future challenges. *Briefings in Bioinformatics*, 4(2):105–123, 2003.

[15] K.-M. Chao and W. Miller. Linear-space algorithms that build local alignments from fragments. *Algorithmica*, 13:106–134, 1995.

[16] T.H. Cormen, C.E. Leiserson, and R.L. Rivest. *Introduction to Algorithms*. MIT Press, Cambridge, MA, 1990.

[17] I. Dagan, M.C. Golumbic, and R.Y. Pinter. Trapezoid graphs and their coloring. *Discrete Applied Mathematics*, 21:35–46, 1988.

[18] A.L. Delcher, S. Kasif, R.D. Fleischmann, and J. Peterson *et al.* Alignment of whole genomes. *Nucleic Acids Res.*, 27(11):2369–2376, 1999.

[19] A.L. Delcher, A. Phillippy, J. Carlton, and S.L. Salzberg. Fast algorithms for large-scale genome alignment and comparison. *Nucleic Acids Res.*, 30(11):2478–2483, 2002.

[20] D. Eppstein, Z. Galil, R. Giancarlo, and G.F. Italiano. Sparse dynamic programming. I: Linear cost functions; II: Convex and concave cost functions. *Journal of the ACM*, 39:519–567, 1992.

[21] S. Felsner, R. Müller, and L. Wernisch. Trapezoid graphs and generalizations, geometry and algorithms. *Discrete Applied Mathematics*, 74:13–32, 1997.

[22] L. Florea, G. Hartzell, Z. Zhang, and G. Rubin *et al.* A computer program for aligning a cDNA sequence with a genomic DNA sequence. *Genome Research*, 8:967–974, 1998.

[23] L.J. Guibas and J. Stolfi. On computing all north-east nearest neighbors in the L_1 metric. *Information Processing Letters*, 17(4):219–223, 1983.

[24] D. Gusfield. *Algorithms on Strings, Trees, and Sequences*. Cambridge University Press, New York, 1997.

[25] A.L. Halpern, D.H. Huson, and K. Reinert. Segment match refinement and applications. In *Proc. Workshop on Algorithms in Bioinformatics*, volume 2452 of *Lecture Notes in Computer Science*, pages 126–139, Berlin, 2002. Springer-Verlag.

[26] M. Höhl, S. Kurtz, and E. Ohlebusch. Efficient multiple genome alignment. *Bioinformatics*, 18:S312–S320, 2002.

[27] X. Huang and W. Miller. A time-efficient, linear space local similarity algorithm. *Advances in Applied Mathematics*, 12:337–357, 1991.

[28] G. Jacobson and K.-P. Vo. Heaviest increasing/common subsequence problems. In *Proc. 3rd Annual Symposium on Combinatorial Pattern Matching*, volume 644 of *Lecture Notes in Computer Science*, pages 52–66, Berlin, 1992. Springer-Verlag.

[29] D.B. Johnson. A priority queue in which initialization and queue operations take $O(\log \log D)$ time. *Mathematical Systems Theory*, 15:295–309, 1982.

[30] D. Joseph, J. Meidanis, and P. Tiwari. Determining DNA sequence similarity using maximum independent set algorithms for interval graphs. In *Proc. 3rd Scandinavian Workshop on Algorithm Theory*, volume 621 of *Lecture Notes in Computer Science*, pages 326–337, Berlin, 1992. Springer-Verlag.

[31] W.J. Kent. BLAT—the BLAST-like alignment tool. *Genome Research*, 12:656–664, 2002.

[32] W.J. Kent and A.M. Zahler. Conservation, regulation, synteny, and introns in large-scale C. briggsae–C. elegans genomic alignment. *Genome Research*, 10:1115–1125, 2000.

[33] E.L. Lawler. *Combinatorial Optimization: Networks and Matroids*. Holt, Rinehart, and Winston, New York, 1976.

[34] D.T. Lee and C.K. Wong. Worst-case analysis for region and partial region searches in multidimensional binary search trees and balanced quad trees. *Acta Informatica*, 9:23–29, 1977.

[35] E.M. McCreight. Priority search trees. *SIAM Journal of Computing*, 14(2):257–276, 1985.

[36] K. Mehlhorn and S. Näher. Bounded ordered dictionaries in $O(\log \log N)$ time and $O(n)$ space. *Information Processing Letters*, 35(4):183–189, 1990.

[37] W. Miller. Comparison of genomic DNA sequences: Solved and unsolved problems. *Bioinformatics*, 17(5):391–397, 2001.

[38] B. Morgenstern, A. Dress, and T. Werner. Multiple DNA and protein sequence align-

ment based on segment-to-segment comparison. *Proc. National Academy of Science USA*, 93:12098–12103, 1996.

[39] B. Morgenstern, K. Frech, A. Dress, and T. Werner. DIALIGN: Finding local similarities by multiple sequence alignment. *Bioinformatics*, 14:290–294, 1998.

[40] E.W. Myers and X. Huang. An $O(n^2 \log n)$ restriction map comparison and search algorithm. *Bulletin of Mathematical Biology*, 54(4):599–618, 1992.

[41] E.W. Myers and W. Miller. Chaining multiple-alignment fragments in sub-quadratic time. In *Proc. 6th ACM-SIAM Symposium on Discrete Algorithms*, pages 38–47, 1995.

[42] F.P. Preparata and M.I. Shamos. *Computational Geometry: An Introduction*. Springer-Verlag, New York, 1985.

[43] S. Schwartz, L. Elnitski, M. Li, and M. Weirauch *et al*. MultiPipMaker and supporting tools: Alignments and analysis of multiple genomic DNA sequences. *Nucleic Acids Res.*, 31(13):3518–3524, 2003.

[44] S. Schwartz, W.J. Kent, A. Smit, and Z. Zhang *et al*. Human-mouse alignments with BLASTZ. *Genome Research*, 13:103–107, 2003.

[45] S. Schwartz, Z. Zhang, K.A. Frazer, and A. Smit *et al*. PipMaker–a web server for aligning two genomic DNA sequences. *Genome Research*, 10(4):577–586, 2000.

[46] S. Shibuya and I. Kurochkin. Match chaining algorithms for cDNA mapping. In *Proc. 3rd Workshop on Algorithms in Bioinformatics*, volume 2812 of *Lecture Notes in Bioinformatics*, pages 462–475, Berlin, 2003. Springer-Verlag.

[47] E. Sobel and M. Martinez. A multiple sequence alignment program. *Nucleic Acids Res.*, 14:363–374, 1986.

[48] P. van Emde Boas. Preserving order in a forest in less than logarithmic time and linear space. *Information Processing Letters*, 6(3):80–82, 1977.

[49] P. van Emde Boas, R. Kaas, and E. Zijlstra. Design and implementation of an efficient priority queue. *Mathematical Systems Theory*, 10:99–127, 1977.

[50] J.-S. Varré, J.-P. Delahaye, and E. Rivals. Transformation distances: a family of dissimilarity measures based on movements of segments. *Bioinformatics*, 15(3):194–202, 1999.

[51] W.J. Wilbur and D.J. Lipman. Rapid similarity searches of nucleic acid and protein data banks. *Proc. National Academy of Science USA*, 80:726–730, 1983.

[52] D.E. Willard. New data structures for orthogonal range queries. *SIAM Journal of Computing*, 14:232–253, 1985.

[53] Z. Zhang, B. Raghavachari, R.C. Hardison, and W. Miller. Chaining multiple-alignment blocks. *J. Compuional Biology*, 1:51–64, 1994.

16

Computational Analysis of Alternative Splicing

Mikhail S. Gelfand

State Scientific Center GosNIIGenetika and
Russian Academy of Sciences

16.1   Introduction

Soon after discovery of splicing [13, 35], it was observed that transcripts of some genes are spliced in different ways [14, 132]. However, until projects of mass sequencing of ESTs started to produce data on composition of total cellular mRNA, alternative splicing was believed to be rather rare, involving about 5% of human genes [143].

Mapping of ESTs to the genomic [113] or mRNA [66, 22] sequences produced an unexpected result: at least one third of human genes had alternatively spliced variants. In subsequent studies these estimates fluctuated between 20% and 60% (Table 16.1). Although the functionality of all of the observed isoforms remains questionable, it is clear that alternative splicing is a major mechanism of generating functional and evolutionary proteome diversity. A major role in studying alternative splicing was played by computational analysis, reviewed in this chapter.

The plan of this chapter is as follows. After a brief biological introduction, we start with description of alternative splicing databases and review of early studies, and then proceed to the analysis of tissue specificity of alternatively spliced isoforms and identification of cancer-

TABLE 16.1   Frequency of Alternative Splicing.

% of alternatively spliced genes	Reference	Comment
5%	[143]	Nobel lecture
5%	[175]	First systematic EST–to–genome analysis
35%	[113]	Pro-EST program; EST/genome alignments
38%	[66],[22]	Based on EST/mRNA alignments
22%	[45]	ISIS database
55%	[80]	TAP program; extrapolation from EST/genome alignments
42%	[117]	HASDB database
59%	[69]	Skipped exons, chromosome 22
∼ 33%	[168]	Human genome paper
59%	[76]	Human genome paper
28%	[38]	AltExtron database
44%	[24]	Approximation from EST/genome alignments (4-34% for other species, dependent on EST coverage).
all?	[81]	Only 17-28% with total minor isoform frequency > 5%
74%	[78]	Exon junction oligonucleotide microarrays
38%	[158]	Only 22% conserved in mouse or supported by 4 or more ESTs
40%	[75]	Full length cDNA/genome alignments, FLcDNAs database
30%	[183]	Mouse, full length cDNA/genome alignments
41%	[53]	Collection of papers about the mouse genome
60%	[184]	Mouse, full length cDNA/genome alignments

Note: By default, human genes are considered.

specific alternative splicing. We then address the dependence of the perceived alternative splicing frequency on the EST coverage and describe computational issues arising from large-scale experiments on alternative splicing. From that we turn to the analysis of functionality of alternative splicing. We consider conservation of alternative isoforms in the human/mouse gene pairs and a possible link between alternative splicing and nonsense-mediated mRNA decay. Further, we present recent results on the impact of alternative splicing on the protein structure. We describe the attempts to analyze the regulation of alternative splicing. The chapter concludes with two general sections on the evolution of alternative splicing and the overall discussion of the role of alternative splicing in the evolution and physiology of eukaryotes.

16.2   Biology of (Alternative) Splicing

Splicing is a process of elimination of introns from transcribed RNA and ligation of remaining exons, leading to the formation of messenger RNA [143]. This process is effected by the spliceosome, a complex consisting of five small nuclear ribonucleoproteins (snRNP) and more than a hundred proteins [131, 187]. The sequence signals involved in this process are, firstly, donor and acceptor splicing sites at the exon–intron and intron–exon boundaries respectively [20, 118], and, secondly, numerous binding sites for additional regulatory proteins, so-called exonic and intronic splicing enhancers and silencers [12, 133, 95, 30].These proteins, belonging to the SR and hnRNP classes, regulate both constitutive and alternative splicing [108, 92, 64]. The process of splicing involves an intricate interaction between these proteins and snRNPs that recognize the splicing sites [67]. Most introns start with GT and end with AG, although there are also rare GC–AG introns and also a special type of intron starting with AT and ending with AC; the latter are spliced by the so-called U12-spliceosome [144, 28, 38]. The U12-spliceosome also splices some GT–AG introns [144].

The main types of elementary splicing alternatives [21] are shown in Figure 16.1. Differential choice of *alternative donor* or *acceptor sites* leads to exon extension or truncation, dependent on what isoform is assumed to be the major one (Figure 16.1-a, b). Complete inactivation of splicing may lead either to *intron retention* (Figure 16.1-e), or to skipping of *cassette exons* (Figure 16.1-c). These four types of events lead to insertions/deletions in

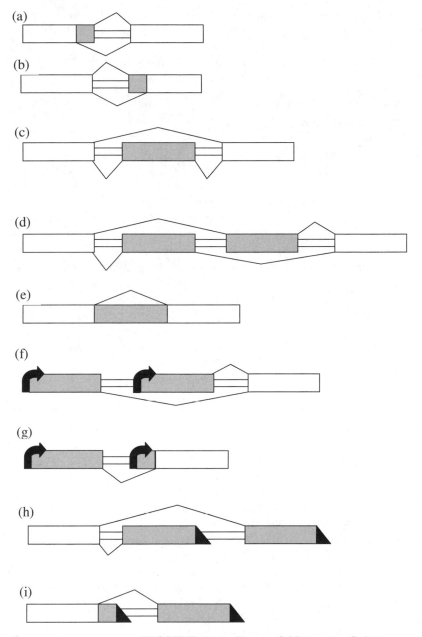

FIGURE 16.1: Types of Alternative Splicing.

the mRNA and, if occurring in the coding region, in the encoded protein. A more complicated case is that of *alternative*, or *mutually exclusive exons* (Figure 16.1-d) that generate different segments in the protein. Alternative splicing can be linked to other types of alternative events. Thus, the start of transcription at alternative promoters leads to alternative initial exons (Figure 16.1-f) or a cassette initial exon *vs.* exon extension alternative (Figure 16.1-g). Similarly, alternative splicing at the last intron may create alternative terminal introns (Figure 16.1-h), whereas competition between splicing and polyadenylation leads to a cassette *vs.* extended terminal exon alternative (Figure 16.1-i). Combinations of elementary events may create a complicated set of alternative isoforms; one rather frequent type of such combination is a cassette exon that has alternative sites as well. There is a clear link between alternative splicing and mutations in splicing sites that also may lead to exon skipping, intron retention, and activation of cryptic (*cf.* alternative) sites [91, 120, 167].

16.3 Large-scale Computational Analysis and Databases on Alternative Splicing

For many years after its discovery, alternative splicing was believed to occur in a minority of cases, up to 5% [143]. Representation of alternative splicing events in the existing databases was not systematic. GenBank [11] and EMBL [94] do not handle alternatively spliced genes in any specific way. Alternatively spliced genes can be found using text search for keywords such as "alternative splicing", "alternative exon", "isoform" or recognized by overlapping mRNA or CDS (coding sequence) descriptors in the feature tables [77]. SWISS-PROT [4] normally describes protein variants produced by alternative splicing using the "VARSPLIC" feature that identifies positions where an alternatively spliced isoform differs from the main one supplemented by the corresponding sequence fragment of the alternative isoform. In some cases alternative isoforms are described by independent SWISS-PROT entries [60]. A special program varsplic.pl has been developed to generate all alternative isoforms, and its output is available for each new release of SWISS-PROT/TrEMBL databases [84]. This is a useful resource, as results of the protein similarity search may depend on the isoform present in the database.

Mass sequencing and analysis of ESTs prompted reappraisal of the prevalence of alternative splicing. In two early studies, ESTs were mapped to human genes [113] or proteins and mRNAs [66, 22]. Both approaches yielded a close estimate: about one third of human genes were found to be alternatively spliced. Further analysis gradually increased the fraction of genes with multiple alternatively spliced isoforms to about 60% (Table 16.1). As all these studies involved automatic processing of large amounts of data, it is not surprising that their results were published not only in the form of numerical estimates, but in most cases were made accessible as databases (Table 16.2).

Most databases are based on mapping of ESTs, mRNAs and proteins to genomic sequences. ESTs were mapped to target genes by BLAST [3], and ESTs arising from one gene were clustered. Spliced alignment of these sequences to the genomic fragment was used for reconstructions of the exon-intron structure. Later, pre-existing clusters of ESTs such as UniGene [189] were used. All such projects involved extensive filtering of possible artifacts [175, 26, 115]. This filtering removes ESTs that arise from unspliced transcripts and contamination of clone libraries by genomic DNA. This makes it difficult to estimate the frequency of intron retention, which cannot be reliably distinguished from the above artifacts. There are also chimeric and aberrant ESTs probably created by recombination during cloning. Such ESTs create intron-like deletions in cDNA that can be recognized by the lack of universal GT–AG dinucleotides at their boundaries. Also, in many such cases the intron-like deletion in a EST sequence aligned to a genomic fragment is flanked by short direct or inverted repeats. EST clusters may be generated by recently duplicated and thus highly similar paralogous genes that could appear as alternatively spliced isoforms. Finally, there may exist unknown systematic errors [151].

Independent alignment of ESTs to the genome may introduce spurious results due to low sequence quality leading to minor differences at exon–intron boundaries. Several programs were developed to find consistent alignments of EST clusters to the genome, while still allowing for alternative splicing [173, 17]. In a pilot study, it was shown that this approach

TABLE 16.2   Alternative Splicing Databases.

Database	Content	Organisms	URL	Reference
EDAS	EST/genome alignments	Human	http://www.ig-msk.ru:8005/EDAS/	[122]
IGMS/ EASED/ ASforms	EST/mRNA alignments	Human (detailed data); mouse,rat cow, *Xenopus*, zebrafish, *Drosophila*, *C. elegans*, *Arabidopsis*	http://eased.bioinf.mdc-berlin.de/, http://eased.bioinf.mdc-berlin.de/igms	[24], [130], [129]
ASD	EST/genome alignments; literature data; regulation	Human,mouse, rat, cow, chicken zebrafish *Drosophila*, *C. elegans*, *Arabidopsis*	http://www.ebi.ac.uk/asd	[165]
PASDB	GenBank and literature data	Plants	http://pasdb.genomics.org.cn	[186]
ASAP/ HASDB	EST/genome alignments	Human	http://www.bioinformatics.ucla.edu/ASAP	[99], [117]
MouSDB	mRNA/genome alignments	Mouse	http://genomes.rockefeller.edu/MouSDB	[184], [53]
ProSplicer	EST/genome and protein/genome alignments	Human	http://140.115.50.96/ProSplicer/	[72]
SpliceNest	EST/genome alignments	Human	http://splicenest.molgen.mpg.de/	[90] [43]
PALS db	EST/mRNA alignments	Human and mouse	http://palsdb.ym.edu.tw/	[73]
Alternative Splicing Supplemental Resources	mRNA/genome alignments	Mouse	http://www.bioinfo.sfc.keio.ac.jp/ research/intron/index.html	[86]
AltExtron	EST/genome alignments	Human	http://www.bit.uq.edu.au/altExtron/, http://www.ebi.ac.uk/~thanaraj/altExtron/	[38]
AsMamDB	Processed GenBank: mRNA/DNA alignments	Human, mouse, rat	http://166.111.30.65/ASMAMDB.html	[77]
STACK	EST/genome alignments	Human	http://www.sanbi.ac.za/Dbases.html	[69]
Intronerator	EST/genome alignments	Nematode *C. elegans*	http://www.cse.ucsc.edu/~kent/intronerator/	[83]
VARSPLIC	Processed SWISS-PROT	All organisms	ftp://ftp.ebi.ac.uk/pub/databases/sp_tr_nrdb, ftp://ftp.expasy.ch/databases/sp_tr_nrdb	[84]
AEDB	Literature data	Human	http://www.ebi.ac.uk/asd/aedb/	[156]

Note: EST/genome implies mRNA/genome.

allows one to find new alternative splicing events supported by multiple ESTs, missed by simple EST–to–genome alignment [17].

After that, the alternative splicing events are classified into elementary alternatives whenever possible. Databases of the last generation, such as EASED and EDAS, incorporate data on expression of observed isoforms and individual alternative events. These data include tissue and stage specificity, as well as the disease state. In addition, functional annotation of the corresponding genes is provided.

However, despite the multiplicity of genome projects, in some cases the genome data for comparison are not available. Indeed, many large-scale sequencing projects involve sequencing of ESTs rather than genomic sequences [15]. Thus there arises the following problem: how to represent the diversity of alternatively spliced isoforms in such cases. One possibility is to use assembly programs borrowed from genome sequencing projects and generate all possible linear isoforms. In particular, this approach was used to create the TIGR gene index [2, 159]. However, as the EST data are local (with the size of a single fragment being 300-500 nt), it does not capture information about long-range dependencies between alternative variants at distant regions. Moreover, it is not immediately clear that such dependencies always exist. Anyhow, this leads to a combinatorial explosion in the number of alternative isoforms. An often cited example is that of the *Drosophila Dscam* gene that may generate up to 38,000 potential isoforms [141, 63]. This may be a likely reason for an EST-based overestimate of the number of human genes [104].

A simpler possibility is to align available ESTs [26]. This avoids the combinatorial explosion, but provides no insight into the structure of the data. Besides, representation of alternating exons and other types of events generating non-alignable segments in mRNA still remains a problem. Still, this is a convenient method when only individual splicing events are studied without the need to reconstruct the overall structure [169]. A procedure allowing one to take into account alternative splicing when using EST data for protein identification from peptide mass fingerprinting was suggested in [106].

To deal with the combinatorial explosion problem, several studies introduced a concept of a *splicing graph* [68, 98, 62, 158]. It is a directed graph in which each vertex corresponds to a position observed in at least one EST, and two vertices are linked by an edge if they are consecutive positions in at least one EST. For convenience, linear chains of non-branching vertices may be collapsed into one vertex. Such a graph may be constructed by multiple alignment of a set of ESTs, or by aligning ESTs to genomic sequences. Each path along this graph represents a possible isoform. Although this representation still does not allow for modeling long-range dependencies, it provides a convenient and intuitively clear representation used in a number of databases.

Importantly, even purely EST-based analysis allows for some insight into the proportion of different elementary alternatives. To do that, one needs to analyze the sequence at the boundaries of alternative regions [86]. Indeed, an overwhelming majority (more than 99%) of introns start at GT and end at AG [27, 28, 163], and there are additional conserved positions in donor and acceptor signals [118, 59]. Internal splicing sites and sites of non-spliced introns are retained in longer isoforms (Figure 16.2), and thus the analysis of candidate sites at the boundaries of an alternative region may tell what type of elementary alternative generated this region.

To analyze the dependencies between elementary alternatives one needs to analyze full-length isoforms. These isoforms can be collected from the literature. Currently three projects, ASDB, ASD/AED and PASDB follow this path; FlyBase contains curated data on the alternative splicing of *Drosophila* genes [57]. Curated literature-based databases have two additional advantages. One of them is the possibility to collect data on regulation, tissue specificity studied in comparable controlled experiments, and cellular localization of resulting proteins. Another advantage is the ability to take into account additional isoforms discovered after publication of the gene. Such isoforms often are not deposited in sequence databases, but are shown in subsequent publications in the form of figures, textual descriptions etc. Analysis of the literature allows one to find such isoforms. As the exhaustive literature search is expensive and time-consuming, there are attempts to apply text mining techniques for finding papers on alternative splicing by analysis of abstracts in the PubMed database, and to link genes, isoforms, and tissue specificity data [142].

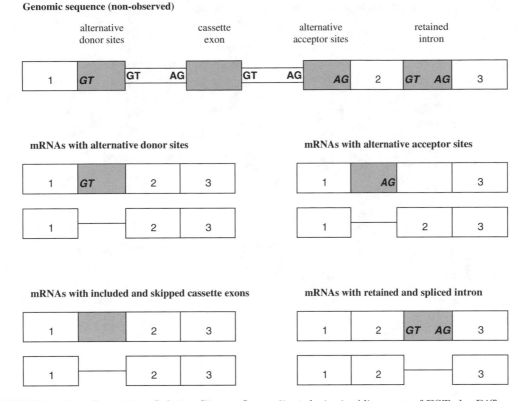

FIGURE 16.2: Remaining Splicing Sites at Insert Boundaries in Alignment of ESTs by Different Types of Alternative Splicing Events.

16.4 Large-scale Experimental Studies

Large-scale experiments aimed at sequencing of full-length cDNAs provide another rich source of data. The FANTOM2/RIKEN project aimed at sequencing mouse RNAs uses a special experimental technique aimed at capturing polyadenylated mRNAs carrying the cap structure at 5′ end [53], whereas the ORESTES approach is based on sequencing EST-s uniformly distributed along mRNA with subsequent mapping to the genome, creating EST contigs, and directed gap closure [29, 137]. Both approaches use hybridization-based normalization strategies to avoid multiple sequencing of abundant transcripts, and thus discriminate against alternatively spliced isoforms. Still, they are able to cover many alternative splicing events, and, as with ordinary ESTs, the number of genes having alternatively spliced isoforms increases with increased coverage [137]. In particular, the estimate of the fraction of alternatively spliced mouse genes based on data generated in the mouse full-length cDNA sequencing project increased in just one year from 30% [183] through 41% [53] to 60% [184] (Table 16.1). A number of new isoforms were identified by alignment of full-length mouse cDNAs [86], although no estimates on the prevalence of alternative splicing were given. Similarly, combination of the data from six projects on sequencing full-length human cDNAs in the framework of the H-Inv project identified alternative splicing isoforms for 40% of human genes. Enrichment strategies increase the number of unique variants (e.g. 40% of alternative exons generated in the ORESTES project do not appear in conventional ESTs) [137], although it is not clear whether these are functional and not

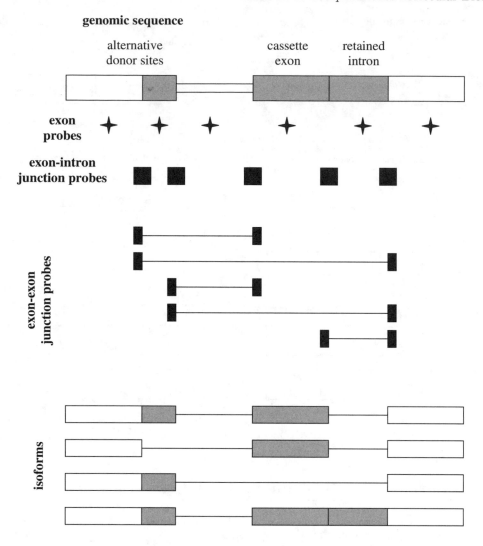

FIGURE 16.3: Microarray-based analysis of alternative splicing. Design of oligonucleotide array assuming complete knowledge of the exon-intron structure and alternative splicing.

results of aberrant splicing, see below.

Another large-scale approach is based on the use of oligonucleotide microarrays. Two types of oligonucleotide probes are used: *exon probes* that hybridize to constant or alternative exons or parts of exons, and *junction probes* that hybridize to exon–exon junction regions. A prerequisite to this analysis is complete description of the exon–intron structure, including all alternative variants (Figure 16.3). An alternative is to create a high-density array covering known mRNAs (isoform 0). At that, skipped exons, alternative exons and internal splicing sites can be observed, as the oligonucleotides occurring within corresponding alternative regions would not hybridize (Figure 16.4, isoforms 1 and 2 respectively), although new exons or alternative splicing sites extending known exons may not be found. However,

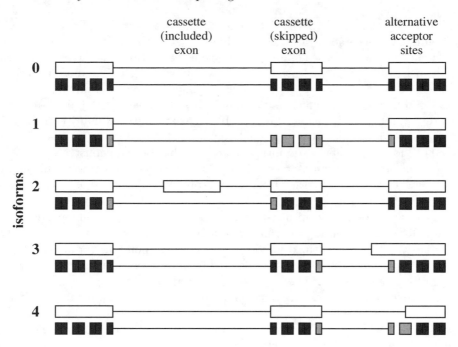

FIGURE 16.4: Microarray-based analysis of alternative splicing. Design of dense oligonucleotide arrays to the main mRNA (isoform 0).

if by chance an oligonucleotide spans an alternative exon–exon junction, even insertion-type events such as additional cassette exons and extending alternative sites might be diagnosed (Figure 16.4, isoforms 1 and 3).

Note that in this case the observed intensities are not uniform, and reflect the level of inclusion of alternative regions and thus the relative fraction of the isoforms in the sampled tissue. Specific algorithms are used to process the data and to estimate the splicing level for all isoforms. One of the first papers specifically aiming at the analysis of alternative splicing was [71]. In that study 1600 rat genes were considered, each represented by twenty 25-mer oligonucleotide probes densely covering the 3' end of the known mRNA, and the expression was studied in ten normal tissues. This resulted in identification of a number of new isoforms confirmed by analysis of available EST data and direct RT–PCR using primer pairs identified as informative in the array experiment.

The exon–exon junction arrays for large-scale analysis of alternative splicing were introduced in [180]. Expression of exons in ten thousand genes was analyzed in 52 tissue samples using ~125,000 exon junction probes [78]. The observed hybridization level was compared to the expected level estimated from the probe response across tissues, and the expression level in a given tissue assuming all exons are present. Tissue-specific alternative splicing was observed for a number of known isoforms; moreover, new alternative splicing events were predicted, when some probes failed to hybridize, indicating that the corresponding splicing sites are not used universally. A fraction of observations was confirmed by RT–PCR. Comparison with EST data demonstrated that arrays seem to be more sensitive in detection of low-level alternative splicing. Further, the comparison of the EST-derived splice junctions and mRNA data from RefSeq demonstrated that 20% exon junctions, not necessarily alternative, are not represented in ESTs, and 11% junctions are present in only one EST, and thus would be deemed unreliable by many analysis tools.

The same techniques can be used to analyze the processes of splicing, processing and decay of mRNAs, as demonstrated in a study of yeast genes [39].

A different approach is based on genome sequence *per se* rather than known exons and introns. In such studies oligonucleotide probes are uniformly distributed along chromosomes, forming so-called *tiling arrays*. Such studies are mainly directed towards identification of new classes of transcripts, such as non-coding RNAs. However, as the distance between adjacent probes and probe length are comparable to exon length, in many cases new isoforms are identified [145, 82]. Sensitivity of relatively long oligonucleotide probes in tiling arrays (up to 60 bp) for detection of exon-intron boundaries has been demonstrated [33]. However, so far only the shortest chromosomes (21 and 22) have been analyzed by this approach.

Since in most cases raw hybridization data represents a mixture of alternative variants produced by different isoforms, special treatment is needed to determine the fraction of each individual isoform, especially in non-trivial cases involving several independent choices of variants in different elementary alternatives. Such a model was proposed in [169]. Observed probe intensities are transformed into fractions of corresponding variants using standard models, whereas isoform frequencies are derived from the elementary variant fractions using a system of linear equations. The approach was validated by comparisons with samples of known isoform composition *in vitro* and the results of RT–PCR applied to genes with known expression patterns *in vivo*.

A database MAASE was developed to simplify the construction of microarray chips for studying alternative splicing and collection of the results [185]. An interactive interface is used to annotate known isoforms combining the data extracted automatically from existing databases and manual analysis of the literature, and then aids in selection of probes.

16.5 Alternative Splicing in Non-mammalian Species

The observed level of alternative splicing is highest in the human genes. Arguably, this is a consequence of higher EST coverage. An attempt to make an independent estimate was made in [24]. The observed prevalence of alternative splicing was computed for comparable numbers of ESTs in several species (human, mouse, rat, cow, *Drosophila*, nematode *C. elegans*, and plant *Arabidopsis*), and the saturation levels were estimated. The level of alternative splicing was comparable in all animal species and slightly lower in *Arabidopsis*.

Alternatively spliced nematode genes are collected in the Intronerator database [83]. The FlyBase database of *Drosophila* genes (release 3.1) [57] contains the data on alternative splicing of 2774 genes (of 13656 genes in the database). Of these, 1815 genes are alternatively spliced in the coding region, whereas 959 are alternatively spliced only in untranslated regions. Most data are derived from ESTs, with only 75 genes with alternative splicing verified by the literature search. A database of alternative splicing in plants, based on literature and EST data, was compiled in [186]. The functional categories most represented in this database were cell growth and enzymatic activity, although the size of the database is rather small and no analysis of statistical significance of this finding was made. About 10% of *Arabidopsis* full-length cDNA were found to be alternative isoforms in a large-scale study[65], although the authors warn that this number should not be extrapolated to the entire genome. Finally, out of approximately 230 intron-containing genes of the yeast *Saccharomyces cerevisiae*, only three are alternatively spliced [63].

16.6 Alternative Splicing, Alternative Promoters, and Alternative Polyadenylation

Alternative splicing is closely related to other types of alternative events, such as alternative starts of transcription and alternative polyadenylation. Indeed, an early study demonstrated that a large fraction of alternative splicing events takes place in non-coding regions and leads to alternative 5' or, less often, 3' mRNA termini [113]. However, it is difficult to exactly estimate the frequency of such events, especially at the 5' end,as this frequency depends on the technique used to generate EST libraries (polyT-primed libraries have low coverage of5'-termini), identification of translation start, and, for so-called full-length libraries, of the completeness of cDNA.Still, the observed trends are the same. Analysis of EST data produced the following breakdown of observed elementary alternatives: 22% in 5-untranslated region, 74% in the coding region, and only 4% in the 3-untranslated region[113]. Meta-analysis of human cDNA sequencing projects demonstrated that 35% of all genes had alternatives in the coding region [75]. Among alternatively spliced transcripts, 48% had alternative 5-ends,62% had internal loops, and 29% had alternative 3-ends.Full-length mouse cDNA had 74% of elementary alternatives overlapping with the coding region [184],whereas ORESTES human cDNAs had 85% of alternative exons in coding regions, compared to 77% alternative exons in a EST sample [137]. Thus, despite some differences, a clearly emerging tendency is higher frequency of alternative splicing at the 5-region compared to the 3-region. One possible explanation for that could be that it allows the cell to use different transcription promoters and thus different regulatory cassettes for the same gene[113]. Thus, instead of a complicated mosaic of regulatory elements in the upstream region of a gene that needs to be expressed in several different conditions, there are several promoters switched on independently, whereas the coding regions is the same and the excessive 5-leader region is removed by splicing. This generates alternatives of the types shown in Figure16.1-f and Figure 16.1-g. On the other hand, analysis of ESTs sequenced from the 3'-termini of mRNAs demonstrated that 29% of human genes have multiple polyadenylation sites [9], and more than 40% of genes have variable termini due to multiple polyadenylation sites and alternative cleavage sites downstream of single polyadenylation site [128]. Moreover, in many cases the choice of a polyadenylation site is tissue- or disease-specific[10]. This analysis is based on a combination of clustering of polyT-primed ESTs [58, 9]with computational identification of candidate polyadenylation sites [61]. Alignment to the genomic sequence is used to filter out ESTs primed at internal polyA runs: unlike polyA tails added to mRNAs during polyadenylation, internal runs are present both in EST and the genomic sequence.

16.7 Tissue-specific and Cancer-specific Isoforms

A natural extension to large-scale analysis of alternative splicing is identification of isoforms specific to tissues, and developmental and disease stages. It has been suggested that a large proportion of alternative splicing events takes place in specific physiological context,such as nervous and immune system [155, 117]. High level of alternative splicing was observed in brain, eye (retina),lymph, as well as testis, muscle and skin, and about 10-30%of alternatively spliced human genes were shown to have tissue-specific isoforms [179]. Clearly, such estimates are very sensitive to the coverage of various tissues by clone libraries. For example, out of 25 exons assumed to be brain-specific in [25], at least 12 turned out to be expressed in other tissues as well [46]. Moreover,absolute tissue specificity is rare [176]. This problem was addressed in a Bayesian statistical approach developed in [179]. It treats the true alternative splicing variant frequency in a given tissue as a hidden variable. The tissue specificity is defined as the difference between the probability that this variant is preferred in a given tissue (more exactly, the conditional probability that its true frequency exceeds 50% given the observed data) and the probability that the variant is preferred in a pool of other tissues. It is a some what indirect measure. A test involving log-odds ratio was developed in [178]. It directly measures the probability that the frequency of one of two alternative variants in one condition (e.g. cancer, see below) is greater than its frequency in a different condition (e.g.,normal). This test can be also applied to analyze tissue-specific splicing.

Tissue-specificity data can be obtained in array experiments, see above. In particular, in[78] it has been demonstrated that the highest frequency of alternative splicing is seen in cell lines, and the overall patterns of alternative splicing are similar in related tissues such as stomach and duodenum, heart and skeletal muscle,as well as in all neuronal tissues. It may be also interesting to identify elementary alternatives with complementary distribution of variants, so that one variant is expressed in one set of tissues, and the other variant in a different, non-overlapping set of tissues. Several such cases were found in [176].However, such analysis is greatly complicated by inhomogeneity of cell types forming most tissue samples. Of major interest are genes having cancer-specific isoforms,especially if there is evidence of medical importance of these isoforms, e.g.[174, 136, 157, 56, 89, 32, 1, 44, 127],see also [178] for more references. Note that cancer specificity in this context should be assumed with some caution.Firstly, it is unlikely that an isoform strictly specific for cancer cells would be evolutionarily stable. More likely, it is expressed in some specific, not yet sampled condition or stage,whereas its appearance in cancer is due to aberrant regulation. There are many examples of such kind in the area of transcriptional regulation, e.g. the so-called cancer/testis antigens, expressed in cancer and germ-line cells [139]. Further, an additional complication is insufficiently detailed and reliable description of many clone libraries [176]. The most frequent type of error is omission of indication to the cancer or cell line origin of a clone library [7]. Manual curation, though labor-consuming, often allows one to resolve this issue [122], (I.Kosmodemyansky, personal communication).It should be interesting also to distinguish between cell lines and primary cancer tissue; no such studies have been reported sofar. Computationally, the analysis of cancer-specific variants is similar to the analysis of tissue specificity, and often such analysis are performed simultaneously. Several cancer-specific isoforms of known genes were identified in [176], and colon cancer-specific isoforms were found in [23]. A z–statistic was applied to identify cancer-specific isoforms,followed by experimental RT–PCR analysis of pairs of tumor and normal samples from several tissues [169]. It turned out that a large fraction of alternative splicing variants were indeed cancer-specific; in addition, in several cases tissue-specific splicing was observed. On the other hand, the increased frequency of non-GT–AG introns in [169] is probably due to contamination, as insufficient filtering criteria were applied[149]. Indeed, it was not clear, whether any of these non-GT–AG introns were supported by experimental analysis in the original paper [169], although some were observed in a follow-up study [170]. The set of genes with cancer-specific isoforms identified in[178] was enriched in tumor suppressors, genes encoding cell cycle, growth and proliferation proteins, genes whose products were involved in transcription and splicing and, more generally,genes encoding DNA- and RNA-binding proteins. This could be due to over-expression of these genes in cancer, leading to increased statistical significance of any trends. However, the results were consistent with mRNA data. Moreover, validation of the EST-derived sets of cancer-specific and normal tissue-specific isoforms by comparison with known mRNAs from GenBank yielded a surprising result: in many cases known mRNAs were the cancer-specific ones,whereas their normal counterparts were identified only by EST analysis. The opposite situation was relatively rare. Further, to avoid

TABLE 16.3   Conservation of Human Alternatives in Mouse and Rat.

Reference	Results
[80]	55% human genes alternatively spliced; 11% alternative splicing patterns observed in mouse ESTs
[123], [121]	17-31% elementary alternatives are non-conserved between human and mouse; 36-55% alternatively spliced genes have species-specific isoforms.
[164]	61% of alternative exon–exon junctions; 74% of constitutive exon junctions conserved in mouse.
[116]	98% constitutive exons, 98% major form cassette exons, 28% of skipped exons conserved between human and mouse.
[134]	Exons observed to be skipped in human and mouse ESTs are 12% of all human skipped exons and 28% of all mouse skipped exons.
[158]	3% retained introns, 38% cassette exons, 8% alternative donor sites, 18% alternative acceptor sites, 32% other alternatives conserved in mouse transcriptome.
[152]	75% cassette exons are not conserved in mouse.

mixing cancer specificity and tissue specificity, tissue-specific samples of cancer and normal origin were compared, and the results were consistent with the data from pooled cancer and normal clone libraries. Conversely, an overwhelming majority of cancer-specific isoforms was supported by evidence from clone libraries both of different tumor origin and from multiple normal tissues. The majority of significant shifts in the isoform distribution were loss (more exactly, strong decrease of frequency) of one variant in cancer (56%) with approximately equal frequency of both variants in the normal sample; 31% of cases demonstrated shift from one prevalent variant to the other one; and only 13% of cases were of the type with a skewed variant frequencies in the normal sample and a more uniform frequencies in cancer. Thus the obtained results were not due to deregulation and loss of specificity in cancer.

16.8   Conservation of Alternative Splicing

Sequencing of additional genomes, especially the mouse genome, created an opportunity to compare the alternative splicing of orthologous genes in the two genomes. Indeed, there were scattered observations of species-specific splicing isoforms (listed in [121], see also [162] about the role of alternative splicing in speciation). However, the extent of non-conservation of alternative splicing was quite surprising: it looks like alternative splicing is much less conserved than the gene complement in general (Table 16.3).

Indeed, 80% of mouse genes form strict pairs of orthologs with human genes [119]. Most of the remaining genes belong to expanded families of paralogs, so that it is impossible to establish 1:1 relation between the human and mouse genes despite clear homology. Not surprisingly, this is characteristic of genes of the reproduction system, pheromones, hormones, the olfactory system, and the immune system. In an early study it was demonstrated that only about 5-14% of the human–mouse gene pairs had different exon–intron structure [8, 119]. On the other hand, a considerable fraction of human genes seem to have isoforms not conserved in the mouse genome, see below. Uneven and incomplete coverage of the human and mouse genomes by ESTs preclude direct comparison of the alternative splicing patterns in the two genomes. Indeed, only 11% of human alternative splicing patterns were observed in mouse ESTs as well [80]. One way to deal with that was to extrapolate the observed fraction of conserved elementary alternatives to the situation of complete coverage [164], an approach reminiscent of the estimates of the prevalence of alternative splicing in different genomes discussed above [24]. Given a number of simplifying assumptions (same tissue specificity of orthologous isoforms, same coverage of tissues by clone libraries, independence of the probability of a gene to be alternative spliced on the expression level, and the independence of the probability of isoform conservation on the expression level), it is possible to derive a formula for the probability of an exon–exon junction (that is, a variant of an elementary alternative) to be conserved and, moreover, to find the confidence interval for this probability. It turned out that 74%(71-78% at the 95% confidence level) of constitutive splice junctions and 61% (47-86% respectively) of alternative splice junctions of human genes are conserved in mouse.Although compared to results of other studies (below) these estimates seem to be somewhat too low, they provide a lower estimate on the conservation of constitutive and alternative splicing. A different approach is to map human isoforms to the mouse genome.At that, if an isoform cannot be expressed because an exon cannot be aligned or splicing sites are destroyed, it is clear that this isoform is not conserved. However, conservation of sequence features does not guarantee existence of the isoform, as other,unknown regulatory elements can be destroyed. Further, it does not allow for the analysis of skipped exons (as the exon will be conserved anyhow because it is included in the longer isoform), intron retention (which in at least one case was proven to be an important diversity-generating mechanism [162]), and alternative splicing in non-coding regions (where the level of sequence conservation is much lower and exons cannot be reliably aligned in many cases). Thus such analysis provides an upper estimate on the conservation of splicing isoforms. Application of this approach to a sample of 166 human–mouse gene pairs demonstrated that 69-76% of human elementary alternatives (mRNA- and EST-derived, respectively) are conserved in mouse, and75-83% of mouse ones, in human. Further, 55% human and36% mouse genes had species-specific isoforms[123, 121]. A similar technique was applied to the analysis of human, mouse,and rat genes [115]. Only cassette exons were considered.These exons were divided into two groups, those present in the major isoform (that is, included in more than 50% ESTs spanning the relevant region) and those present in the minor isoform(skipped in more than 50% ESTs). Mapping of human exons onto mouse and rat genomic sequences was done by BLAST. Similarly to the results of [69], the proportion of genes with major form cassette exons to the genes with skipped exons was approximately 2:1. The levels of inclusion of a given exon inhuman and mouse were significantly correlated (80%). Most importantly, the conservation of the major form exons was the same as of the constitutive exons (98%), whereas as many as 72%of skipped exons turned out to be genome-specific. All types of events were analyzed in [158]. Human and mouse EST libraries were compared. The frequency of genes for which alternative splicing was observed both in human and mouse was 10%. The level of conservation of human alternatives depended on the type of the alternative; variants supported by multiple ESTs were more likely to be conserved. Although the results of this study are not directly comparable to the 11%conservation observed in [80], as it is not clear how to combine the results for different types of alternatives into a single value, it seems that increased EST coverage of the mouse genome leads to increased level of conserved variants. In fact, it could be instructive to apply the same procedure in the opposite direction: consider all mouse alternatives and compute how likely they are to be present in human ESTs. Indeed, the importance of the coverage was demonstrated in[79], where human ESTs were aligned to the mouse genome and vice versa. Human ESTs identify 27% of known mouse alternatives; novel alternatives were found in 51% genes. Conversely, 21% known human alternatives were recovered using mouse ESTs, and new patterns were found in 42% of human genes.In fact, humans ESTs recover more mouse alternatives than mouse mRNAs do, and 60% of EST-derived alternatives. For comparison,40% of mRNA-derived human alternatives were supported by human ESTs. For constitutive splicing events, all these numbers are very close (82-84%) and much higher than for the alternative splices. This again supports the observation that alternative splicing events are much less conserved that constitutive ones.

16.9   Functionality of Alternative Isoforms. Nonsense-mediated Decay

An important question arising from these studies is whether all of the observed isoforms are real. Indeed, given the size of EST datasets, it is possible that some alternative splicing events,especially those supported by unique ESTs, cancer-specific ones, and those non-conserved in mouse, result from aberrant splicing. There is little experimental data about the level of splicing errors in vivo. Up to 2% of aberrant exon skipping in the human HPRT gene was observed in normal tissue, whereas up to 20% aberrant intron retention in the RET gene was observed in cancer [146]. Recently this problem was directly addressed in [152], where it has been suggested that a large fraction of alternative isoforms might not be functional. Indeed, the frequency of frame shifts that result in shortening the protein C-terminus was estimated as19% (with 6% of extensions and overall 46% of C-terminus replacements) [117], 22% [156], 29% (with50% skipped exons retaining the reading frame) [69].In [164], 22% of cassette exons have not lead to frameshifts, nor introduced stop codons. Change of the reading frame due to the use of alternative sites in mouse was estimated as53% for acceptor sites and 35% for donor sites [184], the difference due mainly to a specific case of 3 nt distance between adjacent acceptor sites. A detailed study compared the features of human cassette exons conserved and non-conserved in mouse [152]. The reading frame was retained in 77% of conserved exons, and of these only 5% contained an in-frame stop codon.Thus 73% of such exons did not change the protein C-terminus.For non-conserved cassette exons these numbers were 40% frame retention, 53% of these with open reading frame, and overall only 23% proteins with intact C-termini. However, there are several lines of evidence supporting functionality of observed isoforms. Indeed, at least 47% of observed events were observed more than once, that is, were supported by more than one EST [117]. Similarly, a binomial test was developed in [81] to estimate the frequency of minor variants. It turned out, that although more than 99% of genes with high EST coverage had alternative splicing variants, 17-28% of gene had variants with frequency exceeding 5%; the frequency of reliable intron retention was less than 5%. Very similar conservation rates were observed for elementary alternatives supported by ESTs (69% and 75% in human and mouse, respectively) and mRNAs (resp., 76% and83%) [121]. Of human skipped (minor isoform)exons, 28% were up-regulated in one tissue, whereas 70% of such skipped tissue-specific exons still were not conserved in the mouse genome [116]. The suggestion that many cancer-specific isoforms are due to general deregulation of splicing in cancer contradicts the observation that most changes in isoform frequency observed in cancer are loss of a normal-specific isoform, followed by a switch from a normal-specific to a cancer-specific isoform[178]. This rules out both deregulation of splicing in cancer as a source of new isoforms and the carcinogenic effect of cancer-specific isoforms, and speaks in favor of a tumor suppressor function of normal-specific isoforms, lost in cancer. One possible explanation for the discrepancy between perceived functionality of most isoforms and the fact that most isoforms contain a frameshift could be that truncated proteins encoded by isoforms with alternatives leading to frameshifts or premature stop codons have a dominant negative effect on the main product and serve a regulatory role [93]. Further, isoforms creating premature stop codons (introduced by frameshifts or by in-frame stops in alternative regions) are subject to nonsense-mediated mRNA decay [103]. This process degrades mRNAs where the 3-proximal exon–exon junction is more than 50 nucleotides downstream of the stop codon. It turned out that 35% alternative isoforms suggested by ESTs were candidates for nonsense-mediated decay, whereas only 5% of RefSeq mRNAs belonged to this category. A large fraction of these encoded splicing factors,translation factors, and ribosomal proteins, and in some cases nonsense-mediated decay was shown to regulate gene expression. The authors of [103] have suggested that their results demonstrate that nonsense-mediated decay of alternative isoforms may be a major mechanism of the regulation of gene expression.Subsequent analysis of protein isoforms annotated in Swiss-Prot (more exactly, of corresponding mRNAs)demonstrated that 6% of isoforms corresponding to 7% of genes would be subject to nonsense-mediated decay[70]. Moreover, taking into account nonsense-mediated decay allowed the authors to explain some previously published experimental results. Finally, in plants, incomplete splicing might be associated with gene silencing [111, 65].

16.10 Impact of Alternative Splicing on Protein Structure and Function

Several studies reported relatively high frequency of alternative splicing of genes involved in signal transduction[117, 78] and genes encoding proteins of the nervous and immune systems [117]. Array data show than similar tissues (e.g. stomach and duodenum, skeletal and heart muscle, adult and fetal liver) express similar spectra of alternative splicing isoforms [78]. Analysis of the domain structure of alternatively spliced protein isoforms demonstrated that alternative splicing tends to shuffle domains rather than disrupt domains or change regions between domains [93].The data were compared to several types of random controls aimed at eliminating general correlation of all (not necessarily alternatively spliced) exons and domains as a source of the observed tendency. This was done by fixing the observed domain structure and considering all possible placements of alternatively spliced regions on the protein. Further, those alternative-splicing events that influenced domains and were sufficiently short so as not to disrupt domains completely, were shown to target protein functional sites. Some particular cases were considered in detail in [19]. However, similar analysis of interacting proteins demonstrated lack of correlation between contacting residues and alternatively spliced regions[126]. Analysis of EST-derived isoforms by comparison with PFAM domain annotations demonstrated that alternative splicing frequently targets domains involved in cellular (annexin and collagen), protein-protein (KRAB domains, ankyrin repeats, Kelch domains) and protein-DNA (zinc fingers, SANT domains, homeodomains) interactions [135].However, there was no significant increase in the *rate* of splicing of such domains, and, as noted by the authors, the observation may simply reflect the relative frequency of such domains in the human genome. The same conclusion was reached for mouse genes in [184]. One domain, likely involved in adhesion, and preferentially observed in cancer-specific isoforms, was identified by sequence comparison and named AMOP [37]. Prediction of transmembrane segments in products of alternatively spliced genes showed no significant preference of alternative splicing to target transcript fragments encoding such segments[40]. More exactly, specific targeting was observed only in the case of proteins with single transmembrane segments and likely reflected alternative splicing of signal peptides. Finally, comparison of alternative splicing data and SCOP structural assignments yielded preferential splicing of genes encoding proteins with α/β(alternating α and β secondary structure elements)domains compared to proteins with $\alpha+\beta$ (separatedα and β structural elements) or small domains [75]. However, the details of the normalization procedure were not given, and there remains a possibility that this simply reflects the relative frequency of such domains.

16.11   Evolutionary Origin of Alternative Splicing

An attempt to reconcile the comparative and functional data wasmade in [134], where skipped exons and alternative splicing sites from five genomes (human, mouse, rat, zebrafish*Danio rerio*, and fruit fly *Drosophila*) were considered. For each exon, an attempt was made to identify orthologous exons using the HomoloGene database [172] for vertebrates and BLAST for *Drosophila*. The frequency of frame-preserving(with length divisible by three) constitutive exons was 40% (of all constitutive exons); frame-preserving cassette exons formed 42% of human cassette exons and45% of mouse exons, exactly as expected from the distribution of introns relative to codon positions. However, if an exon was observed to be alternatively spliced in two species, it was frame-preserving in 53-73% of cases (in human-mouse and human-rat pairs). Similarly, higher frequency of frame-preservation was observed for alternative splicing events conserved in zebrafish and *Drosophila* ESTs. A surprising result was higher frequency of frame preservation among minor isoform exons and alternative splicing sites compared to major form and constitutive splicing events. This and earlier[116] studies show very similar behavior of constitutive and major form exons in terms of both frame-preservation and conservation in several genomes, whereas minor form exons are mostly young and more often frame-preserving. The following explanation was suggested: the major form exons are former constitutive exons that encountered competition due to recent activation of cryptic splicing sites. On the other hand, the minor form exons are young, and the fact that they have been fixed means that they produce functional proteins; thus to become fixed, the exon should avoid disrupting the reading frame of downstream exons. Several studies addressed the problem of origin of new isoforms.Analysis of alternative (mutually exclusive) exons demonstrated that at least 9%of them are reliably homologous, that is, arose from tandem exon duplication [87]. This is likely an underestimate,as such exons are often short and thus residual similarity might be unnoticeable or insufficiently significant. Notably, in approximately half of cases the length of alternative regions was not a multiple of 3, and thus both alternative variants could not be present in mRNA without disrupting the reading frame. Thus the isoforms were mutually exclusive immediately after the duplication. When orthologs from several species were known, an upper limit on the time of duplication could be set. It turned out that many duplications are common to mammals, some to mammals and birds, amphibiae, fishes, and even mammals and nematodes. However,some duplications occurred in the last common ancestor of mouse and rat. The number of genes for which such dating could be made is rather small, so no general inferences can be made. Moreover,in some cases the situation may be obscured by secondary loss of alternative exons: one such example is skipped exon of lactadherin that is present in mouse and some other species, but lost in human [18]. A similar study started with identification of all genes with duplicated exons [101]. To avoid artifacts caused by mis-annotation of genes, each exon was compared to adjacent introns, and possible non-functional exons were filtered out based on a number of criteria such as absence of in-frame stop-codons.Exact duplications that likely resulted from genome mis-assembly were also removed. Hypothetical and EST-confirmed duplicated exons had the same statistical features, such as the ratio of non-synonymous and synonymous substitution rates. The fraction of genes with duplicated exons was 11% in human, 7% in fruitfly *D. melanogaster* and 8% in nematode *C. elegans*. In about 60% of cases the length of alternative exons was not a multiple of 3 (close to random expectation), and thus they are likely mutually exclusive exons. On the other hand, in less than 1% of cases there was EST evidence for inclusion of both exons. Thus the vast majority of duplicated exons are likely to be alternative exons. Notably,in some cases the difference between the duplicated exons map to functional sites of proteins and would likely result in changes in the protein function, e.g. by creating dominant negative isoforms with regulatory function (*cf.* discussion of[93] in the previous section). A clear example of independent duplications creating alternative exons is provided by several groups of ion channels in genomes of insects (drosophila and mosquito), vertebrates(human, mouse and fugu, and nematodes *C. elegans* and *C. briggsae*)[41]. In three groups of orthologous ion channels,homologous exons were duplicated independently in these three lineages, creating mutually exclusive pairs or triples of exons.In other members of the family, the corresponding exons are not duplicated and are constitutive. Another source of alternative (in this case, skipped) exons is exonization of non-coding regions. About 5% of internal alternative exons originate from *Alu* sequences, whereas no copies of *Alu* were found in constitutive exons [148]. Most (84%) *Alu*-derived exons in coding regions change the reading frame or introduce in-frame stop codons; they are mainly included in minor isoforms. At that, several different positions in the *Alu* consensus were utilized as splicing sites, both donor and acceptor, mainly in the minus (polyT-containing) *Alu* strand. The two most commonly used positions are two AG dinucleotides that are only two nucleotides apart [102]. If both AG's are present, the upstream one is used unless it is preceded by a purine, thus having a very weak match to the consensus in position +3. In such cases the downstream AG is functional. This observation conforms to the experimental data with mutated sites. Moreover, the balance between competing sites was shown to be essential for alternative splicing, as opposed to constitutive exon skipping or inclusion. An attempt to analyze all types of events leading toinsertions/deletions in proteins was made in [88].To distinguish between insertions and deletions, bacterial and yeast orthologs of alternatively spliced genes as well as protein domain signatures from PFAM and SMART databases were considered as outgroups.The ancestral state was reliably predicted in 73 cases, in which25 of alternatively spliced regions were insertions and 48were deletions. Inserted alternative regions never involved more than one exon. Recently duplicated families of paralogs provide a window through which one can study evolution of alternative splicing at short time periods. One such family is the *MAGE-A* family ofcancer/testis-antigens [139]. This family of at least thirteen genes resulted from multiplication of a common ancestor after the divergence of the primate and rodent lineages. The ancestral gene was generated by retroposition of a processed mRNA[34]. None of the *MAGE-A* genes have introns in the protein-coding region. However, the analysis of the5-untranslated region revealed multiple alternatively spiced introns [5]. Since the genes in this family are sufficiently close to each other to make multiple alignment of non-coding regions possible, one can infer the ancestral state at most alignment positions. After that, one can observe both creation of new exons and splicing sites by mutations activating cryptic sites, and loss of isoforms by mutations disrupting or weakening existing sites. This provides a snapshot of a transitional state in the history of this gene family.

16.12   Regulation of Alternative Splicing

Analysis of regulation of alternative splicing proceeds in two major directions. One of them is comparison of features of constitutive and alternative splicing sites, and the other,analysis of regulatory sites such as splicing enhancers and silencers. The positional nucleotide frequencies of splicing sites at the boundaries of cassette non-conserved exons was claimed to be the same as in constitutive splicing sites [116]. However, other studies reported weaker consensus of alternative sites, compared to constitutive site. Thus, neuron-specific exons were shown to have weaker consensus at positions −3(C) and +4(A) of the donor site and −3(C) (with increased frequency of A) of the acceptor site as well as weaker polypyrimidine tract[155]. A similar trend was observed in a subsequent study of a larger dataset, with difference in the donor sites of cassette and constitutive exons most pronounced in positions+4(A) and +5(G) [156]. Similarly, in mouse,alternative donor sites have weaker positions +4(A) and +5(G)and there are weaker positions −5(Y) and −6(Y) of acceptor sites [184]. Cryptic (minor form) exons tend to have even weaker sites compared to skipped (major form) exons[38]. A special case is that of GC–AG introns. Such introns are rather rare (less than 1% of humanintrons) [27, 28], but much more frequent among sites involved in alternative splicing [163].Constitutive GC donor sites have a stronger match to the consensus in the remaining positions [27]. However, there is a subset of alternative GC donor sites that have a weak match to the consensus [163]. In such introns the corresponding acceptor sites have stronger match to the consensus at positions+1 and +2 compared to acceptor sites of GT–AG introns with weak donor sites or weak polypyrimidine tracts. Thus it seems tobe a specific mechanism facilitating regulation of alternative splicing. Further, there was increased frequency of GC in alternative donor sites. *Vice versa*, 62% of observed GC–AG introns were alternative. Similar analysis performed on the data from nematode *C. elegans* produced somewhat less spectacular results [54]. Most GC–AGintrons (185 of 196 which is less than 1% of all introns in this genome) were not alternatively spliced. Of 26 cases when orthologous introns were available in a related *C. briggsae*genome, only 5 introns retained C at position +2. However, in one of such cases (tenth intron of the *let-2* gene) it was shown that C(+2) is essential for regulated splicing. Thus, as in human, it seems that a subpopulation of GC donor sites indeed are essential for regulation. As more genomes become available,this analysis could be repeated for other pairs of genomes such as two *Drosophila* species or human and mouse. Comparison of oligonucleotide distribution of alternative and constitutive exons produced no consistent pattern. Some purine-rich motifs related to known splicing enhancers were over-represented both in constitutive and alternative exons [156, 184],whereas pyrimidine-rich motifs were over-represented in alternative exons compared to constitutive exons both in human[156] and mouse [184]. The input data for mass analysis of splicing enhancers are SELEX[160, 161], *in vivo* [42] or *in vitro* [166, 108, 140] selection from a pool of random sequences as well as identification of individual binding sites for regulatory proteins [95], in particular by analysis of disease alleles [30, 154]. Then the selected sequences are analyzed in order to determine a common core. Several thus identified signals were purine-rich motifGGGGA/GGAGGA/GGAGA, pyrimidine-rich motif UCUCC/UCUUC/UCCUC,motifs GGACCNG and cCACCc, all of which were similar to naturally occurring enhancers [140]. Recall, however, that splicing enhancers occur not only in alternative, but also in constitutive exons. A SELEX procedure complemented by the Gibbs sampler analysis of selected sites identified binding signals SRSASGA for SF2/ASF,ACDGS for SRp40, and ASCGKM for SRp55; all of these signals differed significantly from signals identified in earlier studies and from known sites in real genomic sequences, indicating that the derived signal might be highly sensitive to the details of the used experimental technique [108]. This makes it difficult to develop methods for *in silico* identification of enhancer sites. On the other hand, some regions known to be involved in regulation of alternative splicing were enriched in candidate enhancers. Further, the observed degeneracy in the enhancer consensuses could answer a functional requirement to reconcile the enhancer sites with the protein-coding message. At that, the functional specificity could be achieved by clustering of multiple sites and co-operative binding of SR-proteins to these sites.Based on this and other studies, a procedure for identification of candidate splicing enhancers for SF2/ASF, SC35, SRp40, and SRp55was developed (ESEfinder, http://exon.cshl.edu/ESE/[31]. ESEfinder analysis can be useful for assessment of the implications of silent mutations in synonymous codon positions; however, as the clustering of candidate sites is not taken into account, it cannot be used for identification of splicing enhancers *de novo*. This program was used to establish that more than half of single-base mutations causing aberrant exon skipping fall into predicted enhancers [107]. A different approach is to start from computational identification of candidate enhancer signals with subsequent experimental verification of obtained results [51]. Known splicing enhancers up-regulate weak splicing sites. Thus each word(hexanucleotide in [51]) can be characterized by its frequency in exons in general, upstream of weak(non-consensus) donor sites and downstream of weak acceptor sites.Hexanucleotides enriched in the vicinity of weak splicing sites,as compared to exonic sequences in general, were clustered by similarity and, based on the above assumption, predicted to be splicing enhancer signals. Notably, there was considerable similarity between acceptor site and donor site enhancers. One of the most prominent signals, GAAGAA, was identified in a number of natural exons and shown to be important for the regulation of splicing by point mutagenesis in mini gene constructs. Again, the degeneracy of the identified signals is so high, that about 10%of all hexanucleotides match the predicted signals, and each human exon contains three to seven candidate sites. However, exons flanked by weak donor and acceptor splicing sites are slightly enriched in candidate enhancer sites, while there was no significant difference between cassette and constitutive exons. Finally, it is possible to take advantage from the fact that while there seems to be no sharp difference between enhancers of constitutive and alternative splicing, some genes are not spliced at all. Indeed, the oligonucleotide composition of these two groups of genes, taking into account the reading frame, was significantly different [55]. However, this difference could not be ascribed to a small group of specific oligonucleotides; rather, it was due to numerous small variations.Further, the list of oligonucleotides with frequencies showing the greatest difference between intron-containing and intronless genes had little intersection with known enhancers. Similarly, a higher fraction of synonymous SNPs (single nucleotide polymorphisms) in single-exon genes compared to intron-containing genes demonstrates that exons in spliced genes contain regulatory elements[109]. In introns, comparison of regions downstream of brain- and muscle-specific and constitutive exons demonstrated over-representation of the UGCAUG hexanucleotide [25].Indeed, in some cases it is known to function as an intronic splicing enhancer [95]. Intron regions adjacent to alternatively exons tend to be more conserved in the human and mouse genes than regions adjacent to the constitutive exons [147]. 77%of conserved alternative exons were flanked on both sides by long conserved intronic sequences, compared to only 17% of constitutive exons. The average length of both upstream and downstream conserved sequences was about 100 nt, whereas the degree of conservation in the upstream introns was slightly higher, 88% compared to 80% in downstream introns, although this might be due to the influence of the polypyrimidine tract in acceptor sites, where the functional pressure decreases the effective alphabet size. The UGCAUG hexanucleotide was over-represented in conserved downstream intronic regions adjacent to alternative exons, but neither downstream of constitutive exons, nor in upstream conserved intronic regions. In most cases(93%) UGCAUG within conserved regions was conserved itself.This coincides with the observation that UGCAUG hexanucleotides downstream of cassette exons tend to be conserved between human and mouse if they occur within conserved regions [46]and demonstrates that such hexanucleotides are likely functionally relevant, but only when placed in suitable sequence context. It is likely that they might be binding sites for a regulatory factor acting in co-operation with additional regulatory proteins. Increased conservation in intronic regions adjacent to skipped exons, alternative donor and acceptor sites was demonstrated also in [158]. Moreover, this increased conservation was observed within exons as well, especially in the regions between alternative splicing sites. Similarly, relative decrease of the fraction of synonymous SNPs near exon termini was demonstrated in[109]. This also might reflect the increasing number of regulatory elements in these regions; however, in this study alternative and constitutive exons were not distinguished. However, given small size of sites regulating splicing, it is difficult to find conserved functional sites on the background of conserved protein-coding message: splicing-related conservation in non-synonymous positions cannot be distinguished from amino acid conservation in the protein, and thus only synonymous positions can be analyzed. The information in pairwise comparisons is clearly insufficient to find conserved regulatory sites within protein-coding regions. However, if more sequences are available,the conservation becomes more significant, and thus highly conserved sites can be analyzed. An analytical theory for such analysis, that is, estimation of nucleotide sequence conservation conditioned on encoded amino acid sequences, was developed in[16]. All types of alternative events (cassette exons and alternative sites) and various types of regions(intronic, exonic/intronic dependent on the isoform, purely exonic) were studied in [138]. One notable observation was observed prevalence of poly-G motifs in the vicinity of alternative splicing. It agrees both with experimental data showing that some such motifs function as intronic splicing enhancers [110], and with early statistical studies that demonstrated elevated frequency of such motifs near intronboundaries[59, 124, 125, 48, 109]and in short introns [110], [105]. Two motifs, one enriched in C and composed of CTCC/CCTCCC repeats, and the other enriched in G and composed of AGGG repeats, were observed upstream and downstream of cassette (skipped) exons in[112], again in agreement with earlier studies where alternatively and constitutively spliced genes were not distinguished [109]. It was suggested that such sites form alternative secondary structure that lead to exon skipping when complementary sites from the upstream and downstream introns form stems with the exon in the loop region.

16.13 Concluding Remarks: Why Is Alternative Splicing So Prevalent?

In the mosaic of observations described above, one can discern several common motifs such as tissue and cancer specificity of alternative isoforms, frame preservation and nonsense-mediated decay, old (conserved) and young (species-specific) elementary alternatives, interplay between alternative splicing and protein structural domains and functional patterns.

Initially most studies addressed these problems separately.Notably, despite differences in the data collection and statistical methods, the obtained results were remarkably consistent. However, the interpretation of the results was difficult as it was not clear whether all observed isoforms and elementary alternatives were indeed functional. Computational studies of the second generation look at several features at once. This uncovered various dependencies between theage of alternatives, their tissue specificity and inclusion level,frame preservation, genome specificity etc. Similarly, modern databases attempt to collect all these data. In fact, it is possible that there exist different subpopulations of alternative splicing events. To uncover such subpopulations one needs to analyze all aspects of the data simultaneously.Frame-preservation, nonsense-mediated decay and protein structure-related features, tissue and cancer specificity, and inclusion levels can be compared for conserved and non-conserved alternative splicing events and isoforms. In addition, for conserved isoforms it might be interesting to compare the ratio of the rate of synonymous and non-synonymous substitutions, Some preliminary observations indicate that there might be positive selection acting on alternative regions [74]. To verify that, one should analyze also the distribution of SNPs in constitutive and alternative regions, but to do that, one needs to restrict the analysis to actually translated regions: here mere conservation of isoforms might be insufficient. A commonly accepted role of alternative splicing is the increase of protein diversity [63]A less appreciated role of alternative splicing is the regulatory one. In addition to the increasing number of individual observations, this role emerges as a likely explanation of some computational observations. In particular, frequent alternative splicing on 5-ends of transcripts can simplify the arrangements of complex transcriptional regulatory cassettes regulating alternative promoters. On the other hand, frame-disrupting alternative splicing events may be related to regulation via nonsense-mediated decay in animals and RNA interference in plants. Finally, alternative splicing is a powerful mechanism of maintaining protein identity. Indeed, a more direct mechanism is available to generate diverse proteins, that is, gene duplication.On the other hand, in many cases the cell needs proteins that are different in some regions but exactly the same in other domains. The most obvious case is that of secreted,membrane-anchored, and cytoplasmic isoforms of some receptors. The recognition domain remains the same, whereas the localization of the protein is determined by the presence of a signal peptide or a transmembrane segment encoded by alternatively spliced parts of the transcript. Another example is provided by isoforms that regulate other isoforms by competitive binding to the same ligand without subsequent functional action. It is clear that it would be difficult to maintain the necessary level of identity in duplicated genes and their concerted evolution by purifying selection, whereas in alternative isoforms it is achieved automatically. Of course, more analyses need to be done before these and other questions are answered. At that, it should be noted that this review covers less than five years of intensive studies. As more data become available from sequencing projects, expression arrays, and proteomic studies, more exact and detailed analyses become possible. In particular, it will be extremely interesting to analyze multiple vertebrate genomes. There already is sufficient data to analyze other groups of genomes: to compare alternative splicing in the genomes of fruit flies *D. melanogaster* and *D. pseudoobscura*, and to compare it with the alternative splicing of the malaria mosquito *Anopheles gambiae*; and the nematodes *Caernorhabditis elegans* and *C.briggsae*. In both cases, several genomes are available,supplemented by more or less extensive EST and cDNA data for one member of the group (*D. melanogaster* and *C. elegans*)respectively. I would not be surprised if such studies are published while this book is completed and printed.

16.14   An Update

Several new studies were published after this chapter was completed. Mainly they continue the trends discussed in the main text, but sometimes add a new twist to the old observations or provide a new angle to look at the data. Two recent reviews are [96] on the evolution of alternative splicing and [100] on the use of microarrays for the analysis of alternative splicing. Application of the PSEP program for gene identification base onEST and comparative genome analysis demonstrated alternative splicing of approximately 75% human genes in standard well-curated datasets (ROSETTA and chromosome 20), with about25% elementary alternatives not conserved between human and mouse. This agrees well with the earlier estimates[36]. The alternative splicing rate in different species defined as the average number of alternative isoforms per gene (as opposed to the fraction of alternatively spliced genes) was measured in[85]. The authors suggested that this number is larger in mammals compared to the *C. elegans* and *Drosophila*.However, as demonstrated in [49], this is likely a result of artifacts in data (using EST contigs instead of raw ESTs). Several studies addressed evolution and origin of alternative splicing. The origins of alternative splicing have been studied in [6], where it has been demonstrated that the information content of donor splicing sites decreases from budding yeast *Saccharomyces cerevisiae* to fission yeast *Schisosaccharomyces pombe* to metazoans. This coincides with increase of the number introns and emergence of alternative splicing. The author suggests that gradual weakening of the donor site consensus and change in the structure of the donor site-U1snRNA allows for regulated splicing, consistent with the observation that alternative sites are generally weaker. Analysis of donor sites generated by exonisation of *Alu*repeats [150] demonstrated that functional sites arise from either C-to-T mutation, creating a GT dinucleotide, or aG-to-A mutation at intron position 3, that strengthens the GC-containing site. The observation that mutation patterns are different in constitutive and cassette exons and in introns adjacent to such exons [102], lead to creation of a program for identification of cassette exons based mainly on comparison of human and mouse orthologs [153]. Another program by the same group uses the support vector machine to combine sequence features for the same recognition task, achieving 50%sensitivity and .5% specificity of discrimination of cassette exons from constitutive ones [47]. Most studies described in the main text centered on finding tissue-specific isoforms. A converse problem was addressed in[181], where tissues expressing the highest fraction of known isoforms were identified. Three clear winners were brain,liver, and testis. At that, brain and testis demonstrated the largest diversity of cassette exons, whereas liver had the most diverse choice of alternative donor and acceptor sites. Expression profiles of splicing factors were considered, and it was demonstrated that the adult liver was an outlier, having the most specific pattern of splicing factor expression. Sequence motifs that could be responsible for alternative splicing in the three tissues with highest isoform diversity were identified and some of them were shown to be similar to known splicing regulatory sites. A program for identification of tissue-specific alternative splicing using microarray expression data was presented in[97]. All types of events, including alternative sites,skipped and mutually exclusive exons, alternative initial and terminal exons could be detected with $75 - 80\%$ validation rate byRT-PCR. The program RESQUE-ESE for identification of candidate regulatory sites, in particular, exonic splicing enhancers, was implemented as a Web server (http://genes.mit.edu/burgelab/rescue-ese/)[52], and applied to several types of analysis. It was demonstrated that exonic splicing enhancer patterns are similar in vertebrates, whereas the intronic signals are different in mammals and fish [182]. This agrees with the fact that SR-proteins,binding to exonic splicing enhancers, are conserved invertebrates, whereas hnRNPs, binding to introns, are more diverse both in domain composition and genomic content. Analysis of human SNPs and comparison with the chimpanzee genome as an outlier allowing for the identification of ancestral alleles demonstrate that mutations disrupting predicted exonic splicing enhancers,especially those close to exon boundaries, are subject to purifying selection [50]. Another group of regulatory sites, exonic splicing silencers, were studied in a combination of computational and experimental (*in vivo* splicing reporter system) approaches [171].Several identified motifs were similar to known binding signals of hnRNPs H and A1. This analysis lead to development of the ExonScan program which simulates splicing by analysis of potential splicing motifs. Analysis of full-length isoforms generated by human and mouse cDNA sequencing projects demonstrated that minor isoforms more often contain premature stop codons that could lead to nonsense-mediated decay (NMD) than major ones (11.1% and 4.7% respectively),and the effect is even more pronounced for genes from the X chromosome, where the overall frequency of premature stop codons is lower compared to autosomes [177]. This suggests mean that NMD reduces selection against aberrant mRNAs that could yield dominant negative phenotypes. Minor, but statistically significant differences between amino acid frequencies in constitutively and alternatively spliced genes were observed in [188]. The value of these observations is not clear, as the direction of change is not only inconsistent between mammals, *C. elegans* and *Drosophila*, but in sometimes even within mammals (human, mouse, rat and bovine). On the other hand, alternatively spliced genes consistently produced longer proteins. It might be interesting to compare amino acid usage in constitutive and alternative regions. Several papers addressed alternative splicing of individual genes or gene families. It has been demonstrated that the expression profiles of splicing factors are different in normal and malignant ovarian tissue, which may explain observed changes in alternative splicing of the CD44 gene in ovarian and breast cancers[154]. High conservation of intronic sites regulating alternative splicing of fibroblast growth receptors from the *FGRF* family in metazoans from sea urchin to mammals was shown in[114].

Acknowledgements

This study was supported by grants from the Ludwig Institute of Cancer Research (CRDF RB0-1268), the Howard Hughes Medical Institute (55000309), the Russian Fund of Basic Research$(04-04-49440)$, and programs "Molecular and Cellular Biology"and "Origin and Evolution of the Biosphere" of the Russian Academy of Sciences.

References

[1] M. Adams, J.L. Jones, R.A. Walker, and J.H. Pringle *et al.* Changes in Tenascin-C isoform expression in invasive and preinvasive breast disease. *Cancer Res.*, 62:3289–3297, 2002.

[2] M.D. Adams, A.R. Kerlavage, R.D. Fleischmann, and R.A.Fuldner *et al.* Initial assessment of human gene diversity and expression patterns based upon 83 million nucleotides of cDNA sequence. *1995*, 377:3–174, 1995.

[3] S.F. Altschul, T.L. Madden, A.A. Schaffer, and J. Zhang *et al.* Gapped BLAST and PSI-BLAST: A new generation of protein database search programs. *Nucleic Acids Res.*, 1997:3389–3402, 1997.

[4] R. Apweiler, A. Bairoch, C.H. Wu, and W.C. Barker *et al.* UniProt: the universal protein knowledgebase. *Nucleic Acids Res.*, 32:D115–D119, 2004.

[5] I.I. Artamonova and M.S. Gelfand. Evolution of the exon-intron structure and alternative splicing of the MAGE-A family of cancer/testis antigens. *J. Mol. Evol.*, 69:620–631, 2004.

[6] G. Ast. How did alternative splicing evolve? *Nature Rev. Genet.*, 5:773–782, 2004.

[7] A.V. Baranova, A.V. Lobashev, D.V. Ivanov, and L.L. Krukovskaya *et al. In silico* screening for tumour-specific expressed sequences in human genome. *FEBS Lett.*, 508:143–148, 2001.

[8] S. Batzoglou, L. Pachter, J.P. Mesirov, and B. Berber *et al.* Human and mouse gene structure: comparative analysis and application to exon prediction. *Genome Res.*, 10:950–958, 2000.

[9] E. Beaudoing, S. Freier, J. Wyatt, and J.M. Claverie *et al.* Patterns of variant polyadenylation signals in human genes. *Genome Res.*, 10:1001–1010, 2000.

[10] E. Beaudoing and D. Gautheret. Identification of alterantive polyadenylation sites and analysis of their tissue distribution using EST data. *Genome Res.*, 11:1520–1526, 2001.

[11] D.A. Benson, I. Karsch-Mizrachi, D.J. Lipman, and J. Ostell *et al.* GenBank: update. *Nucleic Acids Res.*, 32:D23–D26, 2004.

[12] S.M. Berget. Exon recognition in vertebrate splicing. *J. Biol. Chem.*, 270:2411–2414, 1995.

[13] S.M. Berget, C. Moore, and P.A. Sharp. Spliced segments at the 5' terminus of adenovirus 2 late mRNA. *Proc. Natl. Acad. Sci. USA*, 74:3171–3175, 1977.

[14] A.J. Berk and P.A. Sharp. Structure of the adenovirus 2 early mRNAs. *Cell*, 14:695–

711, 1978.

[15] A. Bernal, U. Ear, and N. Kyrpides. Genomes online database (GOLD): a monitor of genome projects world-wide. *Nucleic Acids Res.*, 29:126–127, 2001.

[16] M. Blanchette. A comprarative method for detecting binding sites in coding regions. *Annu. Int. Conf. on Research in Computational Molecular Biology RECOMB'03*, 7:57–65, 2003.

[17] P. Bonizzoni, G. Pesole, and R. Rizzi. A method to detect gene structure and alternative splice sites by agreeing ESTs to a genomic sequence. *Lecture Notes in Bioinformatics (WABI'2003)*, 2812:63–77, 2003.

[18] S. Boue, I. Letunic, and P. Bork. Alternative splicing and evolution. *BioEssays*, 25:1031–1034, 2003.

[19] S. Boue, M. Vingron, E. Kriventseva, and I. Koch. Theoretical analysis of alternative splice forms using computational methods. *Bioinformatics*, 18:S65–S73, 2002.

[20] R. Breathnach and P. Chambon. Organization and expression of eucaryotic split genes coding for proteins. *Annu. Rev. Biochem.*, 50:349–383, 1981.

[21] R.E. Breitbart, A. Andreadis, and B. Nadal-Ginard. Alternative splicing: A ubiquitious mechanism for the generation of multiple protein isoforms from single genes. *Annu. Rev. Biochem.*, 56:467–495, 1987.

[22] D. Brett, J. Hanke, G. Lehmann, and S. Haase *et al.* EST comparison indicates 38% of human mrnas contain possible alternative splice forms. *FEBS Lett.*, 474:83–86, 2000.

[23] D. Brett, W. Kemmner, G. Koch, and C. Roefzaad *et al.* A rapid bioinformatic method identifies novel genes with direct clinical relevance to colon cancer. *Oncogene*, 20:4581–4585, 2001.

[24] D. Brett, H. Pospisil, J. Valcarcel, and J. Reich *et al.* Alternative splicing and genome complexity. *Nature Genet.*, 30:29–30, 2002.

[25] M. Brudno, M.S. Gelfand, S. Spengler, and M. Zorn *et al.* Computational analysis of candidate intron regulatory elements for tissue-specific alternative pre-mRNA splicing. *Nucleic Acids Res.*, 29:2338–2348, 2001.

[26] J. Burke, H. Wang, W. Hide, and D.B. Davison. Alternative gene form discovery and candidate gene selection from gene indexing projects. *Genome Res.*, 8:276–290, 1998.

[27] M. Burset, I.A. Seledtsov, and V.V. Solovyev. Analysis of canonical and non-canonical mammalian splice sites. *Nucleic Acids Res.*, 28:4364–4375, 2000.

[28] M. Burset, I.A. Seledtsov, and V.V. Solovyev. SpliceDB: database of canonical and non-canonical mammalian splice sites. *Nucleic Acids Res.*, 29:255–259, 2001.

[29] A.A. Camargo, H.P. Samaia, E. Dias-Neto, and D.F. Simao *et al.* The contribution of 700,000 ORF sequence tags to the definition of the human transcriptome. *Proc. Natl. Acad. Sci. USA*, 98:12103–12108, 2001.

[30] L. Cartegni, S.L. Chew, and A.R. Krainer. Listening to silence and understanding nonsense: exonic mutations that affect splicing. *Nature Rev. Genet.*, 3:285–297, 2002.

[31] L. Cartegni, J. Wang, Z. Zhu, and M.Q. Zhang *et al.* ESEfinder: a web resource to identify exonic splicing enhancers. *Nucleic Acids Res.*, 31:3568–3571, 2003.

[32] P. Castellani, L. Borsi, B. Carnemolla, and A. Biro *et al.* Differentiation between high- and low-grade astrocytoma using a human recombinant antibody to the extra domain-b of fibronectin. *Am. J. Pathol.*, 161:1695–1700, 2002.

[33] J. Castle, P. Garrett-Engele, C.D. Armour, and S.J. Duenwald *et al.* Optimization of oligonucleotide arrays and RNA amplification protocols for analysis of transcript structure and alternative splicing. *Genome Biol.*, 4:R66, 2003.

[34] P. Chomez, O. De Backer, M. Bertrand, and M. De Plaen *et al.* An overview of the MAGE gene family with the identification of all human members of the family. *Cancer Res.*, 61:5544–5551, 2001.

[35] L.T. Chow, R.E. Gelinas, T.R. Broker, and R.J. Roberts. An amazing sequence arrangement at the 5′ ends of adenovirus 2 messenger RNA. *Cell*, 12:1–8, 1977.

[36] T.J. Chuang, F.C. Chen, and M.Y. Chou. A comparative method for identification of gene structures and alternatively spliced variants. *Bioinformatics*, 20:3064–3079, 2004.

[37] F.D. Ciccarelli, T. Doerks, and P. Bork. AMOP, a protein module alternatively spliced in cancer cella. *Trends Biochem. Sci.*, 27:113–115, 2002.

[38] F. Clark and T.A. Thanaraj. Categorization and characterization of transcript-confirmed constitutively and alternatively spliced introns and exons from human. *Hum. Mol. Genet.*, 11:451–464, 2002.

[39] T.A. Clark, C.W. Sugnet, and Jr M. Ares. Genomewide analysis of mRNA processing in yeast using splicing-specific microarrays. *Science*, 2002:907–910, 2002.

[40] M.S. Cline, R. Shigeta, R.L. Wheeler, and M.A. Siani-Rose *et al.* The effects of alternative splicing on transmembrane proteins in the mouse genome. *Pac. Symp. Biocomput.*, pages 17–28, 2004.

[41] R.R. Copley. Evolutionary convergence of alternative splicing in ion channels. *Trends Genet.*, 20:171–176, 2004.

[42] L.R. Coulter, M.A. Landree, and T.A. Cooper. Identification of a new class of exonic splicing enhancers by *in vivo* selection. *Mol. Cell. Biol.*, 17:2143–2150, 1997.

[43] E. Coward, S.A. Haas, and M. Vingron. Splicenest: visualization of gene structure and alternative splicing based on EST clusters. *Trends Genet.*, 18:53–55, 2002.

[44] M.S. Cragg, H.T. Chan, M.D. Fox, and A. Tutt *et al.* The alternative transcript of CD79b is overexpressed in B-CLL and inhibits signaling for apoptosis. *Blood*, 100:3068–3076, 2002.

[45] L. Croft, S. Schandorff, F. Clark, and K. Burrage *et al.* ISIS, the intron information system, reveals the high frequency of alternative splicing in the human genome. *Nature Genet.*, 24:340–341, 2000.

[46] S. Denisov and M.S. Gelfand. Conservativity of the alternative splicing signal UG-CAUG in the human and mouse genomes. *Biofizika*, 48:30–35, 2004.

[47] G. Dror, R. Sorek, and R. Shamir. Accurate identification of alternatively spliced exons using support vector machine. *Bioinformatics*, page Nov 5, 2004.

[48] J. Engelbrecht, S. Knudsen, and S. Brunak. G+c-rich tract in 5' end of human introns. *J. Mol. Biol.*, 227:108–113, 1992.

[49] W.G. Fairbrother, D. Holste, C.B. Burge, and P.A. Sharp. Reply to [85]. *Nat. Genet.*, 36:916–917, 2004.

[50] W.G. Fairbrother, D. Holste, C.B. Burge, and P.A. Sharp. Single nucleotide polymorphism-based validation of exonic splicing enhancers. *PloS Biology*, 2:e268, 2004.

[51] W.G. Fairbrother, R.-F. Yeh, P. Sharp, and C.B. Burge. Predictive identification of exonic splicing enhancers in human genes. *Science*, 297:1007–1013, 2002.

[52] W.G. Fairbrother, G.W. Yeo, R. Yeh, and P. Goldstein *et al.* Variation in alternative splicing across human tissues. *Nucleic Acids Res.*, 32:W187–W190, 2004.

[53] FANTOM2 Consortium and the RIKEN GSC Genome Exploration Group Phase I & II Team. Analysis of the mouse transcriptome based upon functional annotation of 60,770 full length cDNAs. *Nature*, 420:563–573, 2002.

[54] T. Farrer, A.B. Roller, W.J. Kent, and A.M. Zahler. Analysis of the role of caenorhabditis elegans GC-AG introns in regulated splicing. *Nucleic Acids Res.*, 30:3360–3370,

2002.

[55] A. Fedorov, S. Saxonov, L. Fedorova, and I. Daizadeh. Comparison of intron-containing and intron-lacking human genes elucidates putative exonic splicing enhancers. *Nucleic Acids Res.*, 29:1464–1469, 2001.

[56] C.M. Feltes, A. Kuda, O. Blaschuk, and S.W. Byers. An alternatively spliced cadherin-11 enhances human breast cancer cell invasion. *Cancer Res.*, 62:6688–6697, 2002.

[57] FlyBase Consortium. The flybase database of the drosophila genome projects and community literature. *Nucleic Acids Res.*, 31:172–175, 2003.

[58] D. Gautheret, Poirot. O., F. Lopez, and A. Audic *et al.* Expressed sequence tag (EST) clustering reveals the extent of alternate polyadenylation in human mRNAs. *Genome Res.*, 1998:524–530, 1998.

[59] M.S. Gelfand. Statistical analysis of mammalian pre-mRNA splicing sites. *Nucleic Acids Res.*, 17:6369–6382, 1989.

[60] M.S. Gelfand, I. Dubchak, I. Dralyuk, and M. Zorn. ASDB: database of alternatively spliced genes. *Nucleic Acids Res.*, 27:301–302, 1999.

[61] J.H. Graber, C.R. Cantor, S.C. Mohr, and T.F. Smith. In silico detection of contronl signals: mRNA 3'-end-processing sequences in diverse species. *Proc. Natl. Acad. Sci.*, 96:14055–14060, 1999.

[62] C. Grasso, B. Modrek, Y. Xing, and C. Lee. Genome-wide detection of alternative splicing in expressed sequences using partial order multiple sequence alignment graphs. *Pacific Symp. Biocomput.*, 9:29–41, 2004.

[63] B. Graveley. Alternative splicing: increasing diversity in the proteomic world. *Trends Genet.*, 17:100–107, 2001.

[64] B.R. Graveley. Sorting out the complexity of sr functions. *RNA*, 6:1197–1211, 2000.

[65] B.J. Haas, N. Volofsky, C.D. Town, and M. Troukhan *et al.* Full-length messenger RNA sequences greatly improve genome annotation. *Genome Biol.*, 3:R29, 2002.

[66] J. Hanke, D. Brett, I. Zastrow, and A. Aydin *et al.* Alternative splicing of human genes: more the rule than the exception? *Trends Genet.*, 15:389–390, 1999.

[67] M.L. Hastings and A.R. Krainer. Pre-mRNA splicing in the new millennium. *Curr. Opin. Cell Biol.*, 13:302–309, 2001.

[68] S. Heber, M. Alekseev, S.-H. Sze, and H. Tang *et al.* Splicing graphs and EST assembly problem. *Bioinformatics*, 18:S181–S188, 2002.

[69] W.A. Hide, V.N. Babenko, P.A. van Heusden, and C. Seoighe *et al.* The contribution of exon-skipping events on chromosome 22 to protein coding diversity. *Genome Res.*, 11:1848–1853, 2001.

[70] R.T. Hillman, R.E. Green, and S.E. Benner. An unappreciated role for RNA surveillance. *Genome Biology*, 5:R8, 2004.

[71] G.K. Hu, S.I. Madore, B. Moldover, and T. Jatkoe *et al.* Predicting splice variant from DNA chip expression data. *Genome Res.*, 11:1237–1245, 2001.

[72] H.D. Huang, J.T. Horng, C.C. Lee, and B.J. Liu. ProSplicer: a database of putative alternative splicing information derived from protein, mRNA and expressed sequence tag sequence data. *Genome Biol.*, 4:R29, 2003.

[73] Y.H. huang, Y.T. Chen, J.J. Lai, and S. T. Yang *et al.* PALS db: Putative alternative splicing database. *Nucleic Acids Res.*, 30:186–190, 2002.

[74] K. Iida and H. Akashi. A test of translational selection at 'silent' sites in the human genome: base composition comparisons in alternatively spliced genes. *Gene*, 261:93–105, 2000.

[75] T. Imanishi, T. toh, Y. Suzuki, and C. O'Donovan *et al.* Integrative annotation of 21,037 human genes validated by full-length cDNA clones. *PloS Biology*, 2:0001–

0020, 2004.

[76] International Human Genome Sequencing Consortium. Initial sequencing and analysis of the human genome. *Nature*, 409:860–921, 2001.

[77] H. Ji, Q. Zhou, F. Wen, and H. Xia *et al.* AsMamDB: an alternative splice database of mammals. *Nucleic Acids Res.*, 29:260–263, 2001.

[78] J.M. Johnson, J. Castle, P. Garrett-Engele, and Z. Kan *et al.* Genome-wide survey of human alternative pre-mRNA splicing with exon junction microarrays. *Science*, 302:2141–2144, 2003.

[79] Z. Kan, J. Castle, J.M. Johnson, and N.F. Tsinoremas. Detection of novel splice forms in human and mouse using cross-species approach. *Pacific Symposium in Biocomputing*, 9:42–53, 2004.

[80] Z. Kan, E.C. Rouchka, W.R. Gish, and D.J. States. Gene structure prediction and alternative splicing analysis using genomically aligned ESTs. *Genome Res.*, 11:889–900, 2001.

[81] Z. Kan, D. States, and W. Gish. Selecting for functional alternative splices in ESTs. *Genome Res.*, 12:1837–1845, 2002.

[82] P. Kapranov, S.E. Cawley, J. Drenkow, and S. Bekiranov *et al.* Large-scale transcriptional activity in chromosomes 21 and 22. *Science*, 296:916–919, 2002.

[83] W.J. Kent and A.M. Zahler. The intronerator: exploring introns and alternative splicing in *caenorhabditis elegans*. *Nucleic Acids Res.*, 28:91–93, 2000.

[84] P. Kersey, H. Hermjakob, and R. Apweiler. VARSPLIC: alternatively spliced protein sequences derived from SWISS-PROT and TrEMBL. *Bioinformatics*, 16:1048–1049, 2000.

[85] H. Kim, R. Klein, J. Majewski, and J. Ott. Estimating rates of alternative splicing in mammals and invertebrates. correspondence re [24]. *Nat. Genet.*, 36:915–916, 2004.

[86] H. kochiwa, R. Suzuki, T. Wasio, and R. Saito *et al.* Inferring alternative splicing patterns in mouse from a full-length cDNA library and microarray data. *Genome Res.*, 12:1286–1293, 2002.

[87] F.A. Kondrashov and E.V. Koonin. Origin of alternative splicing by tandem exon duplication. *Hum. Mol. Genet.*, 10:2661–2669, 2001.

[88] F.A. Kondrashov and E.V. Koonin. Evolution of alternative splicing: deletions, insertions and origin of functional parts of proteins from intron sequences. *Trends Genet.*, 19:115–119, 2003.

[89] M. Koslowski, O. Tureci, C. Bell, and P. Krause *et al.* Multiple splice variants of lactate dehydrogenase C selectively expressed in human cancer. *Cancer Res.*, 62:6750–6755, 2002.

[90] A. Krause, S.A. Haas, E. Coward, and M. Vingron. SYSTERS, GeneNest, SpliceNest: exploring sequence space from genome to protein. *Nucleic Acids Res.*, 30:299–300, 2002.

[91] M. Krawczak, J. Reiss, and D.N. Cooper. The mutational spectrum of single base-pair substitutions in mRNA splice junctions of human genes: cases and consequences. *Hum. Genet.*, 90:41–54, 1992.

[92] A.M. Krecic and M.S. Swanson. hnRNP complexes: composition, structure, and function. *Curr. Opin. Cell Biol.*, 11:363–371, 1999.

[93] E.V. Kriventseva, I. Koch, R. Apweiler, and M. Vingron *et al.* Increase of functional diversity by alternative splicing. *Trends Genet.*, 19:124–128, 2003.

[94] T. Kulikova, P. Aldebert, N. Althorpe, and W. Baker *et al.* The EMBL nucleotide sequence database. *Nucleic Acids Res.*, 32:D27–D30, 2004.

[95] A.N. Ladd and T.A. Cooper. Finding signals that regulate alternative splicing in the post-genomic era. *Genome Biol.*, 3:R8, 2002.

[96] L.F. Lareau, R.E. Green, R.S. Bhatnagar, and S.E. Brenner. The evolving roles of alternative splicing. *Curr. Opin. Struct. Biol.*, 14:273–282, 2004.

[97] K. Le, K. Mitsouras, M. Roy, and Q. Wang *et al.* Detecting tissue-specific regulation of alternative splicing as a qualitative change in microarray data. *Nucleic Acids Res.*, 35:e180, 2004.

[98] C. Lee. Generating consensus sequences from partial order multiple sequence alignment graphs. *Bioinformatics*, 19:999–1008, 2003.

[99] C. Lee, L. Atanelov, B. Modrek, and Y. Xing. ASAP: the alternative splicing annotation project. *Nucleic Acids Res.*, 31:101–105, 2003.

[100] C. Lee and M. Roy. Analysis of alternative splicing with microarrays: successes and challenges. *Genome Biol.*, 5:231, 2004.

[101] I. Letunic, R.R. Copley, and P. Bork. Common exon duplication in animals and its role in alternative splicing. *Hum. Mol. Genet.*, 11:1561–1567, 2002.

[102] G. Lev-Maor, R. Sorek, N. Shomron, and G. Ast. The birth of an alternatively spliced exon: 3′ splice-site selection in *alu* exons. *Science*, 300:1288–1291, 2003.

[103] B.P. Lewis, R.E. Green, and S.E. Brenner. Evidence for the widespread coupling of alternative splicing and nonsense-mediated mRNA decay in humans. *Proc. Natl. Acad. Sci. USA*, 100:189–192, 2003.

[104] F. Liang, I. Holt, G. Pertea, and S. Karamycheva *et al.* Gene index analysis of the human genome estimates approximately 120,000 genes. *Nature Genet.*, 25:239–240, 2000.

[105] L.P. Lim and C.B. Burge. A computational analysis of sequence features involved in recognition of short introns. *Proc. Natl. Acad. Sci. USA*, 98:11193–11198, 2001.

[106] F. Lisacek, M. Traini, D. Sexton, and J. Harry *et al.* Strategy for protein isoform identification from expressed sequence tags and its application to peptide mass sequencing. *Proteomics*, 1:186–193, 2001.

[107] H.X. Liu, L. Cartegni, M.Q. Zhang, and A.R. Krainer. A mechanism for exon skipping caused by nonsense or missensenmutations in *brca1* and other genes. *Nature Genet.*, 27:55–58, 2001.

[108] H.X. Liu, M. Zhang, and A.R. Krainer. Identification of functional exonic splicing motifs recognized by individual sr proteins. *Genes Dev.*, 12:1998–2012, 1998.

[109] J. Majewski and J. Ott. Distribution and characterization of regulatory elements in the human genome. *Genome Res.*, 12:1827–1836, 2002.

[110] A.J. McCullough and S.M. Berget. An intronic splicing enhancer binds U1 snRNPs to enhance splicing and select 5′ splice sites. *Mol. Cell. Biol.*, 20:9225–9235, 2000.

[111] M. Metzlaff, M. O'Dell, R. Hellens, and R.B. Flavell. Developmentally and transgene regulated nuclear processing of primary transcripts of chalcone synthase A in petunia. *Plant J.*, 23:63–72, 2000.

[112] E. Miriami, H. Margalit, and R. Sperling. Cosnerved sequence elements associated with exon skipping. *Nucleic Acids Res.*, 31:1974–1983, 2003.

[113] A.A. Mironov, J.W. Fickett, and M.S. Gelfand. Frequent alternative splicing of human genes. *Genome Res.*, 9:1288–1293, 1999.

[114] N. Mistry, W. Harrington, E. Lsda, and E.J. Wagner *et al.* Of urchins and men: Evolution of an alternative splicing unit in fibroblast growth factor receptor genes. *RNA*, 9:209–217, 2004.

[115] B. modrek and C. Lee. A genomic view of alternative splicing. *Nature Genet.*, 30:13–19, 2002.

[116] B. Modrek and C.J. Lee. Alternative splicing in the human, mouse and rat genomes is associated with an increased frequency of exon creation and/or loss. *Nature Genet.*, 34:177–180, 2003.

[117] B. Modrek, A. Resch, C. Grasso, and C. Lee. Genome-wide detection of alternative splicing in expressed sequences of human genes. *Nucleic Acids Res.*, 29:2850–2859, 2001.

[118] S.M. Mount. A catalogue of splice junction sequences. *Nucleic Acids Res.*, 10:459–472, 1982.

[119] Mouse Genome Sequencing Consortium. Initial sequencing and comparative analysis of the mouse genome. *Nature*, 420:520–562, 2002.

[120] K. Nakai and H. Sakamoto. Construction of a novel database containing aberrant splicing mutations of mammalian genes. *Gene*, 141:171–177, 1994.

[121] R.N. Nurtdinov, I.I. Artamonova, A.A. Mironov, and M.S. Gelfand. Low conservation of alternative splicing patterns in the human and mouse genomes. *Hum. Mol. Genet.*, 12:1313–1320, 2003.

[122] R.N. Nurtdinov and I. Kosmodemyansky. The EDAS (EST-derived alternative splicing) database. *Int. Moscow Conf. on Computational Molecular Biology MCCM-B'03*, 1:171–172, 2003.

[123] R.N. Nurtdinov, A.A. Mironov, and M.S. Gelfand. Is alternative splicing of mammalian genes conservative? *Biofizika (Moscow)*, 47:587–594, 2002.

[124] R. Nussinov. (A)GGG(A), (A)CCC(A) and other potential 3′ splice signals in primate nuclear pre-mRNA sequences. *Biochim. Biophys. Acta*, 910:261–270, 1987.

[125] R. Nussinov. Conserved signals around the 5′ splice sites in eukaryotic nuclear precursor mRNAs: G-runs are frequent in the introns and C in the exons near both 5′ and 3′ splice sites. *J. Biomol. Struct. Dynam.*, 6:985–1000, 1989.

[126] M. Offman, R.N. Nurtdinov, M.S. Gelfand, and D. Frishman. No statistical support for correlation between the positions of protein interaction sites and alternatively spliced regions. *BMC Bioinformatics*, 5:41, 2004.

[127] A. Parareda, J.C. Villaescusa, J. Sanchez de Toledo, and S. Gallego. New splicing variants for human tyrosine hydroxylase gene with possible implications for the detection of minimal residue disease in patients with neuroblastoma. *Neurosci. Lett.*, 336:29–32, 2003.

[128] E. Pauws, A.H. van Kampen, van De Graaf S.A., and J.J. de Vijlder. Heterogeneity in polyadenylation cleavage sites in mammalian mRNA sequences: Implications for SAGE analysis. *Nucleic Acids Res.*, 29:1690–1694, 2001.

[129] H. Pospisil, A. Herrmann, R.H. Bortfeldt, and J.G. Reich. EASED: Extended alternatively spliced EST database. *Nucleic Acids Res.*, 32:D70–D74, 2004.

[130] H. Pospisil, A. Herrmann, H. Pankow, and J.G. Reich. A database on alternative splice forms on the integrated genetic map service. *In Silico Biology*, 3:0020, 2003.

[131] J. Rappsilber, U. Ryder, A.I. Lamond, and M. Mann. Large-scale proteomic analysis of the human spliceosome. *Genome Res.*, 12:1231–1245, 2002.

[132] V.B. Reddy, B. Thimmappaya, R. Dhar, and K.L. Subramanian *et al.* The genome of simian virus 40. *Science*, 200:494–502, 1978.

[133] R. Reed. Initial splice-site recognition and pairing during pre-mRNA splicing. *Curr. Opin. Gen. Dev.*, 6:215–220, 1996.

[134] A. Resch, Y. Xing, A. Alekseenko, and B. Modrek *et al.* Evidence for a subpopulation of conserved alternative splicing events under selection pressure for protein reading frame preservation. *Nucleic Acids Res.*, 32:1261–1269, 2004.

[135] A. Resch, Y. Xing, B. Modrek, and M. Gorlick *et al.* Assessing the impact of alternative splicing on domain interactions in the human proteome. *J. Proteome Res.*, 3:76–83, 2004.

[136] H. Saito, S. Nakatsuru, J. Inazawa, and T. Nishihira *et al.* Frequent association of alternative aplicing of NER, a nuclear hormone receptor gene in cancer tissues.

Oncogene, 14:617–621, 1997.

[137] N.J. Sakabe, de Souza J.E., Galante P.A., and P.S. de Oliveira *et al.* ORESTES are enriched in rare exon usage variants affecting the encoded proteins. *C. R. Biol.*, 326:979–985, 2003.

[138] H. Sakai and Maruyama O. Extensive search for discriminative features of alternative splicing. *Pacific Symposium in Biocomputing*, 9:54–65, 2004.

[139] M.J. Scanlan, A.O. Gure, A.A. Jungbluth, and L. J. Old *et al.* Cancer/testis antigens: an expanding family of targets for cancer immunotherapy. *Immunological Rev.*, 188:22–32, 2002.

[140] T.D. Schaal and T. Maniatisclark. Selection and characterization of pre-mRNA splicing enhancers: Identification of novel SR protein-specific enhancer sequences. *Mol. Cell. Biol.*, 19:1705–1719, 1999.

[141] D. Schmucker, J.C. Clemens, H. Shu, and C. A. Worby *et al.* Dscam is an axon guidance receptor exhibiting extraordinary molecular diversity. *Cell*, 101:671–684, 2000.

[142] P.K. Shah, M.A. Andrade, and P. Bork. Text mining for alternative splicing events using support vector machines. *1st Int. Conf. "Functional Genomics and Disease"*, *Prague*, page PT4/185, 2003.

[143] P.A. Sharp. Split genes and RNA splicing. *Cell*, 77:805–815, 1994.

[144] P.A. Sharp and C.B. Burge. Classification of introns: U2-type or U12-type. *Cell*, 91:875–879, 1997.

[145] D.D. Shoemaker, E.E. Schadt, C.D. Armour, and Y.D. He *et al.* Experimental annotation of the human genome using microarray technology. *Nature*, 2001:922–927, 2001.

[146] A. Skandalis, P.J. Ninniss, D. McCormack, and L. Newton. Spontaneous frequency of exon skipping in the human HPRT gene. *Mutat. Res.*, 501:37–44, 2002.

[147] R. Sorek and G. Ast. Intronic sequences flanking alternatively spliced exons are conserved between human and mouse. *Genome Res.*, 2003:1631–1637, 2003.

[148] R. Sorek, G. Ast, and D. Graur. *Alu*-containing exons are alternatively spliced. *Genome Res.*, 12:1060–1067, 2002.

[149] R. Sorek, O. Basechess, and H.M. Safer. Expressed sequence tags: clean before using. correspondence re [169]. *Cancer Res.*, 63:6996, 2003.

[150] R. Sorek, G. Lev-Maor, M. Reznik, and T. Dagan *et al.* Minimal conditions for the exonization of intronic sequences: $5'$ splice site formation in *alu* exons. *Mol. Cell*, 14:221–231, 2004.

[151] R. Sorek and H.M. Safer. A novel algorithm for computational identification of contaminated EST libraries. *Nucleic Acids Res.*, 31:1067–1074, 2003.

[152] R. Sorek, R. Shamir, and G. Ast. How prevalent is functional alternative splicing in the human genome? *Trends Genet.*, 20:68–71, 2004.

[153] R. Sorek, R. Shemesh, Y. Cohen, and O. Basechess *et al.* A non-EST-based method for exon-skipping prediction. *Genome Res.*, 14:1617–1623, 2004.

[154] S. Stamm. Signals and their transcduction pathways regulating alternative splicing: a new dimension of the human genome. *Hum. Mol. Genet.*, 11:2409–2416, 2002.

[155] S. Stamm, M.Q. Zhang, T.G. Marr, and D.M. Helfman. A sequence compilation and comparison of exons that are alternatively spliced in neurons. *Nucleic Acids Res.*, 22:1515–1526, 1994.

[156] S. Stamm, J. Zhu, K. Nakai, and P. Stoilov *et al.* An alternative-exon database and its statistical analysis. *DNA Cell Biol.*, 19:739–750, 2000.

[157] M. Stimpfl, D. Tong, B. Fasching, and E. Schuster *et al.* Vascular endothelial growth factor splice variants and their prognostic value in breast and ovarian cancer. *Clin.*

Cancer Res., 8:2253–2259, 2002.

[158] C.W. Sugnet, W.J. Kent, M. Ares Jr., and D. Haussler. Transcriptome and genome organization of alternative splicing events in humans and mice. *Pacific Symposium in Biocomputing*, 9:66–77, 2004.

[159] G. Sutton, O. White, M. Adams, and A. Kerlavage. TIGR assembler: A new tool for assembling large shotgun sequencing projects. *Genome Science and Technology*, 1:9–19, 1995.

[160] R. Tacke and J.L. Manley. The human splicing factors ASF/SF2 and SC35 possess distinct, functionally significant RNA binding specificities. *EMBO J.*, 14:3540–3551, 1995.

[161] R. Tacke, M. Tohyama, S. Ogawa, and J.L. Manley. Human Tra-2 proteins are sequence-specific activators of pre-mRNA splicing. *Cell*, 93:139–148, 1998.

[162] Y. Terai, N. Morikawa, K. Kawakami, and N. Okada. The complexity of alternative splicing of *hagoromo* mRNAs is increased in an explosively speciated lineage in east african cichlids. *Proc. Natl. Acad. Sci. USA*, 100:12798–12803, 2004.

[163] T.A. Thanaraj and F. Clark. Human GC-AG alternative intron isoforms with weak donor sites show enhanced consensus at acceptor exon positions. *Nucleic Acids Res.*, 29:2581–2593, 2001.

[164] T.A. Thanaraj, F. Clark, and J. Muilu. Conservation of human alternative splice events in mouse. *Nucleic Acids Res.*, 31:2544–2552, 2003.

[165] T.A. Thanaraj, S. Stamm, F. Clark, and J.J. Riethoven *et al.* ASD: the alternative splicing database. *Nucleic Acids Res.*, pages D64–D69, 2004.

[166] H. Tian and R. Kole. Selection of novel exon recognition elements from a pool of random sequences. *Mol. Cell. Biol.*, 15:6291–6298, 1995.

[167] C.R. Valentine. The association of nonsense codons with exon skipping. *Mutat. Res.*, 411:87–117, 1998.

[168] J.C. Venter, M.D. Adams, E.W. Myers, and P.W. Li *et al.* The sequence of the human genome. *Science*, 291:1301–1351, 2001.

[169] Z. Wang, H.S. Lo, H. Yang, and S. Gere *et al.* Computational analysis and experimental validation of tumor-associated alternative RNA splicing in human cancer. *Cancer Res.*, 63:655–657, 2003.

[170] Z. Wang, H.S. Lo, H. Yang, and S. Gere *et al.* Reply to [149]. *Cancer Res.*, 63:6996–6997, 2003.

[171] Z. Wang, M.E. Rolish, G. Yeo, and V. Tung *et al.* Systematic identification and analysis of exonic splicing silencers. *Cell*, 119:831–845, 2004.

[172] D.L. Wheeler, D.M. Church, R. Edgar, and S. Federhen *et al.* Database resources of the national center for biotechnology information: update. *Nucleic Acids Res.*, 32:D35–D40, 2004.

[173] R. Wheeler. A method for consolidating and combining EST and mRNA alignments to a genome to enumerate supported splice variants. *Lecture Notes in Computer Science (WABI'2002)*, 2452:201–209, 2002.

[174] C. J. Wikstrand, L. P. Hale, S. K. Batra, and M. L. Hill *et al.* Monoclonal antibodies against EGFRvIII are tumor specific and reat with breast and lung carcinomas and malignant gliomas. *Cancer Res.*, 55:3140–3148, 1995.

[175] T.G. Wolfsberg and D. Landsman. A comparison of expressed sequence tags (ESTs) to human genomic sequences. *Nucleic Acids Res.*, 25:1626–1632, 1997.

[176] H. Xie, W.Y. Zhu, A. Wasserman, and V. Grebinskiy *et al.* Computational analysis of alternative splicing using EST tissue information. *Genomics*, 80:326–330, 2002.

[177] Y. Xing and C.J. Lee. Negative selection pressure against premature protein truncationis reduced by alternative splicing and diploidy. *Trends Genet.*, 20:472–475,

2004.

[178] Q. Xu and C. Lee. Discovery of novel splice forms and functional analysis of cancer-specific alternative splicing in human expressed sequences. *Nucleic Acids Res.*, 31:5635–5643, 2003.

[179] Q. Xu, B. Modrek, and C. Lee. Genome-wide detection of tissue-specific alternative splicing in the human transcriptome. *Nucleic Acids Res.*, 30:3754–3766, 2002.

[180] J.M. Yeakley, J.B. Fan, D. Doucet, and L. Luo *et al.* Profiling alternative splicing on fiber-optic arrays. *Nature Biotechnol.*, 20:353–358, 2002.

[181] G. Yeo, D. Holste, G. Kreiman, and C.B. Burge. Variation in alternative splicing across human tissues. *Genome Biology*, 5:R74, 2004.

[182] G. Yeo, S. Hoon, B. Venkatesh, and C.B. Burge. Variation in sequence and organization of splicing regulatory elements in vertebrate genes. *Proc. Natl. Acad. Sci. USA*, 101:15700–15705, 2004.

[183] M. Zavolan, S. Kondo, C. Schonbach, and J. Adachi *et al.* Impact of alternative splicing on the diversity of the mRNA transcripts encoded by the mouse transcriptome. *Genome Res.*, 13:1290–1300, 2003.

[184] M. Zavolan, E. van Nimwegen, and T. Gaasterland. Splice variation in muouse full-length cDNAs identified by mapping to the mouse genome. *Genome Res.*, 12:1377–1385, 2003.

[185] C.L. Zheng, T.M. Nair, M. Gribskov, and Y.S. Kwon *et al.* A database designed to computationally aid in experimental approach to alternative splicing. *Pacific Symposium in Biocomputing*, 9:78–88, 2004.

[186] Y. Zhou, C. Zhou, L. Ye, and J. Dong *et al.* Database and analyses of known alternatively spliced genes in plants. *Genomics*, 82:584–595, 2003.

[187] Z. Zhou, L.J. Licklider, S.P. Gygi, and R. Reed. Comprehensive proteomic analysis of the human spliceosome. *Natura*, 419:182–185, 2002.

[188] Y. Zhuang, F. Ma, J. Li-Ling, and X. Xu *et al.* Comparative analysis of amino acid usage and protein length distribution between alternatively and non-alternatively spliced genes across six eukaryotic genomes. *Mol. Biol. Evol.*, 20:1978–1985, 2003.

[189] D. Zhuo, W. Zhao, F. Wright, and H. Yang *et al.* Assembly, annotation, and integration of unigene clusters into the human genome draft. *Genome Res.*, 11:904–918, 2001.

17

Human Genetic Linkage Analysis

Alejandro A. Schäffer
National Institutes of Health

17.1 Introduction

Genetic linkage analysis is a collection of statistical techniques used to identify the approximate chromosomal location of disease-associated genes and other markers of interest. Numerous highly-publicized gene discoveries have included a linkage analysis step. A few examples of such discoveries include: a gene for cystic fibrosis [87, 79], a gene for Huntington's disease [24, 26], a gene for spinal muscular atrophy [9, 56], two genes for hereditary non-polyposis colorectal cancer [71, 22, 57, 70], two genes for hereditary breast and ovarian cancer [29, 61, 99, 98], and four genes for Parkinson's disease [75, 76, 60, 36, 18, 6, 89, 88]. For each gene discovery, I chose only one important paper with pertinent linkage results, although there may have been many papers with linkage results; for example, the entire April 1993 issue of *American Journal of Human Genetics* contains linkage studies of the chromosome 17 region containing the *BRCA1* gene [61]. As can be seen from the above examples, while disease-gene hunting and linkage analysis have been greatly facilitated by the results of the human genome project, linkage analysis was carried out before the genome project started, and continues today after most of the human sequence is complete. Because some genetic linkage analysis computations may take a long time to run, a few computational biologists have been interested in the algorithmic problems that arise in those computations.

In this chapter, I give an overview of human linkage analysis from what I consider to be first principles. My emphasis is on data, practical applications, widely-used software, and algorithms. Readers with more patience or a more statistical orientation will find the books by Ott [69] and Sham [81] more helpful. I was introduced to genetic linkage analysis in

1992 by reading an earlier edition [68], and found it to be easy to read and very helpful.

Genetic linkage analysis and association analysis are essential techniques in a strategy called *positional cloning* that identifies disease-causing genes based on their chromosomal location more so than by the function of the proteins that the genes encode. Even though the human genome sequence was nearly complete in 2004, a list of all genes with functional annotations is far from complete. Moreover, even if one had a list of all genes with all normal functions annotated, the phenotype that results when a gene is mutated may be hard to guess. For example, the normal function of the proteins encoded by the colorectal cancer genes cited above is to participate in repair of DNA mismatches in all cells, which is not a function local to the colorectal organs. Hence, information about the genomic position of a disease-causing gene can be extremely useful in narrowing the list of candidate genes.

The basic steps in identifying a gene by positional cloning are as follows:

1. Identify one or more sets of related patients.
2. Connect patients into families, also called *pedigrees* or *kindreds*.
3. Genotype patients and their families at a coarse set of *markers* throughout the genome.
4. Look for variations that *cosegregate* with the disease by linkage analysis.
5. When linkage is found, narrow the region with more families and/or *markers*.
6. Once the linkage region is small, either find candidate genes and skip to step 9, or clone DNA from the area and put into artificial hosts (vectors).
7. Sequence the DNA in the region.
8. Look for genes (e.g., by computer prediction, exon trapping, cDNA screen, etc.)
9. Screen genes for potential mutations, comparing patients and healthy relatives.
10. Prove mutations are causal.

The focus of this chapter is the data analysis at steps 4 and 5, so I will not discuss the later steps at all, except to say that cosegregation of a DNA variation with disease in families can be an essential component of the "proof of causality" at step 10. To understand the basics of that analysis, one needs to understand first: the biological phenomena of recombination and linkage (discussed in Section 17.2) and what types of markers are used at steps 3 and 5 (discussed in Section 17.3). Steps 1 and 2 are usually not computational, but accurate genealogical databases may be needed (e.g., [2]). Steps 4 and 5 are often called *genome scan* and *fine mapping* respectively. To set up for the next two sections, consider the following three pictures of pedigrees showing what one looks for intuitively when hunting for genetic linkage for a dominant or recessive disease.

To understand pedigree pictures one needs to know the following genetics conventions. Squares are males and circles are females. Two slightly different styles are shown for the lines to connect parents to children. A shaded shape represents an affected (i.e., diseased) individual, while a clear shape represents an unaffected individual. Parents are always drawn above their children. An individual is a *founder* if no parents are shown, and an individual is a *nonfounder* if both parents are shown. By convention, one always shows either zero or two parents, even if the identity of one parent is unknown or that parent is unavailable. One models the disease gene as if it had only two variants: 'h' for healthy and 'd' for disease. Variants in either the disease gene or markers are called *alleles*. The genotype of an individual at a marker is the two alleles. In most examples, I will index the marker alleles by $1, 2, 3, \ldots$, but Section 17.3 explains a few aspects of how the distinct alleles are measured in the laboratory. The *phase* of a genotype indicates which allele (if they are different) came from the father and which allele came from the mother. When genotypes

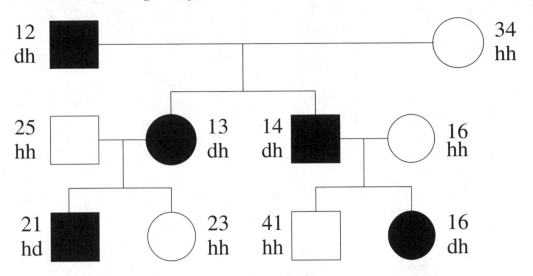

FIGURE 17.1: Dominant inheritance with ideal cosegregation of a marker near the disease locus.

are determined in the laboratory, the phase is unknown. In some cases, the phase can be inferred. Many algorithms in linkage analysis work at their core with genotypes of known phase, and average over the possible phases. An individual (genotype at a marker) is called *heterozygous* if the two alleles are distinct, and *homozygous* if the two alleles are identical.

A disease has a *dominant* inheritance pattern if having a single 'd' allele is sufficient to cause the disease, while a disease has a *recessive* inheritance pattern, if having two 'd' alleles is necessary and sufficient to cause the disease. A probability function that defines the probability of a phenotype conditioned on a genotype is called a *penetrance* function, and I will denote it by pen($x|g$), where x denotes the phenotype and g denotes the genotype. Thus, for a dominant disease, pen(affected | hh) = 0, pen(affected | dh or hd) = pen(affected | dd) = 1. For a recessive disease, pen(affected | hh) = pen(affected | dh or hd) = 0 and pen(affected | dd) = 1. In practice, many linkage analyses use a more equivocal penetrance function; detailed examples of why and how this is done can be found in Chapters 9 and 10 of [86].

In the figures the alleles 'h' and 'd' are shown as if they were known, but in reality they are never known until the disease gene is identified and sequenced. The alleles at the disease locus are inferred from the phenotypes, via the disease status and penetrance function.

Notice that in Figure 17.1, the 1 is always co-inherited with the d, while in Figure 17.2, the 2 and 4 are co-inherited with the d. Notice also that there is an unaffected male in the bottom generation of Figure 17.1 who inherited the 1 allele from his unaffected mother who married in; this unaffected male did not inherit the disease-associated 1 allele from his affected father. When a marker allele is consistently co-inherited with a disease-associated allele along paths in the pedigree, one says that the alleles *cosegregate* with the disease. Figure 17.3 illustrates why inbred populations and pedigrees are popular for studying recessive diseases; in this case the allele 1 and the allele 4 are passed down two distinct paths of inheritance to the affected individuals in the bottom generation, making the affected individuals homozygous at markers flanking the disease gene. One can often detect the disease locus in such pedigrees by looking for regions where the affected individuals are *homozygous* at several markers in a row, but unaffected relatives are not consistently homozygous at those same markers.

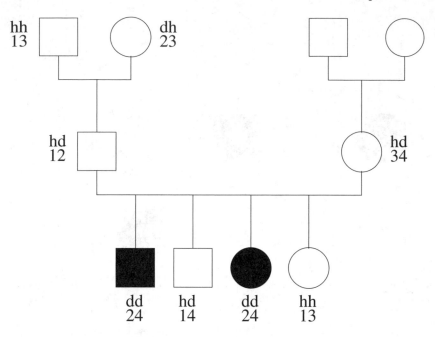

FIGURE 17.2: Recessive inheritance with ideal cosegregation of a marker near the disease locus, but no inbreeding.

Figure 17.4 shows a different representation, called the marriage graph [10], of the pedigree in Figure 17.3. In the marriage graph, the matings are shown as separate nodes of a different type. There is a an edge $i \rightarrow m$ from individual i to mating m, if i is one of the two participants in the mating. There is an edge $m \rightarrow c$, if c is a child produced by the mating. Marriage graphs are often drawn with the directions on edges implicit from top to bottom because the undirected version has some utility as well.

Genetic linkage analysis attempts to compute statistical measures of cosegregation of known marker alleles and putative disease alleles, inferred from the phenotype and penetrance function. As explained in the next section, when such cosegregation is deemed significant, then one may conclude that the putative disease gene is close to the cosegregating marker. Therefore, if one knows the precise chromosomal location of the marker, one may infer an approximate location for the nearby disease gene. In general, one may use more than a single marker within a linkage analysis calculation. Using multiple markers simultaneously can lead to more convincing statistical proof of linkage and in some cases to proof that the gene is likely to lie in between two markers.

The first major distinction between *linkage analysis* and *association analysis* is illustrated by the multiple founders with the allele 1 in Figure 17.1. Linkage analysis looks for the co-inheritance of the allele along paths, while association analysis looks for co-occurrence of the allele. The second major distinction between *linkage analysis* and *association analysis* can be understood intuitively by imagining that one has two pedigrees with the structure of Figure 17.1, except that in the second copy the alleles 1 and 2 are swapped. In linkage analysis, one treats the separate pedigrees separately, so that if marker alleles 1 and 2 were swapped in the second pedigree, both pedigrees would still be indicative of linkage. In association analysis, one is interested in the co-occurrence of a specific marker allele with the disease allele across affected individuals. Thus, to have consistent evidence of association, one would want the same 1 allele to be cosegregating with the d allele in both

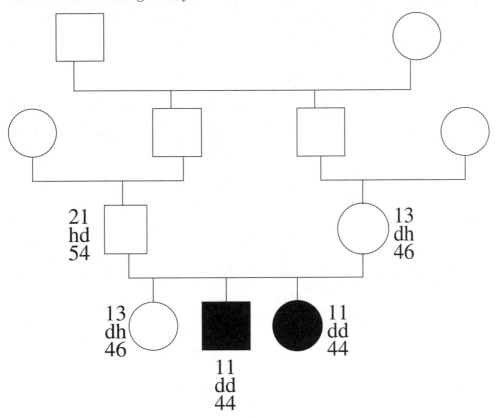

FIGURE 17.3: First-cousin mating leading to a recessive disease in two offspring. Notice that the affected individuals are homozygous for the alleles 1,4 flanking the disease locus.

pedigrees. Association analysis is traditionally done with (unconnected) individuals in a case/control design, but it is also possible to do association analysis on multi-individual pedigrees. See [80] for a survey of association analysis methods and software.

The next section explains why cosegregation of marker and disease alleles occurs when a marker and a disease gene are near each other, and defines proximity in a probabilistic way.

17.2 Chromosomes, Meiosis, and Recombination

The biological basis for genetic linkage analysis is *crossing over* of chromosomes during *meiosis*, through which parents pass DNA to offspring. Before explaining crossing over, I summarize pertinent information about the chromosome content of cells. Most human cells have two copies of every *autosomal* chromosome numbered 1 through 22, in approximate decreasing order by size. For simplicity, all the subsequent descriptions of linkage analysis methods assume autosomal data, but in most cases they have been easily modified to handle X chromosome data. Cells in females have two copies of the X chromosome, while cells in males have one X and one Y. A cell with two copies of every autosome is called *diploid*, while a cell with only one copy is called *haploid*. A cell with more than two copies is called *polyploid*. These terms are also used to describe some organisms in that an organism is called haploid/diploid/polyploid if the vast majority of its cells are

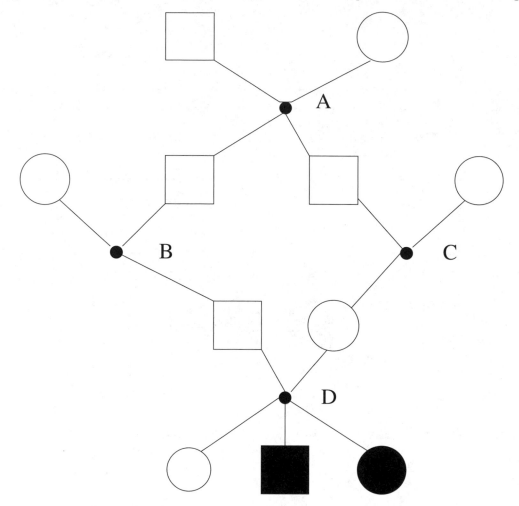

FIGURE 17.4: Marriage graph for the pedigree in Figure 17.3 with genotypes omitted. Each small circle represents a nuclear family; letters are for later usage. Edges are shown as undirected, but may be considered as directed from top to bottom in some usages.

haploid/diploid/polyploid throughout its lifespan. Not all organisms fit neatly into this classification, since some, such as the malaria parasite, change ploidy at different life stages.

One reason I entitled this chapter "Human Genetic Linkage Analysis" is to limit the scope to diploid organisms, such as humans. Much of the content applies to all diploid organisms, including all mammals. However, human studies generally differ from non-human studies in that: 1) mating is close to random and 2) the geneticist has no control over the breeding strategy. The resulting lack of statistical power has to be overcome by collecting more samples or using better methods of data analysis.

Meiosis is the process by which diploid cells divide into haploid gametes, sperm in males and eggs in females. Genetic linkage analysis tries to reverse engineer meiosis computationally to infer what pieces of DNA were transmitted from parent to offspring. As the offspring grow, cells double many times through *mitosis*, during which DNA is copied from parent cells to daughter cells. Maintenance of the two copies of each autosomal piece of DNA and accurate copying of the DNA is important to the health of the cell and its human host. The breast cancer and colorectal cancer gene discoveries highlighted in the opening paragraph illustrate this point. In families with hereditary *BRCA1* or *BRCA2* mutations one copy of the gene has a mutation that is passed from parent to child in *meiosis*, but for breast or ovarian cancer to arise it is necessary that the second copy of the gene be disrupted through an aberrant *mitosis*. The aberrant mitosis could introduce a tiny mutation in a single DNA letter or a large-scale DNA deletion, for example. In families with hereditary nonpolyposis colorectal cancer due to *MSH2* or *MLH1* mutations, local editing errors arise in mitosis at a higher rate than normal. When enough of these errors accumulate in critical genes of a single cell, cancer may follow. Nevertheless, copying errors in mitosis can be quite useful, especially when they occur outside protein-coding genes, as I shall explain in the next section.

During meiosis the four strands of each chromosome come together in "chiasma" enabling the haploid offspring cells to receive a mixture of parental DNA. Genetic linkage analysis is based on a model of how this mixture occurs. I find it useful to explain this model with a medieval metaphor of a monk producing a new copy of a document starting from multiple old editions. Usually the monk will copy the ith and $i + 1^{\text{st}}$ words from the same old edition. If this persisted from the first word to the last, then the new edition would correspond exactly to one of the old ones. However, variation is introduced in two ways:

1. The monk occasionally takes a break; when he resumes, he forgets from which old edition he was copying, and he chooses at random from among the old editions.
2. The monk occasionally makes a mistake in copying, introducing a local change.

In the DNA context, the first source of variation is called *crossing over* and the second source is called *mutation*. When two pieces of DNA, *A* and *B* on the same haploid offspring chromosome derive from *different* chromosomes in the parent, one says that there is a *recombination* between *A* and *B*. Since the parent has two chromosomes, the occurrence of a *recombination* between *A* and *B* implies that an odd number of *crossovers* occurred between the *locus* of *A* and the *locus* of *B*. The process of mutation is necessary to observe recombinations because one can only detect recombinations between loci at which the parental genotypes are heterozygous. Loci at which this property occurs with high probability are sometimes called *markers* and are the topic of the next section.

One says that two loci 1 and 2 are *linked* if the probability of recombination between them $\theta_{12} < 0.5$. The letter θ with various subscripts is almost always used to denote a probability of recombination, and is also called the *recombination fraction*. In disease-gene hunting 1 and 2 would be markers of known location, and one would like to estimate θ_{1D} and θ_{2D} to decide if the disease gene *D* is near those markers. Human marker maps are dense enough that one can expect to have $\theta_{1D} < 0.1$ for monogenic diseases, even for the genome scan at step 4. However, best estimates of θ can be between 0.1 and 0.5 when there are multiple genes involved and and the single locus penetrance model is a poor representation of the inheritance pattern. To publish a statistical proof of linkage, it is not sufficient that the best estimate $\hat{\theta}$ be substantially less than 0.5. Genetic linkage analysis is always done in a framework of statistical hypothesis testing. Therefore, as part of a proof of linkage, one must show that some test statistic for the "alternate hypothesis of linkage" is significantly higher than the same test statistic for the "null hypothesis of no linkage". Much of this chapter is about how to define and compute some commonly used test statistics for genetic linkage analysis.

The rate of crossing over in humans is approximately 1 time per megabase in males and 1.5 times per megabase in females [69]. The average number of crossovers per meiosis varies widely from one mother to another, but not so much from one father to another [40].

The rates vary widely across the genome with the male rate generally higher near the chromosome ends and the female rate higher elsewhere [69]. The analysis of DNA sequence features that may promote or suppress crossing over is an active area of research, but outside the scope of this chapter.

The model of crossing over that I gave above in the monk metaphor is "memoryless", which implies that the process of defining the crossover points along the linear DNA sequence is a Poisson process. The Poisson model is used in most human disease-gene hunting studies, although it is an approximation, and there is ample real data that show it is far from perfect. In Section 17.4, I delineate some models relating distance along the DNA sequence to probability of recombination, and I explain how these models enter into genetic linkage analysis computations.

17.3   Markers Used in Genetic Linkage Analysis

The input data used for linkage analysis computations includes: pedigree structures, phenotypes, and *genotypes* on *polymorphic markers* from the human genome. From now on, I will use the term *marker* to refer to a variable DNA location for which the *alleles* can be determined explicitly in the laboratory, and *locus* to refer to either a marker or a putative disease gene. In a typical linkage analysis computation for hunting a gene, one simultaneously analyzes ($m \geq 2$) loci of which 1 is the putative disease gene, whose position is unknown, and the other ($m - 1$) are markers whose order and approximate location on a *map* are known. I have found it useful to understand some of the laboratory aspects of markers and genotyping. This understanding helps for example in: 1) making sense of diverse data formats used by different laboratories 2) recommending markers for follow-up genotyping in regions that look promising or have large holes and 3) looking for possible laboratory errors in the data.

In mathematical terms, a *marker* is a set of DNA sequences, with 5 pieces: $M = PF_1VF_2R$. I call it a *set* of sequences because the middle part V varies from individual to individual, generating the *polymorphism* that helps track inheritance events. P and R are called the *forward* and *reverse primers*. The sequences P and R are used to specify the conditions for a polymerase chain reaction (PCR) that makes it possible to copy and sequence M in individuals. A frequent source of confusion is that the decomposition of M into parts is not unique. In particular, different primer pairs can be used to sequence the same marker, and different laboratories may use different boundaries for the variable part V. These confusions can make combining genotypes generated in different laboratories most challenging [92]. For this reason, in this section, I will use the term *allele* to refer to either variants of V or variants of M. For example, suppose that the different variants differ by size, then the different sizes in bases of V might be 100, 104, 108, 112, while the corresponding possible sizes of M might be 280, 284, 288, 292.

Some characteristics of a marker useful for linkage analysis are:

1. PCR reaction with primers P and R amplifies exactly one region of the human genome.
2. PCR conditions (e.g., temperature) that allow copying of M are known and work reliably across laboratories.
3. The range of allele sizes and possible alleles are known ahead of time.
4. The probability that a randomly chosen individual has two distinct alleles at M is high.
5. The spontaneous mutation rate in meioses of the sequence V is high enough to

achieve property 4, but low enough that the allele in a child matches one allele in the parent in the vast majority of meioses. Mutation rates between 0.001 and 0.00001 are typically of interest.

A marker that satisfies properties 1 and 2 is often called a *sequence tagged site* (STS), even though it may not be polymorphic. Before the human genome was sequenced extensively, one could not know for certain whether a marker satisfied property 1, but if PCR followed by sequencing consistently generated at most 2 alleles in each individual following rules of inheritance, then it would be assumed to map to a unique site. The probability that an individual is heterozygous (property 4) is called the *heterozygosity* of a marker.

Three commonly used types of markers are *single nucleotide polymorphisms* (SNPs), *variable length tandem repeats* (VLTRs) and *restriction fragment length polymorphisms* (RFLPs). For RFLPs, the variation is due to the presence or absence of a cutter site for a restriction enzyme. Therefore, RFLPs usually have two alleles and the heterozygosity is at most 0.5. VLTRs are variations due to one or more repeated strings in V, such that the number of copies of the repeat(s) varies. *Single tandem repeat polymorphisms* (STPRPs), are a special case of VLTRs in which there is a single substring that is repeated a variable number of times. Many STRPs have been found based on 2,3, or 4 letter repeat units such as CA, AGC, or GATA; such markers are sometimes called *microsatellites*. The CEPH and Marshfield maps described below include thousands of microsatellites with heterozygosities > 0.75. The Marshfield effort also includes recommended screening sets of markers that are coarsely spaced across the genome, with high heterozygosity, and very reliable conditions for PCR. In most cases, microsatellites are outside genes, and the variations are phenotypically irrelevant. However, there is a class of diseases, including Huntington's disease, for which a 3-letter repeat marker (usually CAG, coding for the amino acid glutamine) in a gene can expand to have a pathogenic number of repeat units [26].

SNPs are variations due to substitutions in a single DNA position; since there are 4 DNA letters, a SNP can have up to 4 alleles and heterozygosity up to 0.75. However, most SNPs have only 2 alleles. Most modern linkage studies use STRPs and SNPs. Advantages of SNPs include:

1. SNPs are much more abundant in the human genome.
2. Many genes have SNPs in the promoter or protein coding regions, and these variants may actually be associated with diseases or even causative.

Advantages of STRPs include:

1. STRPs with heterozygosities > 0.75 are abundant [90]
2. Genotyping errors with STRPs are much more likely to lead to violations of the rules of inheritance, and hence be more easily detectable [25]
3. It is currently easier to find STRPs in a given region that have been used in multiple studies.

Above, I alluded to the issues of genotyping errors and rules of inheritance. The pertinent rules of inheritance are that at an autosomal marker, a child receives exactly one allele from the mother and one allele from the father. At an X-chromosome marker, a male child receives exactly one allele from the mother, and a female child receives exactly one allele from each parent. If the genotypes at a marker obey these rules, then they are said to be *consistent*. Consistency is correlated with correctness but neither one implies the other. For example, suppose two parents are heterozygous with alleles 240 244; a child might be reported as homozygous 244 244, and this would be consistent. However, the child may in fact be heterozygous 240 244 and the presence of the 240 allele was missed. Conversely, the child might be reported correctly as heterozygous 240 248 because one of the 244 alleles mutated to a 248 base length allele during meiosis.

When genotypes are not completely filled in, it can be quite challenging computationally to determine whether they are consistent with the rules of inheritance. All of the software packages I describe below indicate correctly when the rules are violated, but information about where the violation(s) occur(s) may be very limited. The best software package for this purpose is PedCheck [65], based on the same algorithmic ideas encoded in the linkage package VITESSE [64] . A particularly useful feature of PedCheck is that when an inconsistent set of genotypes for a single pedigree and marker can be made consistent by a single change, all such single changes are shown along with a likelihood estimate for each change. As indicated above, not all cases of inconsistent genotypes are due to a laboratory error. The variation in the markers arises naturally because of mutations during meiosis, and the mutation rate may be as high as 0.001. Therefore, in any large study one should expect cases where neither of a child's alleles are present in one parent. It is useful to know that most mutations in microsatellites change the number of repeat units by 1. Mutations and genotyping errors that do not violate the rules of inheritance are also important because they often create an apparent recombination, thereby inflating the apparent recombination fraction between markers. There exist several software packages for detecting likely genotype errors and mutations in marker data that is consistent. Two of these packages that attempt to model the sources of error are GenoCheck [19] and a recent version of SIMWALK2 [83].

17.4   Map Functions

One of the primary achievements during the Human Genome Project is the construction of *genetic maps* that describe a sorted order of STS markers along each chromosome, and give distance estimates between markers. Distances on these maps are expressed in Morgans or centiMorgans; 1 Morgan is one expected crossover. For any pair of markers A, B, I will denote the distance in Morgans between A and B as d_{AB}. Some more aspects of marker maps and examples are described in the next section.

Another fundamental quantity in linkage analysis relating two loci A, B is the recombination fraction θ_{AB}. Here I use "loci" rather than "marker", because A or B may represent a putative disease gene. The value θ_{AB} is a unitless probability. Although d is expressed in Morgans and θ is unitless, they can be related; functions that relate the two quantities are called *map functions*. I will mention three commonly used map functions. More material can be found in [69, Sec. 1.5].

The simplest map function used in practice is Morgan's map function: $d_{AB} = \theta_{AB}$. This function cannot work for large values because d is unbounded, while θ (being a probability) must be ≤ 1 always and must be ≤ 0.5 under most circumstances. However, Morgan's function makes some sense for very small values. In general, it is true that $d_{AB} \geq \theta_{AB}$ because the left-hand side counts the number of crossovers, while the right-hand side counts 1 if there are $1, 3, 5, 7, \ldots$ crossovers and 0 otherwise, and then both sides take an average over all meioses. If one assumes that there can be at most 1 crossover between A and B, then the two quantities are equal; this assumption is plausible for very small distances, say under 2cM.

A second widely used map function is Haldane's [28] map function:

$$d_{AB} = -\frac{1}{2}ln(1 - 2\theta_{AB}).$$

This function follows naturally from the assumption that the crossover process is a memoryless Poisson process moving along the chromosome, treated as a one-dimensional space. Geneticists use the term *no interference* to refer to the assumption that the location of the next crossover does not depend on previous crossovers. There is considerable biological data that *in vivo*, interference does occur. The most widely-used function allowing for interference is Kosambi's [41] map function:

$$d_{AB} = \frac{1}{2} arctanh(2\theta_{AB}).$$

The derivation of this function is explained in [81]. Haldane's map function is used in most disease-gene studies, where d_{AB} between adjacent markers is generally $< 0.20M$. In contrast, Kosambi's map function is used for maps, where longer distances must be considered. Other map functions that are of theoretical interest are defined and derived in [81].

As I shall explain below, practical genetic linkage analysis often involves simultaneous consideration of 3 or more loci; an especially important case is the possibility that a gene is located between two markers, as shown in Figure 17.3. For three loci A, B, C in order, it is necessary to compute one of $\theta_{AC}, \theta_{AB}, \theta_{BC}$, given the other two. For Morgan's map function, the relationship is: $\theta_{AC} = \theta_{AB} + \theta_{BC}$. For Haldane's map function, the relationship is: $\theta_{AC} = \theta_{AB} + \theta_{BC} - 2\theta_{AB}\theta_{BC}$. For Kosambi's map function, $\theta_{AC} = (\theta_{AB} + \theta_{BC})/(1 + 4\theta_{AB}\theta_{BC})$. Derivations for the latter two formulas can be found in ftp://fastlink.nih.gov/pub/fastlink/README.mapfun. In [69], Figure 1.4 shows a plot of four map functions, including the three I described, for comparison. It is not necessarily the case that a relation analogous to the above three can be derived for a proposed map function. The theoretical problem of whether a map function is valid for more than two loci is considered in [35, 91], for example.

17.5   Maximum Likelihood in Genetic Linkage Analysis

Many linkage analysis calculations are based on the principle of *maximum likelihood* and on likelihood ratio tests. Typical input to a linkage analysis calculation would include:

1. a pedigree structure K; (I am using K for kindred, so as to reserve P for Probability)
2. loci $l_1, \ldots l_m$, including number and frequency of the alleles;
3. genotypes g_{ij} for some individuals, where i is the index on individuals and j is the index on loci; G_j will denote the vector of all genotypes at locus j.
4. phenotypes x_i for some individuals; X will denote the vector of all phenotypes
5. penetrance functions pen$(x \mid g)$, where different individuals may have different penetrance functions;
6. (when there are at least two markers of known position) recombination fractions $\theta_1, \theta_2, \ldots$ between consecutive markers.

For the simplest case, let us assume there is one putative disease locus l_1 and one marker locus l_2. Then for any fixed recombination fraction θ_1 between the loci, it is possible to define the likelihood L as a function of the input parameters and θ_1. For brevity, I will denote this by $L(K \mid \theta_1)$. One would like to know whether the disease l_1 is *linked* to the marker l_2, which is suggested when the value $\hat{\theta}$ that maximizes the likelihood is much less than 0.5. Formally, one defines the likelihood ratio:

$$\frac{L(K \mid \theta_1 = \hat{\theta})}{L(K \mid \theta_1 = 0.5)} \tag{17.1}$$

that compares the alternate hypothesis of linkage in the numerator to the null hypothesis of no linkage in the denominator. The log base 10 of this ratio (or any other ratio of likelihoods in linkage analysis) is call the *log odds score* or *LOD score* for short. LOD scores are convenient because when the data include multiple pedigrees, the combined likelihood is *product* of single pedigree likelihoods, and the combined LOD score is the *sum* of individual pedigree LOD scores. One can extend the likelihood and the ratio test to use more than 2 loci by allowing multiple recombination fractions in the numerator and the denominator; typically one would use a fixed locus order and fixed recombination values in the denominator, and allow some of these values to vary in the numerator.

For a sequential testing design in which one tests one marker in the genome after another, Morton [62] proposed a LOD score threshold of 3.0 (corresponding to a likelihood ratio of 1000) as high enough to declare linkage, and stop testing markers if the disorder is monogenic. Chotai [11] evaluated this threshold test analytically and by simulation and raised some concerns. To compare the LOD score statistic with other test statistics it would be useful to have a correspondence between LOD scores and the more common measure of statistical significance, p-values. However, the threshold of 3 does not directly correspond to a precise p-value. Under any circumstances, a LOD score of s guarantees a point-wise p-value of 10^{-s} [68]. Ott [68] gives some heuristic arguments that a LOD score of 3 corresponds to a genome-wide false positive rate of approximately 0.05. For traits that appear X-linked based on the inheritance pattern, one may use a LOD threshold of 2.0, since only a small fraction of the genome is considered [86]. Asymptotically, the LOD score multiplied by the constant $2 \ln 10 = 4.6$ is distributed as a 50:50 mixture of a point mass at 0 and a Chi-squared random variable with 1 degree of freedom (df) [81]. This distribution can be used to get asymptotic pointwise p-values, which ignore the usual case that multiple marker loci are tested for linkage to the disease. The empirical significance of any particular LOD score in a specific study can be assessed by simulation as described in Section 17.10.

Lander and Kruglyak [45] questioned the 3.0 LOD score threshold for studies of common, complex diseases, where the correct penetrance function is unclear, pedigrees will usually be small, and the marker set used will be very dense due to uncertainty. Suppose that one is interested in a LOD threshold s, for which the corresponding Chi-quared value $C = (2 \ln 10)s$ has pointwise significance $\alpha_p = 1 - \Phi(\sqrt{(C)})$, where Φ is the standard normal distribution [81]. Based on theoretical work of Feingold et al. [21], Lander and Kruglyak showed that the number of regions where a multilocus LOD score s will achieve a pointwise significance α_p is distributed approximately as a Poisson random variable with mean:

$$\mu = (N + 2MC)\alpha_p,$$

where N is the number of chromosomes, and M is the total length of the genome map in Morgans. Using the above formula, they estimated that the genome wide significance $\alpha_g = 1 - e^{-\mu}$. Plugging in some estimated values for the human genome, they determined that a higher threshold of 3.3 corresponds to the desired genome-wide significance of 0.05. The use of these thresholds is fuzzy in practice because one computes LOD scores for multiple single markers and combinations of markers.

Likelihood ratio tests are used in linkage analysis for purposes other than the basic test of linkage vs. non-linkage. I summarize four such usages.

First, suppose there is a published claim of linkage between disease D and marker l. One is studying another pedigree K with disease D, and one wants to know whether the disease in K is also linked to l. Then one may apply the test in 17.1 restricting the θ in the numerator to very small values, typically < 0.05. Because of this restriction, the LOD score can be negative, and LOD < -2 is considered sufficient evidence to exclude linkage of K to l [86]. Cases where the same apparently monogenic disease is linked to different genomic regions in different pedigrees are common, and this phenomenon is called *locus heterogeneity*. I mentioned colorectal cancer, breast and ovarian cancer, and Parkinson's disease as examples of phenotypes with proven locus heterogeneity in the first paragraph of the Introduction.

Second, suppose one has established linkage of a disease gene to a region containing markers $l_1, l_2, \ldots$, and one wishes to decide where in the region, the gene is most likely to be found. Then one computes the likelihood with different placements of the disease gene relative to the markers. The recombination fraction between consecutive pairs of markers stays fixed, but the recombination fractions between the putative gene and the two closest flanking markers are varied, keeping the fraction between the flanking markers fixed. For this purpose, the equations in the previous section that relate θ_{AC} to θ_{AB}, θ_{BC} are used, with θ_{AC} fixed and the position of the disease (as B) varying. One defines a *support region* in which the LOD score comparing the best position for the disease to any other position in the region is at most s units. A threshold of $s = 1$ is often used to define the support region [86].

Third, suppose one is building a map of markers and one wishes to decide among two locus orders: $l_1, l_2, l_3, \ldots$ and $l_1, l_3, l_2, \ldots$. Then one can determine the recombination fractions that maximize the likelihood of each marker order and do a likelihood ratio test of the two orders. Thresholds of either $s = 1$ or $s = 3$ are used in practice depending on how reliable one wants the marker order to be. There is a tradeoff between being able to order fewer markers with a higher s versus more markers with a lower s.

Finally, one may use likelihood computations to optimize some of the input parameters, other than the recombination fraction. When I defined L, I suppressed all the parameters except the pedigree K treating them as fixed values. In practice, some of the input values may be uncertain. Two common examples are the marker allele frequencies and the penetrance. In this context, one chooses the parameter values that maximize the likelihood, and no likelihood ratio test is used.

17.6 Elston-Stewart Algorithm

The pedigree likelihood can be written as a nested sum and product as described below. If evaluated directly, there are an exponential number of terms. Elston and Stewart [20] found a general method to rearrange the nested computation for some pedigrees that leads to running time polynomial in the number of individuals. Assume that the individuals are numbered $1, \ldots, n$; I will consider different orders depending on pedigree structure. Let $F(i)$ and $M(i)$ be the father and mother of i, if i is not a founder, and 0 otherwise.

I use $P(g_i)$ to denote the probability that i receives genotypes g_i; here g_i denotes multilocus genotype. For founders $P(g_i)$ is the product of the population frequencies of each allele at each locus. For example, suppose that the genotype g_i is 11 at the first locus and 35 at the second locus, with $P(1) = 0.4, P(3) = 0.2, P(5) = 0.1$. If the phases are specified, then $P(g_i) = 0.4 \times 0.4 \times 0.2 \times 0.1 = 0.0032$. If the phases are unknown, then the heterozygous 35 could arise two ways and $P(g_i) = 0.0064$. This definition assumes that allele frequencies within a locus and across loci are independent.

For nonfounders i, $P(g_i) = \text{trans}(g_i \mid g_{F(i)}, g_{M(i)}, \theta)$, which is the (conditional) *transmission probability* for the genotype of a child conditional on the parental genotypes. The algorithm works with phase-known genotypes, so if there are m loci, then each parent can transmit any one of 2^m (not necessarily distinct) combinations of alleles. Let $R_j = 0, 1, 2$ depending on how many recombinations occur between loci j and $j + 1$. Then

$$\text{trans}(g_i \mid g_{F(i)}, g_{M(i)}, \theta) = 4^{-m}\Pi_{1 \leq j < m}\theta_j^{R_j}(1 - \theta_j)^{2 - R_j}.$$

The product can be decomposed into male and female components in case one wants to use distinct male and female θ values.

Using $P(g_i)$ the pedigree likelihood is given by:

$$L(K, X, G, \theta) = \Sigma_{g_1} \cdots \Sigma_{g_n}\Pi_{1 \leq i \leq n}\text{pen}(x_i \mid g_i)P(g_i, \theta). \tag{17.2}$$

Elston and Stewart observed that for some pedigrees the expression 17.2 could be computed efficiently (as a function of n) by factoring out one nuclear family at a time. In a nuclear family with father f, mother m, and children indexed $1, \ldots, c$ the likelihood can be factored as:

$$L = \Sigma_{g_f}\text{pen}(x_f \mid g_f)P(g_f) \times \Sigma_{g_m}\text{pen}(x_m \mid g_m)P(g_m) \times \Pi_{1 \leq i \leq c}\Sigma_{g_i}\text{pen}(x_i \mid g_i)\text{trans}(g_i \mid g_f, g_m, \theta).$$

This factoring can be iterated, using appropriate subexpressions. Suppose f is the only person that connects this nuclear family to the rest of the pedigree. Then

$$P_{\text{cond}}(g_f) = \Sigma_{g_m}\text{pen}(x_m \mid g_m)P(g_m) \times \Pi_{1 \leq i \leq c}\Sigma_{g_i}\text{pen}(x_i \mid g_i)\text{trans}(g_i \mid g_f, g_m, \theta)$$

is the probability for each genotype of the father f conditional on spouse and children. If the children also have descendants, then one uses the more general tail recursion:

$$P_{\text{cond}}(g_f) = \Sigma_{g_m}\text{pen}(x_m \mid g_m)P(g_m) \times \Pi_{1 \leq i \leq c}\Sigma_{g_i}\text{pen}(x_i \mid g_i)\text{trans}(g_i \mid g_f, g_m, \theta)P_{\text{cond}}(g_i). \tag{17.3}$$

For a leaf child i, $P_{\text{cond}}(g_i)$ is initialized to 1 if g_i is possible given the input data, and 0 otherwise.

The above recursion is said to *peel* one nuclear family at a time. More precisely, consider the following algorithm:

1. Order the nuclear families $1, 2, \ldots k$ such that family a comes before family b whenever b has a proper ancestor of a.
2. Apply 17.3 or the corresponding expression when the mother is the connecting parent to nuclear families $1, 2, 3 \ldots$.
3. If family k is the only one with two founders, then sum P_{cond} for the possible genotypes one of the two founder conditioned on the other.

For example, in Figure 17.1 there are 2 nuclear families at the bottom that could be peeled in either order, and the nuclear family at the top would be peeled last.

The above algorithm works for pedigrees in which: 1) there is at most one nuclear family in which there are two founder parents and 2) there are no multiply mated parents. Pedigrees that meet these two conditions are called *simple*. For example, the pedigree in Figure 17.1 is a simple pedigree. Condition 1) is sometimes stated as requiring that each individual have at most 1 pair of grandparents in the pedigree. Condition 2) can be weakened to requiring no multiply-mated nonfounders, and the algorithm still works. Simple pedigrees are quite common in linkage analysis because for a dominant disease, there is typically one affected founder who passed the disease down paths, and in each nuclear family one unaffected parent founder mated with an affected founder. The algorithm was generalized to nonsimple, loopless pedigrees by Lange and Elston [49] by developing an expression analogous to 17.3 for updating the probability of a child's genotype conditional on siblings and parents. This allows one to traverse the pedigree graph in any direction, always peeling off nuclear families that are connected to the rest of the graph by a single individual.

The first production implementation of the generalized Elston-Stewart algorithm was done by Ott in the package LIPED [66], which is still used today. In the process Ott extended the algorithm to handle most looped pedigrees as described in the next paragraph. A pedigree has a loop if and only if the undirected version of the marriage graph has a cycle. For example, the pedigree in Figures 17.3 and 17.4 has a loop involving the nuclear families A, B, D, C. The problem of computing the pedigree likelihood is NP-complete for looped pedigrees [73]. In the example figures the single loop is an *inbreeding loop* because the two parents in family D at the bottom of the loop have a common ancestor (the parents in family A). Loops of other types can arise, for example if two brothers in one nuclear family mate separately with two sisters in another family, and such loops are called *marriage loops*. Some geneticists find it useful to distinguish between inbreeding loops and marriage loops because inbreeding loops typically arise in pedigrees with recessive diseases, where the affected individuals are homozygous for markers near the disease gene. However, it is not possible to clearly define or distinguish inbreeding loops and marriage loops in pedigrees that have many loops of both kinds.

One can make *twin* copies of an individual in the pedigree, so as to break a cycle in the undirected marriage graph. For example, in the pedigree graph of Figure 17.4 one could make either parent in family D into twins. If one chose the father, then one twin would be only the child in family B, while the other would be the child in family D; both copies would be required to have genotypes 21 at the first marker and 46 at the second marker. Suppose individual i is split into individuals i_1 and i_2; usually i_1 retains the parents of i and has no children, while i_2 has all the children of i, but no parents in the pedigree. Only this parent/child split into two individuals is allowed in the widely-used package LINKAGE, but more general splits including allowing more than two copies can speed up the computation [3]. If one forces the resulting copied individuals to have identical genotypes, then the likelihood calculation remains valid. If there are t individuals split into twin copies, the t nested sums for those individuals are moved to the outside in equation 17.2 and they cannot be easily decomposed, so the running time is exponential in t.

Lathrop et al. [52, 54] made an engineering breakthrough by implementing the full multilocus, generalized Elston-Stewart algorithm in LINKAGE [54]. Among the innovations in LINKAGE was the ability to calculate genetic risks based on genotype data and phenotype data on relatives, which is useful in genetic counseling when the causative gene has not yet been found [52]. Another widely-used implementation started in the 1980's is MENDEL [51]. Among the innovations introduced in the first version of MENDEL are dynamic ordering of the nuclear families based on the input data [48] and use of boolean logic and graph algorithms to reduce the list of possible genotypes at each locus [50].

Other major algorithmic improvements include:

1. Take advantage of symmetry in algebraic subexpressions of 17.3 [53].

2. Use sparsity in the arrays storing intermediate terms to replace arithmetic operations with boolean operations [12].

3. Encode possible genotype at each locus separately, and compose the possible multilocus genotypes for an individual on the fly [64].

4. Recode the set of possible alleles inherited by an individual using a "fuzzy"

method to reduce the effective number of possible alleles and hence genotypes at
a locus. [64]

While I have not devoted much space to the algorithmic improvements, they have reduced
the running time of implementations of the generalized Elston-Stewart algorithm by at least
5 orders of magnitude, not even counting the improvements in CPU speeds over the same
time period. The improvements are especially evident when analyzing more loci and looped
pedigrees, although the running time remains exponential both in the number of loops
and the number of loci. The VITESSE package can handle 8-10 loci on medium, loopless
pedigrees [63], while LINKAGE is limited to about 3 loci on the same pedigree. Due to the
concurrent improvement in availability of markers and reliability of maps, 8 loci is far more
than sufficient for most human disease-gene hunting case studies.

17.7   Marker Maps

In 1980, Botstein et al. [7] formally proposed the construction of large-scale marker maps
to be used in disease-gene hunting. The first genome-wide genetic map was published only
seven years later [17], around the same time as numerous higher-density maps of individual
chromosomes. The primary computational problem in genetic map construction is to order
the markers along each chromosome. Efficient marker ordering was greatly facilitated by the
discovery of the Lander-Green algorithm for linkage analysis, described in Section 17.8, and
its implementation in software packages MAPMAKER [47] and CRI-MAP [46]. Another
widely used package for map construction is MULTIMAP [59], which uses CRI-MAP as an
engine to compute likelihoods, but further automates the process of building partial maps
and selecting marker orders. Maps that order markers based on meioses and recombination,
and present intermarker distances in cM, are called *genetic maps*. This nomenclature is to
distinguish them from radiation hybrid and physical maps, which are based on biological
phenomena other than recombination and linkage, and are hence outside the scope of this
chapter.

Genetic map construction is solved using the same maximum likelihood formulation of
linkage analysis that I presented in Section 17.5. However, there are some basic differences
between disease-gene hunting usages and map construction. In disease-gene hunting one
is trying to locate a new locus with respect to an existing map of markers in known order. Thus, the number of loci that need to be simultaneously analyzed is limited. In map
construction one is trying to put m markers into order, and m may be large. In disease-
gene hunting one may be studying a rare disease for which it is desirable to collect one or
more pedigrees with dozens or even hundreds of individuals. For map construction, one
can restrict attention to pedigrees of three generations and of modest size; three-generation
pedigrees are used, so that in the youngest generation it may be possible to identify the
grandparent-of-origin for each allele, and infer phase of genotypes. For disease-gene hunting, the genotype at the disease locus is inferred via the penetrance function, while map
construction usually uses markers for which the genotype and phenotype are identical. For
disease-gene hunting it is necessary to treat pedigrees in which members are not available
for genotyping; this is also done in map construction, but one could reasonably impose the
restriction that all individuals sampled are genotyped at all markers to be ordered.

The Lander-Green algorithm made map construction much easier because it can compute the likelihood of one order of m markers in time that is polynomial in m, although
the time is exponential in the number of individuals. Map construction is still computationally challenging because $m!/2$ different marker orders might have to be considered.
Computational biologists do not necessarily appreciate that there are very time consuming

TABLE 17.1 A Few Markers on Human Chromosome 5 from the Marshfield Genetic Map

Marker	Dname	sex-ave(cM)	female(cM)	male(cM)
AFM028xb12	D5S392	0.00	0.00	0.00
AFMa217zh1	D5S1981	1.72	0.00	3.34
AFMa183wh5	D5S1970	5.43	1.54	9.14
AFM205wh8	D5S417	6.67	2.19	11.07
GATA145D10	D5S2849	7.77	3.28	12.17
AFM336tc1	D5S675	9.41	4.37	14.36

non-computational tasks required in map construction. These include: 1) DNA sequencing to identify markers that meet the criteria discussed in Section 17.3 2) ascertainment of pedigrees 3) genotyping at candidate markers and 4) inter-species hybridization and/or cytogenetic experiments to physically assign markers to chromosomes. Besides the order of markers, and intermarker distances, an important output of some map construction projects is an estimate of frequencies of the observed alleles at each marker. Recall from the previous sections that disease and marker allele frequencies are among the inputs to linkage analysis for disease-gene hunting, and it is often useful to have external estimates not based on the disease pedigrees being analyzed.

Major progress in ascertainment of pedigrees was made under the leadership of Nobel-prize winner Jean Dausset, who set up a foundation Centre d'Etude du Polymorphisme Humain (CEPH) to collect pedigrees, develop markers, and make maps. The most recent of the human maps from the large CEPH effort is described in [16], and it remains widely used. Broman et al. [8] constructed a denser, and more reliable map by genotyping some CEPH pedigrees at lots more markers and developing better marker ordering procedures for short distances. This map is commonly called the "Marshfield map" and can be found at: (`http://research.marshfieldclinic.org/genetics/Map_Markers/maps/IndexMapFrames.html`) Table 17.7 shows a few entries from chromosome 5 from this map. The maps show marker positions in cM from the upper end of the chromosome; these distances can be converted to recombination fractions using map functions. More recently, Kong et al. [40] built an even denser map, called the "deCODE map", using a newly ascertained set of Icelandic families. Now that the genome sequence is nearly complete, one can check marker order in the sequence directly. However, it remains extremely useful to estimate the inter-marker recombination fractions using the published maps.

There is no generally agreed upon algorithm for finding a good ordering of m markers, when m is large, and MAPMAKER and CRI-MAP recommend different heuristic approaches. The MAPMAKER approach is to find a coarse set of f "framework markers" such that: 1) no two of the markers are very close to each other, say at least 10cM apart, 2) the set of markers is much smaller, so many, if not all, of the $f!/2$ orders can be considered and 3) the f markers can be reliably ordered. By "reliably ordered", I mean that the difference in log-likelihood between the best order and the second best order is large, preferably above 3.0. Once a framework map is computed, one can then assign each remaining marker to a place within the framework map and attempt to order all the markers between each pair of framework markers. The main advantage of the framework approach is that a subset of the map is reliably ordered, and in some usages of genetic maps, it is acceptable to use only the framework markers. The main disadvantage is that finding a set of framework markers that meet the three enumerated criteria may be impossible for some data sets.

CRI-MAP and MULTIMAP instead implement iterative algorithms, where one can start with an initial order for some markers, possibly framework markers, and add one marker at a time. These methods have the advantages that they always place all markers, and one can test the reliability of ordering at any stage. A common way to do this is to apply a "flips test" in which each set of k (typical value 5–8) consecutive markers in the current preferred order are permuted in all $k!$ possible ways, leaving the other markers fixed, and the log-likelihoods of each resulting order on all markers computed. The main disadvantage of the iterative approach is that the output map is likely to depend on the order in which the markers are inserted [30].

In the iterative approach it is useful to have a good starting order for at least some markers. The input for this problem is usually taken to be the $\binom{m}{2}$ pairwise LOD scores and best recombination fractions for each pair of markers. Weeks and Lange [93] implemented simulated annealing algorithms for two objective functions: one was the sum of the LOD scores between adjacent markers, and the other was based on least squares fitting for all pairs of markers. In a program called FIRSTORD, Curtis [13] implemented a different heuristic search algorithm to optimize a different objective function based on analogy to minimizing the energy of a collection of springs.

17.8 Lander-Green Algorithm

In 1987, Lander and Green [46] discovered a completely different algorithm for computing multilocus pedigree likelihoods. Their algorithm runs in time that is polynomial in the number of markers m, but exponential in the number of individuals n. The Lander-Green algorithm is ideally suited for computing genetic linkage maps because that can be done with small pedigrees and requires simultaneous analysis of many markers. Indeed, the CRI-MAP and MAPMAKER packages, which both implement variants of the Lander-Green algorithm, have been used to compute many linkage maps for humans and other organisms as well. The initial formulation was primarily for codominant markers, where phenotype = genotype, and there is no need for a penetrance function. I will describe a more general variant of the algorithm developed by Kruglyak et al. [42, 43]. Improved versions of this algorithm are implemented in the widely used software packages GENEHUNTER [43] and Allegro [27].

The critical innovations in the Lander-Green algorithm are to represent the flow of alleles by *inheritance vectors* and to treat the inheritance vectors as hidden states in a Hidden Markov model (HMM). I will use I to denote an inheritance vector; the inheritance vector will usually be associated with a position j, between 1 and m, in an m-locus map, in which I will denote the vector by I_j. I will assume that the recombination fractions between adjacent loci are specified. I reserve positions 0 and $m + 1$ for the situation in which a hypothetical disease gene is a locus outside of the marker map; the disease gene could also be a locus within the map, flanked by two marker loci.

Suppose the pedigree contains n individuals, partitioned into sets F of founders and C for nonfounders (C stands for children) of sizes f and $c = n - f$ respectively. The pedigree contains $2c$ meioses or inheritance events by which the c nonfounders get their genotypes. An inheritance vector I is a 0-1 array of length $2c$ in which the entry $I[k]$ is 0 if that allele was inherited from the grandfather and 1 if that allele was inherited from the grandmother. I will assume that whenever genotypes are known/assigned for founders, they are assigned with phase, so that one knows which allele came from the founder's father and which came from the mother, even though those individuals are not in the pedigree. Given phase-known founder genotypes and an inheritance vector, the phase-known genotypes of all nonfounders are uniquely determined by tracking the meioses down the pedigree. There are 2^{2c} possible inheritance vectors at any locus. Suppose I_j and I_{j+1} are inheritance vectors for adjacent loci, then the positions in which they differ correspond precisely to recombinations. Let

θ_j be the recombination fraction between loci j and $j + 1$. Define a $2^{2c} \times 2^{2c}$ transition matrix T_j indexed by the possible inheritance vectors, such that $T_j[v, w] = \theta_j^d(1 - \theta_j)^{2c-d}$, if the vectors v, w differ in exactly d out of $2c$ entries. Then T_j represents the transition probability matrix in a Markov chain whose states are the inheritance vectors. It is a Hidden Markov model because the inheritance vectors cannot in general be determined uniquely from the available genotypes and phenotypes, but it is possible to infer probabilities on each inheritance vector at each position.

One basic algorithmic problem is to compute probabilities at a single locus that relate genotypes and phenotypes to inheritance vectors. Suppose X_j and G_j represent the vectors of phenotypes and genotypes at locus j, respectively. Then $P(X_j \mid I_j) = \Sigma_{G_j} \text{pen}(X_j \mid G_j)P(G_j \mid I_j)$, where the penetrance function is 1 for markers and provided as input if j is the disease locus. Let G_{Fj} be the founder genotypes at locus j and let G_{Cj} be the nonfounder genotypes. As noted above, if one fully specifies I_j and G_{Fj}, then this determines G_{Cj}, so let that function be $G_{Cj}(G_Fj, I_j)$. One could compute

$$P(G_j \mid I_j) = \Sigma_{G_{Fj}} P(G_{Fj}) \times G_{Cj}(G_Fj, I_j). \tag{17.4}$$

The term

$$P(G_{Fj}) = \Pi_{1 \le i \le 2f} P(a \mid \text{allele assigned to founder position } i \text{ is } a)$$

can be computed from the input marker allele frequencies exactly as in the Elston-Stewart algorithm. I will first explain how to use $P(X_j \mid I_j)$ to compute the likelihood, and then explain how to reduce the number of terms in the sum on the right hand side of equation 17.4..

Using $P(X_j \mid I_j)$ and the transition matrices T_j one can compute the overall likelihood of the data by a chained matrix product. Let 1^{2c} be a vector of all 1's. Let Q_j be a $2^{2c} \times 2^{2c}$ matrix with diagonal entries $P(X_j|I_j)$ and zeroes off the diagonal. Then the likelihood (as a function of phenotypes X, the locus map, and recombination fractions θ_j) is proportional to:

$$1^{2c}Q_1T_1Q_2T_2 \cdots T_mQ_m1^{2c}. \tag{17.5}$$

The above formula is used to compare the likelihood of two maps in genetic map construction or two possible positions of a disease gene in disease gene finding. Specifically, suppose one wishes to compare the hypothesis of linkage with the disease as locus j to the null hypothesis of no linkage. For the likelihood of no linkage treat the disease gene locus as locus 0, with $\theta_0 = 0.5$ (i.e. unlinked to the other loci), and let T_0 represent the transition matrix all of whose entries are 2^{-2c}. Let θ' be the recombination fraction between the markers in positions $j - 1$ and $j + 1$ that flank the putative disease locus, and let T' be the corresponding transition matrix. Then the likelihood ratio test (whose log base 10 is the LOD score) can be computed by using as numerator

$$1^{2c}Q_1T_1Q_2T_2 \cdots Q_{j-1}T_{j-1}Q_jT_jQ_{j+1} \cdots T_mQ_m1^{2c},$$

and as denominator

$$1^{2c}Q_0T_0Q_1T_1Q_2T_2 \cdots Q_{j-1}T'Q_{j+1} \cdots T_mQ_m1^{2c}.$$

In the numerator, there is flexibility as to how to set θ_j and θ_{j+1} subject to the Haldane or Kosambi equations given at the end of Section 17.4. In general, one may wish to consider all possible positions for the disease along the map, so it is most efficient to compute all partial products from the left and the right in the all-locus product equation 17.5 first, and then compose the terms to the left of the disease locus, at the disease locus hypothetical position, and to the right of the disease locus.

One can compute $P(G_j \mid I_j)$ more efficiently in most cases by observing that many assignments of founder genotypes G_{Fj} are not consistent with the input marker data [43]. Moreover, for some inheritance vectors I_j and some founder allele positions, the allele in that position is not passed down to any genotyped individual, so all possible alleles are valid. One can determine which founder allele assignments are consistent with the observed genotypes and a fixed vector I_j by a linear-time graph algorithm. Define a multigraph $H(I_j)$ with $2f$ vertices $v_1, v_2, \ldots v_{2f}$ representing the phased alleles of the founders. Let i be some genotyped individual whose actual unphased alleles are a_1, a_2. Tracing back the inheritance vector, one can determine two founder allele positions $V_1(I_j, i)$ and $V_2(I_j, i)$ such that those are the alleles that must have been passed down to individual i. The entire correspondence between genotyped individuals and founder alleles positions can be determined in one bottom-up traversal of the pedigree graph for each vector I_j. Add to the graph H, the edge $V_1(I_j, i)$—$V_2(I_j, i)$ with label $\{a_1, a_2\}$. Note that the labels are unordered sets. If a vertex v in H has no adjacent edges, then that allele position can be assigned any allele, so the term for position v in $P(G_{F_j})$ is set to 1, and that position is ignored. If there are outgoing edges from v, then the only valid alleles must be in the set labels of all outgoing edges, and that intersection is of size at most 2. If the intersection is empty, then the vector I_j is not consistent with the observed data. If the intersection has a single allele, that allele must be assigned to position v. If there are two choices a_1, a_2, then choose one assignment and that forces the assignment of all vertices in the same connected component as v in the multigraph H, so there at most two possible valid allele assignments for each nontrivial connected component of H. Hence $P(G_{F_j})$ can be expressed as the sum and product of a linear (in number of individuals) number of terms.

Having obtained $P(G_j \mid I_j)$, one can apply Bayes's theorem to compute $P(I_j \mid G_j)$. More generally to compute the conditional probability of an inheritance vector at any position in the map conditional on all the genotype data, one uses the forward-backward dynamic programming algorithm that is standard in HMM theory [77]. Denote this probability for locus j by $P_{\mathrm{cond}}(I_j)$. One advantage of computing $P_{\mathrm{cond}}(I_j)$ conditional on all the data (pedigree, genotypes, phenotypes, recombination fractions) is that one can apply a variety of tests of linkage using this probability, not just the LOD score test. See examples in the next section. The likelihood ratio test for linkage in which the putative disease gene occupies position j can be written as

$$\frac{\Sigma_{I_j} P(X \mid I_j, \theta_{j-1}, \theta j) P_{\mathrm{cond}}(I_j)}{\Sigma_{I_0} P(X \mid I_0, \theta_0 = 0.5) 2^{-2c}}$$

In the denominator, each inheritance vector for the unlinked disease locus has equal prior probability.

Ingenious techniques have been found to speed up the Lander-Green algorithm and these are of at least two different types. There are new algorithmic methods that reduce the number of arithmetic operations to compute the expression in formula 17.5 such as those in [42, 43, 32, 44]. There are also algebraic methods to reduce the dimensionality of the exponential-size inheritance vector space such as those in [43, 27, 58]. The latter can be general or take advantage of pedigree-specific symmetries. For these reasons, I did not give a specific running time for variants of the Lander-Green algorithm. Suffice to say that the recent innovations have enabled one to compute LOD scores instantaneously on pedigrees of modest size that were intractable with the original Lander-Green algorithm, but some real pedigrees are still too large for current implementations.

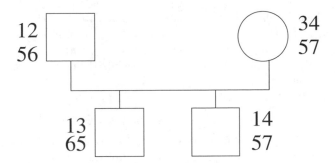

FIGURE 17.5: The siblings share the allele 1 identical by descent. The siblings do not share the allele 5 identical by descent because the left sibling inherited 5 from the mother, while the right sibling inherited 5 from the father.

17.9   Methods That Do Not Use a Penetrance Function Model

Many researchers consider the LOD score method unsuitable for complex traits, which involve multiple genes, because it is impossible to specify an accurate penetrance function. This concern has led to the development of methods that analyze the genotypes of only affected relatives. The intuition is that affected relatives should have marker genotypes that are more similar near disease-associated genes than at random places in the genome. In this section, I introduce a few of the concepts and software packages used for analysis of affected relatives. For the LOD score method, different packages all compute the same test statistic. This is not at all true for affected relative methods. I will mention a few test statistics that are used, but mostly refrain from recommendations.

There is an unfortunate terminology confusion that makes it more difficult than necessary to read the literature on affected relative methods. Many authors refer to methods that use a penetrance function as "parametric" and other methods as "non-parametric". This is incorrect because all methods use parameters such as recombination fractions and allele frequencies. I will instead use the terminology preferred by Robert Elston and Joseph Terwilliger, two linkage analysis luminaries. A method that uses a penetrance function is called *model-based*; other methods are called *model-free*.

The first model-free method was proposed by Penrose [72] and predates the LOD score method by about 20 years. Model-free methods were largely ignored until the late 1970's because early studies focused on diseases for which simple dominant or recessive penetrance functions fit the data well.

Early model-free methods assumed a study design in which clinicians collected nuclear families with at least two affected siblings. Statistical tests for linkage can be based on whether the affected siblings share more alleles than would be expected by chance. To make this more precise, it is useful to distinguish two types of allele sharing. Alleles a and b in two individuals are *identical by state* (IBS) if $a = b$. Alleles a, b are *identical by descent* (IBD) if they are inherited via two paths from the same founder allele. IBD is closely related to the notion of an inheritance vector in the Lander-Green algorithm. Given a pedigree, an assignment of phased genotypes to founders, and an inheritance vector, one can determine whether any two alleles in different individuals are identical by descent. The distinction between IBS and IBD is shown in Figure 17.5.

The advantages of measuring sharing IBS are that: 1) it is computationally trivial for genotyped individuals 2) one can more easily derive asymptotic statistics for how many alleles are shared IBS in large pedigrees [94]. The advantage of measuring sharing IBD is that sharing IBD actually capture the notion of linkage in that the identical by descent alleles should be carrying along the disease associate gene variant, if there is linkage.

Tests of affected sibling pairs (ASP) all compute an estimate of the fractions of sibling pairs $\hat{z}_0$, $\hat{z}_1$, and $\hat{z}_2$ that share 0, 1, or 2 alleles. Using IBD sharing, and under the null hypothesis of no linkage one expects $z_0 = 1/4, z_1 = 1/2, z_2 = 1/4$. Different software packages for ASP linkage tests differ on aspects such as the following:

1. how families with > 2 affected siblings are counted;
2. how pedigrees with multiple nuclear families are counted;
3. how the genotypes of ungenotyped individuals are estimated.

A more fundamental difference is in how to test the estimated proportions $\hat{z}_0$, $\hat{z}_1$, and $\hat{z}_2$ against a null hypothesis. The simplest test, called "the mean test", compares $2\hat{z}_2 + \hat{z}_1$ to its expectation of 1 by a Chi-squared test with 1 degree of freedom (df). More generally, one can compare $\hat{z}_0$, $\hat{z}_1$, and $\hat{z}_2$ to the expected proportions $1/4, 1/2, 1/4$ by a Chi-squared test with 2df. Let $s_{ij}, i = 0, 1, 2$ be the probability that sibling pair j share i alleles IBD and suppose there are N sibships. Risch [78] proposed estimating the $\hat{z}_i$ by maximum likelihood and testing their departure from the null hypothesis by a likelihood ratio test. For the case of sibling pairs the set of formulas is:

$$LR = \Pi_{1 \leq j \leq N}(\hat{z}_0 w_{0j} + \hat{z}_1 w_{1j} + \hat{z}_2 w_{2j})/(0.25w_{0j} + 0.5w_{1j} + 0.25w_{2j}),$$

where the w terms are weights that can be set to make different tests depending on family sizes and mode of inheritance. He called the log 10 of this likelihood ratio a LOD score, but it is not generally equivalent to the model-based LOD score. Knapp et al. [37] showed that the mean test is equivalent to the model-based LOD score method, in that one can define a 1-to-1 correspondence between values of the test statistics, provided the true mode of inheritance is recessive with no phenocopies (penetrance function has values 0, 0, 1). This type of result illustrates why different test statistics on the observed proportions may be preferred for different data sets. Davis and Weeks [15] carried out computational experiments on numerous ASP software packages for realistic complex disease pedigrees.

Holmans [31] proved a triangle constraint:

$$2z_1 \leq 1, (z_1 + z_2) \leq 1, (3z_1/2 + z_2) \geq 1$$

that applies under plausible assumptions about how the disease gene contributes to the disease etiology. So, it is not model-free, but is valid under a wide range of penetrance models. Holmans showed that forcing the maximum likelihood estimates of z_1, z_2, z_3 to lie in the triangle defined by the inequalities leads to a statistically more powerful test. Whittemore and Tu [97] pushed this idea further by asking what is the test that has the best worst-case power (minmax) over all the models allowed by Holmans. This leads to a more general system of algebraic equations whose solution yields the test statistic

$$1.045(1.58 - 2.58\hat{z}_0 - 1.872\hat{z}_1)\sqrt{(N)},$$

where N is the number of affected sibling pairs.

Several different test statistics have been proposed for affected relatives in multi-generation pedigrees; I will mention only three. Weeks and Lange [94] proposed to use IBS status of affected relative genotypes as follows. Let the affected relatives in pedigree p be $A_{p1} \ldots A_{pr}$ and let the two alleles at a marker of A_{pi} be g_{pi1}, g_{pi2}. Let $a =_{IBS} b$ take on the value 1 or 0 depending on whether the alleles a and b are identical by state. Then the affected pedigree member (APM) test statistic for pedigree p is:

$$\Sigma_{A_i,A_j} \Sigma_{k=1,2} \Sigma_{l=1,2} f(g_{pik})(g_{pik} =_{IBS} g_{pjl}),$$

where f is some function of the allele frequency that penalizes more frequent alleles. Weeks and Lange [94] proposed $f(a) = P(a)^{-1/2}$, where $P(a)$ is the input allele frequency of a. The above test statistic has the advantage that for any specific loopless pedigree, one can use variants of the Elston-Stewart algorithm to determine the mean M_p, and the variance V_p for each pedigree, and these in turn can be used to determine the statistical significance of any observed value of the following test statistic for a collection of pedigrees:

$$T = \frac{\Sigma_p W_p (S_p - M_p)}{\Sigma_p W_p^2 V_p},$$

where W_p is some weight function that favors pedigrees with more affected members. The statistic T follows the standard normal distribution asymptotically, so one can assess the significance of a result by table lookup. In practice, one may not know if the number of pedigrees is large enough to use the asymptotic result, so simulation is used, as described in the next section.

Instead of using IBS status one can use IBD status and define:

$$\Sigma_{A_i, A_j} \Sigma_{k=1,2} \Sigma_{l=1,2} f(g_{pik})(g_{pik} =_{IBD} g_{pjl}),$$

When $f(a) = P(a)$, this test statistic is usually denoted $S_{p,\text{pairs}}$ because it sums the shared alleles over all pairs of affected relatives in pedigree p. The statistic $S_{p,\text{pairs}}$ can be computed using the generalized Lander-Green algorithm by noting that for each inheritance vector I, one can determine whether each pair of alleles is identical by descent; let the value of the statistic for a fixed inheritance vector I be denoted by $S_{p,\text{pairs}}(I)$. Then

$$S_{p,\text{pairs}} = \Sigma_I S_{p,\text{pairs}}(I) P(I \mid G, X).$$

Whittemore and Halpern [96] observed that in pedigrees with > 2 affected relatives, it is more indicative of linkage if subsets of size > 2 share the same allele IBD passed down from a single founder. Let sample represent any of the 2^r ways to sample 1 allele from each affected relative. Let $\text{occur}_t(\text{sample})$ denote the number of times the founder allele in position $t = 1, 2 \ldots, 2f$ occurs in sample. The statistic they proposed is

$$S_{\text{all}} = 2^{-r} \Sigma_{\text{sample}} [\Pi_{1 \leq t \leq 2f} (\text{occur}_t(\text{sample})!)].$$

As for S_{pairs}, one computes this by fixing the inheritance vector I, and summing:

$$S_{p,\text{all}} = \Sigma_I S_{p,\text{all}}(I) P(I \mid G, X).$$

For example, consider a pedigree with 3 affected relatives and a dominant inheritance pattern, so that all three affected relatives share 1 allele IBD, and the other alleles are different. Then $S_{\text{pairs}} = 0.75$ because each pair scores 0.25, while $S_{\text{all}} = 1.5$ because the selection of the 3 shared alleles scores 0.75 and each of 3 ways of selecting 2 shared alleles scores 0.25.

A limitation of S_{pairs} and S_{all} is that there is no analytical formula or algorithm to determine their asymptotic distributions. Different versions of GENEHUNTER [43, 39] use different approximations to estimate p-values. One could also use simulation as outlined in the next section.

17.10   Simulation Methods

Simulation can be used in genetic linkage analysis either to assess the statistical power to detect linkage for prospective studies or to assess the statistical significance of possible linkage findings on real data. To illustrate, I will discuss two software packages SLINK [69, 95, 12] and SimIBD [14] that rely on simulation. Another package called SIMLINK [5, 74] is similar in usage and design to SLINK.

The basic paradigm for use of simulation in linkage analysis is:

1. Generate N replicates of the input pedigree(s) filling in genotypes at random.
2. Compute some test statistic S on each replicate, such as its LOD score.
3. If N' replicates have a test statistic better than some threshold T_S, then N'/N is an estimate of the power or significance.

There are some noteworthy distinctions between power estimation and significance testing. Power estimation should be done before any true genotypes are determined in the laboratory, while significance testing is done after linkage analysis of the true data has been done. Power calculations are often done before the DNA samples are collected, and a high estimate of power can be used to justify resources needed to collect and genotype the samples. In power estimation the replicates at step 1 are filled in with the marker data *linked* to the trait locus, with some specified recombination fractions. Because of this, most power estimation is done in a model-based framework with a penetrance function for the trait locus specified and the LOD score as a test statistic. In significance testing, the replicates have the marker data *unlinked* to the trait, so testing can be done either penetrance model-based or model-free.

Because power calculations are done before the genotypes are collected and because they reflect the behavior of "average" markers, one must make some some educated guesses about the upcoming gene hunt. For example, one may represent typical markers with .75 or .80 heterozygosity by using a hypothetical marker with 4 or 5 alleles respectively, and all allele frequencies equal. Also, one has to guess how close the disease gene will be to the nearest marker or flanking markers. One can do a two-stage simulation corresponding to steps 4 and 5 in the gene-hunting procedure summarized in the Introduction. The first stage uses markers spaced as they will be in the genome scan, and a lower threshold for the test statistic. The second stage uses very densely spaced markers and a higher threshold. I typically use 1 as the LOD score threshold in the genome scan and 3 as the LOD score threshold for the fine mapping phase. If 80% of the replicates exceed the thresholds in each stage that is usually considered high enough power to proceed with confidence to collect samples and genotype them.

Suppose one wishes to test the empirical significance of an observed LOD score s. Then one can simulate N replicates using marker spacing and allele frequencies exactly as they occurred in the real data. The threshold in step 3 is the observed score s, rather than any prescribed threshold such as 3 or 3.3. For example, an observed score of 2.92 might turn out to be significant at an empirical genome-wide significance $\alpha < 0.001$ even though it is below 3. This leads to an important difference between power estimation and significance testing. In either setting, N'/N is the maximum likelihood estimate at step 3 of the probability of observing a score above T_S. However, in significance testing, one wants to be extra certain that a score above T_S is unlikely to occur. Therefore it is recommended to generate enough replicates so that the desired α is strictly above the 95% confidence interval of the probability estimate for T_S [68]. In particular, suppose $N' = 0$, that the true LOD score is above any LOD scores observed in the unlinked replicates. Then, it takes $N = 1000$ to get the estimate for $\alpha < 0.001$, but it takes $N \approx 3000$ replicates to get 0.001 above the 95% confidence interval for the estimate [68].

The data set may contain more than $p > 1$ pedigrees and one may wish to consider the possibility that there is locus heterogeneity either for power estimation or for significance testing. One can specify a parameter $0 \le \beta \le 1$ to indicate the fraction of linked replicates.

This can be used to generate a pool of replicates for each pedigree in which the fraction β are expected to be linked to the trait locus and $1 - \beta$ unlinked. Then one can sample one replicate from each pool to get a collection of p pedigrees of which an expected fraction β are linked to the true locus. In practice, one is unlikely to have a good estimate of β and generating N replicates of each pedigree for each possible value of β can be very computationally intensive. Instead one can generate much smaller pools of replicates for each individual pedigree and use bootstrapping to sample with repetition from these to make a p-pedigree compound replicate. This reduces both the computation time to generate the replicates and the time to compute the test statistic. Different sampling and bootstrapping strategies are considered in [4, 85, 84]; a practical example of bootstrapping in significance testing can be found in [34].

The basic computational question in linkage analysis simulation is how can one efficiently generate a single replicate. Consider again the nested sum and product 17.2. With subscript 1 for the outside summation, the summation over the genotypes g_1 occurs last. One can take advantage of the fact that there is flexibility in the numbering of individuals, and in particular which individual gets number 1. The pedigree likelihood can be expressed as

$$\Sigma_{g_1} \left(P(g_1 \mid X, g_2, g_3, \ldots, \theta) \right)$$

In words, it is the sum of the multilocus genotype probabilities of the individual numbered 1, conditioned on all the phenotypes, the recombination fraction(s), and the genotypes of all the other individuals in the pedigree. The genotypes may be known or unknown. By taking the marginal sum over the marker loci, this formulation can be used to make risk calculations that individual 1 has any genotype at the disease locus [52]. This option is offered in LINKAGE/FASTLINK and can be used in genetic counseling for situations where linkage has been established, but the disease gene is unknown and cannot be sequenced for mutations. The above formula can also be used in simulation to present the probability distribution for genotypes of the any individual i, and then one can sample from this distribution at random to fill in a simulated genotype for i. This observation can be made into an iterative algorithm [67]:

1. Order the individuals $1, 2 \ldots, n$. Let R_j be the set of individuals numbered $> j$. Do the next two steps for $i = n, n - 1, \ldots 1$.

2. Treat individual i as if (s)he were number 1 in the pedigree, and use the Elston-Stewart algorithm to carry out the risk calculation for:

$$P(g_i \mid X, g_{R_i}, \theta).$$

   By using g_{R_i} on the right hand side, I mean that the genotypes for whoever is available in R_i have been filled in, while the remaining marker genotypes are blank.

3. Sample a genotype for individual i from the probability distribution generated at step 2.

The above method is implemented in SIMLINK and SLINK. Some of the optimizations to make the Elston-Stewart algorithm faster can be applied, and those in the first version of FASTLINK [12] have been integrated into SLINK. However, most optimizations that depend on reducing the set of possible alleles cannot be applied easily because one wants to sample from the set of all possible alleles.

In practice, one distinguishes pedigree members who are "available" and "unavailable". For example, in the initial pedigree figures, some individuals have no genotypes, indicating that they are unavailable for sampling. When filling in genotypes, only available pedigree members get genotypes filled in, and unavailable individuals get alleles 0 0 at each marker. Usually, all available pedigree members have genotypes zeroed out and then filled in at random in simulation. However, the algorithm outlined above does allow some available pedigree members to have their genotypes fixed as the true genotypes, while other individuals have genotypes filled in differently in each replicate. The SimIBD software package that I describe next takes advantage of this flexibility.

The SimIBD package computes a model-free test statistic and then uses simulation via SLINK as a subroutine to determine the empirical p-value of the test statistic [14]. The test statistic in SimIBD is similar to that in APM, but as the name implies, it attempts to distinguish alleles that are identical by descent from those that are just identical (by state). Like APM, SimIBD works only on pedigrees with at least 2 affected individuals. Let the affected individuals be $A_1, A_2, \ldots$. Let the alleles of A_i be g_{i1} and g_{i2}. Let the specified population frequency of allele g_{ik} be $P(g_{ik})$, and let $f(g_{ik}) = (P(g_{ik}))^{-1/2}$. The SimIBD test statistic for a single pedigree is:

$$\Sigma_{A_i, A_j} \Sigma_{k=1,2} \Sigma_{l=1,2} f(g_{ik}) IBD(g_{ik}, g_{jl}),$$

where $IBD(a_1, a_2)$ is the probability that alleles a_1 and a_2 are identical by descent. This test statistic is similar to that used in APM, but APM uses the indicator function $(G_{ik} = G_{jl})$ instead of the IBD probability. The IBD function can be hard to compute, especially for looped pedigrees that are too large for the Lander-Green algorithm. SimIBD uses a recursive algorithm that is exact in most cases, and approximate in some. I will continue on the theme of approximate algorithms in the next section.

Another major difference between APM and SimIBD is how the genotypes are filled in during the simulation that determines the empirical significance of the test statistic. APM fills in random genotypes for all available individuals. SimIBD keeps the genotypes of the available unaffected individuals fixed at the true genotypes, and fills in random genotypes for the affecteds conditional on the unaffected marker genotypes. The simulation uses the trait values only to determine which individuals keep genotypes fixed or not, because it generates marker genotypes that are *unlinked* to the disease as is always done for significance testing. Keeping the genotypes of the unaffecteds fixed at the true genotypes limits the effects of skewed allele frequencies [96]. Computational experiments show that SimIBD has high power to detect linkage for complex traits [14].

17.11   Some Alternative Data Representations Leading to New Methods

In this section I present a sample of data representation ideas behind three newer software packages MERLIN [1], SIMWALK2 [82], and Superlink [23]. I chose these for two reasons. Firstly, SIMWALK2 and MERLIN are widely used and Superlink shows a lot of potential. Secondly, I wanted to illustrate by example that the use of alternative representations of patterns of inheritance and genotypes can lead to better algorithms and running times. The selections in this section reflect my taste and knowingly omit dozens of other pertinent papers and software packages. A comprehensive list of software packages related to genetic linkage analysis, genetic association analysis, and other problems on pedigrees, such as drawing and database management, can be found at (http://linkage.rockefeller.edu). Many of the entries there include references or links for downloading the software.

The formulas that summarize the Lander-Green algorithm, such as formula 17.5 require computing matrix products and/or sums over all 2^{2c} inheritance vectors. Abecasis et al. [1]

recognized that in typical data sets:

- many inheritance vectors are not consistent with available genotypes, and/or
- due to genotype symmetries or ungenotyped individuals, the likelihood of sub-pedigrees is identical regardless of whether a specific entry in an inheritance vector is 0 (allele came from grandfather) or 1 (allele came from grandmother).

Therefore, in MERLIN, they represent the set of possible inheritance vectors with a compact binary search tree (sometimes called a trie), with a special flag for symmetry between sibling nodes. The root of the tree represents the first bit in possible inheritance vectors. The second level may have two nodes representing vectors that start with 0 or 1; the third level may have four nodes representing vectors that start with 00, 01,10, 11, the fourth level may have up to 8 nodes and so on. Nodes and the subtrees they would root are omitted if any inheritance vector with the corresponding prefix is inconsistent with the available genotypes. If two prefixes $V0$ and $V1$ have the symmetry property that the likelihood summed over all vectors starting $V0$ is identical to the likelihood summed over all vectors $V1$, then the node for $V1$ is made a leaf with a flag indicating it has the same value as the node for $V0$. Symmetry can be detected by rules on the pedigree and genotypes, without actually computing the likelihoods.

As shown in [1], the sparse tree representation leads to some speedups of the Lander-Green algorithm because the number of nodes in the binary tree is often a miniscule fraction of the possible $2^{2c} + 1$. This saves memory leading to faster computation times. It also facilitates the use of sparse matrix algorithms and divide-and-conquer [32] to compute the chained matrix product 17.5.

Another feature of MERLIN is an approximation algorithm in which one requires that there are at most a small constant r recombinations between adjacent loci. This assumption is plausible when the marker map is very dense, as can happen in the use of SNP markers. As a consequence, the transition matrices T_j are necessarily sparse and the matrix chain product 17.5 can be computed more quickly.

The second alternative representation is a graph on the inheritance vectors used as a Markov chain, as implemented in the software package SIMWALK2 [82]. In that paper, the inheritance vectors are themselves represented as graphs called *pedigree descent graphs*, but this is mostly a change of notation, so I will explain the construction of a Markov chain in terms of inheritance vectors directly. The principal advantage of the Markov chain approach is that it gives a structured way to explore the possible inheritance vectors while keeping track of only feasible inheritance vectors that are encountered. Thus, SIMWALK2 goes even farther than MERLIN in reducing the fraction of 2^{2c} possible inheritance vectors about which some information is stored. Therefore, large pedigrees can be analyzed. The tradeoff is that SIMWALK2 results are approximate because they are based on a random sample.

The construction of the Markov chain M starts implicitly with a graph in which there are 2^{2c} vertices representing the inheritance vectors. Inheritance vectors that are not consistent with the available genotypes will be avoided. The next step is to define 3 classes of edges (Sobel and Lange call them transformations) 0,1,2 that define possible direct transitions in the Markov chain.

- Edges of type 0 connect two inheritance vectors if they differ in a single bit position; i.e., these correspond to changing the grandparent origin for a single allele. E.g., there would be an edge 0110—0010 if both vectors are consistent with the available genotypes.
- Edges of type 1 are defined by choosing one individual h, and connecting two

inheritance vectors if they differ exactly on every bit that corresponds to an allele inherited by a child of h from h, and agree on all other bits. This corresponds to flipping the grandparent of origin for every child of h on the allele inherited from h.

- Edges of type 2 are defined by choosing a mated couple h, w, and connecting two inheritance vectors if they differ exactly on every bit that corresponds to an allele inherited from either h or w.

The resulting graph may not be connected (making the Markov chain reducible); to make it connected, Sobel and Lange allow "tunneling edges" that are multi-edge paths composed of basic edges of types 0,1,2. To put probabilities on the Markov chain, there are the edge-type probabilities for traversing an edge of type 0,1,2, and a length probability, defined as $2^{-\ell}$, for taking a multi-edge path of length ℓ in a single Markov chain step. Only edges or multi-edge paths whose endpoints are feasible vectors are considered; but a multi-edge path may traverse through intermediate infeasible inheritance vectors.

The above Markov chain is then used in a Monte Carlo Markov chain algorithm of the proposal form. Each transition can be subdivided into three logical steps.

1. From the current state I_j, one chooses a number of edges ℓ to traverse in a single Markov chain step, and which edges they are by the above probability distributions. This leads to a new state I_j'.

2. Using either the Lander-Green algorithm as described above or a variant of the Elston-Stewart algorithm and Bayes's theorem, one can compute the conditional probabilities $P(G_j \mid I_j)$ and $P(G_j \mid I_j')$ and compare them.

3. If the $P(G_j \mid I_j')$ is higher, then the proposed new state I_j' is automatically accepted as the new current state in the Markov chain traversal. If $P(G_j \mid I_j') < P(G_j \mid I_j)$ then I_j' is chosen as the new state (by generating a random number) with probability $P(G_j \mid I_j')/P(G_j \mid I_j)$.

By allowing transitions to lower probability states, the Markov chain does not get stuck at local maxima. In practice, SIMWALK2 does not compute the new probability $P(G_j \mid I_j')$ from scratch, but uses fast probability update methods that take into account which terms changed as a result of the bit changes between I_j and I_j'.

As in any Markov chain, one can estimate the steady-state probabilities of each inheritance vectors state while doing the traversal. If $S(I_j)$ is the estimated steady-state probability of vector I_j, then the overall pedigree likelihood for the single locus j is $\Sigma_{I_j} S(I_j) P(G_j | I_j)$. SIMWALK2 extends this approach to analyze multiple loci simultaneously including determination of the most likely phase of the genotypes.

The third alternative representation I selected is *Bayesian networks*, also known as *belief networks* or *graphical models*. Computing exact or approximate pedigree likelihoods turns out to be a special case of inference in Bayesian networks. Since powerful, general techniques have been developed by researchers in that domain, representing pedigrees as Bayesian networks has proven useful in practice.

So far as I know, the first researcher to explicitly exploit the Bayesian network representation was Augustine Kong in a 1991 paper [38]. However, the general idea can be seen in a 1978 paper of Cannings et al. [10] providing an alternative formulation of the generalized (to arbitrary pedigrees) Elston-Stewart algorithm. The Bayesian network theme was used further by Jensen and Kong [33] culminating in an MCMC software package (different from SIMWALK2) that uses Gibbs sampling of sets of variables to calculate accurate approximate likelihoods for complex pedigrees. Lauritzen and Sheehan [55] lay out in some detail how Bayesian networks can be applied to genetic linkage analysis and other genetic inference problems.

Becker, Geiger, and I [3] used the Bayesian network approach to address the problem of selecting individuals to break the loops in Ott's extension [66] of the Elston-Stewart algorithm to looped pedigrees. Fishelson and Geiger [23] have pushed the Bayesian network representation much further in a package called Superlink that does exact likelihood computations for complex pedigrees and many loci. The Bayesian network representation allows Superlink to carry an algorithm that is a hybrid between the Elston-Stewart and Lander-Green algorithms, as determined by the input data. For brevity and consistency, I follow the presentation in [23], but this describes an early version, and alternative Bayesian network representations may be useful also.

In general, a Bayesian network is a graph of variables along with conditional probability tables connecting them. The variables in a pedigree likelihood problem can be:

1. the individuals and relationships as represented by the marriage graph,
2. alleles with inheritance information (Superlink uses g_{ijf} g_{ijm} to represent the alleles that individual i inherited at locus j from the father and mother),
3. the phenotypes, as above X_i is the phenotypes of individual i and X the phenotype vector,
4. inheritance vectors, as in the Lander-Green algorithm, and
5. the recombination fractions, $\theta_1, \theta_2, \ldots$ between adjacent loci.

Some of the pertinent probabilities are:

1. the probabilities of transmission from parent to child, as defined in Section 17.6,
2. the penetrance function expressing the probability of the phenotype data X conditional on the genotype data G,
3. rules for recombination that express the probability of changing from an inheritance vector at one locus I_{j-1} to an inheritance vector I_j at the next locus, conditional on the intervening θ_{j-1}, and
4. marker allele frequencies.

Using either the Elston-Stewart approach or the Lander-Green approach, one can write the pedigree likelihood as a nested sum and product of these variables and probabilities. The major virtue of the Bayesian network representation is that one evaluates the nested sum using general techniques that *adapt* to the input data. This allows run-time choices for things such as: the order of the nuclear families, the order of the loci, pruning of infeasible inheritance vectors (as in MERLIN) and infeasible alleles. The algorithms in Superlink use run-time data about the size of conditional probability tables and other intervariable dependencies to dynamically choose an order of evaluation for the pedigree likelihood. Timing data on a couple of sample pedigrees in [23] shows lots of promise. However, experience with more real data is needed to see on which types of inputs the Bayesian network methods work better than the best current implementations of the Elston-Stewart and Lander-Green algorithms.

17.12 Conclusion

I had the good fortune to stumble into the topic of genetic linkage analysis in 1992 on a dare from Robert Cottingham Jr. to speed up and parallelize the LINKAGE package. Since 1992, huge advances have been made by many researchers in genetic linkage analysis methods and software, greatly reducing the frequency with which computational analysis of pedigree data is a time bottleneck in hunting disease-causing genes. I highlighted a few of these advances as well as more fundamental research on some basic algorithms, genetic markers, and marker maps.

Genetic linkage analysis has intrinsic attractions for researchers in genetics, statistics, mathematics, computer science, and other disciplines. Linkage analysis is unusual and extrinsically attractive because it allows computational biologists to participate directly in searching for the subtle DNA differences that distinguish between life and death, and between good health and genetic disease.

Acknowledgements

I thank my past collaborators on linkage analysis methods: Richa Agarwala, Ann Becker, Jeremy Buhler, Robert Cottingham Jr., Alan Cox, Sandhya Dwarkadas, Dan Geiger, Sandeep Gupta, Ramana Idury, Marek Kimmel, Shriram Krishnamurthi, and Willy Zwaenepoel. I thank Leslie Biesecker, Soumitra Ghosh, and Robert Nussbaum for mentoring me on how human genetics research is done in the clinic, in the field, and in the biology laboratory. I thank Richa Agarwala for proofreading this chapter.

References

[1] G. R. Abecasis, S. S. Cherny, W. O. Cookson, and L. R. Cardon. Merlin–rapid analysis of dense gene maps using sparse gene flow trees. *Nature Genetics*, 30:97–101, 2002.

[2] R. Agarwala, L. G. Biesecker, K. A. Hopkins, and C. A. Francomano *et al.* Software for constructing and verifying pedigrees within large genealogies and an application to the Old Order Amish of Lancaster County. *Genome Research*, 8:211–221, 1998.

[3] A. Becker, D. Geiger, and A. A. Schäffer. Automatic selection of loop breakers for genetic linkage analysis. *Human Heredity*, 48:49–60, 1998.

[4] J. Besag and P. Clifford. Sequential Monte Carlo *p*-values. *Biometrika*, 78:301–304, 1991.

[5] M. Boehnke. Estimating the power of a proposed linkage study: A practical computer simulation approach. *American Journal of Human Genetics*, 39:513–527, 1986.

[6] V. Bonifati, P. Rizzu, M. J. van Baren, and O. Schaap *et al.* Mutations in the *DJ-1* gene associated with autosomal recessive early-onset Parkinsonism. *Science*, 299:256–259, 2003.

[7] D. Botstein, R. L. White, M. Skolnick, and R. W. Davis. Construction of a genetic linkage map in man using restriction fragment length polymorphisms. *American Journal of Human Genetics*, 32:314–331, 1980.

[8] K. W. Broman, J. C. Murray, V. C. Sheffield, and R. L. White *et al.* Comprehensive human genetic maps: Individual and sex-specific variation in recombination. *American Journal of Human Genetics*, 63:861–869, 1998.

[9] L. M. Brzustowicz, T. Lehner, L. H. Castilla, and G. K. Penchaszadeh *et al.* Genetic mapping of chronic childhood-onset spinal muscular atrophy to chromosome 5q11.2-13.3. *Nature*, 344:540–541, 1990.

[10] C. Cannings, E. A. Thompson, and M. H. Skolnick. Probability functions on complex pedigrees. *Advances in Applied Probability*, 10:26–61, 1978.

[11] J. Chotai. On the lodscore method in linkage analysis. *Annals of Human Genetics*, 48:359–378, 1984.

[12] R. W. Cottingham Jr., R. M. Idury, and A. A. Schäffer. Faster sequential genetic linkage computations. *American Journal of Human Genetics*, 53:252–263, 1993.

[13] D. Curtis. Another procedure for the preliminary ordering of loci based on two point lod scores. *Annals of Human Genetics*, 58:65–75, 1994.

[14] S. Davis, M. Schroeder, L. R. Goldin, and D. E. Weeks. Nonparametric simulation-based statistics for detecting linkage in general pedigrees. *American Journal of Human Genetics*, 58:867–880, 1996.

[15] S. Davis and D. E. Weeks. Comparison of nonparametric statistics for the detection of linkage in nuclear families: Single marker evaluation. *American Journal of Human Genetics*, 61:1431–1444, 1997.

[16] C. Dib, S. Fauré, C. Fizames, and D. Samson *et al.* A comprehensive genetic map of the human genome based on 5,264 microsatellites. *Nature*, 380:152–154, 1996.

[17] H. Donis-Keller, P. Green, C. Helms, and S. Cartinhour *et al.* A genetic linkage map of the human genome. *Cell*, 51:319–337, 1987.

[18] C. M. Duijn, M. C. Dekker, V. Bonifati, and R. J. Galjaard *et al.* *PARK7*, a novel locus for autosomal recessive early-onset Parkinsonism, on chromosome 1p36. *American Journal of Human Genetics*, 69:629–634, 2001.

[19] M. G. Ehm, M. Kimmel, and R. W. Cottingham Jr. Error detection for genetic data, using likelihood methods. *American Journal of Human Genetics*, 58:225–234, 1996.

[20] R. C. Elston and J. Stewart. A general model for the analysis of pedigree data. *Human Heredity*, 21:523–542, 1971.

[21] E. Feingold, P. O. Brown, and D. Siegmund. Gussian models for genetic linkage analysis using complete high-resolution maps of identity by descent. *American Journal of Human Genetics*, 53:234–251, 1993.

[22] R. Fishel, M. K. Lescoe, M. R. S. Rao, and N. G. Copeland *et al.* The human mutator gene homolog *MSH2* and its association with hereditary nonpolyposis colon cancer. *Cell*, 75:1027–1038, 1993.

[23] M. Fishelson and D. Geiger. Exact genetic linkage computations for general pedigrees. *Bioinformatics*, 18:S189–S198, 2002.

[24] S. E. Folstein, J. A. Phillips III, D. A. Meyers, and G. A. Chase *et al.* Huntington's disease: Two families with differing clinical features show linkage to the G8 probe. *Science*, 229:776–779, 1985.

[25] D. Gordon, S. C. Heath, and J. Ott. True pedigree errors more frequent than apparent errors for single nucleotide polymorphisms. *Human Heredity*, 49:65–70, 1999.

[26] Huntington's Disease Collaborative Research Group. A novel gene containing a trinucleotide repeat that is expanded and unstable on Huntington's disease chromosomes. *Cell*, 72:971–983, 1993.

[27] D. F. Gudbjartsson, K. Jonasson, M. L. Frigge, and A. Kong. Allegro, a new computer program for multipoint linkage analysis. *Nature Genetics*, 25:12–13, 2000.

[28] J. B. S. Haldane. The combination of linkage values and the calculation of distances between the loci of linked factors. *Journal of Genetics*, 8:299–309, 1919.

[29] J. M. Hall, M. K. Lee, B. Newman, and J. E. Morrow *et al.* Linkage of early-onset familial breast cancer to chromosome 17q21. *Science*, 250:1684–1689, 1990.

[30] C. Hitte, T. D. Lorentzen, R. Guyon, and L. Kim *et al.* Comparison of MultiMap and TSP/CONCORDE for constructing radiation hybrid maps. *Journal of Heredity*, 94:9–13, 2003.

[31] P. Holmans. Asymptotic properties of affected-sib-pair linkage analysis. *American Journal of Human Genetics*, 52:362–374, 1993.

[32] R. M. Idury and R. C. Elston. A faster and more general hidden Markov model algorithm for multipoint likelihood calculations. *Human Heredity*, 47:197–202, 1997.

[33] C. S. Jensen and A. Kong. Blocking gibbs sampling for linkage analysis in large pedigrees with many loops. *American Journal of Human Genetics*, 65:885–901, 1999.

[34] T. Kainu, S.-H. H. Juo, R. Desper, and A. A. Schäffer *et al.* Somatic deletions in hereditary breast cancers implicate 13q21 as a putative novel breast cancer susceptibility locus. *Proceedings of the National Academy of Sciences USA*, 97:9603–9608, 2000.

[35] S. Karlin and U. Liberman. Classifications and comparisons of multilocus recombination distributions. *Proceedings National Academy of Sciences USA*, 75:6332–6336, 1978.

[36] T. Kitada, S. Asakawa, N. Hattori, and H. Matsumine *et al.* Mutations in the *parkin* gene cause autosomal recessive juvenile parkinsonism. *Nature*, 392:605–608, 1998.

[37] M. Knapp, S. A. Seuchter, and M. P. Baur. Linkage analysis in nuclear families. 2: Relationship between affected sib-pair tests and lod score analysis. *Human Heredity*, 44:44–51, 1994.

[38] A. Kong. Efficient methods for computing linkage likelihoods of recessive diseases in inbred pedigrees. *Genetic Epidemiology*, 8:81–103, 1991.

[39] A. Kong and N. J. Cox. Allele-sharing models: LOD scores and accurate linkage tests. *American Journal of Human Genetics*, 61:1179–1188, 1997.

[40] A. Kong, D. F. Gudbjartsson, J. Sainz, and G. M. Jonsdottir *et al.* A high-resolution recombination map of the human genome. *Nature Genetics*, 31:241–247, 2002.

[41] D. D. Kosambi. The estimation of map distances from recombination values. *Annals of Eugenics*, 12:172–175, 1944.

[42] L. Kruglyak, M. J. Daly, and E. S. Lander. Rapid multipoint linkage analysis of recessive traits in nuclear families, including homozygosity mapping. *American Journal of Human Genetics*, 56:519–527, 1995.

[43] L. Kruglyak, M. J. Daly, M. P. Reeve-Daly, and E. S. Lander. Parameteric and nonparametric linkage analysis: A unified multipoint approach. *American Journal of Human Genetics*, 58:1347–1363, 1996.

[44] L. Kruglyak and E. S. Lander. Faster multipoint linkage analysis using Fourier transforms. *Journal of Computational Biology*, 5:1–7, 1998.

[45] E. Lander and L. Kruglyak. Genetic dissection of complex traits: guidelines for interpreting and reporting linkage results. *Nature Genetics*, 11:241–247, 1995.

[46] E. S. Lander and P. Green. Construction of multilocus genetic linkage maps in humans. *Proceedings National Academy of Sciences USA*, 84:2363–2367, 1987.

[47] E. S. Lander, P. Green, J. Abrahamson, and A. Barlow *et al.* MAPMAKER: An interactive computer package for constructing primary genetic linkage maps of experimental and natural populations. *Genomics*, 1:174–181, 1987.

[48] K. Lange and M. Boehnke. Extensions to pedigree analysis V. optimal calculation of Mendelian likelihoods. *Human Heredity*, 33:291–301, 1983.

[49] K. Lange and R. C. Elston. Extensions to pedigree analysis. I. Likelihood calculation for simple and complex pedigrees. *Human Heredity*, 25:95–105, 1975.

[50] K. Lange and T. M. Goradia. An algorithm for automatic genotype elimination. *American Journal of Human Genetics*, 40:250–256, 1987.

[51] K. Lange, D. Weeks, and M. Boehnke. Programs for pedigree analysis: MENDEL, FISHER, and dGene. *Genetic Epidemiology*, 5:471–473, 1988.

[52] G. M. Lathrop and J. M. Lalouel. Easy calculations of lod scores and genetic risks on small computers. *American Journal of Human Genetics*, 36:460–465, 1984.

[53] G. M. Lathrop and J.-M. Lalouel. Efficient computations in multilocus linkage analysis. *American Journal of Human Genetics*, 42:498–505, 1988.

[54] G. M. Lathrop, J. M. Lalouel, C. Julier, and J. Ott. Multilocus linkage analysis in humans: Detection of linkage and estimation of recombination. *American Journal of Human Genetics*, 37:482–498, 1985.

[55] S. L. Lauritzen and N. A. Sheehan. Graphical models for genetic analyses. *Statistical Science*, 18:489–514, 2003.

[56] S. Lefebvre, L. Bürglen, S. Reboullet, and O. Clermont *et al.* Identification and characterization of a spinal muscular atrophy-determining gene. *Cell*, 80:155–165, 1995.

[57] A. Lindblom, P. Tannergård, B. Werelius, and M. Nordenskjöld. Genetic mapping of a second locus predisposing to hereditary non-polyposis colon cancer. *Nature Genetics*, 5:279–282, 1993.

[58] K. Markianos, M. J. Daly, and L. Kruglyak. Efficient multipoint linkage analysis through reduction of inheritance space. *American Journal of Human Genetics*, 68:963–977, 2001.

[59] T. C. Matise, M. Perlin, and A. Chakravarti. Automated construction of genetic linkage maps using an expert system (MultiMap): A human genome linkage map. *Nature Genetics*, 6:384–390, 1994.

[60] H. Matsumine, M. Saito, S. Shimoda-Matsubayashi, and H. Tanaka *et al.* Localization of a gene for an autosomal recessive form of juvenile Parkinsonism to chromosome 6q25.2-27. *American Journal of Human Genetics*, 60:588–596, 1997.

[61] Y. Miki, J. Swensen, D. Shattuck-Eidens, and P. A. Futreal *et al.* A strong candidate for the breast and ovarian cancer susceptibility gene *BRCA1*. *Science*, 266:66–71, 1994.

[62] N. E. Morton. Sequential tests for the detection of linkage. *American Journal of Human Genetics*, 7:277–318, 1955.

[63] J. R. O'Connell. Rapid multipoint linkage analysis via inheritance vectors in the Elston-Stewart algorithm. *Human Heredity*, 51:226–240, 2001.

[64] J. R. O'Connell and D. E. Weeks. The VITESSE algorithm for rapid exact multilocus linkage analysis via genotype set-recoding and fuzzy inheritance. *Nature Genetics*, 11:402–408, 1995.

[65] J. R. O'Connell and D. E. Weeks. PedCheck: A program for identification of genotype incompatibilities in linkage analysis. *American Journal of Human Genetics*, 63:259–266, 1998.

[66] J. Ott. Estimation of the recombination fraction in human pedigrees: Efficient computation of the likelihood for human linkage studies. *American Journal of Human Genetics*, 26:588–597, 1974.

[67] J. Ott. Computer-simulation methods in human linkage analysis. *Proceedings National Academy of Sciences USA*, 86:4175–4178, 1989.

[68] J. Ott. *Analysis of Human Genetic Linkage*. The Johns Hopkins University Press, Baltimore and London, 1991. Revised edition.

[69] J. Ott. *Analysis of Human Genetic Linkage*. The Johns Hopkins University Press, Baltimore and London, 1999. Third edition.

[70] N. Papadopoulos, N. C. Nicolaides, Y.-F. Wei, and S. M. Ruben *et al.* Mutation of a *mutL* homolog in hereditary colon cancer. *Science*, 263:1625–1629, 1994.

[71] P. Peltomäki, L. A. Aaltonen, P. Sistonen, and L. Pylkkänen *et al.* Genetic mapping of a locus predisposing to human colorectal cancer. *Science*, 260:810–812, 1993.

[72] L. S. Penrose. The detection of autosomal linkage in data which consist of brothers and sisters of unspecified parentage. *Annals of Eugenics*, 6:133–138, 1935.

[73] A. Piccolboni and D. Gusfield. On the complexity of fundamental computational problems in pedigree analysis. *Journal of Computational Biology*, 10:763–773, 2003.

[74] L. M. Ploughman and M. Boehnke. Estimating the power of a proposed linkage study for a complex genetic trait. *American Journal of Human Genetics*, 44:543–551, 1989.

[75] M. H. Polymeropoulos, J. J. Higgins, L. I. Golbe, and W. G. Johnson *et al.* Mapping of a gene for Parkinson's disease to chromosome 4q21-q23. *Science*, 274:1197–1199, 1996.

[76] M. H. Polymeropoulos, C. Lavedan, E. Leroy, and S. E. Ide *et al.* Mutation in the α-synuclein gene identified in families with Parkinson's disease. *Science*, 276:2045–2047, 1997.

[77] L. Rabiner. A tutorial on hidden Markov models and selected applications in speech recognition. *Proceedings of the IEEE*, 32:257–286, 1989.

[78] N. Risch. Linkage strategies for genetically complex traits III: The effect of marker polymorphism on analysis of relative pairs. *American Journal of Human Genetics*, 46:242–253, 1990.

[79] J. M. Rommens, M. C. Iannuzzi, B. Kerem, and M. L. Drumm *et al.* Identification of the cystic fibrosis gene: Chromosome walking and jumping. *Science*, 245:1059–1065, 1989.

[80] T. G. Schulze and F. J. McMahon. Genetic association mapping at the crossroads: Which test and why? overview and practical guidelines. *American Journal of Medical Genetics*, 114:1–11, 2002.

[81] P. Sham. *Statistics in Human Genetics*. Arnold Publishers, London, 1998.

[82] E. Sobel and K. Lange. Descent graphs in pedigree analysis: Applications to haplotyping, location scores, and marker-sharing statistics. *American Journal of Human Genetics*, 58:1323–1337, 1996.

[83] E. Sobel, J. C. Papp, and K. Lange. Detection and integration of genotyping errors in statistical genetics. *American Journal of Human Genetics*, 70:496–508, 2002.

[84] K. K. Song, D. E. Weeks, E. Sobel, and E. Feingold. Efficient simulation of P values for linkage analysis. *Genetic Epidemiology*, 26:88–96, 2004.

[85] J. D. Terwilliger and J. Ott. A multisample bootstrap approach to the estimation of maximized-over-models lod score distributions. *Cytogenetics and Cell Genetics*, 59:142–144, 1992.

[86] J. D. Terwilliger and J. Ott. *Handbook of Human Genetic Linkage*. The Johns Hopkins University Press, Baltimore and London, 1994.

[87] L.-C. Tsui, M. Buchwald, D. Barker, and J. C. Braman *et al.* Cystic fibrosis locus defined by a genetically linked polymorphic DNA marker. *Science*, 230:1054–1057, 1985.

[88] E. M. Valente, P. M. Abou-Sleiman, V. Caputo, and M. M. K. Muqit *et al.* Hereditary early-onset Parkinson's disease caused by mutations in *PINK1*. *Science*, 304:11158–1160, 2004.

[89] E. M. Valente, A. R. Bentivoglio, P. H. Dixon, and A. Ferraris *et al.* Localization of a novel locus for autosomal recessive early-onset parkinsonism, *PARK6*, on human chromosome 1p35-p36. *American Journal of Human Genetics*, 68:895–900, 2001.

[90] J.L. Weber and P. E. May. Abundant class of human DNA plymorphisms which can be typed using the polymerase chain reaction. *American Journal of Human Genetics*, 44:388–396, 1989.

[91] D. E. Weeks. Invalidity of the Rao map function for three loci. *Human Heredity*, 44:178–180, 1994.

[92] D. E. Weeks, Y. P. Conley, R. E. Ferrell, and T. S. Mah *et al.* A tale of two genotypes: Consistency between two high-throughput genotyping centers. *Genome Research*, 12:430–435, 2002.

[93] D. E. Weeks and K. Lange. Preliminary ranking procedures for multilocus ordering.

Genomics, 1:236–242, 1987.

[94] D. E. Weeks and K. Lange. The affected-pedigree-member method of linkage analysis. *American Journal of Human Genetics*, 42:315–326, 1988.

[95] D. E. Weeks, J. Ott, and G. M. Lathrop. SLINK: a general simulation program for linkage analysis. *American Journal of Human Genetics*, 47:A204(abstr.), 1990.

[96] A. S. Whittemore and J. Halpern. A class of tests for linakge using affected pedigree members. *Biometrics*, 50:118–127, 1994.

[97] A. S. Whittemore and I.-P. Tu. Simple, robust linkage tests for affected sibs. *American Journal of Human Genetics*, 62:1228–1224, 1998.

[98] R. Wooster, G. Bignell, J. Lancaster, and S. Swift *et al.* Identification of the breast cancer susceptibility gene *BRCA2*. *Nature*, 378:789–793, 1995.

[99] R. Wooster, S. L. Neuhausen, J. Mangion, and Y. Quirk *et al.* Localization of a breast cancer susceptibility gene, *BRCA2*, to chromosome 13q12-13. *Science*, 265:2088–2090, 1994.

Haplotype Inference

Dan Gusfield
University of California, Davis

Steven Hecht Orzack
Fresh Pond Research Institute

A "haplotype" is a DNA sequence that has been inherited from one parent. Each human possesses two haplotypes for most regions of the genome. The most common type of variation among haplotypes possessed by individuals in a population is the single nucleotide polymorphism (SNP), in which different nucleotides (alleles) are present at a given site (locus). Almost always, there are only two alleles at a SNP site among the individuals in a population. Given the likely complexity of trait determination, it is widely assumed that the genetic basis (if any) of important traits (e.g., diseases) can be best understood by assessing the association between the occurrence of particular haplotypes and particular traits. Hence, one of the current priorities in human genomics is the development of a full *Haplotype Map* of the human genome [1, 47, 48, 17], to be used in large-scale screens of populations [16, 54]. In this endeavor, a key problem is to infer haplotype pairs and/or haplotype frequencies from genotype data, since collecting haplotype data is generally more difficult than collecting genotype data. Here, we review the haplotype inference problem (inferring pairs and inferring frequencies), the major combinatorial and statistical methods proposed to solve these two problems, and the genetic models that underlie these methods.

18.1 Introduction to Variation, SNPs, Genotypes, and Haplotypes

Now that high-throughput genomic technologies are available, the dream of assessing DNA sequence variation at the population level is becoming a reality. The processes of natural selection, mutation, recombination, gene-conversion, genome rearrangements, lateral gene transfer, admixture of populations, and random drift have mixed and remixed alleles at many loci so as to create the large variety of genotypes found in many populations. The challenge is to find those genotypes that have significant and biologically meaningful associations with important traits of interest. A key technological and computational part of this challenge is to infer "haplotype information" from "genotype information". In this section, we explain the basic biological and computational background for this "genotype to haplotype" problem.

In diploid organisms (such as humans) there are two (not completely identical) "copies" of almost all chromosomes. Sequence data from a single copy is called a haplotype, while a description of the conflated (mixed) data on the two copies is called a genotype. When assessing the genetic contribution to a trait, it may often be much more informative to have haplotype data than to have only genotype data. The underlying data that form a haplotype are either the full DNA sequence in the region, the number of repeats at microsatellite markers, or more commonly the *single nucleotide polymorphisms* (SNPs) in that region. A SNP is a single nucleotide site where more than one (usually two) nucleotides occur with a population frequency above some threshold (often around 5-10%). The SNP-based approach is the dominant one, and high-density SNP maps have been constructed across the human genome with a density of about one SNP per thousand nucleotides [48, 17].

18.1.1 The Biological Problem

In general, it is not easy to examine the two copies of a chromosome separately, and genotype data rather than haplotype data are obtained, although it is the haplotype data that may be of greater use. The data set typically consists of n genotype vectors, each of length m, where each value in the vector is either 0, 1, or 2. The variable n denotes the number of individuals in the sample, and m denotes the number of SNP sites for which one has data. Each site in the genotype vector has a value of 0 (respectively 1) if the associated site on the chromosome has state 0 (respectively 1) on both copies (it is a *homozygous* site); it has a value of 2 otherwise (the chromosome site is *heterozygous*). The goal is to extract haplotype information from the given genotype information.

A variety of methods have been developed and used to do this (e.g., [14, 15, 29, 37, 43, 59, 62, 64, 67, 70]). Some of these methods give very accurate results in some circumstances, particularly when identifying common haplotypes in a population. However, research on haplotype inference continues because no single method is considered fully adequate in all applications, the task of identifying rare haplotypes remains difficult, and the overall accuracy of present methods has not been resolved.

18.1.2 The Computational Problems

The *haplotype inference (HI)* problem can be abstractly posed as follows. Given a set of n genotype vectors, a *solution* to the HI problem is a set of n pairs of binary vectors, one pair for each genotype vector. For any genotype vector g, the associated binary vectors v_1, v_2 must both have value 0 (or 1) at any position where g has value 0 (or 1); but for any position where g has value 2, exactly one of v_1, v_2 must have value 0, while the other has value 1.

A site in g is considered "resolved" if it contains 0 or 1, and "ambiguous" if it contains a 2. If a vector g has zero ambiguous positions, it is called "resolved" or "unambiguous"; otherwise it is called "ambiguous". One can also say that the *conflation* of v_1 and v_2 produces the genotype vector g, which will be ambiguous unless v_1 and v_2 are identical. For an individual with h heterozygous sites there are 2^{h-1} possible haplotype pairs that could underlie its genotype. For example, if the observed genotype g is 0212, then one possible pair of vectors is 0110, 0011, while the other is 0111, 0010. Of course, we want to infer the pair that gave rise to the genotype of each of the n individuals.

A related problem is to estimate the *frequency* of the haplotypes in the sample. We call this the *HF* problem. It is important to note that a solution to the HI problem necessarily solves the HF problem, but the converse is not true.

18.1.3   The Need for a Genetic Model

Non-experimental haplotype inference (the HI and HF problems) would likely be inaccurate without the use of some genetic model of haplotype evolution to guide an algorithm in constructing a solution. The choice of the underlying genetic model can influence the type of algorithm used to solve the associated inference problem.

18.1.4   Two Major Approaches

There are two major approaches to solving the inference problem: combinatorial methods and population-genetic methods. Combinatorial methods often state an explicit objective function that one tries to optimize in order to obtain a solution to the inference problem. Population-genetic methods are usually based on an explicit model of haplotype evolution; the inference problem is then cast as a maximum-likelihood or a Bayesian inference problem. Combinatorial approaches are discussed in Sections 18.2 to 18.4.5, and statistical approaches are discussed in Section 18.5.

18.2   Clark's Algorithm and Other Rule-Based Methods

18.2.1   Introduction to Clark's Algorithm

Clark's algorithm to solve the HI problem [14] has been widely used and is still in use today. The algorithm starts by identifying any genotype vectors with zero or one ambiguous sites, since these vectors can be resolved in only one way. These haplotypes are called the *initial resolved* haplotypes; this method requires that some be derivable from the input vectors (sample). One attempts to resolve the remaining ambiguous genotypes by starting with the initial resolved haplotypes. Clark proposed the following rule that infers a new resolved vector NR from an ambiguous vector A and an already resolved genotype vector R.

The resolved vector R can either be one of the initial resolved haplotypes, or a haplotype inferred by an earlier application of the following **Inference Rule:**

> Suppose A is an ambiguous genotype vector with h ambiguous sites and R is a resolved vector that is a haplotype in one of the 2^{h-1} potential resolutions of vector A. Then infer that A is the conflation of one copy of resolved vector R and another (uniquely determined) resolved vector NR. All of the ambiguous positions in A are set in NR to the *opposite* of the entry in R. Once inferred,

vector NR is added to the set of known resolved vectors, and vector A is removed from the set of ambiguous vectors.

For example, if A is 0212 and R is 0110, then NR is 0011.

When the Inference Rule can be used to infer the vector NR from the vectors A and R, we say that R can be *applied* to resolve A. It is easy to determine if a resolved vector R can be applied to resolve an ambiguous vector A: R can be applied to A if and only if A and R contain identical unambiguous sites.

Clark's entire algorithm for resolving the set of genotypes is to first identify the initial resolved set, and then repeatedly apply the Inference Rule until either all the genotypes have been resolved, or no further genotypes can be resolved. There are important implementation details for Clark's algorithm that need to be specified, and one can choose different ways to do this. Several alternative variations were studied in [64] and the results of that study will be described in Section 18.2.4.

Note that in the application of the Inference Rule, for any ambiguous vector A there may be several choices for vector R, and any one choice can constrain future choices. Hence, one series of choices might resolve all the ambiguous vectors in one way, while another execution involving different choices might resolve the vectors in a different way, or leave ambiguous vectors that cannot be resolved (orphans). For example, consider two resolved vectors 0000 and 1000, and two ambiguous vectors 2200 and 1122. Vector 2200 can be resolved by applying 0000, creating the new resolved vector 1100, which can then be applied to resolve 1122. In this way, one resolves both of the ambiguous vectors and thereby produces the resolved vector set 0000, 1000, 1100 and 1111. But 2200 can also be resolved by applying 1000, creating 0100. At that point, none of the three resolved vectors, 0000, 1000 or 0100 can be applied to resolve the orphan vector 1122.

Clark's method can produce different solutions depending on how the genotype data are ordered, so the problem of choices is addressed in [14] by reordering the data multiple times and running the algorithm on each ordering. The "best" solution among these executions is reported. Of course, only a tiny fraction of all the possible data orderings can usually be tried. We refer to this as the *local inference method*.

Without additional biological insight, one cannot know which solution (or data ordering) is the most accurate. However, simulations discussed in [14] showed that the inference method tended to produce the wrong vectors only when the execution also leaves orphans. The interpretation is that there is some "global" structure to the set of real haplotype pairs that underlie the observed genotypes, so that if some early choices in the method incorrectly resolve some of the genotypes, then the method will later become stuck, unable to resolve the remaining genotypes. Clark recommended that the execution that resolved the most genotypes should be the best and the one most trusted. The efficacy of the local inference method was shown in [14] and in simulations that we have done, when the data are such that there is a unique execution that maximizes the number of resolved genotypes. However, there are also data sets where most, if not all, of the executions resolve all of the genotypes and do so differently. In that case, some other approach must be used. We discuss that situation in Section 18.2.4.

18.2.2   What is the Genetic Model in Clark's Method?

Here we give a partial justification for the Inference Rule described above.

First, note that Clark's method resolves identical genotypes identically, implying the assumption that the history leading to two identical sequences is identical. The genetic model that justifies this is the "infinite sites" model of population genetics, in which only one mutation at a given site has occurred in the history of the sampled sequences [72]. Second, the Inference Rule seems most applicable when it is assumed that the genotypes in the current population resulted from *random mating* of the parents of the current population. The sampled individuals are also drawn randomly from the population, and the sample is small compared to the size of the whole population, so the initial resolved vectors likely represent *common* haplotypes that appear with high frequency in the population.

These two assumptions are consistent with the way in which the method gives preference to resolutions involving two initially resolved haplotypes. Such haplotypes are always queried first in Clark's method. Similarly, the rule gives preference to resolutions involving one initially resolved haplotype as compared to those involving no initially resolved haplotypes. (However, a logically consistent extension of the rule would require use of two initially resolved haplotypes whenever possible, but this is not what Clark's method does).

We can define the "distance" of an inferred haplotype NR from the initial resolved vectors as the number of inferences used on the shortest path of inferences from some initial resolved vector, to vector NR. The above justification for the use of the Inference Rule becomes weaker as it is used to infer vectors with increasing distance from the initial resolved vectors. However, Clark's Inference Rule is justified in [14] by the empirical observation of consistency discussed above. For additional perspective on Clark's method see [41].

18.2.3 The Maximum Resolution Problem

Given what was observed and proposed in [14], the major open algorithmic question is whether *efficient* rules exist to break choices in the execution of Clark's algorithm, so as to maximize the number of genotypes it resolves. This leads to the **Maximum Resolution (MR) Problem** studied in [36, 37]:

> Given a set of genotypes (some ambiguous and some resolved), what execution maximizes the number of ambiguous vectors that can be resolved by successive application of Clark's Inference Rule?

An algorithm to solve the MR problem must take a more *global* view of the data than does the local inference method, in order to see how each possible application of the Inference Rule influences later choices.

We show in [37] that the MR problem is NP-hard, and in fact, Max-SNP complete. However, the MR problem was reformulated as a problem on directed graphs, with an exponential time (worst case) reduction to a graph-theoretic problem that can be solved via integer linear programming. The general idea is to encode all the possible actions of Clark's algorithm as a directed, acyclic graph. In that graph, each node represents a haplotype that could be generated in some execution of Clark's algorithm, and an edge extends from node u to node v if and only if the haplotype at u can be used to resolve some genotype in the data, resulting in the inference of the haplotype at node v. Accordingly, the MR problem can be formulated as a search problem on this graph, and solved using integer linear programming. Computations [37] showed that this approach is very efficient in practice, and that linear programming alone (without explicit reference to integrality) often suffices to solve the maximum resolution problem. However, an alternative modification of Clark's method proved more successful in obtaining more accurate resolutions. We next discuss this modification.

18.2.4   Improving Clark's Method

Computations done on the MR problem suggest that solving it is not a completely adequate way to find the most accurate solutions. One significant problem is that there are often many solutions to the MR problem, i.e., many ways to resolve all of the genotypes. Moreover, while it is clear that Clark's method should be run many times, and this can generate many different solutions, it is not clear how to use the results obtained. In fact, no published evaluations of Clark's method, except for the evaluation in [64], propose an approach to this issue, and almost all have run Clark's method only once on any given data set. This ignores the stochastic behavior of the algorithm, and these evaluations are uninformative. The critical issue in Clark's method is how to understand and exploit its stochastic behavior.

Clark's method is just one specific instantiation of the "rule-based" approach to the HI problem. In this general approach, one starts by enumerating the unambiguous haplotypes in the sample and then proceeds to use these to resolve ambiguous genotypes. However, different variations of the rule-based approach differ in how the list of reference haplotypes is formed and updated during the process of inferral and how the list of ambiguous genotypes is treated. Some of the variations are discussed in [64]; they can differ in the genetic model of haplotype evolution with which they are consistent.

In [64], we examined the performance of several variations of the rule-based method (including Clark's original method), using a set of 80 genotypes at the human APOE locus, of which 47 were ambiguous; each genotype contained nine SNP sites. The real haplotype pairs were experimentally inferred in order to assess the inferral accuracy of each variation (how many inferred haplotype pairs matched the real haplotype pairs). Most variations produced a large number of different solutions, each of which resolved *all* of the 47 ambiguous genotypes. The variability of accuracy among these solutions was substantial, and a solution chosen at random from among the solutions would likely be one with poor accuracy. Hence, an important issue in using rule-based methods, such as Clark's method, is how to exploit the many different solutions that it can produce.

How to Handle Multiple Solutions: The Consensus Approach

The multiplicity of solutions motivates an effort to understand how they can be used so as to provide a single accurate solution.

We found that the following strategy works to greatly improve the accuracy of any of the variations of the rule-based method. First, for the input genotype data, run the algorithm many times (say, 10,000), each time randomizing the order of the input data. In some variations, we also randomize the decisions that the method makes. The result is a set of solutions that may be quite different from one another. Second, select those runs that produce a solution using the *fewest or close to the fewest* number of *distinct* haplotypes; in our analysis of the APOE data, the number of such runs was typically under 100. For this set of runs, record the haplotype pair that was most commonly used to explain each genotype g. The set of such explaining haplotype pairs is called the "consensus" solution. We observed that the consensus solution had dramatically higher accuracy than the average accuracy of the 10,000 solutions. For example, for the APOE data, out of the 10,000 executions of one of the variations, there were 24 executions that used 20 or 21 distinct haplotypes; no executions used a fewer number of haplotypes. The average accuracy of the 10,000 executions was 29 correct haplotype pairs out of the 47 ambiguous genotypes, and the execution with the highest accuracy in the 10,000 had 39 correct pairs. However, the average accuracy of the 24 selected executions was 36, and the consensus solution of those 24 executions had 39 correct pairs. Hence, this simple rule allowed us to generate a single solution that was as good as the most accurate solution out of all 10,000 solutions.

In another variation, the consensus solution had 42 correct pairs, while the average of all the 10,000 solutions had 19 correct pairs. These consensus results compare well with those of other approaches. Multiple executions of the program Phase [70] always produced 42 correct resolutions, whereas the program Haplotyper [62] produced a range of solutions with most getting either 43 or 42 correct, with one solution getting 44 correct, and three solutions getting 37 correct.

We also observed that among the solutions that use the smallest and next-to-smallest number of distinct haplotypes, any haplotype pair that is used with high frequency, say above 85% of the time, was almost always correct. This allows one to home in on those pairs that can be used with high confidence.

18.3 The Pure Parsimony Criterion

18.3.1 Introduction to Pure Parsimony

A different approach to the haplotype inference problem is called the *Pure-Parsimony* approach. To our knowledge, this approach was first suggested by Earl Hubbell, who also proved that the problem of finding such solutions is NP-hard [50]. The Pure-Parsimony problem is:

> Find a solution to the haplotype inference problem that minimizes the total number of distinct haplotypes used.

For example, consider the set of genotypes: 02120, 22110, and 20120. There are solutions for this example that use six distinct haplotypes, but the solution (00100, 01110), (01110, 10110), (00100, 10110) uses only three distinct haplotypes.

Use of such a parsimony criterion is consistent with the fact that the number of distinct haplotypes observed in most natural populations is vastly smaller than the number of possible haplotypes; this is expected given the plausible assumptions that the mutation rate at each site is small and recombinations rates are low. Further, we observed in Section 18.2.4 that the most accurate rule-based solutions were those that inferred a small number of distinct haplotypes.

We note that some authors have described Clark's method [14] as relying on a parsimony criterion for the number of haplotypes used [2, 62], although there is no such criterion in the method and in fact, it rarely produces a most parsimonious solution in our experience (see Section 18.2.4). The inferral program Phase [70] has also been described as relying on a parsimony criterion [23]. However, the complex details of its computation makes it hard to quantify the influence of a parsimony criterion. This makes it difficult to use any of these methods to evaluate the effectiveness of the *parsimony criterion* as an objective function in solving the HI problem.

In [39] we described how to use integer linear programming to compute an HI solution that minimizes the number of distinct haplotypes, i.e., solving the Pure-Parsimony problem. However, the theoretical worst-case running time of this method increases exponentially with the number of genotypes, so empirical studies were undertaken to see if this approach is practical for data sets of current interest in population-scale genomics. The basic approach, combined with additional ideas presented in [39], is practical on moderately sized datasets, and this allows a comparison of the accuracy of solutions based on the Pure-Parsimony criterion and the accuracy of solutions obtained from inferral methods not so based. This is detailed in the next section.

18.3.2   A Conceptual Integer Programming Formulation

We begin by describing a *conceptual* integer-linear-programming solution to the Pure-Parsimony problem. The solution would generally be impractical to use without additional improvements. After describing this solution, we introduce two simple observations that make it practical for data sets of current biological interest.

Let g_i denote the ith genotype input vector, and suppose it has h_i ambiguous sites. There are 2^{h_i-1} pairs of haplotypes that could have generated g_i. We enumerate each one of these pairs, and create one integer programming variable $y_{i,j}$ for each of the 2^{h_i-1} pairs. As we create these y variables, we take note of the haplotypes in the enumerated pairs. Whenever a haplotype is enumerated that has not been seen before, we generate a new integer programming variable x_k for that haplotype. There will only be one x variable generated for any given haplotype, regardless of how often it is seen in a genotype.

What are the linear programming constraints? Consider the following example. For genotype $g_i = 02120$ we enumerate the two haplotype pairs (00100, 01110) and (01100, 00110), and generate the two variables $y_{i,1}$ and $y_{i,2}$ for these pairs. Assuming that these four haplotypes have not been seen before, we generate the four variables x_1, x_2, x_3, x_4 for them. We create the constraint

$$y_{i,1} + y_{i,2} = 1$$

All of the x and y variables can be set only to 0 or 1. Therefore, this inequality says that in a solution, we must select *exactly* one of the enumerated haplotype pairs as the resolution of genotype g_i. Which y variable in this constraint is set to 1 indicates which haplotype pair will be used in the explanation of genotype g_i.

Next, we create two constraints for *each* variable $y_{i,j}$. In our example, these are:

$$y_{i,1} - x_1 \leq 0$$
$$y_{i,1} - x_2 \leq 0$$
$$y_{i,2} - x_3 \leq 0$$
$$y_{i,2} - x_4 \leq 0$$

The first constraint says that if we set $y_{i,1}$ to 1, then we must also set x_1 to 1. This means that if we select the haplotype pair associated with variable $y_{i,1}$ to explain g_i, then we must use the haplotype associated with variable x_1, because that haplotype is one of the pair of haplotypes associated with variable $y_{i,1}$. The second constraint says the same thing for the haplotype associated with variable x_2.

These are the types of constraints that are included in the integer programming formulation for *each* input genotype. If a genotype has h ambiguous sites, then there will be exactly $2^h + 1$ constraints generated for it.

For the objective function, let X denote the set of all the x variables that are generated by the entire set of genotypes. Recall that there is one x variable for each distinct haplotype, no matter how many times it occurs in the enumerated pairs. Then the objective function is:

$$\text{Minimize} \sum_{x \in X} x$$

This function forces the x variables to be set so as to select the *minimum* possible number of distinct haplotypes. Taken together, the objective function and the constraints, along with the restriction that the variables can only be set to 0 or 1, specify an integer-linear-programming formulation whose solution minimizes the number of distinct haplotypes used.

Thus, this formulation solves the "Pure-Parsimony" haplotype problem. This formulation is called the "TIP formulation" in [39].

18.3.3 A More Practical Formulation

For many current data sets (50 or more individuals and 30 or more sites) the large number of constraints generated in the TIP formulation make it impractical to solve the resulting integer program. For that reason, additional ideas are required to make it practical.

The first idea is the following: if the haplotype pair for variable $y_{i,j}$ consists of two haplotypes that are both part of *no* other haplotype pair, then there is no need to include variable $y_{i,j}$ or the two x variables for the two haplotypes in the pair associated with $y_{i,j}$ in the integer program. In this way, we create a "reduced" formulation by removing such y and x variables. This formulation is called the "RTIP" formulation in [39].

The RTIP formulation correctly solves the Pure-Parsimony problem because if there is a genotype vector g such that all associated y variables are removed from the TIP formulation, then there is an optimal solution to the TIP formulation where we arbitrarily choose a permitted haplotype pair for g. Otherwise, there is an optimal solution to the TIP formulation that does not set any of the removed x or y variables to 1. Hence, there is no loss in removing them, and the RTIP formulation will find the same solution that the TIP formulation finds.

This reduced formulation is particularly effective because DNA sequences in populations have generally undergone some amount of recombination, a process that creates two chimeric sequences from two input sequences. Depending on the realized level of recombination in the evolution of the sequences, the reduced formulation can be much smaller (fewer variables and inequalities) than the original original formulation. The reason is that as the level of recombination increases, the number of distinct haplotypes in the sample typically increases and the frequency distribution of the haplotypes becomes more uniform. Therefore, more of the haplotypes appear only in one of the sampled genotypes. These "private" haplotypes are removed in the RTIP formulation. Smaller formulations generally allow integer programming solution codes to run faster.

The reduced formulation preserves the optimal solution to the original formulation. However, if the reduced formulation is created by first creating the original formulation and then removing variables and constraints, the work involved could still make this approach impractical. The following is a more efficient way to create the reduced formulation: let g_i be a genotype vector, and let H_i be the set of haplotypes that are associated with g_i in the original integer programming formulation. Then for any pair of genotypes g_i, g_j, it is easy to identify the haplotypes in $H_i \cap H_j$, and to generate them in time proportional to $m|H_i \cap H_j|$, where m is the length of the genotype vector. Simply scan g_i and g_j left to right; if a site occurs with a value of 1 in one haplotype and 0 in the other, then $H_i \cap H_j = \emptyset$; if a site occurs with a 2 in one vector and a 0 or 1 in the other, then set that 2 to be equal to the other value. Then if there are k remaining sites, where both g_i and g_j contain 2's, there are exactly 2^k distinct haplotypes in $H_i \cap H_j$, and we generate them by setting those k sites to 0 or 1 in all possible ways. The time for this enumeration is proportional to $m|H_i \cap H_j|$. Moreover, each generated haplotype in $H_i \cap H_j$ specifies a haplotype pair that will be included in the reduced formulation, for both g_i and g_j.

Any x variable that is included in the reduced formulation must occur in an intersecting set for some pair of genotypes, and every pair of haplotypes that should be associated with a y variable must also be found while examining some pair of genotypes. Hence, the reduced formulation can be produced very quickly if it is small.

18.3.4   Computational Results

Computations reported in [39] show that a Pure Parsimony solution for problem instances of current interest *can* be efficiently found in most cases. The practicality and accuracy of the reduced formulation depend on the level of recombination in the data (the more recombination, the more practical but less accurate is the method). We show in [39] that the Pure-Parsimony approach is practical for genotype data of up to 50 individuals and 30 sites. Up to moderate levels of recombination, 80 to 95 percent of the inferred haplotype pairs are correct, and the solutions are generally found in several seconds to minutes, except for the no-recombination case with 30 sites, where some solutions require a few hours.

When the recombination rate is low, Pure-Parsimony solutions were generally as accurate as those obtained with the program Phase [70]. However, they become somewhat inferior to Phase solutions when the recombination rate becomes large. Nonetheless, these Pure Parsimony results are a validation of the genetic model implicit in the Pure-Parsimony objective function, for a randomly picked solution would correctly resolve only a minuscule fraction of the genotypes. It is conceivable that a program that adds heuristics to the Pure Parsimony criterion would produce results that are competitive with programs such as Phase.

18.3.5   Further Work on Pure Parsimony

The Pure-Parsimony criterion has also been examined in [74], where a branch-and-bound method, instead of integer programming, was used to solve the problem. More theoretical results on pure parsimony appear in [41, 49, 55, 56]. No computational results on pure parsimony are reported in these papers. The first two papers also presented an integer linear programming formulation of the problem whose size grows polynomially with the size of the input. This is in contrast with the approach in [39] where (in the worst case) the size of the integer program can grow exponentially with the size of the input (although in practice, the growth is more modest).

A Polynomial-size Integer Linear Programming (ILP) Formulation for Pure Parsimony

A different polynomial-size integer linear programming formulation was developed [10] along with additional inequalities (cuts) that decrease the running time needed to solve the integer program. This formulation was also presented in [42] without the additional inequalities, and without computational results.

In this ILP formulation, for each genotype vector i, we create two binary variables (which can take on values 0 or 1 only), $y(2i-1,j)$ and $y(2i,j)$, for each site j in genotype vector i. If site j in genotype i is homozygous with state 0, then we create the constraint:

$$y(2i-1,j) + y(2i,j) = 0$$

If site j in genotype i is homozygous with state 1, then we create:

$$y(2i-1,j) + y(2i,j) = 2$$

If site j in genotype i is heterozygous, then we create:

$$y(2i-1,j) + y(2i,j) = 1$$

For any genotype vector i, the states of the variables $y(2i-1,j)$ and $y(2i,j)$, for all m sites, should define two haplotypes that explain genotype i. The above constraints ensure that any solution to the ILP creates two haplotypes that explain each genotype.

We want to minimize the number of distinct haplotypes used, and the key issue is how to set up constraints to do this. As a first step, let k and $k' < k$ be any two indices between 1 and $2n$, i.e., indices for the $2n$ haplotypes produced by a solution. We will use "haplotype k" to denote the haplotype indexed by k in the solution. For each (k', k) pair, we create the variable $d(k', k)$, which we want to be set to 1 if (but not only if) haplotype k' is different from haplotype k. This is accomplished by creating the following two constraints for each site j:

$$d(k', k) \geq y(k, j) - y(k', j)$$
$$d(k', k) \geq y(k', j) - y(k, j)$$

The variable $d(k', k)$ will be set to 1 if haplotypes k and k' are different, since they will be different if and only if they differ in at least one site j.

We next introduce the variable $x(k)$, for each k from 1 to $2n$, which we want to be set to 1 in a solution, if (but not only if) haplotype k is distinct from all of haplotypes $k' < k$. This is achieved with the following constraint:

$$[\sum_{k'=1}^{i-1} d(k', k)] - i + 2 \leq x(k)$$

To understand this constraint, note that if haplotype k is different from every haplotype $k' < k$, then $\sum_{k'=1}^{i-1} d(k', k) = i - 1$, and so

$$[\sum_{k'=1}^{i-1} d(k', k)] - i + 2$$

will equal one.

With the above constraints, a solution to this integer program specifies a pair of haplotypes that explain the genotypes, where $\sum_{k=1}^{2n} x(k)$ is greater than or equal to the number of distinct haplotypes in the solution. Therefore, by using the objective function

$$Minimize \sum_{k=1}^{2n} x(k),$$

any solution to this integer program will be a solution to the Pure Parsimony problem.

The reader can verify that the number of variables and constraints grows only polynomially with n and m, rather than exponentially (in worst case) as in the TIP and RTIP formulations.

No computation were shown in [42], but extensive computations shown in [10] compared the polynomial-size formulation with the earlier formulation in [39]. Perhaps surprisingly, the exponential-size formulation did not always run slower than the polynomial-size formulation, and there were many cases where the former formulation ran in seconds while the latter formulation took hours (although there were cases where the opposite was observed). Perhaps the reason is that smaller formulation has to computationally discover necessary features of the optimal solution (such as the candidate haplotype pairs) that are explicitly specified in the larger formulation.

Recent Contributions

More recently, a hybrid formulation that combines ideas from [39] and [10] was developed and tested in [11]. The result is an integer programming formulation that again only uses polynomial space (similar to the formulation in [10]), but whose running time in practice is closer to the running time observed with the RTIP formulation, although it is still generally slower than that formulation. The hybrid formulation allows practical computation of problem instances whose RTIP formulation is too large to fit into memory, and whose running time with the formulation from [10] is excessive.

In a somewhat different direction, an approximation algorithm was developed and tested in [49] using Semidefinite Programming Relaxation of an Integer Quadratic Programming formulation of the Pure Parsimony problem. This method was shown to compare well in both speed and accuracy with several other haplotyping methods when applied to simulated and real data sets. Other recent work on Pure-Parsimony includes a heuristic algorithm that builds a solution in a somewhat greedy manner [58].

18.4   Perfect Phylogeny Haplotyping

18.4.1   Introduction to Perfect Phylogeny Haplotyping

As noted earlier, the haplotype inference problem would be impossible to solve without some implicit or explicit genetic assumptions about how DNA sequences evolve. An important set of such assumptions are embodied in the population-genetic concept of a *coalescent* [51, 72]. A coalescent is a stochastic process that provides an evolutionary history of a set of sampled haplotypes. This history of the haplotypes is represented as a directed, acyclic graph, where the lengths of the edges represent the passage of time (in number of generations). In our problems, we ignore time, so we are only concerned with the fact that the history is represented by a directed, acyclic graph. The key observation [51] is that "In the absence of recombination, each sequence has a single ancestor in the previous generation." Hence, if we trace back the history of a single haplotype H from a given individual I, we see that haplotype H is a copy of one of the haplotypes in one of the parents of individual I. It doesn't matter that I had two parents, or that each parent had two haplotypes. The backwards history of a single haplotype in a single individual is a simple path, if there is no recombination. That means the histories of two sampled haplotypes (looking backwards in time) from two individuals merge at the most recent common ancestor of those two individuals.

There is one additional element of the basic coalescent model: the *infinite-sites* assumption (see above). This assumption is justified when the probability of mutation at any given site is small, so that the probability of two or more mutations at a given site can be taken as zero. Hence, the coalescent model of haplotype evolution says that without recombination, the true evolutionary history of $2n$ haplotypes, one from each of $2n$ individuals, can be displayed as a tree with $2n$ leaves, and where each of the m sites labels exactly one edge of the tree.

More formally, if M is a set of binary sequences, and V is a binary sequence that will label the root, the tree displaying the evolution of the haplotypes is called a *perfect phylogeny for M and V* [34, 35]. It is a rooted tree T with exactly $2n$ leaves that obeys the following properties:

1. The root of T is labeled with an m-length binary vector V, which represents the "ancestral sequence", i.e., the ancestral state of each of the m sites.

2. Each of the $2n$ rows labels exactly one leaf of T, and each leaf is labeled by one row.

3. Each of the m columns labels *exactly one* edge of T.

4. Every interior edge (one not touching a leaf) of T is labeled by *at least* one

column.

5. For any row i, the value $M(i,j)$ is unequal to $V(j)$ if and only if j labels an edge on the unique path from the root to the leaf labeled i. Hence, that path, relative to V, is a compact representation of row i.

Often we assume that V is the all-zero vector, but the above definition is more general.

An illustration of a perfect phylogeny and of its use in "association mapping" are presented in [3].

Part of the motivation for the perfect phylogeny model (i.e., coalescent without recombination) comes from recent observations [18, 73] of little or no evidence for recombination in long segments of the Human genome, and the general belief that most SNPs are the result of a mutation that has occurred only once in human history [48].

Formally, the **Perfect Phylogeny Haplotype (PPH) Problem** is:

> Given a set of genotypes, M, find a set of explaining haplotypes, M', which defines a perfect phylogeny.

In the perfect phylogeny model, each genotype vector (from a single individual in a sample of n individuals) was obtained from the mating of two of $2n$ haplotype vectors in an (unknown) coalescent (or perfect phylogeny). In other words, the coalescent with $2n$ leaves is the history of haplotypes in the $2n$ *parents* of the n individuals under study. Those $2n$ haplotypes are partitioned into pairs, each of which gives rise to one of the n genotypes.

So, given a set S of n genotype vectors, we want to find a perfect phylogeny T, and a pairing of the $2n$ leaves of T that explains S. In addition to efficiently finding one solution to the PPH problem, we would like to determine if that is the *unique* solution, and if not, we want to represent the set of *all* solutions, so that each one can be generated efficiently.

18.4.2   Algorithms and Programs for the PPH Problem

The PPH problem was introduced and first solved in [38], where it was explained that after one PPH solution is obtained, one can build an implicit representation of the set of all PPH solutions in $O(m)$ time. The algorithm given in [38] is based on reducing the PPH problem to a well-studied problem in graph theory, called the *graph-realization* problem. The theoretical running time of this initial approach is $O(nm\alpha(nm))$, where α is the inverse Ackerman function, usually taken to be a constant in practice. Hence, the worst-case time for the method is nearly linear in the size of the input, nm. The time for the reduction itself is $O(nm)$, and the graph-realization problem can be solved by several published methods. In [38] we used a graph-realization algorithm (the Bixby-Wagner algorithm) [8] in order to establish the near-linear time bound for the PPH problem. The Bixby-Wagner algorithm is based on a general algorithm due to Löfgren [60], and runs in $O(nm\alpha(nm))$ time. However, the Bixby-Wagner algorithm is difficult to understand and to implement. Accordingly, we implemented a reduction-based approach using a different solution to the graph-realization problem [31]. The resulting program (called GPPH) [13] has a worst-case running time of $O(nm^2)$. Recently, the original reduction-based approach was implemented [57] using a Java implementation of the Bixby-Wagner method [63, 52].

A second program to solve the PPH problem (called DPPH) is based on deeper insights into the combinatorial structure of the PPH problem, rather than on a reduction to the graph-realization problem. The algorithm underlying DPPH was developed in [5]. The running time for the algorithm and program is also $O(nm^2)$, and the algorithm produces a graph that represents all solutions in a simple way. Insights similar to those in [5] were presented in [77], but were not developing into an explicit algorithm for solving the PPH problem.

A third algorithm to solve the PPH problem was developed in [26], and it has been implemented in a program we call BPPH [12]. That algorithm and program also have worst-case running time of $O(nm^2)$, and they can be used to find and represent all solutions.

A Linear-Time Solution

Recently, we developed an algorithm for the PPH problem that runs in $O(nm)$ time, i.e., in *linear* time [22]. The program based on this algorithm is called LPPH. An alternative linear-time algorithm was published in [68]. The results of empirical testing of the first three programs mentioned above can be found in [12]. Some comparisons of LPPH to DPPH (the fastest of the first three) are also shown in [22]. LPPH is significantly faster than DPPH when the number of sites is large. For example, in tests with $n = 1000$ individuals and $m = 2000$ sites, DPPH ran for an average of 467 seconds, while LPPH ran for an average of 1.89 seconds. All four of the PPH programs can be obtained at wwwcsif.cs.ucdavis.edu/~gusfield/.

The conceptual and practical value of a linear-time solution can be significant. Although most current inference problems involve under one hundred sites, where the differences in running time between the programs are not of great practical significance, there are regions of the human genome up to several hundred kilobases long where the SNP states are highly correlated. Such high correlation is called "linkage disequilibrium" (LD), and high LD suggests that little or no recombination has occurred in those regions. Further, there is very little known about haplotype structure in populations of most organisms, so it is too early to know the full range of *direct* application of this algorithm to PPH problems involving long sequences (see [12] for a more complete discussion).

Faster algorithms are of practical value when the PPH problem is repeatedly solved in the inner-loop of an algorithm. This occurs in the inference of haplotype pairs affected by recombination [71], and when searching for recombination hotspots and low-recombination blocks [77]. In both of these cases one finds for every SNP site the longest interval starting at that site for which there is a PPH solution. When applied on a genomic scale (as is anticipated), even a ten-fold increase in speed is important. Moreover, in some applications, one may need to examine *subsets* of the given SNP sites for which there is a PPH solution. This is partly due to small departures from the perfect phylogeny model. It is also motivated by observations of subsets of SNP sites with high pairwise LD, where the sites in the subset are not contiguous in the SNP haplotype, but are are interlaced with other SNP sites which are not in high LD with sites in the subset. Such a subset of SNP sites is called a *dispersed haplotype block*. The lengths of these dispersed haplotype-blocks are not known. When solving the PPH problem repeatedly on a large number of subsets of sites, increased efficiency in the inner loop will be important, even if each subset is relatively small.

The High Level Idea Behind the Linear-Time Solution

In obtaining the linear-time solution [22], we used the general method of Löfgren, but we exploited properties of the PPH problem to obtain a specialized version that is simpler to implement than the Bixby-Wagner graph-realization method. Although there is no explicit mention of the graph-realization problem in [22], in order to develop the intuition behind the method, it is useful to review a bit of the Whitney-Löfgren theory of graph-realization, specialized to the PPH problem.

Let M be an instance of the PPH problem, and let T be a perfect phylogeny that solves the PPH problem for M. Each leaf of T is labeled by one row of M, and each row of M labels *two* distinct leaves of T. We define a *three-partition* of the *edges* of T to be a partition of the edges of T into three connected, directed subtrees T_1, T_2, T_3 of T, such that T_1 is rooted at the root of T, and T_2 and T_3 are rooted at distinct nodes, u, v (respectively) in T_1. Note T_1 might consist only of the root node of T; also note that the *nodes* of T are not partitioned between the three subtrees, and that either, but not both, of u or v might be the root of T. A *three-partition* is *legal* if the set of labels of the leaves in T_2 is identical to the set of labels of the leaves in T_3. Given a legal three-partition, we define a *legal flip* in T of T_2 and T_3 as the following operation: disconnect trees T_2 and T_3 from T, and merge node u in T_2 with node v in T_1, and merge node v in T_3 with node u in T_1.

The application of Whitney's theorem [76] to the PPH problem implies that *every* PPH solution for M can be obtained by a series of legal flips, starting from any PPH solution T, and *every* tree T' created in the series is also a PPH solution for M. Moreover, the number of needed flips is bounded by the number of edges of T. This theorem is the basis for Löfgren's graph-realization algorithm, and the version of the theorem above specializes to a version of Löfgren's algorithm that solves the PPH problem. We will describe this approach for the case when M contains only entries that are 0 or 2. The effect of having no 1 entries is that for every row i, and every PPH solution T, the path in T between the two leaves labeled i must go through the root of T.

We now describe at a high level the approach to solving the PPH problem based on Löfgren's general method for solving the graph-realization problem. Let $T(k)$ be a solution to the PPH problem restricted to the first k rows of M. Let $OLD(k+1)$ be the set of sites that have value 2 in row $k+1$ and have value 2 in some row 1 through k. Each site in $OLD(k+1)$ is an "old" site, and already labels an edge in $T(k)$. Let $N(k+1)$ be the remaining set of sites in row $k+1$, i.e., the "new" sites. Let $M'(k+1)$ be the matrix made up of the first k rows of M, together with a new row created from row $k+1$ by setting to 0 all the entries in $N(k+1)$. Whitney's theorem implies that if there is a PPH solution for $M'(k+1)$, then a PPH solution for $M'(k+1)$ can be obtained by a series of legal flips (each relative to a three-partition) starting from $T(k)$. Löfgren's algorithm finds such a solution $T'(k+1)$ for $M'(k+1)$ by finding an appropriate series of flips. Moreover, there is a series of flips that can find a particular solution $T'(k+1)$, so that the sites in $N(k+1)$ can be added in a single path, at the end of one of the two paths in $T'(k+1)$ that contain the sites of $OLD(k+1)$.

Additional ideas and appropriate data structures are needed to make this approach efficient. A key idea is that when finding a solution $T'(k+1)$, we never allow a flip that is forced to be done later in the opposite direction, and so all the edges incident with the nodes u and v (defined in the three-partition) can be "fixed", thus specifying more of the PPH solution for M (if there is a solution). At each point in the execution of the algorithm, a data structure called a "shadow tree" implicitly represents all possible solutions to the problem seen so far. As the algorithm proceeds, more of the solution becomes fixed, and the shadow tree at the end of the algorithm represents all solutions to the PPH problem.

18.4.3 Uniqueness of the Solution: A Phase Transition

For any given set of genotypes, it is possible that there will be more than one PPH solution. How many individuals should be in the sample so that the solution is very likely to be unique? To answer this question, we did computations that determine the frequency of a unique PPH solution for various numbers of sites and of genotypes [12]. Intuitively, as the ratio of genotypes to sites increases, one expects that the frequency of unique solutions should increase. This was observed, as was a striking *phase transition* in the frequency of unique solutions as the number of individuals grows. In particular, the frequency of unique solutions is close to zero for a small number of individuals, and then jumps to over 90%

with the addition of just a few more individuals. In our computations, the phase transition occurs when the number of individuals is around twenty-five. The phase transition was also found in computations done by T. Barzuza and I. Pe'er [7], although they observed the transition with somewhat fewer individuals than in our computations. These results have positive practical implications, since they indicate that a surprisingly small number of individuals is needed before a unique solution is likely.

18.4.4   Related Models, Results, and Algorithms

The PPH problem has become well-known (see the surveys [9, 40, 41, 42]), and there is now a growing literature on extensions, modifications, and specializations of the original PPH problem [4, 6, 19, 20, 27, 26, 43, 46, 53] and on the PPH problem when the data or solutions are assumed to have some special form [32, 33, 44]. Some of these papers give methods that run in linear time, but only work for special cases of the PPH problem [32, 33], or are only correct with high probability [19, 20]. Some of the papers discuss the problems associated with incomplete or incorrect data, some develop complexity results that limit the extent that one can expect to obtain polynomial-time methods, and some consider different biological contexts that change some of the details of the problem. We will now discuss some of these results.

Papers by [32, 53] showed that the the PPH problem is NP-complete when the data are incomplete. It was established in [4] that the problem of finding a PPH solution that *minimizes* the number of distinct haplotypes it uses is NP-hard. It was also established there that the $O(nm^2)$-time solutions to the PPH problem in [5, 26] are unlikely to be implementable in $O(nm)$ time, even though the same paper shows that if either method could be implemented in $O(nm + m^2)$ time, then the algorithm could be implemented in $O(nm)$ time. The PPH solution in [5] runs in $O(nm)$ time, except for an initial computation that runs in $O(nm^2)$ time but only produces output of size $O(m^2)$. So it seemed attractive to see if that initial computation could be implemented to run in $O(m^2)$ time. The method in [26] contains the same initial computation, and although no explicit algorithm is presented in [77], the ideas there are based on this initial computation. However, we showed in [4] that the initial computational task is equivalent to boolean matrix multiplication. That implies that if the computation could be implemented to run in $O(nm)$ time, then two n by n boolean matrices could be multiplied in $O(n^2)$ time, which is significantly faster than is currently possible.

He and Zelikovsky [46] used linear algebra to find redundancies that can be removed to reduce the number of sites in an instance of the HI problem. This approach is not guaranteed to preserve the set of solutions, but the typical loss of accuracy can vary depending on which specific haplotyping method is used. When tested along with the program DPPH (solving the PPH problem), this approach resulting in little loss of accuracy and a large increase in speed.

In contrast to papers that focus primarily on algorithmic issues related to the PPH problem, several papers discuss variants of the original PPH problem that arise in different biological contexts. The papers [32, 33] considered the PPH problem where the input is assumed to have a row where all the entries have value two. That is, there must be a pair of haplotypes in the solution in which every site is heterozygous. Such a pair is called a "yin-yang" haplotype; they are found in many populations [79]. Hence, in any solution T to a PPH problem of this type, there must be two paths from the root of T that contain all of the sites. Note that there may be rows that are not all-2, but since a solution T must have two directed paths from the root that contain all of the sites, any other haplotype in the solution must be defined by a path in T that forms some initial portion of one of those two paths. Thus, this variant of the PPH problem is called the "Perfect Phylogeny Path Haplotyping" (PPPH) problem. The method in [32] is simple and runs in linear time. The PPPH problem may be related to the classical "consecutive ones" problem. Part of the intuition for this is that the graph-realization problem, which can be viewed as the basis for the PPH problem, is a generalization of the consecutive-ones problem, where the "ones" have to form consecutive intervals in a tree instead of on the line. But the PPPH problem is the PPH problem when restricted to a single path, which can be embedded on a line.

The XOR PPH problem

Suppose that the genotype vector for an individual indicates whether a site is heterozygous or homozygous, but does not indicate the specific state of a homozygous site. Genotype vectors of this type may be cheaper and easier to obtain than those that indicate the specific state at every homozygous site. Such a genotype vector is the XOR (exclusive OR) of the two (0-1)-haplotype vectors [6].

Given a set of n such XOR genotypes, the XOR PPH problem is to find a perfect phylogeny T with $2n$ leaves, and a pairing of the leaf sequences of T, so that the n input genotype vectors result from taking the XOR of each of the n paired leaf sequences. The main result in [6] is that this problem can also be reduced to an instance of the graph-realization problem, as in the original PPH problem, and hence can be solved in $O(\alpha(nm)nm)$ time in theory. As in the PPH problem, initial implementations were based on using a slower and simpler solution to the graph-realization problem, resulting in an $O(nm^2)$-time algorithm for the XOR PPH problem. Just as in the original PPH problem, it is important to assess how many individuals are needed in the sample in order to find a PPH solution (if there is one) that is likely to be unique. Computations were done in [6] to compare the number of needed individuals in the two PPH formulations, and the result is that a high probability of uniqueness is obtained for XOR genotype input using only a few more individuals than with the full PPH genotype input.

There is another interesting and potentially important result in [6] concerning the PPH model and the so-called Tag SNPs in "haplotype blocks". Several recent studies have found long regions in human DNA, called haplotype blocks, where high LD is observed (see [73] for a review). There are other definitions of haplotype blocks that are not explicitly based on LD, such as defining a block to be a region of sufficient length for which only a small number of haplotypes are found among most individuals. (Different instantiations of the words "sufficient", "most", and "few" lead to different precise block definitions and to different methods to recognize blocks or to partition a sequence into blocks.) No matter what the causal basis is for haplotype blocks, they can be exploited to make large-scale genotyping more practical. The high association between the states of SNP sites inside a single haplotype block makes it possible to identify a small number of "Tag-SNP" sites in a set of haplotypes, whose states act as a label for all (or most) of the haplotypes in the sample. In other words, the (0-1) states of the Tag-SNPs for an individual allow one to determine the states of the other SNP sites for that individual. Given the haplotypes in a sample, a smallest set of Tag-SNPs can be found by solving an instance of a "minimal test-set problem", which is easily framed as a set-cover problem. The minimal test-set problem is NP-hard, but is solvable in practice for current data sets (up to a few hundred individuals and one hundred sites). The advantage of knowing a small set of Tag-SNPs is clear: if the haplotypes in yet unstudied individuals are like those in the sampled individuals, one would need to look only at the Tag-SNPs to infer the haplotype pairs of the unstudied individuals.

The definition of a Tag-SNP has been for haplotypes, but it is *genotypes* that will be determined in large screens. Can one find a subset of sites in the genotypes of the sample, so that for any individual in the sample, the values at those genotypes determine the two underlying haplotypes of the individual? One can define and identify "Tag genotype SNPs" as a set of sites that determine the genotype values at the other sites in the sample. But is a set S of Tag genotype SNPs also a set of Tag-SNPs for that underlying haplotype solution in the sample? If so, by knowing the genotype values at S, one would know all the genotype values for the individual, and *also* know the underlying haplotype pair for that individual. It is shown in [6] that a set of Tag genotype SNPs is *not* always a set of Tag haplotype SNPs, but when the data have a PPH solution (or a XOR PPH solution if only XOR genotypes are known) a set of Tag genotype SNPs is also a set of Tag haplotype SNPs. In this case, genotype data in large screens are as useful as haplotype data. This is potentially a very important result.

18.4.5   Near-Perfect Phylogeny

One modification of the PPH problem, called the "imperfect" or "near-perfect" or "almost-perfect" phylogeny haplotyping problem deserves particular attention. This approach was developed in three papers by E. Eskin, E. Halperin, and R.M. Karp [25, 26, 43], and is implemented in a program called HAP [43]. HAP was recently used to infer haplotype pairs and to predict haplotype-blocks, in the largest-yet published study of the patterns of SNP variation in human populations [48].

The main motivation for the near-perfect phylogeny model is the observation that in certain well-studied data sets (e.g., [18]), the common haplotypes (the ones most frequently seen in the sample) fit the perfect phylogeny model, but the full set of haplotypes in the sample do not. Halperin and Eskin [43] stated that "infrequent haplotypes cause the majority of the conflicts with the perfect phylogeny model". They derived this conclusion from studying the data in [18], where very few conflicts remain after the removal of haplotypes that occur in fewer than 5% of the sample, and no conflicts remain after the removal of haplotypes that occur in fewer than 10% of the sample. Thus, the haplotypes fit the perfect-phylogeny model after modifications are made to the data, and are said to fit a "near-perfect-" or "almost-perfect-" phylogeny.

HAP uses this observation to perform haplotype inference in non-overlapping fixed-length intervals of contiguous SNP sites. In each such interval, the program finds a subset of the genotypes (possibly all of them) for which there is a solution to the HI problem that fits a perfect phylogeny. These haplotypes are expected to be the common haplotypes in the population. It then uses these haplotypes to infer haplotype pairs for the remaining genotypes, which may include genotypes initially removed due to missing data. Missing values in a genotype are inferred from the haplotypes in the PPH solution by a maximum-likelihood approach.

We now discuss how HAP finds haplotype pairs for the common haplotypes in a fixed interval. The program derives from an algorithm [25] that solves the original PPH problem. The specific ways that HAP modifies that algorithm and the ways that it modifies the data as the algorithm proceeds have not been completely explained in [43] (HAP contains well over 10,000 lines of code). But the general ideas have been articulated as follows [43]. The PPH algorithm in [25] builds a PPH solution from the root of the tree downward, examining each row of the input in turn. When examining a row, it may learn that in all PPH solutions, a specific pair of sites must be together on a directed path from the root, or it may learn the opposite, that they cannot be together on any directed path from the root. Alternatively, the examination of a row may indicate that there is no PPH solution possible. HAP follows the general outline of this algorithm, but instead of acting under the influence of a single row, it looks at additional rows to see how many rows support the same conclusion (for example, that the edge containing one of the sites should be made an ancestor of the edge containing the other). If only a small number of rows support that conclusion, and the other rows support an alternative conclusion, then the minority action is not taken and the (few) rows supporting the action can be removed (this detail is not explicitly stated in [43]). In this way, a perfect phylogeny is created for a subset of the rows. One of the features of the algorithm in [25] (and other algorithms for the PPH problem) is that it produces an implicit representation of the set of all PPH solutions. In HAP, that representation is used to enumerate all the PPH solutions (for the rows not removed) in order to choose one that best conforms to the assumption that the observed genotypes were created by random mating of the inferred haplotypes, and therefore fits the Hardy-Weinberg equilibrium.

In addition to the assumption of random mating, the implicit model of haplotype evolution embodied in HAP is that the rare haplotypes are created by recent recombinations of a few common haplotypes, or by recent mutations of common haplotypes. That view of haplotype evolution is articulated in [24, 66]. The near-perfect phylogeny model goes one step further, by asserting that the common haplotypes fit a perfect-phylogeny model.

We consider the observation in [43] that well-studied data nearly fit the perfect phylogeny model to be a validation of the original PPH idea. The strict PPH model may be overly-idealized, or too brittle to handle errors in real data, but it is valuable to have a precisely-specified model that leads to an efficiently-solved computational problem that can be used in the core of other more heuristic programs (such as HAP) to handle more complex (and messier) real-world data. An initial effort to more formally specify a model of imperfect phylogeny haplotyping, and to solve the associated computational problems, was recently published in [71].

18.5   Population-Genetic and Statistical Methods

Much of the work described above has been carried out by computer scientists and/or with the methods used by computer scientists. A number of other important approaches to the problem of haplotype inference have originated mainly in the research community of population geneticists and statisticians interested in the assessment and analysis of genetic variation.

Of central historical and scientific note in this regard is the use of maximum likelihood to estimate haplotype frequencies and to infer haplotype pairs. It is straightforward to write down the likelihood function associated with any given sample of individuals if one makes an assumption about the process by which mating occurs within the population. The standard and usually reasonable assumption is that there is a process of random mating among individuals. Given this assumption, one can then derive an explicit likelihood function and the goal is to determine its maximum value [75]. This value will yield estimates of the haplotype frequencies underlying the observed genotype frequencies. Given the haplotype frequencies, one can then determine the most probable pair of haplotypes that underlies any given ambiguous genotype. The main problem in practice then is the derivation of the maximum value of the likelihood function.

For two SNP sites, one can analytically derive the maximum value [65]. This fact has long been known but unfortunately, this approach has almost never been exploited as a means of solution for this important case (exceptions are [21, 78]). Instead, for this case and for the m-site case, the method of expectation-maximization (EM) has been used [28, 45, 61, 67]. The use of this numerical approach to determine the maximum value of the likelihood function has both positive and negative consequences.

On the one hand, it allows the estimation of haplotype frequencies for data sets for which there are few, if any, alternative estimation approaches available. The importance of this can hardly be overestimated. On the other hand, a numerical method such as EM yields only limited information about the overall "dimensionality" of the estimation problem. So, for example, typical use of the EM algorithm does not reveal whether there are multiple peaks on the likelihood surface or whether the surface around a peak is flat or steep. Such information has obvious important implications for the confidence one has in any given result of the application of EM. Examples are shown in Orzack *et al.* [65] of two-site genotype data sets for which the EM algorithm yields misleading results.

In the case of two sites, we recommend use of the analytical method presented in [65]. For data sets with more than two sites, the EM approach is certainly one that should be considered, if one uses it in such a way that the topography of the likelihood surface is at least partially described. The simple way of doing this is by starting the algorithm with many different initial estimates of the unknown haplotype frequencies. If all such different estimates result in the same value of the likelihood function then one can have more confidence that at least there is just one peak. It is not clear with this (or any other method, see below) how to proceed if one has multiple peaks, but at least knowing of their existence is better than proceeding in ignorance.

The likelihood approach can be viewed as a special case of a Bayesian-inference problem. In this case, one has a flat prior, implying that no one prior estimate of haplotype frequencies is better than any other. In the last five years, a number of alternative approaches have been developed in which more informative prior information is incorporated into the estimation procedure, so as to get better estimates of the unknown haplotype frequencies and/or haplotype pairs. Of note in this regard are the programs called Phase [70] and Haplotyper [62]. In the case of Phase, the Bayesian-prior is derived from the infinite-sites model and to this extent, it is a plausible prior to use for many (but not all) data sets. The Gibbs sampler is then used to calculate the posterior distribution from which one can derive estimates of haplotype frequencies and infer haplotype pairs. Further elaboration and discussion of this approach can be found in [59, 69].

In contrast, the other Bayesian method, Haplotyper, uses a prior that is not derived from an explicit population-genetic model; it is consistent with a model of inheritance in which sequences of parents and offspring are potentially independent of one another [69]. The Dirichlet distribution is used here as the sampling distribution. This prior is clearly less biologically meaningful than the prior used in Phase. The Gibbs sampler is also used to calculate the posterior distribution.

What all of these methods have in common is the use of a numerical method to derive estimates of haplotype frequencies and predictions of haplotype pairs. In addition, all of these calculations are stochastic in the sense that one must start each execution with different initial haplotype frequencies (in the cases of Phase and Haplotyper) or should do so (in the case of EM). To this extent, the concern is that the reliability of results derived from any one execution is uncertain. This problem has been recognized by the creators of some of these programs (e.g, [70]) but the resulting implications for how these programs should be used have not been adequately explored. So, typical analyses based on these methods have involved a single execution of a program. How meaningful the associated results are is very unclear; at the very least, it is easy to find genotypic configurations for which different executions of any given method can result in very different estimates of haplotype frequencies (see [65]).

The main problem here is not the discovery of alternative solutions for any given data set. The only requirement is sufficient computing power. Several thousand executions of even the most time-intensive methods that are presently available can be achieved in a few days on the fastest available microprocessors. Accordingly, the central and unresolved problem is what one can make of the possible multiple solutions that one may find for any given data set. One possibility is the use of consensus, just as it was applied in the analysis of multiple solutions stemming from the rule-based algorithms such as Clark's method (as described in Section 18.2.4). Consensus has also been applied in [30]. It is clear that there is great promise to the consensus approach and it may prove widely useful. However, this remains to be seen and additional applications and theoretical analysis are needed. Of course, in any given instance, multiple executions of an inference program may result in the same solution. Such was the case for several thousand Phase analyses of the APOE data set described above (e.g., [64]). However, how typical such monomorphism is remains unknown. Given present evidence and the black-box nature of present stochastic inference methods, we strongly caution against an expectation that the results of any method should be regarded as "true" even if multiple executions result in the same solution. Such confidence can come only from improved understanding of the performance of the algorithms and especially from analyses in which the accuracy of any given method is assessed by a comparison of inferred and real haplotype pairs.

18.6 Going Forward

Our hope is that this review provides a meaningful and stimulating assessment of the present state of the biologically important problem of haplotype inference. While important progress has been made, it is clear that there are substantial questions and issues that remain to be resolved. We hope and expect that further progress will come from the separate and combined efforts of biologists, computer scientists, and statisticians. The interdisciplinary nature of this research effort is testimony to the remarkable state of present research in bioinformatics.

Acknowledgments

Research partially supported by award EIA-0220154 from the National Science Foundation and award P01-AG0225000-01 from the National Institute of Aging.

References

[1] www.hapmap.org.

[2] R. M. Adkins. Comparison of the accuracy of methods of computational haplotype inference using a large empirical dataset. *BMC Genetics*, 5(22), 2004.

[3] D. Altshuler and A. Clark. Harvesting medical information from the Human family tree. *Science*, 307:1052–1053, 2005.

[4] V. Bafna, D. Gusfield, S. Hannenhalli, and S. Yooseph. A note on efficient computation of haplotypes via perfect phylogeny. *Journal of Computational Biology*, 11(5):858-866, 2004.

[5] V. Bafna, D. Gusfield, G. Lancia, and S. Yooseph. Haplotyping as perfect phylogeny: A direct approach. *Journal of Computational Biology*, 10:323–340, 2003.

[6] T. Barzuza, J.S. Beckman, R. Shamir, and I. Pe'er. Computational Problems in perfect phylogeny haplotyping: XOR genotypes and TAG SNPs In *Thirteenth Annual Symposium on Combinatorial Pattern Matching (CPM'04)*, p. 14-31, 2004.

[7] T. Barzuza and I. Pe'er. personal communication.

[8] R. E. Bixby and D. K. Wagner. An almost linear-time algorithm for graph realization. *Mathematics of Operations Research*, 13:99–123, 1988.

[9] P. Bonizzoni, G. Della Vedova, R. Dondi, and J. Li. The haplotyping problem: Models and solutions. *Journal of Computer Science and Technology*, 18:675–688, 2003.

[10] D. Brown and I. Harrower. A new integer programming formulation for the pure parsimony problem in haplotype analysis. *Proceedings of the 2004 Workshop on Algorithms in Bioinformatics*, Springer Lecture Notes in Bioinformatics, LNCS, Vol. 3240 p. 254-265.

[11] D. Brown and I. Harrower. A new formulation for haplotype inference by pure parsimony. Technical report, University of Waterloo, School of Computer Science. Report CS-2005-03.

[12] R.H. Chung and D. Gusfield. Empirical exploration of perfect phylogeny haplotyping and haplotypers. *Proceedings of the 9th International Conference on Computing and Combinatorics COCOON03*. Springer Lecture Notes in Computer Science, Vol. 2697, pages 5–19, 2003.

[13] R.H. Chung and D. Gusfield. Perfect phylogeny haplotyper: Haplotype inferral using a tree model. *Bioinformatics*, 19(6):780–781, 2003.

[14] A. Clark. Inference of haplotypes from PCR-amplified samples of diploid populations. *Molecular Biology and Evolution*, 7:111–122, 1990.

[15] A. Clark, K. Weiss, D. Nickerson, and S. Taylor *et al*. Haplotype structure and population genetic inferences from nucleotide sequence variation in human lipoprotein lipase. *American Journal of Human Genetics*, 63:595–612, 1998.

[16] A. G. Clark. Finding genes underlying risk of complex disease by linkage disequilibrium mapping. *Current Opinion in Genetics & Development*, 13:296–302, 2003.

[17] International HapMap Consortium. HapMap project. *Nature*, 426:789–796, 2003.

[18] M. Daly, J. Rioux, S. Schaffner and T. Hudson *et al*. High-resolution haplotype structure in the human genome. *Nature Genetics*, 29:229–232, 2001.

[19] P. Damaschke. Fast perfect phylogeny haplotype inference. *14th Symposium on Fundamentals of Computation Theory FCT*, 2751:183–194, 2003.

[20] P. Damaschke. Incremental haplotype inference, phylogeny and almost bipartite graphs. *2nd RECOMB Satellite Workshop on Computational Methods for SNPs and Haplotypes*, pages 1–11, 2004.

[21] I. De Vivo, G. S. Huggins, S. E. Hankinson, and P. J. Lescault *et al*. A functional polymorphism in the promoter of the progesterone receptor gene associated with endometrial cancer risk. *Proceedings of the National Academy of Science (USA)*, 99:12263–12268, 2002.

[22] Z. Ding, V. Filkov, and D. Gusfield. A linear-time algorithm for the perfect phylogeny haplotyping problem. *Proceedings of the Ninth Annual International Conference on Computational Biology (RECOMB 2005)*. S. Miyano, J. Mesirov, S. Kasif, and S. Istrail *et al*. (eds). Springer Lecture Notes in Bioinformatics, LNCS Vol. 3500 p. 585-600.

[23] P. Donnelly. Comments made in a lecture given at the DIMACS conference on Computational Methods for SNPs and Haplotype Inference, November 2002.

[24] N. El-Mabrouk and D. Labuda. Haplotype histories as pathways of recombinations. *Bioinformatics*, 20:1836–1841, 2004.

[25] E. Eskin, E. Halperin, and R. M. Karp. Large scale reconstruction of haplotypes from genotype data. *Proceedings of the 7th Annual International Conference on Computational Biology (RECOMB 2003)*. p. 104-113

[26] E. Eskin, E. Halperin, and R. M. Karp. Efficient reconstruction of haplotype structure

via perfect phylogeny. *Journal of Bioinformatics and Computational Biology*, 1:1–20, 2003.

[27] E. Eskin, E. Halperin, and R. Sharan. Optimally phasing long genomic regions using local haplotype predictions. In *Proceedings of the Second RECOMB Satellite Workshop on Computational Methods for SNPs and Haplotypes*, February 2004. Pittsburgh, USA.

[28] L. Excoffier and M. Slatkin. Maximum-likelihood estimation of molecular haplotype frequencies in a diploid population. *Molecular Biology and Evolution*, 12:921–927, 1995.

[29] S. M. Fullerton, A. Clark, C. Sing and D.A. Nickerson *et al.* Apolipoprotein E variation at the sequence haplotype level: implications for the origin and maintenance of a major human polymorphism. *American Journal of Human Genetics*, 67:881–900, 2000.

[30] S. M. Fullerton, A. V. Buchanan, V. A. Sonpar and S. L. Taylor *et al.* The effects of scale: variation in the apoa1/c3/a4/a5 gene cluster. *Human Genetics*, 115:36–56, 2004.

[31] F. Gavril and R. Tamari. An algorithm for constructing edge-trees from hypergraphs. *Networks*, 13:377–388, 1983.

[32] J. Gramm, T. Nierhoff, R. Sharan, and T. Tantau. On the complexity of haplotyping via perfect phylogeny. In *Second RECOMB Satellite Workshop on Computational Methods for SNPs and Haplotypes*, Pittsburgh, USA, February 2004. In Springer Lecture Notes in Bioinformatics, LNCS 2004.

[33] J. Gramm, T. Nierhoff, and T. Tantau. Perfect path phylogeny haplotyping with missing data is fixed-parameter tractable. In *First International Workshop on Parameterized and Exact Computation (IWPEC 2004)*, Bergen, Norway, September 2004. In Springer LNCS 3162, 174-186.

[34] D. Gusfield. Efficient algorithms for inferring evolutionary history. *Networks*, 21:19–28, 1991.

[35] D. Gusfield. *Algorithms on Strings, Trees and Sequences: Computer Science and Computational Biology*. Cambridge University Press, Cambridge, UK, 1997.

[36] D. Gusfield. A practical algorithm for deducing haplotypes in diploid populations. In *Proceedings of 8th International Conference on Intelligent Systems in Molecular Biology*, pages 183–189. AAAI Press, 2000.

[37] D. Gusfield. Inference of haplotypes from samples of diploid populations: complexity and algorithms. *Journal of Computational Biology*, 8(3):305-323, 2001.

[38] D. Gusfield. Haplotyping as Perfect Phylogeny: Conceptual Framework and Efficient Solutions. In *Proceedings of RECOMB 2002: The Sixth Annual International Conference on Computational Biology*, pages 166–175, 2002. Extended Abstract.

[39] D. Gusfield. Haplotype inference by pure parsimony. In E. Chavez R. Baeza-Yates and M. Crochemore, editors, *14th Annual Symposium on Combinatorial Pattern Matching (CPM'03)*, volume 2676, pages 144–155. Springer LNCS, 2003.

[40] D. Gusfield. An overview of combinatorial methods for haplotype inference. In S. Istrail, M. Waterman, and A. Clark, editors, *Computational Methods for SNPs and Haplotype Inference*, volume 2983, pages 9–25. Springer, 2004. Lecture Notes in Computer Science.

[41] B. Halldorsson, V. Bafna, N. Edwards and R. Lipert *et al.* Combinatorial problems arising in SNP and haplotype analysis. In C. Calude, M. Dinneen, and V. Vajnovski, editors, *Discrete Mathematics and Theoretical Computer Science. Proceedings of DMTCS 2003*, volume 2731, pages 26–47. Springer, 2003. Springer Lecture Notes in Computer Science.

[42] B. Halldorsson, V. Bafna, N. Edwards and R. Lipert *et al.* A survey of computational

methods for determining haplotypes. In *Proceedings of the First RECOMB Satellite on Computational Methods for SNPs and Haplotype Inference.* Springer Lecture Notes in Bioinformatics, LNCS, Vol. 2983 p. 26-47, 2003.

[43] E. Halperin and E. Eskin. Haplotype reconstruction from genotype data using imperfect phylogeny. *Bioinformatics*, 20:1842–1849, 2004.

[44] E. Halperin and R. Karp. Perfect phylogeny and haplotype assignment. *Proceedings of The 8th Ann. International Conference Research in Computational Molecular Biology (RECOMB 2004*, pages 10–19. ACM Press, 2004.

[45] M. Hawley and K. Kidd. HAPLO: a program using the EM algorithm to estimate the frequencies of multi-site haplotypes. *Journal of Heredity*, 86:409–411, 1995.

[46] J. He and A. Zelikovsky. Linear reduction for haplotype inference. In *Proc. of 2004 Workshop on Algorithms in Bioinformatics*, Springer Lecture Notes in Bioinformatics, LNCS Vol. 3240. pages 242-253.

[47] L. Helmuth. Genome research: Map of the human genome 3.0. *Science*, 293:583–585, 2001.

[48] D. Hinds, L. Stuve, G. Nilsen, E. Halperin, E. Eskin, D. Gallinger, K. Frazer, and D. Cox. Whole-genome patterns of common DNA variation in three human populations. *Science*, 307:1072–1079, 2005.

[49] Y.T. Huang, K.M. Chao, and T. Chen. An approximation algorithm for haplotype inference by maximum parsimony. *ACM Symposium for Applied Computing SAC'05*, 2005.

[50] E. Hubbell. Personal communication.

[51] R. Hudson. Gene genealogies and the coalescent process. *Oxford Survey of Evolutionary Biology*, 7:1–44, 1990.

[52] S. Iwata. University of Tokyo. Personal Communication.

[53] G. Kimmel and R. Shamir. The Incomplete Perfect Phylogeny Haplotype Problem. *Journal of Bioinformatics and Computational Biology*, 3:1-25, 2005

[54] P.Y. Kwok. Genomics: Genetic association by whole-genome analysis? *Science*, 294:1669–1670, 2001.

[55] G. Lancia, C. Pinotti, and R. Rizzi. Haplotyping populations: Complexity and approximations. Technical Report dit-02-082, University of Trento, 2002.

[56] G. Lancia, C. Pinotti, and R. Rizzi. Haplotyping populations by pure parsimony: Complexity, exact and approximation algorithms. *INFORMS Journal on Computing, Special issue on Computational Biology*, 16:348–359, 2004.

[57] J. Lee. U.T. Austin. Personal Communication.

[58] Z. Li, W. Zhou, X. Zhang, and L. Chen. A Parsimonious tree-grow method for haplotype inference. *Bioinformatics*. 21:3475-3481, 2005.

[59] S. Lin, D. Cutler, M. Zwick, and A. Chakravarti. Haplotype inference in random population samples. *American Journal of Human Genetics*, 71:1129–1137, 2003.

[60] L. Löfgren. Irredundant and redundant boolean branch networks. *IRE Transactions on Circuit Theory*, CT-6:158–175, 1959.

[61] J.C. Long, R.C. William, and M. Urbanek. An E-M algorithm and testing strategy for multiple-locus haplotypes. *American Journal of Human Genetics*, 56:799–810, 1995.

[62] T. Niu, Z. Qin, X. Xu, and J.S. Liu. Bayesian haplotype inference for multiple linked single-nucleotide polymorphisms. *American Journal of Human Genetics*, 70:157–169, 2002.

[63] T. Ohto. An experimental analysis of the Bixby-Wagner algorithm for graph realization problems. *IPSJ SIGNotes ALgorithms Abstract, http://www.ipsj.or.jp/members/SIGNotes/Eng/16/2002/084/article001.html*, 084-001, 2002.

[64] S.H. Orzack, D. Gusfield, J. Olson and S. Nesbitt *et al.* Analysis and exploration of

the use of rule-based algorithms and consensus methods for the inferral of haplotypes. *Genetics*, 165:915–928, 2003.

[65] S.H. Orzack, L. Subrahmanyan, D. Gusfield and S. Lissargue *et al.* A comparison of an exact method and algorithmic method for haplotype frequency inferral. Preprint, 2005.

[66] D. Posada and K. Crandall. Intraspecific gene genealogies: trees grafting into networks. *Trends in Ecology and Evolution*, 16:37–45, 2001.

[67] Z. Qin, T. Niu, and J.S. Liu. Partition-ligation-expectation-maximization algorithm for haplotype inference with single-nucleotide polymorphisms. *American Journal of Human Genetics*, 71:1242–1247, 2002.

[68] R. V. Satya and A. Mukherjee. An Optimal Algorithm for Perfect Phylogeny Haplotyping. *Proceedings of 4th CSB Bioinformatics Conference.* IEEE Press, Los Alamitos, CA, 2005.

[69] M. Stephens and P. Donnelly. A comparison of Bayesian methods for haplotype reconstruction from population genotype data. *American Journal of Human Genetics*, 73:1162–1169, 2003.

[70] M. Stephens, N. Smith, and P. Donnelly. A new statistical method for haplotype reconstruction from population data. *American Journal of Human Genetics*, 68:978–989, 2001.

[71] Y. Song, Y. Wu and D. Gusfield. Haplotyping with one homoplasy or recombination event. Proceedings of Workshop on Algorithms in Bioinformatics (WABI) 2005. Springer, Lecture Notes in Bioinformatics, LNCS Vol. 3692.

[72] S. Tavare. Calibrating the clock: Using stochastic processes to measure the rate of evolution. In E. Lander and M. Waterman, editors, *Calculating the Secrets of Life*. National Academy Press, 1995.

[73] J. D. Wall and J. K. Pritchard. Haplotype blocks and linkage disequilibrium in the human genome. *Nature Reviews*, 4:587–597, 2003.

[74] L. Wang and L. Xu. Haplotype inference by maximum parsimony. *Bioinformatics*, 19:1773–1780, 2003.

[75] B. S. Weir. *Genetic Data Analysis II* Sinauer Associates 1996.

[76] W. T. Whitney. 2-isomorphic graphs. *American Mathematics Journal*, 55:245–254, 1933.

[77] C. Wiuf. Inference of recombination and block structure using unphased data. *Genetics*, 166:537–545, 2004.

[78] R. Y. L. Zee, H. H. Hegener, N. R. Cook, and P. M. Ridker. C-reactive protein gene polymorphisms and the risk of venous thromboembolism: a haplotype-based analysis. *Journal of Thrombosis and Haemostasis*, 2:1240–1243, 2004.

[79] J. Zhang, W. Rowe, A. Clark, and K. Buetow. Genomewide distribution of high-frequency, completely mismatching SNP haplotype pairs observed to be common across Human populations. *American Journal of Human Genetics*, 73:1073–1081, 2003.

V

Phylogenetics

19

An Overview of Phylogeny Reconstruction

C. Randal Linder
The University of Texas at Austin

Tandy Warnow
The University of Texas at Austin

19.1 Introduction

The best evidence strongly supports that all life currently on earth is descended from a single common ancestor. Over a period of at least 3.8 billion years, that single original ancestor has split repeatedly into new and independent lineages (i.e., species), and, on occasion, some of these otherwise independent lineages have come back together to form yet other lineages or to exchange genetic information. The evolutionary relationships among these species is referred to as their "phylogeny", and phylogenetic reconstruction is concerned with inferring the phylogeny of groups of organisms. The ultimate goal is to infer the phylogeny of all life on earth.

Phylogenies are important to biology in many ways. So much so, that phylogenies have become an integral part of much biological research, including biomedical research, drug design, and areas of bioinformatics (such as protein structure prediction and multiple sequence alignment). Accurate phylogenetic reconstructions involve significant effort due to the difficulties of acquiring the primary biological data and the computational complexity of the underlying optimization problems. Not surprisingly, phylogenetic inference is providing interesting and hard problems to the computer science algorithms research community – as witnessed by the three chapters in this volume on novel algorithmic research for phylogeny reconstruction. The limitations of existing phylogenetic reconstruction methods have a direct impact on the ability of systematists (that is, biologists who study the evolutionary history of a group of organisms) to analyze their data with adequate accuracy and efficiency, so that their subsequent scientific inferences are reliable.

The purpose of this chapter is to help computer scientists develop sufficient knowledge and taste in the area of computational phylogenetics, so that they will be able to develop new methods for phylogeny reconstruction that can help the practicing molecular systematist. Understanding the various applications of phylogenies will help the algorithms designer appreciate where errors in phylogeny estimation can be tolerated, and where they will have a more serious impact. Much of our discussion will therefore be from the viewpoint of a molecular systematist, and will (a) elucidate promising areas for additional research, (b) provide a context with which to understand the potential for impact of algorithmic innovations, and (c) give the algorithms research community a better understanding of the strengths and limitations of data collection and analysis in current practice. The final point is important because a better understanding of the type and amount of raw data that biologists can routinely obtain and analyze will help mathematicians and computer scientists in designing methods that are compatible with and take advantage of the types and quantities of data used by biologists. In particular, we will draw attention to the sources of potential error in the primary data, and to those aspects of the input data that have the potential to impact reconstruction methods in significant and potentially different ways.

We therefore begin our discussion in Section 19.2 with an overview of how phylogenies are used in biology, focusing on the questions that can be answered once the phylogeny is obtained. We continue in Section 19.3 with a description of the process a biologist goes through, from the inception of a project to the production of a publishable phylogenetic inference. In Section 19.4 we review each of the steps involved in phylogenetic inference, and discuss their major methodological and algorithmic issues. In Section 19.5 and Section 19.6, we describe advances on two specialized research problems – supertree methods (the subject of one of the chapters in this volume) and gene order phylogenetics (which is discussed in another chapter), respectively. We close in Section 19.7 with some comments about algorithmic research in phylogeny and recommendations for additional reading.

Finally, a caveat. Phylogenies are generally represented as rooted binary trees since speciation events are generally bifurcating, i.e., speciation usually occurs when an ancestral lineage splits into two new, independent lineages. The assumption is that reticulation, i.e., horizontal gene transfer and hybrid speciation, is rare. Nonetheless, reticulation events are known to occur and are fairly common in certain groups of organisms, e.g., hybrid speciation is relatively common in plants [48]. Therefore, the evolutionary history of all life is not properly represented as a tree. Instead, the appropriate graphical model of evolution for all life is a directed acyclic graph (DAG) which we call a "phylogenetic network" [44, 45]. Despite the reality of reticulation, in order to keep the chapter to a reasonable size, we will confine most of our discussion to phylogenetic trees and their reconstruction and will only talk about reticulate evolution insofar as it creates difficulties for tree reconstruction. Readers who want to learn more about reticulate phylogenies may wish to read the tutorial on reticulate evolution at the DIMACS web site for reticulate evolution [38].

19.2   What are Phylogenies Used for in Biological Research?

Phylogenies are reconstructed on the basis of character data, where a "character" is any feature of an organism that can have different states. A typical biological example of a character is a nucleotide position in a DNA sequence, with the character state being the particular nucleotide (A,G,C,T) occupying that position. From a mathematical standpoint, a character is just a function that maps the set of taxa to its set of states. When the set of states is discrete, the character is said to be "qualitative" or "discrete", and when the set of states is continuous, then the character is said to be "quantitative". Molecular phylogenetics research is concerned not only with the evolutionary history of different organisms, but also with how the different characters evolve in the course of that history. Thus, characters are used to infer species trees, but can also be of interest in their own right.

The uses for phylogenies, beyond elucidating the evolutionary relationships of biological species, are many and growing; here we highlight the most common uses and some of the most intriguing.

The most common use of a phylogeny is for a comparative study [5, 30]. A comparative study is one where a particular question is addressed by comparing how certain biological characters have evolved in different lineages in the context of a phylogeny. This information is used to infer important aspects of the evolution of those characters. That statement is vague and general because of the many types of characters studied using a comparative approach. Some examples of areas in which the comparative method can be applied include adaptation, development, physiology, gene function, vaccine design, and modes of speciation. In essence, a comparative approach can be taken any time a biologist wishes to examine a process or aspect of biological organisms that has evolved on a time scale that is greater than the time of an individual species or lineage. An evolutionary perspective, through the use of an accurately estimated phylogeny, makes it possible to ensure that the number of independent data points used in the comparative study is not over estimated, and to determine the order of the events.

Consider the following example, where a biologist is studying the qualitative character *egg color*, and finds that a particular state for that character (e.g., blue egg color) occurs in 50 of the 100 species in a clade (where a "clade" consists of all the descendants from a single node in the tree). The biologist would like to know how many times in the evolutionary history this particular color arose and was lost. Without knowledge of the phylogeny of the clade it is not possible to estimate these numbers, since the pattern itself is consistent with one mutation leading to blue egg color, or even 50 mutations. However, knowing the phylogeny for the clade makes it possible to obtain tighter lower bounds–and to even quantify statistically–the number of times that character has changed state. Hence, these questions can be answered with greater accuracy when the phylogeny is known. Similarly, the biologist might also like to know if the trait is *ancestral* (i.e., that it evolved before the clade of interest) or *derived* (i.e., that it evolved at some point within the clade under study). These questions fall more generally into the question of understanding the evolution of a particular character within a clade of the evolutionary tree for a particular group.

The researcher might also want to know about the rate at which a quantitative character has changed in the clade overall or in different parts of the clade, and for qualitative characters the researcher might like to know the number of times a trait was gained or lost and how those gains and losses are distributed. For both types of traits, the researcher might also like to know if there are correlations between changes in sets of traits or among traits and particular external (environmental) factors. In order to ensure independence between the inferences made in a comparative analysis and the phylogeny that is used to make the inference, biologists usually strive to use different sets of characters for inferring phylogenies and for comparative studies. Comparative studies have become extremely common in biology, and one can easily find examples of them in almost any issue of experimental biological journals, as well as in the general science journals, *Nature* and *Science*.

A second common use of phylogenies is to test biogeographic hypotheses. Biogeography is concerned with the geographical distribution of organisms, extant and extinct. For example, a researcher may be interested in whether a particular group of species has colonized a set of islands a single time or repeatedly. This can be assessed by determining whether all of the species on the island arose from a single most recent mainland common ancestor or whether there are multiple independent mainland ancestors.

Phylogenies can be used to look at the mode and tempo of speciation. Regular bifurcating speciation is hypothesized to occur by several mechanisms. One of these, allopatric speciation, is caused by the geographical separation of a single ancestral species into two geographically isolated groups. Hence, one can look at the geographical distributions of species in a clade to determine whether there is support for allopatric speciation. For example, if the geographical ranges of a set of species is well known, it is sometimes possible to correlate speciation events with large scale geological events such as mountain building or plate tectonics. If there are adequate fossil records or if a molecular clock (the assumption that each character evolves at a rate that is proportional to elapsed time) can reasonably be assumed, one can also compare and contrast the rates of speciation in different clades and in different parts of a clade. However, the assessment of speciation rates is somewhat clouded by the fact that our knowledge of the number of extinct species in a clade is often poor. Since every extinct lineage must also have had a speciation event, it is possible to significantly underestimate the absolute number of speciation events, and their relative proportions between clades. For the same reasons, it is often difficult to use phylogenies to say much about rates of extinction in different clades.

One can also use a phylogeny to attempt to infer the amino acid sequence of extinct proteins. These putative extinct proteins can then be synthesized or an artificial gene coding for them can be produced, and the functional characteristics of the protein that are of interest to the researcher can be tested.

In a more practical vein, phylogenies can be used to track the evolution of diseases, which can, in turn, be used to help design drugs and vaccines that are more likely to be effective against the currently dominant strains. The most prominent example of this use is the flu vaccine, which is altered from year-to-year as medical experts work to keep track of the influenza types most likely to dominate in a given flu season [7].

Finally, phylogenies have even been used in criminal cases, most famously, in a case where a doctor in Louisiana was accused of having deliberately infected his girlfriend with HIV [41]. The phylogenetic evidence featured prominently in the trial and the doctor was ultimately convicted of attempted second degree murder.

The key here is that phylogenies are useful in any endeavor where the historical and hierarchical structure of the evolution of species can be used to infer the history of the point of interest.

19.3   The Steps of a Phylogenetic Analysis

Biologists are reconstructing phylogenies for hundreds of sets of organisms representing all of the major divisions of the "Tree of Life," e.g., bacteria, plants, animals, etc. While that might lead one to think that the practice of inferring phylogenies could be very different depending upon the taxonomic group that is being studied, there are many details and steps in the generation of the primary data that are similar whether the researcher is studying microbes, invertebrates, vertebrates, plants or fungi. And if DNA sequence data are used as the primary data for reconstruction, the steps after purified DNA has been extracted from the group of interest are very similar, if not identical. In this section, we outline the process of inferring a phylogeny, from its conception to the assessment of support for an inferred phylogeny.

Our aim in providing this information is to give developers of phylogenetic methods a feeling for what a biologist must do to produce a phylogeny for a group of organisms. Some steps are relatively easy and not very time consuming; others are rate limiting and have significant impact on the type, quantity and quality of raw data that are available for analysis. A better understanding of how biologists work, what data they can easily augment and what they must struggle to provide, will help researchers to produce methods that will be most useful to biologists.

19.3.1   Designing the study

Before a biologist collects her first organism, the scope and nature of her study must be delineated, and a plan must be put in place to accomplish its goals. The first decision to be made is why the study is being conducted. This decision will determine the taxonomic scope of the project, e.g., a small recently evolved clade (such as a single genus of the great apes), or a clade that encompasses an older group (such as all insects), or even the relationships among the kingdoms of life[1]. However, the broader the taxonomic scope, the more likely a supertree analysis will be needed, for reasons which we will discuss below. In particular, the phylogeny of all life cannot be directly reconstructed. Instead, phylogenies of many subsets of the complete phylogeny are being independently inferred by hundreds of systematists around the world with the goal of ultimately combining the subsets into a single phylogeny for all life. This all-inclusive phylogeny is a major goal of phylogenetic reconstruction.

There are many reasons for wanting to reconstruct phylogenies at all levels, and the reasons for estimating the phylogeny of a particular group can be as varied as the purposes to which a phylogeny can be put (as we discussed in Section 19.2). However, once the taxonomic scope of the work is determined, many of the facets of the study (the sampling scheme, the markers to be used, etc.) will either be determined or at least constrained.

Taxon selection and sampling

Although a phylogenetic reconstruction can occasionally encompass over 10,000 species [61], the majority of studies where the researchers generate the primary data from scratch involve fewer than 100 species. Therefore, unless researchers are working with a small clade of organisms, a sampling scheme needs to be decided upon since it will not be possible to include all of the species in the clade. For example, if a researcher decides to look at floral evolution in a genus of plants with over two hundred described species distributed throughout the globe, several aspects of the study become immediately evident. First, several issues may prevent collection of specimens for every species in the genus: the number of species may be too great to collect and process; the species may be too geographically

---

[1]In traditional taxonomy, organisms are hierarchically grouped according to a standard terminology. From least to most inclusive the hierarchy goes species, genus, family, order, class, phylum, kingdom. Generically, any level can be referred to as a taxon. Ideally each hierarchical level represents a clade of organisms, but there are still many taxonomic groups where it is not yet known if they represent clades, and many organisms have to be reclassified as their phylogenetic relationships become clearer. Since there are at least several million species on earth, the number of taxonomic categories is obviously insufficient to give a name to every level of clades in the phylogeny of life. It is also important to note that the evolutionary depth of a given taxonomic level is not consistent from one group of organisms to the next. For example, *not* all genera represent clades that are approximately 4 million years old.

widespread to allow all collections to be made, even if sympathetic workers in other countries agree to help collect specimens locally; it may be politically infeasible to get collections from some countries; some countries may have highly restrictive laws regarding study of their flora by foreign scientists; and it might not be possible to obtain sufficient funds for all of the travel and permits required. These considerations will necessitate strategic decisions about which taxa are most critical to the study and what sort of sampling scheme will be most likely to yield scientifically valid results, given the goals of the study. It also means that methods designed for inferring phylogenies need to be robust to missing taxa. Note that even if a researcher is able to sample all of the extant taxa for his or her group, there are likely to be some missing taxa due to extinction.

Because our example study is interested in the evolution of characters, the researcher will also have to decide how many individuals from each species will need to be examined to have a statistically valid sample of the range of character variation within and between individual species. Also, if the species are recently evolved and poorly circumscribed taxonomically, the researcher may need to sample several individuals from more than one population of each species to ensure the variation that is used for reconstructing the phylogeny truly represents interspecific variation rather than ancestral variation that has been inherited by several species from a common ancestor. If the species for which a phylogeny will be reconstructed are more distantly related, a smaller number of samples will usually be required from each species. Finally, because in nearly all cases biologists want to reconstruct the rooted phylogeny, the researcher will need to select species to be used as *outgroups*. Outgroup species are ones that are not included in the clade of interest (the ingroup) and which, therefore, can be used to root the clade of interest. This method of rooting works because the common ancestor of the ingroup and the outgroup lies further back in time than the common ancestor of the ingroup. If the outgroup is correctly chosen, and a phylogeny is reconstructed that is correct (as an unrooted tree), then the rooted phylogeny obtained by rooting the reconstructed phylogeny on the edge between the ingroup and the outgroup will be the correct as a rooted phylogeny.

Selecting the outgroup is tricky. The first issue is the obvious one: a taxon may seem to be an outgroup, but it may not be. And if the taxon is not actually an outgroup, the resultant rooting of the reconstructed phylogeny can be incorrect, and hence the inferred order of speciation events will also at least in part be incorrect. Since many of the uses of phylogenies are strongly based upon accurate reconstruction of the order of speciation events, this can have negative consequences for scientific inferences later on. This is not just a theoretical danger: in many cases, work has shown that species that were considered outgroups were misclassified and actually belonged in the ingroup. For example, potatoes and tomatoes were originally thought to be in different genera, but are now known to be close relatives in the same genus.

Since more closely related groups of species are expected to be more similar to one another (because they will have inherited a larger number of unaltered ancestral traits from their most recent common ancestor), it would seem that adding a taxon (as an outgroup) which is most different from the ingroup taxa would be the "right thing" to do; and, indeed, such a choice would avoid the problem of inadvertently picking a member of the ingroup instead of an outgroup. However, this too is potentially problematic. Suppose we take this approach seriously, and we pick taxon A as an outgroup to a set S of taxa. Let us also suppose that A is in fact a true outgroup, and that it is quite dissimilar to every taxon in S. The problem here is that the more dissimilar taxon A is to the taxa of S, the harder it is for any phylogeny reconstruction method to connect the taxon A to the phylogeny on S. That is, evolution is modeled as a random process that is operating on a phylogeny, producing sequences at the nodes of the tree. The more dissimilar A is to S, the closer to random the sequence for A is to the sequences for the taxa in S. The more random these sequences look relative to each other, the more random the placement of the edge that connects A to the rest of the tree. This has the consequence that, once again, the rooting of the resultant phylogeny can be incorrect, and much of the subsequent scientific analysis can be based upon false premises.

Thus, a good outgroup taxon must be dissimilar enough to be definitely an outgroup taxon, but not so dissimilar that the phylogenetic reconstruction method cannot reconstruct the phylogeny correctly. If a molecular clock hypothesis is reasonable for the dataset (including the outgroup), these decisions are much easier to make, but in general this is quite a tricky and difficult problem.

Marker selection

Another critical part of the planning is to choose markers that are appropriate for the study. In this context a "marker" can refer to a particular region of DNA, a protein, a morphological character, or the order of genes on a chromosome. Each type of marker has its own special challenges and techniques. Rather than attempting to provide a comprehensive overview of how all of the different markers are selected and used, we will focus on the most common type in use today, DNA sequences, with occasional comments about the other types.

When a phylogeny is reconstructed for a region of DNA, what is really being reconstructed is the phylogeny of that DNA region, which does not necessarily have to be the same as the phylogeny of the species from which the DNA was taken (see below); this is the classic gene tree/species tree problem [39, 40, 49, 50, 51, 52]. What can a researcher do to increase the probability that their DNA region will produce a gene tree whose evolutionary history is identical to that of the species tree? Achieving this ideal requires picking markers with the following characteristics:

1. *The marker is unrecombined.* When sex cells (eggs, sperm, ovules, pollen, etc.) are produced by normal diploid organisms (organisms that inherit one copy of each nuclear chromosome from their mother and one from their father[2]), they undergo a process called recombination in every generation, which causes individual strands of DNA to be a mixture of two or more different geneological histories. Details about recombination (or more formally "meiotic recombination") are available in any introductory genetics text. What is important here are not the details of how recombination produces DNA sequences having multiple evolutionary histories, but rather the effect this can have on phylogenetic reconstruction. If a recombination event has taken place within a marker used for phylogenetic analysis it will be composed of two different evolutionary histories. If the researcher can determine where the recombination event occurred, then the marker can be broken into these two regions, and each region can be analyzed separately – and each can produce a potentially different gene tree. But since only rarely does a researcher know that a marker has undergone recombination, it is more likely that she will analyze the full DNA sequence without considering the separate histories of the different parts of the sequence. Depending on the reconstruction method used and the relative amounts of data from the

---

[2]The one exception to this rule is the sex chromosomes. The sex that is heterogametic (XY, ZW, etc.) only has one copy of each type, e.g., human males have one copy of the X and one copy of the Y chromosomes.

different histories, analysis of such combined histories can produce a phylogeny that neither reflects the gene trees nor the species tree. To avoid the problem of recombined sequences, in many cases researchers use sequences that are predominantly uniparentally inherited such as mitochondrial and chloroplast sequences. Because these organelles are inherited from only one parent the sequences in these organelles are haploid and do not undergo recombination. This greatly reduces the probability that different genes will have different trees.

2. *The marker is single copy or is subject to rapid concerted evolution.* Another way that gene trees and species trees can differ is the presence of gene duplication, i.e., when a gene has copies at two or more locations (loci) on one or more chromosomes. If a gene is duplicated one or more times before a clade originates and then different copies of the duplications are randomly lost in different lineages ("random assortment"), the leaves of the tree will have different sets of copies of the gene. Depending upon how the systematist analyzes the data (a complicated situation which we will not discuss here - see [38] for more information), this can cause the reconstructed gene tree to differ from the species tree.

   The gene tree is more likely to match the species tree if there is only one copy of a marker that has not undergone duplication and loss in the clade of interest, and for this reason single copy markers are the most desirable in phylogenetics. However, some duplicated genes can be useful in a phylogenetic analysis, provided they undergo "concerted evolution", a process that rapidly homogenizes all the copies of a gene to a single type with the same sequence. As long as concerted evolution is homogenizing the sequences of the copies at a rate significantly faster than the rate of speciation, then the probability that the phylogeny for the marker (the "gene tree") will be identical to that of the species tree will be high. Thus, picking markers with multiple copies can also work, provided that the region has sufficiently rapid concerted evolution. The ribosomal DNA repeat is one such region, and it is broadly used by systematists.

3. *Finally, it is preferable to sequence the same allele of a gene for reconstruction.* Because diploid organisms have two copies of each autosome (non-sex chromosomes), they have two copies (alleles) of every gene at a position (called a "locus") on a chromosome. Although each individual in a species can only have up to two different alleles at a locus, collectively, all the individuals in the species can, and often do, have three or more alleles at that locus. Hence, even single copy genes can experience the same assortment and sampling problems expected with duplicated genes. The only way for a researcher to make sure she is using the same allele for phylogenetic reconstruction is to sequence extensively and determine the phylogenetic relationships among the alleles of all the species in his/her clade. Because this is so much extra work, most systematists focus on organellar DNA sequences or nuclear sequences that undergo rapid concerted evolution, which homogenizes the repeats on both copies of the chromosome.

In addition to the issues surrounding gene trees and species trees, the researcher is also looking for the following characteristics in a DNA region.

1. *The marker is readily amplifiable by polymerase chain reaction (PCR).* At present, almost all DNA sequencing is preceded by amplification of the region to be sequenced using PCR. PCR requires highly conserved sequences at either end of the marker so that "primers" (single stranded DNA sequences that have approximately 18-30 nucleotides) will bind to them for all of the species that will be

sequenced. If a priming sequence for a marker varies significantly within the set of species in the study, it can be very difficult to get DNA sequence data for all the species in a study.

2. *The marker can be sequenced.* Some DNA regions can be amplified but are very difficult to sequence due to repetitive elements in the sequence. This problem can be caused by either very short repeats (one to four nucleotides in length) that cause the DNA polymerase enzyme to stutter and produce different numbers of repeats, or very large repeats that make it difficult to sequence through that region. Modern DNA sequencing methods use primers in a fashion similar to but not identical to PCR. For a given primer only 600-1000 nucleotides can be sequenced. If contiguous repeats are longer than twice this upper limit, it may be impossible to sequence through that repeat region. See Section 19.3.4 for a discussion of the problems with aligning repeats.

3. *The marker evolves quickly enough to distinguish among the most recently evolved species in the ingroup, but does not evolve so quickly that it is either impossible or extremely difficult to infer a reliable multiple sequence alignment* (see Section 19.3.4). Markers that evolve very slowly, such as the ribosomal DNA genes, can be used to reconstruct relationships among organisms from different kingdoms, the broadest traditional taxonomic group, whereas, very rapidly evolving markers, such as mitochondrial sequences in animals, may only be suitable for reconstructing genera, the least inclusive taxonomic group above species. The incomplete taxonomic coverage of all but a few slow evolving DNA regions is part of the reason supertree methods (see Section 19.5) are needed for reconstructing large numbers of species from diverse taxonomic groups. For example, if the ingroup is closely related, rapidly evolving markers that have been vetted by the systematics community for the characteristics enumerated above will be a systematist's first choice, but if these prove to have too little variation to distinguish the species, the researcher may have to invest significant time and resources into developing a new marker or markers. In general, for each kingdom of organisms, there is a fairly small set of DNA regions that are currently considered acceptable for phylogenetic analysis. There are undoubtedly more regions that could be used, but to save time and expense, researchers use the ones that are already developed first.

Having made these critical decisions (and hopefully having secured funding), a systematist can now turn to the next step.

19.3.2 Collecting organisms in the field

Depending on the taxonomic group that is under study, collecting the organisms which will be studied can involve a small number of trips close to the home institution of the researcher or a number of far flung trips to locations that are difficult to access for both geographical and political reasons. Often the researcher will have to obtain permission from one or more governmental agencies to collect specimens. This is especially true in less developed countries where concern about bioprospecting is high. The rules for collecting in different countries are as varied as the countries themselves, and the researcher will have to negotiate the legal web of the countries in which she needs to collect. In some cases, it will be possible to get tissue for DNA for some species by arranging to borrow museum or herbarium collections recently made by other systematists. The older the specimens, the less likely they will have intact DNA.

When the researcher is in the field she must collect and preserve specimens of every species that will be studied. How the specimens are preserved and returned to the lab depends on the taxonomic group studied. For most vertebrates, specimens will either be frozen in the field and then processed back at the lab or placed in a preserving solution. Insects can be easily captured, killed and preserved in the field with particular body parts, e.g., a leg, being harvested for DNA extraction back in the lab. Plants are usually placed in presses in the field to produce dried vouchers that will be kept permanently at a herbarium. At the time of collection, some material (usually leaves or flowers) from the voucher will be quickly dried in silica gel. With plants, the researcher also often has the option of collecting seeds which can later be grown in a greenhouse for vouchering and fresh material for DNA extraction.

For microorganisms or fungi, a researcher might collect from a particular locality and then culture the organisms back in the lab for identification and DNA extraction. However, the vast majority of species in these groups are not culturable. In these cases, the researcher will collect material, e.g., soil or water, from a locality and preserve it until it can be brought back to the lab where molecular techniques can be used to determine the species in the sample.

Collecting is often conducted over a period of several years, so the researcher is at pains to plan well before the trips are made. It may be prohibitively expensive or politically infeasible to return to an area a second time in rapid succession.

19.3.3 In the lab

Once a portion of the species in a study has been collected, work can begin on gathering the primary data for phylogenetic analysis. We focus here on DNA sequence data, researchers might also collect morphological, RNA, protein or gene order data.

For multicellular organisms, generally, a small piece of tissue from the organism is taken through a series of physical and chemical steps to release and purify the DNA from cells and organelles. For single celled organisms, either single species are cultured under sterile conditions and samples are taken from these cultures, or a mixture of many species is extracted simultaneously, e.g., from a soil sample. The steps for extracting and purifying DNA from different groups of organisms differ, but generally, it is easier to get DNA from animals than it is from plants, fungi, and some groups of microbes. Plants and fungi often have secondary compounds that either damage DNA when it is extracted or that co-extract with it, thereby complicating the purification process. Plants, fungi, and some microorganisms also often have either a secondary cell wall or other cell-wall structures that can interfere with DNA extraction. The researcher must often try several different extraction and purification procedures before sufficient quantities of high quality DNA, free of interfering compounds, are reliably obtained. In some cases, months can be spent just on determining an effective extraction and purification procedure.

Once high quality DNA has been obtained, the researcher will usually conduct a preliminary study on a subset of their taxa to determine which markers are likely to have enough informative variation for phylogenetic reconstruction. This study will consist of some taxa that are expected to be closely related and some that are expected to be distantly related, in an attempt to determine whether a given marker has sufficient variation to distinguish among closely related taxa but is not evolving so rapidly that it will be unalignable for the distantly related taxa and the outgroups.

Generally, PCRs will be set up to amplify the region of interest. If the amplifications are successful, they will be purified and sequenced. If they are unsuccessful, the researcher will attempt to determine if there was a problem with the PCR or with the template DNA and will correct the problems with the PCR or try other methods of getting purified DNA, respectively. Although modern DNA sequencing methods make it possible to sequence large quantities of DNA, and the sequence of nucleotides can be called with a fair degree of accuracy by machine when the raw sequence data are of sufficient quality, the automated process is not infallible. Researchers proofread their sequences by eye and resequence regions that are ambiguous or of low quality.

If these preliminary runs indicate the marker is good for the group of interest, the researcher will amplify and sequence the marker for all of the taxa in the study. Because it is often the case that some taxa are more difficult to obtain or extract than others, a researcher often does not complete the sequencing for all of the taxa simultaneously. When this occurs she may perform preliminary phylogenetic analyses on the taxa that are more readily available for sequencing.

Usually, a researcher will sequence two or more markers and then check whether the phylogenetic analyses of each marker produces trees that are topologically identical. Topologically different trees produced by different markers can be caused by several things: gene tree/species tree problems, reticulation, lack of support for parts of the phylogeny, and use of different reconstruction methods for different markers. If a researcher finds that markers and the methods of analysis produce conflicting phylogenies, she will usually take additional steps to determine the source of the conflict. Conflicting tree topologies from multiple markers from the same organelle (or multiple markers from a region experiencing concerted evolution) cannot usually be caused by reticulation or gene tree/species tree problems since the markers usually do not contain multiple evolutionary histories. Therefore, in the absence of compelling evidence that two or more organellar markers have well supported conflicting phylogenies (see Section 19.3.6), it is usually assumed in these cases that conflict is due to particular aspects of the evolutionary history (e.g., evolutionary trees with edges on which very few mutations occur) that make it hard to fully resolve the evolutionary tree. Such conditions usually result in aspects of the reconstructed trees that are not well supported by the data. To solve this problem, the researcher will often sequence more of the organellar genome or the region that is evolving concertedly. On the other hand, if organellar and nuclear or two or more nuclear markers produce well supported conflicting phylogenies, the researcher will have to try to decide whether the cause is gene tree/species tree problems or reticulation. At present, we lack reliable methods for making this judgment solely on the basis of the sequences. However, the researcher may be able to make a judgment using other biological information that we do not discuss here.

19.3.4 Multiple Sequence Alignment

Once the sequence data are available, they need to be put in a multiple sequence alignment (MSA) before a phylogeny reconstruction method can be applied. There are many methods for producing multiple sequence alignments, some of which are quite recently developed, while others (e.g., ClustalW [77]) have been in use for a long time. Many of the most promising MSA algorithms in use are described in this volume, in a chapter on multiple sequence alignment. The focus in that chapter is on MSAs for amino acid sequences, with particular interest in identifying structural features of the proteins. The focus we take this chapter is the use of MSA for phylogeny reconstruction purposes, and thus our discussion will be slightly different.

A multiple sequence alignment of a set S of sequences is defined by a matrix where the rows are the sequences in S, and the entries within each column are "homologous". For phylogenetic reconstruction, this means positional homology, i.e., that all the nucleotides in the same column have evolved from a common ancestor. However, multiple sequence alignments (especially of protein sequences) can also be defined in terms of structural homology, so that columns identify residues that produce identical structural features in the three-dimensional folding of the protein. To a large extent, structural alignments and phylogenetically driven alignments are either the same or very similar, but there can be differences because non-homologous regions in proteins can sometimes evolve to have the same functional/structural form. "Convergent evolution" is the term used to describe this phenomenon, whereby similar characters in different species can evolve from nonhomologous genes or gene regions. Convergent evolution can take place at many levels in organisms, e.g., similar structural features in a protein or the spines on cacti and other desert plants.

The usual procedures for producing a multiple sequence alignment operate by inserting gaps (represented by dashes) into the sequences, so that the final resultant sequences are all the same length. This limitation means that if the sequences submitted to the MSA method do not begin and end at homologous sequence positions, the leading and trailing bases for which at least some of the sequences lack homologous positions will often not be aligned correctly. In some cases this can confuse the alignment algorithms and produce very poor alignments as they attempt to make all of the sequences the same length. To avoid this problem, most researchers trim their sequences to begin and end at what they believe are the same homologous positions before submitting them for multiple alignment.

However, equalizing the length of the sequences is not the objective but rather a feature of the MSA process, as the following discussion should make clear. For example, if we begin with n DNA sequences, with maximum sequence length k, the result of an MSA will be a set of n sequences over the alphabet $\{A, C, T, G, -\}$, each of length $k' \geq k$. These sequences can then be placed in an $n \times k'$ matrix, and hence the correspondence between matrices and multiple alignments.

There are a number of features that are difficult for the current set of methods to handle. The two most prevalent problems are (1) sequences that have diverged so much that similarity is difficult to infer and (2) introduction of large gaps, especially when these are due to different numbers of repeat sequences. When sequences from different taxa have differing numbers of imperfect repeats, the current set of MSA algorithms usually cannot determine which repeats from one sequence should be aligned with which repeats from the other. Consider the simplest example with two sequences. If sequence A has 4 repeats and sequence B has 2 repeats, which of the repeats in A should be aligned with the repeats in B? In some cases, the researcher will simply delete the repeats from his/her analyses. Alternatively, if the researcher thinks the repeats have important phylogenetic information and the repeats are long enough and varied enough, the researcher can try to determine the phylogenetic relationships of the repeats, by producing a MSA consisting of each copy of each repeat from each taxon and then using that MSA as input to a phylogenetic analysis. In this way the researcher can produce a hypothesis of which repeats are homologous (they will appear in clades together) and which are not. The phylogeny can then be used to guide a hand alignment of the putatively homologous repeats within the larger alignment of the marker. An MSA algorithm that could at least partially automate this process would be very useful. Thus, many systematists obtain multiple sequence alignments at least partly by hand, either aligning sequences themselves, or taking the output of some software (e.g. ClustalW [77]) and then modifying the alignment. While this may seem inefficient, the limitations of the current set of MSA algorithms necessitates it.

Finally, after the alignment is obtained, it is often further modified in order to eliminate unreliable sites (columns) or taxa (rows). For example, the systematist may eliminate those columns that contain too many gaps to have confidence in the positional homology of that region of the alignment, or that have a low "score" as computed by ClustalW; she may also elect to eliminate taxa (rows) that contain too many gaps. The objective of these modifications is to reduce the noise in the alignment that arises from poor quality data, since excessive noise (especially due to large numbers of unequal length gaps, which are often introduced in regions that are hard to align well) will result in poorly estimated phylogenies. Consequently, by eliminating the problematic components of the alignment, it may become possible to obtain an accurate reconstruction of the phylogeny on the remaining data.

19.3.5 Phylogenetic reconstruction

After the multiple alignment is obtained and before proceeding to reconstruct the phylogeny, several intermediate steps take place. The first involves deciding how to best analyze datasets which contain multiple markers for the same set of organisms. In this case, the systematist must decide between doing a phylogenetic analysis on a "combined" dataset (obtained by concatenating the individual datasets), or doing phylogenetic analyses on the individual datasets and then comparing the resultant phylogenies. As discussed above, this determination requires ascertaining whether the datasets have the same evolutionary history, so that issues such as lineage sorting or reticulate evolution can be ruled out as causes for making the evolutionary histories different. Methods for determining when it is safe to combine datasets exist, but these methods are not necessarily sufficiently accurate [32]. Most do not even consider whether reticulate evolution is a reasonable explanation for not combining the sets. As pointed out above, sequences from the same uniparentally inherited organelle are generally considered safe to combine, but unless the assumption of uniparental inheritance is explicitly tested for each species–a time consuming and sometimes highly impractical task–combined analyses can be misleading even for these sequences.

The second issue has to do with the choice of phylogenetic reconstruction method. If the systematist wants to use one of the statistical estimation techniques (i.e., Maximum Likelihood or Bayesian MCMC), she needs to decide which stochastic model would be most appropriate for her data. If she is using multiple markers and wishes to perform a combined analysis (combining the datasets into one dataset), then the model selection will in general be different for the different markers, and her phylogenetic analysis will need to be able to maintain these partitions during the inference phase.

Standard stochastic models of evolution

There are many stochastic models of site evolution, most of which have been described in terms of DNA sequence evolution. The models used for DNA sequence evolution do not usually involve bringing in constraints on the evolutionary history that would arise due to structural issues (such as secondary structures for RNA and tertiary structures for regions that code for amino-acid sequences), and so are simpler than stochastic models for either RNA or amino-acid sequences. Stochastic models of DNA sequence evolution used in practice range from the simplest Jukes-Cantor (JC) Markov model, to the fairly complex General Time Reversible (GTR) Markov model. These models describe the evolution of a sequence beginning at the root and evolving down the tree as a sequence of *point mutations*. Thus, no insertions, deletions, or duplications occur, and instead the only changes possible on an edge of the tree are at individual nucleotide positions where the current state of a nucleotide changes to another state. The result of these assumptions is that if the root sequence has length k, then the result of this evolutionary process is that every leaf in the tree is assigned a sequence also of length k. Furthermore, standard models assume that all the positions (sites) within the sequence evolve under identical processes, and independently of each other.

Rate variation across sites

The reader may be aware that it is often assumed that the sites evolve at different rates, with some sites evolving more quickly than others; furthermore, some sites may be *invariable*, and so not be allowed to change at all.[3] However, these observations do not violate the assertion that standard models assume that the sites are evolving identically and independently (*i.i.d.*). That is, when rate variation is incorporated into these models, these rates are assumed to be taken from a distribution (typically a gamma distribution); thus, each site has a probability of being invariable (i.e., having rate 0), but if it is allowed to vary, it then selects its rate from a common distribution, typically the gamma distribution, and then maintains this relative rate on all edges of the tree. Thus, even when sites have variable rates or may be invariable, this way of defining the rates has the consequence that the sites are still evolving under identical and independent processes.

Thus, standard stochastic models of evolution are essentially defined by two considerations: how a random site evolves, since all sites will follow the same model, and then the distribution of rates across sites. Typically these rates are taken from a gamma distribution, and can be incorporated into any single site evolution model. Here we now focus on the single site evolution models.

Different stochastic models of evolution differ substantially with respect to their impact on the resultant phylogenetic analysis, and yet the mathematics involved in the theory underlying the models generally in use in phylogenetics (i.e., from the simplest Jukes-Cantor (JC) model to the complex Generalized Time Reversible (GTR) model) is the same. That is, all of these models are identifiable (the probability distribution a model tree defines on the "patterns" suffices to identify the model tree). To understand how a character defines patterns, suppose that a character has r states and there are n taxa. Then there are r^n possible ways that the character can assign states to the leaves, and these are the "patterns" for the character. Furthermore, the parameter values for the model (such as the substitution probabilities on each edge) determine the probabilities of each pattern occurring. Thus, the model tree itself defines a probability distribution on the r^n possible patterns at the leaves. Saying that a model is identifiable means that knowing the probability distribution of the patterns is sufficient to define the tree. Thus, all the models that are under general consideration, from the JC to the GTR model, are identifiable. However, not all stochastic models are identifiable – the "no-common-mechanism" model [80] is one non-identifiable model. Other models that are not identifiable include standard site evolution models in which sites either do not evolve *i.i.d.*, or have rates of evolution drawn from more complicated distributions than the gamma distribution (see [10, 14, 72, 88]).

Thus, all the standard models currently used in phylogenetic reconstruction are identifiable. Furthermore, methods that have provably good performance under the simplest of these models (i.e., under JC) will also have provably good performance under GTR. In essence, therefore, there is little mathematical difference between any two models in current use in phylogenetic reconstruction.

---

[3]Note that a site that does not vary its state on a particular dataset is said to be "invariant"; however, that does not mean that the site is invariable – rather, its rate of evolution may be so slow as to not exhibit a change on a particular dataset. Thus, saying a site is invariable is a statement about the model, and not about its evolution on a particular dataset.

The Jukes-Cantor (JC) model

The assumption of the JC model which characterizes it is that if a site changes its state, it changes with equal probability to the other states. Hence, in the JC model we can specify the evolutionary process on the tree T by the assigning of substitution probabilities $p(e)$ to each edge e, where $p(e)$ indicates the probability that a site changes on the edge e.

The Generalized Time Reversible (GTR) model

For the GTR model, we do not make the assumption of equiprobable nucleotide substitutions, but we do require that the model be time-reversible. This is a fairly modest assumption, and allows us to model the evolution of a single site with a symmetric 4×4 stochastic substitution matrix M, along with the usual lengths (or substitution probabilities) on the edges. Thus, the GTR model contains the JC model, the Kimura 2-parameter (K2P) model, etc., as special cases, but allows greater complexity by having additional free parameters. It is not the most complex model that has been used to analyze datasets; in particular, the General Markov (GM) model [70] is a model which relaxes the assumption of time-reversibility, while still allowing for identifiability.

Phylogenetic analysis using stochastic models of evolution

Various statistical techniques have been developed which make it possible to select the best fitting model within this spectrum – from JC to GM – for a given DNA sequence dataset [55]. Thus, if a statistical estimation is desired, the first step in a phylogenetic analysis is generally to use one of these statistical tests to select the model under which the data will be analyzed. Once the model is selected, the researcher can then decide how to analyze his/her data – whether with a method that uses the model explicitly, or with one that does not. Phylogeny reconstruction methods come in essentially three flavors: (a) distance-based methods, such as Neighbor Joining [62], which tend to be polynomial time and are very fast in practice; (b) heuristics for either maximum likelihood or maximum parsimony, two hard optimization problems, and (c) Markov Chain Monte Carlo (MCMC) methods. Of these, only distance-based methods are polynomial time; despite this, most systematists prefer to use one of the other types of analyses, because numerous studies (both empirical and simulated) have shown that the other types of methods will often produce better estimates of evolutionary history.

Distance-based methods operate by computing a matrix of pairwise "distances" between the sequences (these are typically not just edit distances, but distances which are supposed to approximate the evolutionary distance, and so are derived from statistically-based distance calculations), and then use only that distance matrix to estimate the tree. Maximum Parsimony is an NP-hard [19] optimization problem in which the tree with the minimum total number of changes is sought (thus, it is the Hamming Distance Steiner Tree problem). Maximum Likelihood (ML) is another NP-hard optimization problem [11], but this problem is defined in terms of an explicit parametric stochastic model of evolution. The optimization problem is then to find the tree and its associated parameters (typically substitution probabilities) that maximizes the probability of the data. The theoretical advantage of Maximum Likelihood over Maximum Parsimony is that it is "statistically consistent" under most models; this means that it is guaranteed to return the correct tree with high probability [9] if the sequences are sufficiently long – something which is not true of Maximum Parsimony [15]. However, ML is even harder in practice than MP, and heuristics for both problems require very substantial amounts of time (weeks or months) for acceptable levels of accuracy on even moderate sized datasets. MCMC methods also explicitly reference a parametric stochastic model of evolution, but rather than trying to solve the ML problem under the model, they perform a random walk through the space of model trees. After some burn-in period, statistics are gathered on the set of trees that are subsequently visited. Thus, the output of an MCMC method is not so much a single tree, but a probability distribution on trees or aspects of evolutionary history.

Thus, the major methods for phylogenetic inference, if run a sufficiently long time to obtain an acceptable level of accuracy, can take weeks, months or longer on large datasets, and even moderate datasets (with just a hundred or so taxa) can take several days. Since there are no well established techniques for determining whether a phylogenetic analysis has run for a sufficient time, most systematists use *ad hoc* methods to determine when to stop.

19.3.6   Support assessment

At the end of the process of reconstructing a phylogeny, a systematist may or may not have a single best reconstruction. This is particularly common for maximum parsimony analyses, where for some datasets there can be thousands of equally good trees. With maximum likelihood this is less likely to happen (because of the real-valued optimization, the optimal solution is more likely to be unique), but it can happen with neighbor joining due to ties during the agglomerative procedure. However, all methods have the potential to produce a set of trees that are very close in score to the best score achievable on the dataset, and which are probably statistically no better. In these cases, the researcher would like to have an objective measure of the support for the best phylogeny (i.e., the one that optimizes the objective criterion, such as MP or ML). In the case where the best tree does not have significant support (such as will happen if the second best trees are not substantially worse than the best tree with respect to the objective criterion), the researcher will still want to know which aspects of the evolutionary history implied by the best tree are reliable, where "reliable" means a measure of how well supported the reconstruction is given the data and the method used. Usually reliability is assessed at the level of individual edges in the tree. Reliability can be addressed through statistical techniques, or through more purely combinatorial or "data-mining" techniques.

The combinatorial approach: consensus techniques

The first step of the combinatorial approach to estimating reliability is to select the profile of trees to evaluate. This profile consists of those trees that are close enough to optimal (with respect to the objective criterion) to be considered equally reliable. From this set, a "consensus" tree will be inferred. Of the many ways of defining consensus trees, the most frequently used consensus methods in systematics are the "strict consensus" and "majority consensus" trees. These are defined in terms of edge-induced bipartitions, in a natural way which we now describe.

Let S be a set of species, T be a tree leaf-labeled by S, and e be an edge in T. The deletion of e from T splits the tree into two pieces, and hence creates a bipartition c_e on the set S of leaves. We can thus identify each edge e in T by the bipartition c_e. Furthermore, the tree T is uniquely identified by its set $C(T) = \{c_e : e \in E(T)\}$, and that set is called the "character encoding" of T. We can now define the strict and majority consensus trees.

Given a collection $\mathcal{T}$ of trees, so $\mathcal{T} = \{T_1, T_2, \ldots, T_k\}$, the tree T_{sc} such that $C(T_{sc}) = \cap_i C(T_i)$ is called the **strict consensus tree**. It is not hard to see that the strict consensus tree always exists, and that is the most refined common contraction of all the trees in $\mathcal{T}$. The tree T_{maj} defined by $C(T_{maj}) = \{c_e : |\{i : c_e \in C(T_i)\}| > k/2\}$ is called the **majority consensus tree**. The majority consensus tree always exists, and it always refines or equals the strict consensus tree. Both the strict consensus and the majority trees can be computed in polynomial time. In practice, the majority consensus tree is more often reconstructed than the strict consensus tree. If desired, other consensus trees can also be computed. For example, instead of picking the tree whose bipartitions appear in more than half the trees, one can pick a different threshold, and every threshold $p > 1/2$ will define a tree that will necessarily exist (if $p < 1/2$ this consensus tree may not exist). See [23, 35, 54, 87] for examples of other consensus methods, and [6] for an overview of consensus methods in terms of their theoretical performance.

Other methods of assessing support are statistical in nature. For MP, ML, or distance-based phylogenetic reconstructions, the most common method of support is a bootstrap analysis, although a jackknife analysis is also sometimes applied. When a Bayesian MCMC approach is used to estimate the tree, posterior probabilities are often estimated.

The Bootstrap

Bootstrap analyses take two basic forms: non-parametric and parametric. A non-parametric bootstrap resamples with replacement each of the positions in the original dataset, creating a new dataset drawn from the same distribution. The researcher then uses the same method by which the phylogeny was reconstructed, and the resulting "bootstrap phylogeny" is compared with the reconstructed phylogeny. Many replicated bootstrap runs are performed and the proportion of times that each edge in the reconstructed phylogeny appears in the bootstrap phylogenies is recorded and interpreted as support for the edge. (In biological papers, edges are frequently referred to as "branches", and so the support of an edge in the tree is called "branch support".)

The interpretation of the non-parametric bootstrap is an important issue, and it is often assumed that the bootstrap support for an edge somehow indicates the likelihood that the estimated edge is correct (that is, that it appears in the true tree). However, this is clearly overly simplified. It is possible for a phylogenetic reconstruction method to return the same tree on all replicated datasets, thereby producing bootstrap proportions that are all 100%, and yet the tree (and hence all its edges) may be incorrect. The standard example for this is the "Felsenstein zone" quartet tree, for which maximum parsimony and UPGMA converge to the wrong tree as the sequence length increases, and thus will give very high bootstrap proportions to the wrong tree; see [15] for this result, and [17] for more about phylogeny reconstruction methods, including UPGMA. (It is also important to remember that the gene tree being estimated in this process can differ from the species tree; this, more general point, has to do with how to interpret *any* phylogenetic analysis.) Thus, strictly speaking, the best way to interpret the bootstrap support for an edge is that it indicates the probability that the edge would continue to be reconstructed if the same phylogenetic estimation procedure were applied to datasets having the same distribution as the original dataset. Thus, when the phylogenetic reconstruction method is statistically consistent for the model, the bootstrap proportion for an edge indicates the strength of support for that edge in the original dataset.

Parametric bootstrapping can only be performed when we assume an explicit model of sequence evolution, such as in ML or distance-based phylogenetic analyses. In this case, we use the original data to estimate the parameters of the stochastic model of evolution on the tree we constructed for the dataset. These parameters will generally include site-specific rates of evolution or the distribution from which the rates of evolution are drawn, substitution probabilities on each edge, and an overall substitution matrix governing the evolutionary process across the tree. Once this model tree is constructed, we can simulate evolution down the model tree, producing datasets of the same length as the original dataset.

We then apply the same method used to construct the original tree, and estimate the tree on these new datasets. As with the nonparametric bootstrap, support is assessed by the proportion of times the edges in the phylogeny reconstructed using the original dataset appear in the parametrically bootstrapped reconstructions.

The Jackknife

Jackknifing involves repeatedly deleting some proportion of the original sites (or in some cases original taxa) at random and then using the same method as was used for the full dataset to reconstruct trees on the reduced dataset. Here again, the aim is to determine the strength of support for the edges the tree constructed using the full dataset by assessing the proportion of times that the edges in the analysis of the full dataset appear in the jackknife reconstructions.

Bayesian MCMC methods

When the phylogenetic reconstruction uses a Bayesian MCMC analysis, the output itself is in the form of an estimation of the statistical support for each of the hypotheses. That is, instead of a single best tree being returned, the output contains a frequency count for each of the trees that is visited after burn-in. These values are then normalized to produce the "posterior probabilities" of the different trees. Just as with the bootstrap and jackknife, the interpretation of these values is complicated and subtle (and like the parametric bootstrap, this interpretation will depend upon issues having to do with the model of evolution used to analyze the data).

Computational issues

Because a single analysis of a large dataset under either maximum likelihood or maximum parsimony (to a reasonable level of accuracy) can take a long time to complete (days, weeks or months), a full bootstrap or jackknife analysis can take prohibitively long. Running the bootstrap analysis using fewer replications, or using faster (and hence less accurate) analyses, changes the outcome, and may therefore produce different estimates of support than would have been obtained if the analyses were correctly repeated. This impacts the accuracy of support estimations obtained using bootstrap or jackknifing. On the other hand, using Bayesian MCMC to produce posterior probabilities does not have the same issue – these values are a natural outcome of the initial analysis.

Interpreting support

In the context of phylogenetic analysis, the features of the evolutionary history for which support is estimated are usually topological – particularly, splits (bipartitions) in the tree. Once the phylogenetic analysis is done and the support values have been estimated, a naive interpretation of support values would suggest that features with high support are likely to be true of the true tree, and that features with low support may not be. However, a more sophisticated user of the support estimation technique understands that these interpretations can only reliably be made when the model that has generated the molecular sequence data is the same as the model used to estimate the tree and subsequently to estimate the support. (Of course, some methods, such as maximum parsimony, for estimating trees and support, are not explicitly based upon models; in this case, we would need to know that the method is reasonably accurate under the generating model.) Thus, except under carefully proscribed circumstances, even high levels of support may not be indicative of true features of evolution, and low levels of support may not be indicative of features unlikely to be true.

19.4   Research Problems in Molecular Phylogenetics

Molecular phylogenetics is concerned with the estimation of phylogenies from molecular (i.e., DNA, RNA, or amino-acid) data, which usually consists of sequences, but for DNA it can also be gene orders. There are many research issues in molecular phylogenetics, each of which could have potentially a large impact on practice in systematics. Some of these issues involve database research, others involve statistical inference (including developing better models of the evolutionary process), and some are algorithmic in nature. Rather than attempting to be comprehensive, in this chapter we will limit ourselves to discussing algorithmic research involved in phylogenetic inference, since the target audience is algorithms researchers in computer science. Also, in order to keep this chapter reasonably moderate in size, we will restrict our discussion to issues involved in reconstructing phylogenetic *trees*, rather than the more general problem of reconstructing phylogenetic networks.

19.4.1   Performance analysis of algorithms

Before discussing research problems in phylogenetics, it is important to discuss how algorithms are evaluated in this community. Algorithm developers are familiar with designing algorithms for numeric or combinatorial optimization problems, like vertex coloring, traveling salesperson, etc. Algorithms for polynomial time problems are generally exact - i.e., they find optimal solutions - and hence they are compared in terms of their running time (whether asymptotic running times or on benchmark datasets). By contrast, if a problem is NP-hard, then algorithms may not be exact, but may instead offer bounded-error guarantees (e.g., obtaining a solution no more than twice the cost of the optimal solution), or simply be designed to find hopefully good local (rather than global) optima. In these cases, algorithms (or heuristics, as they are often called) can be evaluated in terms of the scores they find on benchmark datasets and the time they take to find these scores. Benchmarks can be real datasets (typically from some application domain), or random datasets simulated from some distribution (such as random graphs of some sort). Comparing algorithms for NP-hard problems is thus a bit more complicated than comparing algorithms for polynomial time problems, since there is a trade-off between performance with respect to the optimization criterion and running time. Thus, algorithms for the usual numeric or combinatorial optimization problems can be evaluated both theoretically and empirically, with criteria that include running time and accuracy with respect to the objective criterion.

In this light we now consider phylogenetic estimation. Here, too, we have numeric optimization problems, and for the most part they are hard (either proven to be NP-hard, or conjectured to be so); maximum parsimony and maximum likelihood are two obvious examples. (The evaluation of Bayesian MCMC methods is more complicated, since the output is not a single tree with a score for some objective criterion, but rather a probability distribution on trees. We will discuss the issues in evaluating these methods later in this chapter.)

As with all algorithms for hard optimization problems, methods for MP and ML can be evaluated using the same kind of criteria and methodology as described above. Thus, benchmark datasets can be obtained, some real and some randomly generated, and reconstruction methods can be compared in terms of their accuracy with respect to the referenced objective criterion on these benchmarks. When the method is a local search heuristic (i.e., it keeps on searching for improved solutions), the methods can also be evaluated at different points in time, and their performance can then be measured with respect to how long it takes each method to obtain an acceptable level of accuracy. These studies can help evaluate the performance of phylogenetic reconstruction methods to the extent that the user is interested in solving the particular optimization problem (whether it be MP, ML, or something else).

However, phylogeny reconstruction problems are different from the usual combinatorial optimization problems, in several significant ways. First, we are trying to estimate the "true" tree, not just solve some numeric problem, and so our criteria for success must include how close our reconstructed trees are to the tree that actually generated the data. This statement itself makes it clear that phylogeny reconstruction can be considered a *statistical estimation* problem, whereby we are trying to infer something (the tree) from data generated by a stochastic process (defined by the model tree). Thus, issues such as *how much data does a method need to get an accurate reconstruction with high probability* are just as significant as the running time of the method (if not more so) (see the chapter in this volume on large-scale phylogenetic analysis and [85] for more information on this). These issues can be evaluated theoretically, or in simulation.

Simulation studies

The accuracy of a phylogeny reconstruction method is typically measured *topologically* with respect to the true history. Since the true tree is not usually known on any biological dataset, this accuracy estimation is done using a simulation study, as follows. First, a stochastic model of evolution (such as Jukes-Cantor or the Generalized Time Reversible model) is selected, and a model tree (that is, a rooted tree T along with the parameters necessary to define the evolutionary process) is specified. Then a sequence of some length is placed at the root of the tree T and evolved down the tree according to the specified model. At the end of this process there are sequences at each leaf of the tree, and these can be given as input to a phylogeny reconstruction method (such as neighbor joining, or a heuristic for MP, etc.), thus producing an estimated tree T'. The estimated tree T' is then compared to the model tree T with respect to topological accuracy.

The standard way that trees are compared in the phylogenetic research literature is the Robinson-Foulds (RF) metric [59]. The RF metric between trees is defined in terms of the character encoding of a tree (as described in Section 19.3.6 earlier). Let T be the true tree for a set S of n taxa, and let T' be a reconstructed tree on S. Then $RF(T,T') = \frac{|C(T)-C(T')|+|C(T')-C(T)|}{2(n-3)}$. Note that $0 \leq RF(T,T') \leq 1$, and that $T = T'$ if and only if $RF(T,T') = 0$. RF rates below 10% are generally required, unless the data themselves are so poor that a good estimation of the true tree is unlikely.

The major advantage of using simulations as compared to real data is that for almost all real datasets, it is not possible to know precisely the correct evolutionary history, and those aspects of the evolutionary history that are reliable are also generally easy to infer using any method. Thus, it is not particularly helpful nor straightforward to try to evaluate methods with respect to topological accuracy on real datasets. Since topological accuracy is so important, simulations have become a standard methodology for evaluating the accuracy of phylogenetic reconstruction methods.

There are, however, distinct issues (and disadvantages) in using simulation studies. The most compelling of these issues is that the mathematical models used to define the evolutionary processes are not nearly as complex as those that operate on real organisms and genomes; thus, inferences about accuracy of reconstruction methods must be taken with a certain amount of salt, metaphorically speaking. Additionally, to the extent that simulations are used to evaluate running times for heuristics for hard optimization problems, the landscape produced by these models are also smoother (easier to navigate and find optimal solutions) than the landscapes of real datasets. All in all, however, simulations are important (otherwise we can rarely, if ever, have a real benchmark), and have changed practice within molecular systematics dramatically.

19.4.2   Phylogenetic reconstruction on molecular sequences

Various numeric optimization problems have been formulated for phylogeny reconstruction, of which a few have received significant support by the systematic biology community; these are Maximum Parsimony, Maximum Likelihood, and (in a disguised form) Maximum Integrated Likelihood. While systematists do not generally agree which of these optimization problems is the most appropriate, all of these are of interest to a sizeable community. Unfortunately, these are hard problems (MP and ML provably NP-hard), and so exact solutions are not generally feasible except for sufficiently small datasets.

Heuristic searches for MP and ML

Heuristics for MP and ML (largely based upon hill-climbing) are in very broad use in the systematics user community, and seem to provide quite accurate solutions on small to moderate sized datasets. These heuristics differ from each other in various ways, but most use the same set of transformations for moving from one tree to another. First, a fast method (neighbor joining, or a greedy insertion of taxa into a tree to optimize MP or ML) is used to obtain an initial tree. Then, a neighborhood of the tree is examined, and each tree scored with respect to the objective criterion (MP or ML). If a better tree is found, the search continues from the new tree; otherwise, the current tree is a local optimum and the search may terminate. Some heuristics include additional techniques in order to get out of local optima. The Ratchet [47] is one of the most successful of these techniques for getting out of local optima; it randomly perturbs the sequence data, and then hill-climbs from the current tree (but using the perturbed data to score each visited tree) until a local optimum is found. Then the data are returned to their original values, and the hill-climbing resumes.

It should be clear from the description that these methods may not terminate in any acceptable time period, especially if randomness is included; thus, the systematist must decide when the search has gone on long enough.

The best of these heuristics, as implemented in the popular software packages PAUP* [75], TNT [22], and others, are quite effective at producing good MP analyses on even fairly big datasets (containing a few hundred sequences), provided enough time is allotted. The limit for maximum parsimony analyses using currently available software is probably 1,000 taxa, and maximum likelihood analyses are probably limited to 100 (or fewer) taxa. (Bayesian MCMC methods are reputed to be able to do well on large datasets, but as it is not clear how to evaluate the performance of these methods, this needs additional study.)

However, with lowered costs, automation, and worldwide accumulation of DNA sequence data, systematists now attempt to reconstruct phylogenies on ever larger datasets; many phylogenetic datasets now have easily above a thousand sequences. These datasets are much harder to analyze well in a "reasonable" time (of perhaps a few days or even a few weeks), by comparison to smaller datasets.

Thus, in general, large-scale phylogenetic analysis is quite difficult to do in a reasonable amount of time, and much of the focus of algorithm development (and of our discussion) is on developing better heuristics – ones that can provide sufficient analyses on large datasets in a matter of days rather than months or years. However, since MP and ML searches can take a long time to find optimal solutions, it is also important to be able to assess when it is safe to stop searching for better trees. The current technique is to run the analysis until it seems to have converged. However, it is very difficult to establish convergence, and there is clearly a real possibility that the method has not converged so much as slowed down dramatically with respect to finding improved scores.

Thus, a natural combinatorial optimization problem that is also relevant to practice is to obtain better bounds for MP and ML. If good lower bounds could be obtained, these could be used to evaluate how close to optimal a current best tree is, and that in turn could be used to evaluate whether it was reasonably safe to stop running the heuristic search.

Maximum Parsimony

We begin with Maximum Parsimony, which is the simplest of these optimization problems. Heuristics for this problem have been used to construct perhaps the majority of published phylogenies, and so MP is a major approach to phylogeny estimation. However, optimal solutions to MP may not be correct reconstructions of evolution. There is no guarantee that MP will yield a correct solution, even given infinitely long sequences because MP is not statistically consistent in general. Still, MP is an important problem, and improved algorithms for MP would represent an important advance for computational phylogenetics.

The Maximum Parsimony Search Problem:
- *Input:* Set S of sequences, each of length k, in a multiple alignment
- *Output:* Tree T leaf-labeled by S and with additional sequences, all of length k, labeling the internal nodes of T, so as to minimize

$$\sum_{(u,v)\in E} H(s_u, s_v),$$

where s_u and s_v denote the sequence labeling the nodes u and v, respectively, $H(x,y)$ denotes the Hamming distance between x and y, and E denotes the edge set of T. (The Hamming distance between two sequences is the number of positions in which they differ.)

Maximum Parsimony (MP) is thus the *Hamming distance Steiner Tree problem*. Although MP is NP-hard [19], it, like Steiner Tree problems in general, can be approximated. Also, although finding the best tree is NP-hard, it is possible to score a given fixed tree in linear time (i.e., it is possible to compute sequences at the internal nodes of a given fixed tree so as to minimize the total number of steps in $O(rnk)$ time, where there are n sequences each of length k, each over an alphabet of size r), using the well known Fitch-Hartigan algorithm [18]. Therefore, finding the best MP trees can be solved exactly through techniques such as exhaustive search and branch-and-bound, but such techniques are limited to about 25 or 30 taxa, since the number of trees on n leaves is $(2n - 5)!!$. For larger datasets, heuristic search techniques are used to analyze essentially all datasets of interest to systematists. On moderate sized datasets (up to a few hundred sequences) these heuristics probably work quite well (though there are no theoretical guarantees bounding the error of these heuristics), but much less is understood about their performance on larger datasets, especially datasets with more than a thousand sequences.

Research questions for MP

In addition to the general challenges we discussed earlier in the context of both MP and ML, there are many research questions of particular relevance to MP. One question that is particularly intriguing is to explain, mathematically, why MP is as good as it is. That is, statistical theory has established that MP is not a statistically consistent method for even simple DNA sequence evolution models, and so cannot be guaranteed to reconstruct the true tree (with high probability) even on extremely long sequences. Yet MP's performance in simulation studies (when the model trees are sufficiently large and biological) is clearly not bad. In some cases, it can be better than statistically consistent methods like neighbor joining, and comparable to ML. Why? There must be some theory to explain this phenomenon.

Maximum Likelihood

The usual maximum likelihood problem in phylogenetics is to find a tree and its associated parameters so as to maximize the probability of the observed data. Since stochastic models differ according to the parameters that must be specified (and those that are fixed for the model), the use of a maximum likelihood analysis requires that the stochastic model of evolution already be explicitly specified. While the model choice definitely affects the running time of the software for finding the best ML trees (the more parameters in the model that must be estimated, the more computationally intensive), in essence the mathematics of ML estimation does not change between the simplest model (JC) and the most complex of the standard models (the GTR model). Thus, we will discuss Maximum Likelihood estimation under the JC model.

Recall that, like all of the standard models, the JC model assumes that the sites evolve identically and independently down a tree T, and that the state of each site at the root of T is drawn from a given distribution (usually the uniform distribution or a distribution estimated from the dataset). The model is then simply defined in terms of the evolution of a single site. The main feature of the JC model is that it asserts that if a site changes on an edge e, it changes with equal probability to the remaining states. Thus, the entire evolutionary model can be described by the pair (T, p), where T is a rooted tree, and p is a function $p : E(T) \rightarrow (0, 3/4)$, so that $p(e)$ is the substitution probability on edge e.

Once a model tree (that is, the tree T and its associated parameters as defined by the function p) is specified, it is possible to define the probability $Pr[S|T, p]$ of a given set S of sequences placed at the leaves being generated by the model tree (T, p). Furthermore, this quantity can be calculated in polynomial time, using dynamic programming [16].

The ML score of a fixed tree

Let T be a fixed tree. The score of the tree under ML (for a given model) is defined to be $score_{ML}(T) = sup_p\{Pr[S|T, p]\}$. Note that we have used the supremum, indicated by sup, instead of the maximum. This is because the maximum may not exist, but the supremum will (because the set $\{Pr[S|T, p]\}$ is bounded from above by 1).

Maximum Likelihood for the Jukes-Cantor model:

We now define the ML search problem, again in the context of the Jukes-Cantor model (though the definitions and discussion extend in the obvious way to models with more parameters).

The objective in an ML search is to find the tree with the highest ML score; however, this optimal solution may not exist, because the set is not closed (in the same sense in which we can say there is no largest number in the open interval $(0, 3/4)$). Therefore we will state the ML problem as a decision problem (i.e., does there exist a solution of at least score B?).

- *Input:* A set S of sequences, each of the same length, and a value B.
- *Output:* A model tree (T, p) (where $p : E(T) \rightarrow (0, 3/4)$ defines the substitution probabilities on the edges of T) such that $Pr[S|T, p] \geq B$ (if such a model tree exists); otherwise, *FAIL*.

From a computational viewpoint, ML is very difficult. Solving ML tree under the Jukes-Cantor model is NP-hard [11] (and conjectured NP-hard under the other models), and harder in practice than MP. Worse, even the problem of finding the optimal parameters for a fixed tree is potentially NP-hard, even for trees with only four leaves! Existing approaches to find optimal parameters on a fixed tree, which utilize hill-climbing on the finite dimensional real-parameter space, are not known to solve the problem exactly [69]. Note that the usual exhaustive search strategy isn't feasible, since this optimization is over a continuous space rather than a discrete space.

Research questions for ML

While ML and MP are clearly different, both ML and MP share the same search-space issues, and so heuristics that improve techniques for searching through "treespace" will help speed up both ML and MP analyses. Thus, many of the questions that we discussed in the context of MP apply here as well. However, ML has some additional challenges that are not shared by MP, and which make ML analyses additionally challenging.

The main challenge is computing the ML score of a given tree T; in practice, this amounts to finding parameter settings that will produce (up to some tolerated error) the maximum probability of the data. The computational complexity of this problem is open, and current techniques use hill-climbing strategies which may not find global optima [69]. By contrast, the corresponding problem for MP (computing the "length" of a fixed tree) can be solved in linear time using dynamic programming. Thus, parameter estimation on a given tree is the real bottleneck for ML searches. In fact, this is such a time-consuming step in ML searches, that the popular heuristics for ML do not actually try to find optimal parameters on every tree, but only on some. The cost of computing these optimal parameters is just too high.

Consider then the possibility of *not* computing the optimal parameters. Instead, suppose we could quickly compute an upper bound on the ML score of the tree (that is, the probability of the data under the best possible settings of the parameter values). If we could do this, efficiently, we might be able to speed up solutions to ML. That is, during the heuristic search through treespace, instead of performing the computationally intensive task of computing optimal parameters on a tree we visit, we would simply check that the upper bound we have on its score is at least as big as our current best score. If it is not, we can eliminate this tree from consideration. In this way, we can (rigorously) select those trees which are worth actually spending the time to score exactly, and thus potentially speed up the search.

Closely related to this is the question of simply comparing two trees for their possible scores, rather than scoring either one. Consider the following question:

For fixed phylogenetic trees T_1 and T_2 on set S, is

$$score_{ML}(T_1) > score_{ML}(T_2)?$$

Suppose we could answer this question in a fraction of the time it takes to find the optimal parameters on a tree (i.e. faster than it takes to actually compute $score_{ML}(T)$. In this case, we could also traverse tree space more quickly, and thus get improved solutions to ML.

The challenge in these approaches is to be able to make these comparisons rigorously and efficiently, rather than in an *ad hoc* fashion. In addition, the objective is to obtain an empirical advantage, and not just a theoretical one.

MrBayes, and other Bayesian MCMC methods

It should be clear that Bayesian MCMC methods are not trying to solve maximum likelihood in the sense we have defined, whereby the model tree (that is, the tree with parameter values) is returned that maximizes the probability of the data. However, there is a kind of maximum likelihood problem which Bayesian MCMC methods *can* be used to solve. This problem is the "maximum integrated likelihood" problem, described in [71], and which we now define. Recall that if we are given a set S of sequences and a model tree T with an assignment p of the parameter values to the tree, we can compute $Pr[S|T, p]$ in polynomial time. Thus, for a fixed tree T, we can define the "integrated likelihood" of T to be the *integral* of this quantity, over all the possible parameter settings. In other words, the integrated likelihood of a tree T, which we denote $IL(T)$, is defined by

$$IL(T) = \int Pr[S|T, p] dF(p|T),$$

where $F(p|T)$ is the distribution function of the parameters p on T. In general, $(F(p|T))$ has a probability density function $f(p|T)$, so that we can write this as

$$IL(T) = \int Pr[S|T, p] f(p|T) dp.$$

The tree T which has the maximum possible integrated likelihood value is called the "maximum integrated likelihood tree". The maximum integrated likelihood tree has many desirable properties, some of which are quite surprising [69]. In particular, as Penny and Steel [69] point out, the integrated likelihood of a tree T is proportional to its posterior probability.

Despite the popularity of MrBayes [33] and other Bayesian MCMC methods, not much is known about how to run the methods so as to obtain good analyses, nor about how to evaluate the performance of a Bayesian MCMC method. By contrast, much more is known about MP and ML. Consider, for example, the question of how to evaluate the performance of a heuristic for MP or ML. We can assemble benchmark datasets, and we can analyze each dataset using various heuristics, and record the best score found by each heuristic under various conditions. This is a legitimate way to compare methods, provided that the conditions are identical.

Some systematists use Bayesian MCMC methods as heuristics for ML; rather than using the normal output (i.e., posterior probabilities), they simply return either the most frequently visited tree, or the model tree which had the best likelihood score. If Bayesian MCMC methods are used in this way, then it is reasonable to use the same methodology for evaluating ML methods to evaluate Bayesian MCMC methods. But this is not what Bayesian MCMC methods are really designed for – they produce posterior distributions on trees, not single trees with scores. Therefore, how should we evaluate a Bayesian MCMC method? To do this, we need to know what the "correct" output should be, and failing that, whether one posterior distribution is better than another.

Research questions for Bayesian MCMC methods

The most fundamental problem for Bayesian MCMC is to be able to say what the "correct" output is, and to be able to analytically compute that (even if it would take a long time to obtain an answer). Recall the observation made earlier that the integrated likelihood of a tree T is proportional to its posterior probability, if the MCMC method were to reach the stationary distribution. Hence, if we can calculate the integrated likelihood of each tree exactly, we can actually evaluate the accuracy of a Bayesian MCMC method. The question then becomes: how can we calculate this integral exactly? Once again, this seems difficult, and the problem is that we are trying to calculate something that is in a multi-dimensional continuous space, not a discrete space.

The difficulty in how to evaluate the output of a MCMC method is part of both the appeal and the problem with using MCMC methods – if there is no explicit way to evaluate the quality of the output, then stopping early (and hence finishing quickly) is potentially acceptable. On the other hand, if one wishes to be more conservative about the use of this technology, it becomes necessary to have tools for evaluating how long one should take for the burn-in period, before sampling from what is hoped to be the stationary distribution. Thus, two algorithmic research problems that present themselves in the context of MCMC methods are (1) developing analytical techniques for obtaining bounded-error estimations of the integrated likelihood of fixed trees, and (2) determining when a sufficient amount of time has elapsed so that the burn-in period can be considered complete and the sampling of trees can begin. It is easy to approach these problems using *ad hoc* techniques; the challenge here is (if possible) to develop techniques with a firm theoretical foundation. More generally, of course, designing new MCMC methods with better convergence rates is always beneficial.

19.4.3   Multiple Sequence Alignment (MSA)

MSA remains one of the most significant open problems related to phylogeny estimation, with no really satisfactory software. One of the major challenges for the algorithms designer in developing better MSA methods is that no objective criterion for MSA has been met with general acceptance in the phylogeny research community. Instead, MSA methods (especially MSA methods for protein sequences) are evaluated with respect to accuracy on specific real datasets for which correct structural alignments are known. This makes the development of improved methods for MSA difficult to achieve since the structures of most molecular sequences are not known in advance, and the alignment must be obtained without that knowledge. Furthermore, as noted above, a correct structural alignment may not produce an alignment that maximizes positional homology (i.e., structural alignments need not produce a set of columns where each column has character states that are the result of a common evolutionary history for those character states). Since datasets where analyses based upon these different optimality criteria can lead to different alignments do come up in practice, the current approaches for MSA are inadequate for the purposes of phylogenetic reconstruction.

The rest of this section will describe two numeric optimization problems that have been suggested for MSA. Each of these optimization problems defines the cost of a given multiple alignment on the basis of a set of pairwise alignments. Therefore, we begin by describing how pairwise alignments are scored.

Pairwise alignments

Typically, the cost of a pairwise alignment depends upon the number of each type of substitution, and the number and length of the gaps in the pairwise alignment. The cost of each type of substitution is given by a substitution cost matrix which can be quite arbitrary, although there are standard matrices used in the community. The cost of a gap is more complicated. In general, "affine gap penalties" are the most frequently used. These penalties are of the form $C_0 + C_1(l - 1)$, where l is the length of the gap, and C_0 and C_1 are two positive real numbers. In general, C_1 is much less than C_0, reflecting the model assumption that initiating a gap is harder to do than extending a gap. Note also that if we allowed $C_0 = \infty$, then no gaps would be permitted, and so affine gap penalties can be used to model various model conditions.

Given any such function for the cost of a pairwise alignment, we can then define the obvious optimization problem – given two sequences, find the pairwise alignment of minimum cost. The *pairwise alignment* problem can be solved in polynomial time (standard dynamic programming techniques give an $O(mn)$ time algorithm when the cost function uses an affine gap penalty, where m and n are the lengths of the two sequences).

Consider now a multiple alignment A on the set $S = \{s_1, s_2, \ldots, s_n\}$, and consider two sequences s_i and s_j in S. The pairwise alignment induced by A on sequences s_i and s_j is obtained by examining the alignment A, and restricting the attention to just the ith and j^{th} rows. We denote this induced pairwise alignment by $A(s_i, s_j)$. Then if we are given a cost function $f(\cdot, \cdot)$ on pairwise alignments, we can extend it to any multiple alignment in the obvious way: simply score every induced pairwise alignment, and add up the scores. We formalize this as follows:

Sum-of-Pairs (SOP) alignment
- *Input:* A set $S = \{s_1, s_2, \ldots, s_n\}$, of sequences and a function $f(\cdot, \cdot)$ for computing the cost of a given pairwise alignment between two sequences.
- *Output:* A multiple alignment A on S such that $\sum_{i,j} f(A(s_i, s_j))$ is minimized.

This natural optimization problem is NP-hard to solve exactly [34], but can also be approximated. Despite its natural appeal, it has no demonstrated connection to evolution. Therefore, consider the following alternative way of looking at multiple sequence alignments. This problem (a special case of which was introduced in [64]) can be seen as an extension of the maximum parsimony optimization problem in which we allow for insertions and deletions of substrings during the evolutionary process.

Generalized Tree Alignment (GTA)
- *Input:* A set S of sequences and a function $f(\cdot, \cdot)$ for computing the cost of a given pairwise alignment between two sequence.
- *Output:* A tree T which is leaf-labeled by the set S and with additional sequences labeling the internal nodes of T, so as to minimize

$$\sum_{(v,w) \in E} f(A_{opt}(s_v, s_w)),$$

  where s_v and s_w are the sequences assigned to nodes v and w respectively and E is the edge set of T.

It is easy to see that the Generalized Tree Alignment (GTA) problem is NP-hard, since the special case where gaps have infinite cost (and hence are not permitted) is the maximum parsimony (MP) problem, which is NP-hard. The fixed tree version of GTA is also of interest, and has received as much attention as the Generalized Tree Alignment problem. We now describe this.

Tree Alignment
- *Input:* A tree T leaf-labeled by a set S of sequences and a function $f(\cdot, \cdot)$ for computing the cost of a given pairwise alignment between two sequences.
- *Output:* An assignment of sequences to the internal nodes of T so as to minimize

$$\sum_{(v,w) \in E} f(A_{opt}(s_v, s_w)),$$

  where s_v and s_w are the sequences assigned to nodes v and w respectively and E is the edge set of T.

Unfortunately, Tree Alignment is also NP-hard [81]. Algorithms which return provably optimal solutions for the Tree Alignment problem have been developed for the case where the function $f(\cdot, \cdot)$ uses an affine gap penalty, but these run in $O(c^n k^n)$ time, where c is a constant, k is the maximum sequence length, and n is the number of leaves in the tree [37]; thus, exact solutions to Tree Alignment are computationally infeasible except for extremely small trees with short sequences. Approximation algorithms for the problem have also been developed. One of the simplest of the approximation algorithms is the 2-approximation algorithm in [26], which can be used with arbitrary functions $f(\cdot, \cdot)$ that satisfy the triangle inequality). For the case of affine gap penalties, a PTAS (polynomial time approximation scheme) has also been developed [82]. However, because all the approximation algorithms with good ratios are computationally intensive (even on small datasets!), they are not used in practice. (Gusfield, however, suggests using the 2-approximation in [26] in order to obtain *lower bounds* on achievable alignment costs, rather than to actually estimate a good alignment!)

Heuristics for either the Generalized Tree Alignment problem (in which the tree is not known) or the Tree Alignment problem (when the tree is assumed) have also been developed, and there is still a lively interest in this area (see [21, 26, 37, 56, 65, 57, 82, 86]). However, the performance of these methods is still not well understood, and the standard practice by most systematists is still to use a method such as ClustalW to obtain an alignment, and then to infer a tree on the basis of the alignment.

Maximum likelihood and Bayesian approaches can also be used for phylogenetic multiple sequence alignment, but these require an explicit model of evolution which incorporates insertions and deletions (and perhaps also duplications) as well as site substitutions. Some such models exist, but ML and Bayesian methods based upon these models are extremely computationally intensive, and are unlikely to scale; see [31, 42, 58, 67, 73, 78, 79] for some work in this area.

Research questions for MSA

The main challenge here, from the point of view of phylogenetic estimation, is developing MSA techniques that are appropriate for phylogenetic reconstruction, so that accurate trees can be obtained when the input data are not yet aligned. To establish such a method, however, better models of sequence evolution (ones that include events that make a multiple alignment necessary) need to be developed, so that methods can be tested on simulated data. Such events include duplications of genes, insertions and deletions of DNA regions, and large-scale events such as inversions and transpositions. Realistic simulators should incorporate all these events, while still keeping the flexibility of the standard DNA sequence models which do not enforce molecular clocks or constant rates across sites. No simulator available today has all the flexibility needed to be of real use in testing alignment algorithms. Until we can test methods in simulation, we will not know if trying to optimize the tree length (i.e., trying to solve the Generalized Tree Alignment problem) will produce better trees from unaligned sequences. It is possible that we may need to come up with different optimality criteria in order to best construct trees and alignments simultaneously.

Thus, there are really two main challenges: first, to develop good stochastic models that reflect the properties of real datasets, and then to develop methods for alignment (and perhaps simultaneous alignment and phylogeny reconstruction) that enable accurate phylogenies to be inferred. The general challenge of developing better, more biologically realistic, models of evolution applies to *all* aspects of phylogenetic inference.

19.4.4 Special challenges involved in large-scale phylogenetics

Since most approaches for estimating phylogenies involve solving hard optimization problems, phylogeny reconstruction is generally computationally intensive, and the larger the dataset the more computationally challenging the analysis. This is the obvious challenge in analyzing large datasets. But certain other problems become particularly difficult when large datasets are analyzed. In particular, as mentioned before, assessing confidence in estimated phylogenies using bootstrapping becomes infeasible, unless fast reconstruction methods are used instead of computationally intensive ones.

Another challenge that comes up is storing and analyzing the set of best trees found during an analysis. Even for moderate sized datasets this can be a large number (running in the thousands), and the number of best trees for larger datasets may conceivably run into the millions. How to store these datasets in a space-efficient manner, and so that consensus methods and other datamining techniques can be applied to the set, is still largely an open problem. (See [4] for some progress on this problem.)

Finally, with datasets containing many taxa, the incidence of missing data and difficult multiple sequence alignments increases, thus making the usual approaches to phylogeny estimation difficult. In these cases, the systematist may wish to consider approaches for phylogeny reconstruction which first divide the full data matrix into smaller (probably overlapping) subsets (which may have less missing data, or be easier to align); such subsets should be easier to analyze phylogenetically. These smaller trees on subsets of the taxa may then be used in a *supertree analysis* (the subject of the next section) in order to obtain a phylogeny on the full dataset.

19.5 Special Topic: Supertree Methods

19.5.1 Introduction

Supertree methods attempt to estimate the evolutionary history of a set S of sequences given estimates of evolutionary history for subsets of S. Thus, supertree methods take as input a collection of trees (which may be rooted or unrooted, depending upon the method used to estimate evolution), and they produce a tree on the union of all the input leaf sets. Supertree methods *may* be critical to the inference of the "Tree of Life" (although other approaches do exist, which we discuss below), and for this reason there is an increasing interest in the research community on understanding these methods. See, for example, [3], a volume focusing on supertree methods, their analyses, and discussions of the benefits and pitfalls of these approaches.

Supertree methods can be used in an exploratory fashion, to see (for example) what can be inferred just by combining phylogenies from previously published analyses. However, sometimes a biologist has a dataset that suggests the use of a supertree analysis, due to properties of the data themselves. For example, recall that systematists routinely use multiple markers for phylogenetic reconstruction. Even if the markers are considered to be likely to produce compatible trees (i.e., if reticulation and gene tree/species tree conflicts have been ruled out), when there are enough missing data, each marker may only be relevant for a subset of the taxa. When this happens, each marker may be analyzed separately, with the result being that a different tree is obtained for each marker. Since the trees will not have identical leaf sets, in order to obtain a tree on the entire dataset, a supertree method is then applied. We call this type of use of supertree methods (when only the trees are used to construct the supertree, and not the character matrices as well) *meta-analysis*.

Thus, in some cases the original character datasets are available, but in others only the trees are available. When the original character datasets are available, supertree methods are not the only option – *supermatrix* analyses (where the different matrices for each marker are combined into one data matrix) can also be considered. If the submatrices do not all share the same taxa, then some characters will be missing states for some taxa. In this case, when the supermatrix is created, the missing entries are simply coded as "missing". Phylogeny reconstruction methods are generally adapted for such data, since missing data are fairly common, and so the newly created supermatrix can then be given as input to a phylogenetic reconstruction method. (In this case, however, a reconstruction method may keep track of the way the new supermatrix is composed of submatrices, so that different stochastic models can be used on the different parts of the supermatrix during the estimation of the phylogeny.)

Supertree methods may also be used as part of a divide-and-conquer strategy whereby a dataset is decomposed into smaller, overlapping datasets, trees are reconstructed on each subset, and then the smaller trees are merged into a tree on the full dataset. (See the chapter on large-scale phylogenetic analysis in this volume, and also [60], for more on this kind of application of supertree methods.)

Each of these uses of supertree methods entails somewhat different algorithmic challenges. Supertree methods designed for use in arbitrary meta analyses need to be able to accept arbitrary inputs (and perform well on them). However, supertree methods used in the context of a divide-and-conquer strategy need not be designed to handle (or perform well on) arbitrary inputs, since the input trees can be assumed to have certain overlap patterns (since the divide-and-conquer strategy can produce subset decompositions that are favorable to the supertree method).

The utility of a supertree method must always be considered in comparison to the obvious supermatrix analysis; both approaches have theoretical advantages and disadvantages, and the relative quality of these approaches is not yet known. Note, however, that a supermatrix approach is not always possible; sometimes only the trees are available, and not the original character data.

19.5.2   Tree Compatibility

The most obvious computational problem related to supertree construction is to determine if a collection of trees is compatible, and if so, to construct a supertree consistent with all the input trees. We now make these concepts precise.

Terminology

Let T and T' be two trees on S. Then T is said to *refine* T' if T' can be obtained from T by a sequence of edge-contractions. Thus, every tree refines the star-tree on the same set of leaves, and if T refines T' and both are binary trees (or more simply have the same number of edges), then $T = T'$. Finally, trees T and T' on the same leaf set are *compatible* if there is a tree T'' such that T'' refines both T and T'. Note that if T'' exists, it may be equal to T or T'.

These definitions can be extended to the case where the input trees are not on the same leaf set, by considering trees restricted to subsets of their leaf sets. We can restrict a tree T to a subset A of its leaves in the obvious way: including only the subtree of T connecting the leaves in A and suppressing nodes of degree two. The resultant tree is denoted by T_A. We now define tree compatibility.

Tree compatibility
- **Input:** Set $\mathcal{T} = \{T_1, T_2, \ldots, T_k\}$ of trees on sets $S_1, S_2, \ldots, S_k$, respectively.
- **Output:** Tree T, if it exists, such that for each i, $T|S_i$ refines T_i.

The tree compatibility problem is NP-hard [68], and so is difficult to solve. Furthermore, its relevance to practice is questionable, as almost all phylogenetic analyses have some errors, and so most inputs to a supertree problem will simply not be compatible.

Now consider the case where the input trees are rooted. In this case, we are looking for a rooted supertree which is consistent with all the inputs – thus, the rooted trees must be correct with respect to the location of their roots, as well as topologically. Once again, we can ask if the problem is computationally tractable and relevant to practice. Here, the answers are as follows. First, the rooted compatibility problem is solvable in polynomial time [1]. However, since inputs to the (unrooted) tree compatibility problem are unlikely to be compatible, inputs to the rooted compatibility problem are even less likely to be compatible – therefore, the problem is not particularly relevant to practice.

19.5.3 Matrix Representation Parsimony

Since estimates of evolutionary trees may not be completely correct, supertree methods need to be able to handle incompatibility in their inputs. Fortunately, there are approaches for constructing supertrees from incompatible trees. The most popular method is *Matrix Representation Parsimony* (MRP). This method can be applied to rooted or unrooted trees, and uses maximum parsimony to analyze the data matrix that it creates. The technique used to create the data matrix depends upon whether the trees are rooted or unrooted.

If the trees are unrooted, then each input tree is replaced by a data matrix of partial binary characters on the entire set of taxa, where by "partial binary character" we mean characters whose states are 0, 1, and ?, where the ? indicates that the state is missing for that taxon. An example will make this clear. Let T be one of the trees, and let T have leaf set $A \subset S$. Let $e \in E(T)$ be an edge in T, and let $A_0|A_1$ be the bipartition on A created by deleting e (but not its endpoints) from T. Then e is represented by the partial binary character c_e defined by $c_e(a) = 0$ if $a \in A_i$, $i = 0, 1$, and $c_e(s) =?$ if $s \in S - A$. Note that c_e can be defined in two ways, depending upon the definition of A_0 and A_1; however, this is irrelevant to the computation that ensues. Thus T is replaced by the data matrix on S with character set $\{c_e : e \in E(T)\}$. We represent each tree in the profile by its data matrix, and concatenate all the data matrices.

The same representation is used for rooted trees, except that the characters c_e are defined in such a way that the part of the tree containing the root is assigned the zero state. Also, an additional row consisting of all zeros is added to the data matrix that is created to represent the root, and the trees that are obtained are rooted at this added node. In this way, MRP can be used to analyze either unrooted or rooted trees.

This concatenated data matrix is then analyzed using maximum parsimony, and the output of the maximum parsimony method is returned (this can either be all the optimal trees, or a consensus of the optimal trees, as desired). Note that if all the trees are compatible (meaning that they can be combined into one supertree T without loss of accuracy), then this technique will construct the correct supertree (along with any other supertree which is consistent with all the input trees). However, because MRP involves solving an NP-hard problem, its running time is not generally efficient. Furthermore, most inputs will have some error, and it is not clear how errors in the input trees affect the quality of the MRP tree. Thus, despite the nice theoretical property of MRP, its performance in practice is less clear.

19.5.4   Other supertree methods

Finally, other supertree methods also exist which can handle unrooted or rooted trees which have errors. For example, Gordon's strict consensus supertree [23] method is an interesting method, but not in general use. Another method quite similar to Gordon's strict consensus supertree method is used in the "Disk-Covering Methods" (DCMs) described in this volume. These DCMs are divide-and-conquer methods used to speed up maximum parsimony analyses, as well as improve other phylogenetic reconstruction methods. Both of these supertree methods are guaranteed to solve tree compatibility *if* the input trees are "big enough" (i.e. have enough overlap) and are correct.

19.5.5   Open problems

Despite the potential for supertree methods to be useful, and the interest in them (see the book chapter in this volume on supertrees, and also [3]), little is really known about how well they work on real data analyses. Thus, the main research that needs to be done is to evaluate how well they work in comparison to each other, and also in comparison to supermatrix approaches. In the likely event that current approaches are not able to produce high quality supertrees, new methods should be developed. As a first step, it is likely that new optimization problems should be developed. As these are likely to be NP-hard (as is almost everything in phylogeny estimation), heuristics for these problems will need to be developed.

19.6   Special Topic: Genomic Phylogeny Reconstruction

In the previous section we described phylogenetic inference when the input is a set of aligned molecular (DNA, RNA, or amino-acid) sequences. In this section, we will discuss phylogenetic inference for a different kind of data – whole genomes, in which the information is the order and strandedness of the genes within the genomes.

Just as evolution changes the individual nucleotides within gene sequences, other events also take place which affect the "chromosomal architecture" of whole genomes. Some of these events, such as inversions (which pick up a region and replace it in the same location but in the reverse order, and on the opposite strand) and transpositions (which move a segment from one location to another within the same chromosome) change the order and strandedness of genes within individual chromosomes; others, such as translocations (which move genomic segments from one chromosome to another) duplications, insertions, and deletions can change both the order and also the number of copies of a given gene within a chromosome. Finally, fissions (which split a chromosome into two) and fusions (which merge two chromosomes) change the number of chromosomes within a genome.

All these events are less frequent than individual point mutations, and so have inspired biologists to consider using gene orders as a source of phylogenetic signal in the hope that they might allow evolutionary histories to be reconstructed at a deeper level than is typically possible using molecular phylogenetics. On the other hand, much less is really understood about how genomes evolve, and so the statistical models describing the evolution of whole genomes are not as well developed. Equally problematic is the fact that from a computational standpoint, whole genome phylogenetic analysis is much more difficult than comparable approaches in molecular phylogenetics.

Research in gene order phylogeny has largely focused on two basic approaches: parsimony-style methods that seek to find trees with a minimum total "length", and distance-based methods. Both types of approaches require the ability to compute either edit distances between two genomes (that is, determining the minimum number of events needed to transform one gene order into another) or estimating true evolutionary distances (i.e., estimating the actual number of events that occurred in the evolutionary history between the two gene orders). Algorithms to compute edit distances tend to involve graph-theoretic algorithms (see [27, 28, 29] for initial work in the area) whereas algorithms to estimate true evolutionary distances involve probability and statistics (see [12, 13, 43, 74, 83, 84]).

The estimation of true evolutionary distances (that is, the actual number of events that have occured in the evolutionary history between a pair of gene orders) is directly relevant to phylogeny reconstruction since if this estimation if done sufficiently well and obtained for every pair of chromosomes, then distance-based reconstruction methods (such as neighbor joining) applied to these distances will return accurate trees (the same statement is not true when used with edit distances). However, the inference of these distances requires that a stochastic model for the evolutionary process be given. Typically these algorithms operate by computing one of two standard edit distances on the two chromosomes (either the minimum inversion distance or the breakpoint distance, both of which can be computed in polynomial time and do not depend upon any model assumptions), and then using that value to estimate the number of events that occured in the evolutionary history between the two chromosomes. Since the estimations obtained by these algorithms depends closely on the assumptions of the stochastic model, the more general the model the more accurate (and more generally applicable) the algorithm is likely to be. The algorithms in [12, 13] apply to a model of gene order evolution in which only inversions occur, and the algorithms in [43, 83, 84] apply to a model of gene order evolution in which inversions, transpositions, and inverted transpositions occur. The algorithm in [74] can be used when these events, as well as insertions and deletions, occur.

Note that this last algorithm can analyze datasets which have unequal gene content (i.e., some chromosomes have more than one copy of a gene, while others have only one copy or may even lack any copy of the gene). While some other work has been done for unequal gene content case, the majority of the research has been focused on the equal gene content case. Even for this special case, however, many computational problems are known or conjectured to be NP-hard. For example:

- Computing the inversion distance is solvable in polynomial time [2, 28].
- Computing the transposition distance is of unknown computational complexity.
- Computing the inversion median of three genomes is NP-hard [8].
- Computing the breakpoint median of three genomes is NP-hard (though it reduces to the well studied traveling salesperson problem, and hence can often be solved quickly in practice) [63].

In other words, under even fairly idealized conditions (where the only events are inversions and transpositions), most optimization problems are hard to solve. Heuristics without proven performance guarantees have been developed for these idealized conditions [24], but even these only have good performance under certain conditions.

Research questions in whole genome phylogeny

There are essentially two main obstacles for whole genome phylogeny. The first is that almost all the methods that have been developed (and their stochastic models) assume that all the genomes have exactly one copy of each gene (an assumption that is widely violated), and the second is that despite a fair amount of effort, we still do not have methods that can reliably analyze even moderately large datasets, even under these idealized conditions. Thus, work in both directions needs to be done.

The problem with the assumption that all genomes have one copy of each gene is that events that change gene content (such as insertions, deletions, and duplications) occur in enough datasets to make the current approaches inapplicable without some preprocessing. Thus, in fact, essentially all the datasets that have been analyzed using existing methods that assume equal gene content are first processed to remove duplicate genes. While in some cases this processing is acceptable (and should not change the phylogenetic reconstruction), it is not always clear how to do this rigorously (and in any event, throwing out data is almost always not desirable). Thus, making progress on developing new stochastic models that incorporate these events that change gene content is important, and will allow us to then test the performance of methods that we develop to infer evolutionary history from the full range of data.

Also, the usual stochastic models of gene order evolution make assumptions that all events of the same type are equiprobable, so that (for example) any two inversions have the same probability. However, research now suggests that "short inversions" may have a higher probability than "long inversions". Incorporating these changed assumptions into phylogenetic inference changes the computational problems in interesting ways. For example, instead of edit distances we would have weighted edit distances (so the cost of an event would reflect its probability), and estimations of true evolutionary distances would also need to be changed to reflect the additional complexity of the model. While some progress has been made towards estimating these distances [74], much still needs to be done.

Finally, methods for whole genome phylogeny reconstruction are quite computationally intensive – more so than the corresponding problems for DNA sequence phylogenetics by far. For example, computing the inversion length of a fixed tree on just three leaves can take a long time on some instances! While some progress has been made to provide speed-ups for whole genome phylogeny so that they can analyze large datasets, so far these speed-ups are limited to datasets with certain properties (no long edges, in particular). Therefore, a natural research area is to develop techniques for handling large datasets which do not have as significant limitations as the current set of methods.

19.7 Conclusions and Suggestions for Further Reading

Research into methods for phylogeny reconstruction offers surprisingly deep and interesting challenges to algorithms developers. Yet understanding the data, the methods, and how biologists use phylogenies is necessary in order for the development to be productive. We hope this chapter will help the reader appreciate the difference between pure algorithmic research, and that which could make a tremendous difference to practice.

There is a wealth of books and papers on phylogenetics, from all the different fields (biology, statistics, and computer science). The following list is just a sample of some of these books and papers, that will provide additional grounding in the field of computational and mathematical phylogenetics.

For more information from the perspective of a systematist, see [17, 76]. For books with a greater emphasis on mathematical and/or computational aspects, see [20, 66]. Expositions with a greater emphasis on stochastic models can be found in [25, 46]. Texts that are intermediate between these include [53]. For an on-line tutorial on phylogenetics, see [36].

Acknowledgments

This work was supported in part by the National Science Foundation, the David and Lucile Packard Foundation, and the Institute for Cellular and Molecular Biology at the University of Texas at Austin. Special thanks are due to Martin Nowak, and to the Program for Evolutionary Dynamics at Harvard University which supported both the authors for the 2004-2005 academic year. The authors also wish to thank Erick Matsen, Mike Steel, and Michelle Swenson, for their comments on earlier drafts of this chapter.

References

[1] A. Aho, Y. Sagiv, T. Szymanski, and J. Ullman. Inferring a tree from lowest common ancestors with an application to the optimization of relational expressions. In *Proc. 16th Ann. Allerton Conf. on Communication, Control, and Computing*, pages 54–63, 1978.

[2] D.A. Bader, B.M.E. Moret, and M. Yan. A linear-time algorithm for computing inversion distances between signed permutations with an experimental study. *Journal of Computational Biology*, 8(5):483–491, 2001.

[3] O.R.P. Bininda-Emonds, editor. *Phylogenetic Supertrees: Combining Information to Reveal the Tree of Life*, volume 3 of *Computational Biology*. Kluwer Academics, 2004.

[4] R. Boyer, A. Hunt, and S.M. Nelesen. A compressed format for collections of phylogenetic trees and improved consensus performance. *Proceedings of the 5th Workshop on Algorithmics in Bioinformatics (WABI 2005)*, 2005. Lecture Notes in Bioinformatics, Springer, LNBI 3692.

[5] D.R. Brooks and D.A. McLennan. *Phylogeny, Ecology, and Behavior*. University of Chicago Press, Chicago, 1991.

[6] D. Bryant. A classification of consensus methods for phylogenetics. *DIMACS Series in Discrete Mathematics and Theoretical Computer Science*, 1991.

[7] R.M. Bush, C.A. Bender, K. Subbarao, and N.J. Cox *et al.* Predicting the evolution of human influenza A. *Science*, 286:1921–1925, 1999.

[8] A. Caprara. Formulations and complexity of multiple sorting by reversals. In S. Istrail, P.A. Pevzner, and M.S. Waterman, editors, *Proceedings of the Third Annual International Conference on Computational Molecular Biology (RECOMB-99)*, pages 84–93, 1999.

[9] J.T. Chang. Full reconstruction of Markov models on evolutionary trees: identifiability and consistency. *Mathematical Biosciences*, 137:51–73, 1996.

[10] J.T. Chang. Inconsistency of evolutionary tree topology reconstruction methods when substitution rates vary across characters. *Mathematical Biosciences*, 134:189–215, 1996.

[11] B. Chor and T. Tuller. Maximum likelihood of evolutionary trees is hard. In *Proceedings of the 9th annual international conference on Research in Computational Molecular Biology (RECOMB) 2005*, pages 296–310, 2005.

[12] N. Eriksen. Approximating the expected number of inversions given the number of breakpoints. In *Proceedings of the Workshop on Algorithms for Bio-Informatics (WABI)*, volume 2452, pages 316–330, 2002. Lecture Notes in Computer Science.

[13] N. Eriksen and A. Hultman. Estimating the expected reversal distance after a fixed number of reversals. *Advances of Applied Mathematics*, 32:439–453, 2004.

[14] S.N. Evans and T. Warnow. Unidentifiable divergence times in rates-across-sites models. *IEEE/ACM Transactions on Computational Biology and Bioinformatics*, 1:130–134, 2005.

[15] J. Felsenstein. Cases in which parsimony and compatibility methods will be positively misleading. *Systematic Zoology*, 27:401–410, 1978.

[16] J. Felsenstein. Evolutionary trees from DNA sequences: a maximum likelihood approach. *Journal of Molecular Evolution*, 17:368–376, 1981.

[17] J. Felsenstein. *Inferring Phylogenies*. Sinauer Associates, Sunderland, Massachusetts, 2004.

[18] W. Fitch. Toward defining the course of evolution: minimum change for a specified tree topology. *Systematic Biology*, 20:406–416, 1971.

[19] L. R. Foulds and R. L. Graham. The Steiner problem in phylogeny is NP-complete. *Advances in Applied Mathematics*, 3:43–49, 1982.

[20] O. Gascuel, editor. *Mathematics of Evolution and Phylogeny*. Oxford Univ. Press, 2005.

[21] G. Giribet. Exploring the behavior of POY, a program for direct optimization of molecular data. *Cladistics*, 17:S60–S70, 2001.

[22] P.A. Goloboff. Analyzing large data sets in reasonable times: solution for composite optima. *Cladistics*, 15:415–428, 1999.

[23] A.D. Gordon. Consensus supertrees: the synthesis of rooted trees containing overlapping sets of labeled leaves. *Journal of Classification*, 3:335–348, 1986.

[24] GRAPPA (genome rearrangements analysis under parsimony and other phylogenetic algorithms). http://www.cs.unm.edu/~moret/GRAPPA/.

[25] D. Grauer and W.H. Li. *Fundamentals of Molecular Evolution*. Sinauer Publishers, 2000.

[26] D. Gusfield and L. Wang. New uses for uniform lifted alignments, 1999. DIMACS Series on Discrete Math and Theoretical Computer Science.

[27] S. Hannenhalli and P.A. Pevzner. Towards computational theory of genome rearrangements. *Computer Science Today: Recent Trends and Developments. Lecture Notes in Computer Science*, 1000:184–202, 1995.

[28] S. Hannenhalli and P.A. Pevzner. Transforming cabbage into turnip (polynomial algorithm for sorting signed permutations by reversals). In *Proc. of the 27th Annual Symposium on the Theory of Computing (STOC 95)*, pages 178–189, 1995. Las Vegas, Nevada.

[29] S. Hannenhalli and P.A. Pevzner. Transforming mice into men (polynomial algorithm for genomic distance problem). In *Proc. of the 36 Annual Symposium on Foundations of Computer Science (FOCS 95)*, pages 581–592, 1995. Milwaukee, Wisconsin.

[30] P.H. Harvey and M.D. Pagel. *The Comparative Method in Evolutionary Biology*. Oxford University Press, Oxford, 1991.

[31] I. Holmes and W.J. Bruno. Evolutionary HMMs: a Bayesian approach to multiple alignment. *Bioinformatics*, 17(9):803–820, 2001.

[32] J.P. Huelsenbeck, J.J. Bull, and C.W. Cunningham. Combining data in phylogenetic analysis. *Trends in Ecology and Evolution*, 11(4):152–158, 1996.

[33] J.P. Huelsenbeck and R. Ronquist. MrBayes: Bayesian inference of phylogeny. *Bioinformatics*, 17:754–755, 2001.

[34] W. Just. Computational complexity of multiple sequence alignment with SP-score. *Journal of Computational Biology*, 8(6):615–623, 2001.

[35] S. Kannan, T. Warnow, and S. Yooseph. Computing the local consensus of trees. *SIAM J. Computing*, 27(6):1695–1724, 1995.

[36] J. Kim and T. Warnow. Tutorial on phylogenetic tree estimation, 1999. Presented at the ISMB 1999 conference, available on-line at http://kim.bio.upenn.edu/ jkim/media/ISMBtutorial.pdf.

[37] B. Knudsen. Optimal multiple parsimony alignment with affine gap cost using a phylogenetic tree. In G. Benson and R. D. M. Page, editors, *Workshop on Algorithms for Bioinformatics (WABI)*, volume 2812 of *Lecture Notes in Computer Science*,

pages 433–446. Springer, 2003.

[38] C.R. Linder and B.M.E. Moret. Tutorial on reticulate evolution. Presented at the DIMACS workshop on reticulate evolution, and available online at http://dimacs.rutgers.edu/Workshops/Reticulated_WG/slides/slides.html.

[39] B. Ma, M. Li, and L. Zhang. On reconstructing species trees from gene trees in terms of duplications and losses. In *Proc. 2nd Ann. Int'l Conf. Comput. Mol. Biol. (RECOMB98)*, 1998.

[40] W. Maddison. Gene trees in species trees. *Systematic Biology*, 46(3):523–536, 1997.

[41] M.L. Metzker, D.P. Mindell, X.M. Liu, and R.G. Ptak *et al.* Molecular evidence of HIV-1 transmission in a criminal case. *Proceedings of the National Academy of Sciences of the United States of America*, 99(22):14292–14297, 2002. OCT 29.

[42] I. Miklós, G. A. Lunter, and I. Holmes. A "long indel" model for evolutionary sequence alignment. *Molecular Biology and Evolution*, 21(3):529–540, 2004.

[43] B. Moret, L.S. Wang, T. Warnow, and S. Wyman. New approaches for reconstructing phylogenies based on gene order. In *Proceedings of 9th Int'l Conf. on Intelligent Systems for Molecular Biology (ISMB'01)*, pages 165–173, 2001.

[44] B.M.E. Moret, L. Nakhleh, T. Warnow, and C.R. Linder *et al.* Phylogenetic networks: modeling, reconstructibility, and accuracy. *IEEE/ACM Transactions on Computational Biology and Biocomputing*, 1(1), 2004.

[45] L. Nakhleh, J. Sun, T. Warnow, and C.R. Linder *et al.* Towards the development of computational tools for evaluating phylogenetic network reconstruction methods. In *Proc. 8th Pacific Symp. on Biocomputing (PSB 2003)*, 2003.

[46] M. Nei, S. Kumar, and S. Kumar. *Molecular Evolution and Phylogenetics*. Oxford University Press, 2003.

[47] K.C. Nixon. The parsimony ratchet, a new method for rapid parsimony analysis. *Cladistics*, 15:407–414, 1999.

[48] S.P. Otto and J. Whitton. Polyploid incidence and evolution. *Annual Review of Genetics*, 34:401–437, 2000.

[49] R. Page. Maps between trees and cladistic analysis of historical associations among genes, organisms, and areas. *Systematic Biology*, 43:58–77, 1994.

[50] R. Page. GeneTree: comparing gene and species phylogenies using reconciled trees. *Bioinformatics*, 14(9):819–820, 1998.

[51] R. Page and M. Charleston. From gene to organismal phylogeny: reconciled trees and the gene tree/species tree problem. *Molecular Phylogeny and Evolution*, 7:231–240, 1997.

[52] R. Page and M. Charleston. Reconciled trees and incongruent gene and species trees. In B. Mirkin, F. R. McMorris, F. S. Roberts, and A. Rzehtsky, editors, *Mathematical hierarchies in biology*, volume 37. American Math. Soc., 1997.

[53] R. Page and E. Holmes. *Molecular Evolution: A phylogenetic approach*. Blackwell Publishers, 1998.

[54] C.A. Phillips and T. Warnow. The asymmetric median tree: a new model for building consensus trees. *Discrete Applied Mathematics*, 71:311–335, 1996.

[55] D. Posada and K.A. Crandall. Modeltest: testing the model of DNA substitution. *Bioinformatics*, 14(9):817–818, 1998.

[56] D.R. Powell, L. Allison, and T.I. Dix. Fast, optimal alignment of three sequences using linear gap costs. *Journal of Theoretical Biology*, 207:325–336, 2000.

[57] R. Ravi and J.D. Kececioglu. Approximation algorithms for multiple sequence alignment under a fixed evolutionary tree. *Discrete Applied Mathematics*, 88:355–366, November 1998.

[58] B.D. Redelings and M.A. Suchard. Joint Bayesian estimation of alignment and phy-

logeny. *Systematic Biology*, 54(3):401–418, 2005.

[59] D.F. Robinson and L.R. Foulds. Comparison of phylogenetic trees. *Mathematical Biosciences*, 53:131–147, 1981.

[60] U. Roshan, B.M.E. Moret, T.L. Williams, and T. Warnow. Performance of supertree methods on various dataset decompositions. In O.R.P. Bininda-Emonds, editor, *Phylogenetic Supertrees: Combining Information to Reveal the Tree of Life*, pages 301–328, 2004. Volume 3 of Computational Biology, Kluwer Academics, (Andreas Dress, series editor).

[61] U. Roshan, B.M.E. Moret, T.L. Williams, and T. Warnow. Rec-I-DCM3: A fast algorithmic technique for reconstructing large phylogenetic trees. In *Proc. IEEE Computer Society Bioinformatics Conference (CSB 2004)*, 2004. Stanford University.

[62] N. Saitou and M. Nei. The neighbor-joining method: A new method for reconstructing phylogenetic trees. *Molecular Biology and Evolution*, 4:406–425, 1987.

[63] D. Sankoff and M. Blanchette. Multiple genome rearrangement and breakpoint phylogeny. *Journal of Computational Biology*, 5:555–570, 1998.

[64] D. Sankoff and R.J. Cedergren. Simultaneous comparison of three or more sequences related by a tree. In *Time Warps, String Edits, and Macromolecules: the Theory and Practice of Sequence Comparison*, pages 253–264. Addison-Wesley, 1983. Chapter 9.

[65] B. Schwikowski and M. Vingron. The deferred path heuristic for the generalized tree alignment problem. *Journal of Compututational Biology*, 4(3):415–431, 1997.

[66] C. Semple and M. Steel. *Phylogenetics*. Oxford Series in Mathematics and its Applications, 2004.

[67] A. Siepel and D. Haussler. Phylogenetic hidden Markov models. In R. Nielsen, editor, *Statistical methods in molecular evolution*, pages 325–351. Springer, 2005.

[68] M.A. Steel. The complexity of reconstructing trees from qualitative characters and subtrees. *Journal of Classification*, 9:91–116, 1992.

[69] M.A. Steel. The maximum likelihood point for a phylogenetic tree is not unique. *Systematic Biology*, 43(4):560–564, 1994.

[70] M.A. Steel. Recovering a tree from the leaf colourations it generates under a Markov model. *Applied Mathematics Letters*, 7(2):19–24, 1994.

[71] M.A. Steel and D. Penny. Parsimony, likelihood, and the role of models in molecular phylogenetics. *Molecular Biology and Evolution*, 17(6):839–850, 2000.

[72] M.A. Steel, L.A. Székely, and M.D. Hendy. Reconstructing trees when sequence sites evolve at variable rates. *Journal of Computational Biology*, 1:153–163, 1994.

[73] J. Stoye, D. Evers, and F. Meyer. Rose: generating sequence families. *Bioinformatics*, 14(2):157–163, 1998.

[74] K.M. Swenson, M. Marron, J.V. Earnest-DeYoung, and B.M.E. Moret. Approximating the true evolutionary distance between two genomes. In *Proc. 7th Workshop on Algorithm Engineering and Experiments (ALENEX'05), Vancouver (Canada)*. SIAM Press, 2005.

[75] D. L. Swofford. PAUP*: Phylogenetic analysis using parsimony (and other methods), 1996. Sinauer Associates, Sunderland, Massachusetts, Version 4.0.

[76] D.L. Swofford, G.J. Olsen, P.J. Waddell, and D.M. Hillis. Phylogenetic inference. In D.M. Hillis, C. Moritz, and B.K. Mable, editors, *Molecular Systematics*. Sinauer Associates, Sunderland, Massachusetts, 1996.

[77] J.D. Thompson, D.G. Higgins, and T.J. Gibson. CLUSTAL W: improving the sensitivity of progressive multiple sequence alignment through sequence weighting, position specific gap penalties and weight matrix choice. *Nucleic Acids Research*, 22:4673–4680, 1994.

[78] J.L. Thorne, H. Kishino, and J. Felsenstein. An evolutionary model for maximum likelihood alignment of DNA sequences. *Journal of Molecular Evolution*, 33:114–124, 1991.

[79] J.L. Thorne, H. Kishino, and J. Felsenstein. Inching towards reality: An improved likelihood model of sequence evolution. *Journal of Molecular Evolution*, 34:3–16, 1992.

[80] C. Tuffley and M. Steel. Links between maximum likelihood and maximum parsimony under a simple model of site substitution. *Bulletin of Mathematical Biology*, 59:581–607, 1997.

[81] L. Wang and T. Jiang. On the complexity of multiple sequence alignment. *Journal of Computational Biology*, 1:337–348, 1994.

[82] L. Wang, T. Jiang, and D. Gusfield. A more efficient approximation scheme for tree alignment. *SIAM Journal of Computing*, 30(1):283–299, 2000.

[83] L.S. Wang. Exact-IEBP: A new technique for estimating evolutionary distances between whole genomes. In *Lecture Notes for Computer Sciences No. 2149: Proceedings for the First Workshop on Algorithms in BioInformatics (WABI'01)*, pages 175–188, 2001.

[84] L.S. Wang and T. Warnow. Estimating true evolutionary distances between genomes. In *Proceedings of the Thirty-Third Annual ACM Symposium on the Theory of Computing (STOC'01)*, pages 637–646, 2001.

[85] T. Warnow, B. M. Moret, and K. St. John. Absolute convergence: true trees from short sequences. *Proceedings of ACM-SIAM Symposium on Discrete Algorithms (SODA 01)*, pages 186–195, 2001.

[86] W. Wheeler. Optimization alignment: the end of multiple sequence alignment in phylogenetics? *Cladistics*, 12:1–9, 1996.

[87] M. Wilkinson. Common cladistic information and its consensus representation: reduced Adams and reduced cladistic consensus trees and profiles. *Syst. Biol.*, 43 (3):343–368, 1994.

[88] Z. Yang. Maximum-likelihood estimation of phylogeny from DNA sequences when substitution rates differ over sites. *Molecular Biology and Evolution*, 10:1396–1401, 1996.

20

Consensus Trees and Supertrees

Oliver Eulenstein
Iowa State University

20.1 Introduction

This chapter discusses *supertree methods*, methods that assemble a collection of either rooted or unrooted input trees into one or more trees, called *supertrees* [34]. The strength of supertrees is that they can make statements about branching information that are not contained in a single input tree, but derive from the cumulative branching information of several input trees. Figure 20.1 depicts an example. Typically supertrees are used to describe complex phylogenetic relationships that allow evolutionary biologists to study their implications on the tree of all species — the *Tree of Life*. Thus, ideally, input trees are *species trees*, that describe evolutionary relationships of species. An overview and discussions about the biological relevance of supertree methods can be found in [61, 73, 10, 57, 23, 30, 9].

There is evidence that there are no reasonable supertree methods for unrooted input trees. This evidence is derived from fundamental properties, described by Steel et al. [69], that should be satisfied by any reasonable supertree method. One of these properties is that the method should assemble a supertree that contains all the branching information of the input trees, if they are compatible. Input trees are *compatible*, if there exists a tree, called *parent tree*, that contains each input tree as a homeomorphic subtree. Hence, every supertree method should construct a parent tree for the input trees, if they are compatible. Thus supertree methods need to solve the *compatibility problem*, that is to decide if the

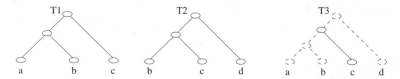

FIGURE 20.1: The supertree $T3$ contains the branching information of the input trees $T1$ and $T2$. The branching information of the dashed subtree tree in $T3$ is not contained in either tree $T1$ or tree $T2$, but derives from the branching information of tree $T1$ and $T2$ together.

input trees are compatible. While the compatibility problem can be solved for rooted input trees in polynomial time using the BUILD algorithm from Aho et al. [3], the problem is **NP**-complete for unrooted input trees [68, 12]. Thus, the time complexity of supertree problems for rooted and unrooted trees are clearly different. An even stronger separation between supertree methods for rooted and unrooted input trees was made by Steel et al. [69]. Steel et al. showed that in contrast to supertree methods for rooted input trees, no supertree method for unrooted input trees exists that satisfies a subset of the fundamental supertree properties. From this result Semple and Steel [63] conclude that there exists no 'reasonable' supertree method for unrooted input trees. Consequently, this chapter will focus on supertree methods for rooted input trees.

A fundamental question in biological classification is how a supertree method should assemble input trees to optimally represent their branching information. Probably the oldest supertree algorithm for rooted input trees is the BUILD algorithm, introduced by Aho et al. [3] in 1976, and that can be used to solve the compatibility problem in polynomial time. While other supertree problems are typically **NP**-complete, there are several modifications of the BUILD algorithm that still run in polynomial time. Semple and Steel [63] introduced a modified version of the BUILD algorithm, the MINCUT SUPERTREE algorithm. While the algorithm does not have an explicitly specified computational problem that it solves, it is designed to handle incompatible input trees and to produce a unique supertree satisfying certain mathematical properties. The MINCUT SUPERTREE algorithm is guided by a local optimization criterion: whenever a conflict is encountered, a minimum amount of incompatible branching information is deleted from the input trees to allow the computation to proceed. Various other polynomial time algorithms are based on the BUILD algorithm (e.g. Consantinescu and Sankoff [22], Bryant and Steel [18], Ng and Wormland [50], Henzinger et al. [41], Böcker et al. [11], Semple and Steel [63], Page [53], Willson [72]). Another group of supertree problems that are typically **NP**-complete, are based on a matrix encoding of the input trees that is then modified by different optimization criteria (e.g. Chen et al. [20], Baum [8], Ragan [59]). The wide variety of existing supertree methods reflects the difficulty to specify criteria to assemble input trees to the needs of evolutionary biologists.

While there is no standard to measure the effectiveness of supertree methods, they can be compared to a special case of supertree methods, called 'consensus tree' methods, that is somewhat better understood. *Consensus tree* methods are supertree methods, where all input trees have the same taxon set and output a unique tree, called *consensus tree*. For consensus tree methods there are natural assembly criteria that are typically defined by set operations and discussed thoroughly in the classification literature. Constraining a supertree method to input trees with the same taxon set allows analyzing the method if it 'preserves' natural assembly criteria of consensus tree methods.

While most supertree methods assume that their input trees are species trees, some methods assemble supertrees from gene trees. Typically the given species trees are directly implied from gene trees that describe evolutionary relationships of genes. However, in many cases the directly implied species trees differ from the real species tree caused by evolutionary mechanisms like gene duplication, lateral gene transfer or lineage sorting [55]. Supertree methods for gene trees reconcile the gene trees to imply (species) supertrees (e.g. [36, 54, 39]).

The chapter is organized as follows. First we introduce basic notations and results in Section 20.2. We survey the axiomatic approach from Steel et al. [69] that shows general and inherent limitations of supertree methods in Section 20.3. In Section 20.4 we first survey basic consensus tree criteria and then define their preservation in supertree methods. Now we are prepared to discuss a selection of three types of supertree methods: supertree methods that are based on the BUILD algorithm in Section 20.5, supertree methods that work with the matrix representation of input trees in Section 20.6, and supertree methods that have gene trees as their input trees in Section 20.7. Concluding remarks are given in Section 20.8.

20.2   Preliminary Notation, Definitions and Results

Here we introduce basic notations and results. For a graph G we refer to its node set by $V(G)$ and to its edge set by $E(G)$. If G is undirected, $deg(v)$ is the degree of a node $v \in V(G)$.

Let T be a tree. If $deg(v) > 1$, then v is an *internal* node, otherwise it is a *leaf*. An edge connecting two internal nodes is an *internal* edge. The set $\mathcal{L}(T)$ denotes the set of all leaf nodes in T, and T is said to be a *graph on* $\mathcal{L}(T)$. An internal node $v \in V(T)$ is a *bifurcation*, if $deg(v) \leq 3$ and a *multifurcation* otherwise. T is *binary*, if every internal node is a bifurcation. A multifurcation in T that is interpreted as an evolutionary polytomy is either 'true' or 'apparent' [44, 66, 53]. The multifurcation is *true* , if three or more lineages diverge from it. The multifurcation is *apparent*, if it represents an unknown binary tree on the set of its adjacent nodes. The graph T is a *(phylogenetic) tree*, if it has no degree two nodes. For convenience the term tree will refer to a phylogenetic tree, unless stated otherwise. A *quartet* is a binary tree with four leaves. For a quartet on $\{a, b, c, d\}$ we write $ab|cd$, if removing the internal edge results in two trees, one on $\{a, b\}$ and one on $\{c, d\}$.

A cycle free graph T is a *rooted (phylogenetic) tree*, if a unique internal node, denoted as $r(T) \in V(T)$, is marked as the *root* of T, and $deg(u) \neq 2$ for any node $u \in (V(T) - \{r(T)\})$. A rooted binary tree T with three leaves is a *triplet*. For a triplet T on $\{a, b, c\}$ we write $bc|a$, if a and $r(T)$ are connected through an edge in T. A set $\mathcal{T}$ of rooted trees is a *profile* and $\mathcal{L}(\mathcal{T}) := \bigcup_{T \in \mathcal{T}} \mathcal{L}(T)$. If $\mathcal{L}(T) = X$, then $\mathcal{T}$ is said to be a *profile on* X. The profile $\mathcal{T}$ is *complete*, if $\mathcal{L}(T) = \mathcal{L}(\mathcal{T})$ for any $T \in \mathcal{T}$. We call $\mathcal{T}$ *binary*, if it contains only rooted binary trees.

Let T be a rooted tree. We define the semi-order $(V(T), \leq)$ where $x \leq_T y$, if y is a node on the path between $r(T)$ and x in T. If $x \leq_T y$ and $x \neq y$ we write $x <_T y$. For convenience we write $\leq$ and $<$, if the tree T is obvious. We call x a *child* of y, if $x < y$ and there exists no node $z \in V(T)$ such that $x < z$ and $z < y$. The set of all children of y is denoted by $Ch(y)$. If $x \leq y$ then x is a *descendant* of y, and y is an *ancestor* of x. If $x < y$, x is a *proper descendant* of y and y is a *proper ancestor* of x.

Let $M \subseteq \mathcal{L}(T)$ be a non-empty set. The *least common ancestor* of a set M in T, denoted as $lca_T(M)$, is the node $y \in V(T)$ such that (i) $x \leq y$ for every $x \in M$ and (ii) there exists no $y' \in V(T)$ such that $y' < y$ and $x \leq y'$ for every $x \in M$. If $M \not\subseteq \mathcal{L}(T)$ we define

$lca_T(M) := NIL.$

Nestings in a tree T are described by the binary relation $<_T$ over the powerset of $\mathcal{L}(T)$, where $A <_T B$ if $lca_T(A) <_T lca_T(B)$. The set A *nests* in the set B with respect to T, if $A <_T B$.

The tree T *constrained by* M, denoted as $T_{|M}$, is the subgraph in T that is induced by the set $\{v \in V(T) \mid \exists x \in M : x \leq v \leq lca_T(M)\}$ with its degree two nodes contracted that are not $lca_T(M)$. The profile $\mathcal{T}$ *constrained by* M is the set $\mathcal{T}_{|M} := \{T_{|M} \mid T \in \mathcal{T}\}$. A tree T' is *displayed* by T, denoted as $T' \leq T$, if T' can be obtained from $T_{|\mathcal{L}(T')}$ by contracting edges. The profile $\mathcal{T}$ is *compatible* if there exists a tree S such that $T' \leq S$ for any $T' \in \mathcal{T}$. If such a tree S on $\mathcal{L}(\mathcal{T}))$ exists, it is called a *parent tree* of $\mathcal{T}$. The *subtree of* T *rooted at* $y \in V(T)$, denoted as $T(y)$, is the subgraph of T that is induced by the set $\{v \in V(T) \mid v \leq y\}$.

The set $c(v) := \mathcal{L}(T(v))$ is a *cluster* in T, for any $v \in V(T)$. If the node $v \in V(T)$ is not a leaf or the root in T, then the cluster $c(v)$ is *proper*. The set $c(T)$ is the set of all proper clusters in T and the set $c(\mathcal{T}) := \bigcup_{T \in \mathcal{T}} c(T)$ is the set of all proper clusters in the profile $\mathcal{T}$. Two sets A and B are not in *conflict* if $A \cap B \in \{\emptyset, A, B\}$. A set $\mathcal{M}$ of sets is *conflict-free*, if no pair $A, B \in \mathcal{M}$ is in *conflict*. For convenience we extend the definition for the least common ancestor to $lca_T(v) := lca_T(c(v))$ for a node $v \in V(T)$.

Let M be a set. A *(directed binary) character over* M is an ordered pair $C := (N, \mathsf{O})$, where $N \subseteq M$ and $\mathsf{O} \subseteq N$. We call N the *taxon set* of C, O the *1-state* of C, and $N - \mathsf{O}$ the *0-state* of C. The character C is *complete*, if $N = M$. A *completion* of C is a complete character $C' = (M, \mathsf{O}')$ such that $\mathsf{O} \subseteq \mathsf{O}'$ and $N - \mathsf{O} \subseteq M - \mathsf{O}'$ (that is, the 1-state and the 0-state of C are contained in the 1-state and 0-state of C', respectively).

Let T be a rooted tree and $C = (N, \mathsf{O})$ be a character. C is *consistent* with T, if and only if T has a cluster X such that $X \cap N = \mathsf{O}$. Character tuple $\mathcal{C}$ is *compatible* if and only if there exists a phylogeny T consistent with C_i for each $i \in \{1, \ldots, r\}$.

The compatibility of a set of characters can be tested in polynomial time [4, 56]. For complete characters, the following result is well known .

THEOREM 20.1 *[Estabrook et al.[28], Gusfield [37]] Character tuple $\mathcal{C}$ is compatible if and only if O_i and O_j are not in conflict for every $i, j \in \{1, \ldots, r\}$.*

20.3 Fundamental Properties of Consensus and Supertree Methods

Supertree methods can be constructed by either (i) devising a method and then determining its mathematical properties (e.g. the MRP and MRF supertree methods in Section 20.6), or (ii) by specifying mathematical properties and then try to design a method that satisfies them [46] (e.g. the MINCUTSUPERTREE algorithm in Section 20.5). For the latter design concept most of the work on properties (axioms) for supertrees can be embedded in the broader framework of Arrow's social choice theory [6]. As a result there are groups of mathematical properties that can not be satisfied together by any supertree method. Here the reader is referred to the work of Day and McMorris [24]. However, independent of the design concept any supertree method should satisfy a set of fundamental properties. As a first step in this direction Steel et al. [69] introduced the following properties that any supertree and consensus tree method should satisfy.

$P1$: The method can be applied to any set of either rooted or unrooted input trees.

$P2$: If the leaves of the input trees are renamed, then the new output trees are the

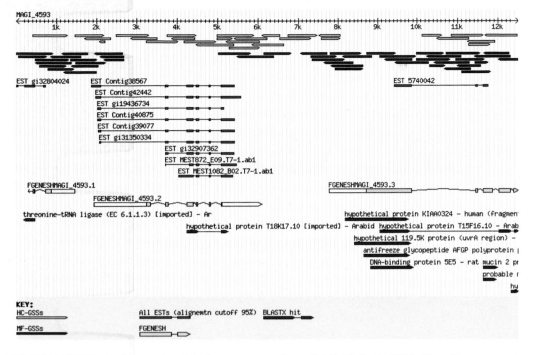

COLOR FIGURE 13.11: Graphical representation of MAGI–3. 1–4593. The first two rows above, as illustrated in the key, correspond to HC and MF sequences, respectively. The next two rows correspond to genes predicted using either EST-based or *ab initio* prediction approaches and include introns. The last row are annotated protein matches. It follows that this single contig shows a case where there are three genes on a genomic island; however, notice how the sampling sources differ in the different intervals above. Some regions are only captured using HC reads (3–5KB) while others are only captured using MF (8—10KB).

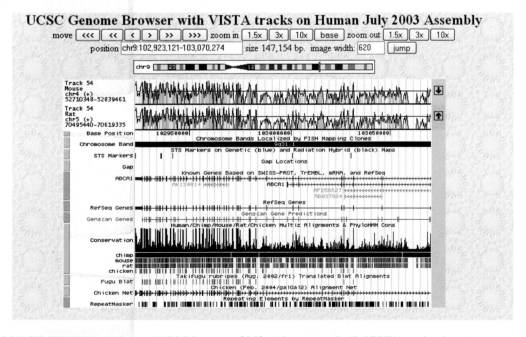

COLOR FIGURE 14.6: UCSC browser [36] with custom-built VISTA tracks showing conservation between the human chr. 9 interval aligned with orthologous mouse and rat sequences.

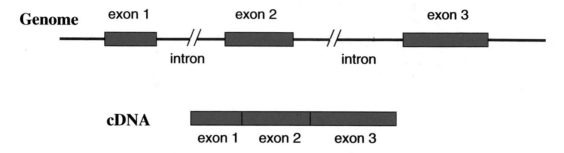

COLOR FIGURE 15.5: cDNA mapped to a genomic sequence.

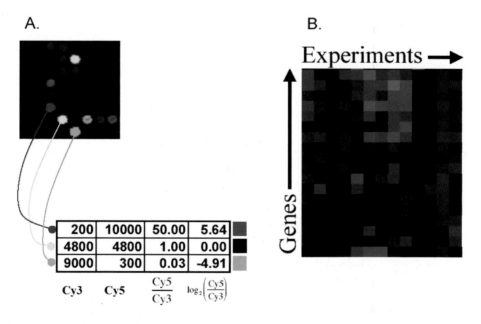

COLOR FIGURE 23.1: Conversion of image data to raw and false color data. A. The intensities in each channel in the scanned image are converted to the raw numeric values, and then used to produce a log ratio. Log ratios above zero are then represented as red, with the intensity being proportional to the value, while green is used to represent negative log ratios. A log ratio of zero is represented by black. B. Data for multiple genes across multiple experiments are represented by block of color, one block per measurement, as suggested in [8].

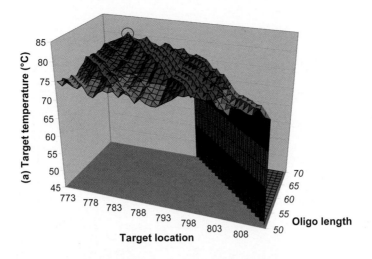

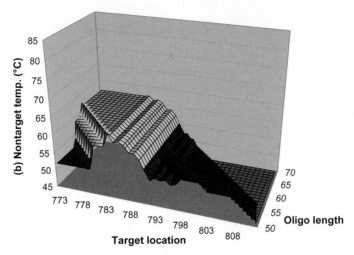

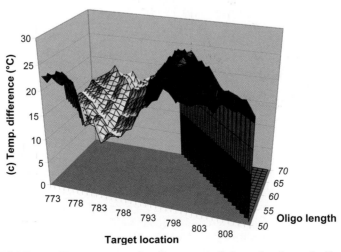

COLOR FIGURE 24.7: Temperature landscapes and the selection of oligo target regions. The (a) target, (b) nontarget and (c) difference melting temperature landscapes are drawn against different sequence locations and oligo lengths. Red circles indicate oligos that might be chosen under different design strategies.

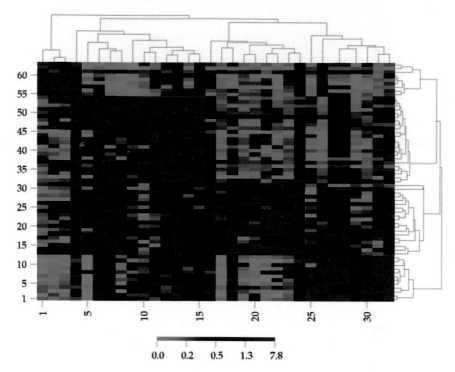

COLOR FIGURE 25.6: Two-dimensional hierarchical clustering of the microarray gene expression data on small round blue cell tumors with selected genes. The dendrograms for gene clusters and sample clusters are shown on top and right of the map, respectively.

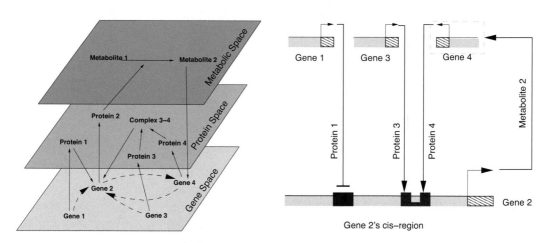

COLOR FIGURE 27.1: A hypothetical gene network. Shown on the left (redrawn from [10]) are the multiple levels at which genes are regulated by other genes, proteins and metabolites. On the right is a useful abstraction subsuming all the interactions into ones between genes only. The cis-regions are shown next to the coding regions, which are marked with pattern and start at the bent arrows. The edges are marked with the name of the molecule that carries the interaction. Some reactions represent trans-factor – DNA binding, happen during transcription, and are localized on the cis-regions. In those cases the corresponding protein-specific binding sites, or cis-elements, on the cis-regions are shown (colored polygons). Otherwise, the interactions can take placeduring transcription or later (e.g. post-translational modications) as maybe the case with Metabolite 2 interacting with Gene 4. The nature of the interactions is inducing(arrow) or repressing (dull end).

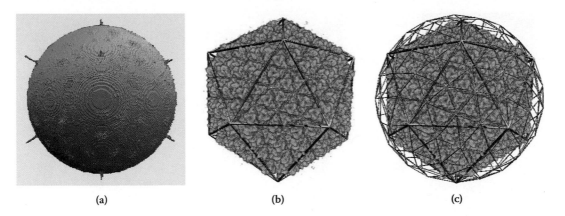

(a)   (b)   (c)

COLOR FIGURE 32.4: Detection of Symmetry axes and construction of global icosahedral symmetry as well as local n-fold symmetry. (a) scoring function. (b) global icosahedral symmetry. (c) local 6-fold symmetry.

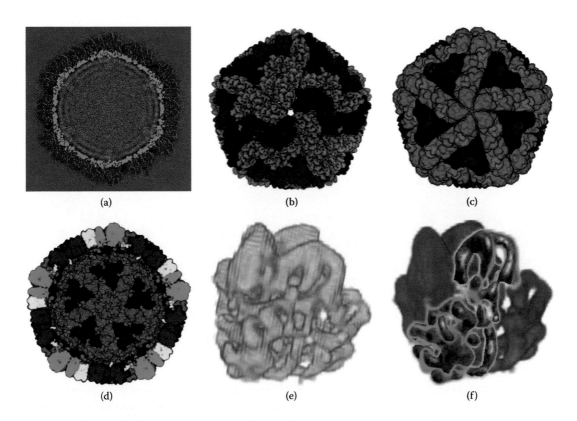

(a)   (b)   (c)

(d)   (e)   (f)

COLOR FIGURE 32.5: Visualization of the architecture of the Rice Dwarf Virus (RDV) 3D map (a) segmented outer and inner icosahedral capsid boundaries (b) segmented asymmetric subunits of the outer capsid (60 subunits in total). Each asymmetric subunit consists of four and one third trimers. (c) & (d) segmented trimeric subunits (260 in total), where (c) shows the view from the outside while (d) shows the view from inside. (e) each segmented trimeric subunit consists of three monomeric sub-subunits. (f) segmented monomeric subunitre presents the 3D density map of a single P8 protein. The RDV 3D map data is courtesy Dr. Wah Chiu, NCMI, BCM, Houston

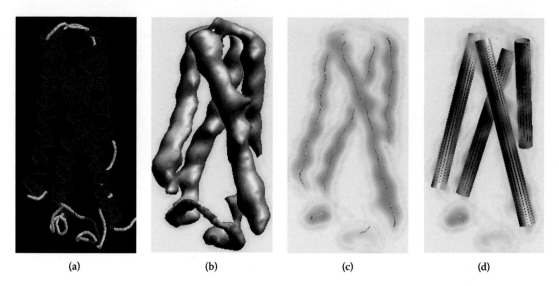

(a)   (b)   (c)   (d)

COLOR FIGURE 32.6: Illustration of secondary structural identification using local structure tensor at critical points of the 3D Map (a) The X-ray atomic structure representation of cytochrome c' (PDB-ID = 1bbh). (b) The volumetric representation of a Gaussian blurred 3D map generated from the X-ray structure (c) The detected skeletons of the 3D map. (d) Four helices are finally constructed fromthe skeletons, while the two on the bottom are discarded as being too small for being helices.

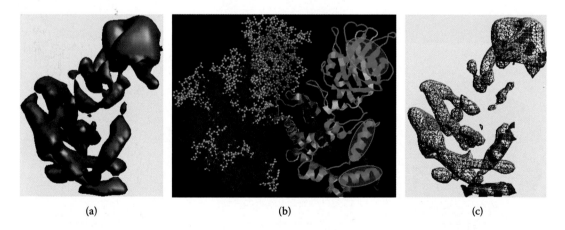

(a)   (b)   (c)

COLOR FIGURE 32.7: Example of structural fitting in the segmented P8 monomeric protein of RDV. (a) P8 monomeric protein iso-surface visualization (b) X-ray atomic structure of the P8 monomeric protein represented in ways of balls & sticks and cartoons. One beta sheet (top) and two alpha helices (middle and bottom) are highlighted and used as a fitting model. (c) By maximizing the correlation between the X-ray atomic model and the 3D map of the P8 monomer, one builds a pseudo-atomic model of the 3D map.

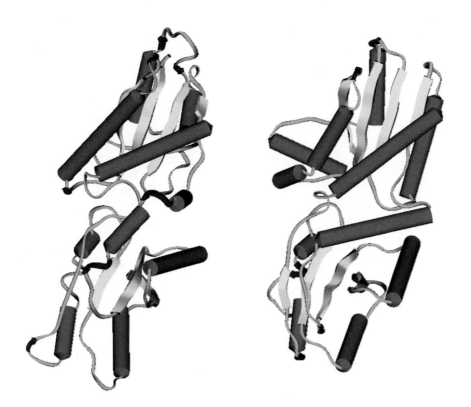

COLOR FIGURE 33.2:   A comparison between the predicted structure (left) and the experimental one (right) for the CASP-3 target t0053. The cylinders indicate α-helices, the strands indicate β-sheets, and the lines indicate loops.

COLOR FIGURE 34.2:   Aquaporin in lipid bilayer membrane, water above and below membrane omitted for clarity.

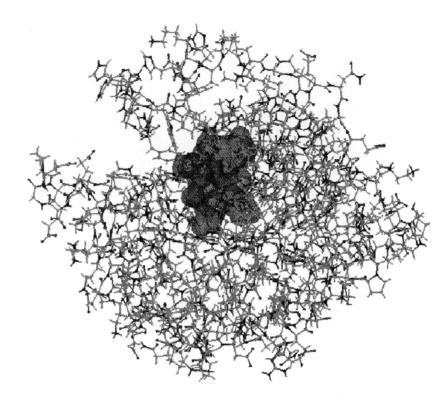

COLOR FIGURE 34.8: Human carbonic anhydrase treated using the mixed *ab initio/* empirical force field based approach described in the text. The full enzyme where in the wire frame represents atomic sites and the blue cloud represents the electron density of the valence electrons associated with "*ab initio* atoms".

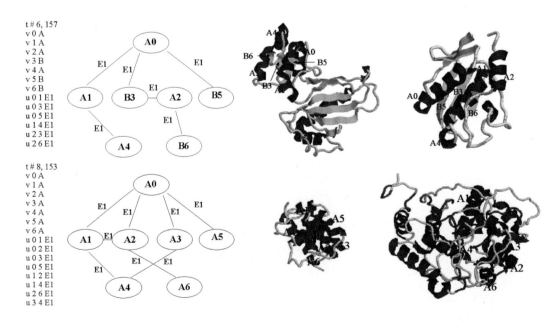

COLOR FIGURE 38.4: **Top:** Alpha-Beta Motif and its Occurrence in PDB files: 1ad2, 1kuh. **Bottom:** All Alpha Motif and its Occurrence in PDB files: 1bg8, 1arv.

old output trees with their leaves renamed.

$P3$: If the input trees are compatible, then every output tree is a parent tree of the input trees.

$P4$: If a leaf is in some input tree, then the leaf is in every output tree.

$P5$: There exists a polynomial time algorithm for the supertree method.

$P6_u$: If a quartet is displayed by a tree in a set of unrooted and compatible input trees, then the quartet is displayed by every output tree.

$P6_r$: If a triplet is displayed by a tree in a set of rooted and compatible input trees, then the triplet is displayed by every output tree.

$P7_r$: If in a set of rooted input trees a triplet $ab|c$ is displayed by a tree and the triplets $ac|b$ and $bc|a$ are not displayed by any tree, then any output tree displays the triplet $ab|c$.

Properties $P1$ and $P2$ seem to be absolutely necessary and the remaining properties are desirable to be satisfied by any supertree method. The following propositions show strong constraints on the ability to assemble input trees into supertrees.

PROPOSITION 20.1 [Steel et al. [69]] There exists no supertree method for unrooted binary input trees that satisfies properties $P1$, $P2$, and $P3$.

PROPOSITION 20.2 [Steel et al. [69]] There exists no consensus tree method for unrooted input trees that satisfies properties $P1$, $P2$, and $P6_u$.

Consequently, there exists no 'reasonable' supertree method for unrooted input trees. On the positive side Proposition 20.1 and Proposition 20.2 do not hold for rooted input trees. Proposition 20.1 does not hold for rooted trees, since the BUILD algorithm, described in Section 20.5, satisfies properties $P1, P2$ and $P3$. Proposition 20.2 does not hold for rooted trees, since the Adams consensus method described in Section 20.4 satisfies properties $P1, P2, P6_r$. However, there is no consensus tree method for rooted supertrees that satisfies all properties.

PROPOSITION 20.3 There exists no consensus tree method for rooted input trees that satisfies properties $P1$, $P2$, and $P7r$.

While no supertree method satisfies $P1, P2$ and $P7r$, a supertree method can maximize the number of uncontradicted triplets in the input trees. An example for such a method is Page's modified MINCUTSUPERTREE algorithm described in Section 20.5.

20.4   Preservation of Consensus Tree Methods

In order to better understand the properties of supertree methods it is natural to investigate if the method for complete input profiles preserves familiar consensus tree properties. Here we first define the preservation of consensus tree methods in supertree methods. We then survey basic consensus tree methods in preparation to analyze their preservation properties in the supertree methods that are discussed in this chapter. For a more complete survey of consensus tree methods the reader is referred to Bryant's work [16, §6.2] and [15].

For a supertree method S let $S(\mathcal{I})$ be the set of output trees of S, for an input profile $\mathcal{I}$.

DEFINITION 20.1 [cluster preservation] Let C be a consensus method and $C(\mathcal{T})$ a resulting consensus tree for a complete profile $\mathcal{T}$. The supertree method S *preserves* C, if $c(\mathcal{T}) \subseteq T$ for any $T \in S(\mathcal{T})$.

DEFINITION 20.2 [nesting preservation] The supertree method S *preserves the nestings* in $\mathcal{T}$, if for any tree $T \in S(\mathcal{T})$ any pair of subsets X, Y of $\mathcal{L}(\mathcal{T})$ where $X <_T Y$ it holds that $X <_{T_i} Y$ for every $i \in \{1, \ldots, |\mathcal{T}|\}$.

Next we survey consensus tree methods for a profile $\mathcal{T}$ of k trees.

20.4.1 Adams Consensus (consensus by nestings)

Likely, the oldest consensus approach for phylogenetic trees is the 'Adams consensus' that was introduced by Adams [1] (see also [2]) in 1972. The *Adams Consensus* $\mathcal{A}(\mathcal{T})$ is a rooted tree over $\mathcal{L}(\mathcal{T})$ that has exactly the nesting relations which are represented in all of the trees in $\mathcal{T}$. That is, $\mathcal{A}(\mathcal{T})$ satisfies the following two properties.

- If $A \subseteq \mathcal{L}(\mathcal{T})$ and $B \subseteq \mathcal{L}(\mathcal{T})$ such that $A <_T B$ for all $T \in \mathcal{T}$, then $A <_{\mathcal{A}(\mathcal{T})} B$.
- If A, B are clusters in $\mathcal{A}(\mathcal{T})$ such that $A <_{\mathcal{A}(\mathcal{T})} B$, then $A <_T B$ for any $T \in \mathcal{T}$.

The Adams Consensus can be computed in polynomial time in the size of the given profile using the algorithm from McMorris et al. [47].

20.4.2 Consensus by cluster occurrence

For $0 \leq l \leq k$ the *consensus rule* $\mathcal{M}_l(\mathcal{T})$ is the set of clusters that appears in at least l of the trees in $\mathcal{T}$. The family of consensus rules has been described axiomatically by McMorris and Neumann [48]. If $l = k$ the consensus rule $\mathcal{M}_l(\mathcal{T})$ is the *strict consensus tree* introduced by Sokal and Rohlf [67]. If $l = \lceil k/2 \rceil + 1$ the consensus rule $\mathcal{M}_l(\mathcal{T})$ is a the *majority rule consensus*, a refinement of the strict consensus tree, introduced by Margush and McMorris [45, 47].

Assuming that cluster can be compared in $O(1)$ time, the following time complexity results are known. Wareham and Day gave an algorithm that computes the majority rule consensus tree in $O(n^2 + k^2 n)$ time [71]. Amenta et al. gave a randomized algorithm to compute the majority consensus tree in time $O(kn)$ [5].

A *semi-strict consensus* or *combinable component* [13, 7] tree of the profile $\mathcal{T}$ is the set of all clusters in trees $T \in \mathcal{T}$ that are not in conflict with any other cluster in $\mathcal{T}$.

20.5   Build Supertree Algorithms

We will survey supertree algorithms that originated from the BUILD algorithm [3]. Therefore we first specify the original objective of the BUILD algorithm, which is based on the following definition of lineage constraints.

DEFINITION 20.3 [lineage constraint] A *lineage constraint* is a binary relation $<$ on ordered pairs of elements: $(i, j) < (k, l)$ is *satisfied* by a rooted tree T, if $\{i, j\} <_T \{k, l\}$. A rooted tree T is *consistent* with a set of lineage constraints S, if every constrain in S is satisfied by T.

The BUILD algorithm constructs a rooted tree that is consistent with a given set of lineage constrains. If no such tree exists the empty tree (represented by NIL) is returned. Given n lineage constrains the BUILD algorithm runs in $O(n^2)$ time [3]. To construct a parent tree for a given profile, the BUILD algorithm is executed on the set of all lineage constrains that are consistent with a tree in the profile.

To derive the supertree methods in this section it is convenient to use a variant of the BUILD algorithm, the BUILD-GRAPH-REPRESENTATION algorithm. Following Semple and Steel [63], Figure 20.2 depicts the BUILD-GRAPH-REPRESENTATION algorithm, which computes a parent tree for a given profile. Note that in case the profile is not compatible the parent tree is empty. The BUILD-GRAPH-REPRESENTATION algorithm associates with a given profile $\mathcal{T}$ the graph $G(\mathcal{T}) = (\mathcal{L}(\mathcal{T}), E_{\mathcal{T}})$, where $e \in E_{\mathcal{T}}$ if there exists a proper cluster $C \in c(\mathcal{T})$ such that $e \subseteq C$. Figure 20.5 shows an example for a graph $G(\mathcal{T})$. We denote the connected components of $G(\mathcal{T})$ by $B(\mathcal{T})$. The correctness of the BUILD-GRAPH-PRESENTATION algorithm follows from the correctness of the BUILD algorithm [3].

FIGURE 20.2: The BUILD-GRAPH-REPRESENTATION algorithm

---

1 BUILD-GRAPH-REPRESENTATION($\mathcal{T}$)
2 **if** $|\mathcal{L}(\mathcal{T})| \leq 2$ **then**
3 $\quad$| return the tree T where $\mathcal{L}(T) = \mathcal{L}(\mathcal{T})$
4 **end**
5 **if** $|B(\mathcal{T})| > 1$ **then**
6 $\quad$**for** *each* $S \in B(\mathcal{T})$ **do**
7 $\quad\quad$| $T_S :=$ BUILD-GRAPH-REPRESENTATION($\mathcal{T}_{|S}$);
8 $\quad\quad$**if** $T_S = Nil$ **then**
9 $\quad\quad\quad$| return Nil;
10 $\quad\quad$**end**
11 $\quad$**end**
12 **else**
13 $\quad$| return Nil;
14 **end**
15 return the tree T where $Ch(r(T)) = \{r(T_S) \mid S \in B(\mathcal{T})\}$;

Other modifications of the BUILD algorithm include the algorithm SUPERB [22] that computes all binary trees consistent with a set of given lineage constrains, and the algorithms ONETREE and ALLTREE [50] that compute for a triplet profile one or all parent trees respectively. A simplified version of ONETREE [18] computes a parent tree in time $O(nm)$ for a set of m triplets over a set of n taxa. Semple and Steel [63] introduced the algorithm MINCUTSUPERTREE that modifies the BUILD algorithm to always return a non-empty tree that contains the nestings common to all trees in a given profile. Page [53] modified the MINCUTSUPERTREE algorithm to optimize the algorithms effectiveness in real applications. Later Semple et al. [62] followed the approach of the BUILD algorithm to include either ancestral divergence dates or a labeling of inner nodes of trees in the profile. In the following we will survey the MINCUTSUPERTREE algorithm and Page's refinement of it.

20.5.1   The MinCutSupertree algorithm

Semple and Steel [63] introduced the algorithm MINCUTSUPERTREE that given a profile $\mathcal{T}$ computes a unique tree, that is denoted by $\mathcal{M}(\mathcal{T})$ and called the *MinCut* supertree. As a major objective of the algorithm, the MinCut supertree contains the nestings that are common to the trees in $\mathcal{T}$.

The MINCUTSUPERTREE algorithm is a modified version of the BUILD-GRAPH- REPRE-SENTATION algorithm. Whenever the BUILD-GRAPH-PRESENTATION algorithm detects a connected component graph $G(\mathcal{T})$ representing incompatibility (in line 13), the MINCUTSU-PERTREE algorithm disconnects this component by deleting a selected set of edges, and thus allows the BUILD-GRAPH-PRESENTATION algorithm to proceed in building a supertree.

To identify the edges to be deleted, each edge $e \in E_{\mathcal{T}}$ is weighted by the number of trees in the profile $\mathcal{T}$ where e is contained in a proper cluster. A natural objective is to disconnect the connected component by deleting edges that cross the minimum cuts in $G(\mathcal{T})$. To return a supertree with the desired nesting property of the MinCut supertree, edges that are supported by each tree in $\mathcal{T}$, denoted by $E_{\mathcal{T}}^{\max}$ need to be exempt from deletion. The edges that are available for deletion are identified through the graph $G(\mathcal{T}) - E_{\mathcal{T}}^{\max}$ that is derived from $G(\mathcal{T})$ as follows. Every maximal subgraph (V, E) in $G(\mathcal{T})$ that contains only edges in $E_{\mathcal{T}}^{\max}$ is collapsed into the node V. If any distinct collapsed nodes V and V' are connected through a non-empty edge set $E' = \{\{v, v'\} \mid v \in V, v' \in V'\}$, then the edges in E' are replaced by the single edge $\{V, V'\}$. The single edge is weighted by the number of trees in the profile $\mathcal{T}$ with a proper cluster that contains the endpoints of at least one edge in the parallel class. Figure 20.3 depicts an example for constructing the graph $G(\mathcal{T}) - E_{\mathcal{T}}^{\max}$.

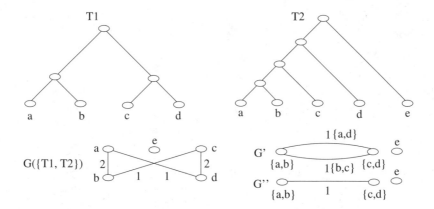

FIGURE 20.3:   The figure depicts how the edges in $E_{\mathcal{T}}^{\max}$ in the graph $G(\mathcal{T})$ for the profile $\mathcal{T} = \{T1, T2\}$ are collapsed. The weight of an edge e in the graph $G(\mathcal{T})$ represents the number of trees in $\mathcal{T}$ in which e is contained in a proper cluster. Thus $E_{\mathcal{T}}^{\max} = \{\{a, b\}, \{c, d\}\}$. Graph G' shows the maximal connected subgraphs that contain only edges from $E_{\mathcal{T}}^{\max}$ collapsed into the nodes $\{a, b\}$ and $\{c, d\}$. In $G'' = G(\mathcal{T}) - E_{\mathcal{T}}^{\max}$ the two 'parallel' edges in G' are replaced by one edge. Only tree $T2$ contains the endpoints of the edge $\{a, b\}$ and $\{c, d\}$. Thus the edge replaced edge is marked by 1.

To disconnect the graph $G(\mathcal{T})$, line 13 is replaced by the line shown in Figure 20.4. After this modification BUILD-GRAPH-REPRESENTATION can not return an empty tree, and

---

FIGURE 20.4: MINCUTSUPERTREE modification

13 remove any edge from $G(\mathcal{T})$, if it is represented by an edge in $G(\mathcal{T}) - E_{\mathcal{T}}^{\max}$ that crosses a minimum-weight cut in $G(\mathcal{T}) - E_{\mathcal{T}}^{\max}$;

---

thus lines $8, 9$ and 10 can be removed from the algorithm. The modified algorithm is the MINCUTSUPERTREE algorithm. Note, that in the original MINCUTSUPERTREE algorithm each tree in the given profile have weights from $\mathbb{Q}^+$ assigned to it, and the edge set $E_{\mathcal{T}}^{\max}$ is defined by the set of edges in $E_{\mathcal{T}}$ weighted by the sum of all weights of trees in $\mathcal{T}$.

Consensus properties of the MinCutSupertree algorithm

Semple and Steel [63] show the following properties for the MINCUTSUPERTREE algorithm. Let $\mathcal{T}$ be a given profile. If $\mathcal{T}$ is compatible, then the MinCut supertree $\mathcal{M}(\mathcal{T})$ is a parent tree of $\mathcal{T}$ (property $P3$ in Section 20.3). Furthermore, the MINCUTSUPERTREE algorithm preserves the nestings shared by the trees in the given $\mathcal{T}$, and consequently the MinCut supertree $\mathcal{M}(\mathcal{T})$ contains the triplets shared by the trees in $\mathcal{T}$ (property $P6_r$ in Section 20.3). By definition the Adams consensus tree $\mathcal{A}(\mathcal{T})$ satisfies this property too, but is not necessarily equal or even comparable with $\mathcal{M}(\mathcal{T})$. However, Semple and Steel showed a strong connection between $\mathcal{A}(\mathcal{T})$ and $\mathcal{M}(\mathcal{T})$. Assuming that the trees in $\mathcal{T}$ have the same taxon set, then either $\mathcal{A}(\mathcal{T}) \leq \mathcal{M}(\mathcal{T})$ or $\mathcal{A}(\mathcal{T})$ and $\mathcal{M}(\mathcal{T})$ are not comparable. As an open problem Semple and Steel [63, p.157] ask, if there exists a modification of the MINCUTSUPERTREE algorithm such that always $\mathcal{A}(\mathcal{T}) \leq \mathcal{M}(\mathcal{T})$. Day and McMorris [24, p.110] conclude from this, that if such a modification could be described axiomatically it would yield a new characterization of the Adams consensus rule.

20.5.2 Page's refinement of the MinCutSupertree algorithm

The refinement of the MINCUTSUPERTREE algorithm from Page [53] allows to display additional nesting information from a profile that is uncontradicted. An edge $e \in E_{\mathcal{T}}$ represent a nesting $e <_T r(T)$ for at least one tree in $\mathcal{T}$. This nesting can be contradicted by some other tree $T' \in \mathcal{T}$ where $e \not<_{T'} r(T')$ and $e \subseteq \mathcal{L}(T')$. If $r(T')$ is a bifurcation or a true multifurcation the nesting is clearly contradicted by $e \not<_{T'} r(T')$. True nestings are rare in practice [66], and Page assumes that multifurcated root nodes are apparent. If $r(T')$ is an apparent multifurcation, it always can be replaced by an unknown rooted binary tree where $e <_{T'} r(T)$. Thus, T' does not make any statement whether the nesting $e <_T r(T)$ is contradiction or not. Hence an edge $e \in E_{\mathcal{T}}$ is *contradicted*, if there exists a bifurcated root $r(T')$ for some tree $T' \in \mathcal{T}$ where $e \not<_{T'} r(T')$. An example for contradicted edges is depicted in Figure 20.5.

Note that the definition for contradicted edges may still contain edges that contradict each other. As an example let $\mathcal{T} = \{a|bc, ab|c, abc\}$, where abc represents the non-binary rooted tree on $\{a, b, c\}$. The edges $\{a, b\}$ and $\{b, c\}$ are elements in $E(\mathcal{T})$. By definition both edges are uncontradicted. However, the edge $\{a, b\}$ assumes that abc is resolved as $c|ab$, while the edge $\{b, c\}$ assumes that abc is resolved as $a|bc$. Thus, for at least one of the edges the nesting is contradicted by the tree abc. Such contradicting edges will be cut by the MINCUTSUPERTREE algorithm.

To refine the MINCUTSUPERTREE algorithm to display more uncontradicted edges , these edges are prevented from being part of any minimum cut. By Proposition 20.2 this is not always possible, since the MINCUTSUPERTREE algorithm satisfies properties $P2$ and $P3$. Therefore, exactly prior to the execution of line 13 in Figure 20.6 the refinement tests if the

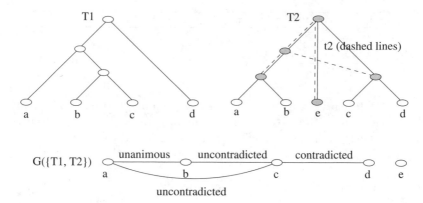

FIGURE 20.5: The graph $G(\mathcal{T})$ associated with the profile $\mathcal{T} = \{T1, T2\}$ is shown. The 'unanimous' edge $\{a, b\}$ is an edge in $E_{\mathcal{T}}^{\max}$, since it is contained in the proper cluster $\{a, b, c\} \in c(T1)$ and the proper cluster $\{a, b\} \in c(T2)$. The edge $\{c, d\}$ is contradicted, since it is not contained in a proper cluster in $c(T1)$ and $r(T1)$ is a bifurcation. The edges $\{a, c\}$ and $\{b, c\}$ are uncontradicted, since the are contained in proper clusters in $c(T1)$ and the apparent multifurcation $r(T2)$ can be replaced by the tree $t2$ (dashed), where both edges are contained in the proper cluster $\{a, b, c, d\}$.

graph $G(\mathcal{T}) - (E_{\mathcal{T}}^{\max})$ can be disconnected by only cutting edges that are contradicted. If not, then the MINCUTSUPERTREE algorithm proceeds. Otherwise, the refinement exempts the set E_u of uncontradicted edges in the graph $G(\mathcal{T}) - (E_{\mathcal{T}}^{\max})$ from being cut by collapsing them. Similar to the graph $G(\mathcal{T}) - (E_{\mathcal{T}}^{\max})$ that results from collapsing edges in $E_{\mathcal{T}}^{\max}$, the graph $G' = (G(\mathcal{T}) - (E_{\mathcal{T}}^{\max})) - E_u$ is constructed. This graph then replaces the graph $G(\mathcal{T}) - (E_{\mathcal{T}}^{\max})$ and the MINCUTSUPERTREE algorithm proceeds to cut edges crossing a minimum cut in G'. The algorithm for the refinement is given below.

FIGURE 20.6: MINCUTSUPERTREE refinement

13(a) determine the set of contradicted edges E_c in $G(\mathcal{T}) - (E_{\mathcal{T}}^{\max})$.;

13(b) **if** *removing E_c from $G(\mathcal{T}) - (E_{\mathcal{T}}^{\max})$ disconnects the graph* **then**
    | $G' := (G(\mathcal{T}) - E_{\mathcal{T}}^{\max}) - E_C$;
else
    | $G' := G(\mathcal{T}) - (E_{\mathcal{T}}^{\max})$;
end

13(c) remove any edge from $G(\mathcal{T})$, if it is represented by an edge in G' that is crossing a minimum-weight cut in G';

The strengths and weaknesses of Page's refinement in practice are discussed in [53].

20.6   Supertree Methods Using Matrix Representations

Several supertree methods take a matrix representation of trees as their input. These methods include 'matrix representation using parsimony' (MRP) and 'matrix representation using flipping' (MRF), which we will survey in this section after we have introduced some basic notation and results.

20.6.1 Notation and basic results

The 'cluster matrix' [14, 8, 59, 58] for a profile encodes each proper cluster in every input tree through a partial binary character. Taxa present in the cluster are scored as a 1-entry, those absent in the cluster are scored as a 0-entry, and those not sampled in the cluster's source tree are scored by a ?-entry. An example for a cluster matrix is depicted in Figure 20.7.

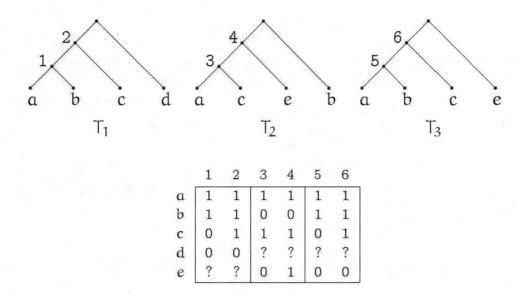

	1	2	3	4	5	6
a	1	1	1	1	1	1
b	1	1	0	0	1	1
c	0	1	1	1	0	1
d	0	0	?	?	?	?
e	?	?	0	1	0	0

FIGURE 20.7: Top: A set of incompatible trees $\mathcal{T} = \{T_1, T_2, T_3\}$. Bottom: $\mathcal{C}(\mathcal{T})$; each column is labeled by the proper cluster to which it corresponds.

If the cluster matrix for a multi set of input trees allows to replace each ?-entry by either a 0- or 1-entry such that there exists a rooted tree that contains the resulting (not necessarily proper) clusters, then the matrix is said to have a 'compatible completion'. Figure 20.8 gives an example for a compatible completion of a cluster matrix. In the following we will give formal definitions and basic results for the matrix representation for a profile $\mathcal{T} = \{T_1, \ldots, T_k\}$.

DEFINITION 20.4 [Complete and incomplete cluster matrix] For $p \in \{1, \ldots, k\}$, let $c(T_P) = \{X_{p1}, \ldots, X_{pq_p}\}$ be the set of proper clusters in the tree T_p. The *cluster matrix* of T_p is the $n \times q_p$ matrix $\mathcal{C}(T_p) := [c_{ij}]$, where

$$c_{ij} := \begin{cases} ? & \text{if } s_i \notin \mathcal{L}(T_p) \\ 1 & \text{if } s_i \in X_{pj} \\ 0 & \text{otherwise} \end{cases}$$

Let $m = \sum_{i=1}^{k} q_i$. The *cluster matrix of the profile* $\mathfrak{T}$ is the $n \times m$ matrix $\mathcal{C}(\mathfrak{T})$ that consists of k blocks of columns, where block p is $\mathcal{C}(T_p)$ (see Figure 20.7). We call the matrix $\mathcal{C}(\mathfrak{T})$ *complete* if it has no ?-entries, and *incomplete* otherwise.

DEFINITION 20.5 [Compatibility and completion of cluster matrices] Let $\mathcal{B} = [b_{ij}]$ be an $n \times m$ complete cluster matrix and let $O_j(\mathcal{B})$ denote the set of row indices i such that $b_{ij} = 1$. The matrix $\mathcal{B}$ is *compatible* if there exists a tree T over the set $\{s_1, \ldots, s_n\}$ such that for every $j \in \{1, \ldots, m\}$, there exists a cluster $X \in c(T)$ such that $X = \{s_i : i \in O_j(\mathcal{B})\}$. Let $\mathcal{C} = [c_{ij}]$ be an $n \times m$ incomplete cluster matrix. A *completion* of $\mathcal{C}$ is an $n \times m$ complete cluster matrix $\mathcal{B} = [b_{ij}]$ such that, for all i, j, $b_{ij} = c_{ij}$ whenever $c_{ij} \neq ?$. $\mathcal{C}$ is said to be *compatible* if it has a compatible completion.

The following corollary follows directly from Theorem 20.1.

COROLLARY 20.1 A complete cluster matrix $\mathcal{B}$ is compatible if and only if for any pair of columns i and j, the sets $O_i(\mathcal{B})$ and $O_j(\mathcal{B})$ are not in conflict.

Thus, a compatible completion of an incomplete cluster matrix $\mathcal{C}$ exists if and only if all ?-entries in $\mathcal{C}$ can be changed to 0-entries or 1-entries such that for any pair of columns i and j, $O_i(\mathcal{B})$ and $O_i(\mathcal{B})$ are not in conflict. Clearly, the profile $\mathfrak{T}$ is compatible if and only if $\mathcal{C}(\mathfrak{T})$ is compatible. Consequently, the BUILD algorithm can be used to decide if there exists a compatible completion of $\mathcal{C}$ in polynomial time. However, there are other polynomial time algorithms that test for the existence of a compatible completion and to construct one, if it exists (e.g. [56]).

20.6.2 MRP **supertrees**

The MRP problem takes the cluster matrix $\mathcal{C}(\mathfrak{T})$ of the profile $\mathfrak{T}$ as input, interprets the clusters in $\mathcal{C}(\mathfrak{T})$ as partial binary characters, and seeks the phylogeny that requires the fewest number of character state transitions, under the parsimony criteria employed (e.g., Dollo, Wagner, irreversible).

While the interpretation of the clusters as characters for the MRP problem seems to be somewhat ad-hoc, the wide availability of parsimony heuristics [32] has made this method for building supertrees the typical choice of phylogeneticists working with real data sets. Since deciding the parsimony problem is NP complete [35], the MRP problem is intrinsically hard to solve.

20.6.3 MRF **supertrees**

The input for the MRF problem is the cluster matrix for a profile $\mathfrak{T}$ of trees. The MRF problem seeks the smallest number of edit steps, called 'flips', that change the given matrix into a compatible one. The resulting matrix then corresponds perfectly to some tree, called a 'MRF supertree'. A 'flip' in a cluster matrix modifies a 1-entry into a 0-entry or vice versa. Figure 20.8 illustrates the notion of the 'flips'. We first describe the MRF problem formally and then survey its theoretical results.

$$\mathcal{C}(\mathcal{T})$$

a	1	1	1	1	1	1
b	1	1	**0**	**0**	1	1
c	0	1	1	1	0	1
d	0	0	?	?	?	?
e	?	?	0	1	0	0

$$\mathcal{C}'$$

1	1	1	1	1	1
1	1	1	1	1	1
0	1	1	1	0	1
0	0	?	?	?	?
?	?	0	1	0	0

$$\mathcal{B}$$

a	1	1	1	1	1	1
b	1	1	1	1	1	1
c	0	1	1	1	0	1
d	0	0	0	0	0	0
e	0	0	0	1	0	0

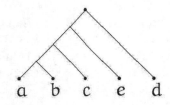

FIGURE 20.8: $\mathcal{C}(\mathcal{T})$ is the cluster matrix for the incompatible set of trees from Figure 20.7. $\mathcal{C}'$ is the compatible cluster matrix that results from flipping the underlined entries of $\mathcal{C}(\mathcal{T})$. $\mathcal{B}$ is a completion of $\mathcal{C}'$; the tree corresponding to $\mathcal{B}$ is also shown. Note that the placement of taxon d as ancestral to taxon e is not supported by the input trees of Figure 20.7; it is simply an artifact of the matrix completion chosen in this example, which is not the only one that would have led to compatibility. Note that the completion of $\mathcal{C}'$ can be interpreted as $? \to 1$ or $? \to 0$ flips with zero cost. Thus, there is no order between flipping and completion

Definitions and Notation

Let $\mathcal{C}$ be a cluster matrix. A *flip* in the matrix $\mathcal{C}$ is the operation of replacing an entry c_{ij}, where $c_{ij} \neq ?$, by its complement. If $c_{ij} = 0$, the flip is called a $0 \to 1$ *flip*, or an *insertion flip*; if $a_{ij} = 1$, the flip is called a $1 \to 0$ *flip*, or a *deletion flip*.

Given is a cluster matrix $\mathcal{C}$, the MRF *problem* is to find the minimum number of flips that convert $\mathcal{C}$ into a compatible matrix $\mathcal{C}'$. If the flips are constrained to be either only insertion or deletion flips the MRF problem is called MRF *insertion problem* or MRF *deletion problem* respectively. A tree T corresponding to $\mathcal{C}'$ is called MRF *supertree*. Note that this tree is not necessarily unique.

Given additionally an integer k, the decision version of the MRF problem is to decide if the matrix $\mathcal{C}'$ can be converted into a compatible matrix by at most k flips. The decision versions of the MRF insertion and MRF deletion problem constrain the flips to be either insertion or deletion flips respectively.

Each MRF problem can be extended to the *weighted* MRF *problem*. A collection of flips is weighted by the sum of their weights. Here flips of matrix elements are weighted by numbers (e.g. to reflect differential node support and confidence from clustering statements). The problem is to find the collection of flips with minimum weight under other collections that convert $\mathcal{C}$ into a compatible matrix $\mathcal{C}'$.

Survey of theoretical results

We survey results on the computational complexity of the MRF problems as well as their consensus properties.

THEOREM 20.2   *[Chen et al. [20]] The decision versions of the* MRF *problem and its restrictions the* MRF *insertion and deletion problem are NP-complete.*

This and other results rely on the graph-theoretic formulation of the MRF decision problem and its variants, which we now summarize.

DEFINITION 20.6   [cluster graph] Let $\mathcal{C}(\mathcal{T})$ be a complete cluster matrix for a profile $\mathcal{T}$. We define the *cluster graph* $G(\mathcal{T})$ of $\mathcal{C}(\mathcal{T})$ to be the bipartite graph (X, Y, E), where $X = \{x_1, \ldots, x_m\}$ represents all columns in $\mathcal{C}(\mathcal{T})$ (these are the proper clusters of trees in $\mathcal{T}$), $Y = \mathcal{L}(\mathcal{T})$ represents the overall taxa set of the rows in $\mathcal{C}(\mathcal{T})$, and $\{x, y\} \in E$ exactly if cluster x contains taxon y.

Corollary 20.1 can be restated in graph-theoretic terms as a forbidden M-graph problem.

DEFINITION 20.7   [M-graph, M-free] Let $G = (X, Y, E)$ be a bipartite graph. A M-*graph* in G is a simple path of length 4, where the degree one nodes are elements in Y. The graph G is M-*free*, if it does not contain an induced M-graph. Figure 20.9 depicts the bipartite graph for a tuple of incompatible complete characters.

	1	2	3	4	5
a	1	1	1	1	0
b	1	1	0	0	0
c	0	1	1	1	0
d	0	0	0	0	1
e	0	0	0	1	1

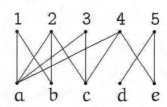

FIGURE 20.9: Left: A tuple of incompatible complete characters. Right: The corresponding character graph; note that the latter contains the induced M-graph defined by the path $\langle b, 2, c, 4, e \rangle$.

THEOREM 20.3   *[Chen et al. [20]] A complete profile $\mathcal{T}$ is compatible if and only if $G(\mathcal{T})$ is M-free.*

If the cluster matrix $\mathcal{C}(\mathcal{T})$ is complete (the profile $\mathcal{T}$ is complete) an edge insertion or deletion in $G(\mathcal{T})$ corresponds to an insertion or a deletion flip in $\mathcal{C}(\mathcal{T})$ respectively. Hence, the MRF decision problems for complete cluster matrices can be stated as the following polynomially equivalent graph modification problem.

DEFINITION 20.8 [M-free graph problems] Given is a cluster graph $G(\mathfrak{T})$ for a complete profile $\mathfrak{T}$ and an integer k. The *M-free graph problem* is to decide if G can be modified into an M-free graph, by at most k deletions or insertions of edges. If only either edge deletions or insertions are allowed the problem is called the *M-free graph insertion problem* or the *M-free graph deletion problem* respectively.

The NP completeness proof for the decision version of the MRF insertion problem is by reduction from the *chain graph completion problem*, studied and proved NP complete by Yannakakis [74].

On the positive side, the MRF problem can be solved approximately within certain guaranteed bounds. A minimization problem P is *approximable within a factor of α*, for some $\alpha \geq 1$, if there exists a polynomial time algorithm A for P such that on any input, the cost c of the solution returned by A is within a factor of α of the cost c^* of the optimum solution; that is, $c/c^* \leq \alpha$. The following theorem relies on some general results on edge modification problems by Natanzon et al. [49].

THEOREM 20.4 *[Chen et al. [20]] Let $\mathfrak{T}$ be a complete profile, and the node degree in X of the cluster graph $G(\mathfrak{T}) = (X, Y, E)$ is at most d (any cluster in each tree in $\mathfrak{T}$ has at most d taxa). The* MRF *problem and the* MRF *deletion problem for the input matrix $G(\mathfrak{T})$ are approximable within a factor of $2d$.*

The graph-theoretic interpretation of clusters is also useful for obtaining an algorithm for the version of the MRF decision problem where a maximum number k of flips is fixed: given a complete profile $\mathfrak{T}$, are at most k flips necessary to make $\mathcal{C}(\mathfrak{T})$ compatible? The next result might be useful when k is small; it implies that the MRF decision problem is *fixed-parameter tractable*, in the sense of Downey and Fellows [26].

THEOREM 20.5 *[Chen et al. [20]] Let $G(\mathfrak{T}) = (X, Y, E)$ be a cluster graph for a complete profile $\mathfrak{T}$, where $x := |X|$ and $y := |Y|$, and $k \in \mathbb{N}$ a fixed number of flips. The* MRF *decision problem can be solved in $O(6^k xy)$ time, and the decision versions of the* MRF *insertion and the* MRF *deletion problem can be solved in $O(2^k xy$ and $O(4^r xy)$ time respectively.*

20.6.4 Consensus properties of MRF and MRP supertrees

On the negative side the MRF and MRP supertree methods do not preserve the majority consensus. The latter property implies that the MRF and MRP supertree methods do not preserve nesting and thus not the Adams consensus [19, 25]. On the positive side, the MRF and MRP supertree methods preserve the semi-strict consensus, and thus the strict consensus. Thus, whenever the input trees are compatible, both any MRF and any MRP supertree is a parent tree of $\mathfrak{T}$ ([17, 19, 25]).

20.7 Supertrees Assembled from Gene Trees

Supertrees describe complex phylogenetic relationships that allow biologists to study their implications on the tree of all species, the Tree of Life. Thus, ideally, input trees for supertree problems are *species trees*, that describe evolutionary relationships of species. Species trees are generally implied from *gene trees*, that represent evolutionary relationships of genes.

Such species trees likely differ from the true species tree, when implied from gene trees that are confounded by a complex history of gene duplication and losses.

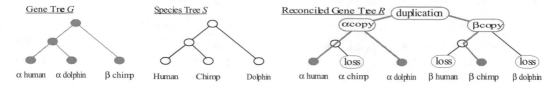

FIGURE 20.10: Reconciling gene tree G with species tree S by the minimum number of duplications and losses $D(G, S)$ (= 1 duplication + 3 losses) needed to achieve agreement between G and S.

Figure 20.10 gives an example, depicting a gene tree G derived from globins of human, dolphin and chimp and the true species tree S of these species. The gene tree G and species tree S differ, because of an ancient gene duplication taking place in the root species of S. Each copy of the duplication develops along the topology of the species tree S and results in the *reconciled gene tree* R. The gene tree G is a homeomorphic subtree of R. Thus duplication is responsible for the incompatibility of trees G and S. The gene duplication could not be detected, because of the leaves in R that are not part of the embedding of G into R.

Gene duplication occurs in prokaryotes [42], but is especially common in eukaryotes, affecting as many as 40% of the genes [60], and leads to the diversification of entire gene families, i.e. globins or rhodopsins, in the same genome [40].

To construct supertrees from gene trees, we need a method to reconcile gene trees. We first describe gene tree reconciliation in Section 20.7.1 and then, in Section 20.7.2, apply this approach to assemble supertrees from gene trees. For convenience, unless noted otherwise, we assume gene and specie trees to be rooted binary trees.

20.7.1   Gene tree reconciliation

In 1979 Goodman et al. [33] introduced the concept *gene tree parsimony* (GTP) of reconciling a gene tree with a species tree. Later this concept was refined and formalized by Page [51], Guigo et al. [36], and Eulenstein [29]. While the literature describes GTP by multi-sets [51, 29], we will introduce GTP through a graph-theoretic approach.

Let G and S be a binary gene and species tree respectively, such that every gene in $\mathcal{L}(G)$ is sampled from exactly one species in $\mathcal{L}(S)$. To represent this association between genes and species we say that G is a gene tree *under* S, if $\mathcal{L}(G) \subseteq \mathcal{L}(S)$. We assume that the gene tree G and the species tree S can only differ because G contains gene duplications. The GTP approach determines a reconciled (gene) tree R, minimal in the number of nodes that can be derived from the species tree S by gene duplications and contains G as a subtree. Trees that are derived from the species tree through zero or more duplications are 'duplication trees'. For example any tree isomorphic to the species tree is a duplication tree, representing a gene in the species $r(S)$ that evolves along the topology of S only through speciation (without duplications). If a gene undergoes a duplication event, it duplicates into two or more copies in the same species. As an example the reconciled tree R in figure 20.10 is a duplication tree whose root $r(R)$ represents a gene that is duplicated in species $r(S)$. We use the duplication function to describes the association of the genes in the duplication tree with their species.

If a gene v undergoes a speciation event this function maps the children of v to the species which are children of the species of v in S. Otherwise v undergoes a duplication event, and thus the duplication function maps the children of v to the species it maps v to. In the following definition duplication trees are defined through the duplication function.

DEFINITION 20.9 [duplication tree] A tree D is a *duplication tree for S*, if there exists a surjective function $d: V(D) \to V(S)$, such that (i) either $d(Ch(v)) = \{d(v)\}$ or $d(Ch(v)) = Ch(d(v))$ for every $v \in V(D)$ and (ii) $d(\mathcal{L}(D)) = \mathcal{L}(S)$. If $d(Ch(v)) = \{d(v)\}$, then v is a *duplication in D* and each node in $Ch(v)$ is a *copy of v*. Otherwise v is a *speciation in D*. The function d is a *duplication function for D*.

Since G and S differ only because of duplications, there exists a duplication tree in which we can embed the gene tree G. Such a duplication tree is called an 'explanation tree', since it explains possible differences between the gene and species tree through its duplications. A 'reconciled tree' is an explanation tree with the minimum number of nodes that explains the difference. We are prepared to give the formal definition of a reconciled tree.

DEFINITION 20.10 [explanation tree, reconciled tree] Let D be a duplication tree for S, and d a duplication function for D. We call D an *explanation tree for G under S*, if there exists a subset $L \subseteq \mathcal{L}(D)$ such that there exists an isomorphism $e: V(G) \to V(D_L)$, called an *embedding of G into R*. A node $v \in V(G)$ is a *duplication in G*, if $e(v)$ is a duplication in D. An explanation tree with the minimum number of nodes is a *reconciled tree*.

THEOREM 20.6 *[Eulenstein [29]]*

1. *The reconciled trees for G under S have a unique topology.*
2. *Let R be a reconciled tree for G under S, and e an embedding of G into R. A node $v' \in V(R)$ is a duplication in R exactly if there exists $v \in V(G)$ such that $e(v) = v'$ where $lca_S(v) = lca_S(c)$ for a child $c \in Ch(v)$. (If G or S are not binary trees the condition is extended by the additional requirement, "or there exist distinct children $a, b \in Ch(v)$ where $lca_S(lca_S(a), lca_S(c(b))) < lca_S(v))$".*

By theorem 20.6 all duplications in a reconciled tree R for G under S are the duplications in G embedded into R, and the duplications in G can be identified through the lca_S mapping from G to S. Thus we can identify the duplications in R through the lca_S relationships independent of the reconciled tree.

DEFINITION 20.11 [duplications] Let G be a gene tree under the species tree S. The set of *duplications* for G under S is $D(G, S) := \{v \in V(G) \mid \exists c \in Ch(g) : lca_S(c) = lca_S(g)\}$.

Let $n := \mathcal{L}(S)$, then $D(G, S)$ can be computed in $O(n)$ on a RAM [75], and in $O(n\alpha(n, n))$, where α is the inverse Ackerman function, on a pointer machine [29]. A reconciled tree R can be constructed from $D(G, S)$ in $\Theta(|V(R)|)$ [29]. Clearly, in the worst case the size of the reconciled tree can be quadratic in the size of S.

The fit of G to S can be measured by the *duplication cost* $c(G, S) := |D(G, S)|$. While the duplication cost satisfies the triangle inequality it is asymmetric [43]. Another measure for the reconciliation adds to the duplication cost the number of maximal unobserved subtrees, called *losses*, in a reconciled tree R for G under S [51]. For example the reconciled tree in Figure 20.10 has three losses, each consisting of a single node. A subtree is *unobserved* if it does not contain any embedded nodes from G. This measure is equivalent [29] to the *mutation cost* [36].

DEFINITION 20.12 [Guigo et al. [36]] Let $g \in V(G)$ and a, b the distinct children of $Ch(g)$, and $p(a, b)$ the number of nodes on the path between $lca_S(a)$ and $lca_S(b)$.

$$l(g) := \begin{cases} 0 & \text{if } lca_S(g) = lca_S(a) = lca_S(b) \\ p(g, a) + 1 & \text{if } lca_S(g) < lca_S(a) \text{ and } lca_S(g) = lca_S(b) \\ p(g, a) + p(g, b) & \text{otherwise.} \end{cases}$$

The *mutation cost* is defined as $l(G, S) := c(G, S) + \Sigma_{g \in V(G)} l(g)$.

Notice that this measure can be determined only from just the gene and species tree.

Software for reconciling gene trees and computing the reconciliation cost includes the phylogenetic software packages COMPONENT [65, 52], FORESTER [76], and Notung [21].

In practice gene trees are often not rooted or not binary. Chen et al. [21] extended the reconciliation of gene trees to unrooted gene trees. The extension of the GTP approach to rooted trees that are not necessarily binary depends on the biological interpretation of a multifurcation in a gene tree [44, 66]. Eulenstein [29] extended the GTP concept from rooted binary trees to rooted gene and species trees with true multifurcations, and showed that gene tree reconciliation can be done in polynomial time.

20.7.2 Supertrees for gene trees

A supertree for a profile of gene trees is a species tree under which the gene trees can be reconciled with the minimum reconciliation cost. We will first define supertree problems for gene trees and then survey results for these problems.

To define supertree problems for gene trees we extend the duplication and mutation cost to a profile of gene trees.

DEFINITION 20.13 Let $\mathcal{G}$ be a profile of gene trees *under* the species tree S, that is $\mathcal{L}(\mathcal{G}) = \mathcal{L}(S)$. The *duplication cost* for G under S is $(\mathcal{G}, S) := \Sigma_{G \in \mathcal{G}} c(G, S)$ and the *optimal duplication cost* for $\mathcal{G}$ is $c^*(\mathcal{G}) := min(\{c(\mathcal{G}, S) \mid S \text{ is a tree over } \mathcal{L}(\mathcal{G})\})$. If $c(\mathcal{G}, S) = c^*(\mathcal{G})$, then S is a *duplication supertree* for $\mathcal{G}$. Similarly the *mutation cost* for $\mathcal{G}$ under S is $l(\mathcal{G}, S) := \Sigma_{G \in \mathcal{G}} l(G, S)$ and the *optimal mutation cost* for $\mathcal{G}$ is $l^*(\mathcal{G}) := min(\{l(\mathcal{G}, S) \mid S \text{ is a tree over } \mathcal{L}(\mathcal{G})\})$. If $l(\mathcal{G}, S) = l^*(\mathcal{G})$, then S is a *duplication-loss supertree* for $\mathcal{G}$.

DEFINITION 20.14 [duplication and duplication-loss problem] Given a profile of gene trees $\mathcal{G}$, the *duplication problem* is to find a *duplication supertree*, and the *duplication-loss problem* is to find a duplication-loss supertree. Given additionally an integer k, the decision version of the duplication and the duplication-loss problem is to decide if there exists a species tree such that $c(\mathcal{G}, S) \leq k$ and $l(\mathcal{G}, S) \leq k$ respectively.

Figure 20.11 shows the profile $\mathcal{G} = \{T_1, T_2, T_3\}$ of pairwise compatible gene trees and the duplication supertree S, which minimizes $D(\mathcal{G}, S)$, by reconciling T_1 with one gene duplication; T_2 and T_3 do not require a gene duplication.

The decision versions of the duplication and the duplication-loss problem are NP complete [43, 31]. The gene duplication problem is fixed parameter tractable, when parameter-

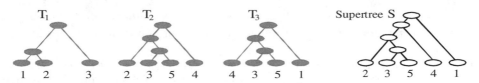

FIGURE 20.11: A gene tree profile $\mathcal{G} = \{T_1, T_2, T_3\}$ and its duplication supertree S.

ized by the number of duplications [70], and the duplication-loss problem is fixed parameter tractable when parameterized by the maximal number of copies that can evolve along a branch in the species tree [39]. Various other parameterizations and restrictions of these problems are shown to be NP complete [43, 31].

Using heuristics, the effectiveness of the duplication loss problem to construct supertrees from gene trees has been evaluated using nuclear gene families. Cotton and Page [23] extracted 118 gene families from HOVERGEN [27] and recovered a nearly classical vertebrate phylogeny. This is in marked contrast to the highly unorthodox trees recovered from analyzes of complete mitochondrial genomes. It also demonstrates the potential of the duplication-loss problem to construct supertrees directly from gene trees, instead from uncertain species trees.

Implying species trees from gene trees is not only complicated by gene duplications, but also by horizontal gene transfer, and lineage sorting [64]. *Horizontal gene transfer* is common in prokaryotes and refers to the copying and insertion of DNA from one genome into another. *Lineage sorting* refers to the persistence of multiple copies of genes through speciation events, followed by their distribution into different descendant lineages. Hallet et al. [38] extended the GTP concept to a combinatorial model that incorporates horizontal gene transfer and gene duplications simultaneously.

20.8 Concluding Remarks

To construct the evolutionary history of life, systematic biologists are faced with the problem of assembling large collections of small phylogenetic trees. This has fostered a strong interest in developing supertree methods, which combine phylogenetic trees into large trees — supertrees. A fundamental issue in developing supertree methods is to resolve conflicting branching information among the input trees.

In this chapter we outlined a variety of supertree methods that resolve conflicting branching information. All presented methods except for the MRP method derive the supertrees from a modification of the input trees that resolves the conflicting branching information. From this resolved branching information supertrees are derived. The MINCUT SUPERTREE methods and the MRF method modify the branching information in the input trees such that they become compatible and output possible parent trees. Both methods differ in the way they modify the branching information. The MINCUT SUPERTREE modifies the branching information to construct one parent tree that satisfies a particular consensus property, which is to preserve the nestings common to the input trees. Note that this is the only consensus property presented in this chapter that applies to input trees with different leaf sets. The MRF method minimizes the overall changes of the branching information. Properties that are satisfied by resulting parent trees are not part of the methods objective. The supertree method for gene trees resolves the conflict in the input trees by reconciling them using an evolutionary model. Reconciling input trees, adds additional subtrees to the input trees and does not modify the original branching information. The objective of reconciling is to minimize the overall number of gene duplications and losses to reconcile the input trees.

Thus the methods can be categorized into two groups by their objectives. One group is specified by an objective that relates only to the input trees. For example the MRF method is designed to correct errors in the input trees and the supertree method for gene trees is designed to correct the gene trees by identifying missing gene duplications that are used to reconcile the input trees. The other group is specified by an objective that the resulting supertree needs to satisfy.

While it might be impossible to determine a supertree objective that performs most effectively in practice, evolutionary biologist have access to supertrees constructed by an increasing number of different objectives and known theoretical properties. Those supertrees can support systematic biologists in their efforts to construct the Tree of Life.

Acknowledgements

The author thanks W. Chang and D. Fernandéz-Baca for careful reading of the manuscript and helpful suggestions. This work was supported in part by the NSF grant EF-0334832.

References

[1] E.N. Adams. Consensus techniques and the comparison of taxonomic trees. *Systematic Zoology*, 21:390–397, 1972.

[2] E.N. Adams. N-trees as nestings: Complexity, similarity, and consensus. *Journal of Classification*, 3:299–317, 1986.

[3] A.V. Aho, J.E. Hopcroft, and J.D. Ullman. On finding lowest common ancestors in trees. *SIAM Journal on Computing*, 1(5):115–132, 1976.

[4] A.V. Aho, Y.Sagiv, T.G. Szymanski, and J.D. Ullman. Inferring a tree from lowest common ancestors with an application to the optimization of relational expressions. *SIAM Journal on Computing*, 10(3):405–421, 1981.

[5] N. Amenta, F. Clarke, and K. St.John. A linear-time majority tree algorithm. In *WABI*, pages 216–227, 2003.

[6] K.J. Arrows. *Social Choice and Individual Values*. Wiley, New York, 1952.

[7] J.P. Barthélemy, F.R. McMorris, and R.C. Powers. Dictatorial consensus functions on n-trees. *Mathematical Social Sciences*, 25(1):59–64, 1992.

[8] B.R. Baum. Combining trees as a way of combining data sets for phylogenetic inference, and the desirability of combining gene trees. *Taxon*, 41:3–10, 1992.

[9] Olaf R.P. Bininda-Emonds, editor. *Phylogenetic supertrees*. Springer Verlag, 2004.

[10] O.R.P. Bininda-Emonds, J.L. Gittleman, and M.A. Steel. The (super) tree of life: procedures, problems, and prospects. *Annual Review of Ecology and Systematics*, 33:265–289, 2002.

[11] S. Böcker, D. Bryant, A.W.M. Dress, and M.A. Steel. Algorithmic aspects of tree amalgamation. *Journal of Algorithms*, 37(2):522–537, 2000.

[12] H.L. Bodlaender, M.R. Fellows, and T.J. Warnow. Two strikes against perfect phylogeny. In *Proc. 19th Internat. Colloq. on Automata, Languages and Programming*, 1992.

[13] K. Bremer. Combinable component consensus. *Cladistics*, 6:369–372, 1990.

[14] D.R. Brooks. Hennig's parasitological method: A proposed solution. *Syst. Zool.*, 30:325–331, 1981.

[15] D. Bryant. A classification of consensus methods for phylogenetics. In *Bioconsensus*. DIMACS-AMS, 163–184.

[16] D. Bryant. *Hunting for trees, building trees andcomparing trees: theory and method in phylogeneticanalysis*. PhD thesis, Dept. of Mathematics, University of Canterbury, 1997.

[17] D. Bryant. A classification of consensus methods for phylogenies. In *Bioconsensus*, volume 61 of *DIMACS: Series in Discrete Mathematics and Theoretic Computer Science*, pages 163–183. American Mathematical Society, Providence, Rhode Island, USA, 2003.

[18] D. Bryant and M. Steel. Extension operations on sets of leaf-labelled trees. *Discrete Applied Mathematics*, 16:425–453, 1995.

[19] D. Chen, L. Diao, O. Eulenstein, and D. Fernández-Baca *et al*. Flipping: A supertree construction method. In *Bioconsensus*, volume 61 of *DIMACS: Series in Discrete Mathematics and Theoretic Computer Science*, pages 135–160. American Mathematical Society, Providence, Rhode Island, USA, 2003.

[20] D. Chen, O. Eulenstein, D. Fernández-Baca, and M.J. Sanderson. Supertrees by flipping. In *Computing and Combinatorics, 8th Annual International Conference, COCOON 2002, Singapore, August 15-17, 2002, Proceedings*, volume 2387 of *Lecture Notes in Computer Science*, pages 391–400. Springer, 2002.

[21] K. Chen, D. Durand, and M. Farach-Colton. Notung: a program for dating gene duplications and optimizing gene family trees. *Journal of Computational Biology*, 7(3/4):429–447, 2000.

[22] M. Constantinescu and D. Sankoff. An efficient algorithm for supertrees. *Journal of Classification*, 12:101–112, 1995.

[23] J. Cotton and R.D.M. Page. Vertebrate phylogenomics: reconciled trees and gene duplications. In *Pacific Symposium on Biocomputing*, pages 536–547, 2002.

[24] W.H.E. Day and F.R. McMorris. Axiomatic consensus theory in group choice and biomathematics. In *Frontiers in Applied Mathematics*, volume 39. Society for Industrial and Applied Mathematics, 2003.

[25] L. Diao, O. Eulenstein, D. Fernández-Baca, and M.J. Sanderson. Consensus properties of MRP supertrees. Technical report, Dept. of Computer Science, Iowa State University, 226 Atanasoff Hall, Ames, IA 50011-1040 USA, 2003.

[26] R.G. Downey and M.R. Fellows. *Parameterized Complexity*. Springer, 1997.

[27] L. Duret, D. Mouchiroud, and M. Gouy. Hovergen: A database of homologous vertebrate genes. *Nucleic Acids Research*, 2:2360–2365, 1994.

[28] G.F. Estabrook, C. Johnson, and F.R. McMorris. An idealized concept of the true cladistic character? *Mathematical Bioscience*, 23:263–272, 1975.

[29] O. Eulenstein. *Predictions of gene-duplications and their phylogenetic development*. PhD thesis, University of Bonn, Germany, 1998. GMD Research Series No. 20 / 1998, ISSN: 1435-2699.

[30] O. Eulenstein, D. Chen, J. G. Burleigh, and D. Fernández-Baca *et al*. Performance of flip supertree construction with a heuristic algorithm. *Systematic Biology*, 53:299–308, 2003.

[31] M. Fellows, M. Hallett, C. Korostensky, and U. Stege. Analogs & duals of the mast problem for sequences & trees. In *European Symposium on Algorithms (ESA), LNCS 1461*, pages 103–114, 1998.

[32] J. Felsenstein. PHYLIP. `http://evolution.genetics.washington.edu/phy-lip.html`.

[33] M. Goodman, J. Czelusniak, G. W. Moore, and A. E. Romero-Herrera *et al.* Fitting the gene lineage into its species lineage. a parsimony strategy illustrated by cladograms constructed from globin sequences. *Systematic Zoology*, 28:132–163, 1979.

[34] A.D. Gordon. Consensus supertrees: The synthesis of rooted trees containing overlapping sets of labelled leaves. *Journal of Classification*, 9:335–348, 1986.

[35] R.L. Graham and L.R. Foulds. Unlikelihood that minimal phylogenies for a realistic biological study can be constructed in reasonable computation time. *Math. Biosci.*, 60:133–142, 1982.

[36] R. Guigó, I. Muchnik, and T.F. Smith. Reconstruction of ancient molecular phylogeny. *Molecular Phylogenetics and Evolution*, 6(2):189–213, 1996.

[37] D. Gusfield. *Algorithms on strings, trees, and sequences.* Cambridge University Press, New York, NY, USA, 1997. ISBN 0 521 58519 8.

[38] M. Hallett, J. Lagergren, and A. Tofigh. Simultaneous identification of duplications and lateral transfers. In *RECOMB*, pages 326–335, 2004.

[39] M.T. Hallett and J. Lagergren. New algorithms for the duplication-loss model. In *RECOMB*, pages 138–146, 2000.

[40] S. E. Henikoff, E. A. Green, S. Pietrokovski, and P. Bork *et al.* Gene families: The taxonomy of protein paralogs and chimeras. *Science*, 278:609–614, 1997.

[41] M. R. Henzinger, V. King, and T. Warnow. Constructing a tree from homeomorphic subtrees, with applications to computational evolutionary biology. *Algorithmica*, 24:1–13, 1999.

[42] I.K. Jordan, K.S. Makarova, J.L. Spouge, and Y.I. Wolf *et al.* Lineage-specific gene expansions in bacterial and archaeal genomes. *Genome Research*, 11:555–565, 2001.

[43] B. Ma, M. Li, and L. Zhang. On reconcstructing species trees from gene trees in term of duplications and losses. In *RECOMB*, pages 182–191, 1998.

[44] W. P. Maddison. Reconstructing character evolution on polytomous cladograms. *Cladistics*, 5:355–377, 1989.

[45] T. Margush and F.R. McMorris. Consensus n-trees. *Bulletin of Mathematical Biology*, 43:239–244, 1981.

[46] F.R. McMorris. On the compatibility of binary qualitative taxonomic characters. *Bulletin of Mathematical Biology*, 39:133–138, 1977.

[47] F.R. McMorris, D.B. Meronk, and D.A. Neumann. *Numerical Taxonomy*, chapter A view of some consensus methods for trees. Springer-Verlag, 1983.

[48] F.R. McMorris and D.A. Neumann. Consensus functions defined on trees. *Mathematical Social Sciences*, 4:131–136, 1983.

[49] A. Natanzon, R. Shamir, and R. Sharan. Complexity classification of some edge modification problems. *Discrete Applied Mathematics*, 113(1):109–128, 2001.

[50] M.P. Ng and N.C. Wormald. Reconstruction of rooted trees from subtrees. *Discrete Applied Mathematics*, 69(1-2):19–31, 1996.

[51] R.D.M. Page. Maps between trees and cladistic analysis of historical associations among genes, organisms, and areas. *Systematic Biology*, 43(1):58–77, 1994.

[52] R.D.M. Page. COMPONENT, 1995. Phylogenetic Program Package, available at *http://taxonomy.zoology.gla.ac.uk/rod/cpw.html*.

[53] R.D.M. Page. Modified mincut supertrees. In D. Gusfield and R. Guigó, editors, *International Workshop, Algorithms in Bioinformatics (WABI)*, volume 2452 of *Lecture Notes in Computer Science*, pages 300–315. Springer Verlag, September 2002.

[54] R.D.M. Page and M.A. Charleston. Reconciled trees and incongruent gene and species trees. In *DIMACS Series in Discrete Mathematics and Theoretical Computer Sciences*, volume 37, 1997.

[55] R.D.M. Page and E.C. Holmes. *Molecular evolution: a phylogenetic approach.* Black-

well Science, 1998.

[56] I. Pe'er, R. Shamir, and R. Sharan. Incomplete directed perfect phylogeny. In R. Giancarlo and D. Sankoff, editors, *Combinatorial Pattern Matching*, volume 1848 of *Lecture Notes in Computer Science*, pages 143–153. Springer-Verlag, 2000.

[57] D. Pisani, A.M. Yates, M.C. Langer, and M.J. Benton. A genus-level supertree of the dinosauria. *Proceedings of the Royal Society of London*, 269:915–921, 2002.

[58] A. Purvis. A modification to Baum and Ragan's method for combining phylogenetic trees. *Systematic Biology*, 44:251–255, 1995.

[59] M.A. Ragan. Phylogenetic inference based on matrix representation of trees. *Molecular Phylogenetics and Evolution*, 1:53–58, 1992.

[60] G.M. Rubin, M.D. Yandell, and J.R. Wortman. Comparative genomics of the eukaryotes. *Science*, 287:2204–2215, 2000.

[61] M.J. Sanderson, A. Purvis, and C. Henze. Phylogenetic supertrees: Assembling the trees of life. *Trends Ecol. Evol.*, 13:105–109, 1998.

[62] C. Semple, P. Daniel, W. Hordijk, and R.D.M. Page *et al.* Supertree algorithms for ancestral divergence dates and nested taxa. Bioinformatics, in press.

[63] C. Semple and M.A. Steel. A supertree method for rooted trees. *Discrete Applied Mathematics*, 105:147–158, 2000.

[64] J. Slowinski and R.D.M. Page. How should species phylogenies be inferred from sequence data? *Systematic Biology*, 105:147–158, 1999.

[65] J.B. Slowinski. Review of Component. *Cladistics*, 9:351–353, 1993.

[66] J.B. Slowinski. Molecular polytomies. *Molecular Phylogenetics and Evolution*, 19:114–120, 2001.

[67] R.R. Sokal and F.J. Rohlf. Taxonomic congruence in the leptopodomorpha re-examined. *Systematic Zoology*, 30:309–325, 1981. And 1981, IBM Watson Research Center. RC no. 8624, 31 pp.

[68] M.A. Steel. The complexity of reconstructing trees from qualitative characters and subtrees. *Journal of Classification*, 9:91–116, 1992.

[69] M.A. Steel, A.W.M. Dress, and S. Böcker. Simple but fundamental limitations on supertree and consensus tree methods. *Systematic Biology*, 49:363–368, 2000.

[70] U. Stege. Gene trees and species trees: The gene-duplication problem is fixed-parameter tractable. In *Proceedings of the 6th International Workshop on Algorithms and Data Structures, LNCS 1663*, Vancouver, Canada, 1999.

[71] H.T. Wareham. An efficient algorithm for computing M_l consensus trees. B.Sc. Honours thesis, Memorial University of Newfoundland, 1985.

[72] S.J. Willson. Reconstruction of additive rooted trees. *http://www.public.iastate.edu-/~swillson/BMB-04-09.pdf*, 2004.

[73] M.F. Wojciechowski, M.J. Sanderson, K.P. Steele, and A. Liston. Molecular phylogeny of the "temperate herbaceous tribes" of papilionoid legumes: a supertree approach. In P. Herendeen and A. Bruneau, editors, *Advances in Legume Systematics*, volume 9, pages 277–298. Royal Botanic Garden, Kew., 2000.

[74] M. Yannakakis. Computing the minimum fill-in is NP-complete. *SIAM Journal on Algebraic and Discrete Methods*, 2(1):77–79, 1981.

[75] L. Zhang. On a Mirkin-Muchnik-Smith conjecture for comparing molecular phylogenies. *Journal of Computational Biology*, 4(2):177–187, 1997.

[76] C.M. Zmasek. FORESTER, 2001. Phylogenetic Program Package, available at *http://www.genetics.wustl.edu/eddy/forester*.

21

Large-scale Phylogenetic Analysis

Tandy Warnow
The University of Texas at Austin

21.1 Introduction

Large-scale phylogeny reconstruction poses several challenges to the algorithms designer. To begin with, the fundamental objective in phylogeny reconstruction is to reconstruct, as accurately as possible, the tree that produced the input dataset (typically aligned biomolecular sequences), rather than to solve any particular numeric optimization problem. Therefore, all reconstruction methods are studied (typically in simulation - see [15, 17, 18] for some examples of simulation studies) with respect to what is called their "topological accuracy" - whereby the reconstructed trees are compared against the model tree and the differences between the branching order in the two trees are quantified. Some reconstruction methods operate in polynomial time and have been shown to perform well with respect to the topological accuracy of the reconstructed trees under many model conditions; however, recent work has shown that for large model trees with high interleaf distances, popular polynomial time methods do not produce trees with acceptable levels of topological accuracy [1, 27, 29, 30, 31]. For this reason, among others, most systematists prefer methods that attempt to solve either maximum parsimony (MP) or maximum likelihood (ML), two NP-hard optimization problems [5, 12]. While many heuristics exist for these optimization problems, it is not clear that these heuristics are able to obtain good enough solutions (for their criteria) on large datasets - that is, they do not scale well. Thus, large-scale phylogenetic analysis is a difficult challenge. (For a more detailed introduction to phylogeny reconstruction, see [16, 22].)

This chapter will present a meta-technique for large-scale phylogenetic analysis which has been shown to improve both types of phylogenetic reconstruction methods. The basic meta-technique first "decomposes" the dataset into overlapping subsets. Trees are then constructed on each of the resultant datasets using a favored phylogenetic reconstruction method; since the subsets overlap, these trees also overlap on their leaf sets. These subtrees can then be merged together, using a preferred "supertree" method, into a tree on the full set of taxa. We call the class of methods using this basic structure "Disk-Covering Methods", or DCMs, for historic reasons. Our research shows that DCMs can make phylogenetic reconstruction more accurate and/or faster, depending upon the particular vulnerabilities of the base method. Thus, a DCM is designed to "boost" the performance of a given "base method". Our current DCMs have focused on two different types of base methods – polynomial time distance-based methods, such as neighbor joining [36], and heuristics for maximum parsimony implemented in software such as PAUP* [41] and TNT (see [13] for a description of TNT). These DCMs differ in their design, because of the particular aspects of the base method. Most noticeably, they use different decomposition strategies. The DCM we use for neighbor joining uses a decomposition that produces very small subproblems with small evolutionary diameters (i.e., interleaf-distances), while the decomposition strategy used in the DCM for maximum parsimony produces larger subproblems, with a maximum subproblem size that is still reasonably large. In addition, the DCM for maximum parsimony is used iteratively within a heuristic search in order to obtain improved results. However, we do not expect our current approaches for improving heuristic maximum parsimony or neighbor joining to be the best that can be achieved, and so new DCMs will continue to be useful for these problems. Furthermore, new DCMs will likely be needed as new problems (such as maximum likelihood or phylogenetic multiple sequence alignment [37]) are considered. Thus, we believe that this type of algorithm design, in which base methods are used on carefully selected subproblems and solutions then merged together, represents a promising approach to phylogeny reconstruction, but that DCMs are really in their infancy. We write this chapter, therefore, in the hope that by presenting both the intuition behind the design strategies and the details involved in designing DCMs for specific base methods, algorithms researchers will be able to design their own DCMs as new challenges arise.

The structure of the chapter is as follows. In Section 21.2 we present the basic issues involved in phylogenetic analysis, including definitions of the basic optimization problems, stochastic models of evolution, statistical aspects of a phylogenetic estimation, and algorithmic issues. In Section 21.3 we present the basic divide-and-conquer strategies we use in our various DCMs. These techniques rely heavily on the theory of triangulated graphs, which we present in Section 21.4. We then discuss general design issues for DCMs in Section 21.5. We then begin our description of a few specific DCMs in order to illustrate these techniques. We begin in Section 21.6 with the DCM we designed for use with maximum parsimony heuristics. We continue in Section 21.7 with a description of DCMs that were designed for use with distance-based methods like neighbor joining. Finally, we conclude in Section 21.8 with a discussion of further research directions, and some specific open problems.

21.2 Phylogenetic Analysis

21.2.1 Stochastic models of sequence evolution

Stochastic models have been proposed for all types of biomolecular sequences, including RNA, DNA, and amino-acid. Of these, DNA have the simplest models, since typically these models do not include any structural constraints, as would be generally true in RNA or amino-acid models. Hence, we will focus here on describing the models that have been proposed for DNA evolution.

A model of DNA sequence evolution must describe the probability distribution of the four states, A, C, T, G, at the root, the evolution of a random site (i.e., position within the DNA sequence) and how the evolution differs across the sites. Typically the probability distribution at the root is uniform (so that all sequences of a fixed length are equally likely). The evolution of a single site on a given edge e is modeled through a collection of parameters - one is the "length" $l(e)$ of the edge (which for many stochastic models is equivalent to the expected number of changes of a random site on that edge) and the other is a substitution probability matrix which determines the probabilities of each substitution of one nucleotide by another in a single substitution. These together can be fully expressed by a single "stochastic substitution matrix," $M(e)$, which for DNA is a 4×4 matrix in which every row sums to 1. Note that the matrix $M(e)$ can have up to 12 free parameters. The simplest such model is the Jukes-Cantor model, with one free parameter, and the most complex is the General Markov model, with all 12 parameters [39].

DEFINITION 21.1 The General Markov (GM) model of single-site evolution is defined as follows.

1. The nucleotide in a random site at the root is drawn from a known distribution, in which each nucleotide has positive probability.
2. The probability of each site substitution on an edge e of the tree is given by a 4×4 stochastic substitution matrix $M(e)$ in which $det(M(e))$ is not 0, 1, or -1.

Note that these models only describe the evolution of a single site down the tree. To model how a sequence evolves we would also need to describe how the different sites evolve. Almost all models of sequence evolution assume that different sites evolve independently (a notable exception is the covarion model [42]), but most phylogenetic analyses are based upon models for which the independence of different sites is assumed. The majority of models used in inference allow for sites to evolve differently, but almost all assume that the differences between sites is limited in the following way. The assumption of how sites differ is expressed by saying that for every site i we have a rate r_i, so that the expected number of changes on the edge e is simply the product of the edge length $l(e)$ with the rate r_i. If all sites evolve under the same rate, then $r_i = r_j$ for all i, j. These $\{r_i\}$ are the "rates-across-sites", with r_i the rate for the ith site. Typically these rates are drawn from some distribution. Note that what this expresses is that if a site i is expected to evolve twice as fast as site j on one edge, then it is expected to evolve twice as fast on every edge – that is, sites speed up or slow down identically under all conditions. While this is not necessarily a reasonable assumption in molecular evolution, it is the underlying assumption of models in practice.

In this chapter, we use the GM model with the assumption that $r_i = r_j$, so as to simplify the analysis. We denote a model tree in the GM model as a pair, $(T, \{M_e \colon e \in E(T)\})$, or more simply as (T, M). We assume that the number of changes of a given site on a given edge obeys a Poisson distribution. For each edge $e \in E(T)$, we define the length $\lambda(e)$ of the edge e to be $-\log|det(M_e)|$. This allows us to define the matrix of leaf-to-leaf distances, $[\lambda_{ij}]$, with $\lambda_{ij} = \sum_{e \in P_{ij}} \lambda(e)$ and where P_{ij} is the path in T between leaves i and j. Note that $[\lambda_{ij}]$ is a symmetric matrix. It is a well-known fact that, given the distance matrix $[\lambda_{ij}]$, it is easy to recover the underlying leaf-labeled tree T in polynomial time.

This general model of site evolution subsumes the great majority of other models examined in the phylogenetic literature, including the Hasegawa-Kishino-Yano (HKY) model, the Kimura 2-parameter model (K2P), the Kimura 3-ST model (K3ST), the Jukes-Cantor model (JC), etc. These models are all special cases of the General Markov model, because they place restrictions on the form of the stochastic substitution matrices (see [25] for more information about stochastic models of evolution).

21.2.2 Statistical performance issues

Once a stochastic model of evolution is stated, it becomes possible to discuss statistical inference under the model, define and develop explicitly statistical estimation methods (such as maximum likelihood), and to ask about the performance of phylogeny reconstruction methods under the model. One of the aspects of performance that is typically considered is whether a reconstruction method is "statistically consistent" under the given model. Put simply, a reconstruction method is statistically consistent under a given model of evolution, if, for all model trees under the model, as the sequence length increases, the probability of reconstructing the model tree goes to 1. This is a mathematical question and establishing statistical consistency requires a mathematical proof. Such proofs have been obtained for many methods, including most distance-based methods (including neighbor-joining [36]) and maximum likelihood (which seeks the model tree that maximizes the probability of the data) under the GM model, and hence under all its submodels.

Another aspect of statistical performance under a model is its "convergence rate", which roughly speaking asks how quickly the error rate in the estimation goes to 0 as a function of the sequence length. In order to make this statement precise, we must provide a definition of topological error. While there are many ways to quantify this, the one that is used most typically in the phylogenetics research community is the Robinson-Foulds [32] rate. This is given as follows.

DEFINITION 21.2 Let T be the true tree, and let T' be an estimated tree, both on the same set of n leaves. For each edge e in T, there is an associated bipartition $\pi(e)$ defined on the leaf set of T which is produced by deleting the edge e from T. We can therefore identify T with its set $C(T) = \{\pi(e) : e \in E(T)\}$. Similarly we can identify the tree T' by its set $C(T')$, defined in the analogous way. The **Robinson-Foulds** distance between T and T' is then the average of $|C(T) - C(T')|/(n-3)$ and $|C(T') - C(T)|/(n-3)$; the first of these two values is called the *missing edge* or *false negative* rate, and the second of these two values is called the *false positive* rate.

These are *rates* because they are divided by $n-3$, where n is the number of leaves in each tree ($n-3$ corresponds to the number of internal edges in a binary tree on n leaves). Error rates that are below 5% are desired, and above 10% are unacceptable - unless the data are just too poor to allow for greater resolution.

Two methods which are both statistically consistent under a model may have very different convergence rates, with one method producing trees with much lower error rates than the other, at most "reasonable" sequence lengths. Thus, the convergence rate of a method is highly significant when it comes to predicting its performance on different datasets.

While statistical consistency is relatively easy to establish (the proofs are not difficult generally), mathematical analyses of the convergence rates of different methods is more difficult (see [1, 8, 9] for some initial results). For this reason, the performance of phylogeny reconstruction methods is typically evaluated in simulation.

21.2.3   Maximum parsimony

We now define the maximum parsimony problem.

- Input: Set S of sequences, each over an alphabet A, and of the same length k.
- Output: a tree T with leaves labeled by S and with internal nodes labeled by other elements of A^k, so as to minimize the total "length" of the tree, which is defined to be $\sum_e H(e)$, where $H(e)$ denotes the Hamming distance between the sequences labeling the different endpoints of the edge e (i.e., $H(e)$ is the number of sites at which the sequences labeling the endpoints of e differ).

Finding the optimal trees under maximum parsimony (MP) is an NP-hard problem, and so hard to solve exactly for datasets beyond about 30 or so taxa (if the tree T is fixed, then the problem of assigning sequences to the internal nodes so as to minimize the total length of the tree is easily solved using dynamic programming in polynomial time [11]). Because MP is important in practice, many heuristics exist which attempt to solve the problem through a combination of hill-climbing and randomization to get out of local optima. These heuristics may actually find optimal solutions, though current approaches do not provide sufficiently good lower bounds to make it possible to assess the degree of suboptimality in a given analysis. Current practice therefore tends to involve running a favored heuristic until it seems that better solutions will not be found. Such analyses can take a few days on moderate sized datasets, to weeks or months on large datasets.

MP is not statistically consistent under the GM model, which means that there are some model trees so that as the sequence length increases, the probability that an exact solution to MP would yield the true tree does *not* provably approach 1 [10]. Even so, MP is a popular method, and many heuristics exist to attempt to solve MP. (For more on MP, see [16].)

21.2.4   Distance-based methods

Distance-based methods operate in two phases: first they use statistically-based techniques to estimate pairwise distances, and then they construct a tree on the basis of these estimated distances. Provided that the appropriate technique is used to estimate pairwise distances, and a good method is used in the second step to construct a tree from the distances, the combined two-phase reconstruction is statistically consistent. The most popular distance-based methods are polynomial-time, and hence these methods are quite attractive.

Of the various distance-based methods, neighbor-joining is probably the most popular; it is statistically consistent under the GM model and performs well in many simulation studies by comparison to other distance-based methods. On the other hand, the only mathematical theory about its convergence rate shows that it can require sequence lengths to be exponential in the maximum leaf-to-leaf distance within the model tree, in order to produce the topologically correct true tree with high probability (see below). Since sequence lengths are generally not extremely long, this is a vulnerability of neighbor joining with respect to large-scale phylogeny reconstruction. (This theoretical statement also holds true for other distance-based methods and so this vulnerability of neighbor joining is not unique.) Simulation studies have verified that neighbor joining's performance degrades as the interleaf-distances increase without a corresponding increase in the sequence length [36], and so from an empirical standpoint this vulnerability seems to be significant. (It is worth noting that all methods seem to degrade in performance with increasing interleaf-distances, with the degradation of neighbor-joining less so than that of some other distance-based methods, but worse - it seems - than some sequence-based methods.)

The basic theorem for the convergence rate of neighbor joining is as follows.

THEOREM 21.1   *[From [1]] Let (T, M) be a General Markov Model tree with n leaves, with $0 < f \leq \lambda_e \leq g < \infty$ for all edges e in T. Let $\epsilon > 0$ be given, and let $\lambda^* = max_{ij}\{\lambda_{ij}\}$. Then there is a constant $C > 0$ such that, if the sequence length exceeds*

$$C \log n e^{O(\lambda^*)}$$

then, with probability at least $1 - \epsilon$, the Neighbor-Joining method recovers the true tree.

Note that $\lambda^* \leq g \cdot diam(T)$, where g is the maximum edge length and $diam(T)$ is the number of edges in the longest path in the tree T (i.e. it is the topological diameter of T), and that $diam(T)$ can be as large as $n - 1$. Thus the sequence length requirement of the neighbor joining method is bounded from above by a function that grows *exponentially* in n, even when g is fixed. Such a method is said to have an *exponential convergence rate*.

21.2.5   Fast-converging methods

Since letting f be arbitrarily small or g be arbitrarily large affects the sequence length requirement, we are interested in developing methods for which polynomially long sequences ensure accuracy under the General Markov model, when both f and g are fixed, but arbitrary. In order to define this concept precisely, we first parameterize the General Markov model.

DEFINITION 21.3 $GM_{f,g}$ contains those $(T, M) \in$ GM for which $f \leq \lambda(e) \leq g$ holds for all edges $e \in E(T)$.

We now define absolute fast convergence:

DEFINITION 21.4   A phylogenetic reconstruction method Φ is *absolute fast-converging (afc)* for the GM model if, for all positive f, g, ε, there is a polynomial p such that, for all (T, M) in the GM model, on set S of n sequences of length at least $p(n)$ generated on T, we have $Pr[\Phi(S) = T] > 1 - \varepsilon$.

Note that method M operates without any knowledge of parameters f or g—or indeed any function of f and g. Thus, although the polynomial p depends upon both f and g, the method itself does not.

There are now several methods which have been proven to be afc (see [6, 7, 30, 43]), and in Section 21.7 we will describe how we derive one such afc method through the use of a DCM. All afc methods, whether implicitly or explicitly, have two steps: first they produce a set of trees, and then they select the best tree from the set. Proofs that the methods are afc then require proving (a) that the first step produces a set that includes the true tree with high probability, given polynomial length sequences, and (b) the second step picks the true tree with high probability, under the assumption that the true tree is in the set and the input sequences are of polynomial length.

21.3   The Basic Divide-and-Conquer Strategy

21.3.1   Introduction

Each DCM is fundamentally based upon a divide-and-conquer strategy, which has the following three phase structure:

- Phase I: Compute a decomposition of the dataset into overlapping subsets, and construct trees on the subsets using the base method.
- Phase II: Use a supertree method to merge the trees on the subsets into a tree on the full dataset.
- Phase III: If the tree obtained in Phase II is not fully resolved (i.e. if the tree is not a binary tree), we resolve it further into a binary tree so that it optimizes the desired objective criterion (e.g., maximum parsimony).

We have designed these phases so that we can guarantee accuracy in the reconstructed tree obtained at the end of the first two phases under certain conditions. These guarantees and other properties of our DCMs rely upon the theory of triangulated graphs which is presented in Section 21.4 below. In general, however, we are motivated not only by theory but also by empirical performance, and so much of our discussion here will attempt to reflect those dual concerns.

21.3.2   Phase I: a brief overview

The main issue in the design of Phase I is the decomposition of the set of taxa into overlapping subsets. Our approach for Phase I is to first construct a triangulated graph (that is, a graph without any simple induced cycle of size four or more) whose vertex set corresponds to the input set of taxa, and then compute a decomposition of the vertices of the triangulated graph (and hence of the set of taxa) into overlapping subsets. We describe how we obtain triangulated graphs from our input sets, and how we decompose these triangulated graphs, in Section 21.4 below.

21.3.3   Phase II: Merging the subtrees

After Phase I is completed, we have a set of trees, one for each of the sets in the decomposition of the set of taxa. These subtrees share taxa in common, and the objective is to merge these subtrees into a tree on the full dataset. We would like this merger to retain accuracy if possible, so that in particular if all the subtrees are correct (meaning that they accurately reflect the true tree restricted to the subset), then the merger of these subtrees should be the true tree. This is the **Subtree Compatibility problem:**

- **Input:** Set $\mathcal{T} = \{T_1, T_2, \ldots, T_k\}$ with T_i an unrooted tree on leaf set S_i.
- **Output:** Tree T on leaf set $S = \cup_i S_i$, if it exists, such that $T|S_i = T_i$.

This problem is NP-hard, as was shown in [38]. Thus, any method for this problem which attempts to retain accuracy is likely to fail under some conditions, or to require exponential time.

The method we use is the *Strict Consensus Merger*, described originally in [19]. This is a polynomial time method which is based upon the theory of triangulated graphs, and has provable accuracy under certain conditions.

Strict Consensus Merger

The Strict Consensus Merger of a set of trees is a technique we developed for use with subtrees obtained through our DCM decompositions. Details of this technique are given in [19, 20, 43]; here we provide a brief description.

The Strict Consensus Merger (SCM) operates by sequentially merging pairs of subtrees until all the subtrees have been merged into a tree on the full set, and the particular order in which the subtrees are merged matters. Given two trees t_1 and t_2 on $S_1 \subseteq S$ and $S_2 \subseteq S$, respectively, SCM operates as follows. First SCM computes $S_1 \cap S_2$, the set of leaves that the two trees share, and considers the two trees restricted to just that common set of leaves. The *strict consensus* of these two subtrees (i.e., the most resolved common contraction of the two trees) is then computed; this constitutes the "backbone" of the resultant tree. The remaining pieces of t_1 and t_2 (on leaf sets $S_1 - S_2$ and $S_2 - S_1$, respectively) are then reattached onto the backbone. If both t_1 and t_2 should contribute pieces to the same edge of the backbone, then that edge is bisected, and all the pieces of both trees are attached to that newly introduced node.

Note the following. First, the SCM of a set of trees can be computed in polynomial time, since the strict consensus of two trees is a linear time operation. Also, the SCM of a set of a set of trees is typically not a binary tree, since any conflict in the trees will result in edge contractions during the merger of the trees together. Finally, even if the set of trees is compatible (meaning that a supertree exists consistent with all the given trees), the SCM of these trees may not produce such a compatible supertree. This last comment is not surprising since the subtree compatibility problem is NP-hard [38], and hence a polynomial time algorithm cannot be expected to solve the problem. On the other hand, we showed in [19] that when the subtrees are "big enough" (a statement we quantify exactly) and the subtrees are compatible, then SCM *does* solve the subtree compatibility problem.

21.3.4 Phase III: Refining trees

The strict consensus merger contracts edges in the subtrees in order to make the subtrees compatible with each other; as a consequence, the tree returned in Phase II is often not fully resolved. We therefore apply techniques for refining the tree obtained in Phase II. A typical approach for this refinement phase is to attempt to find a refinement of the given tree that optimizes some criterion, such as the maximum parsimony criterion, among all refinements. However, such problems tend to be NP-hard (see [4] for this problem when the optimization criterion is maximum parsimony). Heuristics for refining trees so as to optimize maximum parsimony are implemented in the major phylogeny software packages, but are not particularly effective nor fast. Consequently, the optimal tree refinement (OTR) problem is of general importance in phylogeny reconstruction.

21.4 Triangulated Graphs

We now turn to the basic theory of triangulated graphs. The particular properties of triangulated graphs allow us to design these first two phases with provable performance guarantees in terms of accuracy of the reconstructed tree, and in terms of running time. (The interested reader is directed to Golumbic's excellent book [14] from which much of this theory can be obtained.)

21.4.1 Basic material

We begin with a definition.

DEFINITION 21.5 A graph which has no induced simple cycles of length greater than three is a **triangulated graph**.

Triangulated graphs are perfect graphs and are well-studied (see [14] for more on triangulated graphs, and other classes of perfect graphs). The first basic theorem about triangulated graphs is that each such graph has a perfect elimination scheme:

DEFINITION 21.6 A **perfect elimination scheme** for a graph $G = (V, E)$ is an ordering of the vertices $v_1, v_2, \ldots, v_n$, so that for each $i = 1, 2, \ldots, n-1$, $X_i = \Gamma(v_i) \cap \{v_{i+1}, v_{i+2}, \ldots, v_n\}$ is a clique (here $\Gamma(v_i)$ indicates the neighbor set of v_i).

Not only does every triangulated graph have a perfect elimination scheme, but such an ordering can be found in polynomial time. Using the existence of a perfect elimination scheme, the following two theorems (which are the basis for the decomposition of S into overlapping subsets) can also be proved. The first result is as follows:

THEOREM 21.2 *Every triangulated graph $G = (V, E)$ has at most $n = |V|$ maximal cliques, and these can be found in $O(n^2)$ time.*

THEOREM 21.3 *For every triangulated graph $G = (V, E)$, $\exists X \subseteq V$ such that X is a clique and $G - X$ is the disjoint union of components $C_1, C_2, \ldots C_k$. Furthermore, we can find such an X that minimizes $max_i |C \cup X|$ in $O(n^3)$ time, where $n = |V|$.*

A basic aspect of the design of a DCM is producing a triangulated graph. Here we describe two such ways of obtaining triangulated graphs.

21.4.2 Threshold graphs

Let $S = \{s_1, s_2, \ldots, s_n\}$ be a set of taxa, let $[d_{ij}]$ be a distance matrix for the set of taxa, and let q be any non-negative real number. Then the threshold graph for d and q is defined as follows:

DEFINITION 21.7 The threshold graph $TG(d, q)$ has vertex set S and edges (s_i, s_j) such that $d_{ij} \leq q$.

Therefore, if $q \geq \max d_{ij}$ then $TG(d, q)$ is a clique, and for small enough q (for example, $q = 0$), the threshold graph will not be connected.

Before describing how we get triangulated graphs from threshold graphs, we need to define additive distances.

DEFINITION 21.8 An $n \times n$ matrix $[d_{ij}]$ for which there exists a tree T with n labeled leaves, with positive edge-weighting $w : E(T) \to R^+$ so that $d_{ij} = \sum_{e \in P_{ij}} w(e)$ for all i, j (where P_{ij} is the path in T between the leaves i and j) is said to be **additive**.

In [34] we proved the following:

THEOREM 21.4   *Let d be an additive matrix, and let q be a real number. Then $TG(d, q)$ is triangulated.*

21.4.3   Short subtree graphs

Let S be a set of taxa, and let T be a tree leaf-labeled by S, with non-negative edge-weights $w(e)$ assigned to each edge e in T.

DEFINITION 21.9   Let e be an edge of an edge-weighted binary tree T. Let $t_1, t_2, t_3,$ and t_4 be the four subtrees around the edge e (i.e., t_1 through t_4 are the components of $G - \{x, y\}$, where $e = (x, y)$). Let x_i denote those leaves in t_i which are closest to the edge e (using the path lengths defined by the edge-weighting on T). Then the **short subtree around** e is $x_1 \cup x_2 \cup x_3 \cup x_4$.

We now define the short subtree graph.

DEFINITION 21.10   Let T be a tree with leaf set S and edge weighting $w : E(T) \to R^+$. Let G be the graph with vertex set S and edge set E defined by $(s_i, s_j) \in E$ if and only if $\exists e \in E(T)$ such that s_i and s_j are both in the short subtree around e. This is the **short subtree graph** of (T, w), denoted by $SSG(T, w)$.

In [34] we showed the following:

THEOREM 21.5   *Let T be any tree with positive edge-weighting w. Then the short subtree graph defined by T and w is triangulated.*

21.4.4   Decompositions of triangulated graphs

Theorems 21.2 and 21.3 imply two decompositions of the vertex set of a triangulated graph G, as follows:

- **Max-clique decomposition**: Given a triangulated graph G, return the set of maximal cliques of G.
- **Separator-Component decomposition**: Given G, find a clique separator X and compute all the components of $G - X$. Then return the sets of the form $X \cup C$, where C is one of the components of $G - X$.

The first type of decomposition is uniquely determined by the triangulated graph, and for this reason we will refer to it as "*the* max-clique decomposition". However, a triangulated graph can have many different clique separators, each of which thus can define a different decomposition. For this reason, we will refer to any decomposition obtained by choosing a clique separator as "*a* separator-component decomposition". In Theorem 21.3 we showed that picking a clique separator X so as to minimize the maximum size of any created subproblem could be solved in $O(n^3)$ time, but this running time is not always acceptable. Therefore, we may sometimes prefer to use a suboptimal separator-component decomposition, i.e., one that would produce somewhat larger subproblems, if it can be computed faster.

Now consider the difference between the max-clique decomposition and a separator-component decomposition on the same fixed triangulated graph G. It is easy to see that the max-clique decomposition will produce more subproblems (up to n of them, where $n = |V|$), but each subproblem will be smaller (or at least not larger) than the subproblems obtained by the separator-component decomposition. However, the separator-component decomposition will produce potentially only a few subproblems. Furthermore, the pairwise intersections of the subsets produced by the max-clique decomposition can differ significantly (and will even be disjoint in some cases), whereas in any separator-component decomposition all pairwise intersections are the same. These differences will be significant in developing DCMs for different base methods.

21.5 Designing DCMs

21.5.1 Introduction

All of our DCMs use the same three phase structure (although some also use recursion and/or iteration), with the main difference between the DCMs being the decomposition technique. All current DCMs first construct a triangulated graph and then apply either the max-clique or a separator-component decomposition to the graph to obtain subproblems. The combination of base method, choice of triangulated graph, and decomposition technique on that triangulated graph, impact the behavior of the resultant "DCM-boosted" method. For example, methods which will take a long time on big datasets will finish faster on the max-clique decomposition. A more subtle point is the impact of error on subsets – since the technique we use in merging subtrees contracts edges whenever two subtrees disagree, there is a potential for a greater loss of resolution in the max-clique decomposition than in a separator-component decomposition, especially when the separator is small. The difference between using threshold graphs and short subtree graphs is also interesting, but depends as much on the dataset as on the method. Thus, the design of a DCM reflects the particular properties of the base method and of the particular dataset, and only by studying the actual performance of the resultant DCM-boosted methods can we tell which design strategy will be the most beneficial.

We begin this section with a description of how we obtain triangulated graphs from molecular datasets.

21.5.2 Obtaining triangulated graphs from datasets

Threshold graph decompositions.

Threshold graph decompositions can be used on any dataset for which a distance can be defined between each pair of taxa. In molecular sequence datasets, these distances can be Hamming distances, or distances obtained under some statistical estimation procedure selected to match the model of evolution underlying the dataset.

In a threshold graph decomposition, we are given a set S of sequences and a matrix $[d_{ij}]$ of distances on the set S of taxa. To compute a threshold graph we must first select the threshold (the value q). We then construct the threshold graph, $TG(d,q)$ (see Definition 21.7). If $TG(d,q)$ is triangulated, we can compute either the max-clique decomposition, or a separator-component decomposition; however, if $TG(d, q)$ is not triangulated, then we must first triangulate it, by adding edges to $TG(d, q)$ so that the graph is triangulated. However, when we add edges to $TG(d, q)$ we affect the decompositions (either max-clique or separator-component) that we can obtain from the triangulated graph.

In some of our DCMs our objective is to minimize the maximum evolutionary distance within any subproblem, so that in a max-clique decomposition we would wish the cliques to have the smallest maximum distance. This suggests the following objective in the triangulation process: add edges to $TG(d, q)$ so as to minimize the "weight" of the heaviest edge added. This optimization problem is NP-hard, however, and so in our experiments, we have used a greedy triangulation scheme that works reasonably well: compute for each vertex v in the graph the value $W(v) = max\{d_{ij} : \{i, j\} \subseteq \Gamma(v)\}$, where $\Gamma(v)$ denotes the neighbors of the vertex v. Select the vertex v that minimizes $W(v)$ and make it simplicial (i.e., make the neighbors of v into a clique). Recurse on $G - \{v\}$. This approach produces a triangulation of G but may not minimally triangulate the graph G; however, in our experiments this worked quite well. (See also [24] for a polynomial time technique that creates a triangulation with good theoretical properties.)

Thus, threshold graph decompositions have the following structure. They begin with a distance matrix $[d_{ij}]$, and operate as follows:

1. Pick a threshold $q \in \{d_{ij}\}$
2. Construct $TG(d, q)$
3. Add edges to $TG(d, q)$ to triangulate it, producing graph G
4. Compute either the max-clique or a separator-component decomposition from G

Guide tree decompositions:

We now describe how we can obtain guide trees, and from them the triangulated short subtree graph.

The most typical technique for obtaining a guide tree is to use some phylogeny reconstruction method (such as a heuristic for maximum parsimony or maximum likelihood, or perhaps a fast method such as neighbor joining) to obtain an estimation T of the true tree. Given T, we can then use one of many techniques to assign edge lengths. For example, if the set of taxa are biomolecular sequences in a multiple alignment, then we can assign edge lengths to T by using the Fitch-Hartigan dynamic programming fixed-tree maximum parsimony algorithm of [11] to assign sequences to internal nodes, and then use Hamming distances to define edge lengths. We can also use maximum likelihood estimation of edge lengths, which may be more accurate but will take more time than maximum parsimony. In general, however, if T was obtained using a phylogeny reconstruction method, it will typically already have edge lengths (such is the case with the three techniques we mentioned earlier - heuristic MP, heuristic ML, or neighbor joining).

Given the guide tree T with its edge lengths, we then compute the short subtree graph. This is easily done in polynomial time. Once the short subtree graph is obtained, we can compute either the max-clique or a separator-component decompositions on it (since it is already triangulated). As noted before, finding an optimal separator-decomposition - although polynomial time - can be more expensive than desired; consequently faster decompositions based upon clique-separators can also be used. In [33] we showed how we could compute a decomposition we call the "heuristic centroid-edge" decomposition in linear time (this fast running time was accomplished without explicitly constructing the short subtree graph). This decomposition worked very well in practice, as we showed in [33, 34].

21.5.3 Considerations in design strategies

We have described ways we can obtain triangulated graphs, and ways we can decompose a triangulated graph. How do these choices interact with base methods?

Some methods - in particular, distance-based methods like neighbor joining [36], have poor topological accuracy on datasets with large evolutionary diameters, although they are quite fast. These methods would therefore seem to benefit from decompositions that produce the smallest diameter subproblems. Other methods, in particular exhaustive searches for optimal trees under hard optimization criteria, can only realistically handle quite small datasets – maximum parsimony is limited to perhaps 20 or 25 taxa, and maximum likelihood limited to much smaller datasets; these methods would require subproblems to be as small as possible. In a third class are local-search heuristics (like the hill-climbing heuristics used in maximum parsimony searches), which seem not to be impacted by large diameters so much, but are still impacted by dataset size. Understanding the best design strategy for these local-search heuristics is more complicated.

The particular technique used to obtain a triangulated graph also has an impact on the resultant DCM. If we use a guide tree, there is only one triangulated graph that we can obtain, but different guide trees will produce potentially different decompositions. This makes guide tree decompositions useful for heuristic searches for optimal trees under criteria such as maximum parsimony, because as the search finds better solutions, it can become a new guide tree, and a new decomposition can be obtained.

The issues involved in selecting threshold graph decompositions are more complicated. In this case, the distance matrix $[d_{ij}]$ is usually considered fixed, but the threshold can change. If the threshold is too small, the threshold graph will not even be connected, and so the tree on the full dataset cannot be reconstructed from subtrees, even if they are correctly computed. If the threshold is too big, the subproblems become essentially as difficult (almost as large and with almost the same evolutionary diameter) as the full dataset, although correct subtrees on the subproblems would then be likely to define the full tree. Thus, finding the "correct" threshold to use is a difficult problem.

Our experiments with real and simulated data showed us two very interesting things. The first was that the threshold graphs obtained for biomolecular sequence datasets had very large cliques – so large, in fact, that on many datasets the largest subproblems obtained in a threshold graph decomposition were close to the full dataset size [33, 34]. This meant that we'd be analyzing several subproblems of size almost the full dataset size, with the consequence that there was little gained in using a threshold graph decomposition. On the other hand, the subset sizes obtained by using reasonable guide trees (obtained using good heuristic searches for MP, for example) produced much smaller subproblems for the same datasets [33, 34], so that decompositions based upon the short subtree graph were more suited for boosting heuristic searches for maximum parsimony or maximum likelihood.

Other experiments showed that even the best distance-based methods (neighbor joining, for example) had poor topological accuracy on large diameter subproblems, confirming the theory that had been established for the convergence rates (i.e., the sequence length requirements) of these methods [1]. In order to develop methods with provably good convergence rates, we needed to work with threshold graphs, rather than with guide trees. In order to obtain the provable theory, however, we had to take a third approach - rather than selecting a threshold, we compute all possible threshold graphs (one for each threshold that creates a connected graph), and hence we compute $O(n^2)$ trees, one for each threshold graph. We then have to apply yet another step (whereby we obtain a single tree from the $O(n^2)$ trees) in order to return a tree for the input taxa.

Finally, DCMs have also been designed for reconstructing phylogenies on whole genomes (using gene order data) using the GRAPPA software suite (see [28] for a description of GRAPPA and of the DCM-boosting designed for use with GRAPPA techniques). This is empirically a harder problem – datasets above 13 or so genomes cannot be handled by GRAPPA, and so decomposing the dataset into smaller subsets (each subset explicitly limited to at most 13 genomes) is an absolute requirement.

In the next sections we describe the DCMs we used for boosting maximum parsimony heuristics and for use with distance-based methods; readers interested in the use of DCMs for boosting GRAPPA should see [28].

21.6 DCM-boosting Techniques for Maximum Parsimony

21.6.1 Objectives

We begin with a description of our DCMs for maximum parsimony. The main issue confronting maximum parsimony and maximum likelihood heuristics is running time – they take a long time to reach reasonable accuracy on large datasets. In fact, it is not uncommon for phylogenetic analyses to take weeks or months (or years) on large datasets – and with the increasing availability of sequence data, this trend may continue or even get worse. Finding ways to make these analyses much faster is the objective of this algorithm design.

The simplest design of a DCM involves only one application of a base method on a subproblem - so that after the subtrees are constructed and merged together, the analysis stops. When the objective is to solve an optimization problem, however, this no longer makes sense - we will want to continue searching for better trees as long as the data suggests that we may not have found sufficiently good trees. Therefore, we would usually use the output of a DCM as the input to a standard heuristic search, with the hope that the improvement obtained by the DCM would give us a "head start" over standard approaches. In addition, we would like to design DCMs that allow for iterative use, so that instead of switching over to standard heuristic searches we could continue using DCM-boosted heuristics. Thus, iterative-DCMs will be potentially beneficial for maximum parsimony (as well as for other optimization problems).

Thus, DCMs which produce smaller subproblems make a lot of sense, and it would seem that the smallest possible subproblems would be the most desirable. However, in our experiments, the loss of resolution that results from using the max-clique decomposition proved to be more of a problem than the running time used in a separator-component decomposition, because the third phase in which we attempt to resolve the tree optimally with respect to maximum parsimony was too expensive. Furthermore, we observed that the subsets obtained using the threshold graph decomposition produced subproblems that were quite big - 90% of the taxa in the largest subproblem - so that there was little improvement gained in the running time.

For this reason, we developed the short subtree graph decomposition, since these produced maximal cliques that were much smaller than the subproblems we obtained in the threshold graph decompositions. What seems to have worked quite well for maximum parsimony heuristics is to produce a separator-component decomposition on a short subtree graph: these subproblems contain no more than 50-60% of the taxa, and so are substantially smaller than the subproblems obtained using the threshold graph decomposition. The separators we find in these decompositions tend also to be very small – four or five taxa, in fact. Furthermore, because these subproblems overlap only on the separator, very little resolution is lost during the merger of the subtrees into the supertree. Finally, when 50-60% of the taxa is still too large for the base method, we examined recursive uses of the decomposition.

21.6.2   DCM3

Thus, we have several variants of a basic design, which we call DCM3. In its simplest form, we would use some technique to obtain a guide tree, and then compute the short subtree graph from the guide tree. We would then compute some separator-component decomposition based upon the guide tree, thus producing a set of subsets of the original taxa. Following this, we would apply the base method to the subproblems, merge the resultant subtrees using the strict consensus merger, and then refine the tree. In summary, the basic algorithm is as follows.

$DCM3(T_1, w)$.

The input to this routine is a tree, T_1, leaf-labeled by a set S of sequence, and an edge weighting w of T_1.

1. Construct $G = SSG(T_1, w)$, the short subtree graph based upon the guide tree T_1 with edge-weighting w. Compute a separator-component decomposition on G, producing subsets $S_i = C_i \cup X$, where X is the separator and C_i is the ith component of $G - X$.

2. Let t_i be $T_1|S_i$ (that is, t_i is the subtree of T_1 induced by the set S_i of taxa). Use a preferred method to construct trees on each subset, starting with the tree t_i for the subset S_i.

3. Merge the resultant subtrees using the Strict Consensus Merger.

4. Return a refinement of the resultant tree.

Several comments are worth making here. The first is to note that we do not simply apply the favored heuristic to each subproblem, but rather we take advantage of our current best tree (the guide tree, that is) in order to initialize the search at a (hopefully) good tree. That is, we will begin our search for the optimal tree on the subset S_i with the tree t_i.

Secondly, we have left open several steps of the technique: for example, which separator-component decomposition do we compute?, how do we refine the resultant tree?, how exactly do we obtain the guide tree T_1?, and how long do we apply our favored heuristic on each subset S_i? These are all aspects of the design of DCM3 that will depend quite strongly upon the particular base method, as well as the properties of the dataset being analyzed.

Furthermore, we can choose different MP heuristics for each of the steps (one for the initial step where we obtain our tree T_1, and another where we analyze subsets). Also, we can try to optimize the decomposition as described above or just take some reasonably good decomposition, and we can attempt to optimally refine the unresolved tree we get from the Strict Consensus Merger, or we can refine it heuristically (or even randomly). Thus, the particular application of this DCM3 strategy involves decisions at various points.

21.6.3   Iterative-DCM3

The main use we have made of DCM3 is in its iterative form, where we have alternated between the use of a heuristic on the full dataset and the use of the same heuristic (perhaps applied slightly differently) on the subproblems. That is, Iterative-DCM3 follows upon the construction of an initial guide tree (T_1), and has the following steps.

- Repeat until you want to stop:
  - Let $T = DCM3(T_1)$ (i.e. T is the tree obtained by applying DCM3 to the guide tree T_1)

   – Apply your favored heuristic to T until you reach a local optimum, or until you satisfy some stopping rule, and let T_1 be the current best tree.

Note that this use of DCM3 introduces yet another level of flexibility (or ambiguity): now we have three places where we will apply a heuristic for MP to a dataset: the initial stage, where we obtain our first estimate of the optimal tree from scratch, and then we will alternate between applying a heuristic just to subsets, and applying a heuristic to the full set. We may elect to use the same basic heuristic, but apply it with fewer or more iterations depending upon the size of the subset being analyzed. Or, we may change the heuristic we use (and not just vary it by changing the number of iterations) depending upon the size or features of the subset. These issues again will depend upon the features of the dataset and the optimization criterion, as well as upon the desired level of accuracy and/or the amount of time that is available for the analysis.

21.6.4 Recursive DCM3

Recursive DCM3 is a simple modification of DCM3, in which we recursively decompose subsets until we reach a desired subset size. Then we apply the favored heuristic to the subsets (starting, as before, with the subtree induced by the guide tree on each subset), and then merge trees using the strict consensus merger as we go back up the recursion tree. Once all the subtrees are merged into a tree on the full dataset, we then apply the refinement step.

   The main advantage of Recursive-DCM3 is that the subproblems can be made significantly smaller, even with just one or two levels of recursion.

21.6.5 Recursive-Iterative-DCM3

Recursive-Iterative-DCM3 combines both recursion and iteration, so that we iteratively call Recursive-DCM3. This technique has the best performance of the various techniques we examined in our studies, when the base methods were standard heuristics for maximum parsimony.

21.6.6 Experimental results

In this section we summarize the results of fairly extensive studies on real datasets ranging from 1000 sequences up to almost 14,000 sequences, with base methods for maximum parsimony taken from different software packages. An example of one such study is given in Figure 21.6.6.

   We have experimented with several variants of DCMs for use with heuristics for maximum parsimony on a number of very large datasets. These experiments have examined base methods and compared them to their DCM-boosted versions on a number of real datasets. We included DCM2, which is the separator-component decomposition based upon the threshold graph obtained using the smallest threshold that produces a connected graph [20]. In order to compare DCM2 and DCM3 to base heuristics, we used the output of DCM2 and DCM3 as a starting tree for their base heuristics, so that we could explore performance over a longer period of time. Thus, the comparison is made as a function of time – examining the best MP score found at each point in time, over a period of days or weeks (depending upon the dataset). We consider DCM-boosting to be advantageous if we obtain an improvement at every point in time for a considerable length of time, preferably for at least 24 hours. These experiments (detailed in [33, 34]) showed the following:

- The better the base method is, the better the DCM must be in order to obtain an advantage over the base method. Thus, even a poor DCM can improve upon a poor base method, but for the best base methods, we need very good DCMs to obtain an advantage.

- DCM2 (the separator-component decomposition applied to a threshold graph) produces subproblems that are very large, and the decomposition takes a long time. In fact, when we use very good heuristics for MP, DCM2-boosting worsens the performance rather than improving it. Consequently, DCM2 is not helpful for solving maximum parsimony on most datasets we examined, by comparison to good base heuristics.

- DCM3, based upon finding an optimal decomposition but using a random refinement, and then continuing with the base heuristic, did not typically improve the performance of the best base methods, but it also didn't generally make things worse.

- Recursive-DCM3 gave a slight advantage over the best base heuristics, and a somewhat larger advantage over other base methods.

- Iterative-DCM3 gave a somewhat larger advantage over all base methods than Recursive-DCM3.

- Recursive-Iterative-DCM3, using both optimal and heuristic decompositions, gave the largest advantages over even the best base heuristics.

- All advantages obtained by any DCM-boosting technique depended on the difficulty of the dataset and the quality of the base heuristic. Thus, when used with good base heuristics on small to moderately large datasets, the advantage is not always significant. Thus, the main advantage obtained is on the largest and most difficult datasets.

21.7 DCM-boosting Distance-based Methods

21.7.1 Objectives

In this section we describe the DCM we have developed for use with distance-based reconstruction methods, such as neighbor joining (NJ).

The best distance-based methods are very fast, and topologically very accurate even on large datasets as long as the subproblems have small diameter. However, as the diameter increases, their accuracy decreases (see [27, 29, 30, 31]). This empirical observation is supported by the theory that has been established about these methods as described earlier in the chapter.

Thus, the main empirical purpose in devising a DCM for use with neighbor joining is to produce a method that will enable recovering the true tree from shorter sequences (as well as more accurate trees at every sequence length). However, the main theoretical purpose is to use DCM-boosting so as to produce an absolute fast converging (afc) method from NJ (see Definition 21.3). The design of the DCMs for use with NJ (and other distance-based methods) differ slightly depending upon whether the empirical goal or the theoretical objective is more important.

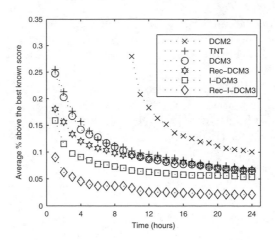

FIGURE 21.1: The performance of different versions of DCM3(TNT) on a dataset of 4114 aligned 16s rRNA Actinobacteria sequences, each of length 1263 (from the Ribosomal Database Project at the University of Michigan). The x-axis reports the best MP score found by each method during the first 24 hours of analysis. The y-axis measures the error rate of each method as a percentage above the current best MP score found on this dataset, where "current best" means the best score found using any method over any amount of time. The base method is the Ratchet heuristic for MP within the TNT software [13]. The recursive-DCM3 code produces subsets of size 1/8th the original dataset size. Each point is the average of 5 runs, and all runs are done on the same machines – 700MHz Pentiums.

21.7.2 $DCM1_{NJ} + SQS$: Designing an afc method using DCM-boosting

In [43] we gave a technique to produce an absolute fast converging method from any method with an exponential convergence rate, such as neighbor joining. Recalling the definition of the threshold graph given in Definition 21.7, the DCM1 decomposition, and the Strict Consensus Merger, the technique is as follows:

$DCM1_{NJ} + SQS$:
- **Phase I:**
    - For each $q \in \{d_{ij}\}$, compute $t_q = DCM1(TG(d,q))$; i.e.,
        * Construct the threshold graph $TG(d,q)$, and triangulate it with a minimum weight triangulation (minimizing the maximum weight of any added edge).
        * Compute the max-clique decomposition of $TG(d,q)$, and construct trees on each maximal clique using neighbor joining.
        * Merge the trees together using the Strict Consensus Merger.
- **Phase II:**
    - Evaluate each tree t_q according to the Short Quartet Support Criterion (SQS) (see [43]).
    - Return the tree with the best score.

We have explicitly referenced the neighbor-joining (NJ) method in this description, but any method could be used in its place as a base method. However, in order to obtain a theoretical guarantee of fast convergence, the base method needs to have a convergence rate that is no worse than exponential in the maximum evolutionary diameter of the model tree (this is true for NJ and most of the other distance-based methods that have been proposed).

Note that this DCM requires that the triangulation of the threshold graph be optimal, which requires therefore solving an NP-hard problem. However, in [24] Jens Lagergren showed that certain polynomial time triangulation techniques also have the desired theoretical properties so that his variant of DCM-boosting would also produce an afc method without requiring an exact solution to the NP-hard problem of minimally triangulating the threshold graph.

21.7.3 Improving the empirical performance of $DCM1_{NJ} + SQS$

$DCM1_{NJ} + SQS$ is designed for theoretical performance, and thus involves more computational time than we would want (in particular, it involves solving a hard computational problem and it produces $O(n^2)$ trees in Phase I). Improving upon the speed of this DCM is a necessary objective if this technique is going to be useful for any real analysis. Furthermore, the specific technique used in Phase II (selecting the tree which optimizes the Short Quartet Support) was also used because of its theoretical properties, but that technique - although theoretically optimal - turns out not to have as good performance (as shown in our simulation studies) as other criteria. Thus, in a later study [30] we explored the empirical consequences of modifying the algorithmic design of $DCM1_{NJ} + SQS$, in order to improve the speed and/or accuracy in simulated and real datasets. We specifically examined a variant where we modified Phase I by examining only ten thresholds (rather than all possible thresholds), and where we used a greedy technique to triangulate the threshold graph (as described earlier in this chapter). We then modified Phase II by selecting the tree that optimized some other criterion (we examined specifically maximum parsimony or maximum likelihood, and found them largely to have the same performance and both superior to SQS). These studies led us to propose (as heuristics) two DCMs for boosting NJ: $DCM1_{NJ} + MP$ and $DCM1_{NJ} + ML$, neither of which has provable theoretical performance, but both of which provide a distinct topological accuracy advantage over NJ, while being still reasonably fast. (Neither is as fast as NJ, but both are fast enough to complete analyses in a few minutes rather than in hours or days, as any serious MP or ML analysis would require.)

21.7.4 Experimental Results

We designed a simulation study to explore the relative performance of our heuristic approximations to $DCM1_{NJ} + MP$ and $DCM1_{NJ} + SQS$, in comparison to NJ and to a provably afc method called HGT+FP (for "Harmonic Greedy Triplets, plus the Four Point Method) [7], a sample of which is shown in Figure 21.7.4. (See [27, 29, 31, 35] for some more experiments.) Our two methods are thus not provably afc but rather only heuristic approximations to $DCM1_{NJ} + SQS$, which is afc.

Our implementation of $DCM1_{NJ} + SQS$ is heuristic because we only examine ten thresholds rather than all possible thresholds, and we do not optimally triangulate the threshold graphs, thus it is not provably afc. Even perhaps more serious, from a theoretical viewpoint, the use of MP as the selection criterion in the second phase of $DCM1_{NJ} + MP$ automatically makes the method not afc and probably not even provably statistically consistent.

However, the difference in performance between the methods is striking. The error rates of NJ rises quite quickly, but the remaining methods have flat error rates in these experiments.

Furthermore, using MP rather than SQS *improves* the performance of $DCM1_{NJ}$, even though from a theoretical perspective it is worse. Finally, although HGT+FP is provably afc its performance is worse than the $DCM1_{NJ}$ methods on all these datasets.

The difference in performance between these methods is due to a combination of factors. NJ's performance problem is due to the fact that it uses all the entries of the matrix, without sufficiently downweighting large entries, and hence has an exponential convergence rate. The difference in performance between HGT+FP and our $DCM1_{NJ}$ methods is largely due to the improvement obtained in practice by using NJ rather than a quartet-based method, which is what HGT+FP essentially uses. The difference in performance between $DCM1_{NJ} + SQS$ and $DCM1_{NJ} + MP$ is more mysterious: why should MP, which is not statistically consistent, outperform SQS as a selection criterion? Once again, the difference may be in the specific design of SQS; it is based upon quartet accuracy, with a particular technique to determine the "correct" tree on each quartet. Although this criterion is theoretically sound, empirically quartet-based methods (such as SQS) are not as accurate as methods that compute trees from all the data. Despite the theoretical guarantees that can be established for quartet methods, they have in general not been able to be as accurate in simulation as the neighbor joining method, as shown in [21].

The lesson from this comparison of distance-based methods is an interesting one, and instructive: while dividing into subproblems can yield improvements in accuracy, precisely how one divides into subproblems is tremendously important.

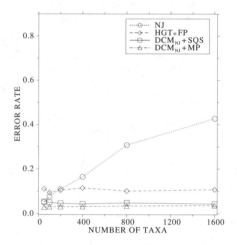

FIGURE 21.2: The performance of $DCM1_{NJ}$ with different techniques used in Phase II, compared to NJ and to another afc method, HGT+FP, as a function of the number of taxa. In this experiment we simulated evolution of sequences with 1000 sites down uniform distribution random trees with branch lengths drawn from the same distribution under the Kimura 2-parameter model [23] of evolution. K2P distances were used as inputs to each method.

21.8   Related Work and Conclusions

DCM-boosting has also been used to improve the speed of software for the inversion and breakpoint phylogeny problems (both NP-hard problems) within the GRAPPA software suite for whole genome phylogeny reconstruction [28]. There are two basic challenges for GRAPPA's phylogenetic analysis: first, because the techniques employed within GRAPPA exhaustively search all trees on each input dataset, GRAPPA is explicitly limited to analyzing datasets below (about) 14 taxa. Secondly, because of its design, GRAPPA is unable to efficiently analyze any dataset where the underlying tree has very large edge lengths. The authors used a DCM in order to address the first of these two challenges; the second remains a problem for any explicit attempt to solve these optimization problems.

The DCM they designed for use with GRAPPA [28] employed many levels of recursion, so that each subproblem was small enough. The specific design of their DCM was similar to the $DCM1$ design, but computed a consensus tree of the different trees (one for each threshold) they obtained, and they studied their DCM-boosting technique in simulation. Their study showed that using DCM-boosting allowed GRAPPA to be applied to datasets with thousands of taxa, in other words a huge improvement in performance.

Future research will explore DCMs for other types of base methods, such as maximum likelihood and phylogenetic sequence alignment [37].

Finally, we note that other divide-and-conquer strategies have been proposed. Some of the most well known are quartet-based methods, such as quartet puzzling [40], quartet-cleaning [3], the Q^* method [2], and the short quartet methods [8, 9], but none of these has been shown to reliably outperform neighbor joining (see [21] for some of these results). On the other hand, methods such as Compartmentalization [26] are also promising, and could be investigated. Research into supertree methods which can construct trees from arbitrarily defined subtrees also needs further investigation.

Acknowledgments

This work was supported in part by the National Science Foundation, the David and Lucile Packard Foundation, the Institute for Cellular and Molecular Biology at the University of Texas at Austin, the Radcliffe Institute for Advanced Study, and the Program for Evolutionary Dynamics at Harvard University.

References

[1] K. Atteson. The performance of the neighbor-joining methods of phylogenetic reconstruction. *Algorithmica*, 25:251–278, 1999.

[2] V. Berry and O. Gascuel. Inferring evolutionary trees with strong combinatorial evidence. In *Proc. 3rd Ann. Int'l Conf. Computing and Combinatorics (COCOON97)*, pages 111–123. Springer Verlag, 1997. in *LNCS* 1276.

[3] V. Berry, T. Jiang, P. Kearney, and M. Li *et al.* Quartet cleaning: improved algorithms and simulations. In *Proc. Europ. Symp. Algs. ESA99*, pages 313–324. Springer Verlag, 1999. in *LNCS* 1643.

[4] M. Bonet, M. Steel, T. Warnow, and S. Yooseph. Faster algorithms for solving parsimony and compatibility. *J. Comput. Biol.*, 5:409–422, 1999.

[5] B. Chor and T. Tuller. Maximum likelihood of evolutionary trees is hard. In *RECOMB*

2005, 2005.

[6] M. Cryan, L. Goldberg, and P. Goldberg. Evolutionary trees can be learned in polynomial time in the two-state general Markov model. In *Proc. IEEE Symp. Foundations of Comput. Sci. (FOCS98)*, pages 436–445, 1998.

[7] M. Csűrös. Fast recovery of evolutionary trees with thousands of nodes. *Journal of Computational Biology*, 9(2):277–297, 2002.

[8] P.L. Erdos, M. Steel, L. Székély, and T. Warnow. A few logs suffice to build almost all trees –I. *Random Structures and Algorithms*, 14:153–184, 1997.

[9] P.L. Erdos, M. Steel, L. Székély, and T. Warnow. A few logs suffice to build almost all trees –II. *Theor. Comp. Sci.*, 221:77–118, 1999.

[10] J. Felsenstein. Cases in which parsimony and compatibility methods will be positively misleading. *Syst. Zool.*, 27:401–410, 1978.

[11] W.M. Fitch. Toward defining the course of evolution: minimum change for a specified tree topology. *Syst. Zool.*, 20:406–416, 1971.

[12] L.R. Foulds and R.L. Graham. The Steiner problem in phylogeny is NP-complete. *Advances in Applied Mathematics*, 3:43–49, 1982.

[13] P.A. Goloboff. Analyzing large data sets in reasonable times: solution for composite optima. *Cladistics*, 15:415–428, 1999.

[14] M. Golumbic. *Algorithmic graph theory and perfect graphs.* Academic Press Inc, 1980.

[15] D.M. Hillis. Inferring complex phylogenies. *Nature*, 383:130–131, 1996.

[16] D.M. Hillis, C. Moritz, and B. Mable. *Molecular Systematics.* Sinauer Pub., Boston, 1996.

[17] J. Huelsenbeck. Performance of phylogenetic methods in simulation. *Syst. Biol.*, 44:17–48, 1995.

[18] J. Huelsenbeck and D. Hillis. Success of phylogenetic methods in the four-taxon case. *Syst. Biol.*, 42:247–264, 1993.

[19] D. Huson, S. Nettles, and T. Warnow. Disk-covering, a fast-converging method for phylogenetic tree reconstruction. *Journal of Computational Biology*, 6:369–386, 1999.

[20] D. Huson, L. Vawter, and T. Warnow. Solving large scale phylogenetic problems using DCM2. In *Intelligent Systems for Molecular Biology*, pages 118–129, 1999.

[21] K. St. John, B.M. Moret, L. Vawter, and T. Warnow. Performance study of phylogenetic methods: (unweighted) quartet methods and neighbor-joining. *J. of Algorithms*, 48(1):173–193, 2003. Also appeared in the Proceedings of 2001 ACM-SIAM Symposium on Discrete Algorithms (SODA 01).

[22] J. Kim and T. Warnow. Tutorial on phylogenetic tree estimation. In *Proc. 7th Int'l Conf. on Intelligent Systems for Mol. Biol. (ISMB99)*, 1999.

[23] M. Kimura. A simple method for estimating evolutionary rates of base substitutions through comparative studies of nucleotide sequences. *J. Mol. Evol.*, 16:111–120, 1980.

[24] J. Lagergren. Combining polynomial running time and fast convergence for the disk-covering method. *Journal of Computer and System Science*, 65(3):481–493, 2002.

[25] W.H. Li. *Molecular Evolution.* Sinauer, Massachuesetts, 1997.

[26] B.D. Mishler. Cladistic analysis of molecular and morphological data. *American Journal of Physical Anthropology*, 94:143–156, 1994.

[27] B.M.E. Moret, U. Roshan, and T. Warnow. Sequence length requirements for phylogenetic methods. In *Proc. 2nd Int'l Workshop Algorithms in Bioinformatics (WABI'02)*, volume 2452 of *Lecture Notes in Computer Science*, pages 343–356. Springer-Verlag, 2002.

[28] B.M.E. Moret, J. Tang, and T. Warnow. Reconstructing phylogenies from gene-content and gene-order data. *Mathematics of Evolution and Phylogeny*, pages 321–352, 2005.

[29] L. Nakhleh, B.M.E. Moret, U. Roshan, and K. St. John *et al*. The accuracy of fast phylogenetic methods for large datasets. In *Proc. 7th Pacific Symp. Biocomputing (PSB'2002)*, pages 211–222. World Scientific Pub., 2002.

[30] L. Nakhleh, U. Roshan, K. St. John, and J. Sun *et al*. Designing fast converging phylogenetic methods. In *Proc. 9th Int'l Conf. on Intelligent Systems for Molecular Biology (ISMB'01)*, volume 17 of *Bioinformatics*, pages S190–S198. Oxford U. Press, 2001.

[31] L. Nakhleh, U. Roshan, K. St. John, and J. Sun *et al*. The performance of phylogenetic methods on trees of bounded diameter. In *Proc. 1st Int'l Workshop Algorithms in Bioinformatics (WABI'01)*, volume 2149 of *Lecture Notes in Computer Science*, pages 214–226. Springer-Verlag, 2001.

[32] D.F. Robinson and L.R. Foulds. Comparison of phylogenetic trees. *Mathematical Biosciences*, 53:131–147, 1981.

[33] U. Roshan. *Algorithmic techniques for improving the speed and accuracy of phylogenetic methods*. PhD thesis, The University of Texas at Austin, 2004.

[34] U. Roshan, B.M.E. Moret, T. Warnow, and T.L. Williams. Rec-I-DCM3: a fast algorithmic technique for reconstructing large phylogenetic trees. In *Proceedings of the IEEE Computational Systems Bioinformatics conference (CSB)*, Stanford, California, USA, 2004.

[35] U. Roshan, B.M.E. Moret, T.L. Williams, and T. Warnow. Performance of supertree methods on various dataset decompositions. In O.R.P. Bininda-Emonds, editor, *Phylogenetic Supertrees: Combining Information to Reveal the Tree of Life*, volume 3 of *Computational Biology*, pages 301–328. Kluwer Academics, 2004. (Dress, A. series ed.).

[36] N. Saitou and M. Nei. The neighbor-joining method: A new method for reconstructing phylogenetic trees. *Mol. Biol. Evol.*, 4:406–425, 1987.

[37] D.D. Sankoff and R.J. Cedergren. Simultaneous comparison of three or more sequences related by a tree. In D. Sankoff and J.B. Kruskal, editors, *Time Warps, String Edits, and Macromolecules: the Theory and Practice of Sequence Comparison*, pages 253–264. Addison-Wesley, Reading, MA, 2003.

[38] M.A. Steel. The complexity of reconstructing trees from qualitative characters and subtrees. *Journal of Classification*, 9:91–116, 1992.

[39] M.A. Steel. Recovering a tree from the leaf colourations it generates under a Markov model. *Appl. Math. Lett.*, 7:19–24, 1994.

[40] K. Strimmer and A. von Haeseler. Quartet puzzling: A quartet maximum likelihood method for reconstructing tree topologies. *Molecular Biology and Evolution*, 13(7):964–969, 1996.

[41] D.L. Swofford. PAUP*: Phylogenetic analysis using parsimony (and other methods), 1996. Sinauer Associates, Sunderland, Massachusetts, Version 4.0.

[42] C. Tuffley and M.A. Steel. Modelling the covarion hypothesis of nucleotide substitution. *Mathematical Biosciences*, 147:63–91, 1997.

[43] T. Warnow, B.M. Moret, and K. St. John. Absolute convergence: true trees from short sequences. *Proceedings of ACM-SIAM Symposium on Discrete Algorithms (SODA 01)*, pages 186–195, 2001.

High-Performance Algorithms for Phylogeny Reconstruction with Maximum Parsimony

David A. Bader
Georgia Institute of Technology

Mi Yan
University of New Mexico

22.1 Introduction

All biological disciplines are united by the idea that species share a common history. These relationships are crucial to the understanding of biological evolution and biological mechanisms in medical and pharmaceutical research. The evolutionary history is usually represented by a phylogeny, an unrooted, binary tree where each leaf represents a species. Phylogeny reconstruction, the problem of inferring the evolutionary relationships from the set of leaves by using sequence (e.g., from the DNA of nuclear or organelle genomes), morphological, or gene-order data, and a plausible model of evolution, is a fundamental problem in computational biology and is increasingly used in drug discovery, epidemiology, and genetic engineering [4]. Unfortunately, most problems in phylogeny reconstruction are proven to be NP-hard problems that can take years to solve on realistic datasets [8, 32]. Despite the large number of available tools and approaches, even moderate-sized datasets can require months or years of computation. Many biologists throughout the world compute phylogenies involving weeks or years of computation without necessarily finding global optima. Certainly more such computational analyses will be needed for larger datasets. The enormous computational demands in terms of time and storage for solving phylogenetic problems can only be met through high-performance computing.

Swofford *et al.* [39] gave an overview of phylogeny reconstruction methods and categorize them into *criteria-based* and *direct* approaches. Criteria-based approaches examine all trees and choose the trees that are optimal according to some criteria, such as Maximum

Likelihood (ML) and Maximum Parsimony (MP); while direct approaches reconstruct a tree directly from a pairwise distance matrix, such as Neighbor-Joining (NJ) [35, 37]. MP is arguably the most widely-used criteria by far, and the experimental studies on sequence data in [34] show that parsimony is hard to beat. The principle of a MP approach is to choose the trees that show the smallest amount of evolutionary change. Since there are $(2n-5)!! = 1 \cdot 3 \cdot 5 \cdots (2n-5)$ unrooted binary trees given n species (e.g., almost 14 billion trees for $n = 13$), it is prohibitively expensive and very time-consuming to examine all trees to obtain the exact optimal trees. Note that we use the term *taxa* in this chapter to represent the taxonomic units such as species contained in the tree. Most researchers focus on heuristic algorithms that examine only a certain number of topologies that are likely to be close to the true tree and choose the best one examined for very large datasets (perhaps examining hundreds to thousands of candidate trees). It may be meaningful to find exact MP solutions for small to moderate sized datasets, or to characterize the quality of approximations from heuristic approaches. Therefore, we investigate new high-performance approaches to find the exact solution for the phylogeny reconstruction problem under maximum parsimony criteria.

Several software packages reconstruct sequence-based phylogeny. The most popular phylogeny software suites that contain parsimony methods are PAUP* by Swofford [38], PHYLIP by Felsenstein [15], and TNT and NONA by Goloboff [19, 31]. We have developed a freely-available shared-memory code for computing MP, that is part of our software suite, GRAPPA (Genome Rearrangement Analysis through Parsimony and other Phylogenetic Algorithms) [28]. GRAPPA was designed to re-implement, extend, and especially speed up the breakpoint analysis (BPAnalysis) method of Sankoff and Blanchette [36]. Breakpoint analysis is another form of parsimony-based phylogeny where species are represented by ordered sets of genes and distances is measured relative to differences in orderings. It is also solved by branch and bound. One feature of our MP software is that it does not constrain the character states of the input and can use real molecular data and also characters reduced from gene-order data such as Maximum Parsimony on Binary Encodings (MPBE) [11].

Scientists study species evolution at two levels: from sequences to whole genomes. Sequence-based evolution has been studied for many years while investigating whole-genome-based evolution is a relatively new area. In the following chapter, we not only investigate new approaches for phylogeny-related problems on sequence data, but also discuss open problems for genome-based phylogeny. In Section 22.2 we describe our work on genome data: the first linear time algorithm to compute the inversion distance between two signed gene order, efficient tree generators, and GRAPPA, software we co-developed for the high-performance reconstruction of evolutionary histories. In Section 22.4 we introduce the main concepts used in sequence-based phylogeny reconstruction. We put forward several new techniques to preprocess the input data before performing combinatorial optimizations such as branch-and-bound (B&B) searches. We also propose a fast character algorithm to compute tree lengths that can be used during a local tree optimization search including B&B, subtree pruning and re-grafting (SPR), and tree bisection and reconnection (TBR) searches. Finally, in Section 22.5 are our conclusions.

22.2 Whole-Genome Based Phylogeny Reconstruction

In whole genome-based phylogeny reconstructions, species are represented by their genomes which are described by a set of chromosomes, each consisting of an ordered set of genes. This type of data (with gene order, orientation, and number) presents new opportunities for discoveries about deep evolutionary rearrangement events. Phylogenetic analysis based on whole genomes and gene-order data is a new and computationally-hard field, and fast and accurate techniques are of utmost concern for large-scale genome-based phylogeny reconstruction.

22.2.1 Incremental Tree Generators

Exploring tree space, whether exhaustively or selectively, requires the efficient generation of tree topologies. Since the tree space for examination is often huge or intractably large, and it is typically impossible to generate and store all of the tree topologies. Instead, one can generate a tree, evaluate it, then generate the next tree. The desired tree generator must be interruptible and restartable at any point. In [40] we give a series of tree generator algorithms that not only meet the above requirement but also allow a stepping feature that enables us to sample the tree space or easily partition the tree space on parallel computers.

Given n taxa, the basic approach to generate all $(2n-5)!!$ tree topologies is illustrated in Fig. 22.1. Tree A is the initial core tree, the unique unrooted binary tree of the first three taxa labeled $1, 2$, and 3. Taxon 4 can be added at any branch of Tree A resulting in the four-taxon trees B, C, and D. In a similar fashion, taxon 5 can be added at any branch of the four-taxon trees to generate the five-taxon trees. For example, from tree C, we can get five five-taxon trees, $C1, C2, C3, C4$, and $C5$. We repeatedly add taxa one by one until all taxa are connected to the tree. An unrooted binary tree T_k with k leaves has $(2k-3)$ edges. Thus, the unrooted binary tree T_{k+1} on $(k+1)$ taxa can be generated from T_k by inserting the $(k+1)^{th}$ taxon onto any edge of T_k.

Let i_k take a value in $\{1, \ldots, k\}$ and represent the position where the new taxon is inserted at each level. Then there is a one-to-one mapping between each tree topology and its unique id, a vector $(i_3, i_5, i_7, \ldots, i_{2n-5})$. We generate each tree T_{id} in the following order. By convention, the digits of the id are ordered from left-to-right going from the least to most significant. The first tree is the one with $id = (1, 1, \ldots, 1)$. During the tree generation, we retain the path from the initial core tree of the first three taxa to the current tree T_m. When generating $T_{(m+s)}$, the tree s steps away from T_m, first we find the nearest common ancestor T^l of T_m and $T_{(m+s)}$, then generate $T_{(m+s)}$ from T^l. By using the pre-order tree encoding, this incremental tree generations leads to a fast tree generator. Assuming any leaf to be a root that has only one left child, each partial tree on the generating path can be represented by its pre-order traversal. After adding one taxon to a partial tree, we create a pair of vertices in the tree, a new internal node bisecting the selected branch and a new leaf. We suggest adding the new taxon as the left child such that this new pair of vertices can be inserted contiguously before any element except the first one in the pre-order encoding.

The stepping feature allows any tree be generated directly. The generator makes it possible to sample the tree space and acquire a gross knowledge of the landscape for large datasets. Partitioning the tree space for processing by multiple processors is straightforward using this tree generator. A parallel version may achieve linear speedup in the number of processors if the computation is fairly-well balanced among the processors.

The similar tree generation mechanism cannot only be applied to the general binary tree generator but can also be applied to the generation of constraint trees, subtree pruning and re-grafting (SPR) trees, and tree bisection and reconnection (TBR) trees [40]. A constraint tree generator produces only unrooted, binary trees that are compatible with a given arbitrary tree and enables one to refine non-bifurcating phylogenies. As mentioned previously, SPR and TBR are techniques to search locally around a good tree [38].

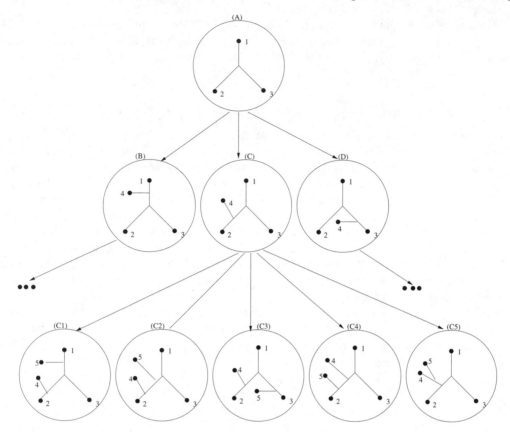

FIGURE 22.1: An example of taxon addition tree generation. At top, tree A is the base three taxa tree. Trees B, C, and D, represent the three four-taxa trees by adding taxon 4 to each of three branches of tree A.

22.2.2   A Linear-Time Algorithm to Compute Inversion Distance

In maximum-parsimony phylogeny reconstruction, one must score trees by computing the tree length, the minimum number of evolutionary events along the edges of the tree. Thus, one subproblem of computing tree length is to compute edge length, the distance between two species. In many cases, the evolutionary process that operates on single-chromosome organisms (such as mitochondria or chloroplasts) consists mostly of inversions of portions of the chromosome. Assuming a fixed set of genes $\{g_1, g_2, \ldots, g_n\}$, each genome is then an ordering (circular or linear) of these genes, with each gene given an orientation that is either positive (g_i) or negative ($-g_i$). Let G be the genome with signed ordering (linear or circular) $g_1, g_2, \ldots, g_n$, then an *inversion* between indices i and j, for $i \leq j$, produces the genome with linear ordering

$$g_1, g_2, \ldots, g_{i-1}, -g_j, -g_{j-1}, \ldots, -g_i, g_{j+1}, \ldots, g_n.$$

The inversion distance between two genomes (two signed permutations of the same set) is then the minimum number of inversions that must be applied to one genome in order to produce the other.

Hannenhalli and Pevzner [20] gave the first polynomial-time algorithm for computing the inversion distance between two signed permutations, as part of the larger task of determining the shortest sequence of inversions needed to transform one permutation into the other. Their algorithm (restricted to distance calculation) proceeds in two stages: in the first stage, the overlap graph induced by the permutation is decomposed into connected components, then in the second stage certain graph structures (hurdles and others) are identified. Berman and Hannenhalli [7] avoided the explicit computation of the overlap graph and gave an $O(n\alpha(n))$ algorithm, based on a Union-Find structure, to find its connected components, where α is the inverse Ackermann function. The bottleneck in their algorithm is to compute the connected components of the overlap graph.

We observe that it is sufficient to know which cycles belong to the same connected component in order to recognize hurdles, and it is not necessary to know the connections between cycles in each connected component. In [6], we introduced the concept of an overlap forest that can be used instead to find the connected components. We describe a linear time algorithm to construct the overlap forest, and thus, have the first linear-time inversion distance calculation between two signed permutations. Our algorithm for computing connected components scans the permutation twice. The first scan sets up a trivial forest in which each node is its own tree, labeled with the beginning of its cycle. The second scan carries out an iterative refinement of this forest, by adding edges and merging trees in the forest. Upon the completion of the algorithm, it is known to which connected component each cycle belongs. Unlike the Union-Find approach, we do not attempt to maintain the forest within certain shape parameters, and this improvement leads to the linear time algorithm. Our algorithm is not only the fastest in theory and in practice, but also is very easy to implement by using a stack.

In [6], we give the results of computational experiments over a large range of permutation pairs produced through simulated evolution. We generated test data according to the evolutionary model of Nadeau and Taylor [29] using five evolutionary rates: 4, 16, 64, 256, and 1024. This rate is the number of evolutionary events applied to the original sequence to generate the two sequences used as the input permutation pairs. We ran experiments on signed permutations of length 10, 20, 40, 80, 160, 320, and 640. Comparing with the Union-Find approach, our experiments show a speedup by a factor of 2 to 5 in the computation of the connected components and by a factor of 1.3 to 2 in the overall distance computation.

22.2.3 GRAPPA

The first heuristic given for reconstructing phylogeny trees from gene order is the breakpoint phylogeny introduced by Sankoff and Blanchette in [36]. They use breakpoint distance as the surrogate of true evolutionary distance such as inversion distance. Later, Blanchette and Sankoff developed BPAnalysis, the software developed to solve the breakpoint phylogeny problem. BPAnalysis (see Fig. 22.2) exhaustively examines every possible tree topology in turn, and, for each topology, generates an optimized set of ancestral genomes that minimize the total breakpoint distance in the tree. For each candidate tree topology, this algorithm solves many instances of the breakpoint median-of-three problem as it searches for a locally optimal tree labeling that minimizes the sum of the tree's edge lengths. The breakpoint median-of-three can be solved by a reduction to the Traveling Salesperson Problem (TSP) where each gene maps to a pair of cities in an interesting twinned-city TSP instance. The TSP instance may be solved exactly or approximately, and its tour gives the gene order and orientation of the median-of-three genome. The computational complexity of the entire algorithm is exponential in each of the number of genomes and the number of genes. Some studies suggest that the breakpoint analysis approach works well for certain datasets, but also find that BPAnalysis is too slow to use on all but very small datasets [11, 12].

for each tree T in the tree space **do**

    Initially label all internal nodes with gene orders;

    Repeat

        for each internal node v, with neighbors A, B, and C, **do**

            Solve the median problem for breakpoints on A, B, C to yield label m;

            if relabeling v with m improves the score of T, **then** do it;

        end

    until no internal node can be relabeled ;

end

FIGURE 22.2: BPAnalysis

22.3 Breakpoint Analysis (BPAnalysis)

With colleagues at the University of New Mexico and the University of Texas at Austin, we designed and implemented an experimental software platform Genome Rearrangement Analysis under Parsimony and other Phylogenetic Algorithms (GRAPPA) [5, 28] for phylogenetic performance studies. GRAPPA performs an exhaustive search under the maximum parsimony criterion and adopts a framework similar to BPAnalysis. In addition, recent versions of GRAPPA now handle inversion phylogeny and unequal gene content. GRAPPA also employs several techniques to make it faster and more flexible:

- **Tree generation** An interruptible and restartable tree generator with a stepwise feature enumerates the trees in amortized constant time and also makes it possible to sample a wide range of tree topologies for large datasets.

- **Tree Labeling** The most time-consuming part of tree labeling is solving the median-of-three problem by using the TSP reduction. We only generate a TSP instance for nodes that saw at least one of their three neighbors relabeled over the last pass.

- **Condensation** In those cases when all genomes in a set contain shared adjacencies, we condense the shared adjacencies: we redefine gene fragments to consist of the longest shared subsequences and replace the original instance by one given in terms of the new gene fragments. Such condensation does not affect labeling or any of the rearrangement based distance measures (breakpoint, inversion, transposition), but decreases the size of the TSP reduction.

- **Approximate TSP solvers** We used the Concorde library [1] for two of our approximate TSP solvers—the chained and the simple versions of the famous Lin-Kernighan heuristics [25]. We also implemented the standard greedy algorithm for TSP.

- **Our exact TSP solver** We implemented a standard include-exclude backtracking search with pruning.

To help improve the overall speed, GRAPPA uses the principle of algorithmic engineering [26], a combination of low-level algorithmic changes, data structure changes, and coding strategies, that combine to eliminate bottlenecks in the code, balance its computational tasks, and make it cache-sensitive. GRAPPA v1.0 includes all of the features of BPAnalysis, but runs about one billion times faster on the IBM/Myrinet 512-processor Los Lobos supercluster at the University of New Mexico to run a complete analysis of the Campanulaceae dataset (13 genomes with 105 genes each).

We ran *GRAPPA* on this system and obtained a 512-fold speed-up (linear speedup with respect to the number of processors): a complete breakpoint analysis (with the more demanding inversion distance used in lieu of breakpoint distance) for the 13 genomes in the Campanulaceae data set ran in less than 1.5 hours in an October 2000 run, for a *million-fold* speedup over the original implementation [3, 4]. Our latest version features significantly improved bounds and new distance correction methods and, on the same dataset, exhibits a speedup factor of *over one billion.* In each of these cases a factor of 512 speed up came from parallelization. The remaining speed up came from algorithmic improvements and improved implementation. The techniques of better bounding and new searching order used in the latest version of GRAPPA make it run much faster [27].

22.4 Sequence-Based Phylogeny Reconstruction

The evolution at DNA or protein level bears little resemblance to the genome level. In sequence-based phylogeny, each species is represented by a string of characters representing sequences of, for example, DNA nucleotides or amino acids. It is usually assumed that each character evolves independently. Researchers have developed a variety of models to study character evolution [39]. Fitch's model [16] is one that is commonly used and allows unordered, multi-state characters, which can transform directly from one state to another. Fitch [16] presented a method to determine the parsimony cost, the minimum number of changes, for a specific tree. His method requires two passes of a rooted binary tree. Although it takes polynomial time to compute the tree cost, there are still $(2n - 5)!!$ trees to be evaluated (where n is the number of species). When n is larger, the branch-and-bound combinatorial optimization technique [21] is often used to prune the search space.

22.4.1 Branch-and-Bound Optimization Approaches

The combinatorial optimization problem is that of maximum parsimony, a *minimization problem* since the most parsimonious tree is that one which has the least number of character changes along its edges. The underlying idea of the branch-and-bound (B&B) technique is the successive decomposition of the original problem into smaller disjoint subproblems until an optimal or all optimal solutions are found. During the B&B search, those subproblems that cannot yield an optimal solution are pruned. A B&B algorithm has four basic rules for a given problem.

- **Branching rule** divides a feasible solution set X into $X_1, X_2, \ldots, X_n$, where $X = \bigcup_{i=1}^{n} X_i$ and $X_i \bigcap X_j = \phi$ for $i \neq j$.
- **Selection Rule** chooses the most promising subproblem for further branching.
- **Elimination Rule** recognizes and eliminates subproblems that cannot yield an optimal solution to the original problem. The most often used rule is called the *lower bound test.* A subproblem Q can be eliminated either when Q has been solved or Q's lower bound is greater than the global upper bound–the best solution obtained so far.
- **Termination Rule** determines whether a feasible solution is optimal.

Next we will discuss the five main aspects that affect the performance of the B&B sequence-based phylogeny reconstruction algorithm: branching scheme, search strategy, lower bounding function, global upper bounding function, and the data structure.

Branching scheme

The branching scheme gives the procedure for decomposing a subproblem in the search s-pace. The branching scheme used in phylogeny reconstruction employs the same mechanism developed for the general tree generator in Section 22.2.1 because the B&B search space is similar to Fig. 22.1. Each node associated with tree T in the B&B search space represents the subproblem to find the most parsimonious tree on all n taxa with strict consensus with the partial tree T. The trees with strict consensus to T are those that can be generated by inserting the remaining species into T. The shape of the B&B search space depends on the addition order of species, which affects the tightness of the lower bound of tree T and thus affects the efficiency of the B&B algorithm.

Search strategy

At any point of the execution, if a subproblem is neither decomposed nor eliminated, we call it an *open problem*. Search strategy decides which of the currently open subproblems will be selected for decomposition. The two strategies most commonly used are *depth-first search (DFS)* which selects the node with longest path to the root and *best-first search (BeFS)* which selects the node with minimum lower bound. DFS is space-saving, while BeFS is more targeted towards a better global upper bound. In the case when the initial global upper bound obtained by heuristic approaches is exactly optimal or very close to the optimal value, there typically is no significant difference in the number of examined subproblems between DFS and BeFS searches, and DFS is then a better search strategy for reasons of space efficiency. Experiments show that the heuristic approaches in Section 22.4.1 can provide a very good solution. Due to the above reasons, we employ DFS as our primary B&B search strategy and for nodes with the same depth we adopt BeFS to break the tie.

Lower bounding of the subproblem

The lower bounding function l associates with each open subproblem p a value v (lower bound) that satisfies the following three conditions:

1. v is less than or equal to the best feasible solution in p;
2. for a leaf of the search space, its lower bound is equivalent to its objective cost; and
3. the function l has nondecreasing values along every path from the root to the leaves of the search space. In other words, the lower bound of a node in the search tree is greater than or equal to the lower bound of its parent.

Ibaraki [22] shows that under the assumption that the dominance test is consistent with lower bound test (it is true in our case), a tight lower bounding function always results in an improvement for DFS and BeFS.

Hendy and Penny [21] used the cost of the associated tree as the lower bound of the subproblem. This traditional approach is straightforward, and obviously, it satisfies the properties of the lower bounding function. However, it is not tight and cannot prune branches very efficiently. Purdom *et al.* used the sum of the single column discrepancy and the cost of the associated tree as the lower bound. For each column (character), the single column discrepancy is the number of states that do not occur among the species in the associated tree but only occur among the remaining species. Purdom's experiments show that it usually reduces the number of decomposed nodes in the search space compared to

Hendy and Penny's lower bounding function [33].

Global upper bounding function

Before starting the B&B search, an initial global upper bound is required. A good upper bound should be somewhat close to the optimal cost in order to work with the lower bound to prune branches efficiently. We investigate two fast heuristic algorithms: neighbor-joining (NJ) [35, 37] and greedy parsimony [13].

Based on the distance matrix, the neighbor-joining algorithm repeatedly pairs two subtrees (at first, a pair of leaves; thereafter, in recursive fashion, entire subtrees), and replaces that pair in further computation with a single artificial taxon representing the subtree, thereby eventually returning a binary tree. The NJ algorithm can always reconstruct the correct tree from an additive distance matrix. However, the distance matrix we use is the minimum distance between sequences and is very likely to underestimate the true distance. Hence, the reconstruction of the correct tree cannot be guaranteed in our case.

The greedy algorithm adopts a strategy that constructs the tree by adding one taxon at a time to the position that yields the best score. Adding taxa in a different order yields different trees. We use the *as is* and *random* option described in [38]. In the *as is* method, the initial core tree is produced by the first three taxa given in the dataset and the following taxon addition is done according to the taxon order in the dataset. In the *random* method, pseudo-random numbers are used to determine the order of taxon addition.

From experiments, we find that the greedy method usually obtains a better tree than NJ. Thus, we use the greedy algorithm with two different addition orders and use the better score as the initial global upper bound.

Data structure for B&B search space

In Fig. 22.3 we illustrate the data structure used to keep open subproblems, where n is the number of species. In the phylogeny reconstruction problem, most of the time is spent on evaluating the tree length of a partial tree, and the choice of different priority queue (PQ) implementations does not make significant difference in performance. So for simplicity, we use a D-heap [23] implementation for the priority queues. A D-heap is organized as an array, using the rule that the first location is the root of the tree, and the locations $2i$ and $2i + 1$ are the children of location i.

22.4.2 Preprocessing of input data

The input data of the sequence-based phylogeny reconstruction is usually a multiple aligned sequence matrix, of which each column represents a character and each row represents the state sequence of a species. We adopt a series of preprocessing steps of the sequence matrix in order to conduct the B&B search efficiently.

Binary encoding of original states

The MP algorithm is dominated by computing the parsimony cost of a tree with Fitch's method. The basic operation of Fitch's method is to compute the *Farris Interval* which is the intersection or union of state sets. Since most computers can perform efficient bitwise logical operations, we use the binary encoding of a state in order to implement intersection and union by bitwise AND and bitwise OR. Each column is encoded independently. There is a one-to-one mapping between the bits of the encoding and the character states. Given a species, if a state is present, then the corresponding bit is set to one, and otherwise it is

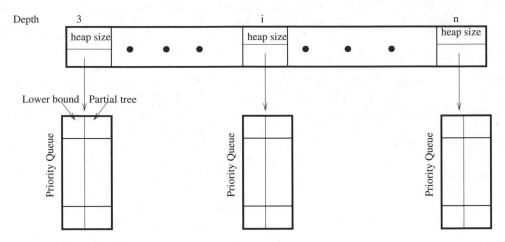

FIGURE 22.3: Data structure to keep the open subproblems.

set to zero.

Deciding the addition order of the species

During the branch and bound process, starting with an initial core tree, the remaining taxa are added one by one as we proceed from the root to the leaves of the search tree. Our experiments show that overall execution time can change drastically depending on the order in which the taxa are added. In this section we discuss the well-known heuristic called the maximum-of-minimum-values (or *max_mini*) algorithm [30] as well as our improved heuristic *max_cross_min*.

The max_mini algorithm proceeds as follows. Starting with initial core tree of three species, for each of the remaining species, we try to insert the species into any edge and find the best position which results the minimum score. Then at the next step, we choose the species that produced the tree with maximum minimum-score to be added at its best position. This process is repeated until the addition order of all of the species is determined.

The max_mini rule is proven to be very efficient and most often used in practice. The intuition behind this rule is that we assume that the greater the score of the best partial tree, the tighter the obtained lower bound. Assuming we have chosen the first i species $\{S_1, \ldots, S_{i-1}, S_i\}$ and the best partial tree T_i, then at the $(i+1)^{th}$ step, for each species S_k from the remaining set $\{S_{i+1}, \ldots, S_n\}$, we must choose the best tree T_{i+1}^k on $\{S_1, \ldots, S_{i-1}, S_i\}$ and S_k. The max_mini rule assumes T_{i+1}^k can be obtained by adding S_k to one branch of T_i. But this speculation is not always true. We have designed a way to improve T_{i+1}^k from the max_mini rule, called *max_cross_mini*.

Suppose the first $(i-1)$ species are $\{S_1, \ldots, S_{i-1}\}$, at the ith step, we get the best tree T_i^k on $\{S_1, \ldots, S_{i-1}\}$ and S_k for $k = i, \ldots, n$. Without loss of generality, we assume T_i^i (that is, $k = i$) has the maximum score and choose to add S_i at step i. T_i^i is T_i in the max_mini rule. But in the max_cross_mini rule, when we add the next $S_{k'}$ at step $(i+1)$, we not only try to insert $S_{k'}$ into each edge of T_i^i, but also try to insert S_i into $T_i^{k'}$. Let the best tree be $T_{i+1}^{k'}$. Similar with max_mini, we select the species k' that gives the tree with maximum minimum-score. Since this rule screens more trees, the best tree obtained this way may be better than max_mini. We compared these rules on 6 datasets and in 4 datasets, the max_cross_mini rule leads to fewer decomposed nodes in the B&B search than the max_mini rule.

Reorder sites

Fitch [17] made a basic classification of sequence sites, the columns of the sequence matrix. At a given site, we say that a state is *non-singleton* if it appears more than once. A site with at most one non-singleton state is a *parsimony uninformative site* since the state changes at such kind of a site can always be explained by the same number of substitutions in all topologies in the following way. If there is one non-singleton state A, we can simply assign all of the internal nodes of the tree topology to be A; if there is no non-singleton state, we can choose any state and assign it to all of the internal nodes. In either case, for any tree topology, the state substitution at this site is equal to the number of states minus one. Since the parsimony uninformative sites do not contribute to the construction of the MP tree, we can ignore them when we evaluate the cost of a tree.

At each level k of the B&B search when we add the kth species, we need to distinguish parsimony informative sites from uninformative sites for the first k species and move all of the parsimony informative sites to a contiguous area in memory in order to evaluate the cost of tree faster. With the addition of species, it is possible for a parsimony uninformative site to turn into a parsimony informative site but the inverse process is not possible. Therefore, we can compute the following by scanning the sequence matrix:

- the level at which a site turns into a parsimony informative site from parsimony uninformative site;
- at each level how many sites turn into parsimony informative sites from parsimony uninformative sites;
- the number of state substitutions on those parsimony uninformative sites;

Then, based on the above information, the targeted position of each site in the new sequence matrix can be computed. Finally, we copy the states of each site from the original sequence matrix to their targeted positions in the new sequence matrix.

22.4.3   Fast algorithms to compute tree length

No matter whether one is performing the local search of heuristic algorithms or the expansion of an active node in a branch-and-bound search, parsimony problems require the evaluation of enormous numbers of trees. Rapid evaluation of a candidate tree generated by stepwise addition or branch swapping (e.g. SPR or TBR) is a crucial factor to the performance of a parsimony program. Although the methods proposed by Farris [14] and Fitch [16] are still considered the basis of the calculation of tree length under Wagner parsimony and Fitch parsimony, respectively, modifications of those basic algorithms can increase efficiency substantially. In the following we will demonstrate an amortized constant time algorithm to compute tree length which is more efficient than Goloboff's method [18] in B&B search since it only makes a single pass in the tree.

First let us see what happens during the stepwise addition. As shown in Fig. 22.4A, a new taxon S is added to an arbitrary branch (U, V) of tree T to obtain a new tree T'. Assume the potential root on (R_{UV}, S) is the calculation root and one post-order pass of T' is sufficient to compute the parsimonious length of T'. The preliminary state set of R_{UV} will not change from T to T'. Our first goal is to pre-process T before the addition of taxa and compute the preliminary state set of R_{UV} for each branch (U, V) of T.

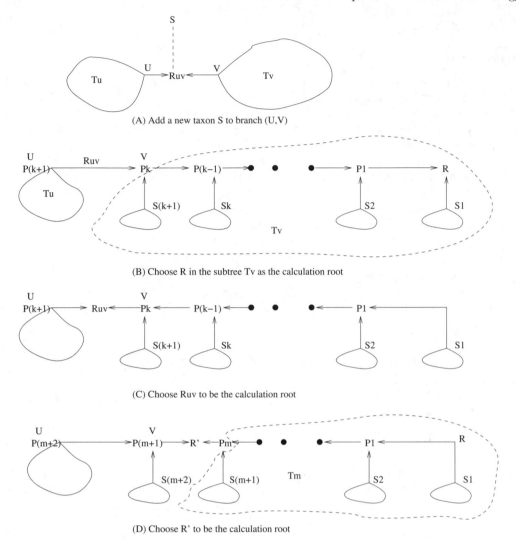

(A) Add a new taxon S to branch (U,V)

(B) Choose R in the subtree Tv as the calculation root

(C) Choose Ruv to be the calculation root

(D) Choose R' to be the calculation root

FIGURE 22.4: Choose different potential root node as the calculation root

Computation of the final states of each potential root node

Without loss of generality, let an arbitrary node R in the subtree T_V be the calculation root of T (see Fig. 22.4B). In this rooted binary tree, V is the parent of U. Let P_{k+1} represent U and P_k represent V, then $(P_{k+1}, P_k, \ldots, P_1, R)$ is the path from P_{k+1} to R. P_{i-1} is the parent of P_i and S_i is the sibling of P_i for $i = 1, \ldots, (k+1)$. (R is the parent of P_1.)

In the post-order traversal of Fitch's method, the preliminary state set of node A comes from the information of all of the nodes in the subtree rooted at A. In Fig. 22.4C, we assume R_{UV} to be the calculation root. Comparing Figs. 22.4B and 22.4C, one can see that the preliminary state sets of each node in $(P_{k+1}, S_{k+1}, \ldots, S_1)$ will be the same whether R or R_{UV} is the calculation root since the subtree rooted at such a node does not change. But the preliminary state set of each node in $(P_k, \ldots, P_1)$ does change with the choice of calculation root and the preliminary state set of P_i depends on that of P_{i-1} and S_{i+1} (see Fig. 22.4C). Therefore, we can first traverse the rooted tree in Fig. 22.4B in post-order and compute the preliminary state set of each node in $(P_{k+1}, S_{k+1}, \ldots, S_1)$, next traverse the tree in pre-order and compute the states sets of each node in $(P_k, \ldots, P_1)$, and then obtain the state set of R_{UV} directly from that of P_{k+1} and P_k.

Assume an arbitrary root for the unrooted binary tree T, then for any branch (U, V), we assume V is the parent of U without loss of generality. Use $\overrightarrow{U}$ to designate the preliminary state set of the root of the subtree T_U which can be obtained from the the post-order traversal of T and use $\overleftarrow{U}$ to designate the preliminary state set of the root of the subtree T_V which can be obtained from the pre-order traversal of T. Since characters are assumed to evolve independently, Algorithm 22.5 describes how two passes of the unrooted binary tree can produce the final states of the potential root node on each branch for one single character.

Computation of tree length in local search

When one searches the neighborhood of a tree by stepwise addition, SPR or TBR, we first use the approach in Section 22.4.3 to compute the tree length of the original tree and the final state set for each potential root node, then for each rearrangement, only a constant time step is required to compute the increase of the tree length from the original tree to the new tree. We explain next the details for different local searches.

B&B Search

Add a new taxon S to tree T with n leaves. Preprocess T as demonstrated in Figure 22.5. Then for each new tree obtained by adding S into edge (U, V), perform Fitch's operations on (S, R_{UV}) to compute the increase of the tree length. The preprocessing needs $(4n - 6)$ Fitch's operations, and each of the $(2n-3)$ new trees takes one Fitch's operation. Therefore, $(6n - 9)$ Fitch's operations are needed to calculate all of the $(2n - 3)$ new trees, and the amortized number of Fitch's operations to evaluate each new tree is 3.

During the B&B search, when we compute the local change between the new tree and the old tree, we actually obtain the preliminary state sets of all internal nodes of the new tree assuming the point of reunion to be the calculation root. If we keep these preliminary state sets, we can save one pass in the preprocessing when adding the next species.

SPR search

In the SPR (subtree pruning and re-grafting) [38] search, a branch of a tree T is cut into two parts: a clipped tree T_c and a target tree T_t. The cutting point of the clipped subtree is then grafted onto each edge of the target tree to produce a new topology. In the preprocessing step, the first pass for T_c and both of the two passes for T_t are required. Let the basic tree length be the sum of the tree lengths of T_c and that of T_t. For each rearrangement by adding the root R_c of T_c into an arbitrary branch (U, V) of T_t, the increase of the tree length from the basic tree length can be obtained by performing Fitch's operation on (R_c, R_{UV}).

TBR Search

In the tree bisection and reconnection (TBR) search, a tree T is cut at an edge and split into two subtrees T_c and T_t. Then the two subtrees are reconnected by joining each pair of edges (one from T_c and the other from T_t) by a new edge. In the preprocessing step, both passes of Algorithm in Figure 22.5 are required for T_c and T_t. Let the basic tree length be the sum of the tree lengths of T_c and T_t. For each rearrangement by connecting branch (U, V) of T_c and (X, Y) of T_t, the increase of the tree length from the original tree length can be obtained by performing Fitch's operation on (R_{UV}, R_{XY}).

Data: T rooted at R.

Result: (1) *cost*, the parsimonious length of T; (2) The final state sets of the potential root on the branch between each node i and its parent.

Begin

    1. $cost = 0$;

    2. Traverse T in post-order.

            foreach internal node i of T **do**

                $\overrightarrow{i}$ = the intersection of the state sets of its two children;

                if $\overrightarrow{i}$ is empty **then**

                    $\overrightarrow{i}$ = the union of the state sets of its two children;

                    $cost = cost + 1$;

                end

            end

    3. Suppose R has two children *lChild* and *rChild*.

        (a) $\overleftarrow{lChild} = \overrightarrow{rChild}$;

        (b) $\overleftarrow{rChild} = \overrightarrow{lChild}$;

        (c) Traverse the subtrees rooted at *lChild* and *rChild* in pre-order, respectively.

            foreach node i whose parent is p and sibling is b **do**

                $\overleftarrow{i}$ = the intersection of $\overleftarrow{p}$ and $\overrightarrow{b}$;

                if $\overleftarrow{i}$ is empty **then**

                    $\overleftarrow{i}$ = the union of $\overleftarrow{p}$ and $\overrightarrow{b}$;

                end

            end

    4. Compute the final state sets of the potential root nodes.

            foreach branch between node i and its parent **do**

                Let the final state set of the potential root node to be the intersection of $\overleftarrow{i}$ and $\overrightarrow{i}$;

                if the intersection is empty **then**

                    Let the final state set to be the union of $\overleftarrow{i}$ and $\overrightarrow{i}$;

                end

            end

End

FIGURE 22.5: Algorithm: pre-process T to get the final states of each potential root node.

22.4.4   Parallel implementation on a symmetric multiprocessor (SMP)

Shared-memory systems, often called symmetric multiprocessors (SMPs), contain from two to hundreds of microprocessors tightly coupled to a shared memory subsystem, running under a single system image of one operating system. For instance, the IBM Power5 p-

Series systems and the Sun Fire Enterprise-class systems all scale from dozens to hundreds of processors in shared-memory images with near-uniform memory access. In addition to a growing number of processors in a single shared memory system, we anticipate the next generation of microprocessors will be "SMPs-on-a-chip". For example, uniprocessors such as the IBM Power4 using simultaneous multithreading (SMT), Sun UltraSparc IV, and Intel Pentium 4 using Hyper-Threading each act like a dual-processor SMP. Future processor generations are likely to have four to eight cores on a single silicon chip. Over the next five to ten years, SMPs will likely become the standard workstation for engineering and scientific applications, while clusters of very large SMPs (with hundreds of multi-core processors) will likely provide the backbone of high-end computing systems.

Since an SMP is a true (hardware-based) shared-memory machine, it allows the programmer to share data structures and information at a fine grain at memory speeds. An SMP processor can access a shared memory location up to two orders of magnitude faster than a processor can access (via a message) a remote location in a distributed memory system. Because processors all access the same data structures (same physical memory), there is no need to explicitly manage data distribution. Computations can naturally synchronize on data structure states, so shared-memory implementations need fewer explicit synchronizations in some contexts. These issues are especially important for irregular applications with unpredictable execution traces and data localities, often characteristics of combinatorial optimization problems, security applications, and emerging computational problems in biology and genomics.

Although the branch-and-bound approach is a very effective technique to solve the exact optimization problem for phylogeny reconstruction when it is applied to large-scale datasets, B&B still requires considerable computation time. In order to improve the performance, we need to utilize the computation power of parallel computers. Since most parallel B&B frameworks [2] are targeted at distributed memory systems, and a symmetric multiprocessor (SMP) has much faster access to their shared memory compared to the message passing between processors on distributed memory systems, we study the parallel B&B algorithms for phylogeny reconstruction with maximum parsimony on Cache-Coherent Uniform-Memory-Access (CC-UMA) SMPs.

Branch-and-bound algorithms can be parallelized at different levels. A low-level parallelization finds maximum concurrency with the computations of each step such as the bounding function, the subproblem selection, node decomposition, and evaluation. A high-level parallelization is such that each processor selects different active nodes then processes the computation associated with its nodes. We focus on the high-level parallelism of the general B&B approach as this coarse-grained parallelism is sufficient for achieving a good load balance among the processors in the system with minimal synchronization. We will tackle a few key issues that affect the performance of the parallel B&B algorithm.

Control scheme for parallel B&B

Since the search space tends to be highly irregular and any static allocation of subtrees to processors is bound to result in significant load imbalance among processors, we need to decide how to coordinate the search subspace of each processor as the computation proceeds. The control scheme of the coordination can be broadly classified into synchronous and asynchronous models.

In the synchronous parallel B&B model [10], the computation is accomplished in consecutive synchronous iterations, and several subproblems are concurrently decomposed during each iteration. In the asynchronous model [9], each processor works at its own pace and does not need to wait at predetermined points for data to become available.

Obviously, the synchronous model guarantees a nearly even load balance between processors. However, the cost of synchronization between processors is still very high on modern computers since in aggregate, processors may sit idle waiting at the synchronization points rather than performing useful work. And for the sequence-based phylogeny reconstruction problem, the computation at each step is low polynomial time and may not be large enough to dominate the overhead of the synchronization. So we prefer to the SPMD asynchronous model that uses the processors more efficiently than the synchronous version.

Data structure for parallel B&B

We use a single shared data structure to hold the priority queues (similar to Fig. 22.3). With the shared data structure, the processors can balance the work fairly evenly and the termination of the algorithm is easy to detect. Since multiple processors access the shared data structure concurrently, the key issue is to avoid contention. Several data structures that allow concurrent operation can be found in the literature, and a comparison between concurrent priority queues is summarized in [24]. However, the comparison is typically conducted using large heaps. While in our serial DFS search, the heap size of each level k is bound by $(2k - 3)$. For such small heaps, the D-heap is simple and efficient. Each heap H_i is protected by a lock $Lock_i$ for $3 \leq i \leq n$. Each processor locks the entire heap H_i whenever it makes an operation on H_i.

In the sequential B&B algorithm, we use DFS strictly meaning that H_i can be accessed only if the heaps at higher levels (H_j for $i < j \leq n$) are all empty. While in the parallel version, in order to minimize contention we allow H_i to be accessed if all of the heaps at higher levels are empty or locked by other processors. When a processor detects that all the heaps are unlocked and empty, no more active nodes exist in the frontier (except those being decomposed by other processors), and this processor can terminate its own execution of the algorithm.

22.4.5   Experimental results

The effectiveness of a phylogenetic reconstruction method depends on the model tree shape, the evolutionary rate, and whether or not the molecular clock hypothesis stands. We use the benchmark collection at (`http://www.lirmm.fr/~ranwez/PHYLO/benchmarks24.html`). Each dataset consists of 24 sequences and the length of DNA sequences is 500. These tests allow comparison on trees whose internal branch lengths are not all equal, and over a wide variety of tree shapes and evolutionary rates.

The experiments of our serial code are carried out on a Sun workstation, whose processor is a 500MHz UltraSparcII and the operating system is Solaris 5.8. We compared the running time between our serial code and PAUP*. Our code uses the max-mini rule as the taxa addition order heuristic and is compiled with option `-xO3 -fast`. We use the `bandb addseq=maxmini` commands for PAUP*. The experimental results are very promising. The results differ for each dataset. Among 20 datasets chosen randomly from 24_tax_500, for 10 datasets our MP implementation, `UNM_MP`, is 1.2-7 times faster than PAUP*, for 5 test instances UNM_MP runs as fast as PAUP*, for 5 other instances UNM_MP is 1.2-2 times slower than PAUP*.

The experiments of our parallel code are run on a Sun E4500, a uniform-memory-access (UMA) shared-memory parallel machine with 14 UltraSparcII 400MHz processors and 14 gigabytes of main memory. Each processor has 16 kilobytes of direct-mapped data (L1) cache and 4 megabytes of external (L2) cache. The test bed uses 200 datasets drawn randomly from the benchmark. Compared with the running time on a single processor, in average we have speedups of 1.92, 2.78, and 4.34, on 2, 4, and 8 processors, respectively. The obtained experimental results show that our strategies on the parallel phylogeny reconstruction problems are efficient.

22.5 Conclusions

We have presented quite a few techniques to compute the exact optimal solutions for phylogeny reconstruction with maximum parsimony. We discussed both the whole-genome-based phylogeny and sequence-based phylogeny. All of the ideas proposed in this chapter are implemented in our serial and parallel code, freely-available as open source code from our web site. For huge datasets, our approaches may successfully serve as base methods for higher-level divide-and-conquer supertree strategies.

Acknowledgments

This work was supported in part by NSF Grants CAREER ACI-00-93039, NSF DBI-0420513, ITR ACI-00-81404, DEB-99-10123, ITR EIA-01-21377, Biocomplexity DEB-01-20709, and ITR EF/BIO 03-31654; and DARPA Contract NBCH30390004.

References

[1] D. Applegate, R. Bixby, V. Chivátal, and W. Cook. CONCORDE: Combinatorial optimization and networked combinatorial optimization research and development environment. http://www.keck.caam.rice.edu/concorde.html.

[2] D.A. Bader, W.E. Hart, and C.A. Phillips. Parallel algorithm design for branch and bound. In H.J. Greenberg, editor, *Tutorials on Emerging Methodologies and Applications in Operations Research*, chapter 5, pages 1–44. Kluwer Academic Press, 2004.

[3] D.A. Bader and B.M.E. Moret. GRAPPA runs in record time. *HPCwire*, 9(47), November 23 2000.

[4] D.A. Bader, B.M.E. Moret, and L. Vawter. Industrial applications of high-performance computing for phylogeny reconstruction. In H.J. Siegel, editor, *Proc. SPIE Commercial Applications for High-Performance Computing*, volume 4528, pages 159–168, Denver, CO, 2001. SPIE.

[5] D.A. Bader, B.M.E. Moret, T. Warnow, and S.K. Wyman *et al.* http://phylo.unm.edu/.

[6] D.A. Bader, B.M.E. Moret, and M. Yan. A linear-time algorithm for computing inversion distance between signed permutations with an experimental study. *Journal of Computational Biology*, 8(5):483–491, 2001.

[7] P. Berman and S. Hannenhalli. Fast sorting by reversal. In D.S. Hirschberg and E.W. Myers, editors, *Proc. 7th Ann. Symp. Combinatorial Pattern Matching (CPM96)*, volume 1075 of *Lecture Notes in Computer Science*, pages 168–185, Laguna Beach, CA, June 1996. Springer-Verlag.

[8] A. Caprara. Formulations and hardness of multiple sorting by reversals. In *3rd Ann. Int'l Conf. Computational Molecular Biology (RECOMB99)*, Lyon, France, April 1999. ACM.

[9] R. Corrêa and A. Ferreira. Modeling parallel branch-and-bound for asynchronous im-

plementation. *DIMACS Series in Discrete Mathematics and Theoretical Computer Science*, 22:45–56, 1995.

[10] R. Corrêa and A. Ferreira. On the effectiveness of synchronous parallel branch-and-bound algorithms. *Parallel Processing Letters*, 5(3):375–386, September 1995.

[11] M.E. Cosner, R.K. Jansen, B.M.E. Moret, and L.A. Raubeson *et al.* An empirical comparison of phylogenetic methods on chloroplast gene order data in Campanulaceae. In D. Sankoff and J. Nadeau, editors, *Comparative Genomics: Empirical and Analytical Approaches to Gene Order Dynamics, Map Alignment, and the Evolution of Gene Families*, pages 99–121. Kluwer Academic Publishers, Dordrecht, Netherlands, 2000.

[12] M.E. Cosner, R.K. Jansen, B.M.E. Moret, and L.A. Raubeson *et al.* A new fast heuristic for computing the breakpoint phylogeny and a phylogenetic analysis of a group of highly rearranged chloroplast genomes. In *Proc. 8th Int'l Conf. Intelligent Systems for Molecular Biology (ISMB00)*, pages 104–115, San Diego, CA, 2000.

[13] R.V. Eck and M.O. Dayhoff. *Atlas of Protein Sequence and Structure*. National Biomedical Research Foundation, Silver Spring, MD, 1966.

[14] J. Farris. Methods for computing wagner trees. *Systematic Zoology*, 34:21–24, 1970.

[15] J. Felsenstein. PHYLIP – phylogeny inference package (version 3.2). *Cladistics*, 5:164–166, 1989.

[16] W.M. Fitch. Toward defining the course of evolution: Minimal change for a specific tree topology. *Systematic Zoology*, 20:406–416, 1971.

[17] W.M. Fitch. On the problem of discovering the most parsimonious tree. *The American Naturalist*, 111(978):223–257, 1977.

[18] P.A. Goloboff. Character optimization and calculation of tree lengths. *Cladistics*, 9:433–436, 1993.

[19] P.A. Goloboff. Analyzing large data sets in reasonable times: Solutions for composite optima. *Cladistics*, 15:415–428, 1999.

[20] S. Hannenhalli and P.A. Pevzner. Transforming cabbage into turnip (polynomial algorithm for sorting signed permutations by reversals). In *Proc. 27th Ann. Symp. Theory of Computing (STOC95)*, pages 178–189, Las Vegas, NV, 1995. ACM.

[21] M.D. Hendy and D. Penny. Branch and bound algorithms to determine minimal evolutionary trees. *Mathematical Biosciences*, 59:277–290, 1982.

[22] T. Ibaraki. The power of upper and lower bounding functions in branch-and-bound algorithms. *Journal of the Operations Research Society of Japan*, 25(3):292–320, 1982.

[23] D. E. Knuth. *The Art of Computer Programming: Sorting and Searching*, volume 3. Addison-Wesley Publishing Company, Reading, MA, 1973.

[24] B. LeCun and C. Roucairol. Concurrent data structures for tree search algorithms. In A. Ferreira and J.D.P. Rolim, editors, *Parallel Algorithms for Irregular Problems: State of the Art*, pages 135–156. Kluwer Academic Publishers, 1995.

[25] S. Lin and B.W. Kernighan. An effective heuristics algorithm for the traveling salesman problem. *Operations Research*, 21:498–516, 1973.

[26] B.M.E. Moret. Towards a discipline of experimental algorithmics. In M.H. Goldwasser, D.S. Johnson, and C.C. McGeoch, editors, *Data Structures, Near Neighbor Searches, and Methodology: Fifth and Sixth DIMACS Implementation Challenges*, volume 59 of *DIMACS Monographs in Discrete Mathematics and Theoretical Computer Science*. American Mathematical Society, 2002.

[27] B.M.E. Moret, J. Tang, L.S. Wang, and T. Warnow. Steps toward accurate reconstruction of phylogenies from gene-order data. *J. Comput. Syst. Sci.*, 65(3):508–525, 2002. Invited, special issue on computational biology.

[28] B.M.E. Moret, S. Wyman, D.A. Bader, and T. Warnow *et al.* A new implementation and detailed study of breakpoint analysis. In *Proc. 6th Pacific Symp. Biocomputing (PSB 2001)*, pages 583–594, Hawaii, 2001.

[29] J.H. Nadeau and B.A. Taylor. Lengths of chromosome segments conserved since divergence of man and mouse. In *Proceedings of the National Academy of Sciences USA*, volume 81, pages 814–818, 1984.

[30] M. Nei and S. Kumar. *Molecular Evolution and Phylogenetics*. Oxford University Press, Oxford, UK, 2000.

[31] K.C. Nixon. The parsimony ratchet, a new method for rapid parsimony analysis. *Cladistics*, 15:407–414, 1999.

[32] I. Pe'er and R. Shamir. The median problems for breakpoints are NP-complete. Technical Report 71, Electronic Colloquium on Computational Complexity, November 1998.

[33] P.W. Purdom Jr., P.G. Bradford, K. Tamura, and S. Kumar. Single column discrepancy and dynamic max-mini optimization for quickly finding the most parsimonious evolutionary trees. *Bioinfomatics*, 2(16):140–151, 2000.

[34] K. Rice and T. Warnow. Parsimony is hard to beat. In *Computing and Combinatorics*, pages 124–133, August 1997.

[35] N. Saitou and M. Nei. The neighbor-joining method: A new method for reconstruction of phylogenetic trees. *Molecular Biological and Evolution*, 4:406–425, 1987.

[36] D. Sankoff and M. Blanchette. Multiple genome rearrangement and breakpoint phylogeny. *Journal of Computational Biology*, 5:555–570, 1998.

[37] J.A. Studier and K.J. Keppler. A note on the neighbor-joining method of Saitou and Nei. *Molecular Biological and Evolution*, 5:729–731, 1988.

[38] D.L. Swofford and D.P. Begle. *PAUP: Phylogenetic analysis using parsimony*. Sinauer Associates, Sunderland, MA, 1993.

[39] D.L. Swofford, G.J. Olsen, P.J. Waddell, and D.M. Hillis. Phylogenetic inference. In D.M. Hillis, C. Moritz, and B.K. Mable, editors, *Molecular Systematics*, pages 407–514. Sinauer, Sunderland, MA, 1996.

[40] M. Yan. *High Performance Algorithms for Phylogeny Reconstruction with Maximum Parsimony*. PhD thesis, Electrical and Computer Engineering Department, University of New Mexico, Albuquerque, NM, January 2004.

VI

Microarrays and Gene Expression Analysis

23

Microarray Data: Annotation, Storage, Retrieval and Communication

Catherine A. Ball
Stanford University

Gavin Sherlock
Stanford University

23.1 Introduction

Thirty years ago, Ed Southern published a method that could be used to detect a particular sequence within a population of nucleic acids attached to a solid support, using a radio-labeled probe [30]. The Southern blot, as it became known, permitted measurement of a nucleic acid's presence and relative abundance. Methods for the screening of clone libraries [10] and the gridding of libraries on filters allowed a larger-scale application of hybridization on filters. These technologies allowed one to detect molecules in a mixed population that hybridized to a single sequence.

The microarray is the high-throughput, miniaturized descendant of the filter-based blots introduced some 30 years ago. Instead of hybridizing a radioactively "known" nucleic acid to a size-sorted population of nucleic acids fixed to a membrane, microarrays use a population of known nucleic acids individually arrayed in known locations on a solid support to report on an unknown population of fluorescently labeled nucleic acids. This simple difference, allows us to determine the presence (and in some cases, the abundance) of thousands of sequences by performing a single hybridization, and has been exploited to great effect. For example, in a pioneering study Alizadeh et al [1] profiled diffuse large B-cell lymphomas, and were able to identify distinct subtypes of the disease with different predicted outcomes, based solely on their molecular signatures. Furthermore, these molecular signatures provided additional prognostic information that was independent of the pre-existing prognostic criteria. It is this ability to gain a global understanding of a biological system that has led to the widespread adoption of the technology.

Two types of microarray were pioneered in the early 1990s. Commercially manufactured microarrays were pioneered by Steve Fodor and colleagues at Affymetrix [9, 21], using photolithography and solid phase chemical synthesis to build short oligonucleotides *in situ* on chips. Around the same time, Patrick Brown and co-workers at Stanford University developed an alternative microarray technology, that involved spotting DNA sequences, such as PCR products or cDNA clones, onto glass microscope slides by capillary action, using a robot with a multi-tipped print head [28, 29]. In the years immediately following, widespread use of microarrays was prevented by the high cost and the degree of expertise required to manufacture the arrays, perform the hybridizations, scan the slides, capture signals from the resulting images and then to analyze the resulting data. In addition, there was no easy-to-use and freely available software for microarray fabrication, image quantification and data analysis. Since that time, microarrays have overcome their "boutique" status and have become widely used for a variety of applications. The widespread manufacture of "home-grown" microarrays was encouraged by the web-based "MGuide" (http://cmgm.stanford.edu/pbrown/mguide/) provided by the Brown lab . The MGuide provided unrestricted access to detailed parts lists and assembly directions for robotic arrayers, software (Joe DeRisi's ArrayMaker) and protocols. Access to microarray technology has been dramatically increased by the widespread access to information necessary to construct microarrays or the ability to simply purchase microarrays and the equipment to perform hybridizations and scans. The concept of mechanically spotting microscopic spots of cDNAs on glass slides has been elaborated so that many companies as well as academic laboratories are manufacturing different types of microarrays. In addition, Mike Eisen, while a postdoctoral researcher with David Botstein and Pat Brown, wrote and released ScanAlyze (for image quantification), Cluster (for data analysis) and TreeView (for data visualization). The release of these software packages made all aspects of dealing with microarray data more accessible to life sciences researchers. Extraction of numerical data from a two color microarray, turning the channel intensities into log ratios, and subsequent visualization of these data using a false color representation is shown in Figure 23.1.

Currently, there are several commercial and "home-grown" microarray platforms in widespread use. Those based on the technology developed by the Brown laboratory still rely on mechanical printing to "print" spots of DNA on an array, and microarrays with up to 60,000 spots have been manufactured using this technique. Affymetrix, using their photolithography method is able to synthesize more than two million 25-mer oligonucleotides on an array as 5-micron features. Agilent uses a technology similar to inkjet printing to synthesize oligonucleotides (typically 60-mers) *in situ* on glass slides, with up to 44,000 features per microarray. Another vendor, Nimblegen arrays uses maskless lithography, a technique which relies on dynamic mirrors to create masks on the fly, to manufacture oligonucleotide arrays, with up to 380,000 oligonucleotides, which can range from 25-mers to 50-mers. As microarrays have become more widely used, there is every reason to expect more companies to throw their hats into the ring.

Regardless of the array platform used, most microarray experiments that assay transcript abundance follow roughly the same experimental design. Populations of mRNA are isolated from experimental samples and a control or reference sample. cDNA is then made from the mRNA, which may be subjected to one or more rounds of linear amplification. The cDNA is labeled by the incorporation of fluorescence-tagged nucleotides, sometimes using an oligo-dT primer, or sometimes using random hexamers, or even a combination of both. When a two-channel platform is used, the experimental and reference samples are labeled with tags that fluoresce at different wavelengths, and these labeled samples are competitively

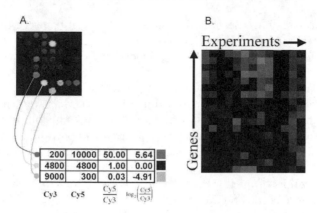

FIGURE 23.1: (See color insert following page 20-4.) Conversion of image data to raw and false color data. A. The intensities in each channel in the scanned image are converted to the raw numeric values, and then used to produce a log ratio. Log ratios above zero are then represented as red, with the intensity being proportional to the value, while green is used to represent negative log ratios. A log ratio of zero is represented by black. B. Data for multiple genes across multiple experiments are represented by block of color, one block per measurement, as suggested in [8].

hybridized to the microarrays in a manner very similar to the original Southern blots. After hybridization, an image for each fluor is created using a scanning laser microscope (typically referred to as the scanner). This image is then processed by software (such as GenePix) to obtain dozens of measurements for each spot on the microarray that can be used to calculate signal, determine data quality and calculate the relative abundance of each transcript between the experimental and control samples. The storage and annotation of these data is the subject of this chapter, while its analysis and processing are covered in subsequent chapters.

While many of the original printed arrays relied on PCR-amplified cDNAs or genomic sequences, more and more array manufacturers (both commercial and academic facilities) are using oligonucleotides for their reporters ('reporter' refers to the nucleic acid that is spotted on a microarray). The use of oligonucleotides has the benefit of providing a lower risk of unwanted cross-hybridization that are possible with longer sequences and the potential to differentiate between alternately spliced transcripts, as well as avoiding costly and time consuming PCR reactions. However, when oligonucleotides are synthesized in situ, there is the possibility of truncated products due to failed synthesis steps, and these may have the potential for non-specific hybridization. For spotted oligonucleotides, the oligos can first be purified based on length, which avoids this problem (though it is more costly).

Microarray technology has been enormously successful in looking at transcript abundance to paint molecular portraits of tumors, tissues, cell cultures and many other experimental systems covering various aspects of fundamental cellular processes (e.g. [7, 13, 25, 32]). However, microarray technology has also been applied to other types of molecular surveys. Array-based comparative genome hybridization (array-CGH or aCGH) has been used to look for copy number changes in genomic DNA in a number of studies, for instance identifying genomic regions of recurrent deletion or amplification in various cancers (e.g. [17, 22, 23]). Others have used microarrays to identify DNA sequences that are immunoprecipitated while cross-linked to DNA-binding proteins. Such chromatin immunoprecipitation (chip-ChIP) experiments have been used to detect targets of transcription factors or other DNA binding proteins on a genome-wide basis (e.g. [11, 16]). Microarrays are now being adapted and used for sequencing, genotyping, identifying SNPs and others applications are sure to arise.

Typically, microarray experiments produce a great deal of data. Not only is the output from a scanned slide a large data set, recording the work that went into performing the experiment in the first place requires storing large quantities of complex data. The output from scanned microarrays includes dozens of measurements pertaining to each spot, each of which can be used either to calculate the signal or the quality of a spot. Other information that must be recorded includes information about the reporter molecules fixed to the surface of the microarray, details of microarray fabrication, conditions of each experiment, quality assessment scores and records of data analysis steps. Successful use of the DNA microarray technology requires that the researcher be able to efficiently retrieve and make sense of this information. Without it, it is almost impossible to track down and troubleshoot problems, such as contamination of samples, or incorrect recording of print information. In addition, unless vital information is comprehensively catalogued, it is impossible for others to replicate the work or assess its quality. Although some researchers use microarrays to perform a limited series of experiments, there are lessons to be learned from those that have been tracking and storing the data for large-scale projects. Many such large-scale projects make use of databases to assist researchers storing and accessing information about many aspects of their experiments and the resulting data. The types of information that should be recorded include:

- Experimental design
- Treatment, isolation and labeling of biosample
- Hybridization conditions
- Scanning parameters
- Image processing/data acquisition procedures
- Data selection, filtering, quality control
- Data analysis methods
- Biological annotation of reporters on microarray

There are three different data storage needs for most microarray studies: methods for tracking the fabrication of the microarrays themselves, methods for storing the data generated from and describing the hybridization of microarrays and finally, methods to archive and make available published microarray data. Most large-scale microarray production facilities use LIM systems (laboratory information management systems) to record the information detailing the manufacture of microarrays. A LIM system is essentially a tracking database that records, for example, how the microarrays were constructed. Results of the hybridization of microarrays are usually stored in a second database that could be referred to as a results database or a research database. Results databases hold more than raw data — they must also contain enough biological annotation to make analysis of the results possible, for in the absence of biological context, expression data are meaningless. While the distinction between LIM systems and results databases seems quite natural, in reality there is significant overlap in the types of information each requires, and it is important that these two separate systems have an interface through which data from one can be connected to data in the other. In many cases, there is considerable duplication of information in LIM systems and in results databases. Consequently, many projects could reasonably make use of a single database that combines a LIM system and a results database. Finally, data that are published need to be made freely available to other researchers so that the conclusions of the original study can be confirmed and so that the data might be used for other research purposes. The databases that provide access to published data can be thought of as archival data repositories.

23.2   Information to be Captured by a LIM System

While researchers purchasing commercially available microarrays have no need to record the details of microarray manufacture, those who are constructing microarrays on-site will need a LIM system. In order to troubleshoot problems and to properly interpret and annotate one's microarray data, it is essential to track data that describes how a microarray is manufactured. For instance, when reporters ('reporter' refers to the nucleic acid that is spotted on a microarray) on a microarray are based on the same sequence yet different results are observed, contamination of one of the reporters might be indicated. Given the tens of thousands of reporter sequences on a typical array, automated approaches are needed to make this situation easy to detect. Conversely, the expression pattern of a given sequence may consistently show close similarity to that of an unrelated sequence. In the absence of tracking information, similar expression patterns would be interpreted as biologically meaningful. However, the ability to track probe information enables a researcher to detect instances where sequences in adjacent wells in a microtiter plate show similar expression patterns. In such a case, it is very likely that one sample has contaminated the other. The ability to detect likely contamination and other procedural problems allows the microarray spots generated from suspect samples to be flagged as unreliable.

An effective microarray LIM system should track procedures that were used to generate sequences spotted on microarrays, should map samples as they are transferred from one microtiter plate to another, should record the identity of the sequences spotted on microarrays as well as their position on the microarray and finally, should store the details of the microarray printing process.

23.2.1   Tracking and preparation of samples

It is essential that information about the sequence identity of reporters and their preparation is stored ('reporter' refers to the nucleic acid that is spotted on a microarray). Access to sequence information about reporters allows them to be properly annotated with biological information, both in the present and in the future event of changes to the genomic sequence record. Reporters are most often PCR products (amplified from cDNAs or from genomic sequences) or oligonucleotides. PCR products can be generated from a variety of templates, such as genomic DNA or cDNAs inserted into vectors. A microarray LIM system must be able to record each step of a sample preparation as well as the sources of templates and primers and their locations in relevant microtiter plates.

23.2.2   Sequences on arrays

An obvious use of a LIM system is to keep track of what is spotted on each array. There are really two components to this. The first is recording what the sequence is supposed to be. In the case of oligonucleotides spotted on a microarray, it could be as simple as recording its sequence. However, if a PCR product is to be spotted, its identity is based on both the template and the primers. In both of these cases, it could be advantageous to record the gene or other genetic element that the spotted sequence is supposed to report. Suspected contamination or PCR failure must be associated correctly with spotted sequences for future reference. The second component is mapping where on a microarray each sequence has been placed.

For example, producing a PCR product from a vector insert requires that the LIM system record the following information (much the same information should be stored for the manufacture of oligonucleotide arrays):

1. The source of clones/bacteria
2. Strains containing plasmids
3. The identity of the plasmid
4. The storage location of the plasmid
5. Any relevant accession numbers or clone ids, or other identifiers pertaining to the insert in the plasmid
6. Identity and source of each primer (usually a common pair of vector primers are used for amplification)
7. Storage locations of each primer
8. Sequences of primers used for amplifying genomic DNA
9. Expected size of the PCR product
10. Identity of the PCR product
11. Expected sequence of the PCR product
12. Storage location of the PCR product
13. Images of any gels used to evaluate the quality of the PCR product
14. The quality of the PCR product, with both respect to whether it is close to the expected size, and/or whether a single, or multiple bands are seen, or indeed no bands are seen
15. Transfer of samples from plate to plate
16. Protocols followed at each step of the production of the microarray

Being able to reconstruct the methods used to generate spots on a microarray is beneficial for troubleshooting. Suspected contamination or incorrect annotation can be verified only if the original samples can all be readily tested.

23.2.3 Microarray printing

The procedural details of microarray fabrication that should be recorded include the arraying machine that was used (if more than one was available), the tip configuration that was employed, and the wash and dry times between each loading of the tips with sample. Most projects record the number of microarrays made, the order of each microarray within the batch (often glass slide microarrays printed at the beginning of a batch may have larger spots than microarrays printed later), the type of slide or surface used, and any pretreatments that might have been applied to the surface prior to printing. In addition, date, time and names of individuals responsible for the print are often useful. Large-scale facilities will also have to track the fates of each microarray (whether given to a researcher or simply dropped and broken) and might even have to track billing and payment records.

23.3 Information to be Stored in a Results Database

Biological samples

A microarray database should allow recording of as much information as possible about the biological samples used for experiments. Such information includes the organism and genotype from which the sample was derived, the organ and/or anatomical derivation of the sample, if appropriate. If possible, these annotations should be from controlled vocabularies, or ontologies so that the descriptions of samples can be consistent, and more easily searched by software, and return more complete results from searches.

Protocols

During the course of a microarray experiment, many manipulations of the biological sample are likely to be carried out, and protocols for these manipulations often need to be accessed by the experimenters themselves, or a third party, to assess exactly how, and in what context the microarray data were generated. The protocols for the extraction of RNA from the biological sample, and the hybridization protocol itself should also be recorded, both for the purposes reproducing the experiment, and for troubleshooting and tracking of any protocol-dependent systematic errors that may have occurred.

Data from the hybridization

Data are obtained by first scanning a hybridized microarray to generate an image of the fluorescent signal. The resulting image is processed to identify spots and to extract data for each spot. Several software packages are available for these tasks and the database a project uses should obviously support data entry from the chosen image analysis program. No matter what microarray platform or data acquisition package is used, the data acquired from a microarray image consist of more than single measurement for each spot, and instead are several dozen values per spot, many of which can be used as quality indicators. Most image analysis programs produce dozens of data metrics per spot, such as total intensity in each channel, background in each channel, regression correlation of all pixels in the spot, etc. While the values used in the ultimate data analysis may simply be the normalized ratios (see below) of background-subtracted signal intensities, many of the other measurements provide information about spot quality (such as regression correlation) and valuable as filters to help researchers select spots that have generated reliable data. If one decides to discard some measurements provided by data acquisition software, one must be well informed about which metrics are least likely to be useful for data selection and quality assessment. Since the analysis of microarray data has so few "Standard Operating Procedures" and we can reasonably expect novel and more sophisticated data analysis procedures to be created, it is probably wise to store as many data columns as possible, if not all of them. However, storage of so many measurements for each spot on a microarray accordingly demands greater storage space.

Normalized data

There are many sources of systematic variation in microarray experiments that affect the measured raw data — normalization is the process used to remove such variation. In the case of two-color microarray experiments, the normalized data values that are most pertinent are normalized log ratios. There are several methods that can be used for data normalization (see [24] for a review), including global mean or median normalization, intensity dependent normalization (using a lowess function [37]), and variance stabilization [12]. In addition, some of these methods may be employed in conjunction with spike-in controls [36], and be applied separately to data from different regions of the microarray to correct for spatial bias. A microarray database must provide one or more of these methods for normalization, as well as to have the ability to store the normalized data.

Data retrieval

Efficient and flexible data queries are what make databases attractive tools for storing microarray data. To be even remotely useful, a microarray database must support a query for data about a single microarray, in a file containing all the results for the array, as well as the associated biological information. Since annotation of the elements on the array is essential for interpretation of the results, it is important to retrieve relevant and up-to-date information describing each of those elements. Another basic requirement is to retrieve specific data fields from multiple microarrays, so that expression of genes in a group of related arrays (for example, arrays that are part of a time series or a study of related tissue samples) can be assayed.

Filtering of data

A microarray database should support filtering of data during retrieval, both on a spot-by-spot basis, and also on a clone-by-clone, oligo-by-oligo or gene-by-gene basis. One should be able to disregard, or filter, data that are generated from spots of low confidence using flexible quality assessment criteria. Some examples of spot metrics that can be used in filtering include the spot's regression correlation, its signal to background ratio or its overall intensity. Additionally, quality measurements from the manufacture of the array (whether a PCR product showed an anomaly or whether the sample might be contaminated) should also be available as metrics to use for data filtering. Examples of LIMS-type data that can be used as filtering metrics include sequence verification, contamination and flag status (indicating that a spot was identified as being suspect by visual inspection or by the data acquisition software package).

When selecting data from multiple arrays it should also be possible to select data for sequences that pass filtering criteria in some percent of arrays in the dataset, or sequences whose signals vary by a certain amount within a set of arrays. Alternatively, a researcher might want to retrieve all data for a subset of genes, clones or oligonucleotides, using a list of genes of interest, or even selecting based on gene annotation.

Modeling of biological samples

An easily overlooked issue in recording microarray data is how to model the reporters on the arrays and how they relate to the biological molecules (often, but not always, genes) that they assay. A simple model would be to simply treat the DNA sequences on the chip as if they were the genes themselves. However most experiments using microarrays require a data model that is more sophisticated. First, such an approach does not allow effective modeling of non-genic sequences (such as intergenic regions), or of large sequences that may span several genes, such as BACs. Second, it is also not adequate to model genes, either. A microarray database must model the sequences using at least two different levels of specificity, first that of the physical sequence that is the reporter on the microarray, and secondly, at the level of the genetic locus to which that reporter maps. It is important to bear in mind that many reporters on a microarray may map to a single gene, and indeed that the mapping between a reporter and a gene may change over time, as our knowledge of genomes, and ability to predict genes within them, becomes more sophisticated. For instance, many sequences (e.g. cDNA clones or oligonucleotides designed using a cDNA clone) may map to a single genetic entity, such as a UniGene cluster. The ability to retrieve individually microarray data as a function of the clones whether or not they map to the same clone is useful to verify the reproducibility of the data from the individual clones. An expression pattern that is significantly different than those of other clones in a UniGene cluster could indicate that clone is contaminated, has been erroneously mapped to a particular UniGene cluster, or may be the result of an alternative splicing event. Alternatively retrieval of data for multiple reporters that map to the same genetic locus as a single, collapsed set of data is also an important function for a microarray database. Collapsing the data as a function of the genetic locus is a means to prevent multiple measurements of the same gene from affecting downstream data analysis. An additional level of identification of the reporters on an array is that of 'instances of a piece of DNA'. For example, a cDNA clone or oligonucleotide may come into a laboratory or array fabrication facility through more than one path, and even though each copy of the clone should ostensibly contain exactly the same sequence, it is prudent to have the ability to track these entities separately in case differences in manufacture or handling cause differences in data obtained from the reporters.

A more sophisticated and much more useful plan would be to map reporter sequences on a microarray not just to the genes they represent, but also to each of the gene's exons or introns. With this information, the data would facilitate comparison of expression patterns of alternative transcripts. As the number of reporters on microarrays increase (particularly those that are fabricated using photolithography, such as Nimblegen and Affymetrix microarrays), it is likely that many array platforms will have enough reporters to assay exon-exon junctions for each exon in each transcript, so as to be able to assay different splice forms, so a data model to take advantage of such a feature of the microarrays would be attractive in a microarray results database.

Biological annotation of sequences

Interpretation of microarray data is impossible without biological information describing the reporters on the array. For this reason, it is essential that a microarray results database links each reporter to accurate and current annotation of the genes to which they map. One factor that complicates this task is that our understanding of genes and their products changes as research uncovers new information. A second complicating factor is that the types of annotation may change. Types and sources of important biological annotation differ with different organisms and with the scientific intent of the experiment. A third, and very challenging factor, is that the actual mapping between a piece of DNA and the gene it is meant to represent can also change. For example, upon each build of the human UniGene clusters, several hundred clones will change their allegiance to different clusters than in the previous build. Thus a microarray database must allow the annotation of entities on chips to be easily updated, flexible and dynamic.

A standard for experimental annotation: MIAME

Independent verification, accurate interpretation and re-use of microarray results require very careful annotation of experiments. Towards this end, the Microarray Gene Expression Data Society (MGED; `www.mged.org`), a grass-roots group of microarray researchers, has developed a standard to define the Minimal Information About Microarray Experiments (MIAME; [6]). The MIAME specification, available at (`www.mged.org`), is not a strict set of requirements, but really a set of guidelines that indicates the types of annotations that should be recorded when microarray data are made publicly available. Since several journals and funding agencies have adopted the MIAME standard for data release, a researcher selecting a microarray results database should check whether the package is MIAME-supportive. The MIAME specification describes experimental annotations that fall into six categories, much of which has been discussed as prerequisites of a good database:

1. **Experimental design:** MIAME allows a group of microarrays to be described in the context of a single experiment. Both the type of experiment can be described (for example, a time series, a comparison of diseased tissue to healthy tissue, or comparison of a wild-type to a mutant) and the relationships of one array to another can be detailed, e.g. an order of the arrays within a series. Experimental factors being assayed, quality control steps and associated publications are part of the experimental design.

2. **Array design:** In addition to a description of the manufacture of the arrays and the microarray platform, this section records details about the features on each spot, such as its sequence. The sections above about LIM systems and biological sequence modeling provide a discussion of the issues associated with recording this type of information.

3. **Samples used, extract preparation and labelling:** In order to understand the experiments performed, information about the origin of each biological sample must be provided as well as details about any procedures or manipulations that were performed using the biological sample. The description of the samples used for hybridization should include topics such as the primary source of the biological sample, the organism from which it was derived, and the protocols used for its preparation, the RNA extraction and subsequent labeling.

4. **Hybridizations procedures and parameters:** This section of MIAME describes the protocol and conditions used for hybridization, blocking and washing, including any post-processing steps.

5. **Measurement data and specifications:** This section of MIAME covers the raw and processed data that make up the actual experimental results. Care should be exercised to record information about the original scans of the array (images), the extracted microarray data based on image analysis and the final data after normalization and consolidation of replicates. Importantly, this section of MIAME specifies that data extraction and processing protocols should be explained in sufficient detail to replicate the data analysis.

A standard for data exchange : MAGE-ML

MAGE-ML is an XML-based markup language that is based on the MicroArray Gene Expression object model (MAGE-OM)indexMAGE-OM developed by members of the MGED society [31]. Expressed in the Universal Modeling Language (UML), MAGE-OM models the entire process of microarray experimentation, from array design, to hybridization and data analysis. Since MAGE-OM permits one to record all MIAME-required annotations, MAGE-ML provides an excellent and comprehensive method to exchange microarray data. To provide information to populate MAGE files, the MAGE Ontology (MO; `http://mged.sourceforge.net/ontologies/`) has been developed. MO provides controlled terms for use within the MAGE-OM. The goal of this ontology is to help different groups to annotate their microarray experiments in consistent and predictable manners so that computational analysis will be easier to accomplish. One obvious advantage of consistent annotation is that data entered into a microarray repositories, such as GEO [5] or ArrayExpress [20], will be more easily queried by investigators.

Tools and analysis packages

A microarray database could simply serve as a data repository, with methods to enter and retrieve the data, as discussed above. However, it is often extremely inconvenient to have to extract data, perhaps re-format it manually and then import it into various analysis tools.

Instead, analytic tools can be associated with the database itself, reducing the manipulation and effort required to explore data sets. An example of a valuable tool would be one to produce graphs of various spot parameters (for example, channel 1 intensity vs. channel 2 intensity), which may be useful for assessing overall array quality. Data analysis tools, such as hierarchical clustering [8], Self Organizing Maps [33] and principal components analysis [2] would be very useful components of a database package. Since there are so many commercial and free stand-alone tools for microarray analysis, it is essential that the database produce data in the correct format for such tools to read.

Archival data repositories

There are currently three archival microarray data repositories. In the US, at the NCBI is the Gene Expression Omnibus (GEO; [5]), which accepts data in multiple formats, including MAGE-ML . In Europe, there is ArrayExpress [20], which is a product of the European Bioinformatics Institute in Cambridge, England, and has more stringent file format and data annotation requirements for data entry. Data can be either directly entered in MAGE-ML format or entered using a web-based interface (MIAMExpress) that will construct MAGE-ML for entry into the database. Finally, in Japan there is Cibex, which is being developed by the DDBJ, though is not yet accepting public data submissions. Despite their geographical separation, one of the stated goals of the three archival repositories to share data in the same spirit as GenBank/EMBL/DDBJ, though no formal accession numbering system of data exchange procedures have yet been formally worked out.

23.4   Freely Available Results Databases

For local installation, a microarray results database can either be purchased from a vendor, such as Rosetta's Resolver, or Iobion's GeneTraffic, or alternatively may be installed from one of several free software projects. While there are several commercial microarray databases available (see [3], for review), their relatively high cost can make their use prohibitive. Below, we present eight freely available results databases, all of which distribute their full source code and schema, permitting customization (or improvement) by enterprising installers, and discuss their advantages and disadvantages, as well as their installation requirements.

23.4.1   BioArray Software Environment (BASE)

BASE [26] is a MIAME-supportive system that provides an integrated framework for storing and analyzing microarray data and related information.

Requirements and Installation: BASE is written in PHP, with some additional C++ code for CPU-intensive tasks, and uses the Open Source MySQL database for data storage. Typically, it is deployed on Linux, though there have also been successful reports of deployment on Sun's Solaris operating system, and MacOSX, and it can be modified for deployment on Windows using Cygwin. The BASE software itself is released under the GNU General Public License, and depends only on Open Source software, such as Linux. Since BASE does not require any software purchase, there is a very low cost for deployment.

Features: BASE is capable of storing data from all aspects of the microarray process, from LIMS data (including tracking clones in microtiter plates, and recording details about microarray printing) to the loading and analysis of results data. Internally, the BASE data model closely resembles the MAGE Object Model [31] and allows users to specify relationships between various aspects of a microarray experiment. For example, which biological samples were hybridized on a given microarray, and data from which hybridizations were grouped together to form an experimental set or series can be easily stored. BASE supports the loading of Array Designs for homemade spotted arrays, as well as for commercial arrays, though currently BASE only supports two-channel arrays, so Affymetrix arrays are not fully supported. BASE has a flexible configuration system, such that importing data from a variety of image extraction packages, such as GenePix and QuantArray is relatively straightforward. BASE also has plug-in software architecture, so external developers could contribute additional software packages, and extracted data are typically produced in a BASEfile format, which can be used by the plug-ins. Several plug-ins are already available, including ones that implement Lowess normalization, and Multi-Dimensional Scaling. BASE also allows data export in a variety of different formats for use in other analysis tools, as well as in MAGE-ML format. BASE also provides tools so users can allow others to view their data, so it can facilitate collaborative research.

Advantages: Completely Open Source solution, with low cost of deployment. MIAME-supportive, able to export MAGE-ML. Flexible data import tool, and ability to share data easily. Able to store at least 3,000 hybridizations, each with 30,000 features. Flexible analysis and filtering pipeline, which can store your parameters for later retrieval. Plug-in architecture that allows new analysis software to be easily added.

Disadvantages: Lack of real support for Affymetrix data. No easy way to transfer data between one BASE instance and another. No MAGE-ML import. Unproven scalability of MySQL may lead to performance issues.

23.4.2 Gecko (Gene Expression: Computation and Knowledge Organization)

Gecko [34] (http://sourceforge.net/projects/geckoe) uses a client server architecture, with a centralized repository that can store tens of thousands of Affymetrix scans, and comes with a suite of analysis tools.

Requirements and Installation: Gecko requires a Sun server running Solaris, and uses Oracle as its RDBMS. The Gecko client currently runs only on Windows, although work is underway to write a Java client, which theoretically could run on any platform.

Features: Gecko features a database, a client tool for accessing the data, and a suite of analysis tools. Currently, the computational engine implements an admirable number of analysis tools, including many two-class comparison tests (Student t-tests, SAM, comparison of variances, Mann-Whitney), as well as multiple-class and multiple-factors tests (one and two-way ANOVA) and the ability to perform contrast calculations. In addition, Gecko has self-organized maps (SOM), average linkage hierarchical clustering, principal component analysis (PCA), multidimensional scaling (MDS) and the ability to build and display correlation or distance matrices, as well as various data transformation tools. Gecko provides a tool they term "Analysis Trees," which enable users to perform and save complex data analysis work flows, which are stored as Directed Acyclic Graphs (DAGs). While Gecko use has thus far been limited to Affymetrix data, it is theoretically able to store two color microarray data as well. While not fully MIAME compliant, Gecko implements some of the MIAME required annotation fields.

Advantages: Scalable to tens of thousands of arrays, with a comprehensive suite of analysis tools, and Analysis Trees that allow users to track and repeat analyses.

Disadvantages: Requires expensive hardware and software for the server, and the client software only runs on Windows. Not fully MIAME compliant, and does not produce MAGE-ML. At the time of writing, only the client software has been made available for download, which significantly diminishes its value.

23.4.3 GeneX and derivatives

GeneX was one of the first Open Source microarray database solutions [19] and was originally developed at the National Center for Genome Research (NCGR), and is now being maintained as GeneX Lite (`http://www.ncgr.org/genex/index.html`). Other groups are also developing two additional variants of GeneX. First, GEOSS (`http://va-genex.sourceforge.net/`), formerly known as GeneX Va [15], is based on the 1.05 release of GeneX, and has additional layers of security to enable data sharing, and also provides support for Affymetrix data. Second, GeneX-2 (`http://genex.sourceforge.net/`) is being developed by a volunteer team that includes several of the original GeneX developers. All three of these GeneX variants provide a freely available database (released under the GNU lesser public license) that uses the free database system (PostgreSQL) to store the data. Commercial installations of GeneX Lite are subject to some restrictions, so license issues should be investigated prior to installation.

Requirements and Installation: GeneX installation requires a Linux machine with Perl installed and an Apache web server. An installation script is used to configure the components of the system.

Features: A Java-based curation tool is provided for formatting datasets for secure upload into the database, and then simple html interfaces are used to retrieve that data. In addition, the GeneX software package provides an Application Programming Interface (API) that allows experienced programmers to make extensions to the interface. There are a number of analytical routines that can be executed with microarray data, such as clustering, multidimensional scaling, cluster validation and Principal Components Analysis, which are provided as separately available add-ons for the database.

Advantages: GeneX can be installed on inexpensive hardware and is relatively easy to install. The flexible data model allows users to store data from different array platforms, such as two-color microarray data, and single-channel Affymetrix data. Some data analysis tools are provided. GeneX is under active development, and thus future improvements can be expected, although this is somewhat mitigated by the fact that the three development tracks are independent.

Disadvantages: GeneX has not yet been demonstrated to scale to hold data for many thousands of experiments, though that of course does not mean it will not scale to that size. GeneX does not allow viewing of proxy images for characterization of array quality.

23.4.4   maxdSQL

maxdSQL (`http://bioinf.man.ac.uk/microarray/maxd/`) is a MIAME-supportive database based on the ArrayExpress schema (which in turn is a direct implementation of the MAGE Object Model).

Requirements and Installation: maxdSQL is implemented entirely using the SQL92 standard, and thus is supported by any database management system that implements that standard. This includes Oracle, PostgreSQL and MySQL. The various tools that comprise the maxd database are written in Java, so theoretically can be run on any platform. There are instructions for Linux, Windows and MacOSX.

Features: Maxd provides a database loading application, maxdLoad2, for the loading of new data into the database. MaxdLoad2 is a standalone Java application that can be run either on the same machine upon which the database resides, or it can also be used on a remote machine. MaxdLoad2 permits users to specify all of the details of their experimental design, closely following the MAGE model in terms of the objects and process they describe. For loading of results data, maxdLoad2 allows users to define the file format of a particular data file, and then save that definition for later reuse. Results data can be loaded in batch. Maxd provides a second application, maxdView, which provides facilities for data visualization, filtering transformation and analysis.

Advantages: maxdSQL is very flexible, and can be deployed on any platform that can support Java, and can use any database that implements the SQL92 standard. This means that an entirely open source implementation can be deployed, that provides a low cost to entry. Data loading is flexible, and many data file formats can be used. Standalone Java applications provide interactive means by which experimental data can be loaded, viewed and analyzed. Data can be output in MAGE-ML, and database fully supports the MIAME standard.

Disadvantages: Scalability not yet proven (though this does not imply that it will not scale).

23.4.5   RNA Abundance Database (RAD)

RAD [18] provides a MIAME-supportive infrastructure for gene expression data management and makes extensive use of ontologies. RAD is part of the more general Genomics Unified Schema (GUS; `www.gus.org`).

Requirements and Installation: Because RAD relies on GUS, it is actually necessary to install GUS. GUS is supported using either Oracle 8i/9i/10g, or PostgreSQL. Perl, PHP and Apache (`www.apache.org`) are required for installation, and there are some additional optional Java modules. GUS installation currently requires a certain level of expertise, but the authors are working on a ready-to-use package.

Features: The RAD Study Annotator records specific details on protocols, biological samples and study designs, which are collected through web-based annotation forms. The RAD Querier provides basic hierarchical clustering tools, plots for quality assessment of single arrays and an in-house algorithm for detecting differentially regulated genes (PaGE), which is also available separately as a Perl program or a Java application (`http://www.cbil.`

`upenn.edu/PaGE/`). For capturing data and meta-data, all microarray platforms and image analysis software are supported. In addition, RAD is being used for CGH, ChIP, and SAGE data. RAD can produce MAGE-ML files for export of data to other databases or software packages. RAD is part of a more general Genomics Unified Schema, which provides a platform to integrate gene and transcript data from a variety of organisms.

Advantages: RAD provides a scalable, web-accessible solution that can accommodate data from several laboratories. The software is provided in an Open Source method. The security features allow fine-tuned access for keeping unpublished data private, sharing data with collaborators and making published data freely available. The methods to store and visualize information about protocols, biological samples and the design of experiments are excellent. RAD can produce MAGE-ML, easing submission of microarray data to public repositories. Since RAD and GUS are being actively developed (indeed, RAD was selected for the microarray data module of the Generic Model Organism Database project (GMOD; `http://www.gmod.org/`), bugs are likely to be fixed and new releases with additional features can be expected.

Disadvantages: Installation and maintenance of RAD can be rather work-intensive, but hardware, software, and personnel requirements are dependent of the scale and scope of the project. RAD (as part of GUS) can be installed on a laptop and maintained by a single computer-savvy student or can be use to support cores and large bioinformatics resources. RAD has limited LIMS features and has limited analysis tools available as part of the package.

23.4.6   Stanford Microarray Database (SMD)

The Stanford Microarray Database [4] is a MIAME-supportive database, which can store results data and some LIMS data, and has data analysis tools available as part of the installation. Its full source code and schema are available from (`http://genome-www.stanford.edu/microarray`)

Requirements and Installation: The Stanford installation of SMD use Oracle 9i on a Sun Microsystems V880 server, running Solaris 2.9, with 8 processors, and 32GB of RAM. While there is no specific reason to use a Solaris system, Oracle updates and bug fixes often appear first on this platform, making it desirable for housing an Oracle database. SMD installation requires the Oracle Enterprise Edition server software, a web server, Perl, and several Perl modules. Although an installer script distributed with the software takes care of many of the steps required to get the system running, it is still not a simple task. Additional details, such as setting up the Oracle instance of the database, and creating all the tables and the relationships between them does require the efforts of a trained database administrator, though all the SQL scripts required to do this are distributed with the SMD package. Making a more easily installed package, with regular, quarterly releases, is an explicit goal of SMD.

Features: SMD incorporates a LIMS tracking system, to track the 96- and 384-well plates that might have been used for printing homemade microarrays, either in people's laboratories, or at the Stanford Functional Genomics Facility (SFGF; `http://www.microarray.org/`). SMD also allows loading of commercial MAGE-ML Array Designs, for example from Agilent Technologies, Affymetrix, or GeneXP Biosciences). Data derived from GenePix, ScanAlyze and SpotReader (Niles Scientific), as well as data extracted by Agilent's Feature

Extraction software from Agilent arrays may be loaded into SMD. In addition, SMD also provides native support for Affymetrix data, allowing users to upload CEL files and d-Chip files. More platforms are added as the SMD user base at Stanford use arrays from new sources - at the time of writing, SMD was adapting software to accept data from arrays manufactured by both Nimblegen (`http://www.nimblegen.com/`) and Combimatrix (`http://www.combimatrix.com/`). Users may enter data derived from any of these packages in either a single or batch mode. For each of the data file types that SMD supports, all data are stored (in many cases several dozen metric per spot), and upon entry, data from two-color microarray experiments may be normalized using either a Global Mean Normalization, or using a Lowess based normalization, either globally, or per print tip. Data retrieval of an arbitrary number of experiments, can be done using complex filters, such that any spot metric, LIMS data, or biological annotation may be used as a filtering criterion, which may be combined together in Boolean queries, using AND, OR or NOT. In addition, SMD has several built-in tools for assessing array quality, experiment reproducibility, and visualizing a representation of the original microarray, as well as tools for downstream analyses. These tools include hierarchical clustering, Self-Organizing Maps, Singular Value Decomposition [2], and imputation of missing data [35]. SMD supports the MIAME standard, and upon publication data can be exported in MAGE-ML for import into the ArrayExpress or GEO data repositories. Finally, SMD has well-developed data access methods so that some data can be restricted to a few close collaborators and other data can be made publicly available. SMD has been successfully used for gene expression, chromatin immunoprecipitation (ChIP), comparative genome hybridization (CGH) and protein microarrays, as well as other applications of microarray technology.

Advantages: SMD is a scaleable solution for storing microarray data from large complex projects — the Stanford installation currently has data from >50,000 microarrays, comprising data from ~1,300,000,000 spots — while a flexible security model allows fine-grained access control to both data and tools. SMD is MIAME-supportive and can export data in MAGE-ML format for direct submission to GEO or ArrayExpress. Several tools are available as part of the package, and software for viewing proxy images of the microarray scans to visually evaluate the quality of the data are also available. Furthermore, SMD dynamically updates annotation of the human, mouse and yeast genes that are represented on the microarrays, and is under active development, so that new features and improved schema and software are regularly available. Finally, SMD has support for both two-color data, and for Affymetrix data.

Disadvantages: As implemented, SMD requires expensive hardware and software, as well as trained staff (at least a database administrator and a programmer/curator) to install it. An offshoot of SMD, the Longhorn Array Database (LAD; [14]), was developed specifically because of these drawbacks of SMD. LAD can be deployed entirely using Open Source software, with its primary platform being PostGreSQL on Linux, and has been shown to be able to store data from several thousand microarrays. As of the time of writing (January 2005), LAD is based on a previous release of SMD, prior to the addition of Affymetrix, Agilent and MAGE-ML support, though it is likely that LAD will update to a newer version of the SMD package.

23.4.7   TM4 and MicroArray DAta Manager (MADAM)

TM4 [27] is apackage that includes a MySQL database and a suite of tools, and is available from The Institute for Genomic Research (TIGR)through (`http://www.tm4.org`). One of the tools associated with TM4, MADAM, provides a graphic user interface for entering data into and retrieving data from the database.

Requirements and Installation: MADAM runs on the Windows 2000/NT/XP systems, as well as Linux, and requires Java v1.4.1 or higher. It uses MySQL for its database.

Features: TM4 with MADAM provides a Java interface for entering and annotating microarray data, in a MIAME-supportive fashion, and provides integrated MAGE-ML export. While MADAM itself does not have any analysis tools, the TM4 package includes MIDAS (Microarray Data Analysis System) and MeV (Multi-experiment viewer). Together, these tools provide the user the ability to carry out various normalizations, quality control steps, data transformations and data analyses on their data, and are tightly integrated with the entire TM4 suite, such that data from one component can easily be loaded into a different component. It is probably fair to say that TM4 contains the most comprehensive set of open source tools available with intuitive graphical interfaces and the system continues to be developed and expanded.

Advantages: Extensive support for various different analysis methods through the integrated tools in the TM4 suite, and completely open source. MIAME supportive annotation and MAGE-ML export.

Disadvantages: MADAM currently only has support for two color microarray experiments and can upload only the ".mev" file format. A utility called ExpressConverter can convert a wide range of formats, including GenePix, ImaGene, ScanArray, ArrayVersion, and Agilent to ".mev" format for loading, but some information in these formats is lost.

23.5 Conclusions

Many researchers conducting experiments using microarray technology are left in a difficult position when it comes time to analyze and interpret their data. While a microarray database is a necessity for handling the quantity of data and the complexity of its annotations, buying a commercial product is costly, and creating and maintaining a local database is no simple proposition. A number of free databases are available for local installation, but none is a "one size fits all" solution. Researchers initiating microarray experiments should carefully consider their database requirements and should plan accordingly to obtain funding. An adequate database solution will significantly ease interpretation, annotation, analysis, communication and publication of microarray experiments. For that reason, database needs should be specified just as explicitly as the experiments that the database will eventually house. Selecting or developing a database to keep data safe and to help researchers retrieve data, keep biological annotations up to date, share data with collaborators, analyze data and publish conclusions should be a high priority for any microarray project.

References

[1] A.A. Alizadeh, M.B. Eisen, R.E. Davis, and C. Ma *et al.* Distinct types of diffuse large B-cell lymphoma identified by gene expression profiling. *Nature*, 403(6769):503–11, 2000.

[2] O. Alter, P.O. Brown, and D. Botstein. Singular value decomposition for genome-wide expression data processing and modeling. *Proc Natl Acad Sci U S A*, 97(18):10101–6, 2000.

[3] P. Anderle, M. Duval, S. Draghici, and A. Kuklin *et al.* Gene expression databases and data mining. *Biotechniques*, pages 36–44, 2003. Suppl.

[4] C.A. Ball, I.A. Awad, J. Demeter, and J. Gollub *et al.* The stanford microarray database accommodates additional microarray platforms and data formats. *Nucleic Acids Res*, (33):D580–2, 2005. Database Issue.

[5] T. Barrett, T.O. Suzek, D.B. Troup, and S.E. Wilhite *et al.* NCBI GEO: mining millions of expression profiles–database and tools. *Nucleic Acids Res*, (33):D562–6, 2005. Database Issue.

[6] A. Brazma, P. Hingamp, J. Quackenbush, and G. Sherlock *et al.* Minimum information about a microarray experiment (MIAME)-toward standards for microarray data. *Nat Genet*, 29(4):365–71, 2001.

[7] J.L. DeRisi, V.R. Iyer, and P.O. Brown. Exploring the metabolic and genetic control of gene expression on a genomic scale. *Science*, 278(5338):680–6, 1997.

[8] M.B. Eisen, P.T. Spellman, P.O. Brown, and D. Botstein. Cluster analysis and display of genome-wide expression patterns. *Proc Natl Acad Sci U S A*, 95(25):14863–8, 1998.

[9] S.P. Fodor, R.P. Rava, X.C. Huang, and A.C. Pease *et al.* Multiplexed biochemical assays with biological chips. *Nature*, 364(6437):555–6, 1993.

[10] M. Grunstein and D.S. Hogness. Colony hybridization: a method for the isolation of cloned DNAs that contain a specific gene. *Proc Natl Acad Sci U S A*, 72(10):3961–5, 1975.

[11] C.T. Harbison, D.B. Gordon, T.I. Lee, and N.J. Rinaldi *et al.* Transcriptional regulatory code of a eukaryotic genome. *Nature*, 431(7004):99–104, 2004.

[12] W. Huber, A. von Heydebreck, H. Sultmann, and A. Poustka *et al.* Variance stabilization applied to microarray data calibration and to the quantification of differential expression. *Bioinformatics*, 18(suppl 1):S96–104, 2002.

[13] V.R. Iyer, M.B. Eisen, D.T. Ross, and G. Schuler *et al.* The transcriptional program in the response of human fibroblasts to serum. *Science*, 283(5398):83–7, 1999.

[14] P.J. Killion, G. Sherlock, and V.R. Iyer. The longhorn array database (LAD): an open-source, MIAME compliant implementation of the Stanford Microarray Database (SMD). *BMC Bioinformatics*, 4(1):32, 2003.

[15] J.K. Lee, T. Laudeman, J. Kanter, and T. James *et al.* GeneX Va: VBC open source microarray database and analysis software. *Biotechniques*, 36(4):634–8, 640, 642, 2004.

[16] T.I. Lee, N.J. Rinaldi, F. Robert, and D.T. Odom *et al.* Transcriptional regulatory networks in *saccharomyces cerevisiae*. *Science*, 298(5594):799–804, 2002.

[17] S.C. Linn, R.B. West, J.R. Pollack, and S. Zhu *et al.* Gene expression patterns and gene copy number changes in *dermatofibrosarcoma protuberans*. *Am J Pathol*, 163(6):2383–95, 2003.

[18] E. Manduchi, G.R. Grant, H. He, and J. Liu *et al.* RAD and the RAD study-annotator: an approach to collection, organization and exchange of all relevant information for high-throughput gene expression studies. *Bioinformatics*, 20(4):452–9, 2004.

[19] H. Mangalam, J. Stewart, K. Zhou, and M. Sclauch *et al.* GeneX: An open source gene expression database and integrated tool set. *IBM Systems Journal*, 40(2):552–569, 2001.

[20] H. Parkinson, U. Sarkans, M. Shojatalab, and N. Abeygunawardena *et al.* ArrayExpress–a public repository for microarray gene expression data at the EBI. *Nucleic Acids Res*, (33):D553–5, 2005. Database Issue.

[21] A.C. Pease, D. Solas, E.J. Sullivan, and M.T. Cronin *et al.* Light-generated oligonu-

cleotide arrays for rapid DNA sequence analysis. *Proc Natl Acad Sci U S A*, 91(11):5022–6, 1994.

[22] J.R. Pollack, C.M. Perou, A.A. Alizadeh, and M.B. Eisen *et al.* Genome-wide analysis of DNA copy-number changes using cDNA microarrays. *Nat Genet*, 23(1):41–6, 1999.

[23] J.R. Pollack, T. Sorlie, C.M. Perou, and C.A. Rees *et al.* Microarray analysis reveals a major direct role of DNA copy number alteration in the transcriptional program of human breast tumors. *Proc Natl Acad Sci U S A*, 99(20):12963–3, 2002.

[24] J. Quackenbush. Microarray data normalization and transformation. *Nat Genet*, (32):496–501, 2002.

[25] D.T. Ross, U. Scherf, M.B. Eisen, and C.M. Perou *et al.* Systematic variation in gene expression patterns in human cancer cell lines. *Nat Genet*, 24(3):227–35, 200.

[26] L.H. Saal, C. Troein, J. Vallon-Christersson, and S. Gruvberger *et al.* BioArray software environment (BASE): a platform for comprehensive management and analysis of microarray data. *Genome Biol*, 3(8):SOFTWARE0003, 2002.

[27] A.I. Saeed, V. Sharov, J. White, and J. Li *et al.* TM4: a free, open-source system for microarray data management and analysis. *Biotechniques*, 34(2):374–8, 2003.

[28] M. Schena, D. Shalon, R.W. Davis, , and P.O. Brown. Quantitative monitoring of gene expression patterns with a complementary DNA microarray. *Science*, 270(5235):467–70, 1995.

[29] D. Shalon, S.J. Smith, and P.O. Brown. A DNA microarray system for analyzing complex DNA samples using two-color fluorescent probe hybridization. *Genome Res*, 6(7):639–45, 1996.

[30] E.M Southern. Detection of specific sequences among DNA fragments separated by gel electrophoresis. *Mol Biol*, 98(3):503–17, 1975.

[31] P.T. Spellman, M. Miller, J. Stewart, and C. Troup *et al.* Design and implementation of microarray gene expression markup language (MAGE-ML). *Genome Biol*, 3(9):RESEARCH0046, 2002.

[32] P.T. Spellman, G. Sherlock, M.Q. Zhang, and V.R. Iyer *et al.* Comprehensive identification of cell cycle-regulated genes of the yeast *saccharomyces cerevisiae* by microarray hybridization. *Mol Biol Cell*, 9(12):3273–97, 1998.

[33] P. Tamayo, D. Slonim, J. Mesirov, and Q. Zhu *et al.* Interpreting patterns of gene expression with self-organizing maps: methods and application to hematopoietic differentiation. *Proc Natl Acad Sci U S A*, 96(6):2907–12, 1999.

[34] J. Theilhaber, A. Ulyanov, A. Malanthara, and J. Cole *et al.* GECKO: a complete large-scale gene expression analysis platform. *BMC Bioinformatics*, 5(1):195, 2004.

[35] O. Troyanskaya, M. Cantor, G. Sherlock, and P. Brown *et al.* Missing value estimation methods for DNA microarrays. *Bioinformatics*, 17(6):520–5, 2001.

[36] J. van de Peppel, P. Kemmeren, H. van Bakel, and M. Radonjic *et al.* Monitoring global messenger RNA changes in externally controlled microarray experiments. *EMBO Rep*, 4(4):387–93, 2003.

[37] Y.H. Yang, S. Dudoit, P. Luu, and D.M. Lin *et al.* Normalization for cDNA microarray data: a robust composite method addressing single and multiple slide systematic variation. *Nucleic Acids Res*, 30(4):e15, 2002.

Computational Methods for Microarray Design

Hui-Hsien Chou
Iowa State University

24.1 Introduction

Microarrays for measuring the expression profile of cells under different developmental or environmental stages are gaining rapid acceptance by biologists as a popular method to conduct whole genome biological studies. This is especially promising given that many genomes of varying sizes have recently been sequenced, including some large genomes like *Drosophila* [1], *Arabidopsis* [31, 35, 7], human [27, 50], mouse [53] and rice [17, 56]. Soon additional large genomes such as maize [13], rat [49], chicken [41] and dog [25] may become available as well.

With the knowledge of the whole genome or a significant portion of its coding region, it is now possible to computationally optimize the design of microarrays to ensure the optimal chance to correctly gather the expression profile deemed highly important for biologists to elucidate gene functions and to eventually better understand life. Given that microarray terminologies are not always consistent in the literature, in this chapter we explicitly use the terms "spot", "probe" or "oligo" to denote the sensor on the microarray surface, and the terms "target", "nontarget" or "sequence" to denote the labeled single-strand DNA or RNA that are converted from extracted cell RNA, commonly just from mRNA.

Currently there are three major formats of microarrays in use, based on spotted cDNA clones [15, 18], lithographically synthesized short oligos (the Affymetrix array) [32, 59], and long oligos either synthesized *in situ* (the Agilent array) or prescribed and spotted by users [21, 9]. See Figure 24.1 for illustrations of the three different types of arrays.

DNA clones are the byproducts of whole genome or EST sequencing projects, therefore they are readily available for DNA microarray manufacturing. After genome assembly or EST clustering, a representative template from the clone library can be computationally

DNA array

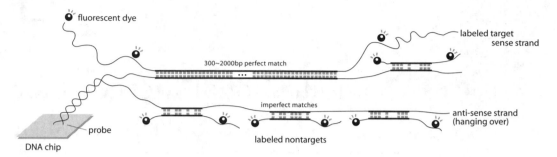

Short oligo (Affymetrix array)

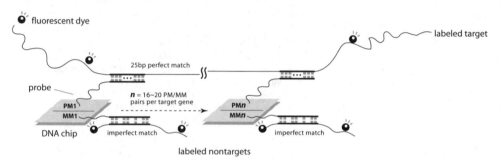

Long oligo (Picky designed)

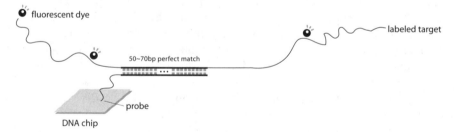

FIGURE 24.1: Illustrations of the three main microarray formats in use today. DNA arrays were made from spotted DNA fragments amplified from clone libraries. Due to their enormous length, uneven melting temperature and the extra anti-sense strands, cross-hybridization with nontargets is the highest with DNA arrays. Short oligos used in Affymetrix arrays were lithographically synthesized directly on the array surface. Due to their shorter length of 25 bp, a set of match/mismatch pairs of oligos is required to uniquely identify the expression level of a single target. For each perfect match (PM) oligo there is a companion mismatch (MM) oligo that is used to gauge the background hybridization level with nontargets. Long synthesized oligo arrays such as those designed by PICKY were computationally optimized that can uniquely identify each labeled target without ambiguity. With a global design approach, theoretically there should not be any cross-hybridization of each probe with nontargets at the optimal experiment temperature determined by the design.

selected to best represent each gene or gene family. Standard PCR amplification procedure can then be followed to manufacture the probe that will be spotted onto the microarray surface. Depending on whether custom primers were used during the amplification process, the DNA probe size can vary from a couple hundred base pairs to thousands of base pairs. DNA arrays usually provide the strongest fluoresce signal due to their long binding site with the targets, however, for the same reason they also suffer the highest cross-hybridization noise due to their inability to differentiate similar genes or gene families sharing long stretch of common subsequences. Since different DNA spots on an array may have very different melting temperatures with their intended targets, it is impossible to obtain an optimal microarray experiment temperature that can successfully separate all probe/target hybridizations from all probe/nontarget hybridizations. The anti-sense strand of each DNA probe may also double the likelihood of cross-hybridization since it is freely available to grab onto labeled nontarget(s). Finally, tracking and maintaining DNA clone libraries may introduce some handling errors into microarray experiment because these libraries were not usually collected for microarray purposes at the beginning.

The Affymetrix lithographically synthesized microarrays are manufactured *in situ* on silicon wafers and are the most popular short oligo (20–25 bp) microarray solution. Since only sequence information is needed to design and manufacture the Affy-arrays, they do not have the clone tracking or PCR synthesis errors that might occur with DNA array manufacturing. Because Affymetrix is the primary vendor of this type of arrays, its operation protocol has been standardized and is much easier to be followed by customers to produce reasonably good results. This is particular beneficial to small labs or new users. Short oligos as used in the Affy-arrays can potentially hybridize to many different genes of a species, therefore multiple oligos must be used as a set to uniquely identify and measure the expression level of each target gene [36]. For each perfect match (PM) oligo in an oligo set, a companion mismatch (MM) oligo is also included in the set to gauge the nontarget background hybridization potential of the PM oligo. The sophisticate Affymetrix microarray data processing software uses data gathered from all PM/MM pairs in an oligo set to computationally determine the actual expression level of each target gene. The silicon wafer masks for manufacturing Affy-arrays are expensive to produce, therefore Affy-arrays cannot be updated as frequently as spotted DNA or long oligo arrays for ongoing genome projects. Affymetrix both designs and manufactures their arrays for its customers, therefore users only need to supply their gene sequences in order to obtain their Affy-arrays. Although general principles of Affymetrix microarrays design can be found in the literature [32, 36], the details remained proprietary to Affymetrix. For these two reasons, this chapter does not attempt to cover Affymetrix microarray design. However, the basic design strategies outlined below are still applicable to similar short oligo array design.

The third type of microarrays uses longer oligos (50–70 bp) as spots. These arrays can be designed by end-users or companies, for all or a subset of the genes of a species, and manufactured in small or large quantities. Since oligos used in these arrays can be individually updated throughout a genome project cycle, they allow quicker incorporation of new gene sequences without incurring a high replacement cost. Due to their longer length and single-strand nature, the long oligos used in this type of arrays can be made very sensitive to their intended targets, therefore often just one oligo is needed to detect a specific gene [40]. Additionally, these oligos can be computationally optimized *together* to achieve a much greater individual specificity and global uniformity, thereby reducing cross-hybridization dramatically. In fact, in theory there should not be any stable cross-hybridization of any designed probe with any nontarget sequence under the globally computed optimal microarray experiment temperature, given that all input sequences are complete and correct. Just like the Affy-array, only sequence data are required to design and manufacture long oligo arrays, therefore clone tracking and PCR errors can be entirely avoided. Finally, optimal long oligo microarray design tools such as PICKY are freely available to academic users to create their own arrays [14]. All these reasons make long oligo microarrays the most suitable ones for studying species under ongoing sequencing projects, or species with lesser commercial interest therefore commercial arrays may not be available for them.

This chapter reviews the fundamental principles of microarray design that are shared by all three types of arrays, then moves on to describe DNA microarray and long oligo microarray design in detail. Emphases are placed more on the design of long oligo microarray because it is currently the most suitable end-user designed array format. Note that this chapter concerns only the computational aspects of microarray design. Microarray experiments are inherently biochemical processes, and there are many other steps such as RNA extraction, labeling, hybridization, and scanning that all have to be conducted correctly for the entire microarray experiment to be successful. Discussions of these non-computational steps are beyond the scope of this chapter, but most of these steps are standardized protocols that can be reliably repeated by end-users if they follow their lab manuals or vendor recommended protocols.

24.2   Basic Design Strategies

Technological advances allow microarray experiments to be conducted at higher throughput and greater convenience, but microarrays must be carefully designed for the data produced from them to be useful. Specifically, spots on the arrays for detecting the expression level of each gene must be unique to a gene (i.e., exhibit high *specificity*), must be able to detect that gene (i.e., exhibit high *sensitivity*), and must function optimally under the same melting temperature and other experiment conditions (i.e., exhibit high *uniformity*). These are fundamental design principles that may not all be achievable by each type of arrays. For example, DNA arrays are the least likely to achieve specificity due to their extended probe length and anti-sense strand, but they have a higher sensitivity level. Short and long oligo arrays can both be computationally designed to be more specific to their intended targets, but it is still relatively more difficult to make all short oligos specific to all genes in a set due to their limited length. Therefore, cross-hybridization level must be estimated using a pair of match/mismatch oligos, and a set of oligo pairs must be used together to determine the actual expression level of a target gene, as in Affymetrix design. On the other hand, the match/mismatch pair strategy does not work for long oligo arrays because at longer length it becomes thermodynamically in-differentiable for a few mismatches during hybridization. Instead, for long oligo arrays, there is a greater chance to achieve specificity through uniqueness in probe selection, so background hybridization does not need to be individually estimated for each oligo probe but can be taken as a global characteristic of the whole array.

Despite the different characteristics of each type of arrays and the different issues that must be addressed in their design, many common computation strategies are still shared among their design. In this section, these common computation strategies are reviewed first. Type specific design strategies will then be discussed when each type of microarray design is discussed.

24.2.1   Sequence level comparison

Since gene sequences to be targeted by microarray experiments are generally known, a lot of design computation can be conducted based on sequence comparison alone. For example,

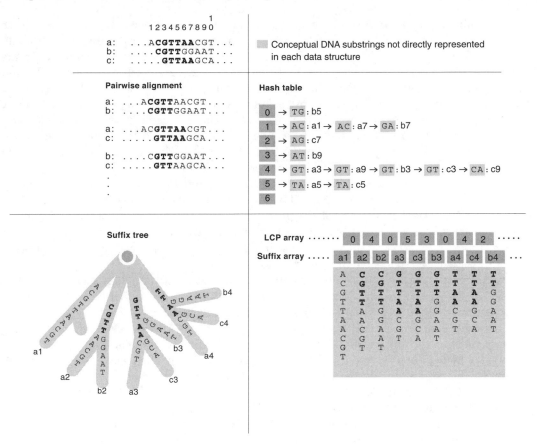

FIGURE 24.2: Popular indexing data structures for quick sequence similarity detection. Illustrations of hash table, suffix tree and suffix array are given for a sample gene set consisting of only three sequences a, b and c. Note that except for pairwise alignments, the DNA fragments shown in each illustration are not explicitly represented in the data structure. Instead, they are indirectly represented by the position information stored in the data structure. Nevertheless, they were explicitly drawn with a grey background to facilitate understanding of these data structures. See the main text for discussions of the indexing data structures.

to achieve probe specificity, a unique region in each gene must be chosen to be targeted by each microarray probe. Although in principle unique regions can be determined simply by running pairwise sequence alignments for all possible pairs of input gene sequences [52], in practice this is too slow. For large gene sets this naive method can take very long time to complete. More efficient indexing methods must be used to achieve a better performance. Three data structures are commonly used for quick sequence similarity identification: hash table, suffix tree and suffix array. See Figure 24.2 for simple illustrations of these data structures when applied to a sample gene set consisting of just three sequences a, b and c. In the following only a brief discussion is given for each indexing method. Interested readers are referred to the references for more detail descriptions.

A hash table [47] is constructed by breaking the input gene sequences into fragments of a certain size (usually 16 bp on 32-bit computers), then converting these fragments using a hashing function to integers no larger than the hash table size. The hash table size is usually a prime number to enhance the hashing effect. The hashed integers then serve as indices into the hash table to add or search for fragment matches. Each hash table entry usually points to a linked list which records sequence fragments that are hashed into the same entry, because different fragments can be hashed into the same hash table entry. To find unique regions in a gene sequence, the sequence is converted into consecutively overlapping fragments of the same size and those fragments were used to search against the hash table. If one or more matches are found in the hash table, the regions from where the matching fragments are derived are then known to be non-unique and should be avoided as the probe target sites. In the example given in Figure 24.2, the three sample sequences are broken down into fragments of 2 bp each, which are then mapped to a hash table of seven entries. To discover sequence similarity, each of the linked list pointed at by each hash table entry must be traversed, e.g., the fragment GT is found along entry 4 at position 3 on all three sample sequences and also at position 9 on sequence a. Note that in practice only fragment positions are recorded in the hash table, not the actual DNA fragments since that can be derived from the positions and the original input sequences. This holds true for the other two data structures suffix tree and suffix array discussed next.

The suffix tree is the fastest indexing data structure to construct and search. As demonstrated in Figure 24.2, each suffix of the input sequences is represented by a path leading from the root of the tree to an external leaf, where the position of the suffix is recorded. If common prefixes are shared between some suffixes, branch nodes are created to accommodate them at the differentiating points, thus to find all other sequences sharing common substrings longer than a certain length with a particular target sequence, one only has to locate all its suffixes on the tree, traverse down the tree with enough length to locate a branch, and then conduct a depth-first search to identify all other suffixes under the same branch. A more elaborate data structure called a suffix link (not shown in the figure) allows jumping from one suffix to the next suffix of the same sequence without having to descent from the root of the tree repetitively, thus dramatically speeding up the search. In the example given in Figure 24.2, only the first few suffixes of each sample sequence are shown on the tree and their common substrings are highlighted in boldface. It is easy to see that sequences a and b share a substring CGTT starting at position 2, for example.

Suffix array is very similar to suffix tree in concept but has a more space-saving data structure that is slightly slower than suffix tree [11, 19]. Suffix array records in alphabetical order all possible suffixes and their locations in the input sequences. The theory of suffix array states that the longest common prefix (LCP) shared by any two non-adjacent suffixes must be equal or shorter than the LCP of any two neighboring suffixes between them in the suffix array [33]. The LCP array can be efficiently computed once the suffix array is constructed. Thus, to determine if a particular gene sequence shares some substrings of certain lengths with the other sequences, we can locate all its suffixes in the suffix array and scan both the left and right sides from each of its suffixes. The other suffixes encountered during the scan indicate shared substrings and their locations in the other sequences can be immediately identified. Once found, regions containing the shared substrings are known to be non-unique and therefore should be avoided as probe target sites. In the example given in Figure 24.2, only the first few suffix positions of each sample sequence are sorted in the suffix array. Their mutual longest common prefix (LCP) are computed. Shared common prefixes of those suffixes are represented by bold type face. To discover if sequence c share any common substring with the other sequences, first we locate the prefix location c3 in the array using an invert suffix array not shown in this figure, then scan toward its right to find a 3 bp overlap with b3, and scan toward its left to find a 5 bp overlap with a3.

All three indexing data structures work efficiently in determining non-unique sequence regions. The hash table method is easier to understand and implement, but its fixed fragment size is less flexible than the suffix tree or suffix array. For example, a 15 bp exact match would not be found if the fragment size was set at 16 bp during hash table construction. Therefore, to ensure the detection of all important matches larger than a minimum fragment size, the hash table fragment size must be set at the minimum, which could result in more hashing collisions, longer linked lists, and less efficient searches. Although the hash table size can be increased to alleviate some of these problems, it is still comparably less flexible than the suffix tree or suffix array.

Although the suffix tree is the quickest data structure to build and search for sequence similarity, it requires a much larger memory footprint, usually in the order of 50 bytes per input DNA base. Therefore, it is not suitable for very large gene sets. For large gene sets suffix array is probably the most suitable data structure, since it requires only about 10 bytes to store each input DNA base, including the suffix array, the invert suffix array and a 2 byte per entry pre-computed LCP array (assuming no common subsequences are longer than 65,536 bp). Once all companion data structures of the suffix array are constructed, it is almost as efficient to search the suffix array as to search the suffix tree. The suffix array can be very flexible in detecting exact matches of arbitrary sizes, depending only on the LCP cutoff value during the left and right scanning from each sequence suffix. The only difficulty to use suffix array is that the algorithm to efficiently construct it is harder to understand and implement. Fortunately, pre-implemented suffix array construction algorithms are readily available on the Internet for use in an array design program [11, 23].

In addition to finding non-unique regions that should be totally avoided, the sequence level comparison can also be used to prioritize the remaining unique regions when they are being considered for probe targeting. Two types of information have been used in oligo design software for prioritizing target regions. The method based on the average longest match length found in each region is easier to compute and can be used directly as the priority score for a region. A more elaborated *average landscape* method can also be used [29, 30] which averages all match lengths in each region and uses that as the priority score for a region. Note that it is relatively easier to find the longest matches than to find all matches for a region. Also, for large genomes there is a greater chance of random short matches among sequences, so averaging short random matches does not help discriminating good and bad target regions. For example, a 3,460 maize gene set is analyzed in Figure 24.3, where the number of gene sequences that cannot have any probe-size region that is free of random short matches to the other sequences is drawn against the match size. When the minimum match size considered is 10 bp or above, none of the sequences can have any unique region, and therefore are all *bad*. Obviously, averaging short matches at this size does not help identifying good and bad sequence regions. If the minimum match size is increased, however, the number of bad sequences drops sharply, and then levels off. This is when randomness ceases, and what remain are significant longest matches that should really be avoided. Therefore, by averaging the longest matches found in each region they can be more correctly prioritized in our opinion.

24.2.2 Thermodynamic calculation

Although sequence level comparison allows non-unique regions to be quickly identified and avoided, it does not automatically suggest the best regions to be targeted by the probes. This conclusion can only be determined by thermodynamic calculations since the biochemical process that drives probe/target or probe/nontarget hybridizations is controlled by thermodynamic free energies, not by sequence similarity per se. Sequence comparison only provides hints that certain sequence regions must be examined for cross-hybridization possibility, i.e., for each candidate region of a sequence that are determined to be relatively

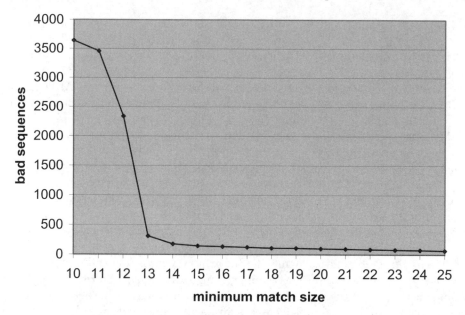

FIGURE 24.3: Random short match statistics. A 3,460 maize gene set is analyzed, where the number of sequences without any probe size region free of matches to the other sequences is drawn against the minimum size of the match considered. Many sequences are totally covered by random short matches and will be considered *bad* if the detectable match size is set too small, but only a few sequences actually have significantly longer matches that should really be avoided as probe target regions.

unique by sequence level comparison, all other sequence regions that are similar to it to some extent must be identified and their thermodynamic characteristics must be compared with the target region. Using the same indexing methods, however, these so called nontargets can all be quickly gathered and their melting temperatures with any oligo candidate targeting the candidate region can then be estimated.

There are several models for estimating the melting temperatures of DNA hybridization. The most popular model is the nearest neighbor (N-N) model [10, 45, 46, 2]. To estimate the melting temperature of a probe with its target, the following equation is used:

$$T_M = \frac{\Delta H}{\Delta S + R\ln(C/4)} + 12.0 \times \log_{10}\left[Na^+\right] - 273.15$$

H and S are the accumulated enthalpy and entropy values based on the sequence content of an oligo and its target region using N-N parameter tables [2, 8], R is the molar gas constant 1.987, C is the molar concentration of total oligonucleotides in the microarray experiment, and $[Na^+]$ is the molar concentration of salt. The oligonucleotide concentration is generally unknown, so a value of $1\times10^{-6}M$ is used by default as suggested in the literature [30, 20, 43]. Salt has a stabilizing effect on oligonucleotide annealing, so a salt concentration term is added to the equation to correct for that temperature shift with a coefficient of 12.0 as suggested by the literature and an oligo vendor [46, 16]. A somewhat higher value of 16.0 were used by others for that term [45, 37]. An example is given in Figure 24.4 to demonstrate how the melting temperature calculation works. Note that there are differences between hybridizations in solution and on a microarray surface because probe oligos are partially

initialization			H, A, T 2300	S 4.1		H, G, C 100	S -2.8

16 Exact Match Values			Left Dangling End			Right Dangling End	
	H	S		H	S		H	S
AA/TT	-7900	-22.2	_A/AT	-700	-0.8	T_/AA	200	2.3
AC/TG	-8400	-22.4	_C/AG	-2100	-3.9	G_/CA	-6300	-17.1
AG/TC	-7800	-21	_G/AC	-5900	-16.5	C_/GA	-3700	-10
AT/TA	-7200	-20.4	_T/AA	-500	-1.1	A_/TA	-2900	-7.6
CA/GT	-8500	-22.7	_A/CT	4400	14.9	T_/AC	600	3.3
CC/GG	-8000	-19.9	_C/CG	-200	-0.1	G_/CC	-4400	-12.6
CG/GC	-10600	-27.2	_G/CC	-2600	-7.4	C_/GC	-4000	-11.9
CT/GA	-7800	-21	_T/CA	4700	14.2	A_/TC	-4100	-13
GA/CT	-8200	-22.2	_A/GT	-1600	-3.6	T_/AG	-1100	-1.6
GC/CG	-9800	-24.4	_C/GG	-3900	-11.2	G_/CG	-5100	-14
GG/CC	-8000	-19.9	_G/GC	-3200	-10.4	C_/GG	-3900	-10.9
GT/CA	-8400	-22.4	_T/GA	-4100	-13.1	A_/TG	-4200	-15
TA/AT	-7200	-21.3	_A/TT	2900	10.4	T_/AT	-6900	-20
TC/AG	-8200	-22.2	_C/TG	-4400	-13.1	G_/CT	-4000	-10.9
TG/AC	-8500	-22.7	_G/TC	-5200	-15	C_/GT	-4900	-13.8
TT/AA	-7900	-22.2	_T/TA	-3800	-12.6	A_/TT	-200	-0.5

Sample Probe Sequence	AATCGTGTCA**T**
Sample Target Sequence	...GCTT**AT**TAGCACAGT**AG**TTAC...

↓

From 16 Exact Match Values Table										
	AA	AT	TC	CG	GT	TG	GT	TC	CA	AT	Sum
H	-7900	-7200	-8200	-10600	-8400	-8500	-8400	-8200	-8500	-7200	-83100
S	-22.2	-20.4	-22.2	-27.2	-22.4	-22.7	-22.4	-22.2	-22.7	-20.4	-225

↓

Add <u>initialization</u> and <u>Dangling End</u> values from the tables above:
Beginning **A**, 2300(H), 4.1(S)   Ending **T**, 2300(H), 4.1(S)
Left Dangling _A/AT, -700(H), -0.8(S) Right Dangling T_/AG, -1100(H), -1.6(S)
Finally, **H = -80300, S = -219.** *Using the N-N formula in the text:*

↓

$$T_m = \frac{-80300}{-219 + (-30.21)} + 12.0 \times \log_{10} 75 \times 10^{-3} - 273.15 = 35.57$$

FIGURE 24.4: Melting temperature calculation example. Probe and target melting temperature can be determined by looking up all the enthalpy and entropy values from the four N-N tables based on sequence content, then adding these values together for use in the equation given in the main text. For example, as calculated, the probe AATCGTGTCAT will hybridize with its target at roughly 35°C.

fixed to the array surface. However, all N-N parameters currently available are measured in solution where oligonucleotides are free to move around, therefore they are used only as the best approximation until surface based parameters become available.

The melting temperature between a probe and its target region can be estimated as illustrated above. In addition, its melting temperatures with all potential nontargets can also be estimated to prevent imperfectly matched cross-hybridizations. Note that perfectly matched nontargets for a candidate probe is impossible since such candidate probes should have already been screened out during the sequence level comparison step. Generally, it is much harder to estimate nontarget melting temperatures precisely given our current limited knowledge of mismatch hybridizations. Although there is only one *perfect* match to an oligo (i.e., its Watson-Crick complement), there are an enormous number of *imperfect* matches between an oligo and its nontargets. Fortunately, very precise nontarget melting temperatures are not necessary when we simply want to know if they can cause cross-hybridizations. Therefore, sequence level comparison can again be used as a guide. For example, any sequence level similarity over 75% may potentially cause problems and must be examined carefully (discussed next). Using one of the indexing methods, all potential nontarget regions can be uncovered and aligned with a probe candidate up to the 75% similarity level using dynamic programming methods. The alignments can then be used to estimate the probe/nontarget melting temperatures using the same equation above but with additional N-N parameters that accommodate simple mismatches, including single base mismatches (e.g. G·A, G·G or G·T) [2, 5, 4, 3, 39], and dangling end and gaps (i.e., bulge or loop) [58, 8]). The calculated melting temperatures of a candidate probe with all its nontargets are to be used to prioritize the probe in the global selection phase.

24.2.3 Additional design criteria

In addition to sequence comparison and thermodynamic calculation, additional criteria for selecting microarray probes must be followed. These criteria are related to the content and morphological shape of the probes. Using long oligo probe as an example, the following conditions are suggested to guarantee good oligo design [32, 21, 30, 40]:

1. Base composition: no single base should make up more than 50% of an oligo.
2. Base distribution: no stretch of a continuous base should exceed 25% of the length of an oligo.
3. GC content: the best is between 30–70%.
4. No secondary structures, i.e., oligo probes should not form dimers and hairpins or attempt to target sequence regions that may form dimers and hairpins under the experiment temperature.
5. Length of exact complementary match to nontargets should ideally be less than 15 bp.
6. Length of overall complementarity match to nontargets should ideally be less than 75%.

Although these conditions are listed separately, enforcing them are not necessarily individual computational steps. For example, condition **2** is implicitly enforced by the other conditions, i.e., a single base region longer than 25% of the oligo size is over 15 bp for a 60 bp oligo, so it cannot be targeted by an oligo by condition **5** if the reverse-complement of each sequence is also considered as a nontarget, which is usually recommended.

24.2.4 Global probe selection

Traditionally, a pipeline of screening is conducted during the probe design process, e.g., BLAST [6] or one of the indexing methods is used to select unique oligo candidates base on similarity level comparison, then MFold [57] or other similar tools are used to estimate thermodynamic properties [54, 37, 43, 51]. Probes are selected when they passed this batch-mode screening. In our opinion, this *batch-mode* design method may not produce the optimal probe sets. The most critical issue of this method is that the size of all oligos and the microarray experiment temperature must be given *a priori* as parameters to get the batch design pipeline started, but our research suggests that these parameters should instead be determined *by* a chosen probe set *after* it has been selected. In addition, similar to others' recent observations [37, 43], we also noticed that the best probe set should allow oligos of varying sizes, i.e., non-uniformity in oligo lengths can achieve greater uniformity in the melting temperature range of all oligos and therefore reduce the chance of cross-hybridization. To sum up, a global step to optimally select the best probe set and to determine the best experimental temperature is essential to achieve the best microarray experiment results.

Therefore, the last step in microarray design is to compare target and nontarget melting temperatures of all probe candidates for all sequences in order to discover an optimal subset that can detect each gene, has the least chance to cross-hybridize to nontargets, shares a uniform temperature range, and maximizes the distance between the lowest target and the highest nontarget melting temperature of the resulted set. Selection of the optimal subset of probe candidates resembles a non-integer knapsack problem [34], which is known to be NP-complete. Fortunately, we can limit ourselves to consider only the best, say, 5 non-overlapping probe candidates of each gene, and we can also use an iterative algorithm to approximate the optimal selection of the probe subset. This iterative algorithm goes as follows. First, the experiment temperature is set to the average mid-temperature among all target and nontarget melting temperatures of all probe candidates. Deviations from this temperature are then computed for each probe candidate. For each gene, their probe candidates up to a user desired number are selected into a new subset based on their deviations from the average mid-temperature. A new average temperature can then be determined from the new subset, and the iteration is repeated until it converges to an optimal subset and an optimal experiment temperature that no longer changes.

24.3 DNA Microarray Design

DNA libraries are usually the byproducts of whole genome or EST sequencing projects, and are readily available for microarray manufacturing. However, DNA microarrays suffer from high cross-hybridization noise due to their double strand probes and their inability to differentiate similar genes or gene families sharing long stretch of common subsequences (see Figure 24.1 on page 24-2 for an illustration). Since different DNA spots can have very different melting temperatures with their intended targets, it is impossible to obtain an optimal experiment temperature that can successfully separate all target hybridizations from all nontarget hybridizations. Also as mentioned earlier, tracking and maintaining DNA libraries can potentially introduce additional errors into the microarray experiments.

After genome assembly or EST clustering, a representative template can be selected for each gene from the library. The selection is usually based on the clone coverage length (the longer the better), and the position of the clone (the closer to the 3′ end of an mRNA, the better). The assembly or clustering overlap information for each gene region or contig is used as the selection input, and it is straightforward to select a representative template for each gene based on the criteria just stated. However, for genome assembly contigs a gene prediction step is usually necessary to target only the exons or 3′ end untranslated region (3′ UTR) of a gene.

Once a template has been selected for a gene, four different methods are possible to manufacture the actual microarray probe from the template. In order of increasing complexity and cost, they are listed below:

- Use the template as the probe, i.e., just spot it with the vector sequence unremoved (not commonly used);
- Use the standard primer pair of the cloning vector to amplify the clone insert from the template and use it as a probe;
- Use one custom primer of the insert and one standard primer of the cloning vector (or the poly-T primer if it is an EST clone) to amplify a specific region closer to the 3′ UTR of the clone insert and use that as a probe (it is hard to come by with a primer pair that works this way); or
- Use a custom primer pair of the insert to amplify a specific region of the clone insert from the template (this is more specific but can be expensive and may not compare favorably to synthesized oligo arrays).

Depending on the library type, some processing may not work. For shotgun library, introns may be contained within the clone insert, therefore only option 4 makes sense to manufacture the microarray probes after gene prediction. However, if custom primer pairs have to be designed in order to amplify probes from templates, it makes more sense to simply synthesize oligo probes and bypass the PCR process altogether, i.e., to use oligo microarrays instead of DNA arrays. On the other hand, EST clone libraries naturally have most of the introns removed from the clone insert, and for higher organisms the 3′ UTR is more unique and more suitable for microarray probe targeting. Therefore, creating cDNA arrays from EST libraries using standard primer pair (i.e. option 2) is economical and easy. In any case, the more specific the selected probes are, the better their specificity and hybridization quality will be, but the cost of custom primer pairs makes option 3 or 4 above less favorable when compared to oligo microarrays, especially if the complete genome sequence is already known.

Since each clone template is amplified separately, it is relatively easy to design custom primers once each template is selected. This job primarily involves thermodynamic consideration only, i.e., within certain desired length or region, a pair of primers must be chosen that are within the same melting temperature range and satisfy the other primer design criteria similar to the oligo design criteria mentioned earlier, e.g., 18–30 bp length, 40–60% GC content, no secondary structure at the melting temperature, and unique to the priming target. The aforementioned basic design strategies can be applied, and a primer design tool like Primer3 [44] is readily available for this task.

Under special situations, custom primer pairs may have to be designed against a whole-genome background in order to amplify specific regions directly from the genome. These situations are generally not related to microarrays. A few examples of why such whole-genome primer design may be necessary are the following:

- For gene rich genome sequencing projects, primer pairs are designed using known assemblies in order to 1) confirm assembly correctness, 2) discover whole region polymorphism, or 3) close physical gaps;
- After a genome sequencing project is finished, using custom whole-genome primer pairs to clone and study genes from closely related species or strands for single nucleotide polymorphism, genotyping or other genetic studies; or
- Simply to uniquely amplify a specific region from a complex genome for various other reasons.

Currently, there are no integrated whole-genome primer pair design software that can directly design the optimal primer pairs based on a complete gene set. Therefore, batch-mode processing must be conducted. In the case of PCR, nontarget mis-priming can happen only when *both* primers in a pair can strongly cross-hybridize to nontargets within about the same distance as the intended target region, be in the proper directions, and be in the same melting temperature range, therefore it is relatively more efficient to design candidate primer pairs first using standard primer design tools like Primer3, then just screen those candidates against the whole genome background using sequence level comparison and thermodynamic calculation.

24.4 Long Oligo Microarray Design

Long oligo microarray design is of high interest and very challenging due to its enormous solution space, e.g., oligos ranging from 50 to 70 bp can all be used, and they all must be unique to their intended targets under a variety of user design parameters and input data characteristics. An oligo design program must be flexible enough to accommodate all reasonable parameters and make minimal assumptions about the input data characteristics, yet still produces computationally optimal probe set for microarray use. As mentioned in Section 24.2.4, traditionally a batch-mode design method has been used for oligo microarray design, but batch-mode design may not find the optimal oligo set because the size of the oligos and the experiment temperature must be given *a priori* as parameters to get a batch design pipeline started. However, these parameters should best be determined *by* a chosen oligo set *after* it has been designed. In this section, we will base our discussion of long oligo microarray design on the newly available tool PICKY, whose integrated design approach allows these two important parameters to be determined by the input data while it selects an optimal probe set [14]. In addition, PICKY is currently the fastest oligo microarray design program, requiring only a few hours to process large gene sets from rice, maize or human.

A comparison of the traditional batch-mode oligo design strategy and the PICKY design strategy is given in Figure 24.5. The most noticeable differences are that in PICKY there are two global steps instead of just one, and the added global step is an iterative step that optimizes the oligo set and determines the best microarray experiment temperature. In a batch-mode design, a global indexing data structure is created first to guide the probe candidate selection and nontarget detection steps. Then, each *fixed-size* probe candidate is aligned with potential nontargets, its melting temperature is estimated, and it is selected based on the deviation of its melting temperature from that of a user preset temperature. In PICKY design, a suffix array is similarly constructed as the first global step to guide the three subsequent screening steps. For each sequence, first its bad regions are screened out, and then its good candidate regions are selected. Both steps use only sequence level comparison and therefore are very fast. The suffix array is also guiding the next step to detect all nontargets that have to be aligned with each candidate region of a sequence in order to obtain a *temperature landscape* of each candidate region. The temperature landscapes will be discussed later in this chapter. Once all processing steps are completed for a sequence, its best *variable* length candidate probes are put into a pool. After all sequences have been processed, the iterative global step is conducted to select the actual probes that will form an optimal oligo set for the microarray. The experiment temperature is finally determined from the chosen optimal set. In the following, the PICKY design steps will be discussed in more detail.

Batch mode strategy

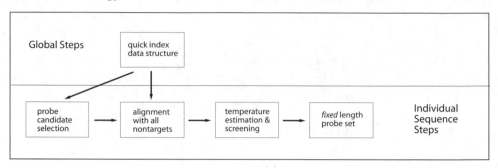

Picky strategy

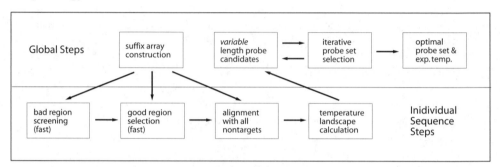

FIGURE 24.5: Comparison of batch-mode and PICKY oligo design strategies. A batch-mode design generally has only one global step to construct an indexing data structure, which is then used to guide the probe candidate selection and the detection of potential nontargets. Fixed length oligos are then selected based on their deviation from a user given microarray experiment temperature. In PICKY design, a suffix array is constructed to similarly guide the subsequent sequence level comparison steps, but there is an additional global step at the end that selects an optimal variable length probe set and actually determines the best microarray experiment temperature.

24.4.1   Picky design parameters

All of the general oligo design criteria previously mentioned in Section 24.2.3 on page 24-10 are considered by PICKY, and most are user adjustable parameters that can be modified depending on the characteristics of the input gene set. In addition to these, PICKY also considers more sophisticated design parameters. PICKY accepts the minimum and maximum oligo lengths instead of a fixed length, and within the specified range it can adjust the length of oligos to achieve greater specificity and uniformity among all oligos. Also, rather than requesting the melting temperature as an input parameter from users, PICKY takes the minimum separation temperature as a parameter, and ranks best oligo candidates by comparing their target and nontarget melting temperature differences. Only oligos that provide at least the minimum separation temperature will be considered in PICKY, and it is their joint temperature ranges that finally determine the optimal temperature suggested by PICKY for microarray experiments. PICKY also handles multiple target and nontarget gene sets, where the nontarget sets are used as a screening background while oligos are being designed for the target sets. This allows, for example, a small budget experiment to study a handful of genes of a large genome and still guarantees the results will be as good as those obtained from a whole genome array.

Among the suggested oligo design criteria for selecting the best oligos on page 24-10, we have found conditions **5** and **6** recommended in Kane *et al.*'s paper [21] to be the most often cited conditions [30, 42, 54]. We call these Kane's first and second conditions. Since these two conditions are only based on sequence similarity, they are not sufficient to determine *good* oligo candidates without in-depth thermodynamic calculation. Nevertheless, they provide the most efficient way to screen out *bad* oligo candidates. The reason is as follows. Although mismatches to nontargets do not ascertain good oligo candidates without thermodynamic calculations, exact matches to nontargets do indeed identify oligos that should be avoided to prevent cross-hybridizations. It will become clear in the following that Kane's two conditions efficiently drive the two initial PICKY steps for screening each sequence.

24.4.2 Suffix array construction and search

The first step in PICKY's oligo computation is to construct a generalized suffix array that can quickly identify all substrings contained in all sequences and their reverse complements. The companion LCP and invert arrays are then computed from the suffix array. With these three data structures, PICKY can quickly determine if a particular gene sequence shares some common substrings of certain lengths with the other sequences. This is conducted by locating all of its suffixes in the suffix array and scanning both the left and right sides of each of its suffixes (cf. Figure 24.2 and the discussion on page 24-6 for details). Such non-unique regions violate Kane's first condition and therefore must be avoided as probe target sites. For large gene sets, due to evolutionary duplications, many gene regions are non-unique and should not be targeted by any oligo probe. PICKY's initial scanning step based on Kane's first condition can quickly identify these *bad* regions and completely avoid any further computation with them. This provides a dramatic speed boost when compared to a batch-mode design method because PICKY can skip all subsequent time-consuming local alignment and thermodynamic calculation steps for many large bad regions without losing any chance of finding the optimal probes. Since a sequence and its own reverse complement are both represented in the suffix array, PICKY scanning steps also detect repetitive, low complexity, self-similar and self-complementary regions. All of these are avoided as probe target regions as well.

The Burkhardt-Kärkkäinen algorithm [11] is used in PICKY after some modification to efficiently construct the generalized suffix array. Although there are several other efficient suffix array construction algorithms, including three linear-time complexity ones [24, 26, 22], we have found through experiments that the Burkhardt-Kärkkäinen algorithm is the quickest and the most memory efficient one in *practice*. The inverse suffix array is linearly constructed from the suffix array, which indexes suffixes in the suffix array from the perspective of each sequence. To avoid string comparison and to speed up suffix array scanning in later steps, the longest common prefix (LCP) array that records the length of the shared prefix of each neighboring suffix pair is also pre-computed using the Kasai *et al.* linear-time algorithm [23]. Altogether, PICKY requires 20 bytes to represent each DNA base in those three arrays. If double-strand screening is turned off, its requirement drops to only 10 bytes per input base.

24.4.3   Local alignment and melting temperature estimation

The next step in PICKY's computation is to find, among the remaining regions of a sequence, the best candidates to be targeted by oligo probes. For each best candidate region, all other sequences that are similar to it to the level of Kane's second condition must be discovered and their thermodynamic properties must be compared to prevent cross-hybridizations. Again using the suffix array, these potential nontargets can be quickly gathered using a smaller exact match size than when checking Kane's first condition. Since suffix array finds only exact matches, rigorous probability models have been established to determine the minimum exact match size that can still guarantee finding a certain similarity level. For example, an exact match of 10 bp guarantees that any 60 bp regions in the data set with a similarity of 90% or higher will be detected using the suffix array, no matter how the mismatch bases are distributed among their alignment [14].

The melting temperatures of all nontargets with any potential oligo targeting the candidate region can then be estimated after an alignment was made with the candidate region using the detected exact matches as alignment *seeds*. This is going to be a slow process using traditional quadratic time dynamic programming algorithm. Therefore, PICKY uses a novel linear time *local* alignment algorithm, which works as follows. Instead of constructing the alignment matrix in a row-by-row or column-by-column fashion as in traditional methods, PICKY interleaves the row and column construction steps. See Figure 24.6 for an illustration. Boundaries are automatically set when the accumulated negative alignment score in a cell falls below a value that prevents any possibility of an *local* alignment having the similarity level set by Kane's second condition to be achievable beyond that cell boundary. For example, for a 75% similarity threshold, for every three base matches there can only be one mismatch, thus if the accumulated mismatches go beyond that threshold, it is impossible to obtain a local alignment that maintains 75% or above similarity. Construction of the alignment matrix in this fashion is therefore bound by a banded region narrowly centered around the starting alignment seed, and it will stop immediately when the upper and lower boundaries converge. Note that the covered region in this alignment matrix does not have to be totally square, i.e., either the row or column construction can be stopped first when the boundary in that direction has been reached.

A very unique strategy in PICKY is that it does not enumerate all possible oligos targeting a candidate region and individually compare them with all potential nontargets as in a batch-mode method. Instead, all nontargets discovered by the suffix array are aligned and compared to the target region just once, and two *temperature landscapes* are derived during this process. One is computed from the target region itself and includes all valid oligos within the allowable length range that are targeting different locations of the region. For example, in Figure 24.7a, the target temperature landscape of a typical candidate region is shown. Oligos targeting between base location 773 to 862 and with variable length 50 to 70 bp can hybridize with this region at different melting temperatures. Only locations targeted by valid oligos have a nonzero temperature, hence temperatures for longer oligos start dropping off beyond location 792 (862-70mer=792) and the last location in this region that can be targeted is 812 (862-50mer=812).

The nontarget temperature landscape for the same region (Figure 24.7b) is computed by aligning each potential nontarget with the target region and estimating its melting temperatures with all oligos targeting the same region. Since the goal is to avoid cross-hybridizations, only the highest nontarget temperature discovered at each location is recorded in the nontarget temperature landscape. Hence, this landscape is relatively flat because the highest temperature found in a location is usually shared by all oligos overlapping the same location. In batch-mode style processing a program might have selected oligos with the highest

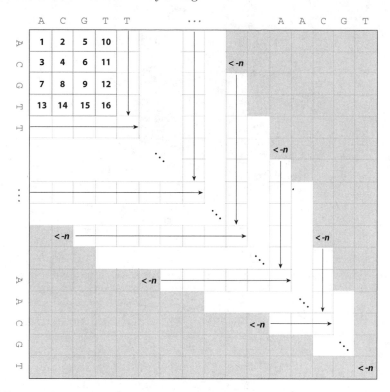

FIGURE 24.6: Linear-time local alignment algorithm. To determine a local alignment in linear-time, one can interleave the row and column construction steps of an alignment matrix in the order indicated by the numbers. The starting boundary of the construction of each row or column is limited by the accumulated negative alignment score boundary, such that when the score is below a certain threshold value $-n$, one can skip those cells above (in column construction) or to the left (in row construction) during the construction process. This maintains a narrow alignment white *band* that grows in linear-time. The alignment stops when the row and column boundaries converge.

target temperatures, e.g., the 70mer oligo targeting location 778 indicated by the red circle in Figure 24.7a. In PICKY's integrated approach, both the target and nontarget landscapes are considered together, and it is their *difference* that finally determines the oligo selection. The temperature difference between the two landscapes is shown in Figure 24.7c. Therefore, the best oligo for this particular region of the sequence should actually be the 64mer targeting location 796 as indicated by the red circle in Figure 24.7c because a greater difference in melting temperature translates to both a greater specificity and a lower background hybridization. The oligo for a gene is later selected among the best non-overlapping oligos found in each candidate region of the gene by further comparing their temperatures. This example also highlights the benefit of varying oligo lengths to achieve a higher difference between target and nontargets melting temperatures as seen in those three figures.

Although the highest temperature in the difference landscape suggests an oligo, it may still be rejected if its value is lower than the minimum temperature separation, i.e., the minimum height of the difference temperature landscape. PICKY's default setting is to ignore oligos whose target and nontarget melting temperature difference is lower than 20°C.

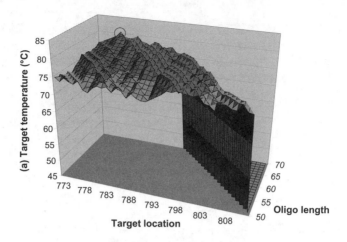

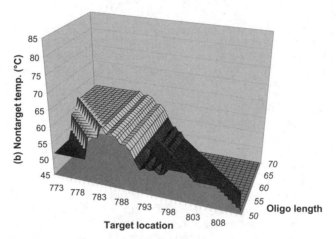

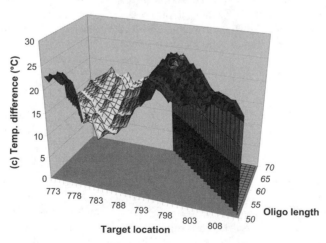

FIGURE 24.7: (See color insert following page 20-4.) Temperature landscapes and the selection of oligo target regions. The (a) target, (b) nontarget and (c) difference melting temperature landscapes are drawn against different sequence locations and oligo lengths. Red circles indicate oligos that might be chosen under different design strategies.

This parameter can be adjusted by the users, and based on our experience 15–20°C seems to be the most suitable range for this parameter. Once an oligo is selected, the other potential oligos overlapping it will not be considered further. Non-overlapping oligos will continue to be selected while they still satisfy the minimum temperature separation. This selection process is independent of the order of the input sequences since genes are considered individually at this stage.

24.4.4 Optimal probe set and experiment temperature determination

The last step is to compare target and nontarget melting temperatures of all probe candidates from all sequences in order to find a subset that can detect each gene, has the least chance to cross-hybridize, shares a uniform temperature range, and maximizes the distance between the lowest target and the highest nontarget melting temperatures of the chosen set. PICKY reports the optimal microarray experiment temperature *after* the probe set has been determined. This differs significantly from most oligo design tools that take the temperature as *input* to screen against probe candidates. PICKY uses an iterative selection algorithm for the optimal subset of probe candidates, which has been outlined in Section 24.2.4 on page 24-10. The oligos designed are then presented in a GUI panel where they can be examined and subsequently saved to files.

It may seem that secondary structure screening is omitted from PICKY steps. That is not the case. Because all sequences and their reverse-complements are represented in the suffix array and are included in cross-hybridization screening, self-similarity tests for secondary structures are conducted alongside mutual similarity tests for cross-hybridizations. Thus, no additional secondary structure screening step is necessary in PICKY. Furthermore, PICKY's inclusion of the reverse-complement of each input sequence ensures that oligos designed will function correctly even in the presence of anti-sense transcripts [28, 48, 12, 38, 55].

References

[1] M.D. Adams, S.E. Celniker, R.A. Holt, and C.A. Evans *et al.* The genome sequence of drosophila melanogaster. *Science*, 287(5461):2185–95, 2000. 0036-8075 Journal Article.

[2] H.T. Allawi and J. SantaLucia Jr. Thermodynamics and NMR of internal G*T mismatches in DNA. *Biochemistry*, 36:10581–10594, 1997.

[3] H.T. Allawi and J. SantaLucia Jr. Nearest neighbor thermodynamic parameters for internal G*A mismatches in DNA. *Biochemistry*, 37:2170–2179, 1998.

[4] H.T. Allawi and J. SantaLucia Jr. Nearest-neighbor thermodynamics of internal A*C mismatches in DNA: Sequence dependence and ph effects. *Biochemistry*, 37:9435–9444, 1998.

[5] H.T. Allawi and J. SantaLucia Jr. Thermodynamics of internal C*T mismatches in DNA. *Nucleic Acids Research*, 26(11):2694–2701, 1998.

[6] S.F. Altschul, W. Gish, W. Miller, and E.W. Myers *et al.* Basic local alignment search tool. *Journal of Molecular Biology*, 215:403–410, 1990.

[7] The Arabidopsis Genome Initiative. Analysis of the genome sequence of the flowering plant *arabidopsis thaliana*. *Nature*, 408:796–815, 2000.

[8] S. Bommarito, N. Peyret, and J. SantaLucia Jr. Thermodynamic parameters for DNA sequences with dangling ends. *Nucleic Acids Research*, 28(9):1929 –1934, 2000.

[9] J.T. Bosch, C. Seidel, S. Batra, and H. Lam *et al.* Validation of sequence-optimized

70 base oligonucleotides for use on DNA microarrays. Technical report, Westburg Company, 2002 2002.

[10] K.J. Breslauer, M. Frank, H. Blocker, and L.A. Marky. Predicting DNA duplex stability from the base sequence. *Biochemistry*, 83:3746–3750, 1986.

[11] S. Burkhardt and J. Krkkinen. Fast lightweight suffix array construction and checking. In R.A. Baeza-Yates, E. Chvez, and M. Crochemore, editors, *14th Annual Symposium, CPM 2003*, volume 2676 of *Lecture Notes in Computer Science*, pages 55–69. Springer, 2003.

[12] G.G. Carmichael. Antisense starts making more sense. *Nat Biotechnol*, 21(4):371–2, 2003.

[13] V.L. Chandler and V. Brendel. The maize genome sequencing project. *Plant Physiol*, 130(4):1594–7, 2002.

[14] H.H. Chou, A.P. Hsia, D.L. Mooney, and P.S. Schnable. Picky: oligo microarray design for large genomes. *Bioinformatics*, 20(17):2893–2902, 2004.

[15] J.L. DeRisi, V.R. Iyer, and P.O. Brown. Exploring the metabolic and genetic control of gene expression on a genomic scale. *Science*, 278(5338):680–6, 1997.

[16] E.J. Devor, L. Huang, and R. Owczarzy. IDT technical bulletins $\hat{A}$ calculation of Tm for oligonucleotides. Technical report, Integrated DNA Technologies, 2002 2002.

[17] S.A. Goff, D. Ricke, T.H. Lan, and G. Presting *et al.* A draft sequence of the rice genome (*oryza sativa l. ssp. japonica*). *Science*, 296(5565):92–100, 2002.

[18] T.R. Golub, D.K. Slonim, P. Tamayo, and C. Huard *et al.* Molecular classification of cancer: Class discovery and class prediction by gene expression monitoring. *Science*, 286(5436):531–537, 1999.

[19] D. Gusfield. *Algorithms on Strings, Trees and Sequences.* Cambridge University Press, Cambridge, United Kingdom, 1997.

[20] L. Kaderali and A. Schliep. Selecting signature oligonucleotides to identify organisms using DNA arrays. *Bioinformatics*, 18(10):1340–1349, 2002.

[21] M.D. Kane, T.A. Jatkoe, Craig R. Stumpf, and J. Lu *et al.* Assessment of the sensitivity and specificity of oligonucleotide (50mer) microarrays. *Nucleic Acids Research*, 28(22):4552–4557, 2000.

[22] J. Karkkainen and P. Sanders. Simple linear work suffix array construction. *Automata, Languages and Programming, Proceedings*, 2719:943–955, 2003. Times Cited: 2 Lecture Notes in Computer Science Article English Cited References Count: 39 Bx39s.

[23] T. Kasai, G. Lee, H. Arimura, and S. Arikawa *et al.* Linear-time longest-common-prefix computation in suffix arrays and its applications. In A. Amir and G.M. Landau, editors, *Combinatorial Pattern Matching, 12th Annual Symposium, Jerusalem, Israel*, Lecture Notes in Computer Science. Springer Verlag, Berlin, 2001.

[24] D.K. Kim, J.S. Sim, H. Park, and K. Park. Linear-time construction of suffix arrays. In R.A. Baeza-Yates, E. Chvez, and M. Crochemore, editors, *14th Annual Symposium, CPM 2003*, volume 2676 of *Lecture Notes in Computer Science*, pages 186–199. Springer, 2003.

[25] E.F. Kirkness, V. Bafna, A.L. Halpern, and S. Levy *et al.* The dog genome: survey sequencing and comparative analysis. *Science*, 301(5641):1898–903, 2003.

[26] P. Ko and S. Aluru. Space efficient linear time construction of suffix arrays. In Ricardo A. Baeza-Yates, Edgar Chvez, and Maxime Crochemore, editors, *14th Annual Symposium, CPM 2003*, volume 2676 of *Lecture Notes in Computer Science*, pages 200–210. Springer, 2003.

[27] E.S. Lander, L.M. Linton, B. Birren, and C. Nusbaum *et al.* Initial sequencing and analysis of the human genome. *Nature*, 409(6822):860–921, 2001. 0028-0836 Journal Article.

[28] B. Lehner, G. Williams, R.D. Campbell, and C.M. Sanderson. Antisense transcripts in the human genome. *Trends Genet*, 18(2):63–5, 2002.

[29] S. Levy, L. Compagnoni, E.W. Myers, and G.D. Stormo. Xlandscapes: The graphical display of word frequencies in sequences. *Bioinformatics*, 14(1):74–80, 1998.

[30] F. Li and G.D. Stormo. Selection of optimal DNA oligos for gene expression arrays. *Bioinformatics*, 17(11):1067–1076, 2001.

[31] X. Lin, S. Kaul, S. Rounsley, and T.P. Shea *et al.* Sequence and analysis of chromosome 2 of the plant arabidopsis thaliana. *Nature*, 402:761–768, 1999.

[32] D.J. Lockhart, H. Dong, M.C. Byrne, and M.T. Follettie *et al.* Expression monitoring by hybridization to high-density oligonucleotide arrays. *Nature Biotechnology*, 14(12):1675–1680, 1996.

[33] U. Manber and E.W. Myers. Suffix arrays: A new method for on-line string searches. *SIAM Journal on Computing*, 22(5):935–948, 1993.

[34] S. Martello and P. Toth. *Knapsack Problems: Algorithms and Computer Implementations.* John Wiley & Sons, New York, NY, 1990.

[35] K. Mayer. Sequence and analysis of chromosome 4 of the plant *arabidopsis thaliana*. *Nature*, 402:769–777, 1999.

[36] R. Mei, E. Hubbell, S. Bekiranov, and M. Mittmann *et al.* Probe selection for high-density oligonucleotide arrays. *Proc Natl Acad Sci U S A*, 100(20):11237–42, 2003.

[37] H. B. Nielsen, R. Wernersson, and S. Knudsen. Design of oligonucleotides for microarrays and perspectives for design of multi-transcriptome arrays. *Nucleic Acids Res*, 31(13):3491–6, 2003.

[38] N. Osato, H. Yamada, K. Satoh, and H. Ooka *et al.* Antisense transcripts with rice full-length cDNAs. *Genome Biol*, 5(1):R5, 2003.

[39] N. Peyret, P.A. Seneviratne, H.T. Allawi, and J. SantaLucia Jr. Nearest-neighbor thermodynamics and NMR of DNA sequences with internal A*A, C*C, G*G, and T*T mismatches. *Biochemistry*, 38:3468–3477, 1999.

[40] A. Relógio, C. Schwager, A. Richter, and W. Ansorge *et al.* Optimization of oligonucleotide-based DNA microarrays. *Nucleic Acids Research*, 30(11):E51, 2002.

[41] C. Ren, M.K. Lee, B. Yan, and K. Ding *et al.* A BAC-based physical map of the chicken genome. *Genome Res*, 13(12):2754–8, 2003.

[42] J.-M. Rouillard, C.J. Herbert, and M. Zuker. Oligoarray: genome-scale oligonucleotide design for microarrays. *Bioinformatics*, 18(3):486–487, 2002.

[43] J.M. Rouillard, M. Zuker, and E. Gulari. Oligoarray 2.0: design of oligonucleotide probes for DNA microarrays using a thermodynamic approach. *Nucleic Acids Res*, 31(12):3057–62, 2003.

[44] S. Rozen and H.J. Skaletsky. Primer3, 1998 1997.

[45] W. Rychlik, W.J. Spencer, and R.E. Rhoads. Optimization of the annealing temperature for DNA amplification in vitro. *Nucleic Acids Research*, 18(21):6409–6412, 1990.

[46] J. SantaLucia Jr., H.T. Allawi, and P.A. Seneviratne. Improved nearest-neighbor parameters for predicting DNA duplex stability. *Biochemistry*, 35:3555–3562, 1996.

[47] R. Sedgewick. *Algorithms in C++ part 1-4: fundamentals, data structures, sorting, searching.* Addison-Wesley, Reading, MA, 1998.

[48] J. Shendure and G.M. Church. Computational discovery of sense-antisense transcription in the human and mouse genomes. *Genome Biol*, 3(9):RESEARCH0044, 2002.

[49] T.J. Summers, J.W. Thomas, S.Q. Lee-Lin, and V.V. Maduro *et al.* Comparative physical mapping of targeted regions of the rat genome. *Mamm Genome*, 12(7):508–12, 2001.

[50] J.C. Venter, M.D. Adams, E.W. Myers, and P.W. Li *et al.* The sequence of the human

genome. *Science*, 291(5507):1304–51, 2001. 0036-8075 Journal Article.

[51] X. Wang and B. Seed. Selection of oligonucleotide probes for protein coding sequences. *Bioinformatics*, 19(7):796–802, 2003.

[52] M.S. Waterman. *Introduction to computational biology: Maps, sequences and genomes*. Chapman & Hall, Florence, KY, 1995.

[53] R.H. Waterston, K. Lindblad-Toh, E. Birney, and J. Rogers *et al.* Initial sequencing and comparative analysis of the mouse genome. *Nature*, 420(6915):520–62, 2002.

[54] D. Xu, G. Li, L. Wu, and J. Zhou *et al.* PRIMEGENS: robust and efficient design of gene-specific probes for microarray analysis. *Bioinformatics*, 18(11):1432–1437, 2002.

[55] R. Yelin, D. Dahary, R. Sorek, and E.Y. Levanon *et al.* Widespread occurrence of antisense transcription in the human genome. *Nat Biotechnol*, 21(4):379–86, 2003.

[56] J. Yu, S. Hu, J. Wang, and G.K. Wong *et al.* A draft sequence of the rice genome (*oryza sativa l. ssp. indica*). *Science*, 296(5565):79–92, 2002.

[57] M. Zuker. Mfold web server for nucleic acid folding and hybridization prediction. *Nucleic Acids Res*, 31(13):3406–15, 2003.

[58] M. Zuker, D.H. Mathews, and D.H. Turner. Algorithms and thermodynamics for RNA secondary structure predictions: A practical guide. In J. Barciszewski and B.F.C. Clark, editors, *RNA Biochemistry and Biotechnology*. Kluwer Academic, Dordrecht, 1999.

[59] F. Zuo, N. Kaminski, E. Eugui, and J. Allard *et al.* Gene expression analysis reveals matrilysin as a key regulator of pulmonary fibrosis in mice and humans. *Proceedings of the National Academic of Sciences*, 99(9):6292–6297, 2002.

25

Clustering Algorithms for Gene Expression Analysis

Pierre Baldi
University of California, Irvine

G. Wesley Hatfield
University of California, Irvine

Li M. Fu
University of Florida, Gainesville

25.1 Problems and Approaches

Differential expression is a useful tool for the analysis of DNA microarray data. However, and in spite of the fact that it can be applied to a large number of genes, differential analysis remains within the confines of the old one-gene-at-a-time paradigm. Knowing that a gene's behavior has changed between two situations is at best a first step. In a cancer experiment, for instance, a significant change could be associated with a direct causal link (activation of an oncogene), a more indirect chain of effects (signalling pathway), a non-specific related phenomena (cell division), or even a spurious event completely unrelated to cancer ("noise").

Most, if not all, genes act in concert with other genes. What DNA microarrays are really after is the *patterns* of expression across multiple genes and experiments. And to detect such patterns, additional methods such as clustering must be introduced. In fact, in the limit, differential analysis can be viewed as a clustering method with only two clusters: change and no-change. Thus, at the next level of data analysis, we want to remove the simplifying assumption that genes are independent and look at their covariance, at whether there exist multi-gene patterns, clusters of genes that share the same behavior, and so forth. While array data sets and formats remain heterogeneous, a key challenge in time is going to be the development of methods that can extract order across experiments, in typical data sets of size $30,000 \times 1,000$ and model, for instance, the statistical distribution of a gene's expression levels over the space of possible conditions. Not surprisingly, conceptually these problems are not completely unlike those encountered in population genetics, such as detecting combinations of SNPs (single nucleotide polymorphisms) associated with complex diseases.

The cluster's boundaries can be very noisy, especially in isolated experiments with low repetition. The key observation, however, is that even in the presence of low-repetition, i.e. highly noisy measurements at the level of individual experiments, complex expression patterns can still be detected robustly across multiple experiments and conditions. Consider, for instance, a cluster of genes directly involved in the cell division cycle and whose expression pattern oscillates during the cycle. For each individual measurement at a given time t, noise alone can introduce distortions so that a gene which belongs to the cluster may fall out of the cluster. However, when the measurements at other times are also considered, the cluster becomes robust and it becomes unlikely for a gene to fall out of the cluster it belongs to at most time steps. The same can be said of course for genes involved in a particular form of cancer across multiple patients, and so forth. In fact, it may be argued that robustness is a fundamental characteristic of regulatory circuits that must somehow transpire even through noisy microarray data.

In many cases, cells tend to produce the proteins they need simultaneously, and only when they need them. The genes for the enzymes that catalyze a set of reactions along a pathway are likely to be co-regulated (and often somewhat co-located along the chromosome). Thus, depending on the data and clustering methods, gene clusters can often be associated with particular pathways and with co-regulation. Even partial understanding of the available information can provide valuable clues. Co-expression of novel genes may provide a simple means of gaining leads to the functions of many genes for which information is not yet available. Likewise, multi-gene expression patterns could characterize diseases and lead to new precise diagnostic tools capable of discriminating, for instance, different kinds of cancers.

Many data analysis techniques have already been applied to problems in this class, including various clustering methods from k-means to hierarchical clustering, principal component analysis, factor analysis, independent component analysis, and self-organizing maps, to name just a few. It is impossible to review all the methods of analysis in detail in the available space and counterproductive to try to single out a "best method" because: (1) each method may have different advantages depending on the specific task and specific properties of the data set being analyzed; (2) the underlying technology is still rapidly evolving; and (3) noise levels do not always allow for a fine discrimination between methods. Rather, we focus on the main methods of analysis and the underlying mathematical background.

Array data is inherently high dimensional, hence methods that try to reduce the dimensionality of the data and/or lend themselves to some form of visualization remain particularly useful. These range from simple plots of one condition versus another, to projection onto lower dimensional spaces, to hierarchical and other forms of clustering. In the next sections, we focus on dimensionality reduction (principal component analysis) and clustering since these are two of the most important and widely used method of analysis for array data. For clustering, we partially follow the treatment in [6]. Clustering methods of course can be applied not only to genes, but also to conditions, DNA sequences, and other relevant data. From array-derived gene clusters it is also possible to go back to the corresponding gene sequences and look, for instance, for shared motifs in the regulatory regions of co-regulated genes.

25.2 Visualization, Dimensionality Reduction, and Principal Component Analysis

The simplest approach to visually explore current array data is perhaps a 2D plot of the activity levels of the genes in one experimental condition versus another. When each experiment is repeated several times, the average values can be used. In such plots, typically most genes are found along the main diagonal (assuming similar calibration between experiments) while differentially expressed genes appear as outliers.

A second more sophisticated approach for dimensionality reduction and visualization is principal component analysis. Principal component analysis (PCA) is a widely used statistical data analysis technique that can be viewed as: (1) a method to discover or reduce the dimensionality of the data; (2) a method to identify new meaningful underlying variables; (3) a method to compress the data; and (4) a method to visualize the data. It is also called the Karhunen-Loéve transform, Hotelling transform, or singular value decomposition (SVD) and provides an optimal linear dimension reduction technique in the mean-square sense.

Consider a set of N points $x_1, \ldots, x_N$ in a space of dimension M. In the case of array data, the points could correspond to genes if the axis are associated with different experiments or to experiments if the axis are associated with the probes/genes. We assume that the x's have already been centered by substracting their mean value, or expectation, $E[x]$. The basic idea in PCA is to reduce the dimension of the data by projecting the x's onto an interesting linear subspace of dimension K, where K is typically significantly smaller than M. The interesting subspace is defined in terms of variance maximization.

PCA can easily be understood by recursively computing the orthonormal axis of the projection space. For the first axis, we are looking for a unit vector u_1 such that, on average, the squared length of the projection of the x's along u_1 is maximal. Assuming that all vectors are column vectors, this can be written as

$$u_1 = \arg \max_{||u||=1} E[(u^T x)^2] \tag{25.1}$$

where u^T denotes transposition, and E is the expected or average value (Figure 25.1). Any vector is always the sum of its orthogonal projection onto a given subspace and the orthogonal complement, so that here $x = (u^T x)u + (x - (u^T x)u)$. The second component maximizes the residual variance associated with $(x - (u^T x)u)$ and so forth. More generally, if the first $u_1, \ldots, u_{k-1}$ component have been determined, the next component is the one that maximizes the residual variance in the form

$$u_k = \arg \max_{||u||=1} E[(u^T (x - \sum_{i=1}^{k-1} (u_i^T x)u_i))^2] \tag{25.2}$$

The principal components of the vector x are given by $c_i = u_i^T x$. By construction, the vectors u_i are orthonormal. In practice, it can be shown the u_i's are the eigenvectors of the (sample) covariance matrix $\Sigma = E[xx^T]$ associated with the K largest eigenvalues and satisfy

$$\Sigma u_k = \lambda_k u_k \tag{25.3}$$

In array experiments, these give rise to "eigengenes" and "eigenarrays" [3]. Each eigenvalue λ_k provides a measure of the proportion of the variance explained by the corresponding eigenvector.

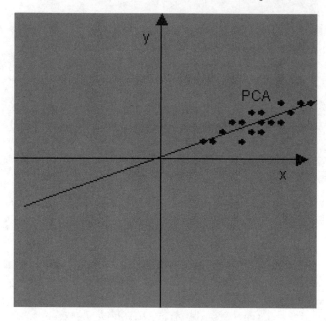

FIGURE 25.1: A two-dimensional set of points with its principal component axis.

By projecting the vectors onto the subspace spanned by the first eigenvectors, PCA retains the maximal variance in the projection space and minimizes the mean-square reconstruction error. The choice of the number K of components is in general not a serious problem–basically it is a matter of inspecting how much variance is explained by increasing values of K. For visualization purposes, only projections onto two- or three-dimensional spaces are useful. The first dominant eigenvectors can be associated with the discovery of important features or patterns in the data. In DNA microarray data where the points correspond to genes and the axis to different experiments, such as different points in time, the dominant eigenvectors can represent expression patterns. For example, if the first eigenvector has a large component along the first experimental axis and a small component along the second and third axis, it can be associated with the experimental expression pattern "'high-low-low." In the case of replicated experiments, we can expect the first eigenvector to be associated with the principal diagonal.

There are also a number of techniques for performing approximate PCA, as well as probabilistic and generalized (non-linear and projection pursuit) versions of PCA [7, 45, 46, 10], and references therein. An extensive application of PCA techniques to array data is described in [3].

Although PCA is not a clustering technique per se, projection onto lower dimensional spaces associated with the top components can help reveal and visualize the presence of clusters in the data. These projections however must be considered carefully since clusters present in the data can become hidden during the projection operation (Figure 25.2). Thus, while PCA is a useful technique, it is only one way of analyzing the data that should be complemented by other methods, and in particular by methods whose primary focus is data clustering.

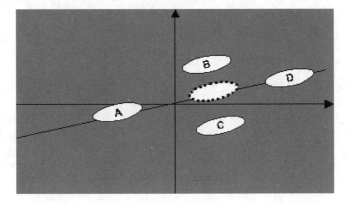

FIGURE 25.2: Schematic representation of four data clusters (A, B, C, and D) in 2D space. Projection of the data onto the first principal component axis results in only three clusters, with clusters B and C projected into the same cluster (enclosed by a dotted line).

25.3 Clustering Overview

Another direction for visualizing or compressing high-dimensional array data is the application of clustering methods. Clustering refers to an important family of techniques in exploratory data analysis and pattern discovery, aimed at extracting underlying cluster structures. Clustering, however, is a "fuzzy" notion without a single precise definition. Dozens of clustering algorithms exist in the literature and a number of *ad hoc* clustering procedures, ranging from hierarchical clustering to k-means have been applied to DNA array data [22, 2, 33, 50, 14, 47, 17, 64]. Because of the variety and "open" nature of clustering problems, it is unlikely that a systematic exhaustive treatment of clustering can be given. However there are a number of important general issues to consider in clustering and clustering algorithms, especially in the context of gene expression.

25.3.1 Data Types

At the highest level, clustering algorithms can be distinguished depending on the nature of the data being clustered. The standard case is when the data points are vectors in Euclidean space. But this is by no means the only possibility. In addition to vectorial data, or numerical data expressed in absolute coordinates, there is the case of relational data, where data is represented in relative coordinates, by giving the pairwise distance between any two points. In many case the data is expressed in terms of a pairwise similarity (or dissimilarity) measure that often does not satisfy the three axioms of a distance (positivity, symmetry, and triangle inequality). There exist situations where data configurations are expressed in terms of tertiary or higher order relationships or where only a subset of all the possible pairwise similarities is given. More importantly, there are cases where the data is not vectorial or relational in nature, but essentially qualitative, as in the case of answers to a multiple-choice questionnaire. This is sometimes also called nominal data. While at the present time gene expression array data is predominantly numerical, this is bound to change in the future. Indeed, the dimension "orthogonal to the genes" covering different experiments, different patients, different tissues, different times, and so forth is at least in part non-numerical. As databases of array data grow, in many cases the data will be mixed with both vectorial and nominal components.

25.3.2   Supervised/Unsupervised

One important distinction amongst clustering algorithms is supervised versus unsupervised. In supervised clustering, clustering is based on a set of given reference vectors or classes. In unsupervised clustering, no predefined set of vectors or classes is used. Hybrid methods are also possible where an unsupervised approach is followed by a supervised one. At the current early stage of gene expression array experiments, unsupervised methods such as k-means and self organizing maps [50] are most commonly used. However supervised methods have also been tried [14, 28], where clusters are pre-determined using functional information or unsupervised clustering methods, and then new genes are classified in the various clusters using a classifier, such as linear and quadratic discriminant analysis, decision trees, neural networks, or support vector machines, that can learn the decision boundaries between data classes. The feasibility of class discrimination with array expression data has been demonstrated, for instance for tumor classes such as leukemias arising from several different precursors [30], and B-cell lymphomas [1] (see also [61, 65]).

25.3.3   Similarity/Distance

The starting point of several clustering algorithms, including several forms of hierarchical clustering, is a matrix of pairwise similarities or distances between the objects to be clustered. In some instances, this pairwise distance is replaced by a distortion measure between a data point and a class centroid as in vector quantization methods. The precise definition of similarity, distance, or distortion is crucial and, of course, can greatly impact the output of the clustering algorithm. In any case, it allows converting the clustering problem into an optimization problem in various ways, where the goal is essentially to find a relatively small number of classes with high intraclass similarity or low intraclass distortion, and good interclass separation. In sequence analysis, for instance, similarity can be defined using a score matrix for gaps and substitutions and an alignment algorithm. In gene expression analysis, different measures of similarity can be used. Two obvious examples are Euclidean distance (or more generally L^p distances) and correlation between the vectors of expression levels. The Pearson correlation coefficient is just the dot product of two normalized vectors, or the cosine of their angle. It can be measured on each pair of genes across, for instance, different experiments or different time steps. Each measure of similarity comes with its own advantages and drawbacks depending on the situation, and may be more or less suitable to a given analysis. The correlation, for instance, captures similarity in shape but places no emphasis on the magnitude of the two series of measurements and is quite sensitive to outliers. Consider, for instance, measuring the activity of two unrelated genes that are fluctuating close to the background level. Such genes are very similar in euclidean distance (distance close to 0), but dissimilar in terms of correlation (correlation close to 0). Likewise, consider the two vectors 1000000000 and 0000000001. In a sense they are similar since they are almost always identical and equal to 0. On the other hand, their correlation is close to 0 because of the two "outliers" in the first and last position.

25.3.4   Number of Clusters

The choice of the number K of clusters is a delicate issue, which depends, among other things, on the scale at which one looks at the data. It is safe to say that an educated partly manual trial-and-error approach still remains an efficient and widely used techniques, and this is true for array data at the present stage. Because in general the number of clusters is relatively small, all possible values of K within a reasonable range can often be tried. Intuitively, however, it is clear that one ought to be able to assess the quality of K from the compactness of each cluster and how well each cluster is separated from the others. Indeed there have been several recent developments aimed at the automatic determination of the number of clusters [47, 53] with reports of good results.

25.3.5 Cost Function and Probabilistic Interpretation

Any rigorous discussion of clustering on a given data set presupposes a principled way of comparing different ways of clustering the same data, hence the need for some kind of global cost/error function that can easily be computed. The goal of clustering then is to try to minimize such function. This is also called parametric clustering in the literature, as opposed to non-parametric clustering, where only local functions are available [12].

In general, at least for numerical data, this function will depend on quantities such as the centers of the clusters, the distance of each point in a cluster to the corresponding center, the average degree of similarity of the points in a given cluster, and so forth. Such a function is often discontinuous with respect to the underlying clustering of the data. Here again there are no universally accepted functions and the cost function should be tailored to the problem, since different cost functions can lead to different answers.

Because of the advantages associated with probabilistic methods and modeling, it is tempting to associate the clustering cost function with the negative log-likelihood of an underlying probabilistic model. While this is formally always possible, it is of most interest when the structure of the underlying probabilistic model and the associated independence assumptions are clear. This is when the additive terms of the cost function reflect the factorial structure of the underlying probabilities and variables. As we shall see this is the case with mixture models, where the k-means clustering algorithm can be viewed as a form of EM (expectation-maximization).

In the rest of this section, we describe in more detail basic clustering algorithms that can be applied to DNA array data, hierarchical clustering, k-means, and self-organizing maps. Many other related approaches, including vector quantization [15], graph methods [47], and factorial analysis can be found in the references.

25.4 Hierarchical Clustering

Clusters can result from a hierarchical branching process. Thus there exist methods for automatically building a tree from data given in the form of pairwise similarities. In the case of gene expression data, this is the approach used in [22].

25.4.1 Hierarchical Clustering Algorithm

The standard algorithm used in [22] recursively computes a dendogram that assembles all the elements into a tree, given the correlation (or distance or similarity) matrix. The algorithm starts by assigning a leaf of the tree to each element (gene). At each step of the algorithm:

- The two most similar elements of the current matrix (highest correlation) are computed and a node joining these two elements is created.
- An expression profile (or vector) is created for the node by averaging the two

expression profiles (or vectors) associated with the two points (missing data can be ignored and the average can be weighted by the number of elements they contain).

- A new smaller correlation matrix is computed using the newly computed expression profile or vector and replacing the two joined elements with the new node.

- With N starting points, the process is repeated at most $N-1$ times, until a single node remains.

This algorithm is familiar to biologists and has been used in sequence analysis, phylogenetic trees, and cluster analysis. As described, it requires $O(N^3)$ steps since for each of the $N-1$ fusions one must search for an optimal pair. An $O(N^2)$ version of the algorithm is described in [23].

In the above algorithm, each node is associated with an expression profile (or feature vector) when created. The similarity/distance between two nodes is computed based on their expression profiles. Alternatively, the similarity/distance between two nodes can be determined using the average linkage method, which takes the average similarity/distance between all possible pairs of elements of the two nodes. Other options exist, such as single linkage and complete linkage [21].

The output of hierarchical clustering is typically a binary tree and not a set of clusters. In particular, it is usually not obvious how to define clusters from the tree since clusters are derived by cutting the branches of the tree at more or less arbitrary points.

25.4.2 Tree Visualization

In the case of gene expression data, the resulting tree organizes genes or experiments so that underlying biological structure can often be detected and visualized [22, 49, 2, 1]. As already pointed out, after the construction of such a dendogram there is still a problem of how to display the result and which clusters to choose. Leaves are often displayed in linear order and biological interpretations are often made in relation to this order, e.g. adjacent genes are assumed to be related in some fashion. Thus the order of the leaves matters.

At each node of the tree, either of the two elements joined by the node can be ordered to the left or the right of the other. Since there are $N-1$ joining steps, the number of linear orderings consistent with the structure of the tree is 2^{N-1}. Computing the optimal linear ordering maximizing the combined similarity of all neighboring pairs seems difficult, and therefore heuristic approximations have been proposed [22]. These approximations weigh genes using average expression level, chromosome position, and time of maximal induction.

More recently, it was noticed in [9] that the optimal linear ordering can be computed in $O(N^4)$ steps, and further improved to $O(N^3)$ steps [8], simply by using dynamic programming, in a form which is essentially the well-known inside portion of the inside-outside algorithm for stochastic context-free grammars [6]. If $G_1, \ldots, G_N$ are the leaves of the tree and ϕ denotes one of the 2^{N-1} possible orderings of the leaves, we would like to maximize the following criterion function

$$\sum_{i=1}^{N-1} C(G_{\phi(i)}, G_{\phi(i+1)}) \tag{25.4}$$

where $G_{\phi(i)}$ is the i-th leaf when the tree is ordered according to ϕ. Let V denote both an internal node of the tree as well as the corresponding subtree. V has two children: V_l on the left and V_r on the right, and four grand-children V_{ll}, V_{lr}, V_{rl}, and V_{rr}. The algorithm

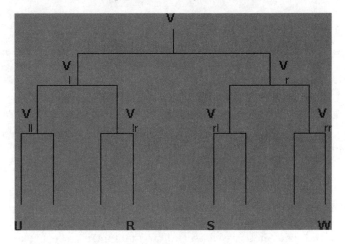

FIGURE 25.3: Tree underlying the dynamic programming recurrence of the inside algorithm.

works bottom up, from the leaves towards the roots by recursively computing the cost of the optimal ordering $M(V, U, W)$ associated with the subtree V when U is the leftmost leaf of V_l and W is the rightmost leaf of V_r (Figure 25.3). The dynamic programming recurrence is given by:

$$M(V, U, W) = \max_{R \in V_{lr}, S \in V_{rl}} M(V_l, U, R) + M(V_r, S, W) + C(R, S) \qquad (25.5)$$

The optimal cost $M(V)$ for V is obtained by maximizing over all pairs U, W. The global optimal cost is obtained recursively when V is the root of the tree, and the optimal tree can be found by standard backtracking. The algorithm requires computing $M(V, U, W)$ only once for each $O(N^2)$ pair of leaves. Each computation of $M(V, U, W)$ requires maximization over all possible $O(N^2)$ (R, S) pairs of leaves. Hence the algorithm requires $O(N^4)$ steps with $O(N^2)$ space complexity, since only one $M(V, U, W)$ must be computed for each pair (U, W) and this is also the size of the pairwise similarity matrix. For some applications, $O(N^4)$ is too slow. A faster algorithm on average is developed in [9] by early termination of search paths that are not promising.

Hierarchical clustering has proved to be a useful for array data analysis in the literature, for instance for finding genes that share a common function [22, 49, 1]. The main clusters derived are often biologically significant and the optimal leaf ordering algorithm can further improve the quality and interpretability of the results [9]. Optimal leaf ordering helps in improving the definition of cluster boundaries and the relationships between clusters.

25.5 K-Means, Mixture Models, and EM Algorithm

25.5.1 K-Means Algorithm

Of all clustering algorithms, k-means [21] is among the simplest and most widely used, and has probably the cleanest probabilistic interpretation as a form of EM (expectation-maximization) on the underlying mixture model. In a typical implementation of the k-means algorithm, the number of clusters is fixed to some value K based, for instance, on the expected number of regulatory patterns. K representative points or centers are initially chosen for each cluster more or less at random. In array data, these could reflect, for instance, regulatory patterns. These points are also called centroids or prototypes. Then at each step:

- Each point in the data is assigned to the cluster associated with the closest representative.
- After the assignment, new representative points are computed for instance by averaging or taking the center of gravity of each computed cluster.
- The two procedures above are repeated until the system converges or fluctuations remain small.

Hence notice that k-means requires choosing the number of clusters and also being able to compute a distance or similarity between points and compute a representative for each cluster given its members.

The general idea behind k-means can lead to different software implementations depending on how the initial centroids are chosen, how symmetries are broken, whether points are assigned to clusters in a hard or soft way, and so forth. A good implementation ought to run the algorithm multiple times with different initial conditions and possibly also try different values of K automatically.

When the cost function corresponds to an underlying probabilistic mixture model [24, 54], k-means is an on-line approximation to the classical EM algorithm [20, 6], and as such in general is bound to converge towards a solution that is at least a local maximum likelihood or maximum posterior solution. A classical case is when Euclidean distances are used in conjunction with a mixture of Gaussian models. A related application to a sequence clustering algorithm is described in [4].

25.5.2 Mixtures Models and EM

To better understand the connection to mixture models, imagine a data set $D = (d_1, \ldots, d_N)$ and an underlying mixture model with K components of the form

$$P(d) = \sum_{k=1}^{K} P(M_k)P(d|M_k) = \sum_{k=1}^{K} \lambda_k P(d|M_k) \tag{25.6}$$

where $\lambda_k \geq 0$ and $\sum_k \lambda_k = 1$ and M_k is the model for cluster k. Mixture distributions provide a flexible way for modeling complex distributions, combining together simple building-blocks, such as Gaussian distributions. The Lagrangian associated with the log-likelihood and the normalization constraints on the mixing coefficients is given by

$$\mathcal{L} = \sum_{i=1}^{N} \log(\sum_{k=1}^{K} \lambda_k P(d_i|M_k)) - \mu(\sum_{k=1}^{K} \lambda_k - 1) \tag{25.7}$$

with the corresponding critical equation

$$\frac{\partial \mathcal{L}}{\partial \lambda_k} = \sum_{i=1}^{N} \frac{P(d_i|M_k)}{P(d_i)} - \mu = 0 \tag{25.8}$$

Multiplying each critical equation by λ_k and summing over k immediately yields the value of the Lagrange multiplier $\mu = N$. Multiplying again the critical equation across by $P(M_k) = \lambda_k$, and using Bayes theorem in the form

$$P(M_k|d_i) = P(d_i|M_k)P(M_k)/P(d_i) \tag{25.9}$$

yields

$$\lambda_k^* = \frac{1}{N} \sum_{i=1}^{N} P(M_k|d_i) \qquad (25.10)$$

Thus the maximum likelihood estimate of the mixing coefficients for class k is the sample mean of the conditional probabilities that d_i comes from model k. Consider now that each model M_k has its own vector of parameters (w_{kj}). Differentiating the Lagrangian with respect to w_{kj} gives

$$\frac{\partial \mathcal{L}}{\partial w_{kj}} = \sum_{i=1}^{N} \frac{\lambda_k}{P(d_i)} \frac{\partial P(d_i|M_k)}{\partial w_{kj}} \qquad (25.11)$$

Substituting Equation 25.9 in Equation 25.11 finally provides the critical equation

$$\sum_{i=1}^{N} P(M_k|d_i) \frac{\partial \log P(d_i|M_k)}{\partial w_{kj}} = 0 \qquad (25.12)$$

for each k and j. The maximum likelihood equations for estimating the parameters are weighted averages of the maximum likelihood equations

$$\partial \log P(d_i|M_k))/\partial w_{kj} = 0 \qquad (25.13)$$

arising from each point separately. As in Equation 25.10, the weights are the probabilities of membership of the d_i in each class.

The maximum likelihood Equations 25.10 and 25.12 can be used iteratively to search for maximum likelihood estimates, yielding also another instance of the EM algorithm. In the E step, the membership probabilities (hidden variables) of each data point are estimated for each mixture component. The M step is equivalent to K separate estimation problems with each data point contributing to the log-likelihood associated with each of the K components with a weight given by the estimated membership probabilities. Different flavors of the same algorithm are possible depending on whether the membership probabilities $P(M|d)$ are estimated in hard or soft fashion during the E step. The description of k-means given above correspond to the hard version where these membership probabilities are either 0 or 1, each point being assigned to only one cluster. This is analogous to the use of the Viterbi version of the EM algorithm for hidden Markov models, where only the optimal path associated with a sequence is used, rather than the family of all possible paths. Different variations are also possible during the M step of the algorithms depending, for instance, on whether the parameters w_{kj} are estimated by gradient descent or by solving Equation 25.12 exactly. It is well known that the center of gravity of a set of points minimizes its average quadratic distance to any fixed point. Therefore in the case of a mixture of spherical Gaussians, the M step of the k-means algorithm described above maximizes the corresponding quadratic log-likelihood and provides a maximum likelihood estimate for the center of each Gaussian component. It is also possible to introduce prior distributions on the parameters of each cluster and/or the mixture coefficients and create more complex hierarchical mixture models.

PCA, hierachical clustering, k-means, as well as other clustering and data analysis algorithms are currently implemented in several publicly (Figure 25.4) or commercially available software packages for DNA array data analysis. It is important to recognize that many software packages will output some kind of answer, for instance a set of clusters, on any kind of data set. These answers should not be trusted always blindly. Rather it is wise practice, whenever possible, to track down the assumptions underlying each algorithm/implementation/package and to run the same algorithms with different parameters, as well as different algorithms, on the same data set, as well as other data sets.

- R Cluster
    - Platform: Web-based
    - Provider: UCI
    - Web URL:www.genomics.uci.edu
    - Functions: Hierarchical clustering, k-means
- EPCLUST (Expression Profiler)
    - Platform: Web-based
    - Provider: European Bioinformatic Institute
    - Web URL: www.ebi.ac.uk/microarray
    - Functions: Hierarchical clustering, k-means
- Cluster and TreeView
    - Platform: Window
    - Provider: Stanford University
    - Web URL: rana.lbl.gov/Eisen/Software.htm
    - Functions: Hierarchical clustering, k-means, PCA, SOM
- Xcluster
    - Platform: Window, Mac, Unix
    - Provider: Stanford University
    - Web URL: genome-www.Stanford.edu/ sherlock/cluster.html
    - Functions: SOM, k-means
- GeneCluster
    - Platform: Window
    - Provider: Whitehead Institute/MIT
    - Web URL: www.genome.wi.mit.edu/MPR
    - Functions: SOM

FIGURE 25.4: Publicly available software for cluster analysis of microarray gene expression data.

25.6   Self-Organizing Maps

The self-organizing map (SOM) [36] is treated as a cluster analysis method. Like k- means, the SOM is a kind of partitional clustering algorithm that partitions the data into a predefined number of clusters. However, unlike k-means, which produces an unorganized or unstructured collection of clusters, the SOM can form a topological map of input distribution.

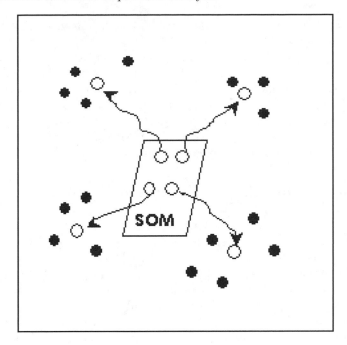

FIGURE 25.5: The formation of a topological map of input distribution (adapted from [50]). Data points and SOM node vectors are represented by solid and blank circles, respectively. The movement trajectories of nodes are indicated by arrows.

The SOM is constructed by choosing an array of nodes (typically, a 2D array). The nodes are mapped into the input feature space, initially at random, and then adjusted iteratively (Figure 25.5). In each iteration, a data point (i.e., the feature vector of an object) P is randomly chosen, and the nodes are moved towards P in the feature space, but only the nearest node (denoted by N_p) and its neighbors in the map can change their positions in the feature space. Through iterations, neighboring nodes in the map evolve into ones with similar feature vectors. As a result, physically adjacent nodes in the map tend to represent related clusters. The correlation between the physical location of nodes in the map and the vectors of the nodes in the feature space is attributed to the adaptive mechanism based on neighborhood. In essence, the SOM transforms the input patterns into a topological map. The SOM has also been conceived as a neural network model, where the vectors associated with the nodes in the map are called the weight vectors. The adaptive formula in each iteration is given by (following [50]):

$$f_{k+1}(N) = f_k(N) + \tau(d(N, N_p), k)(P - f_k(N)) \tag{25.14}$$

In the above equation, $f_k(N)$ denotes the position of node N in the feature space at iteration k; τ is the learning rate that decreases with the distance of node N from N_p and with iteration number k (e.g., τ in the form of a Gaussian function or a window function); N_p is the node closest to the input data point P. What this equation says is that node N_p as well as its neighbor nodes within the limit specified by the function τ is allowed to adjust its vector in the direction of P. By iteratively reducing the learning rate and the radius of the neighborhood, the algorithm converges to a map with optimum clusters and topology, and each node vector settles in the mean of the data points belonging to the node.

The assignment of a data point to a node (cluster) is determined by the shortest distance between the data vector and the node vector.

The SOM lends itself well to exploratory data analysis. It provides a convenient way for data visualization and interpretation and has the advantage of handling data with non-uniformity or irregularities [50]. The landscape of the topological map is psychologically meaningful. On the negative side, there is no theoretical basis for determining the optimal dimensions of the SOM, and it may take a very large number of iterations before convergence (like 20,000 50,000). This technique has been applied to pattern interpretation [50] and class discovery [30] based on microarray gene expression data.

25.7   Cluster Analysis of Microarray Gene Expression Data

With the capability of uncovering the structure and patterns in the data, cluster analysis has become a routine for comprehension and interpretation of complex biological information embedded in microarray gene expression data. Cluster analysis of genome- wide gene expression data offers a system-level exploration of functionally related genes that exhibit similarity or correlation in gene expression across various conditions. Given the gene expression data of a set of tissue samples, cluster analysis may reveal novel class structure.

In general, it is required that the data are normalized and standardized for removing systematic sources of variation and allowing array-to-array comparison before analysis. Cluster analysis software accepts the gene expression data typically represented in the format of a table (or a matrix), where each row, often labeled with a gene name, consists of the expression levels of a particular gene across all experiments, and each column, often labeled with an experiment ID, is formed by all gene expression levels in a particular experiment. Thus, in the data matrix, each entry x_{ij} is the expression level of gene i in experiment j. An experiment here means a hybridization experiment with respect to a condition or a sample. As clustering is a process of grouping objects according to the similarity in their feature descriptions, it is clear that the gene expression data can be clustered in two ways: treating genes as objects and each experiment as a feature, and vice versa, depending on the application objective.

25.7.1   Gene Clustering

Cluster analysis is a crucial step for extracting information from the massive amount of gene expression data. This analysis can generate gene clusters, with each cluster comprising genes whose expressions are correlated across experiments [22]. The co-expression of genes suggests that they are functionally related in the same cellular process and are likely co-regulated. Co-regulated genes here refer to genes regulated by common molecular factors called transcription factors in a mechanism known as transcriptional control. The functions of many uncharacterized genes can be annotated with the functions of respectively co-expressed known genes. The objectives of gene clustering on microarray gene expression data can be boiled down to four essentials: (1) functional organization of genes, (2) interpretation of the cellular status according to the genome-wide expression pattern, (3) functional deduction of unknown genes, and (4) exploration of transcriptional regulation.

Identification of regulatory genes and their target genes is fundamental to building a genetic network essential for understanding basic biological principles. Given microarray gene expression data, cluster analysis can recognize genes whose regulatory regions (cis-regulatory elements) are bound by the same proteins (transcription factors) in vivo. Such a set of co- regulated genes is referred to as a "regulon". Gene regulatory regions are described in more detail in the next section.

Microarray data used for the global analysis of gene function can be collected across multiple growth conditions or collected over a period of time. Time-series microarray gene expression data are particularly useful for studying the dynamics of gene regulation.

Validation of Gene Clusters

Cluster validation is difficult in the absence of formal statistical tests for determining the number of clusters. Statistical bootstrapping has been proposed to assess the reliability of gene clusters [35] in much the same spirit it is used in phylogenetic analysis. In this approach, for example, the match of a gene to a cluster pattern is called 95% stable if this is the case in the actual data clustering and in at least 95% of the bootstrap clusterings.

To evaluate a regulon hypothesis, the upstream regions of co-expressed genes are searched for common motifs, which are the consensus binding sequences (cis-regulatory elements) of the transcription factors. The presence of statistically significant consensus motifs strengthens the belief that the genes are co-regulated. Genes in the same cluster are expected to be involved in the same biological pathway and share similar functional annotations. This can be confirmed by literature and database search.

Assume that cluster analysis arrives at a set of clusters, and there exist known functional categories. The hypergeometric distribution can be used to calculate the chance probability of observing at least k genes from a functional category within a cluster as follows [51]:

$$P = 1 - \sum_{i=0}^{k-1} \frac{\binom{f}{i}\binom{g-f}{n-i}}{\binom{g}{n}} \tag{25.15}$$

where n and f denote the total numbers of genes within the cluster and within the functional category, respectively; and g is the total number of genes in the genome. Consider the yeast genome for example. Since there are about 200 functional categories in the MIPS database [40], only clusters with $P < 0.0003$ for a certain functional category are considered statistically significant at a level of significance of approximately 0.05, adjusted for the multiplicity effect due to 200 categories.

25.7.2 Gene Selection and Filtering

A clustering algorithm relies on a distance or similarity measure that calculates the distance between any two given feature vectors (collections of feature values). The inclusion of irrelevant features in the distance function will lead to imprecise or incorrect distance calculation and impact on clustering. A clustering algorithm could tolerate this problem to some some extent, but the quality of clustering begins to degrade as more irrelevant features get involved. In the context of microarray data clustering, this is not so much a problem in gene clustering as in sample clustering since the dimension of the feature space for gene clustering is considerably smaller than that for sample clustering (e.g., hundreds versus tens of thousands). Data pre-processing for removal of irrelevant genes and selection of relevant genes is thus needed for microarray-based sample clustering. However, this issue may have to be dealt with separately for supervised and unsupervised clustering.

In supervised clustering, samples are associated with class labels so that genes can be selected on the basis of the statistical significance of their differential expression across different classes. Intensive research has resulted in many algorithms for differential gene expression analysis, e.g., [56, 55]. Some algorithms specifically focus on gene selection for discriminant analysis, and tend to select a small set of discriminant genes for predictive or diagnostic purposes, e.g., [31, 52]. Genes selected thereby can be used as a basis to discriminate between classes, and more interestingly, to further analyze the internal structure of the data within classes in an unsupervised manner, leading to possible new subclasses discovered (e.g., [48]).

The approach of using prior class membership information may fail to identify genes related to unknown classes. In unsupervised clustering, this kind of information is unavailable or ignored. Since however, an informative gene is expected to stand out from the background noise and show at least some variation among samples, a simple gene filter can be designed to filter out genes of low expression or low variation. In one study, for example, a gene filter selected genes with signal intensity >1.5-fold over background in both test and reference channels in at least 75% of samples in conjunction with $\geq$ 3-fold variation from the mean in at least two samples [37].

25.7.3 Sample Clustering

Microarray-based gene expression profiling has emerged as a promising approach to disease classification. Taking the same approach to automatic class discovery is even a greater challenge. In both cases, the samples are grouped according to the similarity in their gene expression profiles. It has been demonstrated that this approach can generate clusters, independent of prior biological knowledge, that are consistent with known classes [30]. This demonstration raises the interesting opportunity of class identification in a new domain and subclass distinction within known classes in this approach. This is a significant development since the microarray approach offers a rapid solution in contrast to the traditional typically slow process on this problem. Furthermore, with this approach, the clusters identified can be analyzed directly from the associated gene expression patterns in molecular or clinical perspectives.

Validation of Sample Clusters

Whether putative classes resulting from clustering reflect the true structure in the data and are domain-meaningful can be further evaluated. Basically, a class predictor based on the putative classes derived from one data set is tested on another independent data set. If the initial data set and the independent data set share similar structure, then good predictive performance is expected. However, predictive accuracy cannot be measured for the independent data set in which the samples are not associated with any putative class. Instead, predictive performance can be assessed by the strength of prediction (for instance, in a continuous range between 0 and 1) [30]. High average prediction strength on the independent data suggests the validity of the putative classes.

25.7.4 Two-Dimensional Clustering

A sample cluster is made up of samples sharing similar expression values across genes. Often, a sample cluster is characterized by a subset of genes that are either over-or under-expressed relative to other sample clusters. For example, certain genes are expressed in cancer tissue but not in normal tissue. On the other hand, a gene cluster may be specifically associated with a subset of samples. Two-dimensional clustering on gene expression data combines gene clustering with sample clustering. The graphical display of two-dimensional clustering can reveal the correlation between genes and tissue samples if there is any, whereas either gene clustering or sample clustering alone may not.

A clustering algorithm can be applied to genes or samples or both. If an algorithm performs on these two axes separately, it is called one-way clustering. Such is the case for hierarchical clustering, k-means and self-organizing maps. Some algorithms, however, cluster on both dimensions (gene- and sample-) simultaneously, and are thus called two-way clustering (e.g., [17, 29]). Two-way clustering seeks subsets of the genes and samples so that significant partitions result when one subset is used to cluster the other [29]. These techniques can be found in the references.

25.7.5 Visualization of Gene Expression Data

Hierarchical clustering produces a dendrogram that reveals the structure in the data but does not provide information about the variation with respect to particular features across clusters. When samples are clustered on the basis of their gene expression profiles, visualization of clusters together with their gene expression profiles often permits natural extraction of the correlation information between clusters and gene expression profiles, and thereby gives rise to useful biological insight. The heat map is a 2D grid of color points for representing clustered gene expression data such that each point corresponds to the expression level of a gene in a sample with the color grade indicative of the level of intensity. In two-dimensional hierarchical clustering, it is expedient for data interpretation to order genes and samples so that genes showing a strong correlation across samples appear near each other on the gene tree, and samples with similar gene expression profiles are adjacent on the sample tree, as illustrated by an example in cancer classification [27] (Figure 25.6).

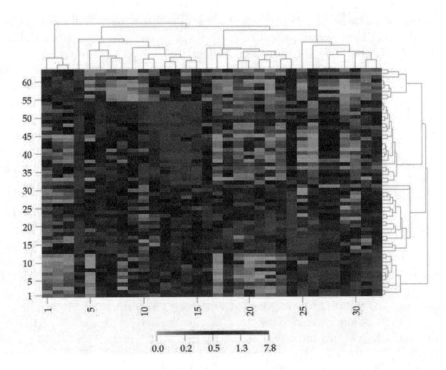

FIGURE 25.6: (**See color insert following page 20-4.**) Two-dimensional hierarchical clustering of the microarray gene expression data on small round blue cell tumors with selected genes. The dendrograms for gene clusters and sample clusters are shown on top and right of the map, respectively.

25.8   DNA Arrays and Regulatory Regions

Another important level of analysis consists in combining DNA array data with DNA sequence data, and in particular with regulatory regions. This combination can be used to detect regulatory motifs, but also to address global questions of regulation. While gene regulatory elements have been found in a variety of regions including introns, distant intragenic regions, and downstream regions, the bulk of the regulation of a given gene is, in general, believed to depend primarily on a more or less extended region immediately upstream of the gene. In [32], for instance, this model was tested on a genomic scale by coupling expression data obtained during oxidative stress response with all pairwise alignments of yeast ORF upstream regions. In particular, it was found that as the difference in upstream regions increases, the correlation in activity rapidly drops to zero and that divergent ORFs, with overlapping upstream regions, do not seem to have correlated expression levels. By and large, however, the majority of current efforts aimed at combining DNA array and sequence data are focused on searching for regulatory motifs.

Several techniques have been developed for the discovery of "significant" patterns from a set of unaligned DNA sequences. Typically these patterns represent regulatory (transcription factor DNA binding sites) or structural motifs that are shared in some form by the sequences. The length and the degeneracy of the pattern are of course two important parameters [44, 43, 41]. Probabilistic algorithms such as EM and Gibbs sampling naturally play an essential role in motif finding, due to both the structural and location variability of motifs [38].

Simple measures of over-representation have also been shown to be effective for detecting such motifs, for instance in sets of gene upstream or downstream [58] regions. While these data mining algorithms can be applied using a purely combinatorial approach to genomic DNA [13, 57], the methods and results can be further refined, and the sensitivity increased, by focusing the search on specific clusters of genes derived from array data analysis, such as clusters of genes that appear to be co-regulated. In addition to regulatory motifs found in the TRANSFAC database [62], these methods can detect novel motifs in the large amounts of more or less unannotated genomic DNA that has become available through genome and other sequencing projects [32, 60, 16, 34, 11].

The basic idea behind these approaches is to compute the number of occurrences of each k-mer, typically for values of k in the range of 3 to 10, within a set of sequences, such as all gene-upstream regions, or all the upstream regions of a particular set of co- regulated genes, and look for k-mers that are over-represented. Over- representation is a statistical concept that can be assessed in a number of different ways and, depending on the problem, a number of points must be carefully considered. These include:

- Regions: How are the upstream, or downstream, regions defined? Do they have fixed length? How are they treated with respect to neighboring genes on each strand and possible overlaps?

- Counts: Are the two strands treated separately or aggregated? It is well known, for instance, that certain regulatory motifs are active regardless of the strand on which they occur and these are better detected if counts on both strands are aggregated. Other motifs are strand-specific.

- Background model: Over-representation must be assessed with respect to a statistical background model. The choice of the background model is critical and non-trivial. In particular the background model cannot be too good otherwise it would predict the counts exactly and therefore would be worthless. Typical models used in the literature are Markov models of various orders, measured on the data or some other reference set. Another possible background model, is to consider the average of single (or multiple) base pair mismatches, i.e. to estimate the counts of a given k-mer using the counts of all the k-mers that differ in one position.

- Statistics: Several statistics can be used to detect significant over-representation from the raw counts, such as ratio, log-likelihood, z-score binomial, t-test, Poisson, and compound Poisson. As in the case of array data, inference based on ratio alone can be tricky, especially for low expected frequencies that can induce false positives (e.g. 1 versus 4 is very different from 1,000 versus 4,000).

- Gene clusters: if the method is applied to the DNA sequences associated with a cluster of genes derived from array data, how is the cluster determined? Are the genes up-or down regulated under a given condition? Etc. Notice also that array data can be used as a filter to detect over-representation before or after the counts, often yielding somewhat different results.

k-mers that are over-represented are of particular interest and have been shown to comprise well-known regulatory motifs (also known as cis-regulatory elements). For instance, when the algorithms are run on the yeast upstream regions using oxidative stress data, one immediately detects the well-known stress element CCCCT [39] and its reverse complement AGGGG, or the YAP1 element TTACTAA and its reverse complement TTAGTAA [63, 25, 19] (See Figure 25.7).

Alignment	Sequence	Distance from ATG	Strand
GATTACTAAG	0	134	1
GCTTACGAAT	1	231	0
GCTTACTAAT	1	256	1
GCTTACTAAT	1	276	0
GCTTAGTAAA	2	171	1
GATTAGTAAT	3	276	1
GATTAGTAAT	3	300	1
GATTAGTAAT	3	312	1
GATTAGTAAT	3	324	1
GATTAGTAAT	3	336	1
GCTGACTAAT	4	331	0
GCTTACTAAT	5	400	1
GATTAATAAT	5	431	1
GCTGACTAAG	6	181	1
ACTTAGTAAT	6	332	0
GATTACTAAT	7	89	1
GCTTAATAAT	8	285	1
GCTTAGTAAT	10	139	1
GCTTACTAAG	10	203	1

FIGURE 25.7: The cis-regulatory element of a group of co-regulated genes identified by the AlignACE program (http://atlas.med.harvard.edu/). The gene sequences (ORFs) are 0:YKL071W, 1:YFL056C, 2:YLL060C, 3:YOL165C, 4:YML116W, 5:YBR008C, 6:YPL171C, 7:YLR460C, 8:YKR076W, 9:YHR179W, 10:YML131W in Saccharomyces cerevisiae. The motif given here satisfies predefined statistical criteria, and is consistent with Yap1-binding sites (TTACTAA or TGACTAA).

In general, however, only a fraction of the putative motifs detected by these techniques nowadays are typically found also in the TRANSFAC [62] data base, or in the current literature, and most must await future experimental verification. In the meantime, over-represented motifs can be further studied in terms of their patterns of localization and co-

9-mer	C0	C1	$27 \times C0/C1$
GCGATGAGC	67	273	6.62
GCTCATCGC	51	262	5.26

FIGURE 25.8: Over-representation of 9-mer GCGATGAGC and its reverse complement across all 500bp gene-upstream regions in yeast. C0 is the total number of occurrences. C1 represents the total number of occurrences of all the 27=3x9 9-mers that differ in only one position from the 9-mer (background model). Under this model, the 9-mer is over 6-fold over-represented.

occurrence within, for instance, upstream regions and/or their DNA structure. Non-uniform patterns of localization, for instance, can be indicative of biological function. To illustrate, consider the over-represented 9- mer GCGATGAGC in yeast (Figure 25.8). When one looks at the 500bp upstream regions of all the genes in yeast, this 9-mer and its reverse complement GCTCATCGC have roughly symmetric distributions with a noticeable peak 50 to 150 bp upstream from the genes they seem to regulate [13, 34]. As far as DNA structure is concerned, it can be analyzed to some extent by using some of the available DNA physical scales [5, 42] (e.g. bendability, propeller twist). Typical over-represented k-mers that have peculiar structural properties include runs of alternating AT which are identical to their own reverse complement and correspond to highly bent or bendable DNA regions (such as the TATA box) or, at the opposite end of the structural spectrum, runs of As or runs of Ts which tend to be very stiff.

All together, these techniques are helping inferential and other data mining efforts aimed at unraveling the "language" of regulatory regions. A somewhat orthogonal approach described in [18] computes for each motif the mean expression profile over a set of array experiments of all the genes that contain the motif in their transcription control regions. These profiles can be useful for visualizing the relationship between the genome sequence and gene expression data, and for characterizing the transcriptional importance of specific sequence motifs.

Detection of gene expression differences, clusters of co-regulated genes, and/or gene regulatory motifs are essential steps toward the more ambitious and long-term goal of inferring regulatory networks on a global scale, or even along more specific sub-components [59, 26, 66] such as a pathway or a set of co- regulated genes.

Acknowledgement

This work has been supported in part by grants from the NIH and NSF to GWH and PB. GWH and PB wish also to thank Cambridge University Press for permission to reuse material from their book DNA Microarrays and Gene Expression–From Experiments to Data Analysis and Modeling.

References

[1] A.A. Alizadeh, M.B. Eisen, R.E. Davis, and C. Ma *et al.* Distinct types of diffuse large B-cell lymphoma identified by gene expression profiling. *Nature*, 403:503–510, 2000.

[2] U. Alon, N. Barkai, D.A. Notterman, and K. Gish *et al.* Broad patterns of gene expression revealed by clustering analysis of tumor and normal colon tissues probed by oligonucleotide arrays. *Proc. Natl. Acad. Sci. USA*, 96:6745–6750, 1999.

[3] O. Alter, P.O. Brown, and D. Botstein. Singular value decomposition for genome-wide expression data processing and modeling. *PNAS*, 97:10101–10106, 2000.

[4] P. Baldi. On the convergence of a clustering algorithm for protein-coding regions in microbial genomes. *Bioinformatics*, 16:367–371, 2000.

[5] P. Baldi and P.-F. Baisnée. Sequence analysis by additive scales: DNA structure for sequences and repeats of all lengths. *Bioinformatics*, 16(10):865–889, 2000.

[6] P. Baldi and S. Brunak. *Bioinformatics: the machine learning approach*. MIT Press, Cambridge, MA, 2001. Second edition.

[7] P. Baldi and K. Hornik. Neural networks and principal component analysis: Learning from examples without local minima. *Neural Networks*, 2(1):53–58, 1988.

[8] Z. Bar-Joseph, E.D. Demaine, D.K. Gifford, and A.M. Hamel *et al.* K-ary clustering with optimal leaf ordering for gene expression data. *Bioinformatics*, 19:1070–1078, 2003.

[9] Z. Bar-Joseph, D.K. Gifford, and T.S. Jaakkola. Fast optimal leaf ordering for hierarchical clustering. *Bioinformatics*, 17(Suppl 1):S22–9, 2001.

[10] C.M. Bishop. Bayesian PCA. In M.S. Kearns, S.A. Solla, and D.A. Cohn, editors, *Advances in Neural Information Processing Systems, volume 11*, pages 382–388. MIT Press, Cambridge, MA, 1999.

[11] M. Blanchette and S. Sinha. Separating real motifs from their artifacts. *Bioinformatics*, 17(Suppl 1):S30–8, 2001.

[12] M. Blatt, S. Wiseman, and E. Domany. Super-paramagnetic clustering of data. *Physical Review Letters*, 76:3251–3254, 1996.

[13] A. Brazma, I.J. Jonassen, J. Vilo, and E. Ukkonen. Predicting gene regulatory elements in silico on a genomic scale. *Genome Research*, 8:1202–1215, 1998.

[14] M.P.S. Brown, W.N. Grundy, D. Lin, and N. Cristianini *et al.* Knowledge-based analysis of microarray gene expression data by using support vector machines. *PNAS USA*, 97:262–267, 2000.

[15] J.M. Buhmann and H. Kuhnel. Vector quantization with complexity costs. *IEEE Transactions on Information Theory*, 39(4):1133–1145, 1993.

[16] H.J. Bussemaker, H. Li, and E.D. Siggia. Building a dictionary for genomes: identification of presumptive regulatory sites by statistical analysis. *PNAS*, 97:10096–10100, 2000.

[17] Y. Cheng and G.M. Church. Biclustering of expression data. In *Proceedings of the 2000 Conference on Intelligent Systems for Molecular Biology (ISMB00), La Jolla, CA*, pages 93–103. AAAI Press, Menlo Park, CA, 2000.

[18] D.Y. Chiang, P.O. Brown, and M.B. Eisen. Visualizing associations between genome sequences and gene expression data using genome-mean expression profiles. *Bioinformatics*, 17(Suppl 1):S49–55, 2001.

[19] S.T. Coleman, E.A. Epping, S.M. Steggerda, and W.S. Moye-Rowley. Yap1p activates

gene transcription in an oxidant-specific fashion. *Mol. Cell. Biol.*, 19:8302–8313, 1999.

[20] A.P. Dempster, N.M. Laird, and D.B. Rubin. Maximum likelihood from incomplete data via the EM algorithm. *Journal Royal Statistical Society*, B39:1–22, 1977.

[21] R.O. Duda and P.E. Hart. *Pattern Classification and Scene Analysis*. John Wiley and Sons, 1973.

[22] M.B. Eisen, P.T. Spellman, P.O. Brown, and D. Botstein. Cluster analysis and display of genome-wide expression patterns. *Proc. Natl. Acad. Sci USA*, 95:14863–14868, 1998.

[23] D. Eppstein. Fast hierarchical clustering and other applications of dynamic closest pairs. *Proceedings of the 9th ACM-SIAM Symp. on Discrete Algorithms*, pages 619–628, 1998.

[24] B.S. Everitt. *An Introduction to Latent Variable Models*. Chapman and Hall, London and New York, 1984.

[25] L. Fernandes, C. Rodrigues-Pousada, and K. Struhl. Yap, a novel family of eight bZIP proteins in Saccharomyces cerevisiae with distinct biological functions. *Mol. Cell Biol.*, 17:6982–6993, 1997.

[26] N. Friedman, M. Linial, I. Nachman, and D. Pe'er. Using Bayesian networks to analyze expression data. *Journal of Computational Biology*, 7:601–620, 2000.

[27] L.M. Fu and C.S. Fu-Liu. Multi-class cancer subtype classification based on gene expression signatures with reliability analysis. *FEBS Letters*, 561(1-3):186–190, 2004.

[28] T.S. Furey, N. Cristianini, N. Duffy, and D.W. Bednarski *et al.* Support vector machine classification and validation of cancer tissue samples using microarray expression data. *Bioinformatics*, 16:906–914, 2000.

[29] G. Getz, E. Levine, and E. Domany. Coupled two-way clustering analysis of gene microarray data. *Proc Natl Acad Sci*, 97(22):12079–84, 2000.

[30] T.R. Golub, D.K. Slonim, P. Tamayo, and C. Huard *et al.* Molecular classification of cancer: class discovery and class prediction by gene expression monitoring. *Science*, 286:531–537, 1999.

[31] I. Guyon, J. Weston, S. Barnhill, and V. Vapnik. Gene selection for cancer classification using support vector machines. *machine learning*, 46:389–422, 2002.

[32] S. Hampson, P. Baldi, D. Kibler, and S. Sandmeyer. Analysis of yeast's ORFs upstream regions by parallel processing, microarrays, and computational methods. In *Proceedings of the 2000 Conference on Intelligent Systems for Molecular Biology (ISMB00), La Jolla, CA*, pages 190–201. AAAI Press, Menlo Park, CA, 2000.

[33] L.J. Heyer, S. Kruglyak, and S. Yooseph. Exploring expression data: identification and analysis of co-expressed genes. *Genome Research*, 9:1106–1115, 1999. in press.

[34] J.D. Hughes, P.W. Estep, S. Tavazole, and G.M. Church. Computational identification of *cis*-regulatory elements associated with groups of functionally related genes in *saccharomyces cerevisiae*. *J. Mol. Biol.*, 296:1205–1214, 2000.

[35] M.K. Kerr and G.A. Churchill. Bootstrapping cluster analysis: assessing the reliability of conclusions from microarray experiments. *Proc Natl Acad Sci*, 98:8961–5, 2001.

[36] T. Kohonen. *Self-Organization, Associative Memory*. Springer-Verlag, New York, NY, 1988.

[37] J. Lapointe, C. Li, J.P. Higgins, and M. van de Rijn *et al.* Gene expression profiling identifies clinically relevant subtypes of prostate cancer. *Proc Natl Acad Sci*, 101:811–6, 2004.

[38] C.E. Lawrence, S.F. Altschul, M.S. Boguski, and J.S. Liu *et al.* Detecting subtle sequence signals: a gibbs sampling strategy for multiple alignment. *Science*, 262:208–14, 1993.

[39] M.T. Martinez-Pastor, G. Marchler, C. Schuller, and A. Marchler-Bauer *et al.* The

saccharomyces cerevisiae zinc finger proteins Msn2p and Msn4p are required for transcriptional induction through the stress-response element (STRE. *EMBO Journal*, 15:2227–2235, 1996.

[40] H.W. Mewes, K. Albermann, M. Bahr, and D. Frishman *et al*. Overview of the yeast genome. *Nature*, 387:7–65, 1997.

[41] G. Pavesi, G. Mauri, and G. Pesole. An algorithm for finding signals of unknown length in DNA sequences. *Bioinformatics*, 17(Suppl 1):S207–14, 2001.

[42] A.G. Pedersen, L.J. Jensen, S. Brunak, and H.H. Staerfeldt *et al*. A DNA structural atlas for *escherichia coli*. *J Mol Biol*, 299:907–30, 2000.

[43] P.A. Pevzner. *Computational Molecular Biology. An Algorithmic Approach*. The MIT Press, Cambridge, MA, 2000.

[44] P.A. Pevzner and S. Sze. Combinatorial approaches to finding subtle signals in DNA sequences. In *Proceedings of the 2000 Conference on Intelligent Systems for Molecular Biology (ISMB00), La Jolla, CA*, pages 269–278. AAAI Press, Menlo Park, CA, 2000.

[45] S. Roweis. EM algorithms for PCS and SPCA. In M.I. Jordan, M.S. Kearns, and S.A. Solla, editors, *Advances in Neural Information Processing Systems, volume 10*, pages 626–632. MIT Press, Cambridge, MA, 1998.

[46] B. Scholkopf, A.J. Smola, and K.R. Mller. Nonlinear component analysis as a kernel eigenvalue problem. *Neural Computation*, 10(5):1299–1319, 1998.

[47] R. Sharan and R. Shamir. CLICK: a clustering algorithm with applications to gene expression analysis. In *Proceedings of the 2000 Conference on Intelligent Systems for Molecular Biology (ISMB00), La Jolla, CA*, pages 307–316. AAAI Press, Menlo Park, CA, 2000.

[48] T. Sorlie, C.M. Perou, R. Tibshirani, and T. Aas *et al*. Gene expression patterns of breast carcinomas distinguish tumor subclasses with clinical implications. *Proc Natl Acad Sci*, 98:10869–74, 2001.

[49] P.T. Spellman, G. Sherlock, M.Q. Zhang, and V.R. Iyer *et al*. Comprehensive identification of cell cycle-regulated genes of the yeast *saccharomyces cerevisiae* by microarray hybridization. *Molecular Biology of the Cell*, 9:3273–3297, 1998.

[50] P. Tamayo, D. Slonim, J. Mesirov, and Q. Zhu *et al*. Interpreting patterns of gene expression with self-organizing maps: methods and application to hematopoietic differentiation. *Proc. Natl. Acad. Sci. USA*, 96:2907–2912, 1999.

[51] S. Tavazoie, J.D. Hughes, M.J. Campbell, and R.J. Cho *et al*. Systematic determination of genetic network architecture. *Nat Genet*, 22:281–5, 1999.

[52] R. Tibshirani, T. Hastie, B. Narasimhan, and G. Chu. Diagnosis of multiple cancer types by shrunken centroids of gene expression. *Proc Natl Acad Sci*, 99:6567–72, 2002.

[53] N. Tishby and N. Slonim. Data clustering by Markovian relaxation and the information bottleneck method. In T. Leen, T. Dietterich, and V. Tresp, editors, *Neural Information Processing Systems (NIPS 2000)*, volume 13. MIT Press, Cambridge, MA, 2001.

[54] D.M. Titterington, A.F.M. Smith, and U.E. Makov. *Statistical Analysis of Finite Mixture Distributions*. John Wiley & Sons, New York, 1985.

[55] O.G. Troyanskaya, M.E. Garber, P.O. Brown, and D. Botstein *et al*. Nonparametric methods for identifying differentially expressed genes in microarray data. *Bioinformatics*, 18:1454–61, 2002.

[56] V.G. Tusher, R. Tibshirani, and G. Chu. Significance analysis of microarrays applied to the ionizing radiation response. *Proc Natl Acad Sci*, 98:5116–21, 2001.

[57] J. van Helden, B. Andre, and J. Collado-Vides. Extracting regulatory sites from the upstream region of yeast genes by computational analysis of oligonucleotide frequen-

cies. *J. Mol. Biol.*, 281:827–842, 1998.

[58] J. van Helden, M. del Olmo, and J.E. Perez-Ortin. Statistical analysis of yeast genomic downstream sequences reveals putative polyadenylation signals. *Nucleic Acids Res.*, 28:1000–1010, 2000.

[59] E.P. van Someren, L.F.A. Wessels, and M.J.T. Reinders. Linear modeling of genetic networks from experimental data. In *Proceedings of the 2000 Conference on Intelligent Systems for Molecular Biology (ISMB00), La Jolla, CA*, pages 355–366. AAAI Press, Menlo Park, CA, 2000.

[60] J. Vilo and A. Brazma. Mining for putative regulatory elements in the yeast genome using gene expression data. In *Proceedings of the 2000 Conference on Intelligent Systems for Molecular Biology (ISMB00), La Jolla, CA*, pages 384–394. AAAI Press, Menlo Park, CA, 2000.

[61] A. von Heydebreck, W. Huber, A. Poustka, and M. Vingron. Identifying splits with clear separation: a new class discovery method for gene expression data. *Bioinformatics*, 17(Suppl 1):S107–14, 2001.

[62] E. Wingender, X. Chen, E. Fricke, and R. Geffers *et al.* The TRANSFAC system on gene expression regulation. *Nucleic Acids Res.*, 29:281–284, 2001.

[63] A.L. Wu and W.S. Moye-Rowley. GSH1 which encodes gamma-glutamylcysteine synthetase is a target gene for YAP-1 transcriptional regulation. *Mol. Cell. Biol.*, 14:5832–5839, 1994.

[64] E.P. Xing, M.I. Jordan, and R.M. Karp. Feature selection for high-dimensional genomic microarray data. In *Proc. 18th International Conf. on Machine Learning*, pages 601–608. Morgan Kaufmann, San Francisco, CA, 2001.

[65] C.H. Yeang, S. Ramaswamy, P. Tamayo, and S. Mukherjee *et al.* Molecular classification of multiple tumor types. *Bioinformatics*, 17(Suppl 1):S316–22, 2001.

[66] A. Zien, R. Kuffner, R. Zimmer, and T. Lengauer. Analysis of gene expression data with pathway scores. In *Proceedings of the 2000 Conference on Intelligent Systems for Molecular Biology (ISMB00), La Jolla, CA*, pages 407–417. AAAI Press, Menlo Park, CA, 2000.

26

Biclustering Algorithms: A Survey

Amos Tanay
Tel-Aviv University

Roded Sharan
Tel-Aviv University

Ron Shamir
Tel-Aviv University

26.1 Introduction

Gene expression profiling has been established over the last decade as a standard technique for obtaining a molecular fingerprint of tissues or cells in different biological conditions [18, 7]. Based on the availability of whole genome sequences, the technology of DNA chips (or microarrays) allows the measurement of mRNA levels simultaneously for thousands of genes. The set (or vector) of measured gene expression levels under one condition (or sample) are called the *profile* of that condition. Gene expression profiles are powerful sources of information and have revolutionized the way we study and understand function in biological systems [1].

Given a set of gene expression profiles, organized together as a *gene expression matrix* with rows corresponding to genes and columns corresponding to conditions, a common analysis goal is to group conditions and genes into subsets that convey biological significance. In its most common form, this task translates to the computational problem known as *clustering*. Formally, given a set of elements with a vector of attributes for each element, clustering aims to partition the elements into (possibly hierarchically ordered) disjoint sets, called clusters, so that within each set the attribute vectors are similar, while vectors of disjoint clusters are dissimilar. For example, when analyzing a gene expression matrix we may apply clustering to the genes (as elements) given the matrix rows (as attributes) or cluster the conditions (as elements) given the matrix columns (as attributes). For reviews on clustering see an earlier chapter in this book. Analysis via clustering makes several a priori assumptions that may not be perfectly adequate in all circumstances. First, clustering can be applied to either genes or samples, implicitly directing the analysis to a particular aspect of the system

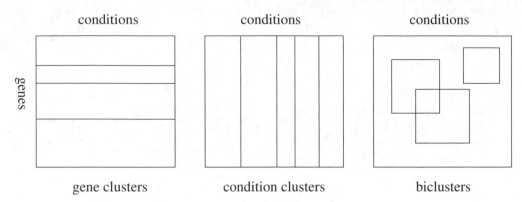

FIGURE 26.1: Clustering and biclustering of a gene expression matrix. Clusters correspond to disjoint strips in the matrix. A gene cluster must contain all columns, and a condition cluster must contain all rows. Biclusters correspond to arbitrary subsets of rows and columns, shown here as rectangles. Note that since gene (condition) clusters are disjoint, the rows (columns) of the matrix can be reordered so that each cluster is a contiguous strip. Similar reordering of rows and columns that shows all the biclusters as rectangles is usually impossible.

under study (e.g., groups of patients or groups of co-regulated genes). Second, clustering algorithms usually seek a disjoint cover of the set of elements, requiring that no gene or sample belongs to more than one cluster.

The notion of a bicluster gives rise to a more flexible computational framework. A *bicluster* is defined as a submatrix spanned by a set of genes and a set of samples (compare Figure 26.1). Alternatively, a bicluster may be defined as the corresponding gene and sample subsets. Given a gene expression matrix, we can characterize the biological phenomena it embodies by a collection of biclusters, each representing a different type of joint behavior of a set of genes in a corresponding set of samples. Note that there are no a-priori constraints on the organization of biclusters and in particular, genes or samples can be part of more than one bicluster or of no bicluster. The lack of structural constrains on biclustering solutions allows greater freedom but is consequently more vulnerable to overfitting. Hence, biclustering algorithms must guarantee that the output biclusters are meaningful. This is usually done by an accompanying statistical model or a heuristic scoring method that define which of the many possible submatrices represent a significant biological behavior. The *biclustering problem* is to find a set of significant biclusters in a matrix.

In clinical applications, gene expression analysis is done on tissues taken from patients with a medical condition. Using such assays, biologists have identified molecular fingerprints that can help in the classification and diagnosis of the patient status and guide treatment protocols [2, 16]. In these studies, the focus is primarily on identifying profiles of expression over a subset of the genes that can be associated with clinical conditions and treatment outcomes, where ideally, the set of samples is equal in all but the subtype or the stage of the disease. However, a patient may be a part of more than one clinical group, e.g., may suffer from syndrome A, have a genetic background B and be exposed to environment C. Biclustering analysis is thus highly appropriate for identifying and distinguishing the biological factors affecting the patients along with the corresponding gene subsets.

In functional genomics applications, the goal is to understand the functions of each of the genes operating in a biological system. The rationale is that genes with similar expression patterns are likely to be regulated by the same factors and therefore may share function. By collecting expression profiles from many different biological conditions and identifying joint patterns of gene expression among them, researchers have characterized transcriptional programs and assigned putative function to thousands of genes [23, 11, 8]. Since genes have multiple functions, and since transcriptional programs are often based on combinatorial regulation, biclustering is highly appropriate for these applications as well.

An important aspect of gene expression data is their high noise levels. DNA chips provide only rough approximation of expression levels, and are subject to errors of up to two-fold the measured value [1]. Any analysis method, and biclustering algorithms in particular, should therefore be robust enough to cope with significant levels of noise.

Below we survey some of the biclustering models and algorithms that were developed for gene expression analysis. Our coverage is not exhaustive, and is biased toward what we believe are the more practical methods. We attempt to cover at least one method from each class of algorithms under development. We do not review methods that are based on extended biological models (e.g., inferring regulation or integrating data types [19, 24]), but focus on algorithms for biclustering per-se. Throughout, we assume that we are given a set of genes V a set of conditions U, and a gene expression matrix $E = (e_{vu})$ where e_{vu} is the expression level of gene v in sample u. We assume that the matrix is normalized, though some of the algorithms below perform additional normalization. A *bicluster* $B = (U', V')$ is defined by a subset of genes $V' \subset V$ and a subset of conditions (or samples) $U' \subset U$. Different algorithmic approaches to the biclustering problem use different measures for the quality of a given biclustering solution. We therefore define the goal function of each algorithm as part of its description.

26.2 Cheng and Church's Algorithm

Cheng and Church were the first to introduce biclustering to gene expression analysis [6]. Their algorithmic framework represents the biclustering problem as an optimization problem, defining a score for each candidate bicluster and developing heuristics to solve the constrained optimization problem defined by this score function. In short, the constraints force the uniformity of the matrix, the procedure gives preference to larger submatrices and the heuristic is a relaxed greedy algorithm.

Cheng and Church implicitly assume that (gene, condition) pairs in a "good" bicluster have a constant expression level, plus possibly additive row and column specific effects. After removing row, column and submatrix averages, the residual level should be as small as possible. More formally, given the gene expression matrix E, a subset of genes I and a subset of conditions J, we define $e_{Ij} = \frac{\sum_{i \in I} e_{ij}}{|I|}$ (row subset average) $e_{iJ} = \frac{\sum_{j \in J} e_{ij}}{|J|}$ (column subset average) and $e_{IJ} = \frac{\sum_{i \in I, j \in J} e_{ij}}{|I||J|}$ (submatrix average). We define the *residue score* of an element e_{ij} in a submatrix E_{IJ} as $RS_{IJ}(i, j) = e_{ij} - e_{Ij} - e_{iJ} + e_{IJ}$ and the *mean square residue score* of the entire submatrix as $H(I, J) = \sum_{i \in I, j \in J} \frac{RS_{ij}^2}{|I||J|}$. The intuition behind this definition can be understood via two examples: a completely uniform matrix will have score zero. More generally, any submatrix in which all entries have the form $e_{ij} = b_i + c_j$ would also have score zero. Given the score definition, the *maximum bicluster problem* seeks a bicluster of maximum size among all biclusters with score not exceeding a threshold δ. The size can be defined in several ways, for example as the number of cells in the matrix ($|I||J|$) or the number of rows plus number of columns ($|I| + |J|$).

The maximum bicluster problem is NP-hard if we force all solutions to be square matrices

```
Cheng-Church(U, V, E, δ):
U : conditions. V : genes.
E : Gene expression matrix.
δ: maximal mean square residue score.
```

Define $e_{Ij} = \frac{\sum_{i \in I} e_{ij}}{|I|}$

Define $e_{iJ} = \frac{\sum_{j \in J} e_{ij}}{|J|}$

Define $e_{IJ} = \frac{\sum_{i \in I, j \in J} e_{ij}}{|I||J|}$

Define $RS_{IJ}(i,j) = e_{ij} - e_{Ij} - e_{iJ} + e_{IJ}$

Define $H(I,J) = \sum_{i \in I, j \in J} \frac{RS_{ij}^2}{|I||J|}$.

Initialize a bicluster (I,J) with $I = U, J = V$.

Deletion phase:
   While $(H(I,J) > \delta)$ do
       Compute for $i \in I$, $d(i) = \frac{1}{|J|} \sum_{j \in J} RS_{I,J}(i,j)$.
       Compute for $j \in J$, $e(j) = \frac{1}{|I|} \sum_{i \in I} RS_{I,J}(i,j)$.
       If $max_{i \in I} d(i) > max_{j \in J} e(j)$ assign $I = I \setminus \{argmax_i(d(i))\}$.
       Else $J = J \setminus \{argmax_j(e(j))\}$

Addition phase:
   assign $I' = I, J' = J$
   While $(H(I',J') < \delta)$ do
       Assign $I = I', J = J'$
       Compute for $i \in U \setminus I$, $d(i) = \frac{1}{|J|} \sum_{j \in J} RS_{I,J}(i,j)$.
       Compute for $j \in V \setminus J$, $e(j) = \frac{1}{|I|} \sum_{i \in I} RS_{I,J}(i,j)$.
       If $max_{i \in I} d(i) < max_{j \in J} e(j)$ assign $I' = I \cup \{argmax_i(d(i))\}$.
       Else $J' = J \cup \{argmax_j(e(j))\}$

Report I, J

FIGURE 26.2: The Cheng-Church algorithm for finding a single bicluster.

$(|I| = |J|)$ or if we use the total number of submatrix cells as our optimization goal (Reductions are from Maximum Balanced Biclique or Maximum Edge Biclique). Cheng and Church suggested a greedy heuristic to rapidly converge to a locally maximal submatrix with score smaller than the threshold. The algorithm (presented in Figure 26.2) can be viewed as a local search algorithm starting from the full matrix. Given the threshold parameter δ, the algorithm runs in two phases. In the first phase, the algorithm removes rows and columns from the full matrix. At each step, where the current submatrix has row set I and column set J, the algorithm examines the set of possible moves. For rows it calculates $d(i) = \frac{1}{|J|} \sum_{j \in J} RS_{I,J}(i,j)$ and for columns it calculates $e(j) = \frac{1}{|I|} \sum_{i \in I} RS_{I,J}(i,j)$. It then selects the highest scoring row or column and removes it from the current submatrix, as long as $H(I,J) > \delta$. The idea is that rows/columns with large contribution to the score can be removed with guaranteed improvement (decrease) in the total mean square residue score. A possible variation of this heuristic removes at each step all rows/columns with a contribution to the residue score that is higher than some threshold.

   In the second phase of the algorithm, rows and columns are being added, using the same scoring scheme, but this time looking for the lowest square residues $d(i), e(j)$ at each move, and terminating where none of the possible moves increases the matrix size without crossing the threshold δ. Upon convergence, the algorithm outputs a submatrix with low mean residue and locally maximal size.

TWOWAY(U, V, E, ALG):
U : conditions. V : genes.
E : Gene expression matrix.
ALG : one-dimensional clustering algorithm. Inputs a matrix and outputs
    significant (stable) clusters of columns or rows.
Initialize a hash table *weight*
Initialize $\mathcal{U}_1 = \{U\}$, $\mathcal{V}_1 = \{V\}$
Initialize $\mathcal{U} = \emptyset$, $\mathcal{V} = \emptyset$
Initialize the sets hierarchy table H_V storing for gene clusters the condition
    subsets used to generate them.
Initialize the sets hierarchy table H_U storing for condition clusters the gene
    subsets used to generate them.
While ($\mathcal{U}_1 \neq \emptyset$ or $\mathcal{V}_1 \neq \emptyset$) do
        Initialize empty sets $\mathcal{U}_2, \mathcal{V}_2$.
        For all $(U', V') \in (\mathcal{U}_1 \times \mathcal{V}_1) \cup (\mathcal{U}_1 \times \mathcal{V}) \cup (\mathcal{U} \times \mathcal{V}_1)$ do
            Run $ALG(E_{U'V'})$ to cluster the genes in V':
                Add the stable gene sets to $\mathcal{V}_2$
                Set $H_V[V''] = U'$ for all new clusters V''.
            Run $ALG(E_{U'V'})$ to cluster the conditions in U':
                Add the stable condition sets to $\mathcal{U}_2$
                Set $H_U[U''] = V'$ for all new clusters U''.
      Assign $\mathcal{U} = \mathcal{U} \cup \mathcal{U}_1$, $\mathcal{V} = \mathcal{V} \cup \mathcal{V}_1$
      Assign $\mathcal{U}_1 = \mathcal{U}_2$, $\mathcal{V}_1 = \mathcal{V}_2$
Report $\mathcal{U}, \mathcal{V}$ and their hierarchies H_U, H_V.

FIGURE 26.3: Coupled two-way clustering.

To discover more than one bicluster, Cheng and Church suggested repeated application of the biclustering algorithm on modified matrices. The modification includes randomization of the values in the cells of the previously discovered biclusters, preventing the correlative signal in them to be beneficial for any other bicluster in the matrix. This has the obvious effect of precluding the identification of biclusters with significant overlaps.

An application of the algorithm to yeast and human data is described in [6]. The software is available at ⟨http://arep.med.harvard.edu/biclustering⟩.

26.3 Coupled Two-way Clustering

Coupled two-way clustering (CTWC), introduced by Getz, Levine and Domany [9], defines a generic scheme for transforming a one-dimensional clustering algorithm into a biclustering algorithm. The algorithm relies on having a one-dimensional (standard) clustering algorithm that can discover significant (termed *stable* in [9]) clusters. Given such an algorithm, the coupled two-way clustering procedure will recursively apply the one-dimensional algorithm to submatrices, aiming to find subsets of genes giving rise to significant clusters of conditions and subsets of conditions giving rise to significant gene clusters. The submatrices defined by such pairings are called *stable submatrices* and correspond to biclusters. The algorithm, which is shown in Figure 26.3, operates on a set of gene subsets $\mathcal{V}$ and a set of condition subsets $\mathcal{U}$. Initially $\mathcal{V} = \{V\}$ and $\mathcal{U} = \{U\}$. The algorithm then iteratively selects a gene subset $V' \in \mathcal{V}$ and a condition subset $U' \in \mathcal{U}$ and applies the one dimensional clustering algorithm twice, to cluster V' and U' on the submatrix $U' \times V'$. If stable clusters are detected, their gene/condition subsets are added to the respective sets $\mathcal{V}, \mathcal{U}$. The process is repeated until no new stable clusters can be found. The implementation makes sure that each pair of subsets is not encountered more than once.

Note that the procedure avoids the consideration of all rows and column subsets, by starting from an established row subset when forming subclusters of established column subsets, and vice versa. The success of the coupled two-way clustering strategy depends on the performance of the given one-dimensional clustering algorithm. We note that many popular clustering algorithms (e.g. K-means, Hierarchical, SOM) cannot be plugged "as is" into the coupled two-way machinery, as they do not readily distinguish significant clusters from non-significant clusters or make a-priori assumption on the number of clusters. Getz et al. have reported good results using the SPC hierarchical clustering algorithm [10]. The results of the algorithm can be viewed in a hierarchical form: each stable gene (condition) cluster is generated given a condition (resp. gene) subset. This hierarchical relation is important when trying to understand the context of joint genes or conditions behavior. For example, when analyzing clinical data, Getz et al. have focused on gene subsets giving rise to stable tissue clusters that are correlative to known clinical attributes. Such gene sets may have an important biological role in the disease under study.

The CTWC algorithm has been applied to a variety of clinical data sets (see, e.g., [17]), the software can be downloaded via the site (`http://ctwc.weizmann.ac.il`).

26.4   The Iterative Signature Algorithm

In the Iterative Signature Algorithm (ISA) [12, 5] the notion of a significant bicluster is defined intrinsically on the bicluster genes and samples — the samples of a bicluster uniquely define the genes and vice versa. The intuition is that the genes in a bicluster are co-regulated and, thus, for each sample the average gene expression over all the bicluster's genes should be surprising (unusually high or low) and for each gene the average gene expression over all biclusters samples should be surprising. This intuition is formalized using a simple linear model for gene expression assuming normally distributed expression levels for each gene or sample as shown below.

The algorithm, presented in Figure 26.4, uses two normalized copies of the original gene expression matrix. The matrix E^G has rows normalized to mean 0 and variance 1 and the matrix E^C has columns normalized similarly. We denote by $e^G_{V'u}$ the mean expression of genes from V' in the sample u and by $e^C_{vU'}$ the mean expression of the gene v in samples from U'. A bicluster $B = (U', V')$ is required to have:

$$U' = \{u \in U : \ |e^G_{V'u}| > T_C \sigma_C\}, V' = \{v \in V : \ |e^C_{vU'}| > T_G \sigma_G\} \qquad (26.1)$$

Here T_G is the threshold parameter and σ_G is the standard deviation of the means $e^C_{vU'}$ where v ranges over all possible genes and U' is fixed. Similarly, T_C, σ_C are the corresponding parameters for the column set V'. The idea is that if the genes in V' are up- or down-regulated in the conditions U' then their average expression should be significantly far (i.e., T_G standard deviations) from its expected value on random matrices (which is 0 since the matrix is standardized). A similar argument holds for the conditions in U'. The standard deviations can be predicted as $\frac{1}{\sqrt{|U'|}}, \frac{1}{\sqrt{|V'|}}$ being a linear sum of $|U'|$ (or $|V'|$) independent standard random variables. Alternatively (and in fact, more practically), the standard deviations can be estimated directly from the data, correcting for possible biases in the statistics of the specific condition and gene sets used. In other words, in a bicluster, the z-score of each gene, measured with respect to the bicluster's samples, and the z-score of each sample, measured with respect to the bicluster's samples, should exceed a threshold.

ISA(U, V, E, V_{in}, T_G, T_C, m, ϵ):
U : conditions. V : genes.
E : Gene expression matrix.
V_{in} : Initial gene set.
T_G, T_C: gene and condition z-score thresholds.
m, ϵ: stopping criteria.
Construct a column standardized matrix E^C.
Construct a row standardized matrix E^G.
Initialize counters $n = 0, n' = 0$.
Initialize the current genes set $V' = V_{in}$
Initialize an empty condition set U'.
While $(n - n' < m)$ do
$\qquad$ Compute $e^G_{V'u} = \frac{1}{|V'|} \sum_{v \in V'} e^G_{vu}$ for $u \in U$.
$\qquad$ $U' = \{u \in U : |e^G_{V'u}| > \frac{T_C}{\sqrt{|V'|}}\}$
$\qquad$ Compute $e^C_{vU'} = \frac{1}{|U'|} \sum_{u \in U'} e^C_{vu}$ for $v \in V$.
$\qquad$ $V'' = V'$
$\qquad$ $V' = \{v \in V : |e^C_{vU'}| > \frac{T_G}{\sqrt{U'}}\}$
$\qquad$ if $(\frac{|V' \setminus V''|}{|V' \cup V''|} < \epsilon)$ then $n' = n$
$\qquad$ $n = n + 1$
Report U', V'

FIGURE 26.4: The ISA algorithm for finding a single bicluster.

As we shall see below, ISA will not discover biclusters for which the conditions (26.1) hold strictly, but will use a relaxed version.

The algorithm starts from an arbitrary set of genes $V_0 = V_{in}$. The set may be randomly generated or selected based on some prior knowledge. The algorithm then repeatedly applies the update equations:

$$U_i = \{u \in U : |e^G_{V_i u}| > T_C \sigma_C\}, V_{i+1} = \{v \in V : |e^C_{vU_i}| > T_G \sigma_G\} \qquad (26.2)$$

The iterations are terminated at step n satisfying:

$$\frac{|V_{n-i} \setminus V_{n-i-1}|}{|V_{n-i} \cup V_{n-i-1}|} < \epsilon \qquad (26.3)$$

for all i smaller than some m. The ISA thus converges to an approximated fixed point that is considered to be a bicluster. The actual fixed point depends on both the initial set V_{in} and the threshold parameters T_C, T_G. To generate a representative set of biclusters, it is possible to run ISA with many different initial conditions, including known sets of associated genes or random sets, and to vary the thresholds. After eliminating redundancies (fixed points that were encountered several times), the set of fixed points can be analyzed as a set of biclusters.

The ISA algorithm can be generalized by assigning weights for each gene/sample such that genes/samples with a significant behavior (higher z-score) will have larger weights. In this case, the simple means used in (26.1) and (26.2) are replaced by weighted means and the algorithm can be represented using matrix operations.

The signature algorithm has been applied for finding cis-regulatory modules in yeast ([12]) and for detecting conserved transcriptional modules across several species ([4]). For software see (http://barkai-serv.weizmann.ac.il/GroupPage/).

26.5   The SAMBA Algorithm

The SAMBA algorithm (Statistical-Algorithmic Method for Bicluster Analysis) [24, 20] uses probabilistic modeling of the data and graph theoretic techniques to identify subsets of genes that *jointly respond* across a subset of conditions, where a gene is termed *responding* in some condition if its expression level changes significantly at that condition with respect to its normal level. Within the SAMBA framework, the expression data are modeled as a bipartite graph whose two parts correspond to conditions and genes, respectively, with edges for significant expression changes. The vertex pairs in the graph are assigned weights according to a probabilistic model, so that heavy subgraphs correspond to biclusters with high likelihood. Discovering the most significant biclusters in the data reduces under this weighting scheme to finding the heaviest subgraphs in the model bipartite graph. SAMBA employs a practical heuristic to search for heavy subgraphs. The search algorithm is motivated by a combinatorial algorithm for finding heavy bicliques that is exponential in the maximum gene degree in the graph.

In the following we describe the probabilistic model used by SAMBA and the theoretical algorithm on which the search method is based. Finally, the full SAMBA algorithm is presented.

Applications of SAMBA for gene expression data are described in [25]. SAMBA was also applied to highly heterogeneous data, including expression, phenotype growth sensitivity, protein-protein interaction and ChIP-chip data [24]. The software is available as part of the Expander package [20, 21].

26.5.1   Statistical Data Modeling

The SAMBA algorithm is based on representing the input expression data as a bipartite graph $G = (U, V, E)$. In this graph, U is the set of conditions, V is the set of genes, and $(u, v) \in E$ iff v responds in condition u, that is, if the expression level of v changes significantly in u. A bicluster corresponds to a subgraph $H = (U', V', E')$ of G, and represents a subset V' of genes that are co-regulated under a subset of conditions U'. The *weight* of a subgraph (or bicluster) is the sum of the weights of gene-condition pairs in it, including edges and non-edges.

Coupled with the graph representation is a likelihood ratio model for the data. Let $H = (U', V', E')$ be a subgraph of G and denote $\overline{E'} = (U' \times V') \setminus E'$. For a vertex $w \in U' \cup V'$ let d_w denote its degree in G. The null model assumes that the occurrence of each edge (u, v) is an independent Bernoulli variable with parameter $p_{u,v}$. The probability $p_{u,v}$ is the fraction of bipartite graphs with degree sequence identical to G that contain the edge (u, v). In practice, one estimates $p_{u,v}$ using a Monte-Carlo process. This model tries to capture the characteristics of the different genes and conditions in the data.

The alternative model assumes that each edge of a bicluster occurs with constant, high probability p_c. This model reflects the belief that biclusters represent approximately uniform relations between their elements. The log likelihood ratio for H is therefore:

$$\log L(H) = \sum_{(u,v) \in E'} \log \frac{p_c}{p_{u,v}} + \sum_{(u,v) \in \overline{E'}} \log \frac{1 - p_c}{1 - p_{u,v}}$$

Setting the weight of each edge (u, v) to $\log \frac{p_c}{p_{u,v}} > 0$ and the weight of each non-edge (u, v) to $\log \frac{1 - p_c}{1 - p_{u,v}} < 0$, one concludes that the score of H is simply its weight.

```
MaxBoundBiClique(U, V, E, d):
Initialize a hash table weight; weight_best ← 0
For all v ∈ V do
      For all S ⊆ N(v) do
            weight[S] ←weight[S]+
                        max{0, w(S, {v})}
            If (weight[S] > weight_best)
               U_best ← S
               weight_best ← weight[S]
Compute V_best = ∩_{u∈U_best} N(u)
Output (U_best, V_best)
```

FIGURE 26.5: An algorithm for the maximum bounded biclique problem.

26.5.2   Finding Heavy Subgraphs

Under the above additive scoring scheme, discovering the most significant biclusters in the data reduces under this scoring scheme to finding the heaviest subgraphs in the bipartite graph. Since the latter problem is NP-hard, SAMBA employs a heuristic search for such subgraphs. The search uses as seeds heavy bicliques and we now present the underlying algorithm to find good seeds. In the rest of the section it will be convenient to assume that the degree of every gene is bounded by d.

Let $G = (U, V, E)$ be a bipartite graph with $n = |V|$ genes. Let $w : U \times V \to \mathcal{R}$ be a weight function. For a pair of subsets $U' \subseteq U, V' \subseteq V$ we denote by $w(U', V')$ the weight of the subgraph induced on $U' \cup V'$, i.e., $w(U', V') = \sum_{u \in U', v \in V'} w((u, v))$. The *neighborhood* of a vertex v, denoted $N(v)$, is the set of vertices adjacent to v in G.

The *Maximum Bounded Biclique* problem calls for identifying a maximum weight complete subgraph of a given weighted bipartite graph G, such that the vertices on one side of G have degrees bounded by d. This problem can be solved in $O(n2^d)$ time (and space) as we show next.

Observe that a maximum bounded biclique $H^* = (U^*, V^*, E^*)$ in G must have $|U^*| \leq d$. Figure 26.5 describes a hash-table based algorithm that for each vertex $v \in V$ scans all $O(2^d)$ subsets of its neighbors, thereby identifying the heaviest biclique. Each hash entry corresponds to a subset of conditions and records the total weight of edges from adjacent gene vertices. The algorithm can be shown to spend $O(n2^d)$ time on the hashing and finding U_{best}. Computing V_{best} can be done in $O(nd)$ time, so the total running time is $O(n2^d)$.

Note that the algorithm can be adapted to give the k condition subsets that induce solutions of highest weight in $O(n2^d \log k)$ time using a priority queue data structure.

26.5.3   The Full Algorithm

Having described the two main components of SAMBA, we are now ready to present the full algorithm, which is given in Figure 26.6. SAMBA proceeds in two phases. First, the model bipartite graph is formed and the weights of vertex pairs are computed. Second, several heavy subgraphs are sought around each vertex of the graph. This is done by starting with good seeds around the vertex and expanding them using local search. The seeds are found using the hashing technique of the algorithm in Figure 26.5. To save on time and space the algorithm ignores genes with degree exceeding some threshold D, and hash for each

SAMBA(U, V, E, w, d, N_1, N_2, k):
U : conditions. V : genes.
E : graph edges. w : edge/non-edge weights.
N_1, N_2 : condition set hashed set size limits. k : max biclusters per gene/condition.
Initialize a hash table *weight*
For all $v \in V$ with $|N(v)| \le d$ do
    For all $S \subseteq N(v)$ with $N_1 \le |S| \le N_2$ do
        $weight[S] \leftarrow weight[S] + w(S, \{v\})$
For each $v \in V$ set $best[v][1\ldots k]$ to the k heaviest sets S such that $v \in S$
For each $v \in V$ and each of the k sets $S = best[v][i]$
    $V' \leftarrow \cap_{u \in S} N(u)$.
    $B \leftarrow S \cup V'$.
    Do {
        $a = argmax_{x \in V \cup U}(w(B \cup x))$
        $b = argmax_{x \in B}(w(B \setminus x))$
        If $w(B \cup a) > w(B \setminus b)$ then $B = B \cup a$ else $B = B \setminus b$
    } **while** improving
    Store B.
Post process to filter overlapping biclusters.

FIGURE 26.6: The SAMBA biclustering algorithm.

gene only subsets of its neighbors of size ranging from N_1 to N_2. The local improvement procedure iteratively applies the best modification to the current bicluster (addition or deletion of a single vertex) until no score improvement is possible. The greedy process is restricted to search around the biclique without performing changes that would eliminate vertices in it or make vertices in it redundant (having a total negative contribution to the bicluster score). To avoid similar biclusters whose vertex sets differ only slightly, a final step greedily filters similar biclusters with more than $L\%$ overlap.

26.6 Spectral Biclustering

Spectral biclustering approaches use techniques from linear algebra to identify bicluster structures in the input data. Spectral biclustering approaches use techniques from linear algebra to identify bicluster structures in the input data. Here we review the biclustering technique presented in Kluger et al. [13]. In this model, it is assumed that the expression matrix has a hidden checkerboard-like structure that we try to identify using eigenvector computations. The structure assumption is argued to hold for clinical data, where tissues cluster to cancer types and genes cluster to groups, each distinguishing a particular tissue type from the other types.

To describe the algorithm, suppose at first that the matrix E has a checkerboard-like structure (see Figure 26.7). Obviously we could discover it directly, but we could also infer it using a technique from linear algebra that will be useful in case the structure is hidden due to row and column shuffling. The technique is based on a relation between the block structure of E and the block structure of pairs of eigenvectors for EE^T and $E^T E$, which we describe next. First, observe that the eigenvalues of EE^T and $E^T E$ are the same. Now, consider a vector x that is *stepwise*, i.e., piecewise constant, and whose block structure matches that of the rows of E. Applying E to x we get a stepwise vector y. If we now apply E^T to y we get a vector with the same block structure as x. The same relation is observed

$$Ex = \begin{bmatrix} 8 & 8 & 7 & 7 & 3 & 3 \\ 8 & 8 & 7 & 7 & 3 & 3 \\ 6 & 6 & 4 & 4 & 5 & 5 \\ 6 & 6 & 4 & 4 & 5 & 5 \end{bmatrix} \begin{bmatrix} a \\ a \\ b \\ b \\ c \\ c \end{bmatrix} = \begin{bmatrix} d \\ d \\ e \\ e \end{bmatrix} = y, E^T y = \begin{bmatrix} 8 & 8 & 6 & 6 \\ 8 & 8 & 6 & 6 \\ 7 & 7 & 4 & 4 \\ 7 & 7 & 4 & 4 \\ 3 & 3 & 5 & 5 \\ 3 & 3 & 5 & 5 \end{bmatrix} \begin{bmatrix} d \\ d \\ e \\ e \end{bmatrix} = \begin{bmatrix} a' \\ a' \\ b' \\ b' \\ c' \\ c' \end{bmatrix} = x'$$

FIGURE 26.7: An example of a checkerboard-like matrix E and the eigenvectors of EE^T and $E^T E$. The vector x satisfies the relation $E^T Ex = E^T y = x' = \lambda x$. Similarly, y satisfies the equation $EE^T y = E\lambda x = \lambda y$.

when applying first E^T and then E (see Figure 26.7). Hence, vectors of the stepwise pattern of x form a subspace that is closed under $E^T E$. This subspace is spanned by eigenvectors of this matrix. Similarly, eigenvectors of EE^T span the subspace formed by vectors of the form of y. More importantly, taking now x to be an eigenvector of $E^T E$ with an eigenvalue λ, we observe that $y = Ex$ is an eigenvector of EE^T with the same eigenvalue.

In conclusion, the checkerboard-like structure of E is reflected in the stepwise structures of pairs of EE^T and $E^T E$ eigenvectors that correspond to the same eigenvalue. One can find these eigenvector pairs by computing a singular value decomposition of E. Singular value decomposition is a standard algebraic technique (cf. [15]) that expresses a real matrix E as a product $E = A\Delta B^T$, where Δ is a diagonal matrix and A and B are orthonormal matrices. The columns of A and B are the eigenvectors of EE^T and $E^T E$, respectively. The entries of Δ are square roots of the corresponding eigenvalues, sorted in a non-increasing order. Hence the eigenvector pairs are obtained by taking for each i the ith columns of A and B, and the corresponding eigenvalue is the Δ_{ii}^2.

For any eigenvector pair, one can check whether each of the vectors can be approximated using a piecewise constant vector. Kluger et al. use a one-dimensional k-means algorithm to test this fit. The block structures of the eigenvectors indicate the block structures of the rows and columns of E.

In the general case, the rows and columns of E are ordered arbitrarily, and the checkerboard-like structure, if E has one, is hidden. To reveal such structure one computes the singular value decomposition of E and analyzes the eigenvectors of EE^T and $E^T E$. A hidden checkboard structure will manifest itself by the existence of a pair of eigenvectors (one for each matrix) with the same eigenvalue, that are approximately piecewise constant. One can determine if this is the case by sorting the vectors or by clustering their values, as done in [13].

Kluger et al. further discuss the problem of normalizing the gene expression matrix to reveal checkerboard structures that are obscured, e.g., due to differences in the mean expression levels of genes or conditions. The assumed model for the data is a multiplicative model, in which the expression level of a gene i in a condition j is its base level times a gene term, which corresponds to the gene's tendency of expression under different conditions, times a condition term, that represents the tendency of genes to be expressed under condition j. The normalization is done using two normalizing matrices: R, a diagonal matrix with the mean of row i at the ith position; and C, a diagonal matrix with the mean of column j at the jth position. The block structure of E is now reflected in the stepwise structure of pairs of eigenvectors with the same eigenvalue of the normalized matrices $M = R^{-1}EC^{-1}E^T$ and M^T. These eigenvector pairs can be deduced by computing a singular value decomposition of $R^{-1/2}EC^{-1/2}$. Due to the normalization, the first eigenvector pair (corresponding to an

Spectral(U, V, E):
U : conditions. V : genes.
$E_{n \times m}$: Gene expression matrix.
Compute $R = diag(E \cdot 1_m)$ and $C = diag(1_n^T \cdot E)$.
Compute a singular value decomposition of $R^{-1/2} E C^{-1/2}$.
Discard the pair of eigenvectors corresponding to the largest eigenvalue.
For each pair of eigenvectors u, v of $R^{-1} E C^{-1} E^T$ and $C^{-1} E^T R^{-1} E$ with the same eigenvalue do:
    Apply k-means to check the fit of u and v to stepwise vectors.
Report the block structure of the p u, v with the best stepwise fit.

FIGURE 26.8: The spectral biclustering algorithm.

eigenvalue of 1) is constant and can be discarded. A summary of the biclustering algorithm is given in Figure 26.8.

The spectral algorithm was applied to human cancer data and its results were used for classification of tumor type and identification of marker genes [13].

26.7   Plaid Models

The Plaid model [14] is a statistically inspired modeling approach developed by Lazzeroni and Owen for the analysis of gene expression data. The basic idea is to represent the genes-conditions matrix as a superposition of *layers*, corresponding to biclusters in our terminology, where each layer is a subset of rows and columns on which a particular set of values takes place. Different values in the expression matrix are thought of as different colors, as in (false colored) "heat maps" of chips. This metaphor also leads to referring to "color intensity" in lieu of "expression level". The horizontal and vertical color lines in the matrix corresponding to a layer give the method its name.

The model assumes that the level of matrix entries is the sum of a uniform background ("grey") and of k biclusters each coloring a particular submatrix in a certain way. More precisely, the expression matrix is represented as

$$A_{ij} = \mu_0 + \sum_{k=1}^{K} \theta_{ijk} \rho_{ik} \kappa_{jk}$$

where μ_0 is a general matrix background color, and $\theta_{ijk} = \mu_k + \alpha_{ik} + \beta_{jk}$ where μ_k describes the added background color in bicluster k, α and β are row and column specific additive constants in bicluster k. $\rho_{ik} \in \{0, 1\}$ is a gene-bicluster membership indicator variable, i.e., $\rho_{ik} = 1$ iff gene i belongs to the gene set of the k-th bicluster. Similarly, $\kappa_{jk} \in \{0, 1\}$ is a sample-bicluster membership indicator variable. Hence, similar to Cheng and Church [6], a bicluster is assumed to be the sum of bicluster background level plus row-specific and column-specific constants.

When the biclusters form a k-partition of the genes and a corresponding k-partition of the samples, the *disjointness constraints* that biclusters cannot overlap can be formulated as $\sum_k \kappa_{jk} \leq 1$ for all j, $\sum_k \rho_{ik} \leq 1$ for all i. Replacing $\leq$ by $=$ would require assignment of each row or column to *exactly* one bicluster. Generalizing to allow bicluster overlap simply means removing the disjointness constraints.

The general biclustering problem is now formulated as finding parameter values so that the resulting matrix would fit the original data as much as possible. Formally, the problem is minimizing

$$\sum_{ij}[A_{ij} - \sum_{k=0}^{K} \theta_{ijk}\rho_{ik}\kappa_{jk}]^2 \tag{26.4}$$

where $\mu_0 = \theta_{ij0}$. If α_{ik} or β_{jk} are used, then the constraints $\sum_i \rho_{ik}\alpha_{ik} = 0$ or $\sum_j \kappa_{jk}\beta_{jk} = 0$ are added to reduce the number of parameters. Note that the number of parameters is at most $k + 1 + kn + km$ for the θ variables, and $kn + km$ for the κ and ρ variables. This is substantially smaller than the nm variables in the original data, if $k << max(n, m)$.

26.7.1 Estimating Parameters

Lazzeroni and Owen propose to solve problem (26.4) using an iterative heuristic. New layers are added to the model one at a time. Suppose we have fixed the first $K - 1$ layers and we are seeking for the K-th layer to minimize the sum of squared errors. Let

$$Z_{ij}^{(K-1)} = A_{ij} - \sum_{k=0}^{K-1} \theta_{ijk}\rho_{ij}\kappa_{jk} \tag{26.5}$$

be the *residual matrix* after removing the effect of the first $K - 1$ layers. In iteration K we wish to solve the following quadratic integer program.

$$\begin{aligned} min \quad & Q^{(K)} = \tfrac{1}{2}\sum_{i=1}^{n}\sum_{j=1}^{p}(Z_{ij}^{(K-1)} - \theta_{ijK}\rho_{iK}\kappa_{jK})^2 \\ s.t. \quad & \sum_i \rho_{iK}^2\alpha_{iK} = 0, \;\; \sum_j \kappa_{jK}^2\beta_{jK} = 0 \\ & \rho_{iK} \in \{0,1\}, \;\; \kappa_{jK} \in \{0,1\} \end{aligned} \tag{26.6}$$

The proposed heuristic method to solve (26.6) is again iterative. To avoid confusion we call the iterations for fixed K *cycles*, and indicate the cycle number by a superscript in parentheses, e.g. $\theta^{(i)}$. The integrality constraints are ignored throughout, and the goal is to solve corresponding relaxation of it. A cycle is done as follows: compute the best values of the θ parameters given fixed ρ and κ values; compute the best values of the ρ parameters given new θ and the old κ values; compute the best values of the κ parameters given the new θ and the old ρ values. In order to avoid "locking in" of the membership variables to 0 or 1, their values are changed only modestly on the first cycle, and they are allowed to become integral only at the final cycle.

The following optimal parameter values in the relaxed version of (26.6) are obtained by using Lagrange multipliers:

$$\mu_K = \frac{\sum_i \sum_j \rho_{iK}\kappa_{jK}Z_{ij}^{K-1}}{(\sum_i \rho_{iK}^2)(\sum_j \kappa_{jK}^2)} \tag{26.7}$$

$$\alpha_{iK} = \frac{\sum_j (Z_{ij}^{(K-1)} - \mu_K\rho_{iK}\kappa_{jK})\kappa_{jK}}{\rho_{iK}\sum_{jK}\kappa_{jK}^2} \tag{26.8}$$

$$\beta_{jK} = \frac{\sum_i (Z_{ij}^{(K-1)} - \mu_K\rho_{iK}\kappa_{jK})\rho_{iK}}{\kappa_{jK}\sum_{iK}\rho_{iK}^2} \tag{26.9}$$

So, in cycle s, we use these equations to update $\theta^{(s)}$ using the old values $\rho^{(s-1)}$ and $\kappa^{(s-1)}$. The values for ρ_{iK} and κ_{jK} that minimize Q are:

$$\rho_{iK} = \frac{\sum_j \theta_{ijK}\kappa_{jK}Z_{ij}^{K-1}}{\sum_j \theta_{ijK}^2\kappa_{jK}^2} \tag{26.10}$$

Plaid(U, V, E, S):
U : conditions. V : genes.
E : Gene expression matrix.
S: maximum cycles per iteration.
Set $K = 0$
adding a new layer:
    K=K+1
    Compute initial values of $\kappa_{jK}^{(0)}, \rho_{iK}^{(0)}$. Set $s = 1$
    While ($s \leq S$) do:
        Compute $\mu_K^{(s)}$, $\alpha_{iK}^{(s)}$, $\beta_{jK}^{(s)}$ using equations (26.7)- (26.9).
        Compute $\kappa_K^{(s)}$ using equations (26.11)
        Compute $\rho_K^{(s)}$ using equations (26.10)
        If $\rho_K^{(s)} > 0.5$ set $\rho_K^{(s)} = 0.5 + s/2S$, else set $\rho_K^{(s)} = 0.5 - s/2S$
        If $\kappa_K^{(s)} > 0.5$ set $\kappa_K^{(s)} = 0.5 + s/2S$, else set $\kappa_K^{(s)} = 0.5 - s/2S$
    If the importance of layer K is non random then record the layer and repeat
    Else exit.
Report layers $1, \ldots, K - 1$.

FIGURE 26.9: The Plaid model algorithm.

$$\kappa_{jK} = \frac{\sum_i \theta_{ijK} \rho_{iK} Z_{ij}^{K-1}}{\sum_i \theta_{ijK}^2 \rho_{iK}^2} \qquad (26.11)$$

At cycle s, we use these equations to update $\rho^{(s)}$ from $\theta^{(s)}$ and $\kappa^{(s-1)}$, and update $\kappa^{(s)}$ from $\theta^{(s)}$ and $\rho^{(s-1)}$. The complete updating process is repeated a prescribed number of cycles.

26.7.2 Initialization and Stopping Rule

The search for a new layer K in the residual matrix $Z_{ij} = Z_{ij}^{(K)}$ requires initial values of ρ and κ. These values are obtained by finding vectors u and v and a real value λ so that $\lambda u v^T$ is the best rank one approximation of Z. We refer the readers to the original paper for details.

Intuitively, each iteration "peels off" another signal layer, and one should stop after $K - 1$ iterations if the residual matrix $Z_{ij} = Z_{ij}^{(K)}$ contains almost only noise. Lazzeroni and Owen define the *importance* of layer k by $\sigma_k^2 = \sum_{i=1}^n \sum_{j=1}^p \rho_{ik} \kappa_{jk} \theta_{ijk}^2$. The algorithm accepts a layer if it has significantly larger importance than in noise. To evaluate σ_k^2 on noise, repeat the following process T times: Randomly permute each row in Z independently, and then randomly permute each column in the resulting matrix independently. Apply the layer-finding algorithm on the resulting matrix, and compute the importance of that layer. If σ_k^2 exceeds the importance obtained for all the T randomized matrices, add the new layer K to the model.

The complete algorithm is outlined in Figure 26.9.

Plaid models have been applied to yeast gene expression data [14]. The software is available at 〈http://www-stat.stanford.edu/~owen/plaid〉.

26.8 Discussion

The algorithms presented above demonstrate some of the approaches developed for the identification of bicluster patterns in large matrices, and in gene expression matrices in particular. One can roughly classify the different methods a) by their model and scoring schemes and b) by the type of algorithm used for detecting biclusters. Here we briefly review how different methods tackle these issues.

26.8.1 Model and score

To ensure that the biclusters are statistically significant, each of the biclustering methods defines a scoring scheme to assess the quality of candidate biclusters, or a constraint that determines which submatrices represent significant bicluster behavior. Constraint based methods include the iterative signature algorithm, the coupled two-way clustering method and the spectral algorithm of Kluger et al. In the first two, we search for gene (condition) sets that define "stable" subsets of properties (genes). In the last, the requirement is for compatibility of certain eigenvectors to a hidden checkboard-like matrix structure.

Scoring based methods typically rely on a background model for the data. The basic model assumes that biclusters are essentially uniform submatrices and scores them according to their deviation from such uniform behavior. More elaborate models allow different distributions for each condition and gene, usually in a linear way. Such are, for example, the Cheng-Church algorithm and the Plaid model and the alternative formulation in [22]. A more formal statistical model for an extended formulation of the biclustering problem was used in [19, 3]. In this family of algorithms a complete generative model including a set of biclusters and their regulation model is optimized for maximum likelihood given the data. Another approach for the modeling of the data is used in SAMBA, where a degree-preserving random graph model and likelihood ratio score are used to ensure biclusters significance.

26.8.2 Algorithmic approaches

The algorithmic approaches for detecting biclusters given the data are greatly affected by the type of score/constraint model in use. Several of the algorithms alternate between phases of gene sets and condition sets optimization. Such are, for example, the iterative signature algorithm and the coupled two-way clustering algorithm. Other methods use standard linear algebra or optimization algorithms to solve key subproblems. Such is the case for the Plaid model and the Spectral algorithm. A heuristic hill climbing algorithm is used in the Cheng-Church algorithm and is combined with a graph hashing algorithm in SAMBA. Finally, EM or sampling methods are used for formulations introducing a generative statistical model for biclusters [19, 3, 22]. The overall picture seems to support a view stressing the importance of statistical models and scoring scheme and restricting the role of the search/optimization algorithm to discovering relatively bold structures. A current important goal for the research community is to improve our understanding of the pros and cons of the various modeling approaches described here, and to enable more focused algorithmic efforts on the models that prove most effective.

26.8.3 Quo vadis biclustering?

Biclustering is a relatively young area, in contrast to its parent discipline, clustering, that has a very long history going back all the way to Aristo. It has great potential to make significant contributions to biology and to other fields. Still, some of the difficulties that haunt clustering are present and are even exacerbated in biclustering: Multiple formulations and objective functions, lack of theoretical and complexity analysis for many algorithms, and few criteria for comparing the quality of candidate solutions. Still, the great potential of the paradigm of biclustering, as demonstrated in studies over the last five years, guarantees that the challenge will continue to be addressed. In time, the concrete advantages and disadvantages of each formulation and algorithm will be made clearer. We anticipate an exciting and fruitful next decade in biclustering research.

Acknowledgments

R. Shamir was supported in part by the Israel Science Foundation (grant 309/02). R. Sharan was supported in part by NSF ITR Grant CCR-0121555. A. Tanay was supported in part by a scholarship in Complexity Science from the Yeshaia Horvitz Association.

References

[1] The chipping forecast II. Special supplement to Nature Genetics Vol 32, 2002.

[2] A.A. Alizadeh , M.B. Eisen, R.E. Davis, C. Ma *et al.* Distinct types of diffuse large B-cell lymphoma identified by gene expression profiling. *Nature*, 403(6769):503–511, 2000.

[3] A. Battle, E. Segal, and D. Koller. Probabilistic discovery of overlapping cellular processes and their regulation. In *Proceedings of the Sixth Annual International Conference on Computational Molecular Biology (RECOMB 2002)*, 2004.

[4] S. Bergman, J. Ihmels, and N. Barkai. Similarities and differences in genome-wide expression data of six organisms. *PLoS*, 2(1):E9, 2004.

[5] S. Bergmann, J. Ihmels, and N. Barkai. Iterative signature algorithm for the analysis of large-scale gene expression data. *Phys Rev E Stat Nonlin Soft Matter Phys*, 67(3 Pt 1):03190201–18, 2003.

[6] Y. Cheng and G.M. Church. Biclustering of expression data. In *Proc. ISMB'00*, pages 93–103. AAAI Press, 2000.

[7] J. DeRisi, L. Penland, P.O. Brown, and M.L. Bittner *et al.* Use of a cDNA microarray to analyse gene expression patterns in human cancer. *Nat Genet*, 14:457–460, 1996.

[8] A.P. Gasch ,M. Huang, S. Metzner, and D. Botstein *et al.* Genomic expression responses to DNA-damaging agents and the regulatory role of the yeast ATR homolog mec1p. *Mol. Biol. Cell*, 12(10):2987–3003, 2001.

[9] G. Getz, E. Levine, and E. Domany. Coupled two-way clustering analysis of gene microarray data. *Proc. Natl. Acad. Sci. USA*, 97(22):12079–84, 2000.

[10] G. Getz, E. Levine, E. Domany, and M.Q. Zhang. Super-paramagnetic clustering of yeast gene expression profiles. *Physica*, A279:457, 2000.

[11] J.D. Hughes, P.E. Estep, S. Tavazoie, and G.M. Church. Computational identification of cis-regulatory elements associated with groups of functionally related genes in *Saccharomyces Cerevisiae*. *J. Mol. Biol.*, 296:1205–1214, 2000. http://atlas.med.harvard.edu/.

[12] J. Ihmels, G. Friedlander, S. Bergmann, and O. Sarig *et al.* Revealing modular organization in the yeast transcriptional network. *Nature Genetics*, 31(4):370–7, 2002.

[13] Y. Kluger, R. Barsi, J.T. Cheng, and M. Gerstein. Spectral biclustering of microarray

data: coclustering genes and conditions. *Genome Res.*, 13(4):703–16, 2003.

[14] L. Lazzeroni and A. Owen. Plaid models for gene expression data. *Statistica Sinica*, 12:61–86, 2002.

[15] W.H. Press, B.P. Flannery, S.A. Teukolsky, and W.T. Vetterling. *Numerical Recipes: The Art of Scientific Computing.* Cambridge University Press, Cambridge (UK) and New York, 2nd edition, 1992.

[16] S. Ramaswamy, P. Tamayo, R. Rifkin and S. Mukherjee *et al.* Multiclass cancer diagnosis using tumor gene expression signature. *Proc. Natl. Acad. Sci. USA*, 98(26):15149–15154, 2001.

[17] T. Rozovskaia, O. Ravid-Amir, S. Tillib, and G. Getz *et al.* Expression profiles of acute lymphoblastic and myeloblastic leukemias with all-1 rearrangements. *Proc Natl Acad Sci U S A*, 100(13):7853–8, 2003.

[18] M. Schena, D. Sharon, R.W. Davis, and P.O. Brown *et al.* Quantitative monitoring of gene expression patterns with a complementary DNA microarray. *Science*, 270:467–470, 1995.

[19] E. Segal, M. Shapira, A. Regev, and D. Pe'er *et al.* Module networks: identifying regulatory modules and their condition-specific regulators from gene expression data. *Nat Genet.*, 34(2):166–76, 2003.

[20] R. Sharan, A. Maron-Katz, N. Arbili, and R. Shamir. EXPANDER: EXPression ANalyzer and DisplayER, 2002. Software package, Tel-Aviv University, http://www.cs.tau.ac.il/~rshamir/expander/expander.html.

[21] R. Sharan, A. Maron-Katz, and R. Shamir. CLICK and EXPANDER: a system for clustering and visualizing gene expression data. *Bioinformatics*, 2003.

[22] Q. Sheng, Y. Moreau, and B. De Moor. Biclustering gene expression data by Gibbs sampling. *Bioinformatics*, 19(Supp 2):i196–i205, 2003.

[23] P.T. Spellman, G. Sherlock,M.Q. Zhang, and V.R. Iyer *et al.* Comprehensive identification of cell cycle-regulated genes of the yeast Saccharomyces cerevisiae by microarray hybridization. *Mol. Biol. Cell*, 9:3273–3297, 1998.

[24] A. Tanay, R. Sharan, M. Kupiec, and R. Shamir. Revealing modularity and organization in the yeast molecular network by integrated analysis of highly heterogeneous genomewide data. *Proc Natl Acad Sci U S A.*, 101(9):2981–6, 2004.

[25] A. Tanay, R. Sharan, and R. Shamir. Biclustering gene expresion data. Submitted for publication, 2002.

27

Identifying Gene Regulatory Networks from Gene Expression Data

Vladimir Filkov
University of California, Davis

27.1 Introduction

The ultimate goal of the genomic revolution is understanding the genetic causes behind phenotypic characteristics of organisms. Such an understanding would mean having a blueprint which specifies the exact ways in which genetic components, like genes and proteins, interact to make a complex living system. The availability of genome-wide gene expression technologies has made at least a part of this goal closer, that of identifying the interactions between genes in a living system, or *gene networks*. Well suited for both qualitative and quantitative level modeling and simulation, and thus embraced by both biologists and computational scientists, gene networks have the potential to elucidate the effect of the nature and topology of interactions on the systemic properties of organisms.

Gene networks can be modeled and simulated using various approaches [9, 17]. Once the model has been chosen, the parameters need to be fit to the data. Even the simplest network models are complex systems involving many parameters, and fitting them is a non-trivial process, known as *network inference*, *network identification*, or *reverse engineering*.

This chapter reviews the process of modeling and inference of gene networks from large-scale gene expression data, including key modeling properties of biological gene networks, general properties of combinatorial models, specifics of four different popular modeling frameworks, and methods for inference of gene networks under those models.

Our knowledge about gene networks, and cellular networks in general, although still limited, has grown in the past decade significantly, due mostly to advances in biotechnology. Some of those known and key properties of gene networks, which can aid in their understanding and modeling, especially with the discrete, combinatorial methods covered in this chapter, are described in Section 27.2. Some properties, like low average connectivity, or the nature of cis-trans interactions during transcription have been used repeatedly in modeling and inference of gene networks.

This chapter is de Some background on the nature of large-scale gene expression experiments together with very short description of methods used for the analysis of the observed data is given in Section 27.3. More detailed descriptions of both the technology and the data analysis methods can be found elsewhere in this book.

Before the actual methods are described in detail, general properties of modeling formalisms which both define and limit them are described in Section 27.4.

Those properties (synchronicity, stochasticity, etc.) are necessary considerations when modeling gene networks and define the resulting models to a large extent.

The four large classes of modeling formalisms covered in this chapter are graph theoretical models in Section 27.5, Bayesian networks in Sec. 27.6, Boolean networks in Sec. 27.7, and Linearized differential equation models 27.8. Together with the models and their properties inference methods and algorithms are presented from the most influential research articles in the area, together with pertinent results on both steady-state and time-course gene expression data. The relationship between model complexity and amount/type of data required versus the quality of the results is underlined.

At the end of the chapter the models and methods for inference are summarized and future directions toward better gene network inference are outlined.

This chapter is not comprehensive with respect to the different frameworks available for modeling gene networks, and more thorough reviews exist in that respect [17]. The emphasis here is on an integrated presentation of the models and the methods for network inference for them. Gene network modeling using *feedback control theory* is presented in another chapter of this book.

27.2   Gene Networks

27.2.1   Definition

Gene regulation is a general name for a number of sequential processes, the most well known and understood being transcription and translation, which control the level of a gene's expression, and ultimately result with specific quantity of a target protein.

A *gene regulation system* consists of genes, cis-elements, and regulators. The regulators are most often proteins, called *transcription factors*, but small molecules, like RNAs and metabolites, sometimes also participate in the overall regulation. The interactions and binding of regulators to cis-elements in the cis-region of genes controls the level of gene expression during transcription. The cis-regions serve to aggregate the input signals, mediated by the regulators, and thereby effect a very specific gene expression signal. *The genes, regulators, and the regulatory connections between them, together with an interpretation scheme form gene networks.*

Depending on the degree of abstraction and availability of empirical data, there are different levels of modeling of gene networks. Figure 27.1 shows a hypothetical gene network together with different levels at which it can be modeled. The particular modeling level depends on the biological knowledge and the available data, as well as the experiment goal, which can be as simple as hypothesis testing, or as complex as quantitative network modeling.

There are a number of gene regulatory networks known in great detail: the lysis/lysogeny cycle regulation of bacteriophage-λ [51], the endomesoderm development network in Sea Urchin [16], and the segment polarity network in Drosophila development [68, 5]. The first

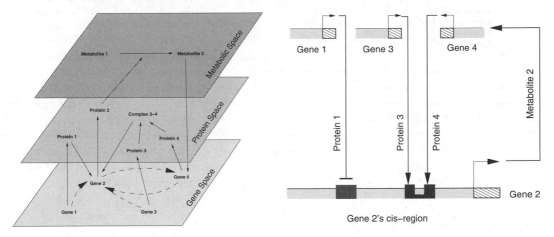

FIGURE 27.1: (See color insert following page 20-4.) A hypothetical gene network. Shown on the left (redrawn from [10]) are the multiple levels at which genes are regulated by other genes, proteins and metabolites. On the right is a useful abstraction subsuming all the interactions into ones between genes only. The cis-regions are shown next to the coding regions, which are marked with pattern fill and start at the bent arrows. The edges are marked with the name of the molecule that carries the interaction. Some reactions represent trans-factor – DNA binding, happen during transcription, and are localized on the cis-regions. In those cases the corresponding protein-specific binding sites, or cis-elements, on the cis-regions are shown (colored polygons). Otherwise, the interactions can take place during transcription or later (e.g. post-translational modifications) as may be the case with Metabolite 2 interacting with Gene 4. The nature of the interactions is inducing (arrow) or repressing (dull end).

two networks are mostly qualitative representations of the relationships among the genes, while the last involves quantitative models. Many other networks are available in some detail, laid out in databases which are publicly available, like KEGG [37] and EcoCyc [38]. Such detailed networks are extracted from the published work of many researchers working on individual links in the networks. With the advent of large-scale technologies in genomics things are becoming faster by orders of magnitude, and there is potential for automating the process of network discovery.

27.2.2 Biological Properties

Substantially more is known about gene regulation and networks today than a decade ago. Some of this knowledge can be used to effectively model gene networks. What follows is a collection of mostly empirical facts which, although few in number, are powerful modeling and design rules.

Topology

The topology of a network defines the connections between nodes, and it can be a starting point for modeling. One of the most important and powerful rules in gene network modeling is that their topology is sparse, i.e. there is a small constant number of edges per node, much smaller than the total number of nodes. This rule is a result of biological observations that genes are regulated by a small constant number of other genes, 2-4 in bacteria [43] and 5-10 in eukaryotes [6]. The sparseness property is often used to prune the search space during network inference, as described later.

More on connectedness, recent studies have shown that the frequency distribution of connectivity of nodes in biological (and other types of naturally arising) networks tends to be longer tailed than the normal distribution [36]. The appropriate distribution seems to belong to a class of power-law functions described by $P(k) = k^{-\gamma}$, where k is the degree of a vertex in the network graph, and γ is some network specific constant. Such networks are called *scale-free* and exhibit several important properties. The first is the emergence of *hubs*, or highly connected nodes, which would have been vanishingly unlikely to appear in a network with normally distributed node degrees. These hubs correspond to highly central nodes in the gene network, i.e. genes that do a large amount of the overall regulation. The second property is that through the hubs, the rest of the nodes are connected by, in the worst case, very short paths, yielding overall short longest paths between nodes (degree of separation, small-world networks).

Transcriptional Control

The cis-regions serve as aggregators of the effects of all transcription factors involved in gene regulation. Through protein-specific binding sites the cis-regions recruit and bring in proximity single or groups of TFs having specific regulatory properties, with the sole purpose of inducing precisely when, where, and at what rate a gene is to be expressed. The hard-coded binding sites, or cis-elements, in the regulatory regions of DNA sequences are in fact the code, or logic, by which input signals are chosen and processed. Very good quantitative examples of such logic processing units are, for example, the *endo16* system in Sea Urchin [15], and the process of lysis/lysogeny in bacteriophage-λ [51].

The range of effects (i.e. output) of the cis-processing logic on the input TF signals has been characterized for very few but telling examples. From them we learn that the cis-function is a multi-valued, complex function of the input concentrations even only for two inputs [55]. However the function becomes simpler, and can be decomposed into linear combinations of independent functional signal contributions, when the functional cis-elements are known (at least when performed over the same conditions, and for genes on the periphery of the network (i.e. ones without feedback input) [77, 76]. The traditional roles of individual or groups of TFs as inducers and suppressors (i.e. activators and inhibitors), with respect to a gene, have been based on the change they cause in expression, and are viable if their effects are strong enough and independent of other TF effects. Those roles have been refined recently by associating function to modules of cis-elements (or equivalently TFs) to include more operators, like enhancers, switches, and amplifiers, which form the repertoire of transcriptional control elements. Some of those have been used to engineer synthetic gene circuits with pre-specified functions [25].

Robustness

Real gene networks are very robust to fluctuations in their parameter values [68, 8], and there is strong indication that only specific topology choices can guarantee such robustness [68, 36]. Insensitivity to variations in molecular concentrations is particularly important during organism development, when things are happening in orchestrated cues around the whole organism at the same time, with the goal of producing the same body plan every time, but under a variety of different conditions [15]. It has also been argued that the process of evolution naturally converges to scale-free design in organized structures [52], and that the scale-freeness ensures the robustness of networks to random topology changes [36].

Noise

Noise is an integral part of gene networks, as they are emerging properties of biochemical reactions which are stochastic by nature [42]. Even small variations in the molecular concentrations during the process of translation can be passed along through the network [65]. The networks control the noise through feedback, although in some cases noise enhances some functional characteristics of the networks [67] and hence may have a role in network evolution. In some cases noise measurements can be used to learn the rates of signal propagation in networks too [48].

27.2.3   Utility

A gene network, at any level of modeling, is a blueprint for understanding the functional cooperativity among genes.

Individually, gene networks are succinct representations of the knowledge about the system studied. For example, they can be used for gene classification, based on the localization of the gene's influence on other genes and the others' influence on them. The regulatory interactions between genes can be studied at large-scales, allowing for example for understanding cascades of gene regulations (e.g during organism development). Another use can be in elucidating the connection between gene regulation and phenotype, again on a systemic scale.

In quantitative settings, they can be used to simulate various scenarios, and even predict future behavior of the system. As such they can serve as in-silico hypothesis generation and testing tools with the potential of running many experiments in a very short time on a desktop computer. Knowing the interacting components can help with identifying molecular targets for specific drugs, or drugs for specific targets. Such knowledge coupled with understanding of the network behavior, can lead to designing controlled systems with potential for producing disease-specific cures and personalized health care solutions.

Having gene networks of multiple organisms will allow for their comparison, yielding understanding of structural and functional changes of the networks with time, i.e. their evolution. Through such network comparisons the different effects of stimuli on organisms can be potentially attributed to differences in their networks.

27.3   Gene Expression: Data and Analysis

The amount of mRNA produced during transcription is a measure of how active or functional a gene is. Gene chips, or microarrays, are large-scale gene expression monitoring technologies, used to detect differences in mRNA levels of thousands of genes at a time, thus speeding up dramatically genome-level functional studies. Microarrays are used to identify the differential expression of genes between two experiments, typically test vs. control, and to identify similarly expressed genes over multiple experiments. Microarray data and analysis methods are described in detail in other chapters of this book. Here we only briefly describe the types of expression data and analysis methods used for gene network inference.

Microarray experiments for our purposes can be classified in two groups: time-course experiments, and perturbation experiments. The former are performed with the goal of observing the change of gene expression with time, and are used for understanding time-varying processes in the cell. The latter are performed to observe the effects of a change or treatment on the cell, and are used for understanding the genetic causes for observed differences between cell/tissue types, or responses of pathways to disruptions.

The processing pipeline of microarray data involves pre-processing the raw data to get a gene expression matrix, and then analyzing the matrix for differences and/or similarities of expression. The gene expression matrix, *GEM*, contains pre-processed expression values with genes in the rows, and experiments in the columns. Thus, each column corresponds to an array, or gene-chip, experiment, and it could contain multiple experiments if there were replicates. The experiments can be time points (for time-course experiments), or treatments (for perturbation experiments). Each row in the matrix represents a *gene expression profile*. An example hypothetical table is given in Table 27.1.

TABLE 27.1   Gene Expression Matrix

	Exp_1	Exp_2	$\cdots$	Exp_m
$Gene_1$	0.12	-0.3	$\cdots$	0.01
$Gene_2$	0.50	0.41		
$\vdots$	$\vdots$	$\vdots$	$\ddots$	
$Gene_n$	-0.02	-0.07		

Gene-chips can hold probes for tens of thousands of genes, whereas the number of experiments, limited by resources like time and money, is much smaller, at most in the hundreds. Thus, the gene expression matrix is typically very narrow (i.e. number of genes, $n \gg$ number of experiments, m). This is known as the *dimensionality curse* and it is a serious problem in gene network inference. Namely, in any complex system, when trying to elucidate the interactions among the state variables it is best to have many more measurements than states, otherwise the system is largely under-constrained and can have many solutions. The inference methods described in Sections 27.5- 27.8 use different ways to address this problem.

Once normalized the data are analyzed for differential expression using statistical tests. The most robust tests allow for replicate measurements, yield a confidence level for each result, and correct for multiple hypothesis testing (i.e. testing if hundreds of genes are differentially expressed). Additionally, to identify functional relationships genes are often combined in groups based on similarities in their expression profiles using a variety of supervised or unsupervised clustering and/or classification techniques [59]. Most of those methods are good for identifying starting points for network inference.

A good source of microarray data is the Stanford Microarray Database (SMD) [29].

27.4   General Properties of Modeling Formalisms

The actual choice of a modeling formalism for a gene network will depend on the type and amount of data available, prior knowledge about the interactions in the network, nature of the questions one needs answered, area of formal training of the modeler, experimental and computational resources, and possibly other study- or organism-specific factors. Here, based on their general properties different models are classified in several groups, which can be used as selection criteria among them.

Physical vs. Combinatorial Models

The most detailed models of any complex dynamical system, like gene networks, are based on differential equations describing the quantitative relationships between the state variables in the system. Although such physical models can be used to run simulations and predict the future behavior of the system, in general any higher level organization is very difficult to obtain from the equations. But, identifying even simple features (e.g. if one variable is responsible for the behavior of another) may be complicated. In addition, physical models typically have many parameters, necessitating large number of experiments to fit them to the data. Some high-level analyses of dynamical systems, like steady state, have yielded promising results (S-systems by Savageau [53]), but lack inference methods.

On the other hand, combinatorial models start from higher level features of the system by defining the characteristics and features of interest, like the important observables, i.e. gene expression levels, and the nature of relationships, like causality for example. A typical representation for such models is a graph of nodes and edges between them from which many important high-level questions can be readily answered (see Section 27.5). Because of the higher level of modeling, the combinatorial models are most often qualitative, and effective methods for their learning or inference exist even for small number of observations (relative to the number of variables). Most models described in this chapter will be combinatorial, except for the linearized model of differential equations.

Dynamic vs. Static Models

Dynamic gene network models describe the change of gene expression in time. Typically, each node in the network is a function that has some inputs, which it aggregates, and produces output based on them. Most dynamic models are based on simplifications of differential rate equations for each node:

$$\frac{dx_i(t)}{dt} = f_i(x_{i_1}(t), x_{i_2}(t), \ldots) \tag{27.1}$$

where the x_i on the left is the observable concentration of gene expression, for gene i, the $x_{i_1}, x_{i_2}, \ldots$ on the right, are concentrations of molecules that anyhow influence x_i's expression, and $f_i(\cdot)$ is the rate function specifying the exact way in which the inputs influence x_i. Dynamic models tend to be more complete than static ones, in that they aim to characterize the interactions among the input signals, and offer quantitative predictions for the observable variables. However, they do require more input data in general because of the large number of parameters to fit, as well as some type of understanding of the nature of the aggregation function. This usually amounts to making simplifying assumptions, e.g. assuming that the interactions between regulators of a gene are limited to being additive. In addition, time is often discretized and the changes are assumed to be synchronous across all variables. Examples of dynamic models covered in this chapter are Boolean networks and linearized differential equations.

Static models do not have a time-component in them. This practically means that static models yield only topologies, or qualitative networks of interactions between the genes, often specifying the nature of the interactions. Static models can be revealing of the underlying combinatorial interactions among genes, a feature most often simplified away in the dynamic models, with the notable exception of the Boolean network model. Examples of static models are: Graph theoretic models, Bayesian networks, and Linear Additive Models (under some modeling assumptions).

The choice between these two modeling paradigms clearly depends on the type and amount of data and experimental setup available, and it often involves understanding and prioritizing the imperatives in the study, for example, the importance of the exact predictions vs. the nature of interactions.

Synchronous Models

Dynamic models are attractive because they offer quantitative predictions. However, many additional assumptions need to be made for the available inference methods to work. One such assumption is synchronous delivery of all signals to the targets or equivalently, updating the states of all genes at the same time. Large-scale gene expression measurements drive these models because all the observables are measured at the same times.

In synchronous models time is discretized to the intervals between consecutive observations. If these time intervals are small enough, then Eq. 27.1 can be approximated as:

$$\frac{x_i(t_{j+1}) - x_i(t_j)}{t_{j+1} - t_j} \approx f_i(x_{i_1}(t_j), x_{i_2}(t_j), \ldots) \qquad (27.2)$$

where t_j and t_{j+1} are two consecutive observation times. Although clearly unrealistic when $t_{j+1} - t_j$ is not small, this approximation is the base for some of the most popular models for gene network inference from expression data, as described in section 27.4.

Finally, reality is obviously asynchronous. However there are no existing effective methods for inferring gene networks solely from expression data for any asynchronous model.

Deterministic vs. Stochastic Models

In deterministic models the expression states of the genes are either given by a formula or belong to a specific class. Measured at two different times or places, while keeping all other parameters the same, a gene's expression would be the same. The precision of the observed expression values, then, depends solely on the experimental setup and technological precision, and can be refined indefinitely with technological advances. The edges stand for relationships which, like the node states, are also deterministic.

Stochastic models, on the other hand, start from the assumption that gene expression values are described by random variables which follow probability distributions. The difference with the deterministic models is fundamental: randomness is modeled to be intrinsic to the observed processes, and thus all things being equal, a gene's expression on two different occasions may be different. Stochastic edges indicate probabilistic dependencies, and their absence may indicate independencies between nodes. They are not easy to interpret in practice.

Stochastic gene network models are especially appropriate for reconstructing gene networks from expression data because of the inherent noise present in them. A fairly general statistical way to accommodate imprecisions and fluctuations in the measurements is to assume that each observed quantity is drawn from an underlying set of values, or a distribution, that the observable variable may take. Then, assessing whether a gene is differentially expressed with respect to another turns into well studied problems of statistical hypothesis testing [59]. In addition to the distribution of a gene's expression one would also like to model the relationships between such distributions as indicators of possible causal relationships between them. One example of a stochastic model of gene networks are the Bayesian Networks, discussed in Section 27.6. Although likely more realistic, stochastic models are more difficult to learn and to interpret.

Expanding Known Networks

Often the goal is to find missing links in a partially known network, or expanding a network around a known core of interacting genes. Although, in general, not necessarily easier to solve than the problem of inferring a network from scratch, expanding a known network is typically a smaller problem and less data might be necessary. In other words, the same amount of data should yield better (i.e. more accurate) or simply more predictions if prior knowledge is available.

27.5   Graph Theoretical Models

Graph theoretical models (*GTMs*) are used mainly to describe the topology, or architecture, of a gene network. These models feature relationships between genes and possibly their nature, but not dynamics: the time component is not modeled at all and simulations cannot be performed. GTMs are particularly useful for knowledge representation as most of the current knowledge about gene networks is presented and stored in databases in a graph format.

GTMs belong to the group of *qualitative network models*, together with the Boolean Network models (Section 27.7), because they do not yield any quantitative predictions of gene expression in the system.

In GTMs gene networks are represented by a *graph* structure, $G(V, E)$, where $V = \{1, 2, \ldots, n\}$ represent the gene regulatory elements, e.g. genes, proteins, etc., and $E = \{(i, j) | i, j \in V\}$ the interactions between them, e.g. activation, inhibition, causality, binding specificity, etc. Most often G is a simple graph and the edges represent relationships between pairs of nodes, although *hyperedges*, connecting three or more nodes at once, are sometimes appropriate. Edges can be directed, indicating that one (or more) nodes are precursors to other nodes. They can also be weighted, the weights indicating the strengths of the relationships. Either the nodes, or the edges, or both are sometimes labeled with the function, or nature of the relationship, i.e. activator, activation, inhibitor, inhibition, etc. (see Figure 27.1). The edges imply relationships which can be interpreted as temporal (e.g. causal relationship) or interactional (e.g. cis-trans specificity).

Many biologically pertinent questions about gene regulation and networks have direct counterparts in graph theory and can be answered using well established methods and algorithms on graphs. For example, the tasks of identifying highly interacting genes, resolving cascades of gene activity, comparing gene networks (or pathways) for similarity correspond to, respectively, finding high degree vertices, topological vertex ordering, and graph iso(homo)-morphisms in graphs.

Inferring gene networks under GTMs amounts to identifying the edges and their parameters (co-expression, regulation, causality) from given expression data. The type of data used directs the approaches taken in the inference. Typically, parsimonious arguments stemming from biological principles are used to restrict the resulting networks. Here, several different approaches are mentioned, using both time-series measurements and perturbation measurements of gene expression.

Inferring Co-expression Networks

Co-expression networks are graphs where edges connect highly co-expressed nodes (genes or gene clusters). Two genes are co-expressed if their expression profiles (rows in the expression matrix) are strongly correlated. The data can be any set of experiments on the genes, possibly a mix of time-series measurements and perturbation data. The goal in this approach is to obtain a graph where the nodes are clusters of genes with indistinguishable expression profiles, and the edges between such clusters are indicative of similarities between clusters.

The process of inference closely follows clustering: given are two-thresholds, within cluster, τ_w, and between cluster, τ_b, a gene expression matrix with n genes and m experiments, and a measure for pair-wise scoring of expression profiles (e.g. correlation coefficient),

$Score(i, j)$. These scores between pairs of genes are used to cluster them in the same cluster based on τ_w, using for example hierarchical clustering [35]. The genes in each cluster are considered indistinguishable for all practical purposes. The gene network is formed by placing edges between clusters of very similar average expression as determined by τ_b, while possibly averaging the profiles within clusters and using those as representative profiles.

The problem with this approach is that resulting edges in the network are difficult to interpret. They identify relationships which follow the specifics of the similarity function, but it is hard to justify any similarity function a priori. When the correlation coefficient is used, for example, the justification is that genes with visually similar expression patterns are likely coexpressed, and possibly regulating each other. When compared to known regulatory pairs such approaches have yielded prediction rates of no more than 20% [22].

Regulatory Interactions from Time-Series Data

Here, given time-series gene expression data the goal is to infer edges as potential regulation relationships between genes. Thus a directed edge (i, j) would imply that gene i regulates gene j. The idea is to consider time-course gene expression experiments and correlate sustained positive and negative changes in the expression levels while incorporating biological considerations.

Chen et al. [12] extended the co-expression networks model by considering a scoring function based on signal, or time-series similarity (instead of correlation at the second clustering step above). The scoring function identified putative causal regulation events of the type: a positive change in the expression of gene i followed by a positive change in the expression of gene j implies a potential regulation (or a joint co-regulation). Such *putative* relationships were consequently pruned by minimizing the overall number of regulators, and consistently labeling nodes as activators or inhibitors. Additional biological considerations were that regulators should either be activators or inhibitors, but not both, and that their number should be small, while the number of regulated nodes should be large. The authors showed that a number of theoretical problems of network inferenc on such labeled graphs are NP-complete, even for networks of very low in-degree (even only 2). By using local search heuristics on more general edge-weighted graphs they maximized the overall regulation while minimizing the number of regulators. Their results yielded networks with interesting biological properties. In a subsequent study they improved on their scoring function for detecting meaningful pairwise regulation signals between genes, and showed that they do significantly better than the correlation measure [22]. This pairwise scoring approach, however, is limited to resolving relationships between pairs of genes, whereas in real networks multiple genes can participate in the regulation.

In a similarly motivated, signal-based approach, Akutsu et al. [4] also considered only qualitative clues in time-series expression data to infer regulation. They define a regulatory relationship between genes to be consistent with the data only when all positive (or negative) changes in the expression of one gene consistently correspond to changes in expression of another gene. In general, their method amounts to solving linear inequalities using *linear programming* methods. The linear inequalities were obtained by estimating the rate (or change) of expression for each gene, from time measurements of gene expression, coupled with linearizing a rate equation 27.1 (as in Section 27.8). This method is comparable to the Linear Additive models' inference, and thus probably has the same limitations, see Sect. 27.8 for more on this. It is notable that their analysis is similar to the qualitative type of steady-state analysis of rate equations in the S-system model [53].

Causal Networks from Perturbation Experiments

Perturbing a gene in a gene network effects all genes downstream of it, but no others. For example, perturbing *Gene 1* in the network in Figure 27.1 can influence *Gene 2* and *Gene 4*, but not *Gene 3*. Thus, in principle, performing gene perturbations is a very good methodology for elucidating causal relationships among genes. The inference problem, then, is to find a network consistent with expression data resulting from genetic perturbations.

A perturbation of gene i in the gene network graph $G = (V, E)$ yields a set of differentially expressed genes $D_{iff}(i) \subseteq V$. This set is a subset of $Reach(i)$–the set of all genes reachable from i. The set $Reach(i)$ includes both the nodes adjacent to i, i.e. the set $Adj(i)$, and the nodes reachable but not adjacent to i in G. As examples of $Adj(i)$ and $Reach(i)$, in the network in Figure 27.1, $Adj(1) = Adj(3) = \{2\}$, $Adj(2) = \{4\}$, $Adj(4) = \{2\}$, and, $Reach(1) = Reach(2) = Reach(3) = Reach(4) = \{2, 4\}$. The graph $G_c = (V, E_c)$, where $E_c = \{(i, j), j \in D_{iff}(i)\}$, for all perturbed nodes i, specifies all perturbation effects, and represents a *causal gene network*. Note that in general G_c has more edges than the gene network graph G.

The inference problem can then be re-stated as: given $D_{iff}(i) \subseteq Reach(i)$ for all $1 \leq i \leq n$, retrieve the graph G. There are, in general, many solutions to this problem, since many different gene networks may yield the same differentially expressed genes under all perturbations. Additional biological assumptions on the nature or number of interactions can be used to resolve the best out of all the graphs consistent with the data.

Wagner [69] considered the case when perturbing a gene changes the expression of all the genes downstream from it, i.e. the observed changes yield the full transitive closure of G, or in other words $D_{iff}(i) = Reach(i)$. (The graph G^* with the same nodes as G and G_c and edges $E^* = \{(i, j), j \in Reach(i)\}$, for all i, is called the *transitive closure* of G (and G_c) [14].) Then, motivated by biological parsimony, he used an additional assumption that the biologically most plausible solution has the minimal number of edges of any graph having the same transitive closure. With that assumption, the inference problem above becomes the well-known problem in graph-theory of *transitive reduction*, which in general is NP-complete [1]. However, if there are no cycles in the graph, there is exactly one transitive reduction of G^*, and it can be obtained from the transitive closure very efficiently [1], also [69] (cyclic graphs can be reduced to acyclic by condensing the cycles into single nodes [69]). The required amount of measurements/experiments to resolve a network with this method is n, i.e. each gene must be perturbed and the effects measured (an experiment consists of measuring the expression of all n genes at a time, i.e. whole-genome microarray experiment). Thus, in the case of yeast, the number of experiments would be ~ 6000, which is still practically infeasible. Biologically, the parsimonious argument of the gene network having the minimal number of relationships is not so easy to justify, especially since for acyclic graphs this implies having no shortcuts (essentially no double paths between two nodes) which in reality is not true (e.g. feed-forward loops are essential in gene networks but because they contain shortcuts are excluded in this model [45]).

Limitations and Extensions

GTMs are very useful for knowledge representation but not simulation. The characteristic of these models is that they only resolve the topology of the networks, and are naturally amenable to inference using graph theoretical arguments. Existing inference methods use parsimonious assumptions about the nature of the networks to reduce the solution space and yield a single solution. The required amount of data points varies between models, and it is between $O(log n)$ for clustering based methods and $O(n)$ for perturbation based methods. GTMs are becoming increasingly important as recent studies indicate that the topology of networks is a determining factor in both re-engineering the network as well as understanding network and organism evolution.

27.6   Bayesian Networks

Bayesian networks are a class of *graphical probabilistic models*. They combine two very well developed mathematical areas: probability and graph theory. A Bayesian network consists of an annotated directed acyclic graph $G(X, E)$, where the nodes, $x_i \in X$, are random variables representing genes' expressions and the edges indicate the dependencies between the nodes. The random variables are drawn from conditional probability distributions $P(x_i|Pa(x_i))$, where $Pa(x_i)$ is the set of parents for each node. A Bayesian network implicitly encodes the *Markov Assumption* that given its parents, each variable is independent of its non-descendants. With this assumption each Bayesian network uniquely specifies a decomposition of the joint distribution over all variables down to the conditional distributions of the nodes:

$$P(x_1, x_2, \ldots, x_n) = \prod_{i=1}^{n} P(x_i|Pa(x_i)) \qquad (27.3)$$

Besides the set of dependencies (children nodes depend on parent nodes) a Bayesian network implies a set of independencies too (see Figure 27.2). This probabilistic framework is very appealing for modeling causal relationships because one can query the joint probability distribution for the probabilities of events (represented by the nodes) given other events. From the joint distribution one can do inferences, and choose likely causalities. The complexity of such a distribution is exponential in the general case, but it is polynomial if the number of parents is bounded by a constant for all nodes.

Learning Bayesian Networks

Given measurements of genome-wide gene expression data the goal is to learn candidate Bayesian networks that fit the data well. A survey of general methods for learning Bayesian networks is given in Heckerman et al. [32]. Here we give a very short overview and point the reader to studies in which Bayesian networks were used to analyze expression data.

Two Bayesian networks are defined to be equivalent, or indistinguishable, if they imply the same set of independencies [24]. From any given data most one can hope to learn is equivalence classes of networks, and not individual networks. The process of learning Bayesian networks from the data is essentially two-fold [23]:

- The first part is *model selection*: Given observed data find the best graph (or model) G of relationships between the variables.

- The second is *parameter fitting*: Given a graph G and observed data find the best conditional probabilities for each node.

Parameter fitting is the easier of the two in general. Given a graph model G, good candidates for the best conditional probability distributions can be found by *Maximum Likelihood Estimation* algorithms (when all nodes are known), or *Expectation Maximization* algorithms (when some nodes are hidden), both well known methods. For model selection, on the other hand, only simple heuristics are known, without solid convergence results, which amount to brute-force search among all graphs.

The gene network inference problem combines both of these: the space of all model graphs is searched, and each candidate model is scored. The highest scoring model is the

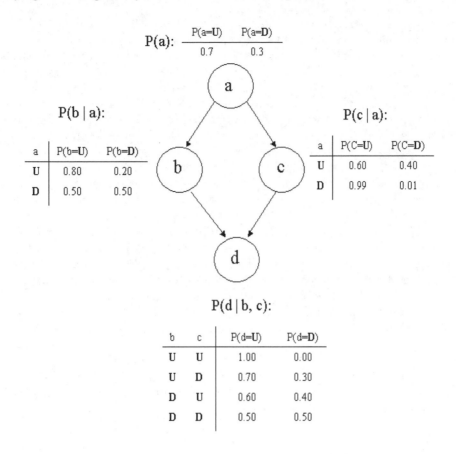

FIGURE 27.2: An Example Bayesian network where the genes can assume discrete states of Up and Down. The probability distributions at the nodes imply the dependencies of nodes only on their parent nodes. The joint probability can be computed directly from the graph, for any instance of the node values. Eg. $P(a = U, b = U, c = D, d = U) = P(a = U)P(b = U|a = U)P(c = D|a = U)P(d = U|b = U, c = D) = 0.7 * 0.8 * 0.4 * 0.7 = 16\%$. The implied independencies in this network are: $I(b; c|a)$ (i.e. b is independent of c given a), and $I(c; b|a)$ (i.e. c is independent of b given a).

best fitting network to the data.

Given a model, its Bayesian score is the posterior probability of the graph G, given the data D:

$$S(G : D) = log P(G|D) = log \frac{P(D|G)P(G)}{P(D)} = log P(D|G) + log P(G) + Const. \quad (27.4)$$

The non-constant terms on the right hand side correspond to the two problems above. Namely, the first term averages the probability of the model G over all possible parametric assignments to it, and it corresponds to the maximum log-likelihood of G given D. The second term average the probability over all possible models G given the data and corresponds to model complexity. The third is a constant independent of the model. Logs are taken to achieve additive independence of the terms above which can be individually estimated.

Finding the best model amounts to optimizing $S(G : D)$, i.e. finding the model and parametric assignment with the best score. To do this efficiently, it is imperative to chose a scoring function which is decomposable to the individual score contributions of each node and for which there are provable guarantees that the highest scoring models are likelier to capture the real network, given the data, than other models. Examples of existing scoring functions with such properties are: the Bayesian Information Criterion (BIC), Akaike Information Criterion (AIC), and Bayesian Dirichlet equivalent (BDe). Even though because of their decomposition property these functions prune the search space, optimizing them over all models (i.e. graphs) is still NP-hard. Thus, heuristics like hill-climbing and simulated annealing are used to efficiently search the space of models by looking at neighboring graphs around a given graph, by adding and deleting edges, or reversing directions of edges.

In addition to the scoring function, other important choices have to be made to get from the observed data of gene expressions to the learned network. These include data discretization into few levels (e.g. -1,0,+1 for under-expression, no expression and over-expression) and choices for priors (e.g. linear, multinomial), and are often related.

In [30] different scoring schemes are illustrated and in [75] various scoring functions, discretizations, and heuristics are compared. The results, although from limited data studies, single out BIC as the score of choice, in conjunction with a three level discretization of the data, and a random-start hill-climbing heuristic for model selection.

The final issue is interpretation. Bayesian networks model the gene expression of each gene as a random variable having a distribution conditional on its parent genes. From that, one wants to infer causal relationships between genes. For that, another assumption, the *local Markov Rule* is needed, which says that variables are independent of all other nodes given their parents. With this assumption, directed edges in the learned equivalence class stand for causal relationships, with the direction indicated by the arrow. If in a class there are undirected edges, then the causality is unknown.

Practical Approaches and the Dimensionality Curse

In practice the available data suffers from the dimensionality curse (see Sec. 27.3) and many different Bayesian networks may fit the data just as well. To lower the number of high-scoring networks, simplifying assumptions regarding the topology of the graph or the nature of the interactions have been used. In addition, instead of trying to learn large-scale networks the focus has been on retrieving features consistently over-represented in high-scoring networks.

In [24] the authors use Bayesian networks to establish regulatory relationships between genes in yeast, based on time-series data of gene expression. To do that, they used several biologically plausible simplifying assumptions. The first one is that the nodes are of bounded in-degree, i.e. the networks are sparse. The second is that the parent and children nodes likely have similar expression patterns, and thus coexpressed pairs of genes might be in regulatory relationships. Both are realistic to a degree, and are meant to reduce the number of potential high-scoring networks. It was found that even with these assumptions the number of high-scoring networks was still high, and instead of full networks subgraphs, or robust features, in common to many high-scoring networks, were considered, like pair-wise relationships for example. Two separate prior distributions were used for the nature of the interactions of the edges at each node (i.e. the combined distributions): a linear Gaussian and a multinomial, requiring different data discretizations each, three levels for the first one and continuous for the second. The results, not surprising, were excellent for the linear prior and mediocre for the multinomial.

A subsequent study by the same group [49], applied Bayesian networks to perturbation gene expression data, with a similar but expanded goal: to identify regulatory relationships and in addition, to predict their nature of activation or inhibition. To do that, they incorporated perturbations into the Bayesian framework. Activations and inhibitions were modeled by requiring that for each change in a regulator's expression the regulated gene must exhibit a change too, much like Akutsu et al. [4] did in their qualitative network model study. The results were, again, conserved features over many runs of the algorithms, and the number of features considered was larger than in the previous study. The resolution of co-regulation was much better than that of correlation methods and yielded subnetworks of known networks automatically from the data.

Although no analysis of the required amount of data was given in either of these studies, the available data is very insufficient. In the first paper, this problem was circumvented by clustering of the data and considering clusters of genes, thus lowering the dimensionality of the original problem. In the second, the results were sub-networks, needing much less data than a genome-scale gene network, specifically those for which there was a lot of support in the expression data.

Extending Bayesian Networks

Bayesian networks offer an intuitive, probabilistic framework in which to model and reason about qualitative properties of gene networks. The gene expression data available is insufficient for full gene network learning, but conserved features over high-scoring models are biologically significant. The Bayesian Network formalism is easily extendable to describe additional properties of the gene networks, like the activatorin the second study above. This implies that other types of data, like regulator candidates, promoter sequence alignment, or TF-DNA binding data, can be used to refine the models, as recent studies have done [54, 7].

The problem with learning of Bayesian networks is combinatorial: if the graph model is not known then the space of all graph models has to be explored. But this space is super-exponential even for directed acyclic graphs and exploring it completely is impossible even with the fastest heuristics. Exploring hierarchical properties of DAGs can help in pruning the search space down to exponential size(from super-exponential), for which, with current technology, exact solutions can be obtained for small networks of $n < 30$ nodes [47]. A promising direction is to restrict the search space further by including additional biological knowledge as it becomes available.

In their basic form, Bayesian networks are qualitative models of acyclic gene networks. There have been efforts to extend them to both include cyclic networks and to model network dynamics by duplicating the nodes in the network. Namely, a state transition of a gene network from time t to time $t + 1$ can be modeled by having two copies of the same Bayesian network, with additional edges between the two. Similar *unrolling* of the network can be used to model cycles. Learning such duplicated networks would require more data in general.

27.7   Boolean Networks

Boolean networks are a dynamic model of synchronous interactions between nodes in a network. They are the simplest network models that exhibit some of the biological and systemic properties of real gene networks. Because of the simplicity they are relatively easier to interpret biologically.

Boolean networks as a biological network modeling paradigm were first used by Kaufmann in the 1970s [39], where he considered directed graphs of n nodes and connectivity, or degree, per node of at most k, called *NK-Boolean networks*. He studied their organization and dynamics and among other things showed that highly-connected NK-networks behave differently than lowly connected ones. In particular, NK-networks of low per-node degree seem to exhibit several of the properties that real life biological systems exhibit, like periodic behavior and robustness to perturbation.

Model and Properties

For completeness, the following summary of Boolean logic is provided. A variable x that can assume only two states or values is called *Boolean*. The values are denoted usually as 0 and 1, and correspond to the logical values **true** and **false**. The logic operators **and**, **or**, and **not** are defined to correspond to the intuitive notion of truthfulness and composition of those operators. Thus, for example x_1 **and** $x_2 = $ **true** if and only if both x_1 and x_2 are **true**. A *Boolean function* is a function of Boolean variables connected by logic operators. For example, $f(x_1, x_2, x_3) = x_1$ **or** (**not** (x_2 **and** x_3)) is a Boolean function of three variables.

A *Boolean network* is a directed graph $G(X, E)$, where the nodes, $x_i \in X$, are Boolean variables. To each node, x_i, is associated a Boolean function, $b_i(x_{i_1}, x_{i_2}, \ldots, x_{i_l}), l \leq n, x_{i_j} \in X$, where the arguments are all and only the parent nodes of x_i in G. Together, at any given time, the states (values) of all nodes represent the *state* of the network, given by the vector $S(t) = (x_1(t), x_2(t), \ldots, x_n(t))$.

For gene networks the node variables correspond to levels of gene expression, discretized to either up or down. The Boolean functions at the nodes model the aggregated regulation effect of all their parent nodes.

The states of all nodes are updated at the same time (i.e. synchronously) according to their respective Boolean functions:

$$x_i(t + 1) = b_i(x_{i_1}(t), x_{i_2}(t), \ldots, x_{i_l}(t)). \tag{27.5}$$

All states' transitions together correspond to a *state transition* of the network from $S(t)$ to the new network state, $S(t + 1)$. A series of state transitions is called a trajectory, e.g. S_5, S_1, S_2 is a trajectory of length 3 in the network in Figure 27.3. Since there is a finite number of network states, all trajectories are periodic. The repeating part of the trajectories are called *attractors*, and can be one or more states long, e.g. S_2 is an attractor in the network in Figure 27.3. All the states leading to the same attractor are the *basin of attraction* [73].

The dynamic properties of Boolean networks make them attractive models of gene networks. Namely, they exhibit complex behavior, and are characterized with stable and reproducible attractor states, resembling many biological situations, like steady expression states. In addition, the range of behaviors of the system is completely known and analyzable (for smaller networks) and is much smaller than that of other dynamic models. In terms of topology, it has been shown that high connectivity yields chaotic behavior, whereas low connectivity leads to stable attractors, which again corresponds well to real biological networks [39].

Reverse Engineering

The goal in reverse engineering Boolean networks is to infer both the underlying topology (i.e. the edges in the graph) and the Boolean functions at the nodes from observed gene expression data.

The actual observed data can come from either time-course or perturbation gene expression experiments. With time-course data, measurements of the gene expressions at two consecutive time points simply correspond to two consecutive states of the network, $S(i)$ and $S(i + 1)$. Perturbation data comes in pairs, which can be thought as the input/output

Identifying Gene Regulatory Networks from Gene Expression Data

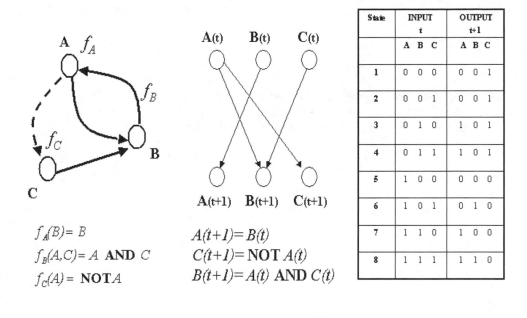

$f_A(B) = B$

$f_B(A,C) = A$ **AND** C

$f_C(A) =$ **NOT** A

$A(t+1) = B(t)$

$C(t+1) =$ **NOT** $A(t)$

$B(t+1) = A(t)$ **AND** $C(t)$

State	INPUT t			OUTPUT t+1		
	A	B	C	A	B	C
1	0	0	0	0	0	1
2	0	0	1	0	0	1
3	0	1	0	1	0	1
4	0	1	1	1	0	1
5	1	0	0	0	0	0
6	1	0	1	0	1	0
7	1	1	0	1	0	0
8	1	1	1	1	1	0

(a) Boolean Network   (b) Wiring Diagram   (c) Truth Table

FIGURE 27.3: An example Boolean network and three possible ways to represent it. The one on the left is a gene network modeled as a Boolean network, in the middle is a wiring diagram obviating the transitions between network states, and on the right is a truth table of all possible state transitions.

states of the network, I_i/O_i where the input state is the one before the perturbation and the output the one after it. In both cases, a natural experimental unit is a pair of consecutive observations of genome-wide gene expression, i.e. states of the network, or a *state transition pair*. The total amount of observed data, m, is the number of available state transition pairs. Perturbation input/output pairs can be assumed to be independent whereas time-course pairs are rarely so. The actual $0, 1$ values are obtained by discretizing the observed continuous expression values, and has to be done with necessary care.

Given the observations of the states of a Boolean network, in general many networks will be found that are consistent with that data, and hence the solution network is ambiguous. There are several variants of the reverse engineering problem: (a) finding a network consistent with the data, (b) finding all networks consistent with the data, and (c) finding the "best" network consistent with the data (as per some pre-specified criteria). The first one is the simplest one and efficient algorithms exist for it, although the resulting network may be very different from the real one. Hence the second problem, which yields all possible solutions, including the real one, although the number of solutions may be very large. The third variant solves the problem of having too many correct solutions by introducing additional assumptions or modeling imperatives. Although very attractive computationally, sometimes the additional criteria may be oversimplifications.

Data Requirement

The reverse engineering problems are intimately connected to the amount of empirical data available. It is clear that, in general, by having more data points the inferred network will be less ambiguous. The amount of data needed to completely determine a unique network is known as the *data requirement problem* in network inference. The amount of data required depends on the sparseness of the underlying topology and the type of Boolean functions allowed. In the worst case, the deterministic inference algorithms need on the order of $m = 2^n$ transition pairs of data to infer a densely connected Boolean Network with general Boolean functions at the nodes [2]. They also take exponential time as the inference problem in this case is known to be NP-complete [4].

If, however, the in-degree of each node is at most a constant k then the data amount required to completely disambiguate a network drops significantly. The lower bound, or the minimum number of transition pairs needed (in the worst case) can be shown to be $\Omega(2^k + k\log n)$ [19, 3]. For the expected number of experiments needed to completely disambiguate a network both randomized and deterministic theoretical results are known, although the latter only for a special case. Akutsu and collaborators have shown [3] that the expected number of pairs is proportional to $2^{2k}k\log n$, with high probability, if the data is chosen uniformly at random. For a limited class of Boolean functions, called *linearly separable* and related to the linear models of Section 27.8, Hertz has shown [33] that $k\log(n/k)$ pairs would suffice to identify all parameters. Empirical results show that the expected number of experiments needed is $O(\log n)$, with the constant before the $\log$ being in the order of $2^k k$ [19, 3], although these studies assumed that the data consists of statistically independent state transition pairs, which is not generally true.

Inference Algorithms

The simplest exhaustive algorithm for inferring Boolean networks consistent with the data is to try out all Boolean functions $b_i(\cdot)$ of k variables (inputs) on all $\binom{n}{k}$ combinations of k out of n genes [3]. For a bounded in-degree network this algorithm works in polynomial time. If all possible assignments for the input state of the network are given (2^{2k} different values), then this algorithm uniquely determines a network consistent with the data. This is of course an extraordinary amount of data which is not available in practice. In [3] the authors show empirically on synthetic data that the expected number of input/output pairs is much smaller, and proportional to $\log n$.

Algorithms that exploit similarity patterns in the data fair much better, on average. Liang et al. [41] used an information theoretic method to identify putative coregulation between genes by scoring the mutual information between gene profiles from the expression matrix. Starting from one regulator per node, their algorithm adds in-edges progressively, trying all the time to minimize the number of edges needed (i.e. variables in the Boolean function at that node) to explain the observed transition pairs. Although in the worst case the data requirement is exponential, their algorithm does much better in practice, demonstrating that $O(\log n)$ experiments usually suffice. A theoretical analysis of the expected number of experiments required to disambiguate a gene network has not been given. Their algorithm worked well in practice for very small k.

The combinatorial approach of Ideker et al [34] also exploits co-expression among genes using steady-state data from gene perturbation experiments. Their algorithm identifies a putative set of regulators for each gene by identifying the differentially expressed genes between all pairs of network states (S_i, S_j), including the wildtype (or baseline) state, S_0. The network states, S_i, correspond to steady-states of gene expression of all the genes in the network following a single gene perturbation. To derive the network topology the authors utilize a parsimony argument, whereby the set of regulators in the network was postulated to be equal to the smallest set of nodes needed to explain the differentially expressed genes between the pairs of network states. The problem thus becomes the classical combinatorial optimization problem of *minimum set covering*, which is NP-complete in general. They solved small instances of it using standard branch and bound techniques. The solutions were graphs, or gene network topologies. To complete the network inference, i.e. to identify the Boolean functions at the nodes, they built truth tables from the input data and the inferred regulators for each node. This procedure does not yield a unique network in general. The authors proposed an information-theoretic approach for predicting the next experiment to perform which best disambiguated the inferred network, based on an information theoretic score of information content. Their results confirmed that the number of experiments needed to fully infer a Boolean network is proportional to *logn*, with double perturbation experiments having better resolving power on average than single perturbation ones.

Limitations and Extensions

Boolean networks make good models for biologically realistic systems because their dynamics resembles biological systems behavior and they are also simple enough to understand and analyze. However, these models are ultimately limited by their definition: they are Boolean and synchronous. In reality, of course, the levels of gene expression do not have only two states but can assume virtually continuous values. Thus discretization of the original data becomes a critical step in the inference, and often reducing the values to two states may not suffice. In addition, the updates of the network states in this model are synchronous, whereas biological networks are typically asynchronous. Finally, despite their simplicity, computationally only small nets can be reverse engineered with current state-of-the-art algorithms.

Boolean network models have been extended in various ways to make them more biologically realistic and computationally more tractable. With the availability of better data and models describing the cis-regulatory control and signal propagation through networks, a number of theoretical models, including *chain functions* [27, 28] and certain *Post classes* of Boolean functions [57], have been proposed to restrict the Boolean functions at the network nodes. In addition to offering more realistic network modeling, these approaches have the computational benefit of significantly pruning the solution space for inference.

Additionally, there have been approaches to introduce stochasticity to these models, through probabilistic Boolean networks [56], related to dynamic Bayesian networks, which further increase their realism.

27.8 Differential Equations Models and Linearization

Differential equations (DE) are the starting point for quantitative modeling of complex systems. DEs are continuous and deterministic modeling formalisms, capable of describing non-linear and emerging phenomena of complex dynamical systems.

DE models of gene networks are based on rate equations, quantifying the rate of change of gene expression as a function of the expressions of other genes (and possibly other quantities). The general form of the equations, one for each of n genes, is:

$$\frac{dx_i}{dt} = f_i(x_{i_1}, x_{i_2}, \ldots, x_{i_l}) \tag{27.6}$$

where each x_j is a continuous function, representing the gene expression of gene j.

Each $f_i(\cdot)$ quantifies the combined effect of its arguments, or regulators, on x_i, and it subsumes all the biochemical effects of molecular interactions and degradation. $\{x_{i_1}, x_{i_2}, \ldots, x_{i_l}\}$, the set of arguments of $f_i(\cdot)$, is a subset of all gene expression functions, $\{x_1, x_2, \ldots, x_n\}$. In the gene network, $f_i(\cdot)$ can be thought of as the function at node i which processes the inputs, $x_{i_1}, x_{i_2}, \ldots, x_{i_l}$ and produces an output rate for gene i.

In addition to the variables, the $f_i(\cdot)$ will include many free parameters whose values must be determined from observed data. Given the $f_i(\cdot)$'s and all their parameters, the dynamics of the network can be approximated even if analytical solutions are unknown, by using various numerical differential equation solvers, or even more specifically, gene network simulator software. Thus, the functions $f_i(\cdot)$, with all parameters fitted, specify the gene network fully.

The specific forms of the node functions $f_i(\cdot)$ come out of biochemical considerations and modeling of the components of the genetic system. In general, these functions are non-linear because, among other things, in reality all concentrations get saturated at some point in time. These functions are usually approximated by sigmoid functions [17]. One such class of functions are the *squashing functions* [70], where $f_i(x(t)) = 1/(1 + e^{-(\alpha_j x(t) + \beta_j)})$. The constants α_j and β_j are gene specific and determine the rapidity of the gene response to the regulation. More complicated, non-linear functions, are also used; for example, Savageau's S-systems, which have some nice representational properties [53].

Identifying a gene network from observed data under this model means estimating (or fitting) the parameters in the functions $f_i(\cdot)$. In general the number of arguments, the functions $f_i(\cdot)$, and their parameters are not known. Given observed gene expression data, the first step to identifying the gene network is to guess or approximate the $f_i(\cdot)$'s. Since the identification process depends solely on the form of the $f_i(\cdot)$'s, the functions are typically linearized, as described below.

The question is how many observations are needed to identify the parameters in the differential equation system? For general differential equations, in a recent work Sontag showed that if the system has r parameters, then $2r + 1$ experiments suffice to identify them [58]. Although it is not immediately obvious how many parameters there are in a gene network, if we assume at most a constant number of parameters per argument in the rate functions, the total number of parameters in a dense network of n nodes is $O(n^2)$, and for a sparse one $O(n)$. A large scale gene expression monitoring technology allows for n observations at a time, which may further lower the number of sufficient experiments.

Linearized Additive Models (LAM)

The simplest interesting form that the $f_i(\cdot)$'s can take are linear additive functions, for which Eq. 27.6 becomes:

$$\frac{dx_i(t)}{dt} = ext_i(t) + w_{i1}x_1(t) + \ldots + w_{in}x_n(t) \tag{27.7}$$

(with possibly some additional linear terms on the right hand side, indicating the degradation rate of gene i's mRNA or environmental effects, which can all be incorporated in the w_{ij} parameters, assuming their influence on x_i is linear [74]). The term $ext_i(t)$ indicates a (possible) controlled external influence on gene i, like a perturbation for example, and is directly observable.

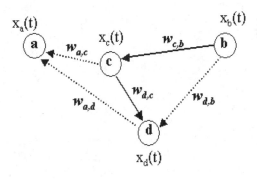

$$x_a(t+1) = x_a(t) + w_{a,c}x_c(t) + w_{a,d}x_d(t)$$
$$x_b(t+1) = x_b(t)$$
$$x_c(t+1) = x_c(t) + w_{c,b}x_b(t)$$
$$x_d(t+1) = x_d(t) + w_{d,c}x_c(t) + w_{d,b}x_b(t)$$

$w_{i,j}$	a	b	c	d
a	0	0	-	-
b	0	0	0	0
c	0	+	0	0
d	0	-	+	0

a) Linear Additive Model

b) Weight Matrix

FIGURE 27.4: Example of a Linear Additive Model for a four node Gene network. The dashed lines are inhibiting relationships, and the full lines are inducing relationships. The weight matrix specifies the existence of relationships between genes and their nature and strength.

The justification for the linear additive model is three-fold. First of all, the w_{ij}'s intuitively relate to the regulatory effect of one gene, j, on another, i, and correspond to the strength of this effect. In the graph representation of the gene network, with genes at the nodes, and edges for each $w_{ij} \neq 0$, the parents of a node are its regulators. The biological interpretation for the weights is that if $w_{ij} > 0$ then gene j induces expression of gene i, and if $w_{ij} < 0$ then gene j represses transcription of gene i. For convenience, we assume $w_{ij} = 0$ if there is no edge from j to i. The second justification for the linearization is that in the immediate neighborhood of any n-dimensional point $(x_1, x_2, \ldots, x_n)$ the surface $f_i(x_1, x_2, \ldots, x_n)$ can be linearized to a plane tangent to it at that point, as in Eq. 27.7, where the coefficients are given by $w_{ij} = \partial f_i / \partial x_j$ [60]. Finally, when the system is observed in a steady-state, or equilibrium, where $dx_i / dt = 0$, and is brought to that steady state slowly, the linear approximation holds.

The reason for linearizing the original system is to turn it into a linear differential equation system, in which the parameters, w_{ij} can easily be fitted to the data using linear algebra methods. From each microarray experiment, be it a time-course, steady-state, or perturbation, one linear equation can be set up for each gene i. Then m such genome-wide microarray experiments would yield nm linear equations. All of those can be written succinctly in a matrix notation as follows.

For experiment l, let $\mathbf{x}^{(l)}$ and $\mathbf{ext}^{(l)}$ be column vectors consisting of the gene expression functions $x_1, \ldots, x_n$, and the external influences on the individual genes, $ext_1, \ldots, ext_n$ respectively. Then, let $\mathbf{X}_{n \times m} = (\mathbf{x}^{(1)}, \ldots, \mathbf{x}^{(m)})$ and $\mathbf{Ext}_{n \times m} = (\mathbf{ext}^{(1)}, \ldots, \mathbf{ext}^{(m)})$, be the matrices consisting of the above vectors for all m experiments. Also, let $\mathbf{W}_{n \times n}$ be the matrix of all weights $w_{ij}, 1 \le i, j \le n$. Then, the system of equations becomes

$$\frac{d}{dt} \mathbf{X}_{n \times m} = \mathbf{W}_{n \times n} \mathbf{X}_{n \times m} + \mathbf{Ext}_{n \times m} \tag{27.8}$$

Here, $\mathbf{X}_{n \times m}$ is the gene expression matrix, consisting of observations of expression for each gene in each experiment, and $\mathbf{Ext}_{n \times m}$ is a matrix of observed external influences. The rate terms on the left hand side of Eq. 27.8, are either known, i.e. observed rates in the experiments, are equal to zero (in the steady state), or are approximated from the available data as $\Delta x_i / \Delta t$, where ideally Δt is very small (in practice that is not the case, and the approximations may be bad).

When dx_i/dt is approximated by $\Delta x_i / \Delta t = (x_i(t+1) - x_i(t))/\Delta t$, the behavior of the network is effectively discretized in time. Then, the next state of a gene's expression can be expressed as a function of the previous states of its regulators, i.e. $x_i(t+1) = \Delta t (x_i(t) + \sum_{j=1}^{n} w_{ij} x_j(t))$. Such models are also known as linear additive, or weight matrix models, and are particularly suited for modeling time-course gene expression experiments.

Gene network identification under LAM

Under the linearized differential equation model network identification amounts to finding a matrix $\mathbf{W}_{n \times n}$ which is consistent with Eq. 27.8, i.e. it is the best solution to that system.

In total, there are n^2 unknowns and nm equations in this system. The existence and uniqueness of solutions to the linear equation system depends on two parameters: m, the number of whole-genome microarray experiments, and $r (\le nm)$ the number of linearly independent equations in the system. Namely, if $r = n^2$ then the system has a unique solution. If $r > n^2$ then the system is over-constrained and there is no solution. However, in practice more than the sufficient number of observations is preferred for better results. A solution for an over-constrained system can be found by performing multiple regression of each gene on the other genes. If $r < n^2$ the system is under-constrained and there are infinitely many solutions, all of which can be expressed in terms of a subset of independent solutions. To address this dimensionality curse, many methods have been used to bridge the gap between r and n^2, resulting in both under-constrained and over-constrained systems.

In practice, the gene expression matrix is very long and narrow, that is, there are typically many more genes than experiments, so $m \ll n$ and thus $r \ll n^2$. Choosing a solution from the infinitely many plausible ones is a non-trivial task. Often a solution with special properties, or a pseudo-inverse is chosen. Such are for example the Moore-Penrose pseudo-inverse, and the Singular Value Decomposition (SVD), and although they rarely give the correct solution [74] they can be used as starting points for generating better solutions.

Additional biological assumptions are needed to prune the search space, of which the network-sparseness assumption has been the most popular one: each gene cannot be regulated by more than a fixed number k of other genes [70, 13, 74]. Chen et al. in [13] translated the problem of finding a solution $\mathbf{W}$ of Eq. 27.8 when the number of nonzero weights w_{ij} for any given i is at most a fixed constant k, into a combinatorial problem called Minimum Weight Solutions to Linear Equations, and showed that it is polynomially solvable in general, although they offered a computationally expensive algorithm.

Collins and collaborators in a series of papers used a system similar to Eq. 27.8 to model gene networks under the sparseness assumption, and offered several methods to identify networks from both time-course and perturbation data under such models. In [74], they used Singular Value Decomposition on time-course experiments to generate an initial solution and then refined it by using an optimization technique called *robust regression*. The solutions were much better than those from using SVD alone. Experimentally, the authors used very fine sampling times in this study which allowed them to approximate the dx_i/dt's very well. This method is particularly well suited for identification of larger networks. In [64] they use the same linear model, but with advanced gene perturbation (over-expression) technology (described in [25]), and measurements at steady-states. The goal there was different from above: based on previous perturbations they wanted to choose the next perturbation to best identify the still unknown parameters in the network. To that end, they used an algorithm which as the next best perturbation chooses genes which have either changed the least in the past perturbations or have most uncertain connections to other genes. However, as a consequence of the steady-state modeling their algorithm is unable to predict w_{ii} for any gene (which is the case in similar methods too, e.g. [18]). In subsequent work Collins and colleagues showed that the linear additive model with steady-state data of targeted perturbations allows for effective identification of both small [26] and large [21] gene networks.

As emphasized above, if only the observed data is used the system of linear equations is mostly under-constrained. However, under some assumptions, the gene expression matrix can be processed to yield an over-constrained system. To identify a small gene network of CNS development regulation in rat, D'Haeseleer and colleagues [20] used cubic interpolation between successive time points of gene expression observations, thereby increasing the total amount of data. They generated just as many interpolated points as needed for the linear system of equations to become over-constrained, which they consequently solved using a least squares optimization. The interpolation approach does well when the phenomena studied have been sampled finely enough to capture the essential changes in gene expression, which cannot be guaranteed in general. In a complementary approach, Someren et al. [66] used clustering of the time-course expression matrix to reduce the dimensionality of **W**. Hierarchical (progressive) clustering was performed until the resulting linear system had the smallest error in explaining the whole data, i.e. was close to being over-constrained. Their approach drew a lot from the success of clustering in identifying coregulated clusters of genes through coexpression, but it has its limitations too: the resulting gene network is a network of gene clusters and not genes, and the interpretation is non-trivial.

Data Requirement

If the weight matrix is dense (i.e. the average connectivity per node in the network is $O(n)$, then $n + 1$ arrays of all n genes are needed to solve the linear system, assuming the experiments are independent (which is not exactly true with time-series data). If instead the average connectivity per node is a fixed constant, k, as is the case in realistic networks, than it can be shown that the number of experiments needed is $O(k \log n)$ [19, 74, 64].

Limitations and Extensions

Linear models yield good, realistic looking predictions when the expression measurements have been performed around a steady state or on a slow changing system. Otherwise, the rates of change of the genes' expressions cannot be estimated well. As with the other methods, the available data is insufficient for large-scale network inference, although the inference methods for this model scale well with the amount of data and should work for larger networks too.

A possible way to extend these models would be to make them hold not just in the vicinity of a steady-state point, but further away too. To do that second order relationships, i.e. $x_i x_j$ product terms, could be considered, although the data requirement may become prohibitive in that case.

27.9   Summary and Comparison of Models

The overall properties of the four models and inference methods are summarized in Table 27.2.

There have been several attempts to compare different network inference approaches based on how well they retrieve a known network [71, 72]. Unfortunately those approaches have compared different modeling strategies without any systematic method behind it. For example Boolean network reverse engineering methods were compared to Bayesian networks etc. The results are therefore not reliable, and by no means definite.

It is more reasonable to compare the efficacy of different network inference methods within the same modeling category. To that end well known real data sets should be used. In addition, results should be reported on much larger sets of artificial data satisfying pre-specified statistical and biological properties, like precise knowledge of what is signal and what noise [44]. Finally, the results of any network inference from public data should be made fully available to the public for future comparison and reference. The trend so far has been of reporting predicted sub-parts of the overall networks, those that had some immediate positive reporting value.

TABLE 27.2   Summary of Model Properties and Inference Methods

	Graph	Bayes	Boolean	LAM
Nodes	Genes	Random Variables	Discretized Expression	Continuous Expression
Edges	Causality	Conditional Dependency	Arguments in Boolean Functions	Arguments in Regulatory Functions
Additional Parameters	Activation, Inhibition	Activation, Inhibition	Boolean Function	Weight Matrix
Properties	Static topology	Static topology, Stochastic	Dynamic, Biologically Realistic	Dynamic, Steady-state realistic
Inference	Biological Parsimony, Optimization	Bayes Scoring, ML, Optimization	Optimization, Information Theory	Linear Regression, Optimization
Data Requirement	$O(logn) - O(n)$	?	$O(2^k klogn)$	$O(knlogn)$

27.10   Future Directions and Conclusions

In this chapter were presented various models, modeling methodologies, and inference methods for reverse engineering gene networks from large-scale gene expression data. In addition to the computational methods, gene network inference involves considerations of type and quantity of data, prior biological knowledge, modeling framework, and experimental technologies. Ultimately, one has to first identify the modeling imperatives (e.g. obtaining the topology only may be enough for some applications but not for others), then compile all available knowledge about the network of interest, decide on what type of data/experiments to use, and only at the end choose the inference method.

The various inference problems presented in this chapter could be used as starting points and ideas to build on; many of them are rather naive although their solutions involve intricate theoretical arguments.

The main challenges in network reconstruction are data related. First of all, typically the currently available data is not sufficient for large-scale network reconstruction (dimensionality curse), and the formal statements of the models are very over-constrained. Second, the question is how much is in the data, i.e. do the experiments capture relevant events or not. For example, in time-series data if the sampling is too coarse the important biological events may happen in-between sampling points. Both of these data related issues will likely be resolved with technological advancements.

Even with ideal gene expression data the inference problems may be difficult to solve because of the large space of solutions. In general, including biological constraints, like the bounded indegree of a node, or restricted cis-functions, narrow down the huge space of possibilities to potentially a manageable one. Such a *model-based* modeling should become the standard as more and more is learned about gene networks.

It is very important to extend the gene network modeling to other levels of physical interactions as in Figure 27.1, especially since some gene–gene interactions cannot be recovered at all from the gene expression data. To do that other types of experiments are needed besides gene expression measurements. Utilizing different types of data together, or data integration, is becoming a hot area of research in functional genomics. The problem is how to put together information from heterogeneous empirical data.

The available large-scale genomic data includes DNA sequences (as most reliable), gene expression data, TF-DNA interaction data (Chromatin Immuno-Precipitation on a chip), protein expressions, and protein-protein interactions (e.g. yeast two-hybrid). Combining these diverse data could potentially reveal gene network interactions beyond what each can individually. For example, promoter DNA regions have been used together with expression data to identify cis-elements in co-regulated genes, starting from either the expression [63], the DNA promoter regions [11], or both [50, 40]. Gene coding DNA together with expression data have been very successfully combined to verify that conserved coding regions across organisms have a conserved function too [61]. TF-DNA interaction data together with expression data has been used to identify modules of functionally related genes [31]. Protein-protein interaction data has been used to refine gene networks estimated from expression data, in a Bayesian setting [46]. Many different types of data were used to identify most of the functional relationships between the genes in yeast [62]. Thus, having more and varied types of data would allow for better, more meaningful predictions of gene networks.

Conclusions

Gene network modeling and inference is a very young research area, mostly because current gene expression data is inadequate for most types of definitive analyses. In addition, most inference methods apply to the simplest models, Boolean networks, and even in those biological over-approximations the algorithms quickly become computationally intractable. In the future, the confluence of better data, through data integration, and powerful inference methods should bring about networks of predictive and comparative value yielding reliable and testable models of systemic gene regulation.

References

[1] A. Aho, M.R. Garey, and J.D. Ullman. The transitive reduction of a directed graph. *SIAM J. Comput.*, 1:131–7, 1972.

[2] T. Akutsu, S. Kuhara, O. Maruyama, and S. Miyano. Identification of gene regulatory networks by strategic gene disruptions and gene overexpressions under a boolean model. *Theor. Comput. Sci.*, 298:235–51, 2003.

[3] T. Akutsu, S. Miyano, and S. Kuhara. Identification of genetic networks from a small number of gene expression patterns under the boolean network model. In *Pac. Symp. Biocomputing*, volume 4, pages 17–28, 1999.

[4] T. Akutsu, S. Miyano, and S. Kuhara. Algorithms for inferring qualitative models of biological networks. In *Pac. Symp. Biocomputing*, volume 5, pages 290–301, 2000.

[5] R. Albert and H.G. Othmer. The topology of the regulatory interactions predicts the expression pattern of the drosophila segment polarity genes. *J. Theor. Biol.*, 223:1–18, 2003.

[6] M.I. Arnone and E.H. Davidson. The hardwiring of development: organization and function of genomic regulatory systems. *Development*, 124:1851–1864, 1997.

[7] Z. Bar-Joseph, G.K. Gerber, T.I. Lee, and N.J. Rinaldi *et al.* Computational discovery of gene modules and regulatory networks. *Nat. Biotech.*, 21(11):1337–1342, 2003.

[8] H. Bolouri and E.H. Davidson. Transcriptional regulatory cascades in development: Initial rates, not steady state, determine network kinetics. *Proc. Natl. Acad. Sci. USA*, 100(16):9371–9376, 2003.

[9] J.M. Bower and H. Bolouri, editors. *Computational modeling of genetic and biochemical networks*. MIT Press, 2001.

[10] P. Brazhnik, A. de la Fuente, and P. Mendes. Gene networks: how to put the function in genomics. *Trends Biotechnol.*, 20:467–472, 2002.

[11] H. Bussemaker, H. Li, and E. Siggia. Regulatory element detection using correlation with expression. *Nat. Genet.*, 27:167–71, 2001.

[12] T. Chen, V. Filkov, and S. Skiena. Identifying gene regulatory networks from experimental data. *Parallel Comput.*, 27(1-2):141–162, 2001.

[13] T. Chen, H.L. He, and G.M. Church. Modeling gene expression with differential equations. *Pac. Symp. Biocomputing*, 4:29–40, 1999.

[14] T.H. Cormen, C.E. Leiserson, and R.L. Rivest. *Intoduction to Algorithms*. MIT Press, 2nd edition, 2001.

[15] E.H. Davidson. *Genomic Regulatory Systems*. Academic Press, 2001.

[16] E.H. Davidson, J.P. Rast, P. Oliveri, and A. Ransick *et al.* A genomic regulatory network for development. *Science*, 295:1669–1678, 2002.

[17] H. de Jong. Modeling and simulation of genetic regulatory systems: a literature review. *J. Comp. Bio.*, 9(1):67–103, 2002.

[18] A. de la Fuente, P. Brazhnik, and P. Mendes. A quantitative method for reverse engineering gene networks from microarray experiments using regulatory strengths. In *2nd International Conference on Systems Biology*, pages 213–221, 2001.

[19] P. D'Haeseleer, S. Liang, and R. Somogyi. Genetic network inference: From co-expression clustering to reverse engineering. *Bioinformatics*, 16:707–26, 2000.

[20] P. D'Haeseleer, X. Wen, S. Fuhrman, and R. Somogyi. Linear modeling of mRNA expression levels during CNS development and injury. *Pac. Symp. Biocomputing*,

pages 41–52, 1999.

[21] D. di Bernardo, T.S. Gardner, and J.J. Collins. Robust identification of large genetic networks. In *Pac. Symp. Biocomputing*, volume 9, pages 486–497, 2004.

[22] V. Filkov, J. Zhi, and S. Skiena. Analysis techniques for microarray time-series data. *J. Comp. Bio.*, 9(2):317–330, 2002.

[23] N. Friedman. Inferring cellular networks using probabilistic graphical models. *Science*, 303:799–805, 2004.

[24] N. Friedman, M. Linial, I. Nachman, and D. Pe'er. Using bayesian networks to analyze expression data. *J. Comp. Bio.*, 7(6):601–620, 2000.

[25] T.S. Gardner, C.R. Cantor, and J.J. Collins. Construction of a genetic toggle switch in *escherichia coli. Nature*, 403:339–342, 2000.

[26] T.S. Gardner, D. di Bernardo, D. Lorenz, and J.J. Collins. Inferring genetic networks and identifying compound mode of action via expression profiling. *Science*, 301:102–105, 2003.

[27] I. Gat-Viks and R. Shamir. Chain functions and scoring functions in genetic networks. *Bioinformatics*, 19(Suppl 1):i108–i117, 2003.

[28] I. Gat-Viks, R. SHAMIR, R.M. KARP, and R. SHARAN. Reconstructing chain functions in genetic networks. In *Pac. Symp. Biocomputing*, volume 9, pages 498–509, 2004.

[29] J. Gollub, C.A. Ball, G. Binkley, and J. Demeter *et al.* The stanford microarray database: data access and quality assessment tools. *Nucleic Acids Res*, 31:94–6, 2003.

[30] A.J. Hartemink et al. Using graphical models and genomic expression data to statistically validate models of genetic regulatory networks. In *Pac. Symp. Biocomputing*, volume 6, pages 422–433, 2001.

[31] A.J. Hartemink, D.K. Gifford, T. Jaakkola, and R.A. Young. Combining location and expression data for principled discovery of genetic regulatory network models. In *Pac. Symp. Biocomputing*, volume 7, pages 437–449, 2002.

[32] D. Heckerman, D. Geiger, and D. Chickering. Learning bayesian networks: The combination of knowledge and statistical data. *Mach. Learn.*, 20:197–243, 1995.

[33] J. Hertz. Statistical issues in reverse engineering of genetic networks. *Pac. Symp. Biocomputing*, 1998. Poster.

[34] T. Ideker, V. Thorsson, and R. Karp. Discovery of regulatory interactions thrugh perturbation: inference and experimental design. In *Pac. Symp. Biocomputing*, volume 5, pages 302–313, 2000.

[35] A. Jain and R. Dubes. *Algorithms for Clustering Data.* Prentice-Hall, Englewood Cliffs NJ, 1988.

[36] H. Jeong, B. Tombor, R. Albert, and Z.N. Oltvai *et al.* The large-scale organization of metabolic networks. *Nature*, 407:651–4, 2000.

[37] M. Kanehisa and S. Goto. *kegg*: Kyoto encyclopedia of genes and genomes. nucleic acid. *Nucl. Acid Res.*, 28:27–30, 2000.

[38] P.D. Karp, M. Arnaud, J. Collado-Vides, and J. Ingraham *et al.* The *e. coli ecocyc* database: No longer just a metabolic pathway database. *ASM News*, 70:25–30, 2004.

[39] S.A. Kauffman. *The Origins of Order: Self Organization and Selection in Evolution.* Oxford University Press, 1993.

[40] M. Lapidot and Y. Pilpel. Comprehensive quantitative analyses of the effects of promoter sequence elements on mRNA transcription. *Nucleic Acids Res.*, 31(13):3824–3828, 2003.

[41] S. Liang, S. Fuhrman, and R. Somogyi. REVEAL, a general reverse engineering algorithm for inference of genetic network architectures. In *Pac. Symp. Biocomputing*,

volume 3, pages 18–29, 1998.

[42] H.H. McAdams and A. Arkin. Stochastic mechanisms in gene expression. *Proc Natl. Acad. Sci. USA*, 94:814–819, 1997.

[43] H.H. McAdams and A. Arkin. Simulation of prokaryotic genetic circuits. *Ann. Rev. Biophys. Biomolecul. Struct.*, 27:199–224, 1998.

[44] P. Mendes, W. Sha, and K. Ye. Artificial gene networks for objective comparison of analysis algorithms. *Bioinformatics*, 19:122–9, 2003.

[45] R. Milo, S. Shen-Orr, S. Itzkovitz, and N. Kashtan *et al.* Network motifs: Simple building blocks of complex networks. *Science*, 298:824–827, 2002.

[46] N. Nariai, S. Kim, S. Imoto, and S. Miyano. Using protein-protein interactions for refining gene networks estimated from microarray data by bayesian networks. In *Pac. Symp. Biocomputing*, pages 336–47, 2004.

[47] S. Ott, S. Imoto, and S. Miyano. Finding optimal models for small gene networks. In *Pac. Symp. Biocomputing*, volume 9, pages 557–567, 2004.

[48] J. Paulsson. Summing up the noise in gene networks. *Nature*, 427:415–418, 2004.

[49] D. Pe'er, A. Regev, G. Elidan, and N. Friedman. Inferring subnetworks from perturbed expression profiles. *Bioinformatics*, 17(Suppl 1):S215–S224, 2001.

[50] Y. Pilpel, P. Sudarsanam, and G. Church. Identifying regulatory networks by combinatorial analysis of promoter elements. *Nat. Genet.*, 29:153–159, 2001.

[51] M. Ptashne. *A Genetic Switch: Phage Lambda and Higher Organisms*. Cell Press and Blackwell Scientific, 2nd edition, 1992.

[52] E. Ravazs, A.L. Somera, D.A. Mongru, and Z.N. Oltvai *et al.* Hierarchical organization of modularity in metabolic networks. *Science*, 297:1551–5, 2002.

[53] M.A. Savageau and P. Sands. Completly uncoupled or perfectly coupled circuits for inducible gene regulation. In E.O. Voit, editor, *Canonical nonlinear modeling: S-system approach to understanding complexity*. Van Nostrand Reinhold, New York, 1990.

[54] E. Segal, M. Shapira, A. Regev, and D. Peer *et al.* Module networks: Identifying regulatory modules and their condition specific regulators from gene expression data. *Nat. Genet.*, 34(2):166–76, 2003.

[55] Y. Setty, A.E. Mayo, M.G. Surette, and U. Alon. Detailed map of a cis-regulatory input function. *Proc. Natl. Acad. Sci. USA*, 100:7702–7707, 2003.

[56] I. Shmulevich, E.R. Dougherty, S. Kim, and W. Zhang. Probabilistic boolean networks: A rule-based uncertainty model for gene regulatory networks. *Bioinformatics*, 18:261–74, 2002.

[57] I. Shmulevich, H. Lähdesmäki, E.R. Dougherty, and J. Astola *et al.* The role of certain post classes in boolean network models of genetic networks. *Proc. Natl. Acad. Sci. USA*, 100:10734–9, 2003.

[58] E.D. Sontag. For differential equations with r parameters, 2r+1 experiments are enough for identification. *J. Nonlinear Sci.*, 12:553–83, 2002.

[59] T. Speed, editor. *Statistical Analysis of Gene Expression Microarray Data*. CRC Press, 2003.

[60] J. Stark, D. Brewer, M. Barenco, and D. Tomescu *et al.* Reconstructing gene networks: what are the limits? *Biochem. Soc. Transact.*, 31:1519–25, 2003.

[61] J. Stuart, E. Segal, D. Koller, and S.K. Kim *et al.* A gene co-expression network for global discovery of conserved genetics modules. *Science*, 302:249–55, 2003.

[62] A. Tanay et al. Revealing modularity and organization in the yeast molecular network by integrated analysis of highly heterogeneous genomewide data. *Proc. Natl. Acad. Sci. USA*, 101(9):2981–2986, 2004.

[63] S. Tavazoie, J.D. Hughes, M.J. Campbell, and R.J. Cho *et al.* Systematic determina-

tion of genetic network architecture. *Nat. Genet.*, 22:281–5, 1999.

[64] J. Tegner, M.K. Yeung, J. Hasty, and J.J. Collins. Reverse engineering gene networks: integrating genetic perturbations with dynamical modeling. *Proc. Natl. Acad. Sci. USA*, 100(10):5944–5949, 2003.

[65] M. Thattai and A. van Oudenaarden. Intrinsic noise in gene regulatory networks. *Proc. Natl. Acad. Sci. USA*, 98(15):8614–8619, 2001.

[66] E.P. van Someren, L.F. Wessels, and M.J. Reinders. Linear modeling of genetic networks from experimental data. In *Proc. ISMB*, pages 355–366, 2000.

[67] J.M. Vilar, H.Y. Kueh, N. Barkai, and S. Leibler. Mechanisms of noise-resistance in genetic oscillators. *Proc. Natl. Acad. Sci. USA*, 99(9):5988–92, 2002.

[68] G. von Dassow, E. Meir, E.M. Munro, and G.M. Odell. The segment polarity network is a robust developmental module. *Nature*, 406:188–92, 2000.

[69] A. Wagner. How to reconstruct a large genetic network from n gene perturbation in n^2 easy steps. *Bioinformatics*, 17:1183–97, 2001.

[70] D.C. Weaver, C.T. Workman, and G.D. Stormo. Modeling regulatory networks with weight matrices. *Pac. Symp. Biocomputing*, 4:112–123, 1999.

[71] L.F. Wessels, E.P. van Someren, and M.J. Reinders. A comparison of genetic network models. In *Pac. Symp. Biocomputing*, volume 6, pages 508–519, 2001.

[72] F.C. Wimberly, C. Glymor, and J. Ramsey. Experiments on the accuracy of algorithms for inferring the structure of genetic regulatory networks from microarray expression levels. *IJCAI 2003 Workshop on Learning Graphical Models for Computational Genomics*, 2003.

[73] A. Wuensche. Genomic regulation modeled as a network with basins of attraction. In *Pac. Symp. Biocomputing*, volume 3, pages 89–102, 1998.

[74] M.K.S. Yeung, J. Tegnär, and J.J. Collins. Reverse engineering gene networks using singular value decomposition and robust regression. *Proc. Natl. Acad. Sci. USA*, 99:6163–6168, 2002.

[75] J. Yu, A.V. Smith, P.P. Wang, and A.J. Hartemink *et al.* Using bayesian network inference algorithms to recover molecular genetic regulatory networks. In *International Conference on Systems Biology*, 2002.

[76] C.-H. Yuh, H. Bolouri, and E.H. Davidson. Genomic *cis*-regulatory logic: experimental and computational analysis of a sea urchin gene. *Science*, 279:1896–1902, 1998.

[77] C.-H. Yuh, J.G. Moore, and E.H. Davidson. Quantitative functional interrelations within the *cis*-regulatory system of the *s. purpuratus endo-16* gene. *Development*, 122:4045–4056, 1996.

28

Modeling and Analysis of Gene Networks Using Feedback Control Theory

Hana El Samad
University of California, Santa Barbara

Mustang Rammish
University of California, Santa Barbara

28.1 Introduction

The Human Genome Project, which sequenced the three billion DNA letters, has resulted in an unprecedented amount of information for scientists to analyze. The ultimate goal is to answer one of biology's next big challenges: how to go from the DNA sequence of a gene, to the structure of the protein for which it encodes to the activity of the protein and its function within the cell, to the tissue and then ultimately to the organism? The answer to this question ultimately involves two central problems. The first is related to the identification of the functional role of a specific gene in the organism; the second is related to the analysis of its interactions within a genetic pathway. Accomplishing this is not a trivial task, especially with the challenges imposed by high dimensionality, uncertainty and complexity of biological systems. As has been frequently suggested, research relying strictly on traditional wet lab or clinical approaches is of prime importance, but is not sufficient by itself to accomplish the goal of using genetic information to understand the functioning of organisms as integrated systems. Therefore, this type of experimental biology should be complemented by the use of mathematical and computational models of the cell and its molecular pathways. These models do not have to be entirely accurate to provide useful insight. Given the enormous increase in genetic and molecular data and with the help of guided experiment, the models will continue to improve to become an essential tool for evaluating hypotheses and suggesting new directions in experimental procedures.

Molecular pathways are a part of a remarkable hierarchy of regulatory networks that operate at all levels of organization. The dynamic character of these pathways and the prevalence of feedback regulation strategies in their operation make them amenable to systematic mathematical analysis using the same tools that have been used with remarkable success in analyzing and designing engineering control systems. The promise of dynamical systems and feedback control theory as an effective tool for the study of biological systems at the molecular level is increasingly being recognized. As a consequence, quantitative tools developed for engineering systems analysis and/or design are being used successfully in the study of systems biology. Indeed, in a recent paper by Hartwell *et al* [34], it has been suggested that ideas borrowed from "synthetic" sciences such as engineering and computer science could be of enormous help in understanding functional biological modules and their interactions. The authors propose that the resulting modular view enables an understanding of the behavior of biological systems that may not be easily attainable from knowledge of the behavior of the underlying molecules.

It is the aim of this article to illustrate this point of view and demonstrate that ideas, methods and principles specifically borrowed from feedback control and dynamical systems theories are necessary in order to fully probe and understand biological complex behavior [80]. We argue this point by providing numerous examples showing the ubiquity of feedback loops in cellular regulatory organization. To this end, we itemize a variety of biological behaviors in terms of the feedback loops that generate these behaviors. We further discuss the use of various ideas from feedback control theory to study these examples and the insight gained in characterizing and understanding the general principles of their functionality and modularity. To set a basis for our analysis, we provide a short review of the mathematical methods that are commonly used in modeling gene networks.

28.2   Gene Regulatory Networks: a Definition

Gene regulatory networks can be broadly defined as groups of genes that are activated by particular signals and stimuli, and once activated, orchestrate their operation to regulate certain biological functions, such as metabolism, development, and the cell cycle. These gene networks are therefore dynamic objects that continuously sense the environment and orchestrate their operation accordingly. The core of this operation lies in the "central dogma" of biology which describes how "operative information" stored in the *DNA* is used to generate "operating elements", mostly proteins. Proteins are produced from an intermediate product, called the *RNA*. First, coding regions of *DNA* (genes) are "transcribed" to synthesize these RNA molecules. Thereafter, proteins are generated through the "translation" of these *RNA* molecules. These proteins, in turn, affect the production of other proteins (or even auto-regulate their own production), or catalyze and regulate reactions responsible for various cellular activities. The organization of these proteins and genetic elements in networks that possess various levels of regulation and feedback can be perceived as a working definition of a gene regulatory network.

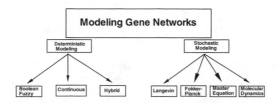

FIGURE 28.1: Modeling approaches to gene regulatory networks.

28.3 Modeling Approaches to Gene Regulatory Networks

Gene regulatory networks can be modeled and analyzed in deterministic or stochastic settings. Within these two broad categories, various approaches can be adopted. We list here a few of these approaches. However, we caution the reader that this list is not in any way exhaustive. It is rather aimed to be a broad account of the most common modeling techniques. A diagrammatic representation of these methods is provided in Figure 28.1.

28.3.1 Deterministic Modeling

Boolean and Bayesian Modeling

Genes can be viewed as logical elements which state is either ON or OFF. Consequently, regulatory control is approximately achieved through some rules that implement logic gates (e.g. AND, OR, NOR) [67, 68]. This is known as the Boolean modeling approach. Boolean modeling has been shown to reproduce qualitatively some of the dynamical behavior of various genetic systems [71, 72]). It has been specifically used to model Drosophila embryogenesis where it yielded precious insight into the genes necessary to simulate embryogenesis in both Drosophila and other insects [10].

Continuous Modeling

Cellular processes are often perceived to be systems of distinct chemical reactions. As in chemical kinetics, these reactions can be described using the laws of mass-action, yielding a set of differential equations (linear or nonlinear) that give the succession of states (usually concentration of species) adopted by the network over time. The equations are usually of the form

$$\frac{dx_i}{dt} = f_i(x), \quad 1 \le i \le n \tag{28.1}$$

where $x = [x_1, ...x_n]'$ is a vector of non-negative real numbers describing concentrations and $f_i : R^n \to R^n$ is a linear or nonlinear function of the concentrations. Ordinary differential equations are arguably the most widespread formalism for modeling gene regulatory networks and their use goes back to the 'operon' model of Jacob and Monod [40] and the early work of Goodwin [31]. Powerful mathematical methods for this approach have been developed, especially in the context of metabolism [11]. These methods are based on different formalisms to specify the f_i functions in order to describe various kinetic interactions. For example, activation of the production of protein x_i by a protein x_j can be described using a function of the form

$$f_i(x_j, \theta_j, m) = \frac{x_j^m}{x_j^m + \theta_j^m}$$

This function is called a *Hill* curve, where the number m (called the Hill constant) indicates the steepness of the curve and θ_j is the threshold for the regulatory influence of x_j on x_i. The sigmoidal shape of the Hill function ($m > 1$) corresponds to experimental data, but other regulatory functions are possible (e.g. step and logoid functions). Due to the importance of the continuous modeling, we illustrate its use through a simple example borrowed from [80]. The example is a sequence of enzyme reactions representing a simplified signal transduction scheme. In this scheme, a substrate S is turned into a product P via an intermediate enzyme-substrate complex SE. The rate of formation of SE is denoted by k_1. Once formed, SE can either dissociate into E and S with a rate constant k_2 or form a product P at a rate k_3. These reactions proceed as follows

$$S + E \underset{k_2}{\overset{k_1}{\rightleftharpoons}} SE \overset{k_3}{\rightarrow} E + P$$

The nonlinear set of differential equations that describes the evolution of species concentrations as dictated by these reactions is

$$
\begin{aligned}
\frac{dS(t)}{dt} &= -k_1 S(t)E(t) + k_2 SE(t) \\
\frac{dE(t)}{dt} &= -k_1 S(t)E(t) + (k_2 + k_3)SE(t) \\
\frac{dSE(t)}{dt} &= k_1 S(t)E(t) - (k_2 + k_3)SE(t) \\
\frac{dP(t)}{dt} &= k_3 SE(t)
\end{aligned}
\tag{28.2}
$$

One should append to these differential equations algebraic constraints describing the relation between total concentration of a certain specie and the different forms it can take. These constraints are often referred to as "mass balance equations". A simple mass balance relation for the example above is one in which the total concentration of enzyme E, E_T, is equal to that of E in its free form in addition to that of E complexed with the substrate. This translates to

$$E(t) + SE(t) = E_T$$

Generic descriptions of the functions generated by various interactions in gene networks have been proposed in the literature. The most influential of these generalizations are the Generalized Mass Action systems (GMA) and Synergetic (S) systems developed by Savageau and various coworkers. GMA and S systems approaches propose the use of power-law functions yielding the following descriptions

$$\frac{dx_i}{dt} = \sum_{k=1}^{r} \alpha_{ik} \prod_{j=1}^{n+m} x_j^{g_{ijk}} - \sum_{k=1}^{r} \beta_{ik} \prod_{j=1}^{n+m} x_j^{h_{ijk}}, \quad i = 1, ..., n \tag{28.3}$$

for GMA systems and

$$\frac{dx_i}{dt} = \alpha_i \prod_{j=1}^{n+m} x_j^{g_{ij}} - \beta_i \prod_{j=1}^{n+m} x_j^{h_{ij}}, \quad i = 1, ..., n \tag{28.4}$$

for Synergetic S systems. g_{ijk} and h_{ijk} are kinetic order parameters for elementary processes corresponding to production and degradation respectively, while α_{ik} and β_{ik} are rate constant parameters for these processes. These descriptions are capable of representing almost any physical system of interest due to the fact that any nonlinear function (that is composite of elementary functions) can be transformed into the power law formalism through an operation called *recasting* [61]. Using this power law formalism, many systemic properties can be derived analytically, including analytical solutions for steady-states (which reduce to conventional linear analysis in logarithmic space) [58], local dynamic behavior through eigenvalue analysis [51], and Hopf Bifurcations [44]. The power law formalism has been successfully applied in the study of various problems, such as the fundamental design principles of gene networks [77, 60], functional effectiveness of different types of coupling in these networks [59], and mathematically controlled comparison [2].

Hybrid Modeling

Combining Boolean and continuous modeling is possible, the result being what is commonly known as hybrid modeling. Boolean logic can be used to represent biochemical processes characterized by sharp thresholds or transients while continuous dynamics are used to model processes that have slower thresholds. Hybrid modeling has been used by McAdams and Shapiro to describe the bacterial λ-phage lysis lysogeny system [48] and Yuh and colleagues [82] to describe the *endo16* developmental gene in sea urchin embryo.

Brief Comparison of Deterministic Modeling Approaches

Boolean, continuous and hybrid approaches have been successfully used in the modeling of various gene regulatory networks. Adopting any of these approaches is very much system dependent, as they all offer advantages and drawbacks. For example, the boolean approach is less computationally intensive than either the continuous or hybrid modeling approaches. However, boolean representations mainly give qualitative information and may even produce erroneous results. For example, mathematical analysis shows that boolean representations can produce steady states that do not exist in the continuous description and periodic solutions that may not correspond to the same periodic solutions in the continuous representation of the same system [30, 4]. Furthermore, spatial dimensions (such as diffusion and transport) cannot be readily incorporated into boolean representations. In contrast, in differential equations descriptions, transport can be incorporated using delays [65, 64], but these equations tend to be computationally expensive. Obviously, one can strike a middle ground by adopting a hybrid modeling approach. Hybrid modeling requires less computational effort than the continuous modeling approach. However, hybrid modeling is still in its infancy, and mathematical results on steady states and other characteristics yielded by the hybrid approach still await further investigation.

28.3.2 The Stochastic Modeling Approach

Gene expression is a "noisy" or stochastic process. Roughly speaking, this noise can come about in two ways. Firstly, the inherent stochasticity in biochemical processes (such as binding, transcription, and translation) generates what is known as "intrinsic noise". Secondly, variations in the amounts or states of cellular components and species that affect those biochemical reactions generate additional fluctuations, termed "extrinsic noise". Intrinsic noise is believed to become especially significant when species are present at low copy numbers [55]. Deterministic modeling does not embody any description of this noise. Hence, alternative stochastic approaches should be used if the stochastic effects are deemed essential in the understanding of the dynamic behavior and performance of a certain genetic system. In this section, we provide a summary of the various stochastic modeling approaches that can be used to this end. We then present a simplified model of the circadian rhythm where stochastic modeling has been proven to be necessary.

Molecular Dynamics Modeling

Molecular dynamics modeling depicts the procedure of producing the exact behavior of chemically reaction systems. Molecular dynamics simulations account for both reacting and non-reacting collisions, in addition to tracking the position and velocity of every molecule in the system. Therefore the information content of this type of modeling includes the temporal and spatial evolution of the system, and as such, it requires a substantial amount of computation and time to simulate even the smallest systems. This approach has mainly been used to study protein folding [63]. In the context of modeling gene networks, it is however often sufficient to track only the reactive collisions under the rationale that non-reacting collisions only serve to stir the system. In this case, and based on the assumption of a uniform spatial distribution, one can go to Master Equation type of modeling.

Master Equation Modeling

The Chemical Master Equation (CME) description accounts for the probabilistic nature of cellular processes. The CME descibes the time evolution of the probability of having a certain number or concentration of molecules, as opposite to deterministic rate equation descriptions of the absolute concentration of these molecules [41, 27]. In the Master Equation, reaction rates are transformed into probability transition rates which can be determined based on physical considerations. The CME can be derived based on the Markov property of chemical reactions. Using this Markov property, one can write the Chapman-Kolmogorov equation, an identity that must be obeyed by the transition probability of any Markov process. Using stationarity and taking the limit for infinitesimally vanishing time intervals, one obtains the Master equation, as the differential form of the Chapman-Kolmogorov equation [41]. Another derivation of the CME based on basic probability and physical principles is given by Gillespie [24]. Here, we give the expression for the CME without proof. The interested reader if referred to [41] or [24] for a more detailed account.

Suppose we are dealing with a chemically reacting system involving N molecular species $S_1, ... S_N$ reacting through M reaction channels $R_1.... R_M$. Let $X(t) = (X_1(t).... X_n(t))$ be the state vector, where $X_i(t)$ is a random number that defines the number of molecules of species S_i in the system at time t. We assume that the system is well stirred and in thermal equilibrium. Under these circumstances, each reaction channel R_k is characterized by a *propensity function* w_k and an N-dimensional *state change vector* $s_k = (s_{1k}.... s_{Nk})$. The vector s_k represents the stoichiometric change of the molecular species by an R_k reaction. Let

$$S = \begin{bmatrix} s_1 & s_2 & . & . & . & . & s_M \end{bmatrix}$$

and

$$W = \begin{bmatrix} w_1 & w_2 & . & . & . & . & w_M \end{bmatrix}^T$$

The Chemical Master Equation written for the evolution of probability in this system is given by:

$$\frac{\partial P(X, t | X_0, t_0)}{\partial t} = \sum_{k=1}^{M} [w_k(X - s_k) P(X - s_k, t | x_0, t_0) - w_k(X) P(X, t | X_0, t_0)]$$

where $P(X, t|X_0, t_0)$ should be interpreted as the probability that at time t, $X(t) = X$ given that $X(t_0) = X_0$ (X and X_0 are integers). To illustrate how one can write the master equation, we give a simple example. Consider a protein existing in two states A or B. This protein can transform from A to B with a transition rate k_1 and from B to A at a rate k_2

$$A \underset{k_2}{\overset{k_1}{\rightleftharpoons}} B$$

For this system,

$$S = \begin{bmatrix} -1 & 1 \\ 1 & -1 \end{bmatrix}$$

and

$$W = \begin{bmatrix} k_1 n_a \\ k_2 n_b \end{bmatrix}$$

where n_a and n_b are the numbers of A and B respectively. The the Master Equation for this system can be written as

$$
\begin{aligned}
\frac{dP(n_a, n_b; t|n_a(t_0), n_b(t_0); t_0)}{dt} &= k_1(n_a + 1)P(n_a + 1, n_b - 1; t|n_a(t_0), n_b(t_0); t_0) \\
&+ k_2(n_b + 1)P(n_a - 1, n_b + 1; t|n_a(t_0), n_b(t_0); t_0) \\
&- (k_1 n_a + k_2 n_b)P(n_a, n_b; t|n_a(t_0), n_b(t_0); t_0)
\end{aligned}
$$

In general, the Chemical Master Equation is not analytically or numerically solvable in any but the simplest cases. Therefore, one has to resort Monte Carlo type simulations that produce a random walk through the possible states of the system under study. Various such methods have been developed, such as StochSim [50] and the Gillespie Stochastic Simulation Algorithm (SSA) [25, 26, 28]. We briefly describe the Gillespie algorithm as the most commonly used representative of these stochastic simulation methods.

The Gillespie Stochastic Simulation Algorithm

The Gillespie Stochastic Simulation Algorithm (SSA) involves the computation of the probability of elementary reactions to occur. The time at which these reactions occur is then determined based on this probability. More specifically, Gillespie proved, that starting at time t, the time τ to the next occurring reaction is the exponentially distributed random variable with mean $\frac{1}{w_0(X)}$. He also proved that the next reaction R_k to occur is the one whose index k is the integer random variable with probability $\frac{w_k(X)}{w_0(X)}$, where $w_0(X)$ is given by

$$w_0(X) = \sum_{k=1}^{M} w_j(X) \tag{28.5}$$

Generating samples of these random variables is then an easy task. For example, one can draw two random number r_1 and r_2 from the uniform distribution in the unit interval, and then take

$$\tau = \frac{1}{w_0(X)} ln \frac{1}{r_1}$$

and $k=$ the smallest integer satisfying $\sum_{j'=1}^{k} w_j'(X) > r_2 w_0(X)$. Based on τ and R_k one can then advance the simulation time by τ, and update the state of the system and repeat until final time or state. The trajectory obtained in this fashion is a stochastic realization based on the description of the Master Equation. The Gillespie stochastic algorithm tracks exactly all the reactions that occur in the system and the species they affect. This often represents a huge computational load which makes these simulations rather prohibitive if the system has species with large numbers of molecules or reactions that evolve at fast time scales. Making the SSA more computationally efficient is the subject of active research. For example, Gibson *et al.* improved the computational and data storage capabilities of the algorithm [23], while Rao *et al.* explored the incorporation of quasi-steady-state assumptions into this stochastic setting [54]. Recently, Rathinam *et al.* devised a "leaping" procedure whereby the algorithm leaps over a number of reactions using preselected τ values [56]. It was argued that this leaping can be safely and conveniently done in dynamically-stiff systems (systems with widely different time scales, the fastest of them being stable).

Langevin Modeling

Langevin modeling was originally devised to incorporate the effects of external noise on a vector process X whose evolution is described by a set of differential equations. In this case, the governing equations are augmented with additive or multiplicative stochastic terms [41]. An example of a Langevin equation is:

$$\dot{X}(t) = f(X) + \Gamma(t)$$

where is $\Gamma(t)$ is Gaussian white noise. However, the use of Langevin type equations in modeling intrinsic noise has not been rigorously justified. Various researchers argue that Langevin modeling could, in some instances, be a good representation of reality. For example, starting from the premises of the Master Equation, Gillespie derived a "Chemical Langevin Equation" valid under some assumptions on the system. This equation takes the form

$$\frac{dX(t)}{dt} \doteq \sum_{k=1}^{M} s_k w_k(X(t)) + \sum_{k=1}^{M} s_k \sqrt{w_k(X(t))}\Gamma_k(t)$$

where s_k, w_k and Γ_k are as defined previously [29]. Furthermore, there has been recently a renewed interest in expansions of the Master Equation which result in a Langevin type equations for the intrinsic noise. More specifically, Elf and coworkers have recently derived a so called Linear Noise Approximation (LNA) [18] of the Master Equation. The LNA is the multidimensional extension of the Van Kampen's system size (Ω-expansion) of the Master Equation [41]. In the Ω-expansion formulation, the master equation is expanded in Taylor series near macroscopic (deterministic) system trajectories or stationary solutions in powers of $1/\sqrt{\Omega}$, where Ω is the system volume. In this expansion, terms of first order in $1/\sqrt{\Omega}$ yield the deterministic rate equations, while terms of second order in $1/\sqrt{\Omega}$ yield the approximate noise equations. The complete exposition of this procedure is given in [18]. We briefly outline this procedure. We start by defining a random vector ξ by the relation $X = \Omega\phi + \Omega^{1/2}\xi$ where ϕ is the macroscopic concentration vector defined by $\phi = \lim_{\Omega \leftarrow \infty} X/\Omega$ (given by the deterministic rate equations). The probability distribution of X is therefore related to that of ξ by

$$P(X,t) = P(\Omega\phi + \Omega^{1/2}\xi, t) = \Pi(\xi, t)$$

One can then differentiate the expression above with respect to time, then expand the propensity functions (which are functions of $X = \Omega\phi + \Omega^{1/2}\xi$) in Taylor series around ϕ, and replace all of the above in the Master Equation (28.5). Ignoring high order terms in this expansion (higher than Ω^0), and matching expressions appearing in powers of Ω, one obtains the following equation for the evolution of the noise *pdf*

$$
\begin{aligned}
\frac{\partial \Pi(\xi, t)}{\partial t} = & -\sum_{i,j} A_{ij} \frac{\partial (\xi_j \Pi(\xi, t))}{\partial \xi_i} \\
& + \frac{1}{2} \sum_{i,j} [BB^T]_{ij} \frac{\partial^2 \Pi(\xi, t)}{\partial \xi_i \partial \xi_j}
\end{aligned}
\tag{28.6}
$$

with $A_{ij} = \sum_{k=1}^M s_{ik} \frac{\partial w_k(\phi)}{\partial \phi_j}$
and $[BB^T]_{ij} = \sum_{k=1}^M s_{ik} s_{jk} w_k(\phi)$. Equation (28.6) is a linear Fokker-Planck Equation with jacobian coefficient matrix $A = S[\frac{\partial W}{\partial x}] \mid_\phi$ and diffusion matrix $D = BB^T \mid_\phi$, where $B = S\sqrt{diag(W(\phi))}$. The stationary solution of (28.6) (with A and D evaluated ϕ^s, the fixed point of ϕ) can be shown to be a multidimensional normal distribution

$$
P(\xi) = ((2\pi)^{N/2} \sqrt{det\Sigma})^{-1} exp(-\xi^T \Sigma \xi / 2)
$$

The covariance matrix $\Sigma = E(\xi \xi^T)$ is given by the solution of the continuous algebraic Lyapunov equation

$$
A \mid_{\phi^s} \Sigma + \Sigma A^T \mid_{\phi^s} + D \mid_{\phi^s} = 0
\tag{28.7}
$$

The covariance matrix of X is given by $C = \Omega \Sigma$. It is worth mentioning here that the Fokker-Planck equation in (28.6) describes an Ornstein-Uhlenbeck process given by the Ito stochastic differential equation (Langevin Equation)

$$
d\xi = A\xi dt + Bd\Lambda(t)
$$

where $d\Lambda(t)$ is a Wiener process in M dimensions.

Fokker-Planck Modeling

The time evolution in the Chemical Langevin Equation induces a similar evolution for the probability distribution of $X(t)$, described by the Fokker-Planck Equation

$$
\begin{aligned}
\frac{\partial P(X, t | X_0, t_0)}{\partial t} \doteq & \sum_{i=1}^N \frac{\partial}{X_i} [(\sum_{j=1}^M \nu_{ij} a_j(X)) P(X, t | X_0, t_0)] \\
& + \frac{1}{2} \sum_{i=1}^N \frac{\partial^2}{\partial X_i^2} [(\sum_{j=1}^M \nu_{ij}^2 a_j(X)) P(X, t | X_0, t_0)] \\
& + \sum_{i,i'=1; i<i'}^N \frac{\partial^2}{\partial X_i \partial X_{i'}} [(\sum_{j=1}^M \nu_{ij} \nu_{i'j} a_j(X)) P(X, t | X_0, t_0)]
\end{aligned}
$$

The Fokker-Planck equation is similar to the Master Equation in that it describes the evolution of a probability distribution of the state $X(t)$. However, $X(t)$ depicts a continuous Markov process instead of a jump (discrete) Markov process as in the Master Equation. Therefore, the Fokker-Planck Equation is sometimes called a diffusion approximation of the Master Equation.

Brief Comparison of Stochastic Modeling Approaches

Biological noise is most accurately captured through the use of the Chemical Master Equation where molecular species (such as proteins, messenger RNA and ribosomes) are modeled as discrete entities. In this framework, and as has been already mentioned, reaction rates are replaced by reaction events which are individually explicitly modeled. From a mathematical perspective, the master equation is simple due to its linearity. It is however not solvable because it is very large (For example, to describe a three-step linear pathway involving one hundred molecules, the master equation requires ten thousand equations to account for each possible combination of molecules [55]). Therefore, it is a much more feasible task to simulate the random evolution of the system using Monte-Carlo techniques as in the Gillespie algorithm [25]. Each run of the algorithm yields a realization of the stochastic process. Therefore, to estimate the statistics of the process, one has to repeat the simulation task many times. The drawback of this procedure is the computational efficiency which rapidly degrades as the complexity of the system increases. Active research is devoted to resolve this issue.

The Master Equation and Fokker-Planck formulations are closely related. One main difference between the two is how the species are represented: In the Fokker-Planck representation, the description is continuous while in the Master Equation the representation is discrete. If the biochemical system involved contains only a few number of reacting molecules, the discrete representation is believed to be more accurate than the continuous representation. Sometimes, working with the Fokker-Planck equation is beneficial in the sense that tools such as sensitivity analysis and bifurcation theory are available. But once again, for systems involving more than a few species (usually ≥ 4), it is impossible to solve the Fokker-Planck equation, even numerically. As an alternative, one can adopt the equivalent Langevin description and again use Monte-Carlo methods to solve the Langevin equation many times in order to estimate the system's statistics.

28.3.3 Stochastic Versus Deterministic

In many cases, the stochastic and deterministic descriptions of a system coincide in the sense that the mean or average behavior of the system can be perceived as accurately captured by the deterministic behavior. Under these circumstance, one can extract a large amount of information about the system by considering the deterministic description (about average behavior, since even in that case, the fluctuations can significant and need to be studied). However, in other cases, deterministic and stochastic behaviors diverge. This can occur when the system is operating near a critical point. In this case, noise can cause the system to undergo a "phase transition" by changing its stability properties. To illustrate this point, we present the example of a simplified model of the circadian rhythm where similar behavior has been observed.

The circadian rhythm is a biological clock that regulates the changes that best suit environmental periodicity, such as daily cycles or light and dark. In [6, 76], a minimal model for the circadian rhythm was presented. The model consists of two genes: one responsible for the synthesis of an activator A and the other for a repressor R. The activator A binds to the A and R gene promoters, increasing their transcription rate. Therefore, A can be thought of as a positive feedback element. R acts as a negative feedback element since it can sequester A (producing the complex C). Sequestered A cannot bind to the gene promoters transcribe for A and R (see Figure 28.2).

These interactions can be described using a set of nonlinear differential. Gillespie's SSA can also be used to provide a stochastic description. Comparing the results of these two

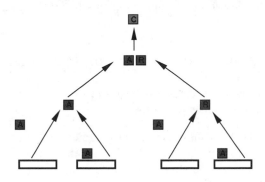

FIGURE 28.2: Simplified model of the circadian oscillator. Adapted from [76]

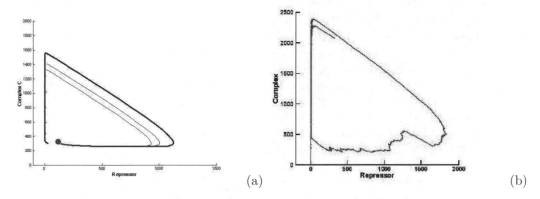

(a)

(b)

FIGURE 28.3: (a) Deterministic simulation results of the circadian oscillator. (b) Stochastic simulation results for the circadian oscillator.

modeling approaches, one observes the following. First, for a certain set of parameters, the stochastic and deterministic time courses coincide. Namely, they both exhibit periodic behavior with the only difference being the presence of random fluctuations in the concentration and period of the stochastic system. However, as parameters are varied, the deterministic description gives rise to a stable steady state, while stable oscillations pertain in the stochastic description (see Figure 28.3). This discrepancy in behavior between stochastic and deterministic can be attributed to the presence of noise. In the deterministic case, the solution, started from any initial conditions, eventually converges to the stable equilibrium and stays there afterwards. In the stochastic case, however, a steady-state distribution rather than a steady state point exists. At steady-state, the process can take any value within this distribution. For some of these values, the trajectory back to steady-state takes the system into a new cycle of oscillations. Therefore, the intrinsic noise eventually initiates a new cycle instead of allowing the time evolution of the system to settle into its stable fixed point. Since Robustness is a feature demanded from these circadian clocks, it can be concluded that noise has increased the reliability of the oscillations. However, the fact remains that the deterministic modeling has been insufficient to account for this observation.

28.4   Computational and Modeling Packages

Many simulation and software packages are being developed for the modeling of gene regulatory networks. Here we list a few of these packages.

- **CellML** CellML is an XML-based make up language to facilitate the integration, storage and and exchange of various biological models. The CellMLTM language is an open standard based on the XML markup language. It enables scientists to share models formulated in various model-building softwares and reuse components from one model in another. CellML includes information about model structure (how the parts of a model are organizationally related to one another) and equations describing the underlying biological processes. It also includes metadata defined as additional information about the model that allows users to search for specific models or model components in a database. CellML is being developed by the Bioengineering Institute at the University of Auckland and affiliated research groups with early support and input from Physiome Sciences Inc. in Princeton, New Jersey [37].

- **E-Cell** E-Cell A software environment for building integrative models based on gene sets. The E-CELL system allows a user to define functions of proteins, protein-protein interactions, protein-DNA interactions, regulation of gene expression and other features of cellular metabolism, as a set of reaction rules. E-CELL simulates cellular functions by numerically integrating the differential equations describing these interactions. The user can observe, through a GUI interafce, dynamic changes in concentrations of chemical species in the cell [74].

- **SBML** SBML is the Systems Biology Makeup Language that is intended to be software-independent. It is a free, open, XML-based format for representing biochemical reaction networks that describe any model component in a biological process, including cell signaling pathways, metabolic pathways, gene regulation, and others [38].

- **Virtual Cell** Virtual Cell is a simulation engine for mammalian cells. It is based on precise measurements of how molecules diffuse and migrate to react with each other. A key feature of the Virtual Cell is that it permits the incorporation of realistic experimental geometries within full 3D spatial models. It enables the formulation of both compartmental and spatial models. Those spatial model could depict either idealized or experimentally derived geometries of one, two or three dimensions [45].

28.5   The Concept of Feedback

In the field of engineering, the demand for robustness and disturbance rejection has long been recognized. Water clocks in 300 BC Alexandria, the Watt Governor in 1788 and the operational amplifiers in the 1920's are classic examples of man-made machines that capitalized on the use of feedback for reliable operation. Present technology is also characterized by the massive use of feedback. "Fly-by-wire" aircrafts, high accuracy positioning robots, automotive cruise control, and chemical process control are some of many examples. The

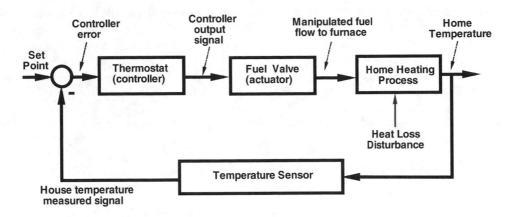

FIGURE 28.4: Diagram showing components of a typical feedback system.

basic idea of feedback is simply to use the current state of a system to make decisions about the course of action for its future. Such a scheme is called "closed loop", while a scheme where no information on the state of the system is used to influence its future operation is called "open loop". An important form of feedback is negative feedback which we illustrate through a common example: Heat Regulation. A simplified block diagram scheme of temperature regulation in a house is shown in Figure 28.4. In this scheme, it is desired to keep the temperature of a house (plant or process) at a certain reference temperature (set point). For this purpose, temperature is measured (sensor) and its deviation from the desired temperature is assessed (error signal). The error signal is then fed to the thermostat (controller) that devises that appropriate control action. The output of the controller is used to operate the heat fuel valve (actuator), therefore generating appropriate actuation signal (fuel to furnace). Errors or deviations from temperature setpoint are hence corrected through the action of this negative feedback.

28.6 Feedback Loops and Their Dynamic Role in Gene Regulatory Networks

Much like technological systems, gene regulatory networks need to operate robustly. Hence, it does not come much as a surprise that regulatory feedback loops are ubiquitously used in gene networks to tackle this robustness demand. In addition to robustness, feedback influences many other dynamical properties in these networks. We review some of these features in the following sections.

28.6.1 Feedback and Steady State Behavior

Often times, the feedback structure in a system dictates its steady state behavior. Here, we focus on the role of feedback in creating monostability, multistability and periodic behavior.

Monostability and Homeostasis

Early mathematical results predicted the important role of autoregulation and feedback in homeostasis (the ability of biological mechanisms to restore their equilibrium in the presence of disturbances [78, 59]), but it was not until recently that this fact has been verified experimentally by Becskei and Serrano [8]. Their elegant experimental setting consisted of a tetR-EGFP (enhanced green fluorescent protein) fusion protein that binds to tetO operator sites in the promoter that drives its own production, thus implementing a negative feedback loop (see Figure 28.5). Using this setup, they demonstrated that autoregulatory negative feedback provides better stability of the homeostatic fixed point than an open loop system, all the while limiting the range over which the concentrations of networks components can fluctuate [8]. We shall return to the mathematical proof of their experimental result in more details later.

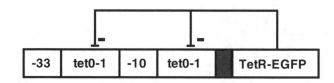

FIGURE 28.5: Simple monostable autoregulatory synthetic gene network. Adapted from [8].

Multistability and the Implementation of Molecular Switches

Positive feedback, could generate multistability [33, 75, 66, 71, 73]. For example, a transcription factor activating its own synthesis (direct positive feedback) or two transcription factors repressing each other (indirect positive feedback) can produce a multistable system [42, 20]. Examples of multistability are abundant in the literature [21, 7]. Here, we give a simple example of multistability in the phage λ system.

The λ bacteriophage system has been thoroughly investigated both experimentally and mathematically [52, 47], and has been used to illustrate the importance of stochasticity in gene regulation [3]. A simplified model for the bacteriophage λ was proposed by Hasty and coworkers [36]. In their model, the gene *cI* expresses the λ repressor CI which dimerizes and binds to DNA as a transcription factor at either of two binding sites, $OR2$ or $OR3$. Binding of this transcription factor to $OR2$ enhances transcription of CI (positive feedback), while binding to $OR3$ represses transcription of CI (negative feedback) (see Figure 28.6(a)). Therefore, the lysis or lysogeny outcome in cells is determined by the specific temporal pattern of CI. For example, lysogeny occurs in cells where, by chance, there is early CI production through the interplay of positive and negative feedback. Due to separation in time scale between all the reactions involved, the description of the system reduces to one differential equation describing the time evolution of repressor concentration.

$$\frac{dx}{dt} = \frac{\alpha x^2}{1 + 2x^2 + 5x^4} - \gamma x + 1$$

where x denotes the concentration of the repressor CI. α and γ are dimensionless parameters which are functions of the original parameters of the system. The steady state concentration of x is dependent on the values of α and γ. For some values of these parameters, the system

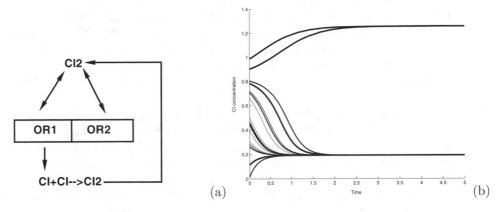

(a)   (b)

FIGURE 28.6: (a) The simplest system that reproduces the dynamics of the lambda bacterio-phage (b) Bistability in the lamdba bacteriophage system. The trajectory of the repressor evolves to one of the two fixed points based on the initial conditions.

FIGURE 28.7: The Repressilator system. Reproduced from [19].

has one fixed point. For another set, the system has three fixed points, one of them being unstable. Depending on the initial concentration for the repressor, the solution will converge to either one of the two stable fixed points. This behavior is shown Figure 28.6(b) where $x(t)$ is shown for different initial conditions.

Stable Oscillations and Limit Cycles

Oscillations in the concentration of some key proteins in the cell are often present to implement clocks that synchronize behavior in response to various growth and adaptation demands (such as day-night cycles). There are various feedback structures that implement such oscillatory behavior.

- **Negative feedback** Negative feedback can be a mediator for oscillatory behavior as shown in the the so-called "Repressilator" [19]. The Repressilator is a bacterial synthetic network consisting of three genes whose product represses each other. The LacI protein represses the promoter of the Tet gene, the TetR protein represses the promoter of the *cI* gene, and the CI protein represses the promoter of the *lac* gene, therefore closing the loop (Figure 28.7). Mathematical analysis of this system showed that oscillations are most likely to occur by the presence of strong promoters coupled to efficient ribosome binding sites, tight transcriptional control, cooperative repression and comparable protein and mRNA decay rates. The experimental system, constructed using these guidelines, produces self sustained oscillations in the concentration of its three proteins: LacI, TetR, and CI.

- **Positive and negative feedback** The interaction of positive and negative feedback can give rise to oscillatory behavior. Roughly speaking, if the positive feedback creates bistability, then the negative feedback drives the system back and forth between the two stable fixed points. An example is that of the Cdc2-cyclin B system in early embryonic cell cycle. Recent experimental and mathematical results have shown that the basic functionality of this oscillator stems from the interplay of positive and negative feedback. Positive feedback implements a toggle switch characterized by biochemical hysteresis (Cdc2 activation system), while a negative feedback loop involving Cdc2 allows sustained oscillations, much like the relaxation oscillator in microelectronics [53].

28.6.2   Feedback and Stability

As hinted to previously, negative feedback loops enhance stability. Let's look back at the system in Figure 28.5 and compare it to an identical system where the negative feedback loop is abolished. We describe the two variations of this system using differential equations, and get the following for the dynamics of the repressor (tetR)

$$f_{unreg} = \frac{dR_{unreg}(t)}{dt} = n\frac{k_pP}{1+k_pP}k_1a - k_{deg}R_{unreg}(t)$$

$$f_{auto} = \frac{dR_{auto}(t)}{dt} = n\frac{k_pP}{1+k_pP+k_rR_{auto}(t)}k_1a - k_{deg}R_{auto}(t) \qquad (28.8)$$

where P is the concentration of the RNA polymerase, k_p and k_r are the binding constants of the polymerase and the repressor respectively, k_1 is the promoter isomerization rate from closed to initiating complex, a the proportionality constant between mRNA and protein, k_{deg} is the degradation rate of the repressor, and n is the gene copy number [79]. Stability (S) in this context is defined as the rate at which the response to a perturbation η from equilibrium decays. We can get the value of S by expanding in Taylor series around R^* (the steady state value of the repressor) such that

$$f(R^* + \eta) = f(R^*) + \eta f'(R^*) + \Theta(\eta^2) \qquad (28.9)$$

Notice that this rate is simply $S = f'(R^*)$. Therefore,

$$S_{unreg} = -k_{deg} \qquad (28.10)$$

$$S_{auto} = n\frac{k_pPk_1ak_r}{(1+k_pP+k_rR^*)^2} - k_{deg} \qquad (28.11)$$

Since $S_{unreg} < S_{auto}$, the stability in the autoregulatory system is higher than that in the unregulated system for all positive values of the parameters and steady states.

28.6.3 Feedback and Sensitivity to Parameter Variations

In engineering, it is generally possible to build robust systems out of imprecise components through the use of feedback. For example, consider the system in Figure 28.8 which consists of the interconnection of a plant K and a controller C. Assume for simplicity that K and C are constant (this could be thought of as the steady-state behavior of a dynamic system after all the transients had settled). The relationship between the output Y and the input R can be easily computed as

$$\frac{Y}{R} = \frac{K}{1 + CK}$$

If $CK \gg 1$ (which ideally can be achieved by the design of the controller C) then $\frac{Y}{R} \simeq \frac{K}{CK} = \frac{1}{C}$. In this case $\frac{Y}{R} \simeq \frac{K}{CK}$ is only dependent on C, which makes it independent of any uncertainties and variations that might occur in the plant K. Such a scheme can be (and is) implemented in gene networks, for example by autogenous regulation of the transcription of a gene by its own product.

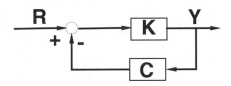

FIGURE 28.8: Elementary feedback system with atucator/system K and controller/sensor C. The goal is for response Y to track reference R, independent of uncertainty and variation in K.

28.6.4 Feedback, Filtering and Noise Resistance

Engineering systems mostly use negative feedback as noise attenuating mechanism. Not surprisingly, molecular networks which must function reliably in the presence of noise, also ubiquitously use feedback. In terms of its filtering properties, a simple negative feedback loop functions as a low-pass filter. More sophisticated types of feedback, such as integral feedback, can shape band-pass filters. Integral feedback is, for example, implemented in bacterial chemotaxis [81](see subsequent sections). The main idea of integral feedback is the use of a negative feedback loops that integrates previous history (keeps a memory) to attenuate high and low frequencies and amplify intermediate frequencies. Feedback is also instrumental in attenuating biochemical intrinsic fluctuations. Results pointing in that direction have been obtained by Thattai and Van Oudenaarden through a simple model for gene expression in prokaryotes [70]. In their example, mRNA molecules are assumed to be synthesized constitutively off a DNA template strand and transcribed at a rate K_R. These molecules are then translated at a rate K_P into proteins. mRNA and proteins are degraded at rates γ_R and γ_P respectively (Figure 28.9). Two scenarios are investigated: an unregulated gene and a gene where the protein regulates its own production in a negative

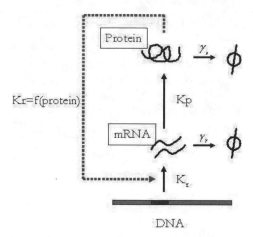

FIGURE 28.9: Simple transcription/translation module. Adapted from [70]

feedback fashion. In the unregulated gene network, K_R is constant whereas in the regulated gene network K_R is taken to be an affine function of P, i.e. $K_R = K_0 - K_1 P$. The statistical moments of this example are solvable analytically from the master equation. The quantity used to quantify the noise rejection properties of this system is the variance of the protein steady state distribution normalized by the mean value of the distribution. This is called the Fano factor f. It is used to characterize the deviation of the protein distribution from Poissonian statistics. This quantity is

$$f^u = \frac{{\sigma_P^u}^2}{<P^u>} = \frac{b}{1+\eta} + 1$$

in the unregulated case and

$$f^r = \frac{{\sigma_P^r}^2}{<P^r>} = (\frac{1-\phi}{1+b\phi})(\frac{b}{1+\eta}) + 1$$

in the regulated case. $\eta = \frac{\gamma_P}{\gamma_R}$, $b = \frac{K_P}{\gamma_R}$, and $\phi = \frac{K_1}{\gamma_P}$. Obviously, $f^u > f^r$, indicating that feedback is indeed an efficient way for intrinsic noise reduction.

28.6.5 Feedback and Transient Response

The work of Monod had suggested that the kinetics of simple transcription units are typically slow [49]. However, it was shown later that many strategies are used to speed up those kinetics. For example, in the regulation of enzyme levels in mammalian tissues, fast dynamics are achieved using large synthesis and degradation rates [62]. Negative feedback can also be used to that effect as first suggested in [59] and proven both mathematically and experimentally in [57]. The experimental setup in [57] consisted of a transcription unit that produces a certain steady-state protein concentration when induced (similar to the simple experimental setup of Becskei and Serrano (see Figure 28.5 and Sections 28.6.1 and 28.6.2). A simplified mathematical representation of this problem takes the form of an ODE for the protein concentraion x

$$\frac{dx(t)}{dt} = A(t) - \alpha x$$

FIGURE 28.10: AND logic gate with memory. Adapted from [35].

$A(t)$ being the production rate and αx the dilution/degradation term (growth rate is $\alpha = ln(2)/\tau$, τ is the cell cycle time). $A(t) = constant$ corresponds to the case where x does not feedback regulate its own production. However, if $A(t)$ is dependent on x (in michaelis-Menten like form, for example), the protein product negatively regulates its own synthesis. The two designs can be built to achieve the same steady state value by tuning the maximal production rate of their promoters, therefore adjusting the production term $A(t)$. To compare the transient performance of both designs, the rise time t_r is adopted. t_r is defined as the time at which x reaches half of its steady state value. For the autoregulated case, $t_r \simeq 0.21\tau$ while for the unregulated case $t_r = \tau$, therefore indicating that feedback was instrumental in speeding up the response of the system.

28.6.6 Feedback and Building Sophisticated Logic Molecular Gates

A classic example of a molecular logic gate is the "AND gate" implemented by the arabinose operon. This operon is only induced if both arabinose and AraC are present. Alternatively, it is in the OFF state if either is absent. Similarly, OR gates can be built. These simple logic gates operate in open loop. The addition of feedback loops can implement complex and sophisticated behavior such as memory. For example, consider the scenario presented in Figure 28.10. Promoter 2 is repressed by either LacI or TetR. At the same time, Promoter 1 directs the transcription of the *lac* and *tet* genes, and is itself induced by the simultaneous presence of the chemical IPTG and anhydrotetracyclin (aTc). This construct implements an AND gate switched ON by the presence of IPTG and aTc and OFF by the absence of either. The memory in this scheme is implemented through the feedback loop from CI to promoter 1. Once the system is in the ON state, the production of LacI and TetR is repressed. Consequently, the system stays in the ON state independently of the subsequent levels of the inducers. By doing so, it remembers that at some point in the past, both inducers were present simultaneously.

28.7 Novel Insights through the Use of Control and Systems Theory: Some Case Studies

28.7.1 Integral Feedback and Chemotaxis

Ideas and concepts stemming from control theory have already been used to delineate the essence of structure and functionality in many examples of gene regulatory networks. The robust perfect adaptation in bacterial chemotaxis is an admirable illustration. Bacterial chemotaxis is the process by which bacteria sense the concentration of chemical species and migrate toward chemoattractants or away from chemorepellants. This behavior is mainly achieved by integrating signals received from the environments (detected through receptors) to modulate the direction of their flagellar rotation. The chemotactic system has long intrigued biologists who accumulated a large body of genetic, structural and biochemical data about this system [9, 32]. It has been established that in bacteria, the chemotactic response is accomplished by signal transmission between two supramolecular entities — the receptor complexes situated at the poles of the cell, and the flagellar motor complexes embedded in the membrane in a random distribution around the cell. The messenger protein *CheY* moves between these two entities, therefore transmitting chemotactic signals from receptors to flagella. The interaction of *CheY* with the flagellar-motor complex increases the probability of changing the flagellar rotation from its default counterclockwise direction (CCW) to clockwise direction (CW). The consequence of CW rotation is abrupt tumbling after which the cell swims in a new direction. The chemotactic system includes cascades of phospho-relay systems consisting of (in addition to *CheY*), five other intracellular proteins: A, B, R, W, Z (see Figure 28.11(a)). Proteins R and B enzymatically add and remove (respectively) methyl groups on the receptor/ligand complex. The phosphorylation level of B dictates its removal activity. This phosphorylation of B is dependent on the histidine kinase A, which is coupled to the receptor via adaptor protein W. Furthermore, protein A mediates the phosphorylation of protein Y, and protein Z removes this phosphorylation. The level of phosphorylated Y affects the cell tumbling frequency through interaction with the flagellar motors. Therefore, the input to this system is usually perceived to be ligand concentration and the output to be the concentration of phosphorylated Y. In a pioneering work, Alon *et al.* [1] demonstrated experimentally that the input-output relationship between ligand and phosphorylated Y concentration possesses a property that Barkai and colleagues had identified as "perfect adaptation" [5]. Upon a change in input (stimulus concentration), perfect adaptation is the process by which the output returns at steady state to its exact prestimulus value, despite the persistence of the input. This property is necessary for the ability of bacteria to respond to stimulus gradients [46]. Although this fact has been long known, an exact understanding of the mechanism that insures this behavior was absent until the elegant work of Yi and coworkers [81]. Their fresh "control engineering" perspective on the problem established the necessity for the presence of an integration process in the chemotactic system in order to accomplish the observed perfect adaptation. Their result was a direct application of a classic control engineering principle: the internal model theorem, due to Francis and Wonham [22]. This theorem states that for a linear system to track a reference signal or robustly reject a disturbance belonging to a class of time functions, it must contain a subsystem which can itself generate all disturbances in this class. In the context of chemotaxis, this result implies that in order to accomplish perfect adaptation after a step increase in ligand concentration, the system **must** possess a subsystem which generates all constant time signals, i.e. an integrator. However, the question remains of how this integration can physically be accomplished by the interaction of molecules and cellular components. To keep our presentation simple, we illustrate the physical implementation of this postulated integral action in a simplified model of chemotaxis in the social amoebea *Dictyostelium discoideum* [39]. As in bacterial chemotaxis, suppose there exists a response regulator (similar to *CheY*) that can exist in one of two states: active (R) and inactive (R^*), such that its total number is constant $R_T = R(t) + R^*(t)$. This regulator is activated through the action of an enzyme A and inactivated through the action of an enzyme I, which are in turn regulated by an external signal S proportional to chemoattractant concentration

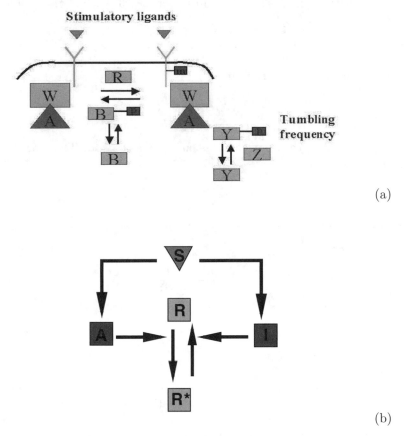

FIGURE 28.11: (a) Schematic illustration of bacterial chemotaxis. Adapted from [81] (b) Simplified model of chemotaxis in the social amoebea *Dictyostelium Discoideum.* Adapted from [39]

(see Figure 28.11(b)). The deterministic rate equations describing these interactions are

$$
\begin{aligned}
\frac{dR^*(t)}{dt} &= -k_{-r}I(t)R^*(t) + k_r A(t)R(t) \\
&= -(k_{-r}I(t) + k_r A(t))R^*(t) + k_r A(t)R_T
\end{aligned}
\tag{28.12}
$$

and

$$
\begin{aligned}
\frac{dA(t)}{dt} &= -k_{-a}A(t) + K_a S(t) \\
\frac{dI(t)}{dt} &= -k_{-i}I(t) + K_i S(t)
\end{aligned}
\tag{28.13}
$$

where $k_{-r}, k_r,\ k_{-a}, k_a, k_{-i}$ and k_i are positive rate constants. We are now in position to analyze this system. When the strength of S increases, stepping from S_0 to S_1 (representing a sharp increase in chemoattractant concentration for example), the enzyme A and I react accordingly. Their time profiles are the solution of (28.13) with $S(t)$ taken as the input

$$A(t) = A_1 K_A + e^{-k_{-a}t}(S_0 - S_1)K_A$$
$$I(t) = I_1 K_I + e^{-k_{-i}t}(S_0 - S_1)K_I \tag{28.14}$$

with $K_A = \frac{k_a}{k_{-a}}$ and $K_I = \frac{k_i}{k_{-i}}$. the ratio $\frac{A(t)}{I(t)}$ is

$$\frac{A(t)}{I(t)} = \frac{K_A}{K_I} \frac{S_1 + e^{-k_{-a}t}(S_0 - S_1)}{S_1 + e^{-k_{-i}t}(S_0 - S_1)}$$

The exponential terms in the above expression decay as $t \to \infty$. Therefore the steady state ratio of A and I is

$$\frac{A(\infty)}{I(\infty)} = \frac{K_A}{K_I} \tag{28.15}$$

Similarly, the steady state concentration of the active regulator R^*, found by setting $\frac{dR^*(t)}{dt} = 0$, is given by

$$R^*(\infty) = \frac{k_r A(\infty)/I(\infty)}{k_{-r} + k_r A(\infty)/I(\infty)} R_T$$

Using (28.15), we get

$$R^*(\infty) = \frac{k_r \frac{K_A}{K_I}}{k_{-r} + k_r \frac{K_A}{K_I}} R_T \tag{28.16}$$

$R^*(\infty)$ is independent of S, indicating that although R^* reacts to the change in chemoattractant (initiating the corresponding course of action for movement toward that chemoattractant), it always returns to its prestimulus value and no residual effect of S pertains. As such, the chemotactic system is ready to react to another gradient change in chemoattractant concentration. The integral action in this system can be better visualized by rewriting (28.12) as

$$\frac{dR^*(t)}{dt} = -(k_{-r}I(t) + k_r A(t)(R^*(t) - S^*(t)) \tag{28.17}$$

with

$$S^*(t) = R_T \frac{\frac{A(t)}{I(t)}}{\frac{k_{-r}}{k_r} + \frac{A(t)}{I(t)}}$$

If $S^*(t)$ is associate with the external disturbance, then the whole system can be seen to be an integral control feedback with time varying gain $k_{-r}I(t) + k_r A(t)$.

We emphasize here that the main contribution of this insight resides in proving the *necessity* aspect for the existence of an integral feedback scheme. The existence of this mechanism is simply *required* to account for the observed adaptation. As a result, known biology can be probed to establish how this integral action is implemented. If the known components cannot account for this action, further guided experiments can be suggested to uncover potential missing parts. The presence of integral feedback in chemotaxis is not a singular occurrence. Indeed, it has been postulated that integral feedback is present at all levels of biological regulation and accounts for various adaptation mechanisms. For example, integral feedback has been identified in the basic hormonal mechanism that underlies the adaptation of the calcium levels in mammals to external disturbances affecting the plasma calcium pools [15].

28.7.2 Robustness and Complexity in the Bacterial Heat Shock Response

High temperatures cause cell proteins to unfold from their normal shapes, resulting in malfunctioning and eventually death of the cell. Cells have evolved gene regulatory mechanisms to counter the effects of heat shock by expressing specific genes that encode heat shock proteins (hsps) whose role is to help the cell survive the consequence of the shock. In *E. coli*, the heat shock (HS) response is implemented through an intricate architecture of feedback loops centered around the σ- factor that regulates the transcription of the HS proteins under normal and stress conditions. The enzyme RNA polymerase (RNAP) bound to this regulatory sigma factor, σ^{32}, recognizes the HS gene promoters and transcribes specific HS genes. The HS genes encode predominantly molecular chaperones (DnaK, DnaJ, GroEL, GrpE, etc.) that are involved in refolding denatured proteins and proteases (Lon, FtsH, etc.) that function to degrade unfolded proteins. At physiological temperatures (30°C to 37°C), there is very little σ^{32} present and hence little transcription of the HS genes. When bacteria are exposed to high temperatures, σ^{32} first rapidly accumulates, allowing increased transcription of the HS genes and then declines to a new steady state level characteristic of the new growth temperature. An elegant mechanism that senses temperature and immediately reacts to its effect is implemented as follows in the bacterial heat shock response. At low temperatures, the translation start site of σ^{32} is occluded by base pairing with other regions of the σ^{32} *mRNA*. Upon temperature upshift, this base-pairing is destabilized, resulting in a "melting" of the secondary structure of σ^{32}, which enhances ribosome entry, therefore increasing the translation efficiency. Indeed the translation rate of the mRNA encoding σ^{32} increases immediately upon temperature increase [69]. Hence, a sudden increase in temperature, sensed through this mechanism, results in a burst of σ^{32} and a corresponding increase in the number of heat shock proteins. This mechanism implements a control scheme very similar to a *feedforward control loop*. With the use of this mechanism, the production of heat-shock proteins is made to be temperature dependent.

The level and activity of σ^{32} are also regulated. In addition to their function in protein folding, the chaperone DnaK are also capable of binding to σ^{32}, therefore limiting its capability to bind to RNA polymerase. Raising the temperature produces an increase in the cellular levels of unfolded proteins that then titrate DnaK/J away from σ^{32}, allowing it to bind to RNA polymerase, resulting in increased trancription of DnaK/J and other chaperones. The accumulation of high levels of HS proteins leads to the efficient refolding of the denatured proteins thereby decreasing the pool of unfolded protein, freeing up DnaK/J to sequester this protein from RNA polymerase. This implements what is referred to as a *sequestration feedback loop*. In this way, the activity of σ^{32} is regulated through a feedback loop that involves the competition of σ^{32} and unfolded proteins for binding with free DnaK/J chaperone pool.

During steady state growth, σ^{32} is rapidly degraded ($t_{1/2} = 1$ minute), but is stabilized for the first five minutes after temperature upshift. The chaperone DnaK and its cochaperone DnaJ are required for the rapid degradation of σ^{32} by the HS protease FtsH. RNAP bound σ^{32} is protected from this degradation. Furthermore, FtsH itself being a product of the heat-shock protein expression, experiences a synthesis rate that is tied to the transcription/translation rate of DnaK/J. Therefore, as protein unfolding occurs, σ^{32} is stabilized by the relief of its sequestration from *DnaK*. However, as more proteins are refolded, and as the number of *FtsH* itself increases, there is a decrease in the concentration of σ^{32} to a new steady state concentration that is dictated by the balance between the temperature-dependent translation of the rpoH mRNA and the level of σ^{32} activity modulated by the hsp chaperones and proteases acting in a negative feedback fashion. In this way, the FtsH

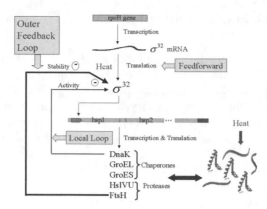

FIGURE 28.12: Biological Block Diagram for the Heat Shock Response

mediated degradation of σ^{32} is feedback regulated. We refer to this as the *FtsH degradation feedback loop*. A biological block diagram of the heat shock response that shows the various regulation mechanisms is shown in Figure 28.12.

Full and reduced order models have been devised to describe the rich dynamics of the bacterial heat shock response [43, 16, 17]. These models have been successful in accurately producing the behavior of the heat shock system. The predictions of the models have been used to mathematically confirm various experimental results (e.g. the absence of translational shutoff for σ^{32} $mRNA$ in the adaptation phase), and suggest new experiments (e.g. temperature downshift, regulation of degradation). The thorough control analysis of the heat shock response also revealed important universal design principle that are likely to be present in other biological system. For example, it has been demonstrated that biological complexity is not a futile outcome of evolution, but rather the result of improvised solutions to achieve various performance objectives, such as robustness, transient response and efficient use of materials [14]. Biological systems share this control design philosophy with engineered systems. To demonstrate this, we start with a minimal system that achieves basic functionality of a heat shock response system. We then add one layer of regulation after another, each time using control engineering intuition to demonstrate how that layer is needed to improve the performance of the overall system with respect to some objective. The simplest functional design that uses σ^{32} to produce heat shock proteins will have the blocks shown in Figure 28.13(a).

In this hypothetical open-loop design strategy, the number of σ^{32} dictates the level of heat shock proteins in the cell. The production of heat-shock proteins can be made to be temperature dependent with the use of the translational control mechanism (feedforward). Figure 28.14 shows the level of σ^{32}, chaperones, and folded proteins associated with this design. In this appealingly simple design, one can observe acceptable steady state levels of folded proteins, both at low and high temperatures. However, this system suffers from a critical shortcoming: the lack of robustness. Indeed, it can be easily shown that the slightest change in the transcription and translation rates results in a corresponding change in the number of heat shock proteins produced. This sensitivity to parametric uncertainty is a well known problem of open loop designs, and is one of the key reasons why feedback control systems are superior to open loop systems. The property of feedback to attenuate this sensitivity has led to the pervasive and extremely successful use of feedback control systems in all engineering disciplines. As in man-made engineering systems, the hyper

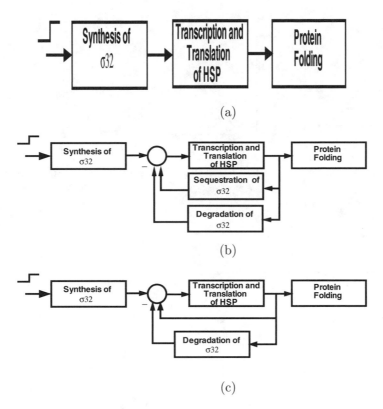

FIGURE 28.13: Hypothetical Design Models for the Heat Shock Response (a) Open loop design with feedforward control that achieves the basic functionality of protein folding (b) closed loop design with sequestration loop that modulates the activity of σ^{32} (c) closed loop design with sequestration loop and degradation loop

sensitivity to parameter variations in the heat shock response seen in the open loop design is circumvented through the use of feedback. This feedback is implemented through the sequestration of free σ^{32} by the chaperones, thus modulating the pool of σ^{32} available for RNA polymerase binding (see Figure 28.13 (b)). In addition to its obvious benefits in reducing parametric uncertainty, the added sequestration loop achieves at least two other functional purposes. Firstly, this loop is instrumental in reducing the delay in the folding of proteins as compared to the open loop case (see Figure 28.15). This effect can be best understood by thinking of the pool of the complex formed by σ^{32} bound to the chaperones as a reservoir for σ^{32} that can be immediately used once the chaperones are recruited to refold denatured proteins. The immediate burst in free σ^{32} results in a faster production of chaperones, and consequently a faster folding of proteins. The use of feedback to improve the transient response is again a common practice in control engineering. Secondly, although the number of chaperones at high temperature in the open loop case is much higher than that with the sequestration loop added, the folded proteins number in the two cases is very

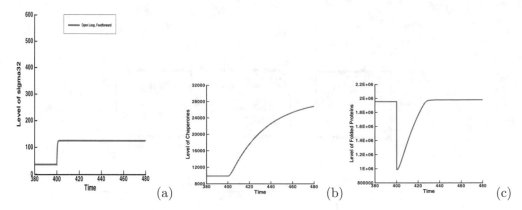

FIGURE 28.14: Levels of σ^{32}, $DnaK$, and unfolded proteins for the open loop design

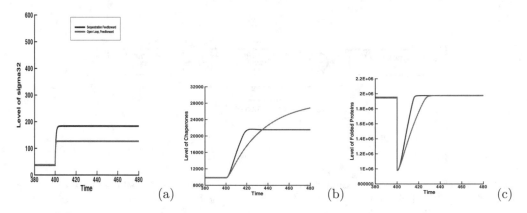

FIGURE 28.15: The levels of σ^{32}, $DnaK$, and unfolded proteins for a design with added sequestration of σ^{32} by chaperone. Parameters are chosen to have same steady-state concentrations of σ^{32} and chaperones in open loop (green) and closed loop with sequestration (blue) at low temperature, and performance is assessed at high temperature

comparable (see Figure 28.15). This is mainly due to the fact that in the closed loop case, the synthesis of proteins is self regulating, and the excess of unneeded heat shock proteins produced in the open loop case is hindered by feeding back a measure of the folding state in the cell (through the $\sigma^{32} - chaperone$ complex) to regulate the pool of free σ^{32}. Hence the unnecessary use of materials and energy that open-loop systems suffer from is completely avoided through the utilization of feedback, which has the advantage of producing the required number of heat-shock proteins thereby avoiding the excessive metabolic burden associated with over-expression of them.

Now, if the speed of the repair response is dependent on the immediate increase in the number of σ^{32}, then slowing down the degradation of σ^{32} after the onset of heat would be also beneficial in that direction. This could be achieved by making the degradation signal of σ^{32} dependent on the protein folding state of the cell. The effect of such an added layer

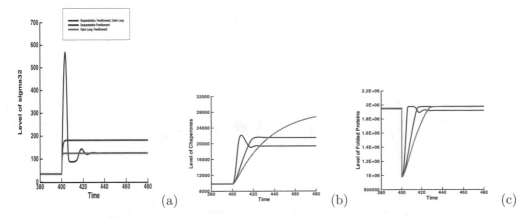

FIGURE 28.16: The levels of σ^{32}, $DnaK$, and unfolded proteins for a design with added sequestration and degradation. Parameters are chosen to have same steady-state concentrations of σ^{32} and chaperones in open loop (green), closed loop with sequestration (blue), and closed loop with sequestration and degradation (red) at low temperature, and performance is assessed at high temperature

of feedback regulation is shown in Figure 28.16 for σ^{32}, $DnaK$, and folded proteins. Notice that the delay in the protein folding process is greatly reduced, while the folding of proteins is only slightly impaired. This comes as a natural consequence of the tighter control over the number of σ^{32}. This is an example of the kind of tradeoffs that usually appear when various performance objectives must be balanced. Yet another objective that is of primary importance in cellular processes is their ability to attenuate undesirable noise and stochasticity. The nature of stochasticity in these processes stems from various sources of uncertainty inside the cell and has been termed intrinsic noise, to differentiate it from extrinsic noise that results from the environment. Various structures and strategies have been identified as leading to intrinsic noise rejection or exploitation mechanisms. For example, the use of feedback has been observed experimentally to attenuate intrinsic cellular noise (see previous sections for more details on the origin of this stochasticity, and various approaches to model it). We briefly describe one simple example in the heat shock system showing the role of feedback in noise rejection. Using the Stochastic Simulation Algorithm of Gillespie, we simulate the system in the presence and absence of the degradation feedback loop. The result of this simulation in shown in Figure 28.17 for two sample paths, one without the degradation loop and the other with that loop in place. It is apparent that this feedback loop is instrumental in reducing the stochastic fluctuations around the heat shock proteins steady-state.

As a conclusion, the detailed scrutiny of the control strategies in the heat shock response has uncovered that the use of elaborate and increasingly sophisticated control mechanisms results in more reliable gene regulatory networks, all the while generating spiraling levels of complexity. This need for robustness is however balanced by constraints resulting from other performance criteria, such as the transient response and the limited cellular energies and materials. This echoes the various tradeoffs considered in the design of engineering systems, and motivates all the more the use of engineering principles for the exploration of biological complexity.

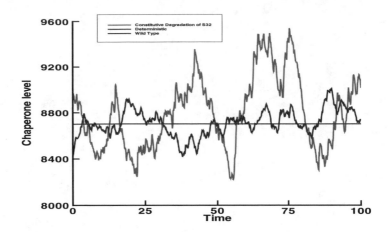

FIGURE 28.17: Stochastic level of chaperones in the presence (green) and absence (red) of the outer degradation feedback loop

28.8 Conclusions

Robust functionality in gene networks is in great part implemented through the use of elaborate regulatory feedback, the outcome being rather complex dynamics that cannot be solely captured through casual intuition alone. Hence the need for theoretical methods and precise mathematical formulations, combined with bench biology experimentation. Ideas borrowed from "synthetic" sciences such as engineering and computer science are increasingly proving to be very powerful tool in understanding functional biological modules and their interactions. In this article, we have presented various examples illustrating the specific use of control theory to that purpose. For a detailed account on the similarities and differences in biology and engineering at the system-level, in addition to engineering results that are most likely to impact biology, the reader is referred [12]. A thorough understanding of these similarities and differences is at the core of "Systems Biology" whose future holds many promises. Deeper understanding of the causes of disease (merely the failure of regulatory networks) and the effects of administration of therapeutic drugs [13] are nothing by a few of the anticipated fruits.

References

[1] U. Alon, M.G. Surette, N. Barkai, and S. Leibler. Robustness in bacterial chemotaxis. *Nature*, 397:168–171, 1999.

[2] R. Alves and M.A. Savageau. Comparing systemic properties of ensembles of biological networks by graphical and statistical methods. *Bioinformatics*, 16(6):527–533, 2000.

[3] A. Arkin, J. Ross, and H. McAdams. Stochastic kinetic analysis of developmental pathway bifurcation in phage λ-infected *escherichia coli* cells. *Genetics*, 149:1633–1648, 1998.

[4] R.J. Bagley and L. Glass. Counting and classifying attractors in high dimensional

dynamical systems. *J. Theor. Biol*, 183:269–284, 1996.

[5] N. Barkai and S. Leibler. Robustness in simple biochemical networks. *Nature*, 387:913–917, 1997.

[6] N. Barkai and S. Leibler. Biological rythms-circadian clocks limited by noise. *Nature*, 403:267–268, 2000.

[7] A. Becskei, B. Seraphin, and L. Serrano. Positive feedback in eukaryotic gene networks: Cell differentiation by graded binary response conversion. *EMBO J.*, 20:2528–2535, 2001.

[8] A. Becskei and L. Serrano. Engineering stability in gene networks by autoregulation. *Nature*, 405:590–593, 2000.

[9] D.F. Blair. How bacteria sense and swim. *Ann. Rev. Microbiol.*, 49:489–522, 1995.

[10] J. Boden. Programming the Drosophila embryo. *J. Theor. Biol.*, 188:391–445, 1997.

[11] A. Cornish-Bowden. *Fundamentals of Enzyme Kinectics*. Portland Press, London, 1995.

[12] M. Csete and J.C. Doyle. Reverse engineering of biological complexity. *Science*, 295:1664–1669, 2002.

[13] E.J. Davidov, J.M. Holland, E.W. Marpleand, and S. Naylor. Advancing drug discovery through systems biology. *Drug Discovery Today*, 8(4):175–183, 2003.

[14] H. El-Samad and M. Khammash. Systems biology: From physiology to gene regulation. *IEE Contr. Syst. Mag.*, 2003.

[15] H. El-Samad, M. Khammash, and J.P. Goff. Calcium homeostasis and parturient hypocalcemia: An integral feedback perspective. *J. Theor. Biol.*, 214:17–29, 2002.

[16] H. El-Samad, M. Khammash, H. Kurata, and J.C. Doyle. Robustness Analysis of the Heat Shock Response in *e. coli*. In *Proceedings of the American Control Conference*, pages 1742–1747, 2002.

[17] H. El-Samad, S. Prajna, A. Papachristodoulou, and M. Khammash *et al.* Model validation and robust stability analysis of the bacterial heat shock response using SOSTOOLS. In *Proceedings of the 42nd IEEE Conference on Decision and Control*, Maui, Hawai, December 2003.

[18] J. Elf and M. Ehrenberg. Fast evaluation of fluctuations in biochemical networks with the linear noise approximation. *Genome Research*, 13:2475–2484, 2003.

[19] M.B. Elowitz and S. Leibler. A synthetic oscillatory network of transcriptional regulators. *Nature*, 403:335–338, 2000.

[20] J.E. Ferrell. Self-perpetuating states in signal transduction: Positive feedback, double negative feedback and bistability. *Curr. Op. Chem. Biol.*, 6:140–148, 2002.

[21] J.E. Ferrell and E.M. Machleder. The biochemical basis for an all-or-none cell fate in *xenopus oocytes*. *Science*, 280:895–898, 1998.

[22] B.A. Francis and W.M. Wonham. The internal model principle for linear multivariable regulators. *Appl. Math. Optim.*, 2:170–194, 1985.

[23] M.A. Gibson and J. Bruck. Exact stochastic simulation of chemical systems with many species and many channels. *J. Phys. Chem.*, 105:1876–1889, 2000.

[24] D. Gillespie. A rigorous derivation of the chemical master equation. *Physica A*, 188:404–425, 1992.

[25] D. T. Gillespie. A general method for numerically simulating the stochastic time evolution of coupled chemical reactions. *J. Comp. Phys.*, 22:403–434, 1976.

[26] D.T. Gillespie. Exact stochastic simulation of coupled chemical reactions. *J. Phys. Chem.*, 81:2340–2361, 1977.

[27] D.T. Gillespie. *Markov Processes: An Introduction for Physical Scientists*. Academic Press, 1992.

[28] D.T. Gillespie. A rigorous derivation of the chemical master equation. *Physica A*,

188:404–425, 1992.

[29] D.T. Gillespie. The chemical Langevin equation. *J. Chem. Phys.*, 113:297–306, 2000.

[30] L. Glass and S.A. Kauffman. The logical analysis of continuous, non-linear biochemical control networks. *J. Theor. Biol.*, 39:103–129, 1973.

[31] B.C. Goodwin. *Temporal Organization in Cells*. Academic Press, New York, 1963.

[32] T.W. Grebe and J. Stock. Bacterial chemotaxis: The five sensors of a bacterium. *Curr. Biol.*, 8:R154–R157, 1998.

[33] J.S. Griffith. Mathematics of cellular control processes. II. positive feedback to one gene. *J. Theor. Biol.*, 20:209–216, 1968.

[34] L.H. Hartwell, J. Hopfield, and A.W. Murray. From molecular to modular cell biology. *Nature*, 81:C47–C52, 1999.

[35] J. Hasty, D. McMillen, and J.J. Collins. Engineered gene circuits. *Nature*, 420:224–230, 2002.

[36] J. Hasty, J. Pradines, M. Dolnik, and J.J. Collins. Noise-based switches and amplifiers for gene expression. *PNAS*, 97:2075–2080, 2000.

[37] W.J. Hedley, M.R. Nelson, D.P. Bullivantand, and P.F. Nielsen. A short introduction to CellML. *Philosophical Transcations of the Royal Society of London Series A-Mathematical Physical and Engineering Sciences*, 359 (1783):1073–1089, 2001.

[38] M. Hucka, A. Finney, H.M. Sauro, and H. Bolouri *et al.* The systems biology markup language (SBML): a medium for representation and exchange of biochemical network models. *Bioinformatics*, 19 (4):524–531, 2003.

[39] P.A. Iglesias. Feedback control in intracellular signaling pathways: Regulating chemotaxis in *dictyostelium discoideum*. *Eur. J. Control*, 9:227–236, 2003.

[40] F. Jacob and J. Monod. On the regulation of gene activity. *Cold Spring Harb. Symp. Quant. Biol.*, 26:193–211, 389–401, 1961.

[41] N.G. Van Kampen. *Stochastic Processes in Physics and Chemistry*. Elsevier Science Publishing Company, 1992.

[42] A. Keller. Model genetic circuits encoding autoregulatory transcription factors. *J. Theor. Biol.*, 172:169–185, 1995.

[43] H. Kurata, H. El-Samad, T.M. Yi, and M. Khammash *et al.* Feedback Regulation of the Heat Shock Response in *e. coli*. In *Proceedings of the 40th IEEE Conference on Decision and Control*, pages 837–842, 2001.

[44] D.C. Lewis. A qualitative analysis of S-systems: Hopf birfucations. In *Canonical Nonlinear Modeling: S-System Approach to Understanding Complexity*, pages 304–344, 1991.

[45] L.M. Loew and J.C. Schaff. The virtual cell: a software environment for computational cell biology. *Trends in Biotechnology*, 19(1):401–406, 2001.

[46] R.M. Macnab and D.E. Koshland. Gradient sensing mechanism in bacterial chemotaxis. *PNAS*, 69:2509–2512, 1972.

[47] H. McAdams and A. Arkin. Stochastic mechanisms in gene expression. In *PNAS*, volume 94, pages 814–819, 1997.

[48] H. McAdams and L. Shapiro. Circuit simulation of genetic networks. *Science*, 269:650–656, 1995.

[49] J. Monod, A.M. Pappenhimer, and G. Cohen Bazire. La cinetique de la biosynthese de la β-galactocidase chez *e. coli* consideree comme fonction de la croissance. *Biochim. Biophys. Acta*, 9:648–660, 1952.

[50] C.J. Morton-Firth and D. Bray. Predicting temporal fluctuations in an intracellular signalling pathway. *J. Theor. Biol.*, 192(1):117–128, 1998.

[51] T.C. Ni and M.A. Savageau. Model assessement and refinement using strategies from biochemical systems theory: Application to metabolism in red blood cells. *J. Theor.*

Biol., 179:329–368, 1996.

[52] M. Plasthne. *A Genetic Switch: Phage λ and Higher Organisms.* Cell Press and Blackwell Scientific Publications, Cambridge, MA, 1992.

[53] J.R. Pomerening, E.D. Sontag, and J.E. Ferrell. Building a cell cycle oscillator: Hysteresis and bistability in the activation of Cdc2. *Nature*, 3:346–351, 2003.

[54] C. Rao and A.P. Arkin. Stochastic chemical kinetics and the quasi-steady-state assumption: Application to the gillespie algorithm. *J. Chem. Phys.*, 118:4999–5010, 2003.

[55] C.V. Rao, D.M. Wolf, and A.P. Arkin. Control, exploitation and tolerance of intracellular noise. *Nature*, 420(6912):231–237, 2002.

[56] M. Rathinam, L. Petzold, and D.T. Gillespie. Stiffness in stochastic chemically reacting systems: The implicit tau-leaping method. To appear J. Chem. Phys.

[57] N. Rosenfeld, M.B. Elowitz, and U. Alon. Negative autoregulation speeds the response time of transcription networks. *J. Molec. Biol.*, 323:785–793, 2002.

[58] M.A. Savageau. Biochemical systems analysis II: The steady-state solution for an n-pool system using a power law approximation. *J. Theor. Biol.*, 25:370–379, 1969.

[59] M.A. Savageau. A comparison of classical and autogenous systems of regulation in inducible operons. *Nature*, 252:546–549, 1974.

[60] M.A. Savageau. Demand theory of gene regulation. I. quantitative development of the theory. *Genetics*, 149(4):1665–1676, 1998.

[61] M.A. Savageau and E.O. Voit. Recasting nonlinear differential equations as S-systems: A canonical nonlinear form. *Math. Biosci.*, 163:105–129, 1987.

[62] R.T. Schimke. On the roles of synthesis and degradation in regulation of enzyme levels in mammalian tissues. *Curr. Top. Cell. Regul.*, 1:77–124, 1969.

[63] J.E. Shea and C.L. Brooks. From folding theories to folding proteins: A review and assessment of simulation studies of protein folding and unfolding. *Ann. Rev. Phys. Chem.*, 52:499–535, 2001.

[64] H. Smith. Monotone semiflows generated by functional differential equations. *J. Diff. Eq.*, 66:420–442, 1987.

[65] H. Smith. Oscillations and multiple steady sates in a cyclic gene model with repression. *J. Math. Biol.*, 25:169–190, 1987.

[66] E.H. Snoussi and R. Thomas. Logical identification of all steady states: The concept of feedback loop characteristic states. *Bull. Math. Biol.*, 55:973–991, 1993.

[67] R. Somogyi and C. Sniegoski. Modeling the complexity of genetic networks: Understanding multigenic and pleitropic regulation. *Complexity*, 1:45–63, 1996.

[68] R. Somogyi and C. Sniegoski. The gene expression matrix: Towards the extraction of genetic network architectures. In *Proceedings of the Second World Congress of Nonlinear Analysis*, 1997.

[69] D.B. Straus, W.A. Walter, and C.A. Gross. The Activity of σ^{32} is Reduced Under Conditions of Excess Heat Shock Protein Production in *escherichia coli*. *Genes & Dev.*, 3:2003–2010, 1989.

[70] M. Thattai and A. Van Oudenaarden. Intrinsic noise in gene regulatory networks. *PNAS*, 98:8614–8619, 2001.

[71] R. Thomas. The role of feedback circuits: Positive feedback circuits are a necessary condition for positive real eigenvalues of the jacobian matrix. *Ber. Busenges Phys. Chem.*, 98:1158–1151, 1994.

[72] R. Thomas and R. D'Ari. *Biological Feedback.* CRC Press, Boca Raton, FL, 1990.

[73] R. Thomas, D. Thieffry, and M. Kauffman. Dynamical behavior of biological regulatory networks- I. biological role of feedback loops and practical use of the concept of the loop-characteristic state. *Bull. Math. Biol.*, 57:247–276, 1995.

[74] M. Tomita, K. Hashimoto, K. Takahashi, and T.S. Shimizu *et al.* E-CELL: Software environment for whole cell simulation. *Bioinformatics*, 15:172–84, 1999.

[75] J. Tyson and H.G. Othmer. The dynamics of feedback control circuits in biochemical pathways. *Prog. Theor. Biol.*, 5:2–62, 1978.

[76] J.M. Vilar, H.Y. Kueh, N. Barkai, and S. Leibler. Mechanisms of noise resistance in genetic oscillators. *PNAS*, 99:5988–92, 2002.

[77] M.E. Wall, W.S. Hlavacek, and M.A. Savageau. Design principles for regulator gene expression in a repressible gene circuits. *J. Molec. Biol.*, 332(4):861–876, 2003.

[78] N. Wiener. *Cybernetics, or Control and Communication in the Animal and the Machine.* John Wiley and sons, New York, 1948.

[79] D.M. Wolf and F.H Eckman. On the relationship between genomic regulatory element organization and gene regulatory dynamics. *J. Theor. Biol.*, 195:167–178, 1998.

[80] O. Wolkenhauer, H. Kitano, and K.H. Cho. Systems biology. *IEEE Contr. Syst. Mag.*, 23(4):38–48, 2003.

[81] T.-M. Yi, Y. Huang, M. Simon, and J.C. Doyle. Robust perfect adaptation in bacterial chemotaxis through integral feedback control. *PNAS*, 97:4649–4653, 2000.

[82] C.H. Yuh, H. Bolouri, and E.H. Davidson. Genomic *cis* regulatory logic, experimental and computational analysis of a sea urchin gene. *Science*, 279:1896–1902, 1998.

VII

Computational Structural Biology

29

Predicting Protein Secondary and Supersecondary Structure

Mona Singh
Princeton University

29.1 Introduction

Proteins play a key role in almost all biological processes. They take part in, for example, maintaining the structural integrity of the cell, transport and storage of small molecules, catalysis, regulation, signaling and the immune system. Linear protein molecules fold up into specific three-dimensional structures, and their functional properties depend intricately upon their structures. As a result, there has been much effort, both experimental and computational, in determining protein structures.

Protein structures are determined experimentally using either x-ray crystallography or nuclear magnetic resonance (NMR) spectroscopy. While both methods are increasingly being applied in a high-throughput manner, structure determination is not yet a straight-forward process. X-ray crystallography is limited by the difficulty of getting some proteins to form crystals, and NMR can only be applied to relatively small protein molecules. As a result, whereas whole-genome sequencing efforts have led to large numbers of known protein sequences, their corresponding protein structures are being determined at a significantly s-lower pace. On the other hand, despite decades of work, the problem of predicting the full three-dimensional structure of a protein from its sequence remains unsolved. Never-theless, computational methods can provide a first step in protein structure determination, and sequence-based methods are routinely used to help characterize protein structure. In this chapter, we review some of the computational methods developed for predicting local

Peptide bond

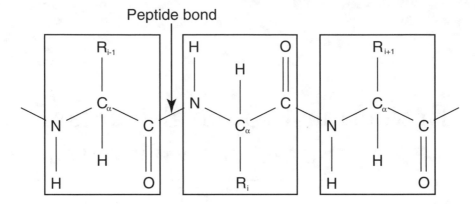

FIGURE 29.1: Proteins are polymers of amino acids. Each amino acid has the same fundamental structure (boxed), differing only in the atoms making up the side chain. Here, the i-th side chain in the protein sequence is designated by R_i. The carbon atom to which the amino group, carboxyl group, and side chain are attached is called the alpha carbon (C_α). Two amino acids $i-1$ and i are linked linearly through a peptide bond between the carboxyl group of amino acid $i-1$ and the amino group of amino acid i; a water molecule is removed in the process of bond formation.

aspects of protein structure.

29.1.1 Background

We begin by giving some introductory background to protein structure; there are many excellent sources for further information (e.g.,[16, 77, 103]).

A protein molecule is formed from a chain of amino acids. Each amino acid consists of a central carbon atom (C_α), and attached to this carbon are a hydrogen atom, an amino group (NH_2), a carboxyl group (COOH) and a *side chain* that characterizes the amino acid. The amino acids of a protein are connected in sequence with the carboxyl group of one amino acid forming a peptide bond with the amino group of the next amino acid (Figure 29.1). Successive bonds make up the protein backbone, and the repeating amino-acid units (also called residues) within the protein consist of both the main-chain atoms that comprise the backbone as well as the side-chain atoms.

There are 20 side chains specified by the genetic code, and each is referred to by a one-letter code. A protein sequence can thus be described by a string over a 20-letter alphabet, and the *primary structure* of a protein refers to the covalent structure specified by its sequence (i.e., Figure 29.1), along with its disulfide bonds. The 20 side chains vary in atomic composition, and thus have different chemical properties. Some side chains are non-polar, or hydrophobic, because of their unfavorable interactions with water. Side chains have many other characteristics, and different side chains are commonly described as being positively charged, negatively charged, polar, small or large. Hydrophobic amino acids include isoleucine (I), leucine (L), methionine (M), phenylalanine (F) and valine (V). Arginine (R) and lysine (K) are positively charged in physiological pH, and aspartic acid (D) and glutamic acid (D) are negatively charged. Polar amino acids include asparagine (N), glutamine (Q) histidine (H), serine (S) and threonine (T). Alanine (A) is a small amino acid that is non-polar. Glycine (G) is the smallest amino acid, with just a hydrogen. Cysteine (C) can take part in disulfide bridges. Proline (P) has the strongest stereochemical constraints, and

FIGURE 29.2: Schematic backbone conformations of an α-helix (left) and a β-sheet (right). An α-helix consists of contiguous amino acid residues. A β-sheet consists of individual β-strands, each of which is made up of contiguous amino acid residues. Here, a 5-stranded β-sheet, without the intervening regions, is shown.

tryptophan (W) and tyrosine (Y) are large, ring-shaped amino acids. There are many other (and sometimes conflicting) ways to classify and describe the amino acids.

The differences in physico-chemical properties of side chains result in the diversity of three-dimensional protein folds observed in nature. In particular, each possible structural conformation brings together a different set of amino acids, and the energy of the conformation is determined by the interactions of the side-chain and main-chain atoms with each other, as well as with solvent and ligands. There are many forces driving protein folding; for water-soluble proteins, the most dominant is the hydrophobic effect, or the tendency of hydrophobic amino acids to avoid water and bury themselves within the core of the protein. Hydrogen bonding, electrostatic interactions and van der Waals forces are also very important.

From a structural perspective, it is useful to think of protein chains as subdivided into peptide units consisting of the main-chain atoms between successive C_α atoms. In protein structures, the atoms in a peptide unit are fixed in a plane with bond lengths and angles similar in all units. Each peptide unit essentially has only two degrees of freedom, given by rotations around its N-C_α and C_α-C bonds. Phi (ϕ) refers to the angle of rotation around the N-C_α bond, and psi (ψ) refers to the angle of rotation around the C_α-C bond. The entire backbone conformation of a protein can thus be specified with a series of ϕ and ψ angles. Only certain combinations of ϕ and ψ angles are observed in protein backbones, due to steric constraints between main-chain and side-chain atoms.

As a result of the hydrophobic effect, the interior of water-soluble proteins form a hydrophobic core. However, a protein backbone is highly polar, and this is unfavorable in the hydrophobic core environment; these main-chain polar groups can be neutralized via the formation of hydrogen bonds. *Secondary structure* is the "local" ordered structure brought about via hydrogen bonding mainly within the backbone. Regular secondary structures include *α-helices* and *β-sheets* (Figure 29.2). A canonical α-helix has 3.6 residues per turn, and is built up from a contiguous amino acid segment via backbone-backbone hydrogen bond formation between amino acids in positions i and $i+4$. The residues taking part in an α-helix have ϕ angles around $-60°$ and ψ angles around $-50°$. Alpha helices vary considerably in length, from four or five amino acids to several hundred as found in fibrous proteins. A β-strand is a more extended structure with 2.0 residues per turn. Values for ϕ and ψ vary, with typical values of $-140°$ and $130°$, respectively. A β-strand interacts via hydrogen bonds with other β-strands, which may be distant in sequence, to form a β-sheet. In parallel β-sheets, the strands run in one direction, whereas in antiparallel sheets, they run in alternating directions. In mixed sheets, some strands are parallel, and some are antiparallel. A β-strand is typically 5–10 residues in length, and on average, there are six strands per sheet. Coil or loop regions connect α-helices and β-sheets and have varying lengths and shapes.

Supersecondary structures, or *structural motifs*, are specific combinations of secondary structure elements, with specific geometric arrangements with respect to each other.[1] Common supersecondary motifs include α-helix hairpins, β hairpins, β-α-β motifs, and coiled coils. Elements of secondary structure and supersecondary structure can then combine to form the full three-dimensional fold of a protein, or its *tertiary structure*. Many proteins exist naturally as aggregates of two or more protein chains, and *quartenary structure* refers to the spatial arrangement of these protein subunits.

29.1.2   Difficulty of general protein structure prediction

Experiments performed decades ago demonstrated that the information specifying the three-dimensional structure of a protein is contained in its amino acid sequence [5, 4], and it is generally believed that the native structure of the majority of proteins is the conformation that is thermodynamically most stable. It is now known that some proteins require specific proteins, or chaperones, to help them fold into their global free-energy minimum. A quantum mechanics treatment to predict structure is intractable for protein sequences, and thus physics-based methods for structure prediction typically use empirical molecular mechanics force fields. In these methods, the system is described as a set of potential energy terms (typically modeling bond lengths, bond angles, dihedral angles, van der Waals interactions and electrostatics), and the goal is to find, for any given protein sequence, the conformation that minimizes the potential energy function (e.g., see [18]). The accuracy of state-of-the-art energy functions, the small energy differences between native and unfolded proteins, and the size of the conformational space that must be searched are all limiting factors in the overall performance of these physics-based methods. In the case where a protein is homologous to another with known structure, the search space is limited, as the homolog provides a template backbone; improved statistical methods for remote homology detection as well as the increasing number of solved protein structures have made such approaches more widely applicable. Purely statistical approaches have also been developed for predicting the tertiary structure of a protein. One such approach is known as *threading* [117, 13, 61, 19], where a sequence is aligned (or "threaded") onto all known backbones using an energy function that is estimated from observed amino acid frequencies in known protein structures. Many modern approaches use a combination of both statistics and physics; for example, in some of the more successful approaches for predicting protein structure, backbone fragments for particular subsequences are sampled from known structures, and then pieced together and evaluated using a molecular mechanics energy function [14]. While there has been much progress in developing computational methods for predicting the three-dimensional structures of proteins, it is clear that the problem is far from being solved (e.g., [92, 72, 2, 118]).

29.1.3   A bottom-up approach

Because of the difficulty of the general protein structure prediction problem, an alternative approach for predicting protein structure is "bottom-up": here, the goal is to focus on specific, *local* three-dimensional structures, and develop specialized computational methods

---

[1]Supersecondary structure is sometimes defined so as to require that the secondary structure units are consecutive in the protein sequence; we do not take that viewpoint here.

for recognizing them within protein sequences. At the most basic level, a protein's secondary structure can be predicted. At the next level, computational methods may be developed to predict local supersecondary structures or structural motifs. Protein structure can also be characterized by identifying portions that are membrane-spanning, or by assessing the solvent accessibility of individual residues, though such subjects will not be reviewed here. By focusing on specific aspects of protein structures, it is possible to develop computational methods that can make high-confidence predictions; these can then be used to constrain methods that attempt to predict tertiary structure. At the same time, one hope is that ultimately it will be possible to build up a "library" of increasingly complex structures that can be recognized via specialized computational methods, and that this library may provide an alternative means for predicting the tertiary structures of proteins.

In the remaining portion of this chapter, we review computational techniques that have been developed for predicting secondary and supersecondary structures. While the most accurate predictions of structure are made by detecting homology to proteins with known structure, we primarily focus on methods that can make predictions even if there are no such homologs. Since there have been hundreds of papers written on predicting the secondary and supersecondary structure of proteins, we will only have a chance to discuss a small subset of the many important papers in the field.

29.2 Secondary Structure

Most commonly, the secondary structure prediction problem is formulated as follows: given a protein sequence with amino acids $r_1 r_2 \ldots r_n$, predict whether each amino acid r_i is in an $\alpha-$helix (**H**), a $\beta-$strand (**E**), or neither (**C**). Predictions of secondary structure are typically judged via the *3-state* accuracy (Q_3), which is the percent of residues for which a method's predicted secondary structure (**H**, **E**, or **C**) is correct. Since residues in known protein structures are approximately 30% in helices, 20% in strands and 50% in neither, a trivial algorithm that always predicts **C** has a 3-state accuracy of 50%. The 3-state accuracy measure does not convey many useful types of information. For example, it does not indicate whether one type of structure is predicted more successfully than another, whether some structure is over- or under- predicted, or whether errors are more likely along the boundaries of secondary structure units than within them. Nevertheless, 3-state accuracy is a concise, useful measure that is frequently used to compare how well different methods perform. Other methods to judge the quality of secondary structure predictions include the Matthews correlation coefficient [87] and measures of how well the predicted secondary structure segments overlap the actual ones [107, 130].

Secondary structural elements are readily evident in the crystal structures of proteins, and are defined operationally based primarily on their hydrogen bonding patterns. Given the 3D atomic coordinates of a protein structure, there are several automated means for extracting secondary structure, including DSSP [62] and STRIDE [44]. The assignment of secondary structure to each amino acid is not completely well-defined, and these two programs differ on approximately 5% of residues (e.g.,see [32]). Both DSSP and STRIDE report detailed descriptions of secondary structure. For example, the DSSP method has eight secondary structure classifications: **H**, α-helix; **E**, β-strand; **G**, 3_{10} helix, a helix with backbone-backbone hydrogen bonds between positions i and $i + 3$; **I**, π-helix, a helix with backbone-backbone hydrogen bonds between positions i and $i+5$; **B**, bridge, a single residue β-strand; **T**, a hydrogen bonded turn; **S**, bend; and **C**, any residue that does not belong to any of the previous seven groups.

There are different schemes for translating the more detailed descriptions given by DSSP and STRIDE into the three broad categories corresponding to helix, sheet and other. One scheme translates all helices (**H**, **G**, and **I**) into **H**, bridges and strands (**E**, **B**) into **E** and every thing else (**T**, **S**, **C**) into **C**. An alternative scheme takes the DSSP categories of **H** and **E** as helix and strand, and maps all other categories into **C**. The reported performance of a secondary structure prediction method can vary depending on which precise translation scheme is used, with the second scheme leading to higher estimates of accuracy [32].

Testing of secondary structure prediction methods has improved over the years. We note that whereas the PDB (the Protein Data Bank of solved structures [11]) contains structures for many very similar sequences, the training set used for estimating parameters should not contain sequences that are too similar to those in the test set. In particular, a protein sequence in the test set should be less than 25–30% similar to any sequence in the training set. Otherwise, reported accuracy is likely to be an overestimate of actual accuracy. Methods are typically tested using N-fold cross-validation, where a dataset is split into N parts. Each part is in turn left out of the training set and performance is judged on it. The performance of the method is the average performance over each left out part.

Early secondary structure prediction methods (such as Chou-Fasman and GOR, described below) have a 3-state cross-validation accuracy of 50–60%. Today's methods have an accuracy of $> 75\%$.

29.2.1 Early Approaches

The earliest approaches for secondary structure prediction considered just single amino acid statistics and properties, and were limited by the small number of proteins with solved structures. While these early methods are not state-of-the-art, they are natural first attempts to the secondary structure prediction problem, and are the basis of many subsequent approaches. Below, we consider three of the most well-known early secondary structure prediction methods.

Chou-Fasman method. One of the first approaches for predicting protein secondary structure, due to Chou and Fasman [27], uses a combination of statistical and heuristic rules. First, using a set of solved protein structures, "propensities" are calculated for each amino acid a_i in each structural conformation s_j, by taking the frequency of a_i in each structural conformation, and then normalizing by the frequency of this amino acid in all structural conformations. That is, if a residue is drawn at random from the space of protein sequences, and its amino acid identity A and structural class S are considered, propensities are computed as $\Pr(A = a_i | S = s_j) / \Pr(A = a_i)$.[2] These propensities capture the most basic concept in predicting protein secondary structure: different amino acids occur preferentially in different secondary structure elements.

Once the propensities are calculated, they are used to categorize each amino acid as either a helix-former, a helix-breaker, or helix-indifferent. Each amino acid is also categorized as either a sheet-former, a sheet-breaker, or sheet-indifferent. For example, as expected, glycine and proline have low helical propensities and are thus categorized as helix-breakers. Then, when a sequence is input, "nucleation sites" are identified as short subsequences with a high-concentration of helix-formers (or sheet-formers). These sites are found with heuristic

---

[2]Sometimes propensities are defined by considering the frequency of a particular structural conformation given an amino acid, and normalizing by the frequency of that structural conformation. These two formulations are equivalent since $\Pr(A = a_i | S = s_j) / \Pr(A = a_i) = \Pr(S = s_j | A = a_i) / \Pr(S = s_j)$.

rules (e.g., "a sequence of six amino acids with at least four helix-formers, and no helix-breakers"), and then extended by adding residues at each end, while maintaining an average propensity greater than some threshold. Finally, overlaps between conflicting predictions are resolved using heuristic rules.

GOR method. The GOR method [47] formalizes the secondary structure prediction problem within an information-theoretic framework. If x and y are any two events, the definition of the information that y carries on the occurrence of event x is [42]:

$$I(x;y) = \log\left(\frac{\Pr(x|y)}{\Pr(x)}\right). \tag{29.1}$$

For the task at hand, the goal is to predict the the structural conformation S_j of residue R_j in a protein sequence, and the GOR method estimates the information that the surrounding "local" 17-long window contains about it:

$$I(S_j; R_{j-8}, \ldots, R_j, \ldots R_{j+8}) = \log\left(\frac{\Pr(S_j|R_{j-8}, \ldots R_j, \ldots, R_{j+8})}{\Pr(S_j)}\right). \tag{29.2}$$

In fact, each structural class x is considered in turn, and the following value, representing the preference for x over all other alternatives $\bar{x}$ is computed:

$$I(S_j = x : \bar{x}; R_{j-8}, \ldots, R_j, \ldots R_{j+8}) = I(S_j = x; R_{j-8}, \ldots, R_j, \ldots, R_{j+8}) - $$
$$I(S_j = \bar{x}; R_{j-8}, \ldots R_j, \ldots, R_{j+8}).$$

To predict residue R_j's structural conformation, these values are computed for all structural states, and the one that has the highest value is taken as the prediction.

Because there are far too many possible sequences of length 17, it is not possible to estimate $\Pr(S_j|R_{j-8}, \ldots R_j, \ldots, R_{j+8})$ with any reliability. Instead, the original GOR method assumes that the values of interest can be estimated using single residue statistics:

$$I(S_j = x; R_{j-8}, \ldots, R_j, \ldots, R_{j+8}) = \sum_{m=-8}^{m=8} I(S_j = x; R_{j+m}), \tag{29.3}$$

where by definition $I(S_j = x; R_{j+m}) = \log(\Pr(S_j = x|R_{j+m})/\Pr(S_j = x)).^3$ $I(S_j = x; R_{j+m})$ represents the information carried by a residue at position $j+m$ on the conformation assumed by the residue at j. If $m \neq 0$, this does not take into account the type of residue at position j, and the intuition is that it describes the interaction of the side chain of residue $j+m$ with the backbone of residue j. For each structural class, this method requires estimating 20×17 parameters.

Lim method. A complicated, stereochemical rule-based approach for predicting secondary structure in globular proteins was developed at about the same time as the statistical methods discussed above. In this method, longer-range interactions between residues are considered. If the protein sequence is $r_1 r_2 \ldots r_n$, then for the i-th residue, the following pairs and triples are considered particularly important for helical regions: (r_i, r_{i+1}), (r_i, r_{i+3}), (r_i, r_{i+4}), (r_i, r_{i+1}, r_{i+4}), (r_i, r_{i+3}, r_{i+4}). Note that residues three and four apart are considered, as they lie on the same face of an α-helix. Similarly, the pair (r_i, r_{i+2}) contains

---

^{3}Note that when $m = 0$, these values are equivalent to taking the log of the Chou-Fasman propensity values.

residues on the same face of a β-strand. Pairs and triplets of particular amino acids are then deemed as compatible or incompatible with helices and strands based on various rules that try to ensure that these residues present a face that allows tight packing of hydrophobic cores. Factors used to determine these rules include each amino acid's size, hydrophobicity, charge, and its ability to form hydrogen bonds. For example, if a protein sequence has hydrophobic residues every three to four residues, this method predicts compatibility with an α-helix, as this would result in one side of the helix being hydrophobic, thus facilitating packing onto the rest of the protein structure.

29.2.2 Incorporating local dependencies

Whereas the first statistical methods for predicting protein secondary structure examined each amino acid individually, later approaches began to consider higher-order residue interactions, either within statistical approaches or via machine learning methods. Reported 3-state accuracies for most of these methods are above 60%.

Information theory approaches. One approach to incorporate higher-order residue interactions is an extension to the original GOR method [48]. The notion of conditional information is helpful here. In particular, $I(x; y_2|y_1)$ is defined as $\log(\Pr(x|y_1, y_2)/P(x|y_1))$. Note that $I(x; y_1, y_2, \ldots, y_n) = I(x; y_1) + I(x; y_2|y_1) + \ldots + I(x; y_n|y_1, y_2, \ldots y_{n-1})$. Instead of the assumption made in equation 29.3, the following assumption is made:

$$I(S_j = x; R_{j-8}, \ldots, R_j, \ldots, R_{j+8}) = I(S_j = x; R_j) + \sum_{m=-8, m \neq 0}^{m=8} I(S_j = x; R_{j+m}|R_j).$$

This formulation incorporates the information carried by the residue at $j + m$ on the conformation of the residue at j, taking into account the type of residue at position j. Note that by changing these assumptions, different pairwise or higher-order residue interactions may be considered. Later versions of the GOR algorithm do precisely this (e.g., see [75]).

Nearest-neighbor approaches. Nearest-neighbor methods classify test instances according to the classifications of "nearby" training examples. In the context of secondary structure prediction, the overall approach is to predict the secondary structure of a residue in a protein sequence by considering a window of residues surrounding it, and finding similar sequence segments in proteins of known structure. The assumption is that short, very similar sequences of amino acids have similar secondary structure even if they come from non-homologous proteins. The known secondary structures of the middle residue in each of these segments are then combined to make a prediction, either via a simple voting scheme or a weighted voting scheme, with segments more similar to the target segment weighed more. Early nearest-neighbor approaches include [94, 78]. Similar segments can be found via sequence similarity, or via structural profiles [13], as in [81].

Neural network approaches. Neural networks provide another means for capturing higher-order residue interactions. They were first applied to predict secondary structure by [101, 55], and some of the most successful modern methods are also based on neural networks (e.g., [105] and its successors).

Because neural nets are widely used in the field of secondary structure prediction, we briefly describe them here. Neural networks, loosely based on biological neurons, are machine learning methods that learn to classify input vectors into two or more categories. Feedforward neural networks consist of two or more connected layers. The first layer is the input layer, and the last layer is the output layer that indicates the predicted category of the input. All other layers are called hidden layers. A simple neural network with no hidden

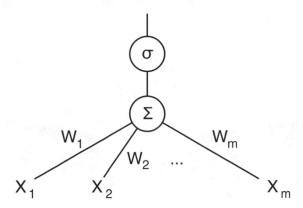

FIGURE 29.3: A simple neural network with no hidden units. There are m inputs $x_1 \ldots x_m$, and the neural net computes a function on these inputs by first calculating $\sum_i w_i x_i$, and then using this as input to an activation function σ.

units is given in Figure 29.3. The inputs can be encapsulated in a vector $\vec{x} = (x_1, \ldots, x_m)^T$, and each of the input edges has a corresponding weight, giving $\vec{w} = (w_1, \ldots, w_m)^T$. Each input is multiplied by the corresponding weight of its edge. Then, the network computes a weighted sum, and feeds it into some activation or continuous threshold function σ. For example, $\sigma(a)$ could be $\frac{1}{1+e^{-a}}$, which is a sigmoidal function with values between 0 and 1.[4] Thus, the function computed by this simple neural network is given by $\sigma(\vec{w} \cdot \vec{x})$, and is essentially linear. In most cases, a neural net must learn the weights from a training set of input vectors $\{\vec{x}_i\}$ where the target value t_i for each is known. For example, in the scenario described, there may be two classes of examples with target values of 0 and 1. Typically, the goal is to find the weights $\vec{w}$ minimizing some error function (e.g., the squared error $E = \sum_i (\sigma(\vec{w} \cdot \vec{x}_i) - t_i)^2)$. Such a $\vec{w}$ can be found via gradient descent.[5] A full-blown neural net is built up from a set of simpler units that are interconnected in some topology so that the outputs of some units become the inputs of other units (e.g., see Figure 29.4). The gradient descent procedure for arbitrary neural networks is implemented via the back-propagation algorithm [109, 108]. While neural nets with multiple layers are not as easy to interpret as those without hidden layers, they can approximate any continuous function $f : \mathcal{R}^m \to \mathcal{R}$ as long as they have a sufficient number of hidden units and at least two hidden layers [33].

The two early neural-network approaches to secondary structure prediction use similar neural network topologies. Holley and Karplus [55] build a neural net that tries to predict the secondary structure of a residue r_j by considering residues $r_{j-8}, \ldots, r_j, \ldots, r_{j+8}$. Each

---

[4]While a strict 0/1 threshold function can also be used, a continuous function is preferred for ease of optimization.

[5]There are many other approaches to find a set of weights that "best" linearly separate two classes. For example, the support vector machine framework (SVM) [119] finds weights so that the margin between the two classes of examples is maximized; that is, an SVM finds the weights by maximizing the distance between the hyperplane specified by the weights and the closest training examples. In the case where the two classes are not linearly separable, the data are typically embedded in a higher dimensional space where they are separable. An alternative approach, linear discriminant analysis, tries to find a set of weights so that when considering $D_x = \vec{w} \cdot \vec{x}$ for all examples $\vec{x}$, these values are as close as possible within the same class and as far apart as possible between classes.

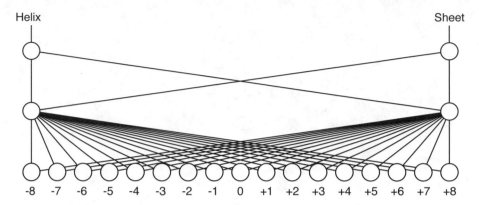

FIGURE 29.4: A neural network topology for predicting secondary structure [55]. To predict
the secondary structure of the middle residue, eight residues on either side are
considered. Each of the 17 input units drawn actually consists of 21 mutually
exclusive binary inputs, one for each possible amino acid, and one used when the
window overlaps the end of the protein sequence. There is one hidden layer with
two units, and an output layer with two units. Here, each node in the hidden
layer and output layer contains both a summation and activation component.
The basic neural network of [101] is similar, but with 13 input units, more hidden
units, and a third output unit corresponding to coil.

of these residues is represented with 21 bits, corresponding to the 20 amino acids and an
extra bit for the case where the window overlaps the beginning or end of the sequence.
Thus, each example is represented by 17×21 bits, with only 17 non-zero entries. The
topology of the neural net has one hidden layer with two nodes, and two output nodes, one
corresponding to helix and the other to sheet (see Figure 29.4). For the training process,
if the middle residue's secondary structure is helix, then the target output has the helix
output set to 1 and the sheet output set to 0. For a new sequence, helix is assigned to four
or more adjacent residues each with helix output value greater than both the sheet output
value and some threshold. Strand is assigned similarly, though only two adjacent strand
residues are required.

Qian and Sejnowski [101] have a slightly different network topology. They consider 13-long
windows, and have three output units (one for each of the three states). More significantly,
they additionally use a cascade of neural networks in order to capture correlations between
secondary structure assignments of neighboring residues. In particular, they show improved
performance by first training a neural net to predict the secondary structure of a central
amino acid, and then taking the outputs for adjacent residues using this trained network
and feeding them into a second network. The input layer of this second network has 13
groups, with three units per group, one for each output unit from the first network.

29.2.3 Exploiting evolutionary information

It is well-known that protein structure is more conserved than protein sequence, and that two
sequences that share more than 30% sequence identity are likely to have similar structures.
Thus, when predicting the secondary structure of a particular protein sequence, predictions
for its homologs may also prove useful. Additionally, conservation evident in multiple
sequence alignments (MSAs) of homologs helps reveal which amino acids are likely to be
functionally or structurally important, and may highlight the characteristic hydrophobic

patternings of secondary structure elements. For example, surface-exposed loop regions that are not important functionally tend to be part of variable regions in MSAs.

A natural first attempt to use homologous proteins in order to improve secondary structure prediction might make predictions for each homolog, and then average (or otherwise combine) these predictions for corresponding amino acids [134, 75]. Alternatively, information from all sequences may be used at once in order to make one set of predictions. This is the approach taken by Rost and Sander [105], and their neural network based program was the first to surpass 70% 3-state accuracy. The use of evolutionary information is critical for the improved performance, and all modern approaches use evolutionary information in making secondary structure predictions.

To make predictions about a single protein sequence, the approach of [105] begins with homologs gathered via database search. These homologs are then aligned in a MSA, and a profile is made. In particular, for each column j in the MSA, the frequency of each amino acid i in the column is computed. To determine the secondary structure of residue r_j, a sequence-to-structure neural network considers a window of 13 residues $r_{j-6} \ldots r_j \ldots r_{j+6}$. For each residue in the window, instead of giving just the identity of this residue as input to the neural net, the frequencies of all amino acids in the corresponding column of the MSA are fed into the neural network. These frequencies encapsulate the evolutionary constraints on each residue. Other features, including overall amino acid composition, are also input to the network. This network has three output units, one for each secondary structure state. Similar to [101], the output of this first level neural network is fed to a structure-to-structure network. Finally, a jury system is used to make the final predictions. Because neural networks are sensitive to topology, the set of training data, the order of training, as well as other parameters, several networks are trained while varying these parameters. The jury system level takes as input the results from each of these nets and averages them. The secondary structure with the highest average score is output as the prediction. A very useful feature of this approach is a per-position reliability index, where higher numbers correspond to more confident predictions. Neural nets have become the most common approach to secondary structure prediction; more recent extensions have included the use of recurrent neural nets to capture non-local interactions [6, 100].

Incorporating evolutionary information into other basic secondary structure prediction methods also results in improved performance, and while it was initially suggested otherwise, it is unlikely that there is some special feature of neural nets that makes them particularly well-suited to predicting secondary structure. For example, similar performance has also been achieved using support vector machines (SVMs) [56]. Adding evolutionary information to nearest-neighbor approaches [110] also performs competitively; here, predictions are made individually for each homolog and then combined. Another MSA approach with similar reported performance [74] uses linear discriminant analysis to combine several predictive attributes. These include: residue propensities, computed as in GOR [47]; distance from the end of the protein sequence; moments of hydrophobicity [39] for each residue under the assumption that it and its three neighboring residues in each direction are in either helices or sheets; whether or not an insertion or deletion is observed in any of the homologs in the MSA; and an entropy-based measurement of residue conservation. Sequence correlations are captured by feeding in the output of the first linear discrimination function into another one, and additionally incorporating smoothing of features over nearby residues, predicted ratios of α-helix and β-strand, and measures of sequence amino acid content.

While most methods incorporate evolutionary information using global MSAs, an alternate method relies solely on pairwise local alignments [45]. A weight is computed for each pairwise alignment based on its score and length. For each residue in the original sequence, the weighted sum over all aligned sequences is computed independently for several propensity values, which are then combined using a rule-based system to make a final prediction. These propensity values are interesting, as several of them try to incorporate non-local interactions [45]. In particular, β-strand hydrogen bonding parallel and anti-parallel propensities (obtained from known structures) are computed between neighboring sequence fragments, and helical hydrogen bonding propensities are computed for fragments by considering residues i and $i + 4$. A propensity concerning β-turn is also used ([58], see below), as well as helical, strand and coil propensities computed using a nearest-neighbor approach.

29.2.4   Recent developments and conclusions

Further improvements in performance have come from better remote homology detection (e.g., using PSI-BLAST [3] or hidden Markov models [64, 65]), and larger sequence databases [31, 104]. For example, Jones [60] obtained better performance than [105] ($> 75\%$ 3-state accuracy) using a similar neural network architecture (without a jury system layer), but where homologs are first detected via PSI-BLAST [3]. PSI-BLAST is an iterative database searching method that uses homologs found in one iteration to build a profile used for searching in the next iteration. The detected homologs are then input into the neural network via the profile provided by PSI-BLAST; this profile incorporates sequence weighting so that several closely-related homologs detected in the database do not overwhelm the contribution of more remote homologs. It is likely that sequence weighting also plays a role in the improved performance of this method, as it has been shown that predictions improve when getting rid of closely related homologs [32].

Several authors have also attempted to predict secondary structure by combining the results of several different programs. For example, Cuff and Barton [32] predict secondary structure by taking the most commonly predicted state by four methods [105, 110, 74, 46], and show a modest improvement in performance. Existing approaches have also been combined using machine-learning methods such as linear discriminant analysis, decision trees and neural nets, and have shown to give upto a 3% improvement in 3-state accuracy over the best individual method [73].

Future evaluation. An important recent development has been to set up continuous evaluation procedures (such as EVA [41]). Protein sequences with newly determined structures are sent to the webservers of the programs being evaluated. In general, evaluation and comparison of methods is often difficult, due to differences in the evaluation methodology and the changing structural databases; thus, a community-wide approach such as this should have great impact on future development of secondary structure prediction methods.

Limitations of secondary structure prediction. In general, it is believed that α-helices are easier to predict than β-sheets. A recent evaluation found that helices were predicted 9.5% more accurately than strands [2]. This may be because the hydrogen bonding patterns for α-helices are among amino acids in close proximity to each other, and those for β-sheets are not. Additionally, shorter secondary structure elements are harder to predict, presumably because the signal is not strong enough from these fragments.

Clearly, protein secondary structure is influenced by both short- and long-range interactions. It has been demonstrated that there are 11-long amino acid sequences that can fold into an α-helix in one context, and a β-sheet in another [91]. However, even assuming that long-range tertiary interactions can be incorporated into secondary structure prediction algorithms, the best possible 3-state accuracy will not be 100%. First, assignment of secondary structure is not always clear even when there is a crystal structure. This is evident from the observed differences between STRIDE and DSSP [32]. Additionally, while secondary structure predictions improve when incorporating evolutionary information, homologous structures do not share identical descriptions of secondary structure assignments [106]. Even when a query sequence can be aligned confidently to a sequence of known structure, the alignment will produce a secondary structure "prediction" with 3-state accuracy of only 88% on average [106]. Accordingly, while secondary structure prediction methods continue to improve, it is unlikely that any method that does not also solve the tertiary structure prediction problem will achieve ideal performance in predicting secondary structure.

29.3   Tight Turns

Tight turns are secondary structure elements consisting of short backbone fragments (no more than six residues) where the backbone reverses its overall direction. Tight turns allow a protein to fold into a compact globular structure, and identifying them correctly in a protein sequence limits the search space of possible folds for the sequence. Tight turns are also important because they are often on the surface of proteins, and thus may play a role in molecular interactions. Tight turns are categorized according to their lengths into $\delta-$, $\gamma-$, $\beta-$, $\alpha-$ and $\pi-$ turns, which consist of two, three, four, five, and six residues respectively.

Computational methods have been developed for recognizing tight turns in protein structures, with most of the work focusing on β-turns, which occur most frequently in protein structures. Approximately one-quarter of all protein residues are in β-turns [62]. A β-turn is defined as four consecutive residues r_i, r_{i+1}, r_{i+2} and r_{i+3}, where the distance between the C_α of residue r_i and the C_α of residue r_{i+3} is < 7 Å, and the central two residues are not helical. These β-turns can be further assigned to one of several (6–10) classes on the basis of the backbone ϕ and ψ angles of residues r_{i+1} and r_{i+2} [120, 80, 103, 58, 25]. The first methods for predicting β-turns focused on identifying which residues take part in β-turns [79, 28], and later methods have additionally attempted to predict the type of β-turn [125]. Some β-turn types show preferences for particular topological environments; for example, type I' and type II' β-turns are preferentially found in β hairpins [112].

As with 3-state secondary structure prediction, methods to predict β-turns fall into two classes: probabilistic methods and machine-learning methods. The earliest probabilistic methods computed the probability that a certain amino acid a_i is located at the j-th position in a β-turn by dividing the number of times the amino acid a_i occurred in the j-th position of a turn by the total occurrences of amino acid a_i [79]. Assuming independence between positions, the probability that a certain 4-long window is an occurrence of a β-turn is calculated by the product of the appropriate four terms, and a cutoff for prediction is chosen. These predictions can be further refined so that a 4-long window that has helical or sheet propensity that is larger than its β-turn propensity is eliminated [28]; structural propensities are defined as in [27]. Modifications of this basic approach to predict turn types include [125, 126, 58].

Other probabilistic methods consider each possibility Ψ (where Ψ can be each type of β-turn as well as non-β-turns) in turn, and compute the probability of observing a particular 4-long window given that it is an instance of Ψ. In particular, given a subsequence $r_1 r_2 r_3 r_4$, it is scored by considered a random subsequence $R_1 R_2 R_3 R_4$ and computing

$$\Pr(R_1 = r_1, R_2 = r_2, R_3 = r_3, R_4 = r_4 | \Psi).$$

The possibility Ψ giving the largest value is taken as the prediction. Assuming that each position is independent of every other, this simplifies to

$$\prod_{i=1}^{i=4} \Pr(R_i = r_i | \Psi).$$

For each type of β-turn, probabilities are estimated from known structures for each of the four positions. Later models [131] consider the spatial arrangement of β-turns and assumed dependencies between the first and fourth position, and the second and third positions:

$$\Pr(R_1 = r_1 | \Psi) \Pr(R_2 = r_2 | \Psi) \Pr(R_3 = r_3 | R_2 = r_2, \Psi) \Pr(R_4 = r_4 | R_1 = r_1, \Psi).$$

Alternate models make the 1st order Markov assumption that all dependencies can be captured by considering adjacent residues [24, 26]:

$$\Pr(R_1 = r_1 | \Psi) \Pr(R_2 = r_2 | R_1 = r_1, \Psi) \Pr(R_3 = r_3 | R_2 = r_2, \Psi) \Pr(R_4 = r_4 | R_3 = r_3, \Psi).$$

The earliest neural network approaches [88] to β-turn prediction take as input a 4-long window of amino acids (each residue is represented with 20 bits), and include a hidden layer. There are four output nodes, two for the most common β-turn classes, one for all other β-turns, and one for non-β-turns. Later approaches subdivide the problem into first predicting whether a window contains a β-turn and then predicting the type of turn [111]. As in neural network based approaches to predicting secondary structure [101, 105], several layers of neural networks are used. In the first, a nine amino acid window is considered. Additionally, for each residue, secondary structure predictions (helix, sheet or other) are considered; inclusion of such predictions improves performance for both neural network [111, 68] and statistical approaches [66] for β-turn prediction. The output for adjacent residues using this neural network are fed into a second structure-to-structure network, along with secondary structure predictions. Predictions are also filtered via a rule-based system. Finally, all data identified by the turn/not-turn networks as possibly taking part in β-turns are input to networks for turn types, with only 4-long amino acid windows considered. When several turn types can be potentially predicted for a particular window, the one with the largest score is taken as the prediction. As with 3-state secondary structure prediction, further improvements in β-turn prediction have been obtained by using evolutionary information, where each sequence position is encoded using a profile describing its amino acid distribution in a MSA [68, 69]. More recently, nearest-neighbor [71] and SVMs [23] have also been applied to predict β-turns.

Predictions of β-turns are not as reliable as 3-state predictions of secondary structure. Approximately 50% of β-turns can be identified with 75% of the sequence fragments predicted as β-turn actually being correct. Overall accuracy of predictions is around 75%; a method that always predicts non-β-turns would have similar accuracy. Furthermore, predictions of β-turn types are only possible for the most frequent turn types.

More recently, attempts have been made to predict γ-turns and α-turns [67, 21, 22]. The computational techniques are very similar to the ones applied to β-turns. Perhaps due to the vastly fewer number of residues taking part in either γ- or α-turns, these methods have only had limited success.

29.4   Beta Hairpins

Beta hairpins are one of the simplest supersecondary structures and are widespread in globular proteins. They consist of short loop regions (or turns) between antiparallel hydrogen bonded β-strands. Typically, the length of these loop regions is eight residues or less, with two residue loops being most common [112, 76]. Correct identification of such structures can significantly reduce the number of possible folds consistent with a given protein, as differing tertiary folds contain different arrangements and numbers of β-strands. As noted in [35, 76], consecutive β-strands in a protein sequence can either form more "local" hairpin structures or "diverge" so that the β-strands may pair with other strands. Methods for predicting β hairpins have just begun to appear, and two recent approaches are based on neural networks.

In the first approach [35], β hairpins are identified by first predicting secondary structure. Each predicted β-coil-β pattern is further evaluated by comparing it to all known β hairpins of the same length. Each comparison between the pattern and a known β hairpin results in 14 scores. These scores are computed based on the compatibility of the predicted secondary structures and solvent accessibilities with those known for the hairpin, and additionally incorporate the segment's turn potential, secondary structure elements' lengths, putative pairwise residue interactions, and pairwise residue contacts. These scores are then fed into a neural network that is trained to discriminate between hairpins and non-hairpins. Finally, all the database matches for a particular β-coil-β segment are evaluated, and if there are more than 10 predictions of a hairpin structure, the segment is predicted as a hairpin.

The second approach [76] incorporates evolutionary information in predicting β hairpins. Homologs are obtained using PSI-BLAST [3], and each position is represented via the underlying profile (as in [60]). Two neural networks are trained, where the first predicts the state of the first residue in a turn, and the second predicts the state of the last residue of the turn. Each neural network predicts whether the residue being considered is the first (or last) residue of a hairpin, a diverging turn, or neither. To predict whether a residue is the start of turn, four residues before it and seven residues after it are considered. Similarly, to predict whether a residue is the end of turn, seven residues before it and four residues after it are considered. Thus, turns up to length eight are completely included in the input window. Each residue in the window is encoded using the appropriate column in the PSI-BLAST profile, as well as three additional parameters corresponding to secondary structure as predicted by [60]. Finally, the per-residue predictions are combined to determine the probability of a particular structure (hairpin turn, other turn, or no turn) starting at residue i and ending at residue j. The authors additionally show that incorporating predictions of hairpins or diverging turns improves their method [113] for tertiary structure prediction.

The performance of the two approaches is not directly comparable, as the first considers hairpins of all lengths, and the second limits itself to hairpins with turn regions of length at most eight. It is likely that longer-range interactions are more difficult to predict. Additionally, the two approaches use different PDB training and testing sets, and report different fractions of β-coil-β patterns that are hairpins (40% vs. 60%). The approach of [35] relies on the correct secondary structure prediction, and thus cannot predict β hairpins whose underlying secondary structure is not predicted correctly. Given an actual turn, the approach of [76] identifies whether it is hairpin or diverging with accuracy 75.9%; a baseline performance of 60% is possible by predicting all turns as hairpin.

29.5   Coiled Coils

The coiled coil is a ubiquitous protein structural motif that can mediate protein interactions. Roughly 5–7% of eukaryotic proteins contain coiled-coil regions. Coiled-coil structures are associated with several cellular functions, including transcription, oncogenesis, cell structure and membrane fusion. Coiled coils consist of two or more right-handed α-helices

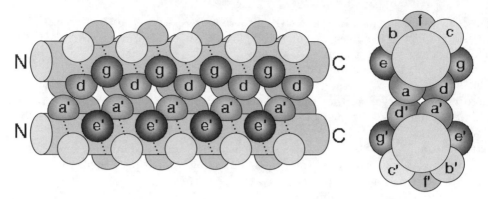

FIGURE 29.5: (a) Side view of a parallel 2-stranded coiled coil. (b) Top view of a parallel 2-stranded coiled coil. The interface between the α-helices in a coiled-coil structure is formed by residues at the core positions **a**, **d**, **e** and **g**. For notational convenience, positions in the two helices are distinguished by the prime notation (e.g., **a** and **a**$'$ are analogous positions in the two helices).

wrapped around each other with a slight left-handed superhelical twist. The helices in a coiled coil may associate with each other in a parallel or anti-parallel orientation, and the sequences making up the helices may either be the same (homo-oligomers) or different (hetero-oligomers). Helices taking part in coiled-coil structures exhibit a characteristic heptad repeat, denoted (**abcdefg**)$_{\mathbf{n}}$, spread out along two turns of the helix (see Figure 29.5). Residues at positions **a** and **d** tend to contain hydrophobic residues, and residues at positions **e** and **g** tend to contain charged or polar residues. The heptad repeat falls 20° short of two complete turns of a regular α-helix, and the supercoiling of the helices maintains that the **a** and **d** positions stay within the core of the structure. Coiled-coil helices pack with each other in a "knobs-into-hole" fashion [30], where a residue in the **a** (or **d**) position is a "knob" that packs into a hole created by four residues on the other α-helix.

Just as secondary structure assignment from known three-dimensional structures is not unambiguous (e.g., [32], and see above discussion), it is non-trivial to determine coiled coils in the set of solved structures. Different researchers may have different opinions on whether a particular structure is a coiled coil or not. The approach of [124] detects coiled coils by searching for knobs-into-holes packing. This approach identifies "true" coiled coils, as well as helical bundle domains where a subset of the helices interact with each other in a knobs-into-holes fashion.

Computational approaches have been developed both for identifying portions of protein sequences that can take part in coiled-coil structures, as well as for predicting specific interactions *between* coiled-coil proteins. While in principle it is possible to identify helices taking part in coiled coils by secondary structure prediction methods, in practice it is more effective to develop specialized methods for recognizing their hallmark heptad repeat. Most of the methods outlined below rely on having databases of known coiled-coil and non-coiled coil sequences. Non-coiled coil databases can be derived from the PDB by excluding potential coiled-coil proteins. Coiled-coil databases are built both from analyzing the PDB, and from including fibrous proteins whose X-ray diffraction patterns reveal coiled-coil structures but do not permit high-resolution structure determination (review, [29]).

29.5.1 Early approaches

The earliest approaches [98, 84] to recognize coiled coils use sequences of known coiled-coil proteins, and construct a 20×7 table tabulating the frequency with which each amino acid is found in each of the seven heptad repeat positions, normalized by the frequency of the amino acid in all protein sequences. These values are very similar to the propensity values computed by the Chou and Fasman approach [27]. For example, for leucine and position **a**, the corresponding entry in the table is the percentage of position **a** residues in the coiled coil database which are leucine, divided by the percentage of residues in all protein sequences that are leucine. For each amino acid in a protein sequence, this approach considers all l-long windows that contain it.[6] Each of the l windows is considered with its first amino acid starting in each of the seven possible heptad repeat positions, and the heptad repeat proceeding uninterrupted in the window. Thus, $7l$ windows are considered for each residue, and each window is scored by taking the product of the propensities for each amino acid (in the appropriate heptad repeat position) in the window. The score for each residue is then the maximum score for any of the windows containing it, and the score for the sequence is the maximum score of any of its residues. Scores are converted to probabilities by approximating both the background and coiled-coil score distributions with Gaussians, and assuming that 1 in 30 residues is in a coiled coil.

This method has also been extended to recognize the "leucine zipper" family of coiled coils found in bZIP transcription factors. The bZIPs are a large family of eukaryotic transcription factors (review, [57]), and their dimerization is mediated by the leucine zipper coiled-coil region. While the tendency is not uniformly true, leucine zippers tend to have leucines in the **d** position of the coiled coil. Early attempts to recognize leucine zippers focused on identifying leucine repeats, but since leucine is the most frequent amino acid, such patterns are frequently found by chance [17]. Both [54] and [12] find leucine zipper proteins by first identifying leucine repeats, and then requiring a coiled-coil prediction by [84]. [54] further uses both disallowed and highly preferred pairs of residues to identify leucine zipper coiled coils. The approach of [12] relaxes the requirement of a strict leucine repeat, and additionally focuses on identifying the short coiled-coil segments found in transcription factors.

29.5.2 Incorporating local dependencies

Subsequent approaches to predicting coiled-coil helices incorporate pairwise frequencies by explicitly considering the problem within a probabilistic framework [7, 10]. This overall framework for coiled-coil prediction is similar to the information theory approaches described above for secondary structure prediction [48]; however, the assumptions used in practice are very different. Here, the goal is to predict whether a subsequence $z = r_1, r_2, \ldots, r_l$ is a coiled coil by estimating $\Pr(z \in C)$, where C is the class of coiled coils [7]. If $X = R_1, R_2, \ldots, R_l$ is a random subsequence selected from the universe of all known protein

---

[6]A typical window length is 28 (four heptads), as it is thought that peptides that can form stable coiled coils in solution should be at least this length. Shorter windows can also be employed; typically, the discriminatory performance of methods deteriorate with shorter window sizes.

sequences, then

$$\begin{aligned}
\Pr(z \in C) &= \Pr(X \in C | X = z) \\
&= \frac{\Pr(X = z | X \in C) \Pr(X \in C)}{\Pr(X = z)} \\
&\propto \frac{\Pr(R_1 = r_1 \wedge \ldots \wedge R_l = r_l | X \in C)}{\Pr(R_1 = r_1 \wedge \ldots \wedge R_l = r_l)}
\end{aligned}$$

Using repeated applications of the definition of conditional probability, this is equal to:

$$\frac{\prod_{i=1}^{l-1} \Pr(R_i = r_i | R_{i+1} = r_{i+1} \wedge \ldots \wedge R_l = r_l \wedge X \in C) \cdot \Pr(R_l = r_l | X \in C)}{\prod_{i=1}^{l-1} \Pr(R_i = r_i | R_{i+1} = r_{i+1} \wedge \ldots \wedge R_l = r_l) \cdot \Pr(R_l = r_l)}. \quad (29.4)$$

To estimate these probabilities, it is necessary to make assumptions. For example, the simplest assumption is that the residues are independent of each other:

$$\Pr(R_i = r_i | R_{i+1} = r_{i+1} \wedge \ldots \wedge R_l = r_l \wedge X \in C) = \Pr(R_i = r_i | X \in C)$$

and

$$\Pr(R_i = r_i | R_{i+1} = r_{i+1} \wedge \ldots \wedge R_l = r_l) = \Pr(R_i = r_i).$$

Simplifying the previous equation with these assumptions gives

$$\Pr(z \in C) \quad \propto \quad \prod_{i=1}^{l} \frac{\Pr(R_i = r_i | X \in C)}{\Pr(R_i = r_i)},$$

and this is equivalent to the approach of [84].

In α-helices, a better assumption might be that a residue in position i is dependent on the next residue in the sequence $i + 1$, as well as on those in positions $i + 3$ and $i + 4$, both of which are on the same face of the helix as position i (see Figure 29.5). This gives the following assumption: $\Pr(R_i = r_i | R_{i+1} = r_{i+1} \wedge \ldots \wedge R_l = r_l \wedge X \in C) = \Pr(R_i = r_i | R_{i+1} = r_{i+1} \wedge R_{i+3} = r_{i+3} \wedge R_{i+4} = r_{i+4} \wedge X \in C)$. However, this would require that $7^4 20^4$ parameters be estimated, and is not feasible in practice. The approach suggested in [7] is to assume that $\Pr(R_i = r_i | R_{i+1} = r_{i+1} \wedge \ldots \wedge R_l = r_l \wedge X \in C)$ can be approximated by some function f (e.g., weighted average, minimum or maximum) over $\Pr(R_i = r_i | R_{i+1} = r_{i+1} \wedge X \in C)$, $\Pr(R_i = r_i | R_{i+3} = r_{i+3} \wedge X \in C)$ and $\Pr(R_i = r_i | R_{i+4} = r_{i+4} \wedge X \in C)$. More generally, if D is the set of dependencies (e.g., for helices, $D = \{1, 3, 4\}$ is the natural set of dependencies), then it is assumed that the probability of interest can be estimated as a function over the corresponding pairwise probabilities. In [10, 8, 115, 114], a geometric average over the pairwise probabilities is used.

The approach outlined above works well if the probabilities are estimated from a database representative of the types of coiled-coil structures that are to be predicted. However, the databases are heavily biased towards certain types of coiled coils. In [8], it is proposed that the basic method be used to iteratively scan a large database of sequences. Initially, the known database is used to estimate the required probabilities. Then, each sequence is scored using the framework described above, and this raw score is converted into a $(0, 1)$ probability p of its being a coiled coil. This probability is computed by fitting a Gaussian to the score distribution. In each iteration of the algorithm, a sequence is chosen with chance proportional to its probability of being coiled coil, and if chosen, its predicted coiled-coil residues will be used in the next iteration of the algorithm to update the probabilities. Single and pairwise frequencies are estimated in a Bayesian manner, with the initial estimates providing the prior. The iterative process continues until it stabilizes. This approach has been successful in identifying coiled-coil-like structures in histidine kinases [115] and viral membrane fusion proteins [114], with crystal structures confirming several novel predictions [133, 85].

Hidden Markov models (HMMs) have also been applied to coiled-coil recognition [9, 37]. (For a general introduction to HMMs, see [38].) These approaches do not require that a fixed-length window be used, and thus may better predict shorter coiled-coil segments. Additionally, for coiled coils longer than a particular window length, HMMs can incorporate longer-range information than window-based approaches. In theory, HMMs permit modeling of interruptions in the heptad repeat pattern; however, in practice, such interruptions are severely penalized.

One HMM approach [37] builds a model of 64 states. There is a background state 0 corresponding to residues that do not take part in coiled coils. The other 63 states are denoted by a group number 1–9 and by a letter that refers to the heptad position. The first four groups model the first four residues in a coiled-coil segment, and the last four groups model the last four residues in a coiled-coil segment. The fifth group models internal coiled-coil residues. Each state corresponding to the same heptad repeat position is given the same emission probabilities. For groups 1–4 and 6–9, transition probabilities are specified to go from group i to group $i + 1$, with deviations from the heptad repeat pattern given some very small (though non-zero) chance. For group 0, self-transitions are allowed, as well as transitions to states in group 1. For state 5, there are transitions between states within this group as well as to states in group 6; in both cases, strong preference is given to transitions maintaining the heptad repeat. For any sequence, the prediction of whether each residue is in a coiled coil or not is given by the most likely state sequence through the HMM, given the sequence.

29.5.3 Predicting oligomerization

Natural coiled coils are known to exist as dimers, trimers, tetramers and pentamers. Attempts to predict oligomeric states of coiled-coil sequences have focused on differentiating between dimeric or trimeric coiled coils. In [128], amino acid frequencies at each heptad repeat position are computed for both dimeric and trimeric coiled coils, and normalized by the frequencies expected by chance. These give dimeric and trimeric propensities for each amino acid/heptad repeat pair. Each coiled-coil segment is then scored by summing the logs of the single frequency dimeric (and trimeric) propensities. Finally, the segment is predicted as dimeric if its dimeric propensity is higher than its trimeric one, and trimeric otherwise.

An alternate approach exploits pairwise residue correlations [127] in predicting oligomerization state. This is a multidimensional scoring approach that uses the framework of [10]. Probabilities are estimated from a dimeric coiled-coil database, and then for $1 \leq d \leq 7$, each subsequence is scored assuming that dependencies exist between residues i and $i + d$. The analogous scores are computed using a trimeric database as well. Finally, a multidimensional score $\vec{s}$ for a subsequence z is converted to a probability of its being a dimeric coiled coil by computing:

$$\frac{\Pr(\vec{s}|z \text{ is dimeric}) \cdot \Pr(\text{dimeric coiled coil})}{\Pr(\vec{s})}.$$

These probabilities are estimated by fitting multivariate Gaussians to the distributions of scores for dimeric coiled coils, trimeric coiled coils and non-coiled coils, and assuming a prior probability of dimeric, trimeric and non-coiled-coil residues. Trimer probabilities are computed similarly.

29.5.4 Structure-based predictions

The approaches outlined above have focused on statistical methods for predicting whether a given sequence takes part in a coiled-coil structure. There has also been work on predicting the high-resolution atomic structures of model coiled-coil systems using molecular mechanics. The earliest such attempts include [132, 121, 36]. In [121], a hierarchical procedure is described to predict the structure of the GCN4 leucine zipper; a backbone root-mean-squared deviation (RMSD) of 0.81 Å is obtained when predicting the dimeric GCN4 leucine zipper. In [36], dimeric and tetrameric variants of GCN4 are considered, and an RMSD of 0.73 Å is obtained for residues in the dimerization interface.

The coiled-coil backbone can be parameterized [30], and [52] show how to exploit this parameterization in order to incorporate backbone flexibility in predicting structures. Coiled-coil backbones can be described by specifying the superhelical radius R_0, the superhelix frequency ω_0, the α-helical radius R_1, the helical frequency ω_1, and the rise per amino acid in the α-helix d. The heptad repeat fixes ω_1 to be $4\pi/7$ radians per amino acid, so that seven residues complete two full turns relative to the superhelical axis, and place every seventh residue in the same local environment. Additionally, it may be assumed that the helices making up the coiled-coil are regular and symmetric, and so d can be fixed to be the rise per amino acid of a regular α-helix (1.52 Å) and R_1 can be fixed to be the C_α radius of a regular α-helix (2.26 Å). The remaining parameters can then be varied to enumerate backbone conformations. Side chains are then positioned on these backbones via energy minimization. This approach has resulted in predictions with root-mean-square deviation from crystal structures of less than 0.6 Å when considering hydrophobic **a** and **d** position residues for three GCN4 variants (2-stranded, 3-stranded and 4-stranded). Additionally, a novel coiled-coil backbone consisting of a *right-handed* superhelical twist and an 11-mer repeat has been designed using the parameterized-backbone approach [51]. A parameterized-backbone approach has also been used to predict the hydrophobic dimerization interface of six designed heterodimeric coiled coils [70], as well as to predict the differences in stabilities of these constructs. In this approach, for each backbone, all near-optimal packings of side chains are identified, and these structures are then relaxed via energy minimization [18] to find the minimum energy backbone and side-chain conformations.

29.5.5 Predicting coiled-coil protein interactions

As outlined above, effective sequence-based prediction methods exist for recognizing single helices that take part in coiled coils. Since coiled coils are made up of two or more helices that interact with each other, a natural next step in predicting their structures is to try to predict which helices are interacting with each other. Since these helices may be in different protein sequences, this begins to address the problem of predicting protein-protein interactions. This is an important problem as protein-protein interactions play a central role in many cellular functions. Furthermore, the difficulty of computationally predicting protein structures suggests a strategy of concentrating first on interactions mediated by specific interfaces of known geometry.

Early approaches towards predicting coiled-coil interaction specificity have counted the number of favorable and unfavorable electrostatic interactions to make some specific predictions about the nature of particular coiled-coil protein-protein interactions [97, 89, 122]; however, it is known that many other factors play a role in coiled-coil specificity (e.g., [96, 83, 53]) and thus such simple approaches are limited in their applicability.

An alternative approach represents coiled coils in terms of their interhelical residue interactions and derives a "weight" that indicates how favorable each residue-residue interaction is [116, 43]. Unlike the other sequence-based approaches outlined in this chapter, this approach uses not only sequence and structural data, but also experimental data. This use of experimental data is critical to its performance. The approach has thus far been applied only to predicting partners for helices taking part in dimeric coiled coils. In dimeric coiled coils, residues at the **a, d, e**, and **g** positions form the protein-protein interface [95, 49] (see Figure 29.5). Experimental studies show that specificity is largely driven by interactions between residues at these core positions (e.g., see [123]). The method further assumes that considering interhelical interactions among these residues in a pairwise manner is sufficient.[7] Based on structural features of the interhelical interface [95, 49] as well as experiments on determinants of specificity (e.g., [96, 83, 122]), the following seven interhelical interactions are assumed to govern partnering in coiled coils:

$$\mathbf{a}_i\mathbf{d}'_i, \ \mathbf{d}_i\mathbf{a}'_{i+1}, \ \mathbf{d}_i\mathbf{e}'_i, \ \mathbf{g}_i\mathbf{a}'_{i+1}, \ \mathbf{g}_i\mathbf{e}'_{i+1}, \ \mathbf{a}_i\mathbf{a}'_i, \ \mathbf{d}_i\mathbf{d}'_i. \tag{29.5}$$

The prime differentiates the two strands and the subscript denotes the relative heptad number (e.g., the first interaction, $\mathbf{a}_i\mathbf{d}'_i$, is between the **a** position in the i-th heptad of one helix and the **d** position in the same heptad of the other helix).

Consequently, each coiled-coil structure is represented as a 2800-dimensional vector $\vec{x}$, the entries of which tabulate the occurrences of amino-acid pairs in the above interactions. Specifically, entry $x_{(p,q),i,j}$ indicates the number of times amino acids i and j appear across the helical interface in positions p and q, respectively.

Scoring framework. For each possible interhelical interaction, the method needs a weight $w_{(p,q),i,j}$ that denotes how favorable the interaction is between amino acid i in position p and amino acid j in position q. A potential coiled coil represented by $\vec{x}$ is then scored by computing $\vec{w} \cdot \vec{x}$ where $\vec{w}$ is a vector of such weights. Initially this weight vector $\vec{w}$ is not known; however, these weights should satisfy certain constraints.

Experimental information on relative coiled-coil stability (e.g, the observation that coiled coil $\vec{x}$ is more stable than coiled coil $\vec{y}$) is used to constrain the weight vector $\vec{w}$ by requiring that

$$\vec{w} \cdot \vec{x} \ > \ \vec{w} \cdot \vec{y}. \tag{29.6}$$

Additionally, sequences known to form coiled coils should score higher than those that do not:

$$\vec{w} \cdot \vec{x} \ > \ 0, \text{ for all coiled coils } \vec{x}, \tag{29.7}$$
$$\vec{w} \cdot \vec{y} \ < \ 0, \text{ for all non-coiled coils } \vec{y}. \tag{29.8}$$

These constraints are similar to those seen most often in machine learning settings.

Finally, knowledge about specific weight elements can be directly incorporated. For example, say it is favorable to have a lysine in a **g** position in one helix with a glutamic acid in

---

[7]It is possible to consider three or more amino acids at a time but this would require a larger coiled-coil database.

the following position **e** in the other helix, but not favorable to have glutamic acid in both these positions (i.e., **g-e** K E is "better than" **g-e** E E). Then the following should be true:

$$w_{(g,e),K,E} > 0, \quad w_{(g,e),E,E} < 0. \tag{29.9}$$

Indexing each constraint with i, the above constraints (equations 29.6–29.9) can be rewritten using vectors $\vec{z}^{(i)}$, such that $\vec{w}$ is constrained to satisfy $\vec{w} \cdot \vec{z}^{(i)} > 0$. Including non-negative slack variables ϵ_i to allow for errors in sequence or experimental data, each constraint can then be relaxed as $\vec{w} \cdot \vec{z}^{(i)} \geq -\epsilon_i$. The goal is to find $\vec{w}$ and $\vec{\epsilon}$ such that each constraint is satisfied and $\sum \epsilon_i$ is minimized. Trade-offs between training and generalization error suggest the approach of support vector machines (SVMs) [119, 20], in which the following quadratic objective function (for some constant C) is minimized, subject to a variation of the previously described set of linear constraints:

$$\frac{1}{2} \parallel \vec{w} \parallel^2 + C(\sum \epsilon_i)$$

subject to

$$\vec{w} \cdot \vec{z}^{(i)} \geq 1 - \epsilon_i \quad \forall i$$
$$\epsilon_i \geq 0 \quad \forall i$$

Differences between this approach and the traditional application of SVMs include constraints on specific elements of the weight vector, and constraints about the relative "score" of different interactions.

This approach has been tested on a near-complete set of coiled-coil interactions among human and yeast leucine zipper bZIP transcription factors [93], and identifies 70% of strong interactions while maintaining that 92% of predictions are correct [43]. Though genomic approaches to predicting protein partners have had some success (e.g., [34, 86, 40, 50, 102, 59, 129]), as have structure-based threading methods [1, 82], the coiled coil is the first interaction interface for which these types of high-confidence, large-scale computational predictions can be made.

29.5.6 Promising future directions

Since secondary structure prediction methods improved considerably by incorporating evolutionary information, the next obvious step in improving recognition of helices taking part in coiled coil structures is to use homologous sequences. For predicting coiled-coil interactions, however, homologous sequences can show very different interaction specificity [93], and thus it is not obvious how to exploit evolutionary information in this context. Additionally, while methods have been developed for predicting whether a coiled coil helix is likely to take part in either a dimeric and trimeric structure [128, 127], there are no methods for predicting higher-order oligomerization states or for predicting whether the helices interact in a parallel or anti-parallel manner. Finally, methods for predicting coiled-coil protein interactions have focused on parallel, 2-stranded coiled coils, and novel approaches are needed for predicting coiled-coil protein interactions more generally.

29.6   Conclusions

In this chapter, we have reviewed the basic computational methods used to predict protein secondary structure, as well as β hairpin and coiled coil supersecondary structures.

Of these problems, secondary structure prediction has been the most widely studied, and almost all successful methods for predicting tertiary structure rely on predictions of secondary structure (e.g.,see [2]). As methods for predicting other types of local structure improve, they are likely to play an increasing role in tertiary structure prediction methods. More recently, effective methods for predicting other types of β-structures, including β-helices [15]and β-trefoils [90], have also been developed, and these types of specialized computational approaches provide a new means for predicting protein tertiary structure. Finally, protein interactions are also mediated by various well-characterized structural motifs (e.g., see [99]), and as demonstrated with the coiled coil, a promising approach for making high-confidence predictions of protein interactions and quartenary structure is to focus first on interactions mediated by specific, local structural interfaces.

Acknowledgments

The author thanks Carl Kingsford, Elena Nabieva and Elena Zaslavsky for helpful discussions, and the NSF for PECASE award MCB–0093399.

References

[1] P. Aloy and R. Russell. Interrogating protein interaction networks through structural biology. *Proceedings of the National Academy of Sciences*, 99:5896+, 2002.

[2] P. Aloy, A. Stark, C. Hadley, and R. Russell. Prediction without templates: New folds, secondary structure, and contacts in CASP5. *Proteins: Structure, Function and Bioinformatics*, 53:436+, 2003.

[3] S. Altschul, T. Madden, A. Schaffer, and J. Zhang *et al.* Gapped BLAST and PSI-BLAST: a new generation of protein database search programs. *Nucleic Acids Research*, 25:3389+, 1997.

[4] C. Anfinsen. Principles that govern the folding of protein chains. *Science*, 181:223+, 1973.

[5] C. Anfinsen, E. Haber, M. Sela, and F. White. The kinetics of formation of native ribonuclease during oxidation of the reduced polypeptide chain. *Proceedings of the National Academy of Sciences (USA)*, 47:1309+, 1961.

[6] P. Baldi, S. Brunak, P. Frasconi, and G. Soda *et al.* Exploiting the past and present in secondary structure prediction. *Bioinformatics*, 15:937+, 1999.

[7] B. Berger. Algorithms for protein structural motif recognition. *Journal of Computational Biology*, 2:125+, 1995.

[8] B. Berger and M. Singh. An iterative method for improved protein structural motif recognition. *Journal of Computational Biology*, 4(3):261+, 1997.

[9] B. Berger and D. Wilson. Improved algorithms for protein motif recognition. In *Symposium on Discrete Algorithms*, pages 58+. SIAM, January 1995.

[10] B. Berger, D. B. Wilson, E. Wolf, and T. Tonchev *et al.* Predicting coiled coils using pairwise residue correlations. *Proceedings of the National Academy of Sciences*, 92:8259+, 1995.

[11] H.M. Berman, J. Westbrook, Z. Feng, and G. Gilliland *et al.* The Protein Data Bank. *Nucleic Acids Research*, 28:235+, 2000.

[12] E. Bornberg-Bauer, E. Rivals, and M. Vingron. Computational approaches to identify leucine zippers. *Nucleic Acids Research*, 26:2740+, 1998.

[13] J. Bowie, R. Luthy, and D. Eisenberg. A method to identify protein sequences that fold into a known three-diensional structure. *Science*, 253:164+, 1991.

[14] P. Bradley, D. Chivian, J. Meiler, and K. Misura *et al.* Rosetta predictions in CASP5: Successes, failures, and prospects for complete automation. *Proteins: Structure, Function and Genetics*, 53:457+, 2003.

[15] P. Bradley, L. Cowen, M. Menke, and J. King *et al.* BETAWRAP: Successful prediction of parallel β-helices from primary sequence reveals an association with many microbial pathogens. *Proceedings of the National Academy of Sciences*, 98:14819+, 2001.

[16] C. Branden and J. Tooze. *Introduction to Protein Structure*. Garland Publishing, Inc., 1999.

[17] V. Brendel and S. Karlin. Too many leucine zippers? *Nature*, 341:574+, 1989.

[18] B. Brooks, R. Bruccoleri, B. Olafson, and D. States *et al.* CHARMM: A program for macromolecular energy, minimization, and dynamics calculations. *Journal of Computational Chemistry*, 4:187–217, 1983.

[19] S. H. Bryant and C. E. Lawrence. An empirical energy function for threading protein sequence through the folding motif. *Proteins: Structure, Function and Genetics*, 16:92–112, 1993.

[20] C. Burges. A tutorial on support vector machines for pattern recognition. *Data Mining and Knowledge Discovery*, 2(2):121+, 1998.

[21] Y.-D. Cai and K.-C. Chou. Artificial neural network model for predicting α-turn types. *Analytical Biochemistry*, 268:407+, 1999.

[22] Y.-D. Cai, K.-Y. Feng, Y.-X. Li, and K. C. Chou. Support vector machine for predicting α-turn types. *Peptides*, 24:629+, 2003.

[23] Y.-D. Cai, X.-J. Liu, Y.-X. Li, and X.-B. Xu *et al.* Prediction of β turns with learning machines. *Peptides*, 24:665+, 2003.

[24] K.-C. Chou. Prediction of β-turns. *Journal of Peptide Research*, 49:120+, 1997.

[25] K.-C. Chou. Prediction of tight turns and their types in proteins. *Analytical Biochemistry*, 286:1+, 2000.

[26] K.-C. Chou and J. Blinn. Classification and prediction of β-turn types. *Journal of Protein Chemistry*, 16:575+, 1997.

[27] P. Chou and G. Fasman. Prediction of protein conformation. *Biopolymers*, 13:211+, 1974.

[28] P. Chou and G. Fasman. Prediction of β-turns. *Biophysical Journal*, 26:367+, 1979.

[29] C. Cohen. Why fibrous proteins are romantic. *Journal of Structural Biology*, 112:3+, 1998.

[30] F. H. C. Crick. The packing of α-helices: simple coiled coils. *Acta Crystallographica*, 6:689, 1953.

[31] J. Cuff and G. Barton. Application of multiple sequence alignment profiles to improve protein secondary structure prediction. *Proteins: Structure, Function and Genetics*, 40:502+, 1999.

[32] J. Cuff and G. Barton. Evaluation and improvement of multiple sequence methods for protein secondary structure prediction. *Proteins: Structure, Function and Genetics*, 34:508+, 1999.

[33] G. Cybenko. Approximation by superpositions of a sigmoidal function. *Mathematics of Control, Signals and Systems*, 2:303+, 1989.

[34] T. Dandekar, B. Snel, M. Huynen, and P. Bork. Conservation of gene order: a fingerprint of proteins that physically interact. *Trends in Biochemical Sciences*, 23(9):324+, 1998.

[35] X. de la Cruz, E. Hutchinson, A. Shepherd, and J. Thornton. Toward predicting

protein topology: an approach to identifying β hairpins. *Proceedings of the National Academy of Sciences*, 99:11157+, 2002.

[36] W. DeLano and A. Brunger. Helix packing in proteins: prediction and energetic analysis of dimeric, trimeric, and tetrameric GCN4 coiled coil structures. *Proteins: Structure, Function and Genetics*, 20:105+, 1994.

[37] M. Delorenzi and T. Speed. An HMM model for coiled-coil domains and a comparison with pssm-based predictions. *Bioinformatics*, 18:617+, 2002.

[38] R. Durbin, S. Eddy, A. Krogh, and G. Mitchison. *Biological sequence analysis: probabilistic models of proteins and nucleic acids*. Cambridge University Press, 2000.

[39] D. Eisenberg, R. Weiss, and T. Terwilliger. The hydrophobic moment detects periodicity in protein hydrophobicity. *Proceedings of the National Academy of Sciences (USA)*, 81:140+, 1984.

[40] A.J. Enright, I. Iliopoulos, N.C. Kyrpides, and C.A. Ouzounis. Protein interaction maps for complete genomes based on gene fusion events. *Nature*, 402:86+, 1999.

[41] V. Eyrich, M. Marti-Renom, D. Przybylski, and M. Madhusudhan *et al.* EVA: continuous automatic evaluation of protein structure prediction servers. *Bioinformatics*, 17:1242+, 2001.

[42] R. Fano. *Transmission of Information*. Wiley, New York, 1961.

[43] J. Fong, A.E. Keating, and M. Singh. Predicting specificity in bZIP coiled-coil protein interactions. *Genome Biology*, 5(2):R11, 2004.

[44] D. Frishman and P. Argos. Knowledge-based secondary structure assignment. *Proteins: Structure, Function and Genetics*, 23:566+, 1995.

[45] D. Frishman and P. Argos. Incorporation of non-local interactions in protein secondary structure prediction from the amino acid sequence. *Protein Engineering*, 9:133+, 1996.

[46] D. Frishman and P. Argos. Seventy-five percent accuracy in protein secondary structure prediction. *Proteins: Structure, Function and Genetics*, 27:329+, 1997.

[47] J. Garnier, D. Osguthorpe, and B. Robson. Analysis and implications of simple methods for predicting the secondary structure of globular proteins. *Journal of Molecular Biology*, 120:97+, 1978.

[48] J. Gibrat, J. Garnier, and B. Robson. Further developments of protein secondary structure prediction using information theory: new parameters and consideration of residue pairs. *Journal of Molecular Biology*, 198:425+, 1987.

[49] J. Glover and S. Harrison. Crystal structure of the heterodimeric bZIP transcription factor c-Fos-c-Jun bound to DNA. *Nature*, 373:257+, 1995.

[50] C. Goh, A. Bogan, M. Joachimiak, and D. Walther *et al.* Co-evolution of proteins with their interaction partners. *J. Mol. Biol*, 299:283+, 2000.

[51] P. B. Harbury, J. J. Plecs, B. Tidor, and T. Alber *et al.* High-resolution protein design with backbone freedom. *Science*, 282:1462+, 1998.

[52] P. B. Harbury, B. Tidor, and P. S. Kim. Predicting protein cores with backbone freedom: Structure prediction for coiled coils. *Proceedings of the National Academy of Sciences*, 92:8408+, 1995.

[53] P. B. Harbury, T. Zhang, P. S. Kim, and T. Alber. A switch between two-, three- and four-stranded coiled coils in GCN4 leucine zipper mutants. *Science*, 262:1401+, November 1993.

[54] J. Hirst, M. Vieth, J. Skolnick, and C. Brooks. Predicting leucine zipper structures from sequence. *Protein Engineering*, 9:657+, 1996.

[55] L. H. Holley and M. Karplus. Protein secondary structure prediction with a neural net. *Proceedings of the National Academy of Sciences (USA)*, 86:152+, 1989.

[56] S. Hua and Z. Sun. A novel method of protein secondary structure prediction with high segment overlap measure: Support vector machine approach. *Journal of Molecular Biology*, 308:397+, 2001.

[57] H. Hurst. Transcription factors 1: bZIP proteins. *Protein Profile*, 2(2):101+, 1995.

[58] E.G. Hutchinson and J. Thornton. A revised set of potentials for β-turn formation in proteins. *Protein Science*, 3:2207+, 1994.

[59] R.H. Jansen, H. Yu, D. Greenbaum, and Y. Kluger *et al.* A Bayesian networks approach for predicting protein-protein interactions from genomic data. *Science*, 302:449+, 2003.

[60] D. Jones. Protein secondary structure prediction based on position-specific scoring matrices. *Journal of Molecular Biology*, 292:195+, 1999.

[61] D. Jones, W. Taylor, and J. Thornton. A new approach to protein fold recognition. *Nature*, 358:86–89, 1992.

[62] W. Kabsch and C. Sander. A dictionary of protein secondary structure. *Biopolymers*, 22:2577+, 1983.

[63] W. Kabsch and C. Sander. How good are predictions of protein secondary structure? *FEBS Lett.*, 155:179+, 1983.

[64] K. Karplus, C. Barret, and R. Hughey. Hidden Markov models for detecting remote protein homologies. *Bioinformatics*, 14:846+, 1998.

[65] K. Karplus, R. Karchin, J. Draper, and J. Casper *et al.* Combining local-structure, fold-recognition, and new fold methods for protein structure prediction. *Proteins: Structure, Function and Genetics*, 53:491+, 2003.

[66] H. Kaur and G. Raghava. An evaluation of β-turn prediction methods. *Bioinformatics*, 18:1508+, 2002.

[67] H. Kaur and G. Raghava. A neural-network based method for prediction of γ-turns in proteins from multiple sequence aligments. *Protein Science*, 12:923+, 2003.

[68] H. Kaur and G. Raghava. Prediction of β-turns in proteins from multiple aligment using neural network. *Protein Science*, 12:627+, 2003.

[69] H. Kaur and G. Raghava. A neural network method for prediction of β-turn types in proteins using evolutionary information. *Bioinformatics*, 20:2751+, 2004.

[70] A.E. Keating, V. Malashkevich, B. Tidor, and P. S. Kim. Side-chain repacking calculations for predicting structures and stabilities of heterodimeric coiled coils. *Proceedings of the National Academy of Sciences*, 98(26):14825+, 2001.

[71] S. Kim. Protein β-turn prediction using nearest-neighbor method. *Bioinformatics*, 20:40+, 2004.

[72] L. Kinch, J. Wrabl, S. Krishna, and I. Majmudar *et al.* CASP5 assessment of fold recognition target predictions. *Proteins: Structure, Function and Bioinformatics*, 53:395+, 2003.

[73] R. King, M. Ouali, A. Strong, and A. Aly *et al.* Is it better to combine predictions? *Protein Engineering*, 13:15+, 2000.

[74] R. King and M. Sternberg. Identification and application of the concepts important for accuracte and reliable protein secondary structure prediction. *Protein Science*, 5:2298+, 1996.

[75] A. Kloczkowski, K.-L. Ting, R. Jernigan, and J. Garnier. Combining the GOR V algorithm with evolutionary information for protein secondary structure prediction from amino acid sequence. *Journal of Molecular Biology*, 49:154+, 2002.

[76] M. Kuhn, J. Meiler, and D. Baker. Strand-loop-strand motifs: prediction of hairpins and diverging turns in proteins. *Proteins: Structure, Function and Bioinformatics*, 54:282+, 2004.

[77] A. Lesk. *Introduction to protein architecture.* Oxford University Press, 2001.

[78] J. Levin, B. Robson, and J. Garnier. An algorithm for secondary structure determination in proteins based on sequence similarity. *FEBS Letters*, 205(2):303+, 1986.

[79] P. Lewis, F. Momany, and H. Scheraga. Folding of polypeptide chains in proteins: a proposed mechanism for folding. *Proceedings of the National Academy of Sciences*, 68:2293+, 1971.

[80] P. Lewis, F. Momany, and H. Scheraga. Chain reversals in proteins. *Biochimica et Biophysica Acta*, 303:211+, 1973.

[81] T. Li and E. Lander. Protein secondary structure prediction using nearest-neighbor methods. *Journal of Molecular Biology*, 232:1117+, 1993.

[82] L. Lu, H. Lu, and J. Skolnick. Multiprospector: an algorithm for the prediction of protein-protein interactions by multimeric threading. *Proteins*, 49(3):1895+, 2002.

[83] K. Lumb and P. S. Kim. A buried polar interaction imparts structural uniqueness in a designed heterodimeric coiled coil. *Biochemistry*, 34:8642+, 1995.

[84] A. Lupas, M. van Dyke, and J. Stock. Predicting coiled coils from protein sequences. *Science*, 252:1162+, 1991.

[85] V. Malashkevich, M. Singh, and P. S. Kim. The trimer-of-hairpins motif in viral membrane-fusion proteins: Visna virus. *Proceedings of the National Academy of Sciences*, 98:8502+, 2001.

[86] E. Marcotte, M. Pellegrini, H. Ng, and D. Rice *et al.* Detecting protein function and protein-protein interactions from genome sequences. *Science*, 285:751+, 1999.

[87] B. Matthews. Comparison of the predicted and observed secondary structure of T4 phage lysozyme. *Biochimica et Biophysica Acta*, 405:442+, 1975.

[88] M. McGregor, T. Flores, and M. Sternberg. Prediction of β-turns in proteins using neural networks. *Protein Engineering*, 2:521+, 1989.

[89] A. McLachlan and M. Stewart. Tropomyosin coiled-coil interactions: Evidence for an unstaggered structure. *Journal of Molecular Biology*, 98:293+, 1975.

[90] M. Menke, E. Scanlon, J. King, B. Berger, and L. Cowen. Wrap-and-pack: A new paradigm for beta structural motif recognition with application to recognizing beta trefoils. In *Proceedings of the 8th Annual International Conference on Computational Molecular Biology*, pages 298+. ACM, 2004.

[91] D. Minor and P. S. Kim. Context-dependent secondary structure formation of a designed protein sequence. *Nature*, 380(6576):730+, 1996.

[92] J. Moult, K. Fidelis, A. Zemla, and T. Hubbard. Critical assessment of methods of protein structure prediction (CASP)-round V. *Proteins: Structure, Function, and Genetics*, 53:334+, 2003.

[93] J. R. S. Newman and A. E. Keating. Comprehensive identification of human bZIP interactions using coiled-coil arrays. *Science*, 300:2097+, 2003.

[94] K. Nishikawa and T. Ooi. Amino acid sequence homology applied to prediction of protein secondary structure, and joint prediction with existing methods. *Biochimica et Biophysica Acta*, 871(1):45+, 1986.

[95] E. O'Shea, J. Klemm, P. S. Kim, and T. Alber. X-ray structure of the GCN4 leucine zipper, a two-stranded, parallel coiled coil. *Science*, 254:539+, October 1991.

[96] E. O'Shea, R. Rutkowski, and P. S. Kim. Mechanism of specificity in the fos-jun oncoprotein heterodimer. *Cell*, 68:699+, 1992.

[97] D. A. D. Parry. Sequences of α-keratin: Structural implication of the amino acid sequences of the type I and type II chain segments. *Journal of Molecular Biology*, 113:449+, 1977.

[98] D. A. D. Parry. Coiled coils in alpha-helix-containing proteins: analysis of residue types within the heptad repeat and the use of these data in the prediction of coiled-coils in other proteins. *Bioscience Reports*, 2:1017+, 1982.

[99] T. Pawson, M. Raina, and P. Nash. Interaction domains: from simple binding events to complex cellular behavior. *FEBS Letters*, pages 2+, 2002.

[100] G. Pollastri, D. Przybylski, B. Rost, and P. Baldi. Improving the prediction of protein secondary structure in three and eight classes using recurrent neural networks. *Proteins: Structure, Function and Genetics*, 47:228+, 2002.

[101] N. Qian and T. Sejnowski. Predicting the secondary structure of globular proteins using neural network models. *Journal of Molecular Biology*, 202(4):865+, 1988.

[102] A. Ramani and E. Marcotte. Exploiting the co-evolution of interacting proteins to discover interaction specificity. *Journal of Molecular Biology*, 327:273+, 2003.

[103] J. Richardson. The anatomy and taxonomy of protein structure. *Advances in Protein Chemistry*, 34:167+, 1981.

[104] B. Rost. Protein secondary structure prediction continues to rise. *Journal of Structural Biology*, 134:204+, 2001.

[105] B. Rost and C. Sander. Prediction of protein secondary structure at better than 70%. *Journal of Molecular Biology*, 232:584+, 1993.

[106] B. Rost, C. Sander, and R. Schneider. PhD: an automatic mail server for protein secondary structure prediction. *Computer Applications in Biosciences*, 10:53+, 1994.

[107] B. Rost, C. Sander, and R. Schneider. Redefining the goals of protein secondary structure prediction. *Journal of Molecular Biology*, 235:13+, 1994.

[108] D. Rumelhart, G. Hinton, and R. Williams. Learning internal representations by error propagation. In D. Rumelhart and J. McClelland, editors, *Parallel Distributed Processing: Explorations in the Microstructure of Cognition*, volume 323, pages 318–362. MIT Press, Cambridge, MA, 1986.

[109] D. Rumelhart, G. Hinton, and R. Williams. Learning representations by back-propagating errors. *Nature*, 323:533+, 1986.

[110] A. Salamov and V. Solovyev. Prediction of protein secondary structure by combining nearest-neighbor algorithms and multiple sequence alignment. *Journal of Molecular Biology*, 247:11+, 1995.

[111] A. Shepherd, D. Gorse, and J. Thornton. Prediction of the location and type of β-turns in proteins using neural networks. *Protein Science*, 8:1045+, 1999.

[112] B. Sibanda and J. Thornton. Beta-hairpin families in globular proteins. *Nature*, 316:170+, 1985.

[113] K. Simons, C. Kooperberg, E. Huang, and D. Baker. Assembly of protein tertiary structures from fragments with similar local sequences using simulated annealing and bayesian scoring functions. *Journal of Molecular Biology*, 268:209+, 1997.

[114] M. Singh, B. Berger, and P. S. Kim. Learncoil-VMF: Computational evidence for coiled-coil-like motifs in many viral membrane-fusion proteins. *Journal of Molecular Biology*, 290:1031+, 1999.

[115] M. Singh, B. Berger, P. S. Kim, and J. Berger *et al.* Computational learning reveals coiled coil-like motifs in histidine kinase linker domains. *Proceedings of the National Academy of Sciences*, 95:2738+, March 1998.

[116] M. Singh and P. S. Kim. Towards predicting coiled-coil protein interactions. In *Proceedings of the 5th Annual International Conference on Computational Molecular Biology*, pages 279+. ACM, 2001.

[117] M. J. Sippl. Calculation of conformational ensembles from potentials of mean force. *Journal of Molecular Biology*, 213:859–883, 1990.

[118] A. Tramontano and V. Morea. Assessment of homology-based predictions in CASP5. *Proteins: Structure, Function, and Genetics*, 53:352+, 2003.

[119] V. Vapnik. *Statistical Learning Theory*. Wiley, 1998.

[120] C. Venkatachalam. Stereochemical criteria for polypeptides and proteins. V. Conformation of a system of three linked peptide units. *Biopolymers*, 6:1425+, 1968.

[121] M. Vieth, A. Kolinski, C. L. Brooks, and J. Skolnick. Prediction of the folding pathways and structure of the GCN4 leucine zipper. *Journal of Molecular Biology*, 237:361+, 1994.

[122] C. Vinson, T. Hai, and S. Boyd. Dimerization specificy of the leucine zipper-containing bZIP motif on DNA binding: prediction and rational design. *Genes and Development*, 7(6):1047+, 1993.

[123] C. Vinson, M. Myakishev, A. Acharya, and A. Mir *et al.* Classification of human bZIP proteins based on dimerization properties. *Molecular and Cellular Biology*, 22(18):6321–6335, 2002.

[124] J. Walshaw and D. Woolfson. Socket: a program for identifying and analysing coiled-coil motifs within protein structures. *Journal of Molecular Biology*, 307:1427+, 2001.

[125] C. Wilmot and J. Thornton. Analysis and prediction of the different types of β-turn in proteins. *Journal of Molecular Biology*, 203:221+, 1988.

[126] C. Wilmot and J. Thornton. β-turns and their distortions: a proposed new nomenclature. *Protein Engineering*, 3:479+, 1990.

[127] E. Wolf, P. S. Kim, and B. Berger. Multicoil: A program for predicting two- and three-stranded coiled coils. *Protein Sci.*, 6:1179+, 1997.

[128] D. Woolfson and T. Alber. Predicting oligomerization states of coiled coils. *Protein Science*, 4:1596–1607, 1995.

[129] H. Yu, N. Luscombe, H. Lu, and X. Zhu *et al.* Annotation transfer between genomes: protein-protein interologs and protein-DNA regulogs. *Genome Research*, 14:1107+, 2004.

[130] A. Zemla, C. Venclovas, K. Fidelis, and B. Rost. A modified definition of Sov, a segment-based measure for protein structure prediction assessment. *Proteins: Structure, Function and Genetics*, 34:220+, 1999.

[131] C.-T. Zhang and K. C. Chou. Prediction of β-turns in proteins by 1–4 and 2–3 correlation model. *Biopolymers*, 41:673+, 1997.

[132] L. Zhang and J. Hermans. Molecular dynamics study of structure and stability of a model coiled coil. *Proteins: Structure, Function and Genetics*, 16:384+, 1993.

[133] X. Zhao, M. Singh, V. Malashkevich, and P. S. Kim. Structural characterization of the human respiratory syncytial virus fusion protein core. *Proceedings of the National Academy of Sciences*, 97:14172+, 2000.

[134] M. Zvelebil, G. Barton, W. Taylor, and M. Sternberg. Prediction of protein secondary structure and active sites using the alignment of homologous sequences. *Journal of Molecular Biology*, 195(4):957+, 1987.

30

Protein Structure Prediction with Lattice Models

William E. Hart
Sandia National Laboratories

Alantha Newman
Massachusetts Institute of Technology

30.1 Introduction

A protein is a complex biological macromolecule composed of a sequence of amino acids. Proteins play key roles in many cellular functions. Fibrous proteins are found in hair, skin, bone, and blood. Membrane proteins are found in cells' membranes, where they mediate the exchange of molecules and information across cellular boundaries. Water-soluble globular proteins serve as enzymes that catalyze most cellular biochemical reactions.

Amino acids are joined end-to-end during protein synthesis by the formation of peptide bonds (see Figure 30.1). The sequence of peptide bonds forms a "main chain" or "backbone" for the protein, off of which project the various side chains. Unlike the structure of other biological macromolecules, proteins have complex, irregular structures. The sequence of residues in a protein is called its *primary structure*. Proteins exhibit a variety of motifs that reflect common structural elements in a local region of the polypeptide chain: α-helices, β-strands, and loops — often termed *secondary structures*. Groups of these secondary structures usually combine to form compact globular structures, which represent the three-dimensional *tertiary structure* of a protein.

The functional properties of a protein depend on its three-dimensional structure. Protein structure prediction (PSP) is therefore a fundamental challenge in molecular biology. Despite the fact that the structures of thousands of different proteins have been determined [10], protein structure prediction in general has proven to be quite difficult. The central dogma of protein science is that the primary structure of a protein determines its

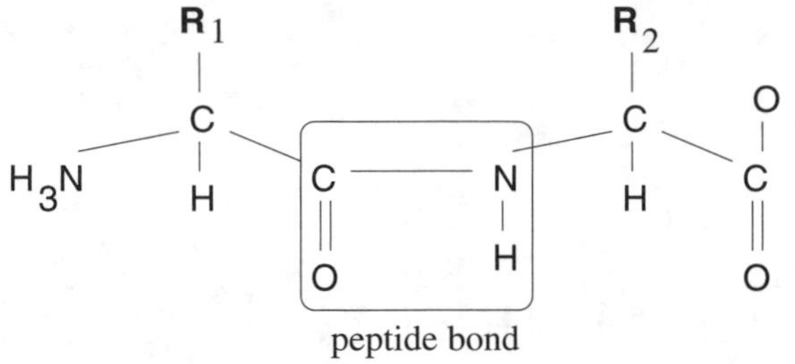

peptide bond

FIGURE 30.1: The peptide bond joining two amino acids when synthesizing a protein.

tertiary structure. Although this is not universally true (e.g. some proteins require *chaperone* proteins to facilitate their folding process), this dogma is tacitly assumed for most of the computational techniques used for predicting and comparing the structure of globular proteins.

Many computational techniques have been developed to predict protein structure, but few of these methods are rigorous techniques for which mathematical guarantees can be described. Most PSP methods employ enumeration or search strategies, which may require the evaluation of exponentially many protein structures. This observation has led many researchers to ask if PSP problems are inherently intractable.

Lattice models have proven to be extremely useful tools for reasoning about the complexity of PSP problems. By sacrificing atomic detail, lattice models can be used to extract essential principles, make predictions, and unify our understanding of many different properties of proteins [18]. One of the important approximations made by lattices is the discretization of the space of conformations. While this discretization precludes a completely accurate model of protein structures, it preserves important features of the problem of computing minimum energy conformations. For example, the related search problem remains difficult and preserves essential features of the conformational space. Consequently, methods that generate low-energy conformations of proteins for lattice models provide insight into the protein folding process.

In this chapter, we review results developed in the past decade that rigorously address the computational complexity of protein structure prediction problems in simple lattice models. We consider analyses of (1) intractability, (2) performance-guaranteed approximations and (3) methods that generate exact solutions, and we describe how the lattice models used in these analyses have evolved. Early mathematical analyses of PSP lattice models considered abstract formulations that had limited practical impact, but subsequent work has led to results that (a) apply to more detailed models, (b) consider lattices with greater degrees of freedom, (c) demonstrate the robustness of intractability and approximability, and (d) solve problems with general search frameworks. Our discussion complements the recent review by Chandru et al. [13], who more briefly survey this literature but provide more mathematical detail concerning some of the results in this area.

We begin by describing the the hydrophobic-hydrophilic model (HP model) [17, 29], which is one of the most extensively studied lattice models. Next, we review a variety of results that explore the possible computational intractability of PSP using techniques from computational complexity theory. These results show that the PSP problem is NP-hard in many simple lattice models, and thus widely believed to be intractable. Because of these hardness results, efficient performance-guaranteed approximation algorithms have been developed for the PSP problem in several lattice models. In particular, many variants of the HP model have been considered, allowing for different degrees of hydrophobicity, explicit side chains and different lattice structures. Finally, we summarize recent efforts to develop exact protein structure prediction methods that provably guarantee that optimal (or near-optimal) structures are found. Although enumerative search methods have been employed for many years, mathematical programming techniques like integer programming and constraint programming offer the possibility of generating optimal protein structures for practical protein sequences.

30.2 Hydrophobic-Hydrophilic Lattice Models

The discretization of the conformational space implicit in lattice models can be leveraged to gain many insights into the protein folding process [18]. For example, the entire conformational space can be enumerated, enabling the study of the folding code. This discretization also provides mathematical structure that can be used to analyze the computational complexity of PSP problems.

A lattice-based PSP model represents conformations of proteins as non-overlapping embeddings of the amino-acid sequence in the lattice. Lattice models can be classified based on the following properties:

1. The *physical structure*, which specifies the level of detail at which the protein sequences are represented. The structure of the protein is treated as a graph whose vertices represent components of the protein. For example, we can represent a protein with a linear-chain structure [18] that uses a chain of beads to represent the amino acids.

2. The *alphabet* of types of amino acids that are modelled. For example, we could use the 20 naturally occurring types of amino acids, or a binary alphabet that categorizes amino acids as hydrophobic (non-polar) or hydrophilic (polar).

3. The *set of protein sequences* that are considered by the model. The set of naturally occurring proteins is clearly a subset of the set of all amino acid sequences, so it is natural to restrict a model to similar subsets.

4. The *energy formula* used, which specifies how pairs of amino acid residues are used to compute the energy of a conformation. For example, this includes contact potentials that only have energy between amino acids that are adjacent on the lattice, and distance-based potentials that use a function of the distance between points on the lattice. Many energy formulas have energy parameters that can be set to different values to capture different aspects of the protein folding process.

5. The *lattice*, in which protein conformations are expressed; this determines the space of possible conformations for a given protein. For example, the cubic and diamond lattices have been used to describe protein conformations (see Figure 30.2).

One of the most studied lattice models is the HP model [17, 29]. This lattice model simplifies a protein's primary structure to a linear chain of beads. Each bead represents an amino acid, which can be one of two types: H (hydrophobic, i.e. nonpolar) or P (hydrophilic, i.e. polar). This model abstracts the hydrophobic interaction, one of the dominant forces in protein folding. Although some amino acids are not hydrophilic or hydrophobic in all contexts, the model reduces a protein instance to a string of H's and P's that represents the pattern of hydrophobicity in the protein's amino acid sequence. Despite its simplicity, the model is powerful enough to capture a variety of properties of actual proteins and has been used to discover new properties. For example, proteins in this model collapse to compact states with hydrophobic cores and significant amounts of secondary and tertiary structure.

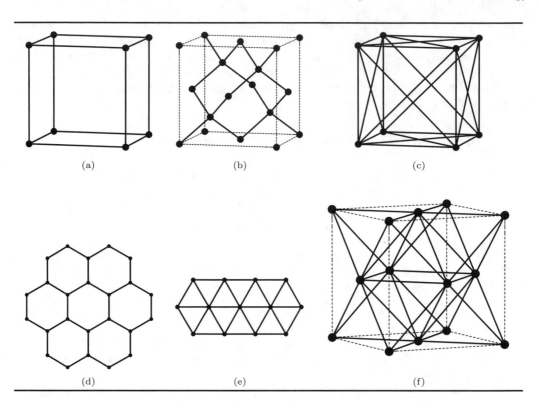

FIGURE 30.2:   Examples of crystal lattices: (a) cubic, (b) diamond, (c) cubic with planar
diagonals, (d) hexagonal, (e) triangular and (f) face-centered-cubic.

For simplicity, we denote H by "1" and P by "0", so the alphabet used in an HP model
is $A = \{0, 1\}$. The set of protein instances typically considered for this model is the set
of binary sequences $\sigma = \{0, 1\}^+$. Each sequence $s \in \sigma$ corresponds to a (hypothesized)
hydrophobic-hydrophilic pattern of a protein sequence. The HP model uses contact energies
between pairs of amino acids: two amino acids can contribute to the protein's energies if
they lie on adjacent points in the lattice. Thus the energy formula used in the HP model
is an energy matrix, $\mathcal{E} = (e(a, b))_{a,b \in A}$, where $e(a, b) = -1$ if $a = b = 1$, and $e(a, b) = 0$
otherwise. The HP model studied by Dill and his colleagues models protein conformations
as linear chains of beads folded in the 2D square or 3D cubic lattices.

Much of our review of the computational complexity of PSP focuses on the HP model,
because it has been so widely studied. Additionally, a variety of extensions of the HP model
have been considered in an effort to make these PSP results more practically relevant. For
example, Agarwala et al. [1] consider an extension of the HP model that allows for various
degrees of hydrophobicity.

More general structures have also been considered than the standard linear-chain model.
One example is a simple side-chain structure that uses a chain of beads to represent the
backbone; amino acids are represented by beads that connect to a linear backbone with
a single edge [11, 25, 28]. Figure 30.3 contrasts the structure of linear and side-chain
conformations in the HP model.

Although most work on the HP model has focused on the 2D square and 3D cubic lattices,

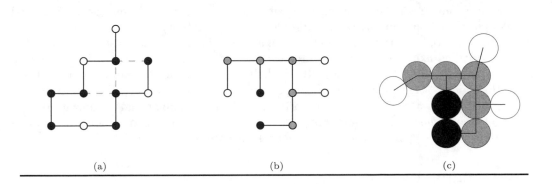

FIGURE 30.3: Illustrations of conformations for: (a) the standard HP model on the square lattice, (b) the HP model with side chains on the square lattice, and (c) the HP tangent spheres model with side chains. Black denotes a hydrophobic amino acid, white denotes a hydrophilic amino acid, and gray denotes a backbone element.

the computational complexity of PSP for the HP model has been studied for a variety of different lattices, including the triangular lattice (see Figure 30.2(e)) [1], the face centered cubic (FCC) lattice (see Figure 30.2(f)) [1, 25], the cubic lattice with diagonal edges on each face (see Figure 30.2(c)) [28], and general crystallographic lattices [27]. Off-lattice variants of the HP model have also been explored by treating a protein structure as a set of connected spheres, with a contact interaction potential that is identical to the standard HP model (see Figure 30.3(c)) [25]. The term off-lattice is used because the protein is not actually folded on a lattice. Conformations on a given lattice can clearly be translated into conformations in this off-lattice model, and near-optimal conformations on triangular and FCC lattices are closely related to near-optimal off-lattice conformations.

30.3 Computational Intractability

Exhaustive search of a protein's conformational space is clearly not a feasible algorithmic strategy. The number of possible conformations is exponential in the length of the protein sequence, and powerful computational hardware would not be capable of searching this space for even moderately large proteins. This observation led Levinthal to raise a question about the paradoxical discrepancy between the enormous number of possible conformations and the fact that most proteins fold within seconds to minutes [36]. While these observations appear contradictory, they can be reconciled by noting that they may simply point to the lack of knowledge that could be used to design an efficient search algorithm (see Ngo et al. [36] for further discussion of this issue). Computational analyses of PSP address this lack of knowledge by providing insight into the inherent algorithmic difficulty of folding proteins.

The native conformation of a protein is the conformation that determines its biological function. Following the thermodynamic hypothesis [19], computational models of protein folding are typically formulated to find the global minimum of a potential energy function. In lattice models, an energy value is associated with every conformation taking into account particular neighborhood relationships of the amino acids on the lattice. Consequently, given a lattice model L and sequence s, the PSP problem is to find a conformation of s in L with minimal energy.

Computational intractability refers to our inability to construct efficient (i.e., polynomial-time) algorithms that can solve a given problem. Here, "inability" refers to both the present state-of-the-art of algorithmic research as well as possible mathematical statements that no such algorithms exist. Customary statements about the intractability of a problem are made by showing that the problem is NP-hard. It is widely believed that a polynomial-time algorithm does not exist for any NP-hard problem, since the class of NP-hard optimization problems includes a wide variety of notoriously difficult combinatorial optimization problems. The best known algorithm for any NP-hard problem requires an exponential number of computational steps, which makes these problems "practically intractable."

30.3.1 Initial Results

PSP has been shown to be NP-hard for various lattice models. Initial intractability analyses of PSP considered models that captured PSP problems in rather limited and unrealistic ways. We survey these analyses and then critique these PSP results in the following two sections.

Fraenkel [20] presents a NP-hardness result for a physical model in which each amino acid is represented as a bead connected to a backbone. The protein must be embedded in a cubic lattice subject to pairwise constraints on the beads, i.e. specified pairs of beads, including pairs of beads on the backbone, are required to be at a fixed distance in the embedding. These specified pairs comprise a *contact graph*. The alphabet consists of three types that represent the charges associated with the amino acids: -1, 0, 1. The model uses a distance-dependent energy formula that computes the product of the charges divided by distance. The energy is the sum over all edges in the contact graph.

Ngo and Marks [35] present a NP-hardness result for a molecular structure prediction problem that encompasses protein structures. This model consists of a chain molecule of atoms that is to be embedded in a diamond lattice. The energy formula is based upon a typical form of the empirical potential-energy function for organic molecules, which is a distance-dependent function.

Paterson and Przytycka [37] present a NP-hardness result for a physical model in which each amino acid is represented as a bead along a chain that is to be embedded in a cubic lattice. A contact energy formula is used, so a pair of amino acids contributes to the conformational energy only if they are adjacent on the lattice. This energy formula has contact energies of one for contacts between identical residues and zero otherwise. The amino acid types in this model are not limited a priori, so instances of this model can represent instances of many specific contact-based PSP problems. However, we note below that this generality is actually a weakness of the model.

Finally, Unger and Moult [43] present a NP-hardness result for a physical model in which each amino acid is represented as a bead along a chain that is to be embedded in a cubic lattice with planar diagonals. The energy formula is a simple form of the empirical potential energy-function for organic molecules, which is a distance-dependent calculation. This NP-hardness result can be generalized to the Bravais lattices (which includes the cubic lattice), as well as the diamond and fluorite lattices [24].

30.3.2 Robust Results

It is difficult to provide strong recommendations for particular PSP formulations because accurate potential energy functions are not known. While various analytic formulations use potentials that capture known features of "the" potential function, the most appropriate analytic formulation of the potential energy for PSP remains an area of active research [15, 44]. Consequently, robust algorithmic results are particularly important for lattice-based PSP models.

Computational robustness refers to the independence of algorithmic results from particular settings. In the context of NP-completeness, robustness refers to the fact that a class of closely related problems can be described, all of which are NP-complete. The members of the class of problems are typically distinguished by some parameter(s) that form a set of reasonable alternate formulations of the same basic problem. Intractability results for PSP can be robust in two different ways [26]. First, an intractability result can be robust to changes in the lattice. The analysis of the PSP problem formulation posed by Unger and Moult [43], which uses a simplified empirical energy potential, can be generalized to show that this PSP problem is NP-hard for any finitely representable lattice [26].

Second, an intractability result can be robust to changes in the energy. Consider a PSP formulation with an objective of the form

$$\sum_{i=2}^{n}\sum_{j=1}^{i-1} C_{s_i,s_j} g\left(|f_i - f_j|\right),$$
(30.1)

where $g : \mathbf{Q} \to \mathbf{R}$ is an energy potential that monotonically increases to zero (in an inversely quadratic fashion) as the distance between amino acids increases. This model can be viewed as a special case of the model examined by Unger and Moult [43], and the class of functions g includes widely used pairwise potential functions like the Lennard-Jones potential. Additionally, the use of the distance $|f_i - f_j|$ makes this energy formulation translationally invariant, which is consistent with practical emperical energy models. For any function g and for an appropriate discretization of the L_2 norm, this PSP problem is NP-hard [26]. Additionally, this result can be generalized to show that this PSP problem is also NP-hard if the protein is modeled with explicit side-chains instead of as a simple linear chain.

30.3.3 Finite-Alphabet Results

A significant weakness of almost all of the models used in these intractability results is that the alphabet of amino acid types used to construct protein sequences is unbounded in size.[1] Let an amino acid *type* be defined by the pattern of interactions it exhibits with all other amino acids. These PSP problems allow for problem instances for which the number of amino acid types are not bounded. For example, a PSP formulation that uses Equation (30.1) allows for $O(n^2)$ amino acid types because the interaction between amino acids i and j is defined in part by the matrix coefficient C_{s_i,s_j}, which can assume any value.

Consequently, the previous models do not accurately model physically relevant PSP problems, for which there are 20 naturally occurring amino acid types. To address this concern, several authors have developed complexity analyses for models with a finite set of amino acids. For example, a PSP problem for which protein sequences are defined from a set of 12 amino acid types and the conformational energy is computed using a contact potential was proved to be NP-hard [2]. Nayak, Sinclair and Zwick [32] consider a string folding problem with a very large alphabet of amino acids, using a technique that "converts" a hardness proof for a model with an unbounded number of amino acids to a hardness proof in a model

---

[1]Fraenkel's model [20] uses a finite number of amino acid types, but it allows the protein chain to be embedded in a lattice without forcing subsequent amino acids to lie in close proximity on the lattice, thereby leading to biologically implausible conformations for certain amino acid sequences.

with a bounded number of amino acids. Crescenzi et al. [16] and Berger and Leighton [9] prove that PSP in the simple HP-model is NP-hard for the 2D square and 3D cubic lattices, respectively.

30.4 Performance-Guaranteed Approximation Algorithms

Performance guaranteed approximation algorithms complement intractability analyses by demonstrating that near-optimal solutions can be efficiently computed. An approximation algorithm has a multiplicative asymptotic approximation ratio of α if the solutions generated by the algorithm are within a factor of α of the optimum. Performance guaranteed approximation methods have been developed for a variety of HP lattice models, as well as some natural generalizations of the HP model.

30.4.1 HP Model

We now take a closer look at some performance guaranteed approximation algorithms that have been developed for the HP model on the 2D square lattice, 3D cubic lattice, triangular lattice and the face-centered-cubic (FCC) lattice [1, 23, 31, 33, 34]. These approximation algorithms take an HP sequence $s \in \{0,1\}^{+}$, and form a conformation on the lattice. Recall that the energy of a conformation is the number of hydrophobic-hydrophobic contacts: hydrophobics (1's) that are adjacent on the lattice but not adjacent on the string.

Square Lattice

The PSP problem in the HP model takes as input an HP sequence S, which can be viewed as a binary string (H=1, P=0). The objective is to find a folding of the string s that forms a self-avoiding walk on a specified lattice and maximizes the number of contacts. Figure 30.4 illustrates an optimal conformation for a binary string on the 2D square lattice (i.e. with the maximum number of contacts). Let $\mathcal{E}[s]$ denote the number of 1's in even positions in the sequence s (even-1's) and let $\mathcal{O}[s]$ denote the number of 1's in odd positions in s (odd-1's). Additionally, let

$$X[s] = \min\{\mathcal{E}[s], \mathcal{O}[s]\}. \tag{30.2}$$

Due to the fact that the square lattice is bipartite, each even-1 in s can have contacts only with odd-1's in s and vice-versa. In any conformation of s on the 2D square lattice, each 1 in the string s that is not in the first or last position on the string can have at most two contacts. Thus, an upper bound on the maximum number of contacts in any conformation of s on the 2D square lattice is:

$$2 \cdot X[s] + 2. \tag{30.3}$$

FIGURE 30.4: An optimal conformation for the string 0010100001011010 on the 2D square lattice. This conformation has four contacts.

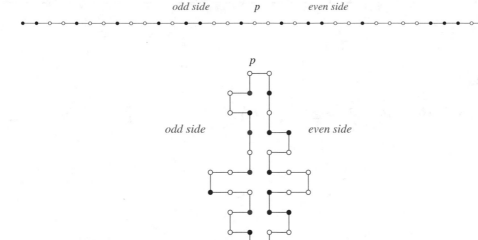

FIGURE 30.5: Illustration of a conformation generated by a simple 1/4-approximation algorithm for the HP model on the square lattice.

The first approximation algorithm developed for the PSP problem on the square lattice has an approximation ratio of 1/4 [23]. For a given sequence s, this algorithm first finds a point p in s such that at least half the odd-1's are in one substring on one side of p (the *odd* substring) and at least half the even-1's are on the other side of p (the *even* substring). Then, the odd substring is embedded in the square lattice such that all odd-1's in the odd substring have the same y-coordinate and the even substring is embedded in a complementary fashion (see Figure 30.5). This conformation yields at least $X[s]/2$ contacts, which is 1/4 of optimal. Mauri, Piccolboni and Pavesi [31] describe an algorithm that also has an approximation ratio of 1/4, which they argue works better in practice.

We now briefly describe how the approximation ratio for this problem can be improved to 1/3 [33]. This approximation algorithm creates "circular" conformations, i.e. it results in foldings in which the end-points of the string occupy adjacent lattice points. For simplicity, we consider even-length sequences s for which $\mathcal{O}[s] = \mathcal{E}[s]$. In the first step of the algorithm, we find a point p such that as we move clockwise in the loop starting at point p, we encounter at least as many odd-1's as even-1's and as we go counter-clockwise, we encounter at least as many even-1's as odd-1's.

Let $B_\mathcal{O}$ be the distance between the first pair of consecutive odd-1's encountered as we go in the clockwise direction starting at point p and let $B_\mathcal{E}$ be the distance between the first pair of consecutive even-1's encountered as we go in the counter-clockwise direction. We sketch the algorithm in Figure 30.6. In cases (a) and (b) of Step 2, we form three contacts and use at most four even- and odd-1's and "waste" at most four even- and odd-1's, i.e. we waste even-1's that occur on the odd side and vice-versa. In cases (c) and (d), we form two contacts and use at most three even- and odd-1's and waste at most three even- and odd-1's. Since there are at most $2\mathcal{O}[s] + 2 = \mathcal{O}[s] + \mathcal{E}[s] + 2$ contacts, this gives a 1/3 approximation ratio.

Step 1:

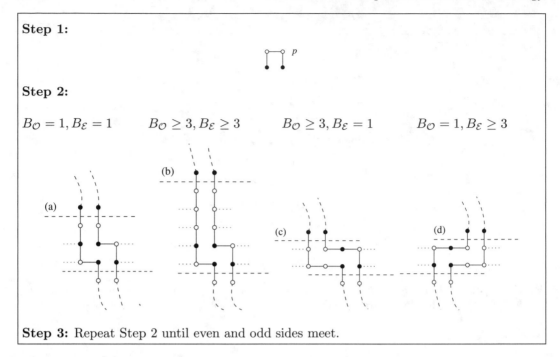

Step 2:

$B_\mathcal{O} = 1, B_\mathcal{E} = 1$ $B_\mathcal{O} \geq 3, B_\mathcal{E} \geq 3$ $B_\mathcal{O} \geq 3, B_\mathcal{E} = 1$ $B_\mathcal{O} = 1, B_\mathcal{E} \geq 3$

Step 3: Repeat Step 2 until even and odd sides meet.

FIGURE 30.6: The steps used in the 1/3-approximation algorithm for the folding problem in the HP model on the square lattice.

Cubic Lattice

In any folding of a sequence s on the 3D cubic lattice, each 1 in the string s that is not in the first or last position can have at most four contacts. Thus, an upper bound on the maximum number of contacts in any conformation of s on the 3D cubic lattice is:

$$4 \cdot X[s] + 2. \tag{30.4}$$

The previously described 1/4-approximation algorithm for the square lattice can be generalized to an approximation algorithm for the problem on the 3D cubic lattice [23]. Suppose the odd side of s has at least k odd-1's and the even side has at least k even-1's, i.e. $k \geq X[s]/2$. Then we can divide the odd side into segments with $\sqrt{k}$ odd-1's and divide the even side into segments with $\sqrt{k}$ even-1's. This approximation algorithm repeats the 2D folding algorithm $\sqrt{k}$ times in adjacent planes, i.e. the first pair of segments is folded in the plane $z = 0$, then next in the plane $z = 1$, etc. In the resulting conformation, each of $X[s]/2 - c\sqrt{X[s]}$ odd-1's has at least 3 contacts for some constant c. Thus, this algorithm has an approximation ratio of $3/8 - \Omega(1/\sqrt{X[s]})$.

Another approximation algorithm, based on different geometric ideas, improves on this absolute approximation guarantee [34]. In this algorithm, the string s is divided into two substrings so that one substring contains at least half the odd-1's and the other substring at least half the even-1's. Each substring is folded along two different diagonals, as shown in Figure 30.7. All but a constant number of odd-1's from the odd substring get three contacts. These geometric ideas can be used to obtain a slightly improved approximation ratio of .37501, which shows that 3/8 is not the best approximation guarantee that can be obtained for this problem, despite the fact that it was the best guarantee known for the past decade.

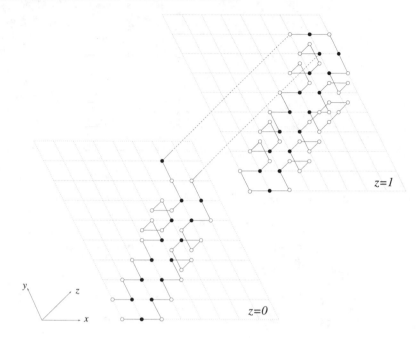

FIGURE 30.7: An illustration of a conformation generated by folding substrings along diagonals of the cubic lattice.

Triangular and FCC Lattices

One undesirable feature of the square lattice is that a contact must be formed between hydrophobics with different parities. There is no such parity restriction in real protein folds. This issue is discussed by Agarwala et al. [1], who suggest that the triangular lattice is more suitable to model protein folding. They give simple 1/2-approximation algorithms and a 6/11-approximation algorithm that uses an improved upper bound. Agarwala et al. generalize these results to a 3D triangular lattice that is equivalent to the FCC lattice, for which they describe an algorithm with an approximation ratio of 3/5.

30.4.2 HP Model with Side-Chains

Performance guaranteed approximation algorithms have also been developed for an HP model that explicitly represents side chains [26, 28]. This lattice model represents the conformation of a protein using a subclass of branched polymers called "branched combs." A homopolymer version of this model was introduced by Bromberg and Dill [11], who argued that linear lattice models fail to capture properties of protein folding, like side chain packing, that affect the stability of the native protein structure. The HP side chain model treats the backbone of the protein as a linear chain of beads. Connected to each bead on the backbone is a bead that represents an amino acid, and each of these side chain beads is labelled hydrophobic or hydrophilic.

Figure 30.3(b) illustrates a conformation of the HP side chain model on the square lattice. Note that there are no interactions between backbone elements and side-chain elements, so the energy of such a conformation is simply the number of contacts between hydrophobic side chains on the lattice. Further, note that adjacent side chains can contribute energy in this model, which is a fundamental difference induced by the branched combs structure.

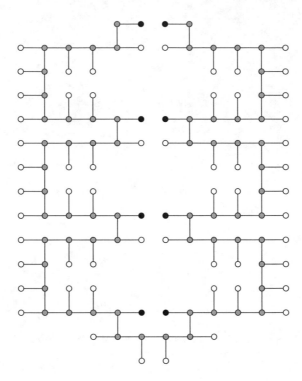

FIGURE 30.8: An illustration of the conformations generated by an approximation algorithm
for the HP side chain model on the square lattice.

Figure 30.8 illustrates the repeated conformational structure produced by an approximation algorithm for the problem on the square lattice [25]. The folding point for this algorithm
is selected in the same manner as for the linear chain model, and thus this structure can
be constructed in linear time. This algorithm guarantees that for a string s, $\lfloor X[s]/4 \rfloor$
hydrophobic-hydrophobic contacts are formed between the two halves of the conformation.
Since each hydrophobic side chain can have at most three contacts, this algorithm has a
$1/12$ approximation ratio.

A similar algorithm for the 3D cubic lattice can also be developed [25]. This approximation algorithm also divides the protein at a folding point, but it then attempts to create a
3D fold with four columns of hydrophobics in the core. Figure 30.9 illustrates the structure
of one of these columns, as well as how the protein sequence forms a hydrophobic core. The
hydrophobic core is formed by threading each half of the protein sequence through the four
columns in an anti-parallel fashion (e.g. up - down - up - down). In this conformation, it
contains at least $4 \lceil X[s]/2 \rceil - 20$ contacts for a sufficiently large sequence s. Since each hydrophobic side chain can have at most five contacts, this algorithm as a $4/10$ approximation
ratio.

These approximation results have been generalized to lattices that do not have the parity
restriction imposed by the cubic lattice: the FCC lattice and the cubic lattice with facial
diagonals (which Heun calls the *extended cubic lattice* (ECL)) [25, 28]. Both of these lattices
allow any hydrophobic amino acids to be in contact with any other hydrophobic amino acid.
Thus if there are $N(s)$ hydrophobic amino acids in a sequence s then we can obtain upper

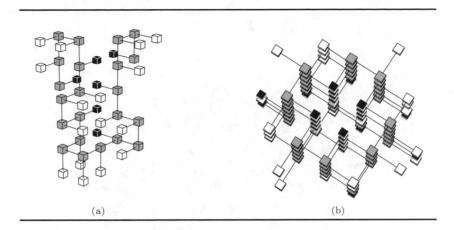

(a)   (b)

FIGURE 30.9:   Illustration of the conformations generated by an approximation algorithm for
the HP side chain model on the cubic lattic: (a) the 3D structure of a single
column, and (b) a perspective of the core generated by interlacing the columns.

bounds of $9N(s)/2$ contacts for the FCC lattice and $7N(s)$ contacts for the ECL.

These approximation algorithms are very similar in that they both place all hydrophobic
side chains in a set of columns, with an algorithm that forms the conformation in a linear
fashion (layer by layer or column by column). The hydrophobic columns form a distinct
hydrophobic core, with an irregular outer layer of hydrophilic side chains. For example,
Figure 30.10 illustrates a conformation generated by an approximation algorithm for the
FCC lattice [25], which generates eight columns of hydrophobics. These tight hydrophobic
cores guarantee that these approximation algorithms have an approximation ratio of $31/36$
on the FCC lattice and $59/70$ on the ECL.

Heun [28] also considers approximation algorithms that are tailored to the characteristics
of sequences commonly found in the SWISS-PROT protein database. Specifically, Heun
considers HP sequences that can be decomposed into blocks of 6 hydrophobics of the form
$\sigma = P^{l_1} H \dots P^{l_6} H$ where

- either there exists $i \in \{2, 3, \dots, 6\}$ such that $l_i = 0$, or
- there exists $i, j \in \{1, 2, \dots, 6\}$, $i \neq j$, such that $l_i + l_j \leq 3$.

Heun notes that over 96% of the sequences in SWISS-PROT can be decomposed into blocks
of 6 hydrophobics with this character, and he describes an approximation algorithm for the
ECL with an approximation ratio of $37/42$.

30.4.3   Off-Lattice HP Model

The HP tangent spheres models are simple PSP models that do not use a lattice but are
analogous to the standard HP model [25]. Because the conformations in these models are
not defined within a lattice, these models are termed off-lattice models. In these models,
the graph that represents the protein is transformed to a set of tangent spheres of equal
radius (or circles in two dimensions). Every vertex in the graph is replaced by a sphere,
and edges in the graph are translated to constraints that force spheres to be tangent in
a conformation (see Figure 30.3(c)). The linear chain model represents the protein as

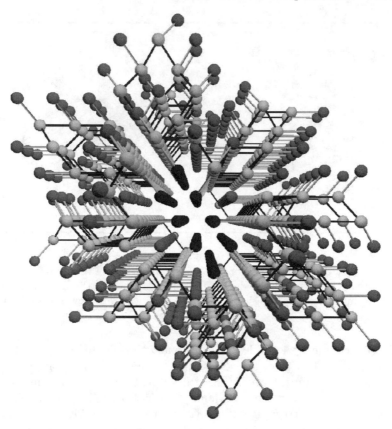

FIGURE 30.10: Illustration of the conformation generated by an approximation algorithm for the HP side chain model on the FCC lattice. The structures generated by Heun's approximation algorithm on the extended cubic lattice have a similar structure, with 10 hydrophobic columns in the core.

a sequence of spheres on a string, consecutive spheres being tangent, which are labelled hydrophobic or hydrophilic. The side chain model represents the backbone as in the linear chain model, but now every sphere in the backbone is tangent to a side chain sphere that models the physical presence of that amino acid's side chain. The side chain spheres are labelled hydrophobic or hydrophilic. A hydrophobic-hydrophobic contact in such a model is obtained when two hydrophobic-spheres are tangent.

The tangent spheres side chain model generalizes the HP model in the sense that for any lattice a conformation on that lattice represents a possible off-lattice conformation. Thus HP tangent spheres models can be analyzed rigorously by transferring algorithmic analyses from various lattice HP-models to the off-lattice setting. In 2D, the maximum number of spheres that can be tangent to a single sphere is 6. Thus a hydrophobic sphere in a linear chain can be tangent to at most 4 other hydrophobic spheres. The arrow-folding algorithm described by Agarwala et al. [1] can be used to construct a conformation (with the linear sphere chain) that has at least $N(s) - 3$ hydrophobic-hydrophobic contacts. Consequently, this algorithm has a 1/4 approximation ratio for the HP tangent spheres model.

To analyze the performance of the HP tangent spheres model in three dimensions, recall that for a set of identical spheres in 3D the maximum number of spheres that can be tangent to a single fixed sphere is 12. This is the so-called the 3D kissing number. From this we can conclude that a hydrophobic sphere in a linear chain can be tangent to only 10 other hydrophobic spheres, and a hydrophobic side chain sphere in a side chain model can be tangent to only 11 other hydrophobic side chain spheres. Thus each hydrophobic sphere in a linear chain can contribute at most 5 contacts and each hydrophobic side chain can contribute at most $11/2$ contacts.

The star-folding algorithm described by Agarwala et al. [1] can be used to construct a FCC conformation (with the linear sphere chain) that has $8N(s)/3$ hydrophobic-hydrophobic contacts (ignoring boundary conditions). Consequently, this algorithm has a $8/15$ approximation ratio for the HP tangent spheres model. Similarly, the approximation for the FCC side chain model [25] can be used to construct a conformation that has at least $31N(s)/8-42$ contacts (for sufficiently long sequences), which yields an algorithm with an approximation ratio of $31/44$ for the HP tangent spheres model with side chains.

30.4.4   Robust Approximability for HP Models on General Lattices

The results that we have surveyed in this section demonstrate that near-optimal protein structures can be quickly constructed for a variety of HP lattice models as well as simple off-lattice protein models. This naturally begs the question of whether approximability is a general property of HP lattice models. Results that transcend particular lattice frameworks are of significant interest because they can say something about the general biological problem with a higher degree of confidence. In fact, it is reasonable to expect that there will exist algorithmic invariants across lattices that fundamentally relate to the protein folding problem, because lattice models provide alternative discretizations of the same physical phenomenon.

Two "master" approximation algorithms have been developed for bipartite and non-bipartite lattices that demonstrate how approximation algorithms can be applied to a wide range of lattices [27]. These master approximation algorithms provide a generic template for an approximation algorithm using only a sublattice called a latticoid, a structured sublattice that in which a skeleton of hydrophobic contacts can be constructed. Further, the analysis of these algorithms includes a complexity theory for approximability in lattices that can be used to transform PSP algorithms in one lattice into PSP algorithms in another lattice such that we can provide a performance guarantee on the new lattice.

Figure 30.11 represents two possible latticoids of the square lattice. The bipartite master approximation algorithm selects a folding point in the same fashion used for the 2D HP model [23], and a hydrophobic core is similarly made by pairing odd and even hydrophobics along two faces of the conformation. The central row in these latticoids indicates the points at which hydrophobic contacts can be made by the master approximation algorithm.

The latticoids in Figure 30.11 can be embedded into a wide range of crystal lattices to provide a performance guaranteed approximation algorithm for the HP model. To illustrate this, consider the diamond lattice, whose unit cell is shown in Figure 30.2(b). Figure 30.12 illustrates how the latticoid in Figure 30.11(a) can be embedded into this lattice to ensure that at least $\lfloor X[s]/4 \rfloor$ hydrophobic-hydrophobic contacts are formed.

The bipartite and non-bipartite master approximation algorithms have performance guarantees for a class of lattices that includes most of the lattices commonly used in simple exact PSP models [27]: square and cubic lattices [18, 22, 39], diamond (carbon) lattices [40], face-centered-cubic lattice [14], and the 210 lattice used by Skolnick and Kolinkski [41]. Additionally, their analysis provides performance guarantees for a wide range of crystallographic lattices: Bravais lattices like the triclinic and triagonal lattices [38], the flourite lattice, 3D close packed lattices, the body centered cubic lattic and the hexagonal lattice.

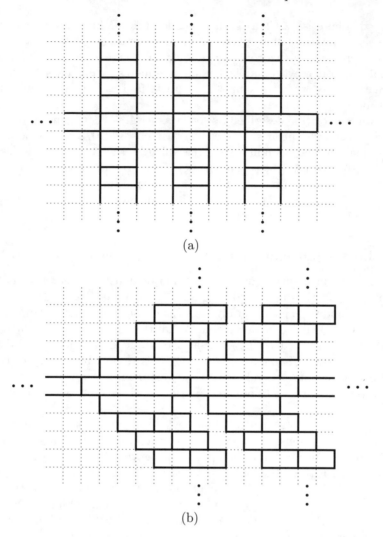

(a)

(b)

FIGURE 30.11: Illustrations of two latticoids of the square lattice. Dark lines indicate edges that are used in some protein conformation and dashed lines indicate remaining edges in the square lattice. The contact edges are the bolded edges in the central horizontal row.

These results demonstrate that approximability is a general feature of HP models on crystal lattices.

30.4.5 Accessible Surface Area Lattice Model

Solvent accessible area (ASA) describes the surface area over which contact between protein and solvent can occur. The concept of the solvent accessible surface of a protein molecule was originally introduced by Lee and Richards [30] as a way of quantifying hydrophobic burial. Subsequently, ASA and similar measures have been integrated into a variety of empirical potentials for PSP. This potential is qualitatively different from the HP model in that it favors hydrophobic burial rather than hydrophobic-hydrophobic interactions.

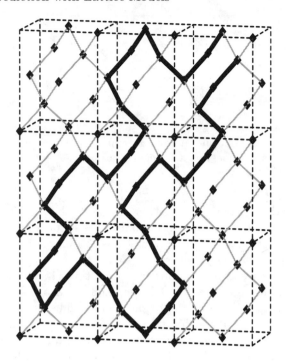

FIGURE 30.12: Illustration of a latticoid embedding into the diamond crystal lattice.

We describe new performance guaranteed approximation algorithms for the ASA lattice model with a linear chain model on the triangular lattice. As with the HP model, in the ASA model, we consider protein sequences $s \in \{H, P\}^+$. On a lattice, the ASA for a protein conformation can be modelled by the number of unoccupied lattice points that are adjacent to hydrophobic amino acids. Since there exist sequences for which the ASA is zero (i.e. all hydrophobics can be buried), it is not possible to develop an approximation algorithm that guarantees a multiplicative approximation ratio. Consequently, we treat this as a covering problem for the hydrophobics in a HP sequence.

Let $\overline{ASA}(s)$ refer to the number of covered hydrophobics in a conformation, which is the value we will attempt to maximize. If a sequence s has $N(s)$ hydrophobics, then $\overline{ASA}(s) \leq 4N(s) + 2$ on the triangular lattice because each amino acid has four neighboring lattice points that are not covered by the chain itself (except for the endpoints). Let $N_{HP}(s)$ denote the number of H-P contacts in a conformation, and let $N_{HH}(s)$ denote the number of H-H contacts. Note that $\overline{ASA}(s) = N_{HP}(s) + 2N_{HH}(s)$, since a single hydrophobic-hydrophobic contact represents the fact that two hydrophobics are being covered. Now consider the conformation of a chain folded back on itself (a simple U-fold). All but 3 hydrophobics in this conformation are guaranteed to have two contacts. Consequently, $N_{HP}(s) + 2N_{HH}(s) \geq 2N(s) - 6$, so an algorithm that generates this conformation has a $1/2$ approximation ratio. A similar analysis applies for a U-fold on the 2D square lattice, so an algorithm that generates that conformation has a $1/2$ approximation ratio.

Now consider the conformation in Figure 30.13, which treats the protein as a circular conformation that is molded into a square shape. If the protein sequence has n amino acids, then approximately $n - 4\sqrt{n}$ amino acids lie strictly within this conformation and are completely

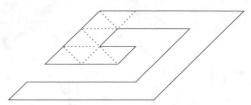

FIGURE 30.13: Illustration of conformations generated by the second approximation algorithm
in the ASA model on a triangular lattice.

buried. Now consider the linear-time algorithm that shifts the protein sequence through the
circular conformation to find the shift that minimizes the number of hydrophobics on the ex-
terior of this conformation. The conformation of this shifted minimal sequence has at most
$4N(s)/\sqrt{n}$ exposed hydrophobics. Thus we have $N_{HP}(s)+2N_{HH}(s) \geq 4N(s)-16N(s)/\sqrt{n}$,
from which it follows that we have an approximation ratio of 1. This implies that asymptotically all but an $o(1)$ fraction of the hydrophobic amino acids are buried in this algorithm.
Note that the conformation in Figure 30.13 can be embedded in the 2D square lattice. A
similar analysis shows that an algorithm that generates this conformation has an approximation ratio of 1. Furthermore, this result naturally generalizes to the 3D cubic and FCC
lattices, since you can create similarly compact structures for which the surface area is
dominated by the volume.

30.5   Exact Methods

Solving PSP problems exactly is an important practical goal because the lowest-energy
structure determines the biological functionality of a protein. Although PSP has been
proven NP-hard for many different lattice models, this does not preclude the development
of practical tools for many protein sequences. Since exhaustive enumeration is clearly not
practical even for relatively small protein sequences, several search techniques have been
developed to solve PSP for simple lattices models. In each of these methods, the lattice
structure is exploited to mathematically limit the search process.

30.5.1   Enumeration of Hydrophobic Cores

Yue and Dill [45] developed the first exact method for exactly finding globally optimal
protein structures on HP lattice models. The surface area of the hydrophobic core is easier
to estimate (given partial information about the final conformation) than the number of
hydrophobic-hydrophobic contacts, and the core surface area and the number of contacts
are related. Yue and Dill developed the constrained hydrophobic core construction (CHCC)
algorithm, which enumerates all possible shapes of the region containing all hydrophobic
amino acids for a sequence. This enumeration of the possible hydrophobic cores is done
so that core shapes with a smaller surface area are enumerated before core shapes with a
larger surface area. For every core shape, CHCC enumerates all positions of the monomers
that fit into the given core shape. CHCC uses some conditions (or constraints) to reduce
the size of the search tree.

The CHCC has been effectively applied to exactly solve PSP problems for HP sequences
with up to 80 amino acids. Perhaps the greatest limitation of this method is that it is specifically tailored for the HP-model on the cubic lattice. Consequently, this basic algorithmic

approach has not been effectively generalized to other simple lattice models for PSP.

30.5.2 Constraint Programming

Backofen et al. [3, 4, 5, 6, 7, 8] provide a declarative formulation of the HP lattice model, which is solved using constraint programming. Constraint programming is a relatively new programming technique that integrates a declarative definition of a problem (e.g. PROLOG) with an inherently concurrent programming paradigm, since all constraints are handled in parallel. The search strategy is not fixed in constraint programming, and systems like Oz [42] offer a flexible environment for defining a search strategy. Constraint programming offers a flexible framework for solving PSP on simple lattice models, and Backofen et al. have described declarative formulations for the HP models on the cubic and FCC lattices, as well as an extended HP model on the cubic lattice.

We illustrate the type of declarative formulation used for constraint programming to define feasible conformations in the cubic lattice. Consider variables X_i, Y_i and Z_i that indicate the position of the i-th amino acid in the lattice. Without loss of generality we can restrict the amino acids with the following constraint:

$$\forall i, X_i \in [1 \ldots (2 \cdot \text{length}(s))] \bigwedge Y_i \in [1 \ldots (2 \cdot \text{length}(s))] \bigwedge Z_i \in [1 \ldots (2 \cdot \text{length}(s))],$$

where length(s) is the length of the HP sequence. We clearly need to satisfy the constraint $\forall i \neq j, (X_i, Y_i, Z_i) \neq (X_j, Y_j, Z_j)$ in a feasible conformation. Additionally, amino acids must be consecutively placed on the lattice. We can enforce this constraint using variables Xdiff$_i$, Ydiff$_i$ and Zdiff$_i$, which represent the difference of the x, y and z coordinates between amino acid i and $i + 1$. The constraints

$$\forall i, \text{Xdiff}_i = |X_i - X_{i+1}|$$
$$\forall i, \text{Ydiff}_i = |Y_i - Y_{i+1}|$$
$$\forall i, \text{Zdiff}_i = |Z_i - Z_{i+1}|$$

define the values of these variables, and the constraint $\forall i, 1 = \text{Xdiff}_i + \text{Ydiff}_i + \text{Zdiff}_i$ ensures that the distance between consecutive amino acids is one.

Backofen et al. apply a search algorithm that is a combination of a branch-and-bound search together with a constrain-and-generate principle, which is common for constraint programming. The branching process selects a variable **var** to branch on and then creates two branches for some value **var**: (1) **var** =: **val**, and (2) **var** $\neq$: **val**. Subsequently, these branches are evaluated using a constraint programming system to evaluate the effected variables according to the constraints, which results in an association of smaller value ranges to some (or many) variables. Further, the search tree may be pruned when an inconsistent conformation is generated. The bounding calculation used in this search requires a problem-specific calculation, based on the feasible domain for a subproblem.

Backofen et al [3, 4, 5, 6, 7, 8] have evaluated constraint programming implementations for HP lattice models using the Oz language [42]. These methods have effectively solved problems of up to 200 amino acids (using pre-calculated hydrophobic cores) within a few seconds. Additionally, these tools have been used to enumerate optimal conformations for the HP cubic model, for which it appears to be more effective than the CHCC algorithm.

30.5.3   Integer Programming

A standard approach for finding exact solutions for hard optimization problems is to model them as integer programs and try to solve these programs to optimality using techniques from the field of integer programming such as branch and bound. Additionally, linear programming relaxations of integer programs often provide efficiently computable non-trivial upper bounds.

Several integer programming formulations have been developed for the PSP problem in the HP model [12, 21, 13]. We illustrate the type of linear constraints used for integer programming to define feasible conformations in the square lattice. Without loss of generality, we can restrict the conformations to lattice points $\mathcal{L} = \{1, 2, \ldots, n^2\}$, such that the coordinates are of the form:

$$y_p = \left\lfloor \frac{p-1}{n} \right\rfloor \text{ and } x_p = p - 1 - n y_p \text{ for } p \in \mathcal{L}.$$

Let $\mathcal{N}(p)$ denote the lattice points adjacent to a point p (whose distance is one away), and let v_{ip} be a binary decision variable that is one if the i-th amino acid is placed at point p on the lattice, and zero otherwise. Now every residue must be placed on a lattice point, which is enforced by the following constraint:

$$\sum_{p \in \mathcal{L}} v_{ip} = 1 \quad , i = 1, \ldots, n.$$

Similarly, each point cannot have more than one amino acid placed at it, which is enforced by the constraint:

$$\sum_{i=1}^{n} v_{ip} \leq 1 \quad , \forall p \in \mathcal{L}.$$

Finally, we can enforce the connectivity between consecutive amino acids with the following two constraints:

$$\sum_{q \in \mathcal{N}(p)} v_{i+1,q} \geq v_{ip} \quad , i = 1, \ldots, n-1, \ p \in \mathcal{L},$$

$$\sum_{q \in \mathcal{N}(p)} v_{i-1,q} \geq v_{ip} \quad , i = 2, \ldots, n, \ p \in \mathcal{L}.$$

These constraints define a convex region that represents valid solutions if we relax the constraint that the $\{v_{ip}\}$ variables are binary. This observation provides a mechanism for computing lower bounds on the minimum energy of a conformation with integer program formulations, for which the lower bound can be computed with linear programming methods.

Linear programming relaxations can provably provide bounds that are at least as strong as the simple combinatorial bound 30.3 and some IP formulations may strengthen this bound even further [12]. Although integer programming formulations have been used to compute such bounds, these formulations can have many variables, which may limit their application to large-scale problems. Additionally, it is not clear whether these integer programming formulations can be used to solve large-scale instances of the PSP problems exactly.

30.6 Conclusions

There are many ways that these analyses and methods for PSP problems can be improved. For example, no intractability analysis has been developed for the HP model on the triangular or FCC lattices. There is wide agreement that these lattices are more practically relevant for PSP because they do not impose the artificial parity constraints found in the square and cubic lattices, so such an intractability analysis would be quite interesting. Similarly, exact methods have not been developed for models like the HP side chain model, which capture greater physical detail. We expect that studies of (near-) optimal conformations in this model would provide significant insight into PSP (e.g. by studying the degeneracy of the optimal solution in these problems).

Improving bounds on lattice models could fundamentally improve our assessment for approximation algorithms. For example, there are strings for which the best conformation on the 2D square lattice achieves only half of the upper bound in Equation 30.3 [33], so this bound is demonstrably weak. However, integer programming formulations may provide a general technique for improving these bounds for specific sequences. The bounds for the HP tangent spheres model might also be improved by generalizing the bound analysis of triangular and FCC lattices. In the triangular and FCC lattices, the bounds on the maximal number of contacts can be tightened by noting that "conflicts" occur between some hydrophobics and non-hydrophobics, thereby limiting the total number of hydrophobic-hydrophobic contacts. However, in 3D it is possible to have 12 spheres touching a given sphere without any pair of them being tangent, so the notion of a "conflict" needs to be generalized in this case to tighten simple upper bounds.

Researchers analyzing PSP in lattice models have increasingly considered detailed models and methods that can be applied to a variety of lattice models. This trend is motivated by the desire to provide robust mathematical insight into protein models that is generally independent of a particular lattice formulation. Analyses that achieve this goal provide greater insight into general PSP complexity, which is not bound by lattice constraints and for which precise empirical energy potentials are not known.

One interesting direction for the analysis of PSP is to consider methods that are tailored to biologically plausible amino acid sequences. Thus we need to develop complexity analyses like Heun's approximation algorithm that is tailored to protein-like sequences. For example, the possible intractability of PSP remains an open question if PSP is restricted in this manner.

Similarly, we expect that methods that can solve more detailed protein models will provide more insight into real protein structures. For example, side chain lattice models are clearly more representative of the structure of actual proteins than linear chain models. However, the analysis of side chain models with variable-size side chains could more accurately capture the complexity of solving side chain packing problems. Additionally, this type of PSP formulation could capture the fact that the hydrophobicity of a side chain is related to its surface area. PSP with variable hydrophobicities has been briefly considered by Agarwala et al. [1], who consider protein structures as linear chains.

Finally, the connection between lattice models and off-lattice models needs to be developed further to more directly impact real-world PSP problems. Performance guaranteed algorithms for the FCC lattice can provide performance guarantees for closely related off-lattice protein models. This is a first step towards a more comprehensive analysis that uses lattice models to provide mathematical insight into off-lattice models. For example, we conjecture that lattice-based search methods like constraint programming can be effectively hybridized with optimizers for standard empirical energy potentials to perform a more effective global search of protein structures.

Acknowledgments

We thank Sorin Istrail for his collaborations on the ASA model. We also thank Edith New-man for her assistance in creating Figure 30.7. This work was performed in part at Sandia National Laboratories. Sandia is a multipurpose laboratory operated by Sandia Corporation, a Lockheed-Martin Company, for the United States Department of Energy under contract DE-AC04-94AL85000. This work was partially funded by the US Department of Energy's *Genomes to Life* program (`www.doegenomestolife.org`), under project "Carbon Sequestration in Synechococcus Sp.: From Molecular Machines to Hierarchical Modeling," (`www.genomes-to-life.org`).

References

[1] R. Agarwala, S. Batzogloa, V. Dančík, and S.E. Decatur *et al.* Local rules for protein folding on a triangular lattice and generalized hydrophobicity in the HP model. *J Comp Bio*, 4(3):276–296, 1997.

[2] J. Atkins and W.E. Hart. On the intractability of protein folding with a finite alphabet of amino acids. *Algorithmica*, 25:279–294, 1999.

[3] R. Backofen. Using constraint programming for lattice protein folding. In R.B. Altman, A.K. Dunker, L. Hunter, and T.E. Klein, editors, *Pacific Symposium on Biocomputing (PSB'98)*, volume 3, pages 387–398, 1998.

[4] R. Backofen. An upper bound for number of contacts in the HP-model on the Face-Centered-Cubic Lattice (FCC). In R. Giancarlo and D. Sankoff, editors, *Proceedings of the 11th Annual Symposium on Combinatorial Pattern Matching*, number 1848 in LNCS, pages 277–292, Montréal, Canada, 2000. Springer-Verlag, Berlin.

[5] R. Backofen. The protein structure prediction problem: A constraint optimisation approach using a new lower bound. *Constraints*, 6:223–255, 2001.

[6] R. Backofen and S. Will. Optimally compact finite sphere packings — hydrophobic cores in the FCC. In *Proc. of the 12th Annual Symposium on Combinatorial Pattern Matching (CPM2001)*, volume 2089 of *Lecture Notes in Computer Science*, Berlin, 2001. Springer–Verlag.

[7] R. Backofen and S. Will. A constraint-based approach to structure prediction for simplified protein models that outperforms other existing methods. In *Proceedings of the Ninetheen International Conference on Logic Programming (ICLP 2003)*, 2003. in press.

[8] R. Backofen, S. Will, and E. Bornberg-Bauer. Application of constraint programming techniques for structure prediction of lattice proteins with extended alphabets. *J. Bioinformatics*, 15(3):234–242, 1999.

[9] B. Berger and T. Leighton. Protein folding in the hydrophobic-hydrophilic (HP) model is NP-complete. *J Comp Bio*, 5(1):27–40, 1998.

[10] H.M. Berman, J. Westbrook, Z. Feng, and G. Gilliland *et al.* The protein data bank. *Nucleic Acids Research*, 28:235–242, 2000. The PDB is at `http://www.rcsb.org/pdb/`.

[11] S. Bromberg and K.A. Dill. Side chain entropy and packing in proteins. *Prot. Sci.*, pages 997–1009, 1994.

[12] R. Carr, W.E. Hart, and A. Newman. Discrete optimization models for protein folding. Technical report, Sandia National Laboratories, 2003.

[13] V. Chandru, A. DattaSharma, and V.S.A. Kumar. The algorithmics of folding proteins

on lattices. *Discrete Applied Mathematics*, 127(1):145–161, Apr 2003.

[14] D.G. Covell and R.L. Jernigan. *Biochemistry*, 29:3287, 1990.

[15] T.E. Creighton, editor. *Protein Folding*. W. H. Freeman and Company, 1993.

[16] P. Crescenzi, D. Goldman, C. Papadimitriou, and A. Piccolboni *et al*. On the complexity of protein folding. *J Comp Bio*, 5(3), 1998.

[17] K.A. Dill. Theory for the folding and stability of globular proteins. *Biochemistry*, 24:1501, 1985.

[18] K.A. Dill, S. Bromberg, K. Yue, and K.M. Fiebig *et al*. Principles of protein folding: A perspective from simple exact models. *Prot. Sci.*, 4:561–602, 1995.

[19] C.J. Epstein, R.F. Goldberger, and C.B. Anfinsen. The genetic control of tertiary protein structure: Studies with model systems. In *Cold Spring Harbor Symposium on Quantitative Biology*, pages 439–449, 1963. Vol. 28.

[20] A.S. Fraenkel. Complexity of protein folding. *Bull. Math. Bio.*, 55(6):1199–1210, 1993.

[21] H.J. Greenberg, W.E. Hart, and G. Lancia. Opportunities for combinatorial optimization in computational biology. *INFORMS Journal of Computing*, 2003. (to appear).

[22] A.M. Gutin and E.I. Shakhnovich. Ground state of random copolymers and the discrete random energy model. *J. Chem. Phys.*, 98:8174–8177, 1993.

[23] W.E. Hart and S. Istrail. Fast protein folding in the hydrophobic-hydrophilic model within three-eighths of optimal. *Journal of Computational Biology*, 3(1):53–96, 1996.

[24] W.E. Hart and S. Istrail. Invariant patterns in crystal lattices: Implications for protein folding algorithms. In *Combinatorial Pattern Matching*, Lecture Notes in Computer Science 1075, pages 288–303, New York, 1996. Springer.

[25] W.E. Hart and S. Istrail. Lattice and off-lattice side chain models of protein folding: Linear time structure prediction better than 86% of optimal. *Journal of Computational Biology*, 4(3):241–259, 1997.

[26] W.E. Hart and S. Istrail. Robust proofs of NP-hardness for protein folding: General lattices and energy potentials. *Journal of Computational Biology*, 4(1):1–20, 1997.

[27] W.E. Hart and S. Istrail. Invariant patterns in crystal lattices: Implications for protein folding algorithms. *Journal of Universal Computer Science*, 6(6):560–579, 2000.

[28] V. Heun. Approximate protein folding in the HP side chain model on extended cubic lattices. *Discrete Applied Mathematics*, 127(1):163–177, 2003.

[29] K.F. Lau and K.A. Dill. A lattice statistical mechanics model of the conformation and sequence spaces of proteins. *Macromolecules*, 22:3986–3997, 1989.

[30] B. Lee and F.M. Richards. The interpretation of protein structures: Estimation of static accessibility. *J Mol Biol*, 55:379–400, 1971.

[31] G. Mauri, A. Piccolboni, and G. Pavesi. Approximation algorithms for protein folding prediction. In *Proceedings of the 10th ACM-SIAM Symposium on Discrete Algorithms, SODA*, pages 945–946, Baltimore, 1999.

[32] A. Nayak, A. Sinclair, and U. Zwick. Spatial codes and the hardness of string folding problems. *J Comp Bio*, pages 13–36, 1999.

[33] A. Newman. A new algorithm for protein folding in the HP model. In *Proceedings of the 13th ACM-SIAM Symposium on Discrete Algorithms, SODA*, pages 876–884, San Francisco, Jan 2002.

[34] A. Newman and M. Ruhl. Combinatorial problems on strings with applications to protein folding. In *Proceedings of the 6th Latin American Theoretical Informatics (LATIN)*, pages 369–378, Buenos Aires, 2004.

[35] J.T. Ngo and J. Marks. Computational complexity of a problem in molecular structure prediction. *Protein Engineering*, 5(4):313–321, 1992.

[36] J.T. Ngo, J. Marks, and M. Karplus. Computational complexity, protein structure

prediction, and the Levinthal paradox. In K. Merz, Jr. and S. Le Grand, editors, *The Protein Folding Problem and Tertiary Structure Prediction*, chapter 14, pages 435–508. Birkhauser, Boston, MA, 1994.

[37] M. Paterson and T. Przytycka. On the complexity of string folding. *Discrete Applied Mathematics*, 71:217–230, 1996.

[38] D.E. Sands. *Introduction to Crystallography*. Dover Publications, Inc., New York, 1975.

[39] E.I. Shakhnovich and A.M. Gutin. Engineering of stable and fast-folding sequences of model proteins. *Proc. Natl. Acad. Sci.*, 90:7195–7199, 1993.

[40] A.j. Sikorski and J. Skolnick. Dynamice Monte Carlo simulations of globular protein folding/unfolding pathways. II. α-helical motifs. *J. Molecular Biology*, 212:819–836, July 1990.

[41] J. Skolnick and A. Kolinski. Simulations of the folding of a globular protien. *Science*, 250:1121–1125, 1990.

[42] G. Smolka. *The Oz programming model*, volume 1000 of *Lecture Notes in Computer Science*, pages 324–343. 1995.

[43] R. Unger and J. Moult. Finding the lowest free energy conformation of a protein is a NP-hard problem: Proof and implications. *Bull. Math. Bio.*, 55(6):1183–1198, 1993.

[44] W.F. van Gunsteren, P.K. Weiner, and A.J. Wilkinson, editors. *Computer Simulation of Biomolecular Systems*. ESCOM Science Publishers, 1993.

[45] K. Yue and K.A. Dill. Sequence-structure relationships in proteins and copolymers. *Phys. Rev. E*, 48(3):2267–2278, 1993.

31

Protein Structure Determination via NMR Spectral Data

Guohui Lin
University of Alberta

Xin Tu
University of Alberta

Xiang Wan
University of Alberta

31.1   Introduction

Protein functions are determined mostly by its three dimensional structure and the structure determination is one of the top challenges in both genomics and proteomics eras. Although having been employed for a long period of time, NMR spectroscopy and X-ray crystallography are still the two main experimental methods for protein structure determination at atomic resolution. It is acknowledged that NMR protein structure determination hasn't been able to achieve the same accuracy as X-ray crystallography does and thus X-ray crystallography remains its dominant position. Nonetheless, NMR spectroscopy complements it in many ways. Typically, structure determination via NMR spectroscopy can provide the three dimensional structure of a protein in solution under nearly physiological conditions along with dynamics information associated with the protein function. Therefore, with the advent of recent innovations such as heteronuclear NMR and cryoprobes [20], NMR spectroscopy is expected to play a more significant role in structural biology, particularly in the high-throughput structure production of the Structural Genomics Initiative [40].

The procedure of protein structure determination through NMR spectroscopy could be roughly partitioned into two stages, the first of which is spectral data analysis and the second is structure calculation under the structural constraints extracted using the analysis results from the first stage. To achieve the high-throughput goal, the spectral data analysis must be automated as it consumes most of the time in the whole procedure. There are a lot of efforts devoted to automate the data analysis in the last two decades. This chapter summarizes the best performed methods to include their key methodologies and their performance guarantees. It also presents some of the most recent developments from our group toward the goal of automated protein structure determination.

The chapter is organized as follows: In the next section, we introduce some basic notions and inherent principles of Nuclear Magnetic Resonance phenomenon, on which the NMR spectroscopy has been developed. Section 31.3 sketches the NMR data acquisition and the Fourier Transformation which transforms the time domain signals into frequency domain signals for later analysis. Section 31.4 summarizes how to identify real spectral peaks from noise peaks. The association of the identified spectral peaks with their host residues in the target protein sequence is given in Section 31.5, where the current methods of choice and our recent developments are introduced. Section 31.6 presents how and what types of the structural constraints are extracted and then put into structure builders for three dimensional structure calculation. We conclude the chapter in Section 31.7.

31.2   Nuclear Magnetic Resonance Phenomenon

Atoms are basic building blocks of matter, and cannot be chemically subdivided by ordinary means. Atoms are composed of three types of particles: protons, neutrons, and electrons. Each proton has a positive charge and each electron has a negative charge, while neutrons have no charge. The number of protons in an atom is the *atomic number*, which determines the type of the atom. Both protons and neutrons reside in the nucleus. A same type of atoms may contain different numbers of neutrons, and they are called *isotopes*.

A nucleus often acts as if it is a single entity with intrinsic total angular momentum I, the nuclear *spin*, which is the overall effect of the imaginary spinning protons and neutrons. Despite many spin-pairing rules, one characteristic is that a nucleus of odd mass number (which is the sum of the numbers of protons and neutrons) will have a half-integer spin and a nucleus of even mass number but odd numbers of protons and neutrons will have an integer spin. For a nucleus of spin I, there are $2I + 1$ spin states (or orientations) ranging from $-I$ to $+I$. In the NMR spectroscopy for protein structure determination, the most important nuclei with spin $I = 1/2$ are ^{1}H (Hydrogen), ^{13}C (Carbon), ^{15}N (Nitrogen), ^{19}F (Fluorine), and ^{31}P (Phosphorus), each of which has two spin states; the nucleus with spin $I = 1$ is deuteron ^{2}H (Hydrogen); and typical isotopes with no spin (i.e., $I = 0$) are ^{12}C, ^{14}N, and ^{16}O (Oxygen).

Nuclear Magnetic Resonance (NMR) is a phenomenon which occurs when the nuclei are immersed in a static magnetic field and are exposed to a second oscillating magnetic field (which is created by radio frequency (r.f.) pulse). In the absence of an external magnetic field, for nuclei of spin I, those $2I + 1$ states are of equal energy. When an external magnetic field is applied, the energy levels split. In an external magnetic field of strength B_0, the spinning rotation axis of a nucleus will *precess* about the magnetic field with angular frequency $\omega_0 = \gamma B_0$, called *Larmor Frequency*, where the *gyromagnetic ratio* γ is different for distinct types of nuclei. For nuclei of spin $I = 1/2$, there will be two possible spinning orientations/states in the external magnetic field, i.e., parallel to the external field (low energy state) and opposite to the external field (high energy state). At the time the external magnetic field is applied, the initial populations of nuclei in the energy levels are determined by thermodynamics, described by the Boltzmann distribution. This means that the low energy level will contain slightly more nuclei than the high level. It is possible to incite the low energy level nuclei into the high energy level with electromagnetic radiation.

In fact, if these aligned nuclei are irradiated with r.f. pulse of a proper frequency, nuclei will spin-flip from low energy state to high energy state or from high energy state to low energy state by absorbing or emitting a quantum of energy, respectively. The frequency of radiation needed is determined by the difference in energy between the two energy levels and when such a spin transition occurs the nuclei are said to be *in resonance* with this radiation. The electromagnetic radiation supplied by the second oscillating magnetic field must be equal to the frequency of the oscillating electric field generated by nucleus precession, which is $\frac{\omega_0}{2\pi}$, because only under that circumstance, the energy needed in resonance can be transferred from electromagnetic radiation to precession nucleus. It is possible that by absorbing energy, the nuclei will reach a state with equal populations in both states. In such a case, the system is *saturated*. If the electromagnetic radiation supply by the second oscillating magnetic field is then switched off, nuclei at the high energy state will be back to the low energy state and the system will return to thermal equilibrium. Such a phase is the *relaxation* process. The relaxation process produces a measurable amount of r.f. signal at the resonant frequency associated with the spin-flip. This frequency is received and amplified to display the NMR signal.

31.2.1 Chemical Shift

The resonance frequencies of individual nuclei are not only relevant to the strength of the external magnetic field B_0 applied on them, but also depend on their local chemical environments. The magnetic field generated by a nucleus itself tends to contradict the effect of the external magnetic field. This contradiction effect is defined as *shielding*. The strength of this shielding effect increases with the electron density. This is called the *Chemical Shift* phenomenon. The actual field present at the nucleus is not B_0 but $B_{\text{local}} = B_0(1 - \sigma)$, where σB_0 is the shielding effect (σ is the shielding factor which is small — typically 10^{-5} for protons and 10^{-3} for other nuclei [7]). Consequently, the Larmor Frequency becomes $\omega_0 = \gamma B_0(1 - \sigma)$. Chemical shift in Parts Per Million (PPM) is defined as

$$\delta = \frac{(\omega_0 - \omega_{\text{reference}}) \times 10^6}{\omega_{\text{reference}}} \approx (\sigma_{\text{reference}} - \sigma) \times 10^6,$$

where $\omega_{\text{reference}}$ is the reference frequency and $\sigma_{\text{reference}}$ is the reference shielding factor. For both protons and carbons, the reference material is often tetramethylsilane $S_i(CH_3)_4$ (TMS). Chemical shift is small but it is a very sensitive probe of the chemical environment of the resonating nucleus, and it is possible that we can distinguish nuclei in different chemical environments using their chemical shift values. Here, chemical environment refers to the interactions between nuclei, including chemical bonds, scalar coupling, dipolar coupling, hydrogen bond, etc. It is observed that even tiny changes of one of these environmental factors will vary the value of chemical shift. Therefore, on the basis of chemical shift phenomenon, we will be able to map spectral peaks back to their host amino acid residues in the target protein sequence through a process called peak assignment, which is one of the key steps in NMR spectroscopy for protein structure determination.

31.2.2 NMR Spectroscopy

Spectroscopy is the study of the interaction of electromagnetic radiation with matter. NMR spectroscopy is the use of the NMR phenomenon to study physical, chemical, and biological properties of matter. As a consequence, NMR spectroscopy finds applications in several areas of science. For example, NMR spectroscopy is routinely used by chemists to study chemical structure using simple one-dimensional techniques. Two and higher dimension-al techniques are used to determine the structure of more complicated molecules. These techniques are improving and are replacing X-ray crystallography for the determination of protein structure.

NMR spectroscopy derives protein structural information by providing a network of dis-tance restraints between spatially close (i.e., < 5Å) hydrogen atoms extracted from the NOEs, dihedral-angle restraints calculated from scalar coupling constants and chemical shifts, and other various geometric restraints including orientation information from the residual dipolar coupling. An NMR structure is typically determined through molecular dynamics (MD) simulation and energy minimization (EM) under the above NMR struc-tural constraints [5, 8, 11, 24].

31.3   NMR Data Acquisition and Processing

A wide variety of NMR instrumentation is available for NMR experiments to produce the data for protein structure determination. The common components of NMR spectrometers are: (a) superconducting magnet for supplying external magnetic field, (b) pulse program-mer and r.f. transmitter to generate and control r.f. pulses, (c) probe for placing the sample in the magnet, (d) receiver for receiving the resulting NMR signals, and (e) computers for data acquisition and processing. Superconducting magnets can provide a wide range of fre-quencies from 60 to 800 MHz. Note that higher frequency implies the higher sensitivity and stability of the NMR spectroscopy, because the differences between the chemical shifts are amplified with the increase of magnetic field strength, meaning better separation between nuclei resonances.

In NMR spectrometers, superconducting magnet provides the external static magnetic field. The transverse magnetic field is generated by a series of r.f. pulses. During the relaxation process of the nuclei in the probe, the time-varying current is amplified and digitized by preamplifier and analog-to-digital converter (ADC), respectively, and then is recorded by the spectrometer. This time domain signal is sent to computer for further processing which transforms the time domain signals into frequency domain signals. The main step of such a processing is the Fourier Transformation, ahead of which multiple processing methods including *zero filling, apodization,* and *linear prediction* are applied to prevent information loss. After Fourier Transformation, a post-processing method *phase correlation* is applied to optimize the appearance of the frequency domain spectrum. The frequency domain signals are the chemical shift values which will be analyzed next.

31.4   NMR Peak Picking

Figure 31.1 shows a sample one dimensional chemical shift spectrum which is a sketch of a proton NMR spectrum for diacetone alcohol molecule [7]. In this spectrum, the x-axis is the chemical shift in ppm and the y-axis is the intensity. The high-resolution peaks can be identified with the functional groups in the molecule: $\delta = 1.23$ for 6 protons in $(CH_3)_2$, $\delta = 2.16$ for 3 protons in $CH_3C=O$, $\delta = 2.62$ for 2 protons in CH_2, and $\delta = 4.12$ for one proton in OH. In the spectrum, the peak at 0ppm is the reference peak and there are some other low intensity peaks which are considered as noise peaks.

Peak picking is a process designed to filter out artificial peaks, to calibrate NMR signal lineshapes, and to recognize the intensity of each peak. For protein structure determination, two and higher dimensional NMR spectra are used, where every axis is the chemical shift in ppm for a certain type of nuclei. Because of strongly overlapping peaks and spectral

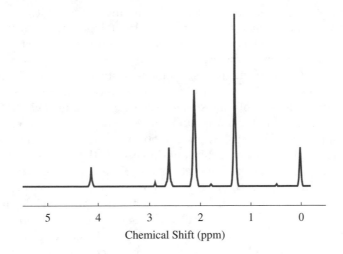

FIGURE 31.1: One dimensional NMR proton spectrum for diacetone alcohol molecule.

distortions due to artificial peaks, robust recognition methods should be developed. In the literature there are a number of existing methods such as neural networks [13, 16], statistical approaches [41, 3], and numerical analysis of various properties of the data points [33, 21]. We remark that even robust methods might fail for NMR signal recognition in complex spectra.

There exists also a number of free software for peak picking process. For example, AU-TOPSY [34] could deal with overlap and deviations from ideal Lorentzian line shape. AT-NOS [30] is mainly developed for automated NOESY peak picking. In more details, AU-TOPSY is an automated peak picking and peak integration method. The essences of this program are the function for local noise level calculation, the use of lineshapes extracted from well-separated peaks for resolving strongly overlapping peaks, and the consideration for symmetry. The key observation used by AUTOPSY is that multidimensional spectra typically contain multiple peaks that have the same lineshape and the same chemical shift in one frequency domain. ATNOS is an automated NOESY peak picking software to extract structural constraints. The input to ATNOS includes target protein sequence, chemical shift lists from peak assignment, and several 2D or 3D NOESY spectra. Current implementation of ATNOS performs multiple cycles of NOE peak identification combining with automated NOE assignment program CANDID [30]. In each cycle, ATNOS performs automated NOE peak picking by NOESY symmetry criterion. By reassessing the NOESY spectra in each cycle of structure calculation, ATNOS enables direct feedback among the NOESY spectra, the NOE assignments, and the protein structure. The new software package RADAR which combines ATNOS and CANDID will be freely available soon [30].

31.5   NMR Peak Assignment

NMR data acquisition and processing and the spectral peak picking require a complete knowledge of the NMR phenomenon. The output from every NMR experiment is a list of spectrum peaks whose quality is mostly dominated by the NMR instrumentation. It should be noted that in practice, the output peak list might still contain artificial peaks and some true peaks might be missing from the list because of data degeneracy. In any case, this list of peaks must be mapped to their host nuclei in the amino acid residues in the target protein sequence in order for the following stage of structural restraint extraction. Such a stage is referred to as *NMR Peak assignment* which usually involves multiple spectra or multiple peak lists. Peak assignment is a crucial stage in the whole process of protein structure determination as a small error in this stage might result in a huge structure gap. The NMR knowledge still plays an essential role in the success of peak assignment, nonetheless, computing techniques come into play with more protein structures being determined via NMR spectroscopy. More importantly, it allows further automation and thus potentially this currently one of the most time consuming steps in NMR structure determination will become high-throughput and thus make the overall structure determination a high-throughput technology.

Peak assignment is to map peaks from multiple spectra to their host nuclei in the target protein sequence. To accomplish the task, two main pieces of information are used. One of them is the nuclear information and the local chemical environmental information associated with the chemical shift values, which is called the *signature information*. In other words, one type of nuclei in a specific local environment will have its chemical shift values in a very narrow range. The other piece of information is from the correlation of multiple spectra. In more detail, the chemical shift of one nucleus will be observed in multiple spectra and thus it might help bridge the peaks. A spin system refers to a set of chemical shifts which are from nuclei residing in a common amino acid residue. One important cross-spectra deduction is that certain pairs of spin systems must be from adjacent amino acid residues in the target protein sequence, which is called the *adjacency information*. This is true because the two and higher dimensional NMR experiments are designed to detect the magnetic interactions among nuclei that are spatially close to each other and thus some peaks (which represent the chemical shifts for the nuclei in the interaction) are intra-residue (nuclei from a common amino acid residue) and some others are inter-residue (nuclei from two adjacent amino acid residues). It should be clear that peak assignment is another phase that detects artificial peaks if they are in conflict with the formed spin systems and their mapped residues.

There are many great efforts devoted to the assignment and several software tools developed (Figure 31.2), some of which are freely available. To name a few, PASTA [35] uses threshold accepting algorithms, GARANT [6, 9] uses genetic algorithm, PACES [17] and MAPPER [26] use exhaustive search algorithms, AutoAssign [51] uses heuristic best-first algorithms, among others.

Every one of these peak assignment approaches takes as input a set of multiple peak lists which come from various NMR experiments. In the following we will be using some of NMR spectra to demonstrate the detailed assignment process. Mathematically, one peak in a list records a vector of chemical shifts corresponding to the nuclei whose magnetic interaction is designed to be captured. For example, in two dimensional HSQC (heteronuclear single quantum correlation) spectrum, every peak contains two entries, one for amide proton (denoted as HN) [1] chemical shift and the other for the directly attached nitrogen (denoted as NH) chemical shift; In three dimensional HNCA spectrum, every peak contains three entries, the first one for NH chemical shift, the second one for carbon alpha (denoted as CA) chemical shift which could be in the same amino acid or in the preceding amino acid, and the third one for HN chemical shift. The peaks from HSQC and HNCA contain

---

[1] In the literature, amide proton is also denoted as H^N; the directly attached nitrogen is also denoted as N^H; the alpha carbon is also denoted as C_α; the beta carbon is also denoted as C_β; and the second backbone carbon is also denoted as C. In this chapter they are denoted as HN, NH, CA, CB and CO, respectively.

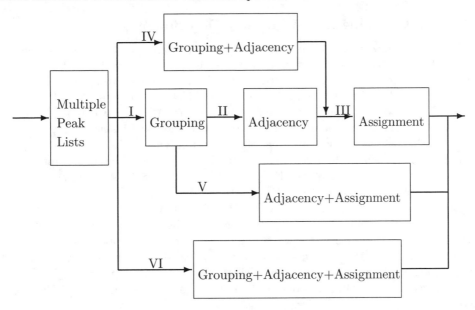

FIGURE 31.2: The flow chart of the peak assignment process: different works assume different starting positions. Phase I includes AutoAssign [51], PASTA [35]; Phase II includes AutoAssign [51], PASTA [35], Random [32]; Phase III includes AutoAssign [51], MAPPER [26], CBM [50]; Phase IV includes SmartNotebook [42]; Phase V includes PACES [17]; Phase VI includes GARANT [6, 9].

chemical shift values for a common nucleus, of which the difference should be within the reading error range, if not identical. On the other hand, every piece of chemical shift has its signature information on which type of amino acid residue the nucleus is in and which type of chemical environment the host amino acid residue is in. By using multiple chemical shifts, such signature information can be intensified. For this reason, almost every peak assignment has a stage call *peak grouping* which essentially groups the chemical shifts for nuclei residing in a common amino acid residue into a vector called *spin system*. For HSQC and HNCA spectra, a spin system would be a vector containing three chemical shifts for HN, NH, and CA (in fact we will not be able to form such spin systems using HSQC and HNCA alone, more details in the following). The spin system signature information is the sum of the signatures of the individual chemical shifts in the spin system. The adjacency information for a pair of spin systems is provided by the inter-residue peaks such that the peak in HNCA spectrum which records NH chemical shift, chemical shift for CA from preceding amino acid residue, and HN chemical shift, since one spin system would contain the NH and HN chemical shifts and the other spin system would contain the CA chemical shift.

As illustrated in Figure 31.2, some peak assignment methods do the peak grouping and adjacency determination first and then move on to the spin system assignment under the adjacency constraints; the others do the peak grouping and adjacency determination and the chained spin system assignment at the same time. We have developed several improved peak assignment methods along both lines and in the following we present some of them.

31.5.1   Peak Grouping and Adjacency Determination

We briefly describe how peaks from multiple spectra are grouped into spin systems and at the same time the adjacency is determined. We go with the ideal case first and then switch to the more complicated real spectral data. As an example, we use three spectra to demonstrate in the ideal case: HSQC, CA(CO)NH, and HNCA. We note that different NMR labs might be able to generate different combinations of NMR spectra, but the grouping and adjacency determination are done in a pretty much the same fashion.

The CA(CO)NH spectrum gives us triples of inter-residue chemical shifts (HN_i, CA_{i-1}, NH_i), where i indexes the residue. The HNCA spectrum gives us triples of inter- and intra-residue chemical shifts (HN_i, CA_{i-1}, NH_i) and (HN_i, CA_i, NH_i). The HSQC spectrum gives us pairs of intra-residue chemical shifts (HN_i, NH_i). Therefore, from these three spectra, we can associate with residue i a triple chemical shifts (HN_i, NH_i, CA_i). Our assignment goal is to identify for each residue its true triple. In the ideal case, the triples can be read out of these three NMR spectra, and the number of triples is equal to the number of residues in the target protein sequence. However, in general, the chemical shifts measured out of one NMR spectrum are different from those measured out of another NMR spectrum. Nonetheless, the difference is very small and we hope we can still be able to extract the triples, besides the existence of noise peaks and missing peaks.

In the following, we describe briefly on identifying the triples, together with the adjacency information between triples.

1. For a pair of chemical shifts (assuming it is associated with the ith residue) in HSQC spectrum, say (HN_i, NH_i), search for one triple in the CA(CO)NH spectrum that shares the HN and NH chemical shifts. Note that there may be none due to noise peaks or missing peaks; there may also be more and when there is more then one triple, we have to identify which one is the true triple. Similarly we expect to find two triples from the HNCA spectrum.

2. Suppose one triple in the CA(CO)NH spectrum is found. Then the CA chemical shift in the triple is from the $(i-1)$th residue. We can denote the triple as (HN_i, CA_{i-1}, NH_i).

3. Suppose two triples in the HNCA spectrum are found. Then these two CA chemical shifts are from the $(i-1)$th residue and from the ith residue. We can denote these two triples as (HN_i, CA_{i-1}, NH_i) and (HN_i, CA_i, NH_i).

4. Since triple (HN_i, CA_i, NH_i) appears in the HNCA spectrum only, we identify it as the triple for ith residue. To identify the triple for $(i+1)$th residue, we try to look for some pair of chemical shifts $(HN_{i'}, NH_{i'})$ from HSQC spectrum such that:

   - $(HN_{i'}, CA_i, NH_{i'})$ appears in the other two spectra,

   which signify the sequential adjacency relationship between two triple spin systems, (HN_i, NH_i, CA_i) and $(HN_{i'}, NH_{i'}, CA_{i'})$. Subsequently, we can assign $i' = i+1$.

This searching process is performed for every pair of chemical shifts in the HSQC spectrum to identify the triple and its adjacent triple. In the case that the chemical shift error is too large to be believed, the identifying process terminates and is re-started on another pair. The results are a set of triples, which are the spin systems, and the adjacency information among these spin systems. In the above ideal case, chemical shifts (and thus peaks) for a same nucleus (a same set of nuclei) are identical across all three spectra. This might not be true in reality as they are observed via different experiments and different references might be used, besides reading errors. The adjacency information connects spin systems into strings, which will be mapped to non-overlapping polypeptides in the target protein sequence. It has been shown [45] that the amount and the quality of adjacency information have significant effects on the assignment results.

As the readers will be seeing in future sections that most peak assignment systems assume the input to be a list of spin systems rather than several peak lists. Furthermore, they assume that the spin systems contain correctly and unambiguously grouped peaks from different spectra. In reality, the peak lists do not have the quality to make the grouping a trivial task, and the quality and quantity of spin systems produced in peak grouping stage have the most significant effect on the assignment. Some existing works (including AutoAssign) address this issue by using an essential the same approach, which considers the HSQC peaks as base peaks and subsequently maps the peaks from other spectra to these base peaks. The NH and HN chemical shift values of mapped peaks must fall within pre-specified match tolerances of base peaks. As the complexity goes high when the length of protein grows, these works depend on many redundant spectra to resolve the ambiguities and missing data. A big drawback of this approach is that it is very sensitive to the pre-specified match tolerance. A high match tolerance will cause more ambiguities that will in turn result in a dramatic increase in search space. A low match tolerance would produce less spin systems for low resolution spectral data, which wouldn't lead to a good assignment.

We propose in the following a new computational model which deals with the real but more complicated case. In this model, peak grouping is formulated as a weighted bipartite matching problem, where one side of vertices represent the peaks from one spectrum and the other side of vertices represent the peaks from another spectrum, and the edge weight measures the probability of mapping them to a common amino acid residue taken as the square root of the product of the differences of HN chemical shifts and NH chemical shifts (and therefore we assume that every peak contains these two chemical shifts). Our goal is to find a minimum weight matching which tells how two spectra can be "merged" into one. Repeating the process will eventually result in one "super" spectra which represents the list of spin systems. Our approach has the advantage that it produces a globally optimal peak grouping without considering the match tolerance and the resolution of spectral data.

31.5.2   Assignment Starting with Spin Systems

Despite a number of efforts devoted to group the peaks from different peak lists into spin systems and at the same time to determine their adjacencies [42], several assignment algorithms start with spin systems without any adjacency information. Rather than determining the adjacency during the grouping, they determine or predict the adjacency information from the spin systems. Such a process is of course easier, since the grouping is assumed. Nonetheless, in some algorithms the adjacency determination is not done alone but combined into the spin system assignment (Figure 31.2).   A few recent works along this line include PACES [17], AutoAssign [51], and Random Graph approach [32].

In essence, PACES represents the adjacency relationships among spin systems as a directed network and enumerates all possible paths in it. Each such path represents a possible string of the involved spin systems which has to be assigned to a polypeptide in the target protein sequence. PACES validates each path by mapping/assigning it to the best possible polypeptide incorporating the spin system signature information. We remark that this approach is only suitable for simple graphs because it is almost impossible to enumerate all paths in a graph with an average out-degree above 2.

AutoAssign combines the adjacency determination and the assignment to validate each other. The combination reduces the total number of possible paths dramatically compared to PACES. In more detail, for each amino acid residue in the target protein sequence, AutoAssign maintains a list of spin systems that the nuclei in this residue may generate. For each pair of spin systems, AutoAssign checks if they reside in the two lists of spin systems from two adjacent residues, respectively. If they do, then the pair is considered as a valid adjacent pair. (At the same time, their assignments could be made if the pair of adjacent residues in the target protein sequence is unique.) As it is obviously seen that the spin system signature information is employed for coming up with the lists of spin systems for every residue. AutoAssign extends this adjacent assignment of two spin systems to more. However, the number of combinations increases exponentially with the length of spin system path increases. Even worse, the ambiguities of the adjacencies among the spin systems also increases the complexity of this approach, even if the list for a residue contains only three spin systems on average. Indeed, AutoAssign requires much redundant information from extra NMR spectra in order to reduce the complexity.

Random graph approach models the adjacency determination again as a directed graph, where vertices represent the spin systems and weighted edges represent their adjacency with probabilities. It starts from an initial set containing all the unambiguous edges and iteratively chooses the edges from the remaining graph with probabilities proportional to their weights. In the real computation, an edge with a high probability doesn't always represent a good adjacency due to the noises and chemical shift degeneracy. On the other hand, the selection of the initial set is crucial to the performance. If the initial set contains some wrong edges, then these wrong edges will lead to the wrong adjacencies in the output.

In summary, researchers have realized that one big issue in adjacency determination is how to identify the correct adjacency out of multiple choices. If the data is of high quality, then the problem might become trivial. In reality, however, this is not the case. Except the works we reviewed in the above, many other approaches assume that a large amount of correct adjacency information could be extracted from the peak lists without much difficulty and focus more on the assignment under the extracted adjacency constraint. We, like the authors of the above works, have a different opinion that peak grouping and adjacency determination is more difficult and once it is done with a certain level of guarantee, then the assignment problem would become trivial. Such an opinion, of course, is based on a certain set of new observations, one of which is the learning of an accurate scoring scheme for quantified spin system signature information. Our simulation study supports the conclusion that when 80% adjacency information could be determined, then the peak assignment problem can be solved efficiently and accurately.

We proposed another way to combine the adjacency determination and the assignment, different from AutoAssign, but similarly assume grouping. The algorithm employs a *best-first* search incorporated with many other heuristics. One of our key observations is that, in practice, a string of connected spin systems typically has a much better score at the correct mapping position in the target protein sequence than almost all the other (incorrect) mapping positions. This stands out quite obviously when the size of the string increases. Such an observation leads to our conclusion that a string of spin systems having an outstanding mapping score has a high probability being correct. In other words, adjacency and assignment support each other. Our algorithm starts with an *Open List* of strings and seeks to expand the string with the best mapping score. The subsequently generated descendant (longer) strings are appended to the Open List only if their (normalized) mapping scores are better than their ancestor's. Another list, *Complete List*, is kept in the algorithm which saves strings not further expandable. At the time Open List becomes empty, high confident strings with their mapping positions are filtered out from Complete List with the conflicts resolved in a greedy fashion. Our preliminary simulation results show that the system outperforms PACES, AutoAssign, and Random Graph Approach significantly, and many instances couldn't be solved by them can be solved by the system.

31.5.3   Scoring Scheme for Signature Information

The spin system assignment problem can be naturally modeled as a weighted bipartite matching problem [50] on two disjoint groups, one group containing spin systems and the other containing a sequence of amino acids, if the adjacency information is not used as matching constraint. Presumably the weight of an edge measures the likelihood of the nuclei in an amino acid residue to generate the set of chemical shifts in the spin system. The matching goal is to find a one-to-one matching between elements of two groups that maximizes the total weight. However, many experimental results show that the quality of assignment from such a weighted bipartite matching is poor because frequently there are multiple amino acid residues of a same type in the target protein sequence. To differentiate the likelihood, the types of local chemical environment are included to provide the spin system signature information. The most commonly employed local chemical environment is the secondary structure which has three types, helix, strand, and coil.

In the rest of this section, we present our work on quantifying the spin system signature information through issuing a score scheme for the association of an arbitrary spin system and an amino acid residue in a certain type of secondary structure. We remark that prior to our work, generally it is assumed that for any type of atom in a combination of an amino acid residue and a secondary structure, the chemical shifts follow a normal distribution. Although well adopted, we justify that such an assumption is a very rough statistics and we conjecture that some other chemical environmental factors might affect the chemical shift values.

Data Collection and Preprocessing

Our training set for score scheme learning consists of the available NMR spectra of proteins collected in BioMagResBank [10]. The chemical shifts stored in the BioMagResBank database cannot be used immediately to do the learning as we want, for two reasons. The first reason is that among the protein sequences collected in the BioMagResBank database, a lot of them are close homologous. "bl2seq" [2] was run on every pair of the 863 sequences collected in the BioMagResBank and suitable clustering is done where each cluster contains sequences sharing at least 50% *modified* sequence identity (which is taken as the maximum number of matched amino acids over all local alignments returned by bl2seq divided by the length of the shorter sequence). This homology filtering gives us at the end 463 clusters and we randomly pick one protein from each cluster for our score scheme training. Secondly, for a single protein, the chemical shifts collected in the BioMagResBank for a few amino acids might be outliers. Since the abnormal behavior of a single outlier may disrupt our scoring scheme, an efficient statistical method, namely "boxplot" [19], was applied to remove the outliers. For example, after the treatment, the accepted range of CA's chemical shifts in our training data set is from 39.93 to 69.80, substantially narrowed from the observed range of CA's chemical shifts across all types of amino acids, i.e. from 4.10 to 85.492.

The chemical shift of a specific atom in a particular type of amino acid is not necessarily a constant, but affected by the local electronic-biochemical environment. For Alanines, the distributions of CA chemical shifts in helices, strands, and loop regions are significantly different, as shown in Figure 31.3. Our score scheme learning accounts for the impact of secondary structure, and uses the ones predicted by the PsiPred program [31, 38] (approximately 78% accuracy).

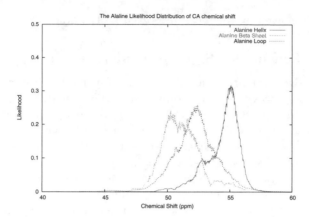

FIGURE 31.3: Alanine CA chemical shifts in different secondary structures.

Learning Scores

Our scoring scheme can take any combination of chemical shifts (the commonly used ones are HN, NH, HA, CA, CB and CO). We present here the scoring scheme learning using the combination of HN, NH, CA, and CB chemical shifts (a portion of the training data we collected in the last step).

(a) **Histogram-Based Score Learning:** For every combination of amino acid aa and secondary structure ss, let $\Pi(aa, ss)$ denote the chemical shift distribution we get out of the training dataset given aa and ss. We do not assume there is any specific pattern that the distribution should follow but use the chemical shift values directly. Let $N(aa, ss)$ denote the total number of chemical shift values collected in the database given aa and ss. For every type of chemical shift, we associate it with an error bound ϵ (which is different for different types of chemical shifts). The error bound ϵ is learned out of the training set such that 10 intervals of length ϵ cover the whole range of the chemical shifts. It's worth pointing out that the error bound maps very well to the reading error collected in BioMagResBank. For every examined quadruple spin system $(\mathrm{HN}_i, \mathrm{NH}_i, \mathrm{CA}_i, \mathrm{CB}_i)$, let $(\mathrm{CA}_i - \epsilon_\alpha, \mathrm{CA}_i + \epsilon_\alpha)$ be the CA chemical shift *window*. Let $N(aa, ss, \mathrm{CA}_i)$ denote the number of CA chemical shift values collected in the database which fall into the window. The *probability* that this examined CA chemical shift is generated from amino acid aa in the secondary structure ss is

$$\mathrm{Prob}(aa, ss, \mathrm{CA}_i) = \frac{N(aa, ss, \mathrm{CA}_i)}{N(aa, ss)}.$$

Similarly we can define the windows for NH and CB chemical shifts. The score for mapping the quadruple $(\mathrm{HN}_i, \mathrm{NH}_i, \mathrm{CA}_i, \mathrm{CB}_i)$ to combination (aa, ss) is

$$\begin{aligned}
&score((\mathrm{HN}_i, \mathrm{NH}_i, \mathrm{CA}_i, \mathrm{CB}_i) \mid (aa, ss)) \\
&= 10 \times \log \Big(\mathrm{Prob}(aa, ss, \mathrm{NH}_i) \times \mathrm{Prob}(aa, ss, \mathrm{CA}_i) \times \mathrm{Prob}(aa, ss, \mathrm{CB}_i) \Big)
\end{aligned}$$

(this number is truncated into an integer).

(b) **Representative-Based Score Learning:** This learning is to apply clustering methods from data mining. Clustering is the process of grouping a set of objects into classes such that objects within a class have high similarity but they are very dissimilar to objects outside of the class. Clustering is an unsupervised learning but with the number of classes pre-specified. The method of our choice is *Ordering Points To Identify the Clustering Structure* (OPTICS) [1] which is a hierarchical density-based clustering method to compute an augmented clustering ordering of objects for automatic and interactive clustering analysis. OPTICS not only discovers the arbitrary-shaped classes but also extracts the intrinsic structure and representatives for every resultant class. In this score scheme learning, every combination (aa, ss) represents a class and thus we have in total 60 classes. Applying OPTICS on the training set we get for each class a set of d representative spin systems.

To estimate the likelihood for any given spin system $v = (\mathrm{HN}_i, \mathrm{NH}_i, \mathrm{CA}_i, \mathrm{CB}_i)$ generated by nuclei in combination (aa, ss), the *tightness* is used which is defined as the sum of the Euclidean distance between the spin system and the d representative spin systems for the combination. In this sense, a spin system is regarded as a point in a four dimensional space. Namely,

$$score(v \mid (aa, ss)) = \sum_{j=1}^{d} ||v, v_j||_2,$$

where $||\cdot, \cdot||_2$ is the Euclidean distance between the two four dimensional points and $v_1, v_2, \ldots, v_d$ are the d representative spin systems for combination (aa, ss). Note that the less the score, the more likely the spin system is generated by nuclei from the combination.

(c) **Multiclass SVM with Error-Correcting Output Codes:** *Support Vector Machine* (SVM) is a binary classification tool. In order to apply it to our case containing 60 classes, we employ a common approach to extend it to a multiclass SVM by combining the outputs of several binary SVMs. In [18], a technique is introduced where *error-correcting codes* are used as a distributed output representation. Error-correcting codes can be regarded as a matrix of binary values such that each row represents a unique coding for a class and each column represents a target binary function associated with a binary classifier. Table 31.1 shows a 6-bit error-correcting output code matrix for 6 classes. In the training

class	Code					
	f_0	f_1	f_2	f_3	f_4	f_5
0	0	0	0	1	0	1
1	1	0	1	0	0	1
2	0	1	1	0	1	0
3	1	0	1	1	0	0
4	0	1	0	0	1	1
5	1	1	0	1	0	0

TABLE 31.1 An example of error-correcting output code matrix.

step, a unique classifier f_i $(i = 0, 1, \ldots, 5)$ is learned for each column. In classification step, a new instance is classified by each classifier and a 6-bit string is produced. This string is compared to each row and the nearest one is chosen as the class this new example belongs to.

Our third score scheme learning employs the multiclass SVM with error-correcting output codes. The code matrix is 60×64 which is generated by using the *Randomized Hill Climbing* algorithm [18]. Each row represents a combination of amino acid and secondary structure, which is a 64-bit code. These 64 SVMs are trained and each of them produces entries in one column in the code matrix. Given a new spin system $v = (\text{HN}_i, \text{NH}_i, \text{CA}_i, \text{CB}_i)$, a 64-bit string is produced by running these 64 SVMs. The hamming distance between it and each of the 60 class strings is taken as the score of mapping v to the combination. Again such a score measures the "distance" rather than likelihood and thus the less the score, the more likely the spin system is generated by nuclei from the combination.

Score Enhancement

One may notice that the distribution of HN chemical shifts was not used in the histogram-based score scheme. The reason is that across 60 *aa* and *ss* combinations, no significant difference is found, and thus adding it into account does not give much useful information. On the positive aspects, there are quite a few special features of the chemical shifts which can be taken advantage to estimate the scores. We point out two in the following: (1) Since there is no CB atom in a Glycine (or GLY, or G), no CB chemical shift should be examined. If a triple does contain a CB chemical shift, then it should not be generated from a GLY. Therefore, we can associate with the mapping a score `minimum`, which tells the assignment algorithm that such a mapping is *illegal*. Note that we only use this way of confident reasoning. For triples not containing a CB chemical shift, they are not necessarily generated from GLYs, due to possible peak missing, which happens over the time. (2) Similarly, since a Proline (or PRO, or P) doesn't have HN atom, a quadruple containing an HN chemical shift gets a score `minimum` when mapping to a PRO.

31.5.4 Assignment Starting with Spin Systems and Adjacency Constraint

Not using the adjacency information to constrain the output matching, the assignment made not much sense in terms of the number of correct mappings are recovered. This was widely true in out extensive simulation study. On the other hand, we have also observed that with an amount of 80% adjacency determined, the optimal constrained matching, employing any one of the above three score schemes, is about perfect meaning all correct mappings are included. We present next the computational model that incorporates both the spin system signature information and adjacency information into assignment, that called *Constrained Bipartite Matching* (CBM) problem. The CBM model is essentially the same as the normal weighted bipartite graph matching problem except that the group of amino acid residues are linearly ordered as they appear in the target protein sequence, and the group of spin systems are partitioned into subsets in each of them the spin systems form into a (directed) string which must be mapped to a polypeptide in the target protein sequence [50]. Theoretically, CBM problem is NP-hard even when the edge weights have only two values (the bipartite graph is complete) [50].

Some heuristics have been proposed very recently, including an exhaustive two-layer search algorithm [50] and some fast approximation algorithms [14]. These heuristics and approximations attempt to find feasible matchings with (approximately) the largest weights, and may work well for the NMR peak assignment. Later on, to overcome the explicit exhaustive search nature in the two-layer algorithm, another heuristics was developed for finding maximum-weight constrained bipartite matchings based on the branch-and-bound technique [36]. The branch-and-bound algorithm uses an efficient (unconstrained) bipartite matching algorithm and the approximation algorithms in [14] to compute necessary lower and upper bounds on optimal solutions to help prune the search tree, and returns an optimal solution, *i.e.* a feasible matching with the largest weight. It runs much faster than the two-layer algorithm. The comparison results for these heuristics and approximations can be found in [15, 36, 50].

MAPPER is a semi-automatic NMR assignment program that also takes in as inputs the spin systems and their adjacencies, but conducts the assignment in a different manner. In more detail, the input to the program consists of the target protein sequence, the spectroscopically assembled short fragments of sequential connected residues, and CA and CB chemical shifts or amino acid type information for each spin system. MAPPER performs first an individual mapping to enumerate all the possible mappings for each fragment; and then performs an exhaustive search for global mapping (i.e., self-consistent mappings of all fragments) to obtain an unambiguous assignment. The global mapping is performed by fragment nested loops, and the forbidden branches of the search tree will be cut as early as possible during the search. The only permissible overlap in global mapping is the overlap between two fragments which share one common residue since the corresponding chemical shift values for the endpoint atoms satisfy a user-defined tolerance.

A New Heuristics

Our assignment algorithm using the CBM formulation can be described as a two-phase procedure: in the first phase, a *greedy filtering* is conducted to select some number of best possible mappings for the identified strings; in the second phase, for every combination of string mappings, an efficient maximum weight bipartite matching algorithm is used to complete the assignment by mapping isolated spin systems to the rest of residues. The algorithm reports the best assignment from all combinations in terms of the assignment confidence — the score. Such a heuristics is fairly intuitive, and is very close to what is currently manually doing in an NMR laboratory. The main difference between the algorithm and manual work is that we employ efficient computational methods to automate the assignment process at a global view, which produce an assignment within seconds on a Pentium IV PC. The global view also helps avoid the tedious "undo-redo" which occurs very often though the manual work.

Greedy Filtering

The greedy filtering process is employed to take advantage of the discerning power of the score schemes. It is expected that with increasing length, the correct mapping positions for strings will stand out significantly. The process first sorts the strings into non-increasing length order, where the length of a string is measured by the number of spin systems therein. Let k be a parameter to bound the number of the best combinations we want to put into the second phase. The greedy filtering process starts with finding the first k best mapping positions for the longest string. This gives the top k combinations, which involve only the longest string at the moment. For every one of the k combinations, the greedy filtering process proceeds to find the first k best mapping positions for the second longest string. In general this will generate in total k^2 combinations involving one more string, and the process only keeps the top k ones, measured by the score, and proceeds to find the k best positions for the third longest string; and so on.

The process is repeated for all strings of length at least L, called *long strings*, where L is a lower bound on the length of strings involved in combinations. The output of the greedy filtering is a set of (at most) k combinations of mapping positions for the long strings. Recall that by the assumption, every identified string should map to a polypeptide in the target sequence and no residue on this polypeptide can be mapped multiple times. Therefore, there might be circumstances that the greedy filtering process terminates at some intermediate combinations, where it fails to find any legal mapping position for the next string.

These two parameters k and L can be tuned suitably to continue the assignment, depending on which algorithms employed in the second phase. In our assignment algorithm, the second phase algorithm is a maximum weight bipartite matching algorithm. Therefore, L is set to 2, meaning that all identified strings should be mapped to non-overlapping polypeptides from the target protein sequence. It worths mentioning that in the *Branch-and-Bound* heuristics studied in [36], L is set to 3, meaning that only those strings of length greater than 2 are long strings and thus should be mapped, and length-2 strings and isolated spin systems is left to be handled by Branch-and-Bound.

In order to find a feasible configuration for k and L, we tested a number of values for k and L and find that for almost all instances all correct mappings are found in the top 6 combinations. That is, $k = 6$ and $L = 4$ are the settings guaranteeing one of the combinations contains the correct mappings for all long strings.

Maximum Weight Bipartite Matching

An instance of the maximum weight bipartite matching problem consists of an edge-weighted bipartite graph $G = (S, R, E)$, where we assume without loss of generality that edge weights are non-negatives. Intuitively, the vertex set S contains all the isolated spin systems left after the greedy filtering (note that $L = 2$); the vertex set R contains all the remaining residues in the target protein sequence. An edge indicates the mapping between a spin system and a residue, where its weight records the confidence of the mapping. The goal of the problem is to compute a matching with maximum weight, corresponding to a partial assignment achieving the maximum confidence for those isolated spin systems. Since we require that all identified strings are mapped to non-overlapping polypeptides in any combination, $|S| = |R|$ is satisfied and the expected matching is perfect (meaning that every spin system is mapped to some residue). This is guaranteed since the generated bipartite graph is complete.

There are various implementations based on efficient algorithms for the maximum weight bipartite matching problem, for example one of the fastest algorithms called CSA developed by Goldberg and Kennedy [22], which is a cost-scaling push-relabel algorithm that finds minimum or maximum weight matchings of maximum cardinality.

31.5.5   Assignment Starting with Peak Lists

Most automated peak assignment programs apply the same general strategies as described in the above to perform peak grouping, adjacency determination, and assignment. Ambiguities arisen at each step are generally resolved within the step. If such ambiguities couldn't be resolved at that moment, then multiple outputs have to be produced and all of them would be considered as candidate inputs to next step. In practice when the spectral data was of low quality, those ambiguities were too complicated to be automatically resolved to produce any output; and manual adjustments had to be done which might require a long time process. During our development, we perceived that a good assignment always comes along with high confident adjacency information which is determined on the basis of correct peak grouping. Consequently, the quality of assignment could be regarded as a means for judging the correctness of resolution to the ambiguities. This observation motivates the fully automated peak assignment to do the three jobs at the same time. The following is a brief description of the system.

(1) All peaks from input spectra are put together to form a super peak list, where suitable

shuffling is required to make the spectra to have the same reference point (NH and HN chemical shifts were employed); (2) A clustering algorithm is applied on the super peak list to generate peak clusters such that peaks within a cluster share close NH and HN chemical shifts, where the number of clusters is set to the estimated number of spin systems using the target protein sequence (note that some different amino acid residues might have close NH and HN chemical shifts and thus multiple spin systems might reside in a cluster); (3) Since we cannot distinguish inter-residue and intra-residue peaks, an *undirected* graph $G = (V, E)$ is defined where each vertex represents a cluster and two vertices are adjacent if they contain close chemical shifts for some nuclei (excluding NH and HN, a tolerance threshold is set); (4) Apply a *best-first* search algorithm which takes in the score scheme learned in the above to find a path cover for graph G; At the same time, the direction of a path will be determined using the spectral nature with the exception that when the direction cannot be determined then two directed copies of it are generated. The output of the search algorithm is a (directed) path cover of G with their mapping positions to the target protein sequence.

We notice that such a system has a strong capability in resolving ambiguities and cross-validation. An existing assignment algorithm GARANT [6, 9] is the most likely one that can be classified into this category. GARANT starts with peak lists in two dimensional COSY and two dimensional NOESY spectra, and uses the knowledge of magnetization transfer pathways as the input. It represents peak assignment as an optimal match between two graphs, of which one is built for expected peaks predicted by combining knowledge of the primary structure and the magnetization transfer paths and the other is for the observed peaks. It employs a genetic algorithm combined with a local optimization routine to find the optimal homomorphism of the graphs of the expected and observed peaks, which is evaluated by a sophisticated statistical score scheme based on *mutual information*.

31.6 Structure Determination

After the peak assignment is done, structural constraints on the target protein structure can be extracted. Some constraints are directly associated with the chemical shift values that were used in the assignment process; while the other should require other NMR experiments to provide. The latter category of structural constraints is correctly associated to the portion of target protein through the peak assignment. We start with the structural constraint extraction and then proceed to secondary structure prediction and tertiary structure calculation. It should be noted that the secondary structure prediction described in the following can be viewed as a dual subject in the above histogram-based score scheme learning, yet it also provides additional structural information using the predicted secondary structure to the next stage of tertiary structure calculation.

31.6.1 Structural Constraint Extraction

Structural constraints refer to the conformations on the target protein structure, which are hidden in the NMR spectral peaks and need to be correctly associated to the portion of the target protein. There are three major types of structural constraints that can be extracted from spectral data, namely, distance constraints, torsion angle constraints, and orientation constraints. Some other additional structural information, including chemical shift values, could be derived to further refine the calculated structure.

NOE-Derived Distance Constraints

Nuclear Overhauser Effect (NOE) is a common phenomenon between pairs of nuclei of any types at a spatial distance within 5Å. NOE-derived distance constraints are the most important source of structural information for protein structure determination. In an NOESY (Nuclear Overhauser Effect Spectroscopy) spectrum, NOE interactions between pairs of nuclei are shown as NOE peaks. Each dimension of the spectrum is the chemical shift of one type of nucleus. For example, a peak at $(4.5\text{ppm}, 4.6\text{ppm})$ in an ^{1}H-^{1}H NOESY spectrum records an interaction between a proton with chemical shift 4.5ppm and another proton with chemical shift 4.6ppm, and its intensity is proportional to the product of the inverse sixth power of the distance between these two protons and a correlation function $f(\cdot)$. It should be noted that the structural constraints for the next step of structure calculation are mostly from ^{1}H-^{1}H NOESY spectrum to provide the distance constraint between a pair of protons, especially for those pairs that are close in space but far away in target protein sequence. For this spectrum, peak intensities are commonly classified into *very strong*, *strong*, *medium*, *weak*, and *very weak* [37] which say that the distance is in the ranges of [2.3Å, 2.5Å], [2.8Å, 3.1Å], [3.1Å, 3.4Å], [3.5Å, 3.9Å], and [4.2Å, 5.0Å], respectively. Besides some modifications that could be applied to resolve the ambiguities, these distance constraints can be directly incorporated into structure calculation such as *Distance Geometry* [5, 29] and *Torsion Angle Dynamics* [24].

J-Coupling-Based Distance and Angle Constraints

J-coupling (or spin-spin coupling) is the interaction between nucleus spins transferred through the electrons of the chemical bonds. There are a few factors in a J-coupling that affect the coupling constant, namely, the nuclei involved, the distance between the two nuclei, the angle of interaction between the two nuclei, and the nuclear spin of the nuclei. Both homonuclear and heteronuclear J-couplings can provide the nucleus distance (the less number of chemical bonds between a pair of nuclei, the stronger the coupling constant is) and the covalent chemical bonds angle (the smaller the angle, the bigger the coupling constant — the *Geminal* coupling or two-bond coupling or ^{2}J coupling). Among them, one of the most employed couplings is *Vicinal* (or three-bond, or ^{3}J) coupling which is dependent upon the dihedral angle θ between the nuclei. Generally, the more eclipsed or antiperiplanar the nuclei the greater the coupling constant and the relationship between dihedral angle and coupling constant is known as the *Karplus* relationship (that is, the coupling constant is $A^2 \cos\theta + B\cos\theta + C$ where constants A, B, C are empirically determined). Obviously, if the involved four nuclei are NH-CA-C-NH, then dihedral angle θ will confine the backbone torsion angles ϕ and ψ as local structural conformation to the next stage of structure calculation.

RDC-Based Orientation Constraints

Residual dipolar couplings (RDCs) constraints are introduced into structure calculation as orientation constraints. Structural information is obtained from RDCs by observing inter-nuclear dipolar interactions.

In solution, proteins are isotropically oriented, and so the inter-nuclear dipolar interactions average to zero and cannot be observed. If proteins are immersed into an anisotropic environment which has different properties in different directions, such as solutions containing phases or bicelles, then dipolar couplings no longer average to zero but produce an observable *Residual Dipolar Coupling* (RDC). The RDC value between two nuclei is a function in θ and ϕ which are cylindrical coordinates describing the orientation of the inter-nuclear vector in the principal axis system of the molecular alignment tensor. More specifically, it is equal to $D_a\left[(3\cos^2\theta - 1) + \frac{3}{2}D_r\sin^2\theta\cos 2\phi\right]$, where D_a is the dipolar coupling tensor and D_r is the rhombicity. Therefore, given the molecular alignment tensor, RDCs provide the orientation of inter-nuclear vectors relative to an external reference frame which is defined in the structure calculation process as an orthogonal axis system [43].

31.6.2   Secondary Structure Prediction

With the structural constraints being correctly associated to their host nuclei and chemical bonds, we may start the three-dimensional structure calculation. As an intermediate step, some research work has been done on target protein secondary structure prediction. Although such a prediction is limited since quite a number of NMR experiments have to be done, and it is very expensive compared to those protein secondary structure predictors not relying on NMR data. Nonetheless, this prediction can be done very fast, compared to three-dimensional structure calculation, and it is very accurate and thus is able to provide more structural constraints to three-dimensional structure calculation.

Most of the secondary structure predictions are done through establishing empirical correlations between protein secondary structure and NMR spectral data. For instance, one correlation between protein secondary structure and NOE-derived distance constraints and J-coupling-based angle constraints has been done in [49, 48]; Other correlations between protein secondary structure and chemical shifts of nuclei such as CA, CB, C, HA, and NH have been found in *Chemical Shift Index* (CSI) [47, 46], which gives an empirically determined table for structural element lookup, and (TALOS) [12], which adopts consensus from a well designed database of 20 proteins with high resolution X-ray structure. After the secondary structure elements for residues in the target protein have been recognized, the constraints on torsion angles will be given as local conformation for three-dimensional structure calculation.

Recall that the predicted target protein secondary structure was used in the peak assignment. However, this step of secondary structure prediction assumes the peak assignment is done. It might look contradictory but keep in mind that this step of secondary structure prediction is only an intermediate step to provide more structural constraints to the next step of three-dimensional structure calculation. Of course, on the other hand, it could also serve as a double-checking step to validate the predicted secondary structure employed in the peak assignment stage.

31.6.3   Three-Dimensional Structure Calculation

With all the structural constraints derived in the above two subsections, we are ready to do the three-dimensional structure calculation. We will introduce four algorithms in the following.

Metric Matrix Distance Geometry

Distance Geometry (DG) is the earliest algorithm used in structure calculation and its underlying principle is that: it is possible to calculate Cartesian coordinates for a set of points in the three-dimensional space if all the pairwise distances are known.

The *Metric Matrix Distance Geometry* [27] uses an $N \times N$ matrix G to solve the structure, where N is the number of atoms in the target protein sequence. Matrix G has only three positive eigenvalues and all the other $N-3$ eigenvalues are zero. Let $r_i, i = 1, 2, \ldots, N$ denote the coordinates of the ith atom in Cartesian three-dimensional space, to be calculated; and D_{ij} denote the Euclidean distance between the ith and the jth atoms. The matrix entry G_{ij} is defined to be the inner product of r_i and r_j, i.e.,

$$G_{ij} = r_i \cdot r_j = \left\{ \begin{array}{ll} \frac{1}{N} \sum_{k=1}^{N} D_{ik}^2 - \frac{1}{2N^2} \sum_{k=1}^{N} \sum_{l=1}^{N} D_{kl}^2, & \text{if } i = j, \\ \frac{1}{2}(D_{ij}^2 - G_{ii} - G_{jj}), & \text{if } i \neq j. \end{array} \right.$$

Theoretically, from all the pairwise distances (i.e., distance matrix D) we will be able to construct metric matrix G and then from G to calculate r_i's,

$$r_i^{\alpha} = \sqrt{\lambda^{\alpha}} e_i^{\alpha}, i = 1, 2, \ldots, N; \alpha = 1, 2, 3,$$

where λ^{α} and e^{α} denote the positive eigenvalues and the corresponding eigenvectors of G, and (r_i^1, r_i^2, r_i^3) is the Cartesian coordinate of the ith atom. In practice, however, not every pairwise distance is available and for most pairs only a range of distance is known. There are a few proposals on how to approximately determine matrix D [29] and afterwards a series of triangle inequality checking have to be done to make sure the resultant matrix G has only three positive eigenvalues [5]. Because of the aforementioned difficulties, DG algorithms are no longer favorable [27]. However, DG algorithms can still be applied to generate starting structures for other better structure calculation algorithms [39, 11, 24].

Variable Target Function Method

Structure calculation in variable target function method is formulated as a target function minimization problem, where the function counts the number of structural constraint violations. The variables in the target function are torsion angles. This says that the degrees of freedom are n torsion angles $\phi_1, \phi_2, \ldots, \phi_n$. The target function $T(\phi_1, \phi_2, \ldots, \phi_n)$ is equal to zero if all the experimentally derived constraints are satisfied. In general, solving a target function starts with small size targets such as intra-residue constraints and then increases the target size in a step-wise fashion up to the whole protein. In other words, local conformations of the protein sequence will be obtained first and the global fold could only be established approaching the end of the calculation.

In the literature there are a number of target functions been employed, among which some representative ones are DISMAN [8] and DIANA [23]. Although having been proven to perform well in determining helical proteins [25], variable target function method is still of low success rate in structure calculation as it is easy to be trapped in local minima [8]. One proposal is to feed in a large number of randomized starting structures in order to generate a group of good structures. On the other hand, compromise will be made between small constraint violation and the computational complexity.

Molecular Dynamics in Cartesian Space

Molecular Dynamics (MD) is a method for simulating the movement of molecular systems. *Simulated Annealing* (SA) is to simulate a slowly cooling process of molecular systems from an extremely high temperature. The method combining MD and SA is called *Molecular Dynamics Simulation* (MDS). The distinctive feature of MDS to other target function minimization methods is the presence of *kinetic energy*, which greatly reduces the probability of being trapped in local minima. Different from the standard MD [44], MDS uses a pseudo potential energy function as the target function.

MDS in Cartesian space defines a system on all the atoms in the target protein sequence. The overall energy function gives a force to each atom which defines an acceleration of the atom and thus a velocity at every time. One of the starting structure for MDS in Cartesian space is provided by Metric Matrix Distance Geometry and the initial coordinates for the atoms are randomly assigned according to some distribution such as Maxwell-Boltzmann distribution in XPLOR [11].

The complete procedure in MDS can be described as follows: (1) initialize structure and velocities for all the atoms; (2) calculate potential energy and from which calculate forces at every atom; (3) calculate the acceleration from force for every atom and update the velocity after time step Δt; (4) update the position (coordinates) after time step Δt for every atom using its velocity; control the temperature for SA by scaling the velocities; (5) repeat the above four steps till the system is equilibrium or the target temperature is reached. Note that multiple starting structures can be used to obtain a group of final structures.

Torsion Angle Dynamics

Torsion Angle Dynamics (TAD) is another molecular dynamics which uses torsion angles as the degrees of freedom rather than the Cartesian coordinates of the atoms, which in fact is the only fundamental difference from MDS in Cartesian space. DYANA [24] is one representative structure calculation program implemented in TAD, which is currently incorporated into software package CYANA [30, 28].

In TAD, a *rigid body* is defined as a collection of atoms whose relative three dimensional positions are unchangeable. These atoms reside in a common amino acid residue. The molecular system of the target protein is represented as a tree structure with a fixed base rigid body (of the N-terminus amino acid residue) and the other n rigid bodies (of the n amino acid residues) connected by n rotatable bonds, where $n + 1$ is the number of amino acid residues in the target protein sequence. These n torsion angles among $n + 1$ rigid bodies are denoted by $\theta_1, \theta_2, \ldots, \theta_n$. For the kth rigid body, there are a number of variables associated with it, $k = 1, 2, \ldots, n$. For example, $\vec{e}_k$ denotes a unit vector along the direction of the bond connecting rigid bodies $k - 1$ and k; and $\vec{r}_k$ is a position vector of the *reference point*, i.e. the endpoint of the bond between rigid bodies $k - 1$ and k. The only movement allowed in this tree structure is the rotation of the bonds. Some incompatible covalent structure of the tree structure, such as closed flexible rings, will be solved by the participation of other methods, such as MDS in Cartesian space.

The structure calculation in TAD is pretty much the same as in MDS, except the involvement of the kinetic energy in the torsion angle accelerations calculation. The complete procedure in TAD can be described as follows: (1) initialize structure by initializing the torsion angles and the torsional velocities, using some distribution; (2) using the torsion angles to calculate the Cartesian coordinates and thus the potential energy; (3) using the torsional velocities and rigid body linear velocities to calculate the kinetic energy; (4) calculate the torsional accelerations using both the potential and kinetic energies according to the Lagrange equations for classical mechanical systems [4]; (5) control the temperature for SA by scaling the velocities; (6) update the torsional velocities at a half time-step using *leap-frog scheme* [27] and update the torsion angles at full time-step; (7) repeat the above four steps till the system is equilibrium or the target temperature is reached. DYANA implements the above recursive process by setting the initial torsional velocities according to a normal distribution with zero mean value and a standard deviation that guarantee some initial temperature. Some experimental results show that DYANA calculates the structure in a reasonable amount of time (in minutes) for target proteins of length under 200.

31.7 Conclusions

This chapter provides a description of the full procedure of protein structure determination via Nuclear Magnetic Resonance spectroscopy. Our intention is to provide computer scientists working in the areas of bioinformatics and structural genomics a global picture of the technology. With this in mind, a number of problem formulations have been provided for every step in the structure determination procedure. In addition, some representative software packages assisting the determination have been briefly introduced. The readers interested in any part of the procedure should look into related references for more detailed description.

Specifically, we put focus on the NMR peak assignment process, whose automation would make the protein structure determination via NMR a high-throughput technology to satisfy the needs from structural genomics. Nonetheless, despite a lot of previous efforts, the automation is still not well solved. Our group has been working toward such a goal and brought out a number of works which partially speed up the peak assignment and make it more accurately by significant percentages. Our current and near future work is to put them together to produce an automated peak assignment pipeline. Many simulation studies are underway, as well as tests on real NMR data.

References

[1] M. Ankerst, M.M. Breunig, H.-P. Kriegel, and J. Sander. OPTICS: Ordering Points To Identify the Clustering Structure. In *Proceedings of ACM SIGMOD'99 International Conference on Management of Data*, pages 49–60, 1999.

[2] S.F. Altschul, T.L. Madden, A.A. Schäffer, and J. Zhang *et al.* Gapped BLAST and PSI-BLAST: a new generation of protein database search programs. *Nucleic Acids Research*, 25:3389–3402, 1997.

[3] C. Antz, K.P. Neidig, and H.R. Kalbitzer. A general Bayesian method for an automated signal class recognition in 2D NMR spectra combined with a multivariate discriminant analysis. *Journal of Biomolecular NMR*, 5:287–296, 1995.

[4] V.I. Arnold. *Mathematical methods of classical mechanics*. Springer, 1978.

[5] W. Braun, C. Bosch, L.R. Brown, and N. Go *et al.* Combined use of proton-proton overhauser enhancements and a distance geometry algorithm for determination of polypeptide conformations. *Biochimica et Biophysics Acta*, 667:377–396, 1981.

[6] C. Bartels, M. Billeter, P. Güntert, and K. Wüthrich. Automated sequence-specific assignment of homologous proteins using the program GARANT. *Journal of Biomolecular NMR*, 7:207–213, 1996.

[7] E.D. Becker. *High Resolution NMR: Theory and Chemical Applications*. Academic Press, 2000.

[8] W. Braun and N. Go. Calculation of protein conformations by proton-proton distance constraints. *Journal of Molecular Biology*, 186:611–626, 1985.

[9] C. Bartels, P. Güntert, M. Billeter, and K. Wüthrich. GARANT – A general algorithm for resonance assignment of multidimensional nuclear magnetic resonance spectra. *Journal of Computational Chemistry*, 18:139–149, 1997.

[10] BioMagResBank. `http://www.bmrb.wisc.edu`. University of Wisconsin. Madison, Wisconsin.

[11] A.T. Brunger. *X-PLOR, Version 3.1. A system for X-ray Crystallography and NMR*. Yale University Press, 1992.

[12] G. Cornilescu, F. Delaglio, and A. Bax. Protein backbone angle restraints from searching a database for chemical shift and sequence homology. *Journal of Biomolecular NMR*, 13:289–302, 1999.

[13] S.A. Corne and P. Johnson. An artificial neural network of classifying cross peaks in two-dimensional NMR spectra. *Journal of Magnetic Resonance*, 100:256–266, 1992.

[14] Z.-Z. Chen, T. Jiang, G.-H. Lin, and J.J. Wen *et al.* Improved approximation algorithms for NMR spectral peak assignment. In *Proceedings of the 2nd Workshop on Algorithms in Bioinformatics (WABI 2002)*, volume 2452 of *Lecture Notes in Computer Science*, pages 82–96. Springer, 2002.

[15] Z.-Z. Chen, T. Jiang, G.-H. Lin, and J.J. Wen *et al.* Approximation algorithms for NMR spectral peak assignment. *Theoretical Computer Science*, 299:211–229, 2003.

[16] E.A. Carrara, F. Pagliari, and C. Nicolini. Neural networks for the peak-picking of nuclear magnetic resonance spectra. *Neural Networks*, 6:1023–1032, 1993.

[17] B.E. Coggins and P. Zhou. PACES: Protein sequential assignment by computer-assisted exhaustive search. *Journal of Biomolecular NMR*, 26:93–111, 2003.

[18] T.G. Dietterich and G. Bakiri. Solving multiclass learning problems via error-correcting output codes. *Journal of Artificial Intelligence Research*, 2:263–286, 1995.

[19] J.L. Devore. *Probability and Statistics for Engineering and the Science.* Duxbury Press, December 1999. Fifth Edition.

[20] A.E. Ferentz and G. Wagner. NMR spectroscopy: a multifaceted approach to macromolecular structure. *Quarterly Review Biophysics*, 33:29–65, 2000.

[21] D.S. Garret, R. Powers, A.M. Gronenborn, and G.M. Clore. A common sense approach to peak picking in two-, three-, and four-dimensional spectra using automatic computer analysis of contour diagrams. *Journal of Magnetic Resonance*, 95:214–220, 1991.

[22] A.V. Goldberg and R. Kennedy. An efficient cost scaling algorithm for the assignment problem. *Mathematical Programming*, 71:153–178, 1995.

[23] P. Güntert, W. Braun, and K. Wüthrich. Efficient computation of three-dimensional protein structures in solution from nuclear magnetic resonance data using the program DIANA and the supporting programs CALIBA, HABAS and GLOMSA. *Journal of Molecular Biology*, 217:517–530, 1991.

[24] P. Güntert, C. Mumenthaler, and K. Wüthrich. Torsion angle dynamics for NMR structure calculation with the new program DYANA. *Journal of Molecular Biology*, 273:283–298, 1997.

[25] P. Güntert, Y.Q. Qian, G. Otting, and M. Muller *et al.* Structure determination of the Antp ($C_{39} \rightarrow$ S) homeodomain from nuclear magnetic resonance data in solution using a novel strategy for the structure calculation with the programs DIANA, CALIBA, HABAS and GLOMSA. *Journal of Molecular Biology*, 217:531–540, 1991.

[26] P. Güntert, M. Salzmann, D. Braun, and K. Wüthrich. Sequence-specific NMR assignment of proteins by global fragment mapping with the program MAPPER. *Journal of Biomolecular NMR*, 18:129–137, 2000.

[27] P. Güntert. Structure calculation of biological macromolecules from NMR data. *Quarterly Reviews of Biophysics*, 31:145–237, 1998.

[28] P. Güntert. Automated NMR protein structure calculation. *Progress in Nuclear Magnetic Resonance Spectroscopy*, 43:105–125, 2003.

[29] T.F. Havel and K. Wüthrich. A distance geometry program for determining the structures of small proteins and other macromolecules from nuclear magnetic resonance measurements of intramolecular ^{1}H-^{1}H proximities in solution. *Bulletin of Mathematical Biology*, 46:673–698, 1984.

[30] T. Herrmann, P. Güntert, and K. Wüthrich. Protein NMR structure determination with automated NOE assignment using the new software CANDID and the torsion angle dynamics algorithm DYANA. *Journal of Molecular Biology*, 319:209–227, 2002.

[31] D.T. Jones. Protein secondary structure prediction based on position-specific scoring matrices. *Journal of Molecular Biology*, 292:195–202, 1999.

[32] C.B. Kellogg, S. Chainraj, and G. Pandurangan. A random graph approach to NMR sequential assignment. In *RECOMB'04*, pages 58–67, 2004.

[33] G.J. Kleywegt, R. Boelens, and R. Kaptein. A versatile approach toward the partially automatic recognition of cross peaks in 2D ^{1}H NMR spectra. *Journal of Magnetic Resonance*, 88:601–608, 1990.

[34] R. Koradi, M. Billeter, M. Engeli, and P. Güntert *et al.* Automated peak picking and peak integration in macromolecular NMR spectra using AUTOPSY. *Journal of Magnetic Resonance*, 135:288–297, 1998.

[35] M. Leutner, R.M. Gschwind, J. Liermann, and C. Schwarz *et al.* Automated backbone assignment of labeled proteins using the threshold accepting algorithm. *Journal of Biomolecular NMR*, 11:31–43, 1998.

[36] G.-H. Lin, D. Xu, Z.Z. Chen, and T. Jiang *et al.* An efficient branch-and-bound algorithm for the assignment of protein backbone NMR peaks. In *Proceedings of the First IEEE Computer Society Bioinformatics Conference (CSB 2002)*, pages 165–174, 2002.

[37] J.L. Markley, A. Bax, Y. Arata, and C.W. Hilbers *et al.* Recommendations for the presentation of NMR structures of proteins and nucleic acids. *Journal of Molecular Biology*, 280:933–952, 1998.

[38] L.J. McGuffin, K. Bryson, and D. T. Jones. The PSIPRED protein structure prediction server. *Bioinformatics*, 16:404–405, 2000.

[39] M. Nilges, G.M. Clore, and A.M. Gronenborn. Determination of three-dimensional structures of proteins from interproton distance data by hybrid distance geometry-dynamical simulated annealing calculations. *FEBS Letters*, 229:317–324, 1988.

[40] *Pilot Projects for the Protein Structure Initiative (Structural Genomics)*. National Institute of General Medical Sciences, Washington, D.C., 1999. `http://www.nih.gov/grants/guide/rfa-files/RFA-GM-99-009.html`.

[41] A. Rouh, A. Louis-Joseph, and J.Y. Lallemand. Bayesian signal extraction from noisy FT NMR spectra. *Journal of Biomolecular NMR*, 4:505–518, 1994.

[42] C.M. Slupsky, R.F. Boyko, V.K. Booth, and B.D. Sykes. SMARTNOTEBOOK: a semi-automated approach to protein sequential NMR resonance assignments. *Journal of Biomolecular NMR*, 27:313–321, 2003.

[43] N. Tjandra, J.G. Omichinski, A.M. Gronenborn, and G.M. Clore *et al.* Use of dipolar ^{1}H-^{15}N and ^{1}H-^{13}C couplings in the structure determination of magnetically oriented macromolecules in solution. *Nature Structural Biology*, 4:732–738, 1997.

[44] W.F. van Gunsteren and H.J.C. Berendsen. Computer simulation of molecular dynamics: methodology, applications and perspectives in chemistry. *Angewandte Chemie International Edition*, 29:992–1023, 1990.

[45] X. Wan, T. Tegos, and G.-H. Lin. Histogram-based scoring schemes for protein NMR resonance assignment. *Journal of Bioinformatics and Computational Biology*, 2:747–764, 2004.

[46] D.S. Wishart and B.D. Sykes. The ^{13}C Chemical-Shift Index: A simple method for the identification of protein secondary structure using ^{13}C chemical-shift data. *Journal of Biomolecular NMR*, 4:171–180, 1994.

[47] D.S. Wishart, B.D. Sykes, and F.M. Richards. The Chemical Shift Index: A fast and simple method of the assignment of protein secondary structure through NMR spectroscopy. *Biochemistry*, 31:1647–1651, 1992.

[48] K. Wüthrich. *NMR of Proteins and Nucleic Acids*. Wiley, John & Sons, New York, 1986.

[49] K. Wüthrich, M. Billeter, and W. Braun. Polypeptide secondary structure determination by nuclear magnetic resonance observation of short proton-proton distances. *Journal of Molecular Biology*, 180:715–740, 1984.

[50] Y. Xu, D. Xu, D. Kim, and V. Olman *et al.* Automated assignment of backbone

NMR peaks using constrained bipartite matching. *IEEE Computing in Science & Engineering*, 4:50–62, 2002.

[51] D.E. Zimmerman, C.A. Kulikowski, Y. Huang, and W.F.M. Tashiro *et al.* Automated analysis of protein NMR assignments using methods from artificial intelligence. *Journal of Molecular Biology*, 269:592–610, 1997.

32

Geometric and Signal Processing of Reconstructed 3D Maps of Molecular Complexes

Chandrajit Bajaj
The University of Texas at Austin

Zeyun Yu
The University of Texas at Austin

32.1 Introduction

Today, hybrid experimental approaches for capturing molecular structures (henceforth, complexes), utilizing cryo-electron microscopy (cryo-EM), electron tomography (ET), X-ray crystallography (X-ray) or nuclear magnetic resonance spectroscopy (NMR) , need to be ably complemented with faster and more accurate computational and geometric processing for ultrastructure elucidation at the best level of resolution that is possible [21].

Electron Microscopy (EM) and in particular single particle reconstruction using cryo-EM, has rapidly advanced over recent years, such that several complexes can be resolved routinely at low resolution (10-20 $\mathring{A}$) and in some cases at sub-nanometer (intermediate) resolution (7-10 $\mathring{A}$) [4]. These complexes provide not only insights into protein and nucleic acid folds, but perhaps even more importantly provide information about how the various structural components interact. There are increasing numbers of molecules where the tertiary or secondary structure of a complex can be fully determined using EM [80]. Often the crystal structures of individual domains or components of these complexes are also known. An emerging trend in these fields is to fit the atomic resolution X-ray crystal structures into the cryo-EM map, to provide a quasi-atomic resolution model of the overall complex, possibly revealing details about molecular interactions within the assembly. In addition, with the increasing capability of determining multiple functional conformers of a complex, there is the promise of studying the dynamics of such interacting systems. The large physical size and complexity of such complexes combined with intermediate to low resolution models, presents challenges for structure to biological function determination.

This chapter reviews some of the crucial three dimensional geometric post-processing once a volumetric cryo-EM map (henceforth a 3D map) has been reconstructed, as essential steps towards an enhanced and automated computational ultrastructure determination pipeline. In particular the paper addresses 3D Map contrast enhancement, filtering, automated structural feature and subunit identification, and segmentation, as well as the development of quasi-atomic models from the reconstructed 3D Map via structure fitting.

32.2   Map Preprocessing

32.2.1   Contrast Enhancement

Many reconstructed 3D Maps, as well as captured 2D EM images, possess low contrast, or narrow intensity ranges i.e small differences between structural features and background densities, thereby making structure elucidation all the more difficult. Image contrast enhancement is a process used to "stretch" the intensity ranges, thereby improving the 2D image or 3D Map quality for better geometric postprocessing such as feature recognition, boundary segmentation, and visualization. The most commonly used methods in the past utilized global contrast manipulation based on histogram equalization [22, 49]. It is however well recognized today that using primarily global information is insufficient for proper contrast enhancement, as it often causes intensity saturation. Solutions to this problem include localized (or adaptive) histogram equalization [7, 56], which considers a local window for each individual image pixel and computes the new intensity value based on the local histogram defined within the local window. A more recently developed technique called the *retinex model* [32], in which the contribution of each pixel within its local window is weighted by computing the local average based on a Gaussian function. A later version, called the *multiscale retinex model* [31], gives better results but is computationally more intensive. Another technique for contrast enhancement is based on wavelet decomposition and reconstruction and has been largely used for medical image enhancement especially digital mammograms [38, 41].

A fast and local method for 2D image or 3D Map contrast enhancement that we have obtained very good success with, is presented in [75]. This is a localized version of classical contrast manipulations [22, 49]. The basic idea of this localized method is to design an adaptive one dimensional transfer function (mapping intensity ranges to intensity ranges) for each individual pixel (2D) or voxel (3D), based on the intensities in a suitable local neighborhood. There are three major steps, which we briefly describe for 2D images as its generalization to 3D Maps is straightforward. First, one computes local statistics (local average, minimum, and maximum) for each pixel using a fast propagation scheme [12, 72]. The propagation rule from a pixel, say, $(m-1, n)$ to a neighboring pixel (m, n) is defined as follows (similar propagation rules exist for other neighbors):

$$\text{lavg}_{m,n} = (1 - C) \times \text{lavg}_{m,n} + C \times \text{lavg}_{m-1,n} \qquad (32.1)$$

where C is called the *conductivity factor*, ranging from 0 to 1. The matrix *lavg* stands for the local average map, initialized with the input image's intensity values. The above propagation rule is sequentially applied in row & column order [12, 72]. In order to compute local min/max maps, some modifications are required for the above propagation scheme. To this end, a *conditional propagation scheme* is introduced in [75]. Assume that *lmin* and *lmax* stand for the local min/max maps, respectively. The *conditional propagation scheme* from $(m-1, n)$ to (m, n) is defined as follows:

$$
\begin{cases}
if(\mathrm{lmin}_{m-1,n} < \mathrm{lmin}_{m,n}) \\
\quad \mathrm{lmin}_{m,n} = (1 - C) \times \mathrm{lmin}_{m,n} + C \times \mathrm{lmin}_{m-1,n} \\
\\
if(\mathrm{lmax}_{m-1,n} > \mathrm{lmax}_{m,n}) \\
\quad \mathrm{lmax}_{m,n} = (1 - C) \times \mathrm{lmax}_{m,n} + C \times \mathrm{lmax}_{m-1,n}
\end{cases}
\tag{32.2}
$$

Once these local statistics are calculated, the second step is to design the 1-dimensional adaptive transfer function, to achieve intensity range stretching on a per pixel basis. Similar to global contrast manipulations, various linear or nonlinear functions can be used here but all such functions should "extend" the narrow range of the local intensity histogram to a much broader range so as to achieve contrast enhancement. In the approach of [75], the transfer function consists of two pieces: a convex curve (for stretching) in the dark-intensity range and a concave curve (for inverse stretching) in the bright-intensity range. The overall transfer function is C^1 continuous. Finally, in the last step, the intensity of each pixel is mapped to a new one using the calculated transfer function. This method inherits the advantages of the three afore-mentioned techniques, namely, global contrast manipulation, adaptive histogram equalization and the retinex model. However, unlike global contrast manipulation, this method is adaptive in the sense that the transfer functions are generally different from pixel to pixel. Also, unlike adaptive histogram equalization, this method considers a weighted contribution of each pixel within a local window. Furthermore, the size of the local window does not need to be pre-specified, due to the conditional propagation scheme used in this approach, which is also a significant difference between this method and the retinex model. Finally, the method of [75] demonstrates a multi-scale property as different choosing different conductivity factors are chosen and used in the propagation scheme. Paper [75], also gives an anisotropic version of the propagation scheme detailed above, and some results are shown in figure 32.1.

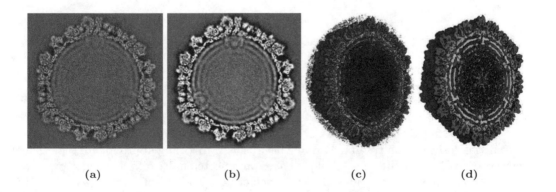

| (a) | (b) | (c) | (d) |

FIGURE 32.1: Anisotropic filtering and contrast enhancement of the Rice Dwarf Virus (RDV). (a) original map (showing only one slice). (b) filtered and enhanced (same slice). (c) original map (volume-rendered). (d) filtered (volume-rendered).

32.2.2 Noise Reduction

Reconstructed 3D Maps are noisy due to both 2D image acquisition as well as computational errors in the 2D to 3D portion of the reconstruction pipeline [21]. Applying 3D
noise reduction techniques on the 3D maps as a pre-processing step, facilitates improved
post-processing feature identification, segmentation and ultra structure determination. Traditional noise reduction filters applied to images include Gaussian filtering, median filtering,
and frequency domain filtering [22]. Most of the recent research however, has been devoted
to local anisotropic filters that operate with a directional bias, and vary in their ability
to reduce noise without blurring the geometric structural features, especially edges and
corners.

Bilateral filtering [5, 15, 16, 59], or sometimes called weighted Gaussian filtering, uses
an additional proximity weighting term to affect quasi-anisotropy. Partial differential equation (PDE) based filtering techniques, known popularly as anisotropic geometric diffusion
[48, 62], differ primarily in the complexity of the local anisotropic modulation. Another
popular anisotropic filtering approach is based on the use of the wavelet transformation
[14]. The basic idea is to identify and zero out wavelet coefficients of a signal that likely
correspond to image noise while maintaining the sharpness of the edges in an image [70].
The development of nonlinear median-based filters in recent years has also produced promising results. One of these filters, the mean-median (MEM) filter [25, 24], behaves differently
from the traditional median filter, and has been shown to preserve fine details of an image
while reducing noise. Among the aforementioned techniques, two noise reduction methods,
namely wavelet filtering [57] as well as non-linear anisotropic diffusion [19], have also been
applied to molecular tomographic imaging data.

An approach we have experimented successfully with on denoising reconstructed 3D maps,
utilizes bilateral pre-filtering [29], coupled to an evolution driven anisotropic geometric
diffusion PDE (partial differential equation) [2]. The PDE model is :

$$\partial_t \phi - \|\nabla\phi\| div \left(D^\sigma \frac{\nabla\phi}{\|\nabla\phi\|} \right) = 0 \qquad (32.3)$$

The efficacy of our method is based on a careful selection of the anisotropic diffusion
tensor D^σ based on estimates of the normal and principal curvature directions of a feature
isosurface (level-set) in three dimensions [2]. The diffusivities along the three independent
directions of the feature boundary, are determined by the local second order variation
of the intensity function, at each voxel. In order to estimate continuous first and second
order partial derivatives, a tricubic B-spline basis is used to locally approximate the original
intensity. A fast digital filtering technique based on repeated finite differencing, is employed
to generate the necessary tri-cubic B-spline coefficients. The anisotropic diffusion PDE is
discretized to a linear system by a finite element approach, and iteratively solved by the
conjugate gradient method.

In Figure 32.1, we show an example of a reconstructed cryo-EM map and the results
of filtering and contrast enhancement. In (a) and (b), only one slice of the 3D map is
illustrated. In (c) and (d), a volume-rendering of the original map is compared to that of
the filtered map.

32.2.3 Gradient Vector Diffusion

In the earlier subsection we considered volumetric filtering in the special context of "critical" feature preservation. For a given volumetric map, the critical features are the essential values that help define the hierarchical structure of a complex. In general these critical features could be points, curves, or surfaces. The critical points of a scalar map can be classified as one of three types: local maxima, local minima, and saddle points of the given scalar function. However, in the context of structure identification, the maximal critical points are of great interest, due to the fact that, in a molecular density map, higher densities imply the existence of more atoms. These critical points can be easily computed from the local maxima of a given scalar map. Since noise is always present in the original maps, a pre-filtering process should be applied. As mentioned in the earlier subsection, a scalar map pre-filter can be either linear or nonlinear. A linear filter (e.g., Gaussian filtering) may destroy some weak features and hence eliminate some critical points. A nonlinear pre-filter [48, 62], however, tends to "deform" a sub-region, yielding many unwanted critical points.

A good alternative is a vector field filtering technique that is based on the diffusion of gradient vectors of the scalar 3D map, from which the afore-mentioned critical points are also easily extracted. In [69], the authors described a diffusion technique to smooth gradient vector fields. The gradient vectors are represented by Cartesian coordinates and a set of partial differential equations (PDEs) are separately applied to each component of the vectors. The equations are linear or isotropic, and therefore inherit the drawbacks of most linear filtering systems. A better way to diffuse a gradient vector field is based on the polar-coordinate representation of the vectors [73, 74]. A drawback of this method is its computational complexity due to the efforts that have to be made to deal with the periodicity of orientation. An improved method is presented in [3, 76], and we provide some details below.

We detect the critical points using a set of anisotropic diffusion equations:

$$
\begin{cases}
\frac{du}{dt} = div(g(\alpha) \cdot \nabla u) \\[2mm]
\frac{dv}{dt} = div(g(\alpha) \cdot \nabla v) \\[2mm]
\frac{dw}{dt} = div(g(\alpha) \cdot \nabla w)
\end{cases}
\tag{32.4}
$$

where (u, v, w) are initialized with the gradient vectors of the original maps. $g(\cdot)$ is a decreasing function and α is the angle between the central vector and its surrounding vectors. For instance, we can define $g(\alpha)$ as follows:

$$
g(\vec{c}, \vec{s}) =
\begin{cases}
e^{\kappa \cdot \left(\frac{\vec{c} \cdot \vec{s}}{\|\vec{c}\| \|\vec{s}\|} - 1 \right)} & \text{if } \vec{c} \neq 0 \ \text{and} \ \vec{s} \neq 0 \\[3mm]
0 & \text{if } \vec{c} = 0 \ \text{or} \ \vec{s} = 0
\end{cases}
\tag{32.5}
$$

where κ is a positive constant; $\vec{c}$ and $\vec{s}$ stand for the central vector and one of the surrounding vectors, respectively.

Once the gradient vector field is generated and diffused, we can define the *critical points* as those where none of the surrounding vectors is pointing away from those points. These critical points shall be frequently used in the following sections dealing with structural feature identification.

To better illustrate the application of the anisotropic gradient vector diffusion technique to accurately extract critical points from a given 3D map, we show cross-sectional two-dimensional (2D) slices (Figure 32.2). The images are from a slice of the herpes virus

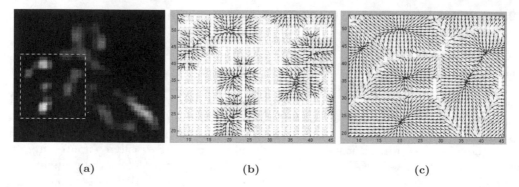

(a)   (b)   (c)

FIGURE 32.2: Illustration of critical point extraction using gradient vector diffusion. (a) one slice of herpesvirus capsid protein, vp5. (b) gradient vector field without diffusion corresponding to the boxed out area in (a). (c) gradient vector field after diffusion (10 iterations) improves the detection of critical points.

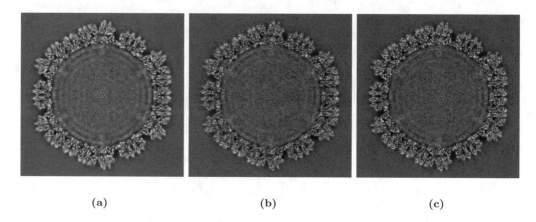

(a)   (b)   (c)

FIGURE 32.3: Illustration of critical point extraction using gradient vector diffusion. (a) one slice (noise reduced) of rice dwarf virus (RDV). (b) after 10 iterations (1214 critical points were extracted). (c) after 30 iterations (781 critical points were extracted). The number of critical points can be further reduced by removing those whose density values are less than a certain threshold.

capsid protein vp5 [81]. For better illustration of vector fields, we only consider a small area as boxed out in Figure 32.2(a). The vector field in Figure 32.2(b) is computed before the vector diffusion. Figure 32.2(c) demonstrates the power of the anisotropic vector diffusion, from which one can easily extract the critical points. Another example with greater detail, is illustrated in Figure 32.3, where one can see that running the vector diffusion with different numbers of iterations can result in multiple levels of critical points.

32.3 Structural Feature Identification

32.3.1 Symmetry Detection

The symmetry of a shape or structure provides fundamental information for shape recognition, and representation. Given the reconstructed 3D map of a large biomolecular complex, one may ask: (1) Does this structure exhibit certain global and local symmetries? (2) If it does, what type of symmetries are present (reflectional, rotational, translational, etc)?

(3) If the symmetry is rotational, what is the folding number and what is the location of the symmetry axis? Past relevant work devoted to answering the above questions in the literature include [13, 40, 43, 55, 58, 71, 79], most of which, however, were applied to simpler inputs, such as a set of points, curves, or polygons.

In many cases, the 3D maps are of spherical viruses, whose protein capsid shells exhibit icosahedral symmetry. In these cases, the global symmetry detection can be simplified to computing the location of the 5-fold rotational symmetry axes, passing through the twelve vertices of the icosahedron, after which the 3-fold symmetry axis for the twenty icosahedron faces and the 2-fold symmetry axis for the thirty icosahedron edges can be easily derived. However local symmetries of the protein arrangement on virus capsid shells are more complicated, exhibiting varied k-fold symmetry and their detection requires a modified correlation based search algorithm explained below [78].

In almost all cases of single particle cryo-EM reconstruction, the origin of the 3D map is identical to the origin of its corresponding icosahedron, as global icosahedral symmetry is utilized in the reconstruction. Given an axis $l_{\theta,\varphi}$ passing through the origin, where θ and φ are defined in a classical way such that $\theta \in [-\pi, \pi]$ and $\varphi \in [-\pi/2, \pi/2]$, a 3D scalar map $f(\vec{r})$ is said to possess a 5-fold rotational symmetry about $l_{\theta,\varphi}$ if the following equation holds:

$$f(\vec{r}) = f(R_{(\theta,\varphi,2\pi/5)} \cdot \vec{r}), \quad \text{for} \quad \forall \vec{r} \tag{32.6}$$

where the 3×3 matrix $R_{(\theta,\varphi,\alpha)}$ is defined as the coordinate transformation that rotates a point counterclockwise about an axis $l_{\theta,\varphi}$ by an angle of α. In particular, the matrix $R_{(\theta,\varphi,\alpha)}$ can be decomposed into five fundamental coordinate transforms.

In order to detect, for example a 5-fold symmetry axis, one can simply correlate the original map with its rotated map and search in the resulting correlation map for peaks [43]. This method has a high computational complexity of $O(NM)$, where N is the number of voxels and M is the number of angular bins. In current applications of icosahedral virus reconstructions at medium resolution, N is roughly 700^3 and M is about 46,000 (a quasi-uniform sampling on the orientation sphere with a radius of 200-voxels). Although a number of techniques can be employed to speed up the search process by reducing the number of the angular bins (e.g., a principal axis method [58] or a coarse to fine hierarchical approach), it is still expensive as N is large. In recent work [77, 78], introduced a method for the fast detection of rotational symmetries, given the fold number. The idea there is to reduce N, the number of voxels to be tested, by restricting the correlation only to a subset of the critical points instead of the entire volume.

An example result of their method is shown in Figure 32.4(a) the scoring function of the outer capsid layer of the rice dwarf virus (RDV) 3D map [80], where one can clearly identify the "peaks" with high contrast. The corresponding 5-fold symmetry axes and the reconstructed icosahedra are shown in Figure 32.4(b). Experiments on this 3D Map data show that the correct symmetry axes could be calculated based only on 23 critical points, in contrast to the total number of 512^3 voxels in the original map (details are given in [77, 78]). The approach has been extended to automatically detect local symmetries, such as the 3- or 6-fold symmetry axes of the RDV map [78]. Figure 32.4(c) demonstrates the detection of the local symmetry axes of the outer capsid layer of RDV.

32.3.2   Boundary Segmentation

Segmentation is a way to electronically dissect significant biological components from a 3D map of a macromolecule, and thereby obtain a clearer view into the macromolecules architectural organization [17]. For instance, it is often helpful to segment an asymmetric

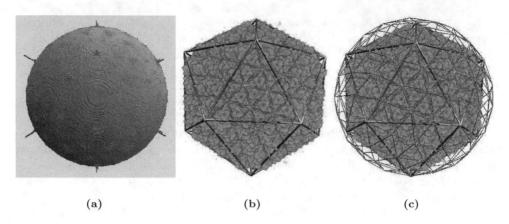

(a)   (b)   (c)

FIGURE 32.4: (See color insert following page 20-4.) Detection of Symmetry axes and
construction of global icosahedral symmetry as well as local n-fold symmetry. (a)
scoring function. (b) global icosahedral symmetry. (c) local 6-fold symmetry.

local subunit out of an icosahedral virus such that further structural interpretation can
be conducted only on the asymmetric unit instead of the entire map without loss of any
structural information. Segmentation of 3D maps is usually carried out either manually
[26, 27, 37, 39, 45] or semi-automatically [3, 20, 60]. Current efforts on the selection and
decomposition of an icosahedral map into its local subunits also relies largely on manual
work with extensive use of a graphical user interface [30, 80]. This manual task can be tedious when the resolution is only marginally high enough to discern the boundaries between
subunits.

Automated segmentation is still recognized as one of the challenge problems in image processing, although various techniques have been proposed for automated or semi-automated
segmentation. Commonly used semi-automatic methods include segmentation based on
edge detection, region growing and/or region merging, active curve/surface motion and
model based segmentation (see for example [54, 74]). In particular, two techniques have
been discussed in detail in the electron tomography community. One is called the *watershed immersion method* [60] and the other is based on normalized graph cut and eigenvector
analysis [20].

Papers [77, 78] present steps towards an automatic approach for asymmetric subunit
detection and segmentation of 3D maps of icosahedral viruses. The approach is an enhanced
variant of the well-known fast marching method [42, 53]. The basic idea of the fast marching
method is that a contour is initialized from a pre-chosen seed point, and the contour is
allowed to grow until a certain stopping condition is reached. Every voxel is assigned with
a value called *time*, which is initially zero for seed points and infinite for all other voxels.
Repeatedly, the voxel on the marching contour with minimal *time* value is deleted from
the contour and the *time* values of its neighbors are updated according to the following
equation:

$$\|\nabla T(\vec{r})\| \cdot F(\vec{r}) = 1 \qquad (32.7)$$

where $F(\vec{r})$ is called the *speed function* that is usually determined by the gradients of the
input maps (e.g., $F(\vec{r}) = e^{-\alpha\|\nabla I\|}$, where $\alpha > 0$ and I is the original map). The updated
neighbors, if they are updated for the first time, are then inserted into the contour. The
traditional fast marching method are designed for a single object boundary segmentation.
In order to segment multiple targets, such as 60-component virus capsids or a 3-component
molecular trimeric subunit, one has to choose a seed for each of the components. However,

assigning only one seed to each component may cause appropriate boundary detection problems, as demonstrated in [78], and hence a *re-initialization* scheme becomes necessary.

The automatic approach of [77, 78] consists of three steps: (1) detection of the critical points; (2) classification of critical points; (3) a multi-seed fast marching method. The technique for (1) the detection of critical points has been briefly described in the earlier subsection on Gradient Vector Diffusion, in this chapter. All the critical points are regarded as seeds in the fast marching method. In general, the number of critical points in a map is much larger than the number of object components of interest. In other words, each component is assigned with a number of seeds instead of just one. Every seed initiates a contour and all contours start to grow simultaneously and independently. Two contours corresponding to the same component merge into a single contour, while two contours corresponding to different components stop on their common boundaries.

The initial classification of critical points as part of step (2) of the algorithm, is crucial in the segmentation of virus 3D maps. The critical points are classified utilizing local or global symmetry and based on their equivalence in terms of the asymmetric components that are to be segmented. Once all the seeds are classified, the above multi-seed variant of the fast marching method is used. First, each component initially possesses multiple seeds and hence multiple initial contours. Second, each marching contour is assigned a membership index based on the classification of seeds and the assignment to components. Once a voxel (volume element of the 3D Map) is conquered by a marching contour, it is assigned with the same index as the marching contour. Third, two marching contours with the same index merge into one when they meet, while two marching contours with different indices stop at their touching boundaries.

The segmentation approach or [77] has been applied to the global asymmetric components dissection of icosahedral virus 3D maps. For viruses with more than 60 subunits that form a quasi-equivalent icosahedron, one additionally needs to incorporate the local symmetry axes of the viruses into the multi-seed classification and segmentation process [78]. Results from the above automatic segmentation technique applied to a reconstructed Cryo-EM 3D Map of the Rice Dwarf Virus (RDV) [80] are shown in Figure 32.5. The RDV has double spherical protein shells (called capsids) with icosahedral symmetry. The first level segmentation is a separation of these two shells from the 3D map (see Figure 32.5 (a)). Next is a segmentation of the asymmetric subunits within each capsid. The sixty asymmetric subunits of the outer capsid viewed from the 5-fold symmetry axis is shown in Figure 32.5(b). Each subunit consists of four and one third trimeric sub-subunits [80]. Figure 32.5 (c) and (d) illustrates the segmented trimers (260 in total), where (c) shows the view from outside while (d) shows the view from the inside. The segmentation shown in (c) and (d) requires the local symmetry detection as shown in Figure 32.4(c) and the algorithm discussed in detail in [78]. Figure 32.5 (e) shows the segmented trimeric subunit consisting of three monomeric units, each of the same protein P8. Figure 32.5(f) shows the P8 protein monomeric unit segmented from the trimeric unit based on local 3-fold symmetry. It is worthwhile pointing out that in the visualization of the segmented trimeric subunits in Figure 32.5(b) only five colors are used to distinguish between sixty subunits, such that any five subunits surrounding the 5-fold symmetry axis would have different colors. In other directions, however, one may see two adjacent subunits having the same color although technically they have different component memberships. One can certainly find a more sophisticated coloring scheme to assure any two adjacent subunits always have different colors. Several more example segmentations for both reconstructed cryo-EM 3D Maps and synthetic 3D maps generated from crystal structure data are given in [78].

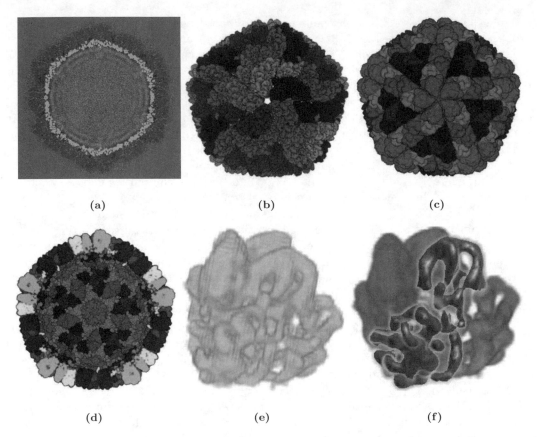

(a)   (b)   (c)

(d)   (e)   (f)

FIGURE 32.5: **(See color insert following page 20-4.)** Visualization of the architecture of
the Rice Dwarf Virus (RDV) 3D map (a) segmented outer and inner icosahe-
dral capsid boundaries (b) segmented asymmetric subunits of the outer capsid
(60 subunits in total). Each asymmetric subunit consists of four and one third
trimers. (c) & (d) segmented trimeric subunits (260 in total), where (c) shows
the view from the outside while (d) shows the view from inside. (e) each segmented trimeric subunit consists of three monomeric sub-subunits. (f) segmented
monomeric subunit represents the 3D density map of a single P8 protein. The
RDV 3D map data is courtesy Dr. Wah Chiu, NCMI,BCM, Houston

32.3.3   Secondary Structure Identification

Although atomic structures are not detectable in reconstructed 3D cryo-EM maps, given
their low feature resolution, it is sometimes feasible to locate secondary structures (alpha
helices and beta sheets) from those maps [9, 80]. An approach for detecting alpha helices in
3D maps has been described in [63], where the alpha helix is modelled with a cylinder (length
and thickness) and the cylinder is correlated with the segmented protein map. Since the
best solution is achieved by exhaustively searching in translation space (3D) and orientation
space (2D), this method is computationally expensive. In addition, this approach is designed
only for alpha helix detection, not for the beta sheets. Another approach, designed for
beta sheet detection, was recently proposed by [34, 35]. This method uses a disk (planar)
model for beta sheets. It inherits the disadvantage of slow computational speed due to the
exhaustive search in both translation and orientation space, and furthermore cannot find
curved beta sheets.

It is of course possible to combine the two methods above to detect both alpha helices and beta sheets, however to detect secondary structures efficiently one must avoiding the exhaustive search in both translation and orientation space. One possible approach is to consider scoring candidate helices/sheets only at the critical points of the 3D Map. This way, the search in translation space can be reduced to a significantly smaller number of locations. In addition, the search in orientation space at each critical point can be further reduced by utilizing the local structure tensor [18, 62]. Given the 3D map $f(x, y, z)$, the gradient tensor is defined as:

$$
G = \begin{pmatrix} f_x^2 & f_x f_y & f_x f_z \\ f_x f_y & f_y^2 & f_y f_z \\ f_x f_z & f_y f_z & f_z^2 \end{pmatrix}
\tag{32.8}
$$

This matrix has only one non-zero eigenvalue: $f_x^2 + f_y^2 + f_z^2$. The corresponding eigenvector of this eigenvalue is exactly the gradient (f_x, f_y, f_z). Therefore, this matrix alone does not give more information than the gradient vector. To make the gradient tensor useful, a spatial average (over the image domain) should be conducted for each of the entries of the gradient tensors, yielding what is called the *local structure tensor*. The averaging is usually based on a Gaussian filter:

$$
T = G_\alpha = \begin{pmatrix} f_x^2 * g_\alpha & f_x f_y * g_\alpha & f_x f_z * g_\alpha \\ f_x f_y * g_\alpha & f_y^2 * g_\alpha & f_y f_z * g_\alpha \\ f_x f_z * g_\alpha & f_y f_z * g_\alpha & f_z^2 * g_\alpha \end{pmatrix}
\tag{32.9}
$$

Here g_α is a Gaussian function with standard deviation α. The eigenvalues and eigenvectors of the structure tensor T indicate the overall distribution of the gradient vectors within the local window, similar to the well-known principal component analysis (PCA). Three typical structures can be characterized based on the eigenvalues [18]. Let the eigenvalues be $\lambda_1, \lambda_2, \lambda_3$ and $\lambda_1 \geq \lambda_2 \geq \lambda_3$. Then we have the following classifications:

1. blobs: $\lambda_1 \approx \lambda_2 \approx \lambda_3 > 0$.

2. lines: $\lambda_1 \approx \lambda_2 >> \lambda_3 \approx 0$.

3. planes: $\lambda_1 >> \lambda_2 \approx \lambda_3 \approx 0$.

For each of the critical points of the 3D map, the structure tensor and its corresponding eigenvalues are calculated. Next, the above criterion based on the eigenvalues of the local structure tensor is computed at each of the critical points to distinguish between alpha helices (line features) and beta sheets (plane features). A critical point classified as an alpha helix, is extended on both sides along the direction of the line structure determined by the local structure tensor, yielding a segment of the median axis of the 3D map. Similarly, for a critical point corresponding to a beta sheet feature, the plane feature is extended yielding a piece of median surface of the density map. Since a true alpha helix or beta sheet may consist of more than one critical point, it is necessary to merge a number of median segments and median surfaces, from which the final alpha helixes and/or beta sheets are constructed.

Figure 32.6 illustrates this approach on a Gaussian blurred map of the X-ray atomic structure of cytochrome c' (PDB-ID = 1bbh). Figure 32.6(a) shows the atomic structure,

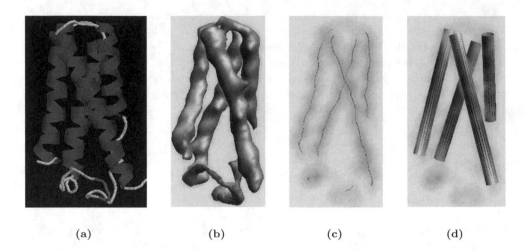

(a)   (b)   (c)   (d)

FIGURE 32.6: **(See color insert following page 20-4.)** Illustration of secondary structural
identification using local structure tensor at critical points of the 3D Map (a) The
X-ray atomic structure representation of cytochrome c' (PDB-ID = 1bbh). (b)
The volumetric representation of a Gaussian blurred 3D map generated from the
X-ray structure (c) The detected skeletons of the 3D map. (d) Four helices are
finally constructed from the skeletons, while the two on the bottom are discarded
as being too small for being helices

consisting of four alpha-helices, visualized as ribbons. The blurred map of this structure is
visualized by contour rendering in Figure 32.6(b). Based on the the critical points of the 3D
map and use of the structure tensor, the skeletons (median segments/planes) are computed
and shown in Figure 32.6(c). From the skeletons, the four alpha helices are constructed
as shown in Figure 32.6(d). Note that two segments of median axes on the bottom are
discarded simply because their lengths are too small to be a true alpha helix.

32.4   Structure Fitting

A primary technique for structure interpretation and molecular model construction is to
attempt to fit a known high-resolution structure (obtained by X-ray or NMR) into a reconstructed 3D density map. This technique is commonly known as *structure fitting* [52].
This technique bridges the resolution gap between low-resolution maps (e.g., lower than
10 Å) [6] and the atomic protein structures (e.g. lower than 3 Å). Figure 32.7 shows an
example of structure fitting between the P8 monomeric protein, segmented from the RDV
3D map [80], and its X-ray atomic structure [47]. Figure 32.7(a) shows the segmented P8
monomeric protein (also see Figure 32.5(f)). The crystal structure of P8 monomer is shown
in Figure 32.7(b), where one beta sheet (top) and two alpha helices (middle and bottom)
are highlighted and used as a high-resolution fitting model. This high-resolution model is
fit against the cryo-EM map of P8 monomer and its best position/orientation within the
cryo-EM map is determined and show in Figure 32.7(c).

There are several papers discussing various techniques on structure fitting. An excellent
review of prior work on this topic is given in [65]. One of the popular methods for volumetric
matching is based on Fourier transforms [44, 46]. The rigid-body fitting can be thought of

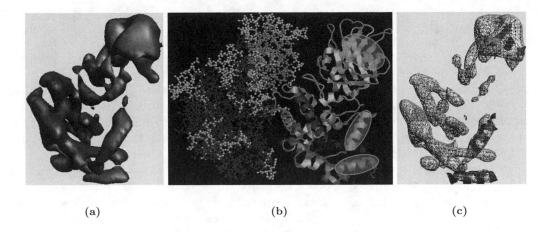

(a)   (b)   (c)

FIGURE 32.7: (See color insert following page 20-4.) Example of structural fitting in the segmented P8 monomeric protein of RDV. (a) P8 monomeric protein iso-surface visualization (b) X-ray atomic structure of the P8 monomeric protein represented in ways of balls & sticks and cartoons. One beta sheet (top) and two alpha helices (middle and bottom) are highlighted and used as a fitting model. (c) By maximizing the correlation between the X-ray atomic model and the 3D map of the P8 monomer, one builds a pseudo-atomic model of the 3D map

as the minimization of the discrepancy between the cryo-EM maps and the atomic structure in Fourier space. The discrepancy is defined as follows:

$$R = \sum_f ||F_{em}(f)| - \lambda |F_{calc}(f, r, t)||^n, \qquad n = 1 \quad or \quad 2 \qquad (32.10)$$

where F_{em} and F_{calc} are the Fourier transforms of the 3D map and the calculated atomic structure (that is, a Gaussian blurred 3D map of the atomic structure). Here r and t stand for rotation and translation parameters, respectively, and both r and t have three degrees of freedom.

Instead of fitting the structures in Fourier space, we can also perform the fitting in the real space [10, 33, 61]. It is known that the minimization of the R factor seen above is equivalent to the maximization of the cross-correlation defined as below:

$$C = \int \rho_{em}(\vec{x}) \rho_{calc}(\vec{x}, r, t) d\vec{r} \qquad (32.11)$$

where ρ_{em} and ρ_{calc} are the twin 3D maps of the cryo-EM and the Gaussian blurred atomic structure. The cross-correlation can be calculated by exhaustive searching with scaling or sampling of the translation (t) and rotation (r) parameters. While the Fast Fourier Transform (FFT) is easily used to speed up the cross-correlation scoring calculation over 3D translations [11, 36], it can also be used to compute the cross-correlation coefficients over rotational parameter (r) space, by first re-expressing the 3D map using trivariate spherical harmonics.

Another improvement on the conventional cross-correlation scoring method is to use a locally defined cross-correlation score [51]. In general, the global correlation method does not exclude the densities in the cryo-EM map that do not correspond to the atomic structure being considered. In addition, maximizing (32.11) often makes the solution "drift" to the highest density region in the cryo-EM map, which, however, does not mean the best-matched region. Hence the normalized and localized method [51] often gives more accurate fitting scores. One disadvantage of this method, however, is that the cross-correlation is conducted in real-space and a six-parameter searching space is considered in [51], resulting in a very slow performance. Recently, Roseman [50] incorporated the fast Fourier transform (FFT) into the local correlation algorithm and applied it to the particle detection in two-dimensional electron micrographs. It was said that the local correlation algorithm together with FFT could be two orders of magnitude faster than the explicit real-space implementation [50]. However, no results have been reported for 3D maps using this fast local correlation algorithm.

The conventional cross-correlation method can also be enhanced by a *contour-based fitting method* [8], in which the correlation coefficient is defined the same as (32.11) except that the Laplacian operator is applied to both maps before the calculation of the cross-correlation. Although this method is called *contour-based fitting*, it is not actually based on the detection of the contours. Due to the Laplacian operator that enhances the edges of both the cryo-EM map and the calculated atomic structure, this method was shown in [8] to give improved results (the resulting correlation map has higher contrast) than the classic cross-correlation method. However, as pointed out in [65], the Laplacian filter may also amplify the noise, which as a result may weaken the performance of this method.

All the above methods for structure fitting are based on cross-correlation between the cryo-EM reconstructions and the calculated atomic structures. A different strategy is based on a data reduction technique. This method has been studied by Wriggers *et al* [68, 67, 66, 64], based on a *vector quantization technique* [23, 28]. The idea of vector quantization is to represent a 3D map with a certain number of vectors (or points in 3D space), from which a weighted graph is constructed. Instead of computing the cross-correlation between the cryo-EM 3D map and the calculated 3D atomic blurred map, one computes a new "difference" function between the two graphs corresponding to the cryo-EM map and the calculated atomic structure map. The "difference" function can be used to search for the best volumetric matching. Although this approach reduces the overall search time for its best match, and it is also possible to extend this to flexible fitting [64, 65], it has two limitations. First, this method requires that the component of the cryo-EM map to be fitted should be isolated from the entire map. Second, the number of vectors must be carefully chosen. A large number of vectors exponentially increases the computational time while a small number of vectors may not be sufficient for perfect alignment and matching of the structural features of the map.

32.5 Conclusion

The field of structural biology is increasingly dependent on computational processing for structural determination of complexes from 3D Maps. Each of the computational structure or ultra-structure elucidation methods that we highlighted above in separate subsections, remains an active area of future research and development, as there is still a ways to go. Furthermore. as the size of these reconstructed 3D EM maps grows, computational processing techniques would need to be developed, working directly from multi-resolution and compressed volumetric representations [1]. Nevertheless, we are optimistic that with progressively better techniques for image acquisition, coupled to efficient map reconstruction, and enhanced computational 3D map processing for structure elucidation, it is only a matter of time until the resolution gap between X-ray structures and cryo-EM structures will be bridged.

Acknowledgements

This work was supported in part by NSF grants INT-9987409, ACI-022003, EIA-0325550, and grants from the NIH 0P20 RR020647, R01 GM074258. We are grateful to Dr. Wah Chiu and his group at the Baylor College of Medicine, for helpful discussions related to this project and for providing us the reconstructed 3D map of the rice dwarf virus.

References

[1] C. Bajaj, J. Castrilon-Candas, V. Siddavanahalli, and Z. Xu. Compressed representations of macromolecular structures and properties. *Structure*, 13(3):463–471, 2005.

[2] C. Bajaj, Q. Wu, and G. Xu. Level-set based volumetric anisotropic diffusion for 3D image denoising. In *ICES Technical Report, University of Texas at Austin*, 2003.

[3] C. Bajaj, Z. Yu, and M. Auer. Volumetric feature extraction and visualization of tomographic molecular imaging. *Journal of Structural Biology*, 144(1-2):132–143, 2003.

[4] T.S. Baker, N.H. Olson, and S.D. Fuller. Adding the third dimension to virus life cycles: three-dimensional reconstruction of icosahedral viruses from cryo-electron micrographs. *Microbiology and Molecular Biology Reviews*, 63(4):862–922, 1999.

[5] D. Barash. A fundamental relationship between bilateral filtering, adaptive smoothing and the nonlinear diffusion equation. *IEEE Trans. on Pattern Analysis and Machine Intelligence*, 24(6).844–847, 2002.

[6] D.M. Belnap, A. Kumar, J.T. Folk, and T.J. Smith *et al*. Low-resolution density maps from atomic models: How stepping 'back' can be a step 'forward'. *Journal of Structural Biology*, 125(2-3):166–175, 1999.

[7] V. Caselles, J.L. Lisani, J.M. Morel, and G. Sapiro. Shape preserving local histogram modification. *IEEE Trans. Image Processing*, 8(2):220–230, 1998.

[8] P. Chacon and W. Wriggers. Multi-resolution contour-based fitting of macromolecular structures. *Journal of Molecular Biology*, 317:375–384, 2001.

[9] W. Chiu, M.L. Baker, J. Wen, and Z.H. Zhou. Deriving folds of macromolecular complexes through electron crymicroscopy and bioinformatics approaches. *Current Opin in Struct Biol*, 12:263–269, 2002.

[10] K. Cowtan. Modified phase translation functions and their application to molecular fragment location. *Acta Crystallography*, D54:750–756, 1998.

[11] R.A. Crowther. The molecular replacement method. pages 173–178. Gordon & Breach, 1972.

[12] R. Deriche. Fast algorithm for low-level vision. *IEEE Trans. on Pattern Recognition and Machine Intelligence*, 12(1):78–87, 1990.

[13] S. Derrode and F. Ghorbel. Shape analysis and symmetry detection in gray-level objects using the analytical fourier-mellin representation. *Signal Processing*, 84(1):25–39, 2004.

[14] D.L. Donoho and I.M. Johnson. Ideal spatial adaptation via wavelet shrinkage. *Biometrika*, 81:425–455, 1994.

[15] F. Durand and J. Dorsey. Fast bilateral filtering for the display of high-dynamic-range images. In *ACM Conference on Computer Graphics (SIGGRAPH)*, pages 257–266, 2002.

[16] M. Elad. On the bilateral filter and ways to improve it. *IEEE Transactions On Image Processing*, 11(10):1141–1151, 2002.

[17] R.J. Ellis. Macromolecular crowding: obvious but underappreciated. *Trends Biochem. Sci.*, 26(10):597–604, 2001.

[18] J.-J. Fernandez and S. Li. An improved algorithm for anisotropic nonlinear diffusion for denoising cryo-tomograms. *J. Struct. Biol.*, 144(1-2):152–161, 2003.

[19] A. Frangakis and R. Hegerl. Noise reduction in electron tomographic reconstructions using nonlinear anisotropic diffusion. *J. Struct. Biol.*, 135:239–250, 2001.

[20] A.S. Frangakis and R. Hegerl. Segmentation of two- and three-dimensional data from electron microscopy using eigenvector analysis. *Journal of Structural Biology*, 138(1-2):105–113, 2002.

[21] J. Frank. *Three-dimensional Electron Microscope of Macromolecular Assemblies.* San Diego: Academic Press, 1996.

[22] R.C. Gonzalez and R.E. Woods. *Digital image processing.* Addison-Wesley, 1992.

[23] R.M. Gray. Vector quantization. *IEEE ASSP Mag.*, pages 4–29, 1983.

[24] A. Ben Hamza, P. Luque, J. Martinez, and R. Roman. Removing noise and preserving details with relaxed median filters. *Journal of Mathematical Imaging and Vision*, 11(2):161–177, 1999.

[25] A.B. Hamza and H. Krim. Image denoising: A nonlinear robust statistical approach. *IEEE Transactions on Signal Processing*, 49(12):3045–3054, 2001.

[26] M.L. Harlow, D. Ress, A. Stoschek, and R.M. Marshall *et al.* The architecture of active zone material at the frog's neuromuscular junction. *Nature*, 409:479–484, 2001.

[27] D. Hessler, S.J. Young, and M.H. Ellisman. A flexible environment for the visualization of three-dimensional biological structures. *Journal of Structural Biology*, 116(1):113–119, 1996.

[28] IEEE. Special issue on vector quantization. *IEEE Transactions on Image Processing*, 5(2), 1996.

[29] W. Jiang, M. Baker, Q. Wu, and C. Bajaj *et al.* Applications of bilateral denoising filter in biological electron microscopy. *J. Struct. Biol.*, 144(1-2):114–122, 2003.

[30] W. Jiang, Z. Li, M.L. Baker, and P.E. Prevelige *et al.* Coat protein fold and maturation transition of bacteriophage P22 seen at subnanometer resolution. *Nature Structural Biology*, 10(2):131–135, 2003.

[31] D.J. Jobson, Z. Rahman, and G.A. Woodell. A multiscale retinex for bridging the gap between color images and the human observation of scenes. *IEEE Trans. Image Processing*, 6(7):965–976, 1997.

[32] D.J. Jobson, Z. Rahman, and G.A. Woodell. Properties and performance of a center/surround retinex. *IEEE Trans. Image Processing*, 6(3):451–462, 1997.

[33] G.J. Kleywegt and T.A. Jones. Template convolution to enhance or detect structural features in macromolecular electron-density maps. *Acta Crystallography, D53*, pages 179–185, 1997.

[34] Y. Kong and J. Ma. A structural-informatics approach for mining b-sheets: locating sheets in intermediate-resolution density maps. *Journal of Molecular Biology*, 332:399–413, 2003.

[35] Y. Kong, X. Zhang, T.S. Baker, and J. Ma. A structural-informatics approach for tracing b-sheets: building pseudo-ca traces for b-strands in intermediate-resolution density maps. *Journal of Molecular Biology*, 339:117–130, 2004.

[36] J.A. Kovacs and W. Wriggers. Fast rotational matching. *Acta Crystallography*, D58:1282–1286, 2002.

[37] J.R. Kremer, D.N. Mastronarde, and J.R. McIntosh. Computer visualization of three-dimensional image data using IMOD. *J Struct Biol*, 116:71–76, 1996.

[38] A.F. Laine, S. Schuler, J. Fan, and W. Huda. Mammographic feature enhancement by multiscale analysis. *IEEE Trans. Medical Imaging*, 13(4):725–738, 1994.

[39] Y. Li, A. Leith, and J. Frank. Tinkerbell-a tool for interactive segmentation of 3D data. *Journal of Structural Biology*, 120(3):266–275, 1997.

[40] G. Loy and A. Zelinsky. Fast radial symmetry for detecting points of interest. *IEEE Trans. on Pattern Analysis and Machine Intelligence*, 25(8):959–973, 2003.

[41] J. Lu, D.M. Healy, and J.B. Weaver. Contrast enhancement of medical images using multiscale edge representation. *Optical Engineering*, 33(7):2151–2161, 1994.

[42] R. Malladi and J.A. Sethian. A real-time algorithm for medical shape recovery. In *Proceedings of International Conference on Computer Vision*, pages 304–310, 1998.

[43] T. Masuda, K. Yamamoto, and H. Yamada. Detection of partial symmetry using correlation with rotated-reflected images. *Pattern Recognition*, 26(8):1245–1253, 1993.

[44] M. Mathieu and F.A. Rey. Atomic structure of the major capsid protein of rotavirus: implication for the architecture of the virion. *EMBO J.*, 20:1485–1497, 2001.

[45] B.F. McEwen and M. Marko. Three-dimensional electron micros-copy and its application to mitosis research. *Methods Cell Biol*, 61:81–111, 1999.

[46] R. Mendelson and E.P. Morris. The structure of the acto-myosin subfragment 1 complex: results of searches using data from electron microscopy and x-ray crystallography. *Proc. Natl. Acad. Sci.*, 94:8533–8538, 1997.

[47] A. Nakagawa, N. Miyazaki, J. Taka, and H. Naitow *et al*. The atomic structure of rice dwarf virus reveals the self-assembly mechanism of component proteins. *Structure*, 11:1227–1238, 2003.

[48] P. Perona and J. Malik. Scale-space and edge detection using anisotropic diffusion. *IEEE Trans. on Pattern Analysis and Machine Intelligence*, 12(7):629–639, 1990.

[49] W.K. Pratt. *Digital Image Processing (2nd Ed.)*. A Wiley-Interscience Publication, 1991.

[50] A. Roseman. Particle finding in electron micrographs using a fast local correlation algorithm. *Ultramicroscopy*, 94:225–236, 2003.

[51] A.M. Roseman. Docking structures of domains into maps from cryo-electron microscopy using local correlation. *Acta Crystallographica, D56*, pages 1332–1340, 2000.

[52] M. G. Rossmann. Fitting atomic models into electron-microscopy maps. *Acta Crystallographica, D56*, pages 1341–1349, 2000.

[53] J.A. Sethian. A marching level set method for monotonically advancing fronts. *Proc. Natl. Acad. Sci.*, 93(4):1591–1595, 1996.

[54] J.A. Sethian. *Level Set Methods and Fast Marching Methods (2nd edition)*. Cambridge University Press, 1999.

[55] D. Shen, H.S. Ip, K.T. Cheung, and E.K. Teoh. Symmetry detection by generalized complex moments: a close form solution. *IEEE Trans. on Pattern Analysis and Machine Intelligence*, 21(5):466–476, 1999.

[56] J.A. Stark. Adaptive contrast enhancement using generalization of histogram equalization. *IEEE Trans. Image Processing*, 9(5):889–906, 2000.

[57] A. Stoschek and R. Hegerl. Denoising of electron tomographic reconstructions using multiscale transformations. *J. Struct Biol*, 120:257–265, 1997.

[58] C. Sun and J. Sherrah. 3D symmetry detection using the extended gaussian image. *IEEE Trans. on Pattern Analysis and Machine Intelligence*, 19(2):164–168, 1997.

[59] C. Tomasi and R. Manduchi. Bilateral filtering for gray and color images. In *1998 IEEE International Conference on Computer Vision*, pages 836–846, 1998.

[60] N. Volkmann. A novel three-dimensional variant of the watershed transform for segmentation of electron density maps. *Journal of Structural Biology*, 138(1-2):123–129, 2002.

[61] N. Volkmann and D. Hanein. Quantitative fitting of atomic models into observed densities derived by electron microscopy. *Journal of Structural Biology*, 125:176–184,

1999.

[62] J. Weickert. *Anisotropic Diffusion In Image Processing*. ECMI Series, Teubner, Stuttgart, ISBN 3-519-02606-6, 1998.

[63] J. Wen, M.L. Baker, S.J. Ludtke, and W. Chiu. Bridging the information gap: computational tools for intermediate resolution structure interpretation. *Journal of Molecular Biology*, 308:1033–1044, 2001.

[64] W. Wriggers and S. Birmanns. Using situs for flexible and rigid-body fitting of multiresolution single-molecule data. *J. Struct. Biol.*, 133:193–202, 2001.

[65] W. Wriggers and P. Chacon. Modeling tricks and fitting techniques for multiresolution structures. *Structure*, 9:779–788, 2001.

[66] W. Wriggers and P. Chacon. Using situs for the registration of protein structures with low-resolution bead models from x-ray solution scattering. *Journal of Applied Crystallography*, 34:773–776, 2001.

[67] W. Wriggers, R.A. Milligan, and J.A. McCammon. Situs: a package for docking crystal structures into low-resolution maps from electron microscopy. *Journal of Structural Biology*, 125:185–195, 1999.

[68] W. Wriggers, R.A. Milligan, K. Schulten, and J.A. McCammon. Self-organizing neural networks bridge the biomolecular resolution gap. *Journal of Molecular Biology*, 284:1247–1254, 1998.

[69] C. Xu and J.L. Prince. Snakes, shapes, and gradient vector flow. *IEEE Trans. Image Processing*, 7(3):359–369, 1998.

[70] Y. Xu, J.B. Weaver, D.M. Healy, and J. Lu. Wavelet transform domain filters: A spatially selective noise filtration technique. *IEEE Trans. Image Processing*, 3(6):747–758, 1994.

[71] R. Yip, W. Lam, P. Tam, and D. Leung. A hough transform technique for the detection of rotational symmetry. *Pattern Recognition Letter*, 15:919–928, 1994.

[72] I.T. Young and L.J. Vliet. Recursive implementation of the gaussian filter. *Signal Processing*, 44:139–151, 1995.

[73] Z. Yu and C. Bajaj. Anisotropic vector diffusion in image smoothing. In *Proceedings of International Conference on Image Processing*, pages 828–831, 2002.

[74] Z. Yu and C. Bajaj. Image segmentation using gradient vector diffusion and region merging. In *Proceedings of International Conference on Pattern Recognition*, pages 941–944, 2002.

[75] Z. Yu and C. Bajaj. A fast and adaptive algorithm for image contrast enhancement. In *Proceedings of International Conference on Image Processing*, pages 1001–1004, 2004.

[76] Z. Yu and C. Bajaj. A segmentation-free approach for skeletonization of gray-scale images via anisotropic vector diffusion. In *Proceedings of 2004 IEEE International Conference on Computer Vision and Pattern Recognition*, pages 415–420, 2004.

[77] Z. Yu and C. Bajaj. Visualization of icosahedral virus structures from reconstructed volumetric maps. *The University of Texas at Austin, Department of Computer Sciences. Technical Report TR-04-10*, 2004.

[78] Z. Yu and C. Bajaj. Automatic ultra-structure segmentation of reconstructed cryoem maps of icosahedral viruses. *IEEE Transactions on Image Processing: Special Issue on Molecular and Cellular Bioimaging*, 14(9):1324–1337, 2005.

[79] K. Yuen and W. Chan. Two methods for detecting symmetries. *Pattern Recognition Letter*, 15:279–286, 1994.

[80] Z.H. Zhou, M.L. Baker, W. Jiang, and M. Dougherty *et al.* Electron cryomicroscopy and bioinformatics suggest protein fold models for rice dwarf virus. *Nature Structural Biology*, 8(10):868–873, 2001.

[81] Z.H. Zhou, M. Dougherty, J. Jakana, and J. He *et al.* Seeing the herpesvirus capsid at 8.5 angstrom. *Science*, 288:877–80, 2000.

33

In Search of Remote Homolog

Dong Xu
University of Missouri-Columbia

Ognen Duzlevski
University of Missouri-Columbia

Xii-Fend Wan
University of Missouri-Columbia

33.1 Introduction

33.1.1 Why remote homolog is interesting

Homolog identification is becoming more and more important in modern biology. Traditional biology studies have been focused extensively on model systems, and these studies provide a tremendous resource to investigate other species. The most used model systems include *E. coli*, budding yeast (*Saccharomyces cerevisiae*), fission yeast (*Schizosaccharomyces pombe*), *Caenorhabditis elegans*, *Drosophila melanogaster*, zebrafish (*Danio Rerio*), Arabidopsis thaliana, and mouse (*Mus musculus*) [7]. Most of the biological knowledge that has been accumulated so far is related to these model organisms. A convenient way to study the functions and structures of a new gene is to identify homologs (evolutionary relationships) in model organisms, from which one can infer structure, function and mechanism of the new gene. Such an approach becomes very powerful nowadays, given the surge in biological sequence data resulting from large-scale sequencing technologies and various genome projects.

Homolog identification can be done through computationally matching a query sequence to similar sequences in the database. However, this matching process is not trivial since two homologous proteins could have been separated a very long time ago in their evolutionary history. Thus, their sequences have diverged substantially and their evolutionary relationship may be very difficult to detect. Such distantly related proteins are called remote homologs. A large proportion, typically 30-40% of the predicted protein coding genes do not have specific function assignments since we cannot relate these proteins to any protein with known function in the database. This is the case even in well-studied model organisms. For example, at present, 2,280 genes have not been annotated with any functions out of 6,324 genes in budding yeast *Saccharomyces cerevisiae* [30]. Many of these genes probably have remote homologs whose functions are well characterized, but existing methods for detecting homology relationships via sequence similarity might not be able to detect such remote homologs.

33.1.2 Homology and evolution

The evolutionary relationship between genes is complicated. When two genes are evolutionarily related (i.e., they are descended from a common ancestor), they are called homologs. Homologs include three different forms (see Figure 33.1), i.e., ortholog, paralog and xenolog [27]. Homologs that have diverged from each other after speciation events are called orthologs. Homologs that have diverged from each other after gene duplication events are called paralogs. Homologs that have diverged from each other after lateral (horizontal) gene transfer events are called xenologs [56]. Homologs only indicate their evolutionary relationship, but they may not have the same function. Dividing genes into groups of orthologs, paralogs and xenologs can help function predictions after the homologs are identified. Orthologs typically preserve the gene functions. Gene duplications and horizontal transfer are frequently accompanied by functional divergence. In many cases, paralogs and also xenologs have related/overlapping, but not identical biological functions [8, 54]. In general, homologs have similar protein structures, but it is not true the other way around. Proteins sharing similar structures but without detectable evolutionary/functional relationship are called analogs.

The evolution connection can go beyond the one-to-one relationship and have mixed homology, as shown in 33.1(b). In this case, a new protein has multiple domains and these domains have homology to domains of different proteins. This happens frequently in evolution when domains are duplicated, inserted or permuted [77]. Such cases can make the remote homology detection even more challenging and lead to misleading computational results. Some computer packages were especially designed to handle such special cases, e.g., DIVCLUS, which detects families of duplication modules from a protein sequence database (http://www.mrc-lmb.cam.ac.uk/genomes)[66]. Some databases use domains as a basis to study the homologous relationship between proteins, e.g. PRODOM [19].

33.1.3 Challenges and progresses in identification of remote homolog

A major challenge for computational identification of remote homolog is the low signal-to-noise ratio. Since remote homologs were separated through a long evolutionary history, similarity due to convergence is generally limited to small regions of genes (signal) and other parts of sequences were diverged too much to have any relationship (noise). Typically, remote homologs cannot be identified through straightforward pairwise sequence comparison using tools such as BLAST [3] or FASTA [69]. When aligning whole sequences or even just domains, most parts of the sequences are forced to align together so that the noise basically buries the signal.

In the past ten years or so, active research has been carried out for remote homologs identification, and some significant advances have been made. New methods for remote homolog identification include sequence comparison, sequence-structure, and structure-structure comparison, as well as other methods that do not depend on protein sequence or structure (e.g., using gene location in the genomic sequence or microarray data). Since

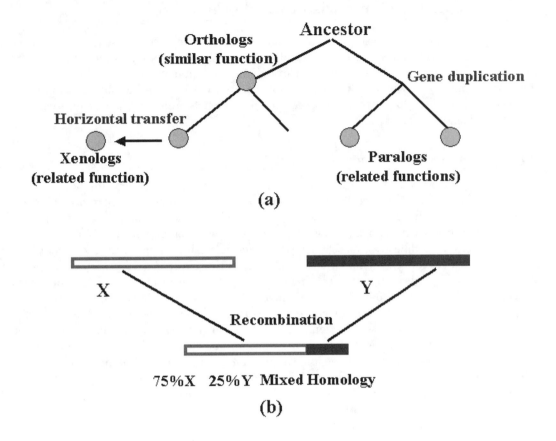

FIGURE 33.1: Evolutionary relationship among proteins.

remote homologs often reveal themselves through evolutionary history, most new methods use the information contained within alignments of multiple closely related sequences (sequence profiles). In addition, as protein three-dimensional structure is more conserved than sequence, protein structure prediction and structure-structure comparison also became a useful tool for identification of remote homologs.

In this chapter, we will provide a review on various methods for identification of remote homologs. Section 33.2 provides an overview on the methods based on protein sequence comparison. Section 33.3 focuses on the methods using sequence-structure comparison, i.e., the threading method. Section 33.4 discusses other computational methods that do not rely on protein sequence for remote-homolog identification. Section 33.5 summarizes the comparisons between different methods. The chapter ends with some discussions in Section 33.6.

33.2   Sequence Comparison Methods

The major methods for identification of remote homologs are based on sequence-sequence comparison. A protein sequence encodes all the information related to its structure and function. Comparing protein sequences through their alignment often provides the most direct relationship between the proteins. Sophisticated methods in identifying remote homologs, such as threading, are often based on the sequence profile resulted from searching a query sequence against a sequence database.

33.2.1   Pairwise sequence alignment

Pairwise sequence alignment is the foundation of all sequence-dependent methods for homolog identification. Although remote homologs are often missed by pairwise sequence comparison, since the method is not sensitive, pairwise sequence comparison is typically the first step in searching for homolog. A pairwise sequence alignment compares two protein sequences according to a match criterion together with an alignment gap penalty for insertion and deletion. The match criterion is generally expressed in a mutation matrix with elements (i, j), describing the preference (score) to replace the amino acid type i with j. The mutation matrices have been developed from the mutation rates found in sequence databases. Most popular mutation matrices are PAM – Percent Accepted Mutation [21] and BLOSUM – BLOcks SUbstitution Matrix [40]. PAM mutation matrix quantifies the amount of evolutionary changes in a protein sequence and defines a PAM unit, which is the amount of evolution that will change, on average, 1% of amino acids in a protein sequence. On the other hand, BLOSUM substitution matrices are derived from observations of the frequencies of substitutions in blocks of local alignments in related proteins. Each BLOSUM matrix is tailored to a particular evolutionary distance. In the BLOSUM62 matrix, for example, the alignment from which scores were derived was created using sequences sharing less than 62% identity. Sequences more identical than 62% are represented by a single sequence in the alignment to avoid redundancy.

The use of a certain matrix depends on different situations. BLOSUM62 is a widely used matrix for searching close homologs. According to some studies [40, 92], BLOSUM matrices have been shown to improve the sensitivity and accuracy of the alignments of homologs significantly over PAM. This is mostly due to the fact that PAM matrices have residue substitution probabilities for distant proteins extrapolated mathematically from rates of closely related sequences, while BLOSUM matrices are based on observed frequencies in real alignments. Furthermore, BLOSUM matrices were simply based on larger datasets used in the experiment. However, in a separate study, it was shown that PAM250 outperformed other matrices using some benchmarks of remote homolog identification [25, 1]. In addition, some variant matrices were also developed in recent years. For example, Probability Matrix from Blocks (PMB) [92] is derived from the Blocks database of protein alignments, and it approximates the residue substitution probabilities as a function of actual evolutionary distance, which is an important characteristic lacking in BLOSUM matrices. PMB was constructed from the original BLOSUM62 matrix and was refined using a Hidden Markov Model. It linked into statistical estimation methods to quantify the level of residual relatedness within a protein sequence database.

Several types of algorithms are used to obtain the optimal or near-optimal alignment given a mutation matrix with gap penalties for the insertions/deletions in the alignment. The first algorithm was developed by [63], who applied the dynamic programming technique to determine the optimal solution for a global alignment between two sequences. This method was further improved by [86] in the aspect that similarity between segments of the two sequences (local alignment) can be identified more efficiently with the guaranteed optimal solution. Several variants of the Smith-Waterman algorithm were developed for fast analysis, such as FASTA [69] and BLAST [3]. These methods use heuristic approaches for identifying regions of similarity between sequences and then further exploring the alignment only between these regions. This heuristics allowed these programs to be much faster than the original Smith-Waterman algorithm while at the same time sacrificing little in terms of accuracy or sensitivity. However, neither BLAST nor FASTA can guarantee the optimal alignment of sequences.

Pairwise sequence comparison has established sound statistical foundation for assessment of homolog identification results [51, 50, 3]. BLAST produces an expectation value (E-value), which indicates the statistical significance of the alignment. The E-value represents the possibility that the alignment is a result of chance, and its value is inversely related to the significance of the alignment. FASTA has a similar measure. It should be noted that the E-value sometimes may be misleading, as alignment significance could be over-estimated [105].

33.2.2 Sequence-profile alignment

Using the sequence profile derived from close homologs of the query protein can significantly increase the sensitivity for remote homolog identification. A profile is a representation of a group of related protein sequences, based on a multiple alignment of these sequences [4]. Once the multiple-sequence alignment is defined, the profile is constructed by counting the number of each amino acid at each position along the alignment. These counts are transformed into probabilities by normalizing the counts using the total number of amino acids and gaps observed at that position. The derived empirical probabilities reflect the likelihood of observing any amino acid k at position i. Since the counts are based on a finite set of sequences, it can happen that not all 20 amino acids are observed at each position. Therefore, pseudo counts are introduced so that no amino acid has a zero probability to occur at position i. Based on the profiles, if an amino acid matches a highly conserved one at a particular location, it will score a high positive result. In contrast, all other amino acids will receive very negative scores for the same position. If the residues at a particular position distribute randomly, their scores in the profile will gravitate around the value of zero.

Typical representatives of methods using sequence-profile alignment are PSI-BLAST [4] and IMPALA [81], where the latter is a modification of the former. PSI-BLAST allows iterative refinement of profiles by way of discovering sequences that are closely related to the profile in one step and incorporating these sequences into the profile for the next step. It uses gapped BLAST (pairwise sequence alignment) to obtain the multiple alignment for a query protein sequence, from which the profile is produced. The profile is compared to the database using a slightly modified BLAST code. The statistical significance of the alignments is then assessed and steps are repeated until convergence or for a user-specified number of times. However, as [5] pointed out, it is important to optimally select the query sequences at each step for the PSI-BLAST search. For example, human experts can recognize undesired proteins and remove them from building the profile for further PSI-BLAST search. The study observed a correlation between the variability in the selected sequences and the ability to sense subtle relationships in following steps. This implies that the level of training of the human operator using the tool would influence the final quality of the results produced. There are alternative approaches for using the sequence profiles based on extensions of the Smith-Waterman algorithm, with significant improvements in the alignment accuracy and the number of detected remote homologs. These algorithms include the Family Pairwise Search (FPS) [36] and Jumping Alignments [88]. FPS is a generalization of the pairwise approach for the case when multiple query sequences are available. The idea is to perform multiple pairwise alignments between the sequences in the database and the sequences in the query set. The resulting scores of the alignment between a database sequence and all sequences in the query set will then be reconciled into an overall score for the given database sequence. The method of Jumping Alignments is a multi-dimensional extension of the Smith-Waterman algorithm for the case of comparing a single sequence to a multiple sequence alignment. The core of it is to compare a query sequence to a "reference" sequence from the multiple alignment. If a "jump" is performed (another sequence is chosen to be the reference), this vertical change is penalized. This implies that the "reference" sequence is allowed to change at a cost. In essence, the algorithm keeps track of all rows of the multiple sequence alignment aligned to each column of the query sequence. The score of this alignment is the standard alignment score (allowing affine gaps and based on a substitution matrix) minus the penalty incurred when changing rows. It can be expressed mathematically in an extension of what is generally known as the recurrence relations for the dynamic programming approach of pairwise alignment of sequences.

33.2.3   Hidden Markov Models

The hidden Markov model (HMM) method [38] is a widely used method to detect remote homologies. This method is based on generalized profiles [14], which have been demonstrated to be very effective in detecting conserved patterns in multiple sequences [46, 22]. The typical profile hidden Markov model is a chain of match, insertion, and deletion nodes, with all transitions between nodes trained to specific probabilities. The single best path through an HMM corresponds to a path from the start state to the end state in which each character of the sequence is related to a successive match or insertion state along that path. In essence a discrete first-order HMM is defined by a set of states S, an alphabet of symbols m (the size of m is 20 in case of identifying protein remote homologs), a probability matrix $P(p_{ij})$, and an emissions matrix $E(e_{ia})$. The system can be in state i at a time, where it has a probability p_{ij} for moving from i to j and probability e_{ia} for emitting a symbol a. Evolution will allow states such as match, insert or delete. In the insert and match states there is an emission while a delete state is quiet. Loops on insertion states allow for expanding gaps. By the very nature of HMMs described above, it is evident that they require a family of related sequences (proteins in our case) to produce the statistical model describing their relatedness and conserved regions. After that, this model can discriminate against each new sequence in a database HMM. A popular implementation of the HMM method is SAM [52], which has been shown to significantly outperform tools like Blast, WU-Blast and Double-Blast in detecting remote homologies.

33.2.4   Profile-profile alignment

Profile-profile comparison provides a sensitive method to detect remote homology. The algorithm for profile-profile comparison is based on comparing numerical alignment profiles typically using the classical dynamic programming approach (other approaches can be co-emission probabilities or Gibbs sampling, for example). It usually consists of four steps:

1. Production of profiles from the multiple sequence alignments between a sequence and its homologs for both the query sequences and each sequence in the search database, the latter of which can be pre-computed. The alignment itself can be performed by using any method discussed above.

2. Apply a weighting scheme for the alignment sequences to avoid the problem of non-uniform distribution of the homologs in the alignment. Usually low weights are assigned to over-represented sequences and high weights to unique ones. A number of variations of weighting have been introduced in the past, e.g. increasing the weights of sequences that are more similar to the query sequence [60].

3. Construct sequence profiles that represent the probabilities for each of the twenty amino acid types at each position in the set. Many times the number of the sequences that are homologous to each other is insufficient to provide a profile based only on the given sequences themselves. In this case, a scheme based on prior probabilities (or expected probabilities) of amino acid occurrences and substitutions is used and a wide number of pseudo-counting schemes can be applied.

4. The final stage of profile-profile alignment is aligning the obtained profiles using gapped alignment with the dynamic programming technique.

Some evaluations show that this method is significantly more sensitive in detecting remote homologs than the sequence-profile based search programs like PSI-BLAST and various H-MM approaches [104, 95]. Sadreyev and Grishin [79] demonstrated their tool COMPASS - "comparison of multiple protein alignments with assessment of statistical significance" significantly improved sensitivity and accuracy against "classic" approaches such as using PSI-BLAST. [104] claimed that the improvement obtained using their method as opposed to PSI-BLAST is comparable to the improvement of PSI-BLAST over regular BLAST. [65] also demonstrated that using profile-profile alignments, a significant (as much as 30%) improvement can be obtained when compared to the sequence-profile approach. On the other hand, [23] concluded that current profile-profile methods improve sensitivity on average by 2% over profile-sequence methods and 40% over sequence-sequence methods.

Since profiles are typically aligned mutually by way of extended or modified dynamic programming algorithm, the choice of the gap-allowing scoring function used in aligning pairs of profile positions is important. Profile columns are vectors of real numbered values. There are different ways to calculate the score of aligning a column of query profile against a column of profiles for a sequence in a search database. A simple scoring scheme is defined as a dot product between the profile vectors. Initial tests among more than twenty various scoring functions have been performed with a general conclusion that performance varied on the order of several percent among different functions [23].

33.2.5 Phylogenetic analysis

To predict/confirm the identified remote homolog and study the evolutionary/functional relationship, phylogenetic analysis can be very useful. Phylogenetic analysis is the process of building a phylogenetic tree for a group of sequences in order to understand their relationship and timing of evolutionary development, as well as possibly infer their function. A phylogenetic tree is composed of branches (or edges), nodes and leafs with each leaf representing a protein or gene (or species). A node in a tree represents a speciation or divergence event and in essence symbolizes ancestry. A branch describes the evolutionary relationship between the nodes and leafs. Trees can be rooted or unrooted. A tree with a root will provide full evolutionary meaning in a historic sense since the root of the tree will signify the common ancestor organism (or gene). An unrooted tree will just provide information on the relationship between the nodes. Since trees can be used to describe species and genes, it is important to note that a speciation event (a node in a species tree) may not coincide with a divergence event in a gene tree.

Phylogenetic trees, in conjunction with other methods, can significantly improve accuracy of remote homolog prediction. For a straightforward implementation [75], a phylogenetic tree is constructed from a given multiple alignment of the family for a sequence-profile alignment. During the search, each database sequence is temporarily inserted into the tree, thus adding a new edge to the tree. Homology between family and sequence is then judged from the length of this edge. A more sophisticated incorporation of phylogenetic information is the Phylogenetic Tree-Based HMMs [72], which can minimize the effect of closely related sequences in the HMM model. Using the posterior probabilities of each amino acid at a particular node in a phylogenetic tree, the sequence profile information can be used more effectively, as it creates an evolutionary representation of the protein family across time. When all is put together, the model can serve as a more accurate representation of the subset of the protein family descended from a particular node.

To make functional predictions based on a phylogenetic tree, it is necessary to first overlay any known functions onto the tree. There are many ways this map can then be used to make functional predictions. First, the tree can be used to identify likely gene duplication events in the past. This allows the division of the genes into groups of orthologs and paralogs. Uncharacterized genes can be assigned a likely function if the function of any ortholog is known. Second, parsimony reconstruction techniques can be used to infer the likely functions of uncharacterized genes by identifying the evolutionary scenario that requires the fewest functional changes over time. The incorporation of more realistic models of functional change (and not just minimizing the total number of changes) may prove to be useful. Using such analysis, the evolutionary and functional relationship between the query protein and a predicted homolog from other methods can be revealed, and as result may help confirm the remote homolog identification.

A number of tools incorporated various ways of constructing trees and overlaying functions onto these trees. Most of the methods come from several basic approaches such as neighbor joining, parsimony, and maximum likelihood. Although a maximum likelihood approach incorporates more biological information into the tree, it is also much slower in constructing the relationships. Tools like Mesquite (`http://mesquiteproject.org`), PAUP (`http://paup.csit.fsu.edu/`), and PHYLIP (`http://evolution.genetics.washington.edu/phylip.html`) can produce trees with different methods, infer consensus trees out of several initial trees and offer other handy ways of manipulating, editing, tracing and viewing the information gained from the data. Such tools will also provide a way to observe similarities between trees or construct fits of one tree over another, which is helpful in identifying functional relationships between sequences.

33.2.6 Other approaches

All the above methods rely on sequence alignment of amino acids, either pairwise or multiple. There are alternative methods that are also based on sequence comparison. These methods may not have good specificity, i.e., the prediction reliability may not be as high as the methods above. However, they may be more sensitive so that they can detect some remote homologs that cannot be identified by the direct sequence alignment of amino acids. Below we summarize five alternative methods:

1. **Homology detection through intermediate sequences** This approach [67] is based on the idea that two sequences might be so divergent that a direct comparison between them will not yield any meaningful results, while both of the sequences may be similar to a third one which will act as a transitive

sequence between the original two. This is the approach implemented in the DOUBLE-BLAST tool, where the BLAST hits of a query protein sequence are used as a new set of queries for the next BLAST run. The work by [67] also suggests that such ISS (Intermediate Sequence Search) approach works better in prediction sensitivity than profile-sequence approaches such as PSI-BLAST. Other implementations of homology detection through ISS are also available [32, 108, 48].

2. **Incorporation of secondary structure information** This approach [31, 33] is based on the fact that secondary structures of proteins are more likely to be conserved than their sequences. In addition, the accuracy of secondary structure prediction has reached as high as 80%. Hence, we can predict the secondary structures for two proteins under alignment, and then use the predicted secondary structures as an additional scoring function. It was found that if the predicted secondary structures of two compared sequences match by more than 50%, then these sequences are more likely to be structurally related (also likely to be homologs). Even when the sequence identity was below 20%, homology could still be detected using the secondary structure comparison [31].

3. **Search remote homolog using hydrophobicity profile** This approach is based on hydrophobicity (measure of preference or absence of preference for water) profile extrapolated from the hydropathy scales of residues along a protein sequence. Hydropathy scale is defined as a measure of hydrophobicity of an amino acid and comes in several different sets. A hydrophobicity profile value at a certain sequence position is obtained by averaging the hydrophobicity scales of several neighboring residues. In some cases, the two remote homologs do not share any significant sequence similarity, but they share similar hydrophobicity profiles. Detecting a similar hydrophobicity profile for the query protein in a protein sequence database can be an alternative approach for possible remote homolog identification. Such strategy is implemented in the Protein Hydrophilicity/Hydrophobicity Search and Comparison Server [71] (`http://bioinformatics.weizmann.ac.il/hydroph/`).

4. **Homolog detection using compositional properties of protein sequence** In this approach, as implemented in PropSearch [42] (`http://www.infobiosud.univ-montp1.fr/SERVEUR/PROPSEARCH/propsearch.html`), a fundamentally different measure of similarity between proteins is used - protein dissimilarity is defined as a weighted sum of differences of compositional (physico-chemical) properties such as singlet/doublet residue composition, molecular weight, and isoelectric point. This approach can use either a single sequence or multiple sequences as a query to the database. In case of multiple sequences being used as a query, they can be reconciled into a consensus sequence describing the "average composition" of the protein family. PropSearch searches do not require alignments and are very fast when scanning a preprocessed database. The searches use reduced information from protein sequence, and hence, more false positives are expected than sequence-alignment methods. Nevertheless, the tool provides a useful alternative for further remote homolog identification when traditional sequence/profile-based methods have failed.

5. **Homologous relationship with frame-shift** This approach [70] is based on the assumption that some nucleotide frameshifts from DNA to RNA to

protein are responsible for the divergence between protein sequences. In these cases, the classical sequence comparison methods, which only consider insertions, deletions and mutations in protein sequences, will not be able to detect such evolutionary relationships. To account for the frame-shifts, sequences were compared using special amino acid substitution matrices for the alternate frames of translation. Such a method provides a sensitive approach for detecting a different type of remote homologs.

33.3 Sequence-Structure Comparison Methods

Since the 3D structures of proteins have been better conserved during evolution than their sequences, predicting protein structures often provides a more sensitive approach to identify distant evolutionary relationship (remote homology) than sequence-comparison methods. Among the protein-structure prediction methods, threading [9, 85, 49, 103] is most suitable for remote homolog identification. The idea of threading was derived from the observations that proteins with no apparent sequence similarity could have similar structural folds and that the total number of different structural folds in nature may be small [24], possibly in the range of a few thousands. Thus, a structure prediction problem can be reduced to a recognition problem, i.e., given a query protein sequence, searching for the most compatible structural fold based on sequence-structure relationships. Sequence-structure relationships include the notion that different amino acids may prefer different structural environments, e.g., a hydrophobic amino acid tends to be in the interior of a globular protein and proline rarely occurs in an α-helix. Once a structural template for the query sequence is identified, the template can serve as a basis for function inference of the query protein, although the template can be an analog of the query protein (i.e., the query and the template do not share the same biological function). In this section, we will first introduce the four components of threading. Then we will discuss how to use the predicted structural template for function inference of the query protein. In the end, we will give an example of remote homolog identification in a large scale, using three cyanobacterial genomes and carboxysomes as examples.

33.3.1 Threading components

A threading method typically consists of four components [87]:

1. a library of representative 3D protein structures for use as templates;
2. an energy function for measuring the fitness between a query sequence and a template structure in the library;
3. a threading algorithm for searching for the lowest energy among the possible alignments for a given sequence-template pair;
4. a criterion for estimating the confidence level of the predicted structure.

The following discussion addresses each aspect in detail.

Fold template library

A fold template library is intended to represent all the experimentally determined protein structures in the database PDB [97]. As many proteins in PDB are similar in sequence and structure, it is not necessary to include all of them in the fold library. Typically, only the representative proteins based on protein structure classification are used. Most template libraries of the existing threading programs are based on three widely used databases of protein structure classifications, i.e., CATH [64], FSSP [45] and SCOP [62]. CATH is a hierarchical classification of protein domain structures. FSSP contains similar information but is based on protein chains rather than domains; in addition, it contains sequence neighbors and multiple structure alignments. SCOP essentially uses a manual procedure. Hence, its classification is probably of higher quality, compared to the other two. However, SCOP has not been updated as frequently as desired simply due to the amount of manual work involved, while FSSP and CATH have been following the PDB updates closely. The classifications for folds by the three databases differ somewhat due to their different classification criteria (e.g., classification on a whole chain or a structure domain) and structure-structure comparison methods. As a result, the number of folds (or unique folds) differs among the three databases. After the templates are selected, some processing is carried out for each template to include derived information from the structure, such as protein secondary structure and solvent accessibility, both of which are needed for threading calculation.

Scoring function

The scoring function describes how favorable an alignment between a query sequence and a template structure is. Threading generally uses knowledge-based scoring functions rather than physical energies, since physical energies are too sensitive to small displacement of atomic coordinates, making them less suitable for threading and too time-consuming for computing. A typical threading scoring function has the following form:

$$S_{total} = S_{mutate} + S_{gap} + S_{single} + S_{pair} \qquad (33.1)$$

The mutation score S_{mutate} describes the compatibility of substituting one amino acid type by another; S_{gap} is the alignment gap penalty; the singleton score S_{single} represents a residue's preference to its local secondary structures (α-helix, β-strand, and loop) and its preference to being in a certain solvent environment (either exposed to solvent or in the interior of the protein); S_{pair} is the pairwise score between spatially close residues that are not neighbors in the protein sequence. The mutation score and the alignment gap penalty are similar to the ones used in sequence alignments. It has been shown that PAM250 is one of the best substitution matrices available for threading [25, 1]. The gap penalty is often a linear function of the gap size, with a penalty for opening a gap and a small penalty for each extension thereafter. Both S_{single} and S_{pair} are typically derived from Boltzmann statistics from a non-redundant protein database [2]. The basic idea is that if an amino acid is frequently observed in the interior of protein structures, a favorable score value will be rewarded when it is aligned to an interior position of a template.

Alignment algorithm

The alignment algorithm in the context of threading means the computational methods to identify a sequence-structure alignment with the best threading score as described in Equation 33.1. If we do not consider the pairwise score, a threading problem is essentially the same as a sequence alignment problem. Such a problem can be solved efficiently by a dynamic programming approach [63, 86]. There are a number of computer programs which essentially use dynamic programming for their threading problem, e.g., 123D [2], TOPIT-S [76] SAS [61], and the UCLA-DOE Structure Prediction Server [25]. An advantage of a threading algorithm without considering pairwise score is its speed. However, without pairwise interactions, the threading accuracy is compromised [100]. Threading with pairwise terms and alignment gaps is generally considered to be a very difficult problem [57].

Two previous existing threading programs with rigorous solutions all have exponential computational complexity [13, 58]. To overcome the computational difficulty, several methods through statistical sampling have been proposed [85, 49, 12, 20]. Such methods do not guarantee to find the globally optimal threading alignment. To resolve this, a unique threading algorithm [103] was developed to solve the globally optimal threading problem efficiently under the assumption that the pairwise term needs to be considered only between spatially close pairs in threading.

Assessment of threading results

A threading score between a query sequence and a template structure may not provide enough information about whether the template is the "correct" fold. This is because the scores are generally not normalized to the same scale. Hence, from the threading scores between a query and a pool of templates, we generally cannot tell if the query's correct fold template is in the pool nor can we always tell which is the correct fold even if it is present. There have been a number of attempts to "normalize" the threading scores so that they can be compared with each other. An early attempt was to use Z-score [28]. There have also been attempts to use the P-value scheme [51, 50] as a way to assign a meaning to a threading score. P-value, which estimates the probability of having a particular alignment score between two random sequences, have been successfully applied to sequence alignment, thanks to Karlin's seminal work on a rigorous model of gapless alignments [51, 50]. Due to the lack of a rigorous model for threading, the P-values are typically estimated through compiling a "large" number of threading scores between a query sequence and a template after randomly shuffling its residues [12]. While some usefulness of the estimated P-value has been demonstrated, the problem of developing a rigorous P-value scheme for threading remains an open challenge. A practical way to "normalize" the threading scores is to feed the threading scores along with various normalization factors such as sequence length to a neural network which has learned to "optimally" combine these factors based on a training set [102].

33.3.2 Identification of remote homolog from structural relationship

Even when the predicted 3D structure has poor quality due to a wrong alignment between the query protein and the template, the identified fold template often represents a remote homolog of the query protein, so that some evolutionary and functional relationship can be inferred between the query and the template. Given that threading often produces inaccurate alignment, it may be more useful in remote homolog identification than in 3D structure prediction.

Although remote homolog may be identified from the predicted structures, a relationship in structure does not guarantee homology relation. The relationship between the proteins can be classified at different hierarchical levels according to structural, functional, and evolutionary relationships. A widely used classification scheme consists of three levels of groups: family, superfamily, and fold, as shown in the SCOP (Structural Classification of Proteins) database [62], which currently has 2327 families, 1294 superfamilies, and 800 fold families (1.65 release, August 1st, 2003). A *family* consists of proteins that have significant sequence identity (often 25% or higher) between each other and share a common evolutionary ancestor (close homolog). Proteins of different families sharing a common evolutionary origin (reflected by their common structural and functional features), typically remote homologs, are placed in the same *superfamily*. Different superfamilies, i.e., analogs, are grouped into a *fold* family if their proteins have the same major secondary structures in the same ar-

Z-score interval	Probability to be correct	Confidence level	Homology
< 6	< 0.3	unlikely	analogs/unrelated
6 - 8	0.35	low	superfamily/analogs
8-10	0.63	medium	superfamily/fold
10-12	0.85	high	superfamily
12-20	0.96	very high	family/superfamily
> 20	> 0.99	certain	family

TABLE 33.1 Interpretation of the Z-scores from PROSPECT. The first column represents the Z-score range. The second column shows the probability of a sequence-template pair sharing the same fold within a certain Z-score range. The third column shows a corresponding qualitative confidence level. The fourth column provides a possible homologous relationship between the query and template protein in terms of the SCOP protein family classification, *family, superfamily*, and *fold* [62].

rangement and with the same topological connections. The structural similarities among proteins of the same fold family (but not the same superfamily) may arise just from the protein energetics favoring certain packing arrangements instead of a common evolutionary origin.

Although proteins with the same fold may not be the homologs, one can suggest a possible homolog of a query sequence from its predicted fold, using the SCOP database. When a predicted structural fold contains multiple superfamilies, it is possible to predict the most likely homolog for the query proteins among all the superfamilies based on threading results. For example, PROSPECT [100] calculates a Z-score that measures the reliability of the structure prediction and the possible homology relationship [53], as shown in Table 33.1. The Z-score is the threading score in standard deviation unit relative to the average of the threading raw score distribution of random sequences with the same amino acid composition and sequence length against the same structural templates. In practice, the average and the standard deviation are estimated by repeated threading between a template and a large number of randomly shuffled query sequences. When the Z-score of the prediction is high, the query and the template are likely to be homologs, and one can simply select the superfamily with the highest Z-score among all the superfamilies in the predicted fold as the (remote) homolog. When the Z-score is low, the predicted fold may not represent a homolog at all.

To further pin down whether a query protein and a predicted template are homologs, one can check functional motifs. If the predicted structure contains a functional motif (conserved residues at a particular position in the 3D structure, not necessarily close to each other on the protein sequence) of a protein in the template, the query protein and the template are probably homologs. [106] have constructed a database of functional motifs for known structures (e.g., EF-hand motif for calcium binding), called SITE. Currently, SITE contains identified functional motifs from about 50% of the SCOP superfamilies. One can also search the predicted protein structure against PROCAT [96], which is a database of 3-D enzyme active site templates. Although the structural motifs are more general, the comparison between the query protein and the template in terms of the motifs depends on the alignment accuracy, which may be different to achieve for threading. Hence, one can also carry out sequence-based motif searches using PROSITE [43], PRINTS [6], and BLOCKS [39]. A good example is the target T0053 of CASP3 [101]. Using PROSPECT, we successfully identified a native-like fold (1ak1) of T0053 in PDB, as shown in Fig. 33.2. T0053 and 1ak1 have only 11.2% sequence identity in the sequence independent structure-structure alignment. Without additional information, it is difficult to determine whether T0053 and 1ak1 are remote homologs. Using the BLOCK search [39], we found that the two proteins share the same sequence block with a conserved active site at His-183 in 1ak1

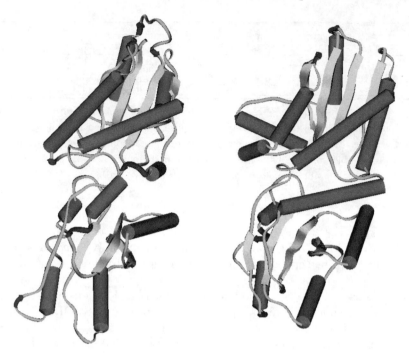

FIGURE 33.2: **(See color insert following page 20-4.)** A comparison between the predicted structure (left) and the experimental one (right) for the CASP-3 target t0053. The cylinders indicate α-helices, the strands indicate β-sheets, and the lines indicate loops.

and His-145 in T0053. This information allowed us to determine that the two proteins are remote homologs. Our prediction turned out to be obvious when the experimental structure of T0053 was determined (PDB code: 1qgo).

33.3.3 Computational Studies of Protein Structures in *Cyanobacteria*

Since the general applicability of threading, one can apply it in conjunction with other computational methods to predict protein structure and identify homologs at the genome scale, as shown by an example in this part for three cyanobacterial genomes; *Synechococcus* sp. WH8102 and two strains of *Prochlorococcus* sp. (MED4 and MIT9313). Using PROSPECT, we performed a global analysis of the structural folds and homologs in the three genomes and a detailed study of several predicted proteins that have been suggested to be essential for the function of carboxysomes, the common microcompartments that are presented in these photosynthetic microorganisms [98].

Overview of three genomes

The cyanobacterial community in the world open oceans is dominated by small unicellular forms of two genera *Synechococcus* and *Prochlorococcus*. Together, these organisms are major primary producers in large oligotrophic central gyres of the world's oceans. Although the two genera are frequently present together, *Synechococcus* is widely distributed and

Species	Synechococcus sp. WH8102	Prochlorococcus sp. MIT9313	Prochlorococcus sp. MED4
Total number of ORFs	2502	2251	1694
Membrane proteins	548 (21.9%)	560 (24.9%)	436 (25.7%)
PSI-BLAST hits	867 (34.7%)	867 (38.5%)	640 (37.8%)
PROSPECT ($z < 20$)	328 (13.1%)	137 (6.1%)	196 (11.6%)
PROSPECT ($12 < z < 20$)	81 (3.2%)	53 (2.4%)	47 (2.8%)
PROSPECT ($10 < z < 12$)	39 (1.6%)	28 (1.2%)	23 (1.4%)
PROSPECT ($8 < z < 10$)	55 (2.2%)	38 (1.7%)	25 (1.5%)
PROSPECT ($6 < z < 8$)	126 (5.0%)	111 (4.9%)	80 (4.7%)
Total number of structural homologs predicted	1496 (59.8%)	1234 (54.8%)	1011 (59.7%)

TABLE 33.2 A summary of predicted structural folds in three cyanobacterial genomes. Membrane proteins are predicted using the SOSUI program [41]. For soluble proteins, PROSPECT was applied to a gene only when it does not have a PSI-BLAST hit. Each row represents the total numbers of structural homologs predicted in three genomes, in a particular range of Z-scores of PROSPCT hits or with PSI-BLAST hits.

dominant in surface water that is rich in nutrients whereas *Prochlorococcus* is limited to $40^{\circ}N - 40^{\circ}S$ latitudes and often found in oligotrophic waters [68]. *Prochlorococcus* sp. MIT9313 is adapted to lower light conditions (at increasing ocean depths) than *Prochlorococcus* sp. MED4. Because regeneration of organic carbon is a critical step in response to anthropogenic inputs of CO_2 into the atmosphere and thus highly relevant to global carbon recycling, a major focus of biological oceanography has been to study, predict, and manipulate the process of carbon fixation in the ocean. The availability of three complete genome sequences (*Synechococcus* sp. WH8102 and two strains of *Prochlorococcus* sp. (MED4 and MIT9313)) enables us to study the global properties of the proteins in these genomes using computational approaches. Such studies will help better understand the structure and function of the proteins encoded by these genomes.

Global analysis of protein structural folds in three genomes

Protein structure predictions were carried out for all the predicted genes in the three genomes using the PROSPECT pipeline. The gene predictions were extracted from ORNL's Genome Channel at (`http://compbio.ornl.gov/channel`). Each genome took about one week for the pipeline to finish all the predictions. These results can be accessed through the Internet at:

$Synechococcus sp. \quad WH8102: \quad http://compbio.ornl.gov/PROSPECT/syn/$
$Prochlorococcus sp. \quad MIT9313: \quad http://compbio.ornl.gov/PROSPECT/pmar\_mit/$
$Prochlorococcus sp. \quad MED4: \quad http://compbio.ornl.gov/PROSPECT/pmar\_med/$

Overall, the PROSPECT pipeline identified structural folds in PDB with reasonable level of confidence (through either PSI-BLAST with E-value less than 10^{-4} or PROSPECT with Z-score 6.0 or above) for 54.8%-59.8% of all the open reading frames (ORFs) in each of the three genomes, as shown in Table 33.2. As indicated in Table 33.1, these predictions can be used to infer remote homologs. Based on the detection of remote homology, we can further predict functional and evolutionary relationships for proteins of unknown functions. Together with annotations of membrane proteins, about 80% of all the ORFs in each genome are characterized.

ORF	Template	Template function	Predicted ORF function
ccmk1	1ris	Ribosomal protein S6	Metallochaperone
csoS2-1	1iir-A	Glycosyltransferase	Structural protein
csoS2-2	1fnf	Fibronectin	Structural protein
csoS3	1qlt-A	Vanillyl-alcohol oxidase	Oxidase
ORFA	1kt9	Diadenosine tetraphosphate hydrolase	Phosphate hydrolase
ORFB	1kt9	Diadenosine tetraphosphate hydrolase	Phosphate hydrolase
ccmK2	1ris	Ribosomal protein S6	Metallochaperone
Or459(MIT9313)	1dcp	Pterin-4a-carbinolamine dehydratase	Dehydratase

TABLE 33.3 Structure, remote homolog (selected template), and function predictions for carboxysome proteins. The table lists the ORF name, the PDB code (and chain name) of the template (remote homolog) used, the function of the template, and the predicted function based on most remote homolog under the context of carboxysome. csoS2-1 and csoS2-2 are the N-terminal domain and C-terminal domain of the csoS2 ORF, respectively. Other than Or459, which is specific to *Prochlorococcus* sp. MIT9313, all others genes are homologous within the three genomes, and hence sharing the same structure template and remote homolog

Computational analysis of predicted carboxysome proteins

The carboxysome is a polyhedral inclusion body found in a variety of microorganisms [15]. The components and the overall structure of the carboxysome is poorly understood although it is known that an enzyme named ribulose1,5, bisphosphate carboxylase/oxygenase (Ru-BisCO) – the major enzyme that converts carbon from an inorganic to organic form – constitutes about 60% of the total carboxysomal proteins [29]. Aiding to the effort of elucidating carboxysome structure and function, we analyzed the automated prediction, as described above, for the proteins found in carboxysome. As shown in Table 33.3, we first identified the structure fold for each ORF, and then found the superfamily with the highest Z-score in the fold as structure template and the remote homolog. Based on the remote homolog, functional information is inferred under the context of carboxysome. For example, a template of csoS2-2 (1fnf) is fibronectin. This fibronectin fragment encompasses the 7th-10th type III repeat, whose presence is found in many proteins with a broad range of biological function, and an RGD motif which is an intergrin-fibronectin interaction site. Other information can be also used for functional inference from remote homology. For example, from PSI-BLAST search, csoS2 are homologous to histones, which also play a role as structural proteins, even though their molecular structure is unknown. Therefore, the C-terminus of csoS2 could play a structural role, or it may be important for promoting protein-protein interactions. Interestingly, it has been reported that antibodies raised against CsoS2 label the edges of the carboxysome of *Thiobacillus neapolitanus*. Combined with the experimental evidence, it may be expected that csoS2 is present in the carboxysome shell.

33.4   Sequence-Independent Methods

Both the sequence-comparison methods and protein-structure predictions for remote-homolog identification use the information from the query-protein sequence. In some cases, the information about a remote homolog is also revealed in other sources, such as structure-structure comparison, evolutionary footprints and gene expression. In this section, we will address these sequence-independent methods for remote homolog identification.

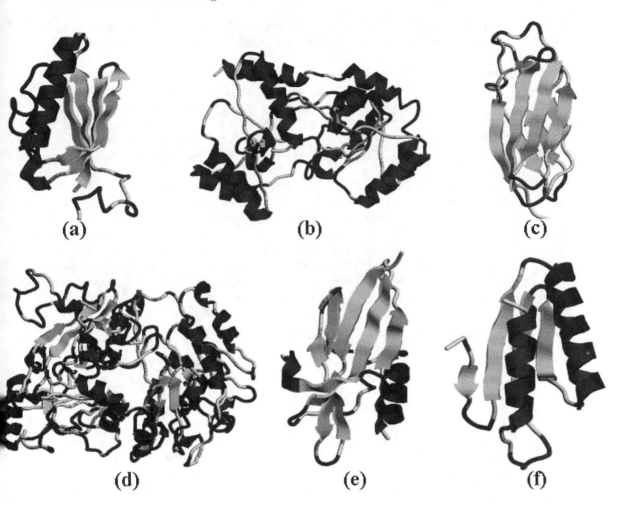

FIGURE 33.3: Predicted structures for (a) ccmk1, (b) csoS2 N-terminal domain, (c) csoS2 C-terminal domain (d) csoS3, (e) ORFA, and (f) Or459 (*Prochlorococcus* sp. MIT9313). (a-e) are 5 ORFs from *Synechococcus* sp. WH8102. (c) shows only one of the repetitive domains of the predicted structure.

33.4.1   Protein structure comparison

When the structure of a protein is known, one can use the structure to identify its homologs through comparing with other known structures. This is similar to the threading method in the sense of using structural information, while the structure-structure comparison is far more reliable than sequence-sequence comparison or sequence-structure comparison in identifying remote homologs. Protein-structure comparison had limited use for remote homolog identification in the past, as there were few protein structures available. With the advent of new technologies such as synchrotron radiation sources and high-resolution nuclear magnetic resonance (NMR), a great number of new protein structures have been determined in recent years. In particular, in the recent effort of structural genomics [107, 10], where protein structures are being determined in large scale, the structures of many proteins were determined without knowing their function. Structure comparison provides a useful tool to identify remote homologs for these proteins, and further predict functions based on the homologs.

Thus, when the structure of a protein is available, one can use the structure to search again the database of known protein structures, i.e., PDB, and the hits with similar structures are potential remote homologs of the query protein. A popular tool for comparing a query protein structure against all the structures in the PDB is the DALI server [44]. A much faster search engine for protein structure comparison, ProteinDBS [83], was developed recently. Based on the alignment between the query protein structure and the hits in the structure database, one may find biologically interesting similarities that are not detectable by sequence comparison or threading. For example, common structural motifs between two aligned structures can be found. Using such information, one can tell whether two proteins of the same fold are remote homologs or merely analogs. Several protein-structure classification databases, such as SCOP, CATH, and FSSP, as discussed in Section 33.3, can facilitate the search for remote homolog using structure comparison. These structural databases provide a useful resource for systematically checking the common features of structural motifs and sequence patterns among proteins in the same superfamily, and these features can help to tell whether two proteins of the same structural fold are homologs or analogs.

33.4.2   Paralog relationship in high-throughput biological data

During evolution, some genes may be duplicated and then diverged (i.e., paralog as defined in section 33.1). The paralogs often have some traces in various high-throughput biological data, including genomic sequence data, gene expression data, and genetic interaction data. Although these traces alone typically are insufficient for predicting paralog relationships, they can help remote paralog identification in conjunction with other methods. In particular, when such traces occur, a paralog prediction from other methods would have an increased confidence level. The following three types of traces can offer some support of possible paralog relationships:

1. **Adjacent genes in genomic sequence.**   Many gene duplications occur in tandem. Hence, it is not surprising that many paralogs were also found in adjacent positions of a genome [18, 94].

2. **Correlated microarray gene expression patterns.**   The neighboring genes due to this duplication mechanism often show similar expression patterns, since these adjacent genes share a single upstream activating sequence in many cases. The correlated gene expression pattern also relates to the distance between the genes on the genomic sequence, as it was found that the expression similarity was correlated to the physical distances in *Saccharomyces cerevisiae* and *Arabidopsis thaliana* [94]. The correlated gene expression pattern among paralogs also extends to orthologs [47].

3. **Genetic interactions based on synthetic lethality screening.** The synthetic lethality screening is a very powerful method for finding "genetic interaction" between gene products [84]. It identifies lethal deletions of two genes at the same time, while either deletion alone is not lethal. A systematic high-throughput synthetic lethal analysis was carried out in yeast *Saccharomyces cerevisiae* for 4700 viable mutants [90]. Between two genes with such a genetic interaction, one may be a backup of the other, and hence, the two genes may be paralogs.

33.5   Assessment of Computational Methods

Given that so many methods are available for remote homolog identification, it is very important to compare these methods based on some benchmark tests. However, such a comparison is not trivial. If we look into any particular paper discussing an individual method, typically the paper shows that the method outperforms others. The results depend on what criteria are used and how the comparisons are performed. At least the following six criteria can be considered when comparing different methods for remote homolog identifications:

1. Sensitivity of remote homolog identification, i.e., how many true remote homologs can we identify as top hits among all remote homologs in the database? For example, if k remote homologs are in the database of a query protein sequence, how many of them rank as top n.

2. Specificity of remote homolog identification, i.e., among the top hits, how many of them represent true homologs? For example, if top n hits in the database are selected for a query protein sequence, how many of them are true homologs?

3. Reliable confidence assessment, i.e., to what extent can the prediction result of homolog identification be trusted, in terms of either probability or Expectation value?

4. Alignment accuracy, i.e., in an alignment between the query protein and the correctly identified remote homolog, how many alignment positions are biologically true? A true biological alignment is typically represented by the structure-structure alignment between the two proteins.

5. Applicability, i.e., what conditions does a method require? For example, for the threading method, it requires that the structure of a remote homolog for the query sequence is available in the database. Some methods do not have explicit requirements, but they tend to work poorly in certain cases, e.g., HMMs do not work well when the query protein does not have close homologs to build profiles.

6. Computational efficiency, i.e., the computing time and memory requirement, and their dependence on the query protein size. This turns out to be very important in practice, especially because many related computations are carried out in large (genome) scale. For example, it is known that many other methods are more accurate in identifying remote homologs, but PSI-BLAST is still the most popular method for remote homolog identification, given that it is very fast and computational time complexity is linear.

Even though there have been constant improvement of methods for discovering remote homologs, there is still much more room for improvement along the six criteria. It has been shown [11] that for close homolog identification (with sequence similarity over 30%), almost all the methods work very well, with insignificant differences for criteria 1-4. However, when predicting remote homologs, none of the methods consistently outperforms others in all of the six criteria. Hence, although PSI-BLAST is the most popular method, many other computational tools are also widely used at the same time.

Some systematic benchmarks between different sequence-comparison methods have been constructed. These comparisons often use SCOP as the gold standard and focus on whether a method can detect remote homologs in the same superfamily but in different families and also how well the sequence alignment compares with the structural alignment. It was found [11] that while comparing sequences below 30% identity (many of them are in the same family), less than 50% of remote homologs could be detected using tools like gapped BLAST, FASTA or the Smith-Waterman SSEARCH. However, even for these identified homologs, the study [11] suggested that the P-values as produced by BLAST seem to underestimate the errors and the alignments are often inaccurate. Another study [80] compared the performance for sequence-alignment accuracy against structure-structure alignment among a pairwise alignment method (BLAST), a sequence-profile method (PSI-BLAST), and an intermediate-sequence-search method (DOUBLE-BLAST). On sequence similarities between 10% and 15%, BLAST, PSI-BLAST, and DOUBLE-BLAST correctly aligned 28%, 40%, and 46% of these sequences, respectively. This indicates that all methods have much room for improvement of alignment accuracy.

Another set of benchmarks came from the protein-structure-prediction community. Although protein structure prediction focuses on structure instead of homology, the dominant method is to identify homologs in the protein structure database PDB and use the homologs as templates to build protein tertiary structures. As a result, the structure-prediction assessment also applies to remote-homolog prediction. To assess objectively the state of the art in prediction tools for protein structures, the computational structural biology community has agreed on an evaluation system called CASP (Community Wide Experiment on the Critical Assessment of Techniques for Protein Structure Prediction). CASP was initiated in 1994 and has been a biannual event since its inception. In each CASP, participants were given tens of protein sequences whose experimental structures were being solved or had been solved but not published. CASP participants then predicted their structures blindly, either in an automated fashion or with manual adjustment. A group of invited assessors evaluated how well each predicted structure matched the experimental structure. At the end of the prediction season, the performance of each team was ranked. The CASP exercises provide an objective way to assess related computational methods, particularly for Criteria 1, 4, and 5. The strengths and weaknesses of each method are often revealed. For example, even though remote homolog identification methods have been consistently improved, the alignment accuracy has very little improvement over the past few years [93, 91]. Two observations from the CASPs are:

1. Manual process (the human knowledge) can help improve prediction significantly,

2. using consensus approach, i.e., to find common hits from different methods, can outperform any individual method substantially [35].

Based on such findings, computational pipelines [98, 82, 34] or expert systems [37] have been developed to incorporate various methods and human knowledge to improve the prediction accuracy. Some hybrid methods using various types of information together were also developed [89, 74].

Other than manual predictions and evaluations in CASP, some fully automated servers for protein structure predictions and evaluations were developed. Such efforts complement CASP and provide useful information for assessments of computational tools themselves (instead of human experts). One of them is CAFASP [26], which was carried out in parallel with CASP, using the same set of prediction targets. The third CAFASP in 2003 showed that several best automated prediction servers using the consensus approaches achieved comparable performance as human CASP predictors. This result shows that significant progress has been achieved in automatic structure prediction. Another automated evaluation server is MaxBench [59](http://www.sanger.ac.uk/Users/lp1/MaxBench/). This system makes it easy for developers to both compare the performance of their methods to standard algorithms and at the same time investigate the results of individual comparisons. Two large-scale evaluation servers using updated PDB entries are LiveBench [78] and EVA [55](http://cubic.bioc.columbia.edu/eva/). The evaluation is updated automatically when sequences of newly available protein structures in PDB are sent to the servers and their predictions are collected. The predictions are then compared to the experimental structures automatically and the results are published on the Web pages. Over time, the two servers have accumulated prediction results for a large number of proteins with various prediction methods and they provide useful information to developers as well as users of these methods.

33.6 Discussions

In summary, significant advances have made for computational identification of protein remote homologs in the past decade. Various methods have pushed our limit to find distantly related homologs that are unidentifiable from simple pairwise sequence comparisons. These methods often utilize the evolutionary information in the sequence database effectively, in particular through building multiple sequence profiles or identifying evolutionary intermediates. Many methods also use protein structural information, including integrating protein secondary structure prediction into the process of sequence comparison, searching through sequence-structure comparisons (threading), and performing structure-structure alignments. More recently, mega-servers using multiple methods to find consensus solutions have been developed. These servers often show significant improvement over any single method. All these developments have a big impact on the field of post-genomic biology, especially for genome annotation, comparative genomics, structural genomics, and functional genomics. Not only computational biologists but also experimentalists benefit tremendously from these tools, which often provide useful information about the structure and function of a protein through its (remote) homologs. The computational results can help develop biological hypothesis for new experiments and also help the interpretation of experimental data.

However, these computational tools may not be used blindly for inferring remote homologies. It should be noted that even when the sequence similarity between two proteins is high, it might not always correspond to homology. There is always a possibility that the sequence similarity is by chance rather than due to biological relationship. When more sensitive methods for remote-homolog identification are used, the confidence level of a comparison result can be low, and it is not rare that false positive predictions are generated. Also, homology may not imply function conservation. Many remote homologs, especially paralogs, have divergent functions, although their functions are often related in a broader category. To best take advantage of the available computational tools and reduce the chance of wrong prediction, it is important to use multiple tools to check for consensus solutions and differences between various results. It is also important to use other computational approaches [99], such as prediction of signal peptide cleavage sites, subcellular localization, protein domain prediction, prediction of transmembrane helices, and sequence motif prediction, to predict the properties and functions of the proteins so that one can better assess the potential homologous and functional relationships between proteins of interest. Furthermore, when high-throughput data (e.g., gene expression data and protein-protein interaction data) are available, it is very useful to utilize these experimental data to confirm and extend the homologous relationship identified from sequence or structure based methods [17, 16]. An example of using various methods in conjunction with homolog identification is described in a recent paper [73]. Finally, additional experiments are generally needed to confirm the predictions.

There are still many challenging problems in remote homolog identification and the related research is very active. More sensitive methods are needed for difficult homolog identifications. Still in many cases, people know a homolog of protein X with well characterized

function in species A should be present in species B, since species B shows the same phenotype related to protein X as does species A. However, current methods may not be sensitive
enough to detect the homolog of protein X in species A. Another challenge is the opposite.
Sometimes there are many homologs detected in species X, but it is unknown which one
represents the true ortholog. Other than the sensitivity issues, current confidence assessment methods of the homolog identification results need further improvement. Some tools,
such as BLAST, PSI-BLAST, and FASTA, have good confidence assessment methods, but
they often overestimate statistical significance. Many other tools have primitive assessment
methods or no assessment at all. As a result, generally, remote homolog identification has
poor prediction specificity, i.e., false positives are frequently predicted. In addition, the
current alignment accuracy between remote homologs is typically poor, and there is much
room for improvement.

Acknowledgments

We would like to thank Drs. Nickolai Alexandrov, Ying Xu, and Frank Larimer for
helpful discussions. This work is supported in part by the US Department of Energy's
Genomes to Life program (http://www.doegenomestolife.org) under project, "Carbon
Sequestration in it Synechococcus Sp.: From Molecular Machines to Hierarchical Modeling"
(www.genomes-to-life.org). It was also supported by the Laboratory Directed Research
and Development Program of Oak Ridge National Laboratory, under Contract DE-AC05-
00OR22725, managed by UT-Battelle, LLC.

References

[1] R.A. Abagyan and S. Batalov. Do aligned sequences share the same fold? *J Mol Biol*, 273(1):355–368, 1997.

[2] N.N. Alexandrov, R. Nussinov, and R.M. Zimmer. Fast protein fold recognition via sequence to structure alignment and contact capacity potentials. *Pac Symp Biocomput*, pages 53–72, 1996.

[3] S.F. Altschul, W. Gish, W. Miller, and E.W. Myers *et al.* Basic local alignment search tool. *J Mol Biol.*, 215(3):403–410, 1990.

[4] S.F. Altschul, T.L. Madden, A.A. Schäffer, and J. Zhang *et al.* Gapped BLAST and PSI-BLAST: a new generation of protein database search programs. *Nucleic Acids Res.*, 25(17):3389–3402, 1997.

[5] L. Aravind and E.V. Koonin. Gleaning non-trivial structural, functional and evo-lutionary information about proteins by iterative database searches. *J. Mol. Biol.*, 287(5):10231040, 1999.

[6] T. K. Attwood, D. R. Flower, A. P. Lewis, J. E. Mabey, S. R. Morgan, P. Scordis, J. N. Selley, and W. Wright. Prints prepares for the new millennium. *Nucleic Acids Res.*, 27(1):220–225, 1999.

[7] C. Bahls, J. Weitzman, and R. Gallagher. Biology's models. *The Scientist*, 17(1), June 2 2003.

[8] N. Baumberger, M. Steiner, U. Ryser, and B. Keller *et al.* Synergistic interaction of the two paralogous arabidopsis genes LRX1 and LRX2 in cell wall formation during root hair development. *Plant J.*, 35(1):71–81, 2003.

[9] J.U. Bowie, R. Luthy, and D. Eisenberg. A method to identify protein sequences that

fold into a known three-dimensional structure. *Science.*, 253(5016):164–170, 1991.

[10] S.E. Brenner. A tour of structural genomics. *Nat Rev Genet.*, 2(10):801–809, 2001.

[11] S.E. Brenner, C. Chothia, and T.J. Hubbard. Assessing sequence comparison methods with reliable structurally identified distant evolutionary relationships. *Proc Natl Acad Sci U S A*, 95(11):6073–6078, 1998.

[12] S.H. Bryant and S.F. Altschul. Statistics of sequence-structure threading. *Curr. Opin. Struct. Biol.*, 5(2):236244, 1995.

[13] S.H. Bryant and C.E. Lawrence. An empirical energy function for threading protein sequence through the folding motif. *Proteins*, 16(1):92112, 1993.

[14] P. Bucher and A. Bairoch. A generalized profile syntax for biomolecular sequence motifs and its function in automatic sequence interpretation. In *ISMB-94*, pages 53–61, Menlo Park, CA, 1994. AAAI/MIT Press.

[15] G.C. Cannon, S.H. Baker, F. Soyer, and D.R. Johnson *et al.* Organization of carboxysome genes in the thiobacilli. *Curr Microbiol.*, 46(2):115–119, 2003.

[16] Y. Chen, T. Joshi, Y. Xu, and D. Xu. Towards automated derivation of biological pathways using high-throughput biological data. In *Proceeding of the 3rd IEEE Symposium on Bioinformatics and Bioengineering*, pages 18–25. IEEE/CS Press, 2003.

[17] Y. Chen and D. Xu. Computational analyses of high-throughput protein-protein interaction data. *Current Protein and Peptide Science.*, 4:159–181, 2003.

[18] B.A. Cohen, R.D. Mitra, J.D. Hughes, and G.M. Church. A computational analysis of whole-genome expression data reveals chromosomal domains of gene expression. *Nat Genet.*, 26(2):183–186, 2000.

[19] F. Corpet, F. Servant, J. Gouzy, and D. Kahn. ProDom and ProDom-CG: tools for protein domain analysis and whole genome comparisons. *Nucleic Acids Res.*, 28(1):267–269, 2000.

[20] O.H. Crawford. A fast, stochastic threading algorithm for proteins. *Bioinformatics*, 15(1):6671, 1999.

[21] M.O. Dayhoff, R.M. Schwartz, and BC. Orcutt. A model of evolutionary change in proteins. matrices for detecting distant relationships. *Atlas of Protein Sequence and Structure(National Biomedical Research Foundation, Washington DC.)*, 5:345358, 1978.

[22] S.R. Eddy, G. Mitchison, and R. Durbin. Maximum discrimination hidden Markov models of sequence consensus. *J Comput Biol.*, 2(1):9–23, 1995. Spring.

[23] R.C. Edgar and K. Sjolander. COACH: profile-profile alignment of protein families using hidden Markov models. *Bioinformatics*, 20(8):13091318, 2004.

[24] A.V. Finkelstein and O.B. Ptitsyn. Why do globular proteins fit the limited set of folding patterns? *Prog Biophys Mol Biol.*, 50(3):171–190, 1987.

[25] D. Fischer, A. Elofsson, J.U. Bowie, and D. Eisenberg. Assessing the performance of fold recognition methods by means of a comprehensive benchmark. In L. Hunter and T. Klein, editors, *Biocomputing: Proceedings of the 1996 Pacific Symposium*, pages 300–318, Singapore, 1996. World Scientific Publishing Co.

[26] D. Fischer, L. Rychlewski, R.L. Jr. Dunbrack, and A.R. Ortiz *et al.* CAFASP3: the third critical assessment of fully automated structure prediction methods. *Proteins*, 53(6):503–516, 2003.

[27] W.M. Fitch. Distinguishing homologous from analogous proteins. *Syst Zool*, 19(2):99–113, 1970.

[28] H. Flockner, M. Braxenthaler, P. Lackner, and M. Jaritz *et al.* Progress in fold recognition. *Proteins*, 23(3):376386, 1995.

[29] D. Friedberg, K.M. Jager, M. Kessel, and N.J. Silman *et al.* Rubisco but not Rubisco

activase is clustered in the carboxysomes of the *cyanobacterium synechococcus sp.* PCC 7942: Mud-induced carboxysomeless mutants. *Mol Microbiol.*, 9(6):1193–1201, 1993.

[30] The Gene Ontology Consortium. Gene ontology: Tool for the unification of biology. *Nature Genetics*, 25:25–29, 2000.

[31] C. Geourjon, C. Combet, C. Blanchet, and G. Deleage. Identification of related proteins with weak sequence identity using secondary structure information. *Protein Sci.*, 10(4):788–797, 2001.

[32] M. Gerstein. Measurement of the effectiveness of transitive sequence comparison, through a third "intermediate" sequence. *Bioinformatics*, 14(8):707–714, 1998.

[33] K. Ginalski, A. Elofsson, D. Fischer, and L. Rychlewski. 3D-Jury: a simple approach to improve protein structure predictions. *Bioinformatics*, 19(8):1015–1018, 2003.

[34] K. Ginalski, J. Pas, L.S. Wyrwicz, and M. von Grotthuss *et al.* ORFeus: Detection of distant homology using sequence profiles and predicted secondary structure. *Nucleic Acids Res.*, 31(13):3804–3807, 2003.

[35] K. Ginalski and L. Rychlewski. Protein structure prediction of CASP5 comparative modeling and fold recognition targets using consensus alignment approach and 3D assessment. *Proteins*, 53(6):410–417, 2003.

[36] W.N. Grundy. Homology detection via family pairwise search. *J Comput Biol.*, 5(3):479–491, 1998.

[37] J.T. Guo, K. Ellrott, W.J. Chung, and D. Xu *et al.* PROSPECT-PSPP: An automatic computational pipeline for protein structure prediction. *Nucleic Acid Research*, 32:W522 – CW525, 2004. Web Server issue.

[38] D. Haussler, A. Krogh, I. S. Mian, and K. Sjölander. Protein modeling using hidden Markov models: Analysis of globins. In *Proceedings of the Hawaii International Conference on System Sciences*, volume 1, pages 792–802, Los Alamitos, CA, 1993. IEEE Computer Society Press.

[39] J.G. Henikoff, S. Henikoff, and S. Pietrokovski. New features of the blocks database servers. *Nucleic Acids Res.*, 27(1):226–228, 1999.

[40] S. Henikoff and J.G. Henikoff. Amino acid substitution matrices from protein blocks. *Proc. Natl. Acad. Sci. U.S.A.*, 89(22):1091510910, 1992.

[41] T. Hirokawa, S. Boon-Chieng, and S. Mitaku. SOSUI: classification and secondary structure prediction system for membrane proteins. *Bioinformatics*, 14(4):378–379, 1998.

[42] U. Hobohm and C. Sander. A sequence property approach to searching protein databases. *J Mol Biol.*, 251(3):390–399, 1995.

[43] K. Hofmann, P. Bucher, L. Falquet, and A. Bairoch. The PROSITE database, its status in 1999. *Nucleic Acids Res.*, 27(1):215–219., 1999.

[44] L. Holm and C. Sander. Protein structure comparison by alignment of distance matrices. *J Mol Biol*, 233(1):123–138, 1993.

[45] L. Holm and C. Sander. Mapping the protein universe. *Science*, 273(5275):595–603, 1996.

[46] R. Hughey and A. Krogh. Hidden Markov models for sequence analysis: extension and analysis of the basic method. *Comput Appl Biosci*, 12(2):95–107, 1996.

[47] J.L. Jimenez, M.P. Mitchell, and J.G. Sgouros. Microarray analysis of orthologous genes: conservation of the translational machinery across species at the sequence and expression level. *Genome Biol.*, 4(1), 2003. R4.

[48] B. John and A. Sali. Detection of homologous proteins by an intermediate sequence search. *Protein Sci.*, 13(1):54–62, 2004.

[49] D.T. Jones, W.R. Taylor, and J.M. Thornton. A new approach to protein fold recog-

nition. *Nature*, 358(6381):86–89, 1992.

[50] S. Karlin and S.F. Altschul. Methods for assessing the statistical significance of molecular sequence features by using general scoring schemes. In *Proc Natl Acad Sci USA*, volume 87, pages 2264–2268, 1990.

[51] S. Karlin, A. Dembo, and T. Kawabata. Statistical composition of high-scoring segments from molecular sequences. *Ann. Statistics*, 18:571–581, 1990.

[52] K. Karplus, C. Barrett, and R. Hughey. Hidden Markov models for detecting remote protein homologies. *Bioinformatics*, 14(10):846–856, 1998.

[53] D. Kim, D. Xu, J.T. Guo, and K. Ellrott *et al.* PROSPECT II: protein structure prediction program for genome-scale applications. *Protein Eng.*, 16(9):641650, 2003.

[54] M.S. Kobor, S. Venkatasubrahmanyam, M.D. Meneghini, and J.W. Gin *et al.* A protein complex containing the conserved Swi2/Snf2-related ATpase Swr1p Deposits Histone Variant H2A.Z into euchromatin. *PLoS Biol*, 2(5):E131, 2004.

[55] I.Y. Koh, V.A. Eyrich, M.A. Marti-Renom, and D. Przybylski *et al.* EVA: Evaluation of protein structure prediction servers. *Nucleic Acids Res*, 31(13):3311–3315, 2003.

[56] E.V. Koonin, K.S. Makarova, and L. Aravind. Horizontal gene transfer in prokaryotes: quantification and classification. *Annu Rev Microbiol*, 55:709–742, 2001.

[57] R.H. Lathrop. The protein threading problem with sequence amino acid interaction preferences is NP-complete. *Protein Eng.*, 7(9):10591068, 1994.

[58] R.H. Lathrop and T.F. Smith. Global optimum protein threading with gapped alignment and empirical pair score functions. *J. Mol. Biol.*, 255(4):641665, 1996.

[59] R. Leplae and T.J. Hubbard. MaxBench: evaluation of sequence and structure comparison methods. *Bioinformatics*, 18(3):494–495, 2002.

[60] M.A. Marti-Renom, M.S. Madhusudhan, and A. Sali. Alignment of protein sequences by their profiles. *Protein Sci.*, 13(4):1071–1087, 2004.

[61] D. Milburn, R.A. Laskowski, and J.M. Thornton. Sequences annotated by structure: a tool to facilitate the use of structural information in sequence analysis. *Protein Eng.*, 11(10):855859, 1998.

[62] A.G. Murzin, S.E. Brenner, T. Hubbard, and C. Chothia. SCOP: a structural classification of proteins database for the investigation of sequences and structures. *J Mol Biol.*, 247(4):536–540, 1995.

[63] S.B. Needleman and C.D. Wunsch. A general method applicable to the search for similarities in the amino acid sequence of two proteins. *J. Mol. Biol.*, 48:443–453, 1970.

[64] C.A. Orengo, A.D. Michie, S. Jones, and D.T. Jones *et al.* Cath–a hierarchic classification of protein domain structures. *Structure*, 5(8):1093–1108, 1997.

[65] A.R. Panchenko. Finding weak similarities between proteins by sequence profile comparison. *Nucleic Acids Res.*, 31:683689, 2003.

[66] J. Park and S.A. Teichmann. DIVCLUS: an automatic method in the GEANFAMMER package that finds homologous domains in single- and multi-domain proteins. *Bioinformatics*, 14(2):144–150, 1998.

[67] J. Park, S.A. Teichmann, T. Hubbard, and C. Chothia. Intermediate sequences increase the detection of homology between sequences. *J Mol Biol.*, 273(1):349–354, 1997.

[68] F. Partensky, W.R. Hess, and D. Vaulot. Prochlorococcus, a marine photosynthetic prokaryote of global significance. *Microbiol Mol Biol Rev.*, 63(1):106–127., 1999.

[69] W.R. Pearson and D.J. Lipman. Improved tools for biological sequence comparison. *Proc. Natl. Acad. Sci. U. S. A.*, 85(8):24442448, 1988.

[70] M. Pellegrini and T.O. Yeates. Searching for frameshift evolutionary relationships between protein sequence families. *Proteins*, 37(2):278–283, 1999.

[71] J. Prilusky, D. Hansen, T. Pilpel, and M. Safran. The protein Hydrophilicity/Hydrophobicity search and comparison server. Technical report, Weizmann Institute of Science, Rehovot, Israel, 1999.

[72] B. Qian and R.A. Goldstein. Detecting distant homologs using phylogenetic tree-based HMMs. *Proteins*, 52(3):446–453, 2003.

[73] K. Qu, Y. Lu, N. Lin, and R. Singh *et al.* Computational and experimental studies on human misshapen/NIK-related kinase MINK-1. *Curr Med Chem*, 11(5):569–582, 2004.

[74] A. Raval, Z. Ghahramani, and D.L. Wild. A bayesian network model for protein fold and remote homologue recognition. *Bioinformatics*, 18(6):788–801, 2002.

[75] M. Rehmsmeier and M. Vingron. Phylogenetic information improves homology detection. *Proteins*, 45(4):360–371, 2001.

[76] B. Rost. TOPITS: threading one-dimensional predictions into three-dimensional structures. *Proc. Int. Conf. Intell. Syst. Mol. Biol.*, 3:314321, 1995.

[77] R.B. Russell and C.P. Ponting. Protein fold irregularities that hinder sequence analysis. *Curr Opin Struct Biol*, 8(3):364–371, 1998.

[78] L. Rychlewski, D. Fischer, and A. Elofsson. LiveBench-6: large-scale automated evaluation of protein structure prediction servers. *Proteins*, 53(6):542–547, 2003.

[79] R. Sadreyev and N. Grishin. COMPASS: a tool for comparison of multiple protein alignments with assessment of statistical significance. *J. Mol. Biol.*, 326(1):317336, 2003.

[80] J.M. Sauder, J.W. Arthur, and R.L. Dunbrack Jr. Large-scale comparison of protein sequence alignment algorithms with structure alignments. *Proteins*, 40(1):6–22, 2000.

[81] A.A. Schäffer, Y.I. Wolf, C.P. Ponting, and E.V. Koonin *et al.* IMPALA: matching a protein sequence against a collection of PSI-BLAST-constructed position-specific score matrices. *Bioinformatics*, 15(12):1000–1011, 1999.

[82] M. Shah, S. Passovets, D. Kim, and K. Ellrott *et al.* A computational pipeline for protein structure prediction and analysis at genome scale. *Bioinformatics*, 19(15):1985–1996, 2003.

[83] C.R. Shyu, P.H. Chi, G. Scott, and D. Xu. ProteinDBS: A real-time retrieval system for protein structure comparison. *Nucleic Acid Research*, 32:W572 – CW575, 2004. Web Server issue.

[84] A.H. Simons, N. Dafni, I. Dotan, and Y. Oron *et al.* Genetic synthetic lethality screen at the single gene level in cultured human cells. *Nucleic Acids Res.*, 29(20):E100., 2001.

[85] M.J. Sippl and S. Weitckus. Detection of native-like models for amino acid sequences of unknown three-dimensional structure in a data base of known protein conformations. *Proteins*, 13(3):258–271, 1992.

[86] T. F. Smith and M. S. Waterman. Identification of common molecular subsequences. *J Mol Biol.*, 147(1):195–197, 1981.

[87] T.F. Smith, C.L. Lo, J. Bienkowska, and C. Gaitatzes *et al.* Current limitations to protein threading approaches. *J Comput Biol.*, 4(3):217–225, 1997.

[88] R. Spang, M. Rehmsmeier, and J. Stoye. A novel approach to remote homology detection: jumping alignments. *J Comput Biol.*, 9(5):747–760, 2002.

[89] C.L. Tang, L. Xie, I.Y. Koh, and S. Posy *et al.* On the role of structural information in remote homology detection and sequence alignment: new methods using hybrid sequence profiles. *J Mol Biol*, 334(5):1043–1062, 2003.

[90] A.H. Tong, M. Evangelista, A.B. Parsons, and H. Xu *et al.* Systematic genetic analysis with ordered arrays of yeast deletion mutants. *Science*, 294(5550):23642368., 2001.

[91] A. Tramontano and V. Morea. Assessment of homology-based predictions in CASP5.

Proteins, 53(6):352–368, 2003.

[92] S. Veerassamy, A. Smith, and E.R. Tillier. A transition probability model for amino acid substitutions from blocks. *J. Comput. Biol.*, 10(6):9971010, 2003.

[93] C. Venclovas. Comparative modeling in CASP5: progress is evident, but alignment errors remain a significant hindrance. *Proteins*, 53(6):380–388, 2003.

[94] W. Volkmuth and N. Alexandrov. Evidence for sequence-independent evolutionary traces in genomics data. *Pac Symp Biocomput*, pages 247–258, 2002.

[95] N. von Ohsen, I. Sommer, and R. Zimmer. Profile-profile alignment: a powerful tool for protein structure prediction. *Pac Symp Biocomput.*, pages 252–263, 2003.

[96] A.C. Wallace, R.A. Laskowski, and J.M. Thornton. Derivation of 3D coordinate templates for searching structural databases: application to Ser-His-Asp catalytic triads in the serine proteinases and lipases. *Protein Science*, 5(6):10011013, 1996.

[97] J. Westbrook, Z. Feng, S. Jain, and T.N. Bhat *et al.* The protein data bank: unifying the archive. *Nucleic Acids Res.*, 30:245–248, 2000.

[98] D. Xu, D. Kim, P. Dam, and M. Shah *et al.* Characterization of protein structure and function at genome scale using a computational prediction pipeline. In J.K. Setlow, editor, *Genetic Engineering, Principles and Methods*, pages 269–293, New York, 2003. Kluwer Academic, Plenum Publishers.

[99] D. Xu, Y. Xu, and E.C. Uberbacher. Computational tools for protein modeling. *Current Protein and Peptide Science*, 1:1–21, 2000.

[100] Y. Xu and D. Xu. Protein threading using PROSPECT: design and evaluation. *Proteins*, 40(3):343–354, 2000.

[101] Y Xu, D. Xu, O.H. Crawford, and J.R. Einstein *et al.* Protein threading by PROSPECT: a prediction experiment in CASP3. *Protein Eng*, 12(11):899–907, 1999.

[102] Y. Xu, D. Xu, and V. Olman. A practical method for interpretation of threading scores: An application of neural network. *Statistica Sinica*, 12:159–177, 2002. special issue in bioinformatics.

[103] Y. Xu, D. Xu, and E.C. Uberbacher. An efficient computational method for globally optimal threading. *J Comput Biol.*, 5(3):597–614, 1998.

[104] G. Yona and M. Levitt. Within the twilight zone: a sensitive profile-profile comparison tool based on information theory. *J Mol Biol*, 315(5):1257–1275, 2002.

[105] G. Yona, N. Linial, and M. Linial. ProtoMap: automatic classification of protein sequences, a hierarchy of protein families, and local maps of the protein space. *Proteins*, 37(3):360–378, 1999.

[106] B. Zhang, L. Rychlewski, K. Pawlowski, and J.S. Fetrow *et al.* From fold predictions to function predictions: automation of functional site conservation analysis for functional genome predictions. *Protein Sci.*, 8(5):1104–1115, 1999.

[107] C. Zhang and S.H. Kim. Overview of structural genomics: from structure to function. *Curr Opin Chem Biol.*, 7(1):28–32, 2003.

[108] J. Zhu, R. Luthy, and C.E. Lawrence. Database search based on bayesian alignment. *Proc Int Conf Intell Syst Mol Biol.*, pages 297–305, 1999.

34

Biomolecular Modeling using Parallel Supercomputers

Laxmikant V. Kalé
University of Illinois

Klaus Schulten
University of Illinois

Robert D. Skeel
Purdue University

Glenn Martyna
IBM TJ Watson

Mark Tuckerman
New York University

James C. Phillips
University of Illinois

Sameer Kumar
University of Illinois

Gengbin Zheng
University of Illinois

34.1 Introduction

The processes of life are carried out at the molecular level by a few distinct categories of biomolecules. The *DNA sequences*, which constitute the genome of an organism, can be thought of as blueprints for making *proteins*, which constitute most of the "machinery of life". The translation from DNA to proteins is carried out via intermediate copies in the form of *RNA sequences*. *Membranes* made up of *lipid bilayers* provide structural separation for cells and other organelles. Short *poly-peptide chains*, made up of the same material as proteins, serve as messengers between cells. Finally, of course, there are a wide variety of molecules being created, transported, and consumed by the organism by means of the proteins.

Among the biomolecules, the proteins are the most diverse group and are responsible for the myriad specific functions carried out by organisms. Proteins carry out their functions inside cells. Embedded in cell membranes, some proteins act as gate-keepers, allowing specific substances to enter or leave the cells selectively, while other proteins act as signaling receptors, recognizing a specific type of signaling agent on the outside of the cell, and then either changing their own (e.g. gatekeeping) behavior or relaying the signal inside of the cell by releasing specific substances. Inside cells, proteins assist in recognizing genes, translating genes to proteins, and catalyzing biochemical reactions, among other activities.

Like DNA molecules, proteins are also polymers: they are made up of one or more chains of amino acids. Typical proteins consist of fifty to several hundred amino acids. Just as the genetic alphabet expressed in the DNA has four letters, a protein chain is made up from an alphabet of size twenty. Indeed, a gene typically corresponds to one protein, and the sequence of amino acids in a protein is determined by the DNA sequence in the corresponding gene, with three DNA bases coding for each residue in the protein sequence. However, unlike the DNA molecule, the 3-dimensional structure of a protein (also called its *conformation*), is strongly determined by its sequence. In most proteins, it is the conformation of the protein, along with the electrostatic properties at its surface resulting from this conformation, that allows it to carry out its biological function.

All amino acids in the polymer have a common part (moiety) and a side chain. The common moiety has a *backbone chain* made up of two carbon atoms and a nitrogen atom, with associated oxygen and hydrogen atoms as shown in Figure 34.1. There are 20 different kinds of side chains that distinguish the 20 amino acids. The atoms between consecutive C_α atoms tend to stay in a single plane. The structural variation arises from the ability of the two consecutive planes to vary their angle around the $N - C_\alpha$ and the $C_\alpha - C$ bonds. (As an aside, these angles are conventionally called ϕ and ψ. Thus the structure of a protein can be almost entirely specified by the successive $\phi - \psi$ angles. The three-dimensional structures of the chain and the side-chains restrict the combination of angles around a C_α atom. The "allowed regions" of the $\phi - \psi$ angles are plotted in the well-known Ramachandran plot for each pair of amino acids.)

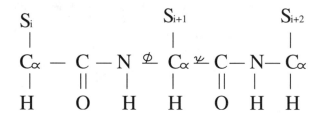

FIGURE 34.1: Amino acids form a polymer

The number of atoms in an amino acid side chain ranges from 1 (in glycine) to the high teens in others. The 19 amino acids, excluding the simple glycine, can be further classified by their electrical charge distributions: 7 uncharged ones, 4 with charged residues, and 8 that have no net charge, but are polarized. The uncharged, nonpolar residues tend to be found away from water (i.e. in the interior regions of the protein), and are called hydrophobic. The others are hydrophillic.

What makes a protein, essentially consisting of one or more long chains, fold itself into a specific structure (conformation)? The chain-forming covalent bonds are strong, and form the backbone of the protein. Independent of the side groups (and therefore independent of the sequence), the amino acids tend to form helices (called α helices) and strands or sheets (called β sheets, formed by two or more portions of the chain lying parallel or antiparallel to each other). Based on the side groups, weaker bonds can form between amino acids that are not next to each other in the chain. In addition, proline, a cyclic amino acid, causes the backbone to bend. The structure is further (and most significantly) determined by the fact that some amino acids are hydrophobic while others are hydrophilic. Water molecules are, of course, abundantly present in a cellular environment. The hydrophobic side groups are repelled by water, and hence cluster in the dry "core" of a folded protein. (Some proteins exist in the aqueous environment of the cell, while others are embedded in the cell membranes, and thus exposed to water molecules only where they emerge from the membrane.) Essentially, a protein folds into a particular shape because that shape presents a minimum free energy configuration.

How do scientists know the structure of a protein? In principle, given the sequence of amino acids, it should be possible to determine its most favorable conformation. But this is an exceedingly difficult computational problem. In current practice, conformations of proteins are determined by crystallizing them and analyzing them by x-ray diffraction.

Even when we know the structure of a protein, and have inferred its function via experimental methods, an important scientific question remains: how exactly does the protein carry out its function? If we understand how this happens, we will have a higher confidence in our guessed function (that indeed it is this particular protein that is responsible for this particular function). Secondly, the details of how this mechanism works may help us understand functions of other similar proteins.

Understanding the relationships between structure and function is often the main objective of molecular dynamics (MD) simulations. For example, an important class of membrane proteins called aquaporins allows the passage of water molecules across cellular membranes in which they are embedded. This would be a simple task, except that energetic processes in the cell are driven by maintaining a voltage gradient across the membrane that would be depleted if protons (positively charged hydrogen atoms lacking their sole electron) leaked through the aquaporin along with water. In many molecules that generate or utilize this voltage gradient, a single-file chain of water molecules will form a proton "wire" in which hydrogen atoms can quantum mechanically jump from one oxygen to the next. MD simulations of aquaporin revealed highly ordered water in the channel with specific charged residues orienting water molecules to prevent proton conduction. (A detailed explanation is in [83].)

FIGURE 34.2: **(See color insert following page 20-4.)** Aquaporin in lipid bilayer membrane, water above and below membrane omitted for clarity.

MD also serves a purpose in the determination of protein structure itself. Crystallographic (or other) data may only determine the structure of a protein at low resolution, leaving several uncertainties. A simulation can be used to refine this structure by starting from the approximate structure and letting the molecule settle into its lower energy conformation in the natural course of simulation.

An MD simulation of a biological system typically involves one or more macromolecules, such as proteins or DNA, and its environment, which consists of a cell membrane made up of lipid bi-layers (in case of membrane proteins), a large number of individual water molecules, and ions. Typically, the number of atoms in a simulation ranges from 10,000 to 100,000 (although simulations with as few as 1,000 atoms or as many as 1,000,000 atoms are sometimes needed).

Cells and their organelles are encapsulated by *membranes* which are made up of lipid molecules. Each lipid has a hydrophillic head and a hydrophobic tail. The lipids aggregate to form two layers, in which the heads form the inner and outer surface while the tails form the hydrophillic core of a membrane.

The computation itself is meant to capture the motions of individual atoms accurately. This motion is determined by Newtonian mechanics. We assume that each atom in the simulation has a fixed residual charge on it. The forces acting on each atom are

- electrostatic forces of attraction or repulsion due to other charged atoms,
- attractive and soft-core repulsive van der Waal's forces due to all other atoms
- forces due to bonds between atoms, which are modeled analogous to springs.

In each time step we need to calculate these forces for each atom and use the net force on each atom to calculate its acceleration, velocity, and new position. Due to the oscillation frequency of the bonds, the time for each step needs to be exceedingly small: 1 femtosecond (10^{-15} s) steps are the norm. The phenomenon of interest may take microseconds or even milliseconds. But, due to computational limitations, it is currently practical to be able to simulate only several tens of nanoseconds.

In most studies, MD is employed as an alternative to Markov Chain Monte Carlo (MCMC) methods, to generate random configurations of the system in accordance with a physically realistic probability distribution, e.g., generating positions x with a probability proportional to the Boltzmann factor $\exp(-U(x)/k_BT)$ where $U(x)$ is the potential energy function, T is temperature, and k_B is Boltzmann's constant. The goal is to explore the configuration space as rapidly as possible, and this is more easily done with MD than with MCMC. In this way the most common structures are found. Also, molecular mechanisms can be inferred from computing relative probabilities of differing values of a *reaction coordinate*, e.g., the progress of an ion through a membrane channel. Relative probabilities are normally described logarithmically as *free energy differences*. In such *sampling* applications of MD it is not necessary, nor probably even desirable, for the motion to be physically realistic as long as the probabilities are accurate. In some studies, however, the computation does aim to calculate realistic trajectories. However, accurate statistics can be gathered from an ensemble of long-time trajectories, e.g., to estimate transition rates.

An MD simulation is computationally intensive: a large number of timesteps are needed — 1 million steps simulate a nanosecond; yet, each timestep may take several seconds to compute on a single processor. This suggests using parallel computers to solve this problem. However, since each step depends on the previous one, parallelism is limited to within a single timestep. To use hundreds or thousands of processors to parallelize each individual timestep, which only represents a few seconds of computation, is the parallelization challenge.

In Section 34.2, we describe this classical molecular dynamics problem in some detail, and summarize several approaches for parallelizing it. We then describe a particular approach, developed by some of the authors, in additional detail to give the reader an idea of issues involved.

The classical picture we painted above breaks down if the phenomenon we are interested in involves making and breaking of bonds. We need a quantum mechanical simulation to understand the phenomenon at the molecular level. We describe approaches for this in Section 34.3. Quantum mechanical simulation is significantly more expensive (by several orders of magnitude) than classical simulation. A hybrid technique (QM/MM) is then used so that a portion of the system can be simulated in quantum mechanical detail, while the atoms surrounding that center are simulated classically. This technique is described in Section 34.4, with a detailed exposition of a particular code developed by several of the authors.

34.2 Classical Molecular Dynamics

Purely classical MD simulations are useful for understanding molecular processes that do not involve chemical reactions, i.e., those that occur entirely through changes in the conformation or aggregation of small numbers of molecules. This approach is particularly useful for biological systems because chemical reactions and electronic excitations in living cells are localized to catalytic sites or photoreceptors. Classical MD is also applicable to the physics of materials such as gels and liquid crystals where properties depend on the shape and conformation of the constituent molecules rather than extended electronic wave functions as in a metal or semiconductor.

Classical MD is based on the heuristic approximation of small-energy molecular deformations and intermolecular repulsion and attraction. The accuracy and usefulness of a simulation depends strongly on efficiency, i.e., can the simulation be performed at all given available resources, how many atoms can be simulated, how much simulated time can elapse, and how many independent sampling runs can contribute statistics. Therefore, the classical force field is designed for speed of evaluation, using as much as possible one term per degree of freedom for bonded (exclusively intramolecular) interactions, and simple pairwise forms for nonbonded (both intra- and intermolecular) interactions.

For simplicity, particularly when dealing with long chain molecules such as proteins and nucleic acids, special intramolecular energy terms are applied additively to every set of up to four atoms separated by three or fewer chemical bonds. The functional form is the simplest available. A *bond* between two atoms is treated as a simple spring with energy $\frac{1}{2}k(d - d_0)^2$. To the *angle* θ between every pair of bonds with a common atom is assigned an energy $\frac{1}{2}k(\theta - \theta_0)^2$. The energy of a *dihedral* torsion angle between the planes of any two angles on opposite ends of a common bond is represented as one or more sinusoids of the form $k \cos n(\phi - \phi_0)$. Finally, if an atom must be restrained to the plane of three atoms to which it is bonded, an *improper dihedral* of the form $\frac{1}{2}k(\phi - \phi_0)^2$ is applied to the angle between the planes of two atoms with the central atom and with the fourth atom. In all cases, the constants k, d_0, θ_0, n, ϕ_0, etc. are specific to the particular combination of atom types (e.g., carbon bonded to two hydrogens and two other single bonds) present in that bond/angle/dihedral/improper. The above bonded terms are the only interactions between directly bonded atoms or between atoms bonded to a common atom; *nonbonded* interactions between such pairs of atoms are said to be *excluded*. All other pairs of atoms in the simulation experience electrostatic, van der Waals, and electronic exclusion interactions. Since the number of pairs of atoms scales as square of the number of atoms in the simulation, nonbonded interactions are normally truncated at a cutoff distance of, e.g., 12 Å. Full electrostatics interactions may be calculated via more efficient algorithms as described below. Each atom in the simulation is assigned a *partial charge* corresponding to the sum of the nuclear charge and the density of the nearby electronic cloud. For example, the oxygen of a water molecule has a charge of -0.834e while each hydrogen bears 0.417e, yielding a neutral molecule. Electrostatic interactions are calculated via the standard Coulomb potential Cq_1q_2/r. The $1/r^6$ van der Waals interaction is combined with atomic repulsion in the easy-to-evaluate Lennard-Jones potential $4\epsilon((\sigma/r)^{12} - (\sigma/r)^6)$ characterized by an attractive well of depth ϵ that ends with a potential of 0 at a distance σ between the two atoms (there is no physical justification for the $1/r^{12}$ form of the repulsive term). The parameters ϵ and σ depend again on the types of the interacting atoms, and are typically derived for interactions between atoms of different types as $\epsilon_{ab} = \sqrt{\epsilon_a \epsilon_b}$ and $\sigma_{ab} = \frac{1}{2}(\sigma_a + \sigma_b)$.

The various constant parameters of the energy functions described above are derived from experimental data on small analogous molecules as well as expensive quantum mechanical calculations. These parameters are typically formulated by a small number of expert groups and then used by an entire community with the expectation that, while not perfect, they are the best available and will yield reasonable results for the typical simulations for which they were developed. These parameters are specified to two or at most three significant digits, and this level of accuracy certainly exceeds the error from the use of the simple functional forms described above. The reader is again reminded that these potentials are a compromise dictated by the need for simulations that are sufficiently long and large to capture interesting phenomena at the expense of detailed accuracy. The transition from the molecular *mechanics* embodied in the above potential function to *dynamics* is simply a matter of combining initial coordinates and velocities for every atom in the simulation with Newton's second law of motion, $F = ma$. Initial coordinates are obtained by combining known structures from crystallography with standard bond lengths and angles, and from minimizing the energy function to eliminate unphysically strong interactions. Initial velocities are chosen at random from the Boltzmann distribution of classical statistical mechanics, $e^{-v^2/2mk_BT}$.

Biological molecules depend for their function and even their stable native structure on a solvent environment of liquid water, or rather a mixture of water, sodium ions, and chloride anions at physiological concentrations. Water is a notoriously complex substance, forming an irregular network of hydrogen bonds with itself and hydrophilic solutes, while avoiding greasy hydrophobic substances such as the tails of lipids. Many biological ion channels have pores that are only wide enough for water molecules to pass through in single file, and chains of water molecules can form "wires" that allow the rapid conduction of protons. For these and other reasons most simulations employ *explicit* solvent, i.e., water and ions are modeled in the same atomic manner as are proteins, lipids, and nucleic acids even though the number of solvent atoms in a simulation may exceed the number of solute atoms by a factor of ten. In order to minimize the amount of solvent in a simulation while avoiding surface effects from a liquid-vacuum boundary, *periodic boundaries* are established for a simulation, wrapping the six faces of a filled simulation cell into a three-dimensional torus. Atoms leaving one side of the cell reappear on the opposite side, and atoms interact with any image of another atom produced by translating its coordinates by integer multiples of the three cell basis vectors.

The temperature of a simulation may be estimated from the atomic velocities via the equipartition theorem of classical statistical mechanics, i.e., that every degree of freedom will have an average energy of $\frac{1}{2}k_BT$. Hence, the temperature of a simulation of N atoms is $(\sum_i m_i v_i^2)/3Nk_BT$. A similar formula (involving also the positions of and forces between atoms) is used to estimate the pressure of a simulation in a periodic cell. Temperature is controlled by modifying the equations of motion to manipulate the atomic velocities either individually via stochastic terms or collectively via instantaneous or gradual rescaling towards a desired average kinetic energy per atom of $\frac{3}{2}k_BT$. Pressure is then additionally controlled by extending the equations of motion to include fluctuations of the periodic cell volume, expanding or contracting the cell under high or low internal pressure, respectively. Pressure control is particularly critical given the incompressibility of liquids and the tendency of water to form "bubbles" of vacuum given an oversized periodic cell.

The basic protocol of a classical MD simulation begins with minimization (of the potential function by adjusting atomic coordinates) so as to eliminate high-energy atomic contacts left over from the assembly of proteins, lipids, solvent, etc. into the final simulated system. While incapable of finding a global energy minimum, or even a nearby local one, this minimization step is needed because of the stability requirements of the explicit integration methods used. When dynamics are begun on the minimized system, the temperature will be found to be near zero and the system must therefore be heated to $300\,\mathrm{K}$ by a sequence of velocity rescalings or reassignments to progressively higher temperatures. When the system has stabilized at the target temperature, more gradual temperature and pressure control algorithms are enabled to allow the cell size to equilibrate. Finally, the simulation is run for an extended period of time to achieve the scientific goals of the simulation.

Running a simulation for an indeterminate period of time, hoping to observe a rare barrier-crossing event is a horrible waste of computer resources. Fortunately, techniques have been developed to force a simulation to follow a given reaction path while monitoring the steering forces required to yield a free energy profile for the transition. It is even possible to calculate the free energy difference of chemical change, e.g., of replacing one ligand directly by another. Finally, conformational exploration may be accelerated by the systematic raising and lowering of temperature as in simulated annealing, or more drastically by the loose coupling of simultaneous replica simulations at a range of temperatures.

34.2.1 Integration Methods

Numerical integrators propagate the state variables of the biological system, e.g., positions and velocities of atoms, as determined by equations of motion. These equations may be either deterministic or stochastic. Deterministic models of molecular dynamics have no dissipation, and due to the enormous number of time steps taken, all approximations must be formulated with enormous care [79]. It is not so much a question of accuracy as ensuring that numerical integrators and fast approximations to forces have the right properties, e.g., energy conservation, among others. It is risky to tamper with the numerics of these algorithms. Stochastic models do have dissipation and are more amenable to standard numerical techniques.

Numerical integrators generate a trajectory $x^n \approx x(n\Delta t)$ where Δt is the size of the time step. The Verlet method:

$$M\frac{1}{\Delta t^2}\left(x^{n+1} - 2x^n + x^{n-1}\right) = F(x^n), \qquad v^n = \frac{1}{2\Delta t}\left(x^{n+1} - x^{n-1}\right), \qquad (34.1)$$

where F is the negative gradient of U, is the most natural approximation to the equations of motion, and fortuituously has all the right properties. This and other typical numerical integrators can be expressed as a mapping from x^n, v^n to x^{n+1}, v^{n+1} of state variable values.

Expressing the Verlet method this way gives the velocity Verlet scheme:

$$x^{n+1} = x^n + \Delta t v^n - \frac{1}{2}\Delta t^2 M^{-1}F(x^n),$$

$$v^{n+1} = v^n + \frac{1}{2}\Delta t M^{-1}F(x^n) + \frac{1}{2}\Delta t M^{-1}F(x^{n+1}),$$

a form with significantly less accumulation of roundoff error than the form obtained by solving eq. (34.1) for x^{n+1}.

The appropriate step size depends on the force term being integrated, and savings in CPU time are possible by using multiple time stepping — different step sizes for different interactions. A suitable scheme for multiple time stepping (MTS) has become popular under the name r-RESPA [85]. As an example, consider the use of two different step sizes having a 3:1 ratio. First split $F = F^{\text{slow}} + F^{\text{fast}}$. Define an (outer) time step of MTS to be 3 Verlet steps, each with step size $\frac{1}{3}\Delta t$:

1. at steps $n = 0, 1, 2, \ldots$, use $F^{\text{slow}} + \frac{1}{3}F^{\text{fast}}$, and
2. at steps $n = \frac{1}{3}, \frac{2}{3}, \frac{4}{3}, \frac{5}{3}, \frac{7}{3}, \frac{8}{3}, \ldots$, use $\frac{1}{3}F^{\text{fast}}$.

This scheme generalizes to more than two different step sizes. To maintain energy conservation, the splitting has to be done at the level of the potential energy function. MTS might employ a short step size for bonded interactions and medium and long step sizes for nonbonded interactions. There is a complication due to the fact that the distance between two nonbonded atoms varies during the course of the simulation. This is handled by splitting each nonbonded potential energy term into a sum of a short-range and a slowly varying long-range part. There is a limit on the longest step size Δt — appreciable energy drift occurs due to 3:1 nonlinear resonance unless Δt is less than a third of the period of the fastest vibration (9 fs) [51]. This limit on Δt rises to one half the period with the use of the mollified impulse MTS method [43].

Larger step sizes are possible if the highest frequencies are removed by introducing rigidity into the molecular model. This can be done for most biomolecular studies with little loss of accuracy. In particular, freezing the lengths of covalent bonds to hydrogens approximately doubles the period of the fastest vibration [76, p. 230]. Rigidity is accomplished by appending a set of constraints $g_i(x) = 0$ to the equations of motions and enforcing these by adding constraint forces $\sum_i \lambda_i \nabla g_i(x)$ to the equations of motion where the λ_i are Lagrange multipliers chosen so as to satisfy the constraints. The Verlet scheme can be extended to handle constraints using the SHAKE [75] discretization:

$$M\frac{1}{\Delta t^2}\left(x^{n+1} - 2x^n + x^{n-1}\right) = F(x^n) + \sum_i \lambda_i^n \nabla g_i(x)$$

where the λ_i^n satisfy the system of equations $g_i(x^{n+1}) = 0$. Commonly, the constraint equations are solved for the Lagrange multipliers iteratively by successive substitution (Newton-Gauss-Seidel to be precise). An accurate solution is needed to avoid energy drift. It is common to use all-rigid models for water and special methods like SETTLE [61] have been developed to solve for the constraints for water non-iteratively.

Typically the computing time is dominated by the evaluation of energies and forces arising from water molecules. This has led to the development of implicit solvent models for water [76]. One such model requires the solution of a nonlinear partial differential equation — the Poisson-Boltzmann equation — for the electrostatic potential. For actual dynamics simulations there is the addition of a friction term and a noise term containing a $3N$ by

$3N$ tensor $D(x)$ to account for contact of the solute with the solvent. The resulting modification of Newtonian dynamics is called *Langevin dynamics*, and a representative temporal discretization is the Brooks-Brünger-Karplus [11] scheme, which appends the terms

$$-k_\mathrm{B}TD(x^n)^{-1}\frac{1}{2\Delta t}(x^{n+1}-x^{n-1})+\sqrt{2}k_\mathrm{B}TD_{1/2}(x^n)^{-\mathsf{T}}\frac{1}{\sqrt{\Delta t}}Z^n$$

to the right-hand side of eq. (34.1). Here Z^n is a collection of independent random numbers from a Gaussian distribution with mean 0 and variance 1 and $D_{1/2}(x)$ is chosen to satisfy $D_{1/2}(x)D_{1/2}(x)^{\mathsf{T}} = D(x)$. There are significant challenges dealing with the inversion and factorization of the diffusion tensor; see, e.g., [6]. A simplification of Langevin dynamics, valid over long enough spatial scales, is to neglect inertia by setting masses to zero, an approximation often called Brownian dynamics. The customary choice of numerical integrator is the Euler(-Maruyama) method, introduced in this context by Ermak and McCammon [28]. This method advances coordinates for a time step Δt by the simple recipe

$$\frac{1}{\Delta t}(x^{n+1}-x^n)=\frac{1}{k_\mathrm{B}T}D(x^n)F(x^n)+\sqrt{2}D_{1/2}(x^n)\frac{1}{\sqrt{\Delta t}}Z^n.$$

For purposes of sampling, one can use nonphysical diagonal diffusion tensors, greatly simplifying the computation. In particular, Langevin dynamics is appropriate for sampling systems in thermal contact with their surroundings (NVT simulations).

Extensions to the equations of motion for the purpose of sampling systems in thermal and mechanical contact with their surroundings (NPT simulations) require careful discretization.

34.2.2   Long Range Forces

Historically, the calculation of nonbonded interactions has been reduced from order N^2 to order N by neglecting all contributions beyond a certain cutoff distance (e.g., 12Å). While this is defensible, in most cases, for the rapidly decaying $1/r^6$ and $1/r^{12}$ terms of the Lennard-Jones potential, the $1/r$ Coulomb potential has a longer effective range and may play a significant role in the function of biomolecular machines through the positioning of charged and polar residues in the native protein structure. Clearly it is possible to specify arbitrary charge configurations for which truncation produces anomolous physical effects, but this does not imply that such situations occur in bone fide simulations.

An effective argument can be made that ions present in solution will be attracted to, migrate near, and thereby effectively screen significant concentrations of positive or negative charge. Therefore, a cutoff distance larger than the effective screening distance should be sufficient to capture scientifically interesting electrostatic effects. This reasoning flows well for the small, globular proteins surrounded on all sides by water and a physiological concentration of ions and anions that were the focus of early MD simulations.

The structure of nucleic acids is more challenging, with the charged DNA or RNA neutralized by a complement of counter-ions. Any DNA-binding protein could be expected to depend on this characteristic charge distribution for adhesion or activation at distances beyond normal ion screening and cutoffs. Similarly, the aligned polar head groups of lipids assembled into a bilayer membrane will induce an electrostatic potential profile across the neutral lipid tails forming the core of the membrane. A protein embedded in a membrane may be sensitive to and rely on this long-range electric field for alignment or function. Therefore, for many biomolecular aggregates the full evaluation of electrostatic interactions is a necessity.

For a simulation without periodic boundary conditions, the calculation of electrostatic interactions between all pairs of atoms is straightforward. For a periodic simulation, however, forces must be evaluated between each atom and all of the infinite periodic images of every other atom. This infinite series is conditionally convergent, and the interaction with each set of periodic images may be summed in order of increasing distance from the central cell. The Ewald method [2, §5.5.2] allows this sum to be performed naively with order N^2 operations, or in a more sophisticated manner with order $N^{3/2}$ at constant accuracy. To each atomic point charge is added a cancelling Gaussian charge distribution, resulting in net interactions that decay exponentially fast (due to the shell theorem of elementary physics) and can be truncated (generally at the same cutoff used for van der Waals interactions) and summed directly with virtually full accuracy. The negative of the cancelling Gaussian distribution (with the same total charge as the original atomic point charge) is then sufficiently slowly varying that its interaction with all periodic images of another such Gaussian may be evaluated as a truncated sum in reciprocal (Fourier) space. The particle-mesh Ewald (PME) method [29] performs the Ewald reciprocal space sum by first spreading each atomic point charge onto (typically a $4 \times 4 \times 4$ portion of) a regular three-dimensional grid (typically of 1Å resolution). A Fourier transform of the charge grid allows the sum over reciprocal vectors can then be accomplished for all pairs of atoms simultaneously through a simple scaling of each element of the transformed grid, after which reversing the Fourier transform results in a potential map from which per-atom energies and forces can be extracted. Biomolecular simulations typically have one atom per $10\,\text{Å}^3$ of volume in the periodic cell, and therefore the number of grid points is roughly ten times the number of atoms. The reciprocal-space sum can then be accomplished in order $N \log N$ operations with the aid of the fast Fourier transform (FFT) algorithm.

In practice, the runtime of the FFT (for which highly tuned libraries are available in abundance) is tiny compared to the cutoff Lennard-Jones and Ewald direct (real-space) force evaluations, or even to the gridding of atomic charges and extraction of forces in the reciprocal-space sum. Therefore, the PME algorithm allows full electrostatics to be evaluated in order N time for any size of simulation contemplated today. In addition, PME allows the cutoff distance for the remaining short-range nonbonded force evaluation to be reduced below what would be reasonable for a simulation with truncated electrostatics, leading to the pleasant result that a simulation with full electrostatics may actually run faster than without. As the long-range electrostatic interations calculated with PME are also comparitively weak and slowly varying, additional performance may be obtained through the use of multiple timestepping as described above.

34.2.3   Parallelization Strategies

The computations in MD simulations stem from calculating forces on each atom in each time step, then integrating these forces to update the position and velocity of each atom. In a parallel computation, the is local and accounts for a small fraction of the computation time.

The force calculations comprise the majority of computational effort. There are two types of force calculations, bonded and non bonded. Forces between bonded atoms are for only two to four atoms total, whereas non bonded forces operate between more atoms and stem from van der Waals and electrostatic forces. As described in Section 34.2.2 computation may be minimized by choosing a cut off radius outside of which the non bonded forces are not calculated. Multiple time-stepping (MTS, see Section 34.2.1) approaches are adopted when the cumulative force of distant atoms cannot be entirely ignored. The other forces must be calculated each time step. Therefore, this discussion will focus on decomposition strategies for the non bonded forces with cut off.

Depending on the size of the cut off radius, non bonded forces consume between 80 and 95 percent of total computation time. The cut off may be as small as 8 Å when using an MTS approach for distant interactions, or range from 12 Å to 18 Å in the non MTS case. The dominance of the non bonded force calculation does not mean the bonded force calculation can be ignored for the purposes of the parallel decomposition strategy. Leaving even as little as 5% of the total calculation purely sequential would limit the maximum obtainable speedup to 20.

The most straightforward parallelization technique is *data replication*. By copying the data for all the atoms to each processor, the forces for any subregion can be independently calculated by any processor. For N atoms and P processors the $O(N)$ forces must be added up across the processors, resulting in a communication time proportional to $N \log P$. The amount of computation is proportional to N (assuming cut off), resulting in N/P computation per processor. Total parallel execution time is the sum of the communication time and computation time. Replicated data leads to a ratio of $P \log P$ as a ratio of communication to computation time, which is entirely independent of the number of atoms N. Thus replicated data is not a scalable strategy, for the fraction of time spent in communication grows with the number of processors. Despite this theoretical limit, replicated data strategies work in practice for tens of processors and the implementation path to parallelize an existing sequential application is straightforward. Replicated data therefore remains in use by many MD applications. Another approach is to use a technique called *atom decomposition* in which the array containing the atoms of the model is partitioned and divided across processors. As the data array is partitioned continuously, atoms close in space may not be nearby in the array. Therefore data from other processors is required, as many as all of them in the worst case. This results in $O(N)$ communication cost which again impedes scalability.

Force decomposition distributes blocks of the sparse force matrix across processors. The $N \times N$ force matrix is divided into blocks of size $(N/\sqrt{P}) \times (N/\sqrt{P})$. A processor needs the coordinates of $2N/\sqrt{P}$ atoms from $\sqrt{P}$ different processors. This leads to a per processor communication cost of $O(N/\sqrt{P})$ and a communication to computation cost ratio of $\sqrt{P}$. Though better than the replicated data method, this fails to achieve ideal scalability because the communication cost rises with the number of processors even if we increase the problem size. It has been shown by Plimpton *et al.* [71] and Hwang *et al.* [41] that this scheme can be used with good speedup for up to hundreds of processors.

The force matrix decomposition can result in a nonuniform distribution of work. This can be redressed by randomizing the sequence of atoms in the array, and thus eliminating all traces of spatial locality [41]. Alternatively one could exploit the spatial locality to co-locate nearby atoms and handle the load imbalance explicitly [71]. The resulting communication costs are smaller than for atom decomposition, but the latter scheme suggests a different technique. *Spatial decomposition* divides the computation by assigning nearby atoms to the same processor. Various techniques fall into three categories:

- Partitioning space into P regular boxes or other regions, one per processor.
- Partitioning space into boxes of fixed size slightly larger than the cut off distance, thereby requiring communication only between neighboring boxes.
- Partitioning space into a very large number of small boxes, thereby requiring each box to communicate with a large number of boxes to cover all atoms within the cut off radius.

In the first case the communication cost is proportional to the surface of the box for sufficiently large N and the computation cost is proportional to the volume, leading to a highly scalable algorithm in theory. Although this works in larger materials science modeling (involving millions of atoms), the number of atoms is not large enough in biomolecular modeling to justify this approximation [45]. Additionally, this method is difficult to employ when the number of processors cannot be factored ($P = L \times M \times N$) into 3 roughly equal integers.

The second case results in each box communicating with a constant 26 neighbor boxes. The number of boxes is now independent of the number of processors. This requires more boxes than processors, but leads to a communication cost of N/P and a constant communication to computation ratio. The problem size can be doubled with a doubling of the number of processors without loss of parallel efficiency. Early versions of NAMD [62] used this technique and achieved good scalability. It is vulnerable to severe load imbalance problems, particularly when simulating non-periodic systems. Recent versions of NAMD use a hybrid combination of force and spatial decompositions, see Section 34.2.4 for details.

The third case results in boxes smaller than the cut off communicating with a larger number of non-neighboring boxes. This was implemented in EulerGromos by Clark [20]. The number of messages (though not the total size of data) can be reduced using a multi-stage algorithm known as "north-south-east-west" or the "shift algorithm" [70, 80]. This produces good speedup if sufficient processors are available.

Another spatial decomposition approach is found in the FAMUSAMM algorithm as implemented in recent versions of EGO [37]. This approach uses hierarchical decomposition of space, based on structural features of biomolecules, to implement a structure-adapted multipole algorithm.

LeanMD is a new experimental prototype [44] implemented in Charm++ which expands on these ideas. It uses a "2-away" or "3-away" strategy where instead of using one box of cut off size with neighbors "1-away", the box is divided into cells where two cells would span the cut off and communicate with all neighbors which are "2-away", or three cells to span the box communicating with all neighbor cells "3-away". This divides the computation into many small objects and communication between them can be optimized by a variety of message consolidation techniques.

Summary of Parallel MD Applications

In addition to NAMD, the biomolecular modeling community sustains a variety of software packages with overlapping core functionality but varying strengths and motivations. For comparison, we select AMBER, CHARMM, GROMACS, NWChem, and TINKER.

AMBER [101] and CHARMM [10] are often considered the standard "community codes" of structural biology, having been developed over many years by a wide variety of researchers. Both AMBER and CHARMM support their own force field development efforts, although the form of the energy functions themselves is quite similar. Both codes are implemented in FORTRAN 77, although AMBER takes the form of a large package of specialized programs while CHARMM is a single binary. GROMACS [49] claims the title of "fastest MD." This can be attributed largely to the GROMOS force field, which neglects most hydrogen atoms and eliminates van der Waals interactions for those that remain. In contrast, the AMBER and CHARMM force fields represent all atoms and new development has centered on increasing accuracy via additional terms. Additional performance on Intel x86 processors comes from the implementation of inner loops in assembly code. GROMACS is implemented in C as a large package of programs and is released under the GNU General Public License (GPL). NWChem [39] is a comprehensive molecular simulation system developed by a large group of researchers at the PNNL EMSL, primarily to meet internal requirements. The code centers on quantum mechanical methods but includes an MD component. Implemented in C and FORTRAN, NWChem is parallelized using MPI and a Global Arrays library[1] which automatically redistributes data on distributed memory machines. Parallel scaling is respectable given sufficient workload, although published benchmarks tend to use abnormally large cutoffs rather than the 12 Å (or PME) typically used in biomolecular simulations. TINKER [72] is a small FORTRAN code developed primarily for the testing of new methods. It incorporates a variety of force fields, in addition to its own, and includes many experimental methods. The code is freely available, but is not parallelized, and is therefore inappropriate for traditional large-scale biomolecular simulations. It does, however, provide the community with a simple code for experiments in method development.

34.2.4 NAMD Structure

We now describe the design and performance of NAMD [69] as a case study. NAMD was designed with three major goals — parallel scalability, maintainability and extensibility. NAMD aimed at utilizing large parallel machines in a scalable manner. Due to the relatively small amount of computation in each timestep of the simulation, effectively parallelizing molecular dynamics simulation is very challenging. Such a strategy needs to generate enough parallelism with fine-grained computation for parallel machines with a large number of processors. Maintainability and extensibility are likewise essential for a parallel molecular dynamics program. A well-motivated application-domain programmer should be able to extend the parallel program to permit novel experiments. NAMD is parallelized using the Charm++ object-oriented parallel language for extensibility as well as the parallel scalability. A novel combination of force and spatial decomposition schemes is deployed, which has been shown to be effective even for very large parallel machines. In this scheme, atoms are partitioned into cubes whose dimensions are slightly larger than the cutoff radius. For each pair of these neighboring cubes, we assign a non-bonded force *compute object*, which can be independently mapped to any processor. Since each cube has at most 26 neighbors, the number of such parallel objects is therefore 14 times $(26/2 + 1$ self-interaction) the number of cubes. For a reasonably large molecular system, this scheme can easily produce a sufficiently large number of parallel calculations for thousands of processors.

For extensibility and modularity, a class hierarchy that incorporates the structure of molecular dynamics was designed. Objects called "patches" represent cubical regions of space, and all the atoms within such a region. Patches are implemented as parallel objects that are distributed to processors. Compute objects signify computation of different kinds of forces on atoms in a set of patches. They are implemented as parallel objects in Charm++ that can be mapped to any processor and are free to migrate among processors. Based on the basic functionality needed in MD computation, a programmer wishing to add features to NAMD can often overload an existing class to add some particular functionality.

The hybrid spatial/force decomposition scheme described above provides the basis the for scalability of NAMD. This scheme requires that a number of entities (multiple patches and force computations) reside on each processor. Instead of a monolithic program that orchestrates all these diverse actions on a single processor, we chose a message-driven object paradigm offered in Charm++. Parallel objects are scheduled based on availability of data needed for their continued execution, and different tasks can interleave based on the availability of data needed. For example, a force computation is scheduled for execution only when all the data it needs are available on the local processor, thus avoiding the possibility that any particular entity will block the processor while waiting for specific data.

---

[1]http://www.emsl.pnl.gov:2080/docs/global/ga.html

The parallel structure of NAMD is shown in Figure 34.3. At the beginning of each timestep, patches send their atom coordinate data to all *compute objects* whose computation depends on them. These compute objects can be pairwise compute objects performing non-bonded force computation; angle compute objects performing various bonded force computation, or PME compute objects performing Particle Mesh Ewald force computation. After compute objects finish their force computation, they send forces back to home patches, which then integrate all the forces to calculate the new atom coordinates.

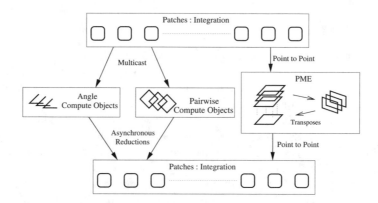

FIGURE 34.3: Parallel structure of NAMD

In order to achieve high scalability over large parallel machines, it is very important to carefully map the computation among processors so that load balance is achieved and communication is minimized. Further, in a long simulation, the force computation tends to change over time when atoms move, leading to new load imbalances. It is clearly impractical to require a programmer to take care of load balance himself manually.

NAMD employs Charm++'s measurement-based load balancing strategy to perform automatic adaptive load balancing. When a simulation begins, only a reasonable distribution is needed. Initially, patches are distributed according to a recursive coordinate bisection scheme. All compute objects are then distributed to a processor owning at least one patch they communicate with. During simulation, Charm++ measures the execution time of each compute object. After the simulation runs for one hundred or so timesteps, the program suspends the simulation to trigger the initial load balancing phase. The Charm++ load balancing module retrieves the object load on each processor, computes an improved load distribution taking into account communication between patches and compute objects, and migrates compute objects to processors to improve load balance. The initial load balancing step is aggressive. It computes a new object-to-processor distribution from scratch with a greedy algorithm. Once a good balance is achieved, atom migration changes load very slowly. Another load balancing phase is only needed after several thousand steps. An aggressive algorithm is not necessary for the subsequent phases, instead, a less expensive refinement scheme that only adjusts load by migrating a few objects from heavily loaded processors to underloaded ones is sufficient.

34.2.5 NAMD Performance

In order to demonstrate the scalability of NAMD for the real problems of biomedical researchers, we have drawn benchmarks directly from simulations being conducted by NIH-funded collaborators. The smaller *ApoA1* benchmark comprises 92K atoms of lipid, protein, and water, and models a high density lipoprotein particle found in the bloodstream. The larger ATP*ase* benchmark consists of 327K atoms of protein and water, and models the F_1 subunit of ATP synthase, a component of the energy cycle in all life. For both benchmarks, short-range nonbonded interactions were cut off at $12\,\text{Å}$ as specified by the CHARMM force field. Full electrostatics interactions were calculated every four timesteps and the PME grid was set at a spacing of approximately $1\,\text{Å}$ $108 \times 108 \times 80$ for ApoA1 and $192 \times 144 \times 144$ for ATPase. Results are shown in Figures 34.5 and 34.4 respectively[2].

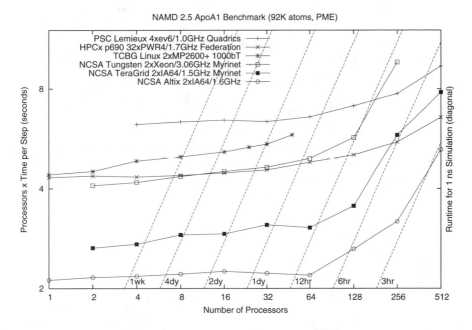

FIGURE 34.4: NAMD Performance on various platforms with ApoA1 benchmark illustrates the portable scalability of NAMD on a variety of platforms employed for production simulations by researchers. Each curve represents total resources (processors multiples by time per step) consumed per step for ApoA1 PME benchmark by NAMD on varying numbers of processors for a specific parallel platform. Perfect linear scaling is a horizontal line. Diagonal scale shows runtime per ns, representing absolute performance — the time to solution as experienced by the user.

NAMD won the Gordon Bell award at 2002 Supercomputing Conference with unprecedented speedup on 3,000 processors on Pittsburgh Supercomputing Center's Lemieux su-

---

[2]A more comprehensive performance comparisons on different platforms can be found at http://www.ks.uiuc.edu/Research/namd/performance.html.

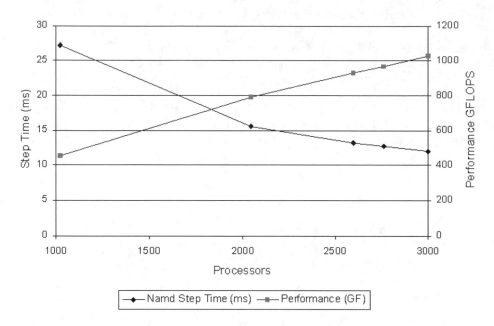

FIGURE 34.5: NAMD Performance on 327K atom ATPase PME benchmark, showing the scalability of the ATPase NAMD simulation to 3000 processors on the PSC Lemieux machine. Both step time and floating point performance are shown in the plot. The best achieved step time is 12ms with a floating point performance of just over one TF.

percomputer[3] with teraflops level peak performance.

34.3   Quantum Mechanical Molecular Dynamics

It is now possible to study novel processes in the condensed phase at an atomistic level of detail using modern theoretical techniques, software and supercomputers to provide new insights into long standing problems in fields ranging from biophysics, condensed matter physics, chemistry and biology. Phenomena such as the self-assembly of biomolecules to form functional nanomachines, the structure and dynamics of the ubiquitous, universal solvent, water, the behavior of minerals in the core of planets and the functionalization of semiconductor surfaces have all been successfully explored [78, 65, 24, 57, 34, 90, 9, 17, 97, 32, 14, 1, 16, 42, 98]. Clearly, a great variety of different physical forces have to be accurately modeled in order to achieve this wide range of applicability. Simulation studies can roughly be divided into two classes, empirical force field based molecular dynamics (MD) calculations described in the previous section [73, 2] and *ab initio* based molecular dynamics simulations [13, 74, 66, 33, 93, 35, 64, 55, 12, 87] to be described here. Although simulation studies performed using empirical potential models have contributed greatly to science

---

[3]750 Quad 1Ghz Compaq Alphaserver ES45 nodes connected by a Quadrics highspeed network.

and engineering, a large number of important processes, particularly those involving bond making and bond breaking, are not treated properly. An empirical force field [52, 18, 21] typically consists of a set of sites, usually the atoms, whose covalent bonding pattern is fixed and special interactions are added to model this pattern (bonds, bends, torsions ...) which in effect prevent the bonding from changing. The sites are assigned charges which are typically fixed. Sites separated by more than 3 covalent bonds then interact in a pairwise fashion via Coulomb's law, a repulsive term with fixed parameters to prevent site interpenetration and an effective dispersion interaction. The parameters in the force field are then fit to describe a small number of high level *ab initio* calculations to set the covalent bonding parameters and a small number of condensed phase experimental data sets to set the remainder of the parameters. The great advantage of the empirical force field approach is its computational simplicity and the speed with which atomic forces can be evaluated, allowing long times and large distance scales to be explored when coupled to MD techniques. However, there are significant disadvantages that need to be discussed.

Fixing the covalent bonding pattern precludes the study of chemistry, biochemistry and geochemistry. Hence, acid-base chemistry cannot be examined, enzyme catalysis cannot be treated nor can the rearrangement of complex minerals under high pressure be studied. There are several ways to lift empirical force fields beyond this limit. Two popular methods are empirical valence bond models [100] and dissociable force fields both of which are beyond the current discussion. While more general, these methods add more empirical parameters to the force field in order to extend it such that some chemistry of interest, for example the breaking of one type of bond, can be described. If another chemical mechanism is present (e.g. other than that hardwired into the more complex description), the investigation will give rise to misleading results.

Similarly, the charge distribution of the molecules comprising a complex manybody system is not fixed. The environment causes the electrons in the molecules to rearrange in a phenomena referred to as manybody polarization [8]. For example, the dipole moment of a typical water molecule in the liquid at T=300K is 1.7 times greater than that of an isolated water molecule due to dipole manybody polarization. Changes also occur to the molecule's quadrupole, octopole electrostatic and higher moments due to appropriately higher order polarization responses. Therefore, in a heterogeneous environment with complex interfaces, for instance a protein in water solution, the solvent molecules at the interface will possess different charge distributions from those in the bulk which possess different charge distributions from those that might penetrate into the interior of a large solute. Indeed, two similar chemical moieties that form separate parts of a large solute will have different charge distributions depending on whether the moiety in question resides on the surface or in the interior of the solute. Neglecting manybody polarization gives rise to incorrect dielectric constants (long range screening of Coulomb interactions), an incorrect number of water molecules around an ion, ions collecting in the bulk as opposed to moving to an interface and a host of other difficulties. As above, empirical force fields can be enhanced with additional terms to treat manybody polarization [82, 81, 23, 15, 53] with the same drawbacks as above. The parameterization is inherently self-limiting and, in present implementations, almost always restricted to the dipole limit (dipole manybody polarization), neglecting the response of higher moments to the environment.

Last, the dispersion or van der Waal's force is the long range attractive interaction that arises between charge distributions comprised of quantum particles, here the valence electrons of the system [40]. If electrons behaved classically, obeying Newton's equations of motion, there would be no dispersion force [40]. Dispersion is not pairwise additive (e.g. when more than three atoms are present). Like polarization, dispersion is a manybody phenomena that depends on environment. Thus, fitting an effective pairwise additive potential to yield correct results for a series of simple condensed phase systems will introduce significant error. For instance, modeling dispersion with an effective pair potential fit to bulk properties of a liquid can underestimate the surface tension by up to 50 percent [4, 5]. Since the behavior of complex interfaces depends on surface tensions, this error can seriously effect important phenomena, particularly self-assembly, which relies on the wetting/dewetting of interfacial structures that is directly controlled by surface tension. Relatively little research has been done to augment standard force fields to account for manybody dispersion. The *ab initio* based molecular dynamics (AIMD) method, a marriage between classical molecular dynamics and *ab initio* electronic structure, was devised [13] to study the complex systems that empirical force field based molecular dynamics fails to describe accurately. In the AIMD method, all the valence electrons of the system are introduced along with the corresponding ions. Ions consist of the core electrons plus the protons/neutrons of each atom treated as a single point particle. (For example, oxygen has 6 valence electrons and 2 core electrons. The charge on the oxygen ion is Z=+6e, and its mass is M=16amu, 8 protons and 8 neutrons). In this way, liquid water consists of 8 electrons, 2 hydrogen ions (protons) and one oxygen ion for every water molecule present in the system. For a fixed configuration of the ions, the quantum mechanical ground state energy of the electrons is determined. Forces on the ions are generated "on the fly" using a potential energy function defined by the Coulomb interaction between the ions and the ground state electronic energy. Given these forces, the ions can be evolved in time using standard MD methods and a new set of forces generated at the new ionic position. In this way, the structure and dynamics of complex systems at finite temperature and pressure can be examined. In a simulation study performed using AIMD, manybody polarization, manybody dispersion and chemical bond making and breaking phenomena are described, perfectly, in principle. Indeed, the method is exact provided the ground state energy is determined correctly, the Born-Oppenheimer (BO) approximation is valid and the ions can be treated as classical point particles. The BO approximation assumes that only the quantum mechanical ground state energy of the electrons is required to describe the system and classical mechanics assumes that the ions obey Newton's equations of motion. Indeed, most chemical, biological and geophysical systems fall within these restrictions. For example, liquid structure [78, 65, 24], acid-base chemistry [57, 34, 90], industrial [9, 17, 97, 32] and biological catalysis [14], as well as geophysical systems [1, 16] have all been successfully treated using AIMD. Examples of systems that violate these requirements include: photochemical reactions important in atmospheric chemistry which violate the BO approximation, and chemical reactions involving light atoms such as hydrogen atoms where the wave-like or quantum nature of the ions becomes important as sometimes occurs in enzymatic reactions. Devising methods to treat these more complex systems is a topic of current research [95, 86, 56, 58].

AIMD, as practically implemented, does not employ an exact solution of the electronic ground state energy. One of the more common methods used to determine the ground state energy is the Kohn-Sham formulation of Density Functional Theory (KS-DFT) [47, 63, 26]. Although KS-DFT is, in principle, exact, the density functional, itself, is not known and approximations are employed. Furthermore, the KS electronic states employed in the KS-DFT formalism are expanded in a finite basis set (as opposed to a complete basis set) which introduces further error. Car-Parrinello AIMD or CPAIMD is a form of AIMD wherein a plane basis set is used to described the KS states and the coefficients of the plane wave basis set are introduced as a set of fictitious dynamical variables which move quickly at low temperature along with the thermal but comparatively slowly evolving ions so as to approximately minimize, for each instantaneous ion configuration, the density functional to produce the ground state electronic energy. The error of a CPAIMD computation is dominated by the functional employed, typically a generalized gradient approximated density functional or a GGA-DFT, given appropriate care is taken. The GGA-DFT class of functional treats manybody polarization and bond making/bond breaking reasonably well but fails to describe dispersion accurately, leading to limitations in the applicability of technique.

Despite its limitations, the CPAIMD method is widely used and most of the AIMD simulations in the scientific literature are, indeed, CPAIMD simulations. Summarizing the discussion above, CPAIMD relies on three fundamental assumptions, the use of classical mechanics to describe ionic motion, the BO approximation and the use of GGA-DFT. However, as in any simulation, the system of interest must be evolved in time long enough and must to be taken large enough that converged results are obtained. That is, appropriate time and length scales must be sampled. Therefore, an additional limitation of the CPAIMD method, is that due to the considerable computational cost of solving the electronic structure problem, it is difficult to reach the time and length scales required to impact science and technology routinely.

The CPAIMD method is a very numerically intensive technique whose serial computational cost scales as the cube of the number of ions in the system, N_I^3. Given CPAIMD's ability to generate new insights into complex systems, it is useful to improve its efficiency. Research is currently being performed both to reduce the scaling with system size and to improve scalar performance through the development of clever new algorithms. However, with the advent of truly massively parallel hardware platforms, such as IBM's BlueGene/L [31], with over 50,000 processors, it is important to increase, significantly, the parallel efficiency of the method. Developing a fine grained parallel CPAIMD algorithm is a non-trivial task due to the CPAIMD's reliance on a plane wave basis set which requires a large number of three dimensional Fast Fourier Transforms (3D-FFT) to be performed. The inherently non-local communication pattern of the 3D-FFT challenges traditional parallelization models, limiting scaling to the number processors less the number of KS electronic states. Recent progress using the Charm++ runtime system and the concept of processor virtualization which it embodies, has, for the first time, yielded a fine parallel CPAIMD framework called leanCP, complementary to the leanMD framework which is similar to NAMD, that exhibits parallel scaling up to processors numbers an order of magnitude greater than the number of KS states [96].

The remainder of this section is organized as follows: In order to better understand the CPAIMD technique and the leanCP framework, the basic formulae underlying the CPAIMD method are presented. Next, the flow chart of the basic algorithm is given followed by a brief description of the multifaceted PINY_MD software package with its basic parallelization scheme. A summary of available CPAIMD freeware is then provided. Last, the leanCP framework is discussed and its parallel performance demonstrated.

34.3.1 Car-Parrinello Molecular Dynamics : CPAIMD

In a CPAIMD simulation, the ground state electronic energy is calculated by minimizing a functional of the electron density following the tenets of KS-DFT[47, 63, 26]. A generalized gradient approximated density functional or a GGA-DFT, is employed because the exact or true functional is not known. The KS electronic states, $\Psi_i(\mathbf{r})$, are used to construct the functional, and are closely related to the electronic states discussed in basic chemistry and physics texts. The symbol $\mathbf{r}$ represents a position in space; remember, electrons are not generally localized at a single point in space and, hence, are described by a function of position. The GGA-DFT contains several terms, the quantum mechanical kinetic energy of non-interacting electrons, the Coulomb interaction between electrons or the Hartree energy, the correction of the Hartree and non-interacting kinetic energy energy to account for the quantum nature of the electrons and their interactions or the exchange-correlation energy, the interaction of the electrons with the ions in the system or the external energy, and finally the interaction of the valence electrons which are treated explicitly and the core electrons which are mathematically removed or the non-local energy. Since only two electrons are permitted to occupy a single electronic state or the electron's satisfy the "Pauli exclusion principle", the electronic states are taken to be "orthogonal", $\int d\mathbf{r}\Psi_i(\mathbf{r})\Psi_j(\mathbf{r}) = 2\delta_{ij}$. The "Fourier expansion coefficients" of the electronic states, $\Psi_i(\mathbf{g})$, or the "expansion coefficients of the states in a plane wave basis set", are used to develop most of the formulas. The symbol $\mathbf{g}$ represents the quantum mechanical momentum, $\mathbf{p} = \hbar\mathbf{g}$, associated with a given Fourier coefficient where $\hbar$ is Planck's constant; remember, electrons do not have a a single well defined momentum and hence are described as a function of momentum or, here, $\mathbf{g}$. The two representations of the states, $\Psi_i(\mathbf{g})$ and $\Psi_i(\mathbf{r})$, are not independent but are related by a "Fourier series". The GGA-DFT is minimized by finding the $\Psi_i(\mathbf{r})$ that allow the functional to take on its lowest possible value subject to the orthogonality constraint at fixed ion positions. If the GGA-DFT was exact, this value would be the ground state energy in the Born-Oppenheimer approximation. Once the GGA-DFT has been minimized, the forces acting on the ions can be computed from the functional and the ion-ion Coulomb interaction (via the negative gradient), and the positions and velocities of the ions evolved in time according to a finite difference solution of Newton's equations of motion. The beauty of the Car-Parrinello method is that these two elements occur simultaneously using a mathematical formulation called an "adiabatic principle". Therefore, the saw tooth nature of the naive method, in which the functional is minimized with the ions fixed, ion forces are determined, and the ions are evolved to the next time step, is neatly avoided.

In the following, the CPAIMD methodology is described, briefly. First, the equations of motion which embody the "adiabatic principle" and permit the naive saw-tooth algorithmic structure to be abandoned, are described. The input to the equations of motion, forces on the electronic states and ions from derived from GGA-DFT are discussed, next. The computational structure of CPAIMD, in serial, is then presented along with a flow chart.

Equations of Motion : CPAIMD

The CPAIMD method is based on an "adiabatic principle" achieved by writing a modified, slightly more complex version of Newton's equations of motion. The Fourier coefficients of the KS electronic states, $\Psi_i(\mathbf{g})$, are introduced as a set of dynamical variables which are assumed to evolve quickly compared to the slowly evolving ions. The ions have a temperature or average kinetic energy that is typical of physical systems at room temperature. The $\Psi_i(\mathbf{g})$ are assigned a "fictitious" temperature that is very cold. In this way, as the ions evolve slowly in time, the $\Psi_i(\mathbf{g})$ adjust quickly because they are "fast" and instantaneously minimize the GGA-DFT because they are "cold". The forces on the ions are, therefore, correctly reproduced, on average, and the Born-Oppenheimer approximation described in the introduction is satisfied to very good approximation. The motion of the $\Psi_i(\mathbf{g})$ in time generated by the modified equations is not physically meaningful nor is their cold temperature which should not be confused with "quantum kinetic energy". The "fictitious" dynamics of the $\Psi_i(\mathbf{g})$ is designed solely to minimize the functional to good approximation. The motion of the ions is physically meaningful and generating this motion is the goal of CPAIMD simulations. In the limit that $\Psi_i(\mathbf{g})$ are permitted to evolve very quickly and are assigned a very low temperature, the exact, but inelegant and computationally inefficient saw-toothed method is reproduced.

The equations of motion that generate the complex motion required to perform CPAIMD simulations studies are

$$\mu(\mathbf{g})\ddot{\Psi}_i(\mathbf{g}) = -\frac{\partial E}{\partial \Psi_i^*(\mathbf{g})} + \sum_j \Lambda_{ij}\Psi_j(\mathbf{g}) = F_{\Psi_i}(\mathbf{g}) + \sum_j \Lambda_{ij}\Psi_j(\mathbf{g})$$

$$M_I\ddot{\mathbf{R}}_I = -\frac{\partial E}{\partial \mathbf{R}_I} = \mathbf{F}_I \tag{34.2}$$

which basically embody the axiom, force equals mass times acceleration. Here, the energy, E, is the electronic energy plus the Coulomb interaction between the ions in the system; the position of the Ith ion is denoted, $\mathbf{R}_I$. The second time derivative is expressed with two dots and $\ddot{\mathbf{R}}_I$ is the acceleration of ion, I. The Λ_{ij} is a set of Lagrange multipliers that enforce the orthogonality condition between the states and $\mu(\mathbf{g})$ is a mass-like parameter (having units of energy$\times$time^2) that controls the time scale of the motion of the expansion coefficients which must be fast compared to the motion of the ions. The initial conditions determine the fictitious temperature of the electronic states which must be very small. The negative derivative of the energy functional with respect to the coefficients of the plane wave expansion must be taken, which is denoted as $F_{\Psi_i}(\mathbf{g})$, the "force" on the coefficients. The equations can be evolved, approximately, in time using the standard integrators of the previous section, Velocity Verlet or Verlet, and the Shake/Rattle procedures to enforce the orthonormality condition which have been expressed as a set of holonomic constraints [91, 92, 88]. **Hence, CPAIMD uses the tools of classical molecular dynamics to perform very elegantly** *ab initio molecular dynamics.*

GGA-Density Functional Theory

Within the KS-DFT formulation of quantum mechanics, the total energy of an N_e electron system in contact with N_I ions at position $\mathbf{R} = \{\mathbf{R}_1 \ldots \mathbf{R}_{N_I}\}$ is given by

$$E[\{\Psi\}, \{\mathbf{R}\}] = -\frac{1}{2}\sum_{i=1}^{N_s}\langle\Psi_i|\nabla^2|\Psi_i\rangle + \frac{1}{2}\int d\mathbf{r}\, d\mathbf{r}'\frac{\rho(\mathbf{r})\rho(\mathbf{r}')}{|\mathbf{r}-\mathbf{r}'|}$$

$$+ \quad E_{xc}[\rho] + E_{ext}[\rho, \{\mathbf{R}\}] + V_{nucl}(\{\mathbf{R}\}) \tag{34.3}$$

where Planck's constant, $\hbar$, and the electron mass, m_e, are both taken to be unity for simplicity or "atomic units" are used. Here $\Psi_i(\mathbf{r})$ is the ith KS electronic state and electron density, $\rho(\mathbf{r})$, is

$$\rho(\mathbf{r}) = \sum_{i=1}^{N_s}|\Psi_i(\mathbf{r})|^2. \tag{34.4}$$

The theory requires the states to satisfy an orthogonality condition of the form

$$\langle\Psi_i|\Psi_j\rangle = \int d\mathbf{r}\Psi_i(\mathbf{r})\Psi_j(\mathbf{r}) = f_i\delta_{ij}. \tag{34.5}$$

where $\sum_{i=1}^{N_s} f_i = N_e$, the number of electrons. Typically, the occupations numbers are $f_i = 2$; that is, the states are doubly occupied by one spin up and one spin down electron following the Pauli exclusion principle. Minimization of the energy functional in Eq. (34.3) with respect to the states subject to the orthonormality condition yields both ground state energy and electron density.

The density functional, Eq. (34.3), consists of several terms. The first term is the quantum kinetic energy of a system of electrons which do not interact, and the second term is the Hartree energy or the Coulomb interaction between electrons in the limit that the quantum nature of the electrons is ignored. The third term, the exchange-correlation functional, $E_{\text{xc}}[\rho]$, which must be approximated [67, 7, 48], accounts for the fact that the electrons indeed both interact and are governed by the quantum mechanical principles. The fourth term is external potential, $E_{\text{ext}}[\rho, \{\mathbf{R}\}]$, which embodies the interaction of the electrons with the ions. Since CPAIMD eliminates the core electrons, the external energy becomes KS state-dependent and takes the form

$$E_{\text{ext}} \quad = \quad E_{\text{ext,loc}}[\rho, \{\mathbf{R}\}] + E_{\text{ext,non-loc}}[\{\Psi\}, \{\mathbf{R}\}] \tag{34.6}$$

where $E_{\text{ext,non-loc}}[\{\Psi\}, \{\mathbf{R}\}]$ [3] takes care of the complexities involved in removing core electrons from the system. It is simply more computationally efficient to treat fewer electrons and introduce more complex terms. As described in basic chemistry and physics texts, the valence electrons determine the chemical behavior of the elements and the periodic table is divided into groups or columns of elements that behave similarly based simply on the number of valence electrons. Thus, removing the core electrons speeds the calculations and, yet, produces accurate results.

Plane Wave Based DFT

In order to minimize the KS-DFT functional, each KS state is expanded in a spherically truncated plane wave basis (Fourier series)

$$\Psi_i(\mathbf{r}) \quad = \quad \sum_{\mathbf{g}} \Psi_i(\mathbf{g}) e^{i\mathbf{g} \cdot \mathbf{r}} \tag{34.7}$$

$$\frac{1}{2}|\mathbf{g}|^2 \quad < \quad E_{\text{cut}}$$

Here, $\Psi_i(\mathbf{g})$ is the plane wave expansion coefficient of the ith state for the plane wave, $e^{i\mathbf{g} \cdot \mathbf{r}}$, characterized by reciprocal lattice vector, $\mathbf{g}$ related to the quantum mechanical momentum, $\mathbf{p} = \mathbf{g}$ (Planck's constant has been set to unity, here). The electrons and the ions are assumed to lie inside a simulation cell or box. Given an simulation box of side L_x, L_y and L_z, $\mathbf{g} = (g_x, g_y, g_z) = (2\pi n_x/L_x, 2\pi n_y/L_y, 2\pi n_z/L_z)$. The truncation is viable because at large lattice vectors, $|\mathbf{g}| \gg 1$, $e^{i\mathbf{g} \cdot \mathbf{r}}$ wildly oscillates and the expansion coefficient $\Psi_i(\mathbf{g})$ approaches zero. Equivalently, it is highly improbable for the electrons to have very large momentum in a physical system. In chemical applications, the KS states can be chosen to be real so that $\Psi_i^*(\mathbf{g}) = \Psi_i(-\mathbf{g})$. Since the KS states are expressed as a linear combination of a finite number of plane-waves, the density can also be so expressed,

$$\rho(\mathbf{r}) \quad = \quad \sum_{\mathbf{g}} \rho(\mathbf{g}) e^{i\mathbf{g} \cdot \mathbf{r}} \tag{34.8}$$

$$\frac{1}{2}|\mathbf{g}|^2 \quad < \quad 4E_{\text{cut}}$$

where $\rho(\mathbf{g})$ are the Fourier or plane wave expansion coefficients of the electron density. Note, density cutoff is 4 times larger than the state cutoff because the density is related to the square of the states. It is useful to define, $\mathbf{G}/2$, the largest reciprocal lattice vector in each direction in the expansion of the density. Of course, $\mathbf{G}/4$ is then the largest reciprocal lattice vector in the expansion of states. The density is always real, $\rho^*(\mathbf{g}) = \rho(-\mathbf{g})$.

The expansion coefficients of the density, $\rho(\mathbf{g})$, can be obtained from the expansion coefficients of the states, $\Psi_i(\mathbf{g})$, exactly, using **3-D FFTs** because the expansion is truncated or finite. A complex-to-real **3-D FFT** of size $\mathbf{G}$ (e.g $\{N \times N \times N\}$ in cubic box) is performed on on each $\Psi_i(\mathbf{g})$ to produce $\Psi_i(\mathbf{r})$ on a discrete real space mesh, the function is squared point by point, $|\Psi_i(\mathbf{r})|^2$, and the result summed over all states to produce the electron density, $\rho(\mathbf{r})$, on the discrete real space mesh (e.g $\{N \times N \times N\}$ in cubic box). The function $\rho(\mathbf{r})$ can then be inverse transformed by a real-to-complex **3-D FFT** to generate $\rho(\mathbf{g})$, exactly. The computational efficiency of the CPAIMD method is due to this clever use of highly optimized **3-D FFT**s.

In order to proceed, each term in the density functional must be expressed in the plane wave basis:

The kinetic energy of non-interacting electrons, depends on the individual electronic states. It can be expressed as

$$E_{kin} = \frac{1}{2}\sum_i \sum_{\mathbf{g}} |\Psi_i(\mathbf{g})|^2 |\mathbf{g}|^2 \qquad (34.9)$$

in the plane wave basis. Remember $\mathbf{g}$ is related to the momentum and E_{kin} is simply related the square of the momentum, $|\mathbf{p}|^2$, as might be expected. The quantum nature of the system is reflected in the fact that more than a single value of the momentum is required to described the kinetic energy of an electron. The orthogonality condition takes the simple form

$$\langle \Psi_i | \Psi_j \rangle = \sum_{\mathbf{g}} \Psi_i(\mathbf{g})\Psi_j^*(\mathbf{g}) = f_i \delta_{ij}. \qquad (34.10)$$

which, again, incorporates the Pauli exclusion principle into the functional.

In a plane wave basis set, the non-local energy is introduced to remove core electrons. The very same notion from basic physics and chemistry is employed. There are two quantum numbers l and m. The quantum number $l = 0$ is associated with "s"-states, $l = 1$ with "p"-states and so on. If core electrons associated with states of type $l = 0$ to $\bar{l}$, are to be removed, the non-local energy, in the Kleinman-Bylander form [46], is given by

$$E_{\mathrm{NL}} = \sum_{i=1}^{N_s} \sum_{I=1}^{n_N} \sum_{l=0}^{\bar{l}-1} \sum_{m=-l}^{l} \mathcal{N}_{Ilm} |Z_{i,I,l,m}|^2 \qquad (34.11)$$

where $\mathcal{N}_{Ilm}$ is a normalization factor, and

$$Z_{i,I,l,m} = \sum_{\mathbf{g}} \Psi_i(\mathbf{g}) e^{i\mathbf{g}\cdot\mathbf{R}_I} h_{Il}(|\mathbf{g}|) Y_{lm}(\theta_{\mathbf{g}}, \varphi_{\mathbf{g}}). \qquad (34.12)$$

In Eq. (34.12), $h_{Il}(|\mathbf{g}|)$ is the lth spherical Bessel function transform of the angular-momentum dependent potential function, $h_{Il}(\mathbf{r})$, describing the interaction that replaces the core electrons of ion, I, with quantum number l and $Y_{lm}(\theta_{\mathbf{g}}, \varphi_{\mathbf{g}})$ is a spherical harmonic. The former function determines the "radial shape" of the interaction that replaces the core electrons while the latter function determines the angular shape.

The local part of the external energy is

$$E_{\mathrm{ext,loc}} = \frac{1}{V}\sum_{\mathbf{g}} \rho^*(\mathbf{g})\sum_I \tilde{\phi}_{\mathrm{loc,I}}(\mathbf{g})S_I(\mathbf{g}) \qquad (34.13)$$

$$S_I(\mathbf{g}) = \exp(-i\mathbf{g}\cdot\mathbf{R}_I)$$

where $\tilde{\phi}_{\mathrm{loc,I}}(\mathbf{g})$ is the Fourier transform of the local interaction, $\phi_{\mathrm{loc,I}}(r)$, between an electron and the Ith ion and $S_I(\mathbf{g})$ is the ion structure factor or the Fourier expansion coefficient of ion, I.

The Hartree energy is given by

$$E_{\text{Hartree}} = \frac{1}{2V} \sum_{\mathbf{g} \neq (0,0,0)} \frac{4\pi}{|\mathbf{g}|^2} |\rho(\mathbf{g})|^2 \qquad (34.14)$$

where V is the volume of the system assuming three dimensional periodic boundary conditions [54, 60]. It is the equivalent to the Coulomb energy of a classical charge density equal to $\rho(\mathbf{r})$.

Both Hartree and the non-interacting kinetic energy must be corrected by the exchange-correlation energy which $E_{\text{xc}}[\rho, \nabla\rho]$ is not known exactly and must, therefore, be approximated. In the generalized gradient approximation, the functional is taken to be of the form

$$E_{\text{xc}}[\rho] = \int d\mathbf{r} \; \rho(\mathbf{r}) \epsilon_{\text{xc}}(\rho(\mathbf{r}), \nabla\rho(\mathbf{r})). \qquad (34.15)$$

In practice, these integrals are evaluated on a set of equally spaced grid points defined by the size of the **3-D FFT** [102], using trapezoidal rule

$$E_{\text{xc}}[\rho] = \Delta^3 \sum_{ijk} \rho(\mathbf{r}_{ijk}) . \epsilon_{xc}(\rho(\mathbf{r}_{ijk}), \nabla\rho(\mathbf{r}_{ijk})) \qquad (34.16)$$

where Δ is the grid spacing and $\rho(\mathbf{r}_{ijk})$ is the electron density at a grid point and $\nabla\rho(\mathbf{r}_{ijk})$ is its gradient [102]. Hereafter the indices, ijk, will be suppressed/understood. More details are provided elsewhere [102].

The negative derivative of each term given above with respect to the coefficients of the plane wave expansion must be taken. These are denoted $F_{\Psi_i}(\mathbf{g})$, the "force" on the coefficients. The forces on the ions, $\mathbf{F}_I$, which have physical meaning only when the density functional is minimized, $F_{\Psi_i}(\mathbf{g}) \equiv 0$, arise from three terms, the local energy external energy, the non-local energy and the ion-ion interaction, $V_{\text{nucl}}(\mathbf{R})$. The ion-ion interaction is general taken to be simply Coulomb's law,

$$V_{\text{nucl}}(\mathbf{R}) = \sum_{\mathbf{S}} \sum_{IJ} {}' \frac{Z_I Z_J}{|\mathbf{R}_I - \mathbf{R}_J + \mathbf{S}\mathbf{h}|}, \qquad (34.17)$$

where $\mathbf{S}$ are the periodic replicas and the prime indicates $I \neq J$ when $\mathbf{S} = 0$. This term is typically evaluated using Ewald summation [38, 22] in periodic systems. The computation of the ion forces takes negligible computer time and will not be discussed further.

Computational Structure of CPAIMD

The two key inputs to a CPAIMD calculation are the ion positions and the plane wave expansion coefficients of the KS-states, the $\Psi_i(\mathbf{g})$ of Eq. (34.7) which are assumed to be orthonormal e.g. satisfy Eq. (34.10). The CPAIMD computation, itself, nicely bifurcates into two branches, a KS state dependent branch and an electron density dependent branch, as given in Fig. 34.6. The non-local pseudopotential energy, Eq. (34.12), and the kinetic energy of the non-interacting electrons, Eq. (34.9) as well as their contribution to $F_{\Psi_i}(\mathbf{g})$, are determined in the left branch.

In order to create the electron density, the states are transformed into real space by **3-D FFT**, squared and summed to form the density, Eq. (34.4), in the right branch. Once the density is formed in real space, the calculation can split again. The Exchange correlation energy is computed using the density and its gradient in real space is computed along with its contribution to $F_{\Psi_i}(\mathbf{g})$. Independently, the density is transformed to g-space by **3-D**

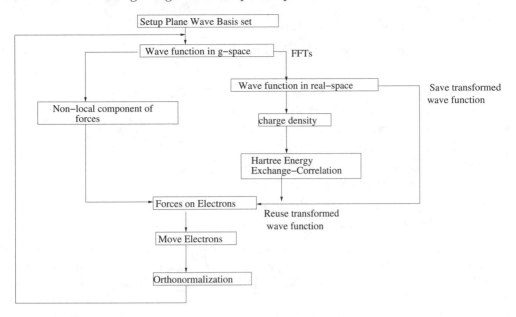

FIGURE 34.6: Schematic flowchart of the implementation of the CP algorithm

FFT and the Hartree and local pseudopotential energy computed using Eq. (34.14) and Eq. (34.14). The contribution of these terms to the force of these terms is computed in g-space, transformed into real space by **3-D FFT** and added into the contribution from the exchange correlation function. This quantity which is referred to as the KS potential is the same size as the density. Each state is then multiplied by the KS potential in real space and a **3-D FFT** is performed to complete the computation of the contribution to the force, $F_{\Psi_i}(\mathbf{g})$, from the Hartree, Exchange-Correlation and External Pseudopotential energies.

When both major branches are complete, the two force contributions are combined. The KS states are, now, evolved using a numerical solver of the CPAIMD equations of motion Eq. (34.2) or as part of an energy minimization procedure at fixed nuclear position. Due to finite time step error in the solvers, the new states will not be perfectly orthonormal. This is adjusted using a variety of methods, Shake/Rattle for CPAIMD, and Grahm-Schmidt or Löwdin techniques for energy minimization.

34.3.2 PINY_MD implementation of CPAIMD

PINY_MD is a distributed memory software package parallelized using the Message Passing Interface (MPI) extension. PINY_MD performs CPAIMD simulation studies, standard MD simulations and QM/MM calculations (see next section) as well as other types of computations that include quantum effects on the nuclei [94]. The PINY_MD implementation of CPAIMD follows Fig. 34.6. Each phase in the diagram is parallelized, in sequence, and the layout of the data arranged to minimize communication between the sequences. This rather basic parallelization scheme leads to an algorithm that scales well when the number of states is less than or equal to the number of physical processors $N_s \leq N_{proc}$. It was found to be sufficient for the types of parallel machines available today (100s of processors) but will not scale on BG/L type systems (except of course for large systems with many states).

34.3.3   Summary of Available Codes

There are a variety of plane wave based density functional theory software packages available on the Web. These include

- PINY_MD (`http://homepages.nyu.edu/~mt33/PINY\_MD/PINY.html`)
- CPMD (`http://www.cpmd.org/`)
- ABINIT (`http://www.abinit.org/`)
- PWSCF (`http://www.pwscf.org/`)
- NWPW/NWCHEM (`http://www.emsl.pnl.gov/docs/nwchem/`)
- DACAPO (`http://www.fysik.dtu.dk/campos/Dacapo/`)
- OCTOPUS (`http://www.tddft.org/programs/octopus/`)
- FHI96MD (`http://www.fhi-berlin.mpg.de/th/fhi96md.html`)

The software packages have various strengths and weakness. Some concentrate on solid-state physics applications, others on chemical applications, and others on applications of time dependent density functional theory (TDDFT) to study excited state chemistry/physics, a topic not discussed covered in this review. The parallel scaling of all, in general, is as in PINY_MD, although efforts are underway in all the groups to improve performance in response to massively parallel computers such as IBM's BG/L.

34.3.4   LeanCP Implementation of CPAIMD

In the LeanCP framework [96], the states in g-space and real-space which are sparse and dense cubes of data, respectively, are each decomposed into planes. The work related to each plane is performed by a Charm++ virtual processor (or chare). In accordance with the philosophy of processor virtualization, the number of virtual processors depends only on issues such as the work to communication ratio but is independent of the total number of physical processors. A collection G holds *objects* $G(i, p)$ corresponding to plane, p, of state, i, in g-space. The plane index, p, is identical to the x coordinate in g-space, g_x, and the object $G(i, g_x)$ houses the coefficients $\Psi_i(g_x, g_y, g_z)$ for all values of g_y and g_z. Similarly, another collection R holds real-space planes $R(i, p)$ corresponding to plane p of state i. However, the axis of decomposition is different for G and R, due to the way the parallel FFTs are implemented (one transpose is required). In addition, there are 1-dimensional chare arrays corresponding to the electron density in real-space, $\rho(\mathbf{r})$ and in g-space, $\rho(\mathbf{g})$ as well as a chare array, $P(i, p)$, associated with computing the non-local pseudopotential interaction, is defined as the $G(i, p)$.

It should be noted that the real-space planes are dense and each state has precisely the same number of planes as the electron density, i.e. N. The g-space planes are sparse and only g-space planes with non-zero elements are included in the calculation. There are roughly twice as many non-zero g-space plane in the reciprocal space representation of the electron density as in the corresponding reciprocal space representation of a state. The non-local pseudopotential chare array is, again, akin to the states. The resultant parallel decomposition of the problem is shown in Fig. 34.7. The algorithm was implemented using the Charm++ runtime system which permits the independent portions of the calculation to be interleaved effectively. For instance, the N_s parallel **3-D FFT**'s of the KS states required to create the electron density are performed, simultaneously. That is, a chare array for each plane of each KS state is launched and as soon as the FFT related computation stage of that plane is completed, and enters a communication phase (a transpose, for example), the computation stage of a different state and plane can be rolled into the idle processor. In the same manner, the non-local pseudopotential chare array is launched simultaneous to the **3-D FFT** computation and is, similarly, rolled into and out of the physical processors. There is a barrier when the forces on the $\Psi_i(g_x, g_y, g_z)$ are collected from the various chare arrays, including the non-local pseudopotential chare array, and the $\Psi_i(g_x, g_y, g_z)$ are evolved to the next time step. Another barrier is required so that states can be reorthogonalized (due to the finite time step error in the numerical solver) before the next time step commences.

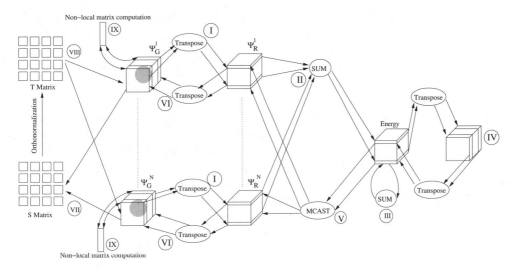

FIGURE 34.7: Parallel structure of our implementation. Roman numbers indicate phases

34.3.5 LeanCP Performance

In order to test the parallel efficiency of the new LeanCP framework, a 32 water molecule system with 128 KS electronic states is examined; this system size has been used in seminal studies published in prestigious journals such as Science and Nature [57, 34, 90]. Using standard parameters (E_{cut}=70Ry), pseudopotentials [84] and exchange correlation functional [7, 48] 32,000 g-space state coefficients per state, 260,000 g-space density coefficients, and a real-space density grid of 100^3 points are employed in the computation.

The scalability of the Charm++based LeanCP CPAIMD framework was studied on PSC-Lemieux, a 750 node, 3000 1Ghz Alpha processor cluster connected via a high speed Quadrics Elan interconnect with $4.5\mu s$ latency. Scaling up to 1500 processors is obtained, over an order of magnitude greater than the number of electronic states (see Table 34.1). Current work involves improving the scalar and parallel efficiency of the LeanCP framework as well as studying a variety of system sizes.

34.3.6 Biological Applications of CPAIMD

The biological applications of pure CPAIMD simulations have had wide impact. Due to the computational cost of the CPAIMD, much of the biological framework must be left out of the simulation. However, interesting biologically relevant systems consisting of small parts of large complex moeities of true interest have been examined. For example, polypeptides,

2 Processors per Node				3 Processors per Node			
P	t(sec)	GFLOPS	Speedup	P	t(sec)	GFLOPS	Speedup
16	13.26	4	16.0	129	2.20	22	96.4
32	6.17	8	34.3	258	1.30	37	163.2
64	3.11	15	68.2	513	0.68	71	312.0
128	2.07	23	102.5	768	0.60	80	353.6
256	1.18	41	179.8	1536	0.40	121	533.1
512	0.65	74	326.4				
1024	0.48	100	442.0				

TABLE 34.1   Execution times on PSC Lemieux, for 128 states. Using more processors, P, per node, N, effects performance adversely as the bandwidth is a limited resource.

chains of amino acids too small to be called proteins, have been studied in water in order to examine the conformations or shapes they take on and inferences drawn to larger systems. Also, reactions involving amino acids, in particular, the acid base chemistry of histidine, a key amino acid in many enzymatic reactions has been studied. More recently, small strands of DNA have been examined. Here, oxidative damage of DNA via radical cation formation is of interest as this is a key element in cancer. Finally, small models of active sites such as the binding pocket of myoglobin, an oxygen carrier similar to hemoglobin, have been studied using approximately 10-20 amino acids as opposed to studying the full system. In the next section, the ability of CPAIMD to interface with molecular mechanics methods to provide more of the biological framework is described.

34.4   QM/MM

34.4.1   Hybrid QM/MM Simulations

Empirical force field models are capable of describing complex systems accurately. In addition, calculations based on these models are quite computationally efficient allowing large systems to be studied at long time scales as the computational effort increases only linearly with system size. However, empirical force field models cannot treat processes involving chemical bond forming and breaking which are of great interest. *Ab initio* methods are, in contrast, more generally applicable. However, *ab initio* methods permit only small systems to be studied at short time scales as the computational effort increases as the cube of the system size and the prefactor is quite large so that even small systems require intensive computation. It is, therefore, useful to consider a combination of the two techniques to study a class of problems where the advantages of the two methods can be exploited.

   An effective way to decompose a physical system in chemistry, physics and biology is to consider part of the system as the reactants/products and the remainder of the system as the "bath" (the environment in which the reactive portion is "bathed"). Typically, the reactants/products are described at a high level of detail while the bath is treated more approximately, even as a simply as a set of harmonic oscillators linearly coupled to reactants/products. Clearly, this basic picture presents an opportunity for *ab initio* or quantum mechanical treatments of the reactive region to be wed to empirical force field or molecular mechanic treatments of the bath to form a hybrid or mixed method often referred to as a QM/MM approach. The QM/MM approach has several advantages. First, the bath can be made quite complex and, indeed, very large. In this way, the fluctuations of the bath that drive the reactants to cross barriers and form products can be studied in detail. Now, if the size of reactive part of the calculation is kept constant and the size of the bath is increased so as to treat the reaction at infinite dilution which is typically desired, the hybrid method scales only linearly with system size. That is, the *ab initio* or reactive part, alone, will be an expensive but fixed part of the calculation, while the computation of the bath scales linearly with size. This assumes that the serial computational cost of determining the interaction of the bath with the reactive part of the system, scales linearly with increasing bath size. Thus, hybrid methods are, in principle, quite desirable. There are several important cases that the division of a system into a reactive region and a bath is desirable. Enzyme catalysis, for instance, can be studied with quite reasonable accuracy by modeling the protein backbone and the water solvent using an empirical force field and treating the valence electrons of the amino acids and the water molecules near the active site, as well as those of the substrate, with *ab initio* techniques [99, 30, 50, 27, 68, 104, 105, 103]. Similarly, organic and inorganic chemical reactions in solution, for example, Diels Alder chemistry, most of the solvent can be modeled using molecular mechanics while the electronic degrees of freedom of the reactants/products and perhaps few solvent molecules near the reactive center can be treated with quantum mechanical methods [99].

It is clear that hybrid-mixed-QM/MM models of chemistry and biochemistry can yield important information about key processes for reduced computational cost. There are two problems that must be resolved. The short range interaction between the *ab initio* region and the empirical region must be treated so that there are no spurious edge effects. There are several techniques that are under development which are quite promising. These include link atoms, pseudobonds and frozen frontier molecular orbitals [30, 104, 68]. However, this portion of the technique is under development and requires more research. Second, the long range interaction of the bath with reactive region must be evaluated in order N computing time where N is the total number of atoms/ions in the simulation.

In previous sections, the plane wave basis set in conjunction with the generalized gradient approximation to density functional theory (GGA-DFT) *ab initio* method was studied in detail [13, 33, 89, 25]. Here, the QM/MM version of this technique will be developed [103]. Applications of the plane wave based, GGA-DFT, QM/MM method is described and then future work on its large scale parallelization is briefly outline.

The QM/MM method : GGA-DFT in a plane wave basis

A plane waves wave basis set naturally possess only a single length scale which has limited the applicability of the method to QM/MM calculations. For instance, let the reactants/products occupy a small simulation cell embedded within a large simulation cell containing the bath. In order to compute the interaction of the electrons of the reactive subsystem with the bath, it is, in principle, necessary to obtain the plane wave expansion coefficients of the electron density in basis spanning the large cell using **same** large cutoff required to describe the rapidly varying electron density in the small cell. Using this simple scheme, memory requirements are prohibitively large and the calculation scales with the cube of the number of particles in the bath (at fixed small cell size). However, such a scheme does allow systems liquids and solids to be studied accurately using standard techniques and clusters, wires and surfaces using recent extensions [54, 60, 59]. In order to remove the difficulties associated with a employing GGA-DFT and a plane wave basis set to examine QM/MM systems, a new dual length scale approach was developed which scales as Nlog N with bath size and has low memory requirement [103].

Two terms in the electron energy have long range character, the Hartree, $E_{\text{Hartree}}[\rho]$, and local pseudopotential energies,

$$E_{\mathrm{H}}[\rho] \;=\; \frac{e^2}{2} \sum_{\hat{\mathbf{S}}} \int_{D(\mathbf{h})} d\mathbf{r} \int_{D(\mathbf{h})} d\mathbf{r}' \; \frac{\rho(\mathbf{r})\rho(\mathbf{r}')}{|\mathbf{r}-\mathbf{r}'+\mathbf{h}\hat{\mathbf{S}}|} \tag{34.18}$$

$$E_{\mathrm{loc}}[\rho] \;=\; \sum_{\hat{\mathbf{S}}} \sum_{I=1}^{N_I} \int_{D(\mathbf{h})} d\mathbf{r} \; \phi_{\mathrm{loc},I}(\mathbf{r}-\mathbf{R}_I+\mathbf{h}\hat{\mathbf{S}})\rho(\mathbf{r}) \tag{34.19}$$

Here, N_I ions/atoms in the system, $\mathbf{R}_I$ is the Cartesian position of the Ith ion/atom, $\mathbf{h}$ is the cell matrix defining the parallelepiped surrounding the system, the $\det \mathbf{h} = V$ is the volume (in a cubic box, the matrix is diagonal, $\mathbf{h}=$diag(L,L,L), and $V = L^3$), and $\hat{\mathbf{S}} = \{\hat{s}_a, \hat{s}_b, \hat{s}_c\}$ is a vector of integers which describe the periodic boundary conditions. Solids and fluids are periodically replicated in all three directions while for clusters, only $\hat{\mathbf{S}} = \{0,0,0\}$ is permitted.

It is clear that the expressions for the Hartree and local external energies possess a single length scale. A second length scale can be introduced using the identity $\mathrm{erf}(\alpha r)+\mathrm{erfc}(\alpha r) = 1$ where $\mathrm{erf}(\alpha r)$ goes to zero at small r and unity at large r while $\mathrm{erf}(\alpha r)$ goes to zero at large r and unity at small r. Inserting the identity yields,

$$
\begin{aligned}
E_{\mathrm{H}}[\rho] \;=\;& \left\{ \frac{e^2}{2} \sum_{\hat{\mathbf{S}}} \int_{D(\mathbf{h})} d\mathbf{r} \int_{D(\mathbf{h})} d\mathbf{r}' \; \frac{\rho(\mathbf{r})\rho(\mathbf{r}')\mathrm{erfc}(\alpha|\mathbf{r}-\mathbf{r}'+\mathbf{h}\hat{\mathbf{S}}|)}{|\mathbf{r}-\mathbf{r}'+\mathbf{h}\hat{\mathbf{S}}|} \right\} \\[2mm]
+\;& \left\{ \frac{e^2}{2} \sum_{\hat{\mathbf{S}}} \int_{D(\mathbf{h})} d\mathbf{r} \int_{D(\mathbf{h})} d\mathbf{r}' \; \frac{\rho(\mathbf{r})\rho(\mathbf{r}')\mathrm{erf}(\alpha|\mathbf{r}-\mathbf{r}'+\mathbf{h}\hat{\mathbf{S}}|)}{|\mathbf{r}-\mathbf{r}'+\mathbf{h}\hat{\mathbf{S}}|} \right\} \\[2mm]
=\;& E_{\mathrm{H}}^{(\mathrm{short})}[\rho] + E_{\mathrm{H}}^{(\mathrm{long})}[\rho]
\end{aligned}
\tag{34.20}
$$

$$
\begin{aligned}
E_{\mathrm{loc}}[\rho] \;=\;& \left\{ \sum_{\hat{\mathbf{S}}} \sum_{I=1}^{N_I} \int_{D(\mathbf{h})} d\mathbf{r} \; \rho(\mathbf{r}) \left[\phi_{\mathrm{loc},I}(\mathbf{r}-\mathbf{R}_I+\mathbf{h}\hat{\mathbf{S}}) + \frac{eq_I\mathrm{erf}(\alpha|\mathbf{r}-\mathbf{R}_I+\mathbf{h}\hat{\mathbf{S}}|)}{|\mathbf{r}-\mathbf{R}_I+\mathbf{h}\hat{\mathbf{S}}|} \right] \right\} \\[2mm]
-\;& \left\{ \sum_{\hat{\mathbf{S}}} \sum_{I=1}^{N_I} \int_{D(\mathbf{h})} d\mathbf{r} \; \rho(\mathbf{r}) \left[\frac{eq_I\mathrm{erf}(\alpha|\mathbf{r}-\mathbf{R}_I+\mathbf{h}\hat{\mathbf{S}}|)}{|\mathbf{r}-\mathbf{R}_I+\mathbf{h}\hat{\mathbf{S}}|} \right] \right\} \\[2mm]
=\;& E_{\mathrm{loc}}^{(\mathrm{short})}[\rho] + E_{\mathrm{loc}}^{(\mathrm{long})}[\rho].
\end{aligned}
\tag{34.21}
$$

The first term in the curly brackets in each equation is short range while the second term is long range. The sum over images can be neglected in the short range terms provided α is selected sufficiently large because the new potential terms vanish exponentially quickly at large distances.

As described above, the electrons are required to be localized in a small region of space that can be surrounded by a small cell, $\mathbf{h}_s$, whose center lies at the point, $\mathbf{R}_c$. The KS states and, electron density are taken to vanish on the surface of $\mathbf{h}_s$. They can therefore be expanded in the plane waves of the small box, $(1/2)|\mathbf{g}|^2 < E_{\mathrm{cut}}^{\mathrm{short}}$, which must be taken rather large, $E_{cut}^{\mathrm{short}} \approx 70$Ry.

The short range components of the Hartree and local pseudopotential energies can be evaluated straightforwardly using the locality assumption,

$$
\begin{aligned}
E_{\mathrm{H}}^{(\mathrm{short})}[\rho] &= \frac{e^2}{2} \int_{D(\mathbf{h_s})} d\mathbf{r} \int_{D(\mathbf{h_s})} d\mathbf{r}' \, \frac{\rho_s(\mathbf{r})\rho_s(\mathbf{r}')\mathrm{erfc}(\alpha|\mathbf{r}-\mathbf{r}'|)}{|\mathbf{r}-\mathbf{r}'|} \\
&= \frac{e^2}{2V_s} \sideset{}{'}\sum_{\hat{\mathbf{g}}_s} \bar{\rho}_s(-\mathbf{g}_s)\bar{\rho}_s(\mathbf{g}_s) \left[\frac{4\pi}{g_s^2}\right]\left[1-\exp\left(-\frac{g_s^2}{4\alpha^2}\right)\right] + \frac{e^2\pi}{2V_s\alpha^2}|\rho_s(0)|^2 \quad (34.22)
\end{aligned}
$$

$$
\begin{aligned}
E_{\mathrm{loc}}^{(\mathrm{short})}[\rho] &= \sum_{J=1}^{N_{I_s}} \int_{D(\mathbf{h_s})} d\mathbf{r}\, \rho_s(\mathbf{r})\left[\phi_{\mathrm{loc},J}(\mathbf{r}-\mathbf{R}_J+\mathbf{R}_c) + \frac{eq_J\mathrm{erf}(\alpha|\mathbf{r}-\mathbf{R}_J+\mathbf{R}_c|)}{|\mathbf{r}-\mathbf{R}_J+\mathbf{R}_c|}\right] \\
&= \frac{1}{V_s} \sideset{}{'}\sum_{\hat{\mathbf{g}}_s} \sum_{J=1}^{N_{I_s}} \bar{\rho}_s^*(\mathbf{g}_s)\exp(-i\mathbf{g}_s\cdot[\mathbf{R}_J-\mathbf{R}_c])\left[\tilde{\phi}_{\mathrm{loc},J}(\mathbf{g}_s) + \frac{4\pi eq_J}{g_s^2}\exp\left(-\frac{g_s^2}{4\alpha^2}\right)\right] \\
&\quad + \frac{1}{V_s} \sum_{J=1}^{N_{I_s}} \bar{\rho}_s(0)\left[\tilde{\phi}_{\mathrm{loc},J}^{(0)} - \frac{eq_J\pi}{\alpha^2}\right]. \quad (34.23)
\end{aligned}
$$

where the J sum runs over the N_{I_s} ions within the small cell, the $\hat{\mathbf{g}}_s$ sum runs over the large reciprocal-space grid of the small cell and $\mathbf{R}_c$ is position of the small cell inside the large. The non-local pseudopotential energy is short range and is assumed to be evaluated within the small cell (only, considering the N_{I_s} ions in the small cell and using the small cell reciprocal space). Similarly, the exchange correlation and the electronic kinetic energies can also be evaluated in the small cell using standard techniques [33, 74]. Again, a large g-space or reciprocal space defined by the cutoff, E_{cut}^{short}, is required to treat the rapid variation of the electron density within the small box.

The long range contributions to Hartree and local pseudopotential energies are obtained by expanding the electron density in the plane waves of the large cell, $\mathbf{g} = \mathbf{h}^{-1}\hat{\mathbf{g}}$,

$$
\begin{aligned}
E_{\mathrm{H}}^{(\mathrm{long})}[\rho] &= \frac{e^2}{2} \sum_{\hat{\mathbf{S}}} \int_{D(\mathbf{h})} d\mathbf{r} \int_{D(\mathbf{h})} d\mathbf{r}' \, \frac{\rho(\mathbf{r})\rho(\mathbf{r}')\mathrm{erf}(\alpha|\mathbf{r}-\mathbf{r}'+\mathbf{h}\hat{\mathbf{S}}|)}{|\mathbf{r}-\mathbf{r}'+\mathbf{h}\hat{\mathbf{S}}|} \\
&= \frac{e^2}{2V} \sideset{}{'}\sum_{\hat{\mathbf{g}}} \bar{\rho}(-\mathbf{g})\bar{\rho}(\mathbf{g})\left[\frac{4\pi}{g^2}\exp\left(-\frac{g^2}{4\alpha^2}\right) + \hat{\phi}^{(\mathrm{screen,Coul})}(\mathbf{g})\right] \\
&\quad + \left(\frac{e^2}{2V}\right)\left[\hat{\phi}^{(\mathrm{screen,Coul})}(0) - \frac{\pi}{\alpha^2}\right]|\bar{\rho}(0)|^2 \quad (34.24)
\end{aligned}
$$

$$
\begin{aligned}
E_{\mathrm{loc}}^{(\mathrm{long})}[\rho] &= -\sum_{\hat{\mathbf{S}}} \sum_{I=1}^{N} \int_{D(\mathbf{h})} d\mathbf{r}\, \rho(\mathbf{r})\left[\frac{eq_I\mathrm{erf}(\alpha|\mathbf{r}-\mathbf{R}_I+\mathbf{h}\hat{\mathbf{S}}|)}{|\mathbf{r}-\mathbf{R}_I+\mathbf{h}\hat{\mathbf{S}}|}\right] \\
&= -\frac{e}{V} \sideset{}{'}\sum_{\hat{\mathbf{g}}} \bar{\rho}^*(\mathbf{g})S(\mathbf{g})\left[\frac{4\pi}{g^2}\exp\left(-\frac{g^2}{4\alpha^2}\right) + \hat{\phi}^{(\mathrm{screen,Coul})}(\mathbf{g})\right] \\
&\quad - \frac{e}{V} \bar{\rho}(0)S(0)\left[\hat{\phi}^{(\mathrm{screen,Coul})}(0) - \frac{\pi}{\alpha^2}\right]. \quad (34.25)
\end{aligned}
$$

where

$$
S(\mathbf{g}) = \sum_I q_I \exp(i\mathbf{g}\cdot\mathbf{R}_I) \quad (34.26)
$$

is the atomic charge density and

$$
\begin{aligned}
\bar{\rho}(\mathbf{g}) &= \int_{D(\mathbf{h})} d\mathbf{r}\, \exp[-i\mathbf{g}\cdot\mathbf{r}]\rho(\mathbf{r}) \qquad\qquad (34.27)\\[2mm]
&= \int_{D(\mathbf{h_s})} d\mathbf{r}_s\, \exp[-i\mathbf{g}\cdot\mathbf{r}_s]\rho(\mathbf{r}_s + \mathbf{R}_c)\\[2mm]
&= \int_{D(\mathbf{h_s})} d\mathbf{r}_s\, \exp[-i\mathbf{g}\cdot(\mathbf{r}_s - \mathbf{R}_c)]\rho_s(\mathbf{r}_s)
\end{aligned}
$$

The integral in Eq. (34.27) can be extended to cover the domain described by the large cell without loss of generality because $\rho(\mathbf{r}_s + \mathbf{R}_c) \equiv 0$ outside of the small cell. Note, $\bar{\rho}(\mathbf{g}) = \bar{\rho}_s(\mathbf{g}_s)$ if $\mathbf{h_s} \equiv \mathbf{h}$ and $\mathbf{R}_c = 0$. The damping factor, $\exp[-g^2/(4\alpha^2)]$, truncates the sum such that only low Fourier coefficients of the electron density on the large cell reciprocal space are required, $E_{cut}^{\text{long}} \ll E_{cut}^{\text{short}}$, and makes the reciprocal or g-space associated with the large box rather small. (Note, the plane wave cutoff of the density is $4E_{cut}^{\text{long}}$ as described in the previous section). The atomic charge density can be evaluated in order $N_I \log N_I$ using Particle Mesh Ewald methods while the sums given in Eq. (34.25) and Eq. (34.24) can be evaluated in order N_I. It remain to compute the $\bar{\rho}(\mathbf{g})$ on the small g-space defined by $E_{cut}^{\text{long}} \ll E_{cut}^{\text{short}}$. Indeed, Euler exponential spline interpolation [19, 77] can be employed to obtain the small $|\mathbf{g}|$ coefficients, $\bar{\rho}(\mathbf{g})$, of an electron density $\rho(\mathbf{r})$ that is assumed to be nonzero only in the small cell described by $\mathbf{h}_s$ in order $N_I \log N_I$, accurately. The discrete but dense real space representation $\rho_s(\mathbf{r})$ generated from the KS states in the small box, is Cardinal B-spline interpolated onto a sparse discrete real space grid which spans the large box defined by $\mathbf{h}$ to generate a new function, $\rho^{(\text{conv})}(\mathbf{r})$. This new function, $\rho^{(\text{conv})}(\mathbf{r})$, is, in turn, transformed into the reciprocal space of the large box via **3-D FFT** where it is multiplied by a **g**-dependent scaling factor derived by Euler to generate a controlled, differentiable approximation to the function of interest, $\bar{\rho}(\mathbf{g})$. Now, $\rho^{(\text{conv})}(\mathbf{r})$ can be evaluated in order $N_{I_s} m^3$ where m is the order of the Cardinal B-spline interpolation and, again, N_{I_s} is the number QM atoms. Thus, the overall computational cost of constructing $\bar{\rho}(\mathbf{g})$ is $N_I \log N_I$ dominated by the **3-D FFT**. The overhead of the calculation is small because the g-space required is small, $E_{cut}^{\text{long}} \ll E_{cut}^{\text{short}}$.

The basic tenets underlying a QM/MM calculation performed using GGA-DFT with a plane wave basis set is now described [103]. Unlike the CPAIMD of the previous section which uses two reciprocal space grids, one for the KS-states and one for the electron density, and one real-space grid, CPAIMD-QM/MM calculations use three reciprocal space grids and two real space grids. The two real-space grids, referred to as the MM-rspace-grid and QM-rspace-grid, respectively, contain the real space representation of the electron density in the large cell or MM-box (**h**) and the electron density in small cell or QM-box (**h**$_s$), respectively. The three reciprocal space grids are as follows: There are the two reciprocal space grids used in standard calculations, the reciprocal space grid for the KS states, QM-KSstate-gspace-grid, and the electron density, QM-e-density-gspace, in the QM-box. The third reciprocal space grid, MM-e-density-gspace, contains the low Fourier components of the electron density in the MM-box. The MM grids contain many fewer elements than the QM grids because $E_{cut}^{\text{long}} \ll E_{cut}^{\text{short}}$.

The short range electron-atom interactions are calculated using the QM-e-density-gspace and the QM-KSstate-gspace-grid. The long range interactions using the MM-e-density-gspace. Briefly,

1. Given N_s KS states on the QM-KSstate-gspace-grid, a set of N_s **3-D FFTs**

is performed in order to generate the states on the QM-rspace-grid. These are squared and used to generate the electron density, in real space, on the QM-rspace-grid.

2. The electron density on the QM-rspace-grid is interpolated onto the MM-rspace-grid using Cardinal B-splines.

3. Two **3-D FFT** of the two representations of the electron density, QM-rspace-grid and MM-rspace-grid, respectively are performed. The g-space representation of the electron density on the QM-e-density-gspace and MM-e-density-gspace grid are, thereby, generated.

4. The calculation now branches as in ordinary CPAIMD. The QM-rspace representation of the density is used to calculate the exchange and correlation potential, while the two reciprocal-space densities are used to calculate the long and short range parts of the Hartree and local pseudopotentials. The long range part includes the interaction of the electrons with the atoms in the MM region of the system. The Kohn-Sham potential is collected.

5. The calculations are then brought back together as in standard CPAIMD. The QM-e-density-gspace representation of the KS potential arising from short range interactions is Fast Fourier transformed back to QM-rspace-grid where it is combined with the exchange and correlation contributions. The MM-e-density-gspace representation of the KS potential arising from long range interactions is transformed back to the MM-rspace-grid and Cardinal B-Spline interpolated back to the QM-rspace-grid and combined with the short range KS contributions. The forces on KS states can then be evaluated in the usual way.

6. The calculation of the intra- and intermolecular MM forces arising from the empirical force field is an independent calculation that can occur before, after, or (preferably) during the dual grid electronic structure calculation.

Modeling the boundary : Pseudobond method

Much like parameterizing an empirical force field, developing a set of pseudopotentials and intermolecular interactions for a mixed *ab-initio*/molecular mechanics system requires melding basic principles with approximations. There is no unique way to treat these interactions and gaining experience through thorough testing for accuracy on various systems is the best way to proceed. Therefore, the following reasonable procedure can be pursued: Define the empirical model consisting of "MM atoms". Define the *ab initio* model consisting of "QM ions" and electrons. Define interactions when the QM/MM moieties are weakly coupled. Define interactions when the QM/MM moieties are strongly coupled.

In order to handle weak coupling between QM/MM and MM regions consider a system consisting of N_i chemically inert molecules and N_a chemically active molecules with $N_i \gg N_a$. That is, the scientist can clearly identify a complete set of molecules that will not undergo chemistry. In this case, it is straightforward to merge QM and MM. Examples include alkali metals in liquid ammonia, e.g. ammonia inert, alkali metals active [24]. Use the empirical model to describe the inert molecules, N_i. Use the *ab initio* model to describe the active molecules, N_a. Allow the ions comprising the active molecules to interact with the inert molecules by Coulombs law (MM atoms in the inert molecules are assigned partial charges). Introduce a short range intermolecular pair potential between the QM ions and the MM atoms in the molecule. Introduce molecular pseudopotential(s) to treat the interaction of the electrons in the system with the inert molecules. In order to fit the molecular pseudopotentials it is useful to invoke a least squares fitting procedure. A one electron "psuedo-atom" or an electron trap that can be moved around the inert molecule is defined and an objective function consisting of deviations from a full QM treatment of the trap plus the molecule and the one electron QM/MM model is minimized,

1. The deviation of the energy,
   E(QM)=E(mol+trap,QM)-E(mol,QM)-E(trap,QM) versus
   E(QM/MM) = E(mol+trap,QM/MM)-E(mol,QM/MM)-E(trap,QM).
2. The deviation of the forces on the atoms in the molecule,
   $-\nabla_I$ E(QM) versus $-\nabla_I$E(QM/MM).

A representative set of trap placements around the molecule is assumed to be included in the fit. Finally, the molecular pseudopotential can be further tuned to work well in the physical situation of interest, e.g. replacing the trap by a more realistic but still small *ab initio* system.

Unfortunately, there are many systems that **cannot** be divided completely into subsystems of chemically active and inactive molecules. For example, biopolymers (proteins, DNA, RNA) have active and inactive areas that are connected together by chemical bonds. Nonetheless, a similar course of action can be applied as in the weakly interacting case. As far as possible, divide the system into inert/active molecules. Next, divide large complex biopolymers into inert/active polymeric subunits or residues. Use the empirical model to describe the inert molecules, N_i and inert residues, R_i. Use the *ab initio* model to describe the active molecules, N_a and active residues, R_a. Allow the ions comprising the active molecules/residues to interact with the inert molecules by Coulombs law (MM atoms in the inert regions are assigned partial charges). Introduce intermolecular pair potentials between the QM ions and the MM atoms in the system. Introduce intramolecular potentials between the QM ions and the MM atoms in the polymeric system as required. Introduce pseudopotential(s) to treat the interaction of the electrons in the system with the inert molecules and residues.

The only difference between the strongly interacting case and the weakly interacting case is that active residues are chemically bonded to inert residues. This is a more delicate case than treating separate molecules. **Chemical bonds must be cut and replaced by pseudopotentials and intramolecular terms**. Therefore, allow a few more residues than necessary to be active so that the effect of the bond-cleavage is down stream from the chemistry. Choose the least polar bond possible as the cleavage site. Fit the pseudopotential(s) or "pseudobonds" to treat the cleavage site! Fit empirical bond, bend, torsion and 1-4 intramolecular terms around the cleavage site to correct for errors in geometry, energetics etc. For a discussion of creating pseudobond-pseudopotentials, themselves, via a more complex least squares procedure to relevant data sets, the reader is referred to [104, 105].

QM/MM using PINY_MD : Model systems and HCA II in water

The order Nlog(N) hybrid empirical force field-*ab initio* dual grid algorithm based on the GGA-DFT *ab initio* technique, a plane wave basis set and pseudobond/molecular pseudopotentials, has been implemented in PINY_MD [94]. In order to demonstrate the accuracy and effectiveness of the dual grid approach three problems have been considered. The first, a Gaussian charge density interacting with a point charge, demonstrates the correctness and accuracy of the dual grid method on an analytically tractable system. The second system, a single *ab initio* water molecule in a bath of empirical model water, can be treated by both a brute force numerical solution and by the novel fast dual grid treatment described above. The third system, an enzyme, Human Carbonic Anhydrase II, surrounded by water for a total of 30,649 atoms with 320 valence electrons of 80 *ab initio* atoms, cannot be treated using brute force, but only by a reduced order method such as that dual grid algorithm.

A comparison of analytical results to a dual grid treatment of a point charge interacting with a Gaussian charge density are shown in Table 34.2. Although the small box containing the Gaussian charge density is treated using a very large plane wave cutoff, $E_{cut}^{(short)}$ =120Ry, the large box can be treated using a very small plane wave cutoff, $E_{cut}^{(long)}$ ≈8Ry without loss of accuracy. Note, the small box is of side $4\mathring{A}$ while the large box is of side $20\mathring{A}$. The point charge can be positioned directly on the small box boundary with effecting the results. Only a relative small Cardinal B-spline interpolation order is required for very high accuracy.

r_0 (Å)	$E_{cut}^{(long)}$ (Rydberg)	m	E_{ext} (Hartree)	ΔE_{ext} (Kelvin)
4	8	4	-0.132293	0
		6	-0.132293	0
6	8	4	-0.088198	1
		6	-0.088198	1
8	8	4	-0.066149	1
		6	-0.066149	1

TABLE 34.2   The interaction of a Gaussian charge density, $\kappa = 3.779454\mathring{A}^{-1}$, with a point charge at distance, r_0 from its center is presented as a function of large cell plane wave cutoff and Cardinal B-Spline interpolation order. The large cell size was fixed at $L_l = 20\mathring{A}$ on edge. The small cell size was fixed at $L_s = 4\mathring{A}$ on edge and the small cutoff was fixed at $E_{cut}^{short} = 120Ry$. The electrostatic division parameter was set to be $\alpha = 6/L_s$ and $\Delta E_{ext} = E_{ext} - E_{ext}^{(exact)}$.

Next, a single *ab initio* water molecule in bath consisting of 31 TIP3P empirical model water molecules is considered. The 8 valence electrons of the *ab initio* molecule interact with the empirical model waters via a molecular pseudopotential. As in the model problem, the small box surrounding the *ab initio* model is treated using a very large plane wave cutoff, E_{cut}^{short}=100Ry but the large box surrounding the full system requires a much more modest number of plane waves to achieve full accuracy $E_{cut}^{long} \approx 8$Ry. The Cardinal B-spline interpolation order required for accuracy is also rather modest.

The HCA-II enzyme solvated by 8,859 waters, for a total of 30,649 atoms is, now, considered (see Fig 34.8). Only the catalytic zinc the side-chains of active site residues and the five water molecules in the active site were described under GGA-DFT (320 valence electrons of 80 atoms). Most of the system is simply the bath and is modeled using CHAR-MM22 [52]. The electronic energy is shown versus the plane wave cutoff of the large cell and the Cardinal B-spline interpolation order in Table 34.4. The results demonstrate that convergenced results can be obtained with a small plane wave cutoff in the big box at small Cardinal B-spline interpolation orders. This indicates that the dual grid method [103] is an attractive way to study QM/MM systems. Future work is described below.

L_s (Å)	E_{cut}^{long} (Rydberg)	m	E_{tot} (Hartree)	ΔE_{tot} (Kelvin)
6	8	6	-20.28133	2000
		8	-20.28134	2000
8	8	6	-20.28718	150
		8	-20.28718	150
9	8	6	-20.28790	-70
		8	-20.28790	-70

TABLE 34.3 The total electronic energy of a single *ab initio* water molecule immersed in a bath of TIP3P molecules as a function of large cell plane cutoff and Cardinal B-spline interpolation order. The large cell size was fixed by the state point, $L_l = 12.43\text{Å}$, on edge. The small cell cutoff was fixed at $E_{cut}^{short}=100\text{Ry}$. The electrostatic division parameter was set to $\alpha = 6/L_s$ and $\Delta E_{tot} = E_{tot} - E_{tot}^{(std)}$ where $E_{tot}^{(std)} = -20.28767$ is the result of a standard calculation with $L_s = L_l = 12.43\text{Å}$.

E_{cut}^{long} (Rydberg)	m	E_{tot} (Hartree)	ΔE_{tot} (Kelvin)
2	6	-2329.34896	32
	8	-2329.34905	3
4	6	-2329.34905	3
	8	-2329.34906	0

TABLE 34.4 The total electronic energy of the active site of HCA II immersed in a bath of TIP3P molecules and CHARMM22 model amino acid residues as a function of large cell plane cutoff and Cardinal B-spline interpolation order. The large cell size is fixed by the state point, 66.7Å, on edge. The small cell size was fixed at 18Å on edge and the small cell cutoff was fixed at $E_{cut}^{short}=70\text{Ry}$. The electrostatic division parameter was set to be $\alpha = 9/L_s$ and the accuracy measure is defined to be $\Delta E_{tot} = E_{tot}(E_{cut}^{long}, m) - E_{tot}(4, 8)$.

Other QM/MM approaches

There are other QM/MM approaches that form important alternatives to the techniques described, in detail, above [99, 30, 50, 68, 27, 104, 105]. First, rather than using a plane wave basis set, a Gaussian basis set can be employed. Second, the electronic structure model need not be GGA-DFT but Hartree Fock, MP2 or even semiempirical methods such as the empirical valence bond model, can be applied. Third, the method by which the long range interaction of the bath with the *ab initio* region need not be treated using the Euler exponential spline technique given here but can equally well be be evaluated using Fast Multipole based methods [36]. Fourth, the pseudobond method can be replaced by the "frozen frontier molecular orbital" method or the "link atom method". The interested reader is encouraged to read the substantial literature on these topics.

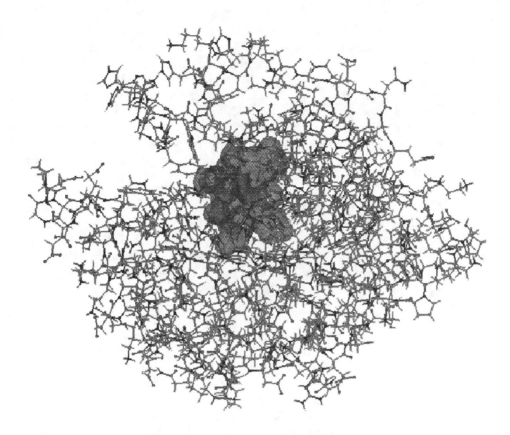

FIGURE 34.8: (See color insert following page 20-4.) Human carbonic anhydrase treated using the mixed *ab initio*/empirical force field based approach described in the text. The full enzyme wherein the wire frame represents atomic sites and the blue cloud represents the electron density of the valence electrons associated with "*ab initio* atoms".

LeanQM/MM : Integrating LeanMD and LeanCP

At present, a fine grain parallel QM/MM framework is not available. Therefore, future work will concentrate on combining the LeanMD framework with the plane wave GGA-DFT based LeanCP framework of the previous sections. On next generation parallel supercomputers such IBM's BG/L, this powerful combination could permit studies of enzyme catalysis with nanoseconds of sampling, allowing for the first time, the effects of the slow fluctuations of the protein backbone on catalytic activity to be accessed. Computational challenges involved in achieving high parallel efficiency for the leanQM/MM framework enabled by the Charm++runtime system include mapping the QM and MM parts to appropriate processors, managing the independent QM and MM communications patterns so they don't interfere and, of course, decomposing the problem so that QM/MM communication is minimized. The opportunity to perform for both exciting computer science and biophysics in this area is clearly available.

34.5 Conclusion

Our knowledge of molecular biology and the machinery of life has been increasing in leaps and bounds. To coalesce this knowledge into a deeper understanding, we need to determine the structure of a multitude of proteins with high resolution, and understand the relationship between their structure and function. Molecular dynamics simulations help further this understanding by allowing us to observe the phenomenon occurring at the scale of picoseconds, and validate our understanding of the basic physical principles embodied in simulations. Simulations based on classical mechanics, with some approximations of the quantum-mechanical "reality" are adequate for many situations; however, for simulations involving making and breaking of bonds, for example, a quantum mechanical simulation is necessary. The Car-Parinello algorithm and the ability to combine classical and quantum models in a single simulation are efficient ways of accomplishing this.

In either case, the computational power needed for carrying out the simulations over an interesting interval of time of the biomolecular phenomena is so large that only parallel computers offer the hope of completing such simulations in a realistic time. Although large parallel computers are available now, it is quite challenging to parallelize the simulations so as to scale to thousands of processors and beyond. This paper presented an overview of strategies aimed at this problem, and presented in some detail the particular strategies the authors have been pursuing.

The future research research agenda in this area will be shaped by two separated but related phenomena. Our capabilities for building faster and larger parallel supercomputers are constantly increasing. This creates newer challenges for parallelizing the simulations. At the same time, as our scientific understanding improves, scientists will pose newer computational problems, and possibly newer algorithms for solving those. Experience from other fields shows that this will typically lead to more efficient but more complex algorithms, which will be further difficult to parallelize. Equipped with the right set of computational tools and techniques, parallel computing experts, numerical analysts and physical scientists can rise to meet this challenge, but only via intense collaborative efforts.

Acknowledgements

The authors would like to thank Eric Bohm, and Chee Wai Lee for their help in preparing this manuscript. Much of the research in the author's research groups reported in this paper was carried out by several additional co-workers, including Emad Tajkhorshid, Yan Shi, Ramkumar Vadali, and Vikas Mehta. The authors would also like to acknowledge support from National Science Foundation (NSF) via grants ITR-0081307 and ITR-0121357, and from National Institutes of Health (NIH) via grant PHS-5P41RR05969-04

References

[1] D. Alfe, M.J. Gillan, and G.D. Price. The melting curve of iron at the pressures of the earth's core from ab initio calculations. *Nature*, 401:462, 1999.

[2] M.P. Allen and D.J. Tildesley. *Computer Simulations of Liquids*. Claredon Press, Oxford, 1989.

[3] G. Bachelet, D. Hamann, and M. Schluter. Pseudopotentials that work. *Phys. Rev. B*, 26:4199, 1982.

[4] J.A. Barker. *J. Chem. Phys.*, 61:3081, 1974.

[5] J.A. Barker. *Phys. Rev. Lett.*, 57:230, 1986.

[6] D.A. Beard and T. Schlick. Inertial stochastic dynamics. I. Long-time-step methods for Langevin dynamics. *J. Chem. Phys.*, 112(17):7313–7322, May 1, 2000.

[7] A.D. Becke. Density-functional exchange-energy approximation with correct assymptotic behavior. *Phys. Rev. A*, 38:3098, 1988.

[8] J.D. Bernal and R.H. Fowler. *J. Chem. Phys.*, 1:515, 1933.

[9] M. Boero, M. Parrinello, and K. Terakura. First principles molecular dynamics study of ziegler-natta heterogeneous catalysis. *J. Am. Chem. Soc.*, 120:2746, 1998.

[10] B.R. Brooks, R.E. Bruccoleri, B.D. Olafson, and D.J. States *et al.* CHARMM: A program for macromolecular energy, minimization, and dynamics calculations. *J. Comp. Chem.*, 4:187–217, 1983.

[11] A. Brünger, C.B. Brooks, and M. Karplus. Stochastic boundary conditions for molecular dynamics simulations of ST2 water. *Chem. Phys. Lett.*, 105:495–500, 1984.

[12] R. Car. Introduction to density-functional theory and ab initio molecular dynamics. *Quant. Struct. Act. Rel.*, 21:97, 2002.

[13] R. Car and M. Parrinello. Unified approach for molecular dynamics and density-functional theory. *Phys. Rev. Lett.*, 55:2471–2474, 1985.

[14] P. Carloni, P.E. Bloechl, and M. Parrinello. Electronic Structure of the Cu,Zn Superoxide dimutase active site and its interactions with the substrate. *J. Phys. Chem.*, 99:1338–1348, 1995.

[15] C.R.A. Catlow, C.M. Freeman, B. Vessal, and S.M. Tomlinson *et al.* Molecular dynamics studies of hydrocarbon diffusion in zeolites. *J. Chem. Soc. Far. Trans.*, 87:1947, 1991.

[16] C. Cavazzoni, G.L. Chiarotti, S. Scandolo, and M. Parrinello *et al.* Superionic and metallic states of water and ammonia at giant planet conditions. *Science*, 283:44, 1999.

[17] J. C. Charlier, A. De Vita, X. Blase, and R. Car. Microscopic growth mechanisms for carbon nanotubes. *Science*, 275:646, 1997.

[18] B. Chen, M.G. Martin, and J.I. Siepmann. Thermodynamic properties of the williams, OPLS-AA and MMFF94 all-atom force fields for normal alkanes. *J. Phys. Chem. B*, 102:2578, 1998.

[19] C.K. Chu. *An Introduction to Wavelets*. Academic Press, Boston,MA, 1992.

[20] T.W. Clark, R.V. Hanxleden, J.A. McCammon, and L.R. Scott. Parallelizing molecular dynamics using spatial decomposition. Technical report, Center for Research on Parallel Comutation, Rice University, P.O. Box 1892, Houston, TX 77251-1892, November 1993.

[21] W.D. Cornell, P. Cieplak, C.I. Bayly, and I.R. Gould *et al.* AMBER95. *J. Am. Chem. Soc.*, 117:5179, 1995.

[22] S.W. deLeeuw, J.W. Perram, and E.R. Smith. Ewald. *Proc. R. Soc. London A*, 373:27, 1980.

[23] Z. Deng, M.L. Klein, and G.J. Martyna. Quantum simulation studies of metal-ammonia solutions. *J. Chem. Phys.*, 100:7590, 1994.

[24] Z. Deng, G.J. Martyna, and M.L. Klein. Structure and dynamics of bipolarons in liquid ammonia. *Phys. Rev. Lett.*, 68:2496–2499, 1992.

[25] M. Diraison, M.E. Tuckerman, and G.J. Martyna. Simulation studies of the structural properties of liquid ammonia by classical, *ab initio* and path integral molecular dynamics. *J. Chem. Phys.*, 110:1096–1103, 1999.

[26] R.M. Dreizler and E.K.U. Gross. *Density Functional Theory*. Springer-Verlag, Berlin/Heidelberg, 1990.

[27] M. Eichinger, P. Tavan, J. Hutter, and M. Parrinello. A hybrid method for solutes in complex solvents: DFT combined with empirical force fields. *J. Chem. Phys.*, 110:10452, 1999.

[28] D.L. Ermak and J.A. McCammon. Brownian dynamics with hydrodynamic interactions. *J. Chem. Phys.*, 69(4):1352–1360, August 15, 1978.

[29] U. Essmann, L. Perera, M.L. Berkowitz, and T. Darden *et al.* Particle mesh ewald. *J. Chem. Phys.*, 103:8577, 1995.

[30] K.P. Eurenius, D.C. Chatfield, B.R. Brooks, and M. Hodoscek. Enzyme mechanism with hybrid quantum and molecular mechanical potentials I. theory. *Int. J. Quant. Chem.*, 60:1189, 1996.

[31] F. Allen *et al.* *IBM Syst. J.*, 40:310, 2001.

[32] G. Galli, R. M. Martin, R. Car, and M. Parrinello. Melting of diamond at high pressure. *Science*, 250:1547, 1990.

[33] G. Galli and M. Parrinello. *Computer Simulation in Materials Science*, 3:283, 1991.

[34] P. L. Geissler, C. Dellago, D. Chandler, and J. Hutter *et al.* Autoionization in liquid water. *Science*, 291:2121, 2001.

[35] M.J. Gillan. *Contemp. Phys.*, 38:115, 1997.

[36] L. Greengard and V. Rokhlin. *Physica Scripta*, 29A:139, 1989.

[37] H. Grubmüller, H. Heller, and P. Tavan. FAMUSAMM: A new algorithm for rapid evaluation of electrostatic interaction in molecular dynamics simulations. *J. Comput. Chem.*, 18:1729–1749, 1997.

[38] J.P. Hansen. *MD simulations of statistical mechanical systems.* North Holland Physics, Amsterdam, 1986.

[39] High Performance Computational Chemistry Group. NWChem, a computational chemistry package for parallel computers, version 4.0.1. http://www.emsl.pnl.gov:2080/docs/nwchem.

[40] J.O. Hirschfelder, C.F. Curtiss, and R.B. Bird. *The Molecular Theory of Gases and Liquids,.* John Wiley and Sons, Inc., New York, NY, 1964.

[41] Y.-S. Hwang, R. Das, J.H. Saltz, and M. Hodoscek *et al.* Parallelizing Molecular Dynamics Programs for Distributed Memory Machines. *IEEE Computational Science & Engineering*, 2(2):18–29, Summer 1995.

[42] C.L. Brooks III, M. Karplus, and B.M. Pettitt. *Proteins: A theoretical perspective.* John Wiley and Sons, NY, 1988.

[43] J.A. Izaguirre, S. Reich, and R.D. Skeel. Longer time steps for molecular dynamics. *J. Chem. Phys.*, 110(20):9853–9864, May 22, 1999.

[44] L.V. Kale, G. Zheng, C.W. Lee, and S. Kumar. Scaling applications to massively parallel machines using projections performance analysis tool. In *Future Generation Computer Systems Special Issue on: Large-Scale System Performance Modeling and Analysis*, number to appear, 2005.

[45] R.k. Kalia, T.J. Campbell, A. Chatterjee, and A. Nakano *et al.* Multiresolution algorithms for massively parallel molecular dynamics simulations of nanostructured material. *Computer Physics Communications*, 128(245), 2000.

[46] L. Kleinman and D.M. Bylander. Efficacious form for model pseudopotentials. *Phys. Rev. Lett.*, 48:1425, 1982.

[47] W. Kohn and L.J. Sham. *Phys. Rev.*, 140:A1133, 1965.

[48] C. Lee, W. Yang, and R.G. Parr. Development of the Calle-Salvetti correlation energy into a functional of the electron density. *Phys. Rev. B*, 37:785, 1988.

[49] E. Lindahl, B. Hess, and D. van der Spoel. GROMACS 3.0: a package for molecular simulation and trajectory analysis. *J. Mol. Mod.*, 2001.

[50] P.D. Lyne, M. Hodoscek, and M. Karplus. A hybrid QM-MM potential employing

HF or DFT methods. *J. Phys. Chem. A*, 103:3462, 1999.

[51] Q. Ma, J. Izaguirre, and R.D. Skeel. Verlet-I/r-RESPA is limited by nonlinear instability. *SIAM J. Sci. Comput.*, 24(6):1951–1973, May 6, 2003.

[52] A. MacKerell Jr., D. Bashford, M. Bellott, and R. L. Dunbrack *et al.* CHARMM22. *J. Phys. Chem. B*, 102:3586, 1998.

[53] P.A. Madden and M. Wilson. *Chem. Soc. Rev.*, 25:339, 1996.

[54] G.J. Martyna and M.E. Tuckerman. A reciprocal space based method for treating long range interactions in ab-initio and force-field-based calculations in clusters. *J. Chem. Phys.*, 110:2810, 1999.

[55] D. Marx and J. Hutter. Ab initio molecular dynamics: Theory and implementation. in *Modern methods and algorithms of quantum chemistry* (J. Grotendorst (Ed.), Forschungszentrum, Juelich, NIC Series), 1:301–449, 2000.

[56] D. Marx and M. Parrinello. Ab initio path integral molecular dynamics. *Z. Phys. B*, 95:143, 1994.

[57] D. Marx, M.E. Tuckerman, J. Hutter, and M. Parrinello. *Nature*, 367:601, 1999.

[58] D. Marx, M.E. Tuckerman, and G.J. Martyna. *Comp. Phys. Comm.*, 118:166, 1999.

[59] P. Minary, J. Morrone, D. Yarne, and M.E. Tuckerman *et al.* A new reciprocal space based treatment of long range interactions on surfaces. *J. Chem. Phys.*, 121:5351, 2004.

[60] P. Minary, M.E. Tuckerman, K.A. Pihakari, and G.J. Martyna. A new reciprocal space based treatment of long range interactions on surfaces. *J. Chem. Phys.*, 116:5351, 2002.

[61] S. Miyamoto and P.A. Kollman. SETTLE: An analytical version of the SHAKE and RATTLE algorithm for rigid water molecules. *J. Comput. Chem.*, 13(8):952–962, 1992.

[62] M. Nelson, W. Humphrey, A. Gursoy, and A. Dalke *et al.* NAMD— A parallel, object-oriented molecular dynamics program. *J. Supercomputing App.*, 1996.

[63] R.G. Parr and W. Yang. *Density Functional Theory of atoms and molecules.* Oxford University Press, Oxford, 1989.

[64] M. Parrinello. *Solid State Commun.*, 102:107, 1997.

[65] A. Pasquarello, I. Petri, P.S. Salmon, and O. Parisel *et al.* First solvation shell of the Cu(II) aqua ion: Evidence for fivefold coordination. *Science*, 291:856, 2001.

[66] M.C. Payne, M. Teter, D.C. Allan, and T.A. Aria *et al.* *Rev. Mod. Phys.*, 64, 2002.

[67] J.P. Perdew and A. Zunger. Self-interaction correction to density functional theory. *Phys. Rev. B.*, 23:5048, 1981.

[68] D.M. Philipp and R.A. Friesner. Mixed ab initio QM/MM modeling using frozen orbitals and tests with analnine peptides. *J. Comp. Chem.*, 20:1468, 1999.

[69] J.C. Phillips, G. Zheng, S. Kumar, and L.V. Kalé. NAMD: Biomolecular simulation on thousands of processors. In *Proceedings of SC 2002*, Baltimore, MD, September 2002.

[70] M.R.S. Pinches, D.J. Tildesley, and W. Smith. Large scale molecular dynamics on parallel computers using the link-cell algorithm. *Molecular Simulation*, 6(1):51, 1991.

[71] S. J. Plimpton and B. A. Hendrickson. A new parallel method for molecular-dynamics simulation of macromolecular systems. *J Comp Chem*, 17:326–337, 1996.

[72] J.W. Ponder and F.M. Richards. An efficient Newton-like method for molecular mechanics energy minimization of large molecules. *J. Comp. Chem.*, 8:1016–1024, 1987.

[73] A. Rahman. Correlations in the motion of liquid argon. *Phys. Rev. A.*, 136:405, 1964.

[74] D.K. Remler and P.A. Madden. *Mol. Phys.*, 70:921, 1990.

[75] J.P. Ryckaert, G. Ciccotti, and H.J.C. Berendsen. Numerical integration of the cartesian equation of motion of a system with constraints: molecular dynamics of n-alkanes. *J. Comput. Phys.*, 23:327–341, 1977.

[76] T. Schlick. *Molecular Modeling and Simulation: An Interdisciplinary Guide*, volume 21 of *Springer Series in Interdisciplinary Applied Mathematics*. Springer-Verlag, New York, 2002.

[77] I.J. Schoenberg. *Cardinal Spline Interpolation*. Society for Industrial and Applied Math, Philadelphia,PA, 1973.

[78] P. L. Silvestrelli and M. Parrinello. Water dipole moment in the gas and liquid phase. *Phys. Rev. Lett.*, 82:3308, 1999.

[79] R.D. Skeel. Integration schemes for molecular dynamics and related applications. In M. Ainsworth, J. Levesley, and M. Marletta, editors, *The Graduate Student's Guide to Numerical Analysis*, volume 26 of *Springer Series in Computational Mathematics*, pages 119–176. Springer-Verlag, 1999.

[80] M.M. Smith. Histone structure and function. *Curr. Opinion Cell Biol.*, 3:429–437, 1991.

[81] M. Sprik. Polarizable water with double nose. *J. Phys. Chem.*, 95:2283, 1991.

[82] F.H. Stillinger and C. David. Dynamics and ensemble averages for the polarization models of molecular interactions. *J. Chem. Phys.*, 71:1647, 1979.

[83] E. Tajkhorshid, P. Nollert, M.O. Jensen, and L.J.W. Miercke *et al.* Global orientational tuning controls the selectivity of the aquaporin water channel family. *Science*, 296:525–530, 2002.

[84] N. Troullier and J.L. Martins. *Phys. Rev. B*, 43:1993, 1991.

[85] M. Tuckerman, B.J. Berne, and G.J. Martyna. Reversible multiple time scale molecular dynamics. *J. Chem. Phys*, 97(3):1990–2001, 1992.

[86] M. Tuckerman, G. Martyna, M.L. Klein, and B.J. Berne. Efficient Molecular Dynamics and Hybrid Monte Carlo Algorithms for Path Integrals. *J. Chem. Phys.*, 99:2796, 1993.

[87] M.E. Tuckerman. Ab initio molecular dynamics: Basic concepts, current trends and novel applications. *J. Phys. Condensed Matter*, 14:R1297, 2002.

[88] M.E. Tuckerman, J. Hutter, and M. Parrinello. *J. Chem. Phys.*, 102:859, 1995.

[89] M.E. Tuckerman, D. Marx, M.L. Klein, and M. Parrinello. On the quantum nature of the shared proton in hydrogen bonds. *Science*, 275:817, 1997.

[90] M.E. Tuckerman, D. Marx, and M. Parrinello. The nature and transport mechanism of hydrated hydroxide ions in aqueous solution. *Nature*, 417:925, 2002.

[91] M.E. Tuckerman and M. Parrinello. *J. Chem. Phys.*, 101:1301, 1994.

[92] M.E. Tuckerman and M. Parrinello. *J. Chem. Phys.*, 101:1316, 1994.

[93] M.E. Tuckerman, P. J. Ungar, T. von Rosenvinge, and M. L. Klein. Ab initio molecular dynamics simulations. *J. Phys. Chem.*, 100:12878, 1996.

[94] M.E. Tuckerman, D.A. Yarne, S.O. Samuelson, and A.L. Hughes *et al.* Exploiting multiple levels of parallelism in molecular dynamics based calculations via modern techniques and software paradigms on distributed memory computers. *Comp. Phys. Comm.*, 128:333, 2000.

[95] J.C. Tully. Molecular dynamics with electronic transitions. *J. Chem. Phys.*, 93:1061, 1990.

[96] R. Vadali, Y. Shi, S. Kumar, and L.V. Kale *et al.* massively parallel cpaimd. *J. Comp. Chem.*, 25:2006, 2004.

[97] A. De Vita, G. Galli, A. Canning, and R. Car. A microscopic model for surface-induced diamond-to-graphite transitions. *Nature*, 379:523, 1996.

[98] A. Warshel. *Computer Modeling of Chemical Reactions in Enzymes and Solutions.*

John Wiley & Sons, 1991.

[99] A. Warshel and M. Levitt. QM/MM first time. *J. Mol. Biol.*, 103:227, 1976.

[100] A. Warshel and R. M. Weiss. *J. Am. Chem. Soc.*, 102:6218, 1980.

[101] P.K. Weiner and P.A. Kollman. AMBER: Assisted model building with energy refinement. A general program for modeling molecules and their interactions. *J. Comp. Chem.*, 2(3):287–303, 1981.

[102] J.A. White and D.M. Bird. Implementation of gradient-corrected exchange-correlation potentials in car-parrinello total-energy calculations. *Phys. Rev. B*, 50:4954, 1994.

[103] D.A. Yarne, M.E. Tuckerman, and G.J. Martyna. A dual length scale method for plane-wave-based, simulation studies of chemical systems modeled using mixed ab initio/empirical force field descriptions. *J. Chem. Phys.*, 115:3531, 2001.

[104] Y. Zhang, T. Lee, and W. Yang. A pseudobond approach to combining quantum mechanical and molecular mechanics methods. *J. Chem. Phys.*, 110:46, 1999.

[105] Y. Zhang, T. Lee, and W. Yang. Free energy calculation on enzyme reactions with efficient iterative procedure to determine minimum energy paths on a combined surface. *J. Chem. Phys.*, 112:3483, 2000.

VIII

Bioinformatic Databases and Data Mining

35

String Search in External Memory: Data Structures and Algorithms

Paolo Ferragina
University of Pisa, Italy

35.1 Introduction

Data should be cheap to store and fast to retrieve or process. Unfortunately there are fundamental reasons why we cannot design a computer memory that is at the same time cheap, compact and fast. No signal can propagate faster than light, and wires reaching every single memory cell would need too much room. As a result, given a storage technology and a desired access latency, there is only a finite amount of data reachable within this time limit. The simplest, and most widely used compromise to escape this problem is the *memory hierarchy.*

Real memory hierarchies on current PCs and workstations are very complex because they consist of multiple levels, all with their own technological specialties: L1 and L2 caches, internal memory, one or more disks, other external storage devices (like CD-ROMs, DVDs and tapes), and memories of multiple hosts over a network. Each of these memory levels has its own cost, capacity, latency, bandwidth and access method. The closer a memory level is to the CPU, the smaller, the faster and the more expensive it is. Currently few nanoseconds suffice to access the processor caches, whereas milliseconds are yet needed to fetch data from disks. Nonetheless, the power of this memory organization is in that it may be able to offer the expected access time of the fastest level while keeping the average cost per memory cell near the one of the cheapest levels, provided that data are properly *cached* and *delivered* to the requiring algorithms.

Virtual memory systems work well if the *working set* of an algorithm (roughly speaking, the set of pages it will reference in the near future) is small and most of it can be stored in internal memory. In this case, the caching and delivering of data is simple and effective.

In many current applications, however, the avalanche of data is not accompanied by its accurate organization over the hierarchical memory, so that the working set is not small and the overall performance results very poor. In fact, the larger is the amount of data to be processed, the wider is the amount of memory needed to store this data, the higher is the number of memory levels involved in the data storage and hence, the more complicated is the design of *algorithms and data structures* that cleverly map these data on the various memory levels to access them efficiently. Neglecting questions pertaining to the cost of memory references in a hierarchical system may even prevent the use of an algorithm on large input data. Engineering research is trying nowadays to improve the input/output subsystem to reduce the impact of these issues, but it is very well known [177] that the improvements achievable by means of a *proper arrangement of data* and a properly *structured algorithmic computation* abundantly surpass the best expected technology advancements. Consequently, the design principles and data structuring tools presented in this chapter should pervade the development of any modern data structure and algorithm devoted to the processing of large (sequence, string) data.

In order to reason about algorithms and data structures for hierarchal memories, we need a *model of computation* that grasps the essence of real situations so that algorithms that are good in the model are also good in practice. Accurate disk models are complex [155], and it is virtually impossible to exploit all the fine points of memory characteristics systematically, either in practice or for algorithmic design. In this chapter we will refer to two models: the *external memory model* [177] and the *cache oblivious model* [70]. The former received much attention in the literature because of its simplicity and reasonable accuracy. Here a computer is abstracted to consist of only *two memory levels*: the internal memory of size M, and the (unbounded) disk memory which operates by reading/writing data by blocks of size B. This view is, however, not limited to identifying the external memory with the disk; in fact, we are actually free to choose any two levels of the memory hierarchy to play the role of the internal and the external memory of the model. The elegant *cache-oblivious* model still assumes a two-level view of the computer memory but, on the other side, it allows to prove results for an unknown multilevel memory hierarchy. Cache-oblivious algorithms know only the existence of the memory hierarchy, but not its parameters which come into play at analysis time. As a consequence such algorithms are designed for a two-level memory hierarchy but *tune automatically* to hierarchies with arbitrarily many memory levels. While these results seem impossible, a recent body of research has developed algorithms and data structures that perform as well, or nearly as well, as standard external-memory solutions which know all about the memory hierarchy (see [49, 109] and references therein).

The issues regarding the design of external-memory algorithms and data structures obviously pertain to the development of efficient computational biology applications. Numerous and large databases holding DNA and protein sequences are now readily available over the Web and are quickly becoming the "lifeblood of molecular biology" [179]. The volume of these data is growing daily and seems to increase significantly: four years ago, Genbank totalled around 10Gb of DNA sequences, now it approaches 40Gb; Swissprot and Trembl stored a smaller volume of protein data that grew significantly in the last years. Whenever a new gene is cloned and sequenced, visiting the appropriate database to *search* for some patterns within its content is the next step for a molecular biologist. The search step is usually carried out by *sequentially scanning* the entire database using a screening approach that identifies the set of desired sequences (e.g. BLAST [5, 4], FASTA [117, 150, 149, 151], FLASH [26], PatternHunter [31]). This approach inevitably suffers from a prolonged response time when dealing with a large amount of database sequences, and its limitations are especially evident when the result set is small compared with the overall database size. Other bio-applications tend to introduce even more stringent performance requirements that lead the *scan-based* algorithms to further reach their limits. Consider, for example, the comparison of an EST database to itself for the purpose of clustering, or the shotgun sequencing where an all-against-all comparison of large amounts of data need to be computed, or finally, the case of a database accessible over the Web that must support several queries per second. Since many bio-queries can be expressed as pattern matching tasks [79], *string-matching data structures* are key tools to support these queries efficiently by reducing the portion of the database to be fully examined. Actually, despite the rapid growth in the database size, the existing biological sequences are relatively stable (most updates concerns the insertion of a new sequence). This further justifies the potential benefit of supporting *searches through a string data structure* since the construction cost can be easily amortized over many searches.

In this chapter with the term *"index"* we mean the class of *full-text indexing data structures*. Namely, string data structures that keep a *compact but detailed picture* of the *full content and structural properties* of the string data they index thus allowing to support powerful search operations, like *substring searches* or statistical, approximated or general fuzzy searches [79]. These data structures, and their several variants, constitute today the "lifeblood" of any current computational biology algorithm. The hierarchical memory setting we consider in this chapter poses many challenges in turning these indexes to be efficient on disk, or multi-level memory, systems. These *I/O-issues* will be the seeds from which the discussion of the next sections will depart from.

Space constraints prevent us to deal with the I/O-issues related to all full-text indexes known in the literature. We content ourselves with three, yet significant, indexes: the well-known Suffix Tree [181, 127] and Suffix Array [121] data structures, as well the less famous String B-tree [59] born within the theoretic string-matching field but currently the best data structure for supporting substring searches on a disk memory. Each of these data structures offers interesting algorithmic and structural specialties that are worth of investigation, especially in the I/O-setting. Therefore we will devote one section to each of them, commenting on the following three main aspects of I/O-computations.

The first aspect concerns with the so called *I/O-bottleneck* in which a naïve index organization might incur. If questions pertaining to the cost of memory references in a hierarchical system are not properly taken into account, it may be the case that index update and query operations might spend most of the time in transferring data to/from the disk (or among the memory levels) with a consequent sensible slowdown of performance. Hence, we will largely comment on the *index scalability* and the *disk consciousness* of index design, nowadays hot research topics. They are important issues because they have been shown to induce a positive effect not limited just to mechanical storage devices, but also to all other memory levels. In this setting we will discuss two main approaches to cope with the I/O-bottleneck: *index packing* and *index design from scratch*. The former approach aims at devising strategies for *packing* known indexes onto disk pages, in order to minimize the I/O-traffic induced by query and update operations executed over them. This is typical of suffix trees and suffix arrays, for which we will resort also to the cache-oblivious model. The latter approach explores innovative indexing data structures which are explicitly designed to take advantage of the blocked access to the disk memory. This is the case of the String B-tree and its variants.

The second aspect concerns with the efficient construction of indexes on very large text collections: *"We have seen many papers in which the index simply 'is', without discussion of how it was created. But for an indexing scheme to be useful it must be possible for the index to be constructed in a reasonable amount of time, ..."* [186]. The construction phase may be, in fact, a bottleneck that can prevent an index to be used even in medium-scale applications [110, 58, 88, 156]. We will discuss state-of-the-art algorithms for index construction, and detail many interesting techniques which have been published in the last two years and have opened new interesting avenues of research.

Starting from the consideration that _"space optimization is closely related to time optimization in a disk memory"_ [106], we will also discuss in detail the third aspect: _space-succinct_ implementation of full-text indexes. Here, data compression appears as an attractive tool because it allows not only to squeeze the space occupancy of an index, but also to improve the computing speed of its operations because they can better use the fast and small memory levels close to CPUs, reduce the disk access time, virtually increase the disk (memory and cache) bandwidth, and come at a negligible cost because of the significant speed of current CPUs. Although it is very well known how compression may operate at the string level (see e.g. [162]), the issues concerned with the compression of indexing data structures is an interesting [33, 135, 132], and recently revitalized topic of research (starting from [64, 77]). Succinct implementations of suffix trees and String B-trees will be discussed in detail, whereas the recent advances onto suffix array compression will be mentioned briefly because they are too complicated to be included in this introductory chapter. Every next section will be concluded with a description of challenging open problems and possible future directions of research.

The present chapter naturally complements Chapter 5, where a RAM-based view of the above full-text indexes has been provided. Additionally to the features highlighted above, our chapter offers some distinguishing features with respect to other surveys present in the literature. It dedicates a special attention to Suffix Arrays because of the recent scientific achievements that have shed new light on some fascinating structural and algorithmic properties of this data structure (Section 35.3). Moreover it presents a deep analysis of I/O issues in disk mapping and engineering of Suffix Trees (Section 35.4). Finally, it details String B-trees and proposes a novel engineering solution that makes them appealing for real sequence databases (Section 35.5). All of these issues and details are conveyed to the reader with the aim of convincing him/her that the text indexing field has grown to such a complicated stage that various issues come into play when studying it: data structure design, database principles, data compression techniques, architectural considerations, caching policies. The expertise nowadays required to design a good index is therefore transversal to many algorithmic fields and much more study on the orchestration of known, or novel, techniques is needed to make progress in this fascinating topic. This chapter illustrates some of the key ideas which should constitute, in our opinion, the current background of every _index designer_. A large, but obviously not complete, literature will accompany our discussion and should be the reference where an eager reader may find further technical details and research hints.

Beyond the exact substring searches

Suffix tree, suffix array and string B-tree are very well-known data structures for supporting _substring searches_, that is, the retrieval of all occurrences in an indexed string of an arbitrary pattern sequence. In biological applications the goal is not only to find _exact_ matches but also to allow some _search fuzziness_ [79, 111, 47, 125]: mismatches, gaps, regular expressions or motif extraction, just to cite a few. In general the sensitivity to detect weak relationships between the query sequence and the indexed sequence is linked to performance, equivalent to the time required to satisfy the query. Known indexes try to balance these factors, however provably efficient performances have still to be achieved [141]. Recently, many interesting results have appeared in the literature shading some new light on this challenging problem. Below we will sketch some of the most promising alternative approaches to _fuzzy searches_ in the external memory setting, and refer the reader to Chapters 6 and 36, and our bibliography for references and research material. At this point, and just for the sake of clarity, we state that most of these sophisticated searches boil down to substring searches, so that full-text indexes play a crucial role in that setting too.

Various similarity measures are known in the literature. Here we concentrate on the *edit distance* between two strings, that is, the number of *character edits* (like insertion, deletion and substitution of single characters) that allow one string to be transformed into the other. This distance model is appropriate in computational genomics [164] and therefore widely used. The edit distance computation between two strings of length n and m requires $O(nm)$ time by using dynamic programming [79].[1] This quadratic bound is unacceptably high in the case of large databases so that the goal of the *fuzzy* indexes is to improve it substantially. Several proposals exist in the literature of computational biology and string matching to design good indexes for efficient *similarity-based searches*. For the sake of clarity and brevity, we classify those proposals into four groups [141, 145] and briefly comment on them.

Backtracking (see e.g. [14, 35, 97, 175]) uses the suffix tree, suffix array or DAWG [43] built over the indexed string in order to factor out its repetitions. A scan-based algorithm over the string is then simulated by backtracking on the data structure. These algorithms take time *exponential* in the pattern length and the number of errors allowed, but turn out to be in many cases independent of the indexed string length. This makes them attractive when searching for short and very similar patterns.

Partitioning (see e.g. [15, 97, 168, 172, 173, 175]) divides the pattern into pieces to ensure that some of these pieces occur exactly into the indexed string (these pieces are usually called q-grams). Any index able to perform exact searches is used to detect the q-gram occurrences which are then checked for a full match with a Dynamic Programming approach [145]. These algorithms work well when q is sufficiently long so as to reduce the *false matches*. This implies that the *error ratio* between number of errors and pattern length should be low. To obtain much faster and/or more efficient filtering we may use a variant of q-grams, called *gapped q-grams*, consisting of a subset of q characters of a random [26] or best chosen [23] arbitrary shape. The first application of those ideas to an external memory setting has been proposed in [22], and we refer the interested reader there for further details and references (see also [47, 48, 111]). Overall the partitioning approach achieves fast delivery of approximate occurrences but it is suitable only for searching *highly similar* sequences.

The third class is a *hybrid* between the other two (see e.g. [140, 143]). The pattern is divided into long pieces that can still contain errors, but they should be very few. These pieces are then searched using backtracking, and the candidate occurrence positions are checked as in the Partitioning method. The hybrid approach is more effective because it can find the good balance between piece length to search and error level allowed. The time complexity is $O(n^\lambda)$ for some positive $\lambda < 1$ that depends on the error level (here n is the database size, or equivalently, the length of the indexed string(s)). This approach tolerates moderate error ratios.

The fourth class is more recent and consists of sophisticated approaches that exploit the *metric properties* of the edit distance and deal with arbitrarily error ratios. Its basic component is the construction of a mapping f (called an *embedding*) which maps any string s into a multi-dimensional vector $f(s)$ such that, for any pair of strings s and s', the l_p

---

[1] Apart from some progress on problem relaxations (see e.g. [40, 137, 139] and references therein), the result of [126] taking $O(nm/\log n)$ time for computing the edit distance between two sequences of length n and m, is the best to date (see also [44]).

distance $||f(s) - f(s')||_p$ is approximately equal to the *edit distance* between s and s'. The approximation factor is called *distortion* of the embedding f. A low-distortion embedding would be very useful because: (1) one could reduce a similarity search between sequences to an analogous problem in normed spaces, thus exploiting many known solutions [92, 112, 113]; (2) if computing $f(s)$ took subquadratic time, then the edit distance could be approximated in subquadratic time as well. Unfortunately little is known on the embedability of the edit distance into normed spaces, apart from some weak lower bounds [8] and some fascinating theoretical achievements [91, 36], whose practical impact is yet to be fully investigated.

To circumvent this lack of "embedding results" various authors tried either to solve a variation of the problem or to impose some relaxations to its performance guarantees. For example, the concept of *block edit distance* has been introduced to model the case of movements of arbitrarily long contiguous blocks of characters in one operation.[2] This metric can be embedded in l_1 with distortion $O(\log \ell \log^* \ell)$, where ℓ is the maximum length of the strings in the database [137, 40, 138, 139]. If we are instead interested in efficient average bounds, the *metric index* proposed in [27] can be a good candidate for efficient complex searches (cfr. [183]). It allows to find the *occ* approximate matches of a pattern $P[1, p]$ as a substring of a text $T[1, n]$ in $O(p \log^2 n + p^2 + occ)$ average time, using $O(n \log n)$ space.

These strictly-algorithmic approaches have been recently complemented with interesting techniques borrowed from the area of signal processing [99, 98] and speech recognition [2]. Their impact on computational biology still needs further validating experiments.

35.2   Background and Terminology

Let $T[1, n]$ be a text string of length n drawn from an alphabet Σ. When not explicitly said Σ is assumed to be of *constant size* ranging from 4 (DNA bases) to 20 (aminoacids) symbols. The *i*th *suffix* T_i of the text T is the substring $T[i, n]$ extending from the *i*th character of T until its end. The operator $\bullet$ is used to denote the concatenation of strings, and $\mathtt{lcp}(\alpha, \beta)$ is used to indicate the *longest common prefix* between the two strings α and β, that is, the largest k such that $\alpha[1, k] = \beta[1, k]$. Since we will deal with rooted trees, the notion of *lowest common ancestor* $\mathtt{lca}(u, v)$ between two nodes u, v will be largely exploited.

Text suffixes play a crucial role in the efficient implementation of (exact) substring searches and therefore in the definition of full-text indexes. Let $P[1, p]$ be a pattern to be searched as a *substring* of the text $T[1, n]$, and let $\mathsf{SUF}(T)$ be the set of all text suffixes *sorted lexicographically*. If P occurs at text position i then $P = T[i, i + p - 1]$; or equivalently, P is a *prefix* of T_i. As a consequence, there is a bijective correspondence between the text suffixes prefixed by P and the pattern occurrences in T. This correspondence leads to two nice observations which are algorithmically useful: (1) the suffixes prefixed by P occur contiguously into $\mathsf{SUF}(T)$, because of its ordering; (2) this contiguous portion occurs after the lexicographic position of P in $\mathsf{SUF}(T)$. These two observations allow us to reduce the *substring-search problem* between P and T to the *prefix-search problem* between P and the ordered string set $\mathsf{SUF}(T)$. This latter problem is intuitively easier to deal with, and an apparent algorithmic solution could be based on binary searching $\mathsf{SUF}(T)$. However the arbitrarily long strings belonging to $\mathsf{SUF}(T)$ pose additional time-efficiency and space-coding problems that make this problem challenging. Actually, all the data structures that we

---

[2] Block edit operations occur as a consequence of large scale inter or intra chromosomal genomic duplication, genome duplications or rearrangements [94]. We cannot allow any block operation because the problem becomes NP-hard [118].

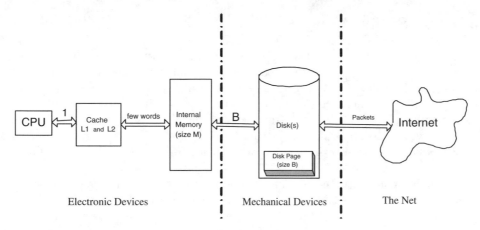

FIGURE 35.1: An example of memory hierarchy of a modern computer.

describe in the present chapter deploy the two basic observations above and differentiate themselves in the way the set $\mathsf{SUF}(T)$ is implemented on the hierarchical memory, and in the way the I/O-conscious prefix-searches are executed over this implementation.

To evaluate the performance of algorithms and data structures in hierarchal memories we will deploy the *external memory model* [177] and the *cache oblivious model* [70]. In the external-memory model the algorithm designer has to worry only about *two memory levels* — internal memory and the disk(s) — and thus only of *two parameters*: disk-block size B and internal-memory size M. Data between the internal memory and the disk are transferred in blocks of size B (called *disk pages*). The algorithm performance is evaluated by counting: (a) the number of disk accesses (hereafter *I/Os*), (b) the internal running time (CPU time), and (c) the number of disk pages occupied by the data structure or used by the algorithm as its working space. This simple model suggests, correctly, that a good external-memory algorithm should exploit both *spatial locality* and *temporal locality*. The former imposes a data organization onto the disk(s) that makes each accessed disk block as much useful as possible; the latter imposes on the algorithms to execute as much useful work as possible onto data fetched in internal memory, before they are written back to disk. These are the *two golden rules* underlying the design of the efficient algorithms and data structures described in the next sections. For the sake of clarity we also remark that the term "I/O" and the *two-level view* must not suggest to the reader that the external memory has to be identified with the disk memory; in fact, we are actually free to choose any two levels of the memory hierarchy for the internal and external memory levels of the model, with their M and B parameters properly set.

The cache oblivious model still assumes a two-level view of the computer memory but its parameters are unknown to the algorithm designer. These parameters come into play only at the time of algorithm analysis. The nice feature of this model is that its algorithms are designed for a two-level memory hierarchy but *tune automatically* to hierarchies with arbitrarily many memory levels [49, 109]. The literature concerning the design of cache-oblivious string data structures is still poor in results. We will deal with the ones pertaining with suffix tree design in the next sections.

35.3   The Suffix Array

The suffix array [121] (or PAT array [73]) is the simplest, the most succinct, easy to code, and elegant data structure currently used to support substring searches. The suffix array of a text $T[1, n]$ is an array SA of n integers in the range $[1, n]$ succinctly representing the string set $\mathsf{SUF}(T)$. Namely, $SA[1]$ is the starting position in T of the smallest suffix in the lexicographic order, $SA[2]$ is the starting position of the second one, and so on. For example, the suffix array for the text $T = \texttt{GAGCT}$ is $SA = [2, 4, 1, 3, 5]$ since $T[2, 5] = \texttt{AGCT}$ is the suffix with the lowest lexicographic rank (hence $SA[1] = 2$), followed by $T[4, 5] = \texttt{CT}$ (hence $SA[2] = 4$), then by $T[1, 5] = \texttt{GAGCT}$ (hence $SA[3] = 1$), and so on.

Notice that SA, coupled with the text T, provides a succinct encoding of the set $\mathsf{SUF}(T)$. Since any integer can be encoded within $O(\log n)$ bits, $O(n \log n)$ bits suffice to store both the suffix array and the text. In practice the integers are limited to be smaller than 2^{32}, hence $5n$ bytes suffice to store SA (4 bytes per suffix pointer) and T (1 byte per character). Notice that $\Omega(n \log n)$ is *not* a lower bound to the storage complexity of SA since $|\Sigma|^n$ is the number of existing suffix arrays for n-long strings drawn from the alphabet Σ, but it is $n! \gg |\Sigma|^n$. Hence the possible instances of $SA[1, n]$ are much fewer than the number of permutations of n integers. It is therefore not surprising mathematically, but challenging algorithmically, that a much succinct [77], or even compressed [68, 64, 75, 76, 69], encodings of SA and T do exist and can be built efficiently [86, 85, 24, 124]; however, we do not go into further details because these results are too involved for this introductory chapter and moreover they are designed so far only for internal-memory use. Hereafter we then assume to encode SA in its raw form using one memory word per integer (i.e. 4 bytes per suffix pointer).

Searching for the *occ* occurrences of an arbitrary pattern $P[1, p]$ into T can be performed via (an indirect) binary search over SA thus taking $O(p \log n + occ)$ time. Actually, the counting of the *occ* occurrences does not need their retrieval and can thus be performed in $O(p \log n)$ time: it suffices to determine the extremes of the subarray containing the suffixes in $\mathsf{SUF}(T)$ prefixed by P. When SA is coupled with information about the longest common prefixes of some text suffixes, taking additional $O(n \log n)$ bits of storage (cfr. [158]), substring searches can be speeded up to take $O(p + \log n)$ time.

Given its simple array structure the mapping of, and the search into, a suffix array on disk seems trivial. Just partition the array in B-sized blocks, and map them onto a contiguous sequence of disk pages. But the coupling of SA and T in the pattern searches poses challenging algorithmic problems whose investigation will start in this section, with simple considerations and solutions, and will lead to precious ideas and non obvious developments in Section 35.5.

35.3.1   I/O issues

Let us start from the previously sketched proposal. Partition $SA[1, n]$ and $T[1, n]$ into disk pages of size B each, allocated contiguously on disk. The binary search, to retrieve the subarray of SA containing suffixes which are prefixed by P, is now executed over the $\Theta(n/B)$ disk pages containing SA. However every binary-search step, they are $\Theta(\log n)$ in the worst case, still needs to access T for retrieving the suffix to be compared with P. Thus a binary-search step takes $O(p/B)$ I/Os. Notice that the suffix retrieval is unavoidable, even if the pattern is assumed to reside in internal memory. As a result the overall substring search cost is $O((p/B) \log n + occ/B)$ I/Os, where the second additive term comes from the cost of retrieving the pattern occurrences via a simple disk scan. Obviously the reduction from p to $\lceil p/B \rceil$ is a small improvement because, in practice, p is usually short compared to the page size B. The use of an additional `lcp` information would achieve $O(p/B + \log n + occ/B)$ I/Os, which is again not enough to call this solution I/O-efficient. Another drawback of this solution resides in its *static nature*: adding to, or removing from, *SA* some suffixes would affect its contiguous storage thus requiring its complete reorganization.

THEOREM 35.1 *A suffix array partitioned among disk pages of size B supports the search for the occ occurrences of a pattern $P[1,p]$ within a text $T[1,n]$ in $O((p/B) \log n + occ/B)$ I/Os. In the case that an additional `lcp` information is provided, the search cost becomes $O((p/B) + \log n + occ/B)$ I/Os. The space occupancy is of $O(n/B)$ disk pages.*

Our second proposal for the disk mapping of *SA* exploits the *not* negligible size M of the internal memory of modern computers. In the case that T fits in internal memory, the random accesses to the text suffixes induced by the binary search over *SA* would be not much of a problem, thus turning the search cost to $O(\log(n/B))$ I/Os. But in the more general situation in which neither *SA* nor T fit into internal memory, only a portion of these arrays could be cached and the on-line nature of the problem does not allow to predict what portions of them would be better to prefetch. A solution to this dilemma is offered by the use of the so called *supra-index*.

A supra-index over *SA* is obtained through a *sampling process* of parameters s and ℓ: copy in internal memory the first ℓ characters of every other s suffix in *SA*. The result is a coarser version of *SA* that can be fit in internal memory given a proper selection of the parameters s and ℓ. The binary search now starts in the supra-index and exploits the available characters to identify, *without executing any I/Os*, a pair of suffixes that delimit the portion of *SA* containing the pattern occurrences. After that, the disk-based binary search is executed over that portion of *SA* thus requiring *some I/Os*. Their actual number depends on the effectiveness of the sampling process with respect to the queried pattern, and this is obviously unpredictable in advance. In practice [13], a 25% to 30% reduction in the search time has been observed. The expensive part of the search still remains the random access to the text suffixes, in this case it may be useful to deviate from standard binary search in order to achieve time reductions close to 60%, as reported in [144]. Although variations on the sampling process may be proposed — dealing with, for example, a non uniform sampling of the suffixes or a different number of characters copied from each suffix — the designed solutions would be yet static and their performance would strongly depend on the searched pattern and the indexed text structure thus resulting, in the worst case, the same as the approach without supra-index.

Nonetheless the sampling idea appears intriguing and worthy of further investigation. In fact, if we choose $s = \Theta(B)$ and $\ell = 0$ (i.e. we do not copy any suffix character), and *iterate* various times the sampling process, we obtain a *hierarchy of supra-indexes* that recalls in its structure a classic *B-tree* [38, 106]. The specialty of this solution is that the keys stored into the B-tree nodes are now the text suffixes ordered lexicographically and indexed via their *constant-sized* pointers. As a result, we have that: (1) every B-tree node can store $\Theta(B)$ suffix pointers, (2) the nodes of a B-tree level provide a paginated supra-index of the suffixes stored in the level below, and (3) the leaf level consists of the paginated *SA*. This data structure uses $O(\log_s n) = O(\log_B n)$ sampling levels and occupies still $O(n/B)$ disk pages, if we assume to fit each B-tree node into one disk page. The search procedure mimics the search into a B-tree. The only specialty is that each visited B-tree node consists of a sampled suffix array, and can be (indirectly) binary searched taking $O((p/B) \log B)$ I/Os. Given that the top-down traversal of the B-tree passes through $O(\log_B n)$ nodes, the overall search cost results $O((p/B) \log n + occ/B)$ I/Os. This approach therefore takes the same I/O-cost of the paginated *SA*, but nonetheless it offers two immediate advantages: it is *dynamic*, as the B-tree does, and its engineering may benefit from the plethora of algorithmic tricks and know-how available for classic B-trees.

An interesting variation on this theme, called the *Prefix B-tree*, has been devised in [17, 20].[3] Here the sampled suffix array present in each B-tree node, is enriched with some characters copied from the indexed suffixes (like the supra-index above). This way the binary search within a B-tree node may benefit from these characters to not access the disk (de-referencing is possibly avoided), and thus save some I/Os. But, on the other side, the choice of how many characters to copy from each indexed suffix is empirical and it induces a decrease in the B-tree fanout (thus augmenting the B-tree height). Various authors, see e.g. [180, 51], have engineered this approach achieving interesting performance in practical situations. And in fact the Prefix B-tree is a remarkable choice in the case of substring (resp. prefix) searches over short patterns (resp. variable-length keys).

35.3.2 Construction

Since the Suffix Array is a sorted sequence of items, namely text suffixes, it is more natural to address its construction by resorting to classical comparison-based sorting algorithms (such as the function QSORT in C, see below), and by specializing the comparison function by means of string comparisons (SUFFIX_CMP below). This is exactly what has been done in [18] for teaching purposes.

COMPARISON_BASED_CONSTRUCTION(char *T, int n, char **SA)

   { for($i = 0$; $i < n$; i ++) $SA[i] = T + i$;

      QSORT(SA, n, sizeof(char *), Suffix_cmp); }

SUFFIX_CMP(char **p, char **q){ return strcmp(*p,*q); }

FIGURE 35.2: An elegant C-coded algorithm to build the Suffix Array.

Although elegant, this algorithm presents various drawbacks which become more apparent in the case of disk-based computations. In fact, QSORT is effective in the presence of virtual memory systems and *atomic* items. But here we are concerned with text suffixes and thus strings of arbitrary lengths, which are represented within *SA* via *indirect pointers*. As a result, every comparison executed by QSORT induces two *random* accesses to the text T and thus probably two *random* I/Os, with a possible subsequent scanning of contiguous disk pages for comparing the corresponding suffix characters. Since the random I/Os are

---

[3]We point out that the original Prefix B-tree has been designed to index *variable-length keys*. The use we make in this chapter is therefore a little bit unusual and borrowed from a solution provided in [59]. We propose this view of the Prefix B-tree because we think that it is an interesting algorithmic solution to the dynamic version of the substring-search problem, and moreover because it is an intermediate step useful to introduce the *String B-tree* data structure in Section 35.5.

costly [155] and, in the best case, we execute $\Theta(n \log n)$ comparisons, each taking $\Theta(n/B)$ sequential I/Os, the overall time cost is dominated by the seek time of the disk with an almost idle CPU (the so called *thrashing*). Actually this is the worst situation one can hope for when designing an external-memory algorithm or, in general, an algorithm working in a hierarchical memory.

THEOREM 35.2   *The* `qsort`*-based algorithm constructs the suffix array of a string* $T[1,n]$ *taking* $O(\lceil \frac{n}{B} \rceil n \log n)$ *I/Os in the worst case. The space usage is* $O(n \log n)$ *bits.*

The previous solution allowed us to point out the two main difficulties arising when we try to build large suffix arrays: (1) we cannot implement suffix comparisons via the direct brute-force scan of their constituting characters, some auxiliary data structures must be built to exploit previously executed comparisons; (2) comparison-based algorithms extended to manage string items are effective when the text length is small but show their limits as soon as the text length grows to medium sizes (i.e. some Megabytes), here radix-based approaches and disk-aware solutions are necessary to not result in many random I/Os.

The last five years have seen a revitalized interest in the design of efficient solutions for suffix array construction. This is due to some interesting uses of this data structure as a basic block of four novel applications: (i) the Burrows-Wheeler compression algorithm [25], which is a practically effective compression tool [165]; (ii) the construction of succinct [77, 157, 158] or compressed [68, 64, 66, 65, 69, 75, 76, 142] indexes, which have among the others interesting biological applications [84, 83, 160]; (iii) the clustering of documents and pages in web applications and search engines [185, 184], and (iv) the extraction of significant patterns in data mining applications [12, 81]. Among the plethora of scientific results devised to efficiently solve the suffix-array construction task, it is possible to identify a taxonomy composed of four main classes:

Tree based. To this class belong algorithms which derive the suffix array from the suffix tree data structure. They deploy the simple fact that the in-order visit of the leaves of the suffix tree actually gives the suffix array pointers. Internal-memory [127, 176] and disk-conscious [55] solutions are known for suffix-tree construction (as well for tree visits), but all of them use an excessive amount of space which goes far beyond the $15n$ bytes for storing and manipulating the tree topology [110].

String comparison. To this class belong comparison-based sorting algorithms specialized to deal with items of variable length. The algorithm of Figure 35.2 belongs to this class, as well the more effective Multikey Quicksort or Ternary Quicksort algorithms [19]. These algorithms sort the suffixes via a direct character-by-character comparison so that they are efficient for strings in which their *longest repeated substring* is short. In the worst case the number of character comparisons may be $\Theta(n^2 \log n)$. A recent development in this class has been achieved by [93, 124, 166] where algorithms exploiting some sophisticated auxiliary data structures have been proposed to skip some suffix comparisons during the ordering process. Indeed [124] presents a deep experimental analysis with respect to the dichotomy cache/internal-memory, considering various CPU architectures. At present the *lightweight* algorithm devised in [124] is the fastest choice if the text fits in internal memory and the space issue is a primary concern.

Incremental construction. The algorithms of this class are mainly based on the *doubling technique* introduced in [102]. They proceed per phases. At the beginning they (radix-)sort the suffixes by their first character, or their first two characters. In each further pass, they double the significant prefix length according to which the suffixes are sorted. This way $O(\log n)$ phases suffice to obtain the sorted sequence of text suffixes. The first algorithm in this class was proposed in the seminal suffix-array paper [121]; subsequently, more engineered and tricky variants have been devised in the scientific literature [116]. In [42] a taxonomy of a large set of doubling-based algorithms has been provided, thus electing [73] as the best choice whenever the text length goes beyond the internal-memory size.

Divide-and-Conquer. In this class fall most of the new proposals appeared in the scientific literature during the last four years. After that the Divide-and-Conquer approach proved useful in [55] for optimally building suffix trees on hierarchical memories (see Section 35.4.2), various researchers addressed the problem of adapting this scheme to the direct construction of suffix arrays. This effort succeeded in the year 2003, when various papers [24, 86, 85, 107, 100, 105] simultaneously devised different solutions based on a Divide-and-Conquer approach. Specifically, [107, 105] proposed linear-time suffix-array construction algorithms for internal memory using $O(n \log n)$ bits of working space. [100] addressed this issue in various models of computations, among which the external-memory model, and found a unifying solution for all of them (actually [24] showed that this approach may achieve $O(n/ \log n)$ bits of additional working space). Finally [85] showed how to achieve both time and space optimality on the unit-cost RAM. We point out that the incremental construction of [73] might be also looked at as a Divide-and-Conquer algorithm in which the partitions are mostly unbalanced. Under this view the Divide-and-Conquer approach results currently the most promising not only from the theoretical point of view, but also in practice; nonetheless, it is premature to be experimentally judged and therefore needs further algorithmic engineering to prove its practical helpfulness.

In the light of these significant progresses in the design of suffix-array construction algorithms, it is currently believed that it is more economical to build the suffix array first, and then derive the suffix tree or other indexing data structures from it (rather than the converse). However, in order to exploit this approach, one needs to compute the additional *longest common prefix* information which is usually stored in an n-length array. Recent papers [24, 103, 100, 122] showed that this computation does not induce any time- or I/O-slowdown to the suffix-array construction process, so that this approach may be much more efficient than managing directly a (suffix) tree topology [110, 158].

Given the importance assumed by suffix-array construction, we decided to describe in this section two algorithms drawn from the Divide-and-Conquer class. The first choice fell onto the *Skew algorithm* of [100]: it is deceptively elegant, implementable with just 50 lines of C++ code, flexible enough to achieve the best theoretical bounds in various models of computations, and one among the algorithms proposed in the year 2003 that solves the long standing open problem of linear-time suffix-array construction in the RAM model. Our second choice fell onto the *BYS algorithm* of [73]: it is the best algorithm for constructing in practice suffix arrays residing on disk, it is extremely simple and admits an efficient theoretical variant [42].

Let us therefore start with the description of the *Skew algorithm*, whose pseudo-code is given in Figure 35.3. The Skew algorithm builds upon the Divide-and-Conquer approach originally proposed in [55] for the construction of suffix trees, here adapted in a novel way to construct suffix arrays. The basic idea is to decompose the construction of a suffix array in three main phases. The first phase is devoted to the (recursive) construction of the suffix

SKEW_ALGORITHM

Phase one:

(1.1)   Construct the string set $\mathcal{S} = \{s_i = T[i, i+2]$ such that $i \bmod 3 \neq 1 \}$.
(1.2)   Radix-sort $\mathcal{S}$ and assign *lexicographic names* s_i' to the strings s_i.
(1.3)   Construct the string $s^{2,0} = [s_i' : i \bmod 3 = 2] \bullet [s_i' : i \bmod 3 = 0]$.
(1.4)   If the s_i' are all distinct, then `return`.
(1.5)   Recursively construct the suffix array $SA^{2,0}$ of the string $s^{2,0}$.

Phase two:

(2.1)   Construct the suffix array SA^1 by stably sorting the entries of $SA^{2,0}$
        that represent the suffixes T_{i+1} with $i \bmod 3 = 1$.

Phase three:

        Merge $SA^{2,0}$ with SA^1 by comparing $T_i \in SA^{2,0}$ with $T_k \in SA^1$ as follows:
(3.1)   If $i \bmod 3 = 2$, compare $\langle T[i], T_{i+1} \rangle$ with $\langle T[k], T_{k+1} \rangle$.
(3.2)   If $i \bmod 3 = 0$, compare $\langle T[i], T[i+1], T_{i+2} \rangle$ with $\langle T[k], T[k+1], T_{k+2} \rangle$.

FIGURE 35.3: The elegant *Skew algorithm* that builds the suffix array of the string $T[1, n]$ within the time- and I/O-complexity of radix sort.

array for the suffixes which start at text positions having the form $3j + 2$ or $3j + 3$, for $j \geq 0$. This suffix array is called $SA^{2,0}$ because the distance from multiple-of-3 positions is 2 or 0 (see Steps (1.1)–(1.5)). The second phase exploits $SA^{2,0}$ to derive the suffix array for the suffixes starting at text positions having the form $3j + 1$, for $j \geq 0$. This other suffix array is called SA^1 (see Step (2.1)). Finally, in the third phase, $SA^{2,0}$ is merged with SA^1 via a linear pass. This pass exploits the carefully designed distribution of the suffixes among the two suffix arrays (see Steps (3.1)–(3.2)). The key difference with the approach in [55] is to recurse on the *two thirds* of the text suffixes instead that on the half of them (see Section 35.4.2). This allows the Skew algorithm to compare in *constant time* the pairs of suffixes involved in the merging process of Steps (3.1)–(3.2), without resorting to the use of tries as occurred in [55]. As a result, the 2/3-recursion changes the constants hidden in the big-O notation but not the overall complexity which is still proportional to the cost of *(radix) sorting* a set of n integers in the range $[1, n]$. Hence it is optimal in various models of computation.

Some more fine comments on the Skew algorithm are in order to better appreciate the elegance of the approach. We invite the reader to follow these comments on the illustrative example of Figure 35.4. First of all, in order to simplify the discussion, we assume that the text string T has multiple-of-three length and number the character positions from 1. If this is not the case, we logically pad T with the special character \$ which is smaller than any other alphabet character. The first phase of the algorithm is the most time consuming. In order to build *recursively* the suffix array $SA^{2,0}$ for the suffixes T_i starting at the text positions $i = 3j + 2$ or $i = 3j + 3$, with $j \geq 0$, the algorithm initially assigns *lexicographic names* s_i' to all substrings of length 3 that start at those positions (in Figure 35.4 these substrings are indicated, above and below, by segments). These names are exploited to construct an auxiliary (shrinked) string $s^{2,0}$, whose length is $\frac{2n}{3}$. $s^{2,0}$ is obtained by justaxposing the string of lexicographic names $[s_i' : i \bmod 3 = 2, \ 1 \leq i \leq n]$ with the string $[s_i' : i \bmod 3 =$

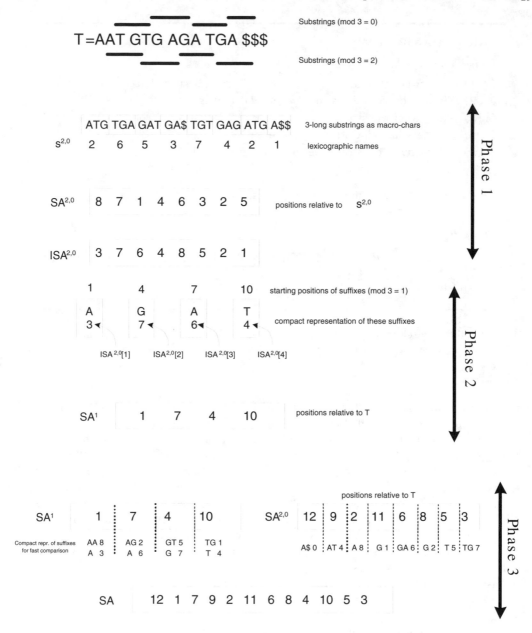

FIGURE 35.4: A running example of the *Skew algorithm* on $T = $ AATGTGAGATGA.

$0,\ 1 \leq i \leq n$]. Notice that the $s entries ensure that the suffixes starting in the first half of $s^{2,0}$ are not influenced by the structure of the second part of that string. The key idea underlying the construction of $s^{2,0}$ is that its suffixes are in a bijective correspondence with the text suffixes we are interested in (i.e. the ones starting at $i = 3j + 2$ or $i = 3j + 3$). Actually, $s^{2,0}[j+1, n/3]$ represents T_{3j+2}; whereas $s^{2,0}[\frac{3j+3+n}{3}, \frac{2n}{3}]$ represents T_{3j+3}. Given this correspondence, the suffix array $SA^{2,0}$ of $s^{2,0}$ actually provides the sorted sequence of text suffixes starting at positions $i \geq 1$ such that $i \bmod 3$ is equal to either 2 or 0. We point out that if the s_i' are all distinct, then we already have the sorted suffixes for the current string $s^{2,0}$, and thus the recursion may be stopped (Step (1.4)).

The correctness of the second phase is simpler to be established. Just observe that we may write $T_{3j+1} = T[3j+1] \bullet T_{3j+2}$. Thus, the ordering of the suffixes starting at positions i such that $i \bmod 3 = 1$ can be derived from the stable sorting of the entries of $SA^{2,0}$ that represent the suffixes T_{3j+2}s via their preceding character $T[3j+1]$. This is the suffix array SA^1 in Figure 35.4.

The Skew algorithm is so simple because the third (merge) phase is also easy. Let $T_i \in SA^{2,0}$ and $T_k \in SA^1$ be two suffixes which are compared during the merging of those suffix arrays. Either $i \bmod 3 = 2$, and thus $i = 3j+2$ for some j; or it is $i \bmod 3 = 0$, and thus $i = 3j+3$ for some j. In the former case, we can write $T_i = T[i] \bullet T_{i+1}$ and $T_k = T[k] \bullet T_{k+1}$, with both suffixes T_{i+1}, T_{k+1} occurring in $SA^{2,0}$, so their order is known. In the latter case, we can write $T_i = T[i] \bullet T[i+1] \bullet T_{i+2}$ and $T_k = T[k] \bullet T[k+1] \bullet T_{k+2}$, with both suffixes T_{i+2}, T_{k+2} occurring in $SA^{2,0}$, so their order is known. As a result, only characters of T and suffixes in $SA^{2,0}$ have to be compared. To this purpose we may compute in linear time the *inverse* array $ISA^{2,0}[j] = i$ iff $SA^{2,0}[i] = j$. Given ISA, each suffix comparison takes constant time.

Overall the execution time of the Skew algorithm is $T(n) = O(n)$ since it obeys the recurrence $T(n) = T(2n/3) + O(n)$. We point out that any step in the Skew algorithm can be simulated via a *sorting* step applied on n atomic items in the range $[1, n]$. As a corollary, the Skew algorithm can be seen as a *reduction* of the suffix-array construction problem to the n-items sorting problem. The latter has been solved optimally in several models of computation, as for example, the external-memory model. Here sorting tuples, lexicographically naming triples of characters and constructing ISA, takes overall $Sort(n) = O(\frac{n}{B} \log_{M/B} \frac{n}{B})$ I/Os by means of one of the many well-known I/O-optimal sorting algorithms [147, 148, 89].

THEOREM 35.3   *The Skew algorithm builds the suffix array of a string $T[1, n]$ in $O(Sort(n))$ I/Os and $O(n/B)$ disk pages. In the case that the alphabet Σ has size polynomial in n, the CPU time is $O(n \log_{M/B} \frac{n}{B})$.*

Except for some very preliminary results [24], an extensive experimental analysis of the Skew algorithm is yet missing. The current winner among the external-memory suffix-array construction algorithms on real-world problems is the one proposed in [73]. It is again a Divide-and-Conquer algorithm with the specialty that the *Divide* step is unbalanced, thus inducing a cubic time complexity (i.e., quadratic number of suffix comparisons). Below we first comment its algorithmic structure, and then discuss its practical efficiency.

Let $\ell < 1$ be a positive constant properly fixed to build the suffix array of a text piece of $m = \ell M$ characters in internal memory. The algorithm computes *incrementally* the suffix array SA in $\Theta(n/M)$ stages (rather than the logarithmic number of stages of the Skew algorithm). At the beginning of stage h, the algorithm maintains the following invariant: *on disk it is stored the array SA^{hm} that contains the sorted sequence of the first hm suffixes of T*. The inner working of the generic hth stage is detailed in Figure 35.5. Actually this stage updates SA^{hm} by inserting into it the text suffixes which start in the substring $T[hm+1, (h+1)m]$. This preserves the invariant above, and so ensures that after all the stages we obtain SA.

Some comments are in order at this point. It is clear that the algorithm proceeds by mainly executing two sequential disk scans: one is performed in Step (1), the other is needed in Step (3) to compute C and merge SA' with SA^{hm}. In particular, Step (3.1) is implemented by scanning rightward the text T (from its beginning) and by computing via a binary search the lexicographic position p_i of each text suffix $T[i, n]$ in SA', with $i \leq hm$.

INCREMENTAL_ALGORITHM

Stage h:

(1) Load the text substring $t = T[hm + 1, (h + 1)m]$ into internal memory.
(2) Build SA' by sorting lexicographically the text suffixes which start in t.
(3) Merge SA' with SA^{hm} as follows:
  (3.1) Compute the counter array $C[1, m + 1]$ storing in $C[j]$ the
          number of suffixes of SA^{hm} that are lexicographically greater than
          the $SA'[j - 1]$-th text suffix and smaller than the $SA'[j]$-th text suffix.
  (3.2) Merge SA' with SA^{hm} via a disk scan, by exploiting the array C.

FIGURE 35.5: The *Incremental algorithm* is effective in practice because it mainly executes sequential disk scans and occupies reduced disk space.

The entry $C[p_i]$ is then incremented to record the fact that $T[i, n]$ lexicographically follows the $SA'[p_i - 1]$-th text suffix, and precedes the $SA'[p_i]$-th text suffix. Array C is then deployed in Step (3.2) to merge the two arrays SA' and SA^{hm}: $C[j]$ indicates how many consecutive suffixes of SA^{hm} lexicographically follow the $SA'[j-1]$-th text suffix and precede the $SA'[j]$-th text suffix. Hence a sequential disk scan, as the one executed in Step (3.2), is enough to build $SA^{(h+1)m}$.

It goes without saying that the algorithm might incur many random I/Os during Steps (2) and (3.1). In both cases we may need to compare a pair of text suffixes which share a long prefix not entirely available in internal memory (i.e., it extends beyond $T[hm+1, (h+1)m]$). In the pathological case that $T = a^n$, the comparison between two text suffixes takes $\Theta(n/B)$ I/Os and thus $O((n^3 \log_2 M)/MB)$ I/Os overall. The total auxiliary disk space used by the algorithm is $4n$ bytes to store SA^{hm} and $8m$ bytes for both C and SA_{int}. The merging step can be easily implemented using some extra space (indeed additional $4n$ bytes are sufficient), or by employing just the space allocated for SA' and SA^{hm} via a more tricky implementation.

Since the worst-case number of total I/Os is cubic, a purely theoretical analysis would classify this algorithm as not much interesting. But there are some considerations that are crucial to look at this algorithm from a different perspective. First of all, we must observe that in practical situations it is very reasonable to assume that each suffix comparison finds in internal memory all the characters needed to compare the two involved suffixes. Consequently, the practical behavior is more reasonably described by the formula: $O(n^2/MB)$ I/Os. Additionally, all I/Os in the analysis above are sequential and the actual number of random seeks is only $O(n/M)$ (i.e., at most a constant number per stage). Consequently, the algorithm takes fully advantage of the large bandwidth of current disks and of the high speed of the current CPUs [155]. Moreover, the reduced working space facilitates the prefetching and caching policies of the underlying operating system and finally, a careful look to the algebraic calculations shows that the constants hidden in the big-O notation are small. [42] has also shown how to make this algorithm no longer questionable from a theoretical viewpoint by proposing a modification that achieves efficient performance in the worst case.

35.3.3 Future directions of research

A frequent statement in text indexing papers and talks is that: *Word-based indexes[4] occupy less space than full-text indexes but they are limited to efficiently support poorer (i.e. word based) search operations.* Such statements have driven many authors to conclude that the increased query power of suffix arrays *has to be paid* by some additional storage. It is nonetheless challenging, from a scientific point of view, to ask ourselves if it is provable that such a tradeoff *does exist* when designing an index. In this context data compression appears as an attractive tool because it allows not only to squeeze the space occupancy but also to improve the computing speed. It is therefore not surprising that IBM has recently installed on the eServers x330 a novel memory chip (based on the Memory eXpansion Technology [90]) that stores data in a compressed form thus ensuring a performance similar to the one achieved by a server with double real memory but, of course, lower cost. All these considerations have raised a renewed interest towards compression techniques within the algorithmic and IR communities.

Compression may of course operate at the text level, or at the index level, or both. The simplest approach consists of compressing the text via a lexicographic-preserving code [87] and then build a suffix array upon it [131]. The improvement in space occupancy is however negligible since the index is much larger than the text. A most promising and sophisticated direction was initiated in [133, 135, 132], and lead [77] to show that a suffix-array implementation does exist that occupies $\Theta(n)$ bits of storage and supports the retrieval of $SA[i]$ in $O(\log^\epsilon n)$ time, where ϵ is an arbitrarily small positive constant. This result has shown that the apparently "random" lexicographic permutation of the text suffixes in suffix arrays can be succinctly coded in optimal space in the worst case [50]. In [157, 158, 159] extensions and variations of this result — e.g. an arbitrary large alphabet or new functionalities — have been considered.

The above index, however, uses space linear in the size of the indexed text even in the case that the text is highly compressible. The first step toward the design of a *compressed full-text index* has been pursued in [64, 68]. The novelty of this approach resides in the careful combination of the Burrows-Wheeler compression algorithm [25] with the suffix array data structure thus obtaining a sort of *compressed suffix array*. More precisely, the index of [64, 68] occupies $5n\,H_k(T) + o(n)$ bits of storage, where $H_k(T)$ is the kth order empirical entropy of the indexed text T. Such index supports the search for an arbitrary pattern $P[1,p]$ as a substring of T in $O(p + occ \log^\epsilon n)$ time.

This line of research seems very promising for computational biology applications, as shown in [84, 83, 160], where compressed suffix arrays have been used to approximately search the human genome in the internal memory of a personal workstation. In this respect, it would be interesting to improve the constants hidden in the big-O notation of the compressed indexes, since they impact their engineering in the case of arbitrarily large alphabets [75, 76, 74, 69, 74, 120, 142]. Experimental results [65, 76, 142] have shown that these compressed indexes are much promising but more extensive tests are needed to investigate their impact especially on biologically motivated applications. Additionally, it would be worth to investigate the *lightweight* construction of the suffix array, because it is true that $5n + o(n)$ bytes [24] seem a small amount of working memory, but this

---

[4]These indexes are also called Inverted Indexes and are widely used in the design of web search engines [182]. They actually consists of a dictionary of words, and for each word, a list of document IDs and positions where that word occurs. The query efficiently supported are restricted to operate on *word boundaries*, which is the typical scenario of web search.

amount may still be too much if the space issue is a primary concern. The recent results in [86, 85, 114] have shown that $\Theta(n)$-bit of working space are enough to optimally build a suffix array, but what about a space complexity that depends on the entropy of the string T (à la [68, 64, 69, 75, 76])? This result would be interesting also in the data compression setting because some effective compressors, like `bzip` [165], exploit a suffix-array construction algorithm to work. Such a type of results could make these algorithms much more space efficient (cfr. [67, 57, 63]).

Finally, various authors have started to freely distribute engineered, and sometimes well documented, versions of their suffix-array construction algorithms [24, 124, 116, 93, 166]. For example in [123] you can download the fastest known internal-memory algorithms to date, whereas for external memory you have to look at [42]. A bigger effort is clearly needed by the theory and the software community to engineer those solutions and thus offer efficient, publicly available, libraries to be used in suffix-array based software systems. This effort is much more precious nowadays than before, in the light of the fascinating results appeared recently in the literature and fully commented above. As it has happened many times in the past, these engineering papers might raise challenging questions for the theory community, thus "closing the cycle" of an effective technology from a stimulating research!

35.4   The Suffix Tree

The suffix tree is the basic and ubiquitous data structure of combinatorial pattern matching because of its elegant and efficient uses in a "myriad" of situations [10, 79]. A suffix tree of a string $T[1, n]$, denoted hereafter by ST_T, is a compressed trie that stores all text suffixes $\mathsf{SUF}(T)$ in a compact form. A compressed trie is a rooted directed tree with exactly n leaves numbered from 1 to n. Each internal node, other than the root, has at least two children and each edge is labeled with a non empty substring of T. No two edges out of a node can begin with the same character, and sibling edges are ordered lexicographically according to that character (see Figure 35.6). The key feature of suffix trees is that, for any leaf i, the concatenation of the edge labels from the root of ST_T to i spells out the text suffix T_i. Hence all the substrings of T, which are $\Theta(n^2)$, are represented in $O(n)$ optimal space by ST_T's structure. Furthermore, the rightward scan of the suffix tree leaves gives the suffix array of T. These two properties allow to look at suffix trees as an index that fully exposes the internal structure of a string in an easier way than suffix arrays do. [5]

An attentive reader may have noticed that the bijection between text suffixes and suffix tree leaves is possible only if it does not exist any text suffix that is the prefix of another one. This is ensured, in case, by appending the endmarker # to the end of T. Figure 35.6 shows an example of a suffix tree.

If we store the edge labels explicitly we incur in a $\Theta(n^2)$ space overhead which makes the data structure unusable even for strings of moderate length. Implementations then usually opt for a compact representation that adds a level of indirection to the retrieval of the edge labels. Indeed, an edge label $T[x, y]$ is represented by the pair $\langle x, y \rangle$ that occupies a

---

[5] Each subtree of the suffix tree corresponds to a subarray of the suffix array, namely the one containing the suffixes descending from that subtree. Any suffix tree traversal operation can then be simulated via a proper binary search of the suffix array, thus introducing a slowdown of $O(\log n)$ time [73]. Recent results [64, 68] have shown that that slowdown can be avoided in some special cases, like the substring search operation.

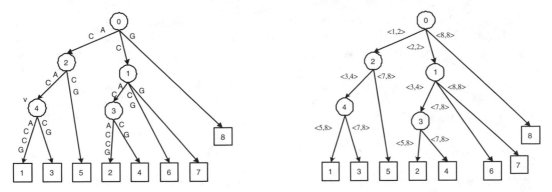

FIGURE 35.6: The suffix tree for the string $T = \texttt{ACACACCG}$ is showed to the left. The endmarker # is not shown. Node v spells out the string $\texttt{ACAC}$. Each internal node stores the length of its associated string, edge labels are explicitly indicated, and each leaf stores the starting position of its corresponding suffix. To the right of the figure, we depict the suffix tree with the edge labels *encoded* by integer pairs.

constant number of memory cells, each of $\Theta(\log_2 n)$ bits (Figure 35.6). In the case of many texts to be simultaneously indexed [80], an edge label is represented by the triple $\langle T, x, y \rangle$. As a result, the overall space occupancy of the suffix tree is $O(n)$, or $O(n \log_2 n)$ bits.

This simple coding trick is not yet enough to turn the suffix tree into an efficient data structure for indexing massive data sets. In fact, a careless pointer-based implementation requires more than $20n$ bytes [121], and more sophisticated solutions [110, 71] achieve an occupancy of at least $12n$ bytes in the worst case, and about $8.5n$ bytes in the average case. For example, a suffix tree built upon 700Mb of DNA sequences may take 40Gb of space [110]. Worse than this, real implementations based on persistent platforms (like PJama [88, 163]) add to these figures a significant space overhead by requiring more than 60 bytes per indexed suffix. As repeatedly observed before, space optimization is closely related to time optimization in a disk-memory system [106], thus the design of *succinct* suffix-tree implementations is a key issue in the indexing of massive textual collections. This topic is an active area of research, full of fascinating solutions which allow to squeeze the suffix tree in a space occupancy much close to the one required by the suffix array (see e.g. [77, 76, 134, 136, 132, 159, 63]). These results, however, are too involved and mainly designed for internal memory use to deserve some room in this introductory chapter. In the following subsections we will detail one specific technique [33] that achieves the best compacting and I/O-results to date.

The search for a pattern string $P[1, p]$ in ST_T consists of a downward traversal starting from the root and proceeding downward as pattern characters are matched against edge labels. Only one path is followed because at each visited node the first characters of the outgoing edges are distinct. If a mismatch character is found during this traversal, the pattern P does not occur in T (case $P = \texttt{CT}$ in Figure 35.6). In the case that the pattern is fully matched (case $P = \texttt{ACA}$ in Figure 35.6), the leaves descending from the final matched edge provide the list of all pattern occurrences in T (leaves descending from v in Figure 35.6). Therefore the cost of pattern searching is $O(p \log |\Sigma| + occ)$ time in the worst case, where Σ is the alphabet and occ is the number of pattern occurrences in T. The factor $O(\log |\Sigma|)$ takes into account the cost of choosing, at each visited node of the downward traversal, the next edge label to match against the pattern.

So far so easy is the design and use of the suffix tree data structure, and actually this information is enough to engineer a good in-memory implementation for it. In fact, in the RAM model every memory access takes $O(1)$ time and retrieves just one item; hence, there is nothing to be "packed". Conversely, various issues deserve attention when dealing with the problem of mapping a suffix tree onto disk (or onto a hierarchical memory, in general). Here, the blocked access to disk pages poses new difficulties that, if underestimated, may even prevent the use of suffix trees on moderately sized data sets. A careful analysis of these issues is presented in the next section.

35.4.1   I/O issues

The first issue to take care of is the *unbalanced tree topology* of suffix trees. It is text dependent because the internal nodes are in correspondence with the repeated substrings of T. Consequently, suffix trees inherit the difficulties pointed out in the literature with regard to paging unbalanced trees on disk. For example [3] showed that paging heuristics based on a Depth-First visit or a Breadth-First visit of the suffix tree can be far from optimality of a $\Omega(\log B/\log\log B)$ factor. Conscious of this difficulty, various authors [37, 101, 136] circumvented the problem above by exploiting a *two-level indexing structure*: one level consists of a suffix tree built on a *sampled subset* of the text suffixes stored in internal memory; the other level is just a suffix array built over all the text suffixes. The sampled suffix tree is used to route the search on a small portion of the suffix array, by exploiting the efficient random-access time of internal memory; an external binary search is subsequently performed on a restricted part of the suffix array thus requiring a reduced number of I/Os [13]. Obviously the choice of the proper suffixes to sample is difficult, and overall the resulting data structures loosely remind the suffix tree and many of its good structural properties (cfr. the supra-index of Section 35.3.1).

The second issue to cope with is related to the *indirect encoding* of the edge labels by integer pairs (or triples, see Figure 35.6). Branching from a node to one of its children requires indeed further I/Os in a disk memory because of the retrieval of the disk pages containing the substring that labels the traversed edge and must be matched against the pattern to be searched. This issue turns the disk mapping of the suffix tree into *not just* a disk-mapping problem of an unbalanced tree.

The third issue is related to a *good engineering and implementation* of the data structure. The encoding of the tree topology and of the edge labels must be as more succinct as possible in order to squeeze them into few disk pages, and thus exploit at the best the caching and prefetching policies of the underlying operating system. Any engineered solution should exploit the interplay between these two types of information — topological versus textual — as well the fact that suffix-tree accesses are constrained to occur via root-to-leaf paths. Another key point to take care in the suffix-tree engineering is the disk-page occupancy. In fact it turns out to be rather complicated to keep it above a given ratio under string insertions or deletions [6, 80, 72]. Moreover, the $\Theta(|\Sigma|)$ fanout of the suffix-tree nodes may prevent the children pointers to be fitted into $O(1)$ disk pages, so requiring a separate B-tree to store them. This may appear not so crucial for biological sequences, being $|\Sigma| \leq 20$, but a character-packing approach for suffix-tree compaction might virtually enlarge Σ thus making this problem evident even in the bio-context. A final observation is that a clever implementation should also take into account the CPU cost of the supported operations because it impacts unfavorably on the (logical) size of the disk pages fetched at each I/O. The lighter is a page processing, the larger a page can be, thus reducing the wasting of disk space because of a possible poor page-fill ratio. This fact constraints the suffix-tree encoding to exploit lighter compression techniques (cfr. [33, 58]). All these specialties and difficulties make not surprising the fact that the suffix-tree engineering is an almost unexplored field [169].

In the rest of this section we deal with all of these three I/O-issues by discussing state-of-the-art solutions for the uneasy problem of suffix-tree mapping onto disk.

Let us start by dealing with the problem of mapping to disk the suffix tree topology, and introduce the so called *tree packing* problem. Here the tree has a fixed topology and this is unbalanced. We point out that the advantage of finding a good tree packing may be unexpectedly large and must be therefore not underestimated. In fact, while balanced trees save a factor $O(\log B)$ when mapped to disk (think at the B-trees), the mapping of unbalanced trees grows with non uniformity and approaches, in the extreme case of a linear-height tree, a saving factor of $\Theta(B)$ over a naïve memory layout. Formally, in the tree packing problem the goal is to find an allocation of the (suffix) tree nodes on the disk pages that *minimizes* either the *total number* of pages loaded in internal memory (i.e. *page faults*) during a pattern search, or the *number of distinct* pages visited (i.e. *working-set size*) during a pattern search. These two parameters model two extreme situations: the case of a one-page internal memory, or the case of an unbounded internal memory. In other words, they model the two cases in which we have either a small buffer or an unbounded buffer to support the pattern searches via root-to-leaf paths. Surprisingly [72], the optimal solution to the tree packing problem is *independent* of the available buffer size because no disk page is visited twice when page faults are minimized or the working set is minimum. Moreover, the optimal solution shows a nice *decomposability property*: the optimal tree packing forms in turn a tree of disk pages. These facts allow to restrict our attention to the page-fault minimization problem, and to the design of recursive approaches to the optimal tree decomposition among the disk pages.

In the literature we find three nice solutions to the tree packing problem of increasing sophistication:

- The first solution [33] operates greedily and bottom-up onto the suffix tree topology by devising a tree packing that *minimizes the maximum number* of page faults executed during a pattern search.

- The second solution [72] assumes a known access distribution over the suffix-tree leaves (nodes), and finds the optimal tree packing via a *dynamic-programming approach*. This solution may be also extended to achieve space optimality, i.e. $\lceil \frac{n}{B} \rceil$ disk-page occupancy, at the cost of only one additional I/O per search.

- The third solution [3] is much sophisticated in that it assumes an access distribution over the suffix-tree leaves and adopts the *cache-oblivious model*. This solution has two nice features: it is within a constant factor of the query performance of the optimal known-block-size layout, although it does not know M and B; and it is computed by deploying one of the two algorithms above [33, 72] as a basic-algorithmic tool. The use of the cache-oblivious model then ensures that it is optimal on every memory hierarchy!

Now we go into the details of each solution. We start with the greedy algorithm of [33] that *minimizes the maximum number of I/Os* executed to visit any root-to-leaf path in the suffix tree (hereafter called the *Min-Max algorithm*). This algorithm proceeds bottom up over the suffix tree, starting with each leaf assigned to its own disk page whose height is set to 1 (see also [39, 129, 128]). Working upward, the algorithm applies the steps illustrated in Figure 35.7 thus producing a Min-Max optimal partitioning of the (binary) tree such that no other partitioning has a smaller root-page height. The proof of optimality parallels the partitioning rules. An attentive reader may notice that these rules may induce a *poor* page-fill ratio. This is true in the worst case, even if there are several changes that can

MIN-MAX ALGORITHM

General step on a (binary) node

(1) If both children have the same page height d:
    (1.1) If the number of nodes in both children's pages is less than B, then
        (1.1.1) Merge the two disk pages and add the current node.
        (1.1.2) Set the height of this new page to d. Exit
    (1.2) Else
        (1.2.1) Close off the pages of the children.
        (1.2.1) Create a new page for the current node and set its height to $d+1$. Exit
(2) Else
    (2.1) Close off the page of the child with the smaller height.
    (2.2) If possible, merge the page of the other child with the current node
        and leave its height unchanged.
    (2.3) Otherwise, create a new page for the current node with height $d+1$
        and close off the child page.

FIGURE 35.7: The *Min-Max algorithm* used to optimally partition an unbalanced tree among disk pages of size B. The tree is assumed binary, by binary encoding each character.

alleviate this problem in real situations:

1. When a page is closed off, scan its children from the smallest to the largest to determine if they can be merged with the parent.
2. Modify the rules to ensure a certain minimum page-fill ratio.
3. Design logical disk pages and pack many of them into one physical disk page; possibly ignore physical page boundaries when placing logical pages onto disk.

Change one should be a part of any implementation of these rules. Change two may result in a non-optimal partition, but should be worthwhile in a practical setting. The last change is interesting but introduces some complications in the management of the external-memory storage. In [33] it was proved the following result:

THEOREM 35.4 *The Min-Max algorithm of Figure 35.7 solves the suffix-tree packing problem in such a way that every root-to-leaf path traverses less than $1 + \lceil \frac{H}{\sqrt{B}} \rceil + \lceil 2 \log_B n \rceil$ disk pages, where H is the height of the suffix tree.*

It is known [174] that H is logarithmic in n with high probability, under very reasonable conditions on the indexed text T. In [33] it has been also experimentally shown that any root-to-leaf path of a suffix tree built over a 500Mb text collection, restricted only to suffixes starting at word boundaries, is mapped to less than 4 disk pages. This is much significant even if we have to still count the I/O-cost for accessing the edge-labels: indeed a suffix array would need $40 \div 50$ I/Os per pattern search. In that paper the problem of dynamically maintaining the tree packing under the insertion/deletion of strings was also addressed by showing that the Min-Max algorithm experiences unfortunately a *low page-fill ratio* (around $35 \div 45\%$).

Usually the distribution of the strings to be searched in ST_T is far from being uniform; in fact it is often skewed towards some root-to-leaf paths that are accessed more frequently than others. In this situation, it would be better to pack the tree nodes in such a way that the more frequently accessed nodes come close to the root page, whereas rarely accessed nodes lie in pages which are far from the root. As we observed above, the advantage of a good tree packing for unbalanced trees under skewed distributions might be close to a factor $\Theta(B)$, but difficult to be devised. In fact an obvious algorithm to solve this problem would be to incrementally grow a root page and repeatedly add the maximum probability node not already packed into that page. When the root page contains B nodes, it is written onto disk and the algorithm lays out the remainder of the tree recursively. Surprisingly enough, the obtained packing is far from optimality of a factor $\Omega(\frac{\log B}{\log \log B})$, but it is surely within a factor $O(\log B)$ from the optimal [3].

The first optimal algorithm for the case of skewed distributions was proposed in [72]. The authors devised a *Dynamic Programming* scheme that optimizes the *expected* number of I/Os incurred by any traversal of a root-to-leaf path. Figure 35.8 illustrates the dynamic-programming computation for the case of a binary tree. The general case of an f-ary tree can be solved by first transforming the tree into a binary one (this is standard!), and by then assigning probability and space-occupancy zero to the nodes added by this transformation. Let us now go into the algorithmic details, by using T to denote the binary tree to be packed, and τ to indicate the optimal tree packing we are searching for. Since initially all pages are isomorphic, we may assume that the root r of T is always mapped to a fixed page $\tau(r) = R$. Consider now the set V of tree nodes that descend from R's nodes but are not themselves in R. Formally, $V = \{v \in T \mid \tau(v) \neq R, \tau(\text{parent}(v)) = R\}$. We observed above that the optimal packing τ induces a tree of disk pages [72]. Consequently, if τ is an optimal packing for T, then τ is an optimal packing for the subtree T_v rooted at any node $v \in V$. This result allows to state a recursive computation for τ that first determines which nodes reside in R, and then continues recursively with all subtrees T_v for which $v \in V$. Dynamic programming provides an efficient implementation of this idea, based on the following definition: An *i-confined* packing of a tree T is a packing in which the page R contains exactly i nodes (clearly $i \leq B$). Now, in the optimal packing τ, the root page R will contain i^* nodes from the left subtree $T_{left(r)}$ and $(B - i^* - 1)$ nodes from the right subtree $T_{right(r)}$, for some i^*. The consequence is that τ is both an optimal i^*-confined packing for $T_{left(r)}$ and an optimal $(B - i^* - 1)$-confined packing for $T_{right(r)}$. This property is at the basis of the Dynamic-Programming rule stated in Figure 35.8 for a generic node v: $A[v, i]$ denotes the cost of an optimal i-confined packing of the subtree T_v. There, the value $w(T_v)$ denotes the probability to access the node v. Rule (1) accounts for the (unbalanced) case in which the i-confined packing is obtained by storing $i-1$ nodes from $T_{left(v)}$ into the v's page; Rule (2) is the symmetric of Rule (1); whereas Rule (3) accounts for the case in which j nodes from $T_{left(v)}$ and $i - j - 1$ nodes from $T_{right(v)}$ are stored into the page of v to form the optimal i-confined packing of T_v. The special case $i = 1$ is given by $A[v, 1] = w(T_v) + A[left(v), B] + A[right(v), B]$. It is easy to check that the algorithm runs in $O(nB^2)$ time, and uses $O(nB)$ space by means of a naïve implementation. Look at [72] for a more space efficient solution.

THEOREM 35.5 *An optimal packing for a f-ary tree of n nodes can be computed in $O(nB^2 \log f)$ time and $O(B \log n)$ space. The packing maps the tree into at most $2\lceil \frac{n}{B} \rceil$ disk pages.*

DYNAMIC-PROGRAMMING COMPUTATION

Compute $A[v, i]$ as $w(v)$ plus the minimum among the following three quantities:

(1) $A[left(v), i - 1] + w(T_{right(v)}) + A[right(v), B]$

(2) $w(T_{left(v)}) + A[left(v), B] + A[right(v), i - 1]$

(3) $\min_{1 \le j < i-1} \{ A[left(v), j] + A[right(v), i - j - 1] \}$

FIGURE 35.8: The *Dynamic-Programming step* used to compute the (page-fault and working-set) optimal packing of a binary tree. Given that $w(v)$ denotes the probability to access the node v, we define the probability to access the subtree T_v as $w(T_v) = \sum_{u \in T_v} w(u)$.

In that elegant paper it is also shown that optimizing both space and I/O performance is NP-complete. Nonetheless an approximation algorithm is additionally provided that uses the *minimum* number of pages to fit the tree nodes, i.e. $\lceil \frac{n}{B} \rceil$ disk pages, but slows down the root-to-leaf tree traversal by (only) *one additional I/O.*

It goes without saying that the packing algorithm above must know the block size B in order to compute the optimal tree packing. The value B is sometimes difficult to establish (think of a software library) and varies according to the disk features. This limitation has been recently overcome in [3] where a general technique, called *Split-and-Refine*, has been devised for converting a family of packing algorithms working with known block-size (like the ones we commented before) into a packing algorithm with *unknown block size*, currently called *cache-oblivious* algorithm. This transformation comes at the cost of a constant-factor increase in the number of (expected) I/Os needed for a tree traversal. But it tunes automatically to arbitrary memory hierarchies with arbitrarily many memory levels.

For the sake of simplicity we describe only the basic ideas underlying that packing algorithm and refer the reader to [3] for technical details. Since we are assuming that the features of the memory hierarchy are unknown, we have to talk about *memory layout* instead of disk mapping, and assume to have an unbounded array of memory cells (it is then the system that will map those cells onto disk pages or other external storage devices). The basic idea of the cache-oblivious layout is to recursively combine optimal layouts for several carefully chosen block sizes. Each layout is computed independently, and the block size is chosen so that the access cost to each of them grows exponentially. The layouts may be radically different; all we know is their order from the coarser (larger B) to the finer (smaller B). To define the layouts recursively we require that the blocks into which they are decomposed, form a *recursive structure*: a block at one level of detail should be made up of *subblocks* at the next finer level of detail. To achieve this, the Split-and-Refine algorithm *refines* a level of detail by *split*ting two nodes in two different subblocks if they occur into different blocks at any coarser level of detail. Given this, each block at any refined level of detail is stored in a contiguous segment of memory (i.e. sequence of disk pages). The subblocks of a block can be stored in any order as long as they are stored contiguously. An elegant, yet sophisticated, expected analysis [3] shows that

THEOREM 35.6   *The Split-and-Refine algorithm produces a cache-oblivious tree layout whose expected I/O-cost for a random root-to-leaf traversal is within a constant multiplicative factor of the optimal.*

Finding a good packing for the suffix tree topology is just the first key ingredient to efficiently use this data structure in a hierarchical memory. In fact, as we stressed before, another key ingredient is the labeling of suffix-tree edges which may consist of *arbitrarily long substrings*. Since the edge labeling is *indirect*, any tree traversal may incur in as many I/Os as there are edges onto the path traversed by a pattern search. This would clearly waste the effort we made above to find an efficient (or, optimal) tree packing. A simple, yet effective, solution to this problem is obtained by adopting the Patricia search method [130, 33] over a suffix tree, and by storing *explicitly the first character* of every edge-label together with its integer pair. Given this additional (and constant space) information, the access to the text can be delayed as long as possible by proceeding initially only over the suffix-tree structure. Namely, the pattern search proceeds blindly by matching just the first character of each traversed edge against the corresponding pattern character, and by assuming that all of the other skipped characters are *magically identical*. As soon as the tree traversal stops, we pick one leaf within the subtree descending from the lastly matched edge and verify that the corresponding suffix is prefixed by the searched pattern. If the match is successful, it is possible to prove that the pattern occurs in T and all the suffixes belonging to that subtree are pattern occurrences; otherwise, we can conclude that the pattern does not occur in T. We observe that, from one side, this approach can exploit any optimal tree packing to obtain an I/O-efficient pattern search on suffix trees, for example it achieves $O(\frac{p}{B} + \frac{H}{\sqrt{B}} + \log_B n)$ I/Os using Theorem 35.4. But, from the other side, this approach seems unusable for more sophisticated operations (like approximate searches) which need to indirectly access the whole edge labels. Here, more research is needed!

We are finally left with the problem of engineering the suffix-tree data structure because, as we repeatedly said in this chapter, space optimization is closely related to performance optimization in a disk memory system [106] and furthermore, we cannot neglect the fact that a space-consuming data structure may become unusable even for moderately sized data. Various papers in the literature dealt theoretically with this problem (see e.g. [134, 136, 50, 64, 68, 77, 75, 69, 76, 142]) or presented heuristics validated through experimentation (see e.g. [7, 9, 13, 46, 71, 110, 128]). In this context, one of the best results to date is the *Compact PAT-tree* [33]. The authors address all I/O-issues by showing that the Compact PAT-tree is a *unifying, elegant and practical* solution to the (static) suffix tree packing problem. The Compact PAT-tree achieves space occupancy close to that of suffix arrays (about $5n$ bytes of disk space) and efficient performance on exact searches (about 5 I/Os on hundreds of Megabytes). Due to space limitations we briefly mention below the main engineering features of this data structure, and refer the greedy reader to [33] and to two other crucial references [110, 71] for other details. We also point out that, unfortunately, there does not exist any publicly available implementation of the Compact PAT-tree, and that this data structure is mainly designed for exact searches, thus it waits for implementation, extensions or new proposals (cfr. [76, 69]).

Given the binary encoding of the alphabet Σ, the Compact PAT-tree is defined as the suffix tree ST_T built over the binary encoding of $T[1, n]$. Each internal node of this suffix tree is therefore binary and its two outgoing edges have a label starting with 0 (the left one) and 1 (the right one), respectively. Each internal node contains a number that denotes the offset of the bit used to distinguish the suffixes which descend from it. In practice, the offset information stored into each node is a *skip* value, one less than the difference between the offset value of the node and its parent. The actual offset is accumulated as the tree is traversed during a search operation. An exact search for a pattern $P[1, p]$ can proceed in the Compact PAT-tree as much like as we described above for the Patricia tree data structure [130]. For engineering reasons, the information stored in the Compact PAT-tree is broken into three categories: the tree structure, the skip values (stored in the internal nodes), and the suffix offsets (stored in the leaves). In what follows we concentrate on the succinct encoding of each class of information, the final result is reported in Theorem 35.7. In order to implement the search operations,[6] the encoding of the tree structure must provide the following functionalities:

- efficient selection of the left and right children of a node;
- support for the inclusion of the skip value in the internal nodes, and the suffix offset in the leaves. Given a node, we need to be able to efficiently determine these values;
- given a node, efficiently retrieve the suffix offset information from some leaf descendent from that node.

In each case we require that the operations be performed in constant time. Finally we want an encoding that is as compact as we can find. The survey papers of [104, 119] present many techniques for binary representations of binary trees that attain $2n$ bits of space occupancy, however none meets the criteria above. The Compact PAT-tree uses a slightly larger encoding developed by [95] that allows the direct implementation of tree traversals on the encoded form of the tree. Specifically, the tree topology is represented as a bit string being the juxtaposition of an *header*, the recursive encoding of the left subtree, and the recursive encoding of the right subtree. The header contains two subfields: a single bit indicating which among the left or right subtree is the smaller, and a prefix coded integer indicating the size of that smaller subtree.

Compressing the skip information in Compact PAT-trees requires an understanding of the distribution of the skip values. In practice, it has been verified [33, 128, 167] that the majority of the skip values are zero and that the probability of higher values decreases geometrically. This distribution leads to a simple encoding of these numbers: just reserve a small fixed number of bits to hold the skip value (usually 6 bits), and introduce an *escape strategy* to deal with the rare cases in which this space is not sufficient. In [136] an intriguing method that avoids even the storage of skip values is presented, but the operations allowed onto the suffix tree are limited.

The suffix offsets take up the bulk of the storage used by the Compact PAT-tree. We might adopt the sophisticated techniques devised in [68, 64, 77, 75, 69, 76], but in order to keep the discussion easier we present the approach proposed in [167]. If k low order bits in the suffix offsets are omitted, nk bits are saved in the final Compact PAT-tree. This change incurs, however, in a 2^k search time (and I/O) slowdown because each offset value needs a search through 2^k bits to locate the exact occurrence of the pattern.

In summary, under the hypothesis that the indexed text is generated by a binary memoryless source, [33] proves the following:

THEOREM 35.7 *The expected size of the Compact PAT-tree is less than $3.5 + \log n + \log\log n + O(\frac{\log\log\log n}{\log n})$ bits per node. The search for an arbitrary pattern $P[1,p]$ takes $O(\frac{p}{B} + \frac{H}{\sqrt{B}} + \log_B n)$ I/Os, where H is the height of the Compact PAT-tree.*

---

[6]We notice that the Compact PAT-tree does not offer the *suffix links*. A suffix link from node u to node v is defined if the string spelled out by u is $a\alpha$ and the string spelled out by v is just α. These links are particularly useful to implement efficiently more complicated searches [79].

35.4.2 Construction

Although the suffix tree data structure is thirty years old, the problem of constructing it efficiently in various models of computations remains an active area of research (see e.g. RAM [28, 52, 85, 108, 127, 176, 181], PRAM [82, 53, 161] and BSP [62, 34] models were investigated). Designing a disk-conscious approach to suffix-tree construction is a challenging problem that has found efficient solutions only in the last years. In fact, almost all previous algorithms inserted one suffix at a time into a growing suffix tree thus exhibiting a marked absence of *locality of reference*. These algorithms elicit many random I/Os when the size of the indexed text is too large to be fit into the internal memory of the computer. This may be obviously a serious problem that, until recently, has prevented this data structure to be built for text collections of even moderate size. The experiments of [33, 110] have shown that classical in-memory approaches need several hours to build a suffix tree lying on disk and, in the case of a 512Mb internal memory, at most 60 million characters could be indexed in reasonable time [110].

In this section we pose attention on the I/O-bottleneck issue arising in the suffix tree construction process, and present three different approaches that carefully structure their pattern of accesses to the disk in order to reduce the number of executed I/Os. Two approaches are mainly theoretical to date, the third one is the best known in the practical setting. A full experimental comparison is still needed in order to establish their comparative performance in the practical setting and, hopefully, achieve a robust suffix-tree construction algorithm for managing Gigabytes of real texts.

- The first algorithm we describe next was the first to achieve I/O-optimality in the external memory setting [55]. It adopted for the first time a *Divide-and-Conquer* approach to suffix tree construction, and showed how to reduce the construction process to external-memory sorting and few low-I/O primitives.

- The second algorithm is the simplest and the most elegant, yet I/O-optimal. It builds the suffix tree indirectly, by exploiting the I/O-effective construction of a suffix array obtained via the Skew algorithm (see Section 35.3). This algorithmic scheme applies successfully also to other indexing data structures, like the String B-tree (see Section 35.5).

- The last algorithm we present achieves the best performance to date in the practical setting [88, 163] and currently is the one to have built the largest suffix tree in reasonable time and internal-memory consumption (i.e. 286Mbps using 2Gbs of internal memory). Surprisingly it is based on the *inefficient* scheme: *one-suffix insertion at a time*; but, it alleviates the I/O-bottleneck of this approach by properly selecting the insertion order of the suffixes and by carefully exploiting the internal memory as a buffer.

The Divide-and-Conquer algorithm of [55] builds the suffix tree ST_T of the string $T[1, n]$ by executing four (macro)steps, detailed in Figure 35.9 and commented below.

It is not difficult to implement the first three steps I/O-efficiently. Actually the first step maps pairs of characters to their lexicographic names via a sorting process. The second step derives the *odd tree* ST_o from $ST_{T'}$ by exploiting the observation that each suffix of T' is indeed a *compacted odd suffix* of T, that is, a suffix starting at an odd position because two characters of T are squeezed into one character of T'. Hence, the lcp of any two suffixes of T' differs from the corresponding lcp in T by at most one unit. Just a character comparison is enough to fix that, and thus all the $O(n)$ nodes in ST_o can be fixed via a batch of $O(n)$ character-comparison queries. Overall the I/O-cost of these two steps is dominated by the sorting process, taking $Sort(n) = O(\frac{n}{B} \log_{M/B} \frac{n}{B})$ I/Os [1, 178]. The third step

DIVIDE-AND-CONQUER ALGORITHM

(1) Construct the string $T'[j] = $ rank of $\langle T[2j], T[2j+1]\rangle$, and recursively compute $ST_{T'}$.

(2) Derive from $ST_{T'}$ the compacted trie ST_o of all suffixes of T beginning at odd positions.

(3) Derive from ST_o the compacted trie ST_e of all suffixes of T beginning at even positions.

(4) Merge ST_o and ST_e into the whole suffix tree ST_T, as follows:

(4.1) Find the anchor pairs and the side trees.

(4.2) For each side tree, find a pair of pull leaves.

(4.3) Overmerge ST_o and ST_e into the tree ST_M.

(4.4) Partially unmerge ST_M to get ST_T.

FIGURE 35.9: The *Divide-and-Conquer algorithm* for suffix tree construction.

has an elegant implementation based on the following observation: *Each suffix starting at an even position (even suffix) is a single character followed by an odd suffix.* Then the lexicographic order of the even suffixes of T can be obtained by stably sorting pairs of the form $\langle T[2i], rank(T_{2i+1})\rangle$, where $rank(T_{2i+1})$ is the lexicographic position of T_{2i+1} among the odd suffixes. Nonetheless, this information is not enough to build the *even tree* ST_e; we further need the lcp-information between pairs of adjacent suffixes. The lcp of two adjacent even suffixes is zero if their first characters do not match, and one plus the lcp of the following odd suffixes otherwise. However, since these odd suffixes may not be adjacent in ST_o, their lcp is computed via a batch of $O(n)$ lca-queries between the corresponding ST_o's leaves. The I/O-cost of this step is still $O(Sort(n))$.

The last merging step is recognized as being difficult, so that the efficiency of the overall approach boils down to the effective implementation of the merging between the odd and even trees. We will only outline it here, and refer the reader to the seminal paper [55] for further technical details and proofs of the observations and properties reported below. For a running example, please have a look at Figure 35.10.

A key concept for the following discussion is the one of *overmerged* tree ST_M. If ST_o and ST_e were *uncompacted tries*, their merging would be simple: it would be enough to perform a *coupled* Depth-First visit along their labeled edges, and merge those edges being equal or split those edges sharing a prefix. The difficulty here is that ST_o and ST_e are compact tries (to occupy linear space), and thus the coupled DFS-visit would execute $\Omega(1)$ I/Os per edge-match test (recall the indirect encoding of the edge labels). In [55] the authors propose an elegant and I/O-optimal solution that proceeds in two steps: (i) it temporarily relaxes the requirement of getting ST_T in one shot, and thus it blindly (over)merges the paths of ST_o and ST_e into the tree ST_M; (ii) it finally re-fixes ST_M by detecting and undoing in an I/O-efficient manner the (over)merged paths to obtain ST_T. The (over)merging approach recalls in some way the Patricia search method used before onto Compacted PAT-trees, in that it (over)merges two edges if their first characters match.

We note that in the final suffix tree ST_T there exists an important subset of nodes, called *odd/even nodes*: They are nodes having both odd and even descendent leaves. Clearly odd/even nodes occur either in ST_o or in ST_e, or in both. The root of ST_T is trivially an odd/even node that occurs in both trees. The construction of ST_M is crucially based on the identification of a *superset* of the odd/even nodes. To this aim, we need to detect *anchor*

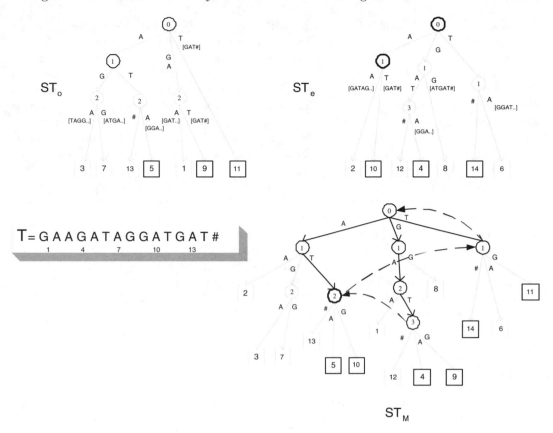

FIGURE 35.10: A running example of the *Divide-and-Conquer algorithm* on the text string $T = \text{GAAGATAGGATGAT}$. Leaves are squares and pull leaves are bold faced. Edge labels are represented with their first character, the remaining string is indicated between square brackets. Bold circles denote the anchor nodes. In ST_M bold paths denote the paths merged by the coupled-DFS. We show via dotted arrows the d-path for the node spelling out the string GAT: this has length three as the string length. So the corresponding node in not unmerged. Actually no overmerge occurs in this example.

pairs and *side trees* in ST_e and ST_o.

DEFINITION 35.1   A pair of nodes $u_o \in ST_o$ and $u_e \in ST_e$ is called an *anchor pair*, if u_o and u_e denote the same string.

All anchor nodes and their ancestors are odd/even nodes. Consequently the part of ST_o and ST_e lying above the anchor nodes is formed by odd/even nodes *only*. The part of ST_o and ST_e below the anchor pairs is structured nicely.

DEFINITION 35.2   A *side tree* is defined to be a subtree of ST_o or ST_e that does not contain an anchor node. A *side tree pair* is defined to be a pair of side trees, one in ST_o and the other in ST_e, such that the parents of their roots form an anchor pair, and the first character on the edge from the anchor pair to the roots is the same.

The nice property about side tree pairs is that the rest of the odd/even nodes in them adhere to a simple pattern: they form two *downward paths* that leave the roots of the side trees and will interdigitate in the final merged path of ST_T. Since path merging can be performed I/O-efficiently, the merging of side tree pairs is reduced to the merging of those downward paths. Actually we first detect a pair of, so called, *pull leaves* descending from those paths, and then merge the paths in the side trees leading to these pull leaves (hence the overmerging). The latter merging step implements a coupled Depth-First merge over paths, hence it is I/O-efficient. In [55] it is shown how to compute the anchor nodes and the pull leaves, and how to merge the part of ST_o and ST_e above the anchor nodes, and how to overmerge the side tree pairs via their pull leaves, in overall $O(Sort(n))$ I/Os.

The last substep (4.4) deals with the detection of some nodes in ST_M that do not occur in ST_T and must be therefore *unmerged*. Since an overmerged node u is a node *uncorrectly forced to be odd/even*, we check its status as follows. Let ℓ_{2j-1} and ℓ_{2i} be a pair of odd and even leaves descending from u in ST_M, and such that $u = \text{lca}(\ell_{2j-1}, \ell_{2i})$. Let us define the pointer $d(u) = \text{lca}(\ell_{2j}, \ell_{2i+1})$. In [55] it is shown that a node u is properly merged if, and only if, the depth of u in ST_M is equal to the depth of u in the tree formed by the d-pointers. Depth information can be computed in $O(Sort(n))$ I/Os, as well the d-tree, via a batch of $O(n)$ lca-queries. As a result, the overall I/O-cost of the Divide-and-Conquer algorithm follows the nice recursive relation $T(n) = T(n/2) + O(Sort(n))$.

THEOREM 35.8   *Given an arbitrary string $T[1,n]$, its suffix tree ST_T can be constructed in $O(Sort(n))$ I/Os, $O(n \log n)$ time, and using $O(n/B)$ disk pages.*

The second algorithm we propose for building large suffix trees is based on the following observation: constructing suffix arrays and suffix trees is equivalent *modulo external-memory sorting or scanning* primitives. In fact, the suffix array can be derived from a rightward scan of the leaves of the suffix tree. The opposite needs little more information, namely the array LCP_T that stores at position i the lcp between the $(i-1)$th and the ith suffix of SA_T. This second proposal builds a suffix tree *indirectly* through the elegant Skew algorithm of [100]. The algorithm is deceptively simple, elegant and I/O-optimal, thus a good candidate to build large suffix trees in the practical setting. The first step of this algorithm consists of computing SA_T and LCP_T in $O(Sort(n))$ I/Os (see Theorem 35.3). After that, the construction of ST_T proceeds by inserting the suffixes of T one at time in lexicographic order, i.e. inserting the leaves in the suffix tree from left to right. A new leaf ℓ_i always becomes the rightmost child of a node on the rightmost path of the tree we are currently building. This path is managed I/O-efficiently by using a stack with the (lastly inserted) leaf ℓ_{i-1} on its top. To insert the next leaf ℓ_i, nodes are popped from the stack until the insertion depth $LCP_T[i]$ is reached. If there is a node u at that depth, ℓ_i is attached to u and pushed onto the stack. If u is absent, then u is created by splitting the edge at that depth, and ℓ_i is attached to u, and both nodes are pushed onto the stack. This preserves the invariant and thus ensures that the overall I/O, time and space complexities are the ones stated in Theorem 35.8.

It is not evident which one of the two construction algorithms above is better in practice. The Divide-and-Conquer algorithm exploits a recursion with parameter $1/2$ but it incurs in a large space-overhead because of the explicit management of the tree topology. The Skew algorithm is more space efficient and clearly more easy to implement, but it exploits a recursion with parameter $2/3$. Experiments are needed to compare their practical performance.

We are left with the description of the best algorithm known for the practical setting. Of course, we may expect this ranking to change in the near future in the light of the active research in this field. In any case, the Incremental algorithm of [88] is an example of good algorithmic engineering that turns an I/O-*inefficient* algorithm into a good one for real data. This algorithm trades the ideal $O(n)$ performance of classical incremental approaches [127, 176, 181] for locality of references on the basis of two decisions: (i) it abandons the use of suffix links (see footnote 6), and (ii) it performs multiple passes over the text T and constructs the suffix tree for a *subrange* of suffixes at each pass (partition). These decisions result in a fan-like tree structure in which partitions can be built independently, and evicted from internal memory as they are completed. The partitions are based on a simple observation: *fix a length q, each q-long string identifies at most one subtree in ST_T*, the one descending from the path spelling out that string. We can thus form the partitions by clustering together the subtrees of more than one q-long string, provided that these trees can be fit into the available internal memory. Actually the authors of [88] propose two approaches to determine this clustering. One way is to fix $q = 3$, count the number of occurrences of any 3-long string within T via a sequential scan, and finally find the best clustering using a bin-packing algorithm. Alternatively, given the pseudo-random nature of DNA that makes ST_T uniformly populated, the authors propose to choose a uniform partition that exploits the lexicographic order among the q-long prefixes, and possibly refine it if much-populated clusters occur. Once the clustering has been computed, we perform as many scanning of T as there are partitions, and for each of them we build the (forest of) subtree(s) for the suffixes belonging to that partition. At the end, all subtrees are put together by exploiting the knowledge about the prefixes that have driven to their construction. We point out that the algorithm is much efficient in practice because of two facts: (1) I/Os are mainly sequential and (2) the random pattern of accesses, incurred by the incremental construction of the subtrees, does not induce many random I/Os because the subtrees are expected to reside in internal memory during their construction. The experiments reported in [88] present a rough implementation of the suffix-tree structure requiring about 65 bytes per indexed suffix! Further compression could be obviously obtained by using techniques similar to those ones proposed for Compact PAT-trees or [110], as well better memory exploitation might be possible to make the algorithm robust against data skewness. Yet other open issues to investigate!

35.4.3 Future directions of research

Because of the recent theoretical [136, 77, 69, 76, 63] and practical achievements [71, 76, 88, 110], we believe that it is no longer the time that "suffix trees are not practical except when the text size to handle is so small that the suffix tree fits in internal memory" [143]. Many sophisticated techniques are around and most of them wait for engineering and experimentation. It is not surprising that [169] claims that suffix trees are the data structure with *the highest need for better implementations*. Hopefully the previous section pointed out some enlightening proposals for achieving that and raised open problems that deserve much attention in the near future.

Before concluding the discussion on suffix trees we would like to address one of the most fascinating issues considered before, that is, the management of *skewed query distributions* on massive data. We observe that, in practice, we have commonly no knowledge about the actual distribution of the queries so that some sort of *self-adjusting strategy* has to be devised. As far as string data structures for RAM are concerned, an optimal solution is known, called the *lexicographic tree* [170]. Recently, this result has been extended to the external-memory setting [32] by proposing a novel self-adjusting index based on a variant of the Skip List data structure [153], called SASL. A technical novelty of SASL is a simple randomized strategy to implement the self-adjusting feature that overcomes the I/O-bottleneck *on expectation* both for the search and the update operations. We will come back to the features of this data structure when dealing with String B-trees (Section 35.5). We wish here to point out only the fact that the incremental construction approach detailed above actually consists of a set of *query/insertion operations* made on strings of variable length (i.e. the text suffixes) with possibly *long-shared prefixes* (that depend on T's structure). In this scenario SASL might turn out to be useful because it could avoid the need of clustering similar suffixes and of scanning multiple times the text T to insert them. An engineering of SASL is therefore worth to be proposed, experimented, and checked against real biological sequences.

35.5   The String B-tree

If we open any textbook on algorithmics we find that an optimal solution to the management of large data sets of *atomic keys* (e.g. numerical values) is provided by the B-tree data structure [38, 106]. And indeed any current DBMS uses B-trees to provide persistent storage capabilities. In Section 35.3 we introduced the *Prefix B-tree* [17] as an I/O-efficient extension of B-trees to the management of *string keys*. This B-tree variant stores a prefix of each key in the B-tree nodes, providing overflow nodes for the remainder. Unfortunately, these overflow nodes not only introduce extra latency into database accesses, but also provide the user with little a priori information about the time required to complete an operation (i.e. bad I/O-performance in the worst case), and finally, they introduce some space overhead that may impact on the practical I/O-performance. In the string setting we are dealing with, the strings are actually suffixes of very long texts so that Prefix B-trees offer very poor performance guarantee. This is the main reason why nobody in the literature has tried to use standard DBMS for string indexing applications!

Despite this scenario, the B-tree scheme is naturally appealing for designing an effective string data structure. This argument has been successfully concretized by the *String B-tree* [59] that overcomes the limitations above working well on any set of arbitrarily long keys. Briefly, the String B-tree is a hybrid data structure that plugs a Patricia tree [130] into the nodes of the B-tree in order to provide a *routing tool* that efficiently drives the string searches and, more importantly, occupies a space proportional to the *number* of indexed strings instead of their total length. This allows us to fit many strings into a single node of the B-tree, independently of their length. As a result, unlike suffix trees and suffix arrays, the String B-tree achieves *optimal I/O-bounds* for searching arbitrary patterns (drawn from unbounded alphabets) and attractive update performance (cfr. Theorems 35.1 and 35.4).[7] In practice it requires a negligible, *guaranteed* number of disk accesses to search for an arbitrary substring of the indexed string, independently of the character distribution [58]. Consequently it solves the long-standing open problem of dynamically managing arbitrary long keys in the worst-case setting.

For the sake of generality, we extend the notation as follows. We denote by Δ a set of arbitrarily long strings which we wish to index by means of the String B-tree data structure. The parameter n denotes their total length. We let $\mathsf{SUF}(\Delta)$ denote the lexicographically ordered set of all suffixes of Δ's strings (cfr. T and $\mathsf{SUF}(T)$ in Section 35.2). It should

---

[7]We remark that the String B-tree supports exact substring searches like suffix trees and suffix arrays. It remains an interesting area of research the resolution of more sophisticated queries, like regular expressions and approximated searches (cfr. [16]).

be clear to the reader that, in order to support substring searches inside the strings of Δ, a data structure must efficiently store $\mathsf{SUF}(\Delta)$. This is what the String B-tree does, with the additional feature that it supports *dynamic* changes to Δ, like insertion and deletion of individual strings. In what follows we will review the basics of String B-trees and discuss an engineered variant worthy of further experimental investigation. For details on the data structure we refer the reader to the seminal paper [59], and to some of its practical variants [21, 58, 154].

The String B-tree built over the set Δ, and denoted hereafter by SB_Δ, has a *structure* similar to the B$^+$-tree [38]: The leaves contain all the indexed keys, while the internal nodes store copies of some keys for routing the subsequent traversals. The definition of *key* is a crucial issue in the context of String B-trees.

DEFINITION 35.3 A key in SB_Δ is a *pointer to a string* of $\mathsf{SUF}(\Delta)$, and therefore it is a pointer to a suffix of some string of Δ.

Consequently SB_Δ indexes *indirectly* $\mathsf{SUF}(\Delta)$. Its *content* is therefore different from that of a Prefix B-tree, since the keys are *pointers to strings* instead of the strings themselves. The order between any two keys is then defined to be the *lexicographic order* among the corresponding pointed strings. The crucial fact about String B-trees is that by storing just the string pointers into their nodes, they allow the indexing of strings of arbitrary length and they are so able to store $\Theta(B)$ keys (pointers to strings) in every disk page.

The mapping of the keys to String B-tree nodes is done as follows. Let us assume that each disk page can contain up to $2b$ keys, where $b = \Theta(B)$ is a parameter depending on the actual space occupancy of a node (this will be discussed in Section 35.5.1). $\mathsf{SUF}(\Delta)$ is partitioned into groups of at most $2b$ strings each, except the last group which may contain fewer strings. Every group is stored into a leaf of SB_Δ in such a way that the left-to-right scanning of these leaves gives the ordered set $\mathsf{SUF}(\Delta)$ (i.e. the suffix array of all suffixes of Δ's strings). Each internal node π has $n(\pi)$ children, with $\frac{b}{2} \leq n(\pi) \leq b$, except the root which has less than b children. Node π also stores the string set $\mathcal{S}_\pi$ formed by copying the leftmost and the rightmost strings contained in each one of π's children. As a result, set $\mathcal{S}_\pi$ consists of $2n(\pi)$ strings. Since the fan-out of each node is $\Theta(B)$, the height of SB_Δ is $O(\log_B n)$. An example of String B-tree is shown in Figure 35.11.

Since the leaves of the String B-tree form a suffix array on $\mathsf{SUF}(\Delta)$, the search for a pattern $P[1,p]$ as a substring of Δ's strings must identify foremost the lexicographic position of P among the string suffixes in $\mathsf{SUF}(\Delta)$, and thus, among the string pointers in the String B-tree leaves. Once this position is known, all the occurrences of P as a substring of Δ's strings are given by the consecutive string suffixes which are stored from that position and have P as a prefix. Their retrieval takes $O((p/B)occ)$ I/Os, in case of a brute-force match between the pattern P and the checked suffixes; or the optimal $O(occ/B)$ I/Os, if some additional information about the $\texttt{lcp}$-length shared by adjacent suffixes is kept into each String B-tree leaf. In the example of Figure 35.11 the search for the pattern $P = \texttt{CT}$ traces a downward path of String B-tree nodes and identifies the lexicographic position of P into the fourth leaf (counting from the left) and before the suffix 42. The pattern occurrences are then retrieved by scanning the String B-tree leaves from that position until the suffix 32 is encountered, because it is not prefixed by P. The positions $\{42, 20, 13, 24, 16\}$ denote the five occurrences of P as a substring of Δ's strings.

Therefore the efficient implementation of substring searches in String B-trees boils down to the efficient routing of the pattern search among the String B-tree nodes. In this respect it is clear that the way a string set $\mathcal{S}_\pi$ is organized, in each traversed node π, plays a crucial

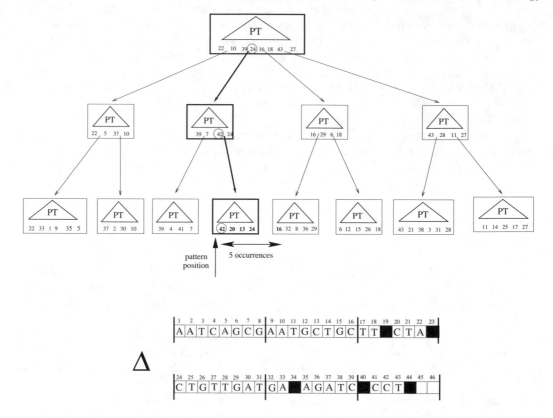

FIGURE 35.11: An illustrative example depicting a *String B-tree* built on a set Δ of DNA sequences. Δ's strings are stored in a file separated by special characters, here denoted with black boxes. Notice that SB_Δ indexes the whole $\mathsf{SUF}(\Delta)$ and thus supports *substring searches*. The triangles labeled with *PT* depict the Patricia trees stored into each String B-tree node. The figure also shows in bold the nodes traversed by the substring search for a pattern $P = \mathtt{CT}$ within Δ's strings. The circled pointers denote the suffixes, one per level, explicitly checked during that search. In the leaf level are indicated the five suffixes prefixed by P, and thus the five positions where P occurs as a substring of Δ's strings.

role. The innovative idea in String B-trees is to use a Patricia tree PT_π to organize the string pointers of $\mathcal{S}_\pi$ [130]. Patricia trees preserve the searching power and properties of compacted tries, although in a reduced space occupancy and reduced I/O-cost, since they defer the access to the disk as long as possible (see Section 35.4). When SB_Δ is traversed downward starting from the root, the traversal is routed by using the Patricia tree PT_π stored in each visited node π. The goal of PT_π is to help finding the lexicographic position of the searched pattern P within the ordered set $\mathcal{S}_\pi$, so that we can detect the next child to proceed in. By assuming that the strings are binary encoded, we can exploit the Patricia search method described for the Compact PAT-trees in Section 35.4, and thus access the disk just when a leaf l of PT_π has been reached. Actually, the difference with that approach is that, now, we need to find the lexicographic position of P, and not just if P is a prefix of the suffix pointed to by that leaf. We solve this problem by comparing the string pointed by l with P in order to determine their longest common prefix. A useful property holds [59]: *the leaf l stores one of the strings in $\mathcal{S}_\pi$ that share the longest common prefix with P.* The

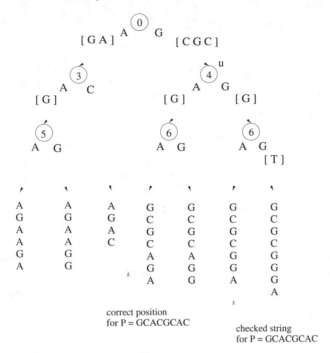

FIGURE 35.12: An example of *Patricia tree* built on a set of $k = 7$ DNA strings. We do not use the binary encoding because the tree is already binary. Each leaf points to one of the k strings; each internal node u (they are at most $k - 1$) is labeled with one integer $len(u)$ which denotes the length of the common prefix shared by all the strings pointed by the leaves descending from u; each edge is labeled with only one character. The characters between square-brackets are not explicitly stored, and denote the other characters labeling an edge.

length ℓ of this common prefix and the mismatch character $P[\ell + 1]$ are used in two ways: first to determine the shallowest ancestor of l spelling out a string longer than ℓ, and then to select the leaf descending from that ancestor which identifies the lexicographic position of P in $\mathcal{S}_\pi$. An illustrative example of a search in a Patricia tree is shown in Figure 35.12.

We remark here that PT_π requires space linear in the *number* of strings of $\mathcal{S}_\pi$, therefore the space usage is independent of their *total length*. Consequently, the number of strings in $\mathcal{S}_\pi$ can be properly chosen to fit PT_π in the disk page allocated for π. An additional nice property of PT_π is that it allows to find the lexicographic position of P in $\mathcal{S}_\pi$ by exploiting the information available in π's page and by fully comparing P with just *one of the strings in $\mathcal{S}_\pi$*. This clearly allows to reduce the number of I/Os needed in the routing step. By counting the number of I/Os required for searching $P[1, p]$, and recalling that Δ's strings have overall length n, we get the I/O-bound $O(\frac{p}{B} \log_B n)$. In fact, SB_Δ has height $O(\log_B n)$, and at each traversed node π we may need to fully compare P against one string of $\mathcal{S}_\pi$ thus taking $O(\frac{p}{B} + 1)$ I/Os.

A further refinement to this idea is possible by observing that we do not necessarily need to compare the two strings, P and the candidate string of $\mathcal{S}_\pi$, starting from their first character. Indeed we can take advantage of the comparisons executed on the ancestors of π in order to skip some character comparisons. An incremental accounting strategy allows to prove that $O(\frac{p}{B} + \log_B n)$ I/Os are indeed sufficient, and this bound is optimal in the case of an unbounded alphabet. A more complete analysis and description of the search and update operations is given in [59] where it is formally proved the following:

THEOREM 35.9   *String B-trees support the search for all the occ occurrences of an arbitrary pattern $P[1, p]$ as a substring of the strings in Δ taking $O(\frac{p+occ}{B} + \log_B n)$ I/Os, where n is the overall length of Δ's strings. The insertion in, or the deletion from, the set Δ of a string of length m takes $O(m \log_B(n + m))$ I/Os. The required space is the optimal $\Theta(\frac{n}{B})$ disk pages.*

Some authors have successfully used String B-trees in other settings: multi-dimensional prefix-string queries [96], conjunctive boolean queries on two substrings [60], dictionary matching problems [61], distributed search engines [62], indexing of XML texts [41] or temporal series [152]. All of these applications show the flexibility of this full-text index, its efficiency in external memory, and foretell engineered implementations. Up to now String B-trees have been confined mainly to the theoretical realm perhaps because of their space occupancy: the best known implementation uses about 12 bytes per indexed suffix [58] (see also [21, 154]). In what follows we try to overcome this limitation by proposing an engineered version of String B-trees suitable for practical implementations.

35.5.1   Engineering

String B-trees have the characteristic that their height decreases exponentially as the node fan-out increases. This value is strictly related to the number of strings contained in each node π (actually it is the half). So that, if the disk page size B increases, we can store more suffixes in $\mathcal{S}_\pi$, and thus we can increase the fan-out of each node π. However, since B is typically chosen to be proportional to the size of a physical disk page, we need a technique that maximizes $|\mathcal{S}_\pi|$ for a fixed B. This is actually the problem solved for the Compact PAT-tree, so that the compaction techniques deployed there might be also used on $\mathcal{S}_\pi$. Rather than do this here, we comment on a more elegant solution that exploits the specialties of the problem we have in our hands. The key observation is that $\mathcal{S}_\pi$ occupies one disk page, so that a CPU cost of $O(|\mathcal{S}_\pi|)$ cannot be considered as a bottleneck, because it is surely dominated by the cost of fetching that page from disk. Hence, what about a slightly slower solution for P's searching within $\mathcal{S}_\pi$, that however offers a more succinct space encoding for that set?

The approach we discuss here surprisingly throws away the Patricia tree topology. It keeps just the string pointers and the offset values stored in the nodes of the Patricia tree, and it is still able to support pattern searches in a constant number of I/Os per visited String B-tree node. As a result, the asymptotic I/O-bounds stated in Theorem 35.9 still hold with a significant space improvement in the constants hidden in the big-O notation. The starting point is the beautiful result of [56] that we briefly recall here.

Let us be given the lexicographically ordered array of pointers to the strings in $\mathcal{S}_\pi$, called A_π, and the array LCP_π of longest-common-prefixes shared by strings adjacent in A_π. We can look at A_π and LCP_π as the sequence of string pointers and offset values encountered during an in-order traversal of PT_π. Substitute then PT_π with A_π and LCP_π into each String B-tree node π. Since A_π is ordered, we might implement the search for $P[1, p]$ via the classical binary search within a logarithmic number of string accesses (see Section 35.3). Instead we deploy here the elegant result of [56] that executes only *one* string access on the disk, $\Theta(p + |\mathcal{S}_\pi|)$ bit comparisons, and one full scan of the arrays A_π and LCP_π. Since these arrays reside in internal memory when the search is performed on π (i.e. their disk page has been fetched) and they consist of *few thousands* of items, a CPU cost linear in their size is negligible with respect to the cost of fetching π's page from disk. Hence this solution is I/O-efficient in that it requires just one sequential string access, it is CPU-efficient because the array-scan can benefit from the reading-ahead policy of the internal cache, and it is space efficient because it avoids the storage of PT_π's topology.

Let us therefore detail the search algorithm which assumes a *binary* pattern P and consists of two phases (see [56] for the uneasy proof of correctness). In the first phase, the algorithm scans rightward the array A_π and inductively keeps x as the position of P in this array (initially $x = 0$). At a generic step i, it computes $\ell = LCP_\pi[i+1]+1$ as the mismatching position between the current string $A_\pi[i]$ and the next string $A_\pi[i+1]$. Notice that the ℓth bit of the string $A_\pi[i]$ is 0, whereas the ℓth bit of the string $A_\pi[i+1]$ is 1 because they are binary and lexicographically ordered. If $P[\ell] = 1$, the algorithm sets $x = i+1$ and increments i; otherwise (it is $P[\ell] = 0$), the algorithm leaves x unchanged and increments i until it is $LCP_\pi[i+1] < \ell$. In this latter case the algorithm is jumping all the succeeding strings which have the ℓth bit set to 1 (since $P[\ell] = 0$). The first phase ends when A_π has been completely scanned. It is possible to prove that $A_\pi[x]$ is one of the strings in S_π sharing the longest common prefix with P. The second phase of the search algorithm initiates by computing the length ℓ' of the longest common prefix between P and the candidate string $A_\pi[x]$. Then, the algorithm starts from position x a backward scanning of A_π if $P[\ell'+1] = 0$, or a forward scanning if $P[\ell'+1] = 1$. This scan searches for the lexicographic position of P in A_π and proceeds until it meets the position x' such that $LCP_\pi[x'+1] < \ell'$. The searched position lies between the two strings $A_\pi[x']$ and $A_\pi[x'+1]$. This is the correct position of P among the strings of S_π.

Notice that the algorithm needs to access the disk just for fetching the string $A_\pi[x]$ and comparing it against P. Hence $O(p/B)$ I/Os suffice to route P through the String B-tree node π. An incremental accounting strategy, as the one devised in [59], allows to prove that we can skip some bit comparisons and therefore require $O(\frac{p+occ}{B} + \log_B n)$ I/Os to search for the occ occurrences of a pattern $P[1,p]$ as a substring of Δ's strings. Preliminary experiments have shown that searching few thousands of strings via this approach needs a negligible time compared to the cost of a single I/O on modern disks. Furthermore, the incremental search strategy allows sometimes to avoid the I/Os needed to access $A_\pi[x]$.

35.5.2   Construction

Given that the leaves of a String B-tree form a paged suffix array on the string set Δ, it is natural to adopt any of the construction algorithms devised for suffix arrays in order to build a String B-tree (see Section 35.3.2). We are left therefore with the problem of building the internal nodes of SB_Δ and we will do it in $O(n/B)$ I/Os.

For the sake of presentation, assume to indicate by SA_i the array of suffix pointers stored in the ith level of SB_Δ (SA_0 is the leaf level) and let LCP_i the corresponding array of lcp information between adjacent strings in SA_i. It goes without saying that SA_i and LCP_i are enough to build the Patricia tree PT_π for the String B-tree nodes at level i, or to build the arrays used in the engineered solution (actually they are the same). To build the next level of SB_Δ, we scan rightward SA_i and take the leftmost string $L(\pi)$ and the rightmost string $R(\pi)$ from each node π. This gives the new array SA_{i+1} whose length is a factor $\Theta(1/B)$ smaller than SA_i. Each pair of adjacent strings is either a $\langle L(\pi), R(\pi)\rangle$ pair or a $\langle R(\pi), L(\pi')\rangle$ pair (derived from consecutive nodes π and π'). In the former case, the lcp of the two strings is obtained by taking the minimum of all the lcp s stored in π; in the latter case, the lcp is directly available in the array LCP_i since $R(\pi)$ and $L(\pi')$ are contiguous there. After that SA_{i+1} and LCP_{i+1} have been constructed, we partition the arrays into disk pages to form a new level of internal nodes of the String B-tree. The process continues for $O(\log_B n)$ iterations until the whole SA and LCP fit into one disk page, in which case the root of the String B-tree is formed and the construction process stops. Since the size of the arrays shrinks by a factor B at each level, the overall I/O-cost is just the cost of scanning the String B-tree leaves, hence $O(n/B)$ I/Os. Preliminary experiments [58] have shown that the time taken to build a String B-tree from its suffix array is negligible with respect to the time taken for the construction of the suffix array itself.

We conclude this section by recalling that one of the key features of String B-trees is their *dinamicity*, which makes them more appealing than suffix trees and arrays. While handling deletions is not really a problem as we have a plethora of tools inherited from standard B-trees [38], implementing the addition of a new string in Δ requires significantly new techniques. This *asymmetry* is better understood if we observe that the insertion of a new string $X[1, x]$ into Δ requires the insertion of all of its x suffixes into the lexicographically ordered set $\mathsf{SUF}(\Delta)$. Since string X can be of few Megabytes (or even more), the *rescanning* of its characters might be a computational bottleneck. On the other hand, the deletion of a string $Y[1, y]$ from Δ consists of a sequence of y standard deletions of pointers to suffixes of Y, and hence it can exploit standard B-tree techniques.

The approach proposed in [59] to avoid the "rescanning" in string insertions is mainly theoretical in its flavor and considers an *augmented* String B-tree where some pointers are added to its leaves. The counterpart for this I/O improvement is that a larger space occupancy is needed and, when rebalancing the String B-tree, the redirection of some of these additional pointers may cause the execution of random I/Os. Therefore, it is questionable if this approach is really attractive from a practical point of view. Starting from these considerations [58] proposed an alternative approach based on a *batched* insertion of the x suffixes of the new string X. This approach exploits the LRU buffering strategy of the underlying operating system and proves to be effective in the case of a large x. In the case of a small x, a different approach must be adopted which is based on the *suffix-array merging* procedure of [73] (see Section 35.3.2): here X plays the role of the new text piece to be indexed, Δ provides the part of the text already indexed, and SB_Δ's leaves provide the suffix array of Δ (the one stored on disk). The merge of SA_X (commonly built in memory) and SB_Δ (and their corresponding LCP arrays) gives the new set of String B-tree leaves. The internal nodes are constructed within $O((n + x)/B)$ I/Os as shown above.

An extensive experimental analysis of these approaches is still needed to validate these appealing theoretical I/O-bounds.

35.5.3   Future directions of research

An important advantage of String B-trees is that they are a variant of B-trees and consequently most of the technological advances and know-how acquired on B-trees can be smoothly applied to them. For example, split and merge strategies ensuring good page-fill ratio, node buffering techniques to speed up search operations, B-tree distribution over multi-disk systems, as well adaptive overflow techniques to defer node splitting and B-tree re-organization, can be applied on String B-trees without any significant modification. Surprisingly enough, there are no publicly available implementations of the String B-tree data structure, whereas some softwares are based on it [21, 41, 96], or some ideas have been tested in [154, 58]. We foretell an engineered and publicly available software library for full-text indexing based on String B-trees. This library should be designed to follow the API of the Berkeley DB [171], thus facilitating its use in well-established applications. The String B-tree could also be adopted in place of the suffix tree for many bio-applications [79], and its support to more sophisticated searches (e.g. regexp or approximate) still waits for theoretical and experimental investigations.

Finally, we wish to make a comment on the case of a *skewed query distribution*. It is apparent that the String B-tree (like the B-tree) is insensible to the frequency with which strings are queried, because string pointers reside on leaves and the String B-tree structure is fixed along the query time. Conversely, the SASL data structure introduced in Section 35.4.3 adjusts itself as the query and update operations are executed. Actually, [32] showed that given k strings S_is of total length n, the data structure SASL built on these strings is able to support a sequence of z searches for $S_{i_1}, S_{i_2}, \ldots, S_{i_z}$ in $O\left(\sum_{j=1}^{z} \left(\frac{|S_{i_j}|}{B} \right) + \sum_{i=1}^{k} \left(n_i \log_B \frac{z}{n_i} \right) \right)$ expected I/Os, where n_i is the number of times the string S_i is queried. If the strings to be indexed in the SASL are taken to be $\mathsf{SUF}(\Delta)$, then the corresponding SASL provides a *self-adjusting version* of the String B-tree. The advantage of SASL is not only theoretical but also practical in that it is simple and, with high probability, significantly better than String B-trees when the sequence of queries is highly skewed and changes over the time. More research is still needed into this fascinating topic which actually introduces a new way of accounting the cost of string searches for the case of a transactional framework, like the one we are faced with genomic databases.

35.6 Final Comments

The reader who has reached this last section has probably grasped a lot of hints and open problems that deserve careful thinking and deep experimental analysis. The latter is actually a crucial point we wish to stress here. As far as we know (and apart of a tentative feasibility project [45]), no public software library does exist that makes available the basic data structures discussed in this chapter in a unifying framework. This would be very valuable because it could allow researchers to "not rediscover the hot water" over and over again, not throw away precious research time to re-implement known things in a bad way, not to perform useless comparisons among known and new solutions, but rather concentrate on significant research and technological improvements.

There are actually many other avenues to investigate in the context of processing large genomic datasets, and here we comment one of them: *compressed full-text indexes* (see e.g. [68, 64, 69, 74, 75, 76, 142]). These indexes allow to fit within (almost) the space needed by the best known compressors both the original text *and* its suffix array (or suffix tree). Although the compression ratios of genomic sequences are yet poor [11, 29, 30, 54, 78, 115, 146]. we believe that these compressed indexes may be useful in the genomic context because they may turn into in-memory some computations which now require the use of the disk. This line of research has been pioneered in the experimental setting by [84, 83, 160] showing that compressed suffix arrays can be used as *filtering data structures* to speed up similarity-based searches on large genomic databases. Actually [160] was able to build a compressed suffix array on the entire human genome within two Gigabytes of internal memory. From the theoretical point of view, it is fascinating the paper [44] that proposed another use of compression for speeding up similarity computations based on a dynamic-programming scheme. We would like to combine these ideas with those developed in [68, 64, 57, 67, 69] in order to reduce the space requirements of these algorithms without impairing their time complexity (which is conjectured in [44] to be close to optimal).

Acknowledgements

I dedicate this chapter to the memory of my father. I thank Valentina Ciriani, Antonio Gullì and Nadia Pisanti for their comments on early versions of this chapter. This work was partially supported by Italian MIUR projects ECD and ALGO-NEXT.

References

[1] A. Aggarwal and J.S. Vitter. The Input/Output complexity of sorting and related problems. *Communications of the ACM*, 31(9):1116–1127, 1988.

[2] A. Aghili, D. Agrawal, and A.E. Abbadi. Filtration of string proximity search via transformation. In *IEEE International Symposium on BioInformatics and Bio-Engineering*, pages 149–157, 2003.

[3] S. Alstrup, M.A. Bender, E.D. Demaine, and M. Farach-Colton *et al.* Efficient tree layout in a multilevel memory hierarchy, 2003. Personal Communication, corrected version of a paper appeared in the *European Symposium on Algorithms 2002*.

[4] S. Altschul, T. Madden, A. Schäffer, and J. Zhang *et al.* Gapped BLAST and PSI-BLAST: a new generation of protein database search programs. *Nucleic Acid Research*, 25:3389–3402, 1997.

[5] S.F. Altschul, W. Gish, W. Miller, and E.W. Myers *et al.* Basic local alignment search tool. *Journal of Molecular Biology*, 215:403–410, 1990.

[6] A. Amir, M. Farach, R. Idury, and J. La Poutré *et al.* Improved dynamic dictionary matching. *Information and Computation*, 119(2):258–282, 1995.

[7] A. Andersson and S. Nilsson. Efficient implementation of suffix trees. *Software–Practice and Experience*, 25(3):129–141, 1995.

[8] A. Andoni, M. Deza, A. Gupta, and P. Indyk *et al.* Lower bounds for embedding of edit distance in normed spaces. In *ACM-SIAM Symposium on Algorithms*, pages 523–526, 2003.

[9] J. Aoe, K. Morimoto, M. Shishibori, and K. Park. A trie compaction algorithm for a large set of keys. *IEEE Transactions on Knowledge and Data Engineering*, 8(3):476–491, June 1996.

[10] A. Apostolico. The myriad virtues of suffix trees. In A. Apostolico and Z. Galil, editors, *Combinatorial Algorithms on Words*, volume 12 of *NATO Advanced Science Institutes, Series F*, pages 85–96. Springer-Verlag, Berlin, 1985.

[11] A. Apostolico and S. Lonardi. Compression of biological sequences by greedy off-line textual substitution. In *IEEE Data Compression Conference*, pages 143–152, 2000.

[12] H. Arimura, J. Abe, H. Sakamoto, and S. Arikawa *et al.* Text data mining: Discovery of important keywords in the cyberspace. In *Kyoto International Conference on Digital Libraries*, pages 121–126, 2000.

[13] R.A. Baeza-Yates, E.F. Barbosa, and N. Ziviani. Hierarchies of indices for text searching. *Information Systems*, 21(6):497–514, 1996.

[14] R.A. Baeza-Yates and G.H. Gonnet. Fast text searching for regular expressions or automaton searching on tries. *Journal of the ACM*, 43(6):915–936, 1996.

[15] R.A. Baeza-Yates and C.H. Perleberg. Fast and practical approximate string matching. *Information Processing Letters*, 59(1):21–27, 1996.

[16] M. Bawa, T. Condie, and P. Ganesan. LSH forest: self-tuning indexes for similarity search. In *International Conference on the World Wide Web*, pages 651–660, 2005.

[17] R. Bayer and K. Unterauer. Prefix B-trees. *ACM Transactions on Database Systems*, 2(1):11–26, 1977.

[18] J. Bentley. *Programming Pearls*. Addison-Wesley, USA, 1989.

[19] J.L. Bentley and M.D. McIlroy. Engineering a sort function. *Software – Practice and Experience*, 23(11):1249–1265, 1993.

[20] P. Bohannon, P. McIlroy, and R. Rastogi. Main-memory index structures with fixed-size partial keys. *SIGMOD Record*, 30(2):163–174, 2001.

[21] P. Bumbulis and I.T. Bowman. A compact B-tree. In *ACM SIGMOD*, pages 533–541, 2002.

[22] S. Burkhard, A. Crauser, H.P. Ferragina, and P. Lenhof *et al.* Q-gram based database searching using suffix array. In *International Conference on Computational Molecular Biology*, pages 77–83, 1999.

[23] S. Burkhardt and J. Kärkkäinen. Better filtering with gapped *q*-grams. *Fundamenta Informaticae*, 23:1001–1018, 2003.

[24] S. Burkhardt and J. Kärkkäinen. Fast lightweight suffix array construction and checking. In *Symposium on Combinatorial Pattern Matching*, volume 2676 of *Lecture Notes in Computer Science*, pages 55–69. Springer-Verlag, 2003.

[25] M. Burrows and D. Wheeler. A block sorting lossless data compression algorithm. Technical Report 124, Digital Equipment Corporation, 1994.

[26] A. Califano and I. Rigoutsos. FLASH: A fast lookup algorithm for string homology. In *International Conference on Intelligent Systems for Molecular Biology*, pages 56–64, 1993.

[27] E. Chávez and G. Navarro. A metric index for approximate string matching. In *Latin American Symposium on Theoretical INformatics*, volume 2286 of *Lecture Notes in Computer Science*, pages 181–195. Springer-Verlag, 2002.

[28] M.T. Chen and J. Seiferas. Efficient and elegant subword tree construction. In *Combinatorial Algorithms on Words*, chapter 12, pages 97–107. NATO ASI Series F: Computer and System Sciences, 1985.

[29] X. Chen, S. Kwong, and M. Li. A compression algorithm for DNA sequences and its applications in genome comparison. In *International Conference on Computational Molecular Biology*, page 107, 2000.

[30] X. Chen, M. Li, B. Ma, and J. Tromp. DNACompress: Fast and effective DNA sequence compression. *Bioinformatics*, 18(12):1696–1698, 2002.

[31] X. Chen, M. Li, B. Ma, and J. Tromp. PatternHunter—fast and more sensitive homology search. *Bioinformatics*, 18:440–445, 2002.

[32] V. Ciriani, P. Ferragina, F. Luccio, and S. Muthukrishnan. Static optimality theorem for external-memory string access. In *IEEE Symposium on Foundations of Computer Science*, pages 219–227, 2002.

[33] D.R. Clark and I. Munro. Efficient suffix trees on secondary storage. In *ACM-SIAM Symposium on Discrete Algorithms*, pages 383–391, 1996.

[34] R. Clifford and M. Sergot. Distributed and paged suffix trees for large genetic databases. In *Symposium on Combinatorial Pattern Matching*, volume 2676 of *Lecture Notes in Computer Science*, pages 70–82. Springer-Verlag, 2003.

[35] A.L. Cobbs. Fast approximate matching using suffix trees. In *Symposium on Combinatorial Pattern Matching*, volume 937 of *Lecture Notes in Computer Science*, pages 41–54. Springer-Verlag, 1995.

[36] R. Cole, L. Gottlieb, and M. Lewenstein. Dictionary matching and indexing with errors and don't cares. In *ACM Symposium on Theory of Computing*, 2004.

[37] L. Colussi and A. De Col. A time and space efficient data structure for string searching on large texts. *Information Processing Letters*, 58(5):217–222, 1996.

[38] D. Comer. Ubiquitous B-tree. *ACM Computing Surveys*, 11(2):121–137, June 1979.

[39] B. Cooper, N. Sample, M.J. Franklin, and G.R. Hjaltason *et al.* A fast index for semistructured data. In *The VLDB Journal*, pages 341–350, 2001.

[40] G. Cormode and S. Muthukrishnan. The string edit distance problem with moves. In *ACM-SIAM Symposium on Discrete Algorithms*, pages 667–676, 2002.

[41] F. Corti, P. Ferragina, and M. Paoli. TReSy: An XML-indexing tool. CRiBeCu – Scuola Normale Superiore (Pisa, Italy), http://www.cribecu.sns.it/, 1999.

[42] A. Crauser and P. Ferragina. A theoretical and experimental study on the construction of suffix arrays in external memory. *Algorithmica*, 32(1):1–35, 2002.

[43] M. Crochemore. Transducers and repetitions. *Theoretical Computer Science*, 45(1):63–86, 1986.

[44] M. Crochemore, G.M. Landau, and M. Ziv-Ukelson. A sub-quadratic sequence alignment algorithm for unrestricted scoring matrices. *SIAM Journal on Computing*, 32(2):1654–1673, 2003.

[45] A. Czumaj, P. Ferragina, L. Gąsieniec, and S. Muthukrishnan *et al.* The architecture of a software library for string processing. In *Workshop on Algorithm Engineering*, pages 166–176, 1997.

[46] W. DeJonge, A.S. Tanenbaum, and R.P. VanDeRiet. Two access methods using compact binary trees. *IEEE Transactions on Software Engineering*, 13(7), 1987.

[47] A.L. Delcher, S. Kasif, R.D. Fleischmann, and J. Peterson *et al.* Alignment of whole genomes. *Nucleic Acid Research*, 27(11):2369–2376, 1999.

[48] A.L. Delcher, A. Phillippy, J. Calton, and S.L. Salzberg. Fast algorithms for large-scale genome alignment and comparison. *Nucleic Acid Research*, 30(11):2478–2483, 2002.

[49] E. Demaine. Cache-oblivious algorithms and data structures. In Gerth Brodal, editor, *Lecture Notes from the EEF Summer School on Massive Data Sets*. Springer-Verlag, 2006.

[50] E. Demaine and A. Lopez-Ortiz. A linear lower bound on index size for text retrieval. In *ACM-SIAM symposium on Discrete algorithms*, pages 289–294, 2001.

[51] G. Diehr and B. Faaland. Optimal pagination of B-trees with variable-length items. *Communications of the ACM*, 27(3):241–247, 1984.

[52] M. Farach. Optimal suffix tree construction with large alphabets. In *IEEE Symposium on Foundations of Computer Science*, pages 137–143, 1997.

[53] M. Farach and S. Muthukrishnan. Optimal logarithmic time randomized suffix tree construction. In *International Colloquium on Automata, Languages and Programming*, volume 1099 of *Lecture Notes in Computer Science*, pages 550–561. Springer-Verlag, 1996.

[54] M. Farach, M.O. Noordewier, S.A. Savari, and L.A. Shepp *et al.* On the entropy of DNA: Algorithms and measurements based on memory and rapid convergence. In *ACM-SIAM Symposium on Discrete Algorithms*, pages 48–57, 1995.

[55] M. Farach-Colton, P. Ferragina, and S. Muthukrishnan. On the sorting-complexity of suffix tree construction. *Journal of the ACM*, 47(6):987–1011, 2000.

[56] D.E. Ferguson. Bit-Tree: a data structure for fast file processing. *Communications of the ACM*, 35(6):114–120, 1992.

[57] P. Ferragina, R. Giancarlo, G. Manzini, and M. Sciortino. Boosting textual compression in optimal linear time. *Journal of the ACM*, 52(4):688–713, 2005.

[58] P. Ferragina and R. Grossi. Fast string searching in secondary storage: Theoretical developments and experimental results. In *ACM-SIAM Symposium on Discrete Algorithms*, pages 373–382, 1996.

[59] P. Ferragina and R. Grossi. The string B-tree: A new data structure for string search in external memory and its applications. *Journal of the ACM*, 46(2):236–280, 1999.

[60] P. Ferragina, N. Koudas, S. Muthukrishnan, and D. Srivastava. Two-dimensional substring indexing. *Journal of Computer System Science*, 66(4):763–774, 2003.

[61] P. Ferragina and F. Luccio. Dynamic dictionary matching in external memory. *Information and Computation*, 146(12), 1998.

[62] P. Ferragina and F. Luccio. String search in coarse-grained parallel computers. *Algorithmica*, 24(3–4):177–194, 1999.

[63] P. Ferragina, F. Luccio, G. Manzini, and S. Muthukrishnan. Structuring labeled trees for optimal succinctness, and beyond. In *IEEE Symposium on Foundations of Computer Science*, 2005.

[64] P. Ferragina and G. Manzini. Opportunistic data structures with applications. In *IEEE Symposium on Foundations of Computer Science*, pages 390–398, 2000.

[65] P. Ferragina and G. Manzini. An experimental study of a compressed index. *Information Sciences: special issue on "Dictionary Based Compression"*, 135:13–28, 2001.

[66] P. Ferragina and G. Manzini. An experimental study of an opportunistic index. In *ACM-SIAM Symposium on Discrete Algorithms*, pages 269–278, 2001.

[67] P. Ferragina and G. Manzini. Compression boosting in optimal linear time using the Burrows-Wheeler transform. In *ACM-SIAM Symposium on Discrete Algorithms (SODA '04)*, pages 648–656, 2004.

[68] P. Ferragina and G. Manzini. Indexing compressed text. *Journal of the ACM*, 52(4):552–581, 2005.

[69] P. Ferragina, G. Manzini, V. Mäkinen, and G. Navarro. An alphabet-friendly FM-index. In *International Symposium on String Processing and Information Retrieval*, volume 3246 of *Lecture Notes in Computer Science*, pages 150–160. Springer-Verlag, 2004.

[70] M. Frigo, C.E. Leiserson, H. Prokop, and S. Ramachandran. Cache-oblivious algorithms. In *IEEE Symposium on Foundations of Computer Science*, pages 285–298, 1999.

[71] R. Giegerich, S. Kurtz, and J. Stoye. Efficient implementation of lazy suffix trees. *Software Practice & Experience*, 33:1035–1049, 2003.

[72] J. Gil and A. Itai. How to pack trees. *Journal of Algorithms*, 32(2):108–132, 1999.

[73] G.H. Gonnet, R.A. Baeza-Yates, and T. Snider. New indices for text: PAT trees and PAT arrays. In B. Frakes and R.A. Baeza-Yates, editors, *Information Retrieval: Data Structures and Algorithms*, chapter 5, pages 66–82. Prentice-Hall, 1992.

[74] Sz. Grabowski, V. Mäkinen, and G. Navarro. First Huffman, then Burrows-Wheeler: an alphabet-independent FM-index. In *International Symposium on String Processing and Information Retrieval*, volume 3246 of *Lecture Notes in Computer Science*, pages 210–211. Springer-Verlag, 2004.

[75] R. Grossi, A. Gupta, and J. Vitter. High-order entropy-compressed text indexes. In *ACM-SIAM Symposium on Discrete Algorithms*, pages 841–850, 2003.

[76] R. Grossi, A. Gupta, and J. Vitter. Indexing equals compression: Experiments on suffix arrays and trees. In *ACM-SIAM Symposium on Discrete Algorithms*, 2004.

[77] R. Grossi and J.S. Vitter. Compressed suffix arrays and suffix trees with applications to text indexing and string matching. In *ACM Symposium on Theory of Computing*, pages 397–406, 2000.

[78] S. Grumbach and F. Tahi. A new challenge for compression algorithms: genetic sequences. *Information Processing and Management*, 30(6):875–886, 1994.

[79] D. Gusfield. *Algorithms on Strings, Trees and Sequences: Computer Science and Computational Biology*. Cambridge University Press, 1997.

[80] D. Gusfield, G.M. Landau, and B. Schieber. An efficient algorithm for the all pairs suffix-prefix problem. *Information Processing Letters*, 41(4):181–185, 1992.

[81] S. Arikawa H. Arimura, H. Sakamoto. Efficient data mining from large text databases. In *Progress in Discovery Science*, pages 123–139, 2002.

[82] R. Hariharan. Optimal parallel suffix tree construction. *Journal of Computer and*

System Sciences, 55(1):44–69, 1997.

[83] J. Healy, E.E. Thomas, J.T. Schwartz, and M. Wigler. Annotating large genomes with exact word matches. *Genome Research*, 13:2306–2315, 2003.

[84] W. Hon, T. Lam, W. Sung, and W. Tse *et al.* Practical aspects of compressed suffix arrays and FM-index in searching DNA sequences. In *Workshop on Algorithm Engineering and Experiments*, pages 31–38, 2004.

[85] W. Hon, K. Sadakane, and W. Sung. Breaking a time-and-space barrier in constructing full-text indices. In *IEEE Symposium on Foundations of Computer Science*, pages 251–260, 2003.

[86] W.K. Hon, T.W. Lam, K. Sadakane, and W.K. Sung. Constructing compressed suffix arrays with large alphabets. In *International Symposium on Algorithms and Comuptation*, volume 2906 of *Lecture Notes in Computer Science*, pages 505–516. Springer-Verlag, 2003.

[87] T.C. Hu and A.C. Tucker. Optimal computer search trees and variable length alphabetic codes. *SIAM Journal of Applied Mathematics*, 21:514–532, 1971.

[88] E. Hunt, M.P. Atkinson, and R.W. Irving. Database indexing for large DNA and protein sequence collections. *The International Journal on Very Large Data Bases*, 11(3):256–271, 2002.

[89] D.A. Hutchinson, P. Sanders, and J.S. Vitter. Duality between prefetching and queued writing with parallel disks. In *European Symposium on Algorithms*, volume 2161 of *Lecture Notes in Computer Science*, pages 62–73. Springer-Verlag, 2001.

[90] IBM Journal on Research and Development. *The Memory eXpansion Technology for xSeries servers*, March 2001.

[91] P. Indyk. Approximate nearest neighbor under edit distance via product metrics. In *ACM-SIAM Symposium on Discrete Algorithms*, pages 646–650, 2004.

[92] P. Indyk and R. Motwani. Approximate nearest neighbors: towards removing the curse of dimensionality. In *ACM symposium on Theory of computing*, pages 604–613, 1998.

[93] H. Itoh and H. Tanaka. An efficient method for in memory construction of suffix arrays. In *Symposium on String Processing and Information Retrieval*, pages 81–88, 1999.

[94] M. Jackson, T. Strachan, and G. Dover. *Human Genome Evolution*. Bios Scientific Publisher, 1996.

[95] G. Jacobson. Space-efficient static trees and graphs. In *IEEE Symposium on Foundations of Computer Science*, pages 549–554, 1989.

[96] H.V. Jagadish, N. Koudas, and D. Srivastava. On effective multi-dimensional indexing for strings. *ACM SIGMOD Record*, 29(2):403–414, 2000.

[97] P. Jokinen and E. Ukkonen. Two algorithms for approximate string matching in static texts. In *Matematical Foundations of Computer Science*, number 520 in Lecture Notes in Computer Science, pages 240–248. Springer-Verlag, 1991.

[98] T. Kahveci and A. Singh. MAP: searching large genome databases. In *Pacific Symposium on Biocomputing*, pages 303–314, 2003.

[99] T. Kahveci and A.K. Singh. Efficient index structures for string databases. In *International Conference on Very Large Data Bases*, pages 351–360, 2001.

[100] J. Kärkkäinen and P. Sanders. Simple linear work suffix array construction. In *International Colloquium on Automata, Languages and Programming*, volume 2719 of *Lecture Notes in Computer Science*, pages 943–955. Springer-Verlag, 2003.

[101] J. Kärkkäinen and E. Ukkonen. Sparse suffix trees. In *International Conference on Computing and Combinatorics*, volume 1090 of *Lecture Notes in Computer Science*, pages 219–230. Springer-Verlag, 1996.

[102] R. Karp, R. Miller, and A. Rosenberg. Rapid Identification of Repeated Patterns in Strings, Arrays and Trees. In *ACM Symposium on Theory of Computation*, pages 125–136, 1972.

[103] T. Kasai, G. Lee, H. Arimura, and S. Arikawa *et al.* Linear-time longest-common-prefix computation in suffix arrays and its applications. In *Symposium on Combinatorial Pattern Matching*, volume 2089 of *Lecture Notes in Computer Science*, pages 181–192. Springer-Verlag, 2001.

[104] J. Katajainen and E. Makinen. Tree compression and optimization with applications. *International Journal of Foundations of Computer Science*, 1(4):425–447, 1990.

[105] D.K. Kim, J.S. Sim, H. Park, and K. Park. Linear-time construction of suffix arrays. In *Symposium on Combinatorial Pattern Matching*, volume 2676 of *Lecture Notes in Computer Science*, pages 186–199. Springer-Verlag, 2003.

[106] D.E. Knuth. *Sorting and Searching*, volume 3 of *The Art of Computer Programming*. Addison-Wesley, Reading, MA, USA, second edition, 1998.

[107] P. Ko and S. Aluru. Linear time construction of suffix arrays. In *Combinatorial Pattern Matching Conference*, volume 2676 of *Lecture Notes in Computer Science*, pages 200–210. Springer-Verlag, 2003.

[108] S. Kosaraju. Real-time pattern matching and quasi-real-time construction of suffix trees. *ACM Symposium on Theory of Computing*, pages 310–316, 1994.

[109] P. Kumar. Cache-oblivious algorithms. In U. Meyer, P. Sanders, and J.F. Sibeyn, editors, *Algorithms for Memory Hierarchies*, volume 2625 of *Lecture Notes in Computer Science*, pages 193–212. Springer-Verlag, 2003.

[110] S. Kurtz. Reducing the space requirement of suffix trees. *Software—Practice and Experience*, 29(13):1149–1171, 1999.

[111] S. Kurtz and C. Schleiermacher. REPuter: Fast computation of maximal repeats in complete genomes. *Bioinformatics*, 15(5):426–427, 1999.

[112] E. Kushilevitz, R. Ostrovsky, and Y. Rabani. Efficient search for approximate nearest neighbor in high dimensional spaces. In *ACM symposium on theory of computing*, pages 614–623, 1998.

[113] E. Kushilevitz, R. Ostrovsky, and Y. Rabani. Efficient search for approximate nearest neighbor in high dimensional spaces. *SIAM Journal on Computing*, 30(2):457–474, 2000.

[114] T.W. Lam, K. Sadakane, W.K. Sung, and S.M. Yiu. A space and time efficient algorithm for constructing compressed suffix arrays. In *International Conference on Computing and Combinatorics*, volume 2387 of *Lecture Notes in Computer Science*, pages 401–410. Springer-Verlag, 2002.

[115] J.K. Lanctot, M. Li, and E. Yang. Estimating DNA sequence entropy. In *ACM-SIAM Symposium on Discrete Algorithms*, pages 409–418, 2000.

[116] N.J. Larsson and K. Sadakane. Faster suffix sorting. Technical Report LU-CS-TR:99-214, Department of Computer Science, Lund University, 1999.

[117] D.J. Lipman and W.R. Pearson. Rapid and sensitive protein similarity searches. *Science*, 227:1435–1441, 1985.

[118] D.P. Lopresti and A. Tomkins. Block edit models for approximate string matching. *Theoretical Computer Science*, 181(1):159–179, 1997.

[119] E. Mäkinen. A survey on binary tree codings. *The Computer Journal*, 34(5):438–443, 1991.

[120] V. Mäkinen, G. Navarro, and K. Sadakane. Advantages of backward searching—efficient secondary memory and distributed implementation of compressed suffix arrays. In *International Symposium on Algorithms and Computation*, Lecture Notes in Computer Science. Springer-Verlag, 2004.

[121] U. Manber and G. Myers. Suffix arrays: a new method for on-line string searches. *SIAM Journal on Computing*, 22(5):935–948, 1993.

[122] G. Manzini. Two space saving tricks for linear time LCP array computation. In *Scandinavian Workshop on Algorithm Theory*, pages 372–383, 2004.

[123] G. Manzini and P. Ferragina. Lightweight suffix sorting home page. `http://www.mfn.unipmn.it/~manzini/lightweight`, 2003.

[124] G. Manzini and P. Ferragina. Engineering a lightweight suffix array construction algorithm. *Algorithmica*, 40:33–50, 2004.

[125] L. Marsan and M.F. Sagot. Algorithms for extracting structured motifs using a suffix tree with an application to promoter and regulatory site consensus identification. *Journal of Computational Biology*, 7:345–360, 2000.

[126] W. J. Masek and M. S. Paterson. A faster algorithm for computing string edit distances. *Journal of Computer System Science*, 20(1):18–31, 1980.

[127] E.M. McCreight. A space-economical suffix tree construction algorithm. *Journal of the ACM*, 23(2):262–272, 1976.

[128] T.H. Merrett and H. Shang. Trie methods for representing text. In *International Conference on Foundations of Data Organization and Algorithms*, volume 730 of *Lecture Notes in Computer Science*, pages 130–145. Springer-Verlag, 1993.

[129] H.W. Mewes and K. Heumann. Genome analysis: Pattern search in biological macromolecules. In *Symposium on Combinatorial Pattern Matching*, volume 937 of *Lecture Notes in Computer Science*, pages 261–285. Springer-Verlag, 1995.

[130] D.R. Morrison. PATRICIA - practical algorithm to retrieve coded in alphanumeric. *Journal of the ACM*, 15(4):514–534, 1968.

[131] E. Moura, G. Navarro, and N. Ziviani. Indexing compressed text. In *South American Workshop on String Processing*, pages 95–111, 1997.

[132] I. Munro. Succinct data structures. In *Workshop on Data Structures*, within the Conference on Foundations of Software Technology and Theoretical Computer Science, pages 1–6, 1999.

[133] I. Munro and V. Raman. Succinct representation of balanced parentheses, static trees and planar graphs. In *Proceedings of IEEE Symposium on Foundations of Computer Science*, pages 118–126, 1997.

[134] I. Munro and V. Raman. Succinct representation of balanced parentheses and static trees. *SIAM Journal on Computing*, 31(3):762–776, 2001.

[135] I. Munro, V. Raman, and S. Srinivasa Rao. Space efficient suffix trees. In *Proceeding of Conference on Foundations of Software Technology and Theoretical Computer Science*, pages 186–195. Springer-Verlag LNCS n. 1530, 1998.

[136] I. Munro, V. Raman, and S. Srinivasa Rao. Space efficient suffix trees. *Journal of Algorithms*, 39(2):205–222, 2001.

[137] S. Muthukrishnan and S.C. Sahinalp. Approximate nearest neighbors and sequence comparison with block operations. In *ACM symposium on Theory of computing*, pages 416–424, 2000.

[138] S. Muthukrishnan and S.C. Sahinalp. Simple and practical sequence nearest neighbors with block operations. In *Symposium on Combinatorial Pattern Matching*, volume 2373 of *Lecture Notes in Computer Science*, pages 262–278. Spring-Verlag, 2002.

[139] S. Muthukrishnan and S.C. Sahinalp. An efficient algorithm for sequence comparison with block reversals. *Theoretical Computer Science*, 321(1):95–101, 2004.

[140] E.W. Myers. A sublinear algorithm for approximate keyword searching. *Algorithmica*, 12(4/5):345–374, 1994.

[141] G. Navarro. A guided tour to approximate string matching. *ACM Computing Surveys*, 33(1):31–88, 2001.

[142] G. Navarro. Indexing text using the Ziv-Lempel trie. *Journal of Discrete Algorithms*, 2(1):87–114, 2004.

[143] G. Navarro and R. Baeza-Yates. A hybrid indexing method for approximate string matching. *Journal of Discrete Algorithms*, 1(1):21–49, 2000.

[144] G. Navarro, E. Barbosa, R. Baeza-Yates, and W. Cunto *et al.* Binary searching with non-uniform costs and its application to text retrieval. *Algorithmica*, 27(2):145–169, 2000.

[145] G. Navarro, E. Sutinen, J. Tanninen, and J. Tarhio. Indexing methods for approximate string matching. *IEEE Data Engineering Bulletin*, 24(4):19–27, 2001.

[146] C.G. Nevill-Manning and I.H. Witten. Protein is incompressible. In *IEEE Data Compression Conference*, pages 257–266, 1999.

[147] M.H. Nodine and J.S. Vitter. Deterministic distribution sort in shared and distributed memory multiprocessors. In *ACM Symposium on Parallel Algorithms and Architectures*, pages 120–129, 1993.

[148] M.H. Nodine and J.S. Vitter. Greed sort: optimal deterministic sorting on parallel disks. *Journal of the ACM*, 42(4):919–933, 1995.

[149] W.R. Pearson. Rapid and sensitive sequence comparison with FASTP and FASTA. *Methods Enzymology*, 183:63–98, 1990.

[150] W.R. Pearson. Flexible sequence similarity searching with the FASTA3 program package. *Methods in Molecular Biology*, 132:185–219, 2000.

[151] W.R. Pearson and D.J. Lipman. Improved tools for biological sequence comparison. *Proc. National Academy of Science USA*, 85:2444–2448, 1988.

[152] C.S. Perng. Indexing temporal series with String B-trees. Manuscript (personal communication), University of California, Los Angeles., 1999.

[153] W. Pugh. Skip Lists: A Probabilistic Alternative to Balanced Trees. *Communications of the ACM*, 33(6):668–676, 1990.

[154] K.R. Rose. Asynchronous generic key/value database. Master's thesis, Massachusetts Institute of Technology, September 2000.

[155] C. Ruemmler and J. Wilkes. An introduction to disk drive modeling. *IEEE Computer*, 27(3):17–29, 1994.

[156] K. Sadakane. A fast algorithms for making suffix arrays and for Burrows-Wheeler transformation. In *IEEE Data Compression Conference*, pages 129–138, 1998.

[157] K. Sadakane. Compressed text databases with efficient query algorithms based on the compressed suffix array. In *International Symposium on Algorithms and Computation*, volume 1969 of *Lecture Notes in Computer Science*, pages 410–421. Springer-Verlag, 2000.

[158] K. Sadakane. Succinct representations of LCP information and improvements in the compressed suffix arrays. In *ACM-SIAM Symposium on Discrete Algorithms*, pages 225–232, 2002.

[159] K. Sadakane. New text indexing functionalities of compressed suffix arrays. *Journal of Algorithms*, 48(2):294–313, 2003.

[160] K. Sadakane and T. Shibuya. Indexing huge genome sequences for solving various problems. *Genome Informatics*, 12:175–183, 2001.

[161] S.C. Sahinalp and U. Vishkin. Symmetry breaking for suffix tree construction. In *ACM Symposium on Theory of Computing*, pages 300–309, 1994.

[162] D. Salomon. *Data Compression: the Complete Reference*. Springer Verlag, 1997.

[163] K.B. Schürmann and J. Stoye. Suffix tree construction for large strings. In *Workshop of Fundamentals of Databases*, 2002.

[164] P.H. Sellers. The theory and computation of evolutionary distances: Pattern recognition. *Journal of Algorithms*, 1(4):359–373, 1980.

[165] J. Seward. BZIP2 home page, 1997. `http://sources.redhat.com/bzip2`.

[166] J. Seward. On the performance of BWT sorting algorithms. In *IEEE Data Compression Conference*, pages 173–182, 2000.

[167] H. Shang. *Trie methods for text and spatial data structures on secondary storage*. PhD thesis, McGill University, 1995.

[168] F. Shi. Fast approximate string matching with q-blocks sequences. In *South American Workshop on String Processing*, pages 257–271, 1996.

[169] S. Skiena. Who is interested in algorithms and why? Lessons from the stony brook algorithm repository. In *Workshop on Algorithmic Engineering*, pages 204–212, 1998.

[170] D.D. Sleator and R.E. Tarjan. Self-adjusting binary search trees. *Journal of the ACM*, 32(3):652–686, July 1985.

[171] Sleepycat Software. The Berkeley DB. `http://www.sleepycat.com/`.

[172] E. Sutinen and J. Tarhio. On using q-gram locations in approximate string matching. In *European Symposium on Algorithms*, volume 979 of *Lecture Notes in Computer Science*, pages 327–340. Springer-Verlag, 1995.

[173] E. Sutinen and J. Tarhio. Filtration with q-samples in approximate string matching. In *Symposium on Combinatorial Pattern Matching*, volume 1075 of *Lecture Notes in Computer Science*, pages 50–63. Springer-Verlag, 1996.

[174] W. Szpankowski. A generalized suffix tree and its (un)expected asymptotic behaviors. *SIAM Journal on Computing*, 22(6):1176–1198, 1993.

[175] E. Ukkonen. Approximate string matching over suffix trees. In *Symposium on Combinatorial Pattern Matching*, volume 684 of *Lecture Notes in Computer Science*, pages 228–242. Springer-Verlag, 1993.

[176] E. Ukkonen. On-line construction of suffix trees. *Algorithmica*, 14:249–260, 1995.

[177] J.S. Vitter. External memory algorithms and data structures: Dealing with MASSIVE DATA. *ACM Computing Surveys*, 33(2):209–271, 2002.

[178] J.S. Vitter and E. Shriver. Algorithms for parallel memory: Two-level memories. *Algorithmica*, 12:110–147, 1994.

[179] M.M. Waldrop. On-line archives let biologists interrogate the genome. *Science*, 269:1356–1358, 1995.

[180] P.J. Weinberger. Unix B-trees. Technical report, AT&T Bell Laboratories, 1995.

[181] P. Weiner. Linear pattern matching algorithm. In *IEEE Symposium on Switching and Automata Theory*, pages 1–11, 1973.

[182] I.H. Witten, A. Moffat, and T.C. Bell. *Managing Gigabytes: Compressing and Indexing Documents and Images*. Morgan Kaufmann Publishers, second edition, 1999.

[183] J. Yang, W. Wang, Y. Xia, and P.S. Yu. Accelerating approximate subsequence search on large protein sequence databases. In *IEEE Computer Society Bioinformatics Conference*, pages 207–218, 2002.

[184] O. Zamir and O. Etzioni. Grouper: a dynamic clustering interface to Web search results. *Computer Networks*, 31(11–16):1361–1374, 1999.

[185] D. Zhang and Y. Dong. Hierarchical, online clustering of web search results. In *International Workshop on Web information and data management*, 2001.

[186] J. Zobel, A. Moffat, and K. Ramamohanarao. Guidelines for presentation and comparison of indexing techniques. *SIGMOD Record*, 25(3):10–15, 1996.

36

Index Structures for Approximate Matching in Sequence Databases

Tamer Kahveci
University of Florida, Gainesville

Ambuj K. Singh
University of California, Santa Barbara

36.1 Why do we Need Index Structures?

Approximate sequence searching is crucial in many problems. Pairwise sequence comparison, multiple sequence alignment, motif finding, shotgun sequence assembly are only a few of countless examples. Hundreds of thousands of approximate sequence search queries are performed daily around the world by scientists. Approximate searches are widely used for evolutionary analysis, identification of coding regions, phylogenetic analysis, structural analysis and classification. Pairwise comparison of sequences is a well studied problem. A number of exhaustive search methods have already been devised to find both local and global alignments such as dynamic programming [48, 56] or finite automata [5]. Why does one need an index structure to find sequence alignments given these powerful tools? In order to understand the need for an index structure, consider the following three examples.

Example 36.1

(Sequence search) One of the elementary problems on sequence databases is searching similarities to a query sequence. This problem involves comparison of two sequences. This type of search is useful for many purposes such as finding related genes, evolutionary analysis, and identification of repeat regions. Exhaustive search methods consider all possible combinations of insertions, deletions and matches/mismatches through dynamic programming. The costs of these methods are determined as the product of the length of the sequences compared. Assume that a user has a query sequence of 10^4 base pairs and searches for similarity in chromosome 22 of *homo sapiens*, one of the shortest chromosomes, that contains $33 \cdot 10^6$ base pairs. Such a comparison requires approximately $33 \cdot 10^{10}$ operations. On a 1 Ghz computer, this search completes in approximately three hours. If the same query is posed on the entire human genome, with more than 3 billion base pairs, then it will require more than $3 \cdot 10^{13}$ operations. In this case, the search takes more than one week. If the search is done against the entire GenBank database, containing 30 billion base pairs (at the end of 2002), the search will complete in more than two months.

Example 36.2

(Multiple alignment) Often, alignment of multiple genomes is needed to understand the structural, functional, and evolutionary relationship across different organisms. Unlike pairwise sequence comparison, multiple alignment brings similar letters of more than two sequences together. This problem is computationally harder than pairwise comparison for it considers all possible combinations of sequences and letters. Assume that each sequence has approximately 10^3 letters. If the multiple alignment involves three sequences, then the alignment involves $O(10^3 \times 10^3 \times 10^3) = O(10^9)$ comparisons, which is still feasible. However, as the number of sequences increases beyond four, multiple alignment using exhaustive search becomes impractical.

Example 36.3

(Shotgun sequencing) Another time-consuming problem on sequence databases is the *shotgun sequencing* problem. This problem requires an all-to-all sequence comparison for repeat and overlap detection. PCAP [26], one of the most recent sequencing tools, assembled the entire mouse genome from 30 Gbp unsequenced database in more than one week using 80 parallel computers. On a single computer, the same problem takes almost two years.

The size of genome databases is increasing exponentially [8]. Statistics show that the size of GenBank has doubled every 15 months. This makes the classic exhaustive search methods impractical due to extensive time requirements. The search time can be reduced by using faster computers or distributing the job to parallel computers. However, this is not feasible for several reasons. First, the speed of computers is growing slower than the size of the sequence databases. Second, many companies or individual researchers prefer to store and search their databases locally for various reasons. Parallel computers are still not affordable for such personal use. The alternative way to reduce the search time is to reduce the amount of data to be searched by ignoring the irrelevant parts of the database for a given query. Usually, only a small percentage of the database matches to a given query segment. If the dissimilar sequences can be determined quickly, the similarity search can avoid inspecting them. A number of index structures have been devised to find such similarities. They can be classified under three categories: k-gram indexing, direct indexing, and vector space indexing. We will discuss the index structures and several tools that employ them for each of these categories.

36.2   K-gram Indexing

One of the features used commonly for sequence searching is a k-gram, also called a k-mer. A k-gram is a sequence of length k, where k is a positive integer. For example, TATGGCAA is an 8-gram. A k-gram is usually considered as the shortest subsequence that should match exactly for meaningful alignments. Such matches are then combined or extended to find longer alignments with mismatches and indels.

A variant of k-gram is the non-sequential k-gram. A non-sequential k-gram is a subsequence of length k', $k' > k$, where $k'-k$ letters are wild cards. For example, TA**GCA*T*AA is an 8-gram with 4 wild cards shown using the symbol *. The wild card letters are ignored when k-grams are matched.

We classify index structures according to how they store k-grams. Throughout this section, we will use sequential k-grams.

36.2.1 Hash tables

One of the most widely used index for sequence search is the *hash table* (also called the *lookup table*). Hash tables enable quick lookup for a set of prespecified sequences. For a given query sequence, once the matches from this set of sequences are determined with the help of a hash table, they are used to find better matches. Both lossless [46] and lossy [38] search tools have been developed with the help of hash tables. Two of the most well known genome search tools that use hash table are FASTA [51] and BLAST [2]. Next, we discuss the hash tables in more detail.

The hash table of a sequence is defined using two parameters: alphabet and word length. Let $\Sigma = \{\alpha_1, \alpha_2, \cdots, \alpha_\sigma\}$ be the alphabet that defines the sequences, where σ is the alphabet size. The ith letter, α_i, is encoded with binary representation of $i-1$ using $\lceil \log_2 \sigma \rceil$ bits. For example, the letters in the DNA alphabet $\{A, C, G, T\}$ are encoded as A $= 00$, C $= 01$, G $= 10$, and T $= 11$. A sequence is encoded as the concatenation of the binary representations of the letters that constitute that sequence. For example, the DNA sequence GGCA is represented as 10100100.

Word length defines the length of the subsequence that will be indexed by the hash table. A *word* is defined as a sequence of length w, where w is the word length. There are σ^w distinct words of length w. This is because each of the w letters can take σ possible distinct values. For a given alphabet Σ and word size w, the hash table for a sequence s is defined as an array H of length σ^w. The entries of this array correspond to the distinct words of length w constructed from Σ (i.e., $H[i]$ stands for the ith word of length w). For all $i \in \{0, \cdots, \sigma^w\}$, $H[i]$ contains the starting positions of all the occurrences of the ith word in s. Figure 36.1 shows the hash table built on a 20 letter DNA sequence for $w = 2$. In this example, the hash table contains 4^2 entries since $\sigma = 4$.

The hash table for s can be constructed by sequentially sliding a window of length k on it. Every positioning of this window corresponds to a k-gram. Sliding this window by one includes a new letter and excludes an existing letter. The index of the new k-gram in the hash table can be determined in $O(k \log_2 \sigma)$ time by removing the preceding $\log_2 \sigma$ bits and appending new $\log_2 \sigma$ bits at the end of the previous k-gram. For example, in Figure 36.1, the first window contains the subsequence AG. The index of this window is 0010. The next window contains GC. Therefore, the preceding two bits (00) for A are removed and two new bits are appended (01) for C. The resulting index is 1001. The total time required to build a hash table on s is

$$O((n - k + 1) \cdot (k \log_2 \sigma)),$$

where n is the length of s, since it contains $n - k + 1$ k-grams.

A hash table enables fast lookup for k-grams. For a given k-gram, its position in the hash table is computed in $O(k)$ time. If the hash table is built on a sequence s, then a given query k-gram matches $(n - k + 1)/\sigma^k$ k-grams of s on the average. This is because s has $n-k+1$ k-grams, and each k-gram matches to a given random k-gram with probability $1/\sigma^k$. Therefore, the total cost of returning the hits for a single k-gram is $O(k + (n - k + 1)/\sigma^k)$. If a given query sequence contains m letters, then looking up all the k-grams of that query sequence costs

$$O((m - k + 1) \cdot (k + (n - k + 1)/\sigma^k)).$$

A good index structure for sequences needs to be dynamic. This is because, as the genomes of new organisms are sequenced, new sequences will be added into existing databases.

```
A   G   C   T   T   T   T   C   A   T   T   C   T   G   A   C   T   G   C   A
0   1   2   3   4   5   6   7   8   9  10  11  12  13  14  15  16  17  18  19
```

```
AA  (0000)
AC  (0001)  ──►14
AG  (0010)  ──►  0
AT  (0011)  ──►  8
CA  (0100)  ──►  7,  18
CC  (0101)
CG  (0110)
CT  (0111)  ──►  2,  11,  15
GA  (1000)  ──►13
GC  (1001)  ──►  1,  17
GG  (1010)
GT  (1011)
TA  (1100)
TC  (1101)  ──►  6,  10
TG  (1110)  ──►12,  16
TT  (1111)  ──►  3,   4,   5,   9
```

FIGURE 36.1: A DNA sequence and the hash table constructed on it when word size is 2. The numbers in parenthesis show the binary code of the corresponding sequence.

Also one may need to modify or remove an existing sequence, or a part of a sequence. Insertions of a new k-gram into an existing hash table is trivial. First, the bit representation of this k-gram is computed in $O(k \log_2 \sigma)$ time. Next, it is appended at the end of the corresponding hash table entry in constant time. In order to remove an existing k-gram, first, its is located by a hash table lookup. Next, it is removed from the hash table. An existing k-gram is modified by a deletion followed by an insertion. Note that modification or removal of an existing letter in a sequence alters k k-grams of that sequence since a letter is a part of k different k-grams. Similarly, insertion of a new letter alters $k-1$ k-grams.

Now, consider the space requirement of the hash tables. Hash table contains σ^k entries. Each of these entries contain one pointer to show the first k-gram for that entry (4 bytes). Each k-gram in the hash table requires one integer to store its location on the sequence and one pointer for the next k-gram (8 bytes). If the hash table is built on more than one sequence simultaneously, then each k-gram also needs to store the identification of its source sequence. Let B be the total number of bytes required for a single k-gram. The total memory consumption of a hash table for a sequence of length n is

$$O(4\sigma^k + B(n - k + 1)).$$

Next we consider some search tools based on hash tables.

FASTP: a case study

FASTP [38], one of the earliest hash table-based sequence search tools, looks for similarities between amino acid sequences. FASTP is a heuristic for finding local alignments. That is, it does not guarantee to return the actual best local alignments. Given two protein sequences q and s of lengths m and n respectively, FASTP runs in 3 phases:

- *Phase 1:* A hash table is created on q for word size of $w = 1$ or 2. Each word of length w in s is extracted by sequentially scanning. Let s_i be the word in s starting at position $i \in [0, n - w]$. The positions of all exact occurrences of s_i in q are found by a hash table lookup. Let j be one such starting position. Then the offset for (s_i, j) pair is computed as $i - j$. The offset can take values anywhere in $[1 - m, n - 1]$. The frequency of an offset shows the number of amino acids that exactly match in the gapless alignment of q and s when s is shifted by that offset. Each offset is scored as the number of matches minus the number of mismatches. The offsets with high scores show the locally similar regions.

- *Phase 2:* The five offsets with highest scores are inspected to locate the beginning and end positions of the alignments. These regions are then re-scored using an amino acid substitution matrix. The default matrix used is PAM250. The score obtained at this phase is called the *initial score*.

- Phase 3: For each of the alignments in Phase 2, an *optimized score* is computed using the Needleman-Wunsch dynamic programming algorithm [48]. The results are reported in decreasing order of initial scores.

FASTA: an improvement over FASTP

Pearson and Lipman [51] developed a more sensitive version of FASTP, called FASTA, which can be used to find similarities between DNA sequences as well as protein sequences. FASTA can also find similar regions between a DNA and a protein sequence by translating the DNA sequence. FASTA has one additional phase, squeezed between the phases 2 and 3 of FASTP, say phase 2.5. Another difference between FASTA and FASTP is that FASTA chooses 10 best alignments at phase 2 instead of five. The additional phase in FASTA is as follows:

- Phase 2.5: Given the beginning and end locations of the alignments that have a score above a certain cutoff, they are checked to see if several of them can be joined together to find better alignments. Joining two alignments incurs a *joining penalty*, which is similar to the gap penalty. FASTA finds the optimal alignment by attaching the alignments with maximal score that are close together.

Other tools based on hash tables

A number of tools employ hash tables for finding sequence similarities. Altschul, Gish, and Miller developed one of the most popular sequence alignment tools, called BLAST [2]. A number of BLAST derivatives have also been developed. Some of them are BL2SEQ [60], PSI-BLAST [3], PHI-BLAST [65], MegaBLAST [66], BLASTZ [54], WU-BLAST (http://blast.wustl.edu), BLAT [32], SSAHA [49], and SENSEI [57]. Unlike these tools, FLASH [14], WABA [34] and PatternHunter [39] build hash table on non-sequential k-grams. GLASS [7] finds initial anchors for global alignment using a fixed value of k for k-grams, and recursively reduces the value of k in order to find the matches between anchors.

The CAP3 algorithm [25], developed by Huang and Madan, uses a hash table for finding repeats and overlaps for sequence assembly. Given a set of nucleotide sequences, called reads, CAP3 creates a single sequence by gluing these sequences with a special letter which is not in the alphabet. Next, a hash table is built on this sequence for its 12-grams. The combined sequence is then sequentially scanned, and the number of exact matches to the 12-grams of each read are accumulated with the help of the hash table. If the number of such matches is greater than a threshold for a letter in a read, then that letter is considered as a repeat. A number of other methods have also been developed for shotgun sequencing. Some of them

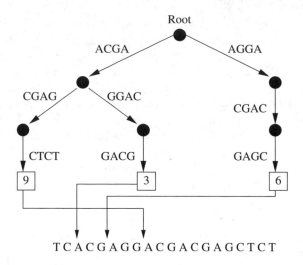

FIGURE 36.2: The ed-tree on a sequence for $k = 12$, $\Delta = 3$, and $H = [4, 4, 4]$.

are GigAssembler [33], RePS [63], Arachne [6, 28], AMASS [35], TIGR assembler [18, 58] Phusion [45], two generations of CAP [23, 24], and JAZZ [4]

36.2.2   ed-trees

The ed-tree enables subsequence searches with the help of k-grams [59]. Similar to BLAST-like searches, the ed-tree runs in two phases. In phase 1, the index is searched for short seeds. In phase 2, the seeds are extended for longer matches. The ed-tree differs from BLAST only in the first phase; the seeds of the ed-tree are longer than that of BLAST. Furthermore, the ed-tree allows some error for seed matches while BLAST's seeds are exact matches. Next, we discuss the construction of the ed-tree.

The ed-tree is defined by three parameters: 1) The gram size, k. 2) The skip interval, Δ. 3) The segment length vector, $H = [h_1, \cdots, h_t]$, where $\sum_{i=1}^{t} h_i = k$, and t is the size of the segment length vector. The value of k defines the length of the k-grams that will be indexed. The skip interval, Δ, defines the locations of the k-grams extracted from the database sequence: Let s be a database sequence. Only the k-grams of s starting at positions $\Delta, 2\Delta, \cdots, \lfloor (|s| - k + 1)/\Delta \rfloor \Delta$ are indexed. The last parameter H defines the partitioning of each k-gram. Each k-gram, x, is partitioned into t non-overlapping subsequences $x_1^H, \cdots$ x_t^H of length $h_1, \cdots, h_t$ respectively. For example, if $x = $ GAATTCGTCGAC is a 12-gram and $H = [3, 5, 4]$, then $x_1^H = $ GAA, $x_2^H = $ TTCGT, and $x_3^H = $ CGAC.

The tree construction algorithm takes the database D, gram size k, skip interval Δ, and the segment length vector H as input. Initially it creates an empty root node. For each sequence in D, the algorithm generates all the k-grams with Δ skips. Later, each gram is partitioned according to H. The partitions are then sequentially inserted into the ed-tree. Figure 36.2 shows an ed-tree constructed on a sequence for $k = 12$, $\Delta = 3$, and $H = [4, 4, 4]$. The typical values of the parameters for the actual ed-trees are $k = 18$, $\Delta = 2$, and $H = [6, 6, 6]$.

The queries on ed-trees use the following observation: $ED(x, y) \geq ||x| - |y||$, where $ED(x, y)$ is the edit distance between sequences x and y. This is true since the optimal alignment of x and y requires at least $||x| - |y||$ indels. Let $x_1^H, \cdots, x_t^H$ and $y_1^{H'}, \cdots, y_t^{H'}$ be the partitioning of x and y respectively such that x_i^H is aligned to $y_i^{H'}$ for $1 \leq i \leq t$

in the optimal alignment of x and y Let $\delta_i = |x_i^H - y_i^{H'}|$, for $1 \leq i \leq t$. It follows that $\mathrm{ED}(x, y) \geq \sum_i |\delta_i|$. The vector $\delta = [\delta_1, \cdots, \delta_t]$ is defined as the length difference vector.

Given a query k-gram, q, and a query radius, r, all possible length difference vectors, δ, for which $\sum_i |\delta_i| \leq r$ are created. Every δ defines a new segment length vector $H' = H + \delta$ that partitions q into $q_1^{H'}, \cdots, q_t^{H'}$. Starting from the root node of the ed-tree, $q_1^{H'}$ is aligned to all the children of the root. If the edit distance, ed, to a child is more than r, then that child is discarded. Otherwise, the radius is refined as $r' = r - ed$ and $q_2^{H'}$ is aligned to all the children of that child. The search proceeds top down similarly for the rest of q. As the value of r decreases, the search becomes faster at the expense of reduced sensitivity. Typically r is set to 2 for good performance and sensitivity.

Some important aspects of ed-trees are: 1) The k-grams are usually longer than that of BLAST. 2) They allow inexact k-gram matches. 3) Only one k-gram out of Δ k-grams is indexed. This reduces the index size at the expense of reduced sensitivity.

The ed-tree is constructed in $O(|D|t/\Delta)$ time, where $|D|$ is the database size. The space consumption of the ed-tree is $O(|D|)$. For a dataset of 2 Gbp, the size of the ed-tree varies from 2.3 to 3.0 GB.

The ed-tree is a dynamic data structure. New k-grams can be inserted in $O(k)$ time. Removal and modification of an existing k-gram is also done in $O(k)$ time.

CHAOS [12] finds local similarities between two sequences using threaded trie, a variation of the ed-tree. Threaded trie has parameters $\Delta = 1$ and $H = [1, 1, \cdots, 1]$. LAGAN [10] and DIALIGN [43, 44] employ CHAOS to find anchors for global alignment. Brudno et al. also use CHAOS to find *glocal alignment* which is a combination of global and local alignments [11].

36.3   Direct Indexing

An important set of index structures indexes sequences directly. Unlike k-gram–based indexes, these techniques index subsequences of varying sizes. As a result of this, longer matches can be found using these index structures. On the other hand, k-gram indexes require an additional step to combine k-gram matches to find longer matches.

36.3.1   Suffix trees

Let $s = s[0]s[1]\cdots s[n]$ be a sequence with n letters, where $s[i]$ is the ith letter, $1 \leq i \leq n$. Let $s[i : j]$ indicate the subsequence of s that starts from position i and ends at position j for $1 \leq i \leq j \leq n$. As the name indicates, the suffix tree of s is a tree structure that contains all suffixes $s[i : n]$. Suffix trees have been useful in sequence searches since they can be used to locate exact subsequence matches quickly.

Suffix trees were first proposed by Weiner [52] under the name *position tree*. McCreight proposed a space efficient technique for the construction of the suffix trees [41]. Later, Ukkonen developed an on-line construction method [62]. A variety of other suffix tree implementations have also been proposed such as implicit suffix tree [62], dynamic suffix tree [15], suffix tree without suffix links [27], string B-tree [17], suffix cactus [31], suffix array [40] and their compressed versions [21]. We will discuss several variations of suffix trees without implementation details. We refer the interested reader to Chapter 35 by Ferragina.

Figure 36.3 shows the suffix tree built on sequence $s = \mathrm{TGAGTGCGA\$}$. A dummy letter, $\$$, marks the end of the sequence. Every path from the root node to a leaf node through the solid arrows defines a suffix of s. The numbers at the leaf nodes show the starting

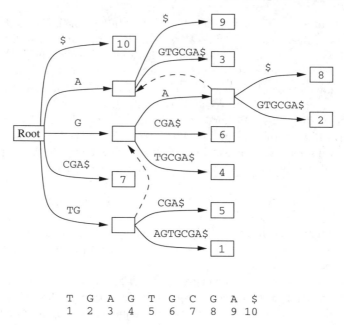

T G A G T G C G A $
1 2 3 4 5 6 7 8 9 10

FIGURE 36.3: Suffix tree built on sequence TGAGTGCGA. The dashed arrows are the suffix links. A dummy letter, $, marks the end of the sequence.

position of the suffix denoted by that leaf. The dashed arrows in the figure are the suffix links. There is a suffix link from an internal node u to another internal node v if u and v are labeled with suffixes $c\alpha$ and α respectively for a given letter c. Suffix links enable faster construction of suffix trees.

Implicit suffix trees [27] do not mark the end of the sequence with a dummy letter. In addition to this, they do not contain the nodes that have only one children node. Instead, each such node collapses with its parent. Figure 36.4 shows the implicit suffix tree for the sequence $s =$ TGAGTGCGA. Compared to the Figure 36.3, the implicit suffix tree in this figure does not have the leaf node 9. Since removal of node 9 leaves its parent with a single child, that parent is replaced with node 3, and the edges A and GTGCGA are appended. Although all the suffixes of s are embedded into the implicit suffix tree, there may not be a leaf node for some of the suffixes. For example, there is no leaf node for the suffix A in Figure 36.4. It is rather detected by the branch that goes from the root node to the node 2.

A suffix cactus [31] reduces the space consumption of the suffix tree by reducing the number of pointers: It collapses the path from a node to a leaf node into a single branch. Each sibling of the nodes on this path creates a new branch. Figure 36.5 depicts the suffix cactus built on the sequence $s =$ TGAGTGCGA.

The naive construction method builds the suffix tree by inserting each suffix into the suffix tree iteratively. This method takes $O(n^2)$ time: $O(m)$ time for suffix of length m, $1 \leq m \leq n$. However, suffix trees can be constructed in $O(n)$ time with a careful implementation [62].

Suffix trees are notorious for their excessive memory usage [47]. Although the space complexity is $O(n)$, the constant in the big-Oh may be large. The size of the suffix tree depends on the alphabet and the distribution of the letters in the database sequence. Kärkkäinen reports 10 bytes per letter [31]. Meek et al. uses 12.5 bytes per letter for SWISSPROT database [42]. Hunt et al. states 21 bytes per letter for DNA databases [27]. Delcher et

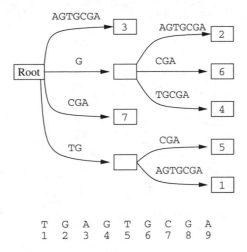

T G A G T G C G A
1 2 3 4 5 6 7 8 9

FIGURE 36.4: Implicit suffix tree built on sequence TGAGTGCGA.

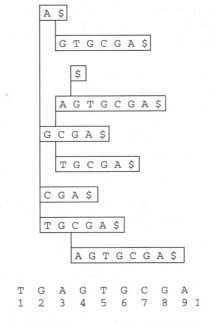

T G A G T G C G A
1 2 3 4 5 6 7 8 91

FIGURE 36.5: Suffix cactus built on sequence TGAGTGCGA. A dummy letter, $, marks the end of the sequence. Rotate the figure by 90° in counterclockwise direction to see why it is called a cactus tree.

al. bounds the memory usage by 37 bytes per letter [16]. Navarro and Baeza-Yates report that the suffix tree is 12 to 70 times larger than the input sequence [47]. Based on these numbers, the suffix tree for a 3 Gbp database, such as the entire human genome takes 30 to 210 GB space. Such an index will be inefficient for main memory search algorithms due to its excessive size.

Suffix tree enables efficient lookup for exact matches of any size. Given a query sequence, q, it can be searched in $O(|q|)$ time by traversing the tree starting from the root. At each

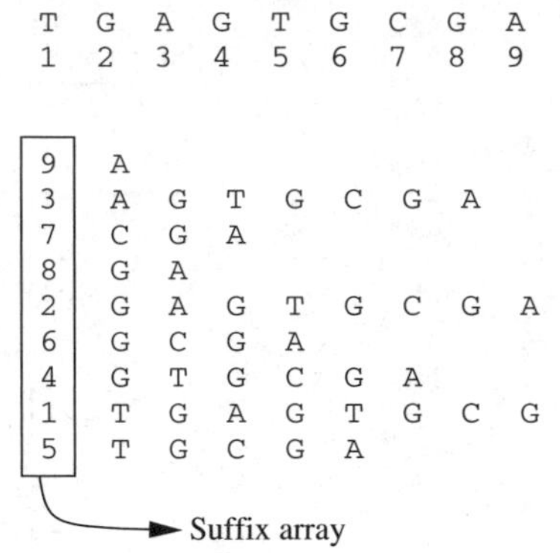

```
T  G  A  G  T  G  C  G  A
1  2  3  4  5  6  7  8  9
```

```
9 │ A
3 │ A  G  T  G  C  G  A
7 │ C  G  A
8 │ G  A
2 │ G  A  G  T  G  C  G  A
6 │ G  C  G  A
4 │ G  T  G  C  G  A
1 │ T  G  A  G  T  G  C  G
5 │ T  G  C  G  A
```

Suffix array

FIGURE 36.6: Suffix array built on sequence TGAGTGCGA.

node, the child node that is labeled with the same letter as that of q is chosen as the next node.

A new sequence, s, can be inserted to a suffix tree in $O(|s|\log(|D|+|s|))$ time, where D is the size of the database that is currently indexed by that suffix tree. An existing sequence, s can be removed from a suffix tree in $O(|s|\log|D|)$ time [15]. However, we are not aware of efficient algorithms for updating, inserting, or deleting individual letters in a sequence that already exists in a suffix tree.

Suffix arrays reduce the space consumption of suffix trees to 5 bytes per letter at the expense of increased search time complexity [40]. The suffix array of a sequence s is an array of integers that shows the alphabetical order of all suffixes of s. Figure 36.6 shows the suffixes of TGAGTGCGA and the suffix array constructed on it. The look up time for a query sequence, q, is $O(|q|\log|D|)$ using suffix arrays.

Next, we discuss a search tool based on suffix trees.

MUMmer: a case study

MUMmer employs suffix trees for global alignment of sequences. Given two sequences x and y, MUMmer aligns them in three steps:

1. *Detecting MUMs:* A MUM (Maximal Unique Match) for x and y is a pair of subsequences (x', y') that exactly match and there is no other matching subsequence pair that contains x' and y' simultaneously. MUMmer first constructs a suffix tree for x. Later, the suffixes of y are also inserted to the same tree. All the MUMs are then detected by traversing this suffix tree.

2. *Finding the backbone of the alignment:* All the MUMs (x', y') are sorted in increasing order of the position of x' in x. Next, the longest sequence of MUMs whose subsequences from x and y are in sorted order are found. These MUMs define the backbone of the alignment.

3. *Closing gaps:* The gaps between consecutive MUMs of the backbone are aligned with the help of the Smith-Waterman method [56].

Other tools that use suffix tree and its variations

A number of tools employ suffix tree and its variations for various sequence problems. Similar to MUMmer, AVID [9] performs global alignment using suffix trees. REPuter [37] uses suffix trees for detecting repeats in a sequence. MGA [22] finds multiMEMs for multiple alignment with the help of suffix trees. A recent paper by Meek et al. [42] uses suffix trees for accurate online searches with short query sequences. QUASAR [13] uses suffix arrays for aligning sequences.

36.3.2   VP-trees

The original VP-tree (Vantage Point tree) [64] indexes the data in general metric space. Şahinalp et al. [53] adapted the VP-tree to sequence databases where the distance is defined as the edit distance or the block edit distance. This technique can also be adapted to other sequence distance functions as long as they are *almost metric*. A distance measure is defined as almost metric if it can be bounded by a metric distance function. VP-tree is a binary tree structure that indexes a database of sequences based on their distances to a vantage point (i.e., a pivot sequence selected from the database). Next, we discuss the construction of VP-trees.

Let $d(x, y)$ be an almost metric distance function, where x and y are two sequences. For example, $d(x, y)$ can be the edit distance between x and y. The tree construction algorithm takes a database of sequences, $D = \{s_1, s_2, \cdots, s_n\}$. It starts by choosing a random vantage sequence s as the root. Later, the median of the distances of the database sequences to s is computed. The database is then partitioned into two equi-sized sets. One of the sets contains the sequences that are closer to s than the median, and the other one contains the rest of the sequences. Each of these sets are then defined as left and right children of the root and recursively partitioned until all the sequences are indexed.

Given a query q, the sequences that are within a distance of r to q are found as follows. First, q is compared to the vantage sequence, s, at the root node. Let M be the median distance of the sequences in the index to the root node.

1. If $d(q, s) \leq r$ then s is inserted to the result set.
2. If $d(q, s) \leq r + M$ then the left child of the root is searched recursively. This is because, by triangle inequality, the distance of a sequence contained in the left sub-tree of the root to the query can be as small as $d(q, s) - M$.
3. If $d(q, s) \geq M - r$ then the right child of the root is searched recursively. This is because, by triangle inequality, the distance of a sequence contained in the right sub-tree of the root to the query can be as small as $M - d(q, s)$.

Depending on the query, either of the following three cases are possible: *Case 1:* Left sub-tree is pruned (q_1 in Figure 36.7(b)). *Case 2:* Right sub-tree is pruned (q_2 in Figure 36.7(b)). *Case 1:* Neither left nor right sub-tree is pruned (q_3 in Figure 36.7(b)). Note that Figure 36.7 is an illustration of the problem drawn on a 2-D plane just for visualizing purposes. The actual sequences are not on a Euclidean space nor the distances are Euclidean as shown in this figure.

There are several important aspects to note about VP-trees. First, the VP-tree is built using sequences. Therefore, the VP-tree enables only global alignment queries. Second, the sequences in the result set are candidates. These sequences still need to be aligned using an existing method such as the dynamic programming technique [48]. Third, the VP-tree can be generalized to higher fanout by using multiple vantage points at each step of the index construction algorithm. Fourth, nearest neighbor queries can also be implemented on the VP-tree by adapting existing methods for multi-dimensional index structures. Fifth, the VP-tree is a static data structure. Therefore, insertions, deletions, and modifications of sequences require reconstruction of the index. However, it can be made dynamic by relaxing the restriction that the VP-tree is balanced.

The VP-tree can be constructed in $O(Tn \log n)$ time where n is the number of sequences

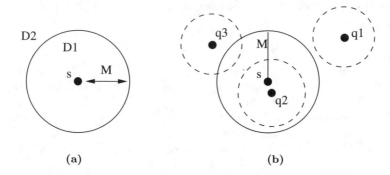

(a)   (b)

FIGURE 36.7: (a) The vantage sequence s and partitioning of the database into two subsets D_1 and D_2. M is the median distance to s. (b) An illustration of the three cases for querying the VP-tree: Prune left sub-tree for q_1, and right sub-tree for q_2. Search both left and right sub-trees for q_3.

in the database, and T is the time required for pairwise sequence comparison. This is because $O(n)$ pairwise sequence comparisons are conducted for each level of the VP-tree, and there are $O(\log n)$ such levels.

36.4   Vector Space Indexing

The index structures we have discussed so far are built directly on the sequences. One disadvantage of such index structures is that they are good for finding exact matches, but inefficient for finding approximate matches. Approximate similarities are usually determined in one of the two ways. 1) Shorter exact matches are combined and extended (e.g. FASTA [51]). 2) A set of inexact neighborhood of the query sequence is generated. Exact matches to this set is then found (e.g. [46]).

A number of index structures have been developed in vector space. These index structures compute approximate similarities directly by mapping sequences or subsequences to vectors in a vector space. The distance between two vectors shows the difference between their sequence counterparts. This enables faster detection of approximate matches.

We will discuss two important index structures for sequences in vector space: SST and the MRS index structure.

36.4.1   SST

Sequence Search Tree (SST) [20] enables near-exact searches for sequence databases, where the distance between two sequences is defined as the edit distance between them. The SST algorithm partitions the database sequences into overlapping subsequences. Later, it maps each of these subsequences to vectors in a multi-dimensional integer space, and builds an index structure using k-means clustering [19]. We discuss each of these steps and how to perform queries in detail next.

Vector space mapping

   The vector space mapping of a database sequence s is determined by three parameters: 1) window size w 2) shift amount Δ and 3) tuple size k. Window size is the length of shortest meaningful query that a user can pose on the database. The window is initially placed at the beginning of s (i.e., the first subsequence is the first w letters of s). The window is then

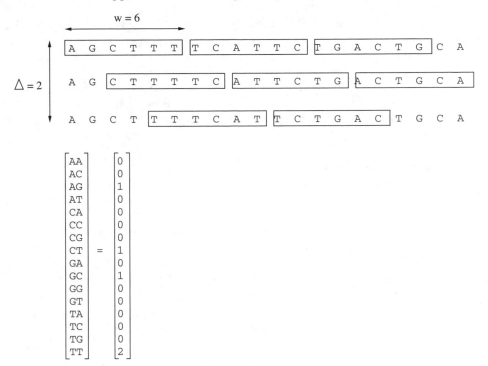

FIGURE 36.8: A DNA sequence and the subsequences extracted from it by the SST algorithm for $w = 6$, and $\Delta = 2$, and the vector computed for the first subsequence for $k = 2$.

shifted by Δ letters to find the next subsequence until the end of s is reached. Figure 36.8 shows a database sequence and the subsequences extracted from it by the SST algorithm for $w = 6$, and $\Delta = 2$. Here, the sequence contains 20 letters. The SST algorithm extracts 8 subsequences from this sequence.

The last parameter, k, determines the size of the vector computed for a subsequence of s. For each subsequence, the number of k-grams of each type is counted for all possible k-grams. There are σ^k such k-grams for alphabet size σ. These counts are then stored in a vector of size σ^k. SST uses this vector to represent each subsequence. For example, Figure 36.8 shows the vector of the first subsequence, AGCTTT, for $k = 2$. Since this is a DNA sequence, the alphabet contains four letters. Thus, the vector size is 4^2. The subsequence contains one AG, CT, and GC, and two TTs.

The L_1 distance between the vectors of two subsequences denotes the number of k-grams that are not common to both of them. Obviously, if two subsequences are equivalent, then their vectors are equal, making the L_1 distance zero. If two sequences differs by only one letter, then the L_1 distance between their vectors differ by at most k since a letter overlaps with up to k k-grams. As the edit distance between two sequences increases, the L_1 distance between their vectors usually increases as well. The SST algorithm is based on this assumption. However, it is also possible that two sequences with a large L_1 distance may have a smaller edit distance. Consider the following example.

Example 36.4

Let $L_1(u, v)$ be the L_1 distance between vectors u and v. Let $\Sigma = \{A, B\}$ be the alphabet.

Let $x = $ AAA, $y = $ ABA, and $z = $ BAB be three sequences defined on Σ. The edit distance between x and y, $ED(x,y)$, is 1, and $ED(y,z) = 2$. This means that y is closer to x than z. Now, let us take a closer look at their vector counterparts. For $w = 3$, $\Delta = 1$, and $k = 2$, the vectors of x, y, and z are $v_x = [2, 0, 0, 0]$, $v_y = [0, 1, 1, 0]$, and $v_z = [0, 1, 1, 0]$ respectively. Here, the vector is computed as [#AAs, #ABs, #BAs, #BBs]. The vector distances are $L_1(x,y) = 4$ and $L_1(y,z) = 0$. This means that y is closer to z than x, contradicting the fact. This example shows that L_1 distance may not reflect the distance between two sequences.

Index structure construction and querying

The vector space mapping of a sequence produces a number of vectors for a given database sequence. These vectors can be considered as points in a σ^k-dimensional integer space. Once the mapping is done, these points can be indexed using any existing multi-dimensional index structure. Here, we will discuss the *k-means clustering* method since the SST algorithm employs it to create a binary tree index structure.

The index construction algorithm takes a set of vectors, D, and the root of the index structure, R, as input. It starts by choosing two centroids randomly. The set D is partitioned into two sets, one for each centroid, by assigning each vector to the set corresponding to the closest centroid. Next the centroids for each of these sets are re-computed as the average of the vectors in these sets. The algorithm updates the sets using the new centroids until the total distance of the vectors in each set to its centroid does not change too much. Once the sets are determined, they form the two children nodes of the binary search tree. Later, each set is partitioned recursively until the set size drops below a threshold.

A query sequence is first divided into subsequences of window size w obtained using a shift amount of $\Delta = w/2$. These query subsequences are mapped to vectors using the same value for k as the SST. Each of the query vectors is then searched on the index structure starting from the root node as follows: The L_1 distances of the query vector to the centroid of the two child node are computed. The child node with the larger L_1 distance is discarded, and the other child node is searched recursively until a leaf node is reached. The subsequences contained in the leaf node are returned as the nearest neighbors of that query subsequence.

The SST algorithm incurs false dismissals for three reasons. First, the L_1 distance between two vectors is not a lower bound to the edit distance (see example 36.4). Second, the search algorithm always chooses the children node whose centroid is closer to the query vector. However, it does not guarantee that all the vectors in that node are closer to the query vector than the vectors in the other node. Third, since the database and query vectors are computed by shifting a window by Δ, $\Delta - 1$ out of Δ subsequences are ignored during similarity search.

The time required to construct the vectors from raw sequences is $O(n)$, where n is the total length of the sequences in the database. The total time to construct the entire tree is $O(Cn \log n)$ if the partitions are of similar size, where C is the number of iterations required to partition a dataset. However, the worst case time complexity is $O(Cn^2)$. This happens when the vector set partitions are unbalanced (e.g., when one partition contains one vector and the other one contains $n - 1$.) The index construction time can be improved by a constant factor by sampling the database.

The look up cost of the SST algorithm is $O(\log n)$ per query vector. This is because the index contains $O(n)$ vectors, and the binary search algorithm eliminates half of the search tree at each step. Let m be the length of the query sequence. The number of query vectors is then $O(m/\Delta)$. Thus, the total time complexity of the index search is $O(m \log n/\Delta)$.

SST is not a dynamic index structure. Insertions, deletions, or modifications require a reconstruction of the entire index structure. However, it can be made dynamic by relaxing the restrictions imposed on the centroids. However, this will further reduce the true positive rate of the index structure.

The database consists $\lceil 1 + (n-w)/\Delta \rceil$ vectors. Each vectors contains σ^k entries. Each entry can be encoded using $\log w$ bits. Thus, the total space needed is $\sigma^k \lceil 1 + (n-w)/\Delta \rceil \log w$ bits. However, this space can be reduced by two observations. First, the entries of the vectors are sparse. Second, many vectors overlap. Using these two observations Giladi et al. [20] reduces the space consumption to two bytes per nucleotide.

36.4.2 Frequency vectors

The MRS (Multi Resolution String) index structure enables both global alignment [29] and local alignment [30] of sequences. It supports both edit distance and BLAST-like scores for similarity. The MRS index structure maps subsequences of the database sequences into an ordered set of points in a vector space for different subsequence lengths. Later, these points are indexed with the help of MBRs (Minimum Bounding Rectangles). We discuss each of these steps and how to run global and local alignment queries on the MRS index structure next.

Vector space mapping

Let s be a sequence from the alphabet $\Sigma = \{\alpha_1, \alpha_2, \cdots, \alpha_\sigma\}$. Let n_i be the number of occurrences of the character α_i in s for $1 \leq i \leq \sigma$. The vector space mapping of s is computed as

$$f(s) = [n_1, n_2, ..., n_\sigma].$$

The vector, $f(s)$ is called the *frequency vector*, of s. For example, let $s = $ AGCTTTTCATTCT-GAC be a DNA sequence. The frequency vector of s is $f(s) = [3, 4, 2, 7]$ ([#As, #Cs, #Gs, #Ts]), since the DNA alphabet contains the letters A, C, G, and T. For simplicity, the letters in the vector are sorted in alphabetical order.

The frequency vectors have three important properties:

1. The size of the frequency vector is equal to the alphabet size. That is, it is independent of the size of the input sequence.

2. The sum of the entries of a frequency vector is equal to the length of the input sequence.

3. An edit operation on the input sequence does not alter its frequency vector too much. We will elaborate on this later.

Properties 1 and 2 follow from the definition of the frequency vectors. The third property is obtained by inspecting the effects of each individual edit operation. Let $f(s) = [v_1, \cdots, v_\sigma]$ be the frequency vector of sequence s.

- Insertion of the letter α_i increases v_i by 1.
- Deletion of the letter α_i decreases v_i by 1.
- Replacement of the letter α_i with α_j decreases v_i by 1 and increases v_j by 1.

Each frequency vector maps to a point in σ dimensional integer space. Each edit operation moves the frequency vector to one of the neighboring points in that space. For example, Figure 36.9(a) shows the movement of the frequency vector for sequence $s = $ YUUUY for three different edit operations (the letters U and Y stand for Purine (A, G) and Pyrmidine (C, T)). Given two sequences x and y, the sequence of edit operations that transforms x into y defines a path between $f(x)$ and $f(y)$ in the frequency space, where each edge on

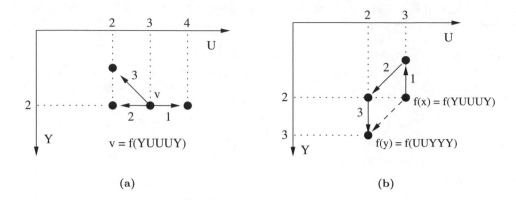

FIGURE 36.9: (a) The frequency vector, v, of the sequence YUUUY and its movement in the frequency space for (1) insertion of a letter U, (2) deletion of a letter U, and (3) replacement of a letter U with letter Y. (b) The solid arrows show a path that translates the frequency vector x to y in three steps. The shortest path is shown by the dashed arrow.

this path is a directed edge from the last visited point to one of its neighbors. For example let $x =$ YUUUY, and $y =$ UUYYY. One possible transformation of x into y is using three operations: $x =$ YUUUY $\rightarrow$ UUUY $\rightarrow$ UUYY $\rightarrow$ UUYYY $= y$. Figure 36.9(b) shows the path for this sequence of edit operations in the frequency domain. The length of a path is defined as the number of edges on that path. Given the frequency vectors, $f(x)$ and $f(y)$, of two sequences, x and y, there are infinitely many paths that move from $f(x)$ to $f(y)$ by visiting neighboring edges. The *frequency distance*, FD$(f(x), f(y))$, between two frequency vectors $f(x)$ and $f(y)$ is defined as the length of the shortest of such paths. The path in Figure 36.9(b) shown by the dashed arrow is the shortest path between the two given frequency vectors. Therefore, FD$(f($YUUUY$), f($UUYYY$)) = 1$.

Two important properties of the frequency distance that follow from its definition are

1. *Lower bounding property:* FD$(f(x), f(y)) \leq$ ED$(x, y), \forall x, y$, where ED(x, y) is the edit distance between x and y. For example, in Figure 36.9(b), ED(x, y) $= 3$ and FD$(f(x), f(y)) = 1$. This is a desirable property. This is because if $r <$ FD$(f(x), f(y))$, for some similarity cutoff r, then $r <$ ED(x, y). Thus, one can conclude that x and y are not similar without actually computing the edit distance between them.

2. *Tightness of the frequency distance:* $\forall x, y, \exists x', y' : f(x) = f(x') \wedge f(y) = f(y') \wedge$ FD$(f(x), f(y)) =$ ED(x', y'). For example, in Figure 36.9(b), $f(x) = [3, 2]$ and $f(y) = [2, 3]$. For $y' =$ YUUYY, $f(y') = f(y)$ and ED$(x, y') = 1$. This property implies that the frequency distance is a tight lower bound to the edit distance.

Figure 36.10 shows the algorithm that computes the frequency distance between two frequency vectors u and v. The algorithm computes the minimum number of increment and decrement operation that need to be applied to u to translate it on to v. Since an increment and a decrement can be overlapped due to a replace type of edit operation, the maximum of these two values is returned as the frequency distance. The frequency distance, FD$(f(x), f(y))$, between two frequency vectors, $f(x)$ and $f(y)$, can be computed in $O(\sigma)$ time. This is much faster compared to the $O(|x| \cdot |y|)$ time required to compute the edit distance between their sequences.

```
/* Let u and v be frequency vectors */
/* Let σ be the number of dimensions */
/* Returns the frequency distance between u and v */
Function FD(u, v)

inc := dec := 0;
For i := 1 to σ
    If uᵢ < vᵢ then
        inc := inc + 1;
    else
            dec := dec + 1;
Return Max{inc, dec};
```

FIGURE 36.10: The function that computes the frequency distance between two frequency vectors u and v.

Index structure construction and querying

Let $S = \{s_1, s_2, ..., s_d\}$ be a database consisting of potentially long sequences from alphabet $\Sigma = \{\alpha_1, \alpha_2, ..., \alpha_\sigma\}$. Let $w_1 = 2^a$ be the length of the shortest possible query sequence. The MRS index structure stores a grid of structures $T_{i,j}$, $a \leq i \leq a + L - 1$, and $1 \leq j \leq d$. The parameter L represents the number of resolution levels in the index structure. Structure $T_{i,j}$ is the index structure for the jth sequence for window size 2^i.

In order to obtain $T_{i,j}$, a window of length $w = 2^i$ is placed at the leftmost point of s_j. Later, this window is slid by one until it reaches to the end of s_j. Each placement of this window produces a subsequence of s_j. The frequency vectors of all those windows are computed. First, the minimum box, called *Minimum Bounding Rectangle* (MBR), that covers the frequency vector of the first subsequence is computed. This box is later extended to cover the frequency vectors of the first c subsequences, where c is the box capacity. Typically, c is set to 1000 to achieve good performance result. (We will later discuss different strategies for choosing box capacity.) After the first c subsequences are indexed, a new MBR is created to cover the next c subsequences. This process continues until all subsequences are transformed. Note that only the lower and higher end points of the MBRs along with the starting locations of the first subsequence contained in that MBR are stored for each MBR. Figure 36.11 shows the construction of the MBRs for a DNA sequence for $w = 16$ and $c = 4$. Sliding the window by one letter includes a new letter to the end of the old window and excludes its first letter. As a result of this, consecutive frequency vectors change only by one in at most two dimensions. Thus, they appear very close in the frequency space. Figure 36.12 shows the first 500 frequency vectors computed from the E.coli bacteria for $w = 128$ in 3-dimensions (number of As, Cs, and Gs).

Let q be a query sequence of length 2^i, where $a \leq i \leq a + L - 1$. Let B be an MBR of the MRS index structure for resolution 2^i. The frequency distance between $f(q)$ and B, $\text{FD}(q, B)$, is defined as the length of the shortest path that translates $f(q)$ into B. From the definition of $\text{FD}(q, B)$, one can prove that if $r \leq \text{FD}(q, B)$ then $r \leq \text{FD}(f(q), f(s)) \leq \text{ED}(q, s)$, $\forall s \in B$.

A range query on the index structure takes a query sequence, q, and a query radius r as input, and returns all the subsequences in the database whose edit distance to q are less than r. The search algorithm runs in two phases:

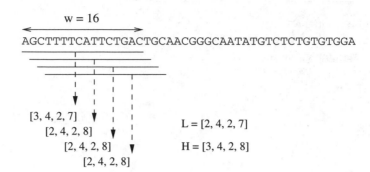

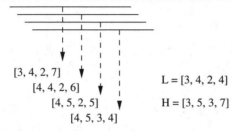

FIGURE 36.11: The construction of the first two MBRs of the MRS index structure for $w = 16$ and $c = 4$ on a DNA sequence.

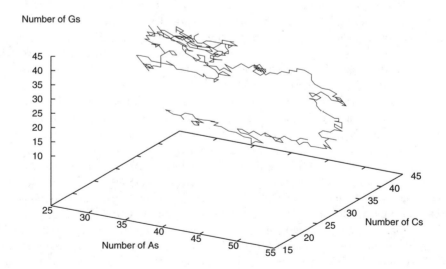

FIGURE 36.12: The path constructed by the first 500 frequency vectors computed from the E.coli bacteria for resolution 128 in 3-dimensions (number of As, Cs, and Gs).

- *Pruning Phase:* The query sequence is partitioned into a number of subqueries at various resolutions available in the MRS index structure. A *partial range query* is performed for each of these subqueries on the appropriate row of the index structure. This is called a partial range query because it only computes frequency distance of the frequency vector of the subquery to the MBRs, not the distance of the query subsequence to the subsequences contained in the MBRs. At the end of this phase, a set of candidate MBRs are determined.

- *Postprocessing Phase:* The database subsequences in the candidate MBRs are read, and their edit distance to the entire query sequence is computed to find the subsequences similar to q.

The second phase can use any lossless sequence alignment method such as Smith-Waterman [56]. We elaborate on the first phase.

Given any query q of length $k2^a$ and a range ϵ, there is a unique partitioning, $q = q_1 q_2 ... q_t$, with $|q_i| = 2^{c_i}$ and $a \le c_1 < ... < c_i \le c_{i+1} \le ... c_t \le a + l - 1$. This partitioning technique chooses the longest possible suffix of q, such that its length is equal to one of the resolutions available in the index, as the last query subsequence. Later, it recursively partitions the rest of the sequence to find the other query subsequences. If the length of the query sequence is not a multiple of 2^a, then the longest prefix of the query whose length is a multiple of the minimum window size can be used. The remaining prefix is ignored.

First, q_1 is searched on row R_{c_1} of the MRS index structure. As a result of this search, a set of MBRs that lie within a distance of $r = \epsilon \times |q|$ from q_1 are returned. Using the frequency distances, $FD(q_1, B)$ to these MBRs, the value of r is refined as $r - FD(q_1, B)$ for each MBR. Later, the second query, q_2, is searched on row R_{c_2} only for the MBRs that are returned using the new value of r. This process continues for the remaining rows R_{c_3} ... R_{c_t}. The final set of MBRs is defined as the candidate MBRs.

A nearest neighbor search can be implemented on the MRS index structure using the range query algorithm. This can be done by exploiting either the two-phase [36] or the multi-phase [55] nearest neighbor strategy.

Some important aspects of the MRS index structure are as follows:

- The range and the nearest neighbor search on the MRS index structure are lossless. That is, the candidate set has 100% recall.

- The searches on the MRS index structure can be parallelized easily since searches on each sequence can be conducted independently.

The time complexity of index construction is $O(L|D|)$. This is because the MRS index structure can be constructed by a single sequential scan of the database sequences, and each letter in the database is counted only twice for each level of the index structure (one when it enters the window and one when it exits the window).

The MRS index stores the following information for each MBR: two $(\sigma - 1)$-dimensional vectors, one for lowest and one for highest coordinates, the id of the sequence to which the MBR belongs and the starting position. Each of the vectors requires $(\sigma - 1)\lceil \log_2(w + 1) \rceil$ bits. The id and the starting position are stored in one word. For example for $w = 128$, one MBR can be stored using 10 bytes. Since there are $O(|D|/c)$ MBRs for each resolution, the total space consumption of the MRS index structure is

$$O((\sigma - 1)\lceil \log_2(w + 1) \rceil L|D|/c).$$

For DNA sequences, the typical values are $c = 1000$ and $L = 4$. In this case, the size of the MRS index structure for a 1 Gbp DNA sequence database is only 10 kB for $w = 128$.

The time for searching the MRS index for candidate MBRs depends on the pruning rate of the index. The query time is

$$O(\frac{\sigma|D|}{c} \sum_{i=1}^{L} \mu_i),$$

where μ_i is the ratio of the number of unpruned MBRs at ith resolution to the number of MBRs at that level.

Insertion of a new letter to the end of a sequence costs $O(L)$ since this can be done by extending the last MBR at each resolution. However, insertion of a new letter into a sequence requires reconstruction of one MBR. Therefore, it costs $O(cL)$ time. Similarly, deletion or modification of an existing letter incurs $O(cL)$ time. Usually, in sequence databases, entire sequences are inserted to or removed from the database. The time requirement for such an operation is $O(L|s|)$, where s is the sequence to be inserted or removed.

Variations of the MRS index structure

Each MBR of the MRS index structure contains a fixed number of frequency vectors, denoted by the box capacity c. The performance of the MRS index structure can be improved by adaptively varying the capacity of each MBR. The motivation behind this is as follows. Let B be an MBR of the MRS index structure obtained by sliding a window from position i to j. If the frequency vector for position $j+1$ is located inside B, then this frequency vector can be inserted into B without increasing its size. Inserting such frequency vectors into B increases the capacity of B without growing it. Thus, the number of MBRs in the index structure is reduced. This reduces the index search time as well as the memory usage. A number of such adaptive strategies can be applied to the MRS index structure:

- *Fixed volume.* Keep adding points to an MBR until its volume exceeds a certain threshold.
- *Fixed density.* Keep adding points until n/V, the number of points divided by the volume of the box, falls below a certain threshold.
- *MHIST–Volume.* Start with one big MBR containing all the points. Find the position to split the MBR such that the sum of the volumes of two new MBRs is minimized, but the points in each MBR still correspond to consecutive positions of the sliding window on the sequence. Keep splitting the MBR with the largest volume to get the desired number of MBRs.
- *MHIST–Density.* Like MHIST–Volume, but split an MBR in to two new MBRs i and j such that the total density, $n_i/V_i + n_i/V_j$, is maximized. Keep splitting the MBR with lowest density.

The frequency vector of a sequence denotes the number of letters of each type in that sequence. However, it does not reflect the distribution of the letters in that sequence. Wavelet decomposition of frequency vectors can be used to extract the distribution of the letters at the expense of increased index space [29]. As the number of wavelet coefficients increases, the original sequence can be reconstructed more precisely from the frequency vector.

The MRS index search returns false positives in the candidate set since there more than one sequence may have the same frequency vector. The number of false positives can be reduced by scaling the frequency distance [50] or by using lossy transformation methods, such as DFT [1]. However, these methods will reduce the sensitivity of the MRS index structure.

Score computation with the MRS index structure

In many biological applications, the similarity between two sequences is defined as the score of their best alignment. The score is defined in terms of four parameters:

1. S_{match}: Score for each matching letter.
2. S_{mismatch}: Penalty for each mismatching letter.
3. $S_{\text{gap-open}}$: Penalty for starting a gap.
4. $S_{\text{gap-extend}}$: Penalty for extending an existing gap by one.

For example, BLAST (for nucleotides) uses a scoring scheme with values $S_{\text{match}} = 1$, $S_{\text{mismatch}} = -3$, $S_{\text{gap-open}} = -5$, and $S_{\text{gap-extend}} = -2$. Usually, the magnitude of $S_{\text{gap-open}}$ is much larger than the rest since the likelihood that a nucleotide is chopped in mutation is very small. This kind of similarity is usually more relevant to biological applications since it reflects the mutation process better than the edit distance. It is also possible to replace S_{match} and S_{mismatch} with a *score matrix*, such as PAM or BLOSUM. A *score matrix* defines the score/penalty of matching two letters for every possible pairs of letters. Frequency vectors can also be employed to find an upper bound on the score of the best alignment of two sequences. We discuss the simpler case of $(S_{\text{match}}, S_{\text{mismatch}})$ next.

Figure 36.13 shows the procedure, $\text{FS}_w(v, B)$, that computes the frequency score between a frequency vector v and an MBR B. This is an upper bound to the score of the best alignment of a query sequence, q, with frequency vector v and a sequence, s, whose frequency vector is in B. The procedure computes the number of increments and decrements needed to translate v into B. The minimum of these two values show the number of mismatches (i.e., modifications) and the difference between them shows the total size of the gap (i.e., indels). The number of matches is found by subtracting the number of mismatches from length of the shorter sequence. The score is found by accumulating the score and penalties for these parameters.

Local alignment with the MRS index structure

Local alignment between two sequences q and s can be carried out using the MRS index structure as follows.

- *Step 1: (Vector space mapping)* Partition q into subsequences of length w, where w is the length of the shortest local match needed. Compute the frequency vector for each of these subsequences. Construct an MRS index on s for resolution w. Figure 36.14 shows the partitioning of q and the MBRs constructed on s.

- *Step 2: (Coarse grain alignment)* Construct a boolean match table: Each row and column corresponds to a page from q and s respectively. Mark the entries $m_{i,j}$ of the match table if the frequency score between a subsequence contained in the ith page of q and a subsequence contained in the jth page of s exceeds a score threshold.

  Conceptually, each sequence is divided into *pages*, i.e., non-overlapping blocks of a fixed size. This is because a page is the minimum I/O unit, i.e. at least one page is read from the disk at a time. Note that there is a many-to-many relationship between boxes and pages: The subsequence corresponding to a box may span more than one page, and (parts of) a page may occur in the subsequences of more than one box.

  The similarity between a pair of pages is defined as the highest frequency score of the (frequency vector, MBR) pair intersecting with those pages. Figure 36.14 depicts the concept of pages and the match table for sequences q and s. Here, the black dots represent the marked entries of the match table. For example, q_1

/* v : σ-dimensional integer point
B : σ-dimensional integer box of lower and
higher coordinates $B.L$ and $B.H$.
w : window size used to construct B */
Procedure $FS_w(v, B)$
1. $inc := dec := sum := 0$;
2. **for** i from 1 to $\sigma - 1$:
        if $v[i] < B.L[i]$ **then**
            $inc \mathrel{+}= B.L[i] - v[i]$;
            $sum \mathrel{+}= B.L[i]$;
        else if $B.H[i] < v[i]$ **then**
            $dec \mathrel{+}= v[i] - B.H[i]$;
            $sum \mathrel{+}= B.H[i]$;
        else
            $sum \mathrel{+}= v[i]$;
3. $ScoreInc := (\min\{sum, w\} - inc) \cdot S_{match} + inc \cdot S_{mismatch}$;
4. $ScoreDec := (\min\{sum, w\} - dec) \cdot S_{match} + dec \cdot S_{mismatch}$;
5. **if** $w < sum$ **then**
        $ScoreInc \mathrel{+}= S_N \cdot (sum - w)$;
    else if $sum < w$ **then**
        $ScoreDec \mathrel{+}= S_N \cdot (w - sum)$;
6. return $\min\{ScoreInc, ScoreDec\}$;

FIGURE 36.13: Procedure $FS_w(v, B)$ for computing the best score of the alignment between a sequence x and a set of sequences $\mathcal{X}$, where v is the frequency vector of s and B is the box that covers the frequency vectors of the sequences in $\mathcal{X}$.

is similar to B_6. Therefore, both $m_{1,3}$ and $m_{1,4}$ are marked.

- *Step 3: (Fine grain alignment)* Align the page pairs whose entries in the match table are marked using an existing alignment tool such as BLAST. Partition the match table if the input sequences are too large to align using the available memory. One strategy for partitioning is as follows: Let r and c be the number of marked rows and columns of the match table. If $r < c$, then the match table is split vertically to obtain a slice big enough to fit into available memory. Otherwise, the match table is split horizontally. This process is carried out recursively for the remaining match table.

36.5   Concluding Remarks and Future Directions

A number of index structures have been proposed to speed up various problems on sequence databases, such as pairwise alignment, multiple alignment, motif finding, shotgun sequencing, and repeat finding. These index structures have been proven to be useful in eliminating less important parts of the database, thus avoiding exhaustive database access. We have discussed these index structures under three categories: k-gram indexing, direct indexing, and vector space indexing.

k-gram indexing enables quick lookups for sequences of length k. Longer matches are

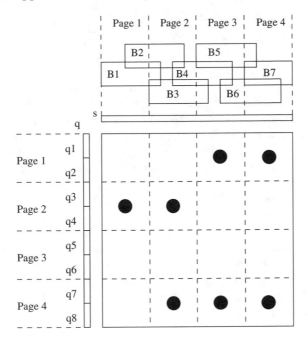

FIGURE 36.14: The match table created on sequences q and s. q is partitioned into 8 subsequences $q_1, \cdots, q_8$. The MRS index structure built on s contains 7 MBRs B_1, $\cdots$, B_7. Both q and s fit into 4 pages. The black dots represent the marked entries of the match table.

found either by extending or stitching these exact matches. Finding inexact matches require additional processing.

Direct indexing is helpful for exact matches of arbitrary lengths. Inexact matches are found either by stitching short exact matches or by finding exact matches to sequences in a specific neighborhood of the query sequence.

Vector space indexing transforms information from sequence domain to a vector space. This enables inexact searches directly on the index structure since the approximate similarities in the vector space can be found by adapting existing search methods for multi-dimensional index structures. Another advantage of vector space indexes such as the MRS index structure is that their sizes are substantially smaller than the other ones. This makes in-memory searches viable for large sequence databases.

A number of open problems remain for indexing sequence databases. Chapter 35 discusses the future directions for suffix tree and its derivatives. Here, we will discuss several open problems for k-grams and vector space indexing.

The speed and the sensitivity of the k-gram–based index structures, depends highly on the choice of k. Large values of k usually improves running time at the expense of reduced sensitivity. Therefore, k must be chosen carefully. Furthermore, the optimal choice of k may also depend on the distribution of the data. For example, if the database contains repeats of length 10, then 11-grams may fail to detect these repeats.

Two other important choices for k-grams are 1) gapped versus ungapped versions of the k-grams, and 2) single versus multiple values of k. These choices also depend on the data distribution.

If the input databases are too large, then each access to a k-gram may cause a page miss. This is because a k-gram may be present at different locations of the database. In this case, every k-gram access incurs a costly random seek. This requires development of novel methods for storing database on disk and in main memory to reduce the amount of random seeks required. Also, new page replacement policies are needed to minimize the number of page faults.

The cost of finding the optimal multiple alignment of more than two sequences is exponential in the number of sequences. Available heuristics, such as CLUSTALW [61], usually employ pairwise alignment to reduce this cost. However, the quality of these methods depend on the order of the sequences aligned. Vector space indexing is shown useful or pairwise sequence comparisons and for searching a query (sub)sequence in a database of sequences. One of the possible future uses of vector space mapping is multiple alignment and motif finding. The data points from more than two sequences that form a cluster in the vector space may define a backbone for the multiple alignment of these sequences, thus reducing the alignment cost.

Existing vector space mapping methods assume that the database is error free. However, the data may be inaccurate due to a number of reasons. One possible extension for vector space mapping methods is the development of statistics–based index structures. This kind of index structures assume that each letter is actually a vector that represents the probability distribution for that letter. Vector space indexing is better suited for this type of databases than k-grams or direct indexing.

Shotgun sequence assembly is one of the hardest problems for sequence databases. Overlay–layout–consensus methods such as PCAP [26] run in three phases. First, low-quality regions at the end of reads are trimmed and the overlaps between reads are computed. Second, reads are joined to form contigs. Finally, a consensus sequence is computed from contigs using multiple alignment. The first phase usually constitutes $\approx 85\%$ of the total running time. Vector space mapping can be utilized to detect overlaps and repeats for shotgun sequence assembly. The overlapping segments traverse similar paths in the vector space, thus, can be used for accelerating the sequence assembly problem.

References

[1] S.A. Aghili, D. Agrawal, and A. El Abbadi. Filtration of string proximity search via transformation. In *IEEE Symp. on BioInformatics and BioEngineering*, pages 149–157, 2003.

[2] S. Altschul, W. Gish, W. Miller, and E.W. Meyers *et al.* Basic local alignment search tool. *Journal of Molecular Biology*, 215(3):403–410, 1990.

[3] S.F. Altschul, T.L. Madden, A.A. Schaffer, and J. Zhang *et al.* Gapped BLAST and PSI-BLAST: A new generation of protein database search programs. *Nucleic Acids Res.*, 25(17):3389–3402, 1997.

[4] S. Aparicio, J. Chapman, E. Stupka, and N. Putnam *et al.* Whole-Genome Shotgun Assembly and Analysis of the Genome of Fugu rubripes. *Science*, 297(5585):1301–1310, 2002.

[5] R. Baeza-Yates and G. Navarro. Faster approximate string matching. *Algorithmica*, 23(2):127–158, 1999.

[6] S. Batzoglou, D.B. Jaffe, K. Stanley, and J. Butler *et al.* ARACHNE: A whole-genome shotgun assembler. *Genome Research*, 12(1):177–189, 2002.

[7] S. Batzoglou, L. Pachter, J.P. Mesirov, and B. Berger *et al.* Human and mouse gene structure: Comparative analysis and application to exon prediction. *Genome Research*, 10(7):950–958, 2000.

[8] D.A. Benson, I. Karsch-Mizrachi, D.J. Lipman, and J. Ostell *et al.* GenBank. *Nucleic Acids Res.*, 28(1):15–18, January 2000.

[9] N. Bray, I. Dubchak, and L. Pachter. AVID: A global alignment program. *Genome Research*, 13(1):97–102, 2003.

[10] M. Brudno, C.B. Do, G.M. Cooper, M.F. Kim, E. Davydov, NISC Comparative Sequencing Program, E.D. Green, A. Sidow, and S. Batzoglou. LAGAN and Multi-LAGAN: Efficient Tools for Large-Scale Multiple Alignment of Genomic DNA. *Genome Research*, 13(4):721–731, 2003.

[11] M. Brudno, S. Malde, A. Poliakov, and C.B. Do *et al.* Glocal alignment: Finding rearrangements during alignment. *Bioinformatics*, 19(90001):54i–62, 2003.

[12] M. Brudno and B. Morgenstern. Fast and sensitive alignment of large genomic sequences. In *Proc. of the IEEE Computer Society Bioinformatic*, 2002.

[13] S. Burkhardt, A. Crauser, P. Ferragina, and H.-P. Lenhof *et al.* q-gram based database searching using a suffix array (QUASAR). In *Research in Computational Molecular Biology*, Lyon, April 1999.

[14] A. Califo and L Rigoutsos. Flash: Fast look-up algorith for string homolgy. *Intelligent Systems for Molecular Biology (ISMB)*, 1993.

[15] Y. Choi and T.W. Lam. Dynamic suffix tree and two-dimensional texts management. *Information Processing Letters*, 61(4):213–220, 1997.

[16] A.L. Delcher, S. Kasif, R.D. Fleischmann, and J. Peterson *et al.* Alignment of whole genomes. *Nucleic Acids Res.*, 27(11):2369–2376, 1999.

[17] P. Ferragina and R. Grossi. The string b-tree: A new data structure for string search in external memory and its applications. *Journal of the ACM*, 46(2):236–280, 1999.

[18] R.D. Fleischmann. Whole-genome random sequencing and assembly of haemophilus influenzae Rd. *Science*, 269(5223):496–498+507–512, 1995.

[19] A. Gersho and R.M. Gray. *Vector Quantization and Signal Compression*. Kluwer Academic Publishers, 1991.

[20] E. Giladi, M.G. Walker, J.Z. Wang, and W. Volkmuth. SST: An algorithm for finding near-exact sequence matches in time proportional to the logarithm of the database size. *Bioinformatics*, 18(6):873–877, 2002.

[21] R. Grossi and J.S. Vitter. Compressed suffix arrays and suffix trees with applications to text indexing and string matching. In *ACM SYMP THEORY COMPUT*, pages 397–406, 2000.

[22] M. Höhl, S. Kurtz, and E. Ohlebusch. Efficient multiple genome alignment. *Bioinformatics*, 18(90001):312S–320, 2002.

[23] X. Huang. A contig assembly program based on sensitive detection of fragment overlaps. *Genomics*, 14:18–25, 1992.

[24] X. Huang. An improved sequence assembly program. *Genomics*, 33:21–31, 1996.

[25] X. Huang and A. Madan. CAP3: A DNA sequence assembly program. *Genome Research*, 9(9):868–877, 1999.

[26] X. Huang, J. Wang, S. Aluru, and S.-P. Yang *et al.* PCAP: A whole-genome assembly program. *Genome Research*, 13(9):2164–2170, 2003.

[27] E. Hunt, M.P. Atkinson, and R.W. Irving. A database index to large biological sequences. In *Proc. of the Conference on Very Large Databases*, pages 139–148, Roma, Italy, September 2001.

[28] D.B. Jaffe, J. Butler, S. Gnerre, and E. Mauceli *et al.* Whole-genome sequence assembly for mammalian genomes: Arachne 2. *Genome Research*, 13(1):91–96, 2003.

[29] T. Kahveci and A. Singh. An efficient index structure for string databases. In *Proc. of the Conference on Very Large Databases*, pages 351–360, Roma, Italy, September 2001.

[30] T. Kahveci and A.K. Singh. MAP: Searching large genome databases. In *Pacific Symposium on Biocomputing*, pages 303–314, January 2003.

[31] J. Kärkkäinen. Suffix cactus: A cross between suffix tree and suffix array. In *CPM*, 1995.

[32] W.J. Kent. BLAT—the BLAST-like alignment tool. *Genome Research*, 12(4):656–664, 2002.

[33] W.J. Kent and D. Haussler. Assembly of the working draft of the human genome with gigassembler. *Genome Research*, 11(9):1541–1548, 2001.

[34] W.J. Kent and A.M. Zahler. Conservation, regulation, synteny, and introns in a large-scale C. briggsae-C. elegans genomic alignment. *Genome Research*, 10(8):1115–1125, 2000.

[35] S. Kim and A.M. Segre. AMASS: A structured pattern matching approach to shotgun sequence assembly. *J. of Comp. Biol.*, 6(2):163–186, 1999.

[36] F. Korn, N. Sidiropoulos, C. Faloutsos, and E. Siegel *et al.* Fast nearest neighbor search in medical databases. In *Proc. of the Conference on Very Large Databases*, pages 215–226, India, 1996.

[37] S. Kurtz and C. Schleiermacher. REPuter: Fast computation of maximal repeats in complete genomes. *Bioinformatics*, 15(5):426–427, May 1999.

[38] D.J. Lipman and W.R. Pearson. Rapid and sensitive protein similarity searches. *Science*, 227(4693):1435–1441, 1985.

[39] M. Ma, J. Tromp, and M. Li. PatternHunter: Faster and more sensitive homology search. *Bioinformatics*, 18(0):1–6, 2002.

[40] U. Manber and E. Myers. Suffix arrays: A new method for on-line string searches. *SIAM Journal on Computing*, 22(5):935–948, 1993.

[41] E.M. McCreight. A space-economical suffix tree construction algorithm. *Journal of the ACM*, 23(2):262–272, 1976.

[42] C. Meek, J.M. Patel, and S. Kasetty. OASIS: An online and accurate technique for local-alignment searches on biological sequences. In *Proc. of the Conference on Very Large Databases*, 2003.

[43] B. Morgenstern. DIALIGN 2: Improvement of the segment-to-segment approach to multiple sequence alignment. *Bioinformatics*, 15(3):211–218, 1999.

[44] B. Morgenstern, K. Frech, A. Dress, and T. Werner. DIALIGN: Finding local similarities by multiple sequence alignment. *Bioinformatics*, 14(3):290–294, 1998.

[45] J.C. Mullikin and Z. Ning. The phusion assembler. *Genome Research*, 13(1):81–90, 2003.

[46] E. Myers. A sublinear algorithm for approximate keyword matching. *Algorithmica*, pages 345–374, 1994.

[47] G. Navarro and R. Baeza-Yates. A hybrid indexing method for approximate string matching. *J. Discret. Algorithms*, 1(1):205–239, 2000.

[48] S.B. Needleman and C.D. Wunsch. A general method applicable to the search for similarities in the amino acid sequence of two proteins. *Journal of Molecular Biology*, 48:443–53, 1970.

[49] Z. Ning, A.J. Cox, and J.C. Mullikin. SSAHA: A fast search method for large DNA databases. *Genome Research*, 11(10):1725–1729, 2001.

[50] Ö. Öztürk and H. Ferhatosmanoğlu. Effective indexing and filtering for similarity search in large biosequence databases. In *IEEE Symp. on BioInformatics and Bio-Engineering*, pages 359–366, 2003.

[51] W.R. Pearson and D.J. Lipman. Improved tools for biological sequence comparison. *Proc. of National Academy of Sciences*, 85:2444–2448, April 1988.

[52] P.Weiner. Linear pattern matching algorithms. *IEEE Symposium on Switching and*

Automata Theory, pages 1–11, 1973.

[53] S.C. Sahinalp, M. Taşan, J. Macker, and Z.M. Özsoyoğlu. Distance based indexing for string proximity search. In *International Conference on Data Engineering*, 2003.

[54] S. Schwartz, Z. Zhang, K.A. Frazer, and A. Smit *et al.* PipMaker—a web server for aligning two genomic DNA sequences. *Genome Research*, 10(4):577–586, April 2000.

[55] T. Seidl and H.P. Kriegel. Optimal multi-step k-nearest neighbor search. In *SIGMOD*, 1998.

[56] T.F. Smith and M.S. Waterman. Identification of common molecular subsequences. *Journal of Molecular Biology*, March 1981.

[57] D.J. States and P. Agarwal. Compact encoding strategies for DNA sequence similarity search. In *ISMB*, 1996.

[58] G. Sutton, O. White, M. Adams, and A. Kerlavage. TIGR assembler: A new tool for asembling large shotgun sequencing projects. *Genome Science and Technology*, 1:9–19, 1995.

[59] Z. Tan, X. Cao, B.C. Ooi, and A.K.H. Tung. The ed-tree: An index for large D-NA sequence databases. In *International Conference on Scientific and Statistical Database Management*, 2003.

[60] T.A. Tatusova and T.L. Madden. BLAST 2 sequences, a new tool for comparing protein and nucleotide sequences. *FEMS Microbiology Letters*, pages 247–250, 1999.

[61] J.D. Thompson, D.G. Higgins, and T.J. Gibson. CLUSTAL W: improving the sensitivity of progressive multiple sequence alignment. *Nucleic Acids Res.*, 22(22):4673–4680, 1994.

[62] E. Ukkonen. On-line construction of suffix-trees. *Algorithmica*, 14:249–260, 1995.

[63] J. Wang, G.K.S. Wong, P. Ni, and Y. Han *et al.* RePS: A sequence assembler that masks exact repeats identified from the shotgun data. *Genome Research*, 12(5):824–831, 2002.

[64] P.N. Yianilos. Data structures and algorithms for nearest neighbor search in general metric spaces. In *ACM-SIAM Symposium on Discrete Algorithms*, pages 311–321, 1993.

[65] Z. Zhang, A.A. Schaffer, W. Miller, and T.L. Madden *et al.* Protein sequence similarity searches using patterns as seeds. *Nucleic Acids Res.*, 26(17):3986–3990, 1998.

[66] Z. Zhang, S. Schwartz, L. Wagner, and W. Miller. A greedy algorithm for aligning DNA sequences. *Journal of Computational Biology*, 7(1-2):203–214, 2000.

37

Algorithms for Motif Search

Sanguthevar Rajasekaran
University of Connecticut

37.1 Introduction

The problem of identifying meaningful patterns (i.e., motifs) from biological data has been studied extensively due to its paramount importance. This problem in general requires finding short patterns of interest from voluminous data. Several variants of this motif search problem have been identified in the literature. In this chapter we survey some of the algorithms that have been proposed for these problems. We are interested in the following three versions:

Problem 1. Input are n sequences of length m each. Input also are two integers l and d. The problem is to find a motif (i.e., a sequence) M of length l. It is given that each input sequence contains a variant of M. The variants of interest are sequences that are at a hamming distance of d from M. (This problem is also known as the *planted (l,d)-motif search problem.*)

Problem 2. The input is a database DB of sequences $S_1, S_2, \ldots, S_n$. Input also are integers l, d, and q. Output should be all the patterns in DB such that each pattern is of length l and it occurs in at least q of the n sequences. A pattern U is considered an occurrence of another pattern V as long as the edit distance between U and V is at most d. We refer to this problem as the *edited motif search problem*.

Problem 3. A pattern is a string of symbols (also called residues) and ?'s. A "?" refers to a wild card character. A pattern cannot begin or end with ?. AB?D, EB??DS?R, etc. are examples of patterns. The length of a pattern is the number of characters in it (including the wildcard characters). This problem takes as input a database DB of sequences. The goal is to identify all the patterns of length at most l (with anywhere from 0 to $\lfloor l/2 \rfloor$ wild card characters). In particular, the output should be all the patterns together with a count of how many times each pattern occurs. Optionally a threshold value for the number of occurrences could be supplied. We refer to this problem as the *simple motifs search problem*.

In this chapter we consider mainly motif search in DNA sequences. These motif search algorithms could be extended to protein sequences as well. Detecting motifs in amino acid sequences requires accounting for the strong, but complex, chemical similarity relationships among the amino acid residues. These complex relationships are quantified and summarized in substitution matrices such as PAM and BLOSUM.

37.2   Algorithms for Planted Motif Search

Though the planted motif search problem (Problem 1) is defined for arbitrary sequences, in the literature it is usually assumed that the input consists of DNA sequences (and hence the alphabet size is 4). We also make this assumption. Numerous papers have been written in the past on the topic of Problem 1. Examples include Bailey and Elkan [4], Lawrence et al. [21], Rocke and Tompa [35]. These algorithms employ local search techniques such as Gibbs sampling, expectation optimization, etc. These algorithms may not output the planted motif always. We refer to such algorithms as *approximation algorithms* . Algorithms that always output the correct answer are referred to as *exact algorithms* .

More algorithms have been proposed for Problem 1 by the following authors: Pevzner and Sze [30], Buhler and Tompa [8]. The algorithm in [30] is based on finding cliques in a graph and the algorithm of [8] employs random projections. These algorithms have been experimentally demonstrated to perform well. These are approximation algorithms as well. Details of these algorithms are given in this chapter.

Algorithms for Problem 1 can be categorized into two depending on the basic approach employed, namely, profile-based algorithms and pattern-based algorithms (see e.g., [31]). Profilebased algorithms predict the starting positions of the occurrences of the motif in each sequence. On the other hand, pattern-based algorithms predict the motif (as a sequence of residues) itself.

Several pattern based algorithms are known. Examples include PROJECTION [8], MULTIPROFILER [19], MITRA [11], and PatternBranching [31]. PatternBranching (due to Price, Ramabhadran and Pevzner [31]) starts with random seed strings and performs local searches starting from these seeds. More details on this algorithm are provided later.

Examples of profile-based algorithms include CONSENSUS [17], GibbsDNA [21], MEME [3], and ProfileBranching [31]. The performance of profile-based algorithms are specified with a measure called "performance coefficient". The performance coefficient gives an indication of how many positions (for the motif occurrences) have been predicted correctly. For the (15, 4) challenge problem, these algorithms have the following performance coefficients (respectively): 0.2, 0.32, 0.14, and 0.57. The run times of these algorithms for this instance are (respectively, in seconds): 40, 40, 5, and 80.

A profile-based algorithm could either be approximate or exact. Likewise a pattern-based algorithm may either be exact or approximate. Algorithms that are exact are also known as *exhaustive enumeration algorithms* in the literature.

Many exact algorithms are known. (See e.g., [23, 7, 12, 37, 38, 40, 42, 32].) However, as pointed out in [8], these algorithms "become impractical for the sizes involved in the challenge problem". Exceptions are the MITRA algorithm [11] and the PMS algorithms of Rajasekaran, Balla and Huang [32]. These algorithms are pattern-based and are exact. MITRA solves for example the $(15, 4)$ instance in 5 minutes using 100 MB of memory [11]. This algorithm is based on the WINNOWER algorithm [30] and uses pairwise similarity information. A new pruning technique enables MITRA to be more efficient than WIN-NOWER. MITRA uses a mismatch tree data structure and splits the space of all possible patterns into disjoint subspaces that start with a given prefix. The same $(15, 4)$ instance is solved in 3.5 minutes by PMS [32].

It is noteworthy here that the profile-based algorithms such as CONSENSUS, GibbsDNA, MEME, and ProfileBranching take much less time for the $(15, 4)$ instance [31]. However these algorithms fall under the approximate category and may not always output the correct answer. Some of the pattern-based algorithms (such as PROJECTION, MULTIPROFILER, and PatternBranching) also take much less time [31]. However these are approximate as well (though the success rates are close to 100%).

Another class of motif finding algorithms, based on the concept of "profiles" can also be found in the literature. Profiles are matrices (also referred to as PSSMs) often used to describe motifs. They are also closely related to Profile-HMMs. Construction of a profile or a profile-HMM is equivalent to capturing the essence of a motif. These algorithms are used in the context of protein sequences as well as nucleotide motifs (see e.g., [15], [16], [39], and [29]).

37.2.1 A Probabilistic Analysis

The problem of planted motif search is complicated by the fact that, for a given value of d, if the value of l is small enough, then the expected number of motifs that occur by random chance could be enormous. For instance, when $n = 20, m = 600, l = 9, d = 2$, the expected number of spurious motifs (that occur in each input sequence at a hamming distance of d) is 1.6. On the other hand for $n = 20, m = 600, l = 10, d = 2$, the expected number of spurious motifs is only 6.1×10^{-8}. A probabilistic analysis to this effect can be conducted as follows.

Let S_k be any input sequence $1 \le k \le n$ and let u be any l-mer. Call the positions $1, l + 1, 2l + 1, \ldots, \lceil \frac{m-l+1}{l} \rceil l + 1$ *special positions*. Probability that u occurs in S_k at a hamming distance of d starting from a specific special position is $p = \binom{l}{d} \left(\frac{3}{4}\right)^d \left(\frac{1}{4}\right)^{l-d}$. Thus, probability that u occurs in S_k starting from at least one of the special positions is $1 - (1 - p)^{m'}$ where $m' = \lceil \frac{m-l+1}{l} \rceil + 1$. As a result, probability that u occurs somewhere in S_k is at least $1 - (1 - p)^{m'}$. This means that the expected number of l-mers that occur in each of the input sequences (at a hamming distance of d) is $\ge 4^l \left[1 - (1 - p)^{m'} \right]^n$.

37.2.2 The WINNOWER and SP-STAR Algorithms

The algorithm of Pevzner and Sze [30] (called *WINNOWER*) works as follows. If A and B are two instances (i.e., occurences) of the motif then the hamming distance between A and B is at most $2d$. In fact the expected hamming distance between A and B is $2d - \frac{4d^2}{3l}$. The algorithm constructs a collection C of all possible l-mers in the input. A Graph $G(V, E)$ is then constructed. Each l-mer in C will correspond to a node in G. Two nodes u and v in G are connected by an edge if and only if the hamming distance between the two l-mers is at most $2d$ and these l-mers come from two different sequences.

Clearly, the n instances of the motif M form a clique of size n in G. Thus the problem of finding M reduces to that of finding large cliques in G. Unfortunately, there will be numerous 'spurious' edges (i.e., edges that do not connect instances of M) in G and also finding cliques is $\mathcal{NP}$-hard. Pevzner and Sze [30] employ a clever technique to prune spurious edges.

Pevzner and Sze [30] observe that the graph G constructed above is 'almost random' and is multipartite. They use the notion of an *extendable clique*. If $Q = \{v_1, v_2, \ldots, v_k\}$ is any clique, node u is called a *neighbor* of Q if $\{v_1, v_2, \ldots, v_k, u\}$ is also a clique. In other words, Q can be extended to a larger clique with the inclusion of u. A clique is called *extendable* if it has at least one neighbor in every part of the multipartite graph G. The algorithm WINNOWER is based on the observation that every edge in a maximal n-clique belongs to at least $\binom{n-2}{k-2}$ extendable cliques of size k. This observation is used to eliminate edges.

WINNOWER is an iterative algorithm where cliques of larger and larger sizes are constructed. The run time of the algorithm is $O(N^{2d+1})$ where $N = nm$. But in practice the algorithm runs much faster. In [30] another algorithm called SP-STAR has also been given. This algorithm is faster than WINNOWER and uses less memory. WINNOWER algorithm treats all the edges of G equally without distinguishing between edges based on similarities. SP-STAR scores the l-mers of C as well as the edges of G appropriately and hence eliminates more edges than WINNOWER in any iteration.

37.2.3   Pattern Branching

A local searching algorithm called *PatternBranching* has been proposed in [31]. If u is any l-mer, then there are $\binom{l}{d}3^d$ l-mers that are at a hamming distance of d from u. Call each such l-mer a *neighbor* of u. One way of solving the planted motif search problem is to start from each l-mer in the input, search the neighbors of this l-mer, score them appropriately and output the best scoring neighbor. There are a total of $n(m-l+1)$ l-mers in the input. Each of these l-mers has $\binom{l}{d}3^d$ neighbors. For each such neighbor, a score can be computed. Having computed the scores of all of these $n(m - l + 1)\binom{l}{d}3^d$ neighbors, the best scoring neighbor is output. A similar approach has been employed by [43, 12].

Let $S = S_1, S_2, \ldots, S_n$ be the collection of n given input sequences. The algorithm of [31] only examines a selected subset of neighbors of any l-mer u of the input and hence is more efficient. For any l-mer u, let $D_i(u)$ stand for the set of neighbors of u that are at a hamming distance of i (for $1 \leq i \leq d$). For any input sequence S_j let $d(u, S_j)$ denote the minimum hamming distance between u and any l-mer of S_j (for $1 \leq j \leq n$). Let $d(u, S) = \sum_{j=1}^{n} d(u, S_j)$. For any l-mer u in the input let $\mathsf{BestNeighbor}(u)$ stand for the neighbor v in $D_1(u)$ whose distance $d(v, S)$ is minimum from among all the elements of $D_1(u)$.

The $\mathsf{PatternBranching}$ algorithm starts from a u, identifies $u_1 = \mathsf{BestNeighbor}(u)$; Then it identifies $u_2 = \mathsf{BestNeighbor}(u_1)$; and so on. It finally outputs u_d. The best u_d from among all possible u's is output.

A pseudocode for this algorithm is given next. In this pseudocode u and u_0 are the same.

Algorithm $\mathsf{PatternBranching}(S, l, d)$

Let M be an arbitrary l-mer;
for each l-mer u in S **do**
    for $j := 0$ **to** d **do**
        if $d(u_j, S) < d(M, S)$ **then** $M := u_j$;
        $u_{j+1} := \mathsf{BestNeighbor}(u_j)$;
output M;

The above algorithm has been shown to perform well in practice [31]. Also, the above algorithm keeps only one best neighbor for each l-mer. Instead, it is possible to keep more than one best neighbors. In other words, BestNeighbor(u_j) could be a set of l-mers instead of a single l-mer.

In [31] a profile-based version of PatternBranching has been given as well. This version is called ProfileBranching.

37.2.4 Random Projection Algorithm

The algorithm of Buhler and Tompa [8] is based on random projections. Let the motif M of interest be an l-mer. Let C be the collection of all the l-mers from all the n input sequences. Project these l-mers along k randomly chosen positions (for some appropriate value of k). In other words, for every l-mer $u \in C$, generate a k-mer u' which is a subsequence of u corresponding to the k random positions chosen. (The random positions are the same for all the l-mers). We can think of each k-mer thus generated as an integer. We group the k-mers according to their integer values. (I.e., we hash all the l-mers using the k-mer of any l-mer as its hash value).

If a hashed group has at least a threshold number s of l-mers in it, then there is a good chance that M will have its k-mer equal to the k-mer of this group. There are $n(m - l + 1)$ l-mers in the input and there are 4^k possible k-mers. Thus the expected number of l-mers that hash into the same bucket is $\frac{n(m-l+1)}{4^k}$. The threshold value s is chosen to be twice this expected value. The value of k is chosen such that $n(m - l + 1) < 4^k$. This ensures that the expected number of random l-mers that hash into the same bucket is less than one. It should also be the case that $k < l - d$. Typical values used for k and s are 7 and 3, respectively.

The process of random hashing is repeated r times (for some appropriate value of r) so as to be sure that a bucket of size $\geq s$ is observed at least once. The value of r can be calculated as follows. Probability p that a given planted motif instance hashes to the planted bucket is given by $\frac{\binom{l-d}{k}}{\binom{l}{k}}$. There are n instances of the planted motif and hence the probability that fewer than s of them hash into the planted bucket is given by $p' = \sum_{i=1}^{s-1} \binom{n}{i} p^i (1 - p)^{n-i}$. As a result, probability that fewer than s instances hash into the planted bucket in each of the r trials is $P = (p')^r$. The value of r is thus $\left\lceil \frac{\log P}{\log p'} \right\rceil$. In [8], a value of 0.05 is used for P.

We collect all the k-mers (and the corresponding l-mers) that pass the threshold and these are processed further to arrive at the final answer M. Processing is done using the expectation maximization (EM) technique of Lawrence and Reilly [22]. The EM formulation employs the following model. Each input sequence has an instance of a length l motif such that these instances are characterized by a $4 \times l$ weight matrix W. In particular, $W[i, j]$ is the probability that base i occurs in position j ($1 \leq i \leq 4$ and $1 \leq j \leq l$). Occurrences of bases in the different positions are independent. Bases for the remaining $m - l$ positions in each sequence are governed by a background distribution B. If S is the set of input sequences then the EM-based technique of [22] determines a weight matrix model W^* that maximizes the likelihood ratio $\frac{\Pr(S|W^*,B)}{\Pr(S|B)}$.

37.2.5 Algorithm PMS1

In this section we present details on the algorithm PMS1 of [32]. Consider the following simple algorithm for the planted motif problem: 1) Let the input sequences be $S_1, S_2, \ldots, S_n$. The length of each sequence is m. Form all possible l-mers from out of these sequences.

The total number of l-mers is $\leq nm$. Call this collection of l-mers C. Let the collection of l-mers in S_1 be C'; 2) Let u be an l-mer in C'. For all $u \in C'$ generate all the patterns v such that u and v are at a hamming distance of d. The number of such patterns for a given u is $\binom{l}{d}3^d$. Thus the total number of patterns generated is $O\left(m\binom{l}{d}3^d\right)$. Call this collection of l-mers C''. Note that C'' contains M, the desired output pattern (assuming that M does not occur in any of the input sequences); 3) For every pair of l-mers (u, v) with $u \in C$ and $v \in C''$ compute the hamming distance between u and v. Output that l-mer of C'' that has a neighbor (i.e., an l-mer at a hamming distance of d) in each one of the n input sequences. The run time of this algorithm is $O\left(nm^2l\binom{l}{d}3^d\right)$. If M occurs in one of the input sequences, then this algorithm will run in time $O(n^2m^2l)$.

Another equally simple algorithm considers every possible l-mer one at a time and checks if this l-mer is the correct motif M. There are 4^l possible l-mers. Let M' be one such l-mer. We can check if $M' = M$ as follows. Compute the hamming distance between u and M' for every $u \in C$. (Note that C is the collection of all possible l-mers in the input sequences.) As a result we can check if M' occurs in each input sequence (at a hamming distance of d). Thus we can identify all the motifs of interest in a total of $O\left(nml4^l\right)$ time. We get the following Lemma.

LEMMA 37.1 We can solve the planted (l, d)-motif problem in $O(nml4^l)$ time.

Algorithm PMS1 is based on sorting and takes the following form.

Algorithm PMS1

1. Generate all possible l-mers from out of each of the n input sequences. Let C_i be the collection of l-mers from out of S_i for $1 \leq i \leq n$.
2. For all $1 \leq i \leq n$ and for all $u \in C_i$ generate all l-mers v such that u and v are at a hamming distance of d. Let the collection of l-mers corresponding to C_i be C'_i, for $1 \leq i \leq n$. The total number of patterns in any C'_i is $O\left(m\binom{l}{d}3^d\right)$.
3. Sort all the l-mers in every $C'_i, 1 \leq i \leq n$ and eliminate duplicates in every C'_i. Let L_i be the resultant sorted list corresponding to C'_i.
4. Merge all the L_i's ($1 \leq i \leq n$) and output the generated (in step 2) l-mer that occurs in all the L_i's.

The following theorem results.

THEOREM 37.1 *Problem 1 can be solved in time* $O\left(nm\binom{l}{d}3^d\frac{l}{w}\right)$ *where w is the word length of the computer. A run time of* $O\left(\left[nm + m\binom{l}{d}^2 3^{2d}\right]\frac{l}{w}\right)$ *is also achievable.*

37.2.6 Algorithm PMS2

An improved algorithm called PMS2 has also been given in [32]. Let M be the planted motif. Note that if M occurs in every input sequence, then every substring of M also occurs in every input sequence. In particular, there are at least $l - k + 1$ k-mers (for $d \leq k \leq l$) such that each of these occurs in every input sequence at a hamming distance of at most d. Let Q be the collection of k-mers that can be formed out of M. There are $l - k + 1$ k-mers in

Q. Each one of these k-mers will be present in each input sequence at a hamming distance of at most d.

In addition, in every input sequence S_i, there will be at least one position i_j such that a k-mer of Q occurs starting from i_j; another k-mer of Q occurs starting from $i_j + 1; \ldots$; yet another k-mer occurs starting from $i_j + l - k$. We can get an l-mer putting together these k-mers that occur starting from each such i_j.

Possibly, there could be many motifs of length k that are in the positions starting from each of $i_j, i_j + 1, \ldots, i_j + l - k$ such that all of these motifs are present in all of the input sequences (with a hamming distance of at most d). Assume that M_{i_j+r} is one motif of length k that starts from position $i_j + r$ of S_i that is also present in every input sequence (for $0 \le r \le l-k$). If the last $k-1$ residues of M_{i_j+r} are the same as the first $k-1$ residues of M_{i_j+r+1} (for $0 \le r \le l-k-1$), then we can obtain an l-mer from these motifs in the obvious way. This l-mer is potentially a correct motif. Also, note that to obtain potential motifs (of length l), it suffices to process one of the input sequences (in a manner described above). Now we are ready to describe the improved algorithm.

There are two phases in the algorithm. In the first phase we identify all $(d+c)$-mers M_{d+c} (for some appropriate value c) that occur in each of the input sequences at a hamming distance of at most d. We also collect potential l-mers (as described above) in this phase. In the second phase we check, for each l-mer M' collected in the first phase, if M' is a correct answer or not. Finally we output all the correct answers.

First we observe that the algorithm PMS1 can also be used for the case when we look for a motif M that occurs in each input sequence at a hamming distance of at most d. The second observation is that if c is large enough then there will not be many spurious hits. A suggested value for c is the largest integer for which PMS1 could be run (without exceeding the core memory of the computer and within a reasonable amount of time).

We present more details on the two phases.

Algorithm PMS2

Phase I

Solve the planted $(d+c, d)$-motif problem on the input sequences (with a hamming distance of $\le d$, using e.g., a modified PMS1). Let R be the set of all motifs found. Let S_k be one of the input sequences. (S_k could be an arbitrary input sequence; it could be chosen randomly as well.) Find all the occurrences of all the motifs of R in S_k (with a hamming distance of up to d). This can be done, e.g., as follows: form all the $(d+c)$-mers of S_k (keeping track of the starting position of each in S_k); For each $(d+c)$-mer $u \in S_k$, find all the $(d+c)$-mers v such that u and v are at a hamming distance of at most d. If R' is the collection of these $(d+c)$-mers, sort R and R' and merge them; and figure out all the occurrences of interest.

Let S_k be of length m. For every position i in S_k, let L_i be the list of all motifs of R that are in S_k (with a hamming distance of $\le d$) starting from position i. Let $\mathcal{A}$ be the l-mer of S_k that occurs starting from position i. Let M_1 be a member of L_i. If M_2 is a member of $L_{i+l-(d+c)}$ such that the last $2(d+c) - l$ characters of M_1 are the same as the first $2(d+c) - l$ characters of M_2, then we could get an l-mer $\mathcal{B}$ by appending the last $l - (d+c)$ residues of M_2 to M_1 (at the end). If the hamming distance between $\mathcal{A}$ and $\mathcal{B}$ is d, then $\mathcal{B}$ is retained as a candidate for the correct motif. We gather all such candidates and check if any of these candidates are correct motifs. Details are given below.

for $i := 1$ **to** $m - l + 1$ **do**
　　for every $u \in L_i$ **do**
　　　　for every $v \in L_{i+l-(d+c)}$ **do**
　　　　　　Let the l-mer of S_k starting from position i be $\mathcal{A}$. If the last
　　　　　　$2(d+c) - l$ residues of u are the same as the first $2(d+c) - l$
　　　　　　residues of v, then form an l-mer $\mathcal{B}$ by appending the last
　　　　　　$l - (d+c)$ residues of v to u. If the hamming distance between
　　　　　　$\mathcal{A}$ and $\mathcal{B}$ is d, then add $\mathcal{B}$ to the list C of candidates.

Phase II

　　for every $v \in C$ **do**
　　　　Check if v is a correct motif in $O(nml)$ time.

For any node u of L_i there can be at most $4^{l-(d+c)}$ candidate motifs. Thus the time
needed to get all the candidate motifs is $O\left(\sum_{i=1}^{m-l+1} |L_i| 4^{l-(d+c)} l\right)$.

We arrive at the following Theorem.

THEOREM 37.2 *Problem 1 can be solved in time*
$O\left(nm \sum_{i=0}^{d} \binom{d+c}{i} 3^i \frac{d+c}{w} + znml + \sum_{i=1}^{m-l+1} |L_i| 4^{l-(d+c)} l\right)$ *where z is the number of potential l-mers collected in the first phase and w is the word length of the computer. If*
$d \leq \lfloor l/2 \rfloor$, *then the run time is* $O\left(mn\binom{d+c}{d} 3^d \frac{d+c}{w} + znml + \sum_{i=1}^{m-l+1} |L_i| 4^{l-(d+c)} l\right)$.

An Alternative Algorithm. We can modify the above algorithm as follows. We first find
the collection R of all the $(d+c)$-mers that are present in every input sequence at a hamming
distance of at most d as before. In the above version, we pick only one sequence S_k and find
all the candidate motifs arising out of S_k. An alternative is to find the candidate motifs
from each sequence and get the intersection of these sets. Let A_i be the set of candidates
from S_i $(1 \leq i \leq n)$. Let $A = \bigcap_{i=1}^{n} A_i$. We output A.

37.2.7 Algorithm PMS3

A third algorithm called PMS3 has been given in [32]. This algorithm enables one to handle
large values of d. Let $d' = \lfloor d/2 \rfloor$. Let M be the motif of interest with $|M| = l = 2l'$ for some
integer l'. Let M_1 refer to the first half of M and M_2 to the second half. We know that
M occurs in every input sequence. Let $S = s_1, s_2, \ldots, s_m$ be an arbitrary input sequence.
Let the occurrence of M (with a hamming distance of d) in S start at position i. Let
$S' = s_i, s_{i+1}, \ldots, s_{i+l'-1}$ and $S'' = s_{i+l'}, \ldots, s_{i+l-1}$.

Then, clearly, either 1) the hamming distance between M_1 and S' is at most d' or 2)
the hamming distance between M_2 and S'' is at most d'. Also, note that in every input
sequence either M_1 occurs with a hamming distance of at most d' or M_2 occurs with a
hamming distance of at most d'. As a result, in at least t' sequences (where $t' = \lceil t/2 \rceil$)
either M_1 occurs with a hamming distance of at most d' or M_2 occurs with a hamming
distance of at most d'. PMS3 exploits these observations. More details can be found in [32].

37.2.8   Saving Memory

The way PMS1 is described, we first form all possible l-mers from out of all the input sequences, generate all relevant neighbors of these l-mers, sort and merge all of them to identify the generated l-mer(s) found in all the sequences. We can modify the algorithm as follows so as to reduce the memory used.

Algorithm PMS1A

Generate all possible l-mers from out of the first input sequence S_1. Let C_1 be the collection of these l-mers. For all $u \in C_1$ generate all l-mers v such that u and v are at a hamming distance of d. Sort the collection of these l-mers and eliminate duplicates. Let L be the resultant sorted collection.

for $i := 2$ **to** n **do**

1. Generate all possible l-mers from out of the input sequence S_i. Let C_i be the collection of these l-mers.
2. For all $u \in C_i$ generate all l-mers v such that u and v are at a hamming distance of d. Let the collection of these l-mers be C_i'.
3. Sort all the l-mers in C_i' and eliminate duplicates. Let L_i be the resultant sorted list.
4. Merge L_i and L and keep the intersection in L. I.e., set $L := L \cap L_i$.

L now has the motif(s) of interest.

37.2.9   Extensions

The planted (l, d)-motif problem as has been defined (in [30] for example) requires discovering a motif M that occurs in every input sequence at a hamming distance of exactly d. Varitations of this problem can be conceived of. We cosider two variants in this section.

Problem 1(a). Input are n sequences each of length m. The problem is to identify a motif M of length l. It is given that each input sequence has a substring of length l such that the hamming distance between this substring and M is **at most** d.

THEOREM 37.3 *Problem 1(a) can be solved in time* $O\left(nm \sum_{i=0}^{d} \binom{l}{i} 3^i \frac{l}{w}\right)$. *If* $d \leq \lfloor l/2 \rfloor$, *then this run time is* $O\left(nm\binom{l}{d} 3^d \frac{l}{w}\right)$.

Proof. An algorithm similar to PMS1 can be devised.

1. Generate all possible l-mers from out of each of the n input sequences. Let C_i be the collection of l-mers from out of S_i for $1 \leq i \leq n$.
2. For all $1 \leq i \leq n$ and for all $u \in C_i$ generate all l-mers v such that u and v are at a hamming distance of at most d. Let the collection of l-mers corresponding to C_i be C_i', for $1 \leq i \leq n$. The total number of patterns in any C_i' is $O\left(\sum_{i=0}^{d} m\binom{l}{i} 3^i\right)$.
3. Sort all the l-mers in every C_i' and eliminate duplicates, $1 \leq i \leq n$. Let L_i be the resultant sorted list corresponding to C_i'.
4. Merge all the L_i's ($1 \leq i \leq n$) and output the generated (in step 2) l-mer that occurs in all the L_i's.

Problem 1(b). Input are n sequences each of length m. The problem is to find all motifs M of length l. A motif M should be output if it occurs in at least ϵn of the input sequences at a hamming distance of d. Here ϵ is a fraction specified as a part of the input. (This variant has been considered in [8]. They use a value of $1/2$ for ϵ).

THEOREM 37.4 *Problem 1(b) can be solved in time* $O\left(nm\binom{l}{d}3^d\frac{l}{w}\right)$.

Proof. The algorithm to be used is the same as PMS1. The only difference is that step 4 now becomes: "Merge all the L_i's ($1 \le i \le n$) and output the generated (in step 2) l-mers that occur in at least ϵn of the L_i's". The run time remains the same asymptotically.

One could also refine Problem 1(b) to look for motifs that occur at a hamming distance of at most d.

37.3   Techniques for Edited Motif Search

In this section we consider edited motif search (Problem 2). Here the input is a database DB of sequences $S_1, S_2, \ldots, S_n$. Input also are integers l, d, and q. The output should be all the patterns in the DB such that each pattern is of length l and it occurs in at least q of the n sequences. A pattern U is considered an occurrence of another pattern V as long as the *edit distance* between U and V is at most d.

An algorithm for the above problem has been given by Sagot [36] that has a run time of $O(n^2ml^d|\Sigma|^d)$ where m is the average length of the sequences in DB and Σ is the alphabet from which the input sequences are generated. It uses $O(n^2m/w)$ space where w is the word length of the computer. This algorithm can be used to solve Problem 1 as well as Problem 2. Consider the case of Problem 1. This algorithm builds a suffix tree on the given sequences in $O(nm)$ time using $O(nm)$ space. Some preprocessing is done on the suffix tree that takes $O(n^2m/w)$ time and $O(n^2m/w)$ space. If u is any l-mer present in the input, there are $O\left(l^d(|\Sigma| - 1)^d\right)$ possible neighbors for u. (A neighbor of u is any word v such that the hamming distance between u and v is d). Any of these neighbors could potentially be a motif of interest. Since there are $O(nm)$ l-mers in the input, the number of such neighbors is $O\left(nml^d(|\Sigma| - 1)^d\right)$. The algorithm of [36], for each such neighbor v, walks through the tree to check if v is a possible answer. This walking step is referred to as 'spelling'. The spelling operation takes a total of $O(n^2ml^d(|\Sigma|-1)^d)$ time using an additional $O(nm)$ space. When employed for solving Problem 2, the same algorithm takes $O(n^2ml^d|\Sigma|^d)$ time.

An algorithm with an expected run time of $O(nm + d(nm)^{1+pow(\epsilon)}\log nm)$ where $\epsilon = d/l$ and $pow(\epsilon)$ is an increasing concave function has been given in [2]. The value of $pow(\epsilon)$ is roughly 0.9 for protein and DNA sequences. This algorithm is also suffix-tree based.

37.3.1   An Algorithm Similar to PMS1

In this section we describe a sorting based algorithm that has the same run time as that of [36]. From hereon the word *occurrence* is used to denote occurrence within an edit distance of d, and the word *presence* is used to denote exact occurrence (i.e., occurrence within an edit distance of zero).

The basic idea behind the algorithm is: We generate all possible l-mers in the database. There are at most mn such l-mers and these are the patterns of interest. For each such l-mer we want to determine if it occurs in at least q of the input sequences. Let u be one of the above l-mers. If v is a string such that the edit distance between u and v is at most d, then we say v is a neighbor of u. We generate all the neighbors of u. For each neighbor v of u we determine a list of input sequences in which v is present. These lists (over all possible neighbors of u) are then merged to obtain a list of input sequences in which u occurs (within an edit distance of d).

Note that if u is an l-mer, then its neighbors will have a length in the interval $[l-d, l+d]$. In other words, there are $(2d+1)$ possible values for the lengths of the neighbors of u. Also note that more than one neighbor of u could have the same length. Corresponding to each r-mer x (where r is an integer in the interval $[l-d, l+d]$) present in the input, we keep a 4-tuple: $(x, SeqNum, Pos, 0)$. Here $SeqNum$ is an index of the input sequence I that x belongs to. There are $O(nmd)$ such 4-tuples. The indexing of the input sequences can be done arbitrarily (e.g., in the order in which they appear in the input). Pos is the starting position of x in I. The fourth entry (0) indicates that x is an r-mer present in one of the input sequences. For every neighbor v of u (u being an l-mer present in the input), we keep a 4-tuple as well: $(v, SeqNum, Pos, 1)$. Here $SeqNum$ is the index of the sequence I that u belongs to and Pos is the starting position of u in I. The fourth entry (1) indicates that this 4-tuple corresponds to a neighbor. We provide details of the algorithm next.

Note that each l-mer present in the input could be represented as a pair $(SeqNum, Pos)$ where $SeqNum$ is the index of the sequence in which the l-mer is present and Pos is the starting position of the l-mer in this sequence. We make use of an array $A[1 : n, 1 : m, 1 : n]$. At the end of the algorithm this array will have the following property: $A[SeqNum, Pos, j] = 1$ if and only if the l-mer $(SeqNum, Pos)$ occurs in the input sequence with index j ($1 \leq j \leq n$).

1. . Generate 4-tuples for all r-mers present in DB where $r \in [l-d, l+d]$. Each of these 4-tuples has a 0 as its fourth entry. Call this collection C. Sort C in lexicographic order and eliminate duplicates among these 4-tuples for which the first two entries are the same to get L_1. Note that the first entries of the 4-tuples in C could be of different lengths. A simple way of sorting these 4-tuples is to group them into $(2d+1)$ groups one corresponding to each possible length of the first entry and handle the groups separately. L_1 has $O(nmd)$ entries. Now sort the 4-tuples of L_1 with respect to their first and fourth entries (in lexicographic order) to get L_2.

2. For every distinct l-mer u present in DB generate all the patterns v such that u and v are at an edit distance of at most d. The number of such neighbors for a given u is $O(l^d |\Sigma|^d)$ (a proof follows). This generation is done using the algorithm of Myers [26]. Form the 4-tuples corresponding to all possible neighbors of all the distinct l-mers in DB. Each of these 4-tuples has 1 as its fourth entry.

3. Sort all the 4-tuples generated in step 2 with respect to their first and fourth entries. The total number of 4-tuples is $O(nml^d |\Sigma|^d)$. Let L_3 be the sorted sequence.

4. Merge L_3 with L_2 to get L_4. We can think of L_4 as consisting of groups where a group has 4-tuples with the same first entry. A group itself can be thought of as consisting of two subgroups. The 4-tuples of the first subgroup have 0 as their fourth entry. The 4-tuples of the second subgroup have 1 as their fourth entry.

> **for** each group G of L_4 **do**
>   Let G_1 and G_2 be the two supgroups of G. Identify the distinct sequence indices (i.e., second entries) in the 4-tuples of G_1. Let these indices be $i_1, i_2, \ldots, i_k$. Note that $k \leq n$.
>   **for** each 4-tuple $(v, SeqNum, Pos, 1)$ in G_2 **do**

$A[SeqNum, Pos, a_j] := 1$, for $1 \leq j \leq k$. (I.e., the l-mer
$(SeqNum, Pos)$ occurs in the input sequences S_{a_j}, for $1 \leq j \leq k$.
Scan through the array A to output the right l-mers. In particular, output
$(SeqNum, Pos)$ if $A[SeqNum, Pos, j]$ is 1 for $1 \leq j \leq n$.

THEOREM 37.5   *The above algorithm runs in time* $O\left(n^2 m l^d |\Sigma|^d\right)$. *The space used is*
$O(nml^d|\Sigma|^d)$. *The space used can be reduced to* $O(nmd + l^d|\Sigma|^d)$.

Proof. The run time of step 1 is $O\left(nmd\frac{l}{w}\right)$ using radix sort algorithm.

Let u be any l-mer. Then the number of patterns v such that the edit distance between u and v is at most d is $O(l^d|\Sigma|^d)$ as argued next. The same fact has been proven by Crochemore and Sagot as well [10]. Let $N(t)$ be the number of patterns obtainable from u by performing t operations (inserts, deletes, and substitutions) on u. The number of patterns of interest is then $\sum_{t=0}^{d} N(t)$. Of the t operations let the number of inserts, deletes, and substitutions be i, del, s, respectively with $i + del + s = t$. For a given choice of i, del, s, it is easy to see that the number of patterns obtainable is $\binom{l+i}{i}\binom{l}{del}\binom{l}{s}|\Sigma|^{s+i}$. As a result, $N(t) \leq \frac{(t+1)(t+2)}{2}\left(\frac{(l+t)e}{t}\right)^t$ using the fact that $\binom{a}{b} \leq \left(\frac{ae}{b}\right)^b$. Finally, summing $N(t)$ over all t's, we see the result (as long as $d \geq 6$).

Thus the generation of all the patterns in step 2 takes $O(nml^d|\Sigma|^d)$ time (using the algorithm in [26]).

In step 3 sorting takes time $O\left(nml^d|\Sigma|^d\frac{l}{w}\right)$.

Merging in step 4 also takes time $O\left(nml^d|\Sigma|^d\frac{l}{w}\right)$. For each 4-tuple of a given G_2 the time spent is $O(k)$ where k is the number of distinct sequence indices in the corresponding G_1. Since $k \leq n$, the total time in processing all the G_2's is $O(n^2 ml^d|\Sigma|^d)$.

These observations prove the theorem (assuming that $\frac{l}{w} = O(n)$).

The above algorithm is simpler than the ones in [36, 2]. The algorithms in [36, 2] employ suffix trees. In comparison the above algorithm uses only arrays. The above algorithm can be expected to perform better than that of [36, 2] in practice.

In practice the above algorithm is expected to run much faster. It is easy to see that the run time of the above algorithm is $O(nml^d|\Sigma|^d z)$ where z is the maximum number of distinct sequence indices in the 4-tuples of any G_1. The expected value of z can be calculated as follows. Let x be any r-mer present in DB. The expected number of sequences that x occurs in is the same as the expected value of z. If I is any input sequence and j is a fixed position in I, probability that x is present in I starting from j is $\left(\frac{1}{4}\right)^r$. Thus, probability that x is present somewhere in I is $\leq m\left(\frac{1}{4}\right)^r$. As a result, the expected number of sequences in which x is present is $\leq nm\left(\frac{1}{4}\right)^r$. Thus the expected value of z is $\leq nm\left(\frac{1}{4}\right)^{l-d}$.

As an example, if $n = 20, m = 600, l - d = 10$, the expected value of z is less than 1.

37.3.2   A Randomized Algorithm

In this section we describe a simple randomized algorithm (due to [33]) that has the potential of performing better than the algorithms of [36, 2]. The algorithms in [36, 2] employ suffix trees and the algorithm to be discussed uses arrays.

Before presenting the randomized algorithm we present a very simple algorithm. The randomized algorithm is based on this simple algorithm. This algorithm works as follows.

1. Generate all possible l-mers in DB. Let the collection of these l-mers be C. There are at most mn elements in C. Duplicates in C could be eliminated by a simple

radix sort.

2. For every l-mer u in C, compute the number of occurrences of u in DB. This can be done in time $O(nmd)$ using the algorithm of Galil and Park [13]. (See also [1, 6, 20, 25, 26, 41]).

Thus we get the following Theorem.

THEOREM 37.6 *Problem 2 can be solved in time $O(n^2 m^2 d)$.*

A Randomized Algorithm. A randomized algorithm can be developed based on the above algorithm.

1. Generate all possible l-mers in DB. Let C the collection of these l-mers. C has at most nm elements.
2. For each element u in C, pick a random sample S_u from DB of $\frac{16\alpha n \ln n}{q}$ sequences where α is the probability parameter (assumed to be a constant). Count the number of occurrences N_u of u in the sample. This will take time $|S_u| m d$ (using the algorithm of Galil and Park [13]) for a single u.
3. For each u in C such that $N_u > 10.34\alpha \ln n$, compute the occurrences of u in the entire input DB. If the number of occurrences of u in DB is q or more, then output u.

THEOREM 37.7 *The above algorithm runs in time $O\left(\frac{n^2 m^2 \log n}{q} d + gmnd\right)$ where g is the number of l-mers that pass the test in step 3. Also, the probability of an incorrect answer is no more than $n^{-\alpha} nm$. The space used is linear in the input size.*

Proof. The run time is easy to see. Note that if an l-mer occurs in less than q input sequences, it will never be output. If an l-mer u occurs in at least q sequences of DB, then the number of occurrences of u in S_u (i.e., the value of N_u) is lower bounded by a binomial random variable with mean $16\alpha \ln n$. An application of the Chernoff bounds (second equation) with $\epsilon = 1/(2\sqrt{n})$ shows that the probability that N_u is less than $10.34\alpha \ln n$ is no more than $n^{-\alpha}$. On the same token, let u' be an l-mer that occurs in at most $(3/8)q$ of the input sequences. The number of occurrences $N_{u'}$ of u' in the sample is a binomial with mean $6\alpha \ln n$. Using Chernoff bounds equation 3 with $\epsilon = 1/\sqrt{2}$, probability that $N_{u'}$ exceeds $10.25\alpha \ln n$ is at most $n^{-\alpha}$.

In summary, if a pattern occurs in q or more input sequences, it will pass the test of step 3 with high probability. Moreover, not many spurious patterns will pass the test of step 3. If a pattern has to pass the test of step 3, then it has to occur in at least $(3/8)q$ of the input sequences (in a high probabilistic sense). Therefore a high probability upper bound on g is the number of patterns that occur in $(3/8)q$ or more of the input sequences. Also note that there at most nm patterns of interest.

Note that this algorithm has the potential of performing better than those of [36, 2], especially for large values of q. When q is large (ϵn for some constant fraction ϵ, for instance), g can be expected to be small and hence the entire run time could be $o(d(nm)^{1+pow(\epsilon)} \log nm)$. Next we show that the expected value of g is very small.

Assume that every residue in each input sequence is picked randomly (from an alphabet of size 4). Let the input consist of n sequences of length m each. Let v be an l-mer and

S_k be any input sequence. Let i be any position in S_k. Probability that v is present in S_k starting from position i is $(1/4)^l$. Thus, the probability that v is present in S_k starting from some position is $\leq (m - l + 1)(1/4)^l$. For every l-mer u there are $\leq cl^d|\Sigma|^d$ (for some constant c) neighbors (as has been shown above). The length of any sych neighbor x is in the interval $[l - d, l + d]$.

In Problem 2, we are supposed to do the following: For every l-mer u in the input, check if it occurs in at least q of the input sequences. Therefore, probability that either u or any of its neighbors is present in S_k is at most $p_1 = (m - l + d + 1)c|\Sigma|^d l^d (1/4)^{l-d}$. As a result, probability that u occurs in q or more of the input sequences is at most $p_2 = \sum_{i=q}^{n} \binom{n}{i} p_1^i (1 - p_1)^{n-i} \leq \sum_{i=q}^{n} \binom{n}{i} p_1^i$. Using Stirling's approximation, we see that $p_2 \leq \frac{2^n \sqrt{2}}{\sqrt{\pi n}} \sum_{i=q}^{n} p_1^i$. By simple arithmetic we see that $p_2 \leq 1/(mn)$ when $q \geq \frac{(n+1) - \log(mn) - (1/2)\log(\pi n)}{2(l-d) - \log(m-l+d+1) - \log c - d\log|\Sigma| - d\log l}$. If $p_2 \leq 1/(mn)$, then the expected number of patterns that occur in q or more of the input sequences is ≤ 1. Thus indeed the value of g will be small even if q is not this high!

As a numerical example, consider the case: $l = 20, d = 2, m = 256, n = 500$. In this case, the condition on q becomes: $q \geq 36$. Also, if α is 4, the probability of an incorrect answer is 2.048×10^{-6}. This is also an upper bound on the expected number of patterns that will be missed by the algorithm! (A pattern is missed by the algorithm if it occurs in at least q of the input sequences but the algorithm fails to detect this).

The above analysis could be tightened further.

37.4   Algorithms for Simple Motif Search

Simple motifs search (Problem 3) takes as input a database DB of n sequences. The goal is to identify all the patterns of length at most l (with anywhere from 0 to $\lfloor l/2 \rfloor$ wild card characters). The output should be all the patterns together with a count of how many times each pattern occurs.

The motif model for Problem 3 has been derived as follows [33]. Rajasekaran, et. al. [33] have generated a list of 312 minimotifs (i.e., motifs of short length) that have defined biological functions. They have used this list to select parameters for a de novo analysis of novel minimotifs in the human proteome. They have chosen to analyze novel motifs with a length (l) of 10 amino acids because 92% of the previously characterized minimotifs in their list are less than 10 amino acids in length. Another reason for choosing a length of 10 amino acids is based on the function of minimotifs. Most minimotifs are in binding domains or substrates of enzymes. The peptide ligand binding surfaces on proteins in the Protein Data Bank is usually no longer than 35 angstroms. A 10 amino acid peptide would achieve a maximum of 35 angstroms in length if it were in a random coil or a beta-sheet structure. Thus, the selection of a length of 10 amino acids is consistent with the length of peptides that interact with binding surfaces on protein domains.

The average minimotif in their list has 2.1 wildcard positions for any amino acid. Wildcards signify any of the 20 amino acids. Since only 13% of minimotifs in their list have more than 50% wild card positions, they chose $l/2$ or 5 wild cards as the maximal number in the algorithm.

In Problem 3 an optional threshold value for the number of occurrences could be supplied. Determining this threshold is a challenging task. One way of determining this threshold is to rank the motifs in the order of the number of their occurrences and choosing certain number of them (either because they are over-represented or because they are under-represented). Another way of determining the threshold is by analyzing the table of occurrences of all the patterns in the database together with a model for the biological sequences under concern.

This differs from the first way in that here the number of occurrences of any motif will be weighted as dictated by the model. A typical value for l is 10. A simple sorting based algorithm for Problem 3 (called Simple Motif Search or SMS) is given in [33]. The run time of this algorithm is $O(l^{l/2}N)$ for a given pattern length l, the number of wild cards being at most $\lfloor l/2 \rfloor$. The number of residues in the database is N.

37.4.1 A Simple Sorting Based Algorithm

The algorithm of Martinez [23] addresses a variant of Problem 3. In particular, the input is just one sequence. The output consists of all repeated patterns. The matches of interest are exact. Even if the input has many sequences, they can be concatenated to get a single sequence.

The algorithm of [23] works as follows. Let $S = x_1, x_2, \ldots, x_n$ be the input sequence. This sequence can be thought of as n sequences where each sequence corresponds to a 'suffix' of S. I.e., S_1 is the same as S; $S_2 = x_2 x_3 \cdots x_n$; and so on. These n sequences are then sorted one residue at a time. At any level of the sorting we have groups of sequences. In particular, after k levels of sorting, two sequences are in the same group, if the first k residues are the same in these two sequences. Sorting at any level is local to groups. A group will not be sorted further if it has only one element.

The expected run time of the above algorithm is $O(n \log n)$ whereas its worst case run time is $\Omega(n^2)$.

The above algorithm can be modified to have an expected run time of $O(n)$ by performing radix sorting with respect to the first $\Omega(\log n)$ residues of the n sequences (see e.g., [18]).

As another variant consider a problem where the input are a sequence S and an integer k. The goal is to report all repeats of length k. This variant can be solved in the worst case in time $O(nk/w)$, w being the word length of the computer as follows. 1) Form all k-mers of S. There are less than n such k-mers; 2) Sort these k-mers lexicographically in time $O(nk/w)$ and 3) Scan through the sorted list to identify the repeats.

Instead of the above algorithm one could also employ a prefix tree or a suffix array to get a run time of $O(n)$. Depending on the underlying constant and the values of k and w, the above algorithm could be faster.

37.4.2 Simple Motif Search (SMS)

As has been pointed out before, we are interested in identifying all the patterns of length at most l (with anywhere from 0 to $\lfloor l/2 \rfloor$ wild card characters). For every pattern, the number of occurrences should be output. How does a biologist identify biologically important patterns? This is a challenging task for biologists and will not be addressed in this chapter.

Define a (u, v)-class as a class of patterns where each pattern has length u and has exactly v wild card characters. For example, GA??C?T belongs to $(7, 3)$-class. Note that there are $\binom{u-2}{v}|\Sigma|^{u-v}$ patterns in a (u, v)-class.

To identify the patterns in a (u, v)-class, we perform $\binom{u-2}{v}$ sorts. More specifically, for each possible placement of v wild card characters (excluding at the end positions) in a sequence of length u, we perform a sorting. As an example, consider a case where $u = 5$ and $v = 2$. There are three possible placements: C??CC, CC??C, and C?C?C, where C corresponds to any residue. Call every placement as a (u, v)-*pattern type*. For every (u, v)-pattern type, we perform the following steps.

Algorithm SMS

For every (u, v)-pattern type **do**

1. If R is a pattern type in (u, v)-class, we generate all possible u-mers in all the sequences of DB. If the sequences in DB have lengths $m_1, m_2, \ldots, m_n$, respectively, then the number of u-mers from S_i is $m_i - u + 1$, for $1 \leq i \leq n$.

2. Sort all the u-mers generated in step 1 only with respect to the non-wild card positions of R. For example, if the pattern type under concern is CC??C?C, we generate all possible 7-mers in DB and sort the 7-mers with respect to positions 1, 2, 5, and 7. Employ radix sort (see e.g., [18]).

3. Scan through the sorted list and count the number of occurrences of each pattern.

The run time of the above algorithm is $O\left(\binom{u-2}{v} N \frac{u}{w}\right)$ for a (u, v)-class, where N is the total number of residues in DB and w is the word length of the computer.

Now we consider the problem of identifying all of the following patterns: The maximum length is 10. Pattern lengths of interest are: 3, 4, 5, 6, 7, 8, 9 and 10. The maximum number of wild cards are 1, 2, 2, 3, 3, 4, 4 and 5, respectively. In other words we are interested in: $(10, 5)$-class, $(10, 4)$-class, $\ldots$, $(10, 1)$-class, $(9, 4)$-class, $(9, 3)$-class, $\ldots$, $(9, 1)$-class, $\ldots$, $(4, 2)$-class, $(4, 1)$-class, and $(3, 1)$-class. Thus the total number of sorts done is

$$\sum_{i=0}^{5}\binom{8}{i} + \sum_{i=0}^{4}\binom{7}{i} + \sum_{i=0}^{4}\binom{6}{i} + \sum_{i=0}^{3}\binom{5}{i} + \sum_{i=0}^{3}\binom{4}{i} + \sum_{i=0}^{2}\binom{3}{i} + \sum_{i=0}^{2}\binom{2}{i} + \sum_{i=0}^{1}\binom{1}{i} = 429.$$

THEOREM 37.8 *SMS algorithm runs in time* $O(l^{l/2} N)$.

37.4.3 Parallelism

SMS is amenable to parallel implementations [33]. One possibility is to partition the number of sorts equally among the processors. For example, if there are two processors, then a reasonable partition is for the first processor to work on a maximum pattern length of 10 and the second processor to work on the remaining maximum pattern lengths. Processor 1 will perform 219 sorts and the second processor will do 210 sorts.

A second possibility is to partition the sequences as equally among the processors as possible and finally merge the occurrence numbers of patterns.

A third possibility is to treat each sort as a job. To begin with all the jobs are available. To begin with, each processor takes up an available job. As soon as a job is taken up (by some processor) it is marked unavailable. When a processor completes its job, it takes up another available job. Computation stops when there are no more available jobs. This third technique has been employed in [33] and the speedups obained are close to linear.

37.4.4 TEIRESIAS Algorithm

The TEIRESIAS algorithm [34] addresses a problem similar to Problem 3. Here we define a *pattern* to be a string of symbols (also called *residues*) and ?'s. A "?" refers to a wild card character. A pattern cannot begin or end with ?. AB?D, EB??DS?R, etc. are examples of patterns. The *length* of a pattern is the number of characters in it (including the wildcard characters). If P is a pattern, any substring of P that itself is a pattern is called a *subpattern* of P. For instance AB is a subpattern of AB?D. A pattern P is called a $< l, W >$ pattern if every subpattern of P of length W or more contains at least l residues. A pattern P' is said to be *more specific* than a pattern P if P' can be obtained from P by changing one or more wild card characters of P into residues and/or by adding one or more residues or wild card characters to the left of P or to the right of P. ABCD and E?AB?D are more specific than AB?D. A pattern P is said to be maximal if there is no pattern P' that is more specific than P and which occurs the same number of times in a given database DB as P. The problem TEIRESIAS addresses takes as input a database DB of sequences, the parameters l, W, and q and outputs all $< l, W >$ maximal patterns in DB that occur in at least q distinct sequences of DB.

The run time of TEIRESIAS is $\Omega(W^l N \log N)$, where N is the size of the database (i.e., the number of characters (or residues) in the database). In this section we describe the TEIRESIAS algorithm in some detail.

The algorithm consists of two phases. In the first phase elementary $< l, W >$ patterns are identified. An elementary $< l, W >$ pattern is nothing but a pattern which is of length W and which has exactly l residues. This phase runs in time $O(NW^l)$ [34].

In the second phase (known as the convolution phase), elementary patterns are combined (i.e., convolved) to obtain larger patterns. For example, AS?TF and TFDE can be combined to obtain AS?TFDE. All the convolved patterns that pass the support level and which are maximal will be output. The run time of this phase is $O(W^l N \log N)$ [34].

A problem similar to the one of [34], as applied to protein sequences, can also be found in the literature (see e.g., [27], [14], [24], [28]). These authors use a variant of the APRIORI algorithm (employed in the field of Data Mining) and incorporate substitution matrix information as part of their detection strategy.

37.5 Random Sampling

Random sampling can be employed to speedup computations. In this section we describe a simple form of sampling (presented in [33]) as it applies to Problem 3.

Let the input database DB consist of the sequences $S_1, S_2, \ldots, S_n$. Assume that the problem is to determine how frequent are certain patterns. In particular, for each pattern we want to determine if the number of sequences in which it occurs is at least q, where q is a given threshold. We can use the following strategy to solve this problem. Randomly choose a sample S' of ϵn sequences (for some appropriate value ϵ), identify the patterns that are frequent in this sample (with an appropriately modified threshold value), and output these patterns.

Analysis. Let U be a pattern that occurs q times in DB. Then the number of occurrences q' of U in S' is Binomially distributed with a mean of ϵq. Using Chernoff bounds, $Pr[q' \leq (1-\alpha)\epsilon q] \leq \exp(-\alpha^2 \epsilon q/2)$. If $q \geq \frac{2\beta \ln n}{\alpha^2 \epsilon}$, then this probability is $n^{-\beta}$. Refer to a probability of $\leq n^{-\beta}$ as *low probability* as long as β is any constant ≥ 1. I.e., if a pattern U occurs q times in DB then its occurrence in S' will be at least $(1-\alpha)\epsilon q$ with high probability. Let $N = \sum_{i=1}^{n} |S_i|$. I.e., N is the number of residues in DB. The total number of patterns that pass the threshold is clearly $< N$. If $q \geq \frac{2}{\alpha^2 \epsilon}(\beta \ln n + \ln N)$, then each pattern that occurs at least q times in DB will occur at least $(1-\alpha)\epsilon q$ times in the sample.

Also, if a pattern U occurs at most $\frac{1-\alpha}{1+\alpha}q$ times in DB, then the expected number of occurrences of U in the sample is $\frac{1-\alpha}{1+\alpha}\epsilon q$. Using Chernoff bounds, this number will be $\leq (1-\alpha)\epsilon q$ with probability $\geq 1 - \exp\left(-\alpha^2 \frac{1-\alpha}{1+\alpha}\frac{\epsilon}{3}q\right)$. Thus if $q \geq \frac{3}{\alpha^2 \epsilon}\frac{1+\alpha}{1-\alpha}(\beta \ln n + \ln N)$, then every pattern that occurs at most $\frac{1-\alpha}{1+\alpha}q$ times in DB will occur at most $(1-\alpha)\epsilon q$ times in S'. We arrive at the following

LEMMA 37.2   Consider the problem of identifying patterns in a database of n sequences. Each pattern of interest should occur in at least q of the input sequences. To solve this problem it suffices to use a random sample of size ϵn and a sample threshold of $(1 - \alpha)\epsilon q$. In this case, with high probability, no pattern that has an occurrence of less than $\frac{1-\alpha}{1+\alpha}q$ in DB will pass the sample threshold, provided $q \geq \frac{3}{\alpha^2 \epsilon} \frac{1+\alpha}{1-\alpha}(\beta \ln n + \ln N)$.

Examples: We present two examples to llustrate the usefulness of sampling. In particular, we want to see how small could ϵ be. For a given n, N, q, we fix suitable values for β and α and use the above constraint on q to evaluate the value of ϵ. The value of α that minimizes $\frac{1}{\alpha^2} \frac{1-\alpha}{1+\alpha}$ is approximately 0.6. Thus we use this value for α. We fix the value of β to be 1.

Consider the case of $n = 1000, m = 200, N = 200000, \alpha = 0.6, \beta = 1$. Here m refers to the length of each sequence. The condition on q becomes: $q \geq \frac{666.6}{\epsilon}$. If $q = 800$, then it means that ϵ could be as small as 0.833.

As another example, consider the case where: $n = 10000, m = 200, N = 2000000, \alpha = 0.6, \beta = 1$. The condition on q becomes: $q \geq \frac{833.25}{\epsilon}$. If $q = 5000$, the value of ϵ could be as small as 0.167.

Appendix A: Chernoff Bounds

A *Bernoulli trial* is an experiment with two possible outcomes viz. *success* and *failure*. The probability of success is p. A binomial random variable X with parameters (n, p) is the number of successes in n independent Bernoulli trials, the probability of success in each trial being p.

We can get good estimates on the tail ends of binomial distributions (see e.g., [9]). In particular, it can be shown that

Lemma A.1. If X is binomial with parameters (n, p), and $m > np$ is an integer, then

$$Pr[X \geq m] \leq (np/m)^m \exp(m - np).$$

Also,

$$Pr[X \leq (1 - \epsilon)np] \leq \exp(-\epsilon^2 np/2)$$

and

$$Pr[X > (1 + \epsilon)np] \leq \exp(-\epsilon^2 np/3)$$

for all $0 < \epsilon < 1$.

37.6   Conclusions

In this chapter we have addressed three versions of the motif search problem. We have also surveyed many deterministic and randomized algorithms that have been proposed for these problems. Developing more efficient algorithms will be an important open problem, given the significance of motif search.

Acknowledgements

The author thanks S. Balla, C.-H. Huang, V. Thapar, M. Gryk, M. Maciejewski, and M. Schiller for many stimulating discussions. This work has been supported in part by the

NSF Grants CCR-9912395 and ITR-0326155.

References

[1] E.F. Adebiyi, T. Jiang and M. Kaufmann, An efficient algorithm for finding short approximate non-tandem repeats, *Bioinformatics* 17, Supplement 1, 2001, pp. S5-S12.

[2] E.F. Adebiyi and M. Kaufmann, Extracting common motifs under the Levenshtein measure: theory and experimentation, *Proc. Workshop on Algorithms for Bioinformatics (WABI)*, Springer-Verlag LNCS 2452, 2002, pp. 140-156.

[3] T.L. Bailey and C. Elkan, Fitting a mixture model by expectation maximization to discover motifs in biopolymers, *Proc. Second International Conference on Intelligent Systems for Molecular Biology*, 1994, pp. 28-36.

[4] T.L. Bailey and C. Elkan, Unsupervised learning of multiple motifs in biopolymers using expectation maximization, *Machine Learning* 21(1-2), 1995, pp. 51-80.

[5] M. Blanchette, Algorithms for phylogenetic footprinting, *Proc. Fifth Annual International Conference on Computational Molecular Biology*, 2001.

[6] M. Blanchette, B. Schwikowski, and M. Tompa, An exact algorithm to identify motifs in orthologous sequences from multiple species, *Proc. Eighth International Conference on Intelligent Systems for Molecular Biology*, 2000, pp. 37-45.

[7] A. Brazma, I. Jonassen, J. Vilo, and E. Ukkonen, Predicting gene regulatory elements in silico on a genomic scale, *Genome Research* 15, 1998, pp. 1202-1215.

[8] J. Buhler and M. Tompa, Finding motifs using random projections, *Proc. Fifth Annual International Conference on Computational Molecular Biology (RECOMB)*, April 2001.

[9] H. Chernoff, A measure of asymptotic efficiency for tests of a hypothesis based on the sum of observations, *Annals of Math. Statistics* 23, 1952, pp. 493-507.

[10] M. Crochemore and M.-F. Sagot, Motifs in sequences: localization and extraction, in *Handbook of Computational Chemistry*, Crabbe, Drew, Konopka, eds., Marcel Dekker, Inc., 2001.

[11] E. Eskin and P. Pevzner, Finding composite regulatory patterns in DNA sequences, *Bioinformatics* S1, 2002, pp. 354-363.

[12] D.J. Galas, M. Eggert, and M.S. Waterman, Rigorous pattern-recognition methods for DNA sequences: Analysis of promoter sequences from *Escherichia coli*, *Journal of Molecular Biology* 186(1), 1985, pp. 117-128.

[13] Z. Galil and K. Park, An improved algorithm for approximate string matching, *SIAM Journal of Computing* 19(6), 1990, pp. 989-999.

[14] Y. Gao, M. Yang, X. Wang and K. Mathee *et al*, Detection of HTH Motifs via Data Mining, *Proceedings of SPIRE99: String Processing and Information Retrieval*, 1999, pp. 63-72.

[15] M. Gribskov, R. Luthy, and D. Eisenberg, Profile anlysis, in R. F. Doolittle (Ed.), *Molecular Evolution: Computer Analysis of Protein and Nucleic Acid Sequences*, volume 183 of *Methods in Enzymology*, Academic Press, 1990, pp. 146-159.

[16] M. Gribskov and S. Veretnik, Identification of sequence pattern with profile analysis, *Methods Enzymol.* 266, 1996, pp. 198-212.

[17] G. Hertz and G. Stormo, Identifying DNA and protein patterns with statistically significant alignments of multiple sequences, *Bioinformatics* 15, 1999, pp. 563-577.

[18] E. Horowitz, S. Sahni, and S. Rajasekaran, *Computer Algorithms*, W.H. Freeman

Press, 1998.

[19] U. Keich and P. Pevzner, Finding motifs in the twilight zone, *Bioinformatics* 18, 2002, pp. 1374-1381.

[20] G.M. Landau and U. Vishkin, Introducing efficient parallelism into approximate string matching and a new serial algorithm, *Proc. ACM Symposium on Theory of Computing*, 1986, pp. 220-230.

[21] C.E. Lawrence, S.F. Altschul, M.S. Boguski and J.S. Liu *et al*, Detecting subtle sequence signals: a Gibbs sampling strategy for multiple alignment, *Science* 262, 1993, pp. 208-214.

[22] C.E. Lawrence and A.A. Reilly, An Expectation Maximization (EM) Algorithm for the Identification and Characterization of Common Sites in Unaligned Biopolymer Sequences, *Proteins: Structure, Function, and Genetics* 7, 1990, pp. 41-51.

[23] H.M. Martinez, An efficient method for finding repeats in molecular sequences, *Nucleic Acids Research* 11(13), 1983, pp. 4629-4634.

[24] K. Mathee and G. Narasimhan, Detection of DNA-binding helix-turn-helix motifs in proteins using the pattern dictionary method, *Methods Enzymol.*, 370, 2003, pp. 250-264.

[25] E.W. Myers, Incremental alignment algorithms and their applications, Technical Report 86-22, Department of Computer Science, University of Arizona, Tucson, AZ 85721, 1986.

[26] E.W. Myers, A sublinear algorithm for approximate keyword searching, Algorithmica 12, 1994, pp. 345-374.

[27] G. Narasimhan, C. Bu, Y. Gao and X. Wang *et al*, Mining Protein Sequences for Motifs, *Journal of Computational Biology*, 9(5), 2002, pp. 707-720.

[28] C.G. Nevill-Manning, T.D. Wu, and D.L. Brutlag, Highly-specific protein sequence motifs for genome analysis, *Proc. Natl. Acad. Sci.* USA, 95, 1998, pp. 5865-5871.

[29] G. Pavesi, G. Mauri, and G. Pesole, In silico representation and discovery of transcription factor binding sites, *Brief Bioinform.*, 5(3), 2004, pp. 217-36.

[30] P. Pevzner and S.-H. Sze, Combinatorial approaches to finding subtle signals in DNA sequences, *Proc. Eighth International Conference on Intelligent Systems for Molecular Biology*, 2000, pp. 269-278.

[31] A. Price, S. Ramabhadran and P.A. Pevzner, Finding subtle motifs by branching from sample strings, *Bioinformatics* 1(1), 2003, pp. 1-7.

[32] S. Rajasekaran, S. Balla, C.-H. Huang, Exact Algorithms for Planted Motif Challenge Problems, *Proc. Third Asia-Pacific Bioinformatics Conference*, Singapore, 2005.

[33] S. Rajasekaran, S. Balla, C.-H. Huang and V. Thapar *et al*, Exact Algorithms for Motif Search, *Proc. Third Asia-Pacific Bioinformatics Conference*, Singapore, 2005.

[34] I. Rigoutsos and A. Floratos, Combinatorial pattern discovery in biological sequences: The TEIRESIAS algorithm, *Bioinformatics*, 14(1), 1998, pp. 55-67. Erratum in: *Bioinformatics* 14(2), 1998, p. 229.

[35] E. Rocke and M. Tompa, An algorithm for finding novel gapped motifs in DNA sequences, *Proc. Second International Conference on Computational Molecular Biology (RECOMB)*, 1998, pp. 228-233.

[36] M.F. Sagot, Spelling approximate repeated or common motifs using a suffix tree, Springer-Verlag LNCS 1380, pp. 111-127, 1998.

[37] S. Sinha and M. Tompa, A statistical method for finding transcription factor binding sites, *Proc. Eighth International Conference on Intelligent Systems for Molecular Biology*, 2000, pp. 344-354.

[38] R. Staden, Methods for discovering novel motifs in nucleic acid sequences, *Computer Applications in the Biosciences* 5(4), 1989, pp. 293-298.

[39] G.D. Stormo, DNA binding sites: representation and discovery, *Bioinformatics*, 16(1), 2000, pp. 16-23.

[40] M. Tompa, An exact method for finding short motifs in sequences, with application to the ribosome binding site problem, *Proc. Seventh International Conference on Intelligent Systems for Molecular Biology*, 1999, pp. 262-271.

[41] E. Ukkonen, Finding approximate patterns in strings, *Journal of Algorithms* 6, 1985, pp. 132-137.

[42] J. van Helden, B. Andre, and J. Collado-Vides, Extracting regulatory sites from the upstream region of yeast genes by computational analysis of oligonucleotide frequencies, *Journal of Molecular Biology* 281(5), 1998, pp. 827-842.

[43] M. Waterman, R. Arratia, and E. Galas, Pattern Recognition in Several Sequences: Consensus and Alignment, *Bulletin of Mathematical Biology* 46, 1984, pp. 515-527.

38

Data Mining in Computational Biology

Mohammed J. Zaki
Rensselaer Polytechnic Institute

Karlton Sequeira
Rensselaer Polytechnic Institute

38.1 Introduction

Data Mining is the process of automatic discovery of valid, novel, useful, and understandable patterns, associations, changes, anomalies, and statistically significant structures from large amounts of data. It is an interdisciplinary field merging ideas from statistics, machine learning, database systems and data-warehousing, high-performance computing, as well as visualization and human-computer interaction. It has been engendered by the phenomenal growth of data in all spheres of human endeavor, and the economic and scientific need to extract useful information from the collected data.

Bioinformatics is the science of storing, extracting, analyzing, interpreting, and utilizing information from biological data such as sequences, molecules, pathways, etc. Genome sequencing projects have contributed to an exponential growth in complete and partial sequence databases. The structural genomics initiative aims to catalog the structure-function information for proteins. Advances in technology such as microarrays have launched the subfields of genomics and proteomics to study the genes, proteins, and the regulatory gene expression circuitry inside the cell. What characterizes the state of the field is the flood of data that exists today or that is anticipated in the future; data that needs to be mined to help unlock the secrets of the cell. Data mining will play a fundamental role in understanding these rapidly expanding sources of biological data. New data mining techniques are needed to analyze, manage and discover sequence, structure and functional patterns/models from large sequence and structural databases, as well as for structure prediction, gene finding, gene expression analysis, biochemical pathway mining, biomedical literature mining, drug design and other emerging problems in genomics and proteomics.

The goal of this chapter is to provide a brief introduction to some data mining techniques, and to look at how data mining has been used in some representative applications in bioinformatics, namely three-dimensional (3D) or structural motif mining in proteins and the analysis of microarray gene expression data. We also look at some issues in data preparation, namely data cleaning and feature selection via the study of how to find normal variation in gene expression datasets. We first begin with a short introduction to the data mining process, namely the typical mining tasks and the various steps required for successful mining.

38.2　Data Mining Process

Typically data mining has the two high-level goals of prediction and description. The former answers the question "what"; while the latter the question "why". That is, for prediction the key criteria is that of accuracy of the model in making future predictions; how the prediction decision is arrived at may not be important. For description, the key criteria is that of clarity and simplicity of the model describing the data, in human-understandable terms. There is sometimes a dichotomy between these two aspects of data mining in the sense that the most accurate prediction model for a problem may not be easily understandable, and the most easily understandable model may not be highly accurate in its predictions. It is crucial that the patterns, rules, and models that are discovered are valid not only in the data samples already examined, but are generalizable and remain valid in future new data samples. Only then can the rules and models obtained be considered meaningful. The discovered patterns should also be novel, and not already known to experts; otherwise, they would yield very little new understanding. Finally, the discoveries should be useful as well as understandable.

38.2.1　Data Mining Tasks

In verification-driven data analysis the user postulates a hypothesis, and the system tries to validate it. The common verification-driven operations include querying and reporting, multidimensional analysis, and statistical analysis. Data mining, on the other hand, is discovery-driven, i.e., it automatically extracts new hypotheses from data. The typical data mining tasks include:

- *Association Rules:* Given a database of transactions, where each transaction consists of a set of items, association discovery finds all the item sets that frequently occur together, and also the rules among them. For example, a rule could be as follows: 90% of the samples that have high expression levels for genes $\{g_1, g_5, g_7\}$ have high expression levels for genes $\{g_2, g_3\}$, and further 30% of all samples support this rule.

- *Sequence Mining:* The sequence mining task is to discover sequences of events that commonly occur together. For example, 70% of the DNA sequences from a family have the subsequence $TATA$ followed by ACG after a gap of, say, 50 bases.

- *Similarity Search:* An example is the problem where we are given a database of objects and a "query" object, and we are then required to find those objects in the database that are similar to the query object. Another example is the problem where we are given a database of objects, and we are then required to find all pairs of objects in the databases that are within some distance of each other. For example, given a query 3D structure, find all highly structurally similar proteins.

- *Deviation Detection:* Given a database of objects, find those objects that are the most different from the other objects in the database, i.e., the outliers. These objects may be thrown away as noise, or they may be the "interesting" ones, depending on the specific application scenario. For example, given microarray data, we might be able to find a tissue sample that is unlike any other seen, or we might be able to identify genes with expression levels very different from the rest of the genes.

- *Classification and Regression:* This is also called supervised learning. In the case of classification, we are given a database of objects that are labeled with predefined categories or classes. We are required to learn from these objects a model that separates them into the predefined categories or classes. Then, given a new object, we apply the learned model to assign this new object to one of the classes. In the more general situation of regression, instead of predicting classes, we have to predict real-valued fields. For example, given microarray gene expression data, with tissues from cancerous and non-cancerous patients, a classification model might be able to predict for a new tissue sample, whether it is cancerous or not.

- *Clustering:* This is also called unsupervised learning. Here, we are given a database of objects that are usually without any predefined categories or classes. We are required to partition the objects into subsets or groups such that elements of a group share a common set of properties. Moreover the partition should be such that the similarity between members of the same group is high and the similarity between members of different groups is low. For example, given a set of protein sequences, clustering can group them into similar (potentially homologous) families.

38.2.2 Steps in Data Mining

Data mining refers to the overall process of discovering new patterns and building models from a given dataset. There are many steps involved in the mining enterprise such as data selection, data cleaning and preprocessing, data transformation and reduction, data mining task and algorithm selection, and finally post-processing and interpretation of discovered knowledge:

- *Collect, clean and transform the target dataset:* Data mining relies on the availability of suitable data that reflects the underlying diversity, order, and structure of the problem being analyzed. Therefore, the collection of a dataset that captures all the possible situations that are relevant to the problem being analyzed is crucial. Raw data contain many errors and inconsistencies, such as noise, outliers, and missing values. An important element of this process is to remove duplicate records to produce a non-redundant dataset. Another important element of this process is the normalization of data records to deal with the kind of pollution caused by the lack of domain consistency.

- *Select features, reduce dimensions:* Even after the data have been cleaned up in terms of eliminating duplicates, inconsistencies, missing values, and so on, there may still be noise that is irrelevant to the problem being analyzed. These noise attributes may confuse subsequent data mining steps, produce irrelevant rules and associations, and increase computational cost. It is therefore wise to perform a dimension reduction or feature selection step to separate those attributes that are pertinent from those that are irrelevant.

- *Apply mining algorithms and interpret results:* After performing the pre-processing steps apply appropriate data mining algorithms – association rule discovery, sequence mining, classification tree induction, clustering, and so on – to analyze the data. After the algorithms have produced their output, it is still necessary to examine the output in order to interpret and evaluate the extracted patterns, rules, and models. It is only by this interpretation and evaluation process that we can derive new insights on the problem being analyzed.

38.3   Mining 3D Protein Data

Protein structure data has been mined to extract structural motifs [29, 22, 31], analyze protein-protein interactions [55], protein folding [65, 40], protein thermodynamic stability [5, 38], and many other problems. Mining structurally conserved motifs in proteins is extremely important, as it reveals important information about protein structure and function. Common structural fragments of various sizes have fixed 3D arrangements of residues that may correspond to active sites or other functionally relevant features, such as Prosite patterns. Identifying such spatial motifs in an automated and efficient way, may have a great impact on protein classification [13, 3], protein function prediction [20] and protein folding [40]. We first describe in detail a graph-based method for mining structural motifs, and then review other approaches.

38.3.1   Modeling Protein Data as a Graph

An undirected graph $G(V, E)$ is a structure that consists of a set of vertices $V = \{v_1, \cdots, v_n\}$ and a set of edges $E \subseteq V \times V$, given as $E = \{e_i = (v_i, v_j) | v_i, v_j \in V\}$, i.e., each edge e_i is an unordered pair of vertices. A *weighted graph* is a graph with an associated weight function $W : E \to \Re^+$ for the edge set. For each edge $e \in E$, $W(e)$ is called the *weight* of the edge e. Protein data is generally modeled as a connected weighted graph where each vertex in the graph may correspond to either a secondary structure element (SSE) (such as α-helix or β-strand) [22, 66], or an amino acid [29], or even an individual atom [31]. The granularity presents a trade-off between computation time and complexity of patterns mined; a coarse graph affords a smaller problem computationally, but runs the risk of oversimplifying the analysis. An edge may be drawn between nodes/vertices indicating proximity, based on several criteria, such as:

- Contact Distance (CD) [29, 66]: Given two amino acids a_i and a_j along with the 3D co-ordinates of their α-Carbon atoms (or alternately β-Carbon), (x_i, y_i, z_i) and (x_j, y_j, z_j), resp., define the Euclidean distance between them as follows:

$$\delta(a_i, a_j) = \sqrt{(x_i - x_j)^2 + (y_i - y_j)^2 + (z_i - z_j)^2}$$

  We say that a_i and a_j are in *contact*, if $\delta(a_i, a_j) \leq \delta^{\max}$, where $\delta^{\max}$ is some maximum allowed distance threshold (a common value is $\delta^{\max} = 7\mathring{A}$). We can add an edge between two amino acids if there are in contact. If the vertices are SSEs then the weight on an edge can denote the number of contacts between the two corresponding SSEs.

- Delaunay Tessellation (DT)[29, 54, 62]: the Voronoi cell of $x \in V(G)$ is given by

$$\mathcal{V}(x) = \{y \in \Re^3 | d(x,y) \leq d(x',y) \ \forall x' \in V(G) \backslash \{x\}\}$$

An edge connects two vertices if their Voronoi cells share a common face.

- Almost Delaunay (AD) [29]: to accommodate for errors in the co-ordinate values of the points in the protein structure data, we say that a pair of points $v_i, v_j \in V(G)$ are joined by an almost-Delaunay edge with parameter ϵ, or AD(ϵ), if by perturbing all points by at most ϵ, v_i, v_j could be made to have Voronoi cells sharing a face.

38.3.2 Graph-based Structural Motif Mining

We now describe our graph-based structural motif mining approach. We represent each protein as an undirected graph; the vertices correspond to the secondary structures, α-helix or β-strands. The vertex numbers increase in the sequence order from N-terminus to C-terminus. The edges correspond to the contacts between two structures. If two SSEs have no contacts between them, there is no edge between them. Each one of the vertices and edges has a label associated with it. The labels on the vertices correspond to the type of secondary structure, and the labels on the edges correspond to the contact types which are defined by the number of contacts between the two SSEs. We use four types of edges $E1$, $E2$, $E3$ and $E4$ depending on the number of contacts between two structures: $E1 \in [1,5), E2 \in [5,10), E3 \in [10,15), E4 \in [15, +\infty)$. The graph representation of the protein 2IGD with contact cutoff $7\mathring{A}$ is shown in Figure 38.1. There are 5 secondary structures in 2igd: $\beta0, \beta1, \alpha2, \beta3, \beta4$ and 7 edges between $\beta1 - \beta0$, $\alpha2 - \beta0$, $\alpha2 - \beta1$, $\beta3 - \beta1$, $\beta3 - \beta2$, $\beta4 - \beta2$, and $\beta4 - \beta3$.

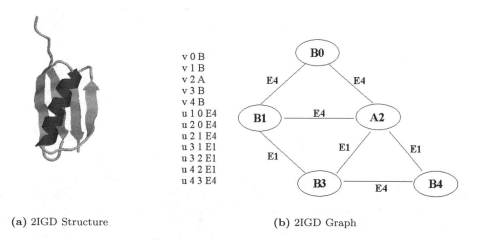

(a) 2IGD Structure   (b) 2IGD Graph

FIGURE 38.1: (a) 3D structure for protein G (PDB file 2IGD, Length 61). (b) Graph for 2IGD: graph input format (left) and graph representation (right). The input format lists the vertices ('v') followed by the edges ('u'). A vertex is a tuple $(v\ n\ l)$, where v denotes a vertex, n gives the vertex number and l its label. An edge is a tuple $(u\ n_1\ n_2\ l)$, where u denotes an edge, n_1, n_2 are the two vertices making the edge, and l is the edge label.

Discovering Frequent Subgraphs

To mine the frequent structural motifs from the database of proteins represented as graphs we used the FSG algorithm for frequent subgraph discovery [44]. The input is a database D of protein graphs and a minimum support threshold σ. The output is the set of all connected subgraphs that occur in at least $\sigma\%$ of the proteins.

There are three considerations when FSG are applied to the database of protein structures. Firstly, we are interested in subgraphs that are composed of all secondary structures in contact, whether or not they are consecutive. A protein sequence or subsequence is naturally connected from N-terminal to C-terminal, thus any two successive structures are connected with each other. Contacts between two non-successive structures can form tertiary structure as well. Secondly, we used labeled graphs, where different graphs can contain vertices and edges with the same label. We classified edges in four types according to the range of contacts. We chose this classification range heuristically to allow us to find patterns containing multiple occurrences of the same structures and contact ranges. A finer classification of edge types will decrease the frequency of common patterns and make the running time extremely slow. Thirdly, we are interested in the subgraphs with at least $\sigma\%$ frequency in the database. This makes sure that the generated subgraphs are the dominant motifs.

FSG uses a level-by-level approach to mine the connected subgraphs. It starts with the single edges and counts their frequency. Those that meet the support threshold are extended by another frequent edge to obtain a set of candidate edge pairs. The counting proceeds by extending a frequent edge-set by one more edge. The process stops when no frequent extension is possible. For a detailed description of FSG see [44].

We ran the FSG algorithm on a dataset of 757 protein graphs obtained from the non-redundant proteins in PDBselect database [27]. The results with different frequency thresholds ranging from 10% to 40%, and with contact cutoff 7Å are shown in Table 38.1. Figure 38.2 shows some mined subgraphs with different number of edges using support 10%.

	10%	20%	30%	40%
1-edge graphs	12	10	8	4
2-edges graphs	34	23	8	6
3-edges graphs	144	42	17	6
4-edges graphs	420	72	22	2
5-edges graphs	1198	142	23	0
6-edges graphs	2920	289	4	0
7-edges graphs	6816	32	0	0
8-edges graphs	14935	114	0	0

TABLE 38.1 Frequent subgraphs.

We mapped the mined graph patterns back to the PDB structure to visualize the mined results. Two such frequent tertiary motifs are shown in Figure 38.3. The top one has 6 edges, and frequency 157. This motif shows four alpha helices and three beta strands. Two proteins where this pattern occurs are also shown: PDB files $1AD2$ and $1KUH$. The bottom motif has 8 edges and frequency 153. It is an all alpha motif. Two occurrences in PDB files $1BG8$ and $1ARV$ are shown. The results obtained from graph mining are quite encouraging. We found that most of the highest scoring patterns match part of the whole protein structures, and we were able to find remote interactions as well.

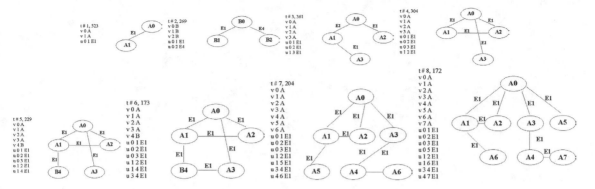

FIGURE 38.2: Highest Frequency Tertiary Motifs: Edge Size 1 to 8. The notation "t #1, 523" in the graph input format denotes edge size = 1, support = 523.

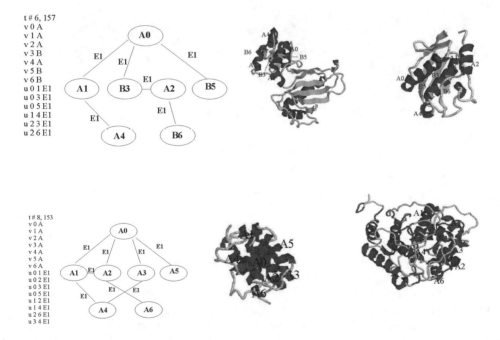

FIGURE 38.3: **(See color insert following page 20-4.)** **Top:** Alpha-Beta Motif and its Occurrence in PDB files: 1ad2, 1kuh. **Bottom:** All Alpha Motif and its Occurrence in PDB files: 1bg8, 1arv.

Comparison with SCOP Database

The SCOP protein database [50], categorizes proteins into all-α, all-β, α and β (interspersed), α plus β (isolated), and multi-domain according to their structure and functional similarity. We applied the graph-mining method to protein families classified in SCOP database. In order to find out whether our method matches with SCOP in mining and categorizing common domains, several protein families were chosen randomly from SCOP and represented as graphs. Within one protein family, we used 100% support to find the largest frequent patterns that appear in every protein in that family.

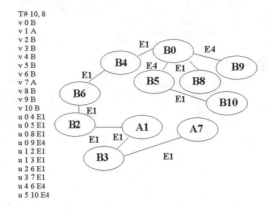

FIGURE 38.4: Largest pattern in 100% of DNA poly.

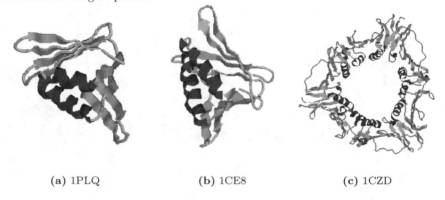

(a) 1PLQ   (b) 1CE8   (c) 1CZD

FIGURE 38.5: Some occurrences of the pattern in PDB files: 1PLQ, 1CE8, 1CZD.

To test retrieval rate, we can create a mixed database from several families within the same superfamily. Mining this database, frequent patterns, could only be found with less than 100% support. If we add to the database some other proteins outside of the superfamily, the maximum size pattern was only found at very small support and it is not frequent any more. This demonstrates that our graph mining method could be used to classify proteins into families based on their shared motifs. For example, consider the DNA clamp superfamily, which is made up of two families: DNA polymerase III, beta subunit and DNA polymerase processivity factor. DNA polymerase III beta subunit family has three E. coli proteins: 2POL, 1JQL and 1JQJ. The DNA polymerase processivity factors family has eight proteins: Bacteriophage RB69 (1B77, 1B8H) and Bacteriophage T4 (1CZD), human herpes virus type 1 (1DML), proliferating cell nuclear antigens (1PLQ, 1PLR, 1AXC, 1GE8). The following pattern, shown in Figure 38.4 and its corresponding occurrences in PDB files, shown in Figure 38.5, appears in every protein of DNA polymerase processivity factors. The graph-based method has the potential to become a powerful tool in structure analysis. It is a rapid, intuitive method to identify and find proteins that display structure similarities at the level of residues and secondary structure elements.

38.3.3   Alternate Approaches for Structural Motif Mining

Another graph based approach to motif discovery was proposed in [29], where they search for frequent subgraphs in proteins belonging to the same structural and functional family in the SCOP database; they propose these subgraphs as family-specific amino acid residue signatures of the underlying family. They represent the protein structures as graphs, using CD, DT and AD representations (see section 38.3.1). They represent each protein graph by an adjacency matrix in which each entry is either a 0 (if there is no edge), a vertex label (if it is a diagonal element), or an edge label (if an edge exists). They define the code of this adjacency matrix as the sequence of the entries in the lower triangular matrix read going left to right and then top to bottom. They use the lexicographic order to impose an ordering on the codes obtained by rearranging the rows of the matrix. Using a novel graph representation such as this, they construct a rooted, ordered tree for each graph and then search for frequent graphs. They then discard those which have a low mutual information score. They conclude that in order to achieve the highest accuracy in finding protein family specific signatures, AD graphs present the best choice both due to their relative computational efficiency and their robustness in taking into account possible experimental errors in determining protein atomic coordinates.

In a different approach, Jonassen et al. [35] represent the neighborhood of each residue r as a neighbor string NS_r of which r is called the *anchor*. NS_r encodes all residues within a d angstrom radius of r. Each residue is encoded by its amino acid type, secondary structure type and a coordinate set (x, y, z) calculated as the mean of the r's side chain atoms. The order of residues in NS_r is governed by their order along the protein's backbone. They then define a packing pattern against which the neighbor strings are matched. A packing pattern consists of a list of elements where each element defines a match set (a set of allowed amino acids), a set of allowed SSE types and one set of coordinates. $NS_r = r_1, \ldots, r_k, \ldots, r_l$, where r_k is the anchor residue, is said to match packing pattern $P = p_1, \ldots, p_l, \ldots, p_n$ if NS_r contains a subsequence $r_{i1}, \ldots, r_{in}$, such that residues have amino acids and SSE types included in the match sets of the corresponding pattern elements and the anchors of the neighbor string and pattern string are aligned. Also, NS_r is said to structurally match P within ϕ, if it is possible to superimpose the coordinates of NS_r onto those of P with a root mean square deviation of maximum ϕ. A neighbor string that structurally matches a packing pattern with a threshold ϕ describes an occurrence of the pattern. A pattern having occurrences in k structures is said to have support k. Thus, they seek packing patterns having support k in a dataset of N structures. For each of the N structures, a packing pattern is generated for the generalization of each neighbor string. There may exist many generalizations of the match sets and hence pruning based on geometrical constraints is used to constrain the length of the neighbor strings. The packing pattern inherits its coordinates and SSE types from the neighbor string. Further, neighbor strings with fewer than 4 elements are discarded. Using depth-first search, they search for all generalizations of a neighbor string having support k moving from simple (short generalizations) to complex (long ones) as the depth of the search increases. For every pattern thus found, they compute a score measuring the pattern's information content divided by its maximum root mean square deviation.

The 3D conformation of a protein may be compactly represented in a symmetrical, square, boolean matrix of pairwise, inter-residue contacts, or "contact map". The contact map provides a host of useful information about the protein's structure. In [28] Hu et al. describe how data mining can be used to extract valuable information from contact maps. For example, clusters of contacts represent certain secondary structures, and also capture non-local interactions, giving clues to the tertiary structure. They focus on two main tasks: 1) Given the database of protein sequences, discover an extensive set of non-local (frequent) dense patterns in their contact maps, and compile a library of such non-local interactions.

2) Cluster these patterns based on their similarities and evaluate the clustering quality. To enumerate all the frequent dense patterns they scan the database of contact maps with a 2D sliding window of a user specified size $W \times W$. Across all proteins in the database, any sub-matrix under the window that has a minimum "density" (the number of contacts) is captured. The main complexity of the method stems from the fact that there can be a huge number of candidate windows. Of these windows only relatively few will be dense, since the number of contacts is a lot less than the number of non-contacts. They propose a fast hash-based approach to counting the frequency of all dense patterns. Finally they use agglomerative clustering to group the mined dense patterns to find the dominant non-local interactions.

Several methods for secondary level motif finding have also been proposed. SPASM can find the motifs consisting of arbitrary main-chain and/or side-chains in a protein database[39]. An algorithm based on subgraph isomorphism was proposed in [49]; it searches for an exact match of a specific pattern in a database. Search for distantly related proteins using a graph to represent the helices and strands was proposed in [42]. An approach based on maximally common substructures between two proteins was proposed in [23]; it also highlights areas of structural overlap between proteins. SUBDUE [15] is an approach based on Minimum Description Length and inexact graph matching for finding patterns in proteins. Another graph based method for structure discovery, based on geometric hashing, was presented in [61]. Most of these methods either focus on identifying predefined patterns in a group of proteins, or find approximate/inexact matches.

38.3.4 Finding Sites of Non-bonded Interaction

Sites of non-bonded interaction are extremely important to find, as they cannot be determined from primary structure and give clues about the higher-order structure of the protein. Side chain clusters in proteins aid in protein folding and in stabilizing the three-dimensional structure of proteins [25]. Also, these sites may be occurring in structurally similar proteins with very low sequence homology [37]. Spectral methods have been gaining favor in finding clusters of non-bonded interaction as they use global information of non-bonded interactions in the protein molecule to identify clusters.

In [11], Brinda et al. construct a graph, where each vertex is a residue and they connect vertices if they are in contact (using cut-off 4.5Å). They represent the protein graph in terms of an adjacency matrix A, where $a_{p,q} = 1/d_{p,q}$, if $p, q \in V(G)$ are connected and $1/100$, otherwise, $d_{p,q}$ = distance between p and q. The degree matrix, D, is a diagonal matrix obtained by summing up the elements of each column. Then, the Laplacian matrix, $L = D - A$, is of dimension $|V| \times |V|$. Using eigen decomposition on L, they get the eigenvalues and the eigenvectors. The Fiedler eigenvector, corresponding to the second lowest eigenvalue, gives the clustering information [24]. The centers of the clusters can be identified from the eigenvectors of the top eigenvalues. The cluster centers identified correspond to the nodes with the highest connectivity (degree) in the cluster, which generally correspond to the geometric center of the cluster. They consider clusters with at least three residues. The residues with the same vector component in the second lowest eigenvalue form a cluster. The residue with the highest magnitude of a vector component in the corresponding top eigenvalue is the center of the cluster.

The α/β barrel proteins are known to adopt the same fold in spite of very low sequence similarity. This could be possible only if the specific stabilizing interactions important in maintaining the fold are conserved in topologically equivalent positions. Such stabilization centers are usually identified by determining the extent of long-range contacts made by the residue in the native structure. In [36], Kannan et al. use a data set of 36 (α/β) barrel proteins having average pair-wise sequence identity less than 10%. They represent each protein by a connected graph using contact distance approach to connect vertices/residues with $\delta = 6.5\text{Å}$. The representation chosen causes high connectivity among the vertices of the graph and hence operating on the Laplacian of the graph is ineffective. Hence they operate on the adjacency matrix corresponding to the proteins. On eigen decomposition, they get a set of eigenvalue-eigenvector pairs. They sort this set based on the eigenvalues. They consider the set of largest eigenvalue-eigenvector pairs. Those residues having a large vector component magnitude in the direction of any of these eigenvectors are believed to belong to the cluster corresponding to that eigenvector. Thus, they cluster the residues. In each cluster, the residue having largest vector component magnitude in the direction of the corresponding eigenvector is the center of that cluster. Using eigenvalue analysis, they infer the degree of each vertex. The residues with the largest degree (typically the cluster centers) correspond to the stabilization centers. They found that most of the residues grouped in clusters corresponding to the higher eigenvalues, typically occur in the strand regions forming the β barrel and were found to be topologically conserved in all 36 proteins studied.

38.4   Mining Microarray Gene Expression Data

High-throughput gene expression has become an important tool to study transcriptional activity in a variety of biological samples. To interpret experimental data, the extent and diversity of gene expression for the system under study should be well characterized. A microarray [59] is a small chip (made of chemically coated glass, nylon membrane or silicon), containing a (usually rectangular) grid into which tens of thousands of DNA molecules (probes) are attached. Each cell of the grid relates to a fragment of DNA sequence.

Typically, two mRNA samples (a test sample and a control sample) are reverse-transcribed into cDNA (targets) and labeled using either fluorescent dyes or radioactive isotopes. They are then hybridized by base-pairing, with the probes on the surface of the chip. The chip is then scanned to read the signal intensity that is emitted from the labeled and hybridized targets. The ratio of the signal intensity emitted from the test target to that emitted from the control target is a measure of the gene expressivity of the test target with respect to the control target.

Typically, each row in the microarray grid corresponds to a single gene and each column, to either an experiment the gene is subjected to, or a patient the gene is taken from. The corresponding gene expression values can be represented as a matrix of real values where the entry (i, j) corresponds to the gene expression value for gene i under experiment j or from patient j. Formally, the dataset is represented as $\mathbf{Y} = \{y(i, j) \in \Re^+ | 1 \leq i \leq n, 1 \leq j \leq m\}$ where n, m are the number of rows (genes) and columns(experiments) and $\Re^+$ is the set of positive real numbers. We represent the vector of expression values corresponding to gene i by $y(i, \cdot)$ and that corresponding to experiment/patient j by $y(\cdot, j)$. A microarray dataset is typically tens of thousands of rows (genes) long and as many as one hundred columns (experiments/patients) wide. It is also possible to think of microarray data as the transpose of $\mathbf{Y}$, i.e., where the rows are experiments and the columns are genes.

The typical objectives of a microarray experiment is to:

1. Identify candidate genes or pathological pathways: We can conduct a microarray experiment, where the control sample is from a normal tissue while the test sample is from a disease tissue. The over-expressed or under-expressed genes identified in such an experiment may be relevant to the disease. Alternatively, a gene A, whose function is unknown, may be similarly expressed with respect to

another gene B, whose function is known. This may indicate A has a function similar to that of B.

2. Discovery and prediction of disease classes: We can conduct a microarray experiment using genes from patients known to be afflicted with a particular disease and cluster their corresponding gene expression data to discover previously unknown classes or stages of the disease. This can aid in disease detection and treatment.

38.4.1 Challenges of Mining Microarray Data

One of the main challenges of mining microarray data is the high dimensionality of the dataset. This is due to the inherent sparsity of high-dimensional space. It has been proven [8], that under certain reasonable assumptions on the data distribution and for a variety of distance functions, the ratio of the distances of the nearest and farthest points to a given point in a high-dimensional dataset, is almost 1. The process of finding the nearest point to a given point is instrumental in the success of algorithms, used to achieve the objectives mentioned above. For example, in clustering, it is imperative that there is an acceptable contrast in distances between points within the same cluster and distances between points in different clusters.

Another difficulty in mining microarray data arises from the fact that there are often missing or corrupted values in the microarray matrix, due to problems in hybridization or in reading the expression values. Finally, microarray data is very noisy. A large number of the genes or experiments may not contribute any interesting information, but their presence in the dataset makes detection of subtle clusters and patterns harder and increases the motivation for highly scalable algorithms.

38.4.2 Association Rule Mining

In order to achieve the objective of discovery of candidate genes in pathological pathways, there has been an attempt to use association rules [2]. Association rules can describe how the expression of one gene may be associated with the expression of a set of genes. Given that such an association exists, one might easily infer that the genes involved participate in some type of gene network.

In order to apply association mining, it is necessary that the data be nominal. In [16], Creighton et al. first discretize the microarray data, so that each $y(a, b)$ is set to $\{high, low, moderate\}$ depending on whether it is up-regulated, down-regulated and neither considerably up nor down regulated, respectively. Then, the data corresponding to each experiment, i.e., $(y(\cdot, a))$ can be thought of as a transaction from the market-basket viewpoint. They then apply the standard Apriori algorithm to find the association rules between the different genes and their expression levels. This yields rules of the form $g_1(\uparrow) \wedge g_3(\downarrow) \Rightarrow g_2(\uparrow)$, which means that if gene g_1 is highly expressed and g_3 is under-expressed, then g_2 is over-expressed. Such expression rules can provide useful insight into the expression networks.

38.4.3 Clustering Algorithms

In order to achieve the objective of discovery of disease classes, there are three types of clustering algorithms:

1. Gene-based clustering: the dataset is partitioned into groups of genes having

similar expression values across all the patients/experiments.

2. Experiment-based clustering: the dataset is partitioned into groups of experiments having similar expression values across all the genes.

3. Subspace clustering: the dataset is partitioned into groups of genes and experiments having similar expression values.

If the algorithm searches for clusters, which have elements which are similar across all dimensions, it can be called a "full dimensional" one.

Similarity Measures

Before delving into the specific clustering algorithms used, we must discuss the measures used to express similarity between the expression values. Although, formulae mentioned in this section describe similarity between rows of the gene expression matrix, they can, without loss of generality be applied, to describe similarity between the columns of the gene expression matrix as well.

One of the most used classes of distance metrics is the L_p distance where

$$||y(a,\cdot) - y(b,\cdot)||_{p\in\Re^+} = \left(\sum_{t=1}^{m} |y(a,t) - y(b,t)|^p \right)^{1/p}$$

In this family, the L_1, L_2 and L_∞ metrics, also called the Manhattan, Euclidean and Chebyshev distance metrics, respectively, are the most studied; the Euclidean distance metric is the most commonly used [21, 47, 12]. However, these measures do not perform too well in high-dimensional spaces. Note that this class of metrics treats all dimensions equally, irrespective of their distribution. This is remedied to some extent, by standardizing the data, i.e. normalizing the data in each row to the range [0,1] having mean 0 and standard deviation 1 [17, 58]. Note that standardization assumes the underlying data is multivariate normal. Alternatively, researchers [18, 19, 12, 63] have used Pearson's correlation coefficient as a similarity measure, where

$$r(y(a,\cdot), y(b,\cdot)) = \frac{\sum_{t=1}^{m}(y(a,t) - \mu(y(a,\cdot)))(y(b,t) - \mu(y(b,\cdot)))}{\sqrt{(\sum_{t=1}^{m}(y(a,t) - \mu(y(a,\cdot)))^2)(\sum_{t=1}^{m}(y(b,t) - \mu(y(b,\cdot)))^2)}}$$

where $\mu(y(a,\cdot)) = \sum_{t=1}^{m} \frac{y(a,t)}{N}$ and $\mu(y(b,\cdot)) = \sum_{t=1}^{m} \frac{y(b,t)}{N}$.

Note that r assumes the two vectors are approximately normally distributed and jointly bivariate normal. In this case there is a strong relationship between r and standardized Euclidean distance, since if $y(a,\cdot)$ and $y(b,\cdot)$ are standardized,

$$||y(a,\cdot) - y(b,\cdot)||_2 = \sqrt{2m(1 - r(y(a,\cdot), y(b,\cdot)))}$$

If $\mu(y(a,\cdot)) = \mu(y(b,\cdot)) = 0$, i.e. the vectors are translated so their mean is 0, then r is identical to the cosine similarity between the vectors, which is a similarity measure known to be highly effective in high-dimensional spaces. Other distance measures tested in microarray data analysis include Kullback-Leibler distance or mutual information [47].

Gene-based Clustering

A number of traditional clustering algorithms like k-means [58] and hierarchical clustering [19] have been used to find full-dimensional clusters in microarray data.

K-means(Y,*sim*,*k*)
1. select k points(genes) from **Y** randomly as cluster centers
2. repeat until convergence
3.   for s=1 to n
4.   assign $y(s, \cdot)$ to the cluster whose center is most similar to it using *sim*
5.   for j=1 to k
6.   recalculate center of cluster c_j as mean of all rows(genes) assigned to it

The k-means algorithm, shown above, is a partitional, iterative clustering algorithm. It takes as input the $n \times m$ dataset **Y**, the similarity measure *sim*, and the number of clusters to mine k. K-means runs very quickly in $O(nmt)$ time (t is the number of iterations, and k is a small constant), but suffers from the disadvantage that it requires a parameter k to be supplied, indicating the number of clusters to be found. This parameter is hard to set. Also, k-means is highly sensitive to noise and outliers, because it assigns every point in the dataset to some cluster. K-means converges to a local optima and theoretical guarantees of its accuracy are yet to be proven. The practical accuracy of k-means is noted to improve considerably, if the initial assignment of points to clusters is not so arbitrary [10].

A Self-Organizing Map (SOM)[43] is based on a single-layered neural network which maps vectors(rows/genes) in the microarray data to a two-dimensional grid of output nodes. Each input node is a row from the microarray dataset and each output node corresponds to a vector in the high-dimensional space.

SOM(Y,*sim, k, g, r*)
1. select k vectors(genes) from **Y** randomly as output nodes
2. repeat until convergence
3.   For s=1 to n
4.   find output node most similar to $y(s, \cdot)$ using *sim*
5.   update all nodes in the r-neighborhood of that output node

As shown above, a SOM trains on the input microarray dataset to adjust its output nodes (line 5), so that they move toward the denser regions of the high-dimensional feature space. The algorithm uses a number of user-specified parameters like the learning rate (g) and the neighborhood size(r). Also, SOM converges slower than k-means but is far more robust than k-means. SOM [56] and related algorithms [26] have been used for gene-based clustering.

Hierarchical clustering has two flavors: agglomerative (bottom-up) and divisive (top-bottom). In agglomerative gene-based clustering (shown below), each gene is initially assigned to its own cluster (line 1). In each iteration the two clusters with the highest inter-cluster similarity (line 3) are merged to form a single one (line 4), until some convergence criterion is satisfied e.g., the desired number of clusters remain, only one cluster remains, etc. The inter-cluster similarity may be computed by a number of methods such as:

- *single linkage*, where $Sim(\mathbf{a}, \mathbf{b}) = \max_{i \in \mathbf{a}, j \in \mathbf{b}} sim(i, j)$.
- *complete linkage*, where $Sim(\mathbf{a}, \mathbf{b}) = \min_{i \in \mathbf{a}, j \in \mathbf{b}} sim(i, j)$.
- *average linkage*, where $Sim(\mathbf{a}, \mathbf{b}) = \frac{\sum_{i \in \mathbf{a}, j \in \mathbf{b}} sim(i,j)}{|\mathbf{a}||\mathbf{b}|}$.
- *average group linkage*, where $Sim(\mathbf{a}, \mathbf{b}) = \frac{\sum_{i \in \mathbf{a} \cup \mathbf{b}, j \in \mathbf{a} \cup \mathbf{b}} sim(i,j)}{|\mathbf{a} \cup \mathbf{b}|^2}$.

Here *Sim* is the inter-cluster similarity of clusters **a** and **b**, each having |**a**| and |**b**| genes assigned to them respectively and *sim* is the similarity measure between two genes. This merging of clusters gives rise to a tree called a dendrogram. This algorithm is greedy and susceptible to noise. Also, it has time complexity $O(n^2 \log n)$ [32] implying it converges slower than K-means.

Agglomerative($\mathbf{Y}$,*Sim*)
1. Assign each $y(1, \cdot), i \in [1, n]$ to its own cluster to form set of clusters $\mathbf{C}$
2. repeat until convergence
3. $\{\mathbf{a}^*, \mathbf{b}^*\} = argmin_{(\mathbf{a},\mathbf{b}) \in \mathbf{C} \times \mathbf{C}} Sim(\mathbf{a}, \mathbf{b})$
4. $\mathbf{a}^* = \mathbf{a}^* \cup \mathbf{b}^*, \mathbf{C} = \mathbf{C} \backslash \mathbf{b}^*$

In divisive clustering, all the genes are initially assigned to the same cluster. Then iteratively, one of the clusters is selected from those existing, and is split into two clusters based on some splitting criterion. This continues, until some convergence criterion is achieved e.g., each gene has its own cluster, desired number of clusters remain, etc.

Experiment-based Clustering

For experiment-based clustering, the dimensionality (i.e. the number of genes) of the space is extremely high. Solutions proposed to remedy the failure of distance metrics in such high-dimensional spaces include, designing new distance metrics [1] and dimensionality reduction [4, 52]. Dimension reduction techniques, such as the Karhunen-Loeve tranformation (KLT), and singular value decomposition (SVD) [4, 52] are applied as a preprocessing step to reduce the number of dimensions prior to the application of a clustering algorithm. In such dimension reduction techniques, the entire database is projected onto a smaller set of new dimensions, called the principal components, which account for a large portion of the variance in the dataset. These new dimensions are mutually uncorrelated and orthogonal. Each of them is a linear combination of the underlying dimensions. Once the data has been projected to a lower-dimensional space, any of the clustering algorithms can be applied.

Subspace Clustering

The strategy of dimension reduction using KLT may be inappropriate as the clusters involving the transformed dimensions may be hard to interpret for the user. Also, data is only clustered in a single subspace. [2] cites an example, in which KLT does not reduce the dimensionality without trading off considerable information, as the dataset contains subsets of points which lie in different and sometimes overlapping lower dimensional subspaces. Hence, the focus of much recent microarray data analysis focuses on subspace clustering [6, 14, 21, 45, 57, 60, 63].

Ben-Dor et al. [6], seek to identify large order-preserving submatrices (OPSMs) in **Y**. A submatrix is order-preserving if there is a permutation of its columns under which the sequence of values in every row is strictly increasing. They prove that the problem is NP-hard. In that paper, they discuss two methods to discover clusters: a complete model and a partial model. The complete model is simply to enumerate every combination, which is unacceptable in reality. The partial model is a stochastic model.

Getz et al. [21] alternate between gene-based and experiment-based clustering. They cluster using super-paramagnetic clustering (SPC), a divisive hierarchical clustering based on the analogy to the physics of inhomogeneous ferromagnets [9], which is robust against noise and searches for "natural" stable clusters. They partition the microarray dataset into gene-based and experiment-based clusters. The gene(experiment)-based clusters specify the group of genes(experiment) which are similar. They then cluster the set of genes reported by each gene-based clustering over the set of experiments specified by each experiment-based cluster. This continues until SPC clustering produces no new robust clusters.

Cheng et al. [14] proposed the biclustering algorithm which seeks to group subsets of microarray data rows and columns having a high similarity score. They use the mean squared residue score as a similarity measure for a cluster. If $O \subseteq [1,n], C \subseteq [1,m]$, the mean square residue of (O,C) is defined as :

$$H(O,C) = \frac{1}{|O||C|} \sum_{i \in O, j \in C} (y(i,j) - \mu_C(y(i,\cdot)) - \mu_O(y(\cdot,j)) + \mu_{O,C}(y(\cdot,\cdot)))$$

where, $\mu_C(y(i,\cdot)) = \frac{1}{|C|} \sum_{j \in C} y(i,j)$, $\mu_O(y(\cdot,j)) = \frac{1}{|O|} \sum_{i \in O} y(i,j)$, $\mu_{O,C}(y(\cdot,\cdot)) = \frac{1}{|O||C|} \sum_{i \in O, j \in C} y(i,j)$. If $H(O,C) \leq \delta$, a user-specified threshold, the cluster (O,C) is retained. This definition imposes a constraint on the variation of gene expression values in the genes in O across the experiments in C. Their algorithm aims to greedily find multiple δ-biclusters, one in each iteration. After discovering a cluster, the values in that cluster are replaced by random data, so that partially overlapping biclusters may be found. However, if clusters naturally overlap, such random data may obstruct the detection of the overlap [60].

The p-clustering algorithm [60] is designed to solve this problem. Unlike biclustering, it is a deterministic algorithm. It defines a pScore of a 2×2 matrix $X = \begin{pmatrix} a & b \\ c & d \end{pmatrix}$ as: pScore$(X) = |(a-b)-(c-d)|$. A sub-matrix (O,C) is a δ-pCluster iff $\forall X_{2\times2} \in (O,C)$ pScore$(X) \leq \delta$, a predefined threshold. The intuition is that δ constrains the variation in the gene expression values $((a-b),(c-d))$, across the conditions (columns in X) over which the genes (rows in X) cluster. They use the implicit recursivity in the definition of a δ-pCluster to design a depth-first algorithm that searches for maximal δ-pClusters. The pCluster algorithm can discover every δ-pCluster in a data set, no matter whether the clusters are overlapping or not. It has time complexity $O(nm^2log(n) + mn^2log(m))$.

38.5   Determining Normal Variation in Gene Expression Data

Having looked at the application of data mining to the problems of motif discovery and microarray gene expression analysis, we now highlight some issues in data pre-processing. We illustrate this with the problem of finding normal variation in gene expression data.

A ubiquitous and under-appreciated problem in microarray analysis is the incidence of microarrays reporting non-equivalent levels of an mRNA or the expression of a gene for a system under replicate experimental conditions. In ideal conditions, the gene expression values for each gene should be the same across all array experiments. But due to the technical limitations the data contains lot of inherent noise, which could also be due to normal variation in expression of the genes across the genetically identical male mice. Our goal is to extract those genes which are contributing to the noise due to their biological variance. This kind of analysis should be done prior to mining, since otherwise, we might come to a wrong conclusion about co-expressed genes or genes that correlate well with a pathological condition.

We try to capture the genes which show variance among the identical mice by trying to eliminate the variations which come in due to experimental errors and fluctuations. We use a very robust method to exclude genes, which would eliminate any considerable variance in the replicates. Our approach is based on the following steps: 1) Calculation of fold-change ratio and discretization of expression levels for each gene, 2) Elimination of experimental noise, 3) Constructing an expression profile for each gene, and 4) Calculating and raking by gene variability via entropy calculation. We describe the steps in more detail below.

38.5.1 Fold-Change Ratio

We assume that we have n genes, in m mice, with r replicates for each mouse, for a given tissue. We denote gene i as g^i. Let S^i_t denote the expression level for gene g^i in the test sample and S^i_r the expression of g^i in the reference microarray samples. We define *fold-change ratio* as the log-odds ratio of the expression intensities of the test sample over the reference sample, given as $\log_2(\frac{S^i_t}{S^i_r})$. To analyze the variability, we discretize the fold-change into k bins ranging from very low expression levels to very high expression levels. The data is normalized in such a way that the median of the deviation from the median was set to the same value for the distribution of all the log-ratios on each array [51]. Similar analysis was done for the Affymetrix data. The raw data containing the intensities was median centered and scaled by the standard deviation. This normalization technique was chosen after experimenting with other methods like linear regression and mean centering. Though none of these methods yielded a normal distribution for the histogram plot of the gene expression values of all clones in a sample, the median centered normalization technique performed the best and also provided a uniform distribution for our binning method.

If the number of bins for expression level discretization is too small or too high, then it leads to problems in analysis. In coarse binning, the information about the values is ignored, and in a very fine binning, the patterns are lost. We tried several values of k and found that $k = 5$ works well. The bin intervals are determined using the uniform frequency binning method. Other popular methods like discriminant discretization, boolean reasoning based and entropy based discretization can be considered [48]. In frequency binning method we discretized the relative expression (fold-change) into 5 levels depending on their expression value. The values of -1.5, -0.5, 0, 0.5, and 1.5 for the fold change ratio were taken as thresholds for very low (VL), low (L), normal (N), high (H), and very high (VH) expression, respectively. That is, $VL \in (-\infty, -1.5]$, $V \in (-1.5, -0.5]$, $N \in (-0.5, 0.5)$, $H \in [0.5, 1.5)$, and $VH \in [1.5, +\infty)$. We use the notation g^i_e to denote the expression level for gene g^i in a given replicate, where $e \in \{VL, L, N, H, VH\}$. Table 38.2 shows an example of the expression of 4 genes in six mice with 4 array replicates for each mouse.

38.5.2 Elimination of Experimental Noise

In order to eliminate the noise due to experimental fluctuations, we process the data taking one mouse at a time. For each mouse the genetic expression signature is obtained and compared across all r replicates. Only those genes which show consistent expression signature in

	Rep1	Rep2	Rep3	Rep4
Mouse1	$\{g^1_{VH}, g^2_{VL}, g^3_{VH}, g^4_L\}$	$\{g^1_{VH}, g^2_{VL}, g^3_{VH}, g^4_N\}$	$\{g^1_{VH}, g^2_{VL}, g^3_{VH}, g^4_N\}$	$\{g^1_{VH}, g^2_{VL}, g^3_{VH}, g^4_N\}$
Mouse2	$\{g^1_{VH}, g^2_{VL}, g^3_L, g^4_N\}$	$\{g^1_{VH}, g^2_{VL}, g^3_L, g^4_N\}$	$\{g^1_{VH}, g^2_{VL}, g^3_L, g^4_N\}$	$\{g^1_{VH}, g^2_{VL}, g^3_H, g^4_N\}$
Mouse3	$\{g^1_{VH}, g^2_N, g^3_{VH}, g^4_N\}$	$\{g^1_{VH}, g^2_N, g^3_{VH}, g^4_N\}$	$\{g^1_{VH}, g^2_N, g^3_{VH}, g^4_e\}$	$\{g^1_{VH}, g^2_N, g^3_{VH}, g^4_L\}$
Mouse4	$\{g^1_{VH}, g^2_N, g^3_{VL}, g^4_L\}$	$\{g^1_{VH}, g^2_N, g^3_{VL}, g^4_N\}$	$\{g^1_{VH}, g^2_N, g^3_{VL}, g^4_N\}$	$\{g^1_{VH}, g^2_N, g^3_{VL}, g^4_N\}$
Mouse5	$\{g^1_{VH}, g^2_H, g^3_L, g^4_L\}$	$\{g^1_{VH}, g^2_H, g^3_L, g^4_L\}$	$\{g^1_{VH}, g^2_H, g^3_L, g^4_L\}$	$\{g^1_{VH}, g^2_H, g^3_L, g^4_L\}$
Mouse6	$\{g^1_{VH}, g^2_H, g^3_{VH}, g^4_L\}$	$\{g^1_{VH}, g^2_H, g^3_{VH}, g^4_L\}$	$\{g^1_{VH}, g^2_H, g^3_{VH}, g^4_N\}$	$\{g^1_{VH}, g^2_H, g^3_{VH}, g^4_N\}$

TABLE 38.2 The gene expression states of 4 genes in 24 (6 mice, 4 replicates) assays, with five possible levels: Very High (VH), High (H), Very Low (VL), Low (L) or Normal (N).

all r replicates are chosen and the ones which show even a slight deviation in any of the replicates are eliminated. This methodology takes a very stringent approach toward eliminating even the slightest errors due to technical noise. One shortcoming of this approach is that it would not eliminate any genes which show high fluctuations in the range $(-0.5, 0.5)$. In our study of normal variance to identify genes which have been falsely reported as differentially expressed, the genes which we might fail to eliminate do not contribute to the databank anyway, because they lie in the normal expression range. So, our approach would eliminate most of the noise which comes due to technical/experimental issues. This operation is done on all m mice, as a result of which we have gene expression signatures in all the mice with minimal experimental noise.

	Gene Expression
Mouse 1 (F_1)	$\{g^1_{VH}, g^2_{VL}, g^3_{VH}\}$
Mouse 2 (F_2)	$\{g^1_{VH}, g^2_{VL}, g^4_{N}\}$
Mouse 3 (F_3)	$\{g^1_{VH}, g^2_{N}, g^3_{VH}\}$
Mouse 4 (F_4)	$\{g^1_{VH}, g^2_{N}, g^3_{VL}\}$
Mouse 5 (F_5)	$\{g^1_{VH}, g^2_{H}, g^3_{L}, g^4_{L}\}$
Mouse 6 (F_6)	$\{g^1_{VH}, g^2_{H}, g^3_{VH}\}$

TABLE 38.3   Gene expressions after elimination of experimental noise.

Table 38.3 illustrates this process on our example data. For example consider Mouse 1. Since gene g^4 is differentially expressed as L in replicate 1, but as N in the other three replicates, we eliminate g^4 from further consideration. The resulting expression signatures for Mouse1 and other mice from our example are shown in Table 38.3.

38.5.3   Gene Expression Profile

Let F_j represent the gene expressions of the j-th mice after the elimination of experimental noise. The F_j's contain the expression level (very high, high, normal, low, very low) information of each gene in each of the m mice in our example. Some values could be missing due to elimination in the first stage. The F_j values, for our example of six mice, are shown in Table 38.3.

From the F_j values we construct a frequency table, which contains the number of occurrences of each gene for each discretized expression level (VH, H, N, L, VL). The frequency of every distinct (gene g^i, expression level e) pair across all F_j, is used to populate the frequency table. The frequency for gene g^i and expression level e is given as $f^i_e = \sum_{j=1}^{m} \delta^i_e(j)$, where m is the number of mice, and $\delta^i_e(j)$ is a characteristic function that notes the presence/absence of gene g^i at level e in mouse j, defined as: $\delta^i_e(j) = 1$, if $g^i_e \in F_j$, and $\delta^i_e(j) = 0$, if $g^i_e \notin F_j$. The frequency table obtained for our example is shown in Table 38.4. As an example, g^2, has expression level VL in mice 1 and 2, level N in mice 3 and 4, and level H in mice 5 and 6. Thus the expression profile for g^2 is given by the vector $(0, 2, 2, 0, 2)$, as shown in the table below.

38.5.4   Entropy-based Variability Ranking

The genes that show presence in more than one discrete level are of interest to us. The frequency table is analyzed further to identify those genes which show considerable variance

	f^i_{VH}	f^i_H	f^i_N	f^i_L	f^i_{VL}
Gene g^1	6	0	0	0	0
Gene g^2	0	2	2	0	2
Gene g^3	3	0	0	1	1
Gene g^4	0	0	1	1	0

TABLE 38.4   Expression Profile: Frequency table for the four genes.

by their presence in more than one state. To capture the variance in a gene's expression level, the entropy measure was used. Entropy gives us the amount of disorder in the expression values of a gene, and thus is a measure of the normal variance, since the noise due to experimental variation is eliminated prior to this step. The entropy measure for a gene g^i is given as follows, $E(g^i) = -\sum_{e=1}^{k} p^i_e \log_2(p^i_e)$, where k is the number of discrete expression levels, and p^i_e is the probability of gene g^i having expression level e, which is given as $p^i_e = \frac{f^i_e}{\sum_{j=1}^{k} f^i_j}$.

By definition of entropy, if a gene has only one expression level (say j), then $p^i_j = 1$ and $E(g^i) = 0$. On the other hand, if a gene has the most variance (i.e., equal occurrence at each expression level), then $P^i_j = 1/k$ for all expression levels j, and $E(g^i) = -\sum_{j=1}^{k} 1/k \log_2(1/k) = -\log_2(1/k) = \log_2(k)$. In our approach genes with entropy 0, i.e., those having no variance in expression across the mice, are discarded, and the remaining genes are ranked in descending order of their entropy (and thus variance). The entropy ranking for the four genes (along with the probability of each expression level) in our example are shown in Table 38.5. Gene g^2 and g^3 are of most interest to us because they show variation in expression states across the six mice. On the other hand gene g^1 is always high in all six mice, showing no variance.

	p^i_{VH}	p^i_H	p^i_N	p^i_L	p^i_{VL}	Entropy
Gene g^2	0	0.33	0.33	0	0.33	1.59
Gene g^3	0.6	0	0	0.2	0.2	1.37
Gene g^4	0	0	0.5	0.5	0	1
Gene g^1	1.0	0	0	0	0	0

TABLE 38.5   Entropy-based gene variability ranking.

38.5.5   Weighted Expression Profiles

In our approach to experimental noise elimination, any gene with varying expression level among the replicates is considered experimental noise, and eliminated. Instead of such a stringent approach, we can choose to retain a gene provided it has the same expression level in a given fraction of the replicates. For instance, gene g^4 has expression level N in three out of the four replicates for Mouse 1. If we set our threshold to 75%, then we would retain g^4_N in the gene expression for Mouse 1 in Table 38.3.

Another approach is to construct a weighted expression signature, as follows: For every gene we record the fraction of the replicates in which it takes a particular value. For instance, for Mouse 1, gene g^1 always takes the value VH, so its weighted expression is

$g^1_{VH(1.0)}$. On the other hand, gene g^4 is N in three and L in one out of the four replicates; we record its weighted expression as $g^4_{N(0.75),L(0.25)}$. We denote by $w^i_e(j)$ the weight of gene g^i at expression level e in Mouse j. Table 38.6 shows the weighted expression signatures for all the six mice (note: if the weight is 1.0 we omit the weight; we write g^1_{VH} instead of $g^1_{VH(1.0)}$).

	Gene Expression
Mouse 1 (F_1)	$\{g^1_{VH}, g^2_{VL}, g^3_{VH}, g^4_{N(0.75),L(0.25)}\}$
Mouse 2 (F_2)	$\{g^1_{VH}, g^2_{VL}, g^3_{H(0.25),L(0.75)}, g^4_N\}$
Mouse 3 (F_3)	$\{g^1_{VH}, g^2_N, g^3_{VH}, g^4_{N(0.25),L(0.75)}\}$
Mouse 4 (F_4)	$\{g^1_{VH}, g^2_N, g^3_{VL}, g^4_{N(0.75),L(0.25)}\}$
Mouse 5 (F_5)	$\{g^1_{VH}, g^2_H, g^3_L, g^4_L\}$
Mouse 6 (F_6)	$\{g^1_{VH}, g^2_H, g^3_{VH}, g^4_{N(0.5),L(0.5)}\}$

TABLE 38.6   Weighted gene expressions.

From the weighted gene expressions, we can construct a weighted profile using the approach in Section 38.5.3. The weighted frequency for gene g^i and expression level e is given as $f^i_e = \sum^m_{j=1} w^i_e(j)$, where m is the number of mice. The weighted frequency table obtained for our example is shown in Table 38.7. As an example, g^4, has expression levels $N(0.75)$ in Mouse 1, $N(1.0)$ in Mouse 2, $N(0.75)$ in Mouse 3 and Mouse 4, and $N(0.5)$ in Mouse 6. Thus $f^4_N = 0.75 + 1.0 + 2 \times 0.75 + 0.5 = 3.75$, and similarly $f^4_L = 2.25$. Thus the weighted expression profile for g^4 is given by the vector $(0, 0, 3.75, 2.25, 0)$, as shown in the table.

	f^i_{VH}	f^i_H	f^i_N	f^i_L	f^i_{VL}
Gene g^1	6	0	0	0	0
Gene g^2	0	2	2	0	2
Gene g^3	3	0.25	0	1.75	1
Gene g^4	0	0	3.75	2.25	0

TABLE 38.7   Weighted Expression Profile.

From the weighted expression profile, we can derive the entropy-based variability ranking for each gene as shown in Table 38.8. Comparing with Table 38.5, we find that g^3 is ranked higher in terms of variability than g^2, but the overall trend is similar.

	p^i_{VH}	p^i_H	p^i_N	p^i_L	p^i_{VL}	Entropy
Gene g^3	0.5	0.04	0	0.29	0.17	1.64
Gene g^2	0	0.33	0.33	0	0.33	1.59
Gene g^4	0	0	0.62	0.38	0	0.95
Gene g^1	1.0	0	0	0	0	0

TABLE 38.8   Entropy-based gene variability ranking.

38.5.6 Application Study

We applied our entropy-based method to detect normal variance in gene expression for the two datasets taken from [51] and [46]. We used three datasets of kidney, liver and testis provided by [51]. Six genetically identical male C57BL6 mice were used to compare the expression values of 5,406 unique mouse genes. Four separate microarray assays were conducted for each organ from each animal, for a total of 24 arrays per organ. Also the dataset of [46] was used which contained the expression values for three mice across the four organs of liver, heart, lung and brain. Each experiment was replicated three times. Affymetrix oligo-chips were used in these experiments.

Using our entropy-based approach, in kidney tissue around 3.5% of the 3088 genes showed considerable variance across the six mice. As reported by [51] we found several immune modulated and stress responsive genes. In liver tissue, 23 out of 2513 genes show significant variation in their expression levels among the six mice. Out of the 3252 genes analyzed in the testis tissue, 63 showed differential expression levels across the six mice. Importantly, many of the genes that we found to vary normally have been reported previously to be differentially expressed because of a pathological process or experimental intervention. One recent study used microarrays to investigate the differential gene expression patterns during pre-implantation mouse development [41]. Rpl12 was reported to be differentially expressed while we found it to be normally varying in the testis tissue. PUFA (polyunsaturated fatty acids) feeding can influence Protein Kinase C (PKC) activity [7]. Itrp1 is another gene which has been reported as differentially expressed [30] in papillary thyroid carcinoma, while we found this gene to be normally varying in kidney tissues. Another study investigated the effects of acetaminophen on gene expression in the mouse liver [53]. Eight of the genes reported to differ in response to acetaminophen, including CisH2, and Hsp40, were genes we found to vary normally.

Principal Component Analysis (PCA) [34] is a classical technique to reduce the dimensionality of the data set by transforming to a new set of variables (the principal components). It has been used in the analysis of gene expression studies. Principal components (PC's) are uncorrelated and ordered such that the k-th PC has the k-th largest variance among all PC's. The k-th PC can be interpreted as the direction that maximizes the variation of the projections of the data points such that it is orthogonal to the first $k-1$ PC's. PCA is sometimes applied to reduce the dimensionality of the data set prior to clustering. Using PCA prior to cluster analysis may aid better extraction of the cluster structure in the data set. Since PC's are uncorrelated and ordered, the first few PC's, which contain most of the variations in the data, are usually used in cluster analysis. Unless external information is available, [64] recommend cautious interpretation of any cluster structure observed in the reduced dimensional subspace of the PC's. They observe no clear trend between the number of principal components chosen and the cluster quality. [33] use PCA analysis for extracting tissue specific signatures.

We used PCA to analyze how well the genes we have extracted capture the normal variance between the mice, eliminating the variance due to any other sources to the maximum possible extent. Projection on to a 3-dimension space (the top three PCs) allows for better visualization of the entire data set. Figure 38.6 shows the arrangement of the samples by plotting them on the principal components derived from: 1) PCA analysis of all the genes in the dataset, and 2) PCA analysis of only those genes which were found to have normal variance across mice. These plots are shown for all the three tissues under study. Kidney and testis show non-random arrangement of the assay points while liver has less discernible patterns. In the case of the kidney tissue for genes with normal variance, the assays arrange into two clusters. One of the clusters has assays which include the replicates from four mice

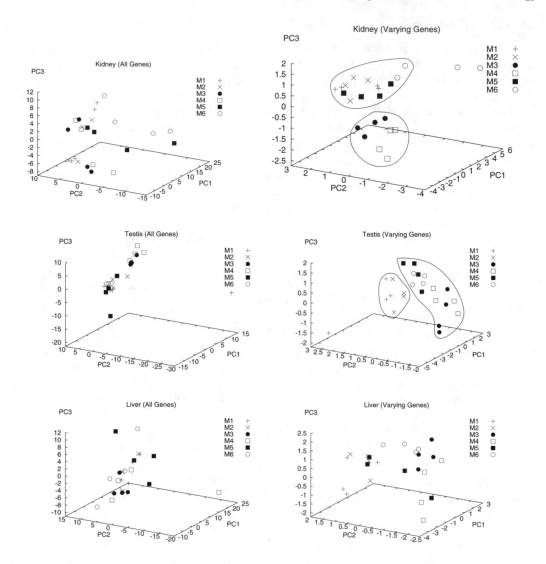

FIGURE 38.6: Principal component analysis of all genes (left column) and genes with normal variance (right column). Three tissues were studied: Kidney (top row), Testis (middle row), and Liver (bottom row). Results for all 6 mice and 4 replicates are shown.

(M1, M2, M5, M6), while the other cluster has mice M3 and M4. This indicates that there is a high similarity among these mice in kidney tissue. In the testis, the first two mice are systematically different from the last four mice. No pattern was observed in liver. PCA can be additionally used as a platform to compare the performance of different methodologies to determine normal variance. The performance can be judged visually on the basis how well the replicates cluster together or measure the goodness of the clusters. We observe that 1) the experimental replicates belonging to any single mice cluster close to each other, and 2) the mice (biological replicates) are also grouped into visible clusters. Pathologically similar mice are clustered together.

38.6   Summary

The goal of this chapter was to provide a brief introduction to some data mining techniques, and to look at how data mining has been used in some representative applications in bioinformatics, namely three-dimensional (3D) or structural motif mining in proteins and the analysis of microarray gene expression data. We also looked at some issues in data preparation, namely data cleaning and feature selection via the study of how to find normal variation in gene expression datasets.

It is clear that data mining is playing a fundamental role in understanding the rapidly expanding sources of biological data. It is equally clear that new data mining techniques are needed to analyze, manage and discover sequence, structure and functional patterns/models from large sequence and structural databases, as well as for structure prediction, gene finding, gene expression analysis, biochemical pathway mining, biomedical literature mining, drug design and other emerging problems in genomics and proteomics.

Acknowledgment

This work was supported in part by NSF CAREER Award IIS-0092978, DOE Career Award DE-FG02-02ER25538, and NSF grant EIA-0103708. We thank Jingjing Hu and Vinay Nadimpally for work on contact map mining and determining normal variations in gene expression data, respectively.

References

[1] C. Aggarwal. Towards systematic design of distance functions for data mining applications. In *9th ACM SIGKDD Conference*, 2003.

[2] R. Agrawal, J. Gehrke, D. Gunopulos, and P. Raghavan. Automatic subspace clustering of high dimensional data for data mining applications. In *ACM SIGMOD Conference*, pages 94–105, 1998.

[3] N. Alexandrov and N. Go. Biological meaning, statistical significance and classification of local spatial similarities in non-homologous proteins. *Protein Sci.*, 3:866–875, 1994.

[4] O. Alter, P. Brown, and D. Botstein. Singular value decomposition for genome-wide expression data processing and modeling. *Proceedings of the National Academy of Sciences*, 97:10101–10106, 2000.

[5] I. Bahar, A. Atilgan, and B. Erman. Direct evaluation of thermal fluctuations in proteins using a single parameter harmonic potential. *Folding and Design*, 2:173–181, 1997.

[6] A. Ben-Dor, B. Chor, R. Karp, and Z. Yakhini. Discovering local structure in gene expression data: The order-preserving submatrix problem. In *6th Annual International Conference on RECOMB*, pages 49–57, 2002.

[7] A. Berger, D.M. Mutch, J. Bruce, and G. Matthew *et al.* Dietary effects of arachidonate-rich fungal oil and fish oil on murine hepatic and hippocampal gene expression. *Lipids in Health and Disease*, 1:2–10, 2002.

[8] K. Beyer, J. Goldstein, R. Ramakrishnan, and U. Shaft. When is nearest neighbors meaningful? In *ICDT Conference*, 1999.

[9] M. Blatt, S. Wiseman, and E. Domany. Data clustering using a model granular magnet. *Neural Computation*, 9:1805–1842, 1997.

[10] P. Bradley and U. Fayyad. Refining initial points for kmeans clustering. In *15th International Conference on Machine Learning*, pages 91–99. Morgan Kaufmann, 1998.

[11] K. Brinda, N. Kannan, and S. Vishveshwara. Analysis of homodimeric protein interfaces by graph-spectral methods. *Protein Engineering*, 15:265–77, April 2002.

[12] Tang C., Zhang L., Zhang A., and Ramanathan M. Interrelated two-way clustering: An unsupervised approach for gene expression data analysis. In *2nd IEEE International Symposium on Bioinformatics and Bioengineering*, pages 41–48, November 2001.

[13] S. Chakraborty and S. Biswas. Approximation algorithms for 3-d common substructure identification in drug and protein molecules. In *Workshop on Algorithms and Data Structures*, pages 253–264, 1999.

[14] Y. Cheng and G. Church. Biclustering of expression data. In *8th International Conference on Intelligent Systems for Molecular Biology*, volume 8, pages 93–103, 2000.

[15] D.J. Cook, L.B. Holder, R. Maglothin S. Su, and I. Jonyer. Structural mining of molecular biology data. *IEEE Engineering in Medicine and Biology*, 20(4):67–74, 2001.

[16] C. Creighton and S. Hanash. Mining gene expression databases for association rules. *Bioinformatics*, 19:79–86, 2003.

[17] F. DeSmet, J. Mathys, K. Marchal, and G. Thijs *et al.* Adaptive quality-based clustering of gene expression profiles. *Bioinformatics*, 18:735–746, 2002.

[18] P. D'haesleer, X. Wen, S. Fuhrman, and R. Somogyi. *Information Processing in Cells and Tissues*. Plenum Press, New York, 1998.

[19] M. Eisen, P. Spellman, P. Brown, and D. Botstein. Cluster analysis and display of genome-wide expression patterns. *Proceedings of the National Academy of Sciences*, 95:14863–8, 1998.

[20] D. Fischer, H. Wolfson, S. Lin, and R. Nussinov. Three-dimensional, sequence order-independent structural comparison of a serine protease against the crystallographic database reveals active site similarities: potential implication to evolution and to protein folding. *Protein Science*, 3:769–778, 1994.

[21] G. Getz, E. Levine, and E. Domany. Coupled two-way clustering analysis of gene microarray data. *Proceedings of the National Academy of Sciences*, 97:12079–12084, October 2000.

[22] H. Grindley, P. Artymiuk, D. Rice, and P. Willet. Identification of tertiary structure resemblance in proteins using a maximal common subgraph isomorphism algorithm. *Journal of Molecular Biology*, 229:707–721, 1993.

[23] H.M. Grindley, P.J. Artymiuk, D.W. Rice, and P. Willett. Identification of tertiary resemblence in proteins using a maximal common subgraph isomorphism algorithm. *J. of Mol. Biol.*, 229(3):707–721, 1993.

[24] K. Hall. An r-dimensional quadratic placement algorithm. *Management Sciences*, 17:219–229, November 1970.

[25] J. Heringa and P. Argos. Side-chain clusters in protein structures and their role in protein folding. *Journal of Molecular Biology*, 220:151–171, 1991.

[26] J. Herrero, A. Valencia, and J. Dopazo. A hierarchical unsupervised growing neural network for clustering gene expression patterns. *Bioinformatics*, 17:126–136, 2001.

[27] U. Hobohm and C. Sander. Enlarged representative set of protein structures. *Protein Science*, 3(3):522–524, 1994.

[28] J. Hu, X. Shen, Y. Shao, and C. Bystroff *et al.* Mining protein contact maps. *2nd BIOKDD Workshop on Data Mining in Bioinformatics*, July 2002.

[29] J. Huan, W. Wang, D. Bandyopadhyay, and J. Snoeyink *et al.* Mining spatial motifs

from protein structure graphs. In *8th Annual International Conference on RECOMB*, 2004.

[30] Y. Huang, M. Prasad, W.J. Lemon, and H. Hampel *et al.* Gene expression in papillary thyroid carcinoma reveals highly consistent profiles. *PNAS*, 98:15044–15049, October 2001.

[31] D. Jacobs, A. Rader, L. Kuhn, and M. Thorpe. Graph theory predictions of protein flexibility. *Proteins: Struct. Funct. Genet.*, 44:150–155, 2001.

[32] A. Jain, M. Murty, and P. Flynn. Data clustering: a review. *ACM Computing Surveys*, 31:254–323, September 1999.

[33] M. Jatin, W. Schmitt, D. Hwang, and L.-L. Hsiao *et al.* Interactive exploration of microarray gene expression patterns in a reduced dimensional space. *Genome Res*, 12:1112–1120, 2002.

[34] I.T. Jolliffe. *Principal Component Analysis, Springer Series in Statistics.* Springer Verlag, New York, 1986.

[35] I. Jonassen, I. Eidhammer, D. Conklin, and W. Taylor. Structure motif discovery and mining the pdb. *Bioinformatics*, 18:362–367, 2002.

[36] N. Kannan, S. Selvaraj, M. Michael Gromiha, and S. Vishveshwara. Clusters in α/β barrel proteins: Implications for protein structure, function and folding: A graph theoretical approach. *Proteins: Struct., Funct., Genet.*, 43:103–112, May 2001.

[37] N. Kannan and S. Vishveshwara. Identification of side-chain clusters in protein structures by graph spectral method. *Journal of Molecular Biology*, 292:441–464, September 1999.

[38] N. Kannan and S. Vishveshwara. Aromatic clusters: a determinant of thermal stability of thermophilic proteins. *Protein Engineering*, 13:753–761, November 2000.

[39] G.J. Kleywegt. Recognition of spatial motifs in protein structures. *J. Mol. Biol*, 285:1887–1897, 1998.

[40] G.J. Kleywegt. Recognition of spatial motifs in protein structures. *Journal of Molecular Biology*, 285:1887–1897, 1999.

[41] M.S.H Ko, J.R. Kitchen, X. Wang, and T.A. Threat *et al.* Large-scale cDNA analysis reveals phased gene expression patterns during preimplantation mouse development. *Development*, 127:1737–1749, 2000.

[42] I. Koch, T. Lengauer, and E. Wanke. An algorithm for finding maximal common subtopologies in a set of protein structures. *J. of Comp. Biol.*, 3(2):289–306, 1996.

[43] T. Kohonen. *Self-Organization and Associative Memory.* Spring-Verlag, Berlin, 1988.

[44] M. Kuramochi and G. Karypis. Frequent subgraph discovery. *1st IEEE Int'l Conf. on Data Mining*, November 2001.

[45] L. Lazzeroni and A. Owen. Plaid models for gene expression data. *Statistica Sinica*, 12:61–86, 2002.

[46] P.D. Lee, R. Sladek, C. Greenwood, and T. Hudson. Control genes and variability: Absence of ubiquitous reference transcripts in diverse mammalian expression studies. *Genome Research*, 12(2):292–297, February 2002.

[47] G. Michaels, D. Carr, M. Askenazi, and S. Fuhrman *et al.* Cluster analysis and data visualization of large-scale gene expression data. In *Pacific Symposium on Biocomputing*, volume 3, pages 42–53, 1998.

[48] H. Midelfart, J. Komorowski, K. Norsett, and F. Yadetie *et al.* Learning rough set classifiers from gene expression and clinical data. *Fundamenta Informaticae*, 53:155–183, November 2002.

[49] E.M. Mitchell, P.J. Artymiuk, D.W. Rice, and P. Willett. Use of techniques derived from graph theory to compare secondary structure motifs in proteins. *J. Mol. Biol.*,

212:151–166, 1990.

[50] A.G. Murzin, S.E. Brenner, T. Hubbard, and C. Chothia. SCOP: a structural classification of proteins database for the investigation of sequences and structures. *J. of Mol. Biol.*, 247:536–540, 1995.

[51] C.C. Pritchard, L. Hsu, J. Delrow, and P.S. Nelson. Project normal: Defining normal variance in mouse gene expression. *PNAS*, 98:13266–13271, 2001.

[52] S. Raychaudhuri, J. Stuart, and R. Altman. Principal components analysis to summarize microarray experiments: Application to sporulation time series. In *Pacific Symposium on Biocomputiing*, pages 455–66, 2000.

[53] T.P. Reilly, M. Bourdi, J.N. Brady, and C.A. Pise-Masison *et al.* Expression profiling of acetaminophen liver toxicity in mice using microarray technology. *Biochem. Biophys. Res. Commun*, 282:321–328, 2001.

[54] R. Singh, A. Tropsha, and I. Vaisman. Delaunay tessellation of proteins. *J. Comput. Biol.*, 3:213–222, 1996.

[55] M. Sternberg, H. Gabb, and R. Jackson. Predictive docking of protein-protein and protein-DNA complexes. *Current Opinion in Structural Biology*, 8:250–256, 1998.

[56] P. Tamayo, D. Solni, J. Mesirov, and Q. Zhu *et al.* Interpreting patterns of gene expression with self-organizing maps: Methods and application to hematopoietic differentiation. *Proceedings of the National Academy of Sciences*, 96:2907–2912, March 1999.

[57] A. Tanay, R. Sharan, and R. Shamir. Discovering statistically significant biclusters in gene expression data. *Bioinformatics*, 18(suppl.1):S136–S144, 2002.

[58] S. Tavazoie, D. Hughes, M. Campbell, and R. Cho *et al.* Systematic determination of genetic network architecture. *Nature Genet*, pages 281–285, 1999.

[59] A. Tefferi, M. Bolander, S. Ansell, and E. Wieben *et al.* Primer on medical genomics. part iii: Microarray experiments and data analysis. *Mayo Clin Proc.*, 77:927–40, September 2002.

[60] H. Wang, W. Wang, J. Yang, and P. S. Yu. Clustering by pattern similarity in large data sets. In *ACM SIGMOD Conference*, 2002.

[61] X. Wang, J.T.L. Wang, D. Shasha, and B.A. Shapiro *et al.* Finding patterns in three-dimensional graphs: Algorithms and applications to scientific data mining. *IEEE Transactions on Knowledge and Data Engineering*, 14(4):731–749, July/August 2002.

[62] L. Wernisch, M. Hunting, and S. Wodak. Identification of structural domains in proteins by a graph heuristic. *Proteins*, 35:338–352, 1999.

[63] J. Yang, W. Wang, H. Wang, and P. Yu. δ-cluster: Capturing subspace correlation in a large data set. In *18th International Conference on Data Engineering*, pages 517–528, 2002.

[64] K.Y. Yeung and W.L. Ruzzo. Principal component analysis for clustering gene expression data. *Bioinformatics*, 17:763–774, 2002.

[65] M. J. Zaki, S. Jin, and C. Bystroff. Mining residue contacts in proteins using local structure predictions. *IEEE Transactions on Systems, Man and Cybernetics – B*, 33(5), October 2003.

[66] M.J. Zaki, V. Nadimpally, D. Bardhan, and C. Bystroff. Predicting protein folding pathways. *12th Int'l Conference on Intelligent Systems for Molecular Biology*, July 2004.

Index